AF393588

ENCYCLOPEDIA OF PHYSICS

EDITED BY
S. FLÜGGE

VOLUME XXIV
FUNDAMENTALS OF OPTICS

WITH 761 FIGURES

SPRINGER-VERLAG
BERLIN · GÖTTINGEN · HEIDELBERG
1956

HANDBUCH DER PHYSIK

HERAUSGEGEBEN VON

S. FLÜGGE

BAND XXIV

GRUNDLAGEN DER OPTIK

MIT 761 FIGUREN

SPRINGER-VERLAG

BERLIN · GÖTTINGEN · HEIDELBERG

1956

ISBN-13: 978-3-642-45851-4   e-ISBN-13: 978-3-642-45850-7
DOI: 10.1007/978-3-642-45850-7

Inhaltsverzeichnis.

Interférences, diffraction et polarisation. Par Maurice Françon, Maître de conférences à la Sorbonne, Professeur à l'Institut d'Optique, Paris (France). (Avec 427 Figures) 171

Determination of the Velocity of Light.

By

E. Bergstrand.

With 31 Figures.

A. Introduction.

1. General survey. The velocity of light propagation plays a dominant role in physics, since light forms an essential part of the electromagnetic wave phenomena. The vacuum value of this velocity is a natural constant of universal importance having more or less direct significance in each process involving energy transport, i.e. where ever anything "happens". Therefore a great number of investigators has worked in that field and tried to determine that velocity, in vacuo and in matter, by direct or by indirect methods. By and by, the subject thus has become rather extensive and still is rapidly growing.

In the present survey of these works the main part is devoted to the determinations in *vacuo* (or in air), beginning with the direct and more classic methods and ending with the new indirect ones, which, due to the advanced modern technics, are of increasing importance. The last part of the survey briefly deals with the experiments on the velocity in matter, including moving bodies, and with some especially connected with the theory of relativity. Describing such experimental work, the basic laws of relativity, such as the law of velocity addition, will be used throughout. The direct experimental verification of these laws may not always be obvious. However, combined with the indirect proofs they form a decisive and reliable complex.

Finally, a word should be said on velocity determinations during the last decades. Michelson's value of 1926 was (299798 ± 4) km./sec. During the following twenty years, five different direct determinations yielded an average value of (299776 ± 4) km./sec., a figure which can be claimed to have been the "official" value during that period. On the other hand, the last results on to 1955 with high probability yield a figure of $(299793 \pm 0,3)$ km./sec. almost generally adopted and close to Michelson's value again. Therefore, describing the preceding determinations, the author has tried to treat sources of error a little more elaboratcly. An attempt has been made to show that these determinations must not be incompatible with the new value, so that we need not be afraid to find varying values of the velocity of light.

2. Definitions of phase and group velocity. Let us start with a simple harmonic oscillation proceeding along the X-axis:

$$y = a \sin (k\,x - \omega\,t + \alpha) \tag{2.1}$$

with an amplitude a and a frequency $\nu = \omega/2\pi$. The phase constant α depends on the choice of the origin of the coordinate x, and of the time t. The wave number k is defined by the relation

$$k = \frac{2\pi}{\lambda_m} \tag{2.2}$$

with λ_m being the wavelength in the substance of propagation.

Phase velocity or wave velocity v in the substance considered will be called the velocity by which a point of constant y is moving. From eq. (2.1) it follows immediately that

$$v = \left(\frac{\partial x}{\partial t}\right)_y = \frac{\omega}{k}.$$
(2.3)

Besides, for a light wave, there holds the equation

$$n = \frac{c}{v}$$
(2.4)

with c being the velocity in vacuo, and n the refractive index of the substance.

Actual light, however, is not strictly monochromatic, but includes several different wavelengths as a consequence of Doppler effects and transient phenomena. These effects, also, make the difference between adjacent wavelengths infinitesimal. Let us take account of the property of variety, first, by simple addition of two wave motions of equal amplitudes [123], i.e.

$$y = a \cdot \sin(k_1 x - \omega_1 t + \alpha_1) + a \cdot \sin(k_2 x - \omega_2 t + \alpha_2),$$
(2.5)

which yields

$$y = 2a \cos\left(\frac{k_2 - k_1}{2} x - \frac{\omega_2 - \omega_1}{2} t + \beta_1\right) \sin\left(\frac{k_2 + k_1}{2} x - \frac{\omega_2 + \omega_1}{2} t + \beta_2\right)$$
(2.6)

where β_1 und β_2 depend upon α_1 and α_2 only.

The sin-factor of eq. (2.6) has a frequency $\frac{\omega_1 + \omega_2}{2}$, and an amplitude varying with a beat frequency $2 \cdot \frac{\omega_2 - \omega_1}{2} = \omega_2 - \omega_1$, where the factor 2 follows from the fact that an amplitude has no sign at all. The wavelength λ_g of the amplitude variation is called the length of a *group* and is, according to definition (2.2):

$$\lambda_g = \frac{1}{2} \cdot \frac{2\pi}{k_2 - k_1} = \frac{\pi}{k_2 - k_1}$$
(2.7)

where again the factor $\frac{1}{2}$ has been added due to the same consideration.

Group velocity, now, will be called the velocity u' of the propagation of a constant amplitude. We apply the operation of eq. (2.3) to the amplitude of eq. (2.6) and get

$$u' = \frac{\omega_2 - \omega_1}{k_2 - k_1}.$$
(2.8)

If both component wavelengths are very close: $\lambda_1 \to \lambda_2 = \lambda_m$, we obtain from eqs. (2.2) und (2.3) the group velocity [see eq. (2.15) also]

$$u = \frac{d\omega}{dk} = v - \lambda_m \frac{dv}{d\lambda_m}.$$
(2.9)

Of course, there is always a great number of different wavelengths [123] which we may consider to form a narrow band. Then, instead of eq. (2.6) we can write

$$y = \int_{k_0 - \varepsilon}^{k_0 + \varepsilon} a(k) e^{i(kx - \omega t)} dk.$$
(2.10)

Each elementary wave has an amplitude $a(k)dk$. Let us write eq. (2.10) in the form

$$y = C e^{i(k_0 x - \omega_0 t)}$$
(2.11)

with an amplitude

$$C = \int_{k_0 - \varepsilon}^{k_0 + \varepsilon} a(k) e^{i[(k - k_0)x - (\omega - \omega_0)t]} dk.$$
(2.12)

This latter quantity will become a constant for such pairs of values (x, t) where the exponential is a constant, i.e. if

$$\frac{dx}{dt} = \frac{\omega - \omega_0}{k - k_0} = \frac{\Delta\omega}{\Delta k} .$$

(2.13)

Again, if Δk is small enough, eq. (2.9) may be applied.

It is not necessary to give a detailed treatment of the general case of an infinitely broad wavelength band where FOURIER integrals have to be used instead of eq. (2.10).

The velocity u of the amplitude is the velocity of energy transport in the light wave. In a number of fundamental experiments, only one of the two quantities, u and v, can be determined by observation. In case it is desirable to obtain the other, too (e.g. in the experiments of Sect. 21), the validity of eq. (2.9) will be of great importance (cf. Sect. 19). Lord RAYLEIGH [1] has been the first to derive it for water waves. He also showed (simultaneously with REYNOLDS, [2]) that the group velocity is closely connected with the velocity of energy transport by using the relation [123]

$$\frac{u}{v} = \frac{S}{E}$$

(2.14)

where u and v are group and phase velocities, S means the energy flow density (passing through unit area in unit time) and E the surplus energy content, caused by the radiation, in a volume of length v and unit cross section.

Introducing into eq. (2.9) the refractive index $n = c/v$, and the vacuum wavelength $\lambda = n\lambda_m$ we obtain

$$\frac{c}{u} = n - \lambda\frac{dn}{d\lambda} = n_g.$$

(2.15)

If $n < 1$, or if the phase velocity $v > c$, the negative quantity $dn/d\lambda$ will become large enough to keep $c/u > 1$ according to the theory of relativity. Close to the limit of anomal dispersion, however, $dn/d\lambda$ becomes positive and u may surpass c. But this does not contradict relativity, because the energy then will be rapidly absorbed (often within a few wavelengths), and eq. (2.15) no longer holds for the velocity of energy transport which still remains below c. In vacuo, $dn/d\lambda = 0$ and $u = v = c$. This readily can be seen from astronomical considerations, e.g. would the light from the eruption of an extragalactic supernova else be spectrally distributed in time.

Some authors define a special signal velocity as the velocity of the centre of the group. This subject has been worked out in more detail by BORGNIS [8]. Outside anomal dispersion signal and group velocities are equal. All direct methods for the determination of the velocity give the group value u and all indirect, except OBOE and SHORAN, the phase velocity v. (Strictly speaking, OBOE and SHORAN, Sect. 16, are direct methods too.)

3. Influence of the atmosphere on the velocity. Eq. (2.15) may be used to calculate the vacuum value from a velocity value determined in air. If the refractive index of air is expanded in the form

$$n_a = A + \frac{B}{\lambda^2} + \frac{C}{\lambda^4} ,$$

(3.1)

then by use of eq. (2.15) we obtain

$$n_g = A + \frac{3B}{\lambda^2} + \frac{5C}{\lambda^4}.$$

(3.2)

(Note: n_g merely expresses the relation c/u, not a real refraction.)

If the value n_g is meant to hold for dry air of 0° C and 760 mm. Hg, we obtain [14] for actual air of temperature t (°C), pressure p (mm. Hg), and humidity e (mm. Hg):

$$n'_g = 1 + \frac{n_g - 1}{1 + \alpha t} \frac{p}{760} - \frac{55 \cdot 10^{-9}}{1 + \alpha t} e, \qquad (3.3)$$

where $\alpha = \frac{1}{273}$. To get n'_a we put n_a for n_g in eq. (3.3). The value of n_a, and almost to the same degree, n_g is known with great accuracy [10], [11], [12], [15], [16], [17]. In ordinary air, the total influence of n_g on the velocity is about 91 km./sec., and the influence of n_a about 88 km./sec. Thus, the use of n_a in the direct determinations in air, causes an error of about 3 km./sec. in the velocity value.

4. Miscellaneous. *α) Influence of intensity.* The intensity of the light has no detectable influence on its velocity. This result is obvious if we regard the intensity as a consequence of the number of independent light quanta.

A possible influence has been checked by Doubt [18] using the Michelson interferometer. For an intensity variation of 1:300000 the change in velocity was less than 6 mm./sec. This result holds also for water and CS_2.

β) Wavelength. In vacuo, there is no influence of the wavelength on the velocity. Otherwise, distant double-stars or other rapidly moving astronomical objects would appear spectrally spread out in space.

γ) Moving light source. No experience exists so far in favour of an influence of the motion of the light source on light velocity. Again, such astronomical objects as the components of a double star, the opposite edges of a rotating and moving star, the surface of a rapidly expanding nova, etc., all would yield peculiar and conflicting phenomena which are not at all observed.

This evidence has only been doubted by la Rosa [23] who based his differing opinion on a special space distribution of the observed double stars. Lots of more recent observations, however, contradict this special distribution.

δ) Strong magnetic field. Banwell and Farr [29] used a Jamin interferential refractometer for the examination of light which had passed a magnetic field of 20000 Gauss. They found a velocity increase of 34 cm./sec. Since the observation error in this experiment was 19 cm./sec. the effect was considered to be doubtful.

ε) Strong electric field. In an experiment by J. Stark [30] a narrow pencil of light passed an inhomogeneous electric field of 1000 kV/cm. A slight deflection of the beam was stated. If we interpret this deflection as a refraction there would be an influence on the velocity. In any case, the effect reported by Stark is almost infinitesimal.

B. Methods of determination in vacuo and in air.

I. Direct methods.

5. Manual method. Galileo (1564 to 1642), the "father of modern physics" was the first who ever seems to have tried to settle the question whether light propagates instantaneously or not. In his "Discorsi" he describes the following experiment: Two observers A and B, a short distance apart from each other, both have a lamp which can be quickly screened off. Right at the instant when B sees A screening off his lamp he does the same. The observer A looks for B screening off the lamp and notes the time elapsed. The experiment then is repeated at a distance of some kilometers, and A checks whether by now the time lag has increased. Since the time difference to be expected ($\sim 10^{-5}$ sec.) is much smaller than the human reaction time, the experiment, of course, must lead to a negative result.

The GALILEO experiment, however, shows already essential points of later terrestrial methods, viz., that the light has to run the distance twice and by the use of a difference of distances the reaction time of the measuring device itself is eliminated. In a modern automatized form the method has reappeared in the beacon-radar!

6. Astronomical method. At the time when the Dane OLE ROEMER (1644 to 1710) was working at Paris, a peculiar regular variation of the revolution period

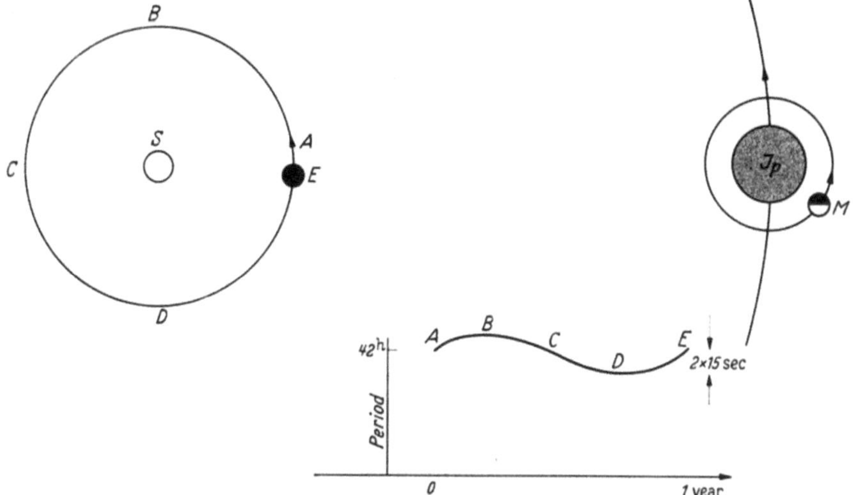

Fig. 1. ROEMER's observations of the Jupiter satellite.

of the innermost of Jupiter's satellites was observed [31], [32], [33]. In August 1676 ROEMER's coworker CASSINI suggested that these variations might be caused by a finite value of the velocity of light. In want of convincing proofs CASSINI very soon gave up his idea and even became its eager opponent. In September of that same year, ROEMER informed the Academy of Sciences of his opinion that the variation depends upon the value of the velocity of light. He predicted an eclipse of the innermost satellite in November to occur 10 minutes later than calculated from the observations of August. This prediction was confirmed and on the 21st of November ROEMER presented his treatise "where he demonstrates that the velocity of light is not instantaneous" (report in manuscript at the Paris observatory).

The following Fig. 1 shows the Earth E in its orbit around the Sun S. $J\!p$ is Jupiter and part of its orbit. The revolving period of Jupiter is 12 years. M is the innermost satellite with a period around the planet of 42 hours. In the position shown in the figure its eclipse is just beginning. Until the next eclipse the Earth has moved a part of its path towards B and the distance between Earth and Jupiter has somewhat increased. The period of M observed on the Earth is the sum of the true period and the travel time of the light for the additional distance. The period between the beginnings (or the ends) of two consecutive eclipses, therefore, will increase up to B where it will become about 15 sec. longer than in the neighbourhood of A. Thus the period observed follows a sine curve. From this effect ROEMER computed the time during which the light crosses the diameter of the Earth's orbit to be 22 min. Accidentally, Roemer's colleagues Piccard and Cassini determined the Earth's radius to 6370 km., and the Sun parallax to 9.3″. From these values the velocity of light was derived to be

$$c = 214300 \text{ km./sec.}$$

The method may be checked with modern values. According to Sampson, eclipses determined photometrically at Harvard Observatory give a traversing time of $16^m 37\overset{s}{.}8$. This value may be combined with Rabe's figure[1] for the radius of the Earth's orbit of 14953×10^4 km. Thus we obtain a velocity value of

$$c = (299840 \pm 60) \text{ km./sec.}$$

The first evidence supporting Roemer's suggestion came in 1727 with Bradley's discovery of the aberration (cf. indirect methods, Sect. 11).

7. The toothed wheel. α) The first terrestrial method to determine the velocity of light was developed by H. L. Fizeau (1819 to 1896). His experiment [34] of 1849 is a decisive step forward in order to use an arrangement the conditions of which are under control. Fig. 2 will facilitate an explanation of his method. L is a light source as point-shaped as possible. After reflection at the semi-transparent

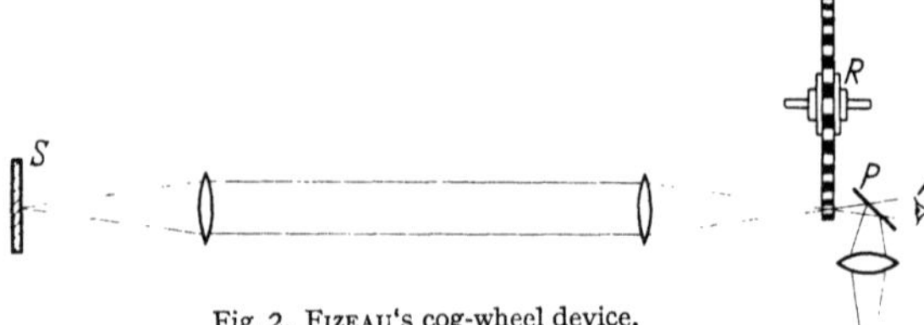

Fig. 2. Fizeau's cog-wheel device.

mirror P an image of L formed by a front lens is projected on the tooth row of the cog wheel R. In Fig. 2 the wheel is seen from the rim. If we rotate the wheel, the cogs cut off the light beam into a series of short flashes which start travelling through the 8633 m. distance to the mirror S where an image is formed by the lenses. The mirror S reflects the light back into its original path. After having passed the openings between the cogs again the light not reflected at P will be observed in A. If the wheel rotates with sufficient speed, a flash passed through an opening after its return from S to R will be stopped by the next cog, now in place of the preceding opening. At constant rotation speed this will occur with every flash and no light will be observed at A. The travel time of the flashes from R to S and back again then is equal to the time needed by a cog to replace its preceding opening. The latter is known from the turn speed and the number of cogs. If we double the turn speed an opening will be in place of the preceding opening, and light will be observed in A. Three times the original speed means darkness again, and so on. In Fizeau's experiment the wheel had 720 cogs and he observed the first eclipse at 12.68 turns per second. Thus knowing the time for the flashes to cover the distance of 2×8633 m. he obtaines a value of

$$c = (315300 \pm 500) \text{ km./sec.}$$

for the velocity of light.

One year later, in 1850, Fizeau, independent of Foucault, used the rotating mirror instead of the toothed wheel as described in Sect. 8.

β) Cornu (1841 to 1902) carefully improved Fizeau's method [37]. He increased the distance to 23000 m. and the turn speed to 1600 turns/sec. Thus, he was able to let 21 cogs pass before return of the light. Cornu's value of 1874 is

$$c = (300030 \pm 200) \text{ km./sec.}$$

---

[1] Astrophs. J. **55**, 112 (1950).

γ) An interesting addition to Fizeau's original device was introduced by Young and Forbes [38]. They used two reflecting systems similar to S in Fig. 2 at distances l_1 and l_2 from the wheel R placed in directions separated sufficiently that the systems did not hide each other. The ratio of l_1 and l_2 was 12:13. The turn speed of R was varied until equal light intensities were observed at A from both systems S. Thus the right turn speed could be defined much more exactly. Unfortunately, there must have been some unknown systematic errors, since the value obtained in 1881 is not very good

$$c = 301\,382 \text{ km./sec.}$$

δ) Perrotin [36] (1845 to 1904) was the last one to use the toothed wheel in 1902. He doubled Cornu's distance to 46 km. and obtained the final value

$$c = (299\,880 \pm 84) \text{ km./sec.}$$

8. Rotating mirrors. α) As early as 1838 Arago [40] suggested the use of Wheatstone's rotating mirror to detect the difference between the velocities of light in air and in water. In 1850, Foucault (1819 to 1868) improved the device by using a concave mirror, and became the first—besides Fizeau's simultaneous and independent work—to determine the velocity in air by this method [39]. The

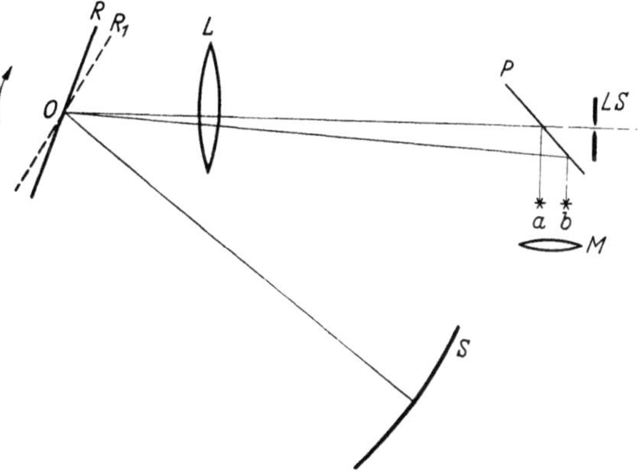

Fig. 3. Foucault's rotating mirror.

principle of the method can be seen from Fig. 3 in which P is a semi-transparent plane mirror, R a plane mirror revolving around an axis O perpendicular to the plane of paper. S is a spherical mirror the center of which lies in O. LS is a slit illuminated by sunlight which is projected on to the mirror S by the lens L. The image of LS on S moves with the rotation of R, but wherever it lies it is always reflected back to O and gives a real image at a (besides the one at LS) which can be observed through a microscope with the front lens M. If the mirror R is put into rapid rotation it turns through the angle ROR' during the time needed by the light to travel through the path OSO. The image of LS then will be shifted to b. From the displacement $a-b$ observed the angle ROR' may be calculated and, the number of revolutions per second being known, we obtain the time for the light passing through OSO. Knowing this latter distance, we readily find the velocity of light. Eventually, the direction of revolution may be reversed and thus the double displacement at a observed.

In Foucault's first experiment a distance OS of 4 m. was used. In 1862 he increased this distance by multiple reflexion to 20 m. The mirror R rotating with 800 turns per second a displacement $a-b$ of 0.7 mm. was obtained. Thus Foucault's best value for the velocity of light in air is

$$c = (298\,000 \pm 500) \text{ km./sec.}$$

β) The original method was improved by Newcomb (1835 to 1909) [41]. Instead of the plane mirror R of Fig. 3 he used a revolving square prism each of the four surfaces of which was a mirror. Combining this with Michelson's improved long distance optics he made a very careful determination in 1882. With a distance of 3720 m. he arrived at a value of

$$c = (299\,860 \pm 30)\ \text{km./sec.}$$

γ) *Modern determinations* by an optical arrangement which was mainly improved by the use of longer focal distances have been done by Michelson (1852 to 1931) [42], [43], [44]. He gained considerably in both light intensity and deviation $a - b$. His first value of 1879 was

$$c = (299\,910 \pm 50)\ \text{km./sec.,}$$

the next one of 1882, using Newcomb's prism, was

$$c = (299\,853 \pm 60)\ \text{km./sec.}$$

Gradually increasing the number of reflecting prism surfaces, in 1926 he arrived at using 3 prisms of glass and 3 of steel having 8, 12, and 16 surfaces. By that time, he increased the distance of the reflecting system S to 35 000 m. Thus his last determinations became the most accurate ones ever done by Foucault's method. Let us, therefore, describe them a little more in detail.

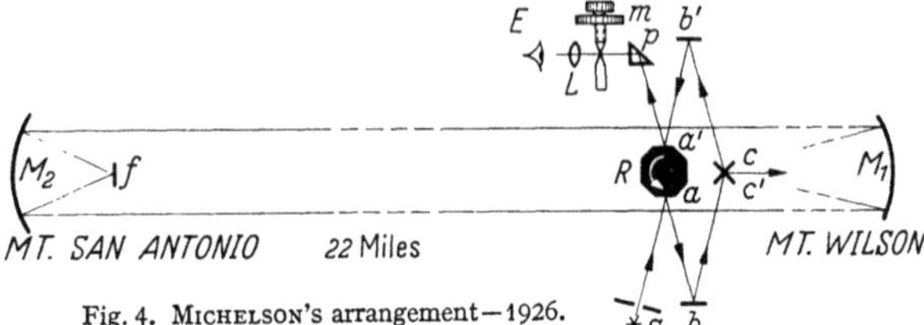

Fig. 4. Michelson's arrangement—1926.

Fig. 4 shows Michelson's famous arrangement on Mt. Wilson [44]. Light from the vertical slit S is reflected at one side (a) of a rotating steel or glass prism R and is then thrown on to the concave mirror M_1 by the fixed plane mirrors b and c. Since the slit S stands in the focus of M_1 a parallel beam, about one meter in width and considerably more in height, is leaving the mirror and passing the 22 mile distance to the reflector system $M_2 f$ on Mt. San Antonio. The light then travels back following its own path. Finally, by M_1 via the fixed mirrors $c' b'$ and the side a' of the prism, the light is focused into a point-like image of that part of the slit-light which is cut out by the mirror M_2. The position of this image-point is determined by means of a lens and a micrometer. The zero-position of the image-point is defined by transmitting light to M_2 while the prism is at rest. Then the prism is made to rotate at a speed just sufficient to allow the next side of the prims to move forward in place of the former side during the travel time of the light to M_2 and back again. The speed is adjusted until the new image point is seen exactly in the former zero-position. Both directions of rotation are used. The velocity of light then can be determined in the usual way from the known distance and the travel time as computed from the speed of rotation.

The distance adopted, 22 miles or 35 km., was determined by triangulation from a 40 km. bas-line the estimated error of which was one part in eleven millions. This high accuracy, however, is of minor importance since the main errors are originated by the triangulation. Between mountain peeks, systematic and masked lateral refraction frequently occurs. Considering these circumstances we can expect the distance value to be correct within an error of 10^{-5} or the velocity within 3 km./sec. The speed of rotation expressed by the number of revolutions is certainly correct within an error of 10^{-6}, but there still remain possible errors due to momentarily non-uniform speed. The rotation was effected by an air blower turbine. Assuming one blower nozzle and the number of blades equal to the number of prism sides, this error may correspond to about 1000 km./sec. It will decrease, say, ten times by increasing the number of blades, another ten times by entirely symmetric construction, five times by using four or more nozzles, and twice by the consecutive use of six different prisms. Thus, we arrive at an error of only 1 km./sec.

Next, we have to consider the influence of atmospheric conditions. MICHELSON used average values of pressure and temperature, based on daily observations at Mt. Wilson Observatory. Errors in pressure are unlikely. In temperature $\pm 5°$ C is ample (gradient: $0.5°$ per 100 m. altitude) and yields ± 1 km./sec. velocity error. There also must have been an influence of the local atmospheric conditions. The expanding air from the blast must have cooled down the turbine close to the prism. We may assume the wind from the rotating prism to have caused a temperature difference of $\pm 0.5°$ C in the light path which is inclined by $6°$ or less with respect to the isotherms. In case of the 12-sided prism there may result a refractive angle of $1''$ corresponding to a velocity error of 2 km./sec. Since the wind participates in rotation, the error can not have been eliminated unless there was complete symmetry on both sides of the prism.

Finally, there is a mean error to be derived from the measurements. MICHELSON computed it from the values obtained for the six different prisms to be ± 4 km./sec. Summing up all sources of error, we thus find the following figures:

Measurements	± 4 km./sec.
Distance	± 3 km./sec.
Atmosphere	± 1 km./sec.
Uneven rotation	± 1 km./sec.
Local atmosphere	± 2 km./sec.

sq. root of sum of squares: ± 6 km./sec. (max. ± 11)

The result as stated by MICHELSON is

$$c = (299\,796 \pm 4) \text{ km./sec.}$$

Curiously enough, MICHELSON seems to have overlooked the correction for group velocity. Thus the result should be increased by 2 km./sec. (thin atmosphere).

δ) MICHELSON took part in planning the next determination with the one-mile evacuated pipe. The observations were carried out by PEASE and PEARSON [46] in 1931 and 1932, mainly after MICHELSON's death. The arrangement is shown in Fig. 5. It was less symmetric than that on Mt. Wilson. Only one prism of 32 sides was used. The outgoing light was reflected at the lower half of prism R and the incoming light at its upper half. By manifold reflection a total effective distance of 10 miles was reached.

If the almost 3000 velocity values measured were distributed entirely at random, there would have been a mean error of only ± 0.3 km./sec. Obviously, this does not account for systematic errors. PEASE and PEARSON reported systematic alterations with time over periods of some months like the following:

1931:	299770
1932:	780, jump to 776, jump to 760.
1932—33:	785, jump to 765, jump to 787.

These values appears as humps on the curve of error distribution in Fig. 6. From this curve we take the humps at 742, 766, 775, 784, 809 and assume the statistical weights proportional to the number of the respective observations. The mean error thus resulting is ± 6 km./sec. [121].

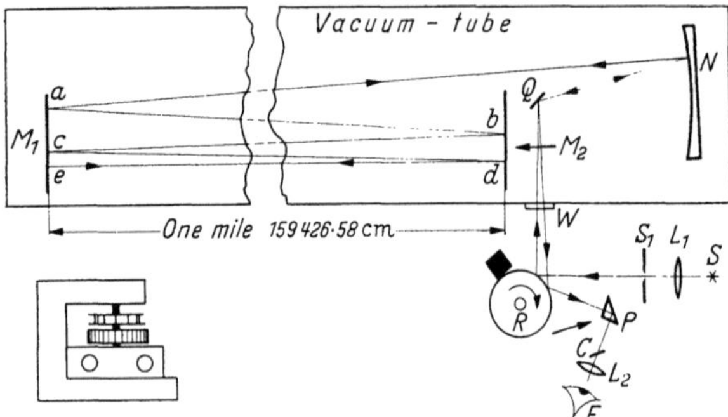

Fig. 5. The evacuated pipe experiment.

There remain those systematic errors which may be estimated. The distance has been measured four times during three years. The maximum difference in velocity corresponding to these different distance values is 2.5 km./sec. However, a mean of ± 2 km./sec. may be ample. The influence of non-uniform rotation should be about three times as large as in the Mt. Wilson experiment due to the smaller angle of rotation when using 32 sides as compared to 12. Assuming a somewhat improved mechanical performance, an error contribution of ± 2 km./sec. may be a fair guess.

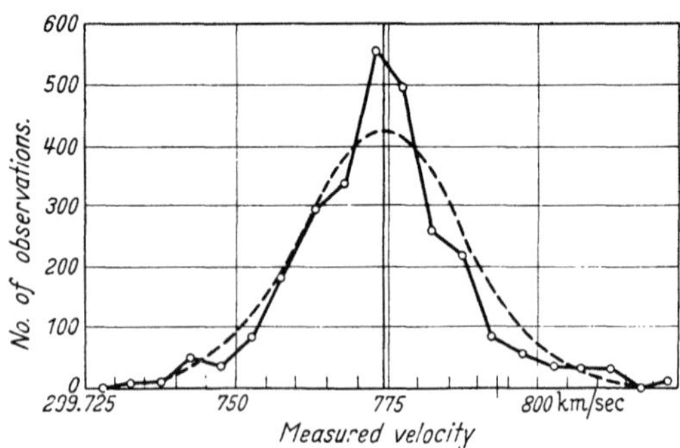

Fig. 6. Error distribution curve of the pipe experiment.

We, then, have to allow for local atmospheric conditions. Since the asymmetry was largeı than that on Mt. Wilson, there may have been an effective temperature difference of $\pm 1°$ C. It is, of course, difficult to estimate its influence on refraction. A possible influence is to be seen in Fig. 5. The arrow pointing on P has the direction of the cool wind if prism R rotates counterclockwise. This wind will be stopped at prism P in a different way if the sense of rotation will be reversed. There also will be caused different ways in which the air sweeps over the light paths by the asymmetric wind arresting yoke (the black square) shown in detail in the bottom left-hand corner of Fig. 5.

The influence of the cool air will be even enlarged if the air used in the turbine is allowed to escape anywhere in the vicinity of the prism such that it may to some extent take part in the rotation. No precautions against these atmospheric influences are mentioned by Pease and Pearson. In any case, there is nothing unreasonable in assuming an effective temperatur difference of 1° C and an inclination of the light beam of 6° or less with respect to the isotherms. This leads to a refractive angle of 2″ or ± 8 km./sec. velocity error contribution. Finally, some mechanical influence of the wind pressure on the prism holder P

is not quite out of the question. A possible torsion of only $0.3''$ would mean 2 km./sec. Summing·up, we find the following error contributions:

Measurements	± 6 km./sec.
Distance	± 2 km./sec.
Uneven rotation	± 2 km./sec.
Local atmosphere	± 8 km./sec.
Wind pressure	± 2 km./sec.
Total	± 11 km./sec. (max. ± 20)

Pease and Pearson give the result

$$c = (299\,774 \pm 11) \text{ km./sec.}$$

It is rather amazing that the pipe experiment gives no smaller error, especially as compared to the Mt. Wilson determination. There are, however, strong reasons to believe that the advantage by the independence of ordinary atmospheric conditions has been overestimated. In the pipe experiment the distance was not known with a convincingly greater accuracy. The main reason for the disappointing results surely was the unavoidable increase in complexity and asymmetry of the device.

9. The Kerr cell. At the same time when the last determinations had been accomplished with rotating mirrors, a new technique was introduced into Fizeau's method using the Kerr cell as an optical shutter of extremely small delay ($< 10^{-8}$ sec.). This improvement, in practice, increased the rapidity about a hundred times.

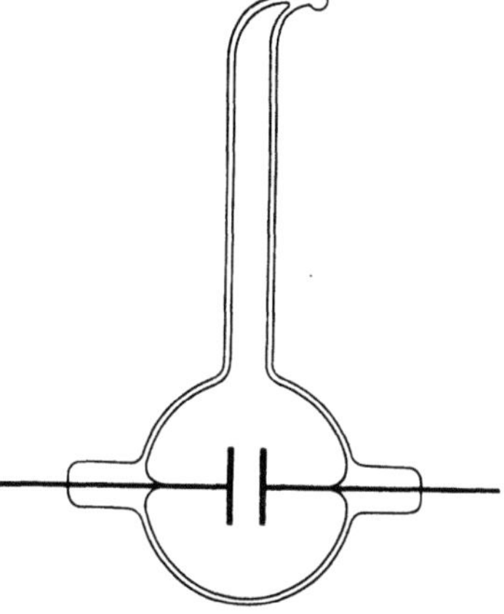

Fig. 7. Kerr cell.

The normal Kerr cell (Fig. 7) consists of a glass vessel with two circular electrodes in it and filled with a liquid of marked dipole structure, usually nitrobenzene. If an electric field is applied between the electrode plates, the molecules of the fluid will, to a certain extent, undergo an orientation in the field direction. The resulting polarization is proportional to the square of the voltage applied. It is limited by the thermal motion. Hence, the Kerr effect will increase at lower temperatures.

Our interest in the Kerr effect is due to the optical double refractive property which the fluid obtains. A light beam polarized by a Nicol prism in a plane of 45° inclination to the direction of the electric field (maximum optical effect) will be more or less elliptically polarized, according to the voltage applied, after passing the Kerr cell. Another Nicol prism, with the polarization plane at 90° with respect to the first one, may be used as analyzer afterwards. The complete arrangement (Fig. 10a) then transmits the light intensity

$$J = J_0 \sin^2 (k V^2) \tag{9.1}$$

with J_0 and k being constants, and V the voltage applied between the plates, eventually $V = E \cdot \sin \omega t$. The constant k depends upon the cell dimensions and the kind of fluid employed. Full intensity of the effect is reached about

10^{-8} sec. after application of the voltage. In Fig. 8 the dependence of the intensity on the voltage applied is shown. Usually, there is applied a steady bias voltage in the middle of the straight part $a-b$ of the curve where the intensity varies linearly with the alternating voltage superimposed.

In the following, more detailed descriptions are given of velocity determinations performed using the Kerr cell, and of the results obtained.

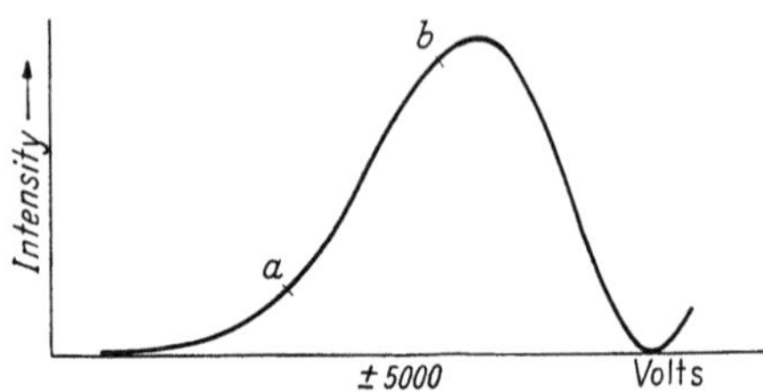
Fig. 8. Intensity of light *vs* applied voltage of Kerr cell.

α) *The* Karolus *and* Mittelstaedt *determination.* After the early work of Gutton (cf. p. 22), Karolus [47] was the first to use the Kerr cell in determinations of the velocity of light in air. In 1925 he started experimental work in that field, finished by his coworker Mittelstaedt [48] in 1929. The device of the latter is shown in Fig. 9. Light from the source LS passes the lens L_1, the Nicol prism N_1, the Kerr cell K_1, the lens L_2, the plane mirror M, the lens L_3, another Kerr cell K_2, another Nicol prism N_2, and finally falls upon the mat glass plate S. The axis of Kerr cell K_2 is oriented perpendicularly to that of cell K_1 so that the cell is seen from top in Fig. 9. Both cells are as similar

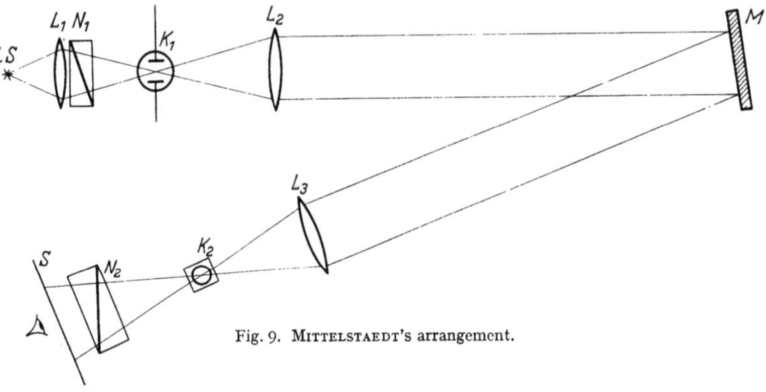
Fig. 9. Mittelstaedt's arrangement.

as possible and connected symmetrically in parallel with a valve transmitter of high frequency voltage. The polarization planes of the two Nicol prisms are at 90° to each other, and at 45° to the direction of the electric field in the Kerr cells. Working with equal cells, and there being the same phase relation between voltage and light in both cells, the ellipticity of light originated in K_1 will be cancelled in K_2 yielding a minimum light intensity on the plate S. This will occur to every light path $K_1 M K_2$ which is an even multiple of $\lambda_a = c_a/n$, c_a being the velocity of light in air and n the frequency of the transmitter. This frequency could be varied from 3 to 7 Mc in order to adjust for minimum light intensity. 6000 volts direct voltage and 2000 volts high frequency voltage were applied to the cell electrodes which were 2.5 mm. apart. The distance to the mirror was about 41.40 m. The real distance, however, was increased to an order of 250 to 330 m. by multiple reflection. This distance could be determined with a mean error of about ± 1 cm.

The frequency was measured in the following way: The high harmonic of an electrically driven tuning-fork was frequency multiplied and compared to the transmitter by means of the beat frequency. The uncertainty was of the order of ± 200 cycles/sec. at the frequency used.

After the KERR cells had been adjusted to cancel the polarization of each other (with only the direct voltage on) the frequency was changed until light minimum was obtained. The minimum was rather flat, probably on account of the inhomogeneity of the electric field. This fact depends on static charges of the nitrobenzene, which accordingly rushes between the electrodes. The light path was reckoned from and to the centres of the KERR cells. The influence on the distance of glass, nitrobenzene and air was calculated according to the ordinary refractive index. The corrections for the media mentioned was respectively 21, 12, and 68 mm. MITTELSTAEDT states the mean of 800 readings to be

$$c = 299\,778 \pm 20 \text{ km./sec.}$$

This value may be increased by 8 km./sec. on account of the group velocity.

MITTELSTAEDT points out that a higher degree of accuracy would be obtainable in the final result by using a longer light path with the consequence of smaller relative errors. The frequency also could be determined more exactly. Thus the method surely allowed for improving the accuracy by a factor 10.

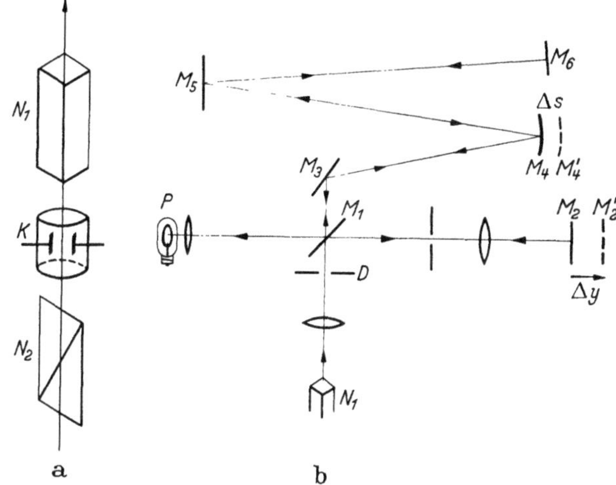

Fig. 10a and b. ANDERSON's method—1940. a KERR cell. b Light paths.

β) ANDERSON's *determination*. Another arrangement was employed by WILMER ANDERSON [49] in 1936. A further development led to his device [50] of 1940 as shown in Fig. 10.

On the left there is the KERR cell with the NICOL's prisms N_1 and N_2. The dependence of the intensity upon the applied voltage was shown in Fig. 8. ANDERSON used a direct bias voltage of 6000 volts and a superimposed high frequency voltage of 2000 volts and 19 Mc/sec. Fig. 10 shows how the light from a high pressure mercury arc of 1000 watts is projected as an image on to the aperture D furnished with a hair cross. M_1 is a semi-transparent mirror, which divides the light into two beams towards the mirrors M_2 and M_4. On its return, the light again passes M_1 and proceeds to the cathode of the multiplying photo-tube P. M_2 can be adjusted until the light from M_2 and M_4 yields a minimum of high frequency variation on the cathode, that is to say, until they have a difference of phase π. Then the mirror M_4 is replaced by M_4'. The light now passes over the additional long path via M_5, M_6, M_5 before it reaches the cathode. As in the former case, the minimum is adjusted by M_2, now for example in the position M_2'. In this way a difference Δ in the distance is obtained, consisting of the path $M_4 - M_5 - M_6 - M_5 - M_4$ plus some small distances Δy and Δs, depending on the exact positions of M_2' and M_4'. The velocity in air will be $c_a = n \cdot \dfrac{\Delta + \Delta y + \Delta s}{N}$, where n is the frequency of the transmitter and N is an integer.

The focal distance of M_4 is chosen to give an image of the hole D in front of P. [121]. This image is reversed. M_4' on the other hand projects an image on the mirror M_6. After the second reflection by M_4' we get a non-reversed image in front of P. An hair cross in D should make it sure that all the images were obtained in the same place. By a small lens in front of the tube the actual image is projected as a spot on the cathode. The diameter of the spot is less than one mm.

Anderson made almost 3000 observations in 22 days in the course of a year. If we consider the daily results and assume random errors, we get $(299\,776 \pm 4)$ km./sec. A closer examination of Anderson's records shows the possibility of systematic errors. However, as there are no distinct clusters of points in the error curve as in the case of the evacuated pipe experiment, we may believe in the ± 4 km./sec. as a mean error from the measurements.

Among the errors that can be estimated, the one due to the frequency can certainly be ignored. The uncertainty of the distance was 2 mm. As the total distance was 172 m. this means ± 3 km./sec. in the velocity value. Anderson considered the transit time of the photo-electrons to be the most significant source of error. In particular, he alluded to that time as depending on the position of the emission point on the cathode. He gave, however, no figures showing

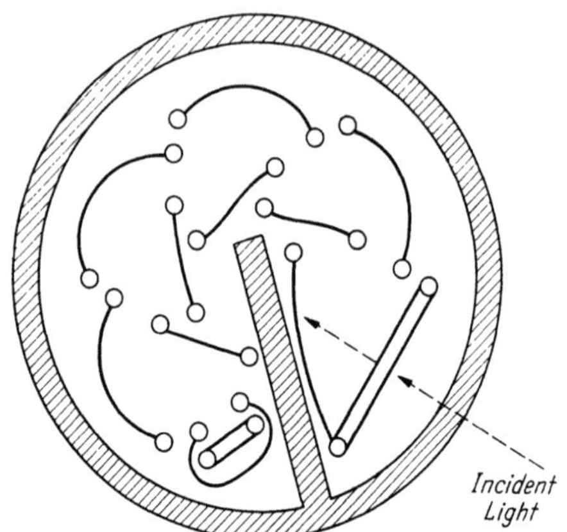

Fig. 12. Arrangement of electrodes in an IP 21 multiplier tube.

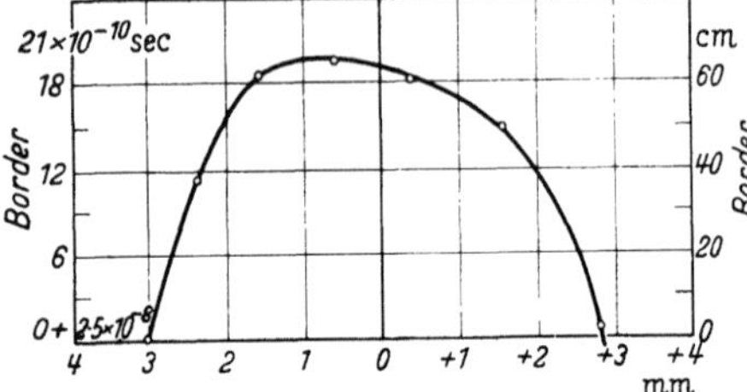

Fig. 11. Dependence of transit time of photo-electrons on the position of the emission point. The cathode borders are at ± 5 mm., not ± 4 as in the figure.

this dependence. Concerning the type of photo-tube, it appears to have been that of the *RCA* series, having nine dynodes. Bergstrand [121] measured the difference in transit time from different positions of the emission point for a specimen of the tube *IP* 21. The voltage across the tube was 800 volts. The result is shown in Fig. 11, where the transit time in seconds (left hand scale), or the corresponding light path in cm. (right hand scale) is drawn vs the horizontal distance in mm. from the centre of the cathode. Probably this state of affairs was unknown to Anderson.

The most natural way of admitting the light to the photomultiplier tube is through the centre of the front aperture, as shown in Fig. 12, hitting the cathode itself one third of the width from the inner edge as indicated by the arrow. Here, a displacement of the emission point of 0.1 mm. corresponds to a change in light path of 3 cm. or, if we consider the inclination of the cathode, 4 cm.

This fact combined with the opposite directions of the images by M_4 and M_4' will arise systematic errors. They will emanate from the hair cross not being exactly in centre of D (0.1 mm.), the light intensity increasing towards the anode in the Kerr cell ($90 \rightarrow 100\%$), the image being chromatic on account of the thickness (5 mm.) of the semitransparent mirror, the image on the cathode not being exactly a point (0.8 mm.). Except for point 4 (smaller than one mm.) no information on these questions is given in Anderson's report. Due to the reversed images the estimated discrepancies are doubled and we get the following possible errors:

Measurements	$\pm$ 4 km./sec.
Distance	$\pm$ 3 km./sec.
Displacement of hair cross	$\pm$ 14 km./sec.
Uneven intensity	$\pm$ 16 km./sec.
Chromatism	$\pm$ 2 km./sec.
Totally	$\pm$ 22 km./sec. (max. $\pm$ 39)

ANDERSON gives the result as

$$c = (299\,776 \pm 14) \text{ km./sec.}$$

γ) HÜTTEL's *determination*. HÜTTEL [*51*], in 1940, used the KERR cell in a new arrangement shown in Fig. 13. On its path from the source L to the photo-tube F the light passes the lenses L_1, L_2, L_3, the NICOL's prisms N_1, N_2,

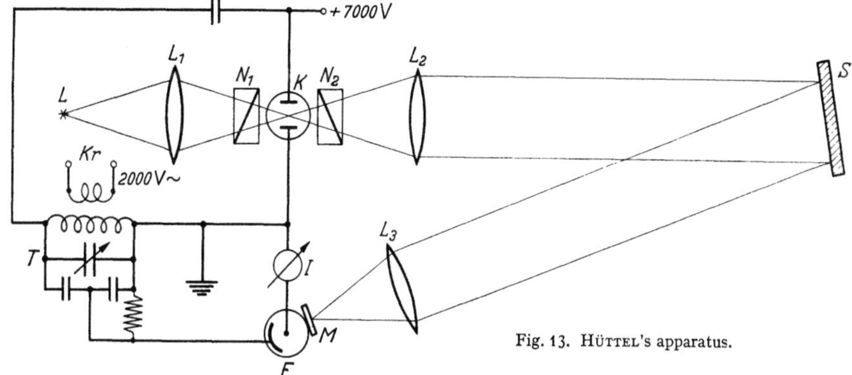

Fig. 13. HÜTTEL's apparatus.

the KERR cell K, the plane mirror S and the mat glass plate M. Besides a direct voltage of 7000 volts the cell K is fed with 2000 volts by the high frequency crystal controlled radio transmitter Kr. The photo-tube F is fed with the same voltage though it is somewhat reduced by a capacitive potentiometer. The resulting photo-current can be read off on the direct current galvanometer I. T is a tuned circuit to pick up the voltage from Kr. The arrangement is further schematized in Fig. 14a, where notations are the same as in Fig. 13.

By successively increasing the distance to the plane mirror S, the active half-cycles of the photo-tube F will be differently timed to the incoming flashes. Then the photo-current I will show a dependence on the distance as is shown in Fig. 14b. HÜTTEL measured the current at a point where the curve had its steepest slope. The current was equal to a certain percentage of that obtained when only the direct voltage was applied. Next the mirror was moved through an even number of cycles to a point at which the percentage was equally large. Then the velocity in air is $c_a = \dfrac{2D}{N} \cdot n$, where D is the displacement of the mirror and n the frequency of Kr. N is an integer. The sensitivity of the method lies in the use of a steep part of the current curve. It is, however, difficult to keep constant the light intensity, the voltages, the temperatures, the photo-sensitivity etc. during the two consecutive readings. As the total light path was only 40 m., systematic errors due to distance would be comparatively large. Therefore the error limits claimed seem to be a little too narrow. HÜTTEL gives the result:

$$c = 299\,771 \pm 10 \text{ km./sec.}$$

δ) Bergstrand's *determination*. This method [52], [53] to some extent is fairly similar to that of Hüttel. From Fig. 14b we see that, if the distance D is zero, we get a maximum of current when high intensity and high tube sensitivity occur simultaneously. Now, supposing we had another apparatus similar to the one mentioned, but with the photo-tube having high sensitivity simultaneously with the minimum of emitted light intensity, we would get a current as in Fig. 14c. Transmitting the currents as in (b) and (c) in opposite directions through a measuring instrument, we can read off the difference as in Fig. 14d. We obtain sharply marked positions of the mirror, where the current is zero and the curve is steep. However, it is difficult to get the two pieces of apparatus entirely similar. On this account, only one is used, working alternately as each of the two. The alternation is simply carried out by replacing the ordinary steady bias voltage on the Kerr cell by a 50-cycle square-

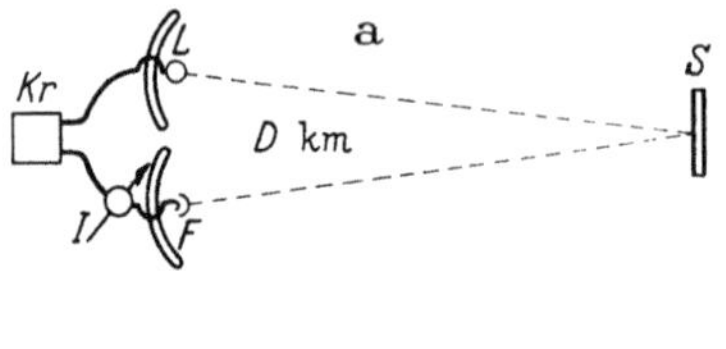

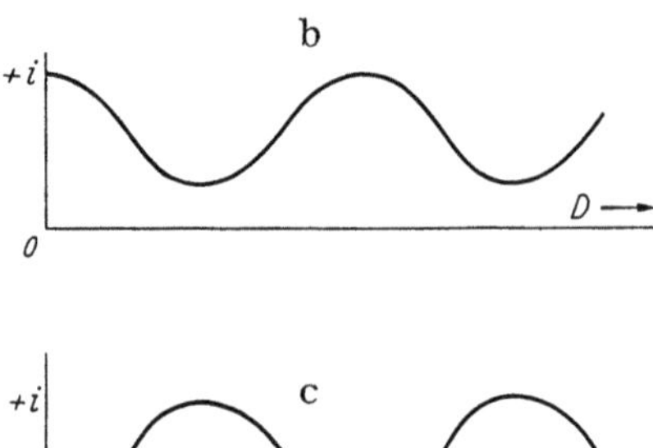

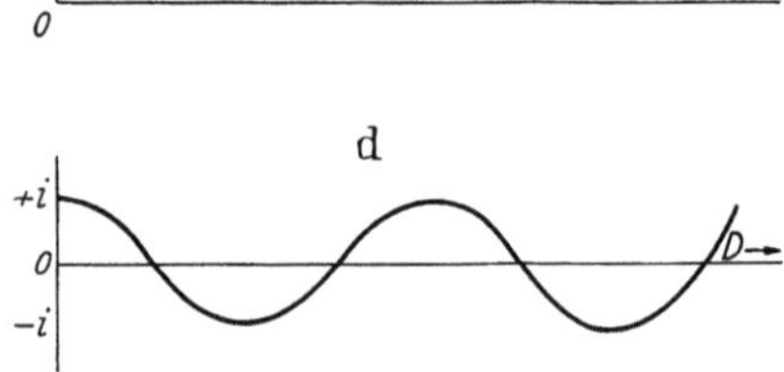

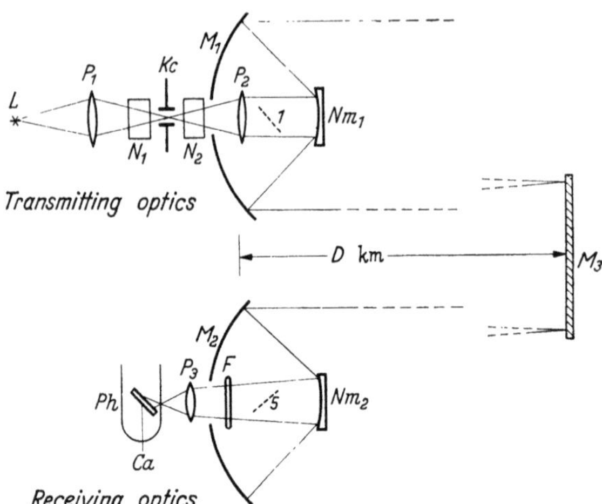

Fig. 14 a—d. Principles employed by Hüttel and Bergstrand.

Fig. 15. Light paths of Bergstrand's device.

shaped one. On account of the 50-cycle voltage the high frequency light variations *change their phase* 100 times a second. Simultaneously the indicating current is applied to the instrument in an alternating manner.

Fig. 15 and 16 will facilitate a description of the apparatur. In Fig. 15 L is the source of light. The intensity is varied by the Kerr cell Kc. The mirror M_1 focusses the light into a beam directed towards the distant plane mirror M_3. By the receiving optics the returning light is thrown on to the cathode of the multiplying photo-tube Ph.

In Fig. 16 the electric circuits are shown, and the light intensity as dependent on the voltage on the Kerr cell again may be taken from Fig. 8. The cell is fed with 2000 volts 8 Mc/sec. from the crystal-controlled transmitter Cr—HA and with 5000 volts 50 c/sec. of rectangular wave shape from the transformer Tr and the neon-tubes G. The anode of the photo-tube Ph is supplied by the 8 Mc from Cr—HA also. By the shifter Sw I the phase of this voltage can be manually reversed. No 50-cycle voltage is supplied to Ph. Due to the photo-current there will arise 50-cycle voltage pulses in the anode resistance 70.

The individual magnitude of every pulse depends on the actual phase difference between the high frequency feeding of the anode and the entering light variations. That phase difference, in turn, depends on the distance to the mirror M_3. The instrument I receives the 50-cycle pulses alternately from the valves 1 and 2, since these valves are alternately blocked every second 50-cycle half-cycle by the voltage from Tr on the grids a and b. By the shifter Sw II the phase of this 50-cycle control voltage can be manually reversed.

The slowly-reacting direct current instrument I shows the difference $i = i_1 - i_2$ of the mean currents i_1 and i_2, delivered respectively by the valves 1 and 2. The current i equals zero every 9th metre (zero-points) of increased distance to the mirror M_3. The use of the shifters Sw I and Sw II prevents electrical asymmetry.

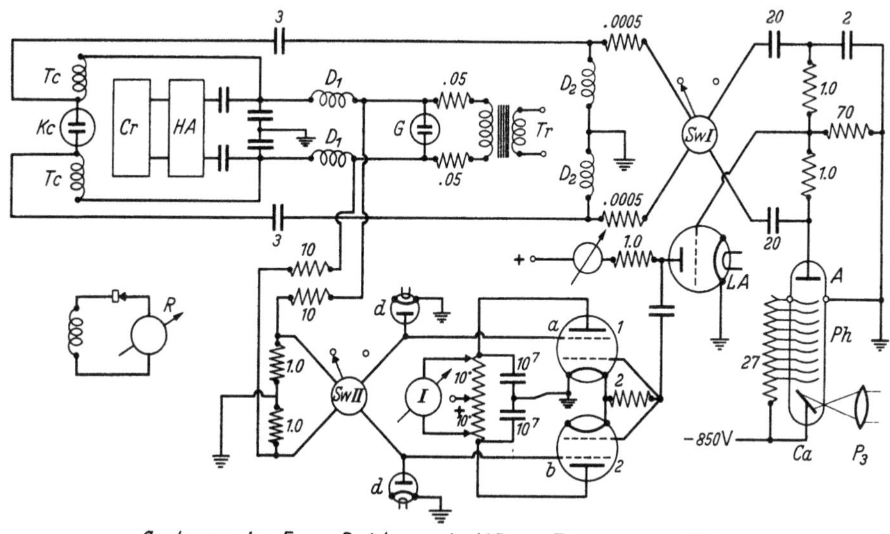

Condensers in pF   Resistances in MΩ   Tube-screens omitted

Fig. 16. Electric circuits of Bergstrand's apparatus.
(The two resistances close to instrument I should read .01 instead of 10·)

Thus, if the distance D to the mirror M_3 is continually increased, the instrument I shows the current as in Fig. 14d. We obtain a series of zero-current points corresponding to certain fixed distances of the mirror M_3. The distances D_N between two zero-points is

$$D_N = \frac{1}{4} N \cdot \frac{c_a}{n},\qquad (9.2)$$

where N is an integer, c_a the velocity in air, and n the frequency of the light variation. The factor $\frac{1}{4}$ comes from the light traveling to and fro and there being two zero-points in each cycle of the curve (d). Thus for air

$$c_a = \frac{4 D_N \cdot n}{N}.\qquad (9.3)$$

The exact setting of a zero-point was effected by a slight and known variation of the frequency. During the measurements the frequency n was checked by the Telegraph Service. The distance D_N was a good Geodetic Survey base line (7 km.). For the transformation of c_a into c for vacuum, eq. (3.3) was used. Temperature and pressure were measured every hour. The temperature gradient above the ground was taken into consideration according to Johnson and Heywood [13]. To obtain the effective colour, the light was dispersed by a prism and the photo-current measured for different wave-lengths. To diminish the influence of a change in colour, a filter was used. The effects of changes in light intensity, supply voltages and temperature were very small or could be estimated. The results are recorded in the following table:

Year	Distance in m.	Weight	Velocity in km./sec.	Mean of year
1947	$11\,025 \pm 0.08$	0.1	$299\,793.9 \pm 2.7$	3.9
1948	$4\,208 \pm 0.01$	0.2	789.3 ± 2.5	
1948	$9\,064 \pm 0.01$	0.6	794.1 ± 1.3	3.1
1949	$1\,762 \pm 0.01$	0.4	792.3 ± 1.5	
1949	$5\,144 \pm 0.01$	1.5	794.0 ± 0.8	
1949 May	$6\,906 \pm 0.0035$	8.0	793.02 ± 0.19	3.1
1949 September	$6\,906 \pm 0.0035$	8.0	793.08 ± 0.20	
1950	$5\,413 \pm 0.0012$	5.0	793.17 ± 0.42	3.2

The value of 1947 was computed in 1950 by using a post-calculated value of the frequency. The limits are derived from the deviations from the mean value. The remaining errors are added in the following table:

Random errors . .	± 150 m./sec.
Refractive index .	$\pm\ 60$ m./sec.
Colour	$\pm\ 60$ m./sec.
Atmosphere . . .	$\pm\ 60$ m./sec.
Base line	± 210 m./sec.
Frequency	$\pm\ 60$ m./sec.
Total	± 285 m./sec. (max. ± 600)

The refractive index for group velocity is slightly less reliable than the ordinary one, which is believed to be better than 10^{-7}. In the base line error, the uncertainty in all steps from the wave length of the red Cd line as a standard of length is accumulated.

Considering the error table we get the result:

$$c = (299\,793.1 \pm 0.3)\ \text{km./sec.}$$

In the form of "geodimeter" the apparatus is employed in several countries for the inverse purpose to determine geodetical lengths up to 40 km. with an accuracy of 2×10^{-6}, using the above value of c. It seems to fit within the error limits of all ordinary triangulation nets.

10. Vibrating quartz. The possibility to use a vibrating piezo-quartz for the determination of the light velocity was first suggested by Kerr and Grant in 1927 [54]. The crystal, activated by an electric field, is placed between two crossed Nicol's prisms. If the beam of light is parallel with the optical axis, the intensity depends on the field voltage in a way [57], which reminds of that for the Kerr cell:

$$J = J_0 \cdot \sin^2 (k \cdot V) \tag{10.1}$$

only instead of V^2 in eq. (9.1) having V. If V is an alternating voltage in resonance with the crystal's natural frequency the constant k increases enormously.

In another method, proposed by Ludwig Bergmann [55] in 1937, the crystal generates a standing supersonic wave in a fluid or a solid or, as a high harmonic, in the crystal itself which forms an intermittent space grating which, by its diffraction, is the essential part of the arrangement.

α) Determination by McKinley. In 1938 (publ. 1950) McKinley [57] used the vibrating quartz in the above mentioned Kerr application. The further arrangement was principally similar to that of Anderson, with the quartz in place of the Kerr cell. The frequency was 8 Mc/sec. and the distance 9 m. The result was stated as

$$c = 299\,780 \pm 70 \text{ km./sec.}$$

The value may be increased by 3 km./sec. for the group correction. The main error is reported to lie in the determination of the distance. There also was some uncertainty as to the correction caused by a lens.

β) Determination by Houstoun. In Fig. 17 the Bergmann application of the vibrating quartz, as used by Houstoun [56] is shown. The crystal is vibrating with a high harmonic fed in by the radio transmitter RT. Light from the slit S is parallelized by the lens L and passes through the semi-transparent mirror

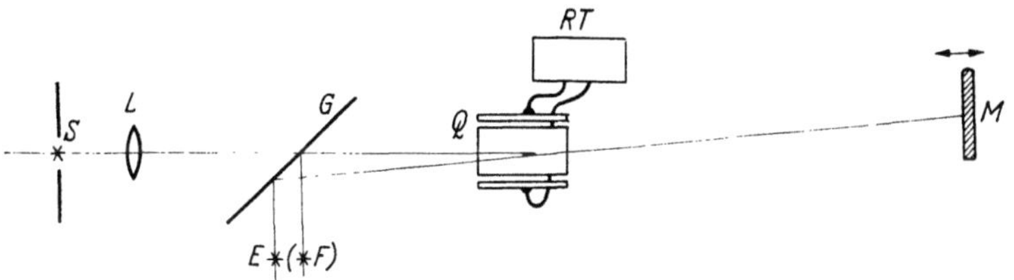

Fig. 17. Crystal light modulator used by Houstoun.

G and the crystal Q. Here the beam of the first order spectrum is intermittently diffracted upwards. After reflection by the mirror M the light travels back on its own path.

If the returning light and the crystal have opposite phase in their vibrating conditions the light goes straight through the crystal and is observed at E. If the light and Q are in phase, the light is diffracted once more and travels mainly to F. In E there will be a minimum of light intensity.

To allow for adjustment to such a minimum, the mirror M is movable. In the same way as all the measurements according to Fizeau's principle do this method yields for air:

$$c_a = \frac{2D}{N} \cdot n, \qquad (10.2)$$

where D is the distance $Q - M$, N is an integer, and n the effective frequency of Q. n is of the order 200 Mc/sec. and $Q - M$ is 40 m.

For vacuum, Houstoun states the mean of 400 readings to be

$$c = (299\,775 \pm 9) \text{ km./sec.}$$

The value for air is noted as 299 698. Therefore the applied air correction seems by 6 to 9 km. too small. Houstoun measured the length from the geometric centre of the crystal. For high harmonic crystals the effective grating centre, however, may well lie 2 mm. outside the geometric centre. Thus a value of ± 18 km./sec. may be more correct for the error limits.

II. Indirect methods.

a) Astronomical methods.

11. Observation of the aberration. *α) Determination by* Bradley (1692 to 1762). In 1725, Bradley [58] tried to detect the annual parallax by observations of the star γ Draconis. He discovered an annual motion of the star, which, however, qualitatively did not agree with that of a parallax. Three years later, Bradley concluded that this very motion occurs because the orbital velocity of the Earth is not negligible in comparison with the velocity of light and that it therefore must be common to all stars. It does, however, not occur in consequence of an "absolute" motion of the Earth, but is only due to the relative motion with its changing direction during the annual revolution [123], [124], [125].

In Fig. 18, S is the star and E the Earth. $x'y'$ is a coordinate system at rest relative to the Earth. The x'-axis lies in the Earth's orbit and the velocity in the orbit as compared to a heliocentric system xy may be v. xy is at rest relative to the Sun and the star. The direction to the star is observed to form the angle α' with the x'-axis. The angle which would be obtained were the Earth at rest in the heliocentric system is denoted by α. Let us put $\alpha'=\alpha-\Delta_c$. The heliocentric velocity of the light is c. Then, using the laws of classical mechanics, its x'-component is

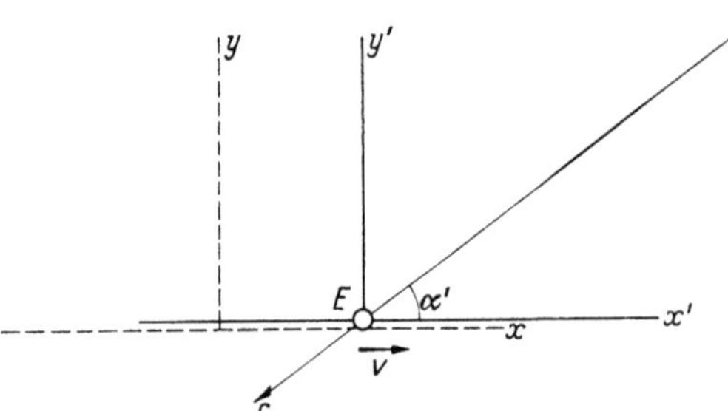

Fig. 18. Reference frames for aberration.

$$u'_x = -(c \cdot \cos \alpha + v), \tag{11.1}$$

and its y'-component

$$u'_y = -c \cdot \sin \alpha. \tag{11.2}$$

Putting $v/c=\beta$, we get

$$\tan \alpha' = \tan(\alpha - \Delta_c) = \frac{\sin \alpha}{\cos \alpha + \beta}. \tag{11.3}$$

By series expansion, considering Δ_c^3 and β^3 as small quantities this yields

$$\Delta_c = \beta \cdot \sin \alpha - \frac{\beta^2}{2} \cdot \sin 2\alpha. \tag{11.4}$$

β is called the aberration constant. A modern value is $\beta = 20''.479 \pm 0''.008$.

Let us now proceed beyond Bradley's original means by using the relativistic Lorentz-transformation connecting the two reference frames x, y and x', y':

$$\left. \begin{aligned} x' &= \frac{x - vt}{\sqrt{1 - \beta^2}}, \\ x' &= y, \\ t' &= \frac{t - \beta \cdot \dfrac{x}{c}}{\sqrt{1 - \beta^2}}. \end{aligned} \right\} \tag{11.5}$$

Now, for an observer on the Earth, $u'_x = \dfrac{dx'}{dt'}$ and $u'_y = \dfrac{dy'}{dt'}$. Using eq. (11.5) we express these components by x, y and use $\dfrac{dx}{dt} = -c \cdot \cos \alpha$ and $\dfrac{dy}{dt} = -c \cdot \sin \alpha$. Thus, we arrive at the relativistically corrected value Δ_r from the equation

$$\frac{u'_y}{u'_x} = \tan \alpha' = \tan(\alpha - \Delta_r) = \frac{\sin \alpha \cdot \sqrt{1 - \beta^2}}{\cos \alpha + \beta} \tag{11.6}$$

which by series expansion renders

$$\varDelta_r = \beta \cdot \sin \alpha - \frac{\beta^2}{4} \cdot \sin 2\alpha, \tag{11.7}$$

where α is the direction to the star.

Using RABE's radius of the Earth's orbit $= 14953 \times 10^4$ km., for the determination of v we find

$$c = (299857 \pm 120) \text{ km./sec.}$$

The error limits are far too wide to allow for a detection of the β^2 term in eq. (11.7).

$\beta)$ *Determination by* AIRY. In 1872 AIRY [59] observed the aberration using a telescope, the tube of which was filled with water. He expected an increased value on account of the lower speed of light in water. If we consider the "dragging" effect of the water according to FIZEAU's formula [eq. (20.3)] and apply v as the speed of the Earth in its orbit, then we obtain an aberration constant identical with that without water. Accordingly, AIRY observed no alteration of the aberration constant.

As in the case of moving matter (see Sect. 20) we can apply the relativistic law for the addition of velocities [124], [125] and get the same result. This seems obvious since FIZEAU's formula is in agreement with relativity. The unchanged aberration is the most natural result also [61]. The observed direction to a star cannot change if the light travels a small part in water which constitutes a part of the tube's optical outfit (water all the way to the star should alter the conditions). Yet, the explanation of this almost selfevident state, requires the "dragging" of the light. Thus, the latter phenomenon appears to be an illusion, merely caused by the relative motion.

b) Electromagnetic methods.

12. Ratio of electro-magnetic units u_m/u_s. The units of the electric and magnetic systems of units are both of them based on COULOMB's law, the connecting link lying in the magnetic effect of the displacement current, which first was expressed in MAXWELL's equations (1864). The propagation speed of any change in the field strenghts occurs in these equations as a constant, which is identical with the ratio between the field units. Provided that the light is an electro-magnetic wave, its velocity should be identical with this ratio and thus equal to the speed of radio waves also.

Denoting electric charge $= Q$, electric field $= E$, capacity $= C$, resistance $= R$, current $= I$ the velocity is given by

$$c = \frac{Q_s}{Q_m} = \frac{E_m}{E_s} = \sqrt{\frac{C_s}{C_m}} = \sqrt{\frac{R_m}{R_s}} = \frac{I_s}{I_m} \tag{12.1}$$

if the same quantity is expressed in units of the electric (s) and magnetic (m) systems.

$\alpha)$ *The experiment of* KOHLRAUSCH *and* WEBER. The first scientists to use the ratio-method were KOHLRAUSCH and WEBER in 1856 [62]. They measured the charge Q_m in a condenser by the discharge through a ballistic galvanometer. To get Q_s the capacity C_s of the condenser was determined by comparison with the capacity of a big sphere with known radius. The potential V_s was determined by transfer of V_s to a small pellet, which thereupon was measured with a COULOMB torsion balance. The result was

$$c = \frac{Q_s}{Q_m} = 310800 \text{ km./sec.}$$

They used the units for current also. The result had its significance by inspiring Maxwell to carry through and publish his theory of the electro-magnetic field.

β) *The experiment of* Maxwell (1831 to 1879). In 1862 Maxwell [63] and Jenkin carried out a measurement using the ratio of capacities and in 1868 Maxwell alone, using the potential, obtained the value

$$c = 284\,300 \text{ km./sec.}$$

γ) *Experiments by* Rosa *and* Dorsey *a. o.* The use of capacities has appeared to be the decidedly most accurate method. It was employed by many investigators, among others J. J. Thomson [71], Abraham [73], Pérot and Fabry [69], Rosa and Dorsey [74]. The two last mentioned authors made the most accurate determination. The capacity C_s was obtained from the dimensions of the condenser and C_m by comparing the discharge current with a battery-current. Likewise, they used a bridge-circuit, where the charge of the condenser was zero-compensated versus a constant battery-current. The result obtained in 1906 has been corrected afterwards by Birge who introduced a more accurate value of the international ohm. Thus Birge reports their value as

$$c = (299\,784 \pm 30) \text{ km./sec.}$$

If the experiment should be repeated by aid of modern means the accuracy would be still better. There is, however, always the limit to which the dimensions of a condenser can be measured.

13. Standing waves in wires and air. High frequency waves, damped (Hertz) or continous, are emitted in each of two adjacent parallel and straight wires, a Lecher system, and run back in the other. The positions of the nodes of the standing wave generated are sharply defined and can be detected by small electric bulbs, Geissler-tubes or thermo-crosses. Small resistance-loops close around the wires in a bolometer-circuit may also have little disturbing influence. The frequency and the distance between two nodes gives the velocity v of propagation along the wires. To get c, the delay due to skin-effect must be determined. The ratio v/v_a, where $v_a =$ speed in air, can be experimentally determined as reported below (Gutton). In earlier measurements by means of Hertz' waves, the frequency often was determined by photographing the spark via a rotating mirror.

α) Blondlot used a Lecher-system in 1891 [77]. He merely short-circuited the wires and determined the resonance positions by observation of the reaction on the spark circuit. The frequency was obtained from the dimensions of the tuned circuit. The result was

$$295\,000 < c < 305\,000 \text{ km./sec.}$$

β) Trowbridge and Duane [78] used a bolometer as detector. Their value was

$$292\,000 < c < 303\,600 \text{ km./sec.}$$

γ) In 1911 Gutton [83] introduced the use of the Kerr effect (see p. 11) in the measurements. His arrangement is shown in Fig. 19.

The spark-circuit of an inductor I feeds the Tesla-transformer T. The established short wave travels out into the wire-system, one half along the branch $OABDC_1$ and the other half along $OFGHKC_2$. C_1 and C_2 are two plate condensers, the dielectricum of which is carbondisulfide. The fluid gets optically double-refractive properties when there is an electric field between the plates

(KERR effect). Entering plane polarized light will then be more or less elliptically polarized. The condenser plate systems are perpendicular to each other. The NICOL's prisms N_1 and N_2 are crossed and form an angle of 45° with the field in the condensers. S is a light source, L a tube for observation. If there is equal direct voltage on the condensers the field of sight in the tube is dark. At alternating voltage the field is still dark if the voltage on C_2 is in phase with the polarization condition of the entering light. This state occurs if the time of propagation along the wire path $OABDC_1$ + the light path $C_1 C_2$ is equal to the running time along $OFGHKC_2$. The loop GHK is adjustable in length to make the field of L dark. Then the difference of wire length between $OABDC_1$ and $OFGHKC_2$ is directly comparable with the optical distance $C_1 C_2$ and the ratio v/v_a as defined above is determined. By inserting different media between C_1 and C_2 the velocity in matter could be obtained.

δ) The ratio v/v_a can also be derived from a theoretical point-of-view as shown by MERCIER [81], [85] in 1921. His formula runs:

$$\frac{v}{v_a} = 1 - \frac{1}{8 \pi R \log D/R \cdot \sqrt{n \cdot \sigma \cdot \mu}}, \quad (13.1)$$

where v and v_a are the velocities along the wires and in air, R the radius of the wires, D the distance between their centres, n the frequency, σ the conductivity and μ the permeability of the wire substance. MERCIER experimentally showed the validity of his formula and also carried out a very careful determination of c. He had the opportunity of using continous waves from a valve-transmitter and the

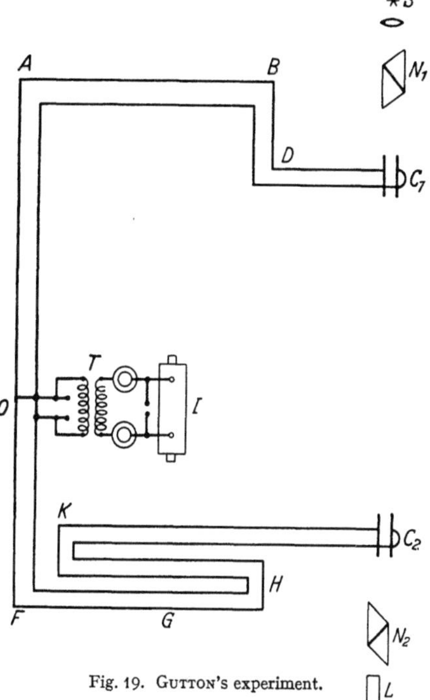

Fig. 19. GUTTON's experiment.

frequency of 46 to 66 Mc/sec. was controlled by the harmonics from a tuning fork. The parallel wires were 11 m. long and tightly stretched at a distance of 2 m. from the floor and neighbouring objects. MERCIER observed resonance by short-circuiting different lengths of the LECHER system. The lengths were measured by an invar tape and this accuracy was of the order of 0.1 mm.

The value of c as corrected by DORSEY was

$$c = (299\,782 \pm 30) \text{ km./sec.}$$

ε) MCLEAN [80] used standing HERTZ waves of $\lambda = 6$ m. originated in air after reflection by a plane metal mirror. The wave-length was obtained from the distance between the knots, detected by a resonator. The frequency was determined by aid of the spark image and a rotating mirror. The result is

$$c = 299\,100 \text{ km./sec.}$$

14. Cavity resonator. During the second World War there was a tremendous development of the shortest radio-wave technics, especially for waves below the meter. The conduction of these super-high frequency currents has been developed in a quite particular manner. We start from a long LECHER-system as the wires a and b in Fig. 20.

Straight across the line we put a bridge, made of two quarterwave long pieces, which are short-circuited in the far end (by the bar c). The entrance impedance is infinite and does not disturb the course in the line. Likewise, we can add another bridge below, d in Fig. 20. If we proceed with the addition of bridges until they are quite close to each other, we get a rectangular tube having the short side c and the long side $\lambda/2$. In reality, the long side must be $\geq \lambda/2$. The tube may be a cylinder as well. If the ends of the tube are short-circuited and the length of it (z-coordinate) is $\lambda_g/2$, where λ_g is the wave length inside the tube, we get a cavity with resonance to oscillation of the said wave length. The exact resonance frequency depends on the number of cycles of oscillation in the other coordinates of the cylindrical tube, that is in azimuthal and radial direction. Thus, to the resonance frequency f is added a suffix ϑ, r, z denoting the mode of the actual oscillation. Sarbacher and Edson have (1943) derived the expression for the frequency as depending on the dimensions of the cylinder:

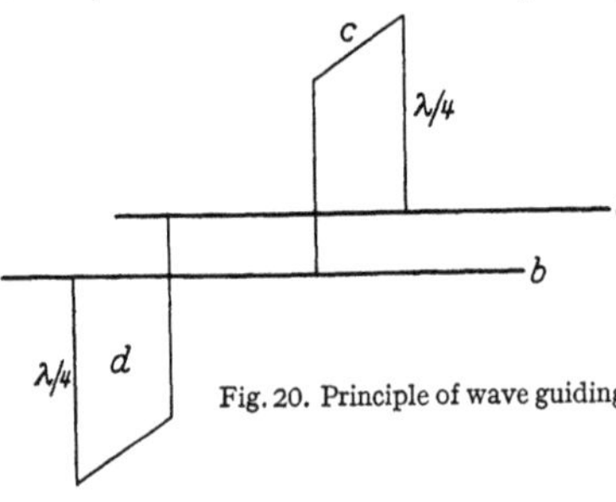

Fig. 20. Principle of wave guiding.

$$f_{\vartheta r z} = v \cdot \sqrt{\left(\frac{k}{\pi D}\right)^2 + \left(\frac{z}{2L}\right)^2}, \qquad (14.1)$$

where v is the wave velocity, D the diameter, L the length and k, in the case of E-modes (electric), is the r-th root of the Bessel eq. $J_\vartheta(x) = 0$ and, for H-modes, $J_\vartheta'(x) = 0$. ϑ, r, z are integers. The cylinder is supposed to have no losses. A finite Q-value lowers the frequency by $f/2Q$, where Q is the measured quality factor. Q is in the region of 15 000. to 50 000. depending on the surface. For an evacuated cavity, where $v = c$, we get

$$c = \frac{f_{\vartheta r z}\left(1 + \dfrac{1}{2Q}\right)}{\sqrt{\left(\dfrac{k}{\pi D}\right)^2 + \left(\dfrac{z}{2L}\right)^2}}. \qquad (14.2)$$

α) *Determination by* Essen *and* Gordon-Smith. In 1947 Essen and Gordon-Smith [86] used the cavity resonator for the determination of c. The method was quite new. The small dimensions of the cavity ($L = 8.5$ cm.) required a very careful performance and accurate determination of the dimensions. A mean was used and the non-uniformity of the dimensions was considered to cause the greatest uncertainty in the determination. With respect to the dimensions, the temperature, of course, also played an important role. The oscillating current penetrates the cylinder wall to a very small depth. It is that part of the wall surface containing residuals from the polishing or oxide by influence of the air. Such contaminations, therefore, increase the resistance, which affects the Q in eq. (14.2). Hence Q must be measured. Here the non-uniformity of the surface somewhat increases the uncertainty of the Q-value obtained. Adding the spread from different modes of vibration to the authors' list of maximum errors, we obtain:

Diff. mod. of vibration	± 1.5 km./sec.
Freq. and setting to resonance . . .	± 1.2 km./sec.
Temperature of resonator	± 0.6 km./sec.
Dimensions	± 0.9 km./sec.
Coupling holes and probes	± 1.8 km./sec.
Non-uniformity of the resonator . .	± 3.0 km./sec.
Q-value.	± 1.5 km./sec.
Total mean error:	± 4.5 km./sec.

The authors use a maximum error of ± 9 km./sec. The result

$$c = (299\,792 \pm 4.5)\ \text{km./sec.}$$

was the most accurate value determined up-to date. It did not agree with the "official" $(299\,776 \pm 4)$ km./sec. This fact required a check determination, if possible of even higher precision.

β) ESSEN's *2nd determination*. In 1949 ESSEN started with an improved apparatus [87], [88], which admitted a variation of the length (L) of the cavity by insertion of a movable plunger. Fig. 21 shows the arrangement schematically. Topmost we see the very cavity C and the plunger P, resting on the exchangeable slip gauge G, the accurate length of which determines the position of the plunge in the cavity. $CV\,129$ is a klystron oscillator, the frequency of which can be varied to fit the actual cavity resonance. Via the coaxial line K the H-field of the cavity is fed by the small loop l. The resonance frequency is detected by aid of the receiver to the right in Fig. 21. The receiver equipment allows for a comparison to a frequency standard. ESSEN writes:

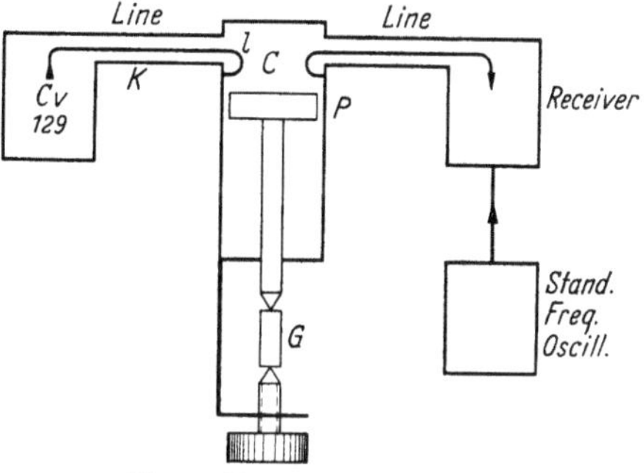

Fig. 21. ESSEN's cavity resonator.

"The principal uncertainties are the value of the diameter and the effect of small imperfections of the cylindrical wall; but by using two different modes of resonance it is possible to measure the effective diameter with a precision of better than 1 part in 10^6, and the length of a set of gauges, which can be checked by interferometer measurements with the same precision, and since, moreover, λ_g is determined by differences, certain small end effects are eliminated and the effects of mechanical and electrical imperfections of the walls and ends of the cylinder are greatly reduced."

In contrast to the earlier resonator, the new one was silverplated inside. The thickness of the Ag layer was $5\,\mu$ or ten times the skin depth. The Q-value of the cavity was 75% of that calculated, surely depending on a slight oxidation by influence of the air on the surface. Q was in the region of 50000.

The gauges G, which determined the different positions of the plunger and the change of L for the cavity, were measured interferometrically after each resonance determination. The angle between the moving direction of the plunger and the length of the gauge in its place, was determined also.

A particular investigation, using the stability of an interference pattern, secured that the lateral displacement of the plunger was of no importance The diameter of the cavity was measured at 78 different positions. Thus the effective diameter was secured to within 10^{-6}.

During the measurement, the whole apparatus was enclosed in an evacuated vessel. The frequencies used were 9000, 9500 and 5960 Mc/sec. The result was

$$c = (299\,792.5 \pm 1)\ \text{km./sec.}$$

γ) *Determination by* HANSEN *and* BOL. HANSEN projected an apparatus similar to that of ESSEN and GORDON-SMITH. After HANSEN's untimely death, BOL [89] concluded the work and his value of 1950 was

$$c = (299\,789.3 \pm 0.4)\ \text{km./sec.}$$

The main difference from ESSEN's method was that the divergence from ideal conditions was calculated rather than eliminated by experimental technique.

With respect to the amount of the skin effect Dayhoff[1] calculates a skin correction of $+8$ km./sec. instead of Hansen and Bol's $+3$ km./sec. In this case the result should be

$$c = 299\,794 \text{ km./sec.}$$

15. Micro-wave interferometry. In 1950 Froome [*90*] used radar waves in the device shown in Fig. 22.

A stabilized klystron emits a wave of $\lambda = 1.25$ cm. through the wave guide towards the hybrid junction T, where it is divided equaly to the left and to the right. After reflection from the plunger P, respectively from the free reflector R the two waves return to T and are again divided into equal parts, now upwards and downwards. The down parts interfere in the guide to the receiver-detector D.

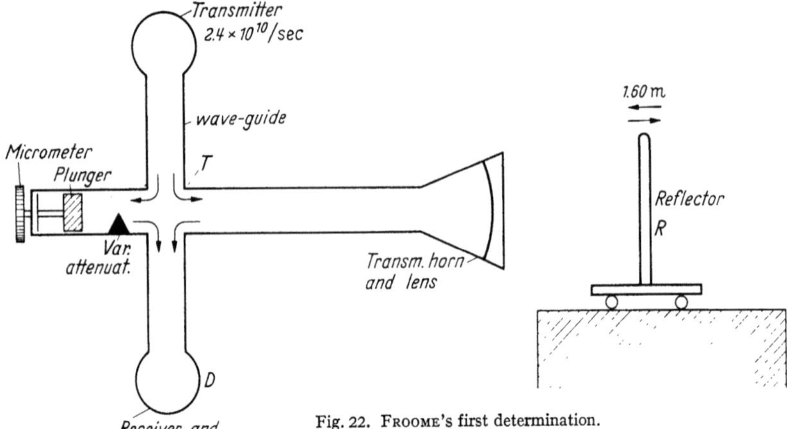

Fig. 22. Froome's first determination.

D shows a minimum of current, when the phase difference is brought to the value π by moving the reflector R. Then, R is moved 1.6 m. through 250 minima, that is 125.5 λ. The value of the displacement thus gives the wave length in air. These 1.6 m. displacements are achieved at different and relatively large distances from the transmitting horn. In this way errors can be eliminated, which depend on the influence of disturbing reflections from the room and of the change of the wave front with the distance from the horn. The frequency being known also, we obtain the velocity for air. After reduction to vacuum the result was:

$$c = (299\,792.6 \pm 0.7) \text{ km./sec.}$$

In the following year, Froome [*91*] was ready with a new improved apparatus as shown in Fig. 23. This new device has been described by Froome as follows:

"The source of micro-waves is a Pound stabilized reflex klystron oscillator of frequency 24 005 Mc/sec., corresponding to a wave-length of about 1.25 cm. The accuracy of frequency measurement is about 1 part in 10^8, and for the method used the reader is referred to the earlier paper" [*90*].

"Energy from the oscillator passes to a hybrid junction ('magic T') which serves as beam divider, from which it passes through to long wave-guide arms to the pair of transmitting horns. The matching stub and the phase-shifter (1) to the left of the beam divider, together with the constant phase auxiliary interferometer (c.p.i.) constitute a device for altering the amplitude of the energy transmitted down this arm without producing a phase displacement. The phase-shifter (2) to the right of the beam divider, together with the variable attenuator, is required in order to adjust and balance the position of the first interference minimum."

---

[1] In a survey for NBS, 1952.

"The movable part of the interferometer consists of a pair of receiving horns mounted on a carriage largely constructed of silica tubes (for thermal stability to minimize the effect of room temperature gradients as the carriage is displaced) and arranged between the transmitting horns. The two received signals are mixed to produce interference and detected by means of a simple superheterodyne arrangement employing the eighth harmonic of a 3000 Mc/sec. klystron local oscillator. This oscillator is unstabilized, and the intermediate frequency amplifier (45 Mc/sec.) consequently has to have a band-width of 2 Mc/sec. The output from this amplifier is rectified and indicated on a milliammeter. An interference minimum is then represented by minimum current through the meter. By such means the carriage position can be set on a minimum to better than 1μ."

"To make a wave-length measurement the micrometer head is set at some convenient reading, the carriage supporting the receiving horns being held against the anvil by means of a weight and pulley arrangement. Phase-shifter (2) and the variable attenuator are adjusted

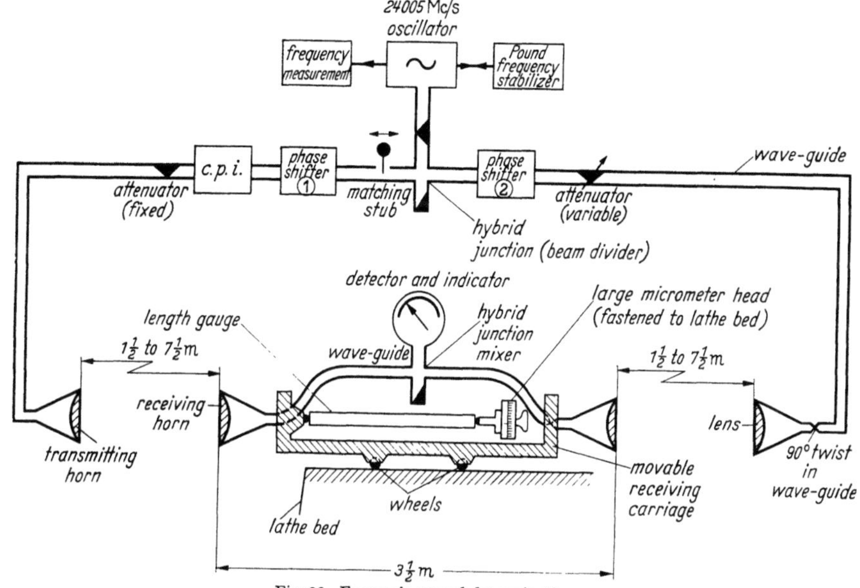

Fig. 23. FROOME's second determination.

to give zero currents in the detector. The carriage is then displaced through 162 minima by means of a meter end-gauge which has a slip gauge wrung on to each face to bring the total length closely equal to 81 wave-lengths. A small movement (e.g. 20μ) of the micrometer suffices to set exactly upon the 163rd minimum. The carriage micrometer has a vernier calibrated to 0.25μ."

"This procedure slightly distorts the table supporting the lathe-bed upon which the carriage runs with the result that although counterbalance weights are used to keep the load upon the floor constant, the lathe-bed and micrometer shift by about 5μ. This movement can be measured by an independently mounted dial-gauge."

"Three series of wave-length measurements were made, each series consisting of ten determinations at different trans-mitting receiving horn separations. The plane surface of the transmitted wave-front of series 1 were obtained by inserting plano-convex lenses made of plastic into all horn mouths. The spherical surfaces of series 2 were obtained by removing these lenses. In series 3 the receiving horns were used without lenses, but the transmitting horns were fitted with a diagonally cut half of a plano-concave lens."

"The horns themselves were not changed throughout the experiments. Their apertures are squares of side 8λ, the distance from mouth to apex being 42λ."

"The ten measurements comprising each series are needed to eliminate the diffraction error. This error always increases with decreasing tranmitter-receiver separation. The plane of polarization of radiation in the left-hand arm of the instrument is arranged to be at 90° to that in the right-hand arm, thus avoiding cross-coupling between the arms."

So far, FROOME's own description. The main improvement in this new apparatus lies in the symmetrical arrangement of the parts. Thus, e.g., the gain in thermal stability was four times that of the earlier instrument. Like its predecessor the new interferometer operates with quasi-spherical waves. Thus the *diffraction errors* must be considered. At great distances between transmitter and receiver, the wave-front may be expressed as a sphere and the interference equation would be:

$$\frac{1}{2} N \lambda = \Delta z + A \left(\frac{1}{z_1} - \frac{1}{z_2} \right) \tag{15.1}$$

where N is the number of interference minima ($= 162$), Δz the displacement of the carriage ($= 101$ cm.), A a constant depending on the actual dimensions, and z the transmitter-receiver distance ($\Delta z = z_1 - z_2$). Multiplying eq. (15.1) by $2nf/N$, where n is the refractive index and f the frequency, we obtain:

$$c = c_m + B \left(\frac{1}{z_1} - \frac{1}{z_2} \right) \tag{15.2}$$

where c is the true phase velocity, c_m the measured velocity, and B is a constant. To make practical use of eq. (15.2) is not very well possible, since the values of z required are too large for the accuracy demanded. Therefore, instead of using eq. (15.1) Froome starts with

$$\frac{1}{2} N \cdot \lambda = \Delta z + A \left(\frac{\delta_1}{z_1} - \frac{\delta_2}{z_2} \right) \tag{15.3}$$

where δ is a correction for departure from the initial or asymptotic spherical form. Froome writes:

"As $z \to \infty$, δ rapidly approaches unity. If, therefore, conditions are chosen so that δ never differs from unity by more than a small percentage (e.g. not more than 5%), and if the experimental results at a number of different $z's$ are fitted to eq. (15.3) by the method of least squares, then the derived value of c is relatively insensitive to the nature of the field assumptions. These are regarded as satisfactory when reasonable agreement is obtained between the least-squares values of A, and the purely theoretical value of A derived on basis of the field assumptions."

"It is seen that this procedure is identical with working a least-squares solution of either of the following equations,

$$\frac{1}{2} \cdot N \cdot \lambda = \Delta z + K (D_{z_1} - D_{z_2}), \tag{15.4}$$

$$c = c_m + \frac{2 K n f}{N} \cdot (D_{z_1} - D_{z_2}). \tag{15.5}$$

Here K is a constant to be determined by the least-squares method; D_z is the additional retardation of the received wave (in addition to z) calculated on basis of the field assumptions. These assumptions are adjusted until the least-squares result gives $K \approx 1$, the best values of c then being obtained. (D_z will, of course, have an inverse variation with z for its asymptotic term.)"

"The three series of experiments described were performed under conditions of large diffraction error, and two of them had bad distorsions deliberately introduced at the mouths of the horns. Thus it is necessary to establish the criteria governing the accurate operation of the interferometer for these instances."

In order to obtain c from the least squares solution of eq. (15.5), Froome derives an expression for D_z.

Besides the diffraction errors we have an error from multiple reflection of microwave energy between transmitting and receiving horns. This error reverses at a displacement of $\frac{1}{4}\lambda$ of the receiving carriage and will thus be eliminated after two wave-length measurements.

Another error is caused by energy scattered from objects of the surrounding room. The error is close to 3 parts in 10^6 for series (1), 6 in 10^6 for series (2) and 12 in 10^6 for series (3). The effect was largely eliminated by making wave-length measurements for a number of slightly different transmitter-receiver distances.

At last, we have a refractive index error. Besides uncertain information on the atmospheric conditions, we have some uncertainty in the index formula. Eq. (3.3) can not be used for 1 cm. waves, particularly on account of the great influence of humidity at these wave lengths. Essen and Froome [15], [16], [17], have given a formula for the regions 9000 and 24000 Mc/sec., which they believe to be accurate to 1 part in 10^7 for dry air as well as for ordinary air.

All errors, to be taken account of, are listed in the following table:

Random scattered errors	± 0.10 km./sec.
Diffraction residual	± 0.09 km./sec.
Length measurement	± 0.09 km./sec.
Meas. of water pressure	± 0.09 km./sec.
Gauge temperature	± 0.06 km./sec.
Gauge coeff. of expansion	± 0.03 km./sec.
Refractive index formula	± 0.03 km./sec.
Constant phase interferometer (c.p.i.)	± 0.03 km./sec.
Air temperature	± 0.02 km./sec.
Barometric pressure	± 0.01 km./sec.
Total:	± 0.2 km./sec. (max. ± 0.55)

The two first errors have been added by the present author. Froome's result is

$$c = (299793.0 \pm 0.3)\ \text{km./sec.}$$

Of the errors listed above, that of the diffraction residual is the most difficult to estimate. The results from the three series of measurements are all consistent to well within their error limits. This gives confidence in the proceeding applied. Froome, however, is planning a further improved equipment which is to work at about 7×10^{10} c./sec. ($\lambda = 4$ mm.) over a path difference incorporating 1000 minima in a much larger room. The random scattered errors and the diffraction errors are expected to be reduced by a factor 10. Then the influence of the refractive index will become the major problem.

16. Oboe, Shoran. During the second world war the Oboe and Shoran system were developed for distance measurements. In Oboe there are two short-wave (~ 0.1 m.) radio ground stations in some distance from each other and with their positions geodetically known. They transmit a continual series of pulses, each 1 μsec. long. An aircraft has a beacon, which captures the pulses and after amplification, re-emits them in a different wave-length. The ground stations receive the returning pulses and their travel times are registered by aid of a cathode-ray tube. The speed of the signals is equal to that of light and the delay in the beacon is known. Then, from the calculated distances, the position of the aircraft was obtained. In contrast to Oboe, Shoran has two beacons on the ground and a radio station on board the aircraft. In the beginning, Anderson's value 299776 km./sec. was used for the speed. However, when applying the methods for geodetical purposes, there occurred always a positive difference in the case of known distances. Essen's value of 299792 km./sec. was more fit for the task.

Jones: In 1947 Jones [92] used the Oboe method at 3000 Mc/sec. in the opposite sense and measured the velocity of light from known distances. The value obtained was

$$c = (299782 \pm 25)\ \text{km./sec.}$$

Aslakson: In 1949 Aslakson [93], [94] calculated c from a great number of 300 Mc/sec. Shoran measurements of first order geodetic distances. His result was

$$c = (299792.3 \pm 2.4)\ \text{km./sec.}$$

In 1951, with improved instruments and further measurements, he obtained a value of

$$c = (299\,794.2 \pm 1.9) \text{ km./sec.}$$

In planned geodetic Shoran projects Aslakson suggests to use 299793.1 km./sec. for the velocity of light.

By these free space radar methods the difficulty is to get sufficiently accurate distances and informations of atmospheric conditions along the pulse-path. Hence, an accuracy of 2 km./sec. seems to be the limit.

17. Rotational Spectrum. The rotational energy levels of a diatomic or linear polyatomic molecule are expressed by

$$h\nu = h\left[B\,J(J+1) - D\,J^2(J+1)^2\right] \tag{17.1}$$

where h is Planck's constant, ν the frequency, $B = \dfrac{h}{8\pi^2 I}$ with I the actual moment of inertia, J the rotational quantum number. The last term describes the frequency shift due to centrifugal distorsion and is small for low numbers of J. ν, B, D are expressed in reciprocal wave lengths (cm.$^{-1}$).

α) In 1952 Rank, Ruth and van der Sluis [95] determined B and D for the ground state by measuring ν of the rotation-vibration bands of HCN in the spectralregion 8000 to 9000 Å. They observed the absorption after a 215 m. of light path through gas at a pressure of from 35 to 70 mm. Hg. They used a plane grating with 15000 lines to the inch in combination with a Fabry-Pérot etalon. Their value of B''_{000} was (1.47830 ± 0.00002) cm.$^{-1}$. Now Nethercot, Klein and Townes had determined the same constant by direct measurement of the frequency of $2B - 4D$, which lies near the upper limit obtainable with precision microwave absorption measurements. The value obtained was $B''_{000} = (4.43159 \pm 0.000025) \cdot 10^{10}$ c/sec. According to the units the velocity of light c is the ratio of the two values or

$$c = (299\,776 \pm 6) \text{ km./sec.}$$

The authors do not preclude the possibility of systematic errors.

β) In 1953 Rank, Shearer and Wiggins [96] applied an improved proceeding, the new method of exact orders[1], to the 002 infrared rotation-vibration absorption band. Instead of glass and brass they used fused quartz and invar in the Fabry-Pérot etalon. Thus the big temperature correction was now considerably reduced. The plates were covered with a dielectrical film of thickness $\lambda/4$. The influence of the phase change on λ in the film was now the greatest uncertainty of the etalon. The absorption was effected by 3 m. of gas at a pressure of 1 to 10 mm. Hg. The value obtained for B_{000} for HCN was 1.478235 cm.$^{-1}$. Nethercot and Klein again made a new microwave determination and the result was $(4.4315800 \pm 0.000001) \cdot 10^{10}$ c/sec. The value of the velocity follows as

$$c = (299\,789.8 \pm 3) \text{ km./sec.}$$

The errors were estimated to be of the order:

Microwave	± 0.1 km./sec.
Abs. value of band origin	± 0.9 km./sec.
Phase change versus λ . .	± 1.0 km./sec.
Statistical error	± 1.0 km./sec.

---

[1] Rank, Shull, Bennett and Wiggins: J. Opt. Soc. Amer. **43**, 693 (1952).

18. Table and best value. In order to facilitate surveying, the results of the determinations reported in this paper are listed in the following table:

Date	Observer	Method[1]	Distance m.	Frequency Mc/sec.	Velocity km./sec.
1676	ROEMER	A	3×10^{11}		214 300
1725	BRADLEY	A			295 000 ± 5000
1849	FIZEAU	T. W.	8633	0.01	315 300 ± 500
1856	KOHLRAUSCH, WEBER	q_s/q_m			310 800
1862	FOUCAULT	R. M.	20	0.001	298 600 ± 500
1868	MAXWELL	e_s/e_m			284 300
1874	CORNU	T. W.	23 000	0.15	300 030 ± 200
1879	MICHELSON	R. M.	700		299 910 ± 50
1881	YOUNG, FORBES	T. W.			301 400
1882	NEWCOMB	R. M.	3720		299 860 ± 30
1882	MICHELSON	R. M.			299 853 ± 60
1891	BLONDLOT	L. W.		10	295 — 305 000
1891	TROWBRIDGE, DUANE	L. W.		5	292 — 303 600
1902	PERROTIN	T. W.	46 000		299 880 ± 84
1906	ROSA, DORSEY	c_s/c_m			299 784 ± 30
1923	MERCIER	L. W.	11	75	299 782 ± 30
1926	MICHELSON	R. M.	35 000	0.004	299 798 ± 4
1929	KAROLUS, MITTELSTAEDT	K. C.	300	3 — 7	299 786 ± 20
1932	(MICHELSON), PEASE, PEARSON	R. M.	15 000	0.02	299 774 ± 11
1940	ANDERSON	K. C.	85	19	299 776 ± 14
1940	HÜTTEL	K. C.	40	10	299 771 ± 10
1947	ESSEN, GORDON-SMITH	C. R.	0.1	3000	299 792 ± 9
1947	JONES	Oboe	70 000	3000	299 782 ± 25
1948	BERGSTRAND	K. C.	9000	8	299 793 ± 2
1949	HOUSTOUN	V. Q.	40	100	299 775 ± 9
1949	ASLAKSON	Shoran	300 000	300	299 792.3 ± 2.4
1950	ESSEN	C. R.	0.1	10 000	299 792.5 ± 1
1950	BERGSTRAND	K. C.	6000	8	299 793.1 ± 0.3
1950	HANSEN, BOL	C. R.	0.1	3000	299 789.3 ± 0.4
1950	MCKINLEY	V. Q.	9	8	299 780 ± 70
1952	ASLAKSON	Shoran	300 000	300	299 794.2 ± 1.9
1952	FROOME	M. I.	2	24 000	299 792.6 ± 0.7
1951	RANK, RUTH, VANDER SLUIS	R. S.		44 000	299 776 ± 6
1953	RANK, SHEARER, WIGGINS	R. S.		44 000	299 789.8 ± 3
1953	FROOME	M. I.	1	24 000	299 793.0 ± 0.3
1955	PLYLER, BLAINE, CONNOR[2]	R. S.			299 792 ± 4
1955	FLORMAN[3]	R. I.	1500	173	299 795.1 ± 3.1

All good determinations of c after 1947 show complete agreement. Contrary to the preceding decades, the methods have been widely differing. This fact will guarantee the stability of the new "high" value. Using the best values from each of the modern methods, of cavity resonator (ESSEN), KERR cell, Shoran, microwave interferometry, rotational spectrum, RF-interferometry, and applying weights inversely as the square roots of errors, we finally obtain for the best value (see [122] also):

$$c = 299793.0 \pm 0.3 \text{ km./sec.}$$

---

[1] In this table the methods are abbreviated by the following symbols: A = astronomical, T. W. = toothed wheel, q_s/q_m etc. = ratio of electrostatic to magnetic units, R. M. = rotating mirrors, L. W. = LECHER wires, K. C. = KERR cell, C. R. = cavity resonator, V. Q. = vibrating quartz, M. I. = microwave interferometer, R. S. = rotational spectrum, R. I. = RF-interferom.

[2] PLYLER, BLAINE and CONNOR: J. Opt. Soc. Amer. **45**, 102 (1955).

[3] FLORMAN: Nat. Bur. Stand. Techn. News Bull. **39**, 1 (1955).

C. Velocity in matter.

19. Stationary matter. One hundred years ago, the velocity of light in dense matter as compared to vacuum (or air) was considered a question of great importance. A correct answer was expected to settle the old dispute between the supporters of Huygens' undulation theory and those of Newton's corpuscular theory of light. According to Huygens, the velocity in matter should be reverse proportional to the refractive index, according to Newton directly proportional to the same number. In the latter case, the refraction was considered as depending upon the attractive forces between matter and the light corpuscles, to whom thus was rendered an increased speed.

Already in 1838, Arago [40] had planned the method later on used by Fizeau and Foucault for a comparison of the velocity in air to that in dense matter.

α) Foucault's arrangement [97] of 1850 is shown in Fig. 24. Parallel light from the slit L passes through a semi-transparent mirror towards the spin axis of the rotating mirror R and is reflected at the spherical mirrors S or S', depending on the position of R, and back again towards R and the semi-transparent plate. During the passage to the mirror S and back again the light travels through a tube with water. Compared to the determination in air

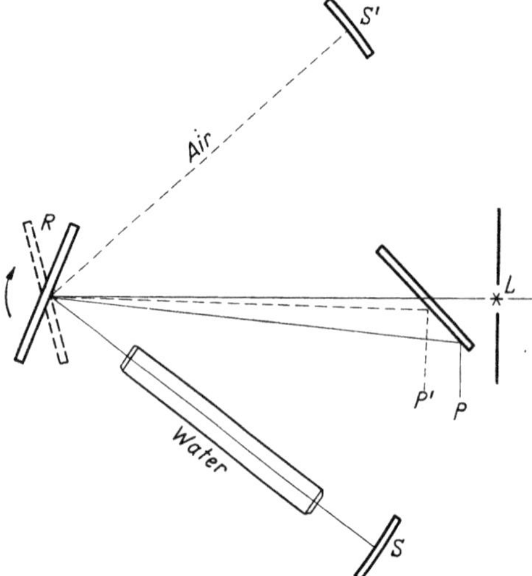

Fig. 24. Foucault's device for measuring the light velocity in water vs air.

(p. 7) there arises a small displacement of the ray, arriving at P or P' depending upon the angle through which the mirror R has rotated during the time for the light to cover the distances RSR or $RS'R$ repectively. The displacement P for water was greater than for air and according to the reversed proportion to index n. The undulation theory thus definitely turned out to be right.

β) In 1885 Michelson [100] carried out a considerably more accurate determination. For the ratio of the velocities in air and water he obtained 1.330, where $n = 1.334$. For carbondisulfide, he found 1.758 with $n = 1.635$. Using white light he got a short spectrum indicating the velocity to depend on colour. In the case of carbondisulfide the result is correct as compared to the group index [eq. (2.15)], in the case of water it is not.

γ) In 1912 Gutton [101] verified the result of Michelson for water. He used the device shown in Fig. 19, placing a tube containing the fluid concerned between C_1 and C_2 and measuring the required prolongation of the system GHK. For water he obtained the ratios 1.32 and 1.36 for respectively yellow and blue light.

δ) Recently, Houstoun [103] carried out a determination for water. Contrary to Michelson, he obtained complete agreement with the group index. The colour variation was also as expected. Principally, his apparatus was similar to the one he used for the air determination (p. 19).

ε) In solid bodies the available light path usually is very short. This may explain that no such measurements were performed in earlier days. Bergstrand [104] in 1954 used a device of some similarity to that of Houstoun. A diagram is shown in Fig. 25. Figures in mm. denote the rough dimensions. By the condenser L_1 an image of the point light source LS is thrown through the semitransparent mirror SM into a 0.7 mm. aperture D_1 in a sooted metal screen S_1. The lens system L_2 parallelizes the light beam passing the vibrating quartz crystal K, the optic axis of which is parallel with the beam of light. The objective L_3 again collects the light to a spot on the sooted screen S_2. By its 150 Mc/sec., the short wave radio transmitter RT keeps the crystal K vibrating on the 209th harmonic. The standing supersonic wave generated forms and deletes, 3×10^8 times per second, an optic space grating, the compressive and dilative plane layers of which create an intermittent image of the light source. This image, a spectrum of first order, is situated 1.5 mm. above the central spot from LS (in practice K is inclined 25′ to give reflexions according to Bragg's law). The spectral image falls into a 0.7 mm. aperture D_2. Hence, beyond the screen S_2 we get a beam, modulated 100% with 300 Mc/sec. For certain purposes a Nicol's prism can be inserted at N. By the lens L_4 the beam is parallelized towards the plane mirror PM. The object for investigation can be placed between L_4 and PM. By the micrometer M the mirror can be displaced in the direction of the beam. Further, the complete micrometer can be moved through certain fixed distances, the magnitudes of which are determined by a couple of iron bars, one of which is denoted B in Fig. 25. By PM the light is thrown back into the opening D_2, passes the crystal for the second time and will again be reproduced, now on the back side of the screen S_1 where the opening D_1 is just in place of the first order spectrum of the returning beam.

If the mirror PM is situated at such a distance from the crystal K that the returning light and K have a difference in phase of $N \cdot \pi$ ($N =$ integer) in their

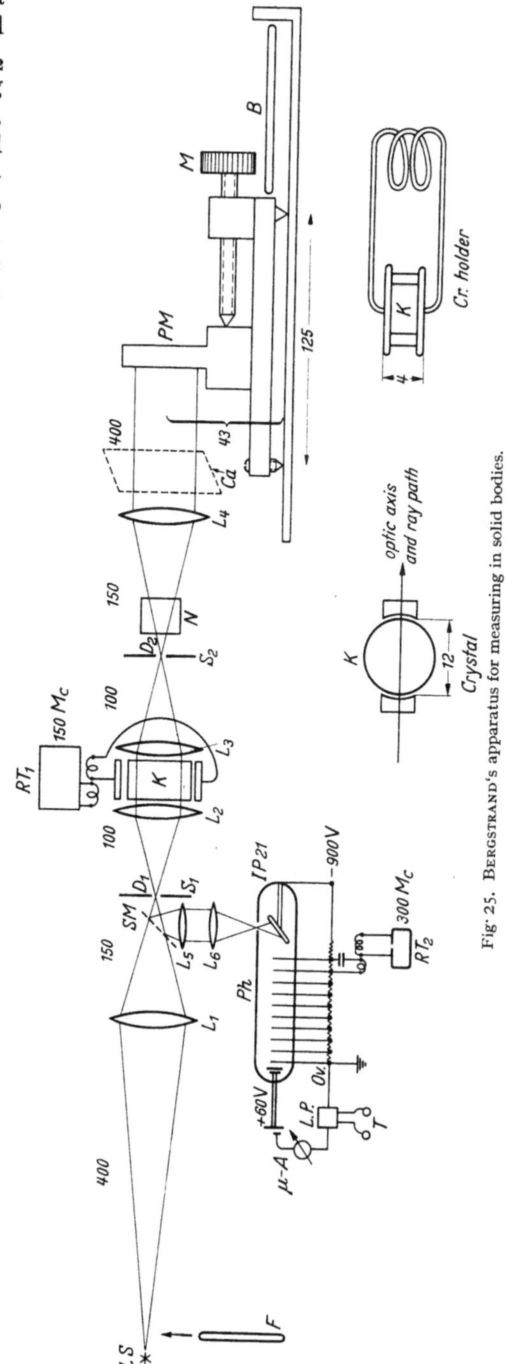

Fig. 25. Bergstrand's apparatus for measuring in solid bodies.

vibrating states, the component 300 Mc will be totally cancelled after this second passage of the aperture D_1. Here the semi-transparent mirror SM directs the beam to the cathode of the photo-multiplier Ph. The transmitter TM_2 produces an alternating voltage between two of the dynodes of Ph. The frequency slightly differs from the 300 Mc/sec. of light variation. Since the current amplification in Ph depends on the dynode voltage, the net result is an audible beat frequency in the telephone T. The harmonics of TM_2 and the light also, give the corresponding acoustic harmonics. They will be stopped by the low-pass filter LP. Thus, a zero-position of the mirror PM occurs at the observed sound minimum in the telephone T. The total photo-current is checked by aid of a µA-meter.

Measuring the crystal Ca the thickness of which is Δ mm. (see Fig. 25), PM was adjusted to zero-position. Then Ca was removed. For the new zero-position the distance to PM now had to be increased by an amount δ, just corresponding to the former delay, caused by Ca. By the prism N the ordinary and extra-ordinary rays were separated. The observed n_g was simply $1 + \dfrac{\delta}{\Delta}$.

With light of 5384 Å the results were: for a glass with $n = 1.519$ and a light path of 45 mm.:

Group index, calculated: $n_g = 1.547$,
Group index, measured $= 1.550 \pm 0.003$.
A calcite crystal with $n_o = 1.6625$, $n_{eo} = 1.4883$:

Angle to optic axis	Ordinary Group index		Extra ordinary Group index	
	Calcul.	Measured	Calcul.	Measured
90°	1.712	1.718	1.511	1.508
57°	1.712	1.719	1.573	1.568
45°	1.712	1.714	1.591	1.595[1]
0°	1.712	1.705	1.712	1.705

To sum up, one may state that, for glass, the investigation has shown the validity of Rayleigh's formula for the group velocity. This formula, also, qualitatively holds for the anisotropic propagation of light energy in a calcite crystal.

20. Moving matter. Using his elastic-solid theory of ether, Fresnel[2] derived a formula for the *dragging* of the ether, which means that moving matter partly takes the light along with it. This formula runs:

$$v' = v \cdot \left(1 - \frac{1}{n^2}\right). \tag{20.1}$$

Thus the velocity of light will change by v', the alteration depending on the velocity v of the substance and on its refractive index n. The bracket in eq. (20.1) is called Fresnel's *dragging coefficient*. (Cf. Sect. 11 β also.)

We assume the direction of v to be equal to that of the propagation of the light. Then, to an observer at rest, the light seems to pass a medium, the number (N_0) of matter particles per unit volume of which is reduced by the rate $\dfrac{v}{c/n} \cdot N_0$. Furthermore, every particle will oscillate slower due to the Doppler effect at

---

[1] 51.7° for the e.o. ray.
[2] Fresnel: Ann. chim. et phys. (2) **9**, 56 (1818).

the same rate. The product of these two effects reduces the matter term of
MAXWELL's displacement equation and, by that, there retarding influence of
the substance on the light velocity, i.e., regarding the dielectric constant, the
refractive index formula $n^2 = 1 + A \cdot N_0 = 1 + r_0$ ($r_0 = \text{const}$) changes into

$$n_v^2 = 1 + \left(1 - \frac{v}{c/n}\right)^2 r_0. \tag{20.2}$$

Noticing the different meaning of v in eq. (2.4) we use the eqs. (2.4) and (20.2)
to derive the resulting light velocity u and obtain

$$u = \frac{c}{n} + v\left(1 - \frac{1}{n^2}\right) \tag{20.3}$$

in accordance with eq. (20.1).Eq. (20.3) is often refered to as "FIZEAU's formula."

A rather complete treatment of the subject, following the dispersion theory
of electromagnetic waves is given e.g. in P. DRUDE's classic ,,Lehrbuch der Optik".
It is, however, unnecessary to go into such highly specialized and unsatisfactory
detail explanations if we immediately turn to the purely kinemativ relativistic
explanation [123], [124], [125]:

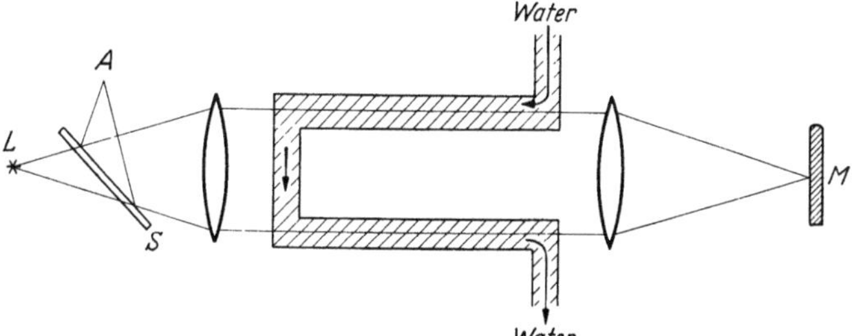

Fig. 26. FIZEAU's experiment with moving matter.

An observer outside the matter (which he finds at a speed v relative to himself)
just has to add the velocity v and that of light inside the matter at rest $v_1 = c/n$,
after the relativistic law of velocity addition

$$u = \frac{v + v_1}{1 + \dfrac{v\,v_1}{c^2}}, \tag{20.4}$$

which yields in this case

$$u = \frac{\dfrac{c}{n} \pm v}{1 \pm \dfrac{v}{nc}} \approx \frac{c}{n} \pm v\left(1 - \frac{1}{n^2}\right), \tag{20.5}$$

in agreement with FRESNEL's formula (20.1). Considering the DOPPLER effect
and its bearing upon $n(\lambda)$, LORENTZ derived the more complete formula

$$u = \frac{c}{n} \pm v\left(1 - \frac{1}{n^2} - \frac{\lambda}{n}\frac{dn}{d\lambda}\right) \tag{20.6}$$

where λ is the wavelength of the light. Eqs. (20.3) and (20.5) differ by omitted
terms, which, due to v/c, are too small for detection. Still, the experimental
corroboration of eq. (20.5) is important as a check of the addition law.

α) In 1859 Fizeau [105] carried out an experiment in order to examine the dragging of the ether. The arrangement is shown in Fig. 26.

Light from the source L passes the upper leg against a stream of water. After reflection by the mirror M it proceeds against the stream in the lower leg. The other part of the light from L first entering the lower leg always travels with the water. Any change in the velocity of the light by the streaming water causes a change of the interference pattern observed at A by aid of the semi-transparent mirror S. Fizeau noted an alteration of the velocity of 0.47 of the water speed. According to Fresnel's formula the figures should be 0.437. The difference surely depends on a wrong value of the water speed, since in the centre of the tube the speed is greater than close to the walls.

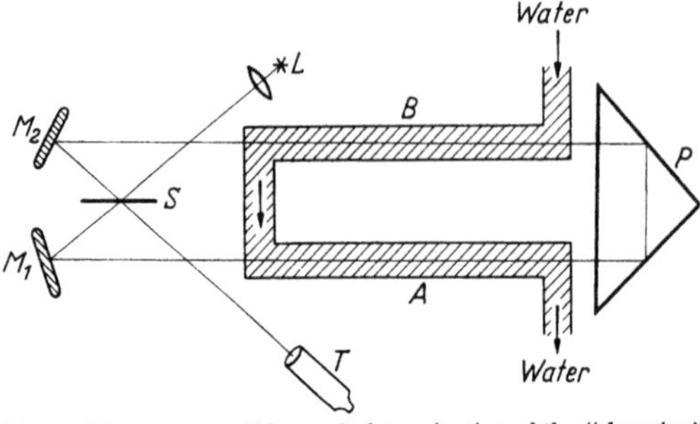

Fig. 27. Michelson and Morley's determination of the "dragging" coefficient.

β) In 1878 Michelson and Morley [107] repeated the Fizeau experiment using an improved arrangement according to Fig. 27.

L is the light source, M_1 and M_2 plane mirrors and P a total-reflecting prism. The semi-transparent mirror S splits up the parallel light beam in two parts, travelling in opposite directions through the water-tubes A and B.

The value obtained, 0.434, is in resonable agreement with Fresnel's coefficient.

γ) Harress in 1912, carried out an experiment with a moving solid body [109]. It was in the form of a horizontal rotating ring of rods of fused quartz. The light was put in and out vertically to reflecting prisms at the centre of the ring. On account of Harress' death in the war the result was not fully definite.

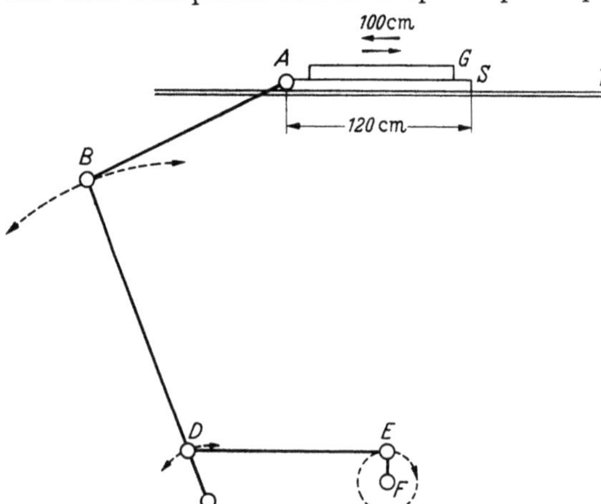

Fig. 28. Zeeman's arrangement for moving solid bodies.

δ) In 1914 to 1920 Zeeman [108] carried out the most careful observations of the dragging coefficient in fluids as well as in solid bodies. He started with water using a device similar to that of Michelson. Special care was taken to get a correct value for the effective velocity of the water. Air-bubbles following the stream were observed by means of a rotating mirror. Thus Zeeman obtained the velocity of the water in the centre of the tube in relation to that calculated from the total flow. Besides, this relation was different for opposite directions of the flow. During the final observations, only a narrow pencil in the centre of the tube was used. The final errors were estimated to be smaller than 0.5% and the result fully confirmed the correctness of the Fresnel formula with the Lorentz *dispersion* term.

For solid bodies ZEEMAN used an apparatus according to Fig. 28.

The slide S is guided on the tracks T. S supports the body G to be investigated. AB, BC, DE, EF are arms which can turn around the joints at A, B, C, D, E, F. C is fixed, and F is the axle of a $3HP$ motor, revolving E in a circular orbit. Thus G will be moved 100 cm. straight back and forth. Within 20 cm. around the middle position the speed of SG is at its maximum, almost constant and at a rate of 900 cm./sec. The body G is a rod of glass or fused quartz, 120 cm. in length.

The required interference fringes are composed of light, one part of which goes through the rod and the other part only through air. The interference pattern is exposed on a photographic plate for 0.01 sec., during the period of constant and maximum speed of the SG-motion. The displacement of SG during the exposure is only 9 cm. There are two instants of exposure: One for SG moving to the right and the second for SG moving to the left. The exposures are released by electric contacts in the orbit of joint E. At a slow motion of G th position of the fringes is seen to change a little, but the change is fully regular with the actual position of the slide S and still at high speed a series of exposures renders sharp pictures of the fringes. The speed of G was optically controlled during the exposures and determined within 1%.

The following example shows the rate of accuracy: At a certain speed, a displacement of the fringes $\Delta = 0.242 \pm 0.004$ has been observed of which figure 0.028 is due to the LORENTZ dispersion term.

The final result agreed fully with the complete FRESNEL formula.

D. Some relativistic experiments.

21. Constancy of c. In the 19th century most theories explaining the propagation of light included a fixed ether, i.e. a medium penetrating everything, even the vacuum. In this ether the light oscillations and the propagation were thought to take place. The existence of the aberration was considered to be a proof that the ether did not follow the Earth in its motion around the Sun. At least at one point of the orbit, the Earth must have a velocity relative to the ether equal to (or larger than) her orbital velocity v. Then, if the relative velocity of light to ether was accepted to be constant, the value of c should be found different in different directions to the Earth's motion. Since, in all terrestrial measurements of c, the light travels a distance forth and back, the value obtained would be close up to the mean c of the velocities $c+v$ and $c-v$ in the two opposite directions. The small difference, build up of second and higher orders of v/c, will, however, vary for different angles to the direction of v.

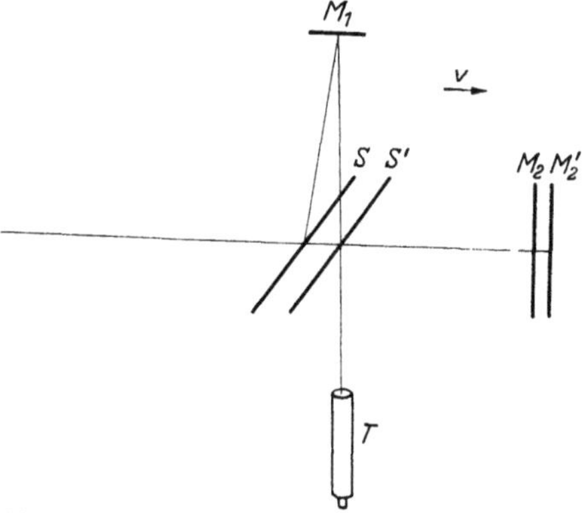

Fig. 29. MICHELSON and MORLEY's ether drift experiment.

$\alpha)$ *Experiment by* MICHELSON *and* MORLEY. In order to test the ether drift MICHELSON and MORLEY [110] started their famous experiment with the MICHELSON interferometer. The observations were performed in July 1881. Their apparatus in principle is shown in Fig. 29.

L emits parallel light towards the semi-transparent plane mirror S. M_1 and M_2 are plane mirrors and T is a tube for observation of the interference fringes, originated by the two light beams traveling the paths $L-S-M_1-T$ and $L-M_2-S-T$. The arrow denotes the direction of the velocity v of the Earth.

The mirrors are adjusted until $S-M_1=S-M_2=d$. During the time used by the light to cover the distance $S-M_2$, the mirror M_2 moves forward to the position M_2'. On its way back, the light hits the mirror S in its new position S'. Hence, the total time for the passage forth and back is

$$t_1 = \frac{d}{c+v} + \frac{d}{c-v} = \frac{2cd}{c^2-v^2}. \tag{21.1}$$

The time t_2 for the other way SM_1S' is

$$t_2 = \frac{2d}{\sqrt{c^2-v^2}}. \tag{21.2}$$

The difference t_1-t_2 multiplied with the velocity c gives the path difference

$$\delta = d\cdot\frac{v^2}{c^2}. \tag{21.3}$$

If we now turn the whole interferometer through 90°, δ will change its sign so that we observe a shift of the fringes corresponding to a total path difference of

$$\varDelta = 2d\cdot\frac{v^2}{c^2}. \tag{21.4}$$

By means of multiple reflection, the distance d was brought to about 11 m. The interferometer was mounted on a large concrete block floating in mercury. During the observations, the whole apparatus was rotated slowly and continuously around a vertical axis.

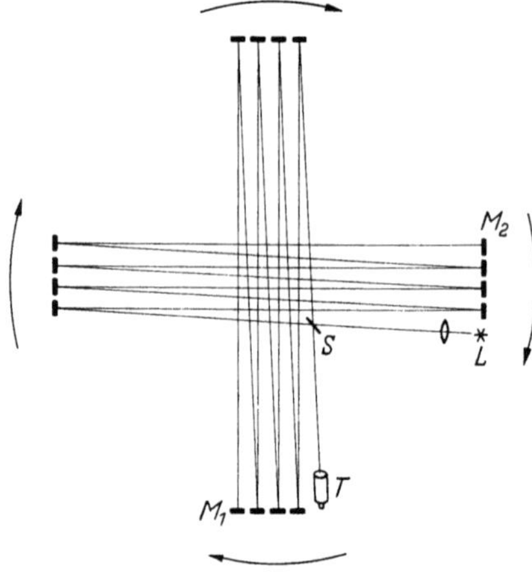

Fig. 30. The light paths in Miller's apparatus.

If we insert the Earth's velocity $v=30$ km./sec. eq. (21.4) yields a displacement of 2.2×10^{-5} cm., that is 0.4 of a fringe for yellow light. The result of the measurements was that there was no shift exceeding the error limits of 0.05 fringe. This result, showing for the first time the independence of light velocity of the observer's frame of reference, was the seed out of which the theory of relativity grew up a score of years later. The experiment has been repeated with the same result by many investigators, with only one exception:

β) *Experiment by* Miller. Miller's investigations [*115*] are very extensive and carefully carried out. His last instrument was of the Michelson type though on a larger scale with a light path of 32 m. (Fig. 30). Measurements were performed at Mt. Wilson Observatory during the years from 1921 to 1926. The experimental conditions were varied in an immense number of ways:

Centre pin fixed or loose, rotation of the instrument in both directions, slow or fast, heavy loads on the arms, the assistant walking in different quadrants, close to or distantly, a great number of temperature tests with heaters in the room, the instrument arms of steel replaced by those of concrete and aluminium, different kind of light sources, e.g. sunlight, etc.

A total of 200000 readings were made and finally there arose a small detectable dependence on the siderial time. The effect was one twentieth of that expected from the Earth's orbital motion. MILLER studied the variation of the effect during the 24 hours of the day at four different epochs of the year. He believes the effect to be present in all similar measurements. In some cases, he has shown the positive effect in measurements by others, but he means that the observational support usually is insufficient to extricate the effect from random errors. With a proposed absolute motion of the solar system of 208 km./sec. towards an apex of $\alpha = 5^h$, $\delta = -70°$ and an observed effect of 5% MILLER obtained the best accordance with his observations. At different epochs, the apex was reported to be:

Epoch	R. A.	Decl.
February	$5^h.2$	$-69°.9$
April	4.8	-70.1
August	4.7	-72.0
September	4.9	-70.2

The result is remarkable in that respect that the derived apex is *unchanged* during the lapse of the year. Whatever the right interpretation of MILLER's observations may be, there is full evidence of a dependence on the siderial time in the tests as they are reported in [115]. MILLER's results caused several investigators to repeat the MICHELSON experiment [112].

γ) *Experiment by* JOOS. With MILLER's experiments in mind, JOOS [113] built his apparatus with extraordinary care in the basement story of the ZEISS works at Jena. The four arms were made of fused quartz and 2 m. in length each. Instead of mercury he used a very great number of springs as strainless support. The whole cross thus was hung up inside a covering air tight metal tube cross, turnable by a motor around its centre on a hollow cylindrical stand. The lengths of the arms were specially examined at different speeds of revolution.

The light path was $\frac{2}{3}$ of MILLER's or 21 m., mainly according to Fig. 31. The final path, however, was directed centrally and the recording camera was placed in the stand under the center of the cross. The light source was a mercury arc, $\lambda = 5461$ Å, which followed the cross. The attendance of the instrument was entirely automatic, e.g. turning, exposures, plate shifts etc. and measurements, carried out during the whole day as by MILLER. The limit of observation was 0.001 λ, corresponding to an "ether-wind" of 1.5 km./sec. In spite of this high accuracy, there were no systematic shifts of the fringes.

The Joos experiment seems to be of higher precision than that of MILLER. Therefore it is to be regretted that it was not repeated in a way following more closely MILLER's leading principle viz. in thin air at great distance from screening or influencing heavy matter. Obviously, the precision would decrease with the metal cover removed. The gain over MILLER's experiment by the automatic attendance should, however, be considerable.

δ) *Experiment by* ESSEN. In 1955 ESSEN [114] employed his cylindrical cavity resonator (Sect. 14) in a new sort of ether-drift experiment. The resonator was used to control the frequency of an oscillator. The cavity cylinder was mounted with its axis horizontal and was rotated continuously in a horizontal plane at a rate of about one turn per minute. In this experiment there should arise a variation proportional to $v^2/2c^2$ of the time for an oscillation to travel to and fro between the end faces of the resonator. The frequency of the controlled oscillator must vary at the same rate. A change in this frequency of 1 part in 10^{10} could be observed. This corresponds to not quite the half of the sensitivity obtained by Joos' device. No positive effect was observed.

22. The relativity of time. If light propagation obeys the laws of the theory of relativity, there should arise a second order effect in the Doppler change of the light frequency. Applying Lorentz transformation on the frequency of the light we get [125]

$$\nu = \nu_0 \cdot \frac{1 \pm v/c}{\sqrt{1 - v^2/c^2}} = \nu_0 \cdot \left(1 \pm \frac{v}{c} + \frac{1}{2} \cdot \frac{v^2}{c^2} \pm \frac{1}{2} \cdot \frac{v^3}{c^3} + \cdots\right). \tag{22.1}$$

ν is the observed frequency, ν_0 is the light frequency for an observer at rest relative to the light source, v is the velocity of the source relative to the observer in the direction of the light propagation. The classic law for a moving source gives:

$$\nu = \frac{\nu_0}{1 \mp v/c} = \nu_0 \cdot \left(1 \pm \frac{v}{c} + \frac{v^2}{c^2} \pm \frac{v^3}{c^3} + \cdots\right). \tag{22.2}$$

As a moving light source, Ives [116] used H_2 and H_3-ions in a canal-ray tube of Dempster type (narrow velocity band) according to Fig. 31.

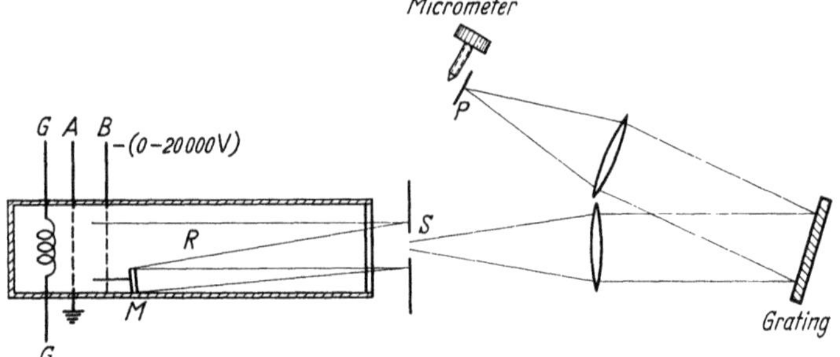

Fig. 31. Ive's proof of the relativity of time.

The ions are produced in an hydrogen arc between the heater G and the perforated electrodes A and B. Between A and B the ions are accelerated by an electric field, thus forming the beam R proceeding with constant speed. Light emitted from these ions will illuminate the slit S. At P the spectral line is visible in a microscope. The displacement of a line can be measured by means of a micrometer. In order to get light from ions moving in the opposite direction Ives placed a slightly concave mirror at M. By M the slit will be illuminated from the "backs" of the ions. At the maximum energy of 20000 volts the H_β-line was displaced by 2 mm. Then v/c is 6×10^{-3}. This first order Doppler effect is superimposed by the second order effect of the rate of 0.005 mm.

The frequencies of light from the direct beam and from that via the mirror are, according to the relativistic eq. (22.1)

$$\nu_1 = \nu_0 \cdot \frac{1 + v/c}{\sqrt{1 - v^2/c^2}} \quad \text{and} \quad \nu_2 = \nu_0 \cdot \frac{1 - v/c}{\sqrt{1 - v^2/c^2}} \tag{22.3}$$

with a mean value of

$$\bar{\nu} = \frac{\nu_0}{\sqrt{1 - v^2/c^2}} = \nu_0 \left(1 + \frac{1}{2} \cdot \frac{v^2}{c^2} + \cdots\right). \tag{22.4}$$

This formula is to be compared with the classical mean of $\nu_0 \cdot \left(1 + \frac{v^2}{c^2} + \cdots\right)$. The difference between the two means is $\left(\frac{1}{2} \nu_0 \frac{v^2}{c^2}\right)$. The position of the observed mean $\bar{\nu}$ was compared to that of the undisplaced line, i.e. of ions at rest. The differences (in Ångström units) obtained are listed in the following table, where V is the applied acceleration voltage and $\Delta\lambda$ denotes the total Doppler shift.

Voltage V	Ion	The ($\gamma_0 v^2/2c^2$) shift as computed		Observed displacement
		from V	from $\Delta\lambda$	
6788	H_3	0.012	0.011	0.011
7780	H_2	0.020	0.020	0.019
9181	H_2	0.024	0.024	0.023
10574	H_2	0.028	0.028	0.027
12100	H_3	0.020	0.020	0.021
13560	H_2	0.035	0.036	0.035
13560	H_3	0.023	0.024	0.022
18350	H_2	0.048	0.047	0.047

References.

Phase and group velocity, energy transport.

[1] RAYLEIGH: Proc. Lond. Math. Soc. **9**, 21 (1877).
[2] RAYLEIGH: Theory of sound, 2nd edit., vol I, p. 301, 475. London 1926.
[3] REYNOLDS: Nature, Lond. **46**, 343 (1877).
[4] GOUY: C. R. Acad. Sci. Paris **91**, 877 (1880); **92**, 34 (1881).
[5] LAUE, v.: Ann. Phys. **18**, 523 (1905).
[6] SOMMERFELD: Phys. Z. **8**, 841 (1907)
[7] SOMMERFELD: Ann. Phys. **44**, 177 (1914).
[8] BORGNIS: Z. Physik **117**, 642 (1941).
[9] SYNGE: Rev. Opt. **31** (1952).

Atmosphere.

[10] PERARD: Trav. Bur. Int. Poids Mes. **1934**.
[11] BARREL and SEARS: Phil. Trans. Roy. Soc. Lond. (I) **1939**.
[12] KÖSTER u. LAMPE: Phys. Z. **1934**.
[13] JOHNSON and HEYWOOD: Met. Off. Geophis. Mem. **9** (1938).
[14] KOHLRAUSCH: Lehrbuch der praktischen Physik. 1935.
[15] ESSEN and FROOME: Proc. Phys. Soc. Lond. B **64**, 862.
[16] ESSEN and FROOME: Nature, Lond. **167**, 512 (1951).
[17] ESSEN: Proc. Phys. Soc. Lond. B **66**, 189 (1953).

Different sorts of influence.
Intensity.

[18] DOUBT: Phys. Z. **5**, 457 (1904).
[19] DOUBT: Phys. Rev. **18**, 129 (1904).

Moving light source.

[20] SITTER, DE: Phys. Z. **14**, 429 (1913).
[21] ZURHELLEN: Astr. Nachr. **198**, 1 (1914).
[22] RITZ: Ann. chim. phys. **13**, 145 (1908).
[23] ROSA, LA: Z. Physik **21**, 339 (1924); **34**, 689 (1925).
[24] SITTER, DE: Bull. Astr. Inst. Netherl. **2**, 121, 163 (1924).
[25] THIRRING: Z. Physik **31**, 133 (1925).
[26] BERNHEIMER: Z. Physik **36**, 302 (1926).
[27] MAJORANA: Phys. Rev. **11**, 411 (1918).
[28] MAJORANA: C. R. Acad. Sci. Paris **165**, 424, (1917); **167**, 71 (1918).

Magnetic field.

[29] BANWELL and FARR: Proc. Roy. Soc. Lond., Ser. A, **175**, 1 (1940).

Electric field.

[30] STARK: Z. Physik **133**, 504 (1952).

Direct methods.
Astronomical.

[31] ROEMER: Mem. Acad. Sci. Paris (I) **10**, 576, 1666—1699 (1733).
[32] ROEMER: J. de Sav. **1676**.
[33] NIELSEN über ROEMER: Nord. Astr. Tidskr. **1944**, Nr. 2.

Toothed wheel.

[34] Fizeau: C. R. Acad. Sci. Paris **29**, 90 (1849)
[35] Fizeau: Pogg. Ann. **79**, 167 (1850).
[36] Perrotin: C. R. Acad. Sci. Paris **131**, 731 (1900); **135**, 881 (1902).
[37] Cornu: C. R. Acad. Sci. Paris **79**, 1361 (1879).
[38] Young and Forbes: Proc. Roy. Soc. Lond. **32**, 247 (1881).

Rotating mirror.

[39] Foucault: C. R. Acad. Sci. Paris **30**, 551 (1850); **55**, 501, 792 (1862).
[40] Arago: C. R. Acad. Sci. Paris **7**, 954 (1838); **30**, 489 (1850).
[41] Newcomb: Naut. Alm. **1885**, 112.
[42] Michelson: Naut. Alm. **1885**, 235.
[43] Michelson: Nature, Lond. **21**, 94.
[44] Michelson: Astrophys. J. **65**, 1 (1927).
[45] Bowie (base line): Astrophys. J. **65**, 14 (1927).
[46] Michelson, Pease and Pearson: Astrophys. J. **82**, 26 (1935).

Kerr cell.

[47] Karolus u. Mitetlstaedt: Phys. Z. **29**, 698 (1928).
[48] Mittelstaedt: Ann. Phys. **2**, 285 (1929).
[49] Anderson: Rev. Sci. Instrum **1937**, 239.
[50] Anderson: J. Opt. Soc. Amer. **31**, 187 (1941).
[51] Hüttel: Ann. Phys. **37**, 365 (1940).
[52] Bergstrand: Ark. Fysik **2**, 119 (1950); **3**, 479 (1951).
[53] Bergstrand: Nature, Lond. **163**, 338 (1949); **165**, 405 (1950).

Vibrating quartz.

[54] Kerr and Grant: Nature, Lond. **120** (1927).
[55] Bergmann: Ultraschall. 1937. English translation: Supersonics, p. 190. New York 1938.
[56] Houstoun: Proc. Roy. Soc. Edinburgh A **61**, 102 (1941); **63**, 95 (1950).
[57] McKinley: J. Roy. Astr. Soc. Canad. **44**, 89 (1950).

Indirect methods.

Aberration.

[58] Bradley: Phil. Trans. Roy. Soc. Lond. **1728**, No. 406.
[59] Airy: Proc. Roy. Soc. Lond. **20**, 35 (1871); **21**, 121 (1873).
[60] Airy: Phil. Mag. (4) **43**, 310 (1872).
[61] Hammer: Phys. Z. **37** (1936).

Ratio of electro-magnetic units.

a) E_m/E_s.

[62] Kohlrausch u. Weber: Elektrodynamische Maßbestimmung, Bd. III, S. 221. 1857.
[63] Maxwell: Phil. Trans. Roy. Soc. Lond. **2**, 643 (1868).
[64] Thomson, N., and King: Rep. Brit. Assoc. **1869**, 434.
[65] McKichan: Phil. Mag. (4) **47**, 218 (1874).
[66] Ayrton and Perry: Electr. Rev. **23**, 337 (1888/89).
[67] Pellat: J. de phys. (2) **10**, 389 (1891).
[68] Hurmuzescu: Ann. chim. phys. (7) **10**, 433 (1897).
[67] Perot and Fabry: Ann. chim. phys. **13**, 404 (1898).

b) $\sqrt{C_s/C_m}$.

[70] Ayrton and Perry: Phil. Mag. (5) **7**, 277 (1879).
[71] Thomson, J. J., and Searle: Phil. Trans. Roy. Soc. Lond. **181**, 593 (1890).
[72] Himstedt: Wied. Ann. **35**, 126 (1888).
[73] Abraham: Ann. chim. phys. (4) **27**, 433 (1892).
[74] Rosa and Dorsey: Bull. Bur. Stand **3**, 433 (1907).
[75] Summary by Abraham: Rapp. prés. au congrès int. de phys. **2**, 247 (1900).

Wires and air.

[76] Hertz: Wied. Ann. **34**, 551 (1881).
[77] Blondlot: C. R. Acad. Sci. Paris **117**, 543 (1893).
[78] Trowbridge and Duane: Phil. Mag. (5) **40**, 211 (1895).
[79] Saunders: Phys. Rev. **4**, 81 (1897).

References. 43

[80] McLean (air): Phil. Mag. (5) **48**, 115 (1899).
[81] Mercier: J. Phys. Radium (6) **5**, 168 (1924).
[82] Summary by Blondlot and Gutton: Rapp. prés au congrès int. de phys. **2**, 268 (1900).

Wires versus air.

[83] Gutton: J. de Phys. (5) **2**, 196 (1912).
[84] Sommerfeld: Ann. Phys. **67**, 223 (1899).
[85] Mercier: Ann. Phys. **20**, 5 (1923).

Cavity resonator.

[86] Essen and Gordon-Smith: Proc. Roy. Soc. Lond., Ser. A **194**, 348 (1948).
[87] Essen: Proc. Roy. Soc. Lond., Ser. A **204**, 260 (1950).
[88] Essen: Nature Lond. **167**, 258.
[89] Bol: Phys. Rev. **80**, 298 (1950).

Microwave interferometry.

[90] Froome: Proc. Roy. Soc. Lond., Ser. A **213**, 123 (1952).
[91] Froome: Proc. Roy. Soc. Lond. Ser. A **223**, 195 (1954).

Oboe, Shoran.

[92] Jones and Cornford: J. Inst. Electr. Engrs. **96**, Pt III, 447 (1949).
[93] Aslakson: Trans. Amer. Geophys. Un. **30**, 475 (1949).
[94] Aslakson: Nature Lond. **168**, 505 (1951).

Rotational spectrum.

[95] Rank, Ruth and Vander Sluis: J. Opt. Soc. Amer. **42**, 693 (1952).
[96] Rank, Shearer and Wiggins: Phys. Rev. **94**, 575 (1953).

Velocity in matter
a) Stationary.

[97] Foucault: C. R. Acad. Sci. Paris **30**, 551 (1850).
[98] Fizeau et Breguet: C. R. Acad. Sci. Paris **30**, 562, 771 (1850).
[99] Fizeau u. Breguet: Pogg. Ann. **82**, 124 (1857).
[100] Michelson: Naut. Alm. **1885**, 235.
[101] Gutton: C. R. Acad. Sci. Paris **152**, 685, 1089 (1911).
[102] Gutton: J. de Phys. **2**, 196 (1912).
[103] Houstoun: Proc. Roy. Soc. Edinburgh A, **62**, 58 (1944).
[104] Bergstrand: Ark. Fysik **8**, 457 (1954).

b) Moving matter.

[105] Fizeau: C. R. Acad. Sci. Paris **33**, 349 (1851).
[106] Fizeau: Pogg. Ann., Erg.-Bd. **3**, 457 (1853).
[107] Michelson and Morley: Sill. J. (3) **31**, 377 (1886).
[108] Zeeman: Proc. Kon. Ned. Ak. Amst. **17**, 445 (1914); **19**, 125 (1917); **29**, 1252 (1922).
[109] Harress: Die Geschwindigkeit des Lichtes in beweglichen Körpern. Diss. Jena 1912.

Relativistic experiments.

[110] Michelson: Amer. J. **22**, 120 (1881); **34**, 333 (1887).
[111] Morley and Miller: Phil. Mag. **8**, 753 (1904); **9**, 680 (1905).
[112] Kennedy: Astrophys. J. **68**, 376 (1928).
[113] Joos: Ann. Phys. **7**, 385 (1930).
[114] Essen: Nature, Lond. **175**, 793 (1955).
[115] Miller: Rev. Mod. Phys. **5**, 203 (1933).
[116] Ives: J. Opt. Soc. Amer. **28**, 215 (1938).

Surveys.

[117] Rapp. prés au congrès internat. de phys., Vol. 2, 1900.
[118] Dorsey: Trans. Amer. Phil. Soc., N.S. (5) **34**, Pt 1, 109 (1944).
[119] Birge: Phys. Soc. Rep. Progr. Phys. **8**, 90 (1941).
[120] Bearden and Watts: Phys. Rev. **81**, 73 (1951).
[121] Bergstrand: Nat. Phys. Lab.s series: Rec. Dev. Standards, p. 75. London 1952.
[122] DuMond and Cohen: Rev. Mod. Phys. **25**, 691 (1953).

Textbooks.

[123] Sommerfeld: Vorlesungen über theoretische Physik. Darmstadt 1950.
[124] Jenkins and White: Fund. of Optics, 2nd ed. 1950.
[125] Bergmann: Intr. to the theory of relat. New York 1948.

Optique géométrique générale.

Par

A. MARÉCHAL.

Avec 208 Figures.

A. Lois générales de l'optique géometrique.

I. Généralités.

1. On peut bâtir l'optique géométrique à partir de diverses propositions qui permettent d'étudier le cheminement des rayons lumineux: lois de DESCARTES ou principe de FERMAT par exemple. Ces diverses propositions équivalentes ne sont en fait que les formes limites que prennent les lois de propagation des ondes électromagnétiques lorsque leur fréquence devient très grande. C'est pourquoi nous nous attacherons tout d'abord à effectuer ce passage à la limite c'est-à-dire à montrer qu'en partant des équations de MAXWELL et en supposant que la fréquence des vibrations tend vers l'infini, on obtient les lois de l'optique géométrique. Nous montrerons ensuite que l'on peut encore retrouver ces résultats en appliquant la méthode de calcul des phénomènes de diffraction au cas où la surface d'onde s'éloigne notablement de la forme sphérique. Nous passerons ensuite à l'étude du principe de FERMAT: partant de son expression générale nous aboutirons tout d'abord à l'équation différentielle des rayons lumineux dans les milieux non homogènes ce qui nous permettra d'en préciser en particulier la courbure; nous étudierons d'autre part les consequences générales les plus importantes du principe de FERMAT, les lois de DESCARTES le théorème de MALUS et la relation de CLAUSIUS. Cette dernière établit l'invariance de l'étendue d'un pinceau lumineux et par conséquent celle de la luminance. Enfin nous terminerons ce chapitre par une comparaison entre l'optique géométrique et l'optique électronique électrostatique en concluant à la possibilité de transposer à cette dernière la plupart des résultats que nous rencontrerons en optique géométrique.

II. L'optique géométrique déduite des équations de MAXWELL.

2. Rappel des relations fondamentales de la théorie électromagnétique. Dans un milieu de constante diélectrique ε, de perméabilité magnétique unitaire les équations de MAXWELL-AMPÈRE et MAXWELL-FARADAY s'écrivent en utilisant les unités électrostatiques pour le champ électrique $\boldsymbol{E}$ et électromagnétiques pour le champ magnétique $\boldsymbol{H}$

$$\left.\begin{array}{l} \dfrac{\varepsilon}{c}\,\boldsymbol{E}' = \operatorname{rot}\boldsymbol{H}, \\[2mm] \dfrac{1}{c}\,\boldsymbol{H}' = -\operatorname{rot}\boldsymbol{E} \end{array}\right\} \tag{2.1}$$

(le signe $'$ désigne bien entendu la dérivée par rapport au temps). Ces deux équations régissent complètement la propagation d'une onde électromagnétique quelconque et dans le but d'étudier le passage à l'optique géométrique il est particulièrement utile de connaître les déplacements de l'énergie électromagnétique au cours du temps; l'énergie localisée dans un colume dv est égale à $dW = \frac{1}{8\pi}(\varepsilon \boldsymbol{E}^2 + \boldsymbol{H}^2)\,dv$ et le flux d'énergie qui sort par seconde d'un volume fermé constant V est égal à

$$-\frac{dW}{dt} = -\frac{d}{dt}\iiint \frac{1}{8\pi}(\varepsilon \boldsymbol{E}^2 + \boldsymbol{H}^2)\,dv = -\frac{1}{4\pi}\iiint(\varepsilon \boldsymbol{E}\cdot \boldsymbol{E}' + \boldsymbol{H}\cdot \boldsymbol{H}')\,dv.$$

On peut tirer $\boldsymbol{E}'$ et $\boldsymbol{H}'$ des équations de Maxwell; on obtient alors

$$-\frac{dW}{dt} = \frac{c}{4\pi}\iiint(\boldsymbol{H}\,\mathrm{rot}\,\boldsymbol{E} - \boldsymbol{E}\,\mathrm{rot}\,\boldsymbol{H})\,dv$$

ou encore puisque l'on a identiquement quels que soient $\boldsymbol{E}$ et $\boldsymbol{H}$

$$\boldsymbol{H}\,\mathrm{rot}\,\boldsymbol{E} - \boldsymbol{E}\,\mathrm{rot}\,\boldsymbol{H} = \mathrm{div}\,(\boldsymbol{E}\times \boldsymbol{H}),$$

$$-\frac{dW}{dt} = \frac{c}{4\pi}\iiint \mathrm{div}\,(\boldsymbol{E}\times \boldsymbol{H})\,dv.$$

On peut poser $\boldsymbol{P} = \frac{c}{4\pi}(\boldsymbol{E}\times \boldsymbol{H})$ et la relation précédente montre (en utilisant l'identité d'Ostrogradsky) que le flux d'énergie qui sort de la surface fermée S devient égal au flux du vecteur $\boldsymbol{P}$ appelé vecteur de Poynting:

$$\boldsymbol{P} = \frac{c}{4\pi}(\boldsymbol{E}\times \boldsymbol{H}). \qquad (2.2)$$

Ce vecteur détermine donc le déplacement de l'énergie: son flux à travers une surface quelconque est constamment égal au flux d'énergie qui sort de cette surface; c'est ce vecteur qui va nous conduire dans l'approximation de l'optique géométrique à la notion de rayon lumineux.

3. Cas des milieux continus non homogènes. Nous allons maintenant chercher une solution «monochromatique» des relations de Maxwell, c'est-à-dire une solution où chacun des champs $\boldsymbol{E}$ et $\boldsymbol{H}$ sera sinusoïdal en fonction du temps. Si f est la fréquence et $\omega = 2\pi f$ la pulsation de la vibration on posera:

$$\left.\begin{array}{l} \boldsymbol{E}(x, y, z, t) = \boldsymbol{E}_0(x, y, z)\,\exp j\omega\left(t - \dfrac{L(x, y, z)}{c}\right), \\[2mm] \boldsymbol{H}(x, y, z, t) = \boldsymbol{H}_0(x, y, z)\,\exp j\omega\left(t - \dfrac{L'(x, y, z)}{c}\right) \end{array}\right\} \qquad (3.1)$$

ainsi les vecteurs $\boldsymbol{E}$, $\boldsymbol{H}$ sont des vecteurs dont le module est un nombre complexe, la valeur physique instantanée étant bien entendu la partie réelle de ce vecteur complexe. Dans les seconds membres des équations on a séparé les vecteurs réels $\boldsymbol{E}_0$, $\boldsymbol{H}_0$, des facteurs de phase variables en fonction du temps et de la position dans l'espace; les fonctions L et L' déterminent seulement les phases locales des vecteurs $\boldsymbol{E}$ et $\boldsymbol{H}$.

Nous allons maintenant examiner les formes limites prises par les relations de Maxwell lorsqu'on y substitue les champs exprimés par les relations (3.1): en appliquant l'identité $\mathrm{rot}\,s\,\boldsymbol{V} = s\,\mathrm{rot}\,\boldsymbol{V} + \mathrm{grad}\,s \times \boldsymbol{V}$ on trouve aisément

$$\mathrm{rot}\,\boldsymbol{E} = \left[\mathrm{rot}\,\boldsymbol{E}_0 - \frac{j\omega}{c}\,\mathrm{grad}\,L \times \boldsymbol{E}_0\right]\exp j\omega\left(t - \frac{L}{c}\right).$$

Lorsque ω devient très grand le premier terme du crochet devient négligeable devant le second; on peut alors porter rot $\boldsymbol{E}$ et rot $\boldsymbol{H}$ dans les équations de Maxwell; on obtient, en supprimant dans les deux membres le facteur commun $\omega \exp j\omega t$ et en posant $k=\omega/c$ la forme limite pour l'une des relations:

$$\varepsilon\,\boldsymbol{E}_0 \exp(-jkL) = -\operatorname{grand} L' \times \boldsymbol{H}_0 \exp(-jkL').$$

Seules les fonctions exponentielles introduisent des quantités complexes et cette identité entraîne nécessairement la condition

$$L(x,y,z) \equiv L'(x,y,z).$$

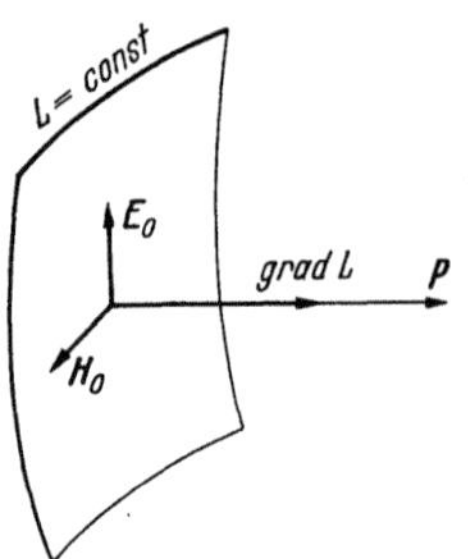

Fig. 1.

Nous appellerons désormais *chemin optique* la fonction unique $L(x,y,z)$ qui détermine la phase commune des deux vecteurs $\boldsymbol{E}$ et $\boldsymbol{H}$.

On obtient finalement comme formes limites des équations de Maxwell:

$$\left.\begin{aligned}\varepsilon\,\boldsymbol{E}_0 &= -\operatorname{grad} L \times \boldsymbol{H}_0,\\ \boldsymbol{H}_0 &= \operatorname{grad} L \times \boldsymbol{E}_0.\end{aligned}\right\} \tag{3.2}$$

On voit que le vecteur grad L va jouer un rôle fondamental dans la propagation; ce vecteur est normal à la surface que nous appellerons désormais *surface d'onde*; les relations de Maxwell imposent que les trois vecteurs $\boldsymbol{E}_0$, $\boldsymbol{H}_0$ et grad L forment un trièdre trirectangle: les vecteurs $\boldsymbol{E}_0$ et $\boldsymbol{H}_0$ sont nécessairement tangents à la surface d'onde et orthogonaux entre eux (Fig. 1).

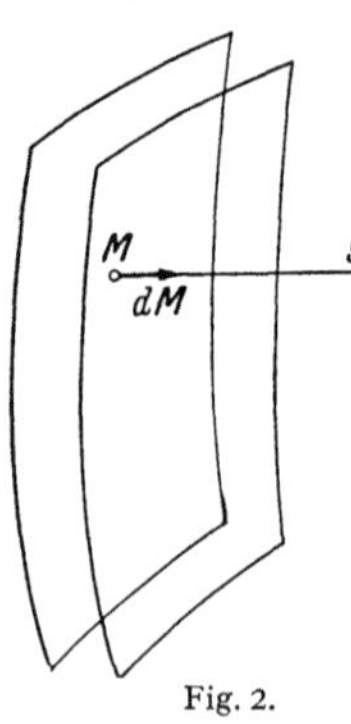

Fig. 2.

Précisons maintenant le module du vecteur grad L. Les deux relations précédentes entraînent les relations suivantes (puisque les vecteurs sont orthogonaux deux à deux)

$$\varepsilon\,|\boldsymbol{E}_0| = |\operatorname{grad} L|\,|\boldsymbol{H}_0|,$$
$$|\boldsymbol{H}_0| = |\operatorname{grad} L|\,|\boldsymbol{E}_0|$$

d'où l'on déduit aisément la relation fondamentale $|\operatorname{grad} L|^2 = \varepsilon$. Nous poserons $\sqrt{\varepsilon}=n$ et nous verrons ultérieurement que cette constante n'est autre que l'indice de réfraction local; on écrira finalement:

$$|\operatorname{grad} L| = n. \tag{3.3}$$

Ceci signifie que si l'on considère deux surfaces d'ondes voisines correspondant aux chemins optiques L et $L+dL$ on doit avoir $dL = \operatorname{grad} L \cdot d\boldsymbol{M} = n\,dl$. *Deux surfaces d'ondes voisines sont séparées par une distance normale dl telle que $n\,dl$ soit constant et indépendant de la position du point M* (Fig. 2).

D'autre part la relation exprimant le champ $\boldsymbol{E}$ peut s'écrire, dans une région peu étendue autour du point considéré M_0

$$\boldsymbol{E} = \boldsymbol{E}_0 \exp j\omega\left(t - \frac{1}{c}\operatorname{grad} L \times (\boldsymbol{M} - \boldsymbol{M}_0)\right)$$

qui est l'équation d'une onde dont la phase se propage dans la direction de grad L avec la vitesse $v = c/|\operatorname{grad} L|$; la vitesse de propagation de la phase est donc $v = c/n$.

Cherchons enfin le vecteur de Poynting qui déterminera la direction de la propagation de l'énergie; il résulte des relations (3.2) et (3.3) que l'on aura

$$|\boldsymbol{H}_0| = n\,|\boldsymbol{E}_0|.$$

Le vecteur $\boldsymbol{P} = \dfrac{c}{4\pi}(\boldsymbol{E} \times \boldsymbol{H})$ est nécessairement colinéaire à grad L (Fig. 1); les valeurs instantanées de $\boldsymbol{E}$ et $\boldsymbol{H}$ seront respectivement $\boldsymbol{E} = \boldsymbol{E}_0 \cos \omega \left(t - \dfrac{L}{c}\right)$ et $\boldsymbol{H} = \boldsymbol{H}_0 \cos \omega \left(t - \dfrac{L}{c}\right)$. Le vecteur $\boldsymbol{P}$ aura donc pour module:

$$|\boldsymbol{P}| = \frac{n\,c\,\boldsymbol{E}_0^2}{4\pi} \cos^2 \omega \left(t - \frac{L}{c}\right).$$

Ce vecteur détermine la direction de propagation de l'énergie, c'est-à-dire la direction du *rayon lumineux*: on voit que les rayons lumineux sont constamment normaux aux surfaces d'ondes et l'on peut résumer ainsi les résultats obtenus:

1. Les champs électriques et magnétiques sont en phase, orthogonaux entre eux et tangents à la surface équiphase (surface d'onde),

2. la phase de la vibration est égale à $-kL$, *la fonction* L *satisfaisant à la condition* $|\operatorname{grad} L| = n = \sqrt{\varepsilon}$,

3. les trajectoires de l'énergie (rayons lumineux) sont normales aux surfaces d'ondes.

Il résulte des deux dernières propositions que le chemin optique L s'exprime aisément si l'on suit un rayon lumineux: on a en effet de façon générale:

$$L = \int \operatorname{grad} L \cdot d\boldsymbol{M},$$

mais le long d'un rayon lumineux le déplacement élémentaire $d\boldsymbol{M}$ est colinéaire à grad L et l'on a dans ces conditions

$$L = \int n\,ds.$$

4. La réfraction sur un dioptre, les lois de DESCARTES-SNELL. Voyons maintenant comment s'exprime la réfraction d'une onde à la surface de séparation de deux milieux d'indices n et n': les conditions de passage relatives aux champs $\boldsymbol{E}$ et $\boldsymbol{H}$ qui se traduisent par la continuité des composantes tangentielles de ces champs exigent en particulier que la phase soit la même des deux côtés de la surface; autrement dit la réfraction n'apporte aucune discontinuité de la fonction L; figurons donc dans les deux milieux dont les indices locaux de part et d'autres du dioptre sont n et n' la surface d'onde correspondant à la valeur L du chemin optique (Fig. 3); appelons $n\boldsymbol{u}$ et $n'\boldsymbol{u}'$ les vecteurs grad L avant et après réfraction: $\boldsymbol{u}$ et $\boldsymbol{u}'$ sont ainsi les vecteurs unitaires portés par les rayons incident et réfracté.

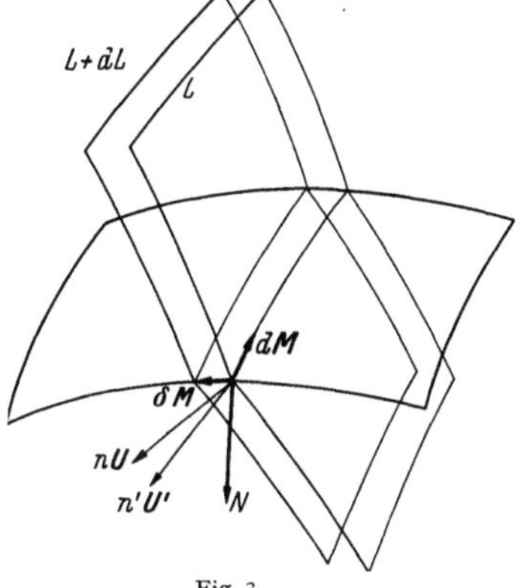

Fig. 3.

Si l'on suit la ligne d'intersection de la surface d'onde avec le dioptre on doit avoir dans les deux milieux en appelant $d\boldsymbol{M}$ le déplacement:

$$n\boldsymbol{u} \cdot d\boldsymbol{M} = n'\boldsymbol{u}' \cdot d\boldsymbol{M} = 0.$$

Ainsi les vecteurs $\boldsymbol{u}$ et $\boldsymbol{u}'$ sont normaux à $d\boldsymbol{M}$. Ceci est également vrai du vecteur unitaire $\boldsymbol{N}$ de la normale au dioptre et l'on en déduit la première loi de DESCARTES:

Le rayon incident, le rayon réfracté et la normale au dioptre sont dans un même plan (plan d'incidence).

Si $\delta\boldsymbol{M}$ est maintenant un déplacement normal à $d\boldsymbol{M}$ dans le plan tangent au dioptre, on doit avoir (Fig. 4):

$$\delta L = \operatorname{grad} L \cdot \delta\boldsymbol{M} = \operatorname{grad} L' \cdot \delta\boldsymbol{M} = n\boldsymbol{u} \cdot \delta\boldsymbol{M} = n'\boldsymbol{u}' \cdot \delta\boldsymbol{M}.$$

On en déduit que $n\boldsymbol{u}$ et $n'\boldsymbol{u}'$ doivent avoir la même projection sur la droite tangente au dioptre contenue dans le plan d'incidence; il en résulte la relation de Descartes-Snell:

$$n \sin i = n' \sin i'. \tag{4.1}$$

L'étude de la réfraction des ondes électromagnétiques de haute fréquence nous amène donc aux lois élémentaires classiques. Nous remarquons d'ailleurs que nous n'avons eu pour cela qu'à considérer les phases des champs électrique ou magnétique; en écrivant d'autre part la continuité des composantes tangentielles nous aurions obtenu les facteurs de réflexion du dioptre exprimés par les relations bien connues de Fresnel.

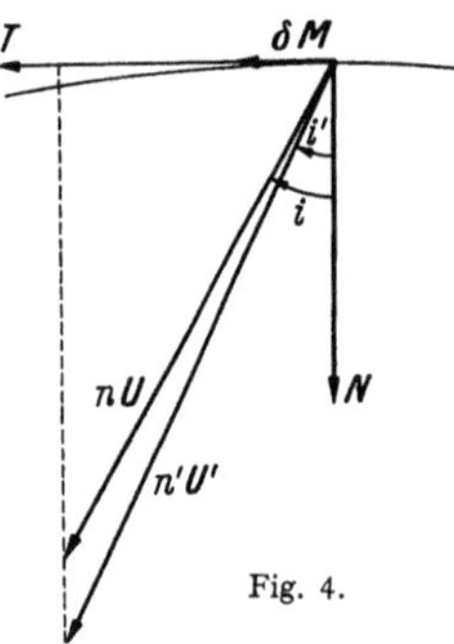

Fig. 4.

Si nous avons montré que la forme limite des équations qui régissent la propagation des ondes conduit à des conclusions simples qui ne sont autres que les lois de l'optique géométrique, il nous est cependant difficile de déduire de là quel sera le degré de validité de l'optique géométrique dans un cas concret fixé à priori: étant donné un instrument dont on connait parfaitement les caractéristiques, avec quelle précision pourra-t-on appliquer cette approximation en un point donné de l'espace? C'est pour aider à résoudre ce problème que nous allons maintenant utiliser une autre méthode basée sur le calcul de l'amplitude vibratoire à l'aide du principe d'Huygens-Fresnel; bien qu'approximative cette méthode nous conduira néamoins à des conclusions intéressantes quant aux limites de validité des lois de l'optique géométrique.

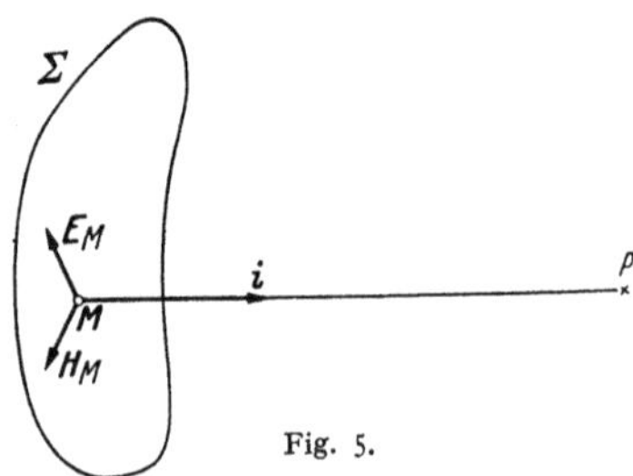

Fig. 5.

5. L'optique géométrique déduite du principe d'Huygens-Fresnel. Nous allons maintenant montrer que le calcul de l'amplitude diffractée en un point de l'expace par une surface d'onde peut conduire dans certaines conditions à une valeur qui représente l'approximation de l'optique géométrique. Le calcul est basé sur l'utilisation du principe d'Huygens-Fresnel, dont on peut donner diverses expres sions; celle que nous utiliserons peut être facilement déduite des relations de Croze et Darmois[1] dans le cas où la distance $r = MP$ du point M de la surface d'onde au point P où l'on veut calculer l'amplitude vibratoire est grande par rapport à la longueur d'onde (Fig. 5)

$$\boldsymbol{E}_p = \frac{jk}{4\pi} \iint_{\Sigma} (\boldsymbol{E}_{Myz} - \boldsymbol{i} \times \boldsymbol{H}_M) \frac{1}{r} \exp(-jkr)\, dS. \tag{5.1}$$

(On suppose $\varepsilon = 1$.) Dans cette relation $\boldsymbol{E}_M$ et $\boldsymbol{H}_M$ sont les champs électrique et magnétique au point M de la surface d'intégration, $\boldsymbol{i}$ le vecteur unitaire porté par MP, $\boldsymbol{E}_{Myz}$ la projection de $\boldsymbol{E}_M$ sur un plan normal à $\boldsymbol{i}$ et dS est l'élément d'aire de la surface d'intégration, qui sera ici une surface d'onde. Si la fréquence des ondes est élevée $\boldsymbol{E}_M$ et $\boldsymbol{H}_M$ seront orthogonaux entre eux et tangents à la surface d'onde Σ. Si d'autre part le point P est situé sur la normale à la surface d'onde dans le sens de la propagation, on aura $-\boldsymbol{i} \times \boldsymbol{H}_M = \boldsymbol{E}_M$ et l'élément d'ampli-

---

[1] F. Croze et G. Darmois: C. R. Acad. Sci. Paris **228**, 824 (1949).

tude diffractée normalement par l'élément dS est

$$d\boldsymbol{E}_p = \frac{jk}{2\pi}\,\boldsymbol{E}_M\,f(r)\,dS.$$

(On a posé ici $f(r) = \dfrac{1}{r}\exp\left(-jkr\right)$.)

Pour une même distance r l'amplitude diffractée dans une direction oblique est toujours inférieure à l'amplitude diffractée normalement puisque le module de $\boldsymbol{E}_{Myz} - \boldsymbol{i}\times\boldsymbol{H}_M$ est maximum lorsque $\boldsymbol{i}$ est normal à $\boldsymbol{E}_M$ et $\boldsymbol{H}_M$.

Il n'est pas aisé de représenter par une construction graphique la sommation de vecteurs élémentaires dont le module est complexe: nous supposerons que l'orientation du vecteur $\boldsymbol{E}_{Myz} - \boldsymbol{i}\times\boldsymbol{H}_M$ varie peu sur la surface d'onde, et nous nous bornerons à représenter dans le plan complexe la formation de l'intégrale qui représente la composante principale du vecteur (par exemple une droite représentant l'orientation du vecteur au point M_0, défini ci-après).

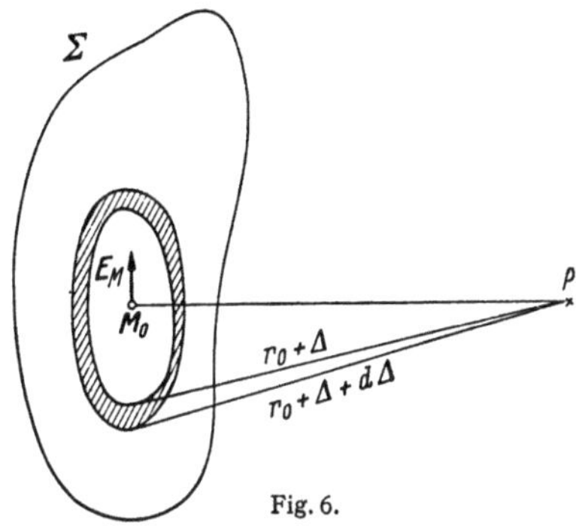

Fig. 6.

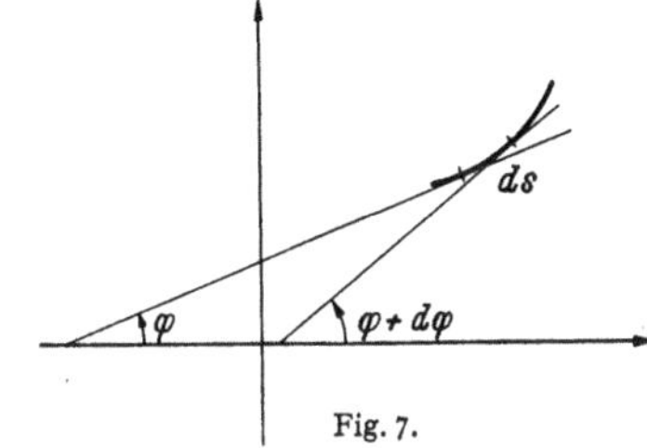

Fig. 7.

Considérons un point P où nous nous proposons de calculer l'amplitude diffractée; soit M_0 un point de la surface d'onde tel que $r = MP$ soit stationnaire en M_0 et égal à r_0: ceci veut dire que PM_0 est *normale à l'onde ou encore que* c'est un *rayon lumineux* (Fig. 6).

Supposons par exemple que PM soit minimum. De P comme centre il est possible de décrire une famille de sphères dont les rayons varient en progression arithmétique de raison $\lambda/2$, ce qui permet de définir sur Σ des zônes sucessives analogues aux zônes de Fresnel: traçons plutôt deux sphères de rayons très voisins $r_0 + \Delta$ et $r_0 + \Delta + d\Delta$. Ces sphères coupent la surface d'onde suivant deux courbes; dS sera l'élément de surface comprise entre ces courbes définies par la condition $r = $ constante; Δ peut être considéré comme une fonction de la surface S ce qui va permettre de représenter la construction de l'intégrale précédente dans le plan complexe. Supposons que r_0 soit un nombre entier de longueurs d'ondes

$$\exp\left(-jkr_0\right) = 1 \quad \text{et} \quad \exp\left(-jkr\right) = \exp\left(-jk\Delta\right)$$

l'élément d'arc de courbe dans le plan complexe est

$$\frac{jk}{2\pi}\,E_0\,\frac{1}{r}\exp\left(-jk\Delta\right)dS.$$

Lorsque la longueur d'onde est petite seule l'exponentielle varie rapidement car $k = 2\pi/\lambda_0$ est grand, alors que r et E_0 varient lentement; d'autre part la courbure de la courbe représentant l'intégrale dans le plan complexe est dans tous les cas (Fig. 7)

$$\frac{d\varphi}{ds} = \frac{-\,d(k\Delta)}{\dfrac{k}{2\pi r}E_0\,ds} = -\frac{2\pi r}{E_0}\frac{d\Delta}{dS}. \tag{5.2}$$

Il nous reste à évaluer $d\Delta/ds$. Si r_1 et r_2 sont les rayons de courbure principaux de la surface d'onde au voisinage de son pôle, la courbe définie par la condition $MP = r_0 + \Delta$ peut être connue en première approximation; par rapport au plan tangent en M_0 à Σ l'équation de la surface est de la forme $z = \dfrac{x^2}{2r_1} + \dfrac{y^2}{2r_2}$ alors que l'équation de la sphère de rayon r_0 est $z = \dfrac{x^2 + y^2}{2r_0}$. Il suffit d'écrire $z - z' = \Delta$ c'est-à-dire

$$\frac{x^2}{2}\left(\frac{1}{r_1} - \frac{1}{r_0}\right) + \frac{y^2}{2}\left(\frac{1}{r_2} - \frac{1}{r_0}\right) = \Delta.$$

Cette courbe est en première approximation une ellipse et sa surface est égale à $\pi a b$, si a et b sont les demi longueurs des axes; on aura ici

$$S = 2\pi \Delta \left[\left(\frac{1}{r_1} - \frac{1}{r_0}\right)\left(\frac{1}{r_2} - \frac{1}{r_0}\right)\right]^{-\frac{1}{2}}$$

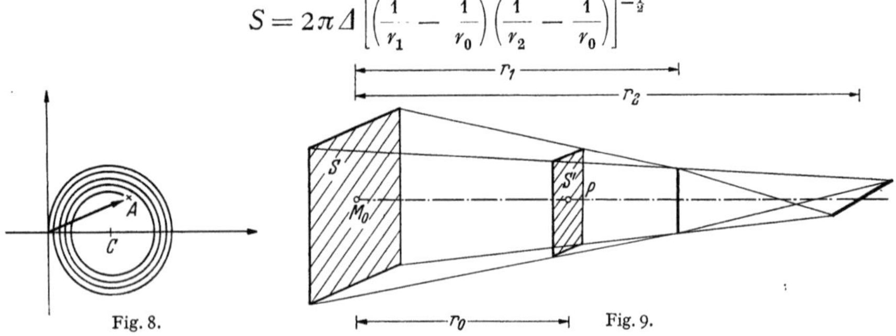

Fig. 8.   Fig. 9.

de sorte que la courbure de la courbe dans le plan complexe sera

$$\frac{r}{E_0}\left[\left(\frac{1}{r_1} - \frac{1}{r_0}\right)\left(\frac{1}{r_2} - \frac{1}{r_0}\right)\right]^{\frac{1}{2}}. \tag{5.3}$$

Lorsque la longueur d'onde est petite il suffit que Δ soit petit lui-même pour introduire des différences de phases appréciables sans que $r = r_0 + \Delta$ varie en valeur relative de façon appréciable: autrement dit la courbure exprimée par (5.3) reste sensiblement constante au début, le rayon de courbure étant

$$\frac{E_0}{\left[\left(\dfrac{r_0}{r_1} - 1\right)\left(\dfrac{r_0}{r_2} - 1\right)\right]^{\frac{1}{2}}}. \tag{5.4}$$

La courbe décrite dans le plan complexe est une spirale tangente en O à l'axe imaginaire (du fait de la présence du facteur j dans la formule (5.1) et le rayon de courbure varie peu au début et est égal à la valeur (5.4). Les zônes de plus en plus éloignées de M diffracteront des éléments de vibration de plus en plus faibles et le rayon de courbure diminue: la spirale tend à se fermer et aboutit finalement à un point extrême A, la vibration résultante étant OA (Fig. 8).

En moyenne la résultante de la vibration est donc de l'ordre de OC, C étant le centre de courbure à la première boucle et l'amplitude obtenue sera égale au rayon de courbure (5.4) c'est-à-dire que l'énergie sera représentée par

$$E_0^2\left[\left(\frac{r_0}{r_1} - 1\right)\left(\frac{r_0}{r_2} - 1\right)\right]^{-1}.$$

Il est facile de montrer que c'est précisemment l'éclairement que l'on doit avoir d'après l'optique géométrique. P est en effet situé à la distance r_0 de la surface d'onde alors que les focales (voir Sect. 65) sont aux distances r_1 et r_2. Si l'on admet que l'énergie se conserve dans le pinceau de rayons, il suffit d'évaluer la surface de section du pinceau en fonction de r et un raisonnement simple montre (Fig. 9) que la section par le plan de front passant par P est

$$S' = S\left[\left(\frac{r_0}{r_1} - 1\right)\left(\frac{r_0}{r_2} - 1\right)\right].$$

On en déduit que l'éclairement doit bien prendre la valeur $E_0^2\, S/S'$ qui est précisément la valeur à laquelle nous sommes parvenus en appliquant le principe d'Huygens-Fresnel.

Quelles conditions doit-on réaliser pour que cette approximation soit valable? Il faut tout d'abord que par le point d'observation P passe un rayon lumineux, (sinon l'approximation de l'optique géométrique indique une énergie nulle) et il faut d'autre part qu'il ne passe pas plus d'un rayon: s'il en passait plusieurs nous aurions en effet à combiner les amplitudes diffractées par les diverses zônes de la surface d'onde entourant les pied de ces rayons en tenant compte des phases et nous observerions des interférences entre ces diverses amplitudes.

Enfin on ne pourrait atteindre effectivement la limite ainsi calculée que si l'on réunissait les deux conditions:

1. La variation Δ de r lorsqu'on parcourt la surface d'onde est importante par rapport à la longueur d'onde (on décrit donc un grand nombre de tours sur la spirale); autrement dit il faut pouvoir décrire sur la surface d'onde un très grand nombre de zônes de Fresnel, ce qui implique que M_0 ne soit pas au voisinage du contour de l'onde ou encore que P ne soit pas au voisinage des limites du pinceau.

2. Il faut encore que la courbure de la spirale diminue notablement lorsque M s'éloigne de M_0; ceci nécessiterait que le champ $\boldsymbol{E}_M$ en M aille en décroissant, ou encore que l'inclinaison de MP sur la normale à l'onde soit importante; cette condition n'est généralement pas réalisée et bien que l'on décrive un grand nombre de tours sur la spirale la résultante obténue est généralement différente de OC (Fig. 8); tout dépend des conditions aux limites et il est possible de préciser l'influence des bords du diaphragme; ainsi van Kampen[1] a montré que les points du contour de l'onde ou Δ est stationnaire diffractent en P une amplitude qui représente la deuxième approximation; nous ne reproduirons pas ici le développement de cette méthode.

III. Le principe de Fermat.

6. Enoncé du principe. Fermat avait pensé que pour aller d'un point a un autre la lumière emprunte le chemin qui correspond au minimum de temps de

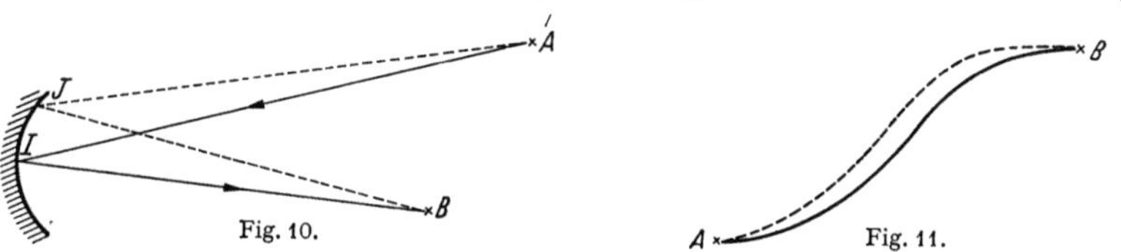

Fig. 10.   Fig. 11.

parcours. Ceci n'est en fait pas exact et pour s'en rendre compte il suffit de considérer l'exemple d'un miroir concave de forte courbure réfléchissant suivant AIB un rayon provenant d'un qui n'est pas l'image de A point éloigné A et aboutissant en un autre point éloigné B (Fig. 10); le chemin réel AIB est certainement plus long que le chemin fictif voisin AJB.

Le temps mis par la lumière pour aller d'un point A à un point B s'exprime par l'intégrale (Fig. 11)

$$T_{AB} = \int_A^B \frac{ds}{v}$$

---

[1] Van Kampen: Physica, Haag **14**, 575 (1949).

si v est la vitesse locale de propagation; mais on a vu que l'on doit avoir $v = c/n$ et l'on peut écrire

$$T_{AB} = \frac{1}{c} \int_A^B n\, ds. \tag{6.1}$$

L'intégrale $L = \int_A^B n\, ds$ s'appelle le *chemin optique* entre A et B le long du chemin envisagé; elle représente le chemin qu'aurait parcouru la lumière dans le vide pendant le temps T qu'elle a mis pour aller de A en B si le chemin considéré est effectivement un rayon lumineux.

On peut énoncer le principe de Fermat sous la forme suivante:

Les rayons lumineux sont les extrêmales du chemin optique.

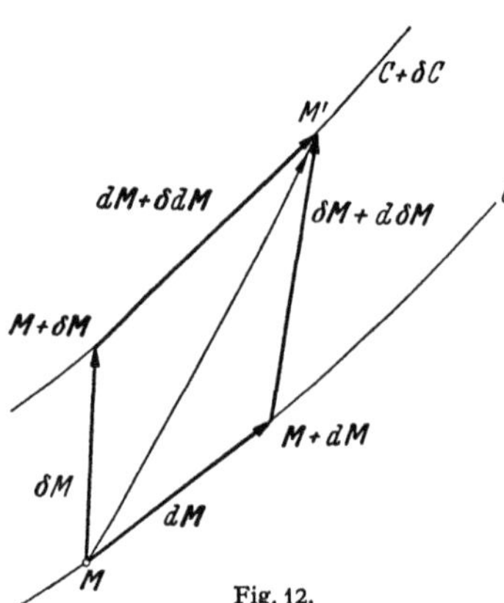

Fig. 12.

Ainsi pour aller de A en B la lumière emprunte un chemin qui correspond à un extrêmum du chemin optique; on écrira donc pour caractériser un rayon lumineux (6.2)

$$\delta \int n\, ds = 0. \tag{6.2}$$

7. Equation des rayons lumineux et courbure. Il nous reste maintenant à déterminer analytiquement les courbes extrêmales qui seront effectivement suivies par la lumière. Pour simplifier l'écriture, nous utiliserons le calcul vectoriel et nous désignerons par $\boldsymbol{M}$ le vecteur joignant un point origine fixe à un point M de la courbe C; soit $\boldsymbol{M} + \delta\boldsymbol{M}$ le vecteur caractérisant un point de la courbe $C + \delta C$ qui résulte d'une déformation de la courbe C (Fig. 12). Nous avons à évaluer la variation de l'intégrale représentant le chemin optique

$$\delta L = \delta \int n\, ds = \int \delta n\, ds + \int n\, \delta ds. \tag{7.1}$$

On a, d'une part,

$$\delta n = \operatorname{grad} n \cdot \delta \boldsymbol{M}$$

et d'autre part,

$$\delta(ds) = \delta(d\boldsymbol{M} \cdot d\boldsymbol{M})^{\frac{1}{2}} = \frac{d\boldsymbol{M} \cdot \delta d\boldsymbol{M}}{ds} = \boldsymbol{T} \cdot \delta d\boldsymbol{M}$$

($\boldsymbol{T}$ est le vecteur unitaire porté par la tangente à la courbe C). Pour transformer la dernière expression, on remarque qu'en égalant les deux expressions du vecteur $\overrightarrow{MM'}$, en évidence sur la Fig. 12, on obtient

$$\delta d\boldsymbol{M} = d\,\delta\boldsymbol{M}$$

et la deuxième intégrale devient

$$\int n\, \delta ds = \int n\, \boldsymbol{T}\, d\,\delta\boldsymbol{M},$$

ce que l'on peut écrire, en intégrant par parties,

$$\left[n\, \boldsymbol{T}\, \delta\boldsymbol{M}\right]_A^B - \int \delta\boldsymbol{M} \cdot \frac{d n\, \boldsymbol{T}}{ds}\, ds = \left[n\, \boldsymbol{T}\, \delta\boldsymbol{M}\right]_A^B - \int_A^B \delta\boldsymbol{M}\left[\boldsymbol{T}\, \frac{dn}{ds} + n\, \frac{d\boldsymbol{T}}{ds}\right] ds.$$

On sait d'autre part que $d\boldsymbol{T}/ds = \boldsymbol{N}/R$ où, $\boldsymbol{N}$ est le vecteur unitaire de la normale principale et R le rayon de courbure de la courbe; on obtient finalement pour variation de l'intégrale L

$$\delta \int_A^B n\,ds = n_B\,\boldsymbol{T}_B\,\delta\boldsymbol{B} - n_A\,\boldsymbol{T}_A\,\delta\boldsymbol{A} + \int_A^B \left(\operatorname{grad} n - \boldsymbol{T}\,\frac{dn}{ds} - n\,\frac{\boldsymbol{N}}{R}\right)\cdot\delta\boldsymbol{M}\,ds. \qquad (7.2)$$

Dans le cas qui nous occupe, les courbes passent constamment par A et B de sorte que les déplacements $\delta\boldsymbol{A}$ et $\delta\boldsymbol{B}$ sont nuls. Pour que cette expression soit nulle d'autre part, quel que soit le déplacement arbitraire $\delta\boldsymbol{M}$ imprimé à chaque point M de la courbe, il est évidemment nécessaire de satisfaire à la relation

$$\operatorname{grad} n = \boldsymbol{T}\,\frac{dn}{ds} + n\,\frac{\boldsymbol{N}}{R}. \qquad (7.3)$$

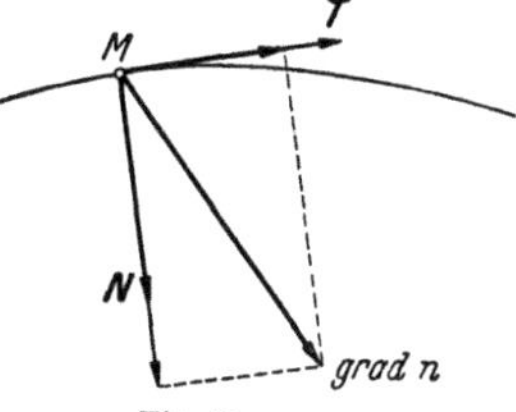

Fig. 13.

Cette relation vectorielle fixe d'abord l'orientation du plan osculateur au rayon lumineux, qui doit nécessairement contenir le vecteur $\operatorname{grad} n$ (Fig. 13); elle détermine d'autre part la courbure du rayon: si $\partial n/\partial N$ est la composante du gradient de l'indice sur la normale principale au rayon lumineux, on devra avoir:

$$n\,\frac{\boldsymbol{N}}{R} = \frac{\partial n}{\partial N}\,\boldsymbol{N}.$$

La courbure devra donc être égale à

$$\frac{1}{R} = \frac{1}{n}\,\frac{\partial n}{\partial N} = \frac{\partial \log n}{\partial N}. \qquad (7.4)$$

Fig. 14.

Ainsi dans un milieu non homogène la courbure des rayons lumineux s'exprime à l'aide de la composante du gradient d'indice sur la normale principale au rayon.

Signalons d'ailleurs un raisonnement simple qui permet de retrouver ce résultat: la figure étant faite dans le plan osculateur aux rayons lumineux considérons deux portions de surface d'ondes très voisines; considérons également deux rayons voisins les perçant respectivement en A, A' et B, B' (Fig. 14); on devra avoir $AA' = v\,dt$ et $BB' = (v+dv)\,dt$, mais le centre de courbure du rayon est la limite du point de rencontre C des deux surfaces d'ondes voisines et l'on aura

$$\frac{AA'}{R} = \frac{BB'}{R - dN}$$

d'où l'on déduit aisément que

$$1 - \frac{dN}{R} = 1 + \frac{dv}{v},$$

mais puisque $dv/v = -dn/n$ on peut encore écrire $dN/R = dn/n$ qui donne effectivement

$$\frac{1}{R} = \frac{1}{n}\,\frac{\partial n}{\partial N}.$$

8. Quelques applications de la courbure des rayons lumineux. Indiquons maintenant quelques exemples d'applications: l'indice de réfraction des gaz suit généralement la loi de GLADSTONE, c'est-à-dire que la quantité $n-1$ est proportionnelle à la densité ϱ du gaz. Si l'on considère l'atmosphère terrestre en équilibre, la densité de l'air va en décroissant avec l'altitude: les rayons lumineux qui nous parviennent des étoiles sont courbés dans le même sens que la surface terrestre et l'étoile parait à l'observateur être plus haute qu'elle n'est située en réalité (Fig. 15). De même, lorsqu'on tente de pointer optiquement la ligne d'horizon, les distances mises en jeu s'évaluent souvent en dizaines de kilomètres, et il est nécessaire de tenir compte de la courbure de rayons qui a encore ici pour effet de déplacer apparemment l'horizon vers le haut (Fig. 16). La courbure des rayons lumineux n'est d'ailleurs pas négligeable par rapport à celle de la terre puisqu'elle est environ 6 à 7 fois plus petite lorsqu'on est à basse altitute.

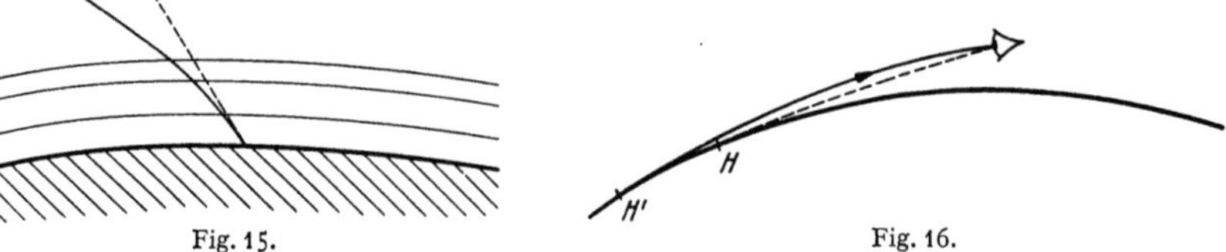

Fig. 15.   Fig. 16.

Lorsque le sol est chauffé par le Soleil et que l'atmosphère n'est plus en équilibre thermique, le phénomène s'inverse, tout au moins pour les rayons qui cheminent près du sol; la décroissance rapide de la température avec l'altitude produit, en supposant la pression à peu près constante, une croissance de la densité et de l'indice. Les rayons se sourbent donc vers le haut et peuvent alors donner

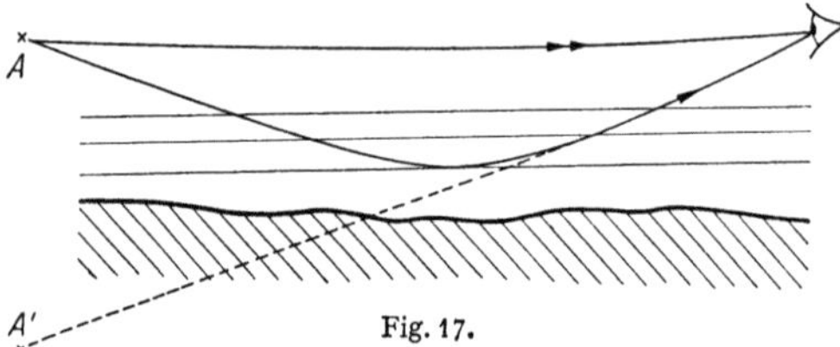

Fig. 17.

lieu à une illusion connue sous le nom de mirage. Admettons que seules les couches d'air situées au voisinage immédiat du sol présentent des variations de température notable avec l'atiltude: le rayon parti du point A et descendant vers le sol peut subir une courbure suffisante pour remonter ensuite vers l'oeil de l'observateur qui perçoit alors deux images du point A, ayant ainsi l'illusion de la présence d'une nappe d'eau qui reflèterait l'objet (Fig. 17).

Admettons par exemple que le gradient de température T soit de l'ordre de $\frac{1}{10}$ de degré par centimètre: si l'on suppose que la pression est sensiblement constante ou, si ϱ est la masse spécifique et h l'altitude

$$\frac{1}{\varrho}\frac{d\varrho}{dh} = -\frac{1}{T}\frac{dT}{dh}.$$

D'après la loi de Gladstone on a d'autre part

$$\frac{1}{n-1}\frac{d(n-1)}{dh} = \frac{1}{\varrho}\frac{d\varrho}{dh}$$

d'où la courbure des rayons

$$\frac{1}{R} = \frac{d\log n}{dh} = \frac{1}{v}\frac{d(n-1)}{dh} = -\frac{n-1}{n-T}\frac{dT}{dh}.$$

Pour

$$n-1 = 3\cdot 10^{-4}, \qquad \frac{dT}{dh} = \frac{1}{10}\ \mathrm{cm^{-1}}, \qquad T = 300°\ \mathrm{K}$$

on trouve

$$\frac{1}{R} = 10^{-7}\ \mathrm{cm^{-1}}.$$

Bien que le gradient de température soit assez faible la courbure peut suffire à produire des mirages: sur une longueur de 1 kilomètre le rayon est dévié de $\frac{1}{100}$ de radian ce qui est très visible.

9. Postulats fondamentaux de l'optique géométrique. Montrons tout d'abord que le principe de FERMAT permet de retrouver le principe de la propagation rectiligne: si nous supposons le milieu homogène, l'indice n est constant, ses dérivées s'annulent et la relation (7.3) indique que la courbure de rayon lumineux est nulle; c'est-à-dire que la lumière se propage en ligne droite. Dans un ensemble de milieux homogènes un rayon lumineux sera composé de segments de droites.

Le principe de FERMAT permet aussi de retrouver le principe du retour inverse: en effet, le chemin optique est le même dans les deux sens puisque l'on change à la fois le sens positif utilisé (sens de propagation de la lumière) et l'ordre des bornes d'intégrations; on trouvera donc les mêmes courbes extrêmales et le lumière suivra le même chemin quel que soit le sens dans lequel elle se propage.

Enfin, le principe de FERMAT englobe les lois de DESCARTES relatives à la réflexion ou à la réfraction sur une surface séparant deux milieux homogènes:

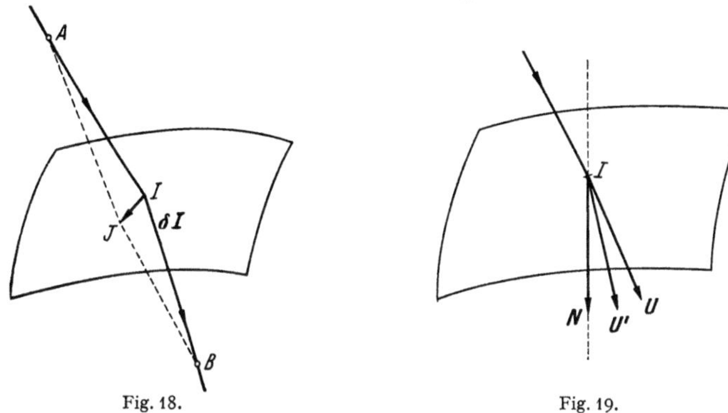

Fig. 18.   Fig. 19.

considérons par exemple un rayon lumineux passant successivement en A dans le premier milieu, au point I sur le dioptre, puis en B dans le deuxième milieu (Fig. 18); la principe de FERMAT indique que le chemin optique doit être stationnaire autour du rayon effectif que nous venons de jalonner. En particulier, déplaçons le point d'iucidence de I en J sur le dioptre et considérons le nouveau chemin (AJB) composé des deux segments AJ et JB; la différence des chemins optiques (AIB) et (AJB) doit être du deuxième ordre par rapport à $\overrightarrow{IJ} = \delta \boldsymbol{I}$. Pour traduire ceci en équations, il nous faut exprimer tout d'abord la différentielle du chemin optique que l'on peut aisément évaluer en utilisant l'expression (7.2) appliquée successivement à AI et IB.

Nous pouvons aussi le trouver à l'aide de considérations géométriques simples.

Appelons $\boldsymbol{u}$ et $\boldsymbol{u}'$ les vecteurs unitaires portés par AI et IB, orientés dans le sens de propagation de la lumière. La différentielle de la distance AI est égale à la projection du déplacement élémentaire $\delta \boldsymbol{I}$ sur AI, c'est-à-dire à

$$\delta(AI) = n\,\boldsymbol{u} \cdot \delta \boldsymbol{I}.$$

On a de même

$$\delta(IB) = -\,n'\boldsymbol{u}' \cdot \delta \boldsymbol{I}.$$

On aura donc

$$\delta(AIB) = (n\,\boldsymbol{u} - n'\boldsymbol{u}') \cdot \delta \boldsymbol{I}.$$

Pour que $\delta(AIB)$ ne soit pas du même ordre que $\delta \boldsymbol{I}$, il faut et suffit que le vecteur $(n\boldsymbol{u} - n'\boldsymbol{u}')$ soit normal à $\delta \boldsymbol{I}$ quel que soit le déplacement de I dans le plan tangent au dioptre; le principe de FERMAT se traduit donc par le fait que le vecteur $n\boldsymbol{u} - n'\boldsymbol{u}'$ est porté par la normale au dioptre. Appelons $\boldsymbol{N}$ le vecteur unitaire de cette normale (Fig.19) et examinons les conséquences ce cette propriété:

1. Les vecteurs u, u' et N sont dans un même plan (première loi de Descartes).

2. Si T est la droite tangente au dioptre et située dans le plan d'incidence, (Fig. 20) les composantes des vecteurs u, u' et N sur cette droite sont respectivement $\sin i$, $\sin i'$, 0; on obtient donc par projection de la condition vectorielle, la relation de la réfraction

$$n \sin i - n' \sin i' = 0.$$

3. Dans le cas de la réflexion, on obtiendra évidemment la relation $i' = -i$.

Les lois de Descartes peuvent donc se déduire du principe de Fermat, et il est évident que le même raisonnement effectué en sens inverse permet de montrer que l'ensemble du principe de la propagation rectiligne et des lois de Descartes entraînent le principe de Fermat, tout an moins en ce qui concerne le cas d'un ensemble de milieux homogènes séparés par des dioptres.

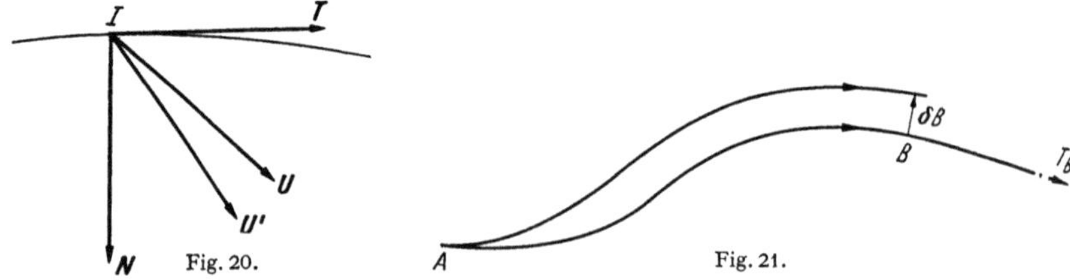

Fig. 20.   Fig. 21.

10. Le théorème de Malus. Montrons que le principe de Fermat entraîne l'existence de surface d'ondes normales aux rayons lumineux. On obtient une surface d'onde en portant sur chacun des rayons issus d'une même source A un chemin optique constant; si l'on considère en particulier deux rayons infiniment voisins (Fig. 21) on peut utiliser l'expression (7.2) de la variation du chemin optique, qui est ici nulle par hypothèse. Le déplacement δA l'est aussi; de plus l'intégrale s'annule le long d'un rayon lumineux; il en résulte que le déplacement δB le long de la surface d'onde est normal au rayon lumineux puisque son produit scalaire par le vecteur T_B doit être nul.

En conclusion, nous voyons que le principe de Fermat suffit à lui seul à définir mathématiquement toutes les hypothèses fondamentales de l'optique géométrique; il est plus général que les lois de Descartes puisqu'il contient en même temps le principe de la propagation rectiligne et la principe du retour inverse. Sa validité s'étend encore au cas des milieux hétérogènes, où l'indice de réfraction varie de façon continue. Il nous faut encore examiner une autre conséquence importante de ce principe.

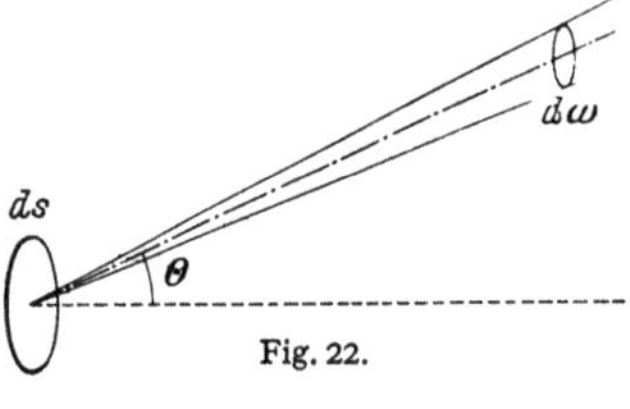

Fig. 22.

11. La relation de Clausius. Cette relation est utile pour étudier la photométrie: elle exprime la conservation de l'étendue d'un pinceau lumineux; si l'on considère un pinceau formé de rayons qui s'appuient sur un élément de surface dS (Fig. 22) et qui sont répartis dans un cône d'angle solide $d\omega$, le rayon moyen de ce cône faisant l'angle ϑ avec la normale à l'élément de surface dS, on appelle *étendue du pinceau* la quantité: $n^2 \, dS \cos \vartheta \, d\omega$.

La relation de Clausius *exprime que l'étendue d'un pinceau se conserve;* si les quantités accentuées sont relatives à l'expace image, on a

$$\boxed{n^2 \, dS \cos \vartheta \, d\omega = n'^2 \, dS' \cos \vartheta' \, d\omega'.} \tag{11.1}$$

Nous ne démontrerons cette relation que dans le cas où le système admet un plan de symétrie que nous prendrons pour plan de la Fig. 23. Il nous faut distinguer deux cas:

1ᵉʳ cas: L'élément dS' considéré dans l'espace image est l'image de l'élément dS situé dans l'espace objet. Nous supposerons que ces éléments sont de petits rectangles; l'élément dS admettra, par exemple, un côté dy dans le plan de figure et un côté dz perpendiculaire à ce plan. Nous supposerons de même que l'angle solide $d\omega$ découpe sur la sphère trigonométrique un rectangle élémentaire dont les côtés correspondent à des angles $d\beta$ et $d\gamma$, l'angle $d\beta$ étant dans le plan de la figure et l'angle $d\gamma$ dans le plan perpendiculaire (Fig. 23).

Exprimons maintenant que tous les rayons issus de A passent par A' (on dit alors qu'il y a *stigmatisme*). Pour chaque rayon, le chemin optique (AA') est stationnaire par rapport aux paramètres (β, γ) fixant l'orientation du rayon; ce chemin optique est donc constant et indépendant du rayon envisagé. Il en

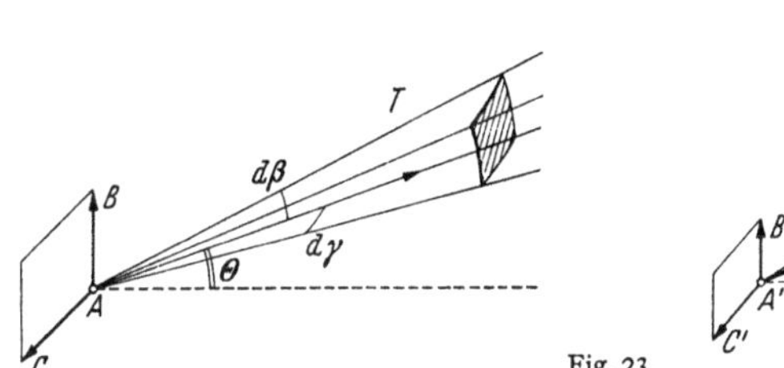

Fig. 23.

est de même de (BB') ou encore de $(BB') - (AA')$. Cette différence de chemins optiques le long de rayons voisins peut s'exprimer facilement à l'aide de la relation (7.2). On aura

$$(BB') - (AA') = n' \overrightarrow{AB'} \cdot \boldsymbol{T}' - n \overrightarrow{AB} \cdot \boldsymbol{T} = n' y' \sin \vartheta' - n y \sin \vartheta$$

qui doit être égal à

$$n' y' \sin (\vartheta' + d\beta') = n y \sin (\vartheta + d\beta);$$

c'est-à-dire que l'on doit avoir

$$n' y' \cos \vartheta' \, d\beta' = n y \cos \vartheta \, d\beta.$$

Le même raisonnement peut être repris pour AC et $A'C'$; on aboutit évidemment à la relation semblable

$$n' z' \, d\gamma' = n z \, d\gamma.$$

(Il n'y a pas d'angles ϑ ou ϑ' à envisager puisque les éléments AC et $A'C'$ sont perpendiculaires aux rayons moyens.)

En faisant le produit des deux dernières égalités, on trouve bien

$$n'^2 \, dS' \cos \vartheta' \, d\omega' = n^2 \, dS \cos \vartheta \, d\omega$$

qui est la relation cherchée.

2ᵐᵉ cas: Il n'y a pas stigmatisme, et les éléments dS, dS' ne sont plus images l'un de l'autre. On prend alors pour élément dS' la surface obtenue en coupant le pinceau des rayons issus de A dans l'angle solide élémentaire $d\omega$. De même, $d\omega'$ sera l'angle solide des rayons qui convergeraient en A' et qui s'appuieraient dans l'expace objet sur la surface dS (Fig. 24). Dans ce cas, par conséquent, dS est proportionnel à $d\omega'$ et $d\omega$ à dS'. On remarque que, si l'on fait varier l'orientation ϑ de l'élément dS, le produit $dS\cos\vartheta$ reste constant puisqu'il représente la projection de dS sur un plan normal au tube de rayons. Nous pouvons

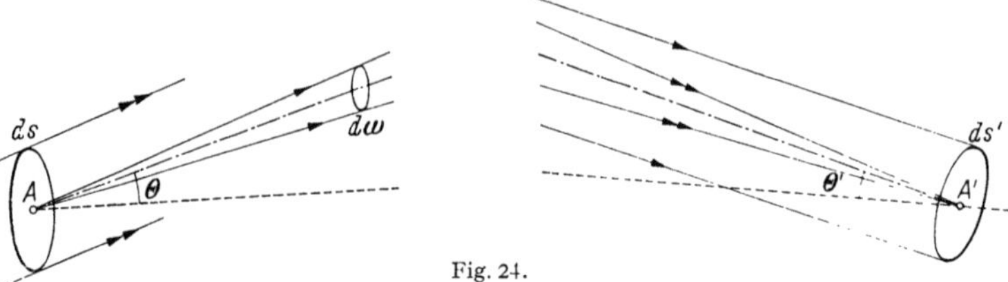

Fig. 24.

donc choisir dès maintenant pour élément dS une section par ce plan normal et annuler ϑ ainsi que ϑ' (Fig. 25).

On remarque aussi que, puisque dS et dS' sont perpendiculaires au rayon moyen, on a $(AB')=(AA')=(A'B)$. En utilisant encore ici la relation (7.2) on écrira d'autre part

$$(BB') - (BA') = n'\,y'\,d\beta',$$

$$(BB') - (AB') = n\,y\,d\beta.$$

En tenant compte des égalités précédentes, on trouve

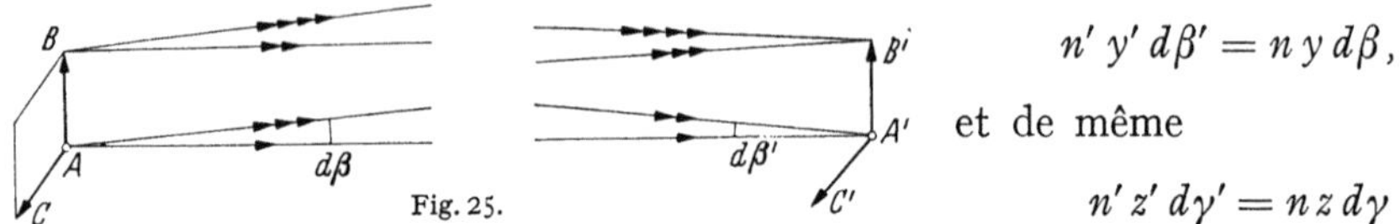

Fig. 25.

$$n'\,y'\,d\beta' = n\,y\,d\beta,$$

et de même

$$n'\,z'\,d\gamma' = n\,z\,d\gamma$$

ce qui nous conduit encore à la relation de Clausius en faisant le produit des deux égalités.

Nous voyons donc que l'invariance de l'étandue d'un pinceau lumineux est une conséquence directe des lois fondamentales de l'optique géométrique: elle ne suppose aucune approximation dans le cas où il n'y a stigmatisme; le plus souvent le stigmatisme n'est qu'approché et les rayons lumineux se concentrent plus ou moins bien autour de l'image: on peut encore montrer que la relation est valable si les aberrations sont assez faibles pour que la diffusion latérale des rayons issus d'un même point de l'objet soit beaucoup plus petite que la surface utilisée dans l'expace image.

12. Application de la relation de Clausius à la photométrie. La relation de Clausius a une conséquence très importante en photométrie: elle impose une limite superieure à la luminance des pinceaux lumineux; si l'on suppose que les milieux objet et image ont le même indice (par exemple celui de l'air) la luminance maximum de l'image est égale à celle de l'objet (théorème de la conservation de la luminance).

Considérons en effet un pinceau dans l'espace objet; le flux transporté est $d\Phi = L\,dS\cos\vartheta\,d\omega$ si L est la luminance de l'objet; le flux qui parvient dans l'espace image est $d\Phi' = \tau\,d\Phi$ où τ est le facteur de transmission de l'instrument, et l'on a d'autre part, en appelant L' la luminance de l'image

$$d\Phi = L'\,dS'\cos\vartheta'\,d\omega',$$

mais l'étendue du pinceau se conserve et l'on a

$$n^2\,dS\cos\vartheta\,d\omega = n'^2\,dS'\cos\vartheta'\,d\omega'.$$

On doit donc avoir finalement

$$\boxed{\frac{L'}{n'^2} = \tau \frac{L}{n^2}.}$$

(12.1)

Dans les meilleurs conditions ($\tau = 1$) on aura $L'/n'^2 = L/n^2$, et cette relation fixe la limite supérieure de la luminance; si l'on suppose en particulier que $n = n'$ la luminance de l'image ne peut en aucun cas dépasser celle de l'objet. Nous rencontrons là une conséquence fondamentale du principe de FERMAT: ce principe entraine l'existence de certains d'invariants (entre l'expace objet et l'expace image) dont la relation de CLAUSIUS est un exemple.

Cette relation exprime l'invariance de l'étendue d'un pinceau (en quelque sorte l'invariance de la quantité de rayons lumineux) et par conséquent la conservation de la luminance. Ce dernier résultat est d'ailleurs très important en thermodynamique: sie l'on forme à l'aide d'un système optique l'image d'une source à la température T, la luminance de l'image ne pourra être supérieure à celle de la source; c'est-à-dire que la température apparente de l'image ne pourra être supérieure à T ou encore qu'il est impossible à l'aide d'un système optique convenable de faire passer de l'énergie de la température T à une température $T' > T$. C'est bien entendu pour établir ce résultat que CLAUSIUS avait démontré tout d'abord la relation exprimant la conservation de l'étendue d'un pinceau.

IV. Optique géométrique et optique électronique.

13. Indiquons pour terminer l'étroite analogie qui existe entre l'optique géométrique et l'optique électronique électrostatique, où l'on utilise exclusivement des champs électriques pour dévier les particules et former des images[1].

Considérons une particule électrisée de masse m, de charge e soumise à un champ électrique $\boldsymbol{E} = -\operatorname{grad} U$ (Fig. 26) l'équation de son mouvement peut s'écrire

$$m \frac{d\boldsymbol{v}}{dt} = -e \operatorname{grad} U$$

Fig. 26. $F = -e \operatorname{grad} U$

(si M est la position de la particule et $\boldsymbol{v}$ son vecteur vitesse). On peut encore écrire cette relation, en mettant en évidence la composante de l'accélération suivant la normale principale à la trajectoire, dont le vecteur unitaire est $\boldsymbol{N}$

$$e \operatorname{grad} U = e \boldsymbol{T} \frac{dU}{ds} - \frac{mv^2}{R} \boldsymbol{N},$$

mais si l'on tient compte de la relation des forces vives

$$\tfrac{1}{2} m v^2 + e U = 0,$$

où U représente le potentiel choisi de façon que $U = 0$ aux points de l'espace où la vitesse est nulle, on pourra écrire

$$\operatorname{grad} U = \boldsymbol{T} \frac{dU}{ds} + \frac{2U}{R} \boldsymbol{N},$$

ce que l'on peut encore écrire en divisant toute la relation par U

$$\operatorname{grad} \log U = \boldsymbol{T} \frac{d \log U}{ds} + \frac{2}{R} \boldsymbol{N},$$

---

[1] Cf. l'article de W. GLASER dans le vol. XXXIII de cette Encyclopédie.

alors que la relation (7.3) relative aux rayons lumineux peut s'écrire

$$\operatorname{grad} \log n = \boldsymbol{T}\, \frac{d \log n}{d s} + \frac{\boldsymbol{N}}{R}.$$

On constate que pour identifier les deux relations, il suffit de prendre

$$\log U = 2 \log n + \text{constante},$$

ou encore

$$U = K n^2. \tag{13.1}$$

Ainsi tous les problèmes d'optique électronique électrostatique peuvent en principe être transposés dans le domaine de l'optique géométrique à condition de prendre pour indice de réfraction $U = n^2$. Ceci n'est d'ailleurs pas étonnant, car il y a identité formelle entre le principe de Fermat et le principe de Maupertuis que l'on peut écrire

$$\delta \int v\, ds = 0$$

à condition de prendre $n = A v$ et en particulier $n = \sqrt{U}$. Dans ces conditions les conclusions générales que nous avons rencontrées ainsi que celles que nous établirons ultérieurement s'appliquent à l'optique électronique; en particulier:

les trajectoires électroniques sont les normales à une famille de surface (surfaces d'ondes) qui satisfont à la condition $|\operatorname{grad} L| = \sqrt{U}$,

l'étendue géométrique d'un pinceau électronique s'écrit sous la forme

$$U\, ds \cos \vartheta\, d\omega.$$

Cette quantité est invariable au cours de la propagation du pinceau et on pourra en déduire comme dans le cas de l'optique classique que la luminance satisfait à la relation

$$\frac{L}{U} = \frac{L'}{U'}.$$

Enfin la formation des images dans l'approximation de Gauss, la classification des aberrations et les relations générales auxquelles elles satisfont pourront se transposer immédiatement à l'optique électronique électrostatique.

Remarquons néanmoins que le potentiel électrique satisfait à la relation de Poisson, qui s'écrit en l'absence de densité électrique

$$\Delta U = 0.$$

Ceci introduit évidemment une limitation aux possibilités de l'optique électronique la répartition des indices dans l'espace ne peut être quelconque et la correction des aberrations ne pourra pas s'effectuer par les mêmes méthodes qu'en optique classique.

B. Recherche du stigmatisme rigoureux.

I. Méthode générale.

14. Les instruments d'optique sont généralement destinés à former des images d'objets étendus. Il est donc indispensable qu'ils fournissent une bonne image de tous les points lumineux qui composent l'objet. Il faut, si l'on en croit l'optique géométrique, que tous les rayons issus d'un point A du plan objet se concentrent effectivement en un point A' du plan image; on dit alors qu'il y a stigmatisme pour le couple de points A et A'. On sera satisfait si le stigmatisme est réalisé pour tous les couples de points conjugués AA' composant l'objet et son image; mais on peut encore poser le problème de façon légèrement différente: on peut d'une part réaliser le stigmatisme regoureux pour un couple de points conjugués particuliers A, A' et s'arranger d'autre part pour que le stigmatisme se maintienne, tout au moins au second ordre près lorsqu'on s'écarte légèrement du couple AA'. On recherchera donc la condition générale pour que le stigmatisme soit stationnaire au voisinage d'un couple de points conjugués A et A'.

Cherchons tout d'abord la condition pour qu'il y ait stigmatisme rigoureux pour deux points conjugués A et A': *il est nécessaire que le chemin optique (AA') soit rigoureusement constant, c'est-a-dire qu'il ne dépende pas du rayon particulier que l'on envisage:* le principe de FERMAT indique en effet que le chemin optique entre A et A' est stationnaire pour chacun des rayons du faisceau; il ne peut donc qu'être constant. En écrivant cette condition, nous allons tout d'abord rechercher les dioptres et miroirs ou les répartitions continues d'indices rigoureusement stigmatiques pour un couple de points A et A' et nous étudierons ensuite les conditions de stigmatisme dans un élément de volume au voisinage de l'axe des instruments de révolution, ce qui nous amenera aux conditions d'ABBE et d'HERSCHEL, dont l'incompatibilité nous fera conclure à l'impossibilité de l'instrument d'optique parfait.

II. Dioptres et miroirs stigmatiques.

15. Condition générale. Si n et n' sont les indices des milieux que sépare le dioptre, la condition de stigmatisme s'écrit (Fig. 27)

$$n\,\overline{AI} + n'\,\overline{IA'} = \text{constante}. \tag{15.1}$$

On évalue les distances algébriques $\overline{AI}$ et $\overline{IA'}$, avant et après réflexion, en utilisant comme sens positif celui de la propagation de la lumière.

Dans le cas de la réflexion, cette condition devient, avec $n = n'$

$$\overline{AI} + \overline{IA'} = \text{constante}. \tag{15.2}$$

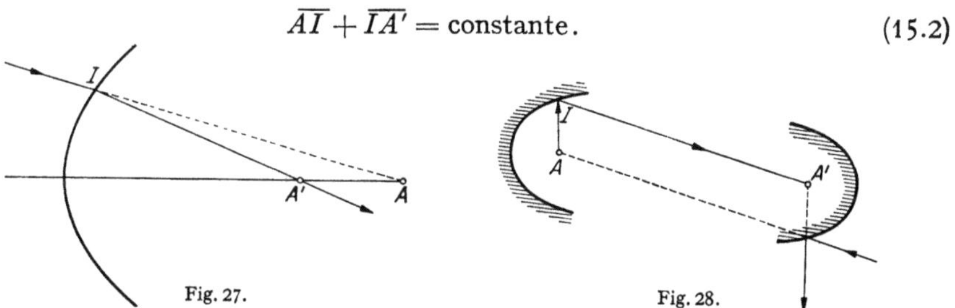

Fig. 27.   Fig. 28.

16. Cas de la réflexion. Nous commencerons par étudier les miroirs stigmatiques definis par la relation (15.2). Il nous faut pour cela distinguer deux cas. Si AI et IA' sont de même signe (A et A' sont de même nature, tous deux réels, ou tous deux virtuels) cette condition définit un elipsoïde de révolution de foyers A et A' (Fig. 28); si, inversement, AI et IA' sont de signes contraires, la différence de leurs valeurs absolues doit rester constante et la surface est un hyperboloïde de foyers A et A' que l'on peut faire fonctionner soit avec objet réel, soit avec objet virtuel (Fig. 29). Bien entendu, lorsque la différence des valeurs absolues s'annule, l'hyperboloïde se réduit à un plan; le miroir plan est d'ailleurs stigmatique pour tous les points de l'espace.

Un cas particulier intéressant est celui où l'objet (ou image) s'en va à l'infini. On ne considère pas alors le chemin optique depuis le point objet A (qui serait infini) mais on le remplace par une surface d'onde; il suffit d'écrire que le chemin optique entre un plan d'onde incident et l'image est constante; la surface d'onde choisie est d'ailleurs arbitraire et la constante l'est également; on peut donc s'arranger pour qu'elle soit nulle; la surface réfléchissante est alors définie par la condition

$$\overline{HI} + \overline{IA'} = \text{constante}$$

où H est le point d'intersection d'un rayon incident et d'une surface d'onde plane particulière, et le lieu de I est le paraboloïde, lieu, des points équidistants du foyer A' et du plan d'onde particulier (Fig. 30). C'est l'objectif parabolique de Newton, encore utilisé en astronomie et dans la construction des projecteurs.

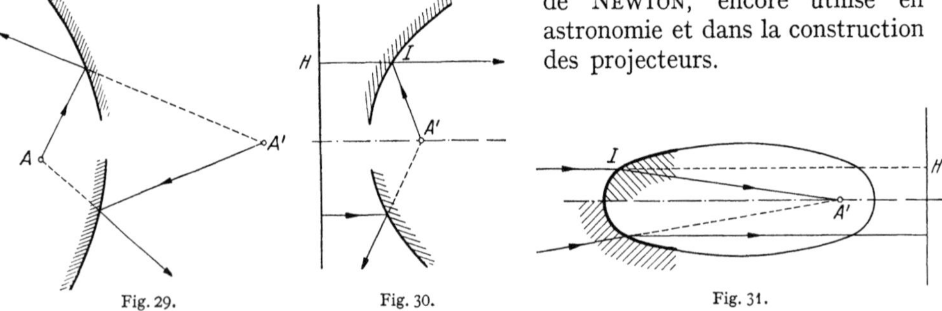

Fig. 29.   Fig. 30.   Fig. 31.

17. Cas de la réfraction. Dans le cas de la réfraction, la condition générale définit une courbe en coordonnées bipolaires appelée ovale de Descartes. Nous nous bornerons à étudier quelques cas particuliers, où la méridienne dégénère en une courbe plus simple.

1. L'objet est à l'infini; on est alors ramené à écrire, en remplaçant l'objet par un plan d'onde arbitraire,

$$n\,\overline{HI} + n'\,\overline{IA'} = \text{constante}.$$

On peut, comme précédemment, annuler la constante, ce qui nous ramène à chercher le lieu des points pour lesquels le rapport des distances à un point A'

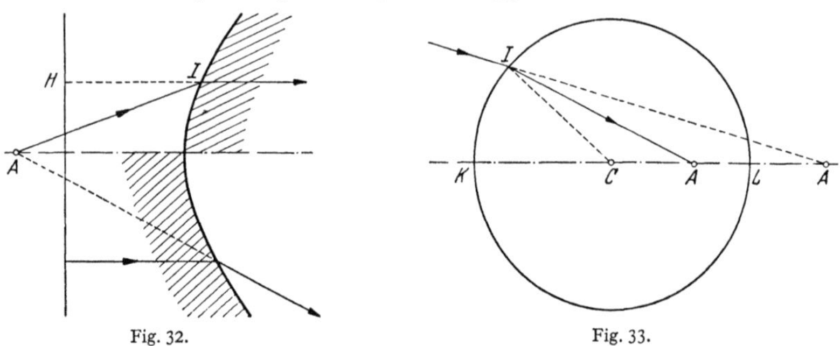

Fig. 32.   Fig. 33.

fixe et à un plan est constant: on sait que c'est une quadrique de révolution engendrée par la rotation d'une conique qui admet A' pour foyer et la trace du plan d'onde particulier dans le plan de figure pour directrice. Sa nature dépend du rapport n'/n et l'on pourrait faire une discussion détaillée des résultats, mais, ces surfaces étant pratiquement très peu utilisées du fait qu'elles ne sont pas aplanétiques, nous nous bornerons à indiquer sur les Figs. 31 et 32 le schéma de fonctionnement de ces diverses surfaces.

2. La constante est nulle; la condition générale s'écrit alors

$$n\,\overline{AI} + n'\,\overline{IA'} = 0$$

et le rapport des longueurs AI et IA' doit rester constant: on sait que le lieu des points I satisfaisant à cette condition est alors une sphère admettant pour diamètre le segment KL qui joint les deux points K et L de la droite AA' qui divisent AA' dans le rapport n'/n (Fig. 33).

Cherchons les abscisses x et x' des points conjugués A et A' par rapport au centre C de la sphère de rayon r; on doit avoir

$$\frac{KA'}{KA} = -\frac{LA'}{LA} = \frac{n}{n'} \quad \text{on encore} \quad \frac{r + x'}{r + x} = -\frac{r - x'}{r - x} = \frac{n}{n'}$$

d'où l'on tire facilement

$$x = \frac{n}{n'}\,r, \qquad x' = \frac{n'}{n}\,r. \tag{17.1}$$

Un dioptre sphérique est donc rigoureusement stigmatique pour les points A et A' situés sur un même diamètre aux distances $\frac{n}{n'}\,r$ et $\frac{n}{n'}\,r$ du centre de la sphère.

On appelle ces points conjugués particuliers, les points de YOUNG, ou encore de WEIERSTRASS. Ces points présentent un très grand intérêt car d'une part le dioptre sphérique est le plus facile à réaliser mais d'autre part nous verrons au Sect. 27 que ces points d'YOUNG sont non seulement stigmatiques mais encore aplanétiques.

III. Surfaces correctrices, retouches optiques.

18. Problème général. On peut envisager le problème suivant: étant donnée une surface d'onde Σ dans l'avant dernier milieu d'un système optique, est-il possible de trouver la forme à donner au dioptre réfringent qui sera la dernière surface de l'instrument pour que l'on obtienne une image stigmatique en un point donné A' situé dans l'air?

Il suffit d'appliquer la méthode générale précédente: on dispose d'une surface d'onde Σ correspondant à un point objet A et l'on sait que par définition le chemin optique (IA) entre A et un point I quelconque de la surface d'onde est constant; il suffira donc d'écrire de plus que $(IA)'$ sera

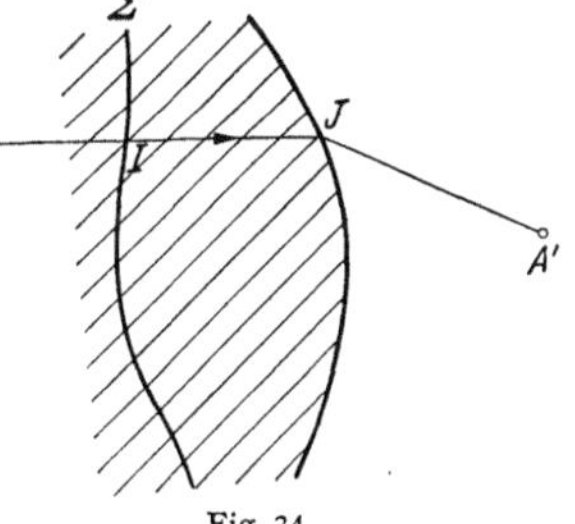

Fig. 34.

constante ce qui définit la forme de la surface (tout au moins à un paramètre près), le point J de la surface (Fig. 34) sera en effet défini par la condition

$$n\,\overline{IJ} + \overline{JK} = \text{constante.}$$

On choisira donc sur chacune des normales à la surface d'onde le point J satisfaisant à cette condition; on pourra en fait trouver une infinité de surface suivant la valeur que l'on choisira pour la constante. En principe il est donc possible de corriger un défaut quelconque de stigmatisme en agissant sur la forme de la dernière surface d'un instrument; il serait facile de montrer plus généralement que l'on peut transformer une surface d'onde incidente Σ en une surface d'onde émergente Σ' données a priori à l'aide d'un dioptre de forme convenable: il suffit que l'on ait (Fig. 35)

$$n\,\overline{IJ} + n'\,\overline{JK} = \text{constante;}$$

on doit donc grouper deux à deux les normales à Σ et Σ' de façon à satisfaire à cette condition; ceci montre finalement qui si Σ est la surface d'onde issue d'un point A, et Σ' la surface d'onde qui donnerait, après traversée de la seconde partie de l'instrument, une image stigmatique A' on peut définir la forme d'une surface réfringente intermédiaire de façon qu'il y ait stigmatisme entre A et A'. On peut être en fait amené à effectuer de faibles retouches locales à la forme d'une surface, mais pour des raisons de commodité on choisit le plus souvent la dernière, nous allons maintenant indiquer comment on peut évaluer l'importance de ces retouches.

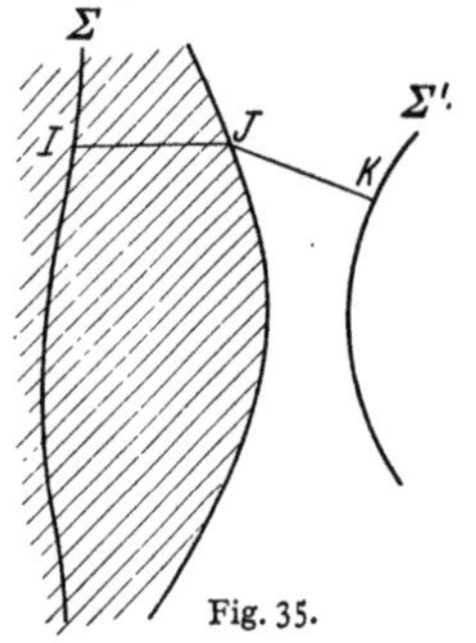

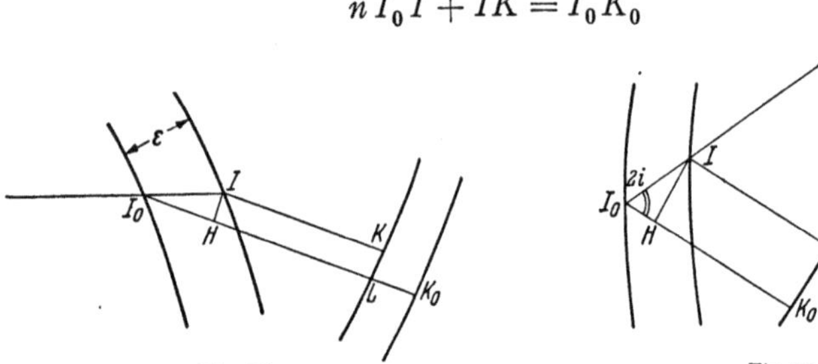

Fig. 35.

19. Retouches locales. Considérons une surface réfringente qui donne une surface d'onde Σ dont la Fig. 36 représente un élément; supposons que la surface d'onde soit légèrement décalée par rapport à sa position théorique Σ_0 qui correspondrait au stigmatisme rigoureux. Soit Δ la distance entre Σ et Σ_0; supposons maintenant que nous enlevions une épaisseur ε de verre d'indice n; le chemin $I_0 I K$ est remplacé par le chemin égal $I_0 K_0$; il est alors aisé de relier ε à Δ

$$n\,I_0 I + IK = I_0 K_0$$

Fig. 36. 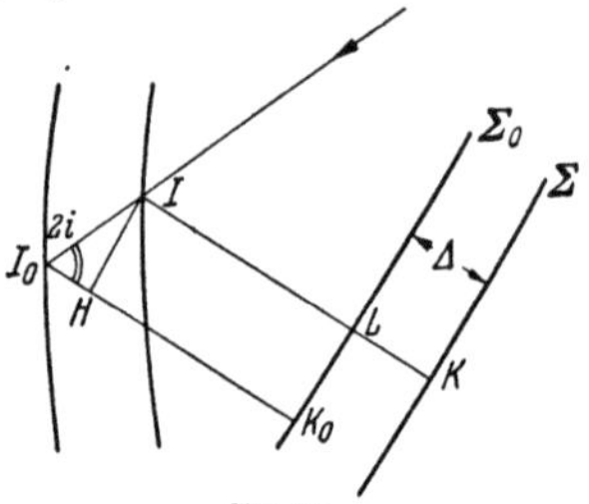   Fig. 37.

c'est-à-dire en projetant encore I en H sur le rayon $I_0 K_0$

$$n\,I_0 I = I_0 H + L K_0,$$

mais

$$I_0 I = \frac{\varepsilon}{\cos i} \quad \text{et} \quad I_0 H = \frac{\varepsilon}{\cos i}\cos(i - r)$$

d'où finalement

$$n\,\frac{\varepsilon}{\cos i} = \frac{\varepsilon}{\cos i}\cos(i - r) + \Delta\,,$$

$$\frac{\varepsilon}{\cos i}\left[n - \cos(i - r)\right] = \Delta\,, \tag{19.1}$$

ce qui détermine la profondeur de la retouche ε à effectuer; le plus souvent les angles i et r sont faibles et l'on aura

$$\varepsilon = \frac{\Delta}{n - 1}\,.$$

Si maintenant la retouche concerne une surface réfléchissante dans l'air, on aura, d'après la Fig. 37

$$IK = I I_0 + I_0 K_0$$

ou encore

$$\Delta = KL = I I_0 + I_0 H = \frac{\varepsilon}{\cos i}(1 + \cos 2i) = 2\varepsilon\cos i,$$

on devra donc avoir

$$\varepsilon = \frac{\varDelta}{2\cos i}.\qquad(19.2)$$

On remarque en conclusion que la précision de la taille d'un miroir doit être environ quatre fois meilleure que celle de la taille d'une surface réfringente: pour le miroir on a en effet $\varepsilon \approx \varDelta/2$ alors que pour le dioptre on a $\varepsilon \approx 2\varDelta$.

Si l'écart $\varDelta$ est fixé, en particulier par des règles de tolérances telles que celles que nous établirons an chapitre F, la tolérance ε sur la teille de la surface du miroir est quatre fois plus petite que celle qui correspond à un dioptre. On verra que dans les très bons instruments cette tolérance est le plus souvent une faible fraction de la longueur d'onde.

IV. Les milieux à variation continue d'indice; applications à certains systèmes optiques pour ondes courtes.

20. Généralités. Examinons maintenant la possibilité d'obtenir le stigmatisme en utilisant une variation continue de l'indice de réfraction; si les études relatives à ce cas ont souvent été du domaine spéculatif du fait des difficultés de réalisation de tels milieux en optique courante, elles ont cependant trouvé récemment d'intéressantes applications dans le domaine des ondes radioélectriques de courtes longueur d'onde (décimètre ou centimètre): il est en effet possible de modifier par des obstacles appropriés la vitesse de propagation de la phase des ondes: c'est ainsi que lorsqu'une onde est diffractée par un trou percé dans une plaque conductrice l'avance de phase de l'onde diffractée par rapport à l'onde incidente dépend du diamètre du trou; on peut alors constituer un milieu à variation d'indice de réfraction en disposant dans des plans parallèles équidistants des plaques conductrices percées de trous de diamètre

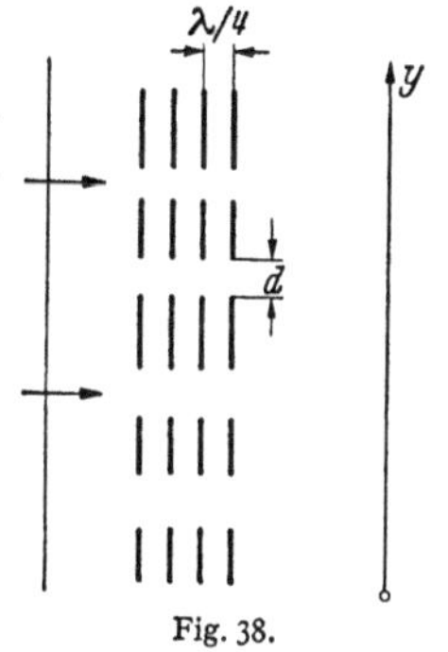

Fig. 38.

variable[1]: la vitesse de phase des ondes dans un tel milieu varie avec y (Fig. 38) si le diamètre D des trous est fonction de y. De tels dispositifs sont employés comme lentilles dans des projecteurs pour ondes courtes. La source S est un cornet terminant un guide d'ondes, les ondes se réfléchissent sur un miroir M et sont ensuite rendues planes au passage à travers une «lentille» constituée en fait par des plaques représentant des trous dont le diamètre dépend de la distance à l'axe de la lentille. On peut régler à volonté la vitesse de progression des ondes dans le milieu diélectrique artificiel que constituent les plaques percées, en choisissant judicieusement la loi de variation du diamètre des trous et obtenir une onde plane à partir d'une onde incidente sensiblement sphérique (Fig. 39).

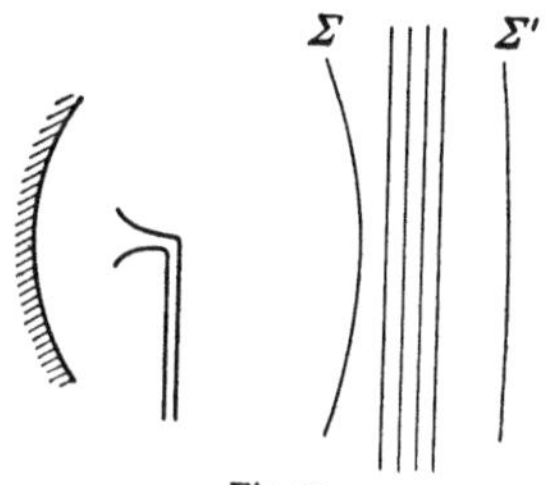

Fig. 39.

Il est bien entendu possible de mettre le problème en équations mais du fait que la longueur d'onde est relativement grande il s'ensuit que la tolérance sur les éventuelles déformations de la surface d'onde émergente est assez importante et il suffit pratiquement de se contenter d'une solution approximative que l'on peut au besoin retoucher à la suite d'une étude expérimentale de la forme de l'onde. Nous allons maintenant nous intéresser au problème particulier du cas où la loi de distribution de l'indice de réfraction présente la symétrie sphérique.

---

[1] J. C. Simon: Thèse Paris 1951.

21. Cas des milieux à symétrie sphérique. Le théorème de Bouguer. On peut se poser le problème suivant: peut-on trouver une distribution continue d'indice de réfraction $n(r)$ ne dépendant que de la distance r à un point fixe O et telle qu'il y ait stigmatisme pour un couple de points donnés A et A'?

Le problème peut être abordé soit en s'appuyant sur les équations des rayons lumineux, soit encore en utilisant le principe de Fermat; nous indiquerons tout d'abord l'existence d'un invariant propre aux milieux à distribution sphérique d'indice.

Théorème de Bouguer: *Dans les milieux à symétrie sphérique la quantité $nr \sin i$ est invariante.*

Dans cette relation r est la distance du centre O, $n(r)$ l'indice de réfraction, i l'angle d'incidence sur le «dioptre» ou encore sur la sphère d'égal indice $n =$

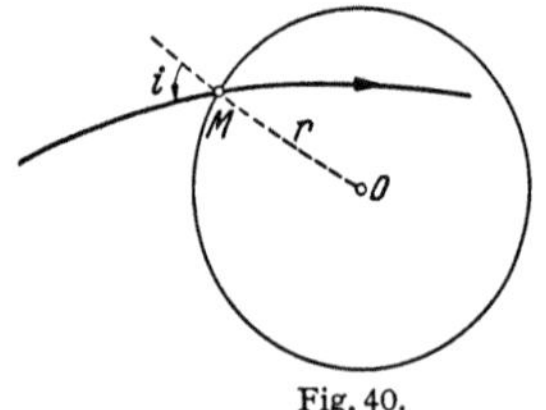

Fig. 40.

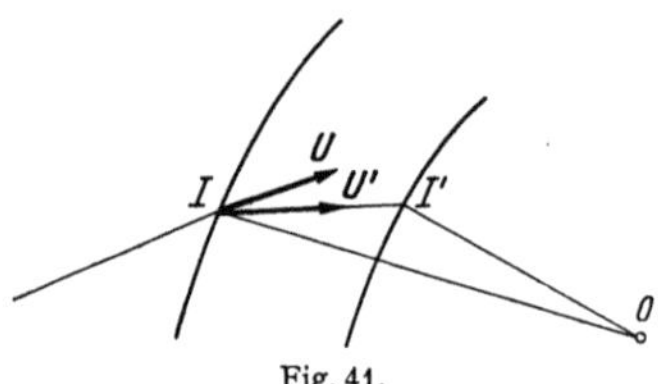

Fig. 41.

constante (Fig. 40). Il est facile de démontrer ce théorème tout d'abord dans le cas des systèmes de dioptres sphériques concentriques séparant des milieux homogènes: si I est un point d'incidence sur le dioptre du fait que OI est la

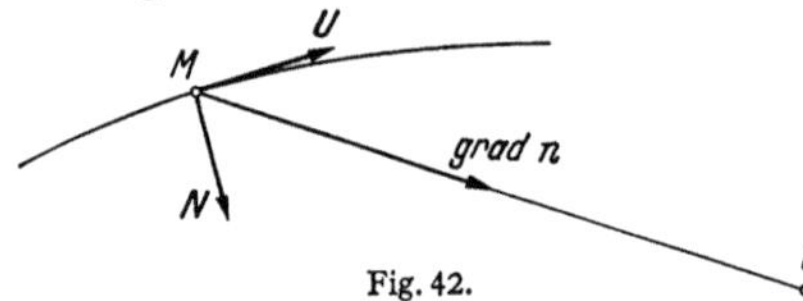

Fig. 42.

normale au dioptre (Fig. 41), les lois de Descartes peuvent se traduire par la relation vectorielle

$$n\,\overrightarrow{OI}\times\boldsymbol{u} = n'\,\overrightarrow{OI}\times\boldsymbol{u}'$$

où $\boldsymbol{u}$ et $\boldsymbol{u}'$ sont les vecteurs unitaires portés par les rayons incident et réfracté. Cette quantité est donc invariante à la réfraction.

Elle est également invariante lorsqu'on traverse l'un des milieux homogènes pour aller de I en I' on aura en effet

$$\overrightarrow{OI'} = \overrightarrow{OI} + \lambda\boldsymbol{u}',$$

et il s'ensuit que

$$n'\,\overrightarrow{OI}\times\boldsymbol{u}' = n'\,\overrightarrow{OI'}\times\boldsymbol{u}'.$$

La quantité ci-dessus est donc invariable quel que soit le nombre de dioptres et l'on remarque aisément que son module est égale à $nr \sin i$, ce qui établit le théorème de Bouguer. Si ce théorème est vrai pour un ensemble de milieux continus séparés par des dioptres il l'est encore dans le cas des distributions continues, que l'on peut obtenir par un passage à la limite du cas précédent. On peut aussi l'établir directement dans ce cas: soit M un point situé sur un rayon lumineux se propageant dans un milieu où l'indice $n(r)$ ne dépend que de la distance $r = OM$ à un point fixe O (Fig. 42): il suffit détablir l'invariance de la quantité

$$\boldsymbol{B} = n\,\overrightarrow{OM}\times\boldsymbol{u}$$

dont la différentielle est

$$d\boldsymbol{B} = \left[\frac{dn}{ds}\,\overrightarrow{OM}\times\boldsymbol{u} + n\boldsymbol{u}\times\boldsymbol{u} + n\,\overrightarrow{OM}\times\boldsymbol{N}\,\frac{1}{R}\right]ds;$$

le second terme est nul et l'on peut écrire en utilisant l'équation du rayon lumineux (7.3)

$$\overrightarrow{OM} \times \left[\boldsymbol{u}\,\frac{dn}{ds} + n\,\frac{\boldsymbol{N}}{R} \right] = \overrightarrow{OM} \times \operatorname{grad} n.$$

Il en résulte immédiatement que $d\boldsymbol{B}=0$ et que $\boldsymbol{B}$ est effectivement invariable.

Le théorème de BOUGUER est en fait l'analogue du théorème de la conservation du moment cinétique ou mécanique, lorsque la particule est soumise à l'action de forces centrales et les démonstrations de ces théorèmes sont évidemment similaires.

22. Equations des rayons lumineux. Si l'on cherche maintenant à déterminer le rayon lumineux par rapport à des coordonnées polaires r et ϑ ayant leur origine en O (Fig. 43) on peut écrire l'invariant de BOUGUER sous la forme suivante

$$B = n\,r \sin i = n\,r\,\frac{r}{\sqrt{r^2 + r'^2}} = n\,r^2 (r^2 + r'^2)^{-\frac{1}{2}}. \quad (22.1)$$

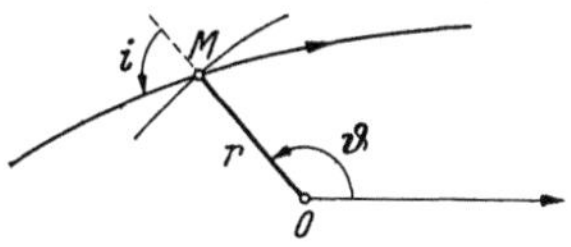

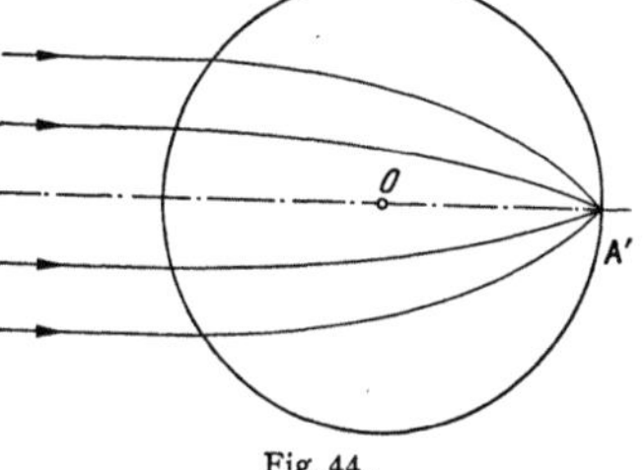
Fig. 43.

Dans cette expression r' désigne la dérivée par rapport à l'angle polaire ϑ et il en résulte que les rayons doivent satisfaire à la relation

$$r^2 + r'^2 = \frac{n^2 r^4}{B^2} \quad \text{ou} \quad \left(\frac{dr}{d\vartheta}\right)^2 = r^2 \left(\frac{n^2 r^2}{B^2} - 1\right)$$

et l'on peut en fait exprimer l'azimut ϑ par une quadrature

$$\vartheta = \int \frac{dr}{r\left[\dfrac{n^2 r^2}{B^2} - 1\right]^{\frac{1}{2}}}. \quad (22.2)$$

Si la loi $n(r)$ est connue on peut ainsi déterminer les rayons lumineux. On peut aussi inversement chercher une distribution $n(r)$ telle que l'on ait stigmatisme pour un couple de points donnés; on arrive alors à une équation intégrale permettant de déterminer $n(r)$; c'est ainsi que R. K. LUNEBERG a proposé comme résultat de calculs que nous ne reproduirons pas ici la loi suivante pour former en un point A' situé à la distance $OA'=1$ (Fig. 44) l'image rigoureusement stigmatique d'un point objet A situé à l'infini

$$\left.\begin{array}{ll} n(r) = \sqrt{2 - r^2} & \text{pour} \quad r < 1, \\ n(r) = 1 & r > 1. \end{array}\right\} \quad (22.3)$$

Fig. 44.

On a donc ainsi un système optique rigoureusement stigmatique et où les variations d'indice sont limités à une sphère donnée. Bien entendu l'image d'un objet étendu situé à l'infini se formera sur une calotte sphérique de rayon unitaire: on verra que dans ces conditions la seule aberration présente est la courbure de champ.

23. Application du principe de FERMAT et utilisation de transformations conformes. On peut encore appliquer directement le principe de FERMAT à la détermination des rayons lumineux: ceux-ci sont, comme on le sait les extrêmales du chemin optique $L = \int n\,ds$.

Il est souvent commode de ramener la recherche d'extrêmales à celle de géodésiques sur une surface simple; supposons par exemple que nous ayons pu exprimer l'élément d'arc du rayon lumineux sous la forme

$$n\,(d\,x^2 + d\,y^2)^{\frac{1}{2}} = (d\,X^2 + d\,Y^2 + d\,Z^2)^{\frac{1}{2}}$$

où l'élément $d\,X, d\,Y, d\,Z$ est tracé sur une surface simple: la recherche des extrêmales de l'intégrale $\int n\,ds = \int d\,S$ se ramène à la recherche des géodésiques (lignes dont la longueur est stationnaire par rapport à toute déformation) sur cette surface. Il est particulièrement indiqué de s'arranger pour que la surface en question soit une sphère: les géodésiques sont alors connues, ce sont le grands cercles de la sphère[1].

Il est possible en particulier d'opérer en deux temps: pour passer de la sphère (de rayon unitaire) à un plan rapporté à des coordonnées $x_1\,y_1$ on peut effectuer une projection stéréographique, qui est une représentation conforme (Fig. 45); il est facile de montrer que l'on a, entre les éléments d'arc sur la sphère $(d\,S)$ et sur le plan $(d\,s_1)$ la relation

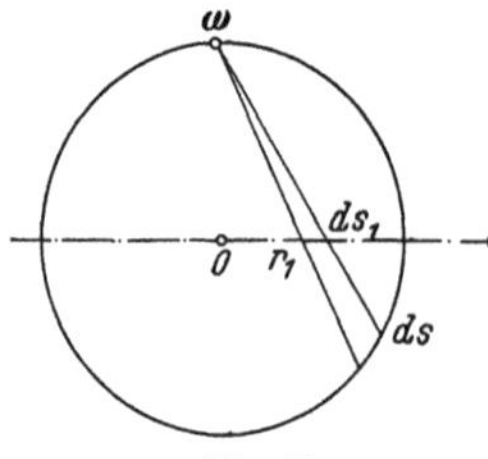

Fig. 45.

$$dS = \frac{2\,d\,s_1}{1 + r_1^2} \quad \text{avec} \quad r_1^2 = x_1^2 + y_1^2.$$

On peut ensuite transformer le plan $x_1 y_1$ sur le plan x, y par une représentation conforme quelconque représentée par la relatiou

$$z_1 = f(z) \quad \text{avec} \quad \begin{cases} z = x + j\,y \\ z_1 = x_1 + j\,y_1, \end{cases}$$

f étant une fonction analytique quelconque. On aura alors

$$ds_1^2 = d\,x_1^2 + d\,y_1^2 = |d\,z_1|^2 = |f'|^2 |d\,z|^2 = f'^2\,d\,s^2$$

et finalement

$$d\,S^2 = \frac{4}{(1 + r_1^2)^2} |f'(z)|^2 d\,s^2. \tag{23.1}$$

Si $f(z)$ est une fonction analytique quelconque on aura donc un moyen simple de trouver les rayons lumineux dans un espace à deux dimensions où l'indice de réfraction est

$$n = \frac{2}{1 + |f(z)|^2} |f'(z)|. \tag{23.2}$$

On remarque alors que les géodésiques de la sphère passant par un point A (qui sont les grands cercles obtenus en coupant la sphère par un plan passant par OA) se recoupent en un point A' diamètralement opposé à A sur la sphère: il en résulte que tout rayon partant du point a correspondant à A dans le plan xy vient passer par une image rigoureusement stigmatique a': on peut donc réaliser de cette façon un instrument stigmatique pour tous les points de l'espace si la répartition envisagée dans le plan rapporté à xy a la symétrie circulaire autour d'un centre O il suffit de généraliser cette répartition à l'espace, ce qui fournit une répartition sphérique rigoureusement stigmatique.

24. L'oeil de poisson de Maxwell. Donnons à titre d'exemple une répartition d'indice qui donne le stigmatisme rigoureux et qui fût suggérée par Maxwell à la suite de l'étude de la structure des yeux de poissons. Faisons pour cela

---

[1] Voir en particulier: G. Toraldo di Francia: On a family of configuration lenses. Optica Acta I **4**, 157 (1955). — R. K. Luneberg: Mathematical Theory of Optics. Providence 1944, p. 213. — R. F. Rinehart: J. Appl. Phys. **19**, 1948, p. 860.

$z_1 \equiv z$ et $f'(z) = 1$. On obtient la répartition

$$n(r) = \frac{2}{1 + r^2}.$$

(24.1)

A un point objet a correspond un point image a' et l'on a le stigmatisme rigoureux entre a et a' (Fig. 46); a et a' sont les points qui correspondent à deux points diamétralement opposés de la sphère S, il en résulte que $Oa \cdot Oa' = 1$ (Fig. 47).

La différence essentielle avec la répartition de LUNEBERG est que l'indice est variable dans tout l'espace: (dans la répartition de LUNEBERG les variation d'indice étaient localisées à l'intérieure de la sphère unitaire).

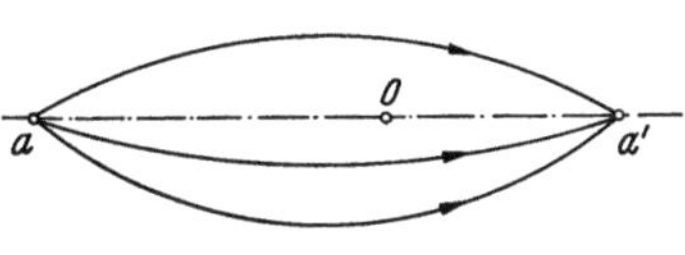

Fig. 46.

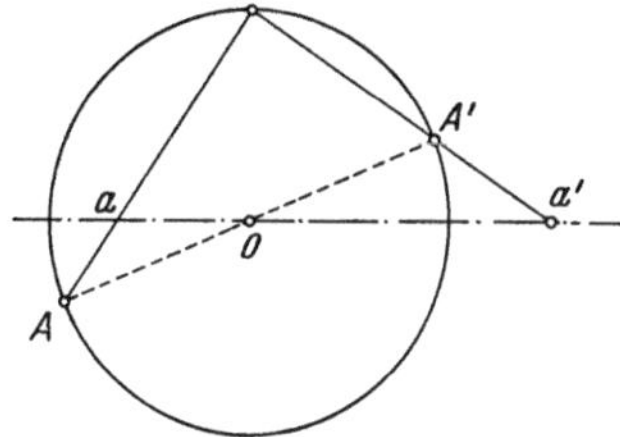

Fig. 47.

25. Les «lentilles de configuration» pour ondes courtes. On sait que l'on peut aisément guider des ondes radioélectriques de faible longueur d'onde en les enfermant entre deux surfaces conductrices rapprochées et sensiblement parallèles (Fig. 48). Admettons que dans ces conditions les ondes cheminent en suivant la surface moyenne: les trajets suivis par l'énergie, c'est-à-dire les rayons lumineux sont alors les géodésiques de la surface car l'indice de réfraction correspondant à la vitesse de propagation de l'onde est constant, tout au moins lorsque la courbure de la surface reste faible (par rapport à l'inverse de la longueur d'onde)

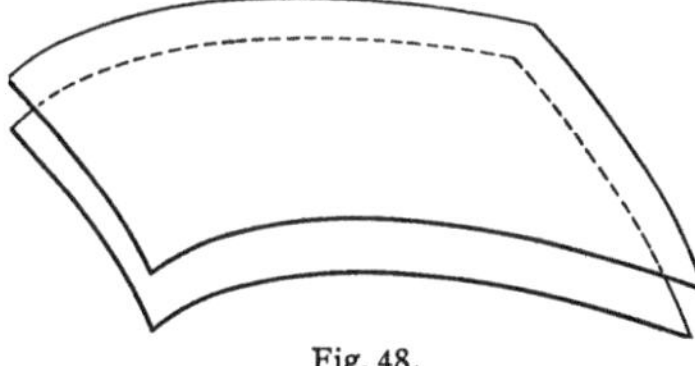

Fig. 48.

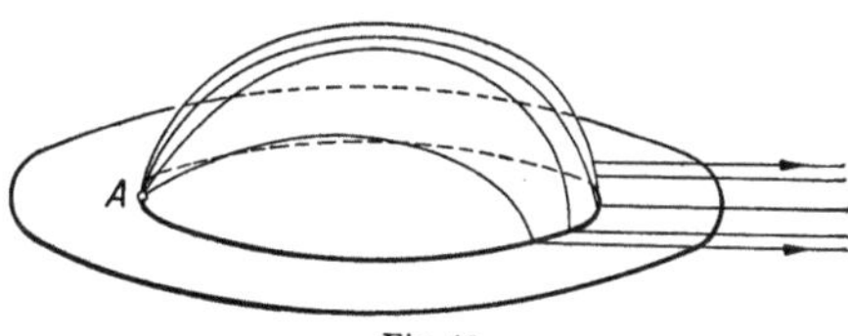

Fig. 49.

On peut alors constituer des systèmes projecteurs pour ondes courtes pouvant servir en particulier à obtenir un faisceau parallèle (ondes planes) à partir d'une source ponctuelle. Ce système peut être utile par exemple pour l'exploration de l'espace à grande vitesse de rotation: on utilise alors un guide dont la surface moyenne est de révolution, et où la source A se déplace rapidement le long d'un parallèle de la surface (Fig. 49).

Voyons maintenant comment on peut déterminer une telle surface: on peut partir d'une répartition plane d'indice de réfraction permettant d'obtenir le stigmatisme: la répartition de LUNEBERG [eq. (22.3)] par exemple. Dans le plan les rayons lumineux sont les extrêmales de l'intégrale $\int n\,ds$ et l'on cherche à faire correspondre point par point la surface de révolution au plan de façon qu'aux rayons lumineux du plan correspondent les géodésiques de la surface: rapportons le plan aux coordonnées polaires r, ϑ alors que la méridienne de la surface de révolution sera definie par la longueur l de l'arc de courbe et par le rayon ϱ (Fig. 50) on devra avoir:

$$n(r)^2 (dr^2 + r^2 d\vartheta^2) = dl^2 + \varrho^2 d\vartheta^2$$

ou encore

$$r\,n(r) = \varrho, \qquad n\,dr = dl. \tag{25.1}$$

On obtient l par intégration de la deuxième équation et la méridienne est alors définie paramétriquement par l et ϱ en fonction de r. On peut quelquefois éliminer r et Rinehart a ainsi obtenu pour équation entre ϱ et l pour la surface associée à la répartition de Luneberg:

$$l = \tfrac{1}{2}\,(\varrho + \text{arc sin}\,\varrho). \tag{25.2}$$

Pour constituer un projecteur on doit associer à la surface une couronne plate et l'on aboutit finalement à une forme de casque: la source est placée sur la ligne anguleuse et les ondes parallèles se propagent dans la région plane (Fig. 49). On doit alort étudier le passage des rayous de la surface convexe à la couronne.

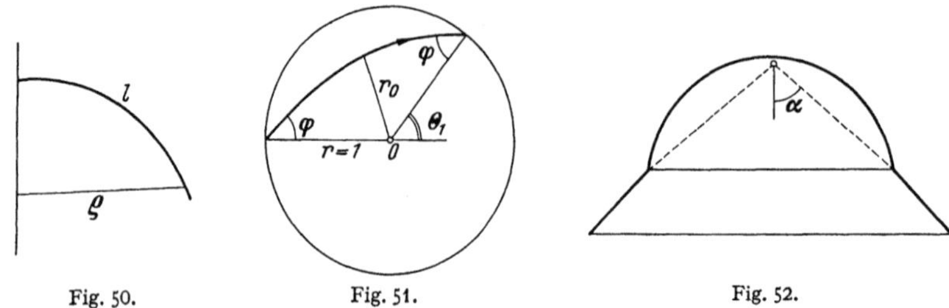

Fig. 50.   Fig. 51.   Fig. 52.

Toraldo di Francia a indiqué l'existence d'une famille de solutions; il commence par étudier les distributions planes à symétrie circulaire limitées au cercle $r=1$ et permettant d'obtenir le stigmatisme; il montre l'existence de la solution générale

$$n^2 r^2 = n^{1/p}\,(2 - n^{1/p}), \tag{25.3}$$

où p est un nombre positif; pour $p=\tfrac{1}{2}$ on obtient la repartition de Luneberg. En utilisant la relation (22.2) qui permet de trouver le rayon lumineux on obtient, en remarquant que $B = \sin\varphi$ (Fig. 51):

$$\vartheta = \sin\varphi \int \frac{dr}{r\,[n^2 r^2 - \sin^2\varphi]^{\frac{1}{2}}}. \tag{25.4}$$

Il existe une valeur r_0 de r pour laquelle le rayon lumineux est perpendiculaire au rayon vecteur: la relation de Bouguer (Sect. 21) donne alors:

$$r_0\,n(r_0) = 1 \tag{25.5}$$

et l'on obtient, en utilisant les notations indiquées sur la Fig. 51

$$\vartheta_1 = \pi - 2\sin\varphi \int_{r_0}^{1} \frac{dr}{r\,\sqrt{n^2 r^2 - \sin^2\varphi}}$$

le résultat de l'intégration (que l'on peut faire en posant $x = n^2 r^2$) s'écrit finalement très simplement

$$\vartheta_1 = 2\,(1 - p)\,\varphi; \tag{25.6}$$

ceci signifie qu'en associant à la surface de révolution une nappe conique d'angle α tel que $\sin\alpha = \dfrac{1}{2\,(1-p)}$ on obtiendra des rayons «parallèles» sur la nappe conique, en ce sens que si l'on développe le cône sur un plan (Fig. 52) les rayons obtenus sur le développement seront parallèles: le long de la ligne de rencontre de la surface et du cône les angles que font les rayons avant et après brisure avec le parallèle d'intersection doivent être égaux (par exemple à φ) et la relation (25.6) est bien la relation que doivent dans ces conditions vérifier les angles ϑ_1 et φ si l'angle α du cône est tel que $\sin \alpha = \dfrac{OP}{O'P'}$ (Fig. 53).

Il reste seulement à rechercher la forme effective des lentilles de «configuration» correspondant au milieu dont l'indice serait donné par la relation (25.2).

On combine pour cela les relations (25.1) et (25.3) et l'on obtient après intégration:

$$l = (1 - p)\,\varrho + p \quad \text{arc} \sin \varrho$$

ce qui définit la courbe méridienne par la longueur

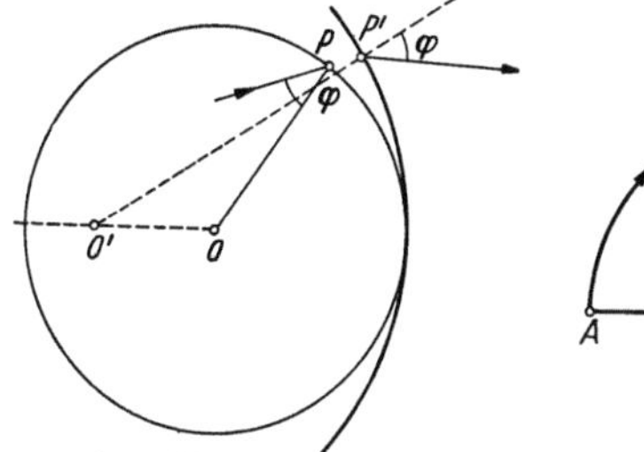

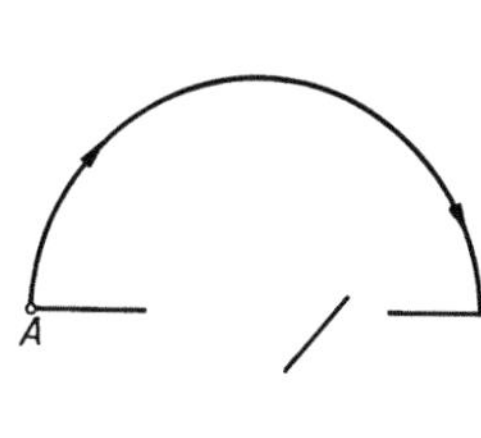

Fig. 53.   Fig. 54.

l de l'arc de courbe en fonction de l'abscisse ϱ. On devra associer à la surface une nappe conique d'ouverture α telle que

$$\sin \alpha = \frac{1}{2(1 - p)}\,.$$

La Fig. 49 représente la solution obtenue pour $p = \frac{1}{2}$ (lentille de RINEHART-LUNEBERG) et la Fig. 54 la solution obtenue opur $p = \frac{3}{2}$. (On doit lui associer en fait un miroir pour dévier les ondes qui sortent de la surface en se dirigeant vers l'axe.)

V. La recherche du stigmatisme dans un élément de volume.

26. Condition générale. Recherchons la condition nécessaire pour que le stigmatisme supposé réalisé pour un couple de points A et A' se maintienne lorsqu'on se déplace de A en B et de A' en B' (Fig. 55) A' étant l'image stigmatique de A on aura $(AIA') = $ constante, on cherche de plus à avoir $(BIB') = $ constante (le point I étant choisi arbitrairement sur le rayon issu de A).

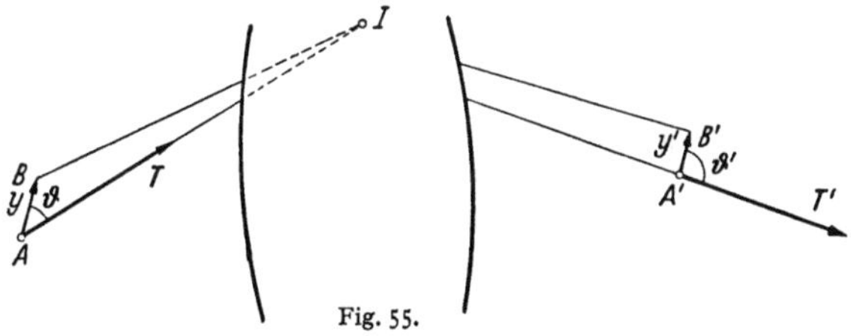

Fig. 55.

En appliquant la relation (7.2) dans laquelle on annule l'intégrale puisque l'on suit un rayon lumineux, on aura

$$(BIB') - (AIA') = n'\,\overrightarrow{A'B'} \times \boldsymbol{T'} - n\,\overrightarrow{AB} \times \boldsymbol{T} \tag{26.1}$$

si l'on appelle y et y' les longueurs AB et $A'B'$, ϑ et ϑ' les angles des vecteurs AB avec $\boldsymbol{T}$ et $A'B'$ avec $\boldsymbol{T'}$, on pourra écrire la condition générale de stigmatisme sous la forme

$$n'\,y'\cos\vartheta' - n\,y\cos\vartheta = \text{constante}. \tag{26.2}$$

Cette quantité devra rester constante quelle que soit l'orientation du rayon dans l'espace objet (qui dépend de deux paramètres) et l'orientation correspondante du rayon dans l'espace image venant passer par le point A'.

Cette condition générale s'applique en toutes circonstances et nous allons en voir deux applications particulièrement utiles.

27. Condition des sinus d'Abbe, Aplanétisme. Considérons en particulier un instrument de révolution parfaitement stigmatique pour un couple de points conjugués A et A' situé sur son axe; cherchons à quelle condition le stigmatisme réalisé pour A et A' se maintient quand on se déplace d'une petite quantité $AB = y$

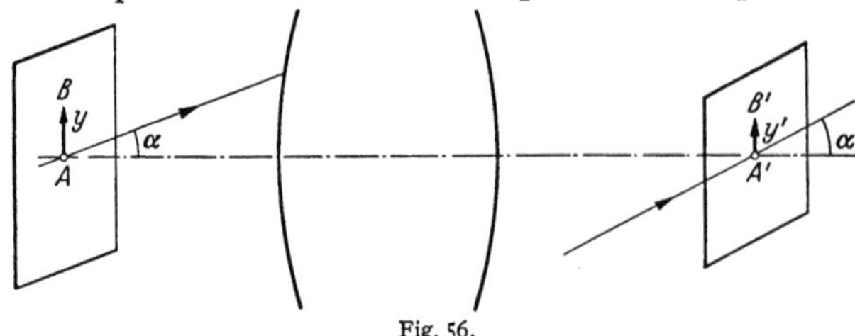

dans le plan de front passant par A l'image de B étant en un point B' situé à la distance $y' = A'B'$ de l'image A' (Fig. 56).

Les quantités y et y' seront bien entendu supposées petites et nous négligerons ultérieurement

Fig. 56.

es quantités du deuxième ordre par rapport à ces infiniments petits principaux; nous aurons ici $\vartheta = \dfrac{\pi}{2} - \alpha$ et $\vartheta' = \dfrac{\pi}{2} - \alpha'$ si α et α' sont les les angles des rayons oncident et émergent avec l'axe (Fig. 56); on devra avoir

$$n' y' \sin \alpha' - n y \sin \alpha = \text{constante}$$

pour déterminer la valeur de la constante, on peut envisager le cas où le rayon chemine suivant l'axe: on a alors $n' y' \sin \alpha' - n y \sin \alpha = 0$ et la condition s'écrit alors

$$\boxed{n y \sin \alpha = n' y' \sin \alpha'.} \qquad (27.1)$$

Cette condition apparait ici comme la condition nécessaire pour que le stigmatisme réalisé pour les points A et A' se maintienne pour des points BB' situés en leur voisinage immédiat dans les plans de front passant par A et A'. On verra ultérieurement que cette condition est suffisante.

Lorsqu'on a réalisé les deux conditions suivantes: a) stigmatisme pour A et A', b) condition des sinus, on dit que l'on a réalisé l'aplanétisme.

Remarquons enfin que la condition des sinus d'Abbe est susceptible d'une représentation géo-

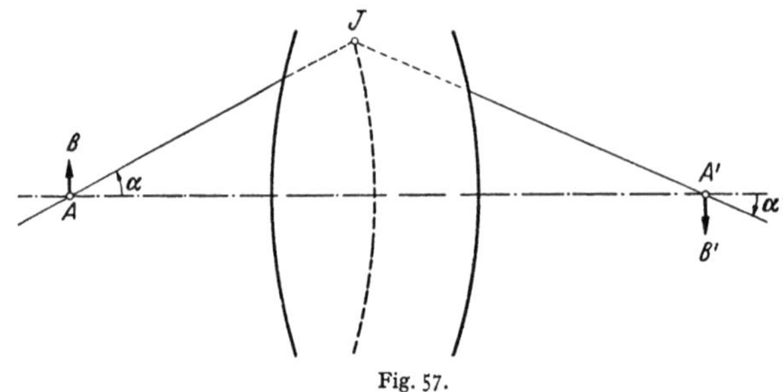

Fig. 57.

métrique simple: si J est le point de rencontre du rayon incident issu de A et du rayon émergent passant en A', le rapport $\dfrac{\sin \alpha'}{\sin \alpha}$ est égal à $\dfrac{JA}{JA'}$ et l'on voit que le point J doit décrire une sphère, lieu des points dont le rapport des distances A et A' est constant (Fig. 57). Le dioptre sphérique fonctionnant pour ses points d'Young satisfait parfaitement à cette condition. Il est donc aplanétique.

28. Condition d'Herschel. Considérons maintenant un instrument stigmatique pour un couple de points A et A' de l'axe et cherchons à quelle condition le stigmatisme se maintient si l'on se déplace de A en B le long de l'axe le point B' étant évidemment également sur l'axe (Fig. 58) soit $d x = AB$ et $d x' = A'B'$ les déplacements infiniments petits.

La condition générale (26.2) s'écrit ici

$$n' \, d\,x' \cos\alpha' - n\,d\,x \cos\alpha = \text{constante}.$$

Pour déterminer la constante, il suffit encore de considérer le cas où le rayon chemine le long de l'axe; on obtient alors: $n'd\,x' - n\,d\,x$ et la condition générale s'écrit finalement

$$n' \, d\,x' \, (1 - \cos\alpha') = n\,d\,x\,(1 - \cos\alpha)$$

ou encore

$$\boxed{n' \, d\,x' \sin^2 \frac{\alpha'}{2} = n\,d\,x \sin^2 \frac{\alpha}{2}.} \qquad (28.1)$$

Cette nouvelle condition est la condition d'HERSCHEL.

29. Incompatibilité des conditions précédentes.
Nous connaissons maintenant les conditions nécessaires pour que le stigmatisme se conserve lorsqu'on se déplace de petites quantités autour des points A et A'; il faut pour cela que $\dfrac{\sin\alpha}{\sin\alpha'}$ soit constant ainsi que $\dfrac{\sin\dfrac{\alpha}{2}}{\sin\dfrac{\alpha'}{2}}$; or,

nous pouvons écrire

$$\frac{\sin\alpha}{\sin\alpha'} = \frac{\sin\dfrac{\alpha}{2}\cos\dfrac{\alpha}{2}}{\sin\dfrac{\alpha'}{2}\cos\dfrac{\alpha'}{2}};$$

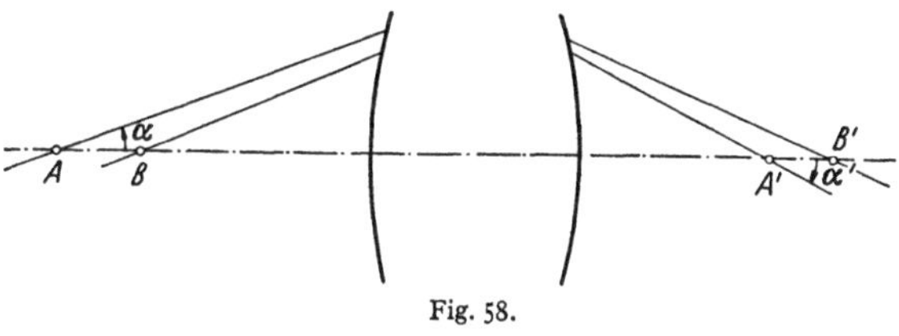

Fig. 58.

pour que les deux conditions soient compatibles, il faudrait que

$$\cos\frac{\alpha'}{2} = \cos\frac{\alpha}{2},$$

c'est-à-dire que $|\alpha| = |\alpha'|$.

Hormis ce cas particulier qui correspond à un grandissement déterminé, égal à ± 1, *les conditions de* HERSCHEL *et d'*ABBE *ne sont pas compatibles:* il sera donc généralement impossible de construire un instrument parfait c'est-à-dire un instrument stigmatique pour un élément de volume entourant le point objet A: le calculateur se bornera donc à satisfaire à la condition qui lui paraîtra la plus utile:

Si l'instrument est destiné à former l'image d'un objet plan perpendiculaire à l'axe, ce qui est le cas le plus fréquent, il sera nécessaire de satisfaire à la condition des sinus, c'est-à-dire que l'instrument devra non seulement être stigmatique mais encore *aplanétique.*

Si, au contraire, l'instrument est destiné à viser un point A mobile sur son axe, on devra satisfaire à la condition de HERSCHEL.

Dans les chapitres relatifs à l'étude des faibles aberrations (Sect. 68) de l'aberration sphérique et de la coma (Sect. 76), nous verrons comment nous pourrions calculer exactement les effets des écarts aux conditions ci-dessus et en particulier les aberrations qui apparaissent lorsqu'on satisfait à la condition d'ABBE et qu'on se déplace le long de l'axe ou au contraire lorsqu'on satisfait à la condition d'HERSCHEL et qu'on se déplace dans un plan de front. On trouve que, si les conditions sont nettement incompatibles lorsque les angles α ou α' sont importants, elles sont pratiquement compatibles lorsque α et α' ne sont pas trop grands.

30. Formes particulières des conditions d'Abbe de Herschel. Il nous paraît utile de signaler ici les formes particulières que prennent les conditions d'Abbe et d'Herschel lorsque l'objet ou l'image s'en vont à l'infini. On sait, que l'on a souvent à étudier des systèmes optiques fonctionnant dans ces conditions et il y a lieu de rechercher comment on doit dans ce cas modifier les conditions

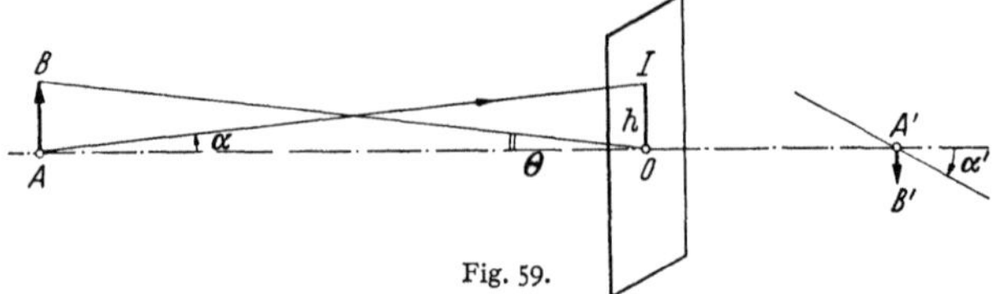

Fig. 59.

précédentes. Nous allons supposer par exemple que l'objet s'en va à l'infini et nous le placerons tout d'abord à une grande distance $x = OA$ de l'instrument.

Traçons alors un rayon faisant un angle α avec l'axe et perçant le plan de front qui passe par l'origine O en un point I tel que $OI = h$ (Fig. 59); l'angle α est petit et l'on peut écrire

$$\sin \alpha \approx \alpha = -\frac{h}{x}$$

si l'objet est vu de O sous un angle ϑ, on aura de même $y' = \vartheta x$ d'où l'on tire

$$n\, y \sin \alpha = -n\, h\, \vartheta,$$

et la condition des sinus devient alors

$$-n\, h\, \vartheta = n'\, y'\, \sin \alpha'. \tag{30.1}$$

Les infiniments petits sont maintenant ϑ et y' qui fixent les grandeurs de l'objet et de l'image; pour que l'instrument soit aplanétique, il faudra que $h/\sin \alpha'$ soit

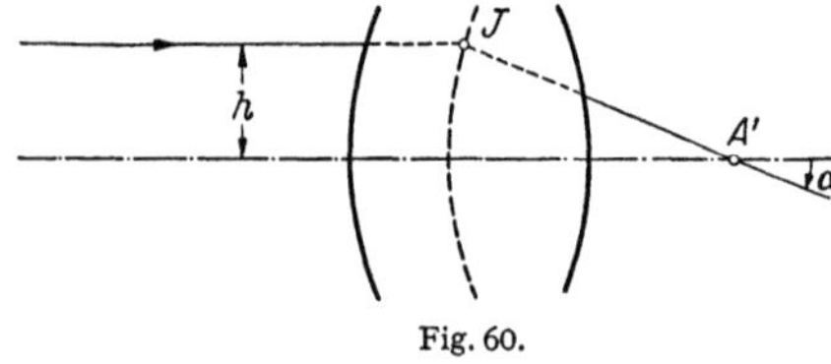

Fig. 60.

constant; si I (Fig. 60) est le point de rencontre du rayon incident et du rayon émergent, on a: $h/\sin \alpha' = IA'$, et le point I devra se déplacer sur une sphère centrée en A'; cette condition n'est évidemment pas satisfaite par les diverses surfaces stigmatiques fonctionnant avec un objet à l'infini étudiées au paragraphes 16 et 17, qui ne peuvent être rigoureusement stigmatiques que lorsque le point objet est sur l'axe, ce qui limite sérieusement leurs applications.

La condition de Herschel peut se transformer par la même méthode: si l'on prend pour variable caractérisant la position de l'objet non pas la distance x, mois son inverse $\xi = \dfrac{1}{x}$, on aura

$$n\, d x \sin^2 \frac{\alpha}{2} = -n\, x^2\, d\xi \left(\frac{\alpha}{2}\right)^2 = -\frac{n\, h^2}{4}\, d\xi \tag{30.2}$$

ou encore

$$-n\, h^2\, \frac{d\xi}{4} = n'\, d x' \sin^2 \frac{\alpha'}{2}.$$

Pour satisfaire à cette condition, il faudra que $h/\sin \dfrac{\alpha'}{2}$ soit constant.

Nous avons, dans ce chapitre, essayé d'obtenir le stigmatisme regoureux à nous pouvons dire en conclusion qu'il est le plus souvent impossible de l'obteni; dans un élément de volume, si l'on considère des rayons assez fortement incliné sur l'axe de l'instrument. Le principe de Fermat nous a conduit, en effet, s des conditions pratiquement incompatibles lorsque les ouvertures angulaires sont importantes. On est amené à étudier maintenant, dans le prochain chapitre, la formation des images dans les instruments de révolution, mais en se limitant à l'étude de la marche des rayons qui cheminent au voisinage immédiat de l'axe, de sorte que les conditions précédentes puissent être approximativement conciliées.

C. Approximation de GAUSS.

I. Définitions, notations et conventions de signes.

31. Nous nous proposons d'étudier maintenant la formation des images dans les systèmes optiques dont les surfaces admettent un axe de révolution commun, en utilitant seulement les rayons lumineux qui cheminent au voisinage de l'axe. Ces systemes peuvent être constitués par un ensemble de surfaces sphériques

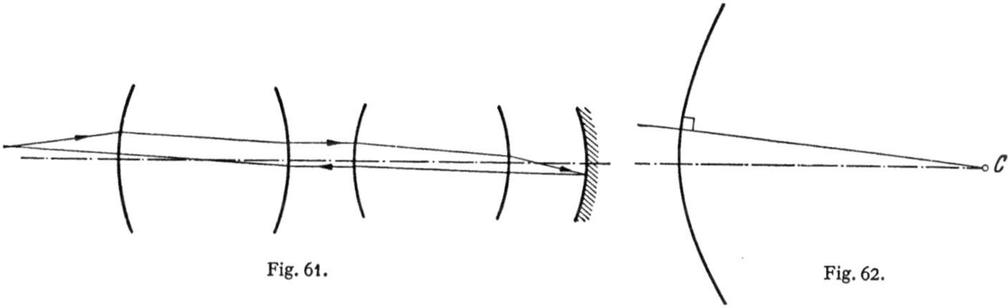

Fig. 61.   Fig. 62.

dont les centres sont alignés, mais on peut, de façon plus générale, envisager le cas de surfaces non sphériques, présentant un axe de révolution commun (Fig. 61).

Nous ferons deux hypothèses:

1. Les rayons lumineux font un petit angle avec l'axe;

2. les angles d'incidence sur les surfaces sont faibles.

On peut alors négliger toutes les quantités du 2ème ordre (ou plus) par rapport à l'ensemble des paramètres angulaires indiqués ci-dessus; autrement dit, on ne conserve dans les expressions des déviations latérales des rayons que les quantités *lineaires* par rapport à ces paramètre: dans le cas des surfaces non sphériques, on peut, bien entendu considérer que la normale passe par le centre de courbure au pôle de la surface (Fig. 62). On dit alors que l'on étudie la marche des rayons satisfaisant à l'approximation de GAUSS ou encore des rayons cheminant dans le domaine *paraxial*. Le caractère linéaire des fonctions à étudier suggéré l'emploi du calcul matriciel; le fonctionnement de l'ensemble de l'instrument peut, en effet, être caractérisé par une matrice égale au produit des matrices correspondant aux divers éléments de l'instrument. Nous ferons en même temps une étude géométrique élémentaire, ce qui nous permettra de mieux comprendre, dans les divers cas particuliers, la marche des rayons.

Au cours de cette étude, nous n'utilisons par les lignes trigonométriques des angles, mais plutôt les angles eux-mêmes de façon à rappeler qu'il est impossible dans l'approximation effectuée de distinguer le sinus de l'angle de sa tangente, ou encore de tout infiniment petit équivalent. Nous verrons d'ailleurs à la fin de ce chapitre (sect. 45) que l'utilisation des tangentes est non seulement arbitraire, mais qu'elle est le plus souvent dangereuse. Il nous faut maintenant préciser les conventions de signes que nous utilisons:

Nous rapportons l'espace-objet à des axes $Oxyz$ dont l'axe Ox coïncide avec l'axe de l'instrument et l'axe Oy est dans le plan de figure. L'espace image est de même rapporté à des axes analogues $O'x'y'z'$ (Fig. 63). Les origines O et O' sont choisies en fonction de la nature du problème à traiter. Enfin les sens positifs seront les mêmes dans l'espace objet et l'espace image (généralement de gauche à droite); ainsi la lumière cheminera avant réflexion sur le miroir M (Fig. 64) dans le sens négatif, elle cheminera après réflexion dans le sens positif: nous devrons donc prendre garde dans l'évaluation des chemins optiques lorsque nous aurons à étudier des réflexions: le chemin optique doit en effet être évalué positivement dans le sens de propagation de la lumière avant et après réflexion.

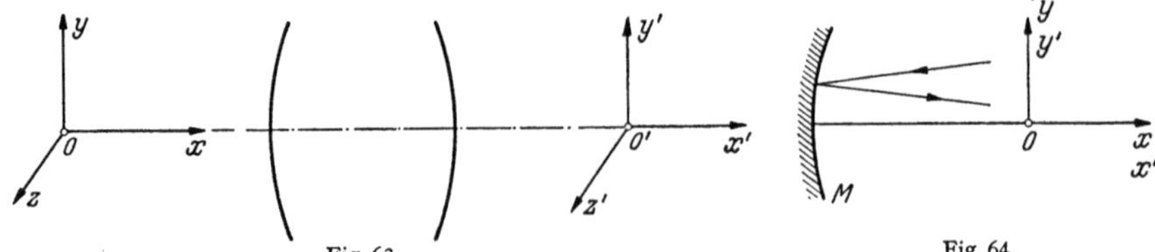

Fig. 63.   Fig. 64.

Il sera d'autre part commode pour éviter d'écrire pour chacune des coordonnées y et z des équations similaires et indépendantes de prendre comme variable caractérisant l'intersection d'un rayon avec un plan de front Oyz la quantité complexe $Z = y + jz$ que nous appellerons affixe de M (Fig. 65): l'écriture d'une relation concernant Z équivaudra à l'écriture de deux relations semblables pour y et z.

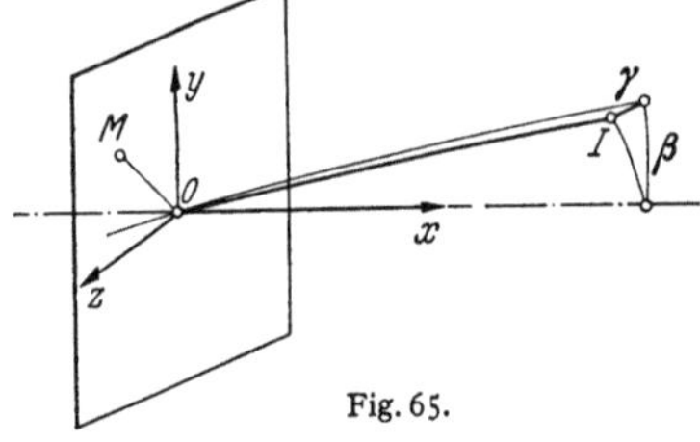

Fig. 65.

Un rayon lumineux sera de même caractérisé par ses cosinus directeurs α, β, γ sur les axes mais dans l'approximation de Gauss β et γ sont petits, de sorte que $\alpha = \sqrt{1 - \beta^2 - \gamma^2}$ pourra être assimilé à l'unité; les seuls paramètres utiles sont donc β et γ. Nous caractériserons l'orientation du rayon par la quantité complexe $\omega = \beta + j\gamma$ que nous appellerons affixe angulaire du rayon. On a représenté sur la Fig. 65 un point M d'affixe Z et un rayon d'affixe angulaire ω (qui est l'affixe de son intersection I avec la sphère unitaire) passant par 0.

Nous nous proposons de montrer tout d'abord que si l'on se donne Z et ω pour un rayon dans l'espace objet, les quantités correspondantes Z' et ω' dans l'espace image s'en déduisent par des transformations linéaires, que l'on représentera commodément par une matrice à coefficients réels.

II. Les éléments constitutifs: le dioptre et le miroir.

32. Le dioptre. Considérons un dioptre dont le rayon de courbure au pôle est $r = SC$, séparant deux milieux d'indices n et n'. Considérons un point I du dioptre (d'affixe Z_0) qui est le point d'impact d'un rayon incident d'affixe angulaire ω. La relation

$$n'\boldsymbol{u}' - n\boldsymbol{u} = (n'\cos i' - n\cos i)\,\boldsymbol{N}$$

permet de connaître le rayon réfracté à condition de connaître tout d'abord la normale: celle-ci a comme affixe angulaire $\omega_N = -Z_0/r$ (Fig. 66); en projetant sur l'axe Oy la relation vectorielle précédente qui exprime la réfraction, on obtient une relation de la forme $n'\beta' - n\beta = (n' - n)\beta_N$ et une relation analogue en projetant sur Oz; ces deux relations peuvent être combinées en utilisant les affixes angulaires sous la forme

$$n'\omega' - n\omega = (n' - n)\,\omega_N. \tag{32.1}$$

Si Z et Z' sont les affixes de points M et M' situés sur les rayons incident et réfracté dans des plans aux distances x et x' on pourra donc écrire (Fig. 67)

$$n'\frac{Z'-Z_0}{x'} - n\frac{Z-Z_0}{x} + (n'-n)\frac{Z_0}{r} = 0. \tag{32.2}$$

On peut donc ainsi caractériser complètement le fonctionnement du dioptre par cette relation; on peut en fait choisir x et x' de façon que Z' ne dépende que de Z, et non de Z_0: tous les rayons issus d'un point B de ce plan, d'affixe Z passeront par le point B' du plan conjugué, d'affixe Z' (Fig. 68); le point B' sera alors

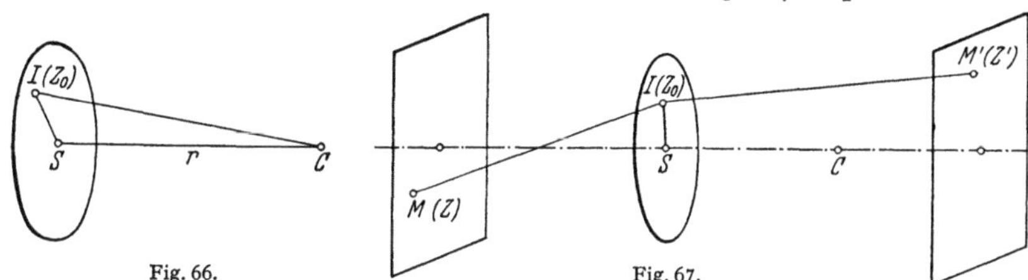

Fig. 66.

Fig. 67.

l'image de B et il y aura stigmatisme dans les limites de l'approximation de Gauss: il suffit pour cela que le coefficient de Z_0 s'annule dans le relation (32.1) c'est-à-dire encore que l'on ait

$$\frac{n'}{x'} - \frac{n}{x} = \frac{n'-n}{r} \tag{32.3}$$

et d'autre part

$$Z' = \frac{n}{n'}\frac{x'}{x}Z. \tag{32.4}$$

On trouve ainsi les formules classiques de conjugaison dans le dioptre sphérique: les plans conjugués sont situés aux abscisses x, x' qui satisfont à la relation (32.3) alors que le grandissement s'exprime par la relation (32.4). On définit en particulier les foyers du dioptre: le foyer objet F dont l'image est à l'infini ($1/x'=0$) et le foyer image F', image d'un point objet à l'infini ($1/x=0$); on posera $f=\overline{SF}$, $f'=\overline{SF'}$, et l'on remarque tout de suite que

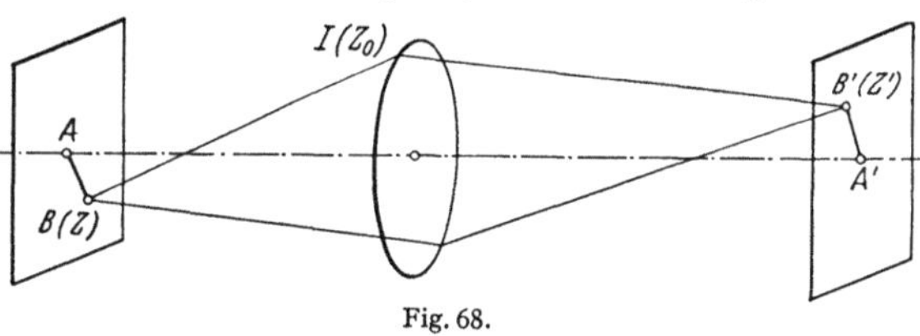

Fig. 68.

$$f = -\frac{n}{n-n'}r, \quad f' = \frac{n'}{n'-n}r, \quad \text{et} \quad \frac{f}{n} = -\frac{f'}{n'}. \tag{32.5}$$

On peut maintenant se poser le problème suivant: étant donné un rayon d'affixe angulaire ω perçant le plan objet au point d'affixe Z comment peut-on connaître les quantités correspondantes dans l'espace image? La relation (32.4) nous fournit Z' en fonction de Z; la relation (32.1) fournit ω', mais nous aurons à y exprimer ω_N, en remarquant que $Z-Z_0=\omega x$

$$\omega_N = -\frac{Z_0}{r} = \frac{\omega_x - Z}{r},$$

et l'on obtient finalement les deux relations (valables seulement lorsque les plans utilisés sont conjugués)

$$\begin{cases} Z' = \dfrac{n}{n'}\dfrac{x'}{x}Z, \\[2mm] \omega' = \left(\dfrac{n}{n'} - 1\right)\dfrac{1}{r}Z + \dfrac{x}{x'}\omega, \end{cases}$$

que l'on peut écrire en posant $g = \dfrac{n}{n'}\dfrac{x'}{x}$

$$Z' = gZ, \qquad \omega' = \left(\frac{n}{n} - 1\right)\frac{1}{r}Z + \frac{n}{n'g}\omega. \qquad (32.6)$$

33. La relation de Lagrange Helmholtz. Considérons un point A de l'axe dont l'image est en A' et d'autre part un rayon passant par A, cheminant dans le plan xoy et faisant avec l'axe l'angle α; soit aussi $y = AB$, la grandeur d'un objet dans le plan «réel» et $y' = A'B' = gy$ celle de l'image (Fig. 69); les relations précédentes s'écrivent, en faisant $Z = Z' = 0$ pour le rayon AIA':

$$\omega' = \alpha' = \frac{n}{n'g}\omega = \frac{n}{n'g}\alpha = \frac{n\,y\,\alpha}{n'\,y'}$$

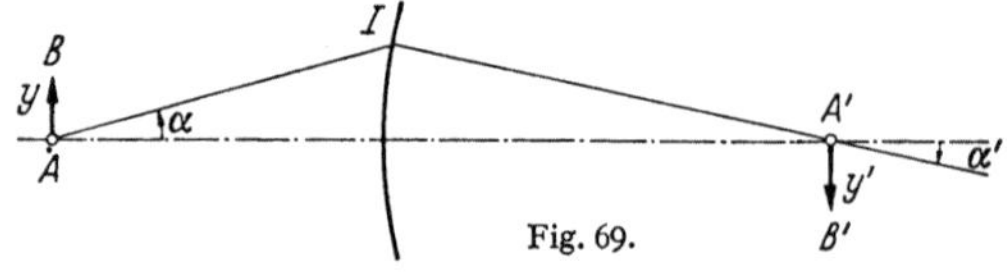

Fig. 69.

d'où l'on déduit que la quantité

$$\boxed{n\,y\,\alpha = n'\,y'\,\alpha'} \qquad (33.1)$$

est invariante entre l'objet et son image[1].

On remarque bien entendu l'analogie avec les résultats obtenus dans l'établissement de la relation de Clausius, ou encore avec la relation des sinus; nous reviendrons ultérieurement sur cette question.

34. L'utilisation des matrices. Si l'on considere un rayon d'affixe angulaire ω, perçant un plan d'abscisse x en un point d'affixe Z on peut se proposer de connaître les quantités correspondantes dans l'espace image, en supposant que le plan de référence d'abscisse x' ne soit pas a priori le plan conjugué du plan d'abscisse x: il suffit d'utiliser les relations (32.1) et (32.2) en y faisant $Z_0 = Z - \omega x$; on trouve alors

$$\omega' = \left(\frac{n}{n'} - 1\right)\frac{Z}{r} + \left(\frac{n}{n'} + \left(1 - \frac{n}{n'}\right)\frac{x}{r}\right)\omega,$$

puis

$$Z' = \left[1 + \left(\frac{n}{n'} - 1\right)\frac{x'}{r}\right]Z + \left[\frac{n}{n'}\frac{1}{x} - \frac{1}{x'} + \left(1 - \frac{n}{n'}\right)\frac{1}{r}\right]x\,x'\,\omega.$$

Les quantités Z' et ω' s'expriment linéairement à l'aide de Z et ω comme on devait s'y attendre et l'on peut exprimer ceci à l'aide de la relation matricielle

$$\begin{vmatrix} Z' \\ \omega' \end{vmatrix} = \begin{vmatrix} 1 + \left(\dfrac{n}{n'} - 1\right)\dfrac{x'}{r} & \dfrac{x\,x'}{n'}\left[\dfrac{n}{x} - \dfrac{n'}{x'} + \dfrac{n'-n}{r}\right] \\ \left(\dfrac{n}{n'} - 1\right)\dfrac{1}{r} & \dfrac{n}{n'} + \left(1 - \dfrac{n}{n'}\right)\dfrac{x}{r} \end{vmatrix} \begin{vmatrix} Z \\ \omega \end{vmatrix} \qquad (34.1)$$

qui se réduit lorsque les plans d'abrcisses x et x' sont conjugués à

$$\begin{vmatrix} Z' \\ \omega' \end{vmatrix} = \begin{vmatrix} \dfrac{n}{n'}\dfrac{x'}{x} & 0 \\ \left(\dfrac{n}{n'} - 1\right)\dfrac{1}{r} & \dfrac{x}{x'} \end{vmatrix} \begin{vmatrix} Z \\ \omega \end{vmatrix}. \qquad (34.2)$$

En posant

$$g = \frac{n}{n'}\frac{x}{x'}, \qquad g_\alpha = \frac{x}{x'}, \qquad \frac{1}{f'} = \left(1 - \frac{n}{n'}\right)\frac{1}{r},$$

---

[1] Lagrange avait en fait écrit cette relation sous la forme $n\,y\tan\alpha = n'\,y'\tan\alpha'$ mais nous préférons ici, pour indiquer qu'elle n'est valable que lorsque les angles sont petits, l'écrire sous la forme précédente, nous verrons d'ailleurs à la Sect. 45 pourquoi il est préférable d'éviter d'écrire la relation utilisont les tangentes.

la matrice «dioptre» s'écrit pour un couple de plans conjugués

$$\begin{vmatrix} g & 0 \\ -\dfrac{1}{f'} & g_\alpha \end{vmatrix}. \tag{34.3}$$

La relation de LAGRANGE HELMHOLTZ s'écrit d'ailleurs $g g_\alpha = n/n'$ et il s'ensuit que les coefficients g et g_α ne sont pas des paramètres indépendants.

Voyons maintenant comment se traduit un simple changement d'origine : si l'on suppose que l'on avance dans le sens des x croissants le plan de référence d'une quantité x' on peut relier les Z', ω' relatifs à la nouvelle position à ceux relatifs à l'ancienne ; il suffit de faire dans la matrice générale (34.1)

$$1/r = 0, \quad n = n' \quad \text{et} \quad x = 0$$

pour obtenir la matrice «translation»

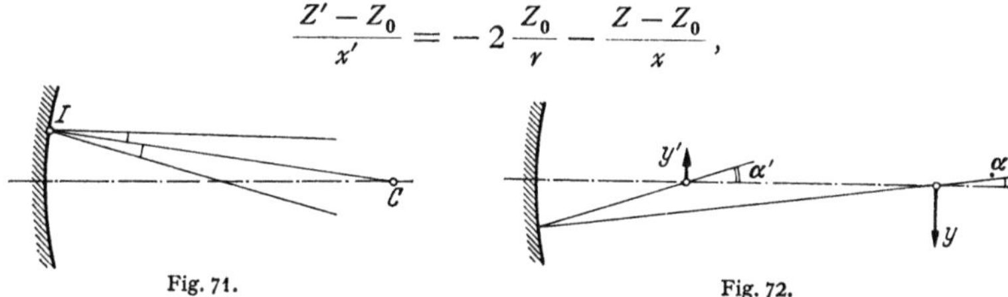

Fig. 70.

$$\begin{vmatrix} 1 & x' \\ 0 & 1 \end{vmatrix} \tag{34.4}$$

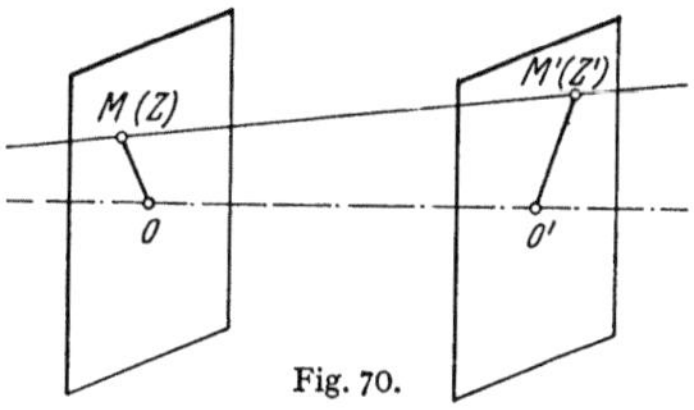

comme on peut le vérifier immédiatement d'ailleurs puisqu'on doit avoir (Fig. 70)

$$Z' = Z + \omega x, \quad \omega' = \omega.$$

35. Le miroir. En suivant la même méthode que pour le dioptre, on écrira pour le miroir la loi de la réflexion sous la forme (Fig. 71)

$$\omega' = 2\omega_N - \omega. \tag{35.1}$$

La relation entre Z, Z' et Z_0 s'écrit alors

$$\frac{Z' - Z_0}{x'} = -2 \frac{Z_0}{r} - \frac{Z - Z_0}{x},$$

Fig. 71.

Fig. 72.

ce qui conduit aux relations de conjugaison et de grandissement

$$\frac{1}{x'} + \frac{1}{x} = \frac{2}{r}, \quad Z' = -\frac{x'}{x} Z. \tag{35.2}$$

On trouve ainsi l'existence de plans conjugués classiques. La relation de LAGRANGE HELMHOLTZ s'écrira dans le cas du miroir (Fig. 72), avec nos conventions de signe :

$$y' \alpha' = - y \alpha. \tag{35.3}$$

Il est aisé d'établir ici encore l'existence de relations matricielles. On trouve successivement dans le cas du miroir

$$\omega' = -\frac{2}{r} Z + \left(\frac{2x}{r} - 1\right)\omega,$$

$$Z' = \left(1 - \frac{2x'}{r}\right) Z + x x' \left(\frac{2}{r} - \frac{1}{x} - \frac{1}{x'}\right)\omega,$$

c'est-à-dire que la relation matricielle générale s'écrira

$$\left|\begin{matrix} Z' \\ \omega' \end{matrix}\right| = \left|\begin{matrix} 1 - \dfrac{2x'}{r} & x\,x'\left(\dfrac{2}{r} - \dfrac{1}{x} - \dfrac{1}{x'}\right) \\ -\dfrac{2}{r} & \left(\dfrac{2x}{r} - 1\right) \end{matrix}\right| \cdot \left|\begin{matrix} Z \\ \omega \end{matrix}\right| \tag{35.4}$$

qui se réduit lorsque les plans d'abscisses x et x' sont conjugués à

$$\left|\begin{matrix} Z' \\ \omega' \end{matrix}\right| = \left|\begin{matrix} g & 0 \\ -\dfrac{1}{f'} & g_\alpha \end{matrix}\right| \left|\begin{matrix} Z \\ \omega \end{matrix}\right| \tag{35.5}$$

en posant

$$g = -\frac{x'}{x}, \qquad g_\alpha = \frac{x}{x'} \quad \text{et} \quad \frac{1}{f'} = \frac{2}{r}.$$

Après avoir vu très rapidement les propriétés des éléments constitutifs des systèmes centrés, nous pouvons maintenant examiner les éléments caractérisant l'ensemble du système.

III. Formation des images dans les systèmes centrés.

36. La matrice caractéristique. Nous avons vu que la réfraction sur un dioptre ou la réflexion sur un miroir, ainsi que le déplacement d'un plan de référence s'expriment à l'aide d'une matrice; on peut donc *a priori* passer de l'espace objet d'un instrument à l'espace image en effectuant un nombre convenable de multiplications de matrices: on peut par exemple écrire la relation matricielle (34.1) qui exprime la réfraction sur un dioptre [ou la relation (35.4) relative à la réflexion sur un miroir] chaque fois que l'on rencontre une surface. Le passage d'une surface à une autre peut être considéré comme une translation du plan de référence que l'on peut traduire encore par la matrice (34.4) ou x' est égal à la distance entre les dioptres; on aura donc à appliquer chaque fois que l'on rencontre une surface la relation (34.1) où l'on ferait $x = x' = 0$, et chaque fois que l'on change de surface la relation (34.4) où x' serait la distance entre surfaces: on obtiendra finalement une relation matricielle globale de la forme

$$\left|\begin{matrix} Z' \\ \omega \end{matrix}\right| = \left|\begin{matrix} A & B \\ B & D \end{matrix}\right| \cdot \left|\begin{matrix} Z \\ \omega \end{matrix}\right| \tag{36.1}$$

mais il est particulièrement utile de faire l'élimination des variables intermédiaires en prenant comme plan de référence dans chaque espace le plan conjugué d'un plan donné dans l'espace objet. Le passage de l'objet à son image dans le premier dioptre s'exprime par la matrice

$$\left|\begin{matrix} g_1 & 0 \\ -\dfrac{1}{f'_1} & \dfrac{n'_1}{n_1}\,\dfrac{1}{g_1} \end{matrix}\right|$$

où f'_1 est la distance focale image du dioptre.

Le passage de l'image fournie par le premier dioptre à l'image fournie par le second s'exprime par la matrice

$$\left|\begin{matrix} g_2 & 0 \\ -\dfrac{1}{f'_2} & \dfrac{n'_2}{n_2}\,\dfrac{1}{g_2} \end{matrix}\right| \cdot$$

on peut montrer aisément que la matrice produit sera du type:

$$\begin{vmatrix} g & 0 \\ -\dfrac{1}{f'} & \dfrac{n'}{n}\,\dfrac{1}{g} \end{vmatrix} \tag{36.2}$$

où $g = g_1,\, g_2,\, \ldots g_n$, et $1/f'$ est un coeffizient caractérisant le système global; ce coefficient peut en effet être déterminé en envoyant dans l'espace objet un rayon parallèle à l'axe ($\omega = 0$) : si Z est l'affixe du point de rencontre avec un plan

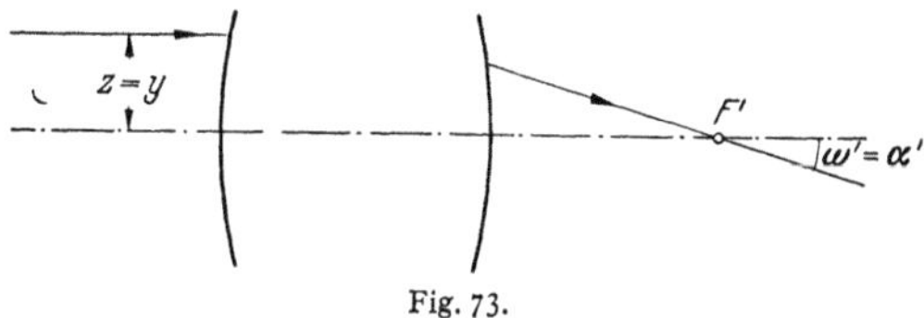

Fig. 73.

quelconque, on aura $\omega' = -Z/f'$ (Fig. 73). Cette quantité ne dépend manifestement pas du couple de plans conjugnés de référence choisis dans l'espace objet et dans l'espace image.

37. La relation de conjugaison avec origines en des point conjugués quelconques. Si l'on choisit comme origines provisoires dans l'espace objet et dans l'espace

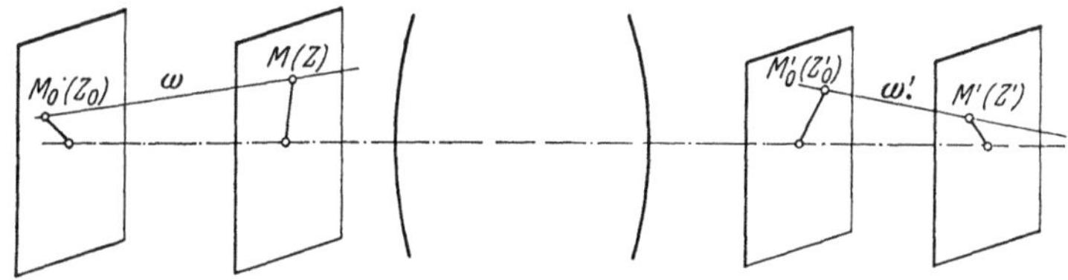

Fig. 74.

image un couple de points conjugués OO' caractérisés par un grandissement transversal g_0 et un grandissement angulaire $g_\alpha = \dfrac{n}{n'g_0}$ on aura

$$Z_0' = g_0 Z_0,$$

$$\omega' = -\frac{Z_0}{f'} + \frac{n}{n'}\,\frac{1}{g_0}\,\omega.$$

Appelons x et x' les abscisses de deux autres points A et A' respectivement par rapport à OO', Z et Z' les affixes correspondantes, on aura, en remplaçant Z_0 par $Z - \omega x$ et Z_0' par $Z' - \omega x'$ (Fig. 74)

$$\left.\begin{aligned} \omega' &= -\frac{Z}{f'} + \left(\frac{n}{n'g_0} + \frac{x}{f'}\right)\omega, \\ Z' &= \left(g_0 - \frac{x'}{f'}\right)Z + \left[\left(\frac{n}{n'g_0} + \frac{x}{f'}\right)x' - g_0 x\right]\omega. \end{aligned}\right\} \tag{37.1}$$

Les deux plans d'abscisses x et x' seront conjugués à condition que Z' ne dépende pas de ω (mais seulement de Z). On doit donc avoir

$$\frac{1}{g_0}\,\frac{n}{x} - g_0\,\frac{n'}{x'} + \frac{n'}{f'} = 0 \tag{37.2}$$

qui est la relation de conjugaison lorsque les origines sont placées en des points conjugués OO' où le grandissement est g_0, le grandissement g étant alors

$$g = g_0 - \frac{x'}{f'}\,. \tag{37.3}$$

Ces deux relations définissent parfaitement la position et la grandeur de l'image d'un objet plan d'abscisse x quelconque. On peut remarquer d'ailleurs que la relation matricielle caractéristique s'écrit

$$\begin{vmatrix} Z' \\ \omega' \end{vmatrix} = \begin{vmatrix} g & 0 \\ -\dfrac{1}{f'} & \dfrac{n}{n'}\dfrac{1}{g} \end{vmatrix} \begin{vmatrix} Z \\ \omega \end{vmatrix}.$$

Le coefficient $-1/f'$ est donc bien invariable lorsqu'on change de couple de plans conjugués, comme on devait s'y attendre d'après la remarque qui termine le Sect. 36.

38. Eléments cardinaux des systèmes centrés à foyers. Nous allons tout d'abord supposer que le coefficient $1/f'$ n'est pas nul; on dira alors que le système est

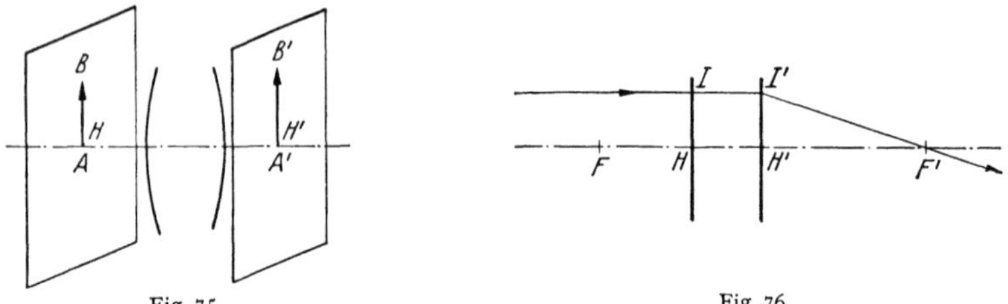

Fig. 75.   Fig. 76.

un système à foyers. On peut dans ces conditions obtenir un grandissement g quelconque fixé a priori: la relation (37.3) fixe en effet x' et la relation (37.2) fixe ensuite x sans ambiguïté; on peut en particulier déterminer les plans pour lesquels le grandissement g est unitaire: les plans conjugués ainsi définis s'appellent *les plans principaux* H, H' du système (Fig. 75).

Dans un système optique à foyers il existe donc un couple de plans particuliers pour lesquels le grandissement transversal est égal à l'unité; la matrice caractéristique s'écrit alors

$$\begin{vmatrix} 1 & 0 \\ -\dfrac{1}{f'} & \dfrac{n}{n'} \end{vmatrix} \tag{38.1}$$

et les relations de conjugaison et de grandissement, si l'on place les origines en HH', prennent la forme

$$\boxed{\dfrac{n'}{x'} - \dfrac{n}{x} = \dfrac{n'}{f'} = C\,,} \qquad \boxed{g = \dfrac{n}{n'}\dfrac{x'}{x}\,.} \tag{38.2}$$

(On pose quelquefois $C = n'/f'$; cette caractéristique du système s'appelle *la convergence du systeme*).

Foyers et plans focaux: les foyers FF' du système sont les conjugués des points à l'infini dans l'espace objet et dans l'espace image (Fig. 76); on les trouve en faisant dans la relation (38.2) soit $1/x = 0$ soit $1/x' = 0$; on aura donc

$$\overline{HF} = f = -\frac{n}{n'}f'\,, \qquad \overline{H'F'} = f'\,. \tag{38.3}$$

Ces longueurs $\overline{HF}$ et $\overline{H'F'}$ sont les longueurs focales du système; on remarque que leur rapport f'/f est égal au rapport des indices des milieux extrêmes changé de signe; on pourra donc écrire la convergence C indifféremment sous les deux formes équivalentes

$$C = \frac{n'}{f'} = -\frac{n}{f}. \qquad (38.4)$$

Considérons maintenant un rayon parallèle à l'axe et perçant le plan principal objet en I. L'image de I est en un point I' du plan principal image, à la même distance de l'axe que le point I. De plus le rayon incident étant parallèle à l'axe il sort du système en passant par le foyer image F'; on peut donc considérer le plan principal image comme le lieu des points tels que I' qui sont l'intersection d'un rayon incident parallèle à l'axe et du rayon émergent correspondant (Fig. 76).

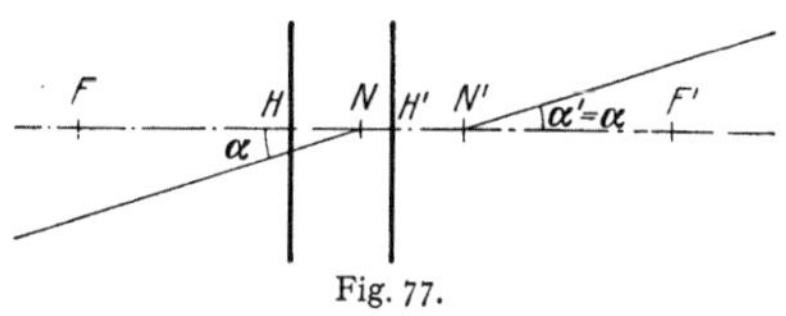
Fig. 77.

Points nodaux N et N': ce sont deux points conjugués tels que le rayon incident et le rayon émergent soient parallèles, ce qui exige que le grandissement angulaire α'/α soit égal à l'unité.

On devra donc avoir d'après la relation de Lagrange Helmholtz $g = n/n'$ et l'on tire alors des relations (38.2)

$$x' = x, \quad \frac{n' - n}{x} = \frac{n'}{f'},$$

ou encore

$$x = x' = \frac{n' - n}{n} f' = \frac{n - n'}{n} f.$$

On aura donc

$$FN = FH + HN = f\left[\frac{n - n'}{n} - 1\right] = f',$$

et de même

$$F'N' = f.$$

Les points nodaux N et N' occupent donc des positions simples par rapport aux foyers et aux points principaux H, H' (Fig. 77).

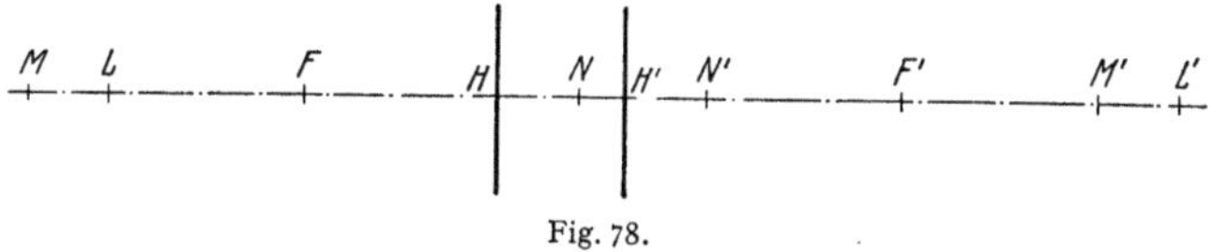
Fig. 78.

Points antiprincipaux et antinodaux: bien que moins utilisés que les éléments cardinaux définis ci-dessus les points antiprincipaux et antinodaux peuvent être utiles dans certains cas: les points antiprincipaux LL' sont définis par la condition que $g = -1$; leurs abscisses par rapport aux points principaux sont respectivement $x = 2f$ et $x' = 2f'$ (Fig. 78).

Les points antinodaux sont de même les points pour lesquels on a $g_\alpha = -1$; on doit alors avoir $x' = -x$ et l'on en déduit facilement que $FM = -f'$ et $F'M' = -f$ (Fig. 78). Les points antinodaux et antiprincipaux sont donc symétriques des points nodaux et principaux par rapport aux foyers.

39. Constructions géométriques de l'image d'un point objet B. On peut maintenant utiliser les résultats précédent pour construire l'image d'un point objet B: on peut tracer en particulier:

Le rayon parallèle à l'axe de l'instrument, qui passe en sortant par le foyer et par le point I' du plan principal image.

Le rayon BF passant par le foyer objet F et par un point J du plan principal objet: il sort parallèlement à l'axe en passant par J.

Le rayon BN qui sort parallèlement à sa direction initiale en passant par le point nodal image N'.

On pourrait encore utiliser d'autres rayons: le rayon BH, qui sort en passant par H' et en faisant un angle que l'on peut connaître en appliquant à H et H' la relation de Lagrange: on aura

$$n\vartheta = n'\vartheta'.$$

La marche des trois rayons principaux est indiquée sur la Fig. 79.

40. Formules de Newton. Plaçons les origines aux foyers: si l'on pose $\Pi = FA$ et $\Pi' = F'A'$ on peut écrire (Fig. 79) les relations

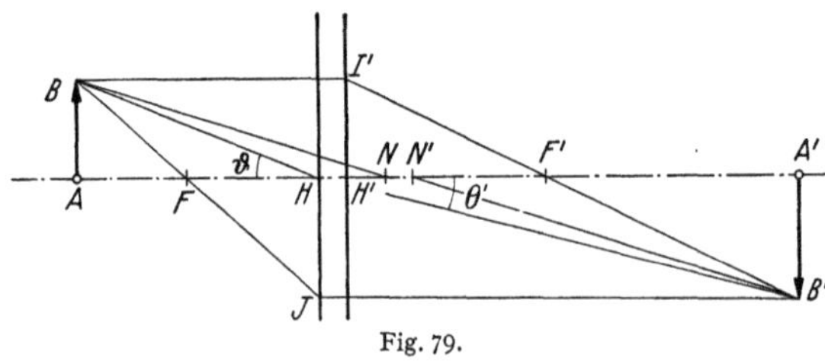

Fig. 79.

$$\frac{y'}{y} = -\frac{f}{\Pi} = -\frac{\Pi'}{f'} \quad (40.1)$$

d'où l'on tire

$$\Pi\Pi' = ff'.$$

Ces formules particulièrement simples sont souvent utilisées dans la pratique; il arrive en effet très souvent que l'objet (ou l'image) soient à grande distance et elles sont particulièrement commodes. Elles montrent que la proximité $1/\Pi$ de l'objet est proportionnelle à la distance du foyer image à l'image.

41. Etude des grandissements. Nous définirons les trois grandisement suivants

$$g = \frac{y'}{y}, \quad g_\alpha = \frac{\alpha'}{\alpha}, \quad g_x = \frac{dx'}{dx}.$$

g_x est le *grandissement longitudinal*, g le *grandissement transversal*, g_α le *grandissement angulaire*, encore appelé *rapport de convergence*.

Nous avons d'après les équations (38.2) (origines en H, H')

$$\left.\begin{aligned} g &= \frac{n}{n'}\frac{x'}{x}, \\ g_\alpha &= \frac{n}{n'}\frac{1}{g} = \frac{x}{x'}, \\ g_x &= \frac{n}{n'}\frac{x'^2}{x^2}. \end{aligned}\right\} \quad (41.1)$$

Ces trois grandissements ne dépendent que du rapport x'/x. En éliminant ce paramètre entre deux des expressions, on obtient une relation permanente entre les deux grandissments correspondants. On trouve

$$g\,g_\alpha = \frac{n}{n'}, \quad g_x\,g_\alpha^2 = \frac{n}{n'}, \quad g_x = \frac{n'}{n}g^2. \quad (41.2)$$

La première de ces relations n'est autre que la relation de Lagrange. Il existe encore la relation suivante entre les trois grandissements

$$g = g_\alpha\,g_x. \quad (41.3)$$

Remarquons enfin que, lorsqu'on utilise les foyers pour origines, on a

$$g_x = \frac{d\Pi'}{d\Pi} = -\frac{\Pi'}{\Pi}, \quad \text{puisque} \quad \Pi\Pi' = \text{constante.}$$

42. Associations de systèmes centrés. Etant donnés deux systemes S_1, S_2 de convergences C_1, C_2, on se propose de déterminer les éléments du système centré S formé par leur association. La détermination des foyers est immédiate: le foyer image F' est le conjugué de F_1' à travers S_2, le foyer objet le conjugué de F_2 à travers S_1. Si Δ est la distance $\overline{F_1'F_2}$ (Fig. 80), on a en appliquant les formules de NEWTON

$$\Delta \cdot \overline{F_1 F} = f_1 f_1'$$

et de même

$$-\Delta \cdot \overline{F_2 F'} = f_2 f_2',$$

équations qui fixent les positions de F et F'.

Pour caractériser complètement le système global il suffit d'effectuer la multiplication de deux matrices: nous allons le faire principalement dans le but de

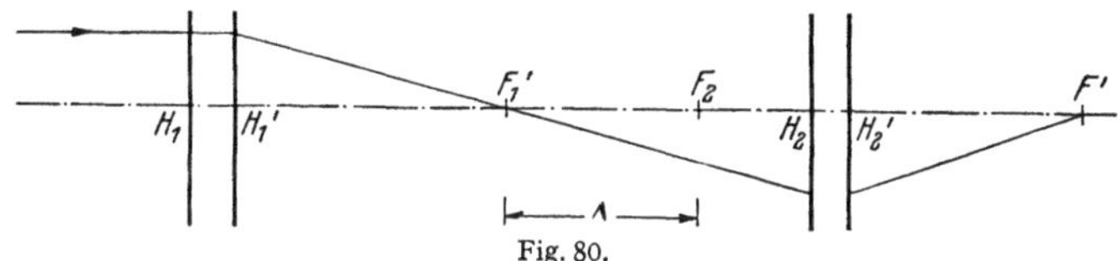

Fig. 80.

connaître la distance focale de l'association. Si g_1 et g_2 sont les grandissements obtenus dans les systèmes S_1, S_2 pour une position particulière de l'objet, f_1' et f_2' les distances focales, f' la distance focale de l'ensemble, on aura à faire le produit de matrices:

$$\begin{vmatrix} g_2 & 0 \\ -\dfrac{1}{f'_2} & \dfrac{n_2}{n_2' g_2} \end{vmatrix}, \quad \begin{vmatrix} g_1 & 0 \\ -\dfrac{1}{f_1'} & \dfrac{n_1}{n_1' g_1} \end{vmatrix}.$$

En remarquant que $n_1' = n_2$, on trouve comme matrice produit

$$\begin{vmatrix} g_1 g_2 & 0 \\ -\dfrac{g_1}{f_2'} - \dfrac{n_2}{n_2' g_2 f_1'} & \dfrac{n_1}{n_2' g_1 g_2} \end{vmatrix}.$$

On aura donc

$$\frac{1}{f'} = \frac{g_1}{f_2'} + \frac{n_2}{n_2' g_2 f_1'},$$

mais les formules de NEWTON indiquent que $g_1 = -\dfrac{\overline{F_1' A'}}{f_1'}$ et $g_2 = -\dfrac{f_2}{\overline{F_2 A'}}$, et l'on trouve finalement:

$$f' = \frac{f_1' f_2'}{\Delta} \tag{42.1}$$

où Δ est l'intervalle $\overline{F_1' F_2}$.

On peut encore exprimer la convergence du système en fonction des convergences C_1 et C_2 des systèmes composant l'ensemble et de l'intervalle $e = \overline{H_1' H_2}$ entre les plans principaux dans l'espace intermédiaire; on aura $\Delta = e + f_2 - f_1'$ et en remplaçant Δ par cette valeur dans la relation (42.1), on trouve

$$\boxed{C = C_1 + C_2 - \frac{e}{N} C_1 C_2} \tag{42.2}$$

où N est l'indice du milieu intermédiaire. Cette relation, due à GULLSTRAND est très souvent utilisée.

43. Les systèmes afocaux. L'étude des associations de systemes centrés nous amène maintenant à examiner le cas particulier des systèmes afocaux, où l'image d'un objet à l'infini est également située à l'infini.

Reprenons pour cela la matrice générale (36.2) en supposant maintenant $1/f' = 0$; on aura simplement pour un couple de points conjugués

$$Z' = gZ, \qquad \omega' = \frac{n}{n'g}\,\omega,$$

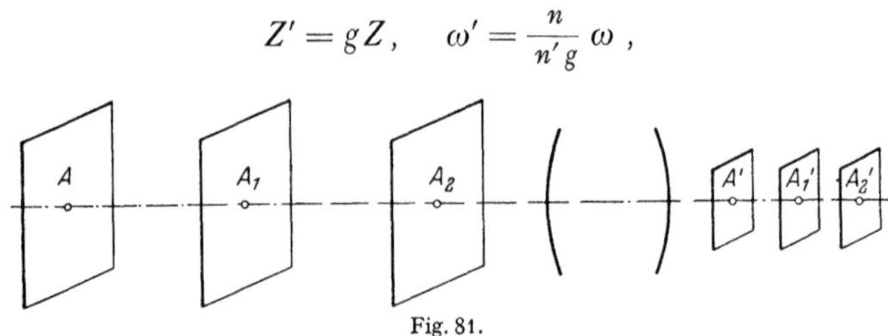

Fig. 81.

et l'on peut utiliser la même méthode qu' à la Sect. 37 pour rechercher la position des plans conjugués; on aura en déplaçant l'origine d'une distance x dans l'espace objet, d'une distance x' dans l'espace image

$$\left.\begin{array}{l} Z' = gZ + \left(\dfrac{n}{n'g}\,x' - g\,x\right)\omega, \\[2mm] \omega' = \dfrac{n}{n'g}\,\omega. \end{array}\right\} \tag{43.1}$$

Ainsi Z' ne dépendra que de Z si l'on a

$$x' = \frac{n'}{n}\,g^2\,x. \tag{43.2}$$

Le déplacement de l'image est cette fois proportionnel à celui de l'objet; si l'objet va à l'infini, l'image y va également ce qui justifie la dénomination de système afocal; on aura de plus

$$Z' = gZ \tag{43.3}$$

qui indique que le grandissement transversal est constant; on aura d'ailleurs

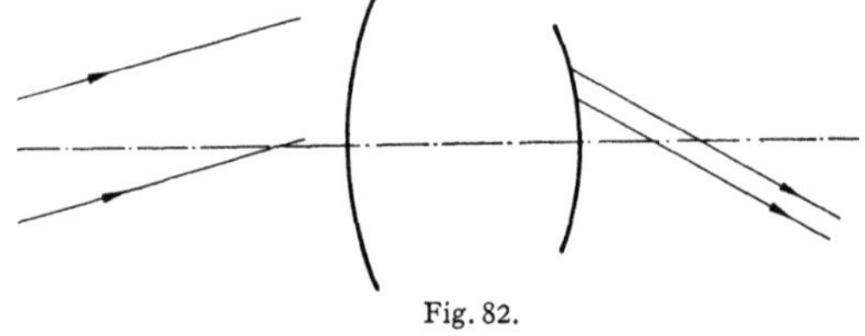

Fig. 82.

$$g_\alpha = \frac{n}{n'g} = \text{constante}.$$

Ainsi dans un système afocal tous les grandissements sont constants. On a représenté sur la Fig. 81 quelques positions de plans conjugués. Le grandissement angulaire étant constant, un faisceau de rayons parallèles dans l'espace objet donnera un faisceau de rayons également parallèles dans l'espace image (Fig. 82).

En conclusion la formation des images dans les systèmes afocaux est remarquablement simple, la relation de conjugaison est linéaire et les trois grandissement sont constants.

44. Les lentilles. Nous appliquerons maintenant la théorie générale des systèmes centrés au cas simple des lentilles; nous supposerons que les milieux extrêmes sont constitués par l'air. Nous désignerons par N l'indice du verre; nous appellerons d'autre part r_1 et r_2 les deux rayons de courbure des dioptres. (Fig. 83). Nous étudierons tout d'abord le cas général des lentilles épaisses, dont l'épaisseur sera désignée par e; d'après la relation (38.3) le rapport des distances focales sera égal à -1 et l'on aura

$$f' = -f.$$

Cherchons tout d'abord la convergence de la lentille: la formule de GULLSTRAND nous donne ici

$$C = C_1 + C_2 - \frac{e}{N} C_1 C_2$$

où C_1 et C_2 sont les convergences des dioptres qui limitent la lentille; on aura pour le premier dioptre

$$C_1 = \frac{N}{S_1 F_1'} = \frac{N-1}{r_1}$$

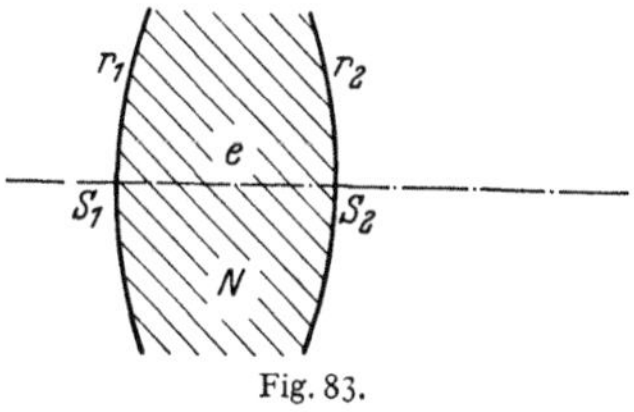

et de même pour le second

$$C_2 = \frac{1-N}{r_2}$$

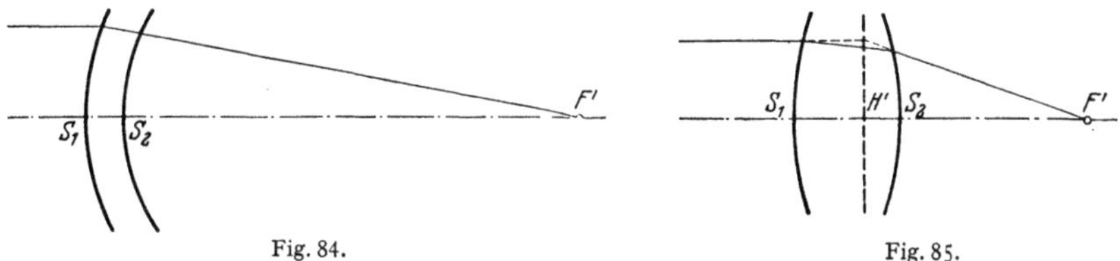
Fig. 83.

d'où pour la lentille

$$C = (N-1)\left(\frac{1}{r_1} - \frac{1}{r_2}\right) + \frac{e(N-1)^2}{N r_1 r_2}. \qquad (44.1)$$

La convergence dépend en particulier de l'épaisseur; ainsi si $r_1 = r_2$, ce qui annule le premier terme on obtient toujours une lentille convergente (Fig. 84). Cherchons

Fig. 84.

Fig. 85.

maintenant les foyers: ce sont les images dans la lentille d'un point objet situé à l'infini; on aura donc dans les deux dioptres:

$$\frac{N}{x_1'} = \frac{N-1}{r_1},$$

$$\frac{1}{x'} = \frac{N}{x_1'-e} + \frac{1-N}{r_2}$$

d'où l'on déduit la distance

$$S_2 F' = x' = \frac{1}{C}\left[1 - \frac{N-1}{N}\frac{e}{r_1}\right]. \qquad (44.2)$$

On en déduit d'ailleurs la position du plan principal image (Fig. 85), dont l'abscisse est

$$S_2 H' = x' - f' = \frac{1-N}{NC}\frac{e}{r_1}. \qquad (44.3)$$

On aurait de même dans l'espace objet:

$$S_1 H = \frac{1-N}{NC}\frac{e}{r_2}.$$

Remarquons d'ailleurs que la position des plans principaux peut se retrouver aisément par la considération du centre optique: soit O le point qui divise le segment $S_1 S_2$ dans le rapport r_1/r_2; O est le centre d'homothétie pour les deux dioptres sphériques et il résulte que tout rayon passant par ce point donne avant et après réfraction deux rayons parallèles (Fig. 86), car les angles d'incidence sur les dioptres sont identiques. Si les rayons sont peu inclinés sur l'axe (domaine paraxial) on peut ainsi définir les points nodaux NN' qui coïncident avec les points principaux HH' puisqu'ici

$$f = f'.$$

On aura

$$\frac{OS_1}{r_1} = \frac{OS_2}{r_2} \quad \text{avec} \quad OS_2 - OS_1 = e$$

d'où

$$S_1O = \frac{e\,r_1}{r_1 - r_2}, \qquad S_2O = \frac{e\,r_2}{r_2 - r_1}.$$

H' sera par exemple le conjugué de O dans le dioptre de rayon r_2 et l'on retrouverait facilement les valeurs indiquées ci-dessus. Nous représenterons sur la Fig. 87 la position des plans principaux dans quelques cas.

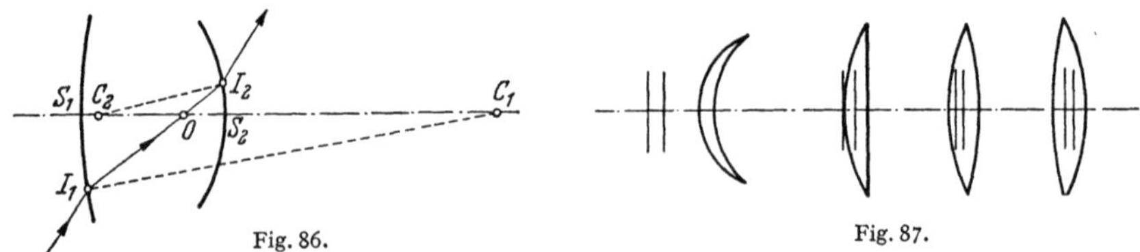

Fig. 86.   Fig. 87.

IV. L'homographie optique; impossibilité du système optique parfait.

45. Dans l'approximation de GAUSS chaque plan de l'espace objet admet un plan conjugué dans lequel les images qui se forment sont semblables à l'objet. On peut alors imaginer un instrument d'optique idéal qui permettrait de prolonger ces propriétés hors de l'approximation de GAUSS; on pourrait demander à cet instrument de donner d'un objet plan de dimensions étendues une image stigmatique et semblable à l'objet quelle que soit la position du plan objet. Il est facile de voir en fait que cet instrument idéal est impossible.

Considérons en effet tout d'abord le couple de plans conjugués passant par A et A', puis un autre couple de plans conjugués passant par A_1 et A_1'. L'image $A_1'B_1'$ d'un objet A_1B_1 devant être semblable à l'objet, on devra avoir

$$\frac{A_1'B_1'}{A_1B_1} = \text{constante}$$

ou encore

$$\frac{AA_1 \tan \alpha_1}{A'A_1' \tan \alpha_1'} = \text{constante},$$

d'où

$$\frac{\tan \alpha_1}{\tan \alpha_1'} = \text{constante}$$

mais l'instrument devant être aplanétique pour tout couple de plans on devra avoir en fait, d'après la condition d'ABBE,

$$\frac{\sin \alpha_1}{\sin \alpha_1'} = \text{constante}.$$

On voit donc qu'il est impossible de concevoir un instrument parfait du fait de l'impossibilité de concilier la relation utilisant les tangentes et le principe de FERMAT, exprimé par la condition d'ABBE [sauf dans le cas particulier où $|\alpha'| = |\alpha|$: il s'agit alors d'un système afocal de grandissement unitaire]; on a d'ailleurs déjà vu qu'il était impossible d'obtenir l'aplanétisme et la stabilité du stigmatisme le long de l'axe (incompatibilité des conditions d'ABBE et d'HERSCHEL). Dans la construction d'un instrument on se bornera donc à corriger les aberrations qui sont les plus gênantes dans les conditions d'emploi de l'instrument, mais un instrument d'excellente qualité ne pourra être stigmatique pour une position de l'objet très différente de celle pour laquelle il a été construit. C'est pourquoi on ne peut en particulier espérer une bonne précision dans les mesures de grandeurs caractéristiques (grandissements, distances focales etc.) si ces dernières ne sont pas faites dans les plans conjugués pour lesquels le système optique permet d'obtenir le stigmatisme.

D. Le champ.

46. Définitions. Un instrument d'optique ne fournit pas une image de n'importe quel point de l'espace: il faut que le point lumineux envisagé puisse envoyer dans l'instrument des rayons qui soient susceptibles de le traverser sans être arrêtés par les diverses montures de lentilles, diaphragmes etc.: pour une position déterminée de plan image (par exemple le plan d'une plaque photographique) seule la portion centrale du plan conjugué dans l'espace objet sera visible: c'est ce que nous appelons *le champ latéral*. De plus si le point objet se déplace légèrement en profondeur, son image ne cesse pas immédiatement d'être nette: on peut se déplacer légèrement tout en obtenant encore une image acceptable; la portion de l'espace que l'on peut ainsi décrire est *le champ en profondeur*. Nous étudierons tour à tour ces deux caractéristiques.

I. Le champ latéral.

47. Le problème du champ. Nous allons etudier le probleme suivant: étant donné un point B dans l'espace objet de l'instrument et un rayon issu de B, dans quelles conditions ce rayon pourra-t-il traverser tout l'instrument sans être arrêté par les divers diaphragmes ou montures de lentilles? L'étude de l'approximation de GAUSS nous fournit une solution approchée de ce problème: nous pouvons en effet déterminer les positions et les grandeurs des images des divers diaphragmes et montures dans l'espace objet: si le rayon issu de B passe à l'intérieur de toutes ces images (Fig. 88) il traversera les diveres lentilles et diaphragmes ans être arsrêté; bien entendu, certaines de ces images pourront être situées avant l'objet.

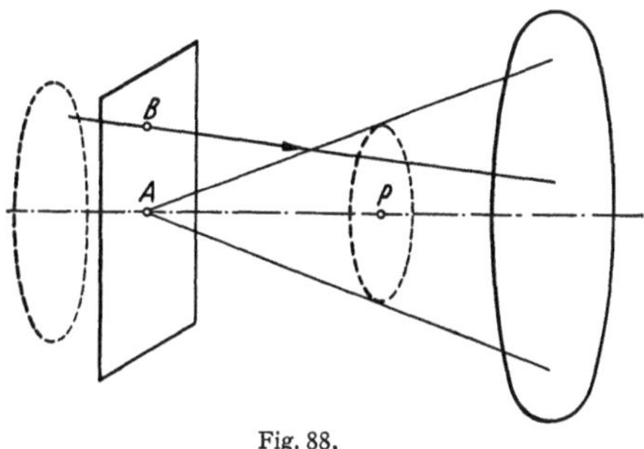

Fig. 88.

La solution théorique de ce problème est donc simple, tout au moins lorsqu'on se limite à l'approximation de GAUSS. Mais il nous faut maintenant étudier plus en détail la limitation des faisceaux.

48. La pupille. Plaçons tout d'abord le point objet au point A, sur l'axe de l'instrument. Le cône des rayons issus de A est manifestement limité par celle des images qui est vue de A sous l'angle le plus petit (Fig. 88). Cette image porte le nom de *pupille d'entrée P*, elle est conjuguée dans l'espace-objet d'un certain diaphragme (ou monture de lentille) qui limite en fait l'ouverture du faisceau issu de A dans un espace quelconque de l'instrument. Ce diaphragme réel s'appelle la *pupille de l'instrument* et son conjugué dans l'espace image porte le nom de *pupille de sortie*. Bien entendu, la pupille de sortie est vue du centre A' de l'image sous un angle plus petit que n'importe laquelle des images des autres diaphragmes. On peut rechercher la pupille de l'instrument en opérant soit dans l'espace objet, soit dans l'espace image ou encore dans un espace intermédiaire quelconque; il suffit pour cela de connaître toutes les images des divers diaphragmes dans l'espace où l'on se propose d'étudier le champ.

Signalons ici une convention relative à l'ouverture de la pupille des objectifs photographiques: on caractérise le diamètre de la pupille d'entrée par le rapport N de la distance focale à ce diamètre; ce rapport est appelé nombre d'ouverture

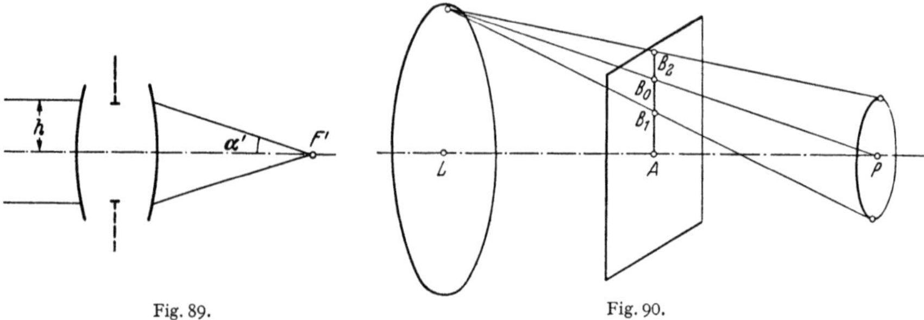

Fig. 89.   Fig. 90.

et on peut encore l'exprimer en fonction de l'ouverture angulaire du faisceau dans l'espace image (Fig. 89): on a en effet, d'après l'équation (30.1) où l'on fait $n = n'$:

$$\sin \alpha' = -\frac{h\vartheta}{y'} = \frac{h}{f'} = \frac{1}{2N}$$

on encore:

$$N = \frac{1}{2\,|\sin \alpha'|}\,.$$

49. La lucarne. Nous allons poursuivre cette étude en nous plaçant dans l'espace objet: déplaçons le point dans le plan de l'objet et plaçons-le en un point source B voisin de A: tous les rayons qui traversent la pupille réussissent encore à traverser tous les diaphragmes. Si l'on éloigne encore le point B dans le plan de front, il arrive un moment où le cône de rayons de sommet B_1 vient toucher le bord de l'image d'un autre diaphragme (Fig. 90). Ce diaphragme s'appelle *la lucarne* et ses conjugués dans l'espace objet et l'espace image s'appellent respectivement *lucarne d'entrée* et *lucarne de sortie*. Si l'on représente la trace dans le plan de la lucarne d'entrée du cône de rayons issus de B, et s'appuyant sur le contour de la pupille d'entrée, on constate que lorsque B et en B_1 le cercle projection de la pupille est tangent au bord de la lucarne et là lucarne n'arrête encore aucun rayon; si B dépasse la position B_1, le faisceau de rayons est partiellement arrêté par la lucarne et il ne subsiste du faisceau que la partie commune aux deux cercles (Fig. 91). Lorsque B est en B_0, le rayon qui passe par le centre de la pupille atteint le bord de la lucarne et l'on a déjà perdu plus de la moitié de la surface de section du faisceau; enfin on pourra arriver à une occultation complète du faisceau à partir du point B_2. On voit donc que les points B_1 et B_2 jouent des rôles particuliers: lorsqu'on se déplace dans le champ le faisceau est progressivement occulté à partir de B_1 et disparait complètement en B_2. Le cercle de rayon AB_1 est appelé champ de pleine lumière, celui de rayon AB_0 le champ moyen, et la zone comprise entre les cercles de rayons AB_0 et AB_2 est appelée champ de contour. On évite le plus souvent d'utiliser le champ de contour qui est trop peu lumineux, et le champ est généralement limité au voisinage du champ moyen par un diaphragme convenable placé dans le plan d'une image intermédiaire. Suivant les conditions d'emploi de l'instrument on tolèrera une occultation plus ou moins importante de la pupille. C'est ainsi que dans les instrument destinés à l'observation visuelle nocturne on limite le champ au champ de pleine lumière, alors que pour les instruments diurnes on peut tolérer une baisse très importante de la clarté. D'autre part ce seront le plus souvent les aberrations

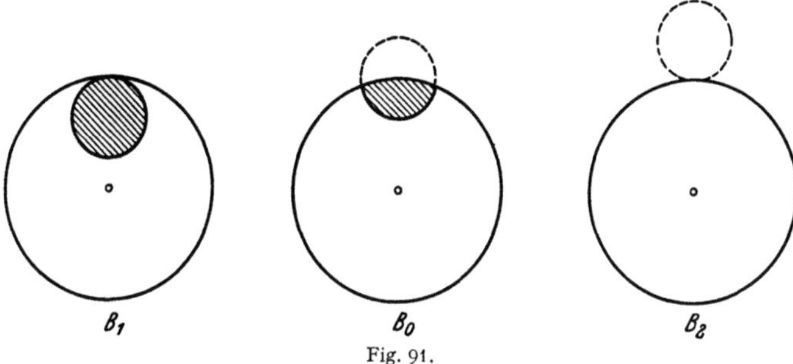

Fig. 91.

qui limiteront le champ acceptable d'un instrument: les aberrations sont en effet convenablement réduites dans une portion du plan image, qui est pratiquement utilisable, mais elles deviennent très rapidement intolérables au delà d'une valeur maximum du champ: ce sera pratiquement ce champ de bonne qualité qui sera le champ de l'instrument.

II. La profondeur de champ et la latitude de mise au point.

50. Il nous faut maintenant savoir de quelle quantité le point objet A peut se déplacer autour de sa position théorique A_0 sans que l'image cesse d'être nette dans le plan conjugué A_0' (Fig. 92). Deux cas sont à distinguer:

1. L'approximation de l'optique géométrique est valable: c'est-à-dire qu'on ne cherche pas à obtenir du système optique tout

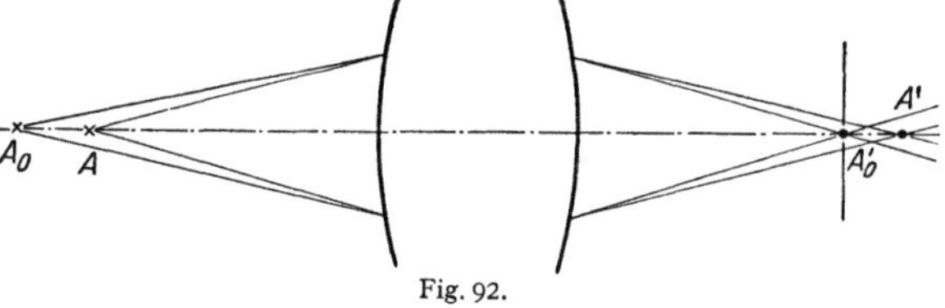

Fig. 92.

son pouvoir séparateur: ce sera le case de l'objectif photographique, car la structure granulaire de l'émulsion photographique limite le pouvoir séparateur linéaire à une distance le plus souvent supérieur à $\frac{1}{100}$ mm., nettement plus grande que les dimensions de la tache de diffraction fournie par l'objectif. Dans ce cas il suffit que la section circulaire du faisceau qui converge vers A' ne dépasse pas le grain de la plaque; on écrira donc:

$$\varepsilon \cdot 2\alpha' < d$$

où ε est le déplacement de l'image, α' la demi ouverture du cône de rayons (Fig. 92) et d le diamètre de tache tolérable, déterminée par la structure de l'émulsion. Pour déterminer les limites entre lesquelles peut se déplacer l'objet il est souvent commode d'utiliser les formules de NEWTON, surtout lorsque l'objet est situé à grande distance; sie Π_0' et Π' sont les distances de A_0' et A' au foyer image, Π_0 et Π les quantités correspondantes dans l'espace objet on aura:

$$|\Pi_0' - \Pi'| < \frac{d}{2\alpha'},$$

ou encore

$$\left| \frac{1}{\Pi} - \frac{1}{\Pi_0} \right| < \frac{d}{2\alpha' f^2} \, .$$

Cette formule limite donc les variations de la proximité de l'objet.

2. Dans le cas des instruments d'excellente qualité (microscope, viseurs, etc.) on peut se demander quel est le déplacement de l'objet qui produira une baisse sensible de la qualité de l'image, du fait du léger défaut de mise au point. Quelle sera inversement la précision avec laquelle un instrument permet de pointer sur un plan objet déterminé? Le problème ne peut évidemment pas être traité dans le cadre de l'optique géométrique et il est indispensable de faire intervenir la diffraction; nous utiliserons ici un résultat qui sera justifié ultérieurement: la qualité de l'instrument est acceptable tant que la surface d'onde Σ (Fig. 93) centrée sur l'image A' ne s'écarte pas de plus de $\lambda/4$ de la sphère S centrée sur A_0' (règle de Lord Rayleigh).

On aura dans ces conditions à écrire que les deux sphères S et ε ne sont pas distantes de plus de $\lambda/4$ au bord, ce que l'on peut traduire comme suit:

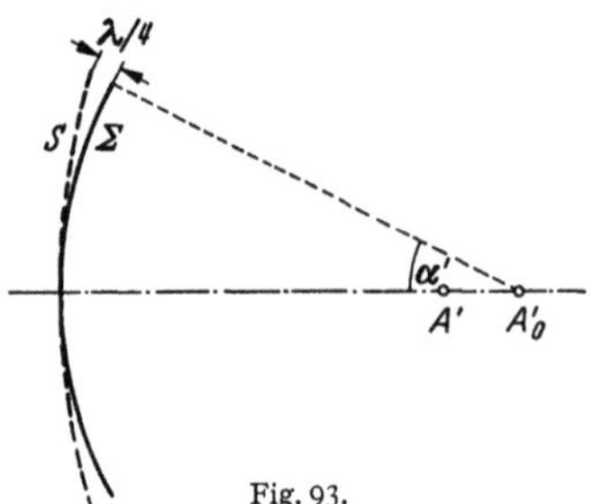

Fig. 93.

$$\frac{h^2}{2R} - \frac{h^2}{2(R+\varepsilon)} < \frac{\lambda}{4} \, ,$$

ou encore, puisque ε' est petit

$$\frac{\varepsilon' h^2}{2R^2} < \frac{\lambda}{4} \, ,$$

ce qui donne comme tolérance

$$\varepsilon' < \frac{\lambda}{2\alpha'^2} \, . \tag{50.1}$$

(α' étant le demi angle d'ouverture).

Dans le cas où l'objet est situé à faible distance, il est convenable d'exprimer la tolérance ε' en fonction de l'angle d'ouverture α dans l'espace objet; si g_α est le grandissement angulaire, on aura:

$$\varepsilon' < \frac{\lambda}{2 g_\alpha^2 \alpha^2} \, ,$$

mais on a vu que (en supposant $n = n'$) $g_x \, g_\alpha^2 = 1$, et l'on trouve ainsi

$$\varepsilon = \frac{\varepsilon'}{g_x} = \frac{\lambda}{2\alpha^2} \, . \tag{50.2}$$

La précision des pointés dans l'espace objet est inversement proportionnelle au carré de l'ouverture angulaire α; ainsi un microscope «à sec» d'ouverture angulaire $\sin \alpha = 0{,}3$ permet d'obtenir une précision de l'ordre de 5λ.

Les bons observateurs arrivent à faire une mise au point avec une bien meilleure précision mais le défaut de mise au point ci-dessus est nettement perceptible à un observateur non exercé par le fait que le contraste de l'image obtenue est sensiblement affecté: nous l'admettrons donc comme valeur approximative de la latitude de mise au point l'expression (50.1).

Dans le cas des instruments visant à une distance importante (viseurs dioptriques, par exemple) on peut se servir encore de l'expression (50.2) mais il est commode d'écrire (Fig. 94) $\alpha = h/x$ et d'exprimer la variation de la proximité $\xi = 1/x$; on aura:

$$\varepsilon = x^2 \, d\xi = \frac{\lambda}{\dfrac{2h^2}{x^2}}$$

d'où l'on tire la précision longitudinale du pointé exprimée en proximité:

$$d\xi = \frac{\lambda}{2\,h^2}.$$

Ainsi un viseur dont l'objectif a un diamètre de 20 mm permet d'avoir une précision de pointé de l'ordre de:

$$d\xi = 0,3 \cdot 10^{-5}\,\text{mm}^{-1} = 0,3 \cdot 10^{-2} \quad \text{dioptrie.}$$

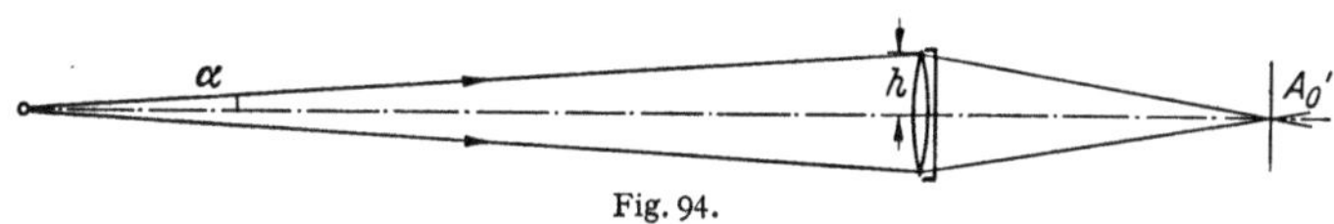

Fig. 94.

A une distance de 1 mètre on fera donc le pointé avec une précision de l'ordre de 2 mm. Un bon observateur obtiendra d'ailleurs une bien meilleure précision.

E. Les aberrations.

I. Généralités.

51. Les chapitres précédents ont été consacrés à l'étude de la formation des images dans les systèmes optiques centrés, en nous limitant néanmoins à l'approximation de Gauss.

Les instruments réels doivent en fait avoir une pupille assez grande pour que la quantité de lumière qui les traverse ne soit pas négligeable et qu'ils aient une clarté acceptable; il doivent de plus avoir un champ latéral aussi étendu que possible. Ils ne peuvent donc être caractérisés complètement par leurs propriétés paraxiales: les angles d'incidence sur les surfaces peuvent être importants et il nous faut étudier comment cheminent les rayons qui s'écartent notablement de l'approximation de Gauss.

Nous avons, de plus, envisagé jusqu'ici l'utilisation de lumière monochromatique; on utilise en fait le plus souvent des lumières complexes (très souvent la lumière blanche) et la dispersion des verres est alors responsable de légères variations de position et de grandeur des images: les défauts ainsi produits sont les aberrations chromatiques, que nous allons étudier tout d'abord: elles apparaissent en effet déjà dans l'approximation de Gauss et les résultats simples obtenus dans le domaine paraxial pourront être utilisés en première approximation. Nous passerons ultérieurement à l'étude des aberrations géométriques.

II. Aberrations chromatiques.

52. Aspect expérimental. Prenons une lentille mince convergente de grande distance focale (par exemple 1 mètre) et d'ouverture modérée (ne dépassant pas par exemple le $\frac{1}{10}$ de la distance focale). Essayons de former, à l'aide de cette lentille, l'image d'un point A très éloigné situé sur l'axe de la lentille (Fig. 95). On constate que si A émet de la lumière blanche on ne parvient pas à obtenir sur un écran une image satisfaisante par suite de la présence d'irisations au bord de l'image: la lentille décompose la lumière et donne naissance à une infinité d'images échelonnées le long de l'axe dont chacune correspond à une longueur d'onde déterminée. On constate en effet qu'en coupant le faisceau par un écran situé en E_1 on obtient un halo rougeâtre au bord du faisceau, en E_2 le halo rouge entoure une image bleue; entre E_2 et E_3 se forment les diverses images allant du bleu au rouge; en E_3 l'image rouge est bordée d'un halo bleu; enfin en E_4 le halo borde la trace du faisceau; dans cette expérience la tache image conserve cependant la symétrie circulaire, et l'on dit alors que la lentille présente une *aberration chromatique longitudinale*.

Si maintenant, le point objet est un point B hors de l'axe de l'instrument et si l'on diaphragme par une pupille située à une certaine distance devant la lentille, on constate que l'image de B n'a plus la symétrie circulaire, comme l'avait l'image de A; aux phénomènes précédents vient s'ajouter un échelonnement latéral des diverses radiations qui donne naissance à un petit spectre étalé sur l'écran; ceci subsiste si l'on réduit la pupille à un trou de faibles dimensions: la lentille fonctionne alors comme un prisme qui disperserait les différentes radiations incidentes (Fig. 96). On observe ainsi une aberration chromatique latérale.

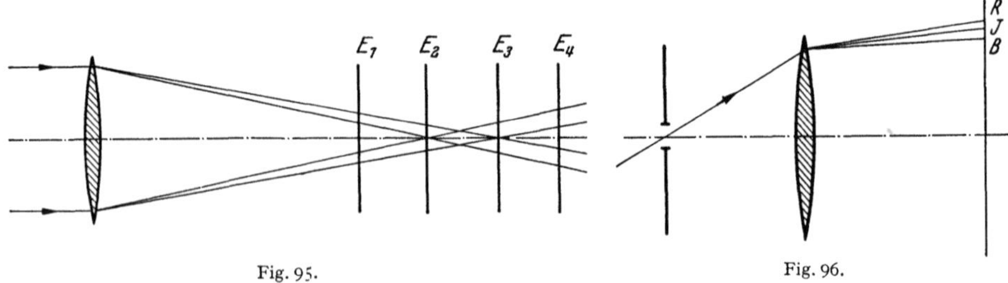

Fig. 95.   Fig. 96.

En résumé l'image d'un point situé sur l'axe est irisée mais conserve une symétrie de révolution alors que l'image d'un point situé hors de l'axe peut ne plus posséder qu'un plan de symétrie. On voit que les variations d'indice de réfraction du verre se manifestent de diverses façons et nous allons entreprendre l'étude des aberrations qui en résultent.

53. Dispersion des verres. Il faut d'abord indiquer comment on caractérise la dispersion d'un verre: il est commode de choisir un indice de référence correspondant à une radiation peu éloignée du maximum de sensibilité du récepteur, c'est ainsi que pour l'oeil on choisit d'habitude la raie D du sodium. Cette raie est en fait un doublet ($\lambda = 5890$ et 5896 Å) et il serait plus convenable de choisir une raie simple situé de préférence plus près du maximum de sensibilité de l'oeil (5550 Å).

La disperion pourra être caractérisée d'autre part à l'aide de la différence des indices pour deux radiations encadrant la radiation moyenne, et pour lesquelles la sensibilité du récepteur est notablement plus faible que la sensibilité maximum (par exemple le $\frac{1}{10}$). On utilise traditionnellement les raies C et F de l'hydrogène, qui ont respectivement pour longueurs d'ondes 6563 et 4861 Å; la dispersion du verre peut être caractérisée par la différence des indices $n_F - n_C$ mais on est amené à utiliser en fait le paramètre

$$\nu = \frac{n_D - 1}{n_F - n_C}. \tag{53.1}$$

Ce paramètre est inversement proportionnel à la dispersion et il varie en gros de 65 pour les verres peu dispersifs (borosilicate crowns ordinaires) à 35 ou moins pour les verres au plomb dont la dispersion croit avec la teneur en plomb (flints).

On peut distinguer dans les verres couramment utilisés en optique deux catégories: a) les verres «anciens» pour lesquels la disperion est pratiquement fonction de l'indice: dans un graphique (Fig. 97) représentant (conformément à l'habitude des opticiens) l'indice du verre en fonction du paramètre ν porté de droite à gauche en abcisse, on trouve les verres anciens dans une région très étroite (hachurée) qui va des crowns aux flints extradenses. b) Les verres «nouveaux» qui contiennent souvent de la baryte et qui ont permis d'étendre notablement la carte des verres: les verres au baryum permettent en particulier d'atteindre des indices élevés en maintenant la dispersion à une valeur relativement faible.

Enfin l'emploi de certains cristaux naturels ou artificiels permet encore d'étendre le choix des matières réfringentes utilisables en optique. Les plus utilisés sont indiqués également sur la Fig. 97.

Dans les limites de l'approximation de GAUSS nous déterminerons les variations de la position et de la grandeur de l'image d'un objet AB perpendiculaire à l'axe.

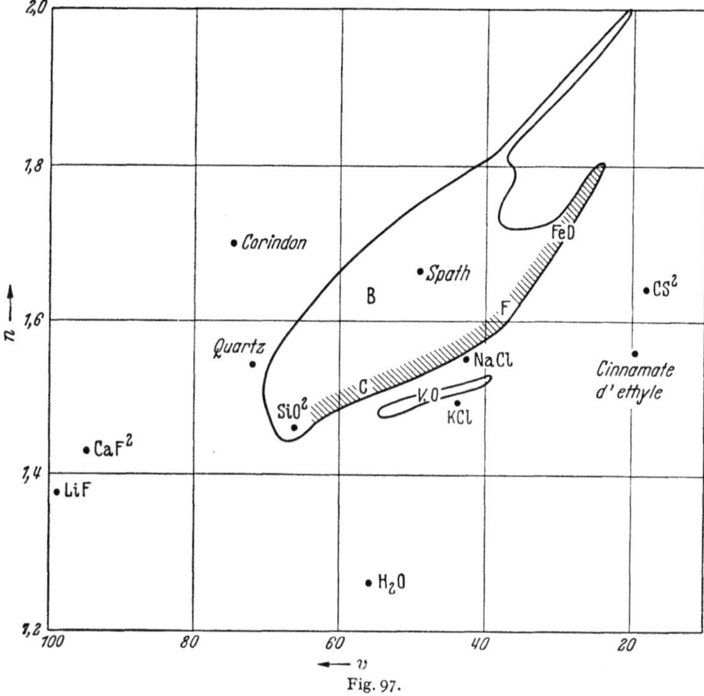

Fig. 97.

Nous n'envisagerons par convention que deux longueurs d'onde, par exemple celles des radiations C et F et nous rapporterons les aberrations à l'image obtenue pour la raie C: les variations d'indices sont alors positives pour les longueurs d'onde inférieures à 6563 Å qui représentent la plus grande partie du spectre visible.

54. Chromatisme longitudinal. Nous examinons à titre d'exemple le cas d'une lentille mince convergente mais il est évident que les résultats s'appliquent à un système centré quelconque.

Soit A un point objet, situé sur l'axe et A' son image. Pour chaque longueur d'onde on a

$$\frac{1}{x'} = \frac{1}{x} + D,$$

ce qui donne par différentiation

$$-\frac{dx'}{x'^2} = dD = \left(\frac{1}{r} - \frac{1}{r'}\right) dn = \frac{D}{n-1} dn$$

et, si nous utilisons les raies C et F,

$$-\frac{dx'}{x'^2} = \frac{D}{v}. \tag{54.1}$$

dx' est donc de signe contraire à celui de D: pour une lentille convergente dx' est négatif (Fig. 98), l'image bleue A'_b est toujours avant l'image rouge A'_r et si nous coupons le faisceau par un plan passant par le point I la section du faisceau sera bordée de rouge; en J au contraire le faisceau sera bordé de bleu. Calculons la diffusion obtenue dans le plan de A'_r du fait de l'aberration chromatique de position; prenons pour cela des axes $A'_r x' y' z'$. Si le rayon chemine dans le plan

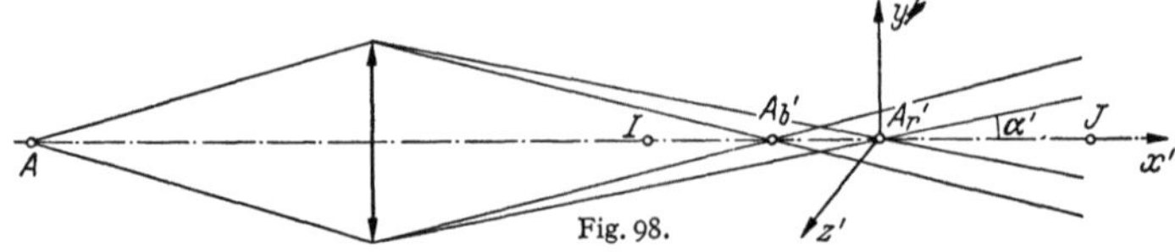

Fig. 98.

méridien contenant $A'_r y'$ (que l'on prend pour plan du tableau), les coordonnées de l'intersection du rayon bleu avec le plan $A'_r y' z'$ seront

$$dy' = -\alpha' \, dx'; \quad dz' = 0.$$

Si, maintenant, le point M n'est plus dans le plan $A'_r y' x'$ mais dans un plan méridien faisant l'azimut φ avec ce dernier on aura manifestement

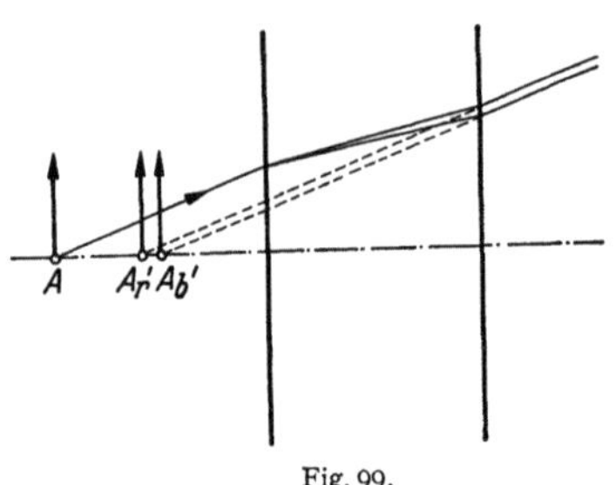

Fig. 99.

$$dy' = -\alpha' \, dx' \cos\varphi; \quad dz' = -\alpha' \, dx' \sin\varphi.$$

Les résultats précédents s'appliquent immédiatement à un système centré quelconque: néanmoins le signe de l'aberration chromatique longitudinale dépend à la fois de la constitution de l'instrument et aussi de la position de l'objet; si dx' est négatif, comme dans la lentille mince convergente, on dit qu'il y a souscorrection; si dx' est positif on dit au contraire qu'il y a sur-correction.

Signalons l'exemple de la lame à faces parallèles. On a vu que la distance de l'objet à l'image est égale à:

$$AA' = \frac{n-1}{n} e$$

e étant l'épaisseur de la lame, n son indice; l'aberration chromatique de la lame isolée sera

$$dx' = d\left(\frac{n-1}{n}\right) e = \frac{dn}{n^2} e > 0.$$

Une lame à face parallèles présente donc une sur-correction chromatique (Fig. 99); l'image rouge est située avant l'image bleue.

55. Chromatisme de grandeur et chromatisme latéral. De même que la position de l'image d'un point A est déplacé de A'_r en A'_b lorsqu'on passe de la raie C à la raie F, la grandeur de l'image d'un objet AB varie avec la longueur d'onde. Envisageons par exemple le cas d'une lentille mince; le grandissement et

$$g = \frac{x'}{x}$$

et l'on a, dans le cas où l'objet n'est pas entaché lui-même de chromatisme

$$\frac{dg}{g} = \frac{dx'}{x'}.$$

Sur la Fig. 100 l'image bleue est plus petite que l'image rouge on dit alors que le système présente une *aberration chromatique de grandeur*.

Dans le cas d'un système centré quelconque, le calcul de la différence de grandeur des deux images est plus compliqué et nous l'effectuerons ultérieurement. Signalons seulement que δg est nul pour une lame à faces parallèles; l'image est en effet toujours égale à l'objet quelle que soit la longueur d'onde (Fig. 99).

En fait, le chromatisme de grandeur ne se manifeste pas directement à l'observateur; celui-ci utilise en effet un écran (passant par exemple par A'_r) ou encore il observe l'image à l'oeil nu (lorsque celle-ci est éloignée). Calculons la diffusion de l'image de B' dans le plan image de référence (plan de l'image rouge); la tache obtenue pour les rayons bleus n'est autre

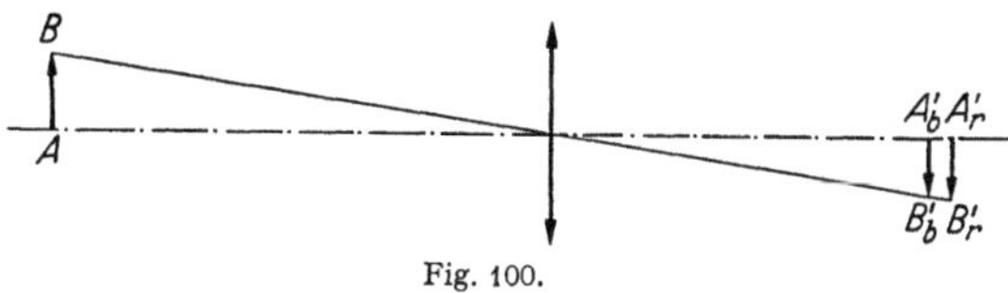

Fig. 100.

que la projection conique de la pupille de sortie de B'_b comme sommet (Fig. 101): ce sera un cercle ayant le même rayon que le cercle de diffusion relatif à A'_r mais dont le centre C' sera sur la droite $P'B'_b$. Il ne coïncide généralement pas avec B'_r de sorte que les différentes radiations ne donnent pas naissance à des taches concentriques. On dit alors que l'image présente du *chromatisme latéral ou apparent*.

Calculons la distance $B'_r C'$ caractérisant l'importance de cette aberration; p' sera désormais l'abscisse de la pupille de sortie par rapport au dernier dioptre et x' l'abscisse de l'image.

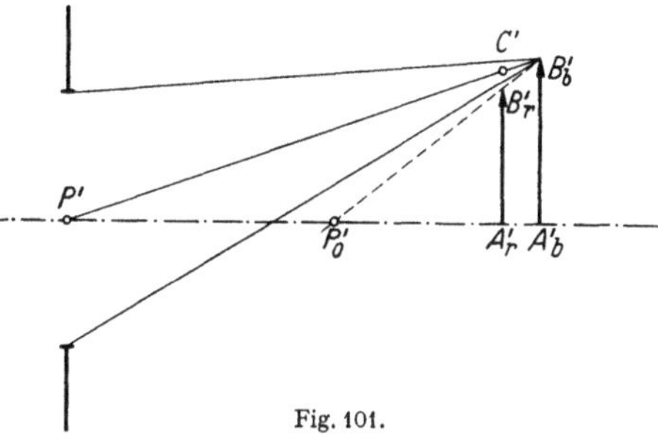

Fig. 101.

$$A'_r C' - A'_r B'_r = A'_b B'_b \frac{P'A'_r}{P'A'_b} - y' = y' \frac{1 + \dfrac{dg}{g}}{1 + \dfrac{dx'}{x' - p'}} - y' = y'\left(\frac{dg}{g} - \frac{dx'}{x' - p'}\right).$$

La diffusion par rapport à l'image rouge B'_r sera donc obtenue finalement, en ajoutant les effets du chromatisme longitudinal et du chromatisme latéral,

$$\left.\begin{aligned} dy' &= -\alpha'\, dx' \cos\varphi + y'\left(\frac{dg}{g} - \frac{dx'}{x' - p'}\right), \\ dz' &= -\alpha'\, dx' \sin\varphi. \end{aligned}\right\} \tag{55.1}$$

Pour chaque longueur d'onde la tache obtenue est encore circulaire et égale à celle que l'obtenait en A', mais les divers cercles ne sont plus concentriques: leurs centres sont à une distance de B'_r égale à

$$y'\left(\frac{dg}{g} - \frac{dx'}{x' - p'}\right);$$

cette distance caractérise le chromatisme latéral.

Pour éviter le chromatisme latéral, il faut satisfaire à la relation

$$\frac{dg}{g} = \frac{dx'}{x' - p'},$$

ce qui assigne à la pupille une position bien déterminée. Ceci est d'ailleurs évident géométriquement: pour annuler le chromatisme latéral on placera le centre P' de la pupille en P'_0 sur la droite $B'_r B'_b$ (Fig. 101). Cette condition est satisfaite par une lentille mince jouant elle-même le rôle de pupille (Fig. 100); dans ce cas, bien que les images n'aient pas même grandeur leurs projections sur l'écran sont cependant superposées.

Si le chromatisme de position est nul les relations (55.1) montrent que le *chromatisme latéral se confond avec le chromatisme de grandeur et ne dépend plus de la position de la pupille*; les images $A'_b B'_b$ et $A'_r B'_r$ sont alors situées dans le même plan; ce résultat est à rapprocher de ceux que nous obtiendrons ultérieurement dans l'étude de l'influence de la position de la pupille sur les aberrations géométriques (voir Sect. 88).

56. Recherche de l'achromatisme. Nous avons vu qu'un lentille mince est entachée de diverses aberrations chromatiques; essayons de la remplacer par deux lentilles accolées de puissance D_1 et D_2 et voyons s'il est possible de supprimer le chromatisme de système. Si l'on réussit à obtenir un système mince dont la

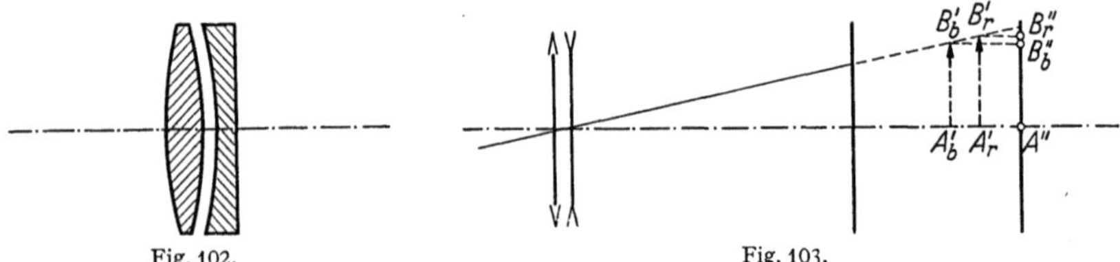

Fig. 102.   Fig. 103.

puissance D soit indépendante de la longueur d'onde, on aura supprimé le chromatisme de position et de grandeur. Il suffit d'écrire pour cela

$$dD = dD_1 + dD_2 = 0$$

ou encore, en faisant intervenir les dispersions des verres et en adjoignant l'équation de convergence du système, on a

$$\frac{D_1}{\nu_1} + \frac{D_2}{\nu_2} = 0,$$

ce qui conduit aux valeurs suivantes des puissances des deux lentilles composantes:

$$\left.\begin{aligned} D_1 &= D\,\frac{\nu_1}{\nu_1 - \nu_2}, \\ D_2 &= D\,\frac{\nu_2}{\nu_2 - \nu_1}. \end{aligned}\right\} \tag{56.1}$$

Supposons par exemple $D > 0$ et $\nu_1 > \nu_2$; pour ne pas être conduit à des valeurs trop grandes de D_1 et D_2, il faut que la différence $\nu_1 - \nu_2$ ne soit pas trop petite: la première lentille est le plus souvent en crown peu dispersif et la deuxième lentille en flint; dans ces conditions, on aura $D_1 > 0$ et $D_2 < 0$, c'est-à-dire que pour obtenir un objectif achromatique convergent, on associe une lentille convergente en verre peu dispersif (crown le plus souvent) à une lentille divergente en verre dispersif (flint le plus souvent).

Les objectifs doublets construits sur ce principe portent le nom d'objectifs astronomiques (Fig. 102); ceux d'entre eux qui présentent la même courbure pour les faces en regard des deux lentilles peuvent être collés; ils prennent alors le nom d'objectifs de Clairaut.

En fait, on a souvent besoin de placer derrière l'objectif une lame à faces parallèles, ou un système équivalent (dans les jumelles à prisme par exemple);

si l'on veut que l'image définitive soit exempte de chromatisme axial, il faut laisser à l'objectif une sous-correction convenable, pour compenser la surcorrection de la lame; la distance des deux images objectives devra être égale à l'aberration chromatique longitudinale de la lame, soit

$$e\,\frac{dn}{n^2}\,,$$

mais il subsistera alors un chromatisme de grandeur donnant naissance à du chromatisme latéral (Fig. 103): l'image rouge sera plus grande que l'image bleue; ce défaut peut heureusement être corrigé par l'oculaire.

L'achromatisme de position peut encore être obtenu par un système de deux lentilles séparées par un intervalle d'air, mais il est alors impossible d'obtenir l'achromatisme de grandeur un rayon passant par le centre de la lentille convergente n'est pas dispersé par celle-ci mais la lentille divergente le disperse nécessairement: l'image bleue est cette fois plus grande que l'image rouge (Fig. 104).

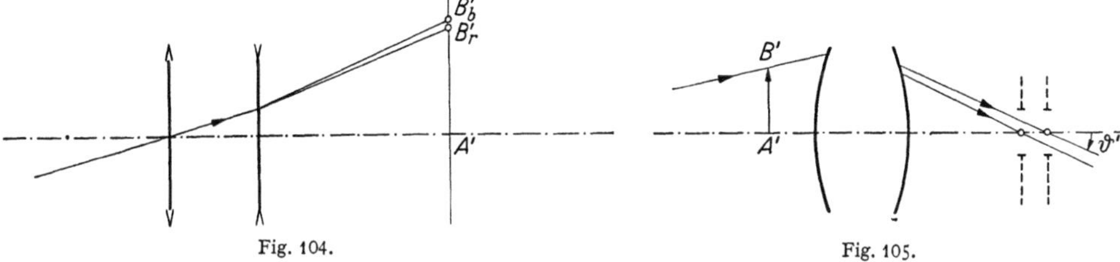

Fig. 104.   Fig. 105.

57. Achromatisme apparent dans un oculaire doublet. Recherchons à quelle condition un oculaire constitué par deux lentille de puissance D_1 et D_2 séparées par un intervalle peut fournir, d'un objet parfaitement achromatique, une image à l'infini présentant l'achromatisme apparent. La condition s'écrit simplement $d\vartheta' = 0$ (Fig. 105).

Supposons d'abord que la pupille d'entrée de l'oculaire soit très éloignée; le rayon moyen est alors parallèle à l'axe dans l'espace objet et, si y est la grandeur de l'objet, D la convergence de l'oculaire, l'angle sous lequel on aperçoit l'image sera

$$\vartheta' = -\,y\,D\,.$$

Pour que $d\vartheta' = 0$, il suffit que $dD = 0$, et en dérivant la formule de GULLSTRAND

$$D = D_1 + D_2 - e\,D_1 D_2$$

on obtient la condition

$$\frac{f_1}{v_2} + \frac{f_2}{v_1} - e\left(\frac{1}{v_1} + \frac{1}{v_2}\right) = 0$$

qui s'écrit, lorsque $v_1 = v_2$

$$f_1 + f_2 - 2e = 0. \tag{57.1}$$

Remarquons que l'objet peut ne pas être rigoureusement au foyer objet, le raisonnement suppose seulement que la pupille d'entrée est è l'infini.

Si maintenant la pupille n'est plus à l'infini mais si on la suppose parfaitement achromatique ainsi que l'objet AB, on peut appliquer la relation de LAGRANGE aux pupille (Fig. 106) en remarquant que ϑ ne dépend pas de la longueur d'onde:

$$\varrho\,\vartheta = \varrho'\,\vartheta'\,.$$

Pour que $d\vartheta' = 0$, il suffit que le grandissement ϱ'/ϱ soit constant; on voit que l'achromatisme apparent est alors équivalent à l'achromatisme de grandeur des pupilles (Fig. 106). Dans la pratique, la pupille sera souvent à assez grande distance de l'oculaire et si, en toute rigueur, on doit écrire l'achromatisme des

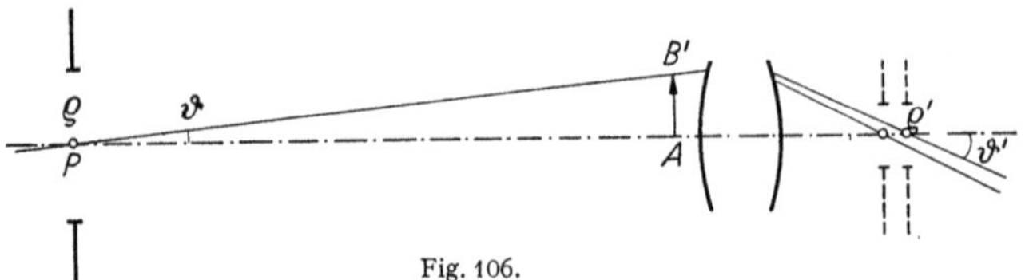

Fig. 106.

pupilles, la relation (57.1) fournira une solution approchée du problème de l'achromatisme apparent.

58. Spectre secondaire. Voyons maintenant si l'achromatisme obtenu par les moyens précédents est parfaitement satisfaisant; examinons par exemple le cas des objectifs astronomiques: la puissance du système est égale pour toute longueur d'onde à

$$D = D_1 + D_2$$

et, en mettant en évidence les constantes géométriques caractérisant les lentilles,

$$K_1 = \frac{1}{r_1} - \frac{1}{r_1'}, \qquad K_2 = \frac{1}{r_2} - \frac{1}{r_2'}$$

où r_1, r_1', r_2, r_2' sont les rayons de courbure des lentilles. On a alors

$$D = K_1\,(n_1 - 1) + K_2\,(n_2 - 1).$$

La puissance D sera rigoureusement constante et l'achromatisme sera parfait s'il existe une relatiou linéaire entre les indices n_1 et n_2 des lentilles composantes.

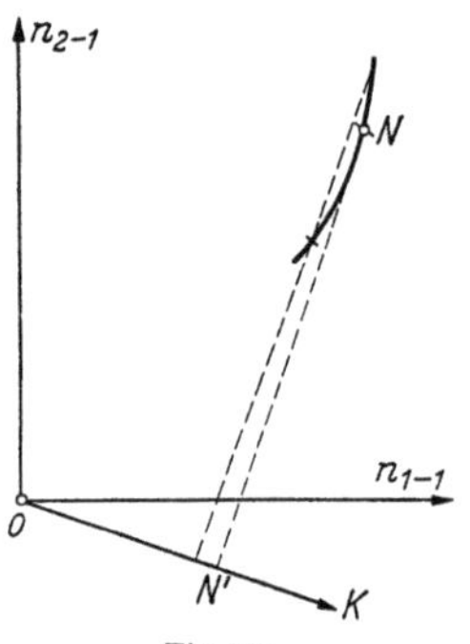

Représentons, dans un système d'axes rectangulaires, les variations de $n_2 - 1$. en fonction de $n_1 - 1$; les verres réels ne conduisent jamais à une droite, mais à une courbe dont la concavité est tournée vers le haut si n_2 est relatif au verre le plus dispersif (Fig. 107).

Fig. 107.

Fig. 108.

Plaçons en K le point de coordonnées K_1, K_2; l'expression algébrique de la puissance D montre qu'elle est égale au produit scalaire $\overrightarrow{OK} \cdot \overrightarrow{ON}$. Si N' est la projection de N sur la droite OK, cette puissance varie comme ON'. Si l'on a amené les puissances pour les raies C et F à être égales, on voit que D passe par un maximum entre les raies C et F et décroit ensuite au delà de F. En conséquence, la distance focale du système passe par un minimum et la disposition des foyers est représentée conventionnellement sur la Fig. 108 par une courbe à minimum, allongée le long de l'axe.

On n'a donc pas réussi à supprimer complètement les aberrations chromatiques: le résidu obtenu porte le nom de spectre secondaire. En coupant le faisceau par un écran, comme au Sect. 52, on obtiendra des colorations particulières dues au mélange des différentes radiations: la superposition des radiations violettes et rouges donne naissance à une teinte pourpre qui borde le faisceau lorsque l'écran est en I; en J au contraire, on obtiendra une teinte jaune verdâtre.

Indiquons maintenant l'ordre de grandeur du spectre secondaire pour un objectif astronomique fait avec des verres du type aucien, l'objet étant supposé à l'infini:

1/2000 de la distance focale entre les raies D et F (ou D et C);

2/1000 de la distance focale entre les raies F et G.

Remarquons que la position du point K (Fig. 107) reste arbitraire et, en le déplaçant légèrement, on peut faire coïncider deux radiations arbitrairement choisies: pour l'observation visuelle on fait généralement coïncider les raies C et F (le maximum de sensibilité de l'oeil se situant entre les deux), on dit alors qu'on a réalisé l'achromatisme visuel.

Dans les objectifs photographiques anciens, à mise au point visuelle, il y avait lieu de faire coïncider les images relatives aux radiations efficaces pour l'oeil (5550 Å) et celles pour la plaque photographique; on choisissait par exemple les raies D et G'; on dit alors qu'on a réalisé l'achromatisme actino-visuel.

Si enfin on veut réduire le spectre secondaire dans le domaine des radiations actiniques on peut faire coïncider par exemple, les raies G' (4341 Å) et Hg (4047 Å); on obient ainsi l'achromatisme actinique pur. On voit que la correction des aberrations chromatiques dépend tout d'abord de l'utilisation de l'appareil.

59. Apochromatisme. On peut maintenant se demander si l'on ne peut pas parvenir à réduire l'étendue du spectre secondaire. On peut tout d'abord essayer de faire un choix plus judicieux du couple de verres que l'on emploie; malheureusement ceci ne permet pas d'améliorer beaucoup le résultat; il est difficile d'arriver à réduire notablement le spectre secondaire par un choix convenable des verres: il reste en général de l'ordre de grandeur des valeurs indiquées ci-dessus. On

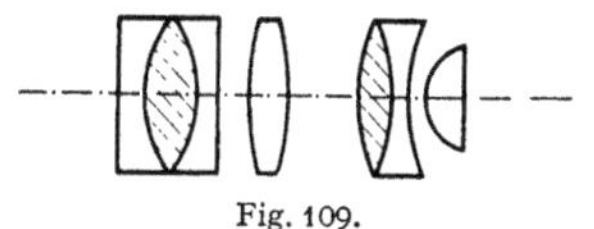

Fig. 109.

peut alors penser à se donner des paramètres supplémentaires: en associant par exemple une troisième lentille, on peut écrire à priori que les plans focaux coïncident non plus pour deux mais pour trois radiations; on peut alors déterminer la convergence des lentilles composantes à l'aide d'un système de trois équations linéaires à trois inconnues; malheureusement ce procédé conduit à des convergences très élevées, qui nécessitent des courbures importantes et rendent difficile la correction des aberrations géométriques: le résultat ne sera donc pas intéressant.

Dans le calcul des objectifs de microscope, on parvient à réduire à la fois le spectre secondaire à un spectre tertiaire et aussi à obtenir l'aplanétisme pour au moins deux radiations: on appelle alors ces objectifs, objectifs apochromatiques; ce résultat ne peut en fait être obtenu qu'à condition d'introduire dans l'objectif au moins un élément constitué par une substance dont l'allure de dispersion est nettement différente de celle du verre; on utilise par exemple la fluorine (fluorure de calcium CaF_2) que l'on sait maintenant produire sous forme de cristaux artificiels; la Fig. 109 représente à titre d'exemple la coupe d'un objectif de microscope du type apochromatique, où deux lentilles sout constituées par de la fluorine. Il est à noter d'ailleurs que dans ces objectifs la correction du chromatisme axial est excellente mais qu'il subsiste un chromatisme de grandeur qui peut être appréciable: on le compense en fait à l'aide d'un oculaire spécialement adapté (oculaire compensateur).

III. Les aberrations géométriques.

a) Etude expérimentale.

60. L'aberration sphérique. Essayons tout d'abord de former l'image d'un point éloigné sur l'axe d'une lentille simple de grande ouverture: utilisons par exemple une lentille de 10 cm. de longueur focale et dont le diamètre soit de l'ordre de 5 cm., présentant une face plane à la lumière incidente (Fig. 110).

Nous constatons qu'il n'existe pas de position de l'écran pour laquelle l'image se réduise à un point; si l'écran est d'abord placé dans le plan de l'image paraxiale, les rayons provenant du centre de la lentille viennent bien se concentrer en F', image de Gauss, mais ceux qui ont traversé la lentille à une distance notable de l'axe produisent autour de F' un halo circulaire diffus. Si, maintenant, on coupe le faisceau par un plan situé en avant du foyer on obtient les aspects indiqués sur la Fig. 110; lorsque le plan de section n'est pas trop éloigné de F', la tache présente un point brillant central entouré aussi d'un anneau brillant dont le diamètre augmente avec la distance de l'écran à F'. L'ensemble du phénomène peut d'ailleurs être mis en évidence si on utilise la diffusion produite par une fumée blanche très fine (chlorure d'ammonium par exemple). On peut alors observer une nappe de surface vivement éclairée, cette nappe est une zone de concentration particulière et nous verrons qu'elle est l'enveloppe des rayons lumineux. On lui donne le nom de surface caustique à cause de la grande quantité d'énergie qui est concentrée en son voisinage.

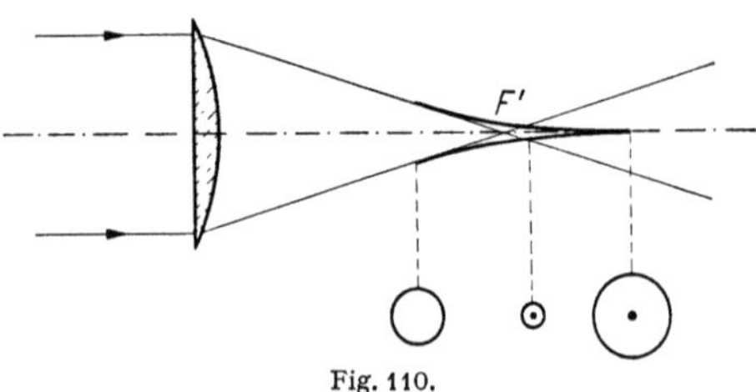

Fig. 110.

La surface caustique se compose en fait de la nappe que nous venons d'étudier, que l'on nomme nappe tangentielle, et d'une autre nappe, dégénérée ici en un fragment de droite (nappe sagittale). On constate en effet que la lumière se concentre également sur une partie de l'axe de l'instrument et tout se passe comme si chaque zone de la lentille donnait naissance à un foyer particulier dont la position dépend du diamètre de la zone utilisée. Cette aberration porte le nom d'aberration sphérique ou aberration de sphéricité.

61. Point-objet à faible distance de l'axe. La coma. Les instruments d'optique de bonne qualité ne présentent pas d'aberration sphérique, mais il peut apparaître de nouveaux défauts lorsqu'on quitte l'axe de l'instrument. Utilisons par exemple un objectif de Clairaut d'assez grande ouverture, ne satisfaisant pas à la condition des sinus, ou encore un miroir parabolique. L'aspect de l'image d'un point objet éloigné B situé au voisinage de l'axe est très particulier: si l'on se place dans le plan de l'image paraxiale, la tache obtenue est symétrique par rapport à l'axe $A'y'$, elle présente une forme d'aigrette (Fig. 111) ou encore de queue de comète, aussi appelle-t-on cette aberration coma ou aigrette. Les dimensions de la tache augmentent d'ailleurs avec la distance du point objet à l'axe et elles varient encore plus vite avec l'ouverture de la pupille de l'instrument.

Si l'on isole maintenant, par un diaphragme annulaire, les rayons qui traversent une même zone de l'objectif et qui formeraient une nappe conique de sommet B' en l'absence d'aberration, on peut observer sur en écran mobile en profondeur les diverses courbes indiquées sur la Fig. 111. Sur ces courbes les points numérotés correspondent aux rayons issus du point de même numéro sur la pupille. On voit que la courbe présente d'abord une cavité, puis un point de rebroussement et devient une courbe à point double qui se transforme en un cercle décrit deux fois lorsqu'on passe dans le plan de l'image paraxiale. Lorsqu'on continue à déplacer l'écran, on obtient à nouveau une courbe à point double, un point de rebroussement, etc. L'aspect de l'aberration est, cette fois, à peu près symétrique de part et d'autre de l'image paraxiale B'.

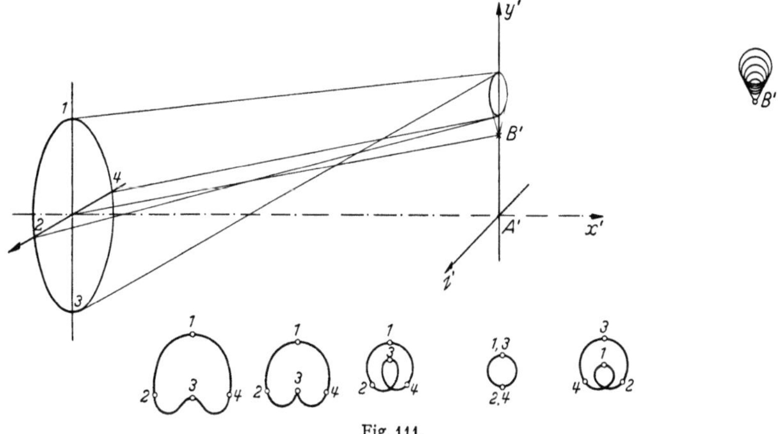

Fig. 111.

La tache comatique est constituée par l'ensemble des cercles doubles provenant des différentes zones de l'objectif ils sont homothétiques par rapport au foyer paraxial (Fig. 112) et le théorie nous montrera que, du point B' on voit tous ces cercles sous un angle de 60°.

En fait, il est rare que l'on observe la coma à l'état pur, elle se superpose souvent à l'aberration sphérique ou encore à une nouvelle aberration que nous allons étudier: l'astigmatisme.

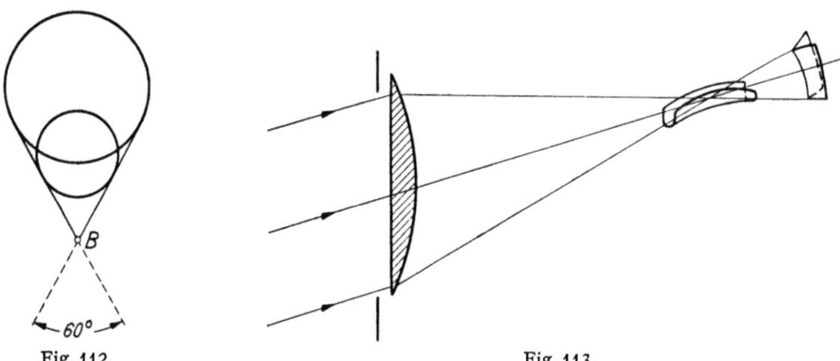

Fig. 112.   Fig. 113.

62. L'astigmatisme. Focales et caustiques. Utilisons à nouveau la lentille simple de la Sect. 60 et éloignons notablement l'objet de l'axe de la lentille. On constate que la caustique se modifie sensiblement, mais elle présente encore deux nappes distinctes issues respectivement des nappes sagittale et tangentielle de l'aberration sphérique. Les rayons lumineux sont tangents à ces deux nappes et nous rencontrons ici le cas général de la caustique d'un faisceau lumineux qui n'est autre que l'enveloppe (à deux nappes) des rayons lumineux. Nous supposerons d'abord que l'ouverture de la pupille est petite et que le faisceau émergent présente une faible ouverture angulaire (Fig. 113). On peut alors constater la présence de deux aires d'amincissement qui sont les éléments de surface caustique situés au voisinage du rayon moyen.

Lorsque l'ouverture angulaire est très petite (cas du pinceau lumineux) les petits morceaux de caustique peuvent être assimilés à leurs plans tangents; l'un d'eux est nécessairement confondu avec le plan de symétrie, l'autre est perpendiculaire à ce plan (Fig. 114). Si l'on coupe le faisceau par un écran perpendiculaire au rayon moyen, on pourra trouver deux positions de l'écran pour lesquelles la section du pinceau se réduit pratiquement à un petit segment de droite; les deux segments ainsi obtenus sont les *focales* du pinceau; l'une d'elles est perpendiculaire au plan de la figure: c'est la *focale tangentielle T*; l'autre est contenue dans ce plan et s'appelle *focale sagittale S*. On dit que le pinceau présente de l'astigmatisme, la distance d'astigmatisme étant égale à la distance des focales T et S.

Coupons le faisceau à grande distance des focales; la trace du faisceau est sensiblement circulaire, puis le cercle s'aplatit peu à peu en une ellipse qui se réduit à une droite lorsqu'on passe en T; on obtient ensuite une ellipse dont le

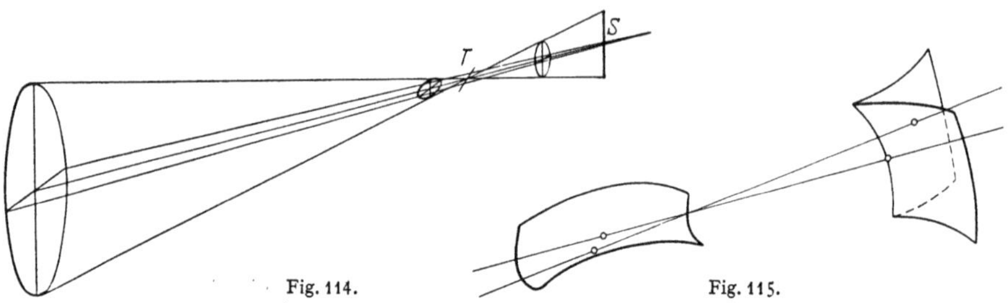

Fig. 114.   Fig. 115.

grand axe est parallèle à la focale tangentielle qui devient un cercle lorsqu'on est sensiblement à égale distance de T et de S. Les dimensions de la tache sont alors minima et ce cercle s'appelle le cercle de moindre diffusion. Le grand axe de l'ellipse est ensuite parallèle à la focale sagittale et l'ellipse se réduit encore à un segment lorsqu'on passe en S. Les différentes sections obtenues sont alors indiqués en perspective sur la Fig. 114.

Donnons maintenant à la pupille une plus grande ouverture; on voit alors apparaitre autour de T et de S deux éléments du surface caustique dont les dimensions vont en croissant avec le diamètre de la pupille. Ces deux nappes sont tangentes en T et en S au rayon moyen et leurs plans tangents sont manifestement rectangulaires puisqu'ils contiennent respectivement les focales T et S. La théorie nous montrera l'existence de focales orthogonales sur tous les rayons du faisceau même s'ils ne sont pas contenus dans le plan méridien de symétrie. La caustique apparait alors comme le lieu des focales des différents rayons du faisceau de même qu'il existe deux focales distinctes, la caustique comprendra deux nappes auxquelles tous les rayons seront tangents (Fig. 115) et les plans tangents aux deux nappes de la caustique aux points de contact d'un même rayon seront toujours perpendiculaires.

Reprenons maintenant l'expérience de la Fig. 114 et prenons encore un objet plan très éloigné; pour chaque point de l'objet, on obtient deux focales T et S qui décrivent, lorsque le point-objet se déplace, deux surfaces de révolution tangentes au plan focal image de la lentille; ainsi l'image d'un objet plan étendu est constituée, si la pupille est petite, par deux surfaces courbes qui sont les surfaces focales (tangentielle et sagittale). L'image présentant les dimensions minima se forme sensiblement à égale distance des deux focales (cercle de moindre diffusion) et le lieu en est une surface courbe. La meilleure image pour un objet ponctuel n'est donc pas appliquée sur un plan de front, on dit qu'on a de la *courbure de champ* (Fig. 116).

Cette aberration subsiste d'ailleurs si l'on supprime l'astigmatisme. Plaçons pour cela la pupille avant la lentille (à une distance voisine du tiers de la distance

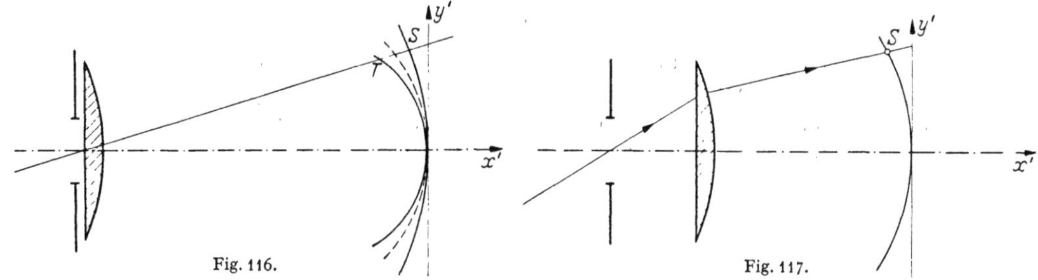

Fig. 116.　　Fig. 117.

focale); on constate que l'astigmatisme est supprimé, mais cependant l'image ne se forme pas sur un plan de front et la surface image S conserve une courbure notable (Fig. 117.).

63. La distorsion. Utilisons comme loupe une lentille plan convexe et observons un quadrillage régulier situé dans son plan focal objet; l'apparence de l'image est celle indiquée sur la Fig. 118. La quadrillage est déformé, les images des droites éloignées du centre du champ présentent une courbure notable; on dit que l'image a de la *distorsion*. Cette déformation subsiste bien entendu si l'on remplace le pupille naturelle de l'oeil de l'observateur par une pupille artificielle

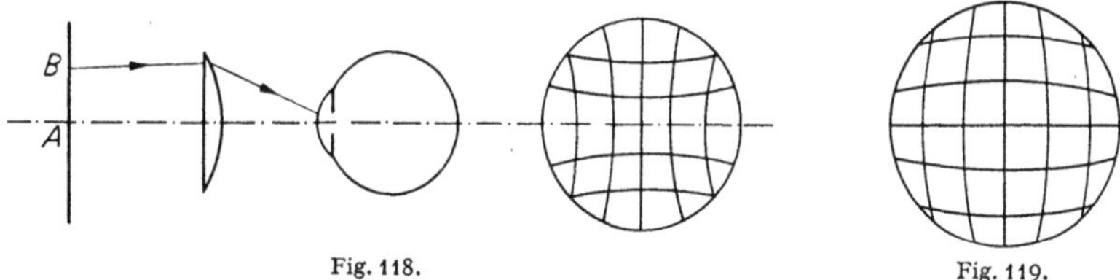

Fig. 118.　　Fig. 119.

de très petites dimensions; cette aberration n'est pas relative au défaut de stigmatisme du faisceau, elle affecte seulement la position du rayon moyen et la forme de l'image d'un objet étendu.

La distorsion indiquée sur la Fig. 118 est la *distorsion en coussinet* ou en *croissant*. Dans d'autres cas, la déformation se produit en sens contraire, la distorsion obtenue est dite en *barillet* (Fig. 119).

Les diverses aberrations que nous venons de décrire n'affectent pas uniquement les lentilles minces: on les rencontre dans tous les systèmes centrés et nous allons en aborder maintenant l'étude théorique.

b) Etude théorique des caustiques et focales.

64. Caustiques. Lorsqu'il n'y a pas stigmatisme, les rayons lumineux dans l'espace image ne viennent plus se concentrer en un point. Ils forment un faisceau de droites normales à une surface d'onde Σ, ou plutôt à une infinité de surfaces d'ondes que l'on peut déduire de l'une quelconque d'entre elles en menant sur chacun des rayons lumineux une longueur constante l (Fig. 120) en faisant varier $l = II'$, la surface d'onde Σ' viendra s'appliquer successivement sur toutes les surfaces d'ondes relatives au milieu image, qui forment ainsi une famille de surfaces parallèles.

Lorsqu'il y a stigmatisme, les surfaces d'onde sont des sphères concentriques; lorsqu'il n'y a pas stigmatisme ce sont des surfaces quelconques parmi lesquelles nous choisirons une surface particulière Σ; les rayons lumineux apparaissent ainsi comme une famille de droites dépendant de deux paramètres (par exemple deux des coordonnées de leur intersection par Σ); d'une façon générale, une famille de courbes à deux paramètres admet une enveloppe; les rayons lumineux resteront donc tangents à une surface au voisinage de laquelle l'énergie lumineuse viendra se concentrer; cette enveloppe est la surface caustique. Nous allons montrer que cette surface admet deux nappes distinctes en étudiant les positions des éléments obtenus au voisinage d'un rayon particulier.

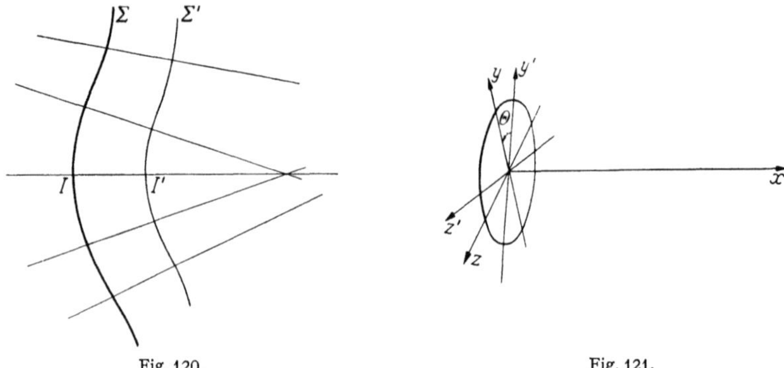

Fig. 120.   Fig. 121.

65. Focales. Soit O un point de Σ, Ox le rayon lumineux (Fig. 121), Oy' et Oz' deux axes rectangulaires tangents à Σ et formant avec O un trièdre trirectangle; l'équation de la surface au voisinage de O peut s'écrire, en limitant le développement de Taylor de $x(y', z')$ aux termes du deuxième ordre,

$$x = a y'^2 + 2b y' z' + c z'^2 + \varepsilon_3 \tag{65.1}$$

ou ε_3 désigne un résidu du troisième degré au moins en y' et z'.

Montrons d'abord que, par une rotation convenable des axes autour de Ox, on peut obtenir un développement par rapport aux nouvelles variables y et z ne contenant pas de termes en yz; (directions principales de la surface) les formules de transformation sont:

$$\begin{cases} y' = y \cos \vartheta - z \sin \vartheta, \\ z' = y \sin \vartheta + z \cos \vartheta. \end{cases}$$

Il suffit de porter ces valeurs dans (65.1) et de développer pour voir que le terme en yz sera nul si $(c - a) \tan 2\vartheta + 2b = 0$ ce qui fixe une série de valeurs de ϑ définies à $k\pi/2$ près k étant un entier.

Les directions rectangulaires ainsi définies sont les directions principales de la surface Σ au point O. Plaçons les axes Oy et Oz suivant ces directions et appelons R et R' les rayons de courbure des sections de Σ par les plans xOy et xOz; l'équation (65.1) s'écrira sous la forme:

$$x = \frac{y^2}{2R} + \frac{z^2}{2R'} + \varepsilon_3.$$

Cherchons maintenant la normale à Σ; ses paramètres directeurs seront avec la même approximation, $1, -\left(\dfrac{y}{R} + \varepsilon_2\right)$ et $-\left(\dfrac{z}{R'} + \varepsilon_2\right)$ le symbole ε_2 désignant des résidus quelconques du deuxième ordre; ses équations seront, en appelant m le paramètre dont dépend le point,

$$\left.\begin{aligned}
X &= \frac{y^2}{2R} + \frac{z^2}{2R'} + \varepsilon_3 + m \\[2mm]
Y &= y - \left(\frac{y}{R} + \varepsilon_2\right) m \\[2mm]
Z &= z - \left(\frac{z}{R'} + \varepsilon_2\right) m.
\end{aligned}\right\} \tag{65.2}$$

Cette droite dépend des deux paramètres, y et z; son enveloppe s'obtient en écrivant

$$\begin{vmatrix} X'_m & X'_y & X'_z \\ Y'_m & Y'_y & Y'_z \\ Z'_m & Z'_y & Z'_z \end{vmatrix} = 0$$

ce qui conduit à une équation de la forme:

$$\left(1 - \frac{m}{R}\right)\left(1 - \frac{m}{R'}\right) + \varepsilon_1 = 0. \tag{65.3}$$

Remarquons d'abord que, lorsque $y = z = 0$, ε_1 s'annule et l'on a deux racines $m = R$ et $m = R'$.

Les racines d'une équation étant en général des fonctions continues des coefficients, on peut écrire les racines sous la forme

$$m = R + \varepsilon_1; \quad m = R' + \varepsilon_1$$

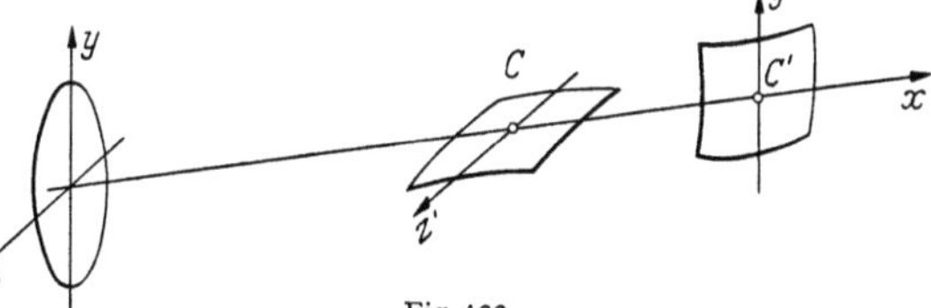

Fig. 122.

où les ε_1 désignent des polynômes commençant par un binôme du premier degré en y et z; on aboutit alors aux deux solutions:

$$\begin{aligned}
X &= R + \varepsilon_1 & \qquad X &= R' + \varepsilon_1 \\[2mm]
Y &= \varepsilon_2 & \qquad Y &= y\left(I - \frac{R'}{R}\right) + \varepsilon_2 \\[2mm]
Z &= z\left(I - \frac{R}{R'}\right) + \varepsilon_2 & \qquad Z &= \varepsilon_2.
\end{aligned} \tag{65.4}$$

Les équations (65.4) montrent qu'au voisinage du rayon particulier issu de O la caustique se décompose en deux nappes tangentes à Ox aux points C et C', centres de courbure des sections de la surface par xOy et xOz (Fig. 122). La première nappe admet en C le plan tangent à xOz, la seconde est tangente en C' à xOy. Ces résultats étant valables quel que soit le rayon choisi pour axe Ox, on en déduit que la surface caustique se décompose en deux nappes distinctes, lieux des points C et C', les plans tangents en C et C' étant rectangulaires, et parallèles aux directions principales de la surface d'onde au point utilisé.

Coupons maintenant le pinceau par un écran perpendiculaire au rayon moyen passant par C: on obtient une section d'un élément de caustique que l'on peut assimiler à un petit segment de la droite Cz'. Dans les mêmes conditions on obtiendrait en C' un petit élément de la droite $C'y'$. La théorie nous montre ainsi l'existence des focales.

Si l'écran ne passe plus par C ou C', les équations (65.2) nous permettent de trouver la forme de la section du pinceau. Supposant par exemple que le pinceau soit limité à un cercle de rayon a dans le plan yOz (Fig. 123) on obtiendra les équations d'un rayon en écrivant que les projections de ce rayon sur les plans xOy et xOz passent respectivement par C et C'. Ces équations s'écriront:

$$X = m, \qquad Y = \cos\varphi\left(I - \frac{m}{R}\right), \qquad Z = a\sin\varphi\left(I - \frac{m}{R'}\right).$$

Il en résulte que les sections du pinceau par un plan d'abscisse m seront généralement des ellipses, sauf si $\dfrac{2}{m} = \dfrac{1}{R} + \dfrac{1}{R'}$, c'est-à-dire (lorsque $R - R'$ est faible devant R ou R') si $m = \frac{1}{2}(R + R')$ on obtient alors le cercle de moindre diffusion. On peut en effet vérifier que, pour cette position de l'écran, la section du pinceau

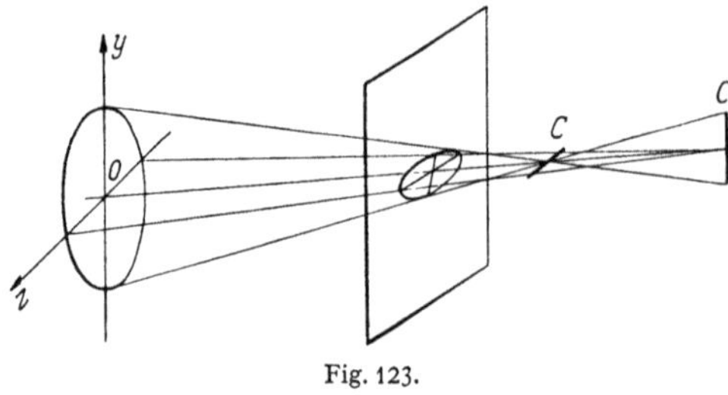

Fig. 123.

est la plus réduite dans l'ensemble des directions y et z.

Les focales précédemment définies (qui sont perpendiculaires à Ox) s'appellent focales de Sturm, mais nous devons remarquer qu'elles ne jouissent d'aucune propriété particulière. Si nous coupions le pinceau par une plan quelconque passant par C, les conclusions seraient identiques, on trouverait que le pinceau s'appuie avec la même approximation que précédemment sur des morceaux de droites contenues dans le plan xOy, mais non perpendiculaires à Ox.

En résumé les focales sont les sections des plans tangents à la caustique par le plan de l'écran. Parmi ces sections, les focales de Sturm sont le plus souvent utilisées, mais elles ne présentent pas d'avantage théorique sur les focales obliques.

c) Propriétés générales des faibles aberrations géométriques.

66. La méthode des chemins optiques. Lorsque les aberrations d'un instrument d'optique sont faibles, il est commode de les étudier en calculant l'écart entre la surface d'onde et une sphère centrée sur l'image de Gauss; l'application du principe de Fermat et du théorème de Malus permet en effet d'établir aisément une relation entre cet écart et les variations du chemin optique entre la source et le centre de la sphère; on peut d'ailleurs calculer pratiquement ce chemin optique en ajoutant les termes d'aberrations relatifs aux diverses surfaces réfringentes ou réfléchissantes composant l'instrument et déterminer ensuite les déplacements latéraux des rayons par les formules différentielles (69.1). Nous allons établir ces diverses propriétés élémentaires, dont nous verrous ultérieurement les applications. Nous éviterons en fait de développer sous une forme très générale l'étude des chemins optiques (théorie de l'iconale) en nous limitant aux propriétés aisément utilisables.

67. Quelques propriétés des faibles aberrations. Considérons une surface d'onde Σ s'écartant peu d'une sphère S et soit $\varDelta = IM$ l'écart entre S et Σ (Fig. 124) que nous appellerons écart normal.

Nous supposerons que $\varDelta$ peut s'écrire sous la forme $\varDelta = \varDelta_0 m$ où $\varDelta_0$ représente l'allure des variations de $\varDelta$, et m est un paramètre fixant l'importance des aberrations, dont nous préciserons la nature dans chaque cas particulier (par exemple dans le cas de la coma, ce sera la grandeur de l'image $A'B'$.

Nous nous proposons d'étudier quelques propriétés particulières que l'on peut établir lorsque m est assez petit pour que l'on puisse négliger les quantités de l'ordre de m^2 ou plus.

1. $\varDelta$ ne varie pas si l'on se déplace sur la surface d'onde d'un infiniment petit de l'ordre de m.

On peut écrire en effet, s étant l'élément d'arc représentant le déplacement sur Σ,

$$\varDelta' = \varDelta + \frac{\partial \varDelta}{\partial s}\, ds + \cdots;$$

mais $\partial \varDelta / \partial s$ est, comme $\varDelta$, un infiniment petit du premier ordre; si ds le déplacement est aussi du premier ordre, on a $\varDelta' \approx \varDelta$.

2. *Théorème de* GOUY, *invariance de $\varDelta$ le long d'un rayon lumineux.*

Considérons une surface d'onde quelconque incidente sur un système optique et la surface d'onde correspondante dans l'espace image (en traits interrompus

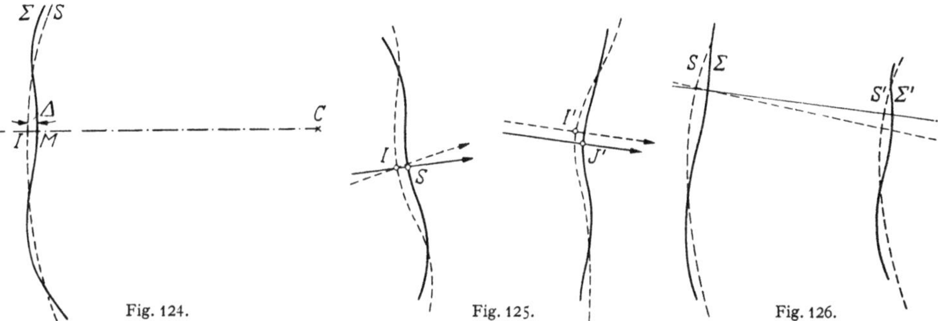

Fig. 124.   Fig. 125.   Fig. 126.

sur la Fig. 125); si l'on applique à la surface d'onde incidente une petite déformation $\varDelta$, il en résulte une déformation $\varDelta' = n\varDelta/n'$ de la surface d'onde émergente.

La relation (7.2) permet en effet d'écrire ici puisque par hypothèse $II' = JJ'$:

$$0 = n'\, \boldsymbol{T}'\, \overrightarrow{I'J'} - n\, \boldsymbol{T}\, \overrightarrow{IJ},$$

et il est évident sur la Fig. 125 que cette relation s'écrit:

$$n'\varDelta' = n\varDelta.$$

Les quantités telles que $n\varDelta$ ou $n'\varDelta'$ seront encore appelées *chemins optiques aberrants* lorsque la surface de référence sera une sphère.

Si l'on envisage en particulier le cas d'un instrument composé de divers éléments (dioptres, lentilles, miroirs) entachés chacun de faibles aberrations, le théorème de GOUY permet d'établir immédiatement *qu'on obtient le chemin optique aberrant $n'\varDelta'$ relatif à l'ensemble de l'instrument en ajoutant les chemins optiques aberrants des divers éléments.* Il suffit, en effet, de supposer d'abord tous les dioptres parfaitement stigmatiques, puis de faire intervenir à tour de rôle les chemins optiques aberrants de chacun d'eux.

Nous pouvons choisir pour surfaces d'ondes particulières deux sphères concentriques S, S' dans le milieu image. Soient Σ, Σ' deux surfaces d'ondes aberrantes voisines respectivement de S et de S' et séparées par le même chemin optique (Fig. 126), on aura cette fois $\varDelta = \varDelta'$ et l'on en conclut que $\varDelta$ se conserve au cours de la propagation dans un même milieu; il ne dépend que de la direction du rayon lumineux.

68. Expression de Δ en fonction des variations de chemins optiques. Le chemin optique aberrant est égal aux variations de chemin optique entre la source et le centre C de la sphère de référence.

Nous appellerons chemin optique L entre le point objet A et le centre C le chemin optique obtenu entre la source A et le pied H de la perpendiculaire abaissée de C sur le rayon peu aberrant $L = (AH)$ (Fig. 127).

Soient M_0 un point commun à Σ et S, CK la perpendiculaire abaissée de C sur le rayon passant par M_0, L_0 le chemin optique (AK); on peut écrire

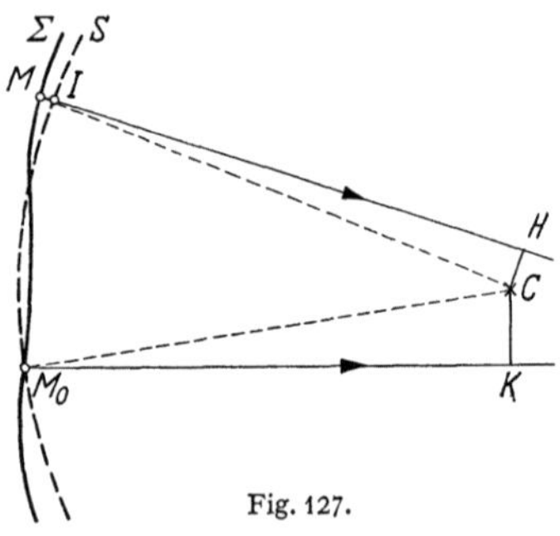

Fig. 127.

$$L = (AM) + n'(\overline{MI} + \overline{IH}) = (AM) + n'(R - \Delta),$$

$$L_0 = (AM_0) + n'\overline{M_0K} = (AM) + n'R$$

d'où l'on déduit la relation:

$$n'\Delta = L_0 - L. \qquad (68.1)$$

Cette relation sera extrêmement utile pour calculer les aberrations d'un système optique quelconque; supposons en effet que nous connaissions les marches approximatives des rayons lumineux, l'incertitude sur la position des rayons pouvant être caractérisée par un infiniment petit ε; le calcul des quantités L_0 et L évaluées le long de ces rayons permettra de connaître Δ avec une erreur de l'ordre de ε^2 et d'étudier par conséquent les aberrations dont l'ordre d'infinitude (en ce qui concerne Δ) est inférieur à ε^2. C'est ainsi que le calcul du chemin optique le long des rayons paraxiaux permettra de connaître les aberrations dites du troisième ordre, ou encore les aberrations chromatiques en différentiant par rapport aux indices de réfraction (voir Sect. 86).

69. Relations entre Δ et les aberrations transversales. Nous rapporterons la surface d'onde à un système d'axes rectangulaires $Cx'y'z'$ et nous définirons un point M de cette surface par l'azimut φ du plan méridien qui le contient et par l'angle α' de la droite MC et de Cx' (Fig. 128). Nous nous proposons d'établir par une méthode géométrique les relations qui lient les coordonnées y', z' de l'intersection P du rayon MP et du plan $y'Cz'$ à l'écart Δ (les coordonnées y' et z' sont, bien entendu, des quantités de l'ordre de m). Nous supposerons d'abord $\varphi = 0$ et nous tracerons au voisinage de M de petits éléments des lignes σ_1 et σ_2 qu'on obtient en coupant Σ par un plan méridien ($\varphi = $ constante) et par un cône de révolution ($\alpha' = $ constante). Le rayon lumineux MP est perpendiculaire à ces lignes et les angles ε_1 et ε_2 marqués sur la figure ne sont autres que les pentes que présente la surface d'onde par rapport à la sphère de centre C le long des lignes σ_1 et σ_2; ε_1 et ε_2 sont donc les dérivées partielles de Δ par rapport à des déplacements effectués le long de σ_1 ou σ_2. On pourra écrire, R étant le rayon de la sphère,

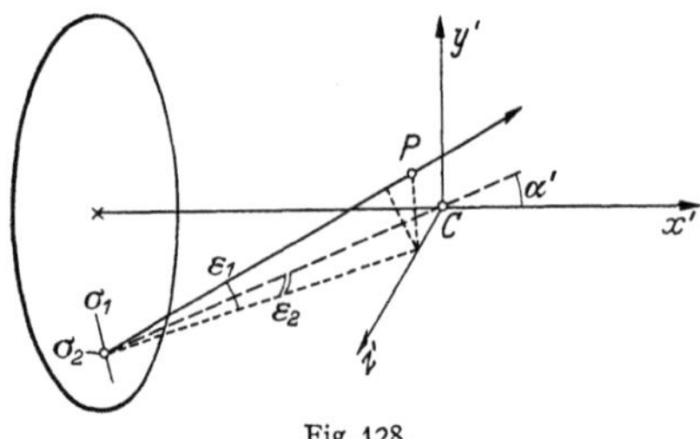

Fig. 128.

$$y'\cos\alpha' = R\,\varepsilon_1 = R\,\frac{\partial\Delta}{\partial\sigma_1},$$

$$z' = R\,\varepsilon_2 = R\,\frac{\partial\Delta}{\partial\sigma_2},$$

mais

$$d\sigma_1 = R\,d\alpha' \quad \text{et} \quad d\sigma_2 = R\sin\alpha'\,d\varphi.$$

On en tire immédiatement les relations

$$y' = \frac{1}{\cos\alpha'}\frac{\partial\Delta}{\partial\alpha'}; \quad z' = \frac{1}{\sin\alpha'}\frac{\partial\Delta}{\partial\varphi}.$$

Si maintenant l'azimut φ est quelconque, il suffit de faire tourner le vecteur CP de l'angle A, ce qui donne:

$$\left.\begin{aligned} y' &= \frac{\cos\varphi}{\cos\alpha'}\frac{\partial\Delta}{\partial\alpha'} - \frac{\sin\varphi}{\sin\alpha'}\frac{\partial\Delta}{\partial\varphi}, \\ z' &= \frac{\sin\varphi}{\cos\alpha'}\frac{\partial\Delta}{\partial\alpha'} + \frac{\cos\varphi}{\sin\alpha'}\frac{\partial\Delta}{\partial\varphi}. \end{aligned}\right\} \tag{69.1}$$

On constate, que conformément ou théoréme de Gouy le rayon R de la sphère de référence n'intervient pas dans ces formules. D'autre part, la propriété démontrée à la Sect. 67 nous autorise à y remplacer au besoin α' ou φ par une quantité n'en différant que d'un infiniment petit du premier ordre.

Ces formules serviront soit à calculer les coordonnées de l'intersection d'un rayon aberrant par le plan de l'image (cas de la coma), soit inversement à déterminer Δ par intégration si l'on connait y' et z' (étude expérimentale des aberrations).

70. Fonctions iconales et formules fondamentales. Les formules (69.1) sont en fait susceptibles d'être écrites plus simplement si l'on n'est pas amené par la nature du problème à utiliser obligatoirement les coordonnées sphériques α' et φ. On peut même définir de façon précise le chemin optique L de façon que les formules obtenues ne supposent a priori aucune approximation mais qu'elles résultent directement du principe de Fermat.

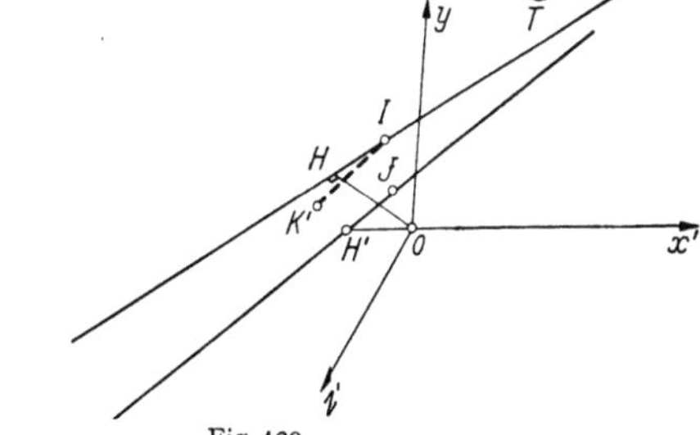

Fig. 129.

Considérons (Fig. 129) un instrument devant lequel nous plaçons une source lumineuse A. L'orientation des rayons qui cheminent dans l'espace image peut être caractérisée par les cosinus directeurs α', β', γ' dont deux suffiront en fait à fixer l'orientation; nous choisirons pour cela β' et γ'. Si de l'origine O on abaisse la perpendiculaire OH sur le rayon, on définit ainsi le chemin optique (AH) en fonction de l'orientation du rayon; nous appellerons *fonction iconale* $L(\beta',\gamma')$ le chemin optique AH.

Montrons maintenant que la connaissance de la fonction iconale permet de déterminer l'intersection I du rayon avec le plan $O'y'z'$: considérons un rayon ayant une orientation légèrement différente, caractérisée par les paramètres $\beta'+d\beta'$ et $\gamma'+d\gamma'$ et perçant le plan $O'y'z'$ en J (Fig. 129). Pour aller de A en I on peut suivre le rayon effectif, qui correspond au chemin optique $L+n'HI$ ou encore suivre le chemin très voisin fictif $AK'I$, $K'I$ étant parallèle à $H'J$ et $H'K'$ perpendiculaire à $H'J$, mais on a $(AK')=(AH')=L+dL$. On écrira donc $dL+d(HI)=0$, mais on a:

$$HI = \overrightarrow{OI}\cdot\boldsymbol{T} = \beta'\gamma' + \gamma'z'$$

d'où l'on tire par exemple:

$$\frac{\partial L}{\partial \beta'}\, d\beta' + y'\, d\beta' = 0.$$

On aura donc les deux relations fondamentales:

$$\boxed{y' = -\,\frac{\partial L}{\partial \beta'} \qquad z' = -\,\frac{\partial L}{\partial \gamma'}} \tag{70.1}$$

Ces relations, souvent appelées relations fondamentales de l'iconale montrent que si la fonction iconale L est connue on connait complètement la marche des rayons lumineux. On remarque d'ailleurs que l'établissement de ces relations n'implique aucune approximation; elles sont valables quelle que soit l'importance de la déformation de la surface d'onde.

Les fonctions iconales L sont encore fonctions de la position du point A: on peut par exemple caractériser complètement l'instrument par la fonction générale de quatre variables $L(y, z, \beta', \gamma')$ où y et z sont les coordonnées de A. Le calcul de ces fonctions n'est en fait pas aisé et on ne les utilisera que dans l'approximation précédente, où les aberrations sont supposées faibles. On peut alors écrire les formules précédentes sous la forme:

$$y' = \frac{\partial \varDelta}{\partial \beta'}\,; \qquad z' = \frac{\partial \varDelta}{\partial \gamma'} \tag{70.2}$$

qui sont bien entendu équivalentes aux formules (69.1) en coordonnées rectangulaires.

d) Classification des aberrations géométriques des instruments de révolution.

71. Classification ordinaire. Soit $AB = y$ un petit objet perpendiculaire à l'axe dans l'espace objet, $A'B'$ l'image paraxiale de AB; si g est le grandissement obtenu dans le domaine paraxial, la grandeur de l'image sera définie par l'équation

$$y' = A'B' = g\,y.$$

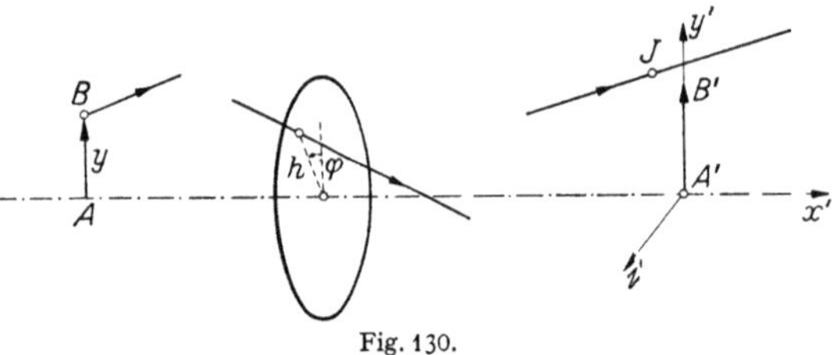

Fig. 130.

Si nous nous écartons maintenant du domaine paraxial, les rayons issus de B ne viendront plus passer par B', mais viendront couper le plan de front $A'y'z'$ en un point J voisin de B' (Fig. 130); nous appellerons dy', dz' les coordonnées de l'intersection J du rayon par rapport à B', c'est-à-dire que dans les axes $A'y'z'$ les coordonnées de J seront $y'+dy', dz'$. Le vecteur $\overrightarrow{B'J}$ sera appelé *aberration transversale*.

Les composantes dy', dz' de l'aberration transversale dépendent de y' et du point I où le rayon a percé la pupille. Le diaphragme réel qui joue le rôle de pupille peut être situé dans l'espace objet ou l'espace image, ou dans un milieu intermédiaire quelconque, nous choisirons dans son plan des coordonnées polaires h et φ; le rayon lumineux sera parfaitement défini si l'on connait B (c'est-à-dire y ou y') et le point I où il perce le plan du diaphragme (c'est-à-dire h et φ). Imaginons maintenant que nous fassions tourner toute la figure d'un demi-tour autour de xx', ce qui revient à changer y' en $-y'$ et h en $-h$; il est évident que

dy', dz' changent aussi de signe; on en déduit que dy' et dz' sont des fonctions impaires par rapport à l'ensemble des variables h et y'; leurs développements en série ne comprendront donc que des termes de degré impair.

Les termes de degré 1 seront proportionnels à h ou à y' mais l'étude de l'approximation de Gauss a montré que, lorsque h et y' sont petits, les rayons viennent sensiblement passer par l'image de Gauss B'. Ces termes sont donc nuls (ils ne le seraient pas si l'on n'avait pas choisi comme origine l'image paraxiale B').

Les termes du degré 3 constituent les aberrations dites du troisième ordre. Nous montrerons que:

Le terme en h^3 caractérise l'aberration sphérique du 3ème ordre,

$h^2 y'$ caractérise la coma du 3ème ordre,

$h y'^2$ caractérise l'astigmatisme et la courbure de champ du 3ème ordre,

y'^3 caractérise la distorion.

Plus généralement on peut classer les termes d'aberrations dans un tableau à double entrée ou figurent en coordonnées les puissances respectives de h et y.

On apellera de façon générale termes

d'aberration sphérique ceux qui ne dépendent pas de y' (1re colonne);

de *coma* ceux qui sont proportionnels à y' (2e colonne);

d'astigmatisme et de *courbure* ceux qui sont proportionnels à h (2e ligne);

de *distorsion* ceux qui ne dépendent pas de h (1re ligne).

Tableau 1.

	y'^0	y'^1	y'^2	y'^3	y'^4	y'^5
h^0				y'^3		y'^5
h^1			$h y'^2$		$h y'^4$	
h^2		$h^2 y'$		$h^2 y'^3$		
h^3	h^3		$h^3 y'^2$			
h^4		$h^4 y'$				
h^5	h^5					

72. Classification de Nijboer. Le tableau 1 permet d'établir une classification des aberrations transversales. On peut encore les classer en étudiant la déformation de la surface d'onde. Une méthode particulièrement utile pour l'étude des phénomènes de diffraction a été proposée par Nijboer: elle consiste à développer en série de Fourier l'écart Δ entre la surface d'onde et une sphère centrée en B'. Comme Hamilton l'avait déjà montré, Δ n'est pas une fonction quelconque de h, y' et φ: l'instrument est de révolution et Δ est déterminé si l'on se donne h^2 et y'^2 (c'est-à-dire les valeurs absolues de h et y') ainsi que le produit scalaire $\overrightarrow{PI} \cdot \overrightarrow{A'B'} = h y' \cos \varphi$; on peut donc développer Δ sous forme d'une série triple par rapport aux variables h^2, y'^2 et $h y' \cos \varphi$; c'est-à-dire que les termes sont de la forme:

$$h^{2k} y'^{2l} (h y' \cos \varphi)^n = h^{2k+n} y'^{2l+n} \cos^n \varphi,$$

mais on sait que $\cos^n \varphi$ est une combinaison linéaire de $\cos n\varphi$, $\cos(n-2)\varphi$, $\cos(n-4)\varphi$ etc. et les termes peuvent alors s'écrire:

$$h^{2k+n} y'^{2l+n} \cos (n - 2i) \varphi.$$

Posant $m = n - 2i$; $p = k + i$; $q = l + i$, la forme générale devient:

$$h^{2p+m} y'^{2q+m} \cos m \varphi$$

et l'on peut ainsi préciser comment chaque terme dépend de la variable φ, ce que ne nous permettait pas la classification précédente. On remarque d'ailleurs qu'il est nécessaire d'utiliser trois paramètres (p, q, m) pour effectuer une classification complète.

On peut remarquer que cette classification permet en somme de développer Δ en série de Fourier; les termes indépendants de φ dans le développement ($m=0$) sont par exemple l'aberration sphérique et la courbure de champ; les termes en $\cos \varphi$ contiennent en particulier la coma et la distorsion, les termes en $\cos 2\varphi$ l'astigmatisme etc. Cette classification permet évidemment de dénombrer les termes d'aberrations. On verra ultérieurement que les termes de degré n par rapport à l'ensemble des variables h et y' dans Δ donnent naissance aux termes de degré $n-1$ dans l'expression de l'aberration transversale; c'est ainsi qu'aux termes dits du 3ème ordre dans l'aberration transversale correspondent des termes du 4ème degré dans Δ; on peut alors aisément les écrire tous:

$m=0$	h^4		$h^2 y^2$		y^4
$m=1$		$h^3 y' \cos \varphi$		$h\,y'^3 \cos \varphi$	
$m=2$			$h^2 y^2 \cos 2\varphi$		

En fait, le terme en y^4 qui ne correspond à aucune variation de Δ sur la pupille ne donne naissance à aucune aberration. Il subsiste donc 5 termes indépendants d'aberration dites du 3ème ordre.

Les termes correspondants aux aberrations du 5ème ordre sont de même:

$m=0$	h^6	$h^4 y'^2$	$h^2 y'^4$	y'^6
$m=1$	$h^5 y' \cos \varphi$	$h^3 y'^3 \cos \varphi$	$h y'^5 \cos \varphi$	
$m=2$		$h^4 y'^2 \cos 2\varphi$	$h^2 y'^4 \cos 2\varphi$	
$m=3$		$h^3 y'^3 \cos 3\varphi$		

Il subsiste 9 termes indépendants (celui en y'^6 ne comptant pas). On peut montrer de façon générale que les termes d'aberration d'ordre $2n+1$ (en ce qui concerne les aberrations latérales) sont au nombre de:

$$\frac{(n+2)(n+3)}{2} - 1.$$

On est amené en fait à classer comme suit les termes les plus usuels (nous en verrons ultérieurement la justification):

aberration sphérique: termes indépendants de y' ($q=m=0$);

coma: termes proportionnels à y' ($q=0, m=1$) et l'on voit que le chemin optique aberrant contient alors nécessairement $\cos \varphi$;

courbure de champ: termes proportionnels à h^2 et indépendants de φ ($p=1, m=0$);

astigmatisme: termes proportionnels à h^2 et à $\cos 2\varphi$ ($p=0, m=2$);

distorsion: termes proportionnels à h et à $\cos \varphi$ ($p=0, m=1$).

e) L'aberration sphérique.

73. Relations générales. Dans les classifications précédentes, nous avons eppelé termes d'aberrations phérique ceux qui ne dépendent pas de y'; ils doivent donc exister déjà pour les points conjugués A et A' situés sur l'axe de révolution.

Lorsque l'ouverture de l'instrument est faible, le rayon dans l'espace image passe par le point A', image de Gauss de A; mais lorsqu'on s'éloigne du domaine paraxial, le rayon vient couper l'axe en un point L qui peut être différent de A' (Fig. 131); nous poserons $A'L=l$ et nous appellerons cette longueur *l'aberration sphérique longitudinale.*

Cette quantité est manifestement une fonction paire de h, qui s'annule lorsque $h=0$; son développement en série ne contiendra donc que des termes en h^2, h^4 etc; pour plus de commodité nous caractériserons la position du rayon par l'angle α' qu'il fait dans l'espace image avec l'axe $A'x'$; le calcul des combinaisons optiques

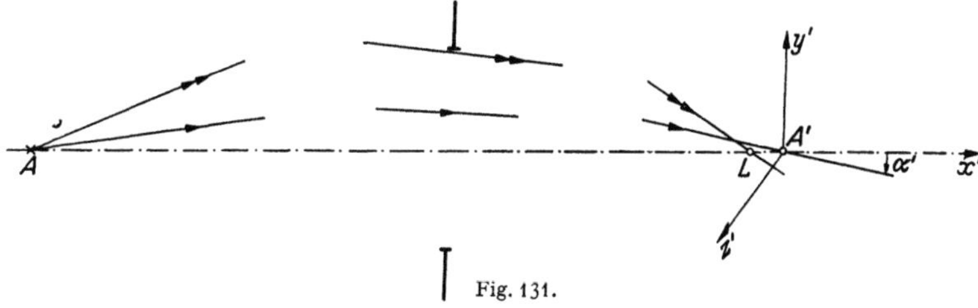

Fig. 131.

conduit en fait à utiliser $\sin\alpha'$ et nous effectuerons les développements en séries par rapport à ce paramètre. Il sera donc commode de développer sous la forme:

$$l = a\sin^2\alpha' + a'\sin^4\alpha' + \cdots$$

lorsqu'on est dans le plan méridien ($\varphi=0$). Les composantes dy', dz' de la diffusion latérale du rayon sont évidemment,

$$dy' = LA'\tan\alpha' = -l\tan\alpha',$$
$$dz' = 0$$

et, si le rayon chemine dans un plan d'azimut φ:

$$dy' = -l\tan\alpha'\cos\varphi,$$
$$dz' = -l\tan\alpha'\sin\varphi.$$

Ces relations vont nous permettre de déterminer la forme de la surface d'onde et de calculer l'écart Δ, dans l'hypothèse où l'aberration est faible.

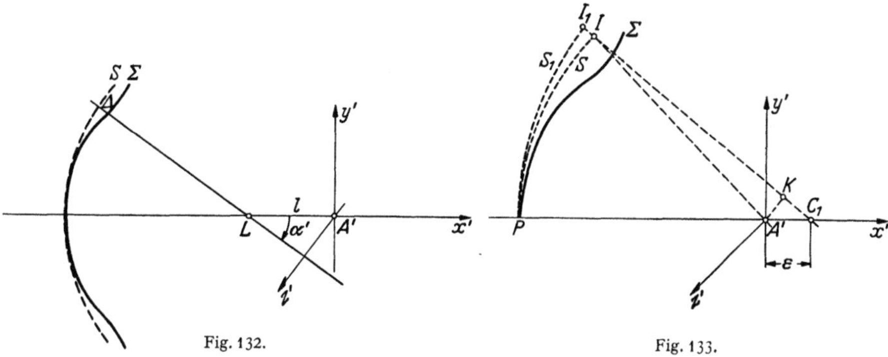

Fig. 132.   Fig. 133.

Choisissons pour sphère de référence la sphère S centrée en A' (Fig. 132). Les formules (69.1), dans lesquelles on fera $\varphi=0$ et $\partial\Delta/\partial\varphi=0$, donnent

$$\Delta = \int_0^{\alpha'} (-l\tan\alpha')\cos\alpha'\,d\alpha' = \int_0^{\alpha'} -l\sin\alpha'\,d\alpha' = \int_0^{\alpha'} l\,d(\cos\alpha'). \qquad (73.1)$$

Cherchons aussi l'écart que présenterait la surface d'onde par rapport à une sphère S_1 ayant don centre C_1 à une distance $\overline{A'C_1}=\varepsilon$ du point A'; il suffit pour cela de déterminer la distance normale II_1, entre les deux sphères (Fig. 133):

si K est la projection de A' sur $I_1 C_1$ on a:

$$PA' = IA' \quad \text{et} \quad PC_1 = I_1 C_1$$

d'où

$$\varepsilon = PC_1 - PA' = I_1 C_1 - IA' \approx I_1 I + KC_1$$

c'est-à-dire

$$I_1 I = \varepsilon (1 - \cos \alpha').$$

On en déduit que l'écart $\varDelta_1$ par rapport à la sphère S_1 est égal à:

$$\varDelta_1 = \varDelta + I_1 I = \varDelta + \varepsilon (1 - \cos \alpha'). \tag{73.2}$$

La caustique est évidemment de révolution autour de xx' et nous pouvons en déterminer la méridienne: elle s'obtiendra en cherchent l'enveloppe des rayons lumineux situés dans un plan méridien particulier. La recherche analytique de cette courbe nous fournira le lieu de la focale T perpendiculaire au plan méridien (focale tangentielle). Il suffit pour cela de dériver l'équation du rayon

$$y' = [x' - l (\sin \alpha')] \tan \alpha'$$

par rapport au paramètre α' pour obtenir le point de contanct avec la nappe tangentielle. Nous le ferons dans l'étude de l'aberration du 3ème ordre.

Pour rechercher la focale située dans le plan méridien (focale sagittale), il suffit de remarquer que tous les rayons lumineux rencontrent l'axe xx' ce qui permet d'affirmer que la focale sagittale S est constituée par un petit segment de l'axe; la nappe correspondante de la caustique est exceptionnellement dégénérée en un segment de droite: c'est la partie de l'axe décrite par le point L lorsque α' varie.

74. L'aberration du troisième ordre.

Pour étudier l'aberration sphérique du 3ème ordre, nous allons limiter le développement de l à son premier terme:

$$l = a \sin^2 \alpha'$$

ou encore avec la même approximation,

$$l = a \alpha'^2, \tag{74.1}$$

et les relations (69.1) deviennent

$$\left. \begin{array}{l} dy' = - a \alpha'^3 \cos \varphi, \\ dz' = - a \alpha'^3 \sin \varphi. \end{array} \right\} \tag{74.2}$$

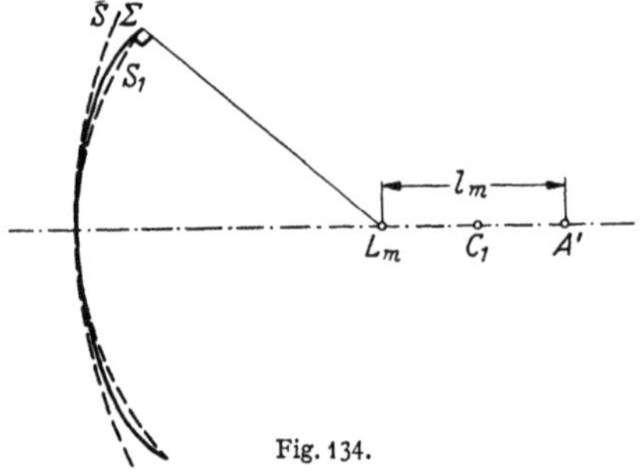

Fig. 134.

On trouve d'autre part, en appliquant la relation (73.1) l'expression de la déformation de la surface d'onde:

$$\varDelta = - \frac{a \alpha'^4}{4}. \tag{74.3}$$

La surface d'onde $\varDelta$ ne s'écarte de S que par un terme du 4ème ordre par rapport à l'ouverture: elle est sur-osculatrice à S en P (Fig. 134).

Si on rapporte $\varSigma$ à une sphère S_1 dont le centre est à distance ε de A', on aura [relation (73.2)]:

$$\varDelta_1 = \varDelta + \frac{\varepsilon}{2} \alpha'^2. \tag{74.4}$$

On verra ultérieurement, dans l'étude des effets combinés de la diffraction et des aberrations géométriques, que l'on est amené à rechercher un défaut de mise au point tel que $\Delta_1 = 0$ au bord, c'est-à-dire, en appelant $\alpha'm$ la valeur maximum de α':

$$- \frac{a\,\alpha'^4 m}{4} + \varepsilon \frac{\alpha'^2 m}{2} = 0$$

ou encore $\varepsilon = \tfrac{1}{2} a\alpha_m'^2$; la sphère S_1 doit donc être centrée au milieu du segment $A'L_m$, L_m étant la position particulière de L obtenue lorsque $\alpha' = \alpha'_m$ (Fig. 134).

L'écart Δ_1 prend alors la forme:

$$\Delta_1 = - \frac{a\,\alpha'^4}{4} + \frac{a\,\alpha_m'^2\,\alpha'^2}{4} = \frac{a\,\alpha'^2}{4}\left(\alpha_m'^2 - \alpha'^2\right).$$

Δ_1 est nul pour $\alpha' = 0$ et α'_m, mais sa valeur absolue passe par un maximum lorsque

$$\alpha'^2 = \frac{\alpha_m'^2}{2}$$

et l'on a alors

$$\Delta_1 = \frac{a\,\alpha_m'^4}{16}.$$

La valeur maximum de Δ_1 est donc (en module) quatre fois plus faible que celle de Δ (qui est égale à $-\tfrac{1}{4} a\,\alpha_m'^4$).

Pour rechercher la caustique, nous partirons de l'équation du rayon lumineux:

$$y' = \alpha'\left(x' - a\alpha'^2\right).$$

On obtient immédiatement les coordonnées du point de contact T en dérivant l'équation précédente par rapport au paramètre α'; on trouve:

$$x' = 3\,a\alpha'^2; \quad y' = 2\,a\alpha'^3.$$

La courbe ainsi déterminée admet un rebroussement pour $\alpha' = 0$ ce qui donne à la nappe tangentielle de la caustique la forme d'un filet à papillons (Fig. 135).

Nous aurons l'occasion de calculer quelques valeurs de l'aberration sphérique de systèmes très simples (miroirs, lentilles minces) lorsque nous traiterons les tolérances relatives aux aberrations.

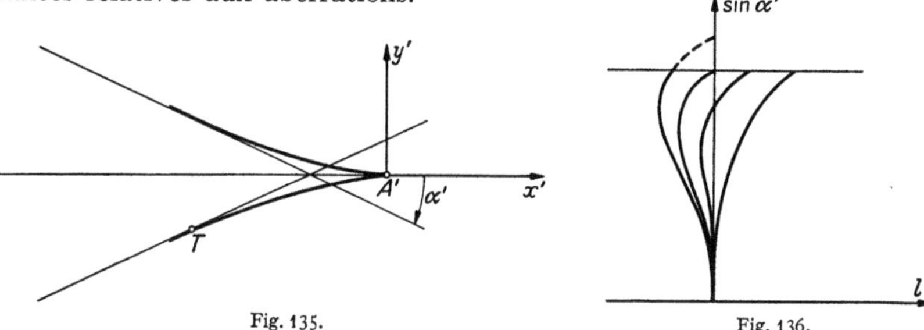

Fig. 135.   Fig. 136.

75. L'aberration sphérique repliée. On dit qu'on a «corrigé» l'aberration sphérique lorsque la courbe qui représente l'aberration longitudinale est repliée (Fig. 136); il faut pour cela que le terme d'aberration du 3ème ordre soit suffisamment réduit pour être compensé par les termes d'ordre supérieur, dont nous ne retiendrons que le terme du 5ème ordre. On écrira donc:

$$l = a \sin^2\alpha' + a' \sin^4\alpha'. \tag{75.1}$$

Les instruments pour lesquels une telle représentation de l'aberration est convenable sont ceux pour lesquels l'ouverture n'est pas exceptionnellement grande (puisqu'on néglige les termes du 6ème ordre et au-dessus). L'étude de cas pratiques montre que l'on peut le plus souvent remplacer dans ces conditions $\sin \alpha'$ par α' sans que la validité des résultats en soit pratiquement affectuée.

l peut s'annuler pour la valeur particulière α'_0 donnée par la relation $a + a' \sin^2 \alpha'_0$; il est commode d'utiliser pour variable le rapport

$$v = \frac{\sin^2 \alpha'}{\sin^2 \alpha'_m}$$

ou α'_m est l'ouverture maximum de l'instrument.

Si l'on pose

$$v_0 = \frac{\sin^2 \alpha'_0}{\sin^2 \alpha_m},$$

l'aberration prend la forme

$$l = A\, v\, (v - v_0).$$

On aura d'autre part, lorsque l'ouverture n'est pas très grande,

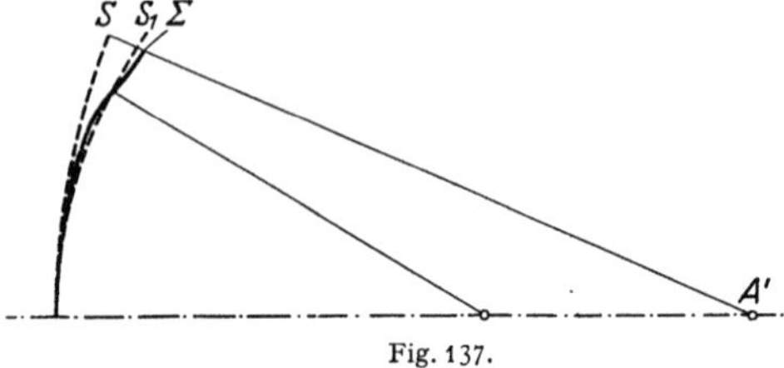

Fig. 137.

$$\Delta = - \frac{a\, \alpha'^4}{4} - \frac{a'\, \alpha'^6}{6} \qquad (75.2)$$

et, en utilisant encore les paramètres v et v_0:

$$\Delta = a'\, \alpha'^6_0 \left[\frac{v_0 v^2}{4} - \frac{v^3}{6} \right]. \qquad (75.3)$$

La surface d'onde aura la forme indiquée sur la Fig. 137; comme dans le cas de l'aberration du 3ème ordre, on peut trouver une sphère de référence, telle que S_1, par rapport à laquelle l'écart Δ_1 sera beaucoup plus faible que Δ.

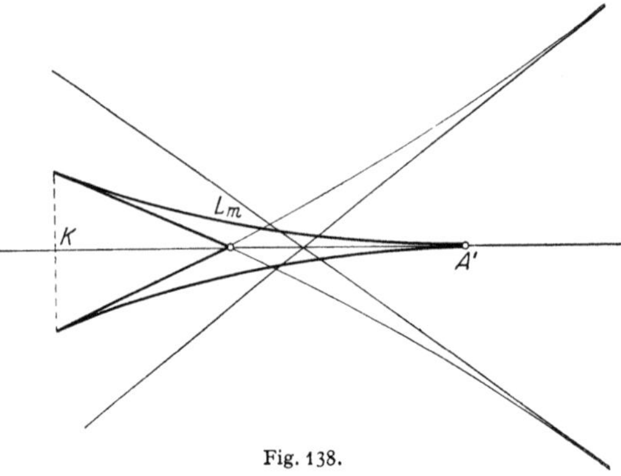

Fig. 138.

La nappe tangentielle de la caustique présentera un point de rebroussement en A' et une arête de rebroussement circulaire dans le plan de front passant par K (Fig. 138). Les phénomènes de diffraction viendront, en fait, modifier la forme de la tache image et la caustique géométrique restera généralement invisible, mais on observe cependant un anneau brillant lorsqu'on coupe la caustique par un plan de front tel que K.

Nous n'avons pas à étudier les procédés pratiques qui permettent de replier l'aberration sphérique; disons simplement qu'il faut réduire suffisamment le coefficient «a» pour que les termes du 3ème ordre et du 5ème ordre puissent se compenser au voisinage du bord de l'ouverture. On associera, pour cela, des lentilles convergentes sous-corrigées et des divergentes sur-corrigées jusqu'à ce que ce coefficient soit suffisamment petit; ce sera le cas des objectifs astronomiques constitués par deux lentilles simples minces, que l'on peut parfois coller (objectifs de Clairaut).

Remarquons qu'en faisant varier légèrement l'une des caractéristiques de l'instrument, on pourra donner au coefficient a des valeurs voisines de $a = 0$. La courbe représentant l pourra alors évoluer comme l'indique la Fig. 136 et l'on pourra faire varier la hauteur de correction α_0'.

Dans certains instruments très ouverts (objectifs de microscope, par exemple), on parvient à replier deux fois la courbe d'aberration sphérique mais il devient difficile de vouloir représenter ici l'aberration par les premiers termes d'un développement en série; l'aberration sera considérée comme une fonction empirique n'ayant plus d'expression algébrique simple.

f) La coma.

76. Cas des instruments stigmatiques. Etude de la coma pure. Nous devons rechercher les termes d'aberration qui sont proportionnels à la grandeur y de l'objet AB (ou à la grandeur de son image paraxiale y'). Pour cela nous supposerons y petit et nous négligerons les infiniment petits de l'ordre de y^2.

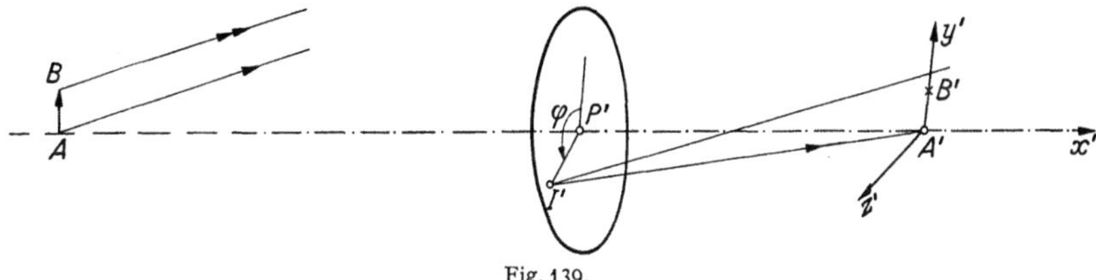

Fig. 139.

Supposons d'abord l'instrument parfaitement corrigé de l'aberration sphérique pour les points A et A' ce qui entraine que le chemin optique (AA') soit constant. D'après la propriété établie à la Sect. 68, le calcul des variations du chemin optique (BB') nous permettra de connaître la forme de la surface d'onde. Reprenons ce calcul en utilisant la même méthode qu'à la Sect. 27 pour l'établissement de la condition des sinus.

Soit $AI'A'$ le chemin suivi par un rayon lumineux (I' est par exemple le point où le rayon perce le plan de la pupille de sortie), évaluons le chemin optique $BI'B'$ le long d'un rayon qui ne passe pas effectivement par B' (Fig. 139) mais simplement en son voisingage; le point I' sera défini par les angles habituels α' et φ.

Nous allons récrire la relation (26.1) en utilisant les cosinus directeurs du rayon AI qui sont $\cos\alpha$, $\sin\alpha\cos\varphi$, $\sin\alpha\sin\varphi$ et les composantes de l'objet $AB\,(0, y, 0)$; on aura donc :

$$n\, y \cos\vartheta = n\, y \sin\alpha\,\cos\varphi$$

et une quantité analogue dans l'espace image; on aura donc :

$$L - L_0 = (BIB') - (AIA') = (n'\, y' \sin\alpha' - n\, y \sin\alpha)\cos\varphi$$

ce qui nous fournit immédiatement l'expression de $\varDelta$ d'après la relation (68.1) :

$$\boxed{\varDelta = (n\, y \sin\alpha - n'\, y' \sin\alpha')\,\frac{\cos\varphi}{n'}} \tag{76.1}$$

qui est l'expression de l'écart $\varDelta$ dans le cas de la coma.

On constate d'abord que si la condition des sinus $\dfrac{\sin\alpha}{\sin\alpha'} = $ constante est satisfaite, on a $\varDelta = 0$ et il n'y a pas de coma. La condition des sinus est donc non seulement nécessaire mais aussi suffisante pour qu'il n'y ait pas de coma.

Si maintenant la condition des sinus n'est pas satisfaite, nous allons exprimer Δ en fonction de l'écart à cette condition: le rapport $\dfrac{n \sin \alpha}{n' \sin \alpha'}$ tend, lorsque α et α' sont très petits, vers le grandissement $g_y = y'/y$; on a en effet, d'après la relation de Lagrange, $\dfrac{n\,\alpha}{n'\,\alpha'} = \dfrac{y'}{y}$ lorsque α et α' sont petits.

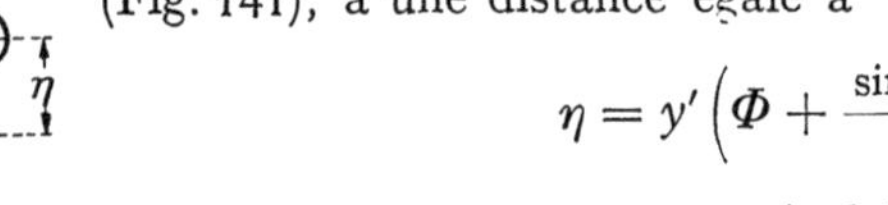

On pourra donc écrire:

$$G = \frac{n \sin \alpha}{n' \sin \alpha'} = g\left[I + \Phi(\sin \alpha')\right]. \qquad (76.2)$$

La fonction Φ représente la variation relative du rapport G, c'est-à-dire l'écart à la condition des sinus. Cette fonction est paire et s'annule pour $\alpha' = 0$.

En remplaçant, dans l'équation (76.1), $n \sin \alpha$ par sa valeur tirée de la relation précédente, on obtient immédiatement:

$$\Delta = y' \, \Phi(\sin \alpha') \sin \alpha' \cos \varphi. \qquad (76.3)$$

Il est facile d'interpreter géométriquement cette expression dans le cas où l'ouverture α' est petite: on peut obtenir la surface d'onde entachée de coma à partir de la surface sphérique de référence en faisant tourner autour d'un axe parallèle à la direction Z' chacune des zônes de la sphère, la rotation étant égale à $\dfrac{y' \, \Phi(\sin \alpha')}{P'A'}$ (Fig. 140). Dans ces conditions, l'écart Δ varie bien comme $\cos \varphi$. Les relations (69.1) permettent de déterminer les composantes des déplacements latéraux des rayons; on trouve

Fig. 140.

$$\boxed{\begin{aligned} d y' &= y' \left[\Phi(\sin \alpha') + \tfrac{1}{2} \sin \alpha' \, \Phi'(\sin \alpha') (1 + \cos 2\varphi)\right], \\ d z' &= \tfrac{1}{2} y' \sin \alpha' \, \Phi'(\sin \alpha') \sin 2\varphi. \end{aligned}} \qquad (76.4)$$

Ces relations montrent que, lorsque le rayon lumineux décrit une fois un cercle de la pupille ($\alpha' =$ constante, φ variable), son intersection par le plan image décrit deux fois un cercle dont le centre est situé sur $A'y'$ (Fig. 141), à une distance égale à

$$\eta = y' \left(\Phi + \frac{\sin \alpha' \, \Phi'}{2}\right) \qquad (76.5)$$

du point B' et son rayon est égal à

$$\varrho = y' \frac{\sin \alpha' \, \Phi'}{2}. \qquad (76.6)$$

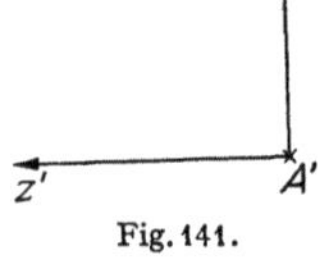

Fig. 141.

Ce cercle est appelé cercle comatique. Lorsque α' varie, η et ϱ varient, le cercle comatique se déplace dans le plan $y'A'z'$ en engendrant ainsi une tache de diffusion que l'on appelle tache comatique; bien entendu, cette tache admet $A'y'$ pour axe de symétrie.

Nous allons maintenant envisager deux cas particuliers importants:

77. Coma du troisième ordre. On ne considère que le premier terme du développement en série de Δ, en écrivant

$$\Phi(\sin \alpha') = b(\sin^2 \alpha') \qquad (77.1)$$

d'où l'on tire,

$$\Delta = b y' \sin^3 \alpha' \cos \varphi \qquad (77.2)$$

et les composante du déplacement latéral du rayon sont:

$$dy' = by' \sin^2\alpha' (2 + \cos 2\varphi) \left.\right\}$$
$$dz' = by' \sin^2\alpha' \sin 2\varphi \qquad \left.\right\} \qquad (77.3\,\text{a})$$

ou encore

$$\eta = 2by' \sin^2\alpha', \left.\right\}$$
$$\varrho = by' \sin^2\alpha'. \qquad \left.\right\} \qquad (77.3\,\text{b})$$

On remarque que la distance η est constamment égale au double du rayon ϱ du cercle comatique; celui-ci reste donc tangent à deux droites passant par B' et faisant chacune un angle de 30° avec $B'y'$; lorsque α' est petit, le cercle est très voisin de B' et, si α' augmente, le rayon du cercle varie comme $\sin^2\alpha'$ la tache comatique est constituée par un ensemble de cercles homothétiques tangents aux deux droites précédentes (Fig. 142). Nous avons vu à la Sect. 61 que les images entachées de coma présentent effectivement cet aspect.

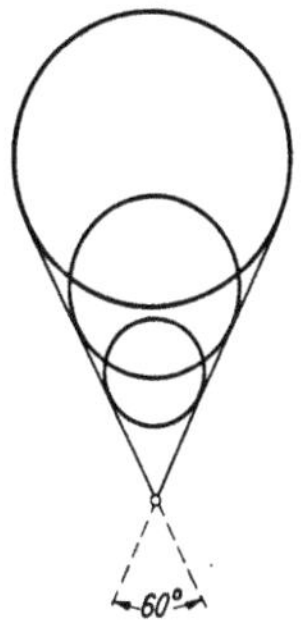

L'étendue totale de cette tache est égale à $3\,\varrho_m$ en appelant ϱ_m le rayon du cercle comatique relatif au bord de la pupille

$$\varrho_m = by' \sin^2\alpha'_m.$$

60°

Fig. 142.

ϱ_m/y' peut d'ailleurs se calculer facilement: en appelant $\delta G_m/G$ la variation relative maximum du rapport G, on a

$$\frac{\delta G_m}{G} = \Phi(\sin\alpha'_m) = b \sin^2\alpha'_m$$

d'où l'on tire:

$$\frac{\varrho_m}{y'} = \frac{\delta G_m}{G}. \qquad (77.4)$$

Cette relation très simple montre que le rapport du rayon du cercle comatique à la grandeur de l'image est égale à la variation relative du rapport

$$G = \frac{n \sin\alpha}{n' \sin\alpha'}.$$

Cette propriété sera très utile dans la pratique du calcul pour déterminer l'étendue de la tache comatique. On peut aussi remarquer que le rayon du cercle comatique est lié à l'écart Δ par la relation: $\Delta = \varrho \sin\alpha' \cos\varphi$.

L'étude analytique de la caustique relative à la coma du 3ème ordre est assez longue et la caustique obtenue est complexe; elle présente deux nappes symétriques par rapport au plan $B'y'z'$, chacune possédant dans le plan $B'x'y'$ une arête de rebroussement; ces nappes se coupent (sans se raccorder) suivant les deux droites à 60° passant par B'. Nous ne l'étudierons pas en détails.

78. Coma «corrigée». Pour atténuer la coma, on cherchera à réduire les variations du rapport G et l'on peut, en particulier (comme dans le cas de l'aberration sphérique), obtenir que la courbe représentant la fonction $\Phi(\sin\alpha')$ soit repliée pour une certaine valeur de l'ouverture (Fig. 143 a). On peut alors représenter convenablement $\Phi(\sin\alpha')$ par les deux premiers termes de son développement en série. On écrira:

$$\Phi(\sin\alpha') = b' \sin^2\alpha' (\sin^2\alpha' - v_0 \sin^2\alpha').$$

b' représente le coefficient de coma du 5ème ordre, v_0 caractérise l'ouverture pour laquelle on a réalisé la correction

$$\sin^2 \alpha_0' = v_0 \sin^2 \alpha_m'.$$

Les calculs sont semblables à ceux qu'on effectuait pour la coma du 3ème ordre; on obtient successivement:

$$\Delta = b' y' \sin^3 \alpha' (\sin^2 \alpha' - v_0 \sin^2 \alpha_m') \cos \varphi,$$

$$\eta = b' y' (3 \sin^4 \alpha' - 2 v_0 \sin^2 \alpha_m' \sin^2 \alpha'),$$

$$\varrho = b' y' (2 \sin^4 \alpha' - v_0 \sin^2 \alpha_m' \sin^2 \alpha').$$

Lorsque α' est petit, les termes du 3ème ordre sont prépondérants et l'on a $\eta/\varrho \approx 2$; le cercle comatique est sensiblement tangent aux droites à 60° issues

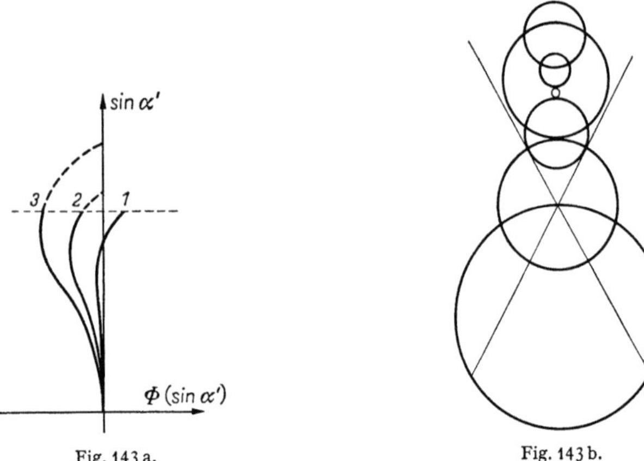

Fig. 143 a.   Fig. 143 b.

de B'; lorsque $\sin \alpha'$ croit, ϱ et η passent à tour de rôle par un maximum (en valeur absolue) puis s'annulent; ϱ s'annule d'abord, lorsque

$$\sin^2 \alpha' = \tfrac{1}{2} v_0 \sin^2 \alpha_m'$$

η s'annule ensuite lorsque

$$\sin^2 \alpha' = \tfrac{2}{3} v_0 \sin \alpha_m'.$$

Si l'on suppose $v_0 = 1$ la tache comatique aura l'aspect indiqué sur la Fig. 143 b; on voit que les cercles comatiques ne restent plus tangents au droites à 60°, et lorsque $\sin \alpha' = \sqrt{v_0 \sin \alpha_m'}$ le cercle obtenu a des dimensions importantes. On aura avantage à éviter la présence de cercles de grand rayon en limitant l'ouverture α_m' de l'instrument à une valeur inférieure à celle pour laquelle il y a correction de la condition des sinus ($\sin \alpha' = v_0 \sin \alpha_m'$), c'est-à-dire qu'on fera $v_0 > 1$ et la courbe représentant Φ (ou G) devra avoir l'allure de la courbe 2 de la Fig. 143 a. En fait, la coma obtenue correspondra à de faibles valeurs de Δ et l'on doit faire intervenir la théorie de la diffraction pour rechercher la valeur optimum de v_0. Nous verrons que cette théorie conduit, dans le cas des faibles aberrations, à la valeur optimum $v_0 = 1,2$ lorsque l'objet est un point isolé.

79. Coma en présence d'aberration sphérique; isoplanétisme. Les exemples qui précèdent concernent le cas où l'aberration sphérique a été parfaitement corrigée, c'est-à-dire où il y a stigmatisme lorsque le point objet A est sur l'axe; si l'on s'écarte alors de l'axe et si l'on ne satisfait pas à la condition des sinus, il apparait de la coma pure. Pratiquement il subsiste souvent des résidus d'aberration sphérique et l'on doit étudier maintenant la coma obtenue en présence d'aberration sphérique et les modifications que l'on doit apporter dans ce cas á la condition des sinus.

Montrons tout d'abord par un raisonnement géométrique simple que la position de la pupille influe sur la valeur de la coma. Supposons que pour une position P_0' de la pupille l'instrument ne présente pas de coma: ceci signifie que l'image obtenue en B' est de même nature que celle obtenue en A', c'est-à-dire que l'on a uniquement de l'aberration sphérique, qui peut être caractérisée par une caustique dont la pointe est en B' (Fig. 144).

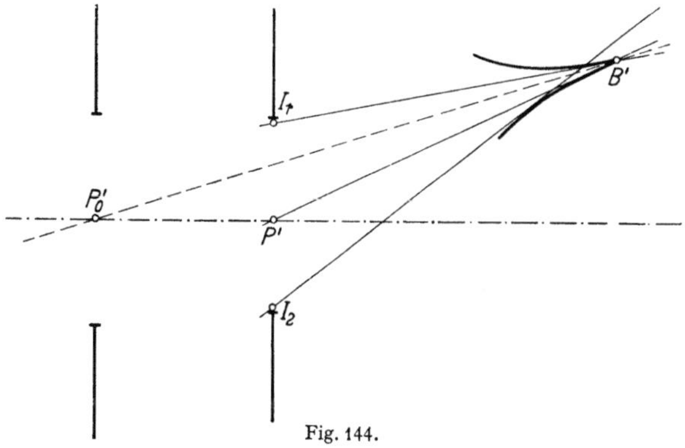

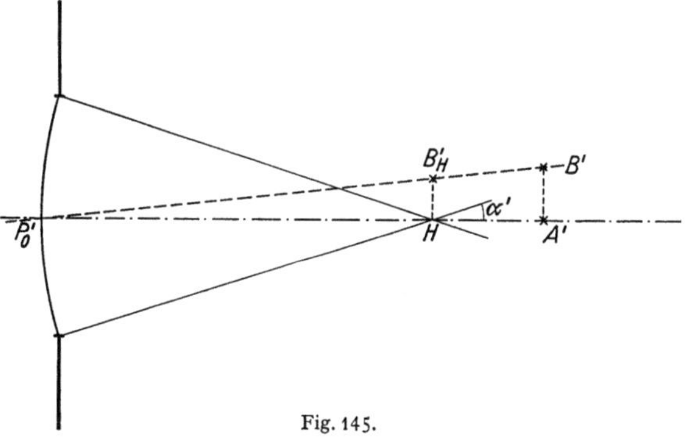

Fig. 144.

Supposons maintenant que la pupille soit déplacée de P_0' en P'; bien que la caustique reste la même le faisceau ne présente pas la même symétrie que la caustique; il en résulte que si l'on prend trois rayons dont l'un (passant par P') correspond à une ouverture de pupille nulle et les autres (passant par I_1 et I_2) correspondent à des ouvertures symétriques de la pupille ces derniers correspondant aux ouvertures $+\alpha'$ et $-\alpha'$ ne se rencontrent pas sur le rayon moyen; autrement dit, il existe un terme d'aberration d'ordre pair par rapport à α', qui n'est autre que la coma (la coma du 3ème ordre donnera un terme en α'^2). On voit qu'en présence d'aberration sphérique le coefficient de coma dépend de la position de la pupille; il est souvent possible de trouver une position P_0' de la pupille telle que la coma soit nulle (on verra que ceci est toujours possible dans le cadre des aberrations du 3ème ordre).

Fig. 145.

Lorsque l'on a ainsi supprimé la coma on dit qu'il y a *isoplanétisme* et la position correspondante P_0' de la pupille est appelée *pupille isoplanétique*.

Cherchons la condition générale d'isoplanétisme: considérons pour cela un instrument entaché d'aberration sphérique.

Soit $A'H$ l'aberration sphérique longitudinale (Fig. 145), α' l'angle de convergence du cône de rayons passant en H. Reprenons ici le calcul de la Sect. 76; si l'on définit la position d'un point B_H' à la distance $y_H' = HB_H'$ du point H par la condition:

$$n\, y \sin \alpha - n'\, y_H' \sin \alpha' = 0 \tag{79.1}$$

le chemin optique jusqu'en B_H' sera indépendant de φ pour l'ouverture considérée; c'est-à-dire que le point B_H' sera tel que sa distance à la surface d'onde issue de B sera constante si l'on suit le cercle de la surface d'onde correspondant à l'ouverture α'. Cette propriété permet de connaitre la surface d'onde. Pour qu'il y ait isoplanétisme il suffit que les divers points B'_H définis pour chaque ouverture α' soient alignés sur une droite; si P'_0 est le point ou cette droite coupe l'axe on obtiendra ainsi une caustique admettant $P_0 B'$ pour axe; il faudra d'autre part que le faisceau soit diaphragmé par la pupille tracée en trait plein sur la Fig. 145, qui sera la pupille isoplanétique.

On peut bien entendu traduire algébriquement la condition géométrique précédente: il faut que l'on ait, si D est la distance $P'_0 A'$:

$$\frac{y'_H}{D + l} = \text{constante}.$$

Si l'on tire maintenant y'_H de la relation (79.1) et que l'on appelle g le grandissement paraxial qui est la limite du rapport $\dfrac{n \sin \alpha}{n' \sin \alpha'}$ lorsque α et α' sont petits, on peut encore écrire la condition d'isoplanétisme sous la forme suivante (condition de Staeble et Lihotzky):

$$\frac{n \sin \alpha}{n' \sin \alpha'} = g\left(1 + \frac{l}{D}\right). \tag{79.2}$$

Ainsi lorsque la pupille est à distance finie ($1/D$ non nul) la condition de suppression de la coma n'est pas la condition d'Abbe. Il faut en fait que le rapport $\dfrac{\sin \alpha}{\sin \alpha'}$ varie suivant une loi imposée.

Dans le cadre des aberrations du 3ème ordre on pourra écrire:

$$\frac{n \sin \alpha}{n' \sin \alpha'} = g\left(1 + b\,\alpha'^2\right) \quad \text{et} \quad l = a\,\alpha'^2.$$

La position de la pupille isoplanétique sera donc défine par la condition:

$$b = \frac{a}{D}.$$

g) L'astigmatisme et la courbure de champ.

80. Généralités. Nous avons appelé, à la Sect. 71, termes d'astigmatisme et de courbure de champ les termes qui, dans les développements de $d y'$ et $d z'$, sont proportionnels à l'ouverture de la pupille. Ils doivent donc se manifester déjà lorsque la pupille a un faible diamètre. Dans ce cas, le faisceau se réduit à un pinceau dont toutes les propriétés sont connues, si l'on connait la position de ses focales; les termes d'aberration que nous recherchons sont donc déterminés par les positions des focales; par raison de symétrie l'une d'elles sera perpendiculaire au plan méridien de symétrie: on l'appelle focale tangentielle; l'autre sera contenue dans ce plan: c'est la focale sagittale (Fig. 146).

Donnons-nous les abscisses t', s' de ces focales par rapport au plan de l'image de Gauss et soient h' et φ les coordonnées polaires de l'intersection du rayon par le plan de la pupille de sortie. Si nous projetons la figure sur les deux plans rectangulaires $x' A' y'$, $x' A' z'$ (Fig. 146), nous aurons en projection sur le plan $x' A' y'$ (triangles semblables ayant T pour sommet commun),

$$\frac{d y'}{h' \cos \varphi} = \frac{B' T}{P' T} \approx \frac{t'}{x' - p'}$$

et, sur l'autre projection,

$$\frac{d z'}{h' \sin \varphi} \approx \frac{s'}{x' - p'},$$

mais, on a, d'autre part: $h' = -\alpha'(x' - p')$, d'où l'on déduit les expressions suivantes:

$$\left.\begin{aligned} dy' &= -\alpha' t' \cos \varphi, \\ dz' &= -\alpha' s' \sin \varphi. \end{aligned}\right\} \qquad (80.1)$$

Si le plan $A'y'z'$ est le plan de l'image paraxiale, l'astigmatisme est nul lorsque l'objet AB est petit (domaine de GAUSS) et les deux focales viennent alors se confondre en B'; chacune des focales décrit une surface de révolution autour de l'axe, tangente en A' au plan $A'y'z'$. Ce sont la surface focale sagittale et la surface focale tangentielle (Fig. 147).

Les distances t' et s' seront des fonctions paires de y' dont les développements en série débutent par des termes en y'^2, on aura donc:

$$dy' = -\alpha' \cos \varphi \, (a_1 y'^2 + a_4 y'^4 + \cdots),$$
$$dz' = -\alpha' \sin \varphi \, (b_2 y'^2 + b_4 y'^4 + \cdots).$$

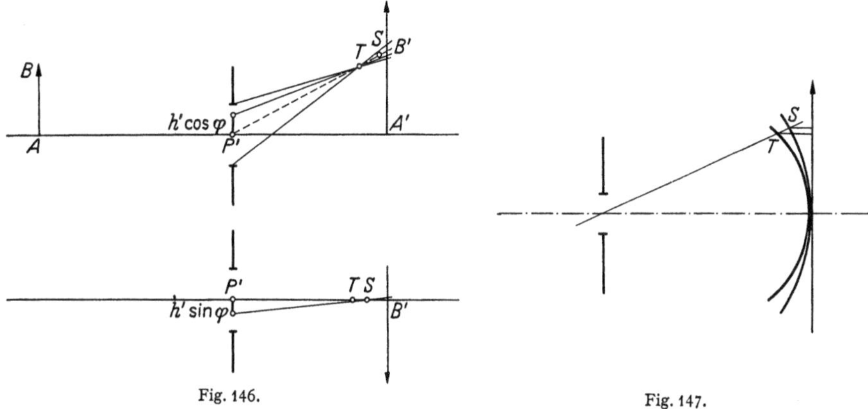

Fig. 146.   Fig. 147.

En limitant ces expressions aux termes du 3ème ordre, on ne conservera que les termes en y'^2 que l'on peut encore exprimer en fonction des courbures C_T et C_S des surfaces focales en leur pôle A'; on écrira:

$$t' = C_T \frac{y'^2}{2}; \qquad s' = C_S \frac{y'^2}{2}$$

d'où, dans le domaine du 3ème ordre:

$$dy' = -\alpha' \cos \varphi \, \frac{c_T \, y'^2}{2},$$
$$dz' = -\alpha' \sin \varphi \, \frac{c_S \, y'^2}{2}.$$

On voit que ces aberrations sont liées (en ce qui concerne les termes du 3ème ordre) seulement aux courbures C_S et C_T des surfaces focales. Le cercle de moindre diffusion décrit aussi une surface de révolution dont la courbure est égale à la moyenne des courbures précédentes; cette courbure

$$C = \frac{c_T + c_S}{2}$$

est la courbure de champ que nous avons déjà rencontré à la Sect. 62; c'est la courbure que présente le lieu de la meilleure image de l'objet.

81. Aberrations du troisième ordre. Courbure des surfaces focales et chemin optique aberrant. Le calcul des aberrations du 3ème ordre montrera que les courbures C_S et C_T des surfaces focales s'expriment comme suit

$$C_S = \mathscr{D} - \mathscr{A},$$
$$C_T = \mathscr{D} - 3\mathscr{A},$$

où $\mathscr{D}$ est une constante appelée courbure de Petzval dont la valeur ne dépend ni de la position des plans conjugués, ni de la position de la pupille. Si l'on considère l'instrument comme un ensemble de lentille épaisses, $\mathscr{D}$ peut se mettre sous la forme:

$$\mathscr{D} = - \Sigma \frac{C_0}{n}$$

où C_0 serait la puissance de la lentille que l'on aurait réduite à une épaisseur nulle en maintenant constantes les courbures de ses faces:

$$C_0 = (n-1)\left(\frac{1}{r} - \frac{1}{r'}\right).$$

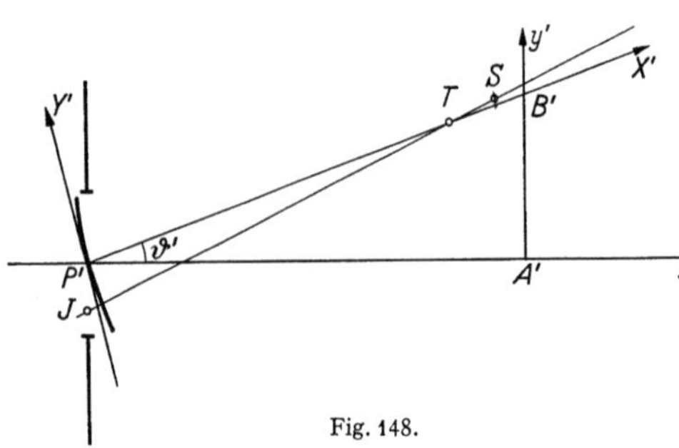

Fig. 148.

Dans le cas d'un miroir, la contribution à la somme de Petzval sera $+ C$, elle est donc de signe contraire à celle des lentilles de même convergence.

D'autre part, $\mathscr{A}$ sera un coefficient caractérisant l'astigmatisme de l'instrument, qui dépend à la fois de la position de l'objet et de celle de la pupille.

Représentons la pupille de sortie et un fragment de la surface d'onde qui admet pour rayons de courbure principaux $P'S$ et $P'T$ (Fig. 148). Prenons des axes $P'X'Y'Z'$, $P'X'$ étant confondu avec le rayon moyen et $P'Y'$ étant contenu dans le plan de la figure.

L'équation de Σ sera (si on la limite à ses termes du 2ème ordre):

$$X' = \frac{Y'^2}{2\,\overline{P'T}} + \frac{Z'^2}{2\,\overline{P'S}},$$

tandis que l'équation de la sphère centrée en B' serait:

$$X'_S = \frac{Y'^2 + Z'^2}{2\,\overline{P'B'}}.$$

On obtient donc l'écart Δ par différence et, si ϑ' est l'angle de champ, on a:

$$\Delta = X' - X'_s = -\frac{1}{2}\left[\frac{Y'^2\,\overline{B'T}}{\overline{P'B'^2}} + \frac{Z'^2\,\overline{B'S}}{\overline{P'B'^2}}\right] = -\frac{1}{2}\frac{\cos\vartheta'}{\overline{P'A'^2}}\left[t'\,Y'^2 + s'\,Z'^2\right].$$

Si l'on veut maintenant caractériser la position du rayon par les coordonnées h' et φ de son intersection J par le plan de la pupille, on a à des infiniment petits du 2ème ordre près,

$$Y' = h'\cos\varphi\cos\vartheta',$$
$$Z' = h'\sin\varphi$$

et, si l'on porte dans l'expression de Δ, on obtient après réduction:

$$\Delta = -\frac{\alpha'^2}{2}\left[t'\cos^2\varphi\cos^3\vartheta' + s'\sin^2\varphi\cos\vartheta'\right].$$

En se limitant aux aberrations du 3ème ordre, on peut écrire:

$$t' = C_T \frac{y'^2}{2}, \qquad s' = C_S \frac{y'^2}{2}$$

et, en remarquant que $\cos \vartheta'$ n'introduit dans $\varDelta$ que des termes en $\alpha'^2 y'^4$ au moins, on écrira:

$$\varDelta = - \frac{\alpha'^2 y'^2}{4} (C_T \cos^2 \varphi + C_S \sin^2 \varphi).$$

Utilisons comme paramètres le coefficient de courbure de champ $C_m = \frac{1}{2}(C_T + C_S)$ et le coefficient d'astigmatisme $\mathcal{A} = \frac{1}{2}(C_S - C_T)$ on a:

$$C_S = C_m + \mathcal{A}, \qquad C_T = C_m - \mathcal{A}.$$

On aboutit finalement à l'expression:

$$\varDelta = - \frac{\alpha'^2 y'^2}{4} (C_m - \mathcal{A} \cos 2\varphi) \qquad (81.1)$$

où les termes correspondant à la courbure de champ et à l'astigmatisme sont séparés.

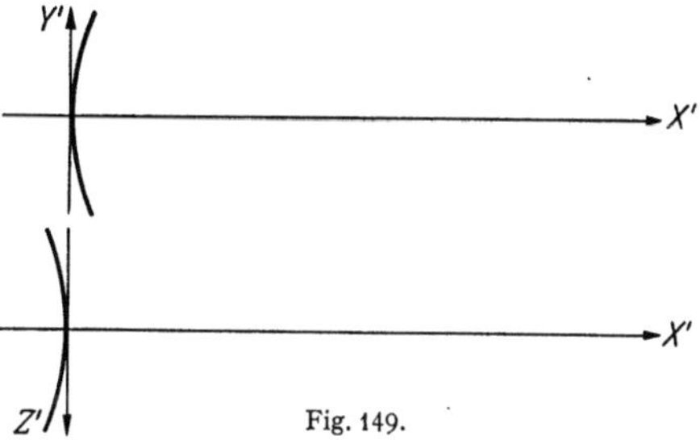

Fig. 149.

On voit que la courbure de champ introduit un terme proportionnel à α'^2, que l'on peut assimiler à un défaut de mise au point. Remarquons cependant que ce défaut de mise au point est d'autre partproportionnel à y'^2.

L'astigmatisme introduit au contraire un terme en $\cos 2\varphi$: imaginons, par exemple, que l'image soit très éloignée et que la sphère S soit pratiquement un plan. Les sections de la surface d'onde $\varSigma$ par les plans $X'P'Y'$ et $X'P'Z'$ présentent des courbures égales en valeur absolue mais de signes contraires, comme l'indique la Fig. 149. La surface $\varSigma$ est une surface gauche et, lorsqu'on décrit un cercle de la pupille, $\varDelta$ change de signe quatre fois: on s'explique ainsi que le terme d'astigmatisme varie comme $\cos 2\varphi$ et l'on justifie maintenant la dénomination indiquée à la Sect. 72.

82. Calcul de la position des focales. Nous allons pour terminer établir les formules très générales qui permettent de connaître les positions des focales d'un pinceau réfracté en fonction de celles relatives au pinceau incident. Considérons un pinceau incident caractérisé par les abscisses T, S des focales, tombant sous l'incidence i sur un dioptre. Pour déterminer la position et l'orientation des focales dans l'espace image nous chercherons la forme d'une surface d'onde réfractée, dont les lignes de courbure détermineront l'orientation des focales, les valeurs des courbures permettant de connaître T', S'. Nous pouvons pour cela rechercher les équations de la surface d'onde incidente, de la surface d'onde réfractée et du dioptre dans le but d'écrire que le chemin optique est constant entre les deux surfaces d'ondes (en fait ce chemin optique est identiquement nul); il suffit d'ailleurs de conserver dans ces calculs les termes de courbure (c'est-à-dire les termes du second degré) ce qui permettra certaines approximations.

Prenons des axes $Oxyz$ (Fig. 150) tels que Oyz soit le plan tangent au dioptre et Oxy le plan d'incidence. Soit AOA' le rayon moyen, i et i' les angles d'incidence et de réfraction, r_1 et r_2 les rayons de courbure principaux du dioptre au point O; l'équation du dioptre sera, si on la limite à ses termes du 2ème ordre:

$$x = \frac{1}{2} \left[\frac{(y \cos \vartheta + z \sin \vartheta)^2}{r_1} + \frac{(z \cos \vartheta - y \sin \vartheta)^2}{r_2} \right], \qquad (82.1)$$

où ϑ est l'angle que fait l'une des lignes de courbure avec le plan d'incidence.

Soit OX le rayon moyen incident (situé dans le plan xOy), OY et Oz les autres axes du trièdre. Pour plus de généralité nous supposerons que les focales T et S ne sont pas respectivement parallèles à Oz et OY mais qu'on les a fait tourner d'un angle α; l'équation de la surface d'onde incidente sera ainsi:

$$X = \frac{1}{2}\left[\frac{(Y\cos\alpha + Z\sin\alpha)^2}{T} + \frac{(Z\cos\alpha - Y\sin\alpha)^2}{S}\right] \tag{82.2}$$

et de même pour la surface d'onde réfractée dans des axes similaires:

$$X' = \frac{1}{2}\left[\frac{(Y'\cos\alpha' + Z'\sin\alpha')^2}{T'} + \frac{(Z'\cos\alpha' - Y'\sin\alpha')^2}{S'}\right]. \tag{82.3}$$

Il reste à évaluer la distance MI entre le point M de la surface d'onde et le point I du dioptre situés sur le même rayon. On peut écrire avec une approximation

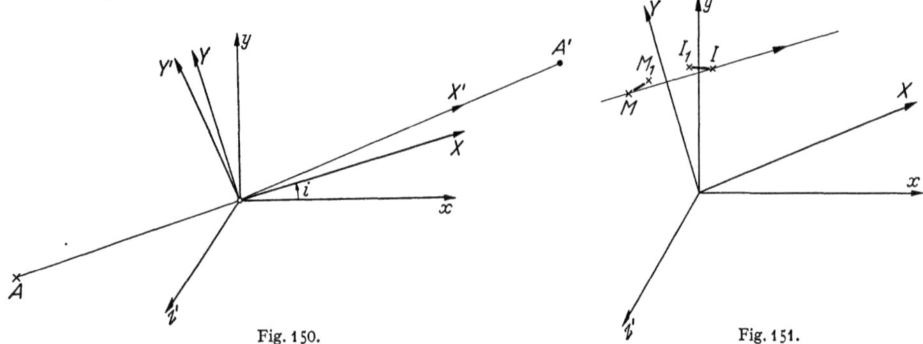

Fig. 150.   Fig. 151.

suffisante (Fig. 151) en remarquant que $MM_1 = X$, $II_1 = x$, MM_1 et II_1 étant respectivement perpendiculaires à YOz et yOz

et de même
$$\left.\begin{array}{l} MI = y\sin i - X + x\cos i, \\ M'I = y\sin i' - X' + x\cos i'. \end{array}\right\} \tag{82.4}$$

Il est commode de tout exprimer à l'aide des coordonnées y, z de I en écrivant (avec une approximation également suffisante):

$$Y = y\cos i, \quad Z = z = Z', \quad Y' = y\cos i'.$$

Il suffit alors d'écrire que $n\overline{MI} + n'\overline{IM'} = 0$ (le chemin optique entre les deux surfaces d'onde est nul puisqu'elles se rencontrent en O). On obtient immédiatement pour les termes du second ordre:

$$n(x\cos i - X) - n'(x\cos i' - X') = 0.$$

Il reste à ordonner cette expression et à annuler respectivement les termes en y^2, en yz et en z^2 pour obtenir les trois conditions:

$$\left.\begin{aligned}
&n\left[\left(\frac{\cos^2\vartheta}{r_1} + \frac{\sin^2\vartheta}{r_2}\right)\cos i - \frac{\cos^2 i\cos^2\alpha}{T} - \frac{\cos^2 i\sin^2\alpha}{S}\right] \\
&\quad = n\left[\left(\frac{\cos^2\vartheta}{r_1} + \frac{\sin^2\vartheta}{r_2}\right)\cos i'' - \frac{\cos^2 i'\cos^2\alpha'}{T'} - \frac{\cos^2 i'\sin^2\alpha'}{S'}\right], \\
&n\cos i\left[\sin 2\vartheta\left(\frac{1}{r_1} - \frac{1}{r_2}\right) - \sin 2\alpha\left(\frac{1}{T} - \frac{1}{S}\right)\right] \\
&\quad = n'\cos i'\left[\sin 2\vartheta\left(\frac{1}{r_1} - \frac{1}{r_2}\right) - \sin 2\alpha'\left(\frac{1}{T'} - \frac{1}{S'}\right)\right], \\
&n\left[\left(\frac{\sin^2\vartheta}{r_1} + \frac{\cos^2\vartheta}{r_2}\right)\cos i - \left(\frac{\sin^2\alpha}{T} + \frac{\cos^2\alpha}{S}\right)\right] \\
&\quad = n'\left[\left(\frac{\sin^2\vartheta}{r_1} + \frac{\cos^2\vartheta}{r_2}\right)\cos i - \left(\frac{\sin^2\alpha'}{T'} + \frac{\cos^2\alpha'}{S'}\right)\right].
\end{aligned}\right\} \tag{82.5}$$

Dans ces expressions n, n', i, i'. ϑ, α, T, S, r_1, r_2 sont connus; on peut en déduire α', T', et S'. On peut en particulier tirer successivement des trois équations les quantités suivantes:

$$\left.\begin{aligned}
\frac{\cos^2\alpha'}{T'} + \frac{\sin^2\alpha'}{S'} &= A, \\[4pt]
\sin 2\alpha'\left[\frac{1}{T'} - \frac{1}{S'}\right] &= B, \\[4pt]
\frac{\sin^2\alpha'}{T'} + \frac{\cos^2\alpha'}{S'} &= C.
\end{aligned}\right\} \qquad (82.6)$$

On aura donc

$$\cos^2\alpha'\left[\frac{1}{T'} - \frac{1}{S'}\right] = A - C$$

qui donne, en la combinant avec la seconde équation

$$\tan 2\alpha' = \frac{B}{A - C},$$

relation qui fixe l'orientation des focales sur le rayon réfracté; leurs abscisses T' et S' seront ensuite obtenues aisément. Le cas où le plan d'incidence est un plan de symétrie conduit à des résultats plus simples: si $\alpha = \vartheta = 0$ on aura d'après la relation (82.5) $\alpha' = 0$, et l'on a alors pour fixer les valeurs de T' et S':

$$\boxed{\begin{aligned}
\frac{n'\cos^2 i'}{T'} - \frac{n\cos^2 i}{T} &= \frac{n'\cos i' - n\cos i}{r_1}, \\[6pt]
\frac{n}{S'} - \frac{n}{S} &= \frac{n'\cos i' - n\cos i}{r_2}.
\end{aligned}} \qquad (82.7)$$

Ces relations sont souvent utilisées par les calculateurs opticiens pour déterminer de proche en proche la position des focales. Il serait facile de montrer que dans le cas de la réflexion on obtiendrait des formules analogues: dans le cas où il y a un plan de symétrie, on aurait par exemple:

$$\frac{1}{T'} + \frac{1}{T} = \frac{2}{r_1\cos i}, \qquad \frac{1}{S'} + \frac{1}{S} = \frac{2\cos i}{r_2}.$$

h) La distorsion.

83. Généralités. Conditions d'orthoscopie. Nous avons appelé termes de distorsion des aberrations latérales dans la classification l'ensemble des termes qui ne dépendent pas de h: ils sont proportionnels à y^3, y'^5 etc. ... Cette aberration doit donc exister déjà lorsqu'on fait $h = 0$, c'est-à-dire lorsqu'on réduit le faisceau à son rayon moyen; elle n'affectera pas la qualité de l'image, mais seulement sa position.

Soit Π le centre du diaphragme qui joue le rôle de pupille; le rayon moyen du pinceau passe en Π et coupe l'axe dans l'espace objet en P, dans l'espace image en P' (Fig. 152). Les angles de champs ϑ, ϑ' pouvant être importants, P et P' ne seront peut-être pas les images paraxiales de Π, mais pourront être entachés d'aberration sphérique pupillaire qui aura pour effet de faire varier les positions de P et P' en fonction de ϑ (ou ϑ').

Le rayon moyen coupe le plan de front passant par A' en un point situé sur $A'y'$, mais ce point ne coïncide pas nécessairement avec l'image paraxiale B' définie par la relation $\overline{A'B'} = g\overline{A\,B}$ où g représente encore le grandissement paraxial.

Pour supprimer la distorsion, c'est-à-dire pour que le grandissement $\dfrac{A'D'}{AB}$ obtenu soit constant, on est amené à écrire :

$$\frac{\overline{P'A'}\tan\vartheta'}{\overline{PA}\tan\vartheta} = \text{constante}. \tag{83.1}$$

Si cette condition est réalisée, l'image est rigoureusement semblable à l'objet et l'on dit qu'il y a orthoscopie.

On utilise quelquefois une forme simplifiée de la relation (83.1) proposée par l'astronome Airy :

$$\frac{\tan\vartheta'}{\tan\vartheta} = \text{constante},$$

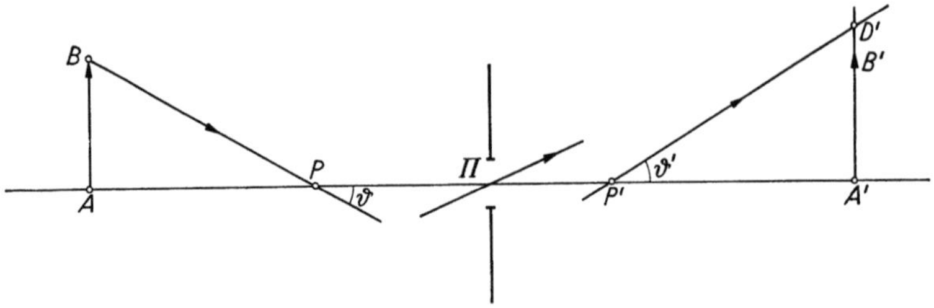

Fig. 152.

relation qui n'entraine l'orthoscopie qu'à la condition que le rapport $\overline{P'A'}/\overline{PA}$ soit aussi constant ; c'est le cas :

1. Si l'aberration sphérique pupillaire est négligeable ; P et P' sont alors des points fixes et le rapport $\overline{P'A'}/\overline{PA}$ est constant lorsqu'on se déplace dans le champ ;

2. si l'objet et l'image sont suffisamment éloignés des pupilles pour que $\overline{P'A'}$ et $\overline{PA}$ soient très grands ; les variations relatives de $\overline{P'A'}$ et $\overline{PA}$, dues aux déplacements des points P et P', sont alors négligeables ; c'est le cas des instruments afocaux pour lesquels la condition d'Airy suffit à exprimer l'orthoscopie.

84. Etude algébrique. Nous allons définir une fonction $f(y')$ caractérisant la distorsion, en écrivant :

$$A'D' = y'\left[1 + f(y')\right].$$

$f(y')$ sera nécessairement une fonction paire qui s'annule lorsque $y' = 0$. Le grandissement effectif sera :

$$\frac{A'D'}{AB} = \frac{y'\left[1 + f(y')\right]}{y} = g\left[1 + f(y')\right].$$

Si, d'autre part, on envisage un petit objet rectangulaire de côtés dy et dz et admettant l'un de ses sommets en B (Fig. 153), l'image qu'en donnera l'instrument sera sensiblement rectangulaire, mais les côtés du rectangle seront :

$$d(A'D') = d\left[g\,y\,(1 + f(y'))\right] = g\left[1 + f(y') + y f'(y')\right]dy,$$

$$A'D'\,d\varphi = A'D'\,\frac{dz}{y} = g\left[1 + f(y')\right]dz.$$

On voit sur ces expressions que les deux grandissements obtenus ne sont pas égaux ; si l'objet est un carré $(dy = dz)$ son image sera un rectangle ; l'instrument a produit une déformation que nous allons préciser dans le cas de la distorsion du 3ème ordre. On peut écrire dans ce cas $f(y') = d \cdot y'^2$ où d est un coefficient constant.

On a alors, pour expression du déplacement latéral du rayon,

$$d\,y' = B'D' = d \cdot y'^{3}.$$

Supposons $d>0$ et prenons pour objet une droite ne passant pas par A sur laquelle nous marquons un certain nombre de points $B_1, B_2, B_3, \ldots$; déterminons leurs images paraxiales $B_1', B_2', B_3', \ldots$; ces images sont évidemment alignées puisque l'image paraxiale est semblable à l'objet; en fait, la distorsion déplace les images effectives de B_1' en D_1', de B_2' en D_2' etc. (Fig. 154). La distance $d\,y'$ allant rapidement en croissant avec la distance $y' = A'B'$, l'image obtenue n'est pas une droite mais une courbe convexe vers A', c'est ainsi que, si l'angle $B_1'A'B_3'$ est égal à $\pi/4$, on a $\overline{A'B_3'} = \overline{A'B_1'}\,\sqrt{2}$ et $\overline{B_3'D_3'} = \overline{B'D_1'}\,2\,\sqrt{2}$. L'image d'un quadrillage aura l'aspect de la Fig. 118; on obtient une distorsion en coussinet; si au contraire, on a $\alpha >0$ la distorsion est en barillet.

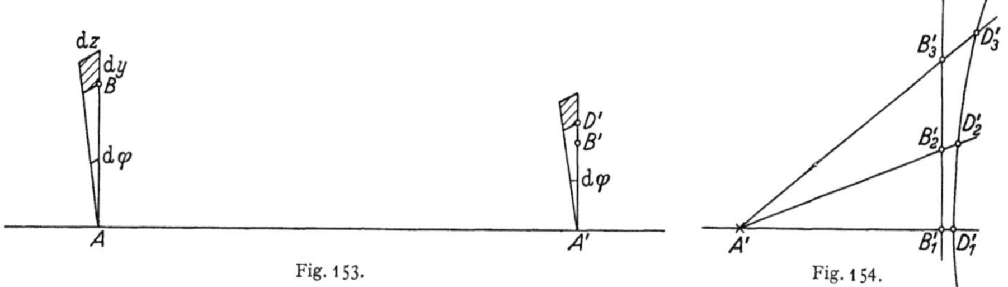

Fig. 153.   Fig. 154.

Lorsqu'on corrige la distorsion d'un instrument, on peut arriver à réduire le terme du 3ème ordre à une faible valeur, de façon qu'il soit compensé, pour une certaine grandeur de l'image, par les termes d'ordre supérieur. La courbe représentant la fonction est alors repliée et la valeur de l'aberration est réduite. Il n'est cependant pas très intéressant de rechercher quelle serait la forme de l'image d'une droite, car l'aberration n'est généralement plus apparente à l'oeil; elle peut néanmoins être gênante dans certains instruments de haute précision tels que les appareils de photogrammétrie ou les projecteurs de profils et seules des mesures de précision permettront de la déceler et de l'étudier.

i) Calcul des aberrations chromatiques et des aberrations géométriques du troisième ordre.

85. La méthode des chemins optiques. Des techniques variées peuvent être utilisées pour exprimer les aberrations chromatiques ou géométriques en fonction des courbures, épaisseurs, proximités de l'objet et de l'image caractérisant le système optique. On peut étudier la réfraction des rayons sur les diverses surfaces et effectuer des développements en série des paramètres d'orientation et de position des rayons lumineux, ce qui permet d'obtenir les expressions des aberrations des divers ordres; cette technique est assez facile à mettre en oeuvre pour le calcul des aberrations chromatiques ou des aberrations géométriques du 3ème ordre; elle présente néanmoins quelques difficultés et, en particulier, l'élimination des paramètres intermédiaires, relatifs à chacune des surfaces de l'instrument, peut paraître artificielle. Les résultats obtenus dans l'étude des faibles aberrations (Sect. 68) nous fournissent au contraire une méthode rationnelle de calcul que nous allons maintenant étudier.

Supposons que nous connaissions la marche approximative R' d'un rayon lumineux R partant d'un point-objet A et passant au voisinage de l'image théorique A', ainsi que la position approximative Σ' d'une surface d'onde Σ dans l'espace-image (Fig. 155). Appelons ε un infiniment petit pouvant caractériser l'incertitude sur la position du rayon (par exemple, la distance entre R et R'), et ε' un infiniment petit pouvant caractériser l'incertitude sur la position de Σ'. La méthode consiste à préciser par approximations successives les positions du rayon et de la surface d'onde en calculant le chemin optique entre la source et la surface d'onde le long de R'. D'après le principe de Fermat, ce calcul sera exact à des termes en ε^2 près et il nous permettra de corriger ε' si ce dernier est d'un ordre d'infinité inférieur à celui de ε^2. Ainsi, le calcul du chemin optique le long des rayons paraxiaux permet de connaître les aberrations dites du 3ème ordre: en effet, les rayons paraxiaux diffèrent des rayons effectifs par des distances qui sont au moins du 3ème ordre par rapport à l'ensemble des variables d'ouverture et de champ h et y et le calcul du chemin optique sera exact au 6ème ordre près, c'est-à-dire qu'il permet de connaître les chemins

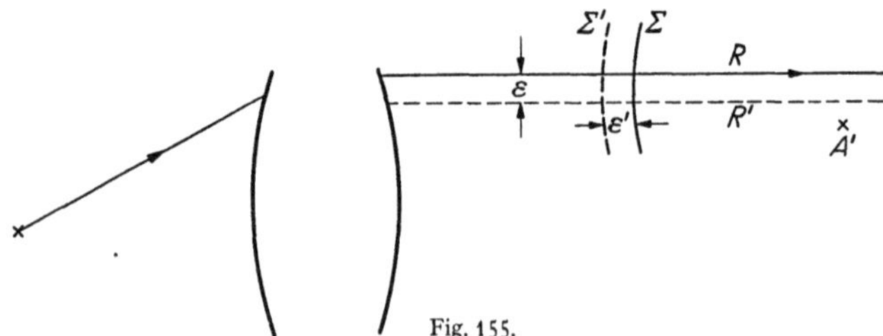

Fig. 155.

optiques aberrants du 4ème ordre, qui correspondent aux aberrations dites du 3ème ordre, comme on l'a vu successivement pour les diverses aberration (voir Sect. 73 à 84).

Il est ensuite possible, en tenant compte des résultats de ce premier calcul, de corriger la marche des rayons à travers le système en ne commettant plus sur leur position qu'une erreur du 5ème ordre au moins; on peut alors calculer les chemins optiques aberrants des 6ème et 8ème ordres (aberrations des 5ème et 7ème ordres), mais ce calcul est très pénible dans le cas général et nous nous bornerons ici au calcul théorique des aberrations du 3ème ordre; la même méthode nous permettra d'ailleurs de connaître les aberrations chromatiques du système, en prenant cette fois pour infiniment petit du 1er ordre les dispersions des différents verres du système optique.

86. Calcul des aberrations chromatiques. Considérons un dioptre rapporté à des axes $OXYZ$; soient $B(x, y, z)$ le point-objet, $B'(x', y', z')$ l'image, r le rayon de courbure, n et n' les indices. L'équation du dioptre, limitée aux termes du 4ème ordre en Y et Z, s'écrit

$$X = \frac{Y^2 + Z^2}{2r} + \frac{\lambda + 1}{8r^3}(Y^2 + Z^2)^2 \qquad (86.1)$$

où λ est un paramètre permettant de tenir compte d'une déformation éventuelle de la surface $\lambda = 0$ pour la sphère et $\lambda = -1$ pour le paraboloïde. On doit calculer le chemin optique $(BB') = n\,\overline{BI} + n'\,\overline{IB'}$, où I est un point du dioptre (Fig. 156); on peut exprimer L sous la forme:

$$L = n\left[(X - x)^2 + (Y - y)^2 + (Z - z)^2\right]^{\frac{1}{2}} + \left.\begin{array}{l} \\ + n'\left[(X - x')^2 + (Y - y')^2 + (Z - z')^2\right]^{\frac{1}{2}}, \end{array}\right\} \qquad (86.2)$$

les racines à prendre devant être respectivement des signes de $X - x$ et $x' - X$.

Développons cette expression jusqu'au 4ème ordre par rapport à l'ensemble des variables y, z, Y, Z, en remarquant que la quantité figurant dans les premiers crochets s'écrit en tenant compte de l'équation du dioptre

$$x^2\left[1 - \frac{Y^2 + Z^2}{r\,x} + \frac{(Y-y)^2 + (Z-z)^2}{x^2} + \frac{(Y^2+Z^2)^2}{4r^3 x^2}\,(r + \overline{\lambda + 1}\,x)\right].$$

En appliquant d'abord le développement de

$$(1 + \varepsilon)^{\frac{1}{2}} = 1 + \frac{\varepsilon}{2} - \frac{\varepsilon^2}{8}$$

limité provisoirement au terme $\varepsilon/2$, on trouve aisément

$$L = -\,n\,x\left[1 - \frac{Y^2+Z^2}{2x}\left(\frac{1}{r} - \frac{1}{x}\right) - \frac{y\,Y + z\,Z}{x^2} + \frac{y^2 + z^2}{2x^2}\right]$$
$$+ n'\,x'\left[1 - \frac{Y^2+Z^2}{2x'}\left(\frac{1}{r} - \frac{1}{x'}\right) - \frac{y'\,Y + z'\,Z}{x'^2} + \frac{y'^2 + z'^2}{2x'^2}\right].$$

B' étant l'image paraxiale de B, L ne doit pas contenir de termes des 1er ou 2ème degrés en Y et Z ce qui amène aux relations paraxiales:

$$Q_x = n\left(\frac{1}{r} - \frac{1}{x}\right) = n'\left(\frac{1}{r} - \frac{1}{x'}\right);$$
$$n\,\frac{y}{x} = n'\,\frac{y'}{x'}, \qquad n\,\frac{z}{x} = n'\,\frac{z'}{x'}.$$

La quantité Q_x est l'invariant paraxial de réfraction.

Sans restreindre la généralité du calcul, on peut maintenant supposer que B et B' sont dans le plan méridien de symétrie: $z = z' = 0$.

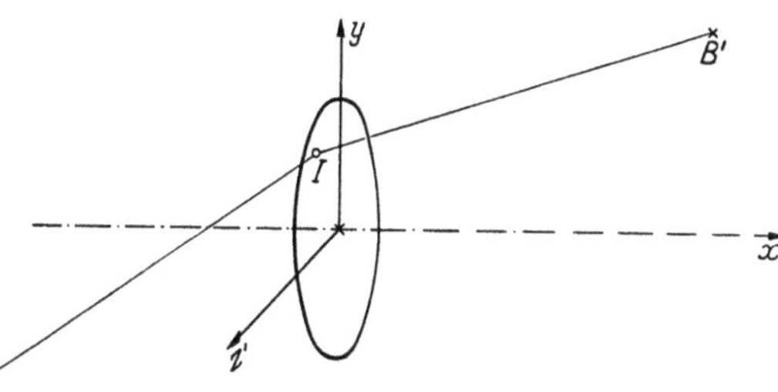

Fig. 156.

On obtient les conditions d'achromatisme complet en écrivant que $\delta L = 0$ pour l'ensemble de l'instrument; les chemins optiques aberrants étant additifs pour les différentes surfaces, on aura à écrire

$$\left.\begin{aligned}n'\,\varDelta = \sum \frac{Y^2 + Z^2}{2}\left[\left(\frac{1}{r} - \frac{1}{x'}\right)\delta n' - \left(\frac{1}{r} - \frac{1}{x}\right)\delta n\right] + \\ + \sum Y\left[\frac{y'\,\delta n'}{x'} - \frac{y\,\delta n}{x}\right] = 0.\end{aligned}\right\} \tag{86.3}$$

Il est utile en fait d'exprimer le chemin optique aberrant $n'\varDelta$ en fonction des variables pupillaires d'une part, et de la grandeur de l'objet ou de l'image d'autre part. Il nous faut donc exprimer Y et Z en fonction des coordonnées de l'intersection du rayon avec le plan de la pupille d'entrée par exemple que nous rapporterons à des coordonnées polaires h et φ [h est égal à $\alpha(p - x)$ si p et x sont les abscisses de P et A]. Si h_j est la hauteur d'incidence du rayon issu de A (marqué d'une flèche qur la Fig. 157) sur le dioptre d'ordre j, on posera

$$\frac{h_j}{h} = H_j.$$

D'autre part le rayon moyen BP du faisceau issu de B perce le plan de P suivant l'angle de champ ϑ et le dioptre d'ordre j en un point M_j dont l'ordonnée est proportionnelle à ϑ; cette ordonnée sera prise égale à

$$-n\,\vartheta\,k_j = -\frac{n\,y}{x-p}\,k_j$$

ce qui définit un coefficient k_j caractérisant l'ordonnée de M_j. On aura dans ces conditions:

$$\left.\begin{aligned}Y_j &= h_j\cos\varphi - \frac{n\,y}{x-p}\,k_j,\\ Z_j &= h_j\sin\varphi,\end{aligned}\right\} \tag{86.4}$$

d'où l'on tire

$$Y_j^2 + Z_j^2 = h_j^2 - 2h_j\frac{n\,k_j}{x-p}\,y\cos\varphi + \frac{n^2\,k_j^2}{(x-p)^2}\,y^2.$$

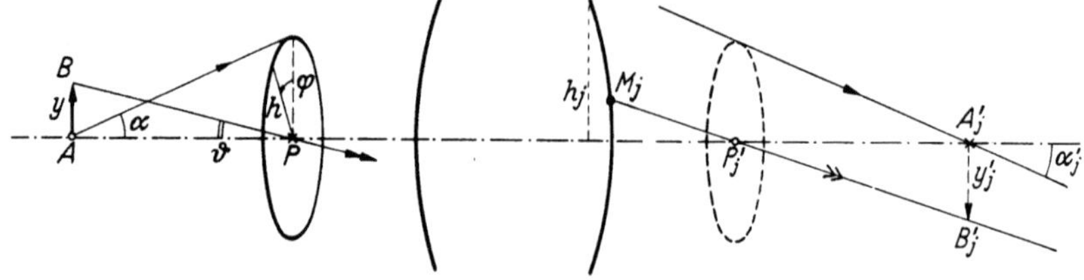

Fig. 157.

Le chemin optique aberrant $n'\Delta = L_0 - L$ s'écrit alors

$$\left.\begin{aligned}n'\Delta &= \frac{h^2}{2}\sum\frac{h_j^2}{h^2}\,Q_x\left(\frac{\delta n'}{n'}-\frac{\delta n}{n}\right) -\\ &\quad - \frac{n\,h\,y\cos\varphi}{x-p}\left[\sum\frac{h_j}{h}\,k_j\,Q_x\left(\frac{\delta n_j'}{n_j'}-\frac{\delta n_j}{n_j}\right) - \left(\frac{\delta n'}{n'}-\frac{\delta n}{n}\right)\right];\end{aligned}\right\} \tag{86.5}$$

expression où apparaissent les sommes bien connues des calculateurs C_x et C_y; on peut écrire

$$n'\Delta = \frac{h^2}{2}\,C_x - \frac{n\,h\,y}{x-p}\cos\varphi\left[C_y - \left(\frac{\delta n'}{n'}-\frac{\delta n}{n}\right)\right].$$

On reconnait dans le premier terme l'influence d'un défaut de mise au point (chromatisme axial) égal, d'après la relation (73.2) à

$$\delta x' = -\frac{2\Delta}{\alpha'^2} = -\frac{h^2(x'-p')^2}{n'\,h'^2}\,C_x;$$

le second correspond à un décalage latéral de l'image (chromatisme latéral) égal à

$$\left[C_y - \left(\frac{\delta n'}{n'}-\frac{\delta n}{n}\right)\right]y'.$$

Si l'on applique à l'expression de Δ les relations (69.1) en faisant, $\alpha' = -h'/(x'-p')$, on obtient les aberrations chromatiques transversales:

$$\delta y' = -\frac{h^2(x'-p')}{n'\,h'}\,C_x\cos\varphi + \left[C_y - \left(\frac{\delta n'}{n}-\frac{\delta n}{n}\right)\right]y',$$

$$\delta z' = -\frac{h^2(x'-p')}{n'\,h'}\,C_x\sin\varphi.$$

87. Calcul des aberrations géométriques du troisième ordre; sommes de Seidel. Il suffit, pour les calculer, de poursuivre le développement du chemin optique aberrant $n'\Delta$ jusqu'aux termes de 4ème degré. On trouve alors, en éliminant les termes en y^4 qui ne correspondent à aucune aberration puisqu'ils ne varient pas sur la surface de la pupille:

$$n' \Delta = \frac{(Y^2 + Z^2)^2}{8} \left[Q_x^2 \left(\frac{1}{n' x'} - \frac{1}{n x} \right) + \lambda \frac{n' - n}{r^3} \right] +$$
$$+ \frac{Y(Y^2 + Z^2)}{2} Q_x \frac{n' y'}{x'} \left(\frac{1}{n' x'} - \frac{1}{n x} \right) - \frac{Y^2 + Z^2}{4} Q_x \frac{n'^2 y'^2}{x'^2} \left(\frac{1}{n'^2} - \frac{1}{n^2} \right) +$$
$$+ \frac{Y^2}{2} \frac{n'^2 y'^2}{x'^2} \left(\frac{1}{n' x'} - \frac{1}{n x} \right) - \frac{Y}{2} \frac{n'^3 y'^3}{x'^3} \left(\frac{1}{n'^2} - \frac{1}{n^2} \right). \qquad (87.1)$$

On applique encore les transformations (86.4) qui conduisent aux expressions suivantes, où l'on ne conserve que les termes qui font intervenir h_j,

$$(Y_j^2 + Z_j^2)^2 \approx h_j^4 - \frac{4n}{x - p} h_j^3 k_j y \cos \varphi +$$
$$+ 2 \frac{n^2}{(x - p)^2} h_j^2 h_j^2 y (2 + \cos 2\varphi) - \frac{4n^3}{(x - p)^3} h_j k_j^3 y^3 \cos \varphi,$$

$$(Y_j^2 + Z_j^2) Y_j \approx h_j^3 \cos \varphi - \frac{n}{x - p} h_j^2 k_j y (2 + \cos 2\varphi) +$$
$$+ \frac{3n^2}{(x - p)^2} h_j k_j^2 y^2 \cos \varphi,$$

$$Y h_j^2 \approx h_j^2 \frac{1 + \cos 2\varphi}{2} - \frac{2n}{x - p} h_j k_j y \cos \varphi.$$

On voit aussi (voir Fig. 157) que:

$$\frac{n_j' y_j'}{x_j'} = - n_j' y_j' \frac{\alpha_j'}{h_j} = - \frac{n y \alpha}{h_j} = \frac{n h y}{h_j (x - p)},$$

et l'on trouve l'expression du chemin optique aberrant où interviennent les cinq sommes classiques de Seidel. On peut poser pour simplifier l'écriture

$$S = Q_x^2 \left(\frac{1}{n' x'} - \frac{1}{n x} \right) + \lambda \frac{n' - n}{r^3} \quad \text{et} \quad \delta = \frac{1}{n' x'} - \frac{1}{n x}.$$

Le résultat s'écrit, en fonction de h et φ (variables d'ouverture) et de $\vartheta = \dfrac{y}{x - p}$ (variable de champ):

$$n' \Delta = \frac{1}{8} h^4 \sum \left(\frac{h_j}{h} \right)^4 S$$
$$- \frac{n h^3 y \cos \varphi}{2(x - p)} \sum \left[\left(\frac{h_j}{h} \right)^3 k S - \left(\frac{h_j}{h} \right)^2 Q_x \delta \right]$$
$$+ \frac{n^2 h^2 y^2}{2(x - p)^2} \sum \left[\left(\frac{h_j}{h} \right)^2 k_j^2 S - 2 \frac{h_j}{h} k_j Q_x \delta + \delta - \frac{1}{2r} \left(\frac{1}{n'} - \frac{1}{n} \right) \right]$$
$$+ \frac{n^2 h^2 y^2 \cos 2\varphi}{4(x - p)^2} \sum \left[\left(\frac{h_j}{h} \right)^2 k_j^2 S - 2 \frac{h_j}{h} k_j Q_x \delta + \delta \right] \qquad (87.2)$$
$$- \frac{n^3 h y^3 \cos \varphi}{2(x - p)^3} \sum \left[\frac{h_j}{h} k_j^3 S - 3 k_j^2 Q_x \delta + 2 \frac{h}{h_j} k_j \delta - \right.$$
$$\left. - \frac{h}{h_j} k_j \left(Q_x - \frac{h}{h_j k_j} \right) \left(\frac{1}{n'^2} - \frac{1}{n^2} \right) \right].$$

Rappelons que dans ces expressions le coefficient Q_x est l'invariant de réfraction sur le dioptre $Q_x = n \left(\frac{1}{r} - \frac{1}{x} \right)$ qui est lié dans le domaine paraxial à l'angle d'incidence du rayon issu de A par la relation

$$n\,i = n'\,i' = h\,Q_x.$$

Il serait donc facile d'exprimer au besoin les sommes de Seidel en fonction des angles d'incidence[1].

88. Influence de la position de la pupille sur les aberrations du troisième ordre.

Nous avons déjà vu à la Sect. 55 l'influence de la position de la pupille sur le chromatisme latéral; pour étudier son influence sur l'ensemble des aberrations nous pourrions partir des sommes de Seidel mais en fait il n'est pas besoin de connaître les contributions individuelles des différents dioptres mais de se donner seulement les valeurs des coefficients d'aberrations dans l'espace image et d'étudier par une étude géométrique l'influence de la position de la pupille

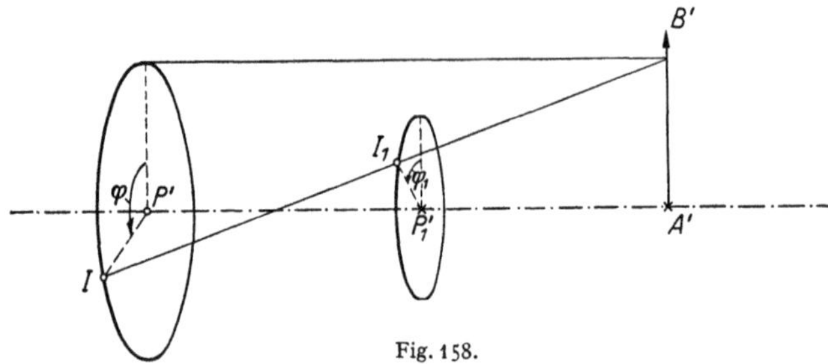

Fig. 158.

de sortie. On peut écrire en rassemblant les résultats obtenus dans les chapitres concernant les divers termes d'aberrations (Sect. 73 à 84)

$$\varDelta = - \frac{a\,\alpha'^4}{4} + b\,\alpha'^3\,y'\cos\varphi - \frac{\alpha'^2\,y'^2}{4}\,(c - \mathcal{A}\cos 2\varphi) + d\,\alpha'\,y'^3\cos\varphi. \quad (88.1)$$

Déplaçons maintenant le centre de la pupille de P' en P'_1 (Fig. 158), définissons les coordonnées polaires α'_1, φ_1 permettant de décrire la surface de cette nouvelle pupille et essayons de connaître les coefficients $a_1, b_1, c_1, \mathcal{A}_1, d_1$ du nouveau développement de $\varDelta$:

$$\varDelta = - \frac{a_1\,\alpha_1'^4}{4} + b_1\,\alpha_1'^3\,y'\cos\varphi_1 - \frac{\alpha_1'^2\,y'^2}{4}\,(c_1 - \mathcal{A}_1\cos 2\varphi_1) + d_1\,\alpha_1'\,y'^3\cos\varphi_1. \quad (88.2)$$

Cette expression n'est fonction que des coordonnées paraxiales du rayon envisagé et il suffit d'écrire que les points I, I_1 et B' sont alignés, en les caractérisant par leurs coordonnées paraxiales

$$I\begin{cases} p' = A'P' \\ p'\,\alpha'\cos\varphi \\ p'\,\alpha'\sin\varphi \end{cases} \qquad I_1\begin{cases} p'_1 = AP''_1 \\ p'_1\,\alpha'_1\cos\varphi_1 \\ p'_1\,\alpha'_1\sin\varphi_1 \end{cases} \qquad B'\begin{cases} 0 \\ y' \\ 0. \end{cases}$$

En posant

$$\pi' = \frac{1}{p'}, \quad \pi'_1 = \frac{1}{p'_1},$$

on aura

$$\left.\begin{aligned} \alpha'\cos\varphi &= \alpha'_1\cos\varphi_1 + y'\,(\pi' - \pi'_1), \\ \alpha'\sin\varphi &= \alpha'_1\sin\varphi_1. \end{aligned}\right\} \quad (88.3)$$

---

[1] Voir en particulier H. H. Hopkins: Wave theory of aberrations. Oxford: Clarendon Press 1950.

Ces relations permettent d'exprimer toutes les quantités qui interviennent dans l'expression (88.1); on trouve successivement:

$$\left.\begin{aligned}
\alpha'^2 &= \alpha_1'^2 + 2\alpha_1' y' \cos\varphi_1 (\pi' - \pi_1') + y'^2 (\pi' - \pi_1')^2, \\
\alpha'^2 \cos 2\varphi &= \alpha_1'^2 \cos 2\varphi_1 + 2\alpha_1' y' \cos\varphi_1 (\pi' - \pi_1') + y'^2 (\pi' - \pi_1')^2, \\
\alpha'^3 \cos\varphi &= \alpha_1'^3 \cos\varphi_1 + \alpha_1'^2 y' (\pi' - \pi_1') (2 + \cos 2\varphi_1) \\
&\quad + 3\alpha_1' y'^2 (\pi_1 - \pi_2') \cos\varphi_1 + y'^3 (\pi' - \pi_1')^3, \\
\alpha'^4 &= \alpha_1'^4 + 4\alpha_1'^3 y' \cos\varphi_1 (\pi' - \pi_1') + 2\alpha_1'^2 y'^2 (\pi - \pi_1')^2 \times \\
&\quad \times (2 + \cos 2\varphi_1) + 4\alpha_1' y'^3 (\pi' - \pi_1')^3 \cos\varphi_1 + y'^4 (\pi - \pi_1')^4.
\end{aligned}\right\} \quad (88.4)$$

On peut ensuite porter ces valeurs dans l'équation (88.1) et identifier les termes obtenus à ceux de l'expression (88.2). On trouve ainsi les valeurs des nouveaux coefficients d'aberrations en fonction des anciens:

$$\left.\begin{aligned}
a_1 &= a, \\
b_1 &= b + a(\pi_1' - \pi'), \\
c_1 &= c + 8b(\pi_1' - \pi'') + 4a(\pi_1' - \pi')^2, \\
\mathcal{A}_1 &= \mathcal{A} - 4b(\pi_1' - \pi') - 2a(\pi_1' - \pi')^2, \\
d_1 &= d + \frac{c - \mathcal{A}}{2}(\pi_1' - \pi') + 3b(\pi_1' - \pi')^2 + a(\pi_1' - \pi')^3.
\end{aligned}\right\} \quad (88.5)$$

La première équation exprime que l'aberration sphérique ne dépend pas de la position de la pupille pour une ouverture angulaire donnée (ce qui était évident);
on voit ensuite que le coefficient de coma est une fonction linéaire de la proximité de la pupille, que les coefficients d'astigmatisme et de courbure de champ des fonctions du 2ème degré, le coefficient de distorsion en polynôme du 3ème degré. Ces résultats sont susceptibles d'une interprétation simple que nous allons maintenant donner. Remarquons d'abord que la position

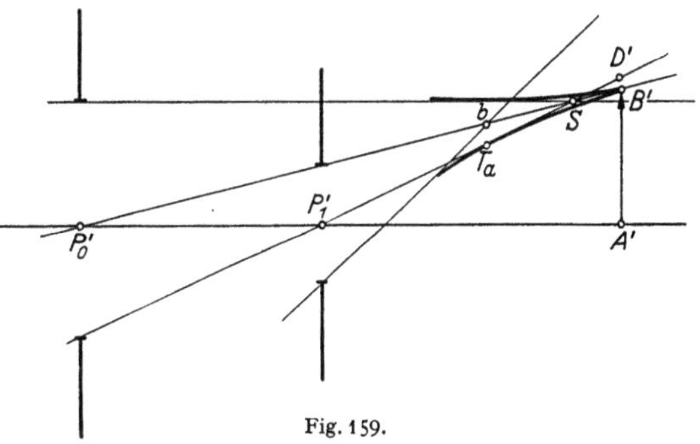

Fig. 159.

de la pupille n'influe évidemment pas sur la position des rayons ni par conséquent sur la forme de la caustique; en déplaçant la pupille, on déplace seulement le rayon moyen du faisceau des rayons utilisés, la caustique restant immobile.

Considérons un instrument présentant de l'aberration sphérique: la deuxième des relations (88.5) indique qu'en plaçant convenablement la pupille il est toujours possible de s'arranger pour que le coefficient b_1 soit nul comme on l'a déja on à la Sect. 79: on a alors obtenu dans le cadre des aberrations du 3ème ordre l'isoplanétisme. Supposons d'autre part que pour cette position de la pupille l'astigmatisme soit négligeable, la caustique, centrée sur le rayon moyen issu de P_0' a la forme d'une caustique d'aberration sphérique (Fig. 159); si nous déplaçons la pupille de P_0' en P_1' le faisceau perd la symétrie de révolution et n'admet plus qu'un plan de symétrie; la coma apparait et nous pouvons la caractériser par le fait que deux rayons provenant d'un même cercle de la pupille et issus de deux point opposés de ce cercle dans le plan de symétrie ne se coupent plus sur le rayon moyen du faisceau; leur intersection est en b sur la Fig. 159, et la distance

ab caractérise la coma (le point a étant confondu avec le point de contact T du rayon issu de P_1' avec la caustique). La variation de la coma sera évidemment proportionnelle au coefficient d'aberration sphérique a.

On interprète de façon analogue les variations des coefficients $\mathcal{A}$ et c; supposons par exemple que l'astigmatisme soit nul lorsque la pupille est en P_0' (P étant la pupille isoplanétique); la caustique conservera la forme d'une caustique d'aberration sphérique, même si l'objet s'éloigne assez notablement de l'axe. Lorsqu'on déplace la pupille, l'astigmatisme apparaît et la distance d'astigmatisme sera TS (Fig. 159). Cette distance passe par un minimum nul lorsque P_1' vient en P_0', et les variations du coefficient d'astigmatisme $\mathcal{A}$ s'expriment en effet dans ce cas par la relation

$$\mathcal{A} = -\,2\,a\,(\pi_1' - \pi')^2.$$

Enfin, les variations de distorsion s'interprètent aisément: lorsqu'on déplace la pupille, le rayon moyen reste tangent à la section de la caustique par le plan

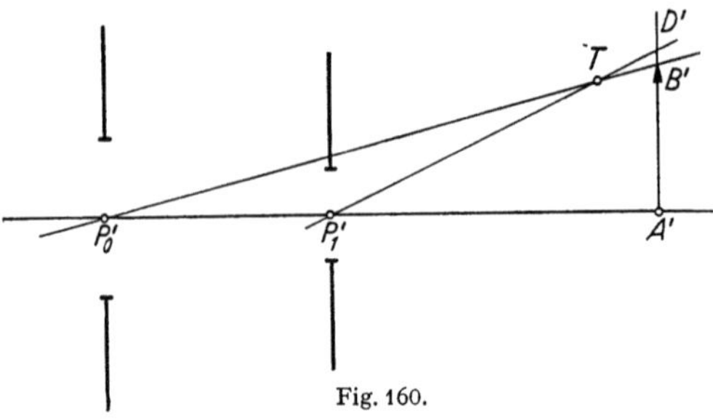

Fig. 160.

de figure et le point D' se déplace en conséquence (Fig. 159).

Si les aberrations se réduisent à l'astigmatisme, le rayon tourne simplement autour de la focale tangentielle située en T et les variations de distorsion peuvent alors se calculer très simplement; le coefficient d devient d'ailleurs une fonction linéaire de $\pi_1' - \pi'$ (Fig. 160).

Signalons enfin que les formules (88.5) permettent de déterminer les différentes formes de caustiques que l'on peut obtenir, dans le cadre des aberrations du 3ème ordre; une discussion rapide permet d'établir que les caustiques possibles se réduisent à quatre:

1. caustique d'aberration sphérique pure;
2. caustique de coma pure;
3. superposition d'aberration sphérique et d'astigmatisme;
4. astigmatisme pur (caustique dégénérée en deux focales).

Les cas 2. et 3. nécessitent un peu de calcul pour déterminer la forme des caustiques obtenues; nous ne les effectuerons pas ici.

89. Variations des aberrations du troisième ordre avec la position de l'objet. Nous venons d'étudier comment les aberrations du 3ème ordre varient en fonction de la position de la pupille et de donner une interprétation géométrique des résultats obtenus. Cette étude n'a cependant pas de nombreuses applications dans la pratique du calcul; lorsqu'un instrument a été convenablement calculé, les aberrations du 3ème ordre sont le plus souvent très réduites; on s'applique en effet à leur donner de faibles valeurs de façon qu'elles compensent au mieux les aberrations d'ordre supérieur. Si l'on déplace maintenant le plan objet, les termes du 3ème ordre peuvent reprendre des valeurs relativement importantes et masquer pratiquement les autres ordres. Il est donc intéressant de savoir comment varient les aberrations du 3ème ordre avec la position de l'objet, afin de prévoir quels seront les déplacements maxima que l'on peut faire subir au plan objet sans que l'image cesse d'être bonne: on peut penser, en particulier, à appliquer les résultats à l'objectif photographique auquel on demande de fournir de bonnes images pour diverses mises au point, ou encore à l'objectif de microscope pour savoir de quelle quantité on peut faire varier la position de l'image, ou encore la distance de l'objectif à l'oculaire.

Considérons la pupille d'entrée de l'instrument (Fig. 161), dans laquelle le point d'impact I du rayon sera supposé être fixe et à la distance h de l'axe. Soient AB la position d'un objet caractérisé par les paramètres y et Ψ, $A_1 B_1$ la position d'un autre objet tel que le point B_1 soit sur le rayon BI. On suppose connues les aberrations du 3ème ordre lorsque l'objet est en AB, et l'on peut écrire, en utilisant des coefficients a_p, b_p, $\mathcal{C}_p$, $\mathcal{A}_p$, d_p relatifs à la *pupille d'entrée* et aux paramètres qui y sont attachés,

$$n'\Delta = a_p h^4 + b_p h^3 \vartheta \cos\Psi + \mathcal{C}_p h^2 \vartheta^2 + \mathcal{A}_p h^2 \vartheta^2 \cos 2\Psi + d_p h \vartheta^3 \cos\Psi. \quad (89.1)$$

Il reste à évaluer les variations du chemin optique $(B_1 I I' B_1')$ en fonction de celles de $BII'B')$; on aura:

$$\left. \begin{aligned} n'\Delta_1 - n'\Delta &= (B_1 PP' B_1') - (B_1 II' B_1') - (B PP' B') + (B II' B') \\ &= - n'(B'I' - B'P' - B_1'I' + B_1'P') + \\ &\quad + n(BI - BP - B_1 I + B_1 P) + (PP')_{\vartheta_1} - (PP')_{\vartheta}. \end{aligned} \right\} \quad (89.2)$$

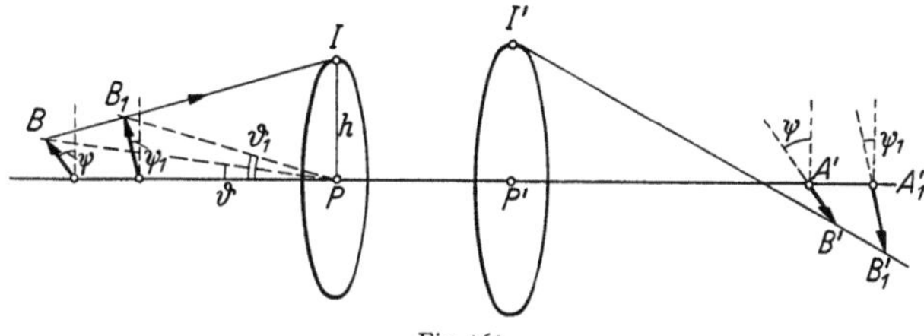

Fig. 161.

Exprimons que B, B_1 et I sont alignés, ainsi que B', B_1', I'; on aura, en posant $x = \overline{PA}$ et $X = 1/x$, des relations analogues pour $A_1 B_1$

$$\left. \begin{aligned} \vartheta \cos\Psi &= \vartheta_1 \cos\Psi_1 + h(X - X_1), \\ \vartheta \sin\Psi &= \vartheta_1 \sin\Psi_1. \end{aligned} \right\} \quad (89.3)$$

Les coefficients $a_p b_p \ldots$ pourraient, bien entendu, être exprimés en fonction de $a, b \ldots$ ou encore des sommes de Seidel, mais ceci est parfaitement inutile pour les applications.

On en déduit aisément, en se limitant aux termes linéaires en $X - X_1$, ce qui signifie que nous n'envisageons que de faibles déplacements de l'objet ou de l'image,

$$\left. \begin{aligned} \vartheta^2 &= \vartheta_1^2 + 2h\vartheta_1(X - X_1)\cos\Psi_1, \\ \vartheta^2 \cos 2\Psi &= \vartheta_1^2 \cos 2\Psi_1 + 2h\vartheta_1(X - X_1)\cos\Psi_1, \\ \vartheta^3 \cos\Psi &= \vartheta_1^3 \cos\Psi_1 + h\vartheta_1^2(X - X_1)(2 + \cos 2\Psi_1). \end{aligned} \right\} \quad (89.4)$$

Il reste à exprimer les différences telles que $BI - BP$, avec la précision du 4ème degré par rapport à h et ϑ (ou ϑ_1)

$$BI - BP = [x^2 + (\vartheta x \cos\Psi - h)^2 + (\vartheta x \sin\Psi)^2]^{\frac{1}{2}} - [x^2 + \vartheta^2 x^2]^{\frac{1}{2}};$$

on trouve, en développant les racines carrées jusqu'au 4ème degré,

$$BI - BP = -\frac{h^2}{2} X + h\vartheta \cos\Psi + \frac{1}{8} h^4 X^3 - \frac{1}{2} h^3 \vartheta X^2 \cos\Psi +$$
$$+ \frac{1}{4} h^2 \vartheta^2 X (2 + \cos 2\Psi) - \frac{1}{2} h \vartheta^3 \cos\Psi.$$

En appliquant les relations (89.3) et (89.4), on trouve aisément,

$$(BI - BP) - (B_1 I - B_1 P)$$
$$= \left[\frac{h^2}{2} - \frac{h^4 X^2}{8} + \frac{h^3 \vartheta X \cos \Psi}{2} - \frac{h^2 \vartheta^2}{2} - \frac{h^2 \vartheta^2}{4} \cos 2\Psi\right](X - X_1).$$

Il est alors possible d'exprimer le second membre de l'équation (89.2) en remarquant que le théorème de Gouy (Sect. 67) appliqué à deux surfaces d'ondes sphériques s'exprime par la relation

$$n\, h^2 (X - X_1) = n'\, h'^2 (X' - X_1').$$

On aboutit à l'expression suivante:

$$\left. \begin{aligned} n'(\Delta_1 - \Delta) &= n h^2 (X - X_1)\left[\frac{h'^2 X'^2 - h^2 X^2}{8} - \right. \\ &\left. - \frac{\cos \Psi}{2}(h'\vartheta' X' - h\vartheta X) + \frac{2 + \cos 2\Psi}{4}(\vartheta'^2 - \vartheta^2)\right] + (PP')_{\vartheta_1} - (PP)_\vartheta. \end{aligned} \right\} \quad (89.5)$$

Les derniers termes ne font intervenir que l'aberration sphérique pupillaire; si a' est un coefficient caractérisant cette aberration, on aura

$$(PP')_{\vartheta_1} - (PP')_\vartheta = \frac{a'_p}{4}(\vartheta_1^4 - \vartheta^4).$$

L'identification des termes dans l'équation (89.5) conduit aux résultats définitifs:

$$\left. \begin{aligned} a_{1p} &= a_p + \left[b_p + \frac{n}{8h^2}(\alpha'^2 - \alpha^2)\right](X - X_1), \\ b_{1p} &= b_p + \left[2(\mathcal{C}_p + \mathcal{A}_p) - \frac{n}{2h\vartheta}(h'\vartheta' X' - h\vartheta X)\right](X - X_1), \\ \mathcal{C}_{1p} &= \mathcal{C}_p + \left[2d_p + \frac{n}{2\vartheta^2}(\vartheta'^2 - \vartheta^2)\right](X - X_1), \\ \mathcal{A}_{1p} &= \mathcal{A}_p + \left[d_p + \frac{n}{4\vartheta^2}(\vartheta'^2 - \vartheta^2)\right](X - X_1), \\ d_{1p} &= d_p - a'_p\ (X - X_1). \end{aligned} \right\} \quad (89.6)$$

On constate tout d'abord une différence fondamentale avec les relations (88.5): si l'on suppose l'instrument parfaitement corrigé pour une position de l'objet, tous les coefficients $a_p, b_p\ldots$ étant nuls, le déplacement de l'objet aura cependant pour effet de faire apparaître de la courbure, de l'astigmatisme, de la coma et de l'aberration sphérique, ainsi que de la distorsion si l'aberration sphérique pupillaire n'est pas nulle.

On peut néanmoins éviter l'apparition de ces aberrations dans certains cas particuliers:

Il n'apparaît pas d'aberration sphérique si $\alpha'^2 = \alpha^2$ (points nodaux ou antinodaux). On a en effet supposé qu'il n'y avait pas de coma c'est-à-dire que la condition d'Abbe était satisfaite. Pour satisfaire encore à la condition de Herschel, il est nécessaire que $\alpha' = \pm\alpha$ (voir Sect. 29).

Il n'apparaît pas de coma si $h'\vartheta' X' = h\vartheta X$ ou encore $\vartheta\alpha = \vartheta'\alpha'$ et, en tenant compte de la relation de Lagrange appliquée successivement à AA' et PP',

$$\frac{n'^2}{n^2} g_A g_P = 1\,;$$

dans le cas où $n = n'$, cette condition devient $g_A g_P = 1$, c'est-à-dire que les grandissements objectif et pupillaire doivent être inverses.

Il n'apparaît pas de courbure ou d'astigmatisme si $\vartheta'^2 = \vartheta^2$, c'est-à-dire si les pupilles sont placées aux points nodaux ou antinodaux.

Nous allons maintenant montrer comment ces relations permettent de prévoir les variations d'aberrations dans deux cas concrets.

α) Objectif photographique. Considérons un objectif de longueur focale image f' (Fig. 162), dont l'ouverture est caractérisée par le nombre $N = f'/2h$ de sorte que l'ouverture angulaire dans l'espace image est $\sin \alpha' = 1/2N$; supposons encore que les pupilles sont dans les plans principaux; il ne peut apparaître dans ces conditions que de l'aberration sphérique ou de la coma; on admettra aussi que l'objectif est parfaitement corrigé lorsque l'objet est à l'infini; si l'objet se rapproche à la distance $x = -pf'$ (p étant un nombre positif assez grand), c'est-à-dire à la proximité $X_1 = -1/pf'$, il apparaît de l'aberration sphérique et de la coma caractérisées par les écarts suivants, que l'on obtient en faisant $h = h'$, $\vartheta = \vartheta'$, $X = 0$, $X' = 1/f'$, $\alpha = 0$ dans les relations (89.6)

$$\varDelta_{\mathrm{sph}} = \frac{h^2}{8 p f'} \sin^2 \alpha',$$

$$\varDelta_{\mathrm{coma}} = \frac{h^3}{2\, p f'^2}\,\vartheta,$$

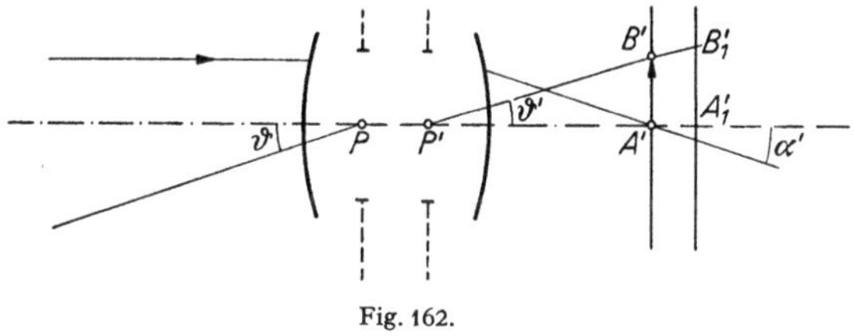

Fig. 162.

et l'on peut constater que l'aberration sphérique correspondra à une sous-correction, alors que la coma sera interne. Calculons les dimensions des taches de diffusion produites par ces aberrations.

1. Aberration sphérique: l'aberration longitudinale sera $l = \dfrac{4\varDelta}{\alpha'^2} = \dfrac{h^2}{2 p f'}$. Dans le plan de l'image paraxiale on a un cercle de rayon $\delta y' = 4\varDelta/\alpha'$, mais si l'on suppose que la diffraction n'intervient pas le cercle de diffusion minimum, obtenu dans le plan de mise au point pour lequel le cône des rayons marginaux coupe la nappe sagittale de la caustique, admet un rayon égal au quart et un diamètre égal à la moitié de la valeur précédente, soit $2\varDelta/\alpha'$; on aura donc une tache dont l'étendue sera au minimum de

$$\frac{h^2}{4 p f'}\, \sin \alpha' = \frac{f'}{32\, p\, N^3}\,.$$

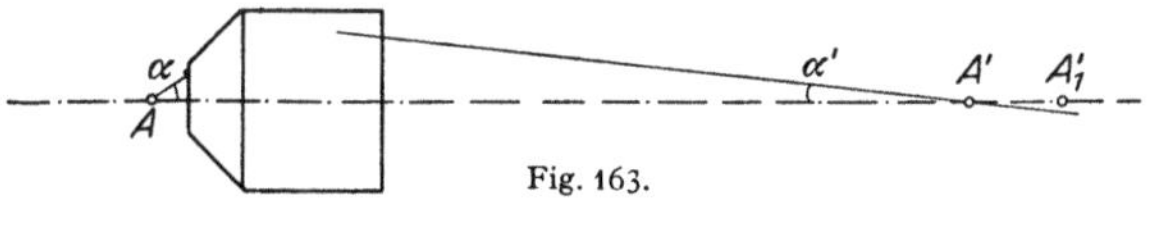

Fig. 163.

2. Coma: l'étendue totale de la tache sera $3\,\varDelta/\alpha'$ (Sect. 77), soit encore

$$\frac{3}{2}\,\frac{f'\alpha'^2\vartheta}{p} = \frac{3 f'\vartheta}{8\, p\, N^2}\,.$$

Choisissons, à titre d'exemple, un objectif de longueur focale $f' = 50$ mm., ouvert à $f/1,5$, devant photographier un objet à 1 mètre ($p = 20$), le demi-angle de champ étant $\vartheta = 1/4$ de radian. Les diffusions obtenues sur la plaque seront, pour l'aberration sphérique: $\dfrac{50}{32 \times 20 \times 1,5^3} = 0,02$ mm., pour la coma: $\dfrac{3 \times 50}{8 \times 20 \times 1,5^2 \times 4}$ $= 0,1$ mm.

Cette dernière peut être gênante, alors que l'aberration sphérique ne le sera pas dans la plupart des cas.

β) Objectif de microscope. Les objectifs de microscope sont bien corrigés de l'aberration sphérique et de la coma pour une position déterminée de l'objet et de l'image (Fig. 163), définie pratiquement par la longueur de tube, qui fixe la distance entre l'objectif et l'oculaire. Cherchons quelles seraient les valeurs de l'aberration sphérique et de la coma obtenues lorsqu'on fait varier la longueur du tube de dx'; on supposera que la pupille de sortie de l'objectif est située dans le plan focal image, de sorte que $X=0$; le produit $h\vartheta$ restant borné (et égal à $-y\alpha$) lorsque la pupille s'en va à l'infini, la quantité $h\vartheta x$ s'annulera.

La qualité optique d'un objectif de microscope devant être excellente, il est indiqué de caractériser les aberrations par les chemins optiques aberrants; on a respectivement pour l'aberration sphérique et pour la coma, en remplaçant $n h^2(X-X_1)$ par $h'^2(X'-X_1')$, en appelant g le grandissement de l'objectif et en utilisant le paramètre $\sin\alpha$:

$$\Delta_{\text{sph}} \approx -\frac{h'^2}{8}(\sin^2\alpha - \alpha'^2)\frac{dx'}{x'^2} \approx \frac{-\alpha'^2\sin^2\alpha\,dx'}{8} \approx -\frac{n^2\sin^4\alpha}{8g^2}dx',$$

$$\Delta_{\text{coma}} \approx -\frac{h'^2}{2}h'\vartheta' X'\frac{dx'}{x'^2} = \frac{\alpha'^3}{2}\vartheta'\,dx' = \frac{(n\sin\alpha)^3\vartheta'\,dx'}{2g^3}.$$

On constate que, si l'on allonge le tube $(dx'>0)$, il apparaît une surcorrection sphérique $(\Delta<0)$ et une coma externe. Il est utile de calculer numériquement les valeurs de Δ obtenues dans divers cas concrets correspondant à des objectifs courants, pour $dx'=20$ mm., ce qui correspond en gros à la différence maximum de longueurs de tube entre divers constructeurs et en admettant une image $y'=5$ mm., soit $\vartheta'\approx 3\cdot 10^{-2}$.

Type objectiv	Grandissement	$n\sin\alpha$	Δ (sph) μ	Δ (coma) μ
Apochromat à sec	$\times 10$	0,3	0,2	$8\cdot 10^{-3}$
Achromat à sec	$\times 40$	0,65	0,3	$1,2\cdot 10^{-3}$
Apochromat à sec	$\times 40$	0,95	1,25	$4\cdot 10^{-3}$
id. à sec	$\times 60$	0,95	0,55	$1,2\cdot 10^{-3}$
id. Immersion	$\times 90$	1,3	0,4	$1\cdot 10^{-3}$

On constate qu'il est indispensable de respecter la distance objectif-oculaire avec une précision de l'ordre de quelques millimètres si l'on veut que l'image soit satisfaisante, surtout si l'on utilise un objectif à grande ouverture et à grandissement relativement faible; il est souhaitable en effet dans un microscope que la déformation de la surface d'onde ne dépasse pas $\lambda/4$ d'aberration sphérique au foyer paraxial (c'est-à-dire $\lambda/16$ au meilleur foyer); la coma n'est, par contre, pas gênante.

90. Etude de la forme de la surface d'onde dans quelques cas simples. Pour terminer ce chapitre sur les aberrations nous indiquerons la forme que prend la surface d'onde dans quelques cas d'aberrations simples; nous nous aiderons pour cela de figures d'interférences obtenues par la méthode de Twyman Green. Cette méthode est souvent utilisée pour connaître avec une bonne précision la forme de la surface d'onde fournie par un instrument et juger ainsi de sa qualité. On utilise pour cela un interféromètre de Michelson que l'on modifie de la façon suivante (Fig. 164): l'un des miroirs plans est remplacé par l'ensemble du système S à étudier et d'un miroir parfaitement sphérique M_1 dont le centre est situé très sensiblement au foyer de S. Si la surface d'onde fournie par S est rigoureusement sphérique, elle s'applique sur M_1 et l'on obtiendra à nouveau, après une deuxième traversée de S en sens inverse, une surface d'onde plane qui donnera naissance, par interférence avec la surface d'onde provenant du deuxième bras de l'interféromètre, à un système de franges rectilignes et équidistantes. Les franges pourront être supprimées si les deux surfaces d'ondes planes sont rigoureusement appliquées l'une sur l'autre. Si S présente des aberrations, elles se manifesteront par l'apparition de franges qui représenteront simplement la distance entre la surface d'onde et le miroir sphérique; du fait de l'autocollimation le chemin optique aberrant est doublé et l'on écrira, si k est le nombre de franges observées à partir d'une origine située par exemple au centre de la pupille

$$\Delta = k\,\frac{\lambda}{2}. \qquad (90.1)$$

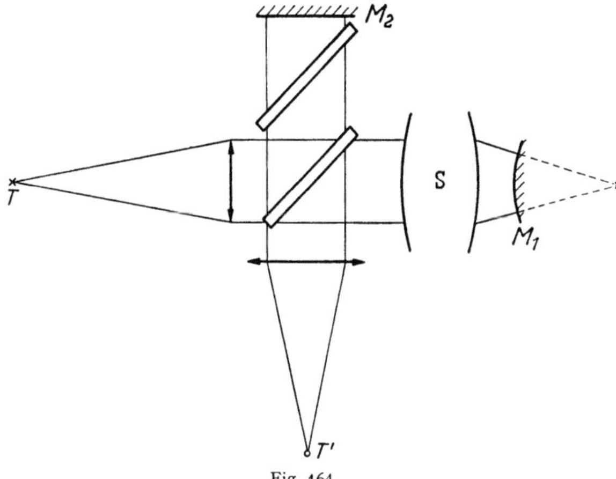

Fig. 164.

On voit que cette méthode présente le grand avantage de fournir immédiatement la forme complète de la surface d'onde et d'apprécier l'importance de ses déformations.

Cherchons maintenant quel sera l'effet d'un déplacement longitudinal ou latéral du centre C du calibre de référence: ceci revient à changer de sphère de référence pour étudier la surface d'onde. (On a déjà vu à la Sect. 73 l'effet d'un déplacement longitudinal.) Soit $C(0, 0, 0)$ le centre de l'ancienne sphère de référence et $C_1(\varepsilon, \varrho_1 \cos \varphi_1, \varrho_1 \sin \varphi_1)$ celui de la nouvelle que l'on rapporte à des coordonnées cylindriques $\varepsilon, \varrho_1, \varphi_1$ ayant leur origine en C (Fig. 165). Traçons de C_1 pour centre une sphère S_1 qui passe par le pôle P de la sphère S et considérons un rayon lumineux perçant S en I, S_1 en I_1. Les coordonnées sphériques de I, par rapport à C, sont R, α', φ.

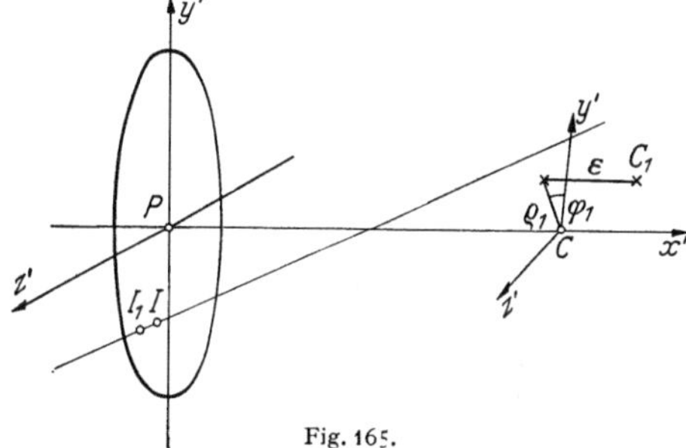

Fig. 165.

On a manifestement:

$$PC_1 - PC = I_1C_1 - IC.$$

Le premier membre s'évalue en projetant CC_1 sur PC, le second en projetant I_1I et CC_1 sur IC dont les cosinus directeurs sont $\cos \alpha'$, $\sin \alpha' \cos \varphi$ et $\sin \alpha' \sin \varphi$; on obtient

$$\varepsilon = I_1 I + \varepsilon \cos \alpha' + \varrho_1 \sin \alpha' \cos (\varphi - \varphi_1),$$

c'est-à-dire que, si Δ est l'écart normal par rapport à la sphère S, l'écart Δ_1 par rapport à S_1 sera

$$\Delta_1 = \Delta + \varepsilon(1 - \cos \alpha') - \varrho_1 \sin \alpha' \cos (\varphi - \varphi_1). \qquad (90.2)$$

Pratiquement, l'angle d'ouverture α' est souvent faible et l'on peut, avec une approximation suffisante, écrire

$$\Delta_1 = \Delta + \varepsilon\,\frac{\alpha'^2}{2} - \varrho_1 \alpha' \cos (\varphi - \varphi_1). \qquad (90.3)$$

Dans l'interféromètre de TWYMAN GREEN il sera facile de produire soit un déplacement longitudinal, soit un déplacement latéral; on voit qu'ils se traduisent respectivement par:

1. Un terme de défaut de mise au point proportionnel au carré de la distance au centre P.

2. Un terme de déplacement latéral du centre de la sphère, qui est une forme linéaire par rapport aux coordonnées Y' et Z' sur la pupille.

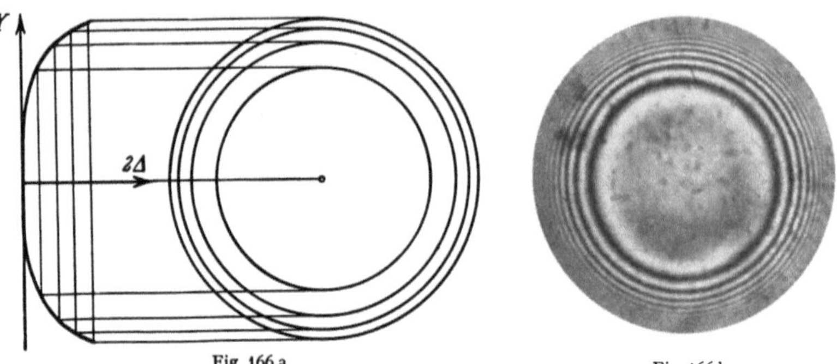

Fig. 166 a.   Fig. 166 b.

Nous allons maintenant examiner quelques cas particuliers:

α) *Effet d'un défaut de mise au point.* L'équation (90.1) donne pour ordre d'interférence

$$k = \frac{\varepsilon \alpha'^2}{\lambda} = \frac{\varepsilon (Y'^2 + Z'^2)}{\lambda p'^2}.$$

Les franges sont les cercles

$$Y'^2 + Z'^2 = \text{constante}$$

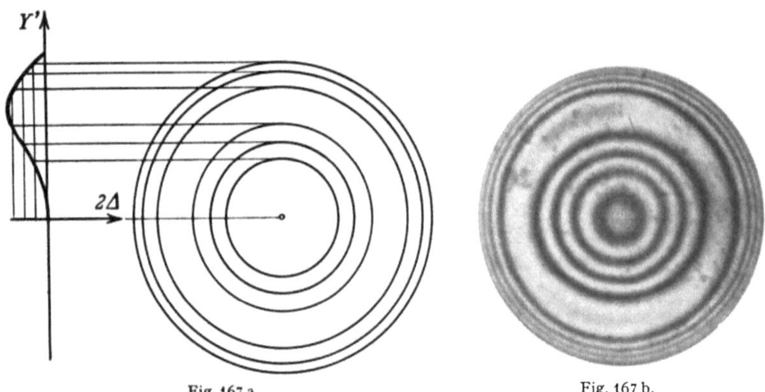

Fig. 167 a.   Fig. 167 b.

et la loi d'espacement de ces cercles est identique à celle des anneaux de Newton; les franges sont d'autant plus serrées que le défaut de mise au point est plus important.

β) *Aberration sphérique du 3ème ordre.* Nous supposerons tout d'abord que la mise au point est faite sur le foyer paraxial; on a alors

$$\Delta = - \frac{a \alpha'^4}{4} = - \frac{a}{4} \frac{(Y'^2 + Z'^2)^2}{p'^4}$$

et la construction géométrique de la position des franges peut s'effectuer en traçant la courbe $\Delta(Y')$ pour $Z' = 0$ comme l'indique la Fig. 166a; les franges sont analogues aux anneaux de Newton, mais elles sont beaucoup plus serrées au bord; la Fig. 166b en est une photographie.

Si l'on suppose maintenant que la mise au point soit effectuée en un point distant de ε du foyer, on aura

$$\Delta_1 = -\frac{a}{4}\frac{(Y'^2+Z'^2)^2}{p'^4} + \frac{\varepsilon(Y'^2+Z'^2)}{2p'^2}.$$

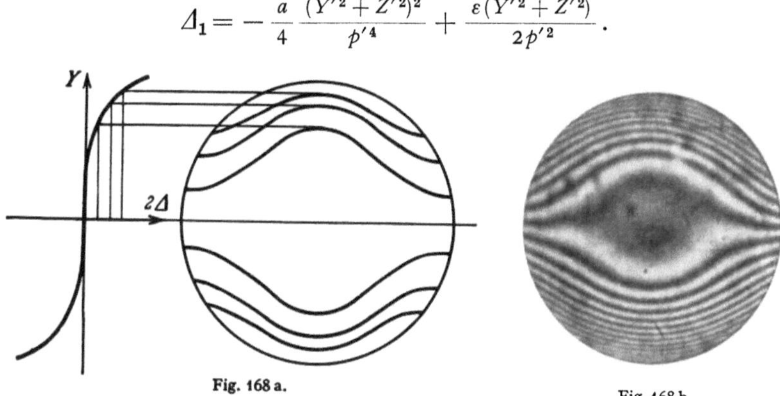

Fig. 168 a.

Fig. 168 b.

Nous avons choisi sur la Fig. 167 le cas où Δ_1 est nul pour l'ouverture maximum. Nous avons vu à la Sect. 74 que ceci correspond à la mise au point au milieu de la nappe sagittale de la caustique.

γ) *La coma.* Nous nous bornerons au cas de la coma du 3ème ordre caractérisée par l'expression suivante de Δ

$$\Delta = b\,y'\,\alpha'^3\cos\varphi = \frac{b\,y'}{p'^3}\,Y'(Y'^2+Z'^2);$$

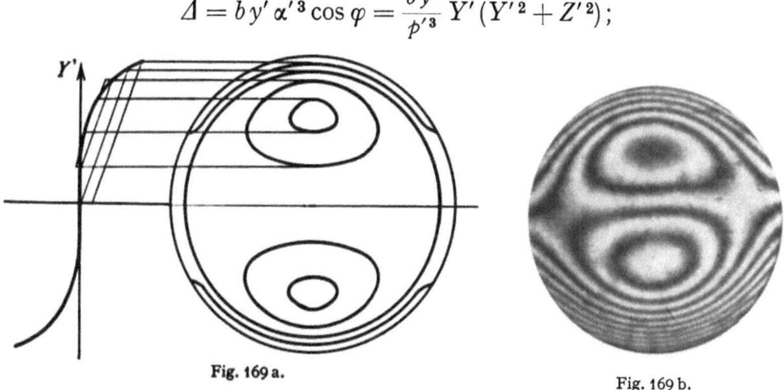

Fig. 169 a.

Fig. 169 b.

les franges sont les courbes $Y'(Y'^2+Z'^2)=$ constante, et leur aspect est indiqué sur la Fig. 168. On peut encore effectuer une construction qui permet de connaître les positions des intersections des franges par le plan méridien $Z'=0$: on trace pour cela la courbe $2\Delta(Y')$ pour $Z'=0$ et l'on prend encore son intersection par une série de droites distantes de λ (Fig. 168a). On peut aussi étudier l'effet d'un décalage latéral du centre de la sphère de référence, dans le plan méridien $Z'=0$. On écrit, pour tenir compte de ce décalage,

$$\Delta_1 = (b\,y'\,\alpha'^3 - \varrho_1\alpha')\cos\varphi = \frac{b\,y'\,Y'(Y'^2+Z'^2)}{p'^3} - \varrho_1\frac{Y'}{p'},$$

ce qui a pour effet d'ajouter un terme linéaire en Y' et de modifier les courbes représentant $2\Delta_1$ en fonction de Y' comme l'indique la Fig. 169. On peut remarquer que, dans le cas de la Fig. 169, Δ_1 est nul le long d'un cercle particulier de la pupille: pour connaître l'ouverture α' qui lui correspond, il faut écrire

$$b\,y'\alpha'^3 - \varrho_1\alpha' = 0, \quad \text{ou} \quad \varrho_1 = b\,y'\alpha'^2,$$

c'est-à-dire que le centre C_1 de la sphère de référence est situé précisément au point du cercle comatique le plus proche du foyer paraxial: on a vu en effet que le rayon du cercle comatique est égal à $b\,y'\,\alpha'^2$. et que distance de son centre au foyer paraxial est $2\,b\,y'\,\alpha'^2$.

δ) *L'astigmatisme.* Il est intéressant d'étudier l'aspect des franges obtenues en présence d'astigmatisme pour diverses mises au point. On peut, pour cela, supposer que l'origine soit, par exemple, au milieu de l'intervalle qui sépare les deux focales, dont les abscisses seront t et s $(s = -t)$. L'expression de Δ, que nous avons établie à la Sect. 81 s'écrit

$$\Delta = \frac{t}{2p'^2}\,(Z'^2 - Y'^2),$$

et, si l'on effectue la mise au point C_1, d'abscisse ε, on aura

$$\Delta = \frac{t}{2p'^2}\,(Z'^2 - Y'^2) + \frac{\varepsilon}{2p'^2}\,(Y'^2 + Z'^2) = \frac{1}{2p'^2}\,[(\varepsilon - t)\,Y'^2 + (\varepsilon + t)\,Z'^2].$$

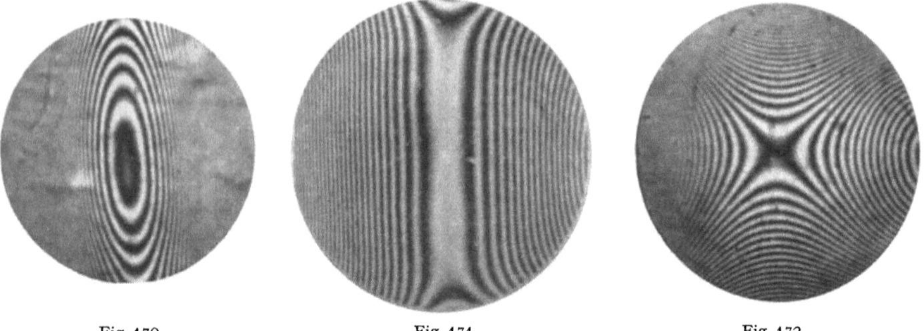

Fig. 170.   Fig. 171.   Fig. 172.

D'une façon générale, les franges sont des coniques admettant OY' et OZ' pour axes et le numéro d'ordre de la frange varie comme le carré de la distance au centre, comme dans le cas du défaut de mise au point. Leur nature dépend des valeurs relatives *de* $\varepsilon + t$ et $\varepsilon - t$. c'est-à-dire de la position de C_1 par rapport aux focales:

a) Si C_1 est à l'extérieur de l'intervalle qui sépare les focales, $\varepsilon - t$ et $\varepsilon + t$ sont de même signe; les franges sont elliptiques et ressemblent à des anneaux de Newton auxquels on aurait fait subir une anamorphose (Fig. 170).

b) Si C_1 est sur une focale (par exemple sur la focale tangentielle), l'ordre d'interférence est donné par la relation

$$k = \frac{(\varepsilon + t)\,Z'^2}{\lambda\,p'^2} = \frac{2t\,Z'^2}{\lambda\,p'^2};$$

les franges sont alors parallèles à OY' et sont analogues à celles que l'on obtiendrait entre un cylindre de faible courbure et un plan amenés au contact suivant une génératrice. Nous remarquons en passant que ces franges sont perpendiculaires à la direction de la focale (Fig. 171).

c) Si C_1 est à l'intérieur de l'intervalle des deux focales, les franges sont hyperboliques (Fig. 172), les asymptotes faisant un angle ϑ donné par la relation

$$\mathrm{tg}^2\vartheta = -\,(t - \varepsilon)/(t + \varepsilon).$$

L'étude rapide des aberrations nous a permis de préciser les divers termes d'aberrations, d'étudier leurs caractères généraux, de calculer les aberrations du 3ème ordre, etc. Nous ne donnerons ici aucune indication sur les méthodes de réduction des aberrations dans le calcul des instruments d'optique; ces méthodes sont en effet très variées et dépendent dans une assez large mesure de l'instrument considéré; on peut dire néanmoins qu'après la réduction des termes d'aberration du 3ème ordre elles se terminent par un calcul numérique de la marche de certains rayons qui permet d'effectuer des retouches finales aux caractéristiques de l'instrument (courbures des surfaces, distances, indices). Nous n'étudierons pas non plus ici les aberrations d'excentrement qui apparaissent lorsque les divers éléments d'un système optique ne sont pas centrés: le calcul de ces aberrations est surtout destiné à déterminer les tolérances de montage des instruments et est plus intéressant pour le technicien que pour le physicien. Nous aborderons maintenant l'étude de la formation des images d'objets étendus, en nous attachant surtout à déterminer l'influence des aberrations sur le contraste des images, ce qui nous amènera à définir en particulier des tolérances.

F. Contraste des images; influence des aberrations.

91. Généralités. Dans les chapitres précédents nous avons étudié la formation de l'image d'un point isolé dans le cadre de l'optique géométrique.

Néanmoins l'utilisateur d'un instrument s'intéressera le plus souvent à la nature de l'image d'un objet complexe, composé d'un ensemble de plages lumineuses de luminance variable (quelquefois limitées par des discontinuités) de points, de lignes etc. La qualité de l'image pourra alors être caractérisée par son contraste, c'est-à-dire par l'importance des variations relatives d'éclairement dans l'image: chacun sait en effet que lorsqu'on passe de l'objet à l'image les contours sont adoucis, les discontinuités sont remplacées par des variations continues plus ou moins étendues, enfin certains détails de petites dimensions disparaissent. Il est donc utile d'étudier maintenant la formation des images d'objets étendus. Nous suivrons pour cela une méthode analogue à celle utilisée en acoustique ou en électricité: grâce à la transformation de FOURIER on décomposera l'objet en une infinité de répartitions sinusoïdales dont l'amplitude dépend de la période p de la sinusoïde envisagée ou plutôt de son inverse $1/p$ que nous appellerons la «fréquence spatiale». Le rôle de cette décomposition est particulièrement important lorsqu'on étudie le fonctionnement d'un appareil où l'on effectue un balayage, par exemple les appareils de télévision, où les fréquences des diverses composantes des courants ou tensions électriques correspondront aux fréquences spatiales de l'objet exploré.

La qualité de l'image globale dépend bien entendu de la nature de la tache image d'un point isolé; mais il est souvent plus commode d'étudier la loi de filtrage des fréquences spatiales, qui est, comme nous allons le voir la transformée de FOURIER de la répartition des éclairements dans la tache image d'un point isolé.

Lorsque les aberrations géométriques sont parfaitement supprimées, on obtient une tache de diffraction d'AIRY dans le cas très fréquent où la pupille est circulaire. Si au contraire les aberrations sont importantes les résultats de l'optique géométrique sont applicables et l'on peut déduire la qualité de l'image de la diffusion latérale des rayons autour de l'image paraxiale. Mais souvent les aberrations géométriques sont assez réduites pour que la déformation de la surface d'onde Δ soit de l'ordre de grandeur de la longueur d'onde ou souvent moins. On peut alors étudier à l'aide de développements en séries l'influence des faibles aberrations sur le contraste des images, ce qui nous permettra de déterminer en particulier des tolérances sur les aberrations et de préciser la vieille règle du quart de longueur d'onde due à Lord Rayleigh. Il sera possible en fait d'étudier, dans le cadre de la théorie de la diffraction, les pertes de contraste des images d'objets types tels que points noirs, lignes noires, bords de plages, etc. provoquées par la présence de faibles aberrations. Il nous faut tout d'abord établir certains résultats généraux qui serviront de base à l'étude de l'influence des aberrations.

I. La formation des images d'objets étendus.

92. Méthode générale. Considérons un objet étendu caractérisé par une répartition de luminance $O(y, z)$; chaque petit élément de l'objet donne dans l'espace image de l'instrument une petite tache image dont la superposition forme l'image globale; nous ferons ici deux hypothèses:

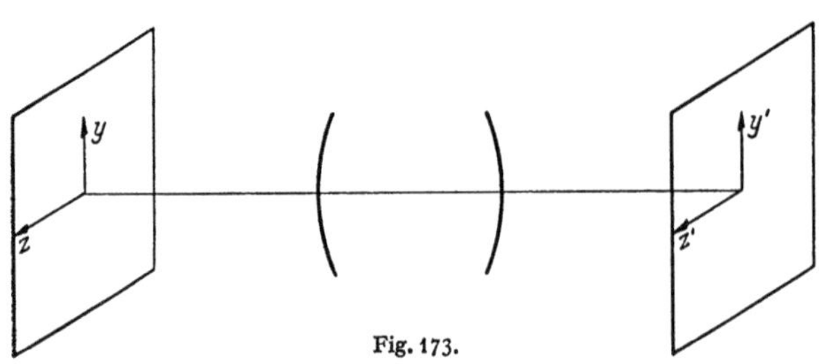

Fig. 173.

a) Les divers éléments de l'objet émettent des vibrations incohérentes entre elles: ainsi pour connaître l'image il suffira d'ajouter les éclairements produits par les divers éléments de l'objet (dans le cas de l'éclairage cohérent il faudrait au contraire ajouter les amplitudes complexes).

b) La loi de répartition des éclairements dans les taches images des divers points de l'objet est stationnaire: ceci signifie que les aberrations varient relativement lentement ou encore que pour observer leurs variations dans le champ il faut se déplacer de quantités importantes par rapport aux dimensions de la tache de diffusion qui représente l'image d'un point: cette hypothèse est généralement vérifiée dans les instruments courants.

Dans ces conditions on peut définir la tache de diffusion qui représenterait l'image d'un point lumineux placé à l'origine des coordonnées dans l'espace objet par une fonction $D(y', z')$. Si maintenant on place un point objet en un point de coordonnées y et z son image sera représentée par la fonction $D(y' - g y, z' - g z)$ si g est le grandissement de l'instrument (Fig. 173). On peut choisir les unités dans les plans objet et image de façon que l'on ait $g = 1$ et l'image d'un élément, $d y\, d z$ de l'objet sera alors:

$$O(y \cdot z)\, d y\, d z \times D(y' - y, z' - z).$$

L'image globale est constituée par la sommation de ces divers éléments et l'on aura pour répartition:

$$I(y', z') = \iint O(y, z)\, D(y' - y, z' - z)\, d y\, d z. \tag{92.1}$$

Ainsi une image optique est liée à l'objet et à la tache de diffusion par une relation de *convolution*. On sait que cette relation se traduit très simplement dans le langage de la transformation de Fourier; considérons en effet les transformées de Fourier à deux variables des fonctions précédentes définies par les relations:

$$\left.\begin{aligned}
i(\mu', \nu') &= \iint I(y', z') \exp\left[j\, 2\pi(\mu'\, y' + \nu'\, z')\right] d y'\, d z', \\
o(\mu', \nu') &= \iint O(y, z) \exp\left[j\, 2\pi(\mu'\, y + \nu'\, z)\right] d y\, d z, \\
d(\mu', \nu') &= \iint D(y', z') \exp\left[j\, 2\pi(\mu'\, y' + \nu'\, z)\right] d y'\, d z'.
\end{aligned}\right\} \tag{92.2}$$

On peut remplacer I dans la première des relations ci-dessus par son expression (92.1); on aura:

$$i(\mu', \nu') = \iint \exp\left[j\,2\pi(\mu'\,y' + \nu'\,z')\right] \left[\iint O(y,z)\,D(y'-y, z'-z)\,dy\,dz\right] dy'\,dz'.$$

Il suffit de prendre pour variables $y' - y = Y$ et $z' - z = Z$ qui remplaceront les variables y' et z', et de grouper les termes:

$$i(\mu', \nu') = \iint \exp\left[j\,2\pi(\mu'Y + \nu'Z)\right] D(Y,Z)\,dY\,dZ \iint \exp\left[j\,2\pi(\mu'\,y + \nu'\,z)\right] O(y,z)\,dy\,dz$$

ce qui met en évidence le résultat bien connu:

$$i(\mu', \nu') = o(\mu', \nu')\,d(\mu', \nu'). \tag{92.3}$$

Puisque l'image optique s'obtient par convolution de l'objet et de la tache de diffusion, sa transformée de FOURIER et le produit des transformées de l'objet et de la tache de diffusion. Ainsi pour passer de l'objet à l'image on peut commencer par effectuer sur l'objet la transformation de FOURIER, ce qui donne la fonction $o(\mu', \nu')$: cette opération revient en fait à représenter l'objet par la superposition d'une infinité de composantes sinusoïdales dont l'amplitude dépend de la fréquence, c'est-à-dire des deux composantes μ' et ν': considérons par exemple une composante sinusoïdale isolée, de période p et orientée dans la direction ϑ elle sera représentée par une fonction de la forme:

$$A \cos\left(2\pi\,\frac{y \cos\vartheta + z \sin\vartheta}{p} - \varphi\right)$$

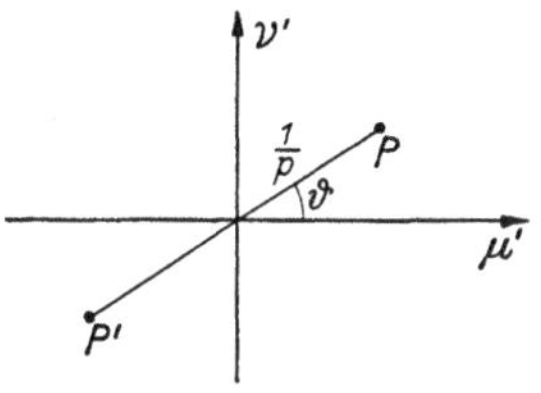

que l'on peut encore écrire:

$$\frac{A}{2}\exp(-j\varphi)\exp\left(j\,2\pi\frac{y\cos\vartheta + z\sin\vartheta}{p}\right) + \frac{A}{2}\exp(j\varphi)\exp\left(-j\,2\pi\frac{y\cos\vartheta + z\sin\vartheta}{p}\right).$$

Fig. 174.

On aura ainsi en fait deux composantes complexes dont les fréquences ont respectivement pour composantes $\mu' = \pm\dfrac{1}{p}\cos\vartheta$, $\nu' = \pm\dfrac{1}{p}\sin\vartheta$ représentables par les deux points P et P' du plan des fréquences (Fig. 174). Les modules $OP = OP'$ des fréquences sont bien égaux à $1/p$. Ayant ainsi décomposé l'objet en une infinité de composantes sinusoïdales (ou exponentielles imaginaires) on effectue ensuite un filtrage des amplitudes des composantes (le filtre caractérisant l'instrument n'est autre que la fonction $d(\mu', \nu')$ et l'on obtient la transformée de FOURIER de l'image; on met donc en évidence le fait que le passage d'un objet à son image optique est un filtrage des fréquences spatiales. Nous allons en donner un exemple simple qui nous domera l'occasion de montrer la possibilité de produire l'inversion des contrastes.

93. Image d'une mire périodique dans un instrument mal mis au point. Considérons une «mire de FOUCAULT» constituée par des traits alternativement noirs et blancs. Elle peut être représentée par la série de FOURIER:

$$O(y) = 1 + \frac{4}{\pi}\left[\sin 2\pi\frac{y}{p} + \frac{1}{3}\sin 3\,\frac{2\pi y}{p} + \cdots\right]$$

ou encore:

$$O(y) = 1 - \frac{2j}{\pi}\sum_{-\infty}^{+\infty}\frac{\exp j\,(2n+1)\,2\pi\dfrac{y}{p}}{2n+1}.$$

Ainsi la transformée de Fourier de l'objet ne présente que les fréquences 0, $1/p$, $\pm 3/p$ etc. Nous la représentons sur la Fig. 175 par un diagramme en perspective où l'on figure dans un plan complexe d'abscisse μ' la valeur de l'amplitude complexe.

La tache de diffusion dans un instrument mal mis au point, et où la diffraction n'intervient pas, n'est autre que la projection de la pupille de sortie de l'instrument sur le plan de la plaque. Nous supposerons pour simplifier que la pupille a la forme d'une fente et que l'image ainsi obtenue sur la plaque est une bande (ou

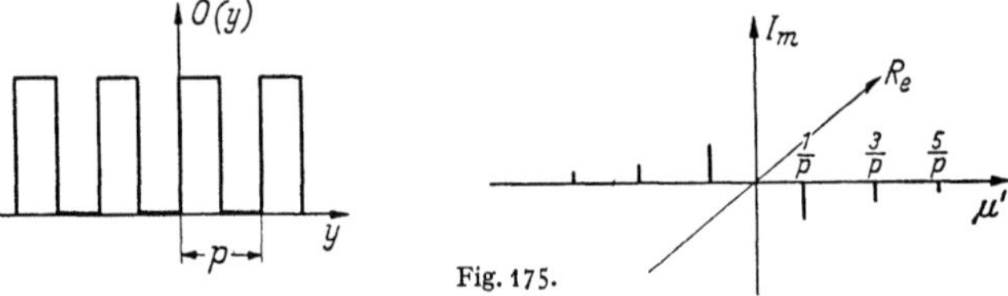

Fig. 175.

un rectangle) de largeur a. La transformée de Fourier d'une fonction nulle à l'extérieur de l'intervalle $\pm a/2$ et égale à $1/a$ à l'intérieur de cet intervalle est de la forme:

$$d(\mu') = \frac{\sin \pi \mu' a}{\pi \mu' a}$$

et ses variations sont représentées également sur la Fig. 176. Ainsi les diverses composantes de l'objet vont être filtrées par la fonction $d(\mu')$: on obtient en

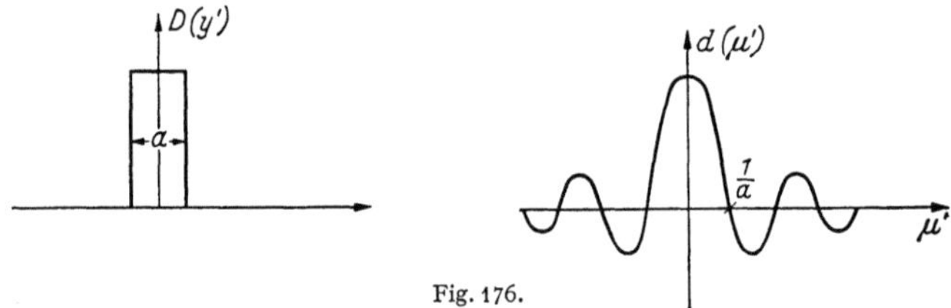

Fig. 176.

fait la transformée de Fourier de l'image en faisant le produit $o(\mu')\,d(\mu')$ et l'image peut s'écrire sous la forme:

$$I(y') = 1 - \frac{2j}{\pi}\left[d\left(\frac{1}{p}\right)\exp\left(j\,\frac{2\pi y'}{p}\right) - d\left(-\frac{1}{p}\right)\exp\left(-j\,\frac{2\pi y'}{p}\right) + \cdots\right]$$

$$= 1 + \frac{4}{\pi}\left[d\left(\frac{1}{p}\right)\sin 2\pi\,\frac{y'}{p} + \frac{1}{3}\,d\left(\frac{3}{p}\right)\sin 3\,\frac{2\pi y'}{p} + \cdots\right].$$

Prenons tout d'abord un défaut de mise au point assez faible: a est petit et la fréquence $\mu' = 1/a$ pour laquelle l'instrument est imperméable et assez grande; il en résulte que $d(1/p)$, $d(3/p)$ sont voisins de l'unité; l'image est assez voisine de l'objet mais néanmoins les contours seront arondis du fait de la disparition d'harmoniques élevés. Si maintenant le défaut de mise au point augmente il ne subsiste pratiquement que le terme fondamental: les harmoniques sont pratiquement éliminés, l'image est pratiquement sinusoïdale et son amplitude est $\frac{4}{\pi}\,d\left(\frac{1}{p}\right)$.

Lorsque le défaut de mise au point est tel que $a = p$ la modulation disparaît: on a en effet $d(1/p) = 0$ et l'image est une répartition uniforme d'éclairement (tous les harmoniques disparaissent en effet en même temps que le fondamental). Si maintenant on augmente le défaut de mise au point la valeur de la fréquence

$\mu' = 1/a$ qui correspond au premier zéro de $d\,(1/p)$ diminue et $d\,(1/p)$ devient négatif: la modulation de l'image reparaît, mais cette fois avec une invérsion de contraste: les minima de l'image correspondent aux maxima de l'objet (Fig. 177); on pourra bien entendu produire un défaut de mise au point plus important et le contraste changera de signe comme $d\,(1/p)$ en même temps qu'il diminuera en valeur absolue.

Ainsi peut on interpréter les inversions de contraste souvent observées en pratique pour les objets présentant une structure périodique. La Fig. 178 représente l'image d'une mire à traits légèrement convergents où ce phénomène est nettement mise en évidence.

En conclusion on voit qu'il existe une grande analogie entre le fonctionnement d'un instrument d'optique qui peut être complètement caractérisé par une loi de filtrage des fréquences spatiales, et les instruments électroacoustiques où les fréquences sonores

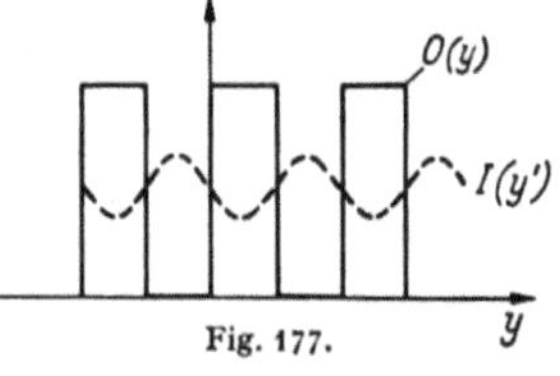

Fig. 177.

sont transmises suivant une loi qui dépend de l'instrument. Il est à prévoir qu'en optique comme en acoustique ou en électricité il y aura intérêt à rechercher une loi de filtrage aussi uniforme que possible. Nous allons voir maintenant que ce n'est pas le cas de l'instrument stigmatique limité par la diffraction, où la loi de filtrage est du type passe bas à bande de fréquence limitée.

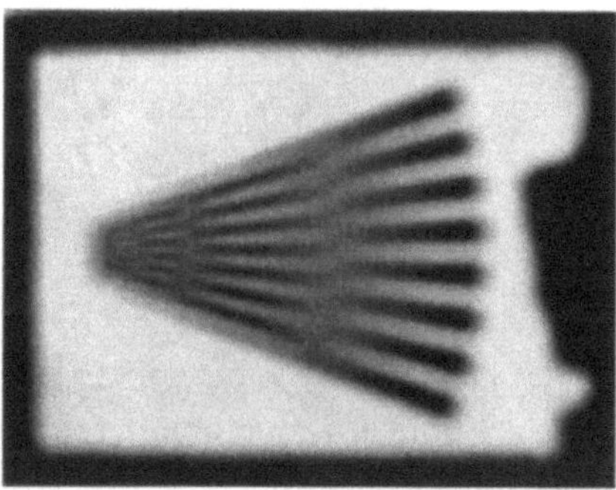

Fig. 178.

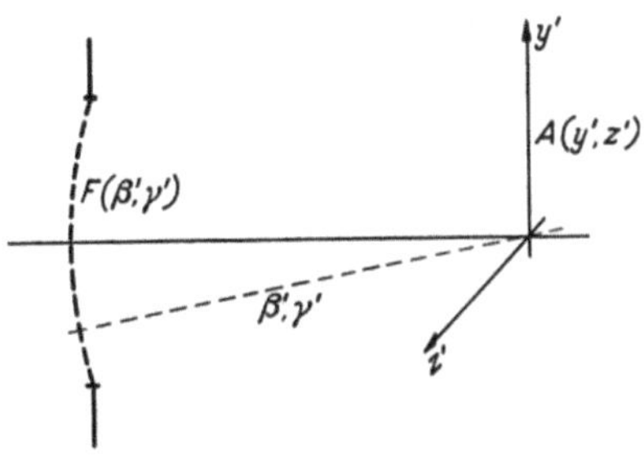

Fig. 179.

94. La loi de filtrage des fréquences spatiales dans les instruments stigmatiques. Considérons maintenant un instrument parfaitement stigmatique: on sait que la tache image est une tache de diffraction où la répartition de l'énergie peut être caractérisée par une fonction $D\,(y', z') = |A\,(y' \cdot z')|^2$, la fonction complexe. A représentant la répartition des amplitudes vibratoires. On sait aussi que la fonction A s'exprime à l'aide d'une transformation de FOURIER à partir de la distribution des amplitudes $F\,(\beta', \gamma')$ sur une sphère centrée en O et limitée au contour de la pupille (Fig. 179); on a en effet:

$$A\,(y', z') = \frac{j\,R}{\lambda} \iint F\,(\beta', \gamma')\, \exp\left[-j\,k\,(\beta'\,y' + \gamma'\,z')\right]\, d\beta'\, d\gamma' \qquad (94.1)$$

et réciproquement

$$F\,(\beta', \gamma') = -\frac{j}{\lambda\,R} \iint A\,(y', z')\, \exp\left[j\,k\,(\beta'\,y' + \gamma'\,z')\right]\, dy'\, dz' \qquad (94.2)$$

où R est le rayon de la sphère de référence, β' et γ' des coordonnées angulaires permettant de décrire la pupille et k la quantité $\dfrac{2\pi}{\lambda}$.

Il est facile d'utiliser les propriétés de la transformation de Fourier pour rechercher en particulier la transformée de la fonction $D(y', z') = |A(y', z')|^2$ qui s'écrit :

$$d(\mu', \nu') = \iint A(y', z') \, A^*(y', z') \exp\left[j\,2\pi(\mu'\,y' + \nu'\,z')\right] dy'\,dz'. \qquad (94.3)$$

Remplaçons $A(y', z')$ par son expression tirée de (94.1) :

$$d(\mu', \nu') = \frac{jR}{\lambda} \iint A^*(y', z') \exp\left[j\,2\pi(\mu'\,y' + \nu'\,z')\right] \times$$
$$\times \left\{ \iint F(\beta', \gamma') \exp\left[-j\,k(\beta'\,y' + \gamma'\,z')\right] d\beta'\,d\gamma' \right\} dy'\,dz',$$

ou encore, en intervertissant l'ordre des intégrations :

$$\frac{jR}{\lambda} \iint F(\beta', \gamma') \left\{ \iint A^*(y', z') \exp\left[-j\,k(\beta' - \mu'\,\lambda, \gamma' - \nu'\,\lambda)\right] dy'\,dz' \right\} d\beta'\,d\gamma'$$

et en appliquant la deuxième relation (94.1), on peut encore écrire :

$$d(\mu', \nu') = R^2 \iint F(\beta', \gamma') \, F^*(\beta' - \mu'\,\lambda, \gamma' - \nu'\,\lambda) \, d\beta'\,d\gamma'; \qquad (94.3')$$

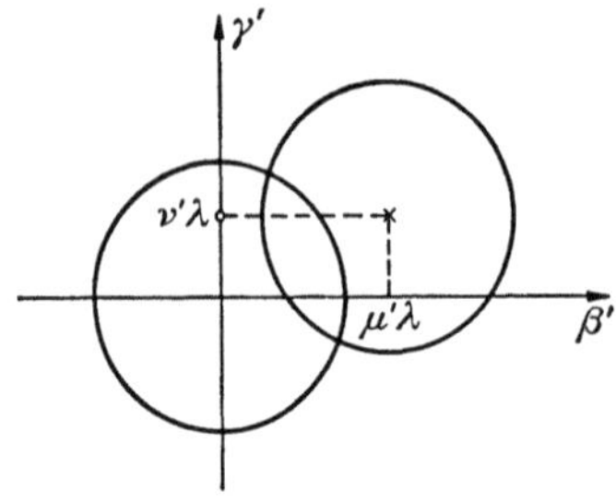

Fig. 180.

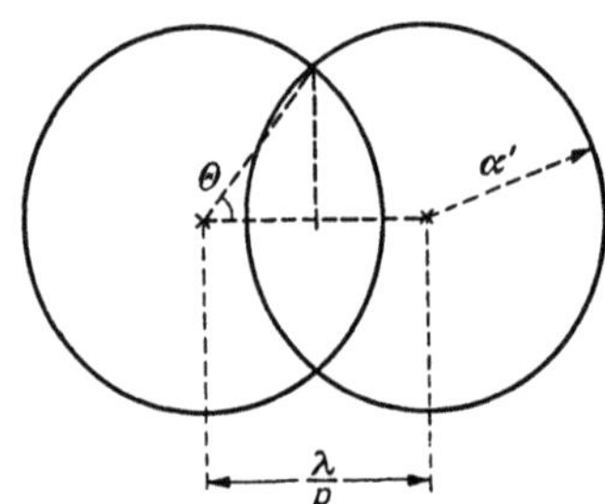

Fig. 181.

c'est-à-dire que l'analyse harmonique de la fonction $D(y', z')$ est représentée par la fonction d'autocorrélation de la répartition des amplitudes complexes $F(\beta', \gamma')$ sur la sphère de référence.

Considérons le cas particulier de l'instrument stigmatique : $F(\beta', \gamma')$ peut être considérée comme constant et égal à l'unité sur la sphère de référence (qui est une surface d'onde) ; si l'on représente dans un plan le contour de la pupille, dans un système de coordonnées β', γ' qui représentent en fait des angles on peut dire que la loi de filtrage des fréquences spatiales est donnée par la surface commune aux deux contours qui limitent les fonctions $F(\beta', \gamma')$ et $F(\beta' - \mu'\,\lambda, \gamma' - \nu'\,\lambda)$ (Fig. 180).

Dans le cas particulier où la pupille est circulaire la transformée $d(\mu', \nu')$ présente évidemment la symétrie de révolution et se réduit à une fonction d'une seule variable $d\left(\sqrt{\mu'^2 + \nu'^2}\right)$ dont on peut déterminer la méridienne : il suffit pour cela d'évaluer la surface commune à deux cercles de rayon dont les centres sont décalés d'une distance égale à $\sqrt{\mu'^2 + \nu'^2}\,\lambda$. Il suffit de supposer par exemple $\nu' = 0$ et de discuter en fonction de $\mu' = \dfrac{1}{p}$ si l'on fait apparaître la période p de la composante sinusoïdale envisagée ; la surface commune peut aisément être évaluée en fonction du paramètre ϑ (Fig. 181) ; on a d'une part pour définir ϑ :

$$\cos\vartheta = \frac{\lambda\mu'}{2\alpha'} = \frac{\lambda}{2p\,\alpha'} \qquad (94.4)$$

et d'autre part l'expression du facteur de modulation M, représenté par la surface commune aux deux cercles (en prenant comme unité la surface de l'un des cercles):

$$M = \frac{2}{\pi}\left[\vartheta - \sin\vartheta\cos\vartheta\right]. \tag{94.5}$$

Les variations de M en fonction de $\cos\vartheta$ (qui est proportionnel à la fréquence spatiale) sont représentées sur la Fig. 182 et le Tableau 2 ci-dessous en donne quelques valeurs.

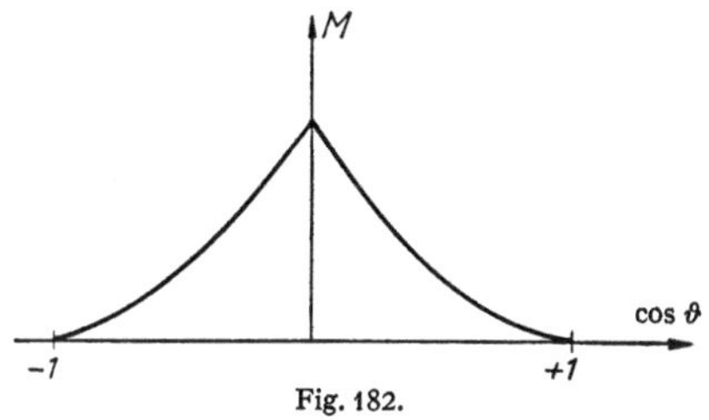
Fig. 182.

Tableau 2.

$\cos\vartheta$	M	$\cos\vartheta$	M
0	1	0,6	0,2848
0,1	0,8729	0,7	0,1881
0,2	0,7471	0,8	0,1041
0,3	0,6238	0,9	0,0374
0,4	0,5046	1	0
0,5	0,3910		

Ainsi on voit qu'un instrument stigmatique établit un filtrage des fréquences du type passe bas, où le facteur de transmission décroît régulièrement jusqu'à une fréquence limite; il est particulièrement intéressant de préciser cette fréquence limite on plus exactement la période $1/p$ qui lui correspond; on doit avoir lorsque les deux cercles deviennent tangents:

$$\cos\vartheta = 1 \quad \text{ou:} \quad \boxed{p_0 = \frac{\lambda}{2\alpha'}}. \tag{94.6}$$

On voit que le période limite est un peu inférieure au rayon de la tache de diffraction qui est égal comme on sait à:

$$1{,}22\,\frac{\lambda}{2\alpha'}.$$

Etant donné un objet périodique il est facile de déduire de là la nature de son image: considérons par exemple une mire constituée par de fins traits blancs de largeur ε équidistants (à la distance p); on peut représenter une telle mire par le développement:

$$O(y) = \frac{\varepsilon}{p} + \frac{2}{\pi}\sum_1^\infty \frac{1}{n}\sin\frac{n\pi\varepsilon}{p}\cos\frac{2n\pi y}{p}$$

et l'image sera:

$$I(y') = \frac{\varepsilon}{p} + \frac{2}{\pi}\sum_1^\infty \frac{1}{n}\sin\frac{n\pi\varepsilon}{p}\,M\left(\frac{n\lambda}{2p\alpha'}\right)\cos\frac{2n\pi y'}{p}. \tag{94.7}$$

Lorsque la fréquence $1/p$ est petite, $\dfrac{\lambda}{2p\alpha'}$ est faible et la fréquence fondamentale est transmise, ainsi que les quelques premiers harmoniques, mais lorsque p diminue la fréquence fondamentale augmente et on arrive rapidement à éliminer complètement les harmoniques; ainsi l'image d'un objet périodique dans un instrument stigmatique devient rigoureusement sinusoîdale aussitôt que $\dfrac{1}{p} > \dfrac{\lambda}{4\alpha'}$ (Fig. 183); on peut alors écrire puisque ε/p est petit:

$$I(y') = \frac{\varepsilon}{p}\left[1 + 2M\left(\frac{\lambda}{2p\alpha'}\right)\cos\frac{2\pi y'}{p}\right]. \tag{94.8}$$

Il est d'autre part facile de connaître le contraste de l'image qui doit être supérieur au minimum de contraste perceptible pour qu'il y ait résolution. Si le récepteur est l'oeil le minimum de contraste perceptible est très faible: on peut donc résoudre des structures périodiques dont la période est très proche de la période limite p_0; si le minimum de contraste est par exemple de 2/100 on peut écrire:

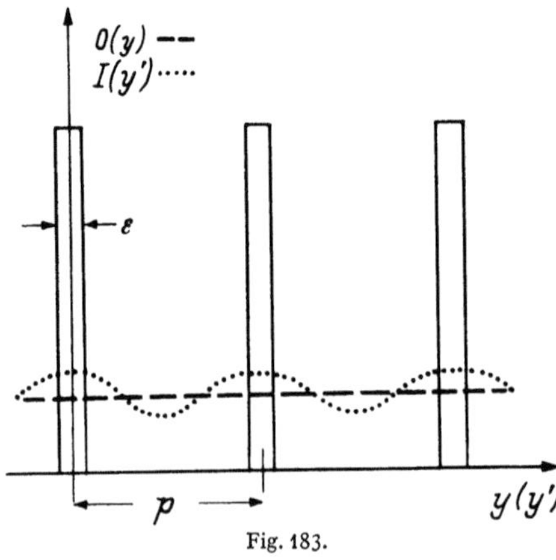

Fig. 183.

$$2M\left(\frac{\lambda}{2p\alpha'}\right) = \frac{1}{100}$$

(la variation d'éclairement entre les maxima et les minima sera en effet de 2/100) et l'on en tire:

$$M\left(\frac{\lambda}{2p\alpha'}\right) = \frac{1}{200};$$

c'est-à-dire que $\cos\vartheta$ sera très voisin de l'unité; au voisinage du point limite on peut d'ailleurs utiliser la formule approchée:

$$M = \frac{4}{3\pi}\left[2\left(1-\cos\vartheta\right)\right]^{\frac{3}{2}}$$

qui donne ici: $\cos\vartheta = 1 - 0{,}025$; c'est-à-dire encore: $p = 1{,}025\,p_0$. On peut donc s'approcher de la fréquence limite à quelques centièmes près.

On note d'autre part que la courbe de filtrage des fréquences ne présente pas de changement de signe: il n'y aura donc pas d'inversion de contraste.

La connaissance de la courbe de filtrage des fréquences peut aider à la recherche de l'image d'un objet quelconque; on se bornera à mentionner quelques exemples simples qui seront étudiés plus en détails dans le chapitre relatif à l'influence des aberrations.

Contraste de l'image d'un point noir de surface s (Fig. 184): $\dfrac{s\,\omega}{\lambda^2} = \dfrac{\pi s\alpha'^2}{\lambda^2}$.

Contraste de l'image d'une ligne noire de largeur ε: $\dfrac{16}{3\pi}\dfrac{\varepsilon\alpha'}{\lambda}$.

Pente de l'image d'un bord de plage (Fig. 185) $\dfrac{16}{3\pi}\dfrac{\alpha'}{\lambda}$.

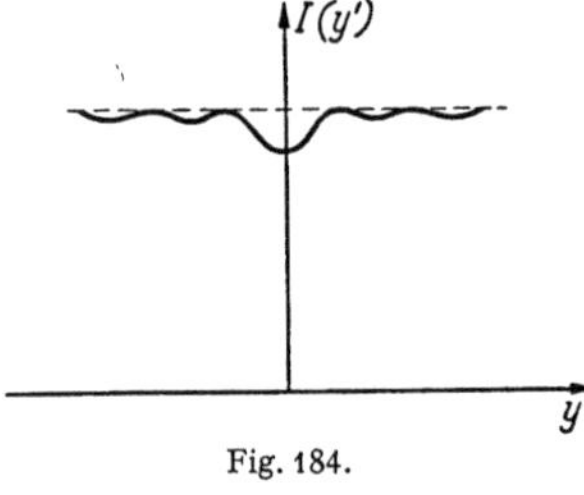

Fig. 184.

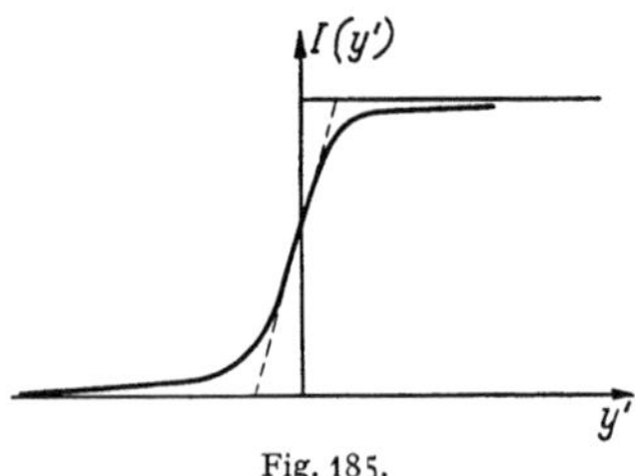

Fig. 185.

Signalons enfin que la limitation de la bande de fréquences spatiales transmissibles influe sur la quantité d'information contenue dans une image optique: il résulte en effet du fait que la bande de fréquences est limité à $\pm 2\alpha'/\lambda$ que l'on peut connaître l'image optique si l'on connaît l'éclairement en des points disposés régulièrement suivant un quadrillage dont le côté serait $\lambda/4\alpha'$ (théorème d'interpolation, dont on trouvera en particulier la démonstration et l'utilisation dans les travaux de A. Blanc-Lapierre); ainsi la répartition des éclairements dans une image ne dépend pas d'une infinité de paramètres, comme la répartition des luminances dans l'objet, qui est parfaitement arbitraire en chacun des points de l'objet: pour fixer la nature de l'image il suffit de connaître la valeur de l'éclairement en un nombre fini de points (Fig. 186); si S est la surface de l'image on ne dispose en fait pour la caractériser que de $16\alpha'^2 S/\lambda^2$ paramètres arbitraires.

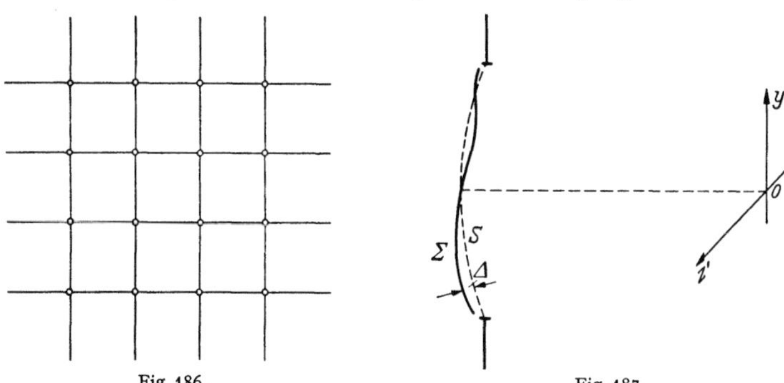

Fig. 186.   Fig. 187.

II. Influence des aberrations sur le contraste.

95. L'instrument d'optique aberrant peut être caractérisé par la déformation de la surface d'onde $\varDelta(\beta', \gamma')$ dans l'espace image: on a vu que chaque type d'aberration apportait une déformation que l'on introduira en temps utile dans l'expression de $\varDelta$ mais il nous faut tout d'abord écrire les relations générales

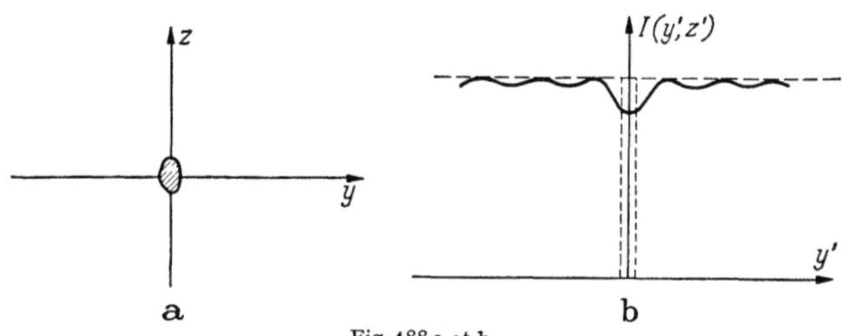

a   b

Fig. 188 a et b.

qui permettent l'étude de la formation des images: Sur une sphère centrée en O (Fig. 187) la répartition des amplitudes vibratoires est représentée par $F(\beta', \gamma') = \exp\left[jk\varDelta(\beta', \gamma')\right]$ (avec $k = 2\pi/\lambda$), fonction qui caractérise complètement les aberrations de l'instrument.

L'amplitude $A(y', z')$ dans la tache de diffraction est égale à

$$A(y', z') = \frac{jR}{\lambda} \iint F(\beta', \gamma') \exp\left[-jk(\beta' y' + \gamma' z')\right] d\beta' d\gamma' \qquad (95.1)$$

et la répartition de l'énergie est représentée par:

$$D(y', z') = |A(y', z')|^2. \qquad (95.2)$$

Dans ces conditions l'image d'un petit point noir de surface s (Fig. 188a) placé à l'origine des coordonnées au milieu d'un fond uniforme de luminance unitaire $[O(y, z) = 1$ sauf à l'origine, à l'intérieur de la surface $s]$ sera [en appliquant la relation (92.1)]:

$$I(y', z') = \iint D(y' - y, z' - z)\, dy\, dz - s\, D(y', z'). \qquad (95.3)$$

On doit donc retrancher d'un fond uniforme représenté par le premier terme la fonction $s D(y', z')$: l'allure de la répartition des éclairements est représentée sur la Fig. 188 b et en tenant compte des propriétés générales de la transformation de Fourier on trouve [en combinant par exemple les relations (94.2) et (94.3) et en faisant $\mu' = \nu' = 0$]

$$\iint D(Y, Z)\, dY\, dZ = R^2 \iint |F(\beta', \gamma')|^2\, d\beta'\, d\gamma' = R^2 \omega \qquad (95.4)$$

si ω est l'angle solide du cône de rayons qui viennent former l'image; la répartition des éclairements dans l'image d'un point est finalement:

$$I(y', z') = R^2 \omega - s D(y', z')$$

et le contraste s'exprime, en appelant D_M la valeur maximum de $D(y', z')$ par:

$$\gamma = \frac{s D_M}{R^2 \omega}.$$

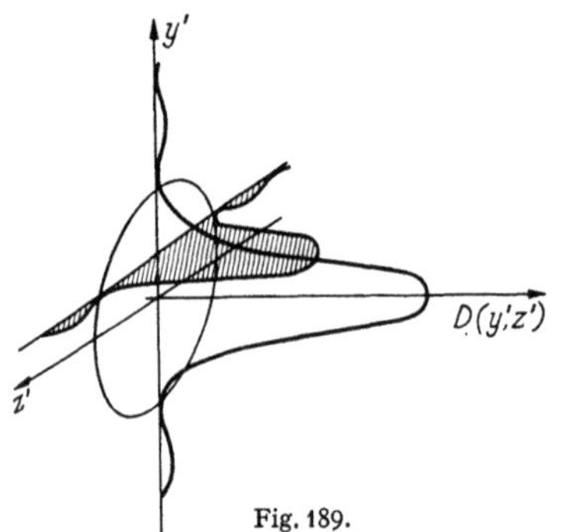

Fig. 189.

On voit donc que le contraste de l'image d'un point noir pourra être caractérisé par la valeur maximum de la fonction $D(y', z')$.

Dans le cas de l'instrument stigmatique on aurait d'ailleurs, en prenant $F(\beta', \gamma') = 1$ à l'intérieur de la pupille

$$D_M = D(0, 0) = |A(0, 0)|^2 = \frac{R^2 \omega}{\lambda^2}$$

et l'on retrouve ainsi l'expression déjà indiquée précédemment.

Considérons maintenant une ligne claire de largeur ε placée suivant l'axe Oz: en utilisant toujours l'expression (95.1), on trouve:

$$I(y') = R^2 \omega - \varepsilon \int\limits_{-\infty}^{+\infty} D(y', z' - z)\, dz.$$

Posons $S(y') = \int D(y', Z)\, dZ$; $S(y')$ représente la surface de section du «solide de diffraction» par le plan d'ordonnée y' (Fig. 189); le contraste de l'image sera manifestement égal à $\dfrac{\varepsilon S_M}{R^2 \omega}$ si S_M représente la valeur maximum de $S(y')$.

Exprimons maintenant la fonction $S(y')$:

$$S(y') = \int A(y', z')\, A^*(y', z')\, dz', \qquad (95.5)$$

mais on peut écrire

$$A(y', z') = \frac{jR}{\lambda} \int \exp(-j k \gamma' z') \left\{ \int F(\beta', \gamma') \exp(-j k \beta' y')\, d\beta' \right\} d\gamma'$$

qui montre que $A(y', z')$ peut être considéré comme la transformée de Fourier à une variable de

$$f(\gamma', y') = \frac{jR}{\lambda} \int F(\beta', \gamma') \exp(-j k \beta' y')\, d\beta'. \qquad (95.6)$$

En appliquant un théorème bien connu de la transformation de Fourier, on aura:

$$S(y') = \int A(y', z')\, A^*(y', z')\, dz' = \lambda \int f(\gamma', y')\, f^*(\gamma', y')\, d\gamma'. \qquad (95.7)$$

Inversement si l'on considère maintenant une ligne noire de largeur ε son image sera représentée par la fonction:

$$I(y') = R^2 \omega - \varepsilon S(y')$$

et le contraste sera $\dfrac{\varepsilon S_M}{R^2 \omega}$ où S_M désigne la valeur maximum de $S(y')$.

Cette expression s'applique bien entendu au cas particulier de l'instrument stigmatique à pupille circulaire: on aura dans ce cas en prenant

$$F(\beta',\gamma') = \begin{cases} 1 \ \ \text{si}\,\beta'^2 + \gamma'^2 < \alpha'^2, \\ 0 \ \ \text{si}\,\beta'^2 + \gamma'^2 > \alpha'^2. \end{cases}$$

Le maximum de $S(y')$ a lieu dans ce cas lorsque $y' = 0$:

$$f(\gamma',0) = \frac{2jR}{\lambda}\sqrt{\alpha'^2 - \gamma'^2} \quad \text{et} \quad S(0) = \frac{4R^2}{\lambda^2}\int(\alpha'^2 - \gamma'^2)\,d\gamma' = \frac{16R^2}{3\lambda}\alpha'^3$$

d'où le contraste déjà indiqué à la Sect. 94:

$$\gamma = \frac{16}{3\pi}\frac{\varepsilon\alpha'}{\lambda}.$$

Le contraste d'un bord de plage peut également s'exprimer à l'aide $S(y')$: l'image du bord du demi plan $y > 0$ peut en effet être représentée par

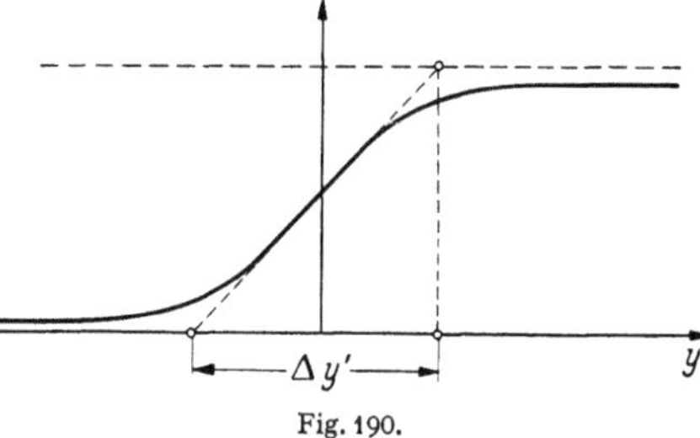

Fig. 190.

$$I(y') = \int_0^\infty\!\!\int D(y'-y,z'-z)\,dy\,dz = \int_0^\infty S(y'-y)\,dy = \int_{-\infty}^{y'} S(Y)\,dY$$

et la pente de la courbe $I(y')$ est ainsi $S(y')$ (Fig. 190).

Si l'on considère enfin un objet périodique son contraste s'exprimera aisément en fonction de la courbe de transmission des fréquences spatiales que traduit la relation (94.2). On pourra finalement caractériser le contraste par les intégrales suivantes, effectuées à l'intérieur du contour de la pupille:

Point: $\qquad\qquad\qquad |\int F(\beta',\gamma')\exp[-jk(\beta'y'+\gamma'z')]\,d\beta'\,d\gamma'|^2,$

Ligne noire, bord de plage: $\int|\int F(\beta',\gamma')\exp(-jk\beta'y')\,d\beta'|^2\,d\gamma',$ $\qquad$ (95.8)

Structure périodique: $\qquad \int F(\beta',\gamma')\,F^*(\beta'-\lambda\mu',\gamma'-\lambda\nu')\,d\beta'\,d\gamma'$

étant entendu que l'on recherche pour les deux premières expressions les valeurs maxima en fonction de y' et z' pour la première, de y' pour la seconde.

Il reste maintenant à exprimer ces intégrales lorsque la surface d'onde est déformée par les aberrations classiques, que nous écrirons sous la forme:

$$\Delta = dh^2 + s_1h^4 + s_2h^6 + Kh\cos(\varphi - \Phi) + (c_1h^3 + c_2h^5)\cos\varphi + ah^2\cos2\varphi \quad (95.9)$$

où h et φ sont des coordonnées polaires sur la pupille de sortie, telles que $h=1$ au bord. On aura ainsi:

$$h = \frac{1}{\alpha'}\sqrt{\beta'^2 + \gamma'^2}$$

d caractérise le défaut de mise au point (ou la courbure de champ);

s_1 caractérise l'aberration sphérique du 3ème ordre;

s_2 caractérise l'aberration sphérique du 5ème ordre;

K caractérise le basculement latéral de la sphère de référence dans la direction Φ;

c_1 caractérise la coma du 3ème ordre;

c_2 caractérise la coma du 5ème ordre;

a caractérise l'astigmatisme.

Le terme contenant K peut remplacer en fait le terme $\beta'y' + \gamma'z'$ qui permet de calculer l'amplitude en un point mobile dans le plan de front y', z'.

On pourrait tout d'abord montrer rapidement que le contraste des images ne change pas si l'on change Δ en $-\Delta$: le contraste est donc une fonction paire de l'aberration envisagée et nous nous bornerons tout d'abord aux termes quadratiques par rapport aux coefficients s_1, s_2, c_1, c_2, a qui indiqueront aisément l'influence de faibles aberrations.

96. Influence des faibles aberrations sur l'image d'un point. Il est particulierement aisé dans ce cas de donner une expression simple de la baisse d'éclairement au centre de la tache de diffraction (ou en son voisinage); on a, en posant

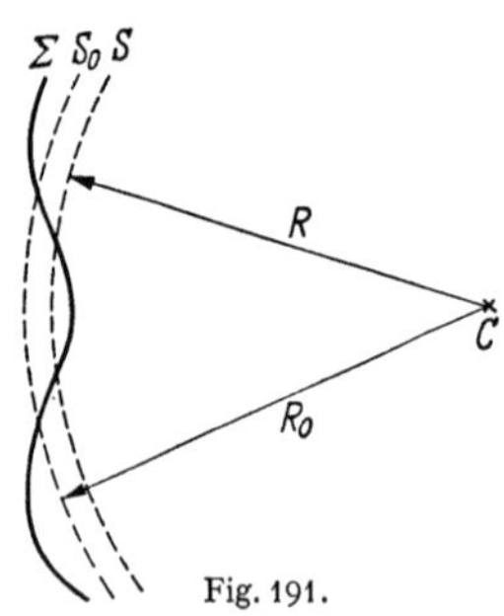

Fig. 191.

$$\Delta' = \Delta - \beta' y' - \gamma' z',$$

$$A(y', z') = \frac{jR}{\lambda} \iint \exp(jk\Delta')\, d\beta'\, d\gamma'$$

et si Δ' n'est pas trop grand on peut écrire:

$$\exp(jk\Delta') = 1 + jk\Delta' - \frac{k^2\Delta'^2}{2}.$$

Il est commode de poser

$$ds = \frac{1}{\pi\alpha'^2}\, d\beta'\, d\gamma'$$

(ainsi ds est l'élément de surface d'onde lorsqu'on prend pour unité la surface de la pupille); on a finalement:

$$D(y', z') = \frac{R^2}{\lambda^2}\left[1 + jk \iint \Delta'\, ds - \frac{k^2}{2} \iint \Delta'^2\, ds\right]\left[1 - jk \iint \Delta'\, ds - \frac{k^2}{2} \iint \Delta'^2\, ds\right]$$

ou encore:

$$\boxed{\frac{\lambda^2}{R^2} D(y', z') = 1 - k^2\left[\overline{\Delta'^2} - (\overline{\Delta'})^2\right]}$$

ou l'on désigne par $\overline{\Delta'}$ et $\overline{\Delta'^2}$ les valeurs moyennes des quantités correspondantes.

Cette relation nous permet d'évaluer aisément la baisse relative d'éclairement au voisinage du centre de la tache de diffraction en présence de faibles aberrations.

Assurons nous tout d'abord que cette expression ne dépend pas du rayon R de la sphère de référence: si l'on ajoute une constante a à Δ' on aura en effet:

$$\overline{(\Delta' + a)^2} - (\overline{\Delta' + a})^2 = \overline{\Delta'^2} + 2a\overline{\Delta'} + a^2 - (\overline{\Delta'} + a)^2 = \overline{\Delta'^2} - (\overline{\Delta'})^2.$$

On peut en particulier choisir le rayon R_0 de la sphère de référence de telle sorte que

$$\overline{\Delta_0'} = \overline{\Delta'} + R_0 - R = 0$$

et l'on reconnaît que l'expression du contraste s'écrit alors

$$\boxed{1 - k^2\, \overline{\Delta_0'^2}} \tag{96.1}$$

où Δ_0' est l'écart par rapport à la sphère de rayon R_0 tel que $\overline{\Delta_0'} = 0$ (Fig. 191); cette sphère peut être appelée «sphère moyenne», elle laisse entre la surface d'onde Σ et elle-même le même volume de part et d'autre.

*On voit que la baisse d'éclairement par rapport au centre de la tache d'*Airy *n'est pas directement liée à la valeur maximum de l'écart Δ' (regle de* Rayleigh*) mais plutôt à l'écart quadratique moyen de la surface d'onde par rapport à une sphère convenable.*

On peut alors formuler une règle générale de correction des aberrations si l'on admet, en suivant l'exemple de Lord Rayleigh qu'une baisse d'éclairement de 20% n'affecte pas sensiblement la qualité; on écrira:

$$k^2\,\overline{\varDelta_0'^2} \leq 0,2 \tag{96.2}$$

ou encore

$$\overline{\varDelta_0'^2} \leq \frac{\lambda^2}{180}. \tag{96.3}$$

Une étude plus précise montre d'ailleurs que l'éclairement est certainement supérieur à $\left(1-\dfrac{k^2}{2}\,\overline{\varDelta_0'^2}\right)^2$, ce qui permet d'être sûr que lorsqu'on satisfait à la tolérance ci-dessus la baisse de contraste n'excède pas 20%.

Ainsi pour que l'éclairement au centre de la tache de diffraction ou le contraste de l'image d'un point noir ne décroissent pas de plus de 20%, il suffit que le carré moyen de l'écart de la surface d'onde par rapport à une sphère convenable n'excède pas $\lambda^2/180$.

Ce résultat permet évidemment d'étudier les variations d'éclairement au voisinage du «centre» d'une surface d'onde quasi sphérique: le point de l'espace où l'éclairement est maximum est évidemment celui pour lequel $\overline{\varDelta_0'^2}$ est maximum. On a donc ainsi la possibilité de trouver le véritable «foyer» de l'onde, c'est-à-dire le point où l'on obtient le maximum d'énergie. Après avoir déterminé la position du foyer F on

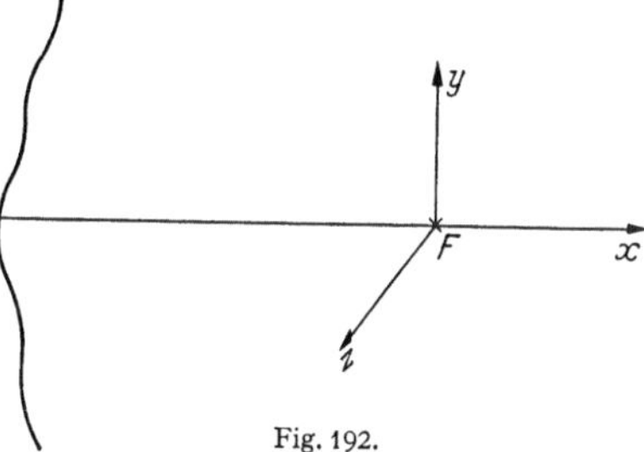

Fig. 192.

peut également étudier les variations d'éclairement autour du foyer, en particulier en fonction d'un déplacement longitudinal suivant Ox (Fig. 192) ou latéral. Nous nous bornerons ici à étudier l'influence des diverses aberrations, ce qui nous donnera d'ailleurs l'occasion de rechercher la position du foyer physique.

Reprenons pour cela la relation (95.9) et calculons les quantités $\overline{\varDelta'}$ et $\overline{\varDelta'^2}$ en faisant $\varDelta' = \varDelta$ puisque le déplacement latéral du centre de la sphère de référence peut être obtenu à l'aide du paramètre K. Il suffit pour cela de prendre $ds = \dfrac{1}{\pi}\,h\,dh\,d\varphi$ et d'effectuer l'intégration entre 0 et 1 pour h, et de 0 à 2π pour φ. On trouve alors l'expression suivante pour l'éclairement relatif:

$$\left.\begin{aligned}
1 - \frac{4\pi^2}{\lambda^2}\Bigg[& \frac{d^2}{12} + \frac{ds_1}{6} + \frac{4}{45}s_1^2 + \frac{3}{20}ds_2 + \frac{1}{6}s_1 s_2 + \frac{9}{7\times 16}s_2^2 + \\
& + \frac{K^2}{4} + \frac{KC_1}{3} + \frac{C_1^2}{8} + \frac{KC_2}{4} + \frac{C_1 C_2}{5} + \frac{C_2^2}{12} + \frac{A^2}{6}\Bigg].
\end{aligned}\right\} \tag{96.4}$$

97. Discussion des résultats. Tolérances. On peut tirer de cette expression des valeurs des tolérances relatives à toutes les catégories d'aberrations que nous avons envisagées.

α) *Défaut de mise au point.* On doit avoir

$$\frac{4\pi^2 d^2}{12\lambda^2} < 0,2$$

soit encore

$$d \leq \frac{\lambda}{4}.$$

On retombe ici sur la règle de Lord RAYLEIGH: dans un instrument stigmatique fournissant une surface d'onde sphérique centrée en un point O, on peut encore mettre au point en un point O' à condition que la distance entre la surface d'onde

Σ et la sphère de référence centrée en O' ne s'écartent pas au bord de plus de $\lambda/4$ (Fig. 193). D'après la relation (74.4), il suffit que le défaut de mise au point ε soit tel que $\dfrac{\varepsilon \alpha'^2}{2} \leq \dfrac{\lambda}{4}$ ou encore $\varepsilon < \dfrac{\lambda}{2\alpha'^2}$.

On peut également appliquer cette relation dans l'espace objet et elle nous fournit la valeur de la précision avec laquelle il est possible d'effectuer un pointé longitudinal. On voit ainsi que la précision des pointés longitudinaux varie comme le carré de l'ouverture de l'instrument.

β) *Aberration sphérique.* Considérons tout d'abord l'aberration sphérique du 3ème ordre, caractérisée par le coefficient s_1; il est manifeste sur l'expression (96.4) que l'on aura avantage à rechercher la meilleure mise au point, c'est-à-dire celle pour laquelle

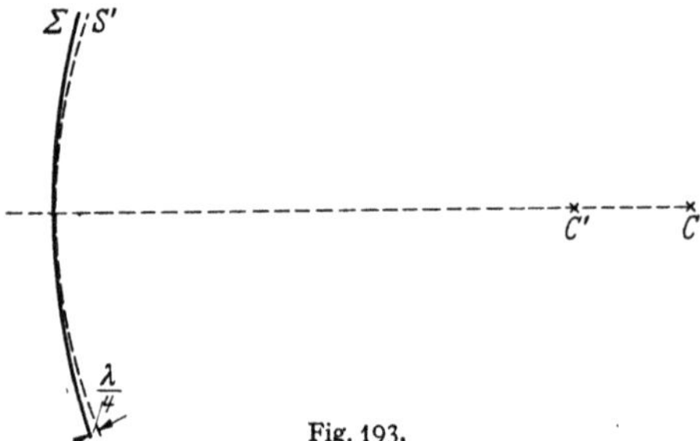

Fig. 193.

$$\frac{d^2}{12} + \frac{d s_1}{6} + \frac{4}{45} s_1^2$$

est minimum. Ceci à lieu lorsque $d = -s_1$, c'est-à-dire lorsque la mise au point est faite au centre de la sphère S' qui coupe la surface d'onde d'aberration sphérique au bord (Fig. 194).

En se reportant á la Sect. 74 on trouverait aisément que la mise au point est alors faite au milieu du segment d'aberration sphérique longitudinale (celui-ci est en fait complètement invisible lorsque l'aberration est faible).

En écrivant la tolérance générale pour cette mise au pount particulière, on trouve alors

$$s_1 < 0{,}95\, \lambda \approx \lambda.$$

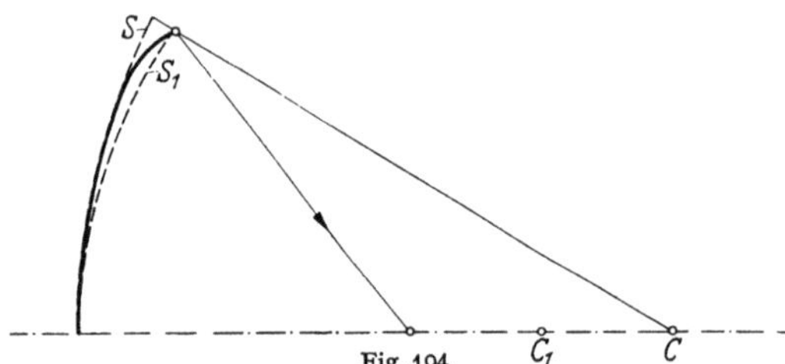

Fig. 194.

Ainsi la déformation de la surface d'onde par rapport à la sphère S centrée au foyer paraxial ne doit pas excéder une longueur d'onde; on trouverait d'ailleurs que dans ces conditions l'écart maximum par rapport à la sphère S' n'excède pas $\lambda/4$. L'aberration longitudinale maximum est alors égale (d'après les relations de la Sect. 74) à

$$l = \frac{4\lambda}{\alpha'^2}.$$

Dans le cas de l'aberration sphérique du 5ème ordre le problème se pose le plus souvent de la façon suivante: les termes d'aberration du 3ème ordre étant réduits à une très faible valeur, il subsiste une aberration du 5ème ordre qui est pratiquement difficile à réduire. On peut chercher à savoir quel résidu d'aberration du 3ème ordre il convient de laisser subsister pour atténuer au mieu l'effet du terme du 5ème ordre, de plus la mise au point peut également être choisie au mieux de sorte que l'on doit simplement rechercher le minimum du polynôme en d, s_1, s_2 de la relation (96.4), par rapport à d et s_1: en trouve qu'il fait $d = \frac{3}{5} s_2$ et $s_2 = -\frac{3}{2} s_1$; on trouve alors en écrivant la tolérance générale:

$$s_2 \leq 3{,}75\, \lambda.$$

Bien entendu, l'écart de la surface d'onde par rapport à la sphère optimum restera très inférieur à la longueur d'onde. On pourra finalement écrire que la déformation de la surface d'onde par rapport au foyer paraxial s'exprime par:

$$\Delta = 3{,}75\,\lambda\left[-\tfrac{3}{2}\,h^4 + h^6\right]$$

et l'on peut aisément déduire de cette relation l'aberration longitudinale: on trouve que l'aberration sphérique longitudinale s'annule pour l'ouverture maximum ($h=1$); ainsi la correction optimum de l'aberration sera obtenue lorsque l'aberration longitudinale marginale est nulle (Fig. 195); on montrerait dans les mêmes conditions que la valeur maximum de l'aberration longitudinale est $\dfrac{6\lambda}{\alpha'^2}$.

γ) *Coma.* Comme pour l'aberration sphérique nous pouvons rechercher le foyer physique de l'onde cette fois en utilisant le paramètre K qui traduit un déplacement latéral du centre de la sphère de référence à l'intérieur de la tache géométrique de coma, d'ailleurs complètement masquée par la diffraction.

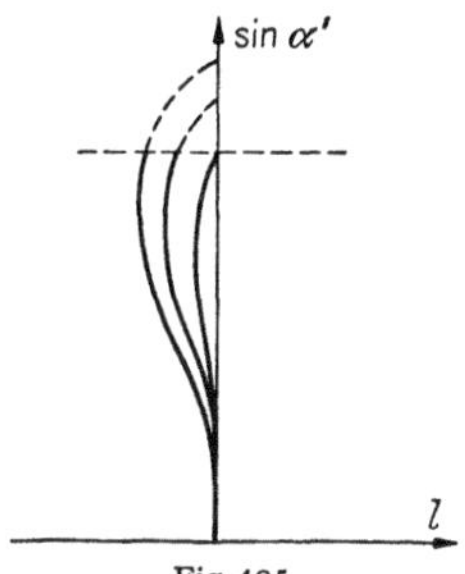

Fig. 195.

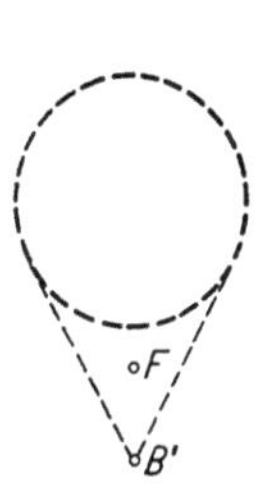

Fig. 196.

Fig. 197.

Pour la coma du 3ème ordre on trouve que le minimum du polynôme a lieu lorsque $K=-\tfrac{2}{3}\,C_1$: c'est-à-dire que le foyer est situé au centre d'une sphère S' que l'on peut connaître aisément; on trouve que ce foyer est situé aux $\tfrac{2}{3}$ de l'étendue totale de la tache comatique (Fig. 196) qui est d'ailleurs complètement masquée par la diffraction. Appliquant maintenant la tolérance générale, on trouve

$$C_1 \leq 0{,}6\,\lambda.$$

Ainsi la déformation maximum de la surface d'onde par rapport à une sphère centrée sur l'image paraxiale ne doit pas excéder 0,60 mais bien entendu l'écart par rapport à la sphère S', centrée au foyer physique F de l'onde est nettement plus faible (Fig. 197).

On peut d'ailleurs écrire encore la tolérance en fonction de l'écart à la condition des sinus: on a vu que l'on a

$$\Delta = y'\sin\alpha'\,\Phi(\sin\alpha'),$$

ou Φ est l'écart relatif à la condition des sinus; on pourra donc écrire pour la tolérance à l'écart à la condition des sinus:

$$\Phi(\sin\alpha') < \frac{0{,}6\,\lambda}{y'\alpha'}.$$

Lorsque l'on cherche à compenser au mieux la coma du 5ème ordre par un résidu de coma du 3ème ordre et un basculement éventuel de la sphère de référence, on devra chercher à rendre minimum le polynome en K, C_1, C_2, par rapport à K et C_1; on trouve:

$$C_1 = -\frac{6}{5}\,C_2 \quad \text{et} \quad K = \frac{3}{10}\,C_2;$$

d'autre part la valeur maximum de C_2 est 2,5 λ. On peut d'ailleurs traduire ceci par une tolérance sur l'écart maximum par rapport à la condition des sinus: après quelques calculs simples on trouve que la courbe (Fig. 198) représentant $\Psi(\sin\alpha')$ doit être repliée pour une ouverture légèrement supérieure à l'ouverture maximum (dans le rapport $\sqrt{1,2}$) et que l'écart maximum ne doit pas dépasser $\dfrac{0,9\,\lambda}{y'\alpha'}$.

δ) *Astigmatisme.* On trouve dans ce cas

$$\frac{a^2}{6} < \frac{\lambda^2}{180} \quad \text{soit à peu près} \quad a < 0,17\,\lambda.$$

On voit que la tolérance sur l'astigmatisme est nettement plus sévère que pour les autres aberrations: cette aberration est généralement représentée effectivement

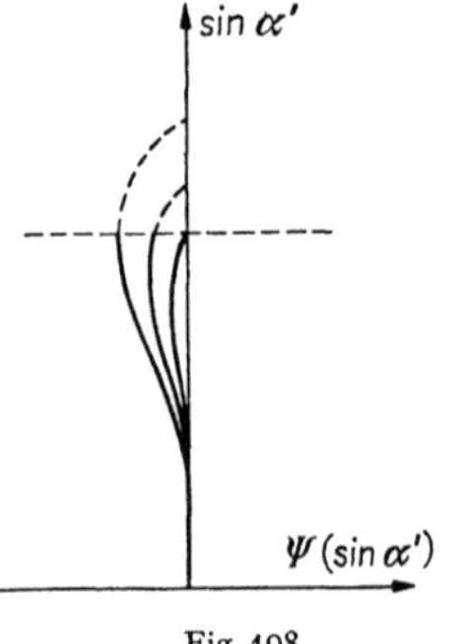

Fig. 198.

par un terme en h^2 difficile à compenser et il s'ensuit qu'il sera difficile dans ce cas de satisfaire à la tolérance ci-dessus.

ε) *Ensemble des aberrations.* On remarque sur l'expression (96.4) que les aberrations forment des groupes indépendants: aberration sphérique, coma, astigmatisme; il en résulte que, si l'on définit pour chacun une tolérance spécifique valable lorsque cette aberration est seule, on pourra formuler une tolérance générale sur l'ensemble des aberrations sous la forme suivante:

$$\left(\frac{S}{S_M}\right)^2 + \left(\frac{C}{C_M}\right)^2 + \left(\frac{A}{A_M}\right)^2 \le 1,$$

où S_M, C_M, A_M sont les tolérances spécifiques relatives aux types d'aberration sphérique, de coma et d'astigmatisme envisagés. Il serait aisé de montrer que le défaut de mise au point est en fait indépendant de l'aberration sphérique si celle-ci est définie par rapport au foyer physique de l'onde, de sorte que l'on peut encore ajouter à la some précédente $\left(\dfrac{D}{D_M}\right)^2$ moyennant certaines précautions.

98. Influence des faibles aberrations sur la qualité de l'image d'une ligne ou d'un bord de plage; tableau des tolérances. Dans le cas de l'image d'une ligne, on utilise d'une part la deuxième des expressions (95.8) en effectuant les mêmes approximations que dans le cas du point, et d'autre part les expressions (95.9) où l'on remplacera dans les termes en C_1, C_2 et a le paramètre φ par $\varphi - \Phi$ de façon à pouvoir faire varier l'orientation des aberrations dissymétriques (coma et astigmatisme), par rapport à celle de la ligne (supposée parallèle à l'axe Oz, comme à la Sect. 95).

On obtient alors l'expression suivante pour la variation du contraste en présence de faibles aberrations:

$$1 - \frac{3\pi^2}{4\lambda^2}\,[0{,}339\,s_1^2 + 0{,}305\,s_2^2 + 2\times0{,}320\,d\,s_1 + 2\times0{,}285\,d\,s_2 + 2\times0{,}317\,s_1 s_2 +$$
$$+ \sin^2\Phi\,(0{,}01655\,C_1^2 + 0{,}03868\,C_1 C_2 + 0{,}02392\,C_2^2) +$$
$$+ \cos^2\Phi\,(0{,}652\,K^2 + 0{,}856\,K C_1 + 0{,}636\,K C_2 + 0{,}318\,C_1^2 +$$
$$+ 0{,}506\,C_1 C_2 + 0{,}210\,C_2^2) + 0{,}325\,(a\cos 2\Phi + d)^2 + 0{,}812\,a^2\sin^2 2\Phi].$$

On peut bien entendu faire à ce propos une discussion analogue à celle de paragraphe précédent: on trouve des résultats très voisins, mais il est cependant à remarquer que l'orientation de la ligne par rapport aux aberrations a une influence; l'ensemble des tolérances sur le point et la ligne sont rassemblés sur le Tableau 3.

Tableau 3. *Aaleurs de diverses aberrations produisant une perte de contraste de 20%.*

		Eclairage incohérent			Eclairage cohérent				
		point	ligne		point noir	ligne noire		ligne lumineuse	
Défaut de mise au point	$D\ll$	$0,25\ \lambda$	$0,29\ \lambda$		$0,166\ \lambda$	$0,22\ \lambda$		$0,24\ \lambda$	
Aberr. sph. du 3ème ordre — isolée	$S\ll$	$0,25\ \lambda$	$0,28\ \lambda$		$0,22\ \lambda$	$0,30\ \lambda$		$0,27\ \lambda$	
combinée avec un défaut de mise au point	$D=$	$-S$	$-0,98\ S$		$-3/4\ S$	$-5/7\ S$		$-6/7\ S$	
	$S\ll$	$0,95\ \lambda$	$1,04\ \lambda$		$0,9\ \lambda$	λ		$0,92\ \lambda$	
Aberr. sph. repliée (3ème et 5ème ordre) et défaut de mise au point	$v_0=$	1	1		$8/9$	$0,85$		$0,91$	
	$S'\ll$	$3,75\ \lambda$	$4\ \lambda$		$4\ \lambda$	$4,5\ \lambda$		$3,7\ \lambda$	
			$\Phi=0$	$\Phi=\pi/2$		$\Phi=0$	$\Phi=\pi/2$	$\Phi=0$	$\Phi=\pi/2$
Coma du 3ème ordre — isolée	$C\ll$	$0,20\ \lambda$	$0,2\ \lambda$	$0,86\ \lambda$	$0,28\ \lambda$	$0,265\ \lambda$	∞	$0,19\ \lambda$	∞
combinée à un décalage latéral	$K=$	$-2/3C$	$-2/3C$	$?$	$-2/3C$	$-3/5C$	$?$	$-3/5C$	$?$
	$C\ll$	$0,60\ \lambda$	$0,58\ \lambda$	$0,86\ \lambda$	$0,85\ \lambda$	$0,66\ \lambda$	∞	$0,47\ \lambda$	∞
Coma repliée (3ème et 5ème ordre) et décalage latéral	$v_0=$	$6/5$	$6/5$	$1,17$	$6/5$	$10/9$	$?$	$10/9$	$?$
	$C'=$	$2,5\ \lambda$	$2,35\ \lambda$	$3,1\ \lambda$	$3,5\ \lambda$	$2,7\ \lambda$	∞	$1,85\ \lambda$	∞
Astigmatisme — $\Phi=0$	$A\ll$	$0,17\ \lambda$			$0,25\ \lambda$				
$\Phi=\pi/2$ $D\mp A\ll$			$0,29\ \lambda$			$0,22\ \lambda$		$0,24\ \lambda$	
$\Phi=\pi/4$	$A\ll$	$0,17\ \lambda$	$0,18\ \lambda$		$0,25\ \lambda$	∞		∞	

11*

On remarque tout d'abord dans ce tableau que les tolérances relatives à la ligne obtenues pour $\Phi=0$ dans le cas de la coma sont très voisines de celles relatives au point: dans ce cas la tache géométrique serait étendue dans la direction de l'axe Oy', c'est-à-dire perpendiculairement à la ligne. Dans le cas où la tache géométrique serait au contraire parallèle à la ligne ($\Phi=\pi/2$) les tolérances sont plus importantes ce qui signifie que la coma a moins d'influence.

Le cas de l'astigmatisme doit être précisé: si la ligne est parallèle à l'une des focales d'astigmatisme ($\Phi=0$ ou $\pi/2$) on obtient le même contraste que pour un instrument stigmatique à condition de mettre au point sur la focale convenable la tolérance se réduit alors à une tolérance de défaut de mise au point par rapport à cette focale (terme en $D \pm A$).

Si maintenant on prend $\Phi=\pi/2$ (la ligne objet fait un angle de 45° avec chacune des focales) on obtient une tolérance sur l'astigmatisme seul, plus sévère que dans le cas précédent.

On trouvera sur la partie droite du tableau les tolérances correspondantes relatives au cas de l'éclairage cohérent, que nous donnons ici, à titre d'information: les calculs relatifs à l'éclairage cohérent sont analogues à ceux que nous avons esquissés pour l'éclairage cohérent; on doit cependant utiliser les amplitudes complexes et non les carrés des modules, ce qui conduit d'ailleurs à des calculs plus simples.

99. Le cas des structures périodiques. On peut encore étudier l'influence de faibles aberrations sur le contraste des images d'objets périodiques: il suffit d'utiliser la troisième relation (95.8) et de la développer avec les mêmes approximations que précédemment; ce travail a été effectué par W. H. Steel qui a obtenu une expression générale du contraste en fonction des aberrations (supposées faibles) et de la fréquence spatiale.

Nous poserons pour caractériser la fréquence spatiale $\omega = \cos\vartheta = \dfrac{\lambda}{2p\alpha'}$ et $\varepsilon = \sin\vartheta$.

Posant d'autre part

$$P_1 = 3\,(\vartheta - \sin\vartheta\cos\vartheta) = 3\,(\vartheta - \varepsilon\omega),$$
$$P_2 = P_1 - 2\varepsilon^3\omega,$$
$$P_3 = P_2 - \tfrac{8}{5}\varepsilon^5\omega,$$
$$P_4 = P_3 - \tfrac{48}{35}\varepsilon^7\omega,$$

on obtient l'expression suivante du facteur de transmission de l'instrument aberrant, où le premier terme et relatif á l'instrument parfait:

$$\frac{2}{3\pi}P_1 + \{$$

$$-D^2 \quad \times \tfrac{16}{3}\pi\omega^2\left[(1 + 4\omega^2)\,P_3 - \tfrac{32}{5}\varepsilon^7\omega\right]$$

$$-2DS_1 \times \tfrac{64}{9}\pi\omega^2\left[(1 + 12\omega^2 + 12\omega^4)\,P_3 - \tfrac{24}{5}\varepsilon^7\omega\,(3 + 4\omega^2)\right]$$

$$-S_1^2 \quad \times \tfrac{32}{9}\pi\omega^2\left[(3 + 68\omega^2 + 192\omega^4 + 96\omega^6)\,P_3 - \tfrac{48}{5}\varepsilon^7\omega\,(7 + 28\omega^2 + 16\omega^4)\right]$$

$$-2DS_2 \times \tfrac{8}{3}\pi\omega^2\left[(3 + 72\omega^2 + 240\omega^4 + 128\omega^6)\,P_3 - \tfrac{16}{5}\varepsilon^7\omega\,(21 + 104\omega^2 + 64\omega^4)\right]$$

$$-2S_1 S_2 \times \tfrac{32}{15}\pi\omega^2\left[(6 + 225\omega^2 + 1320\omega^4 + 2000\omega^6 + 640\omega^8)\,P_3 -\right.$$
$$\left. - \tfrac{16}{5}\varepsilon^7\omega\,(57 + 497\omega^2 + 920\omega^4 + 320\omega^6)\right]$$

$$-S_2^2 \quad \times \tfrac{16}{15}\pi\omega^2\left[(15 + 828\omega^2 + 7960\omega^4 + 24\,000\omega^6 + 23\,040\omega^8 + 5120\omega^{10})\,P_3 -\right.$$
$$\left. - \tfrac{64}{35}\varepsilon^7\omega\,(315 + 4578\omega^2 + 17\,298\omega^4 + 19\,040\omega^6 + 4480\omega^8)\right]$$

$$- B^2 \cos^2 \vartheta \times \tfrac{16}{3}\pi\omega^2 P_1$$
$$-2 BC_1 \cos^2\vartheta \times \tfrac{16}{3}\pi\omega^2 \left[(1 + 4\omega^2) P_1 - 8\,\varepsilon^3\omega\right]$$
$$- C_1^2 \cos^2 \vartheta \times \tfrac{8}{3}\pi\omega^2 \left[(3 + 34\omega^2 + 32\omega^4) P_1 - \tfrac{2}{5}\varepsilon^3\omega\,(129 + 206\omega^2)\right]$$
$$- C_1^2 \sin^2 \vartheta \times \tfrac{8}{9}\pi\omega^2 \left[(1 + 6\omega^2) P_4 - \tfrac{288}{55}\varepsilon^9\omega\right]$$
$$- A^2 \sin^2 2\vartheta \times \tfrac{8}{3}\pi\omega^2 P_2\}.$$

Cette expression complexe peut en fait plus aisément être utilisée si l'on calcule pour diverses valeurs de la fréquence spatiale réduite ω les coefficients des divers termes d'aberrations; on obtient le tableau suivant:

Tableau 4. *Pupille sans obturation. Image de la mire: coefficients des aberrations dans l'expression du contraste.*

Aberration	Fréquence spatiale $\omega = \cos\Theta$								
	0,1	0,2	0,3	0,4	0,5	0,6	0,7	0,8	0,9
Constante	0,8729	0,7471	0,6238	0,5046	0,3910	0,2848	0,1881	0,1041	0,0374
$- D^2$	0,5517	1,4761	2,1018	2,1995	1,8326	1,2196	0,61197	0,19508	0,02201
$- 2 D S_1$	0,6272	1,4752	1,9358	1,9822	1,7224	1,2642	0,72892	0,27376	0,03682
$- S_1^2$	0,8000	1,6395	1,9499	1,9108	1,6918	1,3435	0,87853	0,38583	0,06166
$- 2 D S_2$	0,6089	1,3046	1,6459	1,6831	1,4941	1,1453	0,71166	0,29911	0,04659
$- 2 S_1 S_2$	0,8279	1,5411	1,7512	1,6971	1,5165	1,2424	0,86685	0,42329	0,07809
$- S_2^2$	0,8922	1,5049	1,6294	1,5552	1,3954	1,1706	0,86433	0,46627	0,09897
$- C_1^2 \cos^2\vartheta$	0,2234	0,4773	0,5381	0,4467	0,3021	0,1726	0,0816	0,0275	0,00364
$- C_1^2 \sin^2\vartheta$	0,0854	0,2096	0,2691	0,2487	0,1778	0,09736	0,03767	0,00823	0,00049
$- A^2 \sin^2 2\vartheta$	0,3281	1,0536	1,8238	2,3620	2,4987	2,1940	1,5459	0,7769	0,1839

On peut faire à propose de ces résultats une remarque générale: les aberrations de toutes sortes ont une influence relativement importante sur le contraste lorsque l'on considère les fréquences réduites intermédiaires (ω de l'ordre de 0,5), mais leur influence est très faible pour les basses fréquences et encore plus faible pour les fréquences voisines de la fréquence limite: la courbe de filtrage d'un instrument aberrant présentera l'allure de celle de la Fig. 199: elle se rapproche de celle de

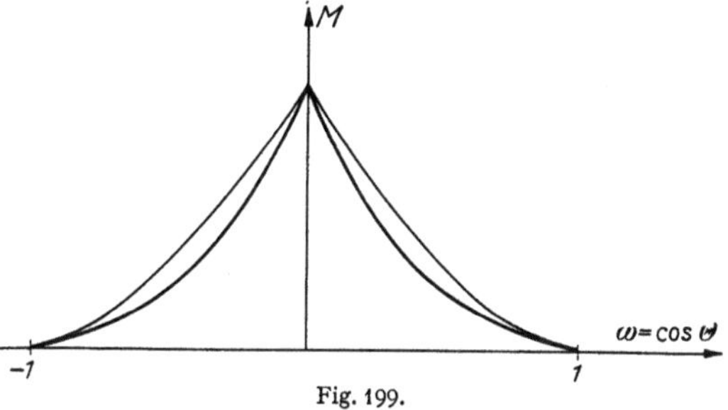

Fig. 199.

l'instrument parfait au voisinage de $\omega = 0$ et de $\omega = 1$ mais s'en écarte sensiblement pour les fréquences intermédiaires. On peut considérer en particulier l'influence d'un défaut de mise au point égal à $\lambda/4$; on trouve les facteurs de transmission suivants (représentés au Tableau 5).

Tableau 5. *Influence d'un defaut de mise au point égal à $\lambda/4$ sur le facteur de transmission.*

$\omega =$	0	0,1	0,2	0,3	0,4	0,5	0,6	0,7	0,8	0,9	1
Instrument parfait	1	0,8729	0,7471	0,6238	0,5046	0,3910	0,2848	0,1881	0,1041	0,0374	0
Défaut de mise au point ($\lambda/4$)	1	0,8384	0,6548	0,4924	0,3671	0,2765	0,2086	0,1499	0,0919	0,0360	0
Perte relative de contraste	0	0,04	0,12	0,21	0,27	0,29	0,27	0,20	0,12	0,04	0

Il en résulte que si l'on détermine par la méthode utilisée à la Sect. 94 la limite de résolution de l'instrument aberrant, on trouvera qu'elle est extrêmement voisine de celle de l'instrument parfait puisqu'au voisinage du point limite les deux courbes sont pratiquement confondues; si l'on veut au contraire détecter dans un instrument la présence de faibles aberrations il faudra utiliser des fréquences intermédiaires (ω voisin de 0,5); on pourra alors mesurer la baisse de contraste produite par les aberrations par un procédé photométrique convenable, mais si le contraste de l'objet est élevé le contraste de l'image sera toujours surabondant pour obtenir la résolution. On sera loin de la limite de résolution et les mesures de contraste seront indispensables. On peut néanmoins produire des variations de limites de résolution si l'on utilise comme objets des tests de faible contraste: le contraste de l'image pourra alors être voisin du minimum perceptible et la présence de faibles aberrations se traduira par une chute de la fréquence résoluble (pour un contraste objet donné).

On voit que la théorie de la diffraction permet de préciser l'influence des aberrations sur la qualité des images optiques: à cette occasion nous sommes sortis du cadre de l'optique géométrique. Nous avons obtenu des résultats utiles à l'opticien en déterminant en particulier des tolérances. Il serait souhaitable de donner maintenant des exemples d'applications mais nous nous bornerons à en indiquer un dans le prochain chapitre qui nous servira de conclusion.

G. Conclusion.

100. Dans les chapitres qui précèdent nous nous sommes efforcés d'étudier les principes généraux de l'optique géométrique en s'attachant à en indiquer autant que possible quelques aspects modernes.

Pour conclure cet article nous donnerons maintenant un aperçu très rapide de certains progrès réalisés dans le domaine instrumental.

De nombreux systèmes optiques ont été proposés dans ces dernières décades. Beaucoup représentent un progrès sensible principalement quant à la luminosité ou à l'étendue du champ utilisable; la découverte de ces nouvelles solutions a parfois été due à l'amélioration des moyens de calcul: on dispose maintenant de machines automatiques très rapides qui permettent d'effectuer une analyse complète des aberrations du système et de rechercher dans une famille de solutions voisines celle qui correspond à la correction optimum des aberrations. Les progrès les plus importants ont néanmoins été dus à la mise en oeuvre de nouvelles conceptions, de nouvelles méthodes de correction de l'ensemble des aberrations. Nous étudierons à titre d'exemple l'utilisation de systèmes correcteurs associés à un miroir sphérique.

101. L'utilisation des correctrices, les objectifs de Schmidt et de Bouwers-Maksutof. Ces objectifs nouveaux sont nés principalement des besoins de l'astronomie et nous nous placerons dans ce cadre pour étudier les mérites des solutions modernes.

Un objectif astronomique destiné à la photographie doit tout d'abord présenter un chromatisme très réduit: cette condition rend difficile l'utilisation de systèmes dioptriques si l'on veut atteindre une distance focale importante et suggère l'emploi de miroirs. Le miroir sphérique présente de l'aberration sphérique: si h est la hauteur d'incidence sur le miroir on a (Fig. 200)

$$CH = \frac{R}{2\cos\dfrac{\alpha'}{2}} = \frac{f}{\cos\dfrac{\alpha'}{2}} \approx f\left(1 + \frac{\alpha'^2}{8}\right).$$

L'aberration sphérique longitudinale est ainsi

$$l = f\,\frac{\alpha'^2}{8}\,. \tag{101.1}$$

A cette aberration longitudinale correspond une déformation marginale Δ de la surface d'onde (Fig. 201) égale, d'après les relation de la Sect. 74, à:

$$\Delta = l\,\frac{\alpha'^2}{4} = f\,\frac{\alpha'^4}{32}\,. \tag{101.2}$$

D'après le tableau, la tolérance applicable à cette aberration est de l'ordre de λ; celle-ci est rapidement atteinte et les ouvertures obtenues sont pratiquement

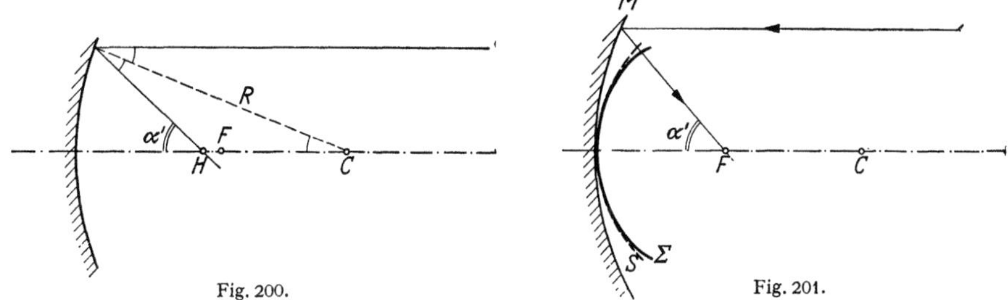

Fig. 200.   Fig. 201.

trop faibles: si l'on prend $f = 10$ mètres et $\Delta = \lambda = 0,5\,\mu$ on trouve que l'ouverture α' ne peut dépasser $3,5 \cdot 10^{-2}$ ce qui est manifestement très insuffisant ($f/14$ environ). On remarque néanmoins que le miroir sphérique diaphragmé dans le plan de son centre ne présente pas de coma ni d'astigmatisme: le faisceau

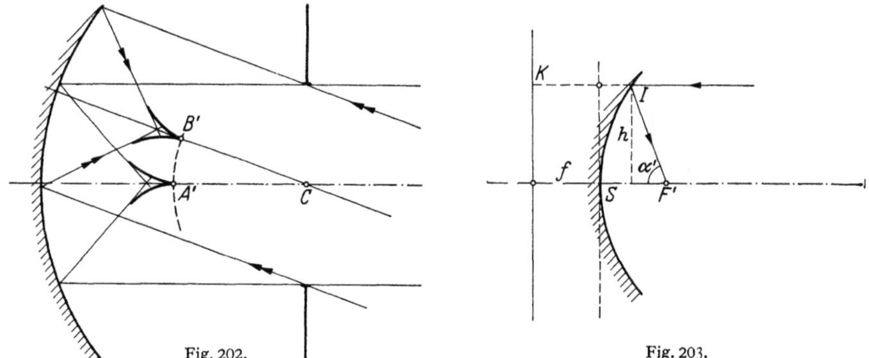

Fig. 202.   Fig. 203.

oblique (Fig. 202) donne naissance en effet à une caustique d'aberration sphérique centrée sur l'axe secondaire CB' identique à celle que l'on obtenait pour A' et, si le faisceau est diaphragmé symétriquement par rapport à l'axe CB', on n'aura pas de coma ni d'astigmatisme. Le miroir sphérique diaphragmé en son centre ne présente donc que de l'aberration sphérique et de la courbure de champ: l'image obtenue pour une petite ouverture vient en effet se peindre sur une surface sphérique de rayon f, centrée en C (Fig. 202).

Le paraboloïde de Newton ne présente pas d'aberration sphérique mais sa coma est très gênante: reprenant les notations des Sects. 76 et 77 on peut écrire (Fig. 203) pour rechercher l'écart à la condition des sinus:

$$\frac{h}{\sin\alpha'} = IF' = KI = f + \frac{h^2}{4f} = f\left[1 + \frac{\alpha'^2}{4}\right]\,.$$

Ainsi l'écart à la condition des sinus peut être caractérisé, dans le domaine du ème ordre par la fonction

$$\Phi(\sin\alpha') = \frac{\alpha'^2}{4}$$

et la déformation de la surface d'onde sera, d'après la relation (76.3):

$$\Delta = y' \frac{\alpha'^3}{4}. \tag{76.3}$$

Les tolérances figurant au Tableau 3 nous permettent de calculer la valeur maximum du champ utilisable; on écrira donc $\Delta < 0,6\,\lambda$; prenant l'exemple du grand télescope du Mont Palomar, on aura avec $\alpha' = 1/6,6$ et $\lambda = 0,5\,\mu$:

$$y' < \frac{4\Delta}{\alpha'^3} = \frac{4 \times 0,6\,\lambda}{\alpha'^3} = 0,35 \text{ mm.},$$

soit encore un champ angulaire total de

$$\frac{0,7 \text{ mm.}}{16 \text{ mm.}} = 0,4 \cdot 10^{-4} \text{radian.}$$

Ce champ est donc très petit et l'instrument serait inutilisable si l'on ne lui avait adjoint un système dioptrique correcteur placé au voisinage de l'image (correcteur de Ross).

On obtient cependant une bien meilleure correction de l'ensemble des aberrations si l'on revient au miroir sphérique, à condition de corriger au péalable son aberration sphérique par un dispositif additionnel convenable.

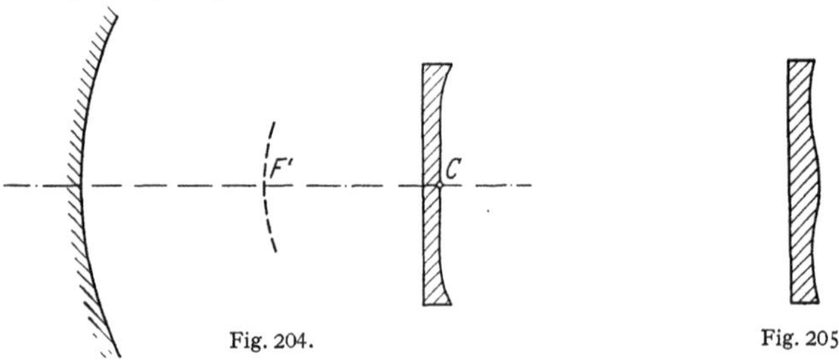

Fig. 204.   Fig. 205.

B. Schmidt a proposé de corriger l'aberration sphérique du miroir à l'aide d'une lame correctrice située dans le plan du centre: il suffit pour cela que les variations d'épaisseur e de la lame soient telles qu'elles compensent le chemin optique aberrant: il faut que l'on ait d'après l'équation (101.2), en appelant δe la variation d'épaisseur de la lame:

$$\delta e\,(n-1) = \frac{h^4}{32 f^3}$$

et l'on est ainsi conduit à donner à la lame la forme indiquée sur la Fig. 204.

On peut en fait donner à la lame une lègère convergence dans le but de diminuer l'épaisseur de matière à enlever: ceci revient à faire former l'image non pas au foyer paraxial mais en un point voisin; on a vu à la l'on met au point au milieu du segment d'aberration sphérique longitudinale, la déformation de la surface d'onde peut s'écrire sous la forme:

$$\Delta' = f\alpha'^2\,(\alpha'^2 - \alpha_m'^2).$$

On donnera alors à la lame une forme indiquée sur la Fig. 205 définie par la relation:

$$\delta e\,(n-1) = \frac{h^2}{32 f^3}\,(h^2 - h_m^2)$$

qui passe par un maximum égal à $\dfrac{h_m^4}{128\,f^3}$ pour $h^2 = \dfrac{h_m^2}{2}$. Prenons l'exemple de la chambre de Schmidt de l'observatoire du Mont Palomar: l'ouverture est $\dfrac{f}{2,5}$ soit $\dfrac{h_m}{f} = \dfrac{1}{5}$ et l'on aura avec $f = 3$ mètres:

$$\delta e\,(n-1) = 0,04 \text{ mm.}$$

La variation d'épaisseur δe sera de l'ordre de 0,08 mm., ce qui est remarquablement petit.

Si maintenant l'on envoie un faisceau légèrement incliné sur l'axe la correction de l'aberration sphérique se maintient, tout au moins en première approximation: l'analyse des chemins optiques montre qu'il n'apparait pas en tous cas

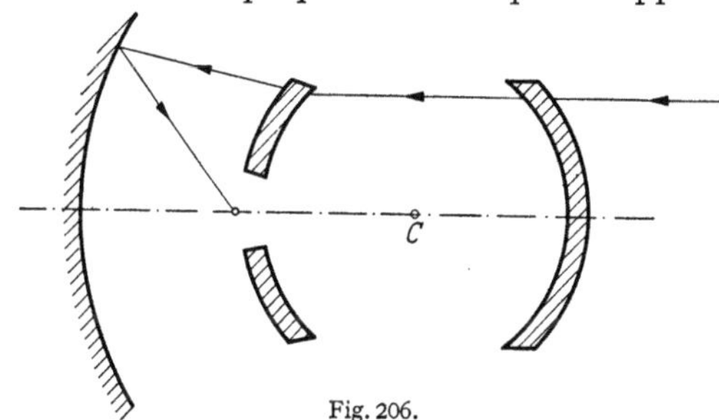

Fig. 206.

de terme de coma du 3ème ordre et l'image obtenue est sensiblement stigmatique mais elle vient se former en fait sur une sphère centrée sur C: le défaut le plus important est donc la courbure de champ.

Ce genre de système est très employé pour obtenir des champs angulaires importants tout en conservant une grande luminosité: la chambre de Schmidt

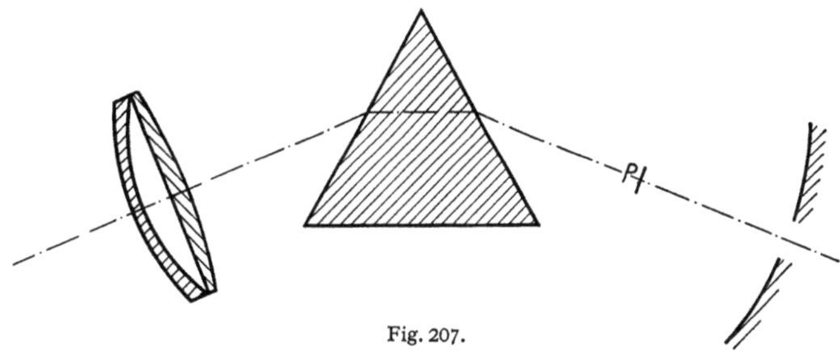

Fig. 207.

du Mont Palomar qui a une ouverture de $f/2,5$ et une distance focale de 3 mètres couvre un champ d'environ 9 degrés.

On voit donc que ceci représente un progrès considérable en particulier par rapport au télescope parabolique de Newton.

On a proposé d'autres systèmes correcteurs possibles: Bouwers et Maksukof ont suggéré l'emploi de ménisques concentriques au miroir et caculés de façon qu'ils corrigent l'aberration sphérique (Fig. 206). La parfaite symétrie sphérique du système nous assure que la seule aberration présente sera la courbure de champ: l'image se peint sur une sphère concentrique à l'ensemble. Arnulf a utilisé dans un objectif de spectrographe un doublet correcteur (Fig. 207) Baker, Linfoot et Slevogt ont suggéré d'associer une lame correctrice à un système de Cassegrain (Fig. 208) ce qui permet d'éviter la courbure de champ des systèmes précédents (qui ne comprennent qu'un seul miroir).

On voit en conclusion qu'il est possible de profiter de la symétrie du miroir sphérique et de son isoplanétisme lorsque la pupille est dans le plan du centre pour concevoir des systèmes dont les aberrations de champ sont remarquablement réduites, il suffit pour cela de corriger l'aberration sphérique du miroir par un système qui peut présenter lui-même la symétrie sphérique (BOUWERS, MAKSUKOF) ou encore dont les propriétés soient suffisamment stationnaires en fonction de l'inclinaison des faisceaux.

Nous terminerons en indiquant la possibilité d'effectuer le transport d'une image optique le long d'un trajet qui peut être assez lumineux grâce à l'utilisation de fines fibres: si l'on considère une tige cylindrique de verre on sait que la lumière peut y être conduite grâce aux réflexions totales successives (principe de la fontaine lumineuse); si l'on utilise un faisceau de fibres de verre de très

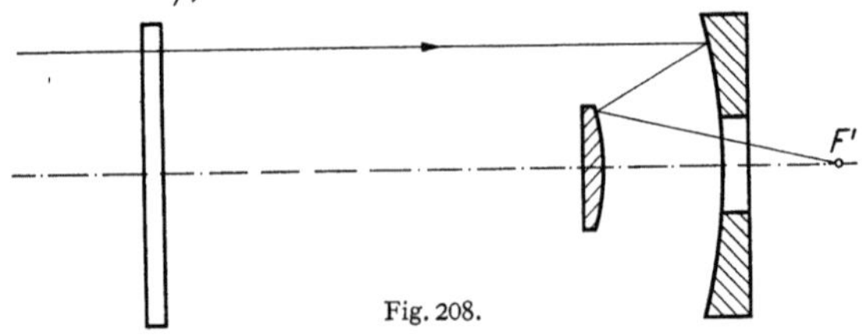

Fig. 208.

petit diamètre on peut ainsi conduire une image optique d'une extrémité à l'autre du faisceau en obtenant un pouvoir de résolution qui est défini par le diamètre des fibres. De tels systèmes peuvent être particulièrement utiles dans les endoscopes (A. C. S. VAN HEEL et H. H. HOPKINS).

Bibliographie.

BEREK, M.: Grundlagen der praktischen Optik. Berlin: W. de Gruyter 1930.
BORN, M.: Optik. Berlin 1933.
BLANC LAPIERRE, A.: Ann. Inst. H. Poincaré **13**, 245 (1953) et Optica Acta II, **1**, 1 (1955).
BLEIN, J.: Optique géométrique. Paris: Gaston Doin 1913.
BOUTRY, G. A.: Optique instrumentale. Paris: Masson & Cie. 1946.
BOUWERS, A.: Achievements in Optics. Amsterdam: Elsevier 1946.
CHRÉTIEN, H.: Calcul des Combinaisons Optiques. Paris: Rev. Opt. 3e éd. 1938.
CONRADY, M.: Applied Optics and Optical design. Oxford: Oxford Univ. Press 1929.
DANJON, A., et A. COUDER: Lunettes et Télescopes. Paris: Rev. Opt. **1935**.
DIMITROFF, G. Z., and J. G. BAKER: Telescopes and accessories. Philadelphia: P. Blakiston Son & Co. 1945.
DUFFIEUX, P. M.: L'intégrale de FOURIER et ses applications à l'optique. Rennes 1946.
FLEURY, P., A. MARÉCHAL et Mme C. ANGLADE: La théorie des images optiques (Réunions organisées en 1946). Paris: Rev. Opt. **1949**. — Réunions d'opticiens. Rev. Opt. **1950**.
FRANÇON, M., et A. MARÉCHAL: Influence de la diffraction. A paraître aux editions de la Revue d'Optique Paris.
HERZBERGER, M.: Strahlenoptik. Berlin: Springer 1931.
HOPKINS, H. H.: Wave Theory of Aberrations. Oxford: Clarendon Press 1950.
KÖNIG, A.: Handbuch der Experimentalphysik, Bd. XX/2.
LINFOOT, E. H.: Recent Advances in Optics. Oxford: Clarendon Press 1955.
LUNEBERG, R. K.: Mathematical Theory of Optics. Providence 1944.
MARÉCHAL: A., Imagerie Géométrique. Aberrations. Paris: Rev. Opt. **1952**.
MARTIN, L. C.: An introduction to applied optics. London: Isaac Pittman 1930.
NIJBOER, B. R. A. Thése Gröningen 1942.
PICHT, J.: Optische Abbildung. Braunschweig 1931.
Lord RAYLEIGH: Scientific Rapert, Vol. I, p. 435.
STEEL, W. H.: Thèse Paris 1952 et Rev. Opt. **1953**.
TORALDO DI FRANCIA, G.: Onde elettromagnetiche. Bologne: Zanichelli 1953.
ZERNIKE, F.: Physica, Haag **1**, 689 (1934).
Optical Image Evaluation (Symposium held in October 1951). Washington: N.B.S. 1954. — Optical Instruments (Proceedings of the London Conference). London: Chapman 1951.
Problems in contemporary Opties. Istituto nazionale di Ottica, Arcetri-Firenze 1956.
Quelques publications périodiques concernant l'optique géométrique et instrumentale: *Journal of the Optical Society of America*; *Optica Acta* (Revue Européenne); *Optik*; *Ottica*; *Proceedings of the Physical Society*; *Revue d'Optique*.

Interférences, diffraction et polarisation.

Par

M. Françon.

Avec 427 Figures.

A. Interférences.

I. Introduction sur les vibrations.

a) Définitions.

1. Radiation simple. Période, vitesse et longueur d'onde. L'élément simple d'un rayonnement est la radiation simple ou radiation monochromatique. Les radiations simples sont en nombre infini et forment une série continue.

Dans cet ensemble, un petit groupe impressionne notre oeil en nous donnant la sensation de lumière. L'ensemble de ces radiations «visibles» se prolonge des deux côtés par des radiations très étendues et auxquelles notre oeil n'est pas sensible.

L'étude des phénomènes d'interférence montre qu'une radiation simple est une perturbation périodique qui se propage. Or, toutes les fois qu'il y a propagation d'un phénomène périodique, trois éléments caractéristiques sont à considérer:

la période T ou son inverse la fréquence $N = \dfrac{1}{T}$,

la vitesse de propagation V,

la longueur d'onde $\lambda = V T = \dfrac{V}{N}$.

La longueur d'onde représente la période dans l'espace. Si on considère des temps $T, 2T, 3T \ldots$ le système se retrouve dans le même état. De même si on considère dans l'espace des points distants de $\lambda, 2\lambda, 3\lambda \ldots$ le système se retrouve dans la même position.

La longueur d'onde est accessible à des mesures très précises, elle peut s'exprimer en différentes unités: le micron ou 10^{-3} mm. désigné par la lettre μ, le millimicron ou 10^{-6} mm. désigné par l'abréviation mμ, l'angström Å ou $10^{-4}\,\mu$.

Les fréquences sont des nombres très grands et d'un maniement incommode. Aussi les remplace-t-on par ce qu'on appelle les «nombres d'ondes par unité de longueur dans le vide» $\nu_0 = 1/\lambda_0 = N/c$, c étant la vitesse de la lumière et λ_0 la longueur d'onde dans le vide.

Ainsi la radiation de longueur d'onde $\lambda_0 = 6\,000$ Å $= 0{,}6\,\mu$ a pour période $T = 2 \cdot 10^{-5}$ sec, pour fréquence $N = 0{,}5 \cdot 10^{15}$ sec^{-1} et pour nombre d'ondes au centimètre $\nu_0 = 17\,000$ cm^{-1}.

2. Chemin optique. On est souvent amené à considérer le temps que met la lumière pour parcourir la distance qui sépare deux points. On peut relier ce temps à une longueur convenablement choisie. Soient A et A' deux points distants de la longueur x et $t = x/V$ le temps que met la lumière pour parcourir la distance x

à la vitesse V. Si n est l'indice du milieu, on a $V = c/n$ et

$$t = \frac{n\,x}{c}.\tag{2.1}$$

Posons $l = nx$, l est la longueur du chemin optique parcouru par la lumière : c'est la longueur que parcourerait la lumière dans le vide pendant le même temps. Dans beaucoup de cas, la lumière traverse plusieurs milieux successifs et parcourt un chemin x_1 dans un premier milieu d'indice n_1, un chemin x_2 dans un second milieu d'indice n_2 etc. La longueur totale du chemin optique est

$$l = n_1\,x_1 + n_2\,x_2 + \cdots = \sum n\,x.\tag{2.2}$$

L'indice peut varier d'une façon continue. C'est par exemple le cas qui se présente pour la lumière émise par une étoile et qui traverse l'atmosphère terrestre. L'indice n est alors une fonction de x et on aura pour chemin optique total l'intégrale $l = \int n\,dx$ prise entre des limites convenables.

3. Classification des radiations. Dans la série continue des radiations simples, seules des raisons de commodité de technique ou d'usage font établir des divisions

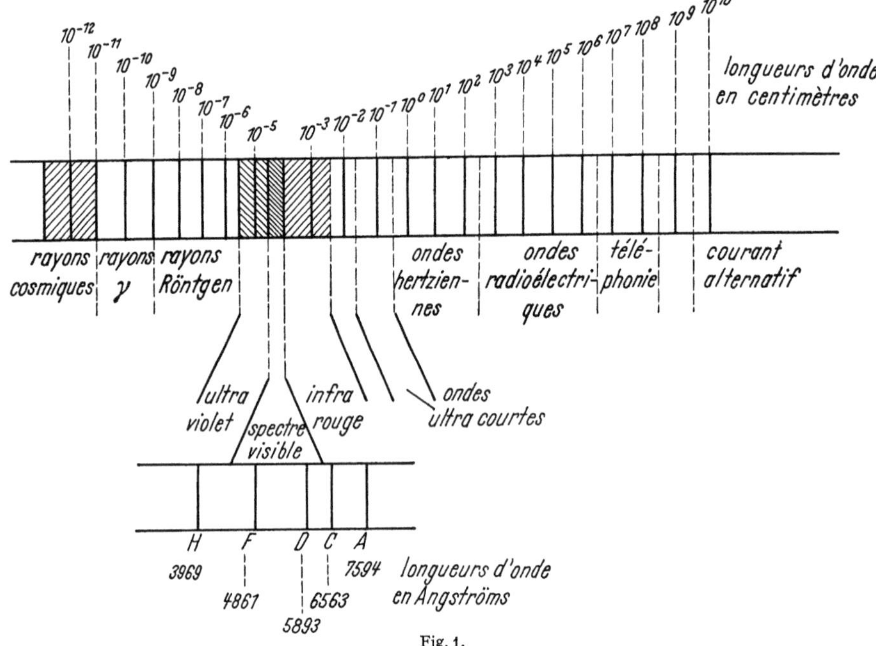

Fig. 1.

dans cet ensemble. Ces divisions sont basées sur les considérations suivantes : aucun milieu n'est transparent pour toutes les radiations d'où la nécessité d'employer telle ou telle substance pour les appareils d'optique avec lesquels on cherche à étudier un groupe de radiations. Ces appareils ne pourront donc être sensibles qu'à un certain domaine de radiations. Enfin, il faut employer des sources différentes pour obtenir avec une intensité suffisante des radiations de fréquences très différentes.

La Fig. 1 donne un schéma de la classification des radiations et montre la faible place occupée par les radiations visibles. Ce groupe s'étend de $0,4\,\mu$ (extrême violet) à $0,8\,\mu$ (extrême rouge) environ. Notons que vers $0,35\,\mu = 3500\,\text{Å}$

le verre cesse d'être transparent, sauf en lames minces et ne peut plus être utilisé pour la construction de prismes. Vers $0,29\,\mu = 2900$ Å se trouve la fin du spectre solaire. Vers $0,20\,\mu = 2000$ Å les plaques photographiques ordinaires cessent d'être utilisables à cause de l'opacité de la gélatine. Vers $0,18\,\mu = 1800$ Å le quartz cesse d'être transparent mais la fluorine l'est encore jusqu'à 1200 Å. A partir de cette longueur d'onde, aucun milieu, si ce n'est le vide, ne peut transmettre de radiation.

Du côté des radiations de grandes longueurs d'onde, le verre est pratiquement opaque à partir de 3 ou $4\,\mu$. On peut alors explorer l'infra-rouge en employant successivement des prismes de sel gemme, de fluorine, de chlorure de potassium dont les bandes d'absorption ne se confondent pas.

4. Vibration lumineuse. Dans l'étude des phénomènes d'interférences et de diffraction, nous emploierons généralement l'expression de «vibration lumineuse» sans préciser la nature de la variable qui la représentera.

La vibration lumineuse en un point de l'espace est représentée par un vecteur ayant ce point pour origine: l'extrémité de ce vecteur décrit une certaine courbe dans un plan perpendiculaire à la direction de propagation et sa projection sur un axe

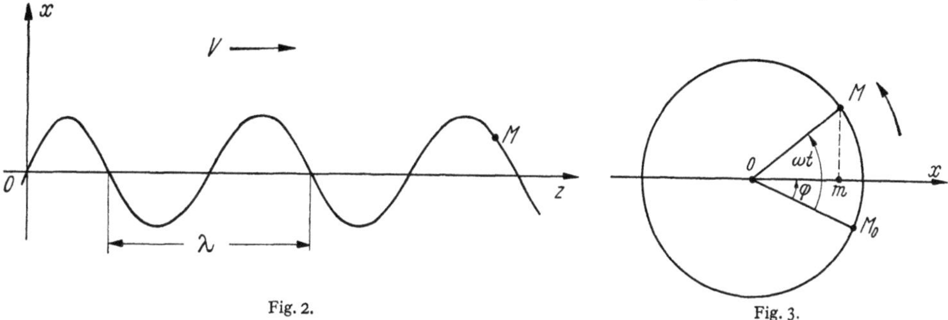

Fig. 2.   Fig. 3.

est une fonction périodique du temps. On sait que toute fonction périodique de période T peut être représentée par une somme de fonctions sinusoïdales de périodes T, $T/2$, $T/3$, ... et de phases φ_1, φ_2 φ_3 ...

Les phénomènes d'interférence et de diffraction conduisent à admettre que la fonction périodique représentant la vibration est la plus simple de toutes: c'est une fonction sinusoïdale.

L'expression de la vibration lumineuse x le long d'un faisceau de rayons parallèles est une fonction sinusoïdale du temps que l'on peut écrire pour le point O

$$x = a \cos \frac{2\pi t}{T} = a \cos 2\pi N t = a \cos \omega t \tag{4.1}$$

où a est l'amplitude de la vibration et $\omega = 2\pi N$ la pulsation. Au point M d'abscisse z (Fig. 2) la vibration est à l'instant t ce qu'elle était au point origine O au temps $t - \dfrac{z}{V}$, on a donc

$$x = a \cos \frac{2\pi}{T}\left(t - \frac{z}{V}\right) = a \cos 2\pi \left(\frac{t}{T} - \frac{z}{V}\right) = a \cos (\omega t - \varphi). \tag{4.2}$$

$\varphi = 2\pi z/\lambda$ est la phase de la vibration en M ou encore la différence de phase entre la vibration en M et la vibration en O. $\lambda = \lambda_0/n$ est la longueur d'onde dans le milieu d'indice n où se propage la vibration. On peut représenter cinématiquement la vibration en associant à l'expression $x = a \cos (\omega t - \varphi)$ le vecteur tournant OM (Fig. 3) de module a dont la projection OM sur l'axe des x a pour abscisse $x = a \cos (\omega t - \varphi)$. La phase φ détermine la position M_0 de M à l'origine des temps.

Il pourra être avantageux par la suite d'utiliser la notation imaginaire dans laquelle on représente la vibration par l'affixe du point M

$$a\,[\cos(\omega t - \varphi) + j\sin(\omega t - \varphi)] = a\,e^{j(\omega t - \varphi)} = a\,e^{j\omega t} \times e^{-j\varphi}. \tag{4.3}$$

Le facteur $e^{j\omega t}$ indique la nature ondulatoire de la lumière et se trouve en facteur dans tous les calculs. On peut en faire abstraction et représenter simplement la vibration par l'expression $a\,e^{-j\varphi}$, l'amplitude réelle de la vibration étant le module de l'amplitude imaginaire $a\,e^{-j\varphi}$.

Quant à l'éclairement en un point, il sera proportionnel au carré de l'amplitude a en ce point.

La lumière naturelle émise par une source isotrope possède la symétrie de révolution, ses propriétés sont les mêmes dans tous les azimuts. On est amené à supposer que pendant le temps nécessaire à une impression visuelle ou photographique, un très grand nombre de perturbations se produisent dans la source, chacune d'elles modifiant au hasard l'orientation et la phase de la vibration émise.

b) Composition des vibrations.

5. Composition de deux vibrations parallèles. Soit à composer les deux vibrations sinusoïdales de même période

$$x_1 = a\,e^{j\varphi_1}, \qquad x_2 = b\,e^{j\varphi_2}.$$

La vibration résultante somme de ces deux vibrations est représentée par la fonction

$$x = x_1 + x_2 = a\,e^{j\varphi_1} + b\,e^{j\varphi_2} = A\,e^{j\Phi}. \tag{5.1}$$

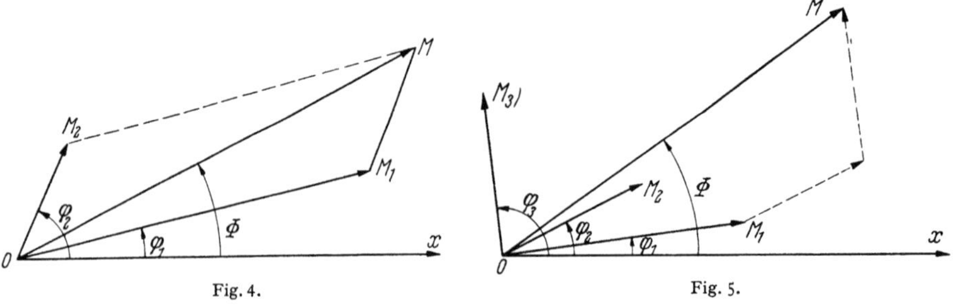

L'amplitude A de la vibration résultante est donnée par le module de cette expression. Le carré A^2 du module est égal au produit de l'expression (5.1) par la quantité conjuguée:

$$\begin{aligned} A^2 &= (a\,e^{j\varphi_1} + b\,e^{j\varphi_2})(a\,e^{-j\varphi_1} + b\,e^{-j\varphi_2}), \\ A^2 &= a^2 + b^2 + 2\,a\,b\cos(\varphi_1 - \varphi_2). \end{aligned} \tag{5.2}$$

La phase Φ de la vibration résultante sera donnée par

$$\tan\Phi = \frac{a\sin\varphi_1 + b\sin\varphi_2}{a\cos\varphi_1 + b\cos\varphi_2}. \tag{5.3}$$

Ces relations conduisent à la construction de Fresnel.

A partir d'un axe Ox quelconque (Fig. 4) construisons le vecteur OM_1 de module a_1 qui fait avec l'axe Ox l'angle φ_1. Construisons de même le vecteur OM_2 de module a_2 et faisant l'angle φ_2 avec Ox. On voit immédiatement que leur somme géométrique OM représente la vibration résultante. L'amplitude A de la vibration résultante est égale à la longueur OM de la Fig. 4 et sa phase Φ est l'angle de OM et Ox. La considération du triangle OM_1M permet de retrouver immédiatement l'expression (5.2).

D'autre part, les projections de OM sur Ox et sur l'axe perpendiculaire sont respectivement $a\cos\varphi_1 + b\cos\varphi_2$ et $a\sin\varphi_1 + b\sin\varphi_2$, et on retrouve bien pour l'angle Φ de OM avec Ox la valeur donnée par la relation (5.3).

La construction précédente ou règle de Fresnel s'étend à l'addition d'un nombre quelconque de vibrations de même période et de même direction: l'amplitude de la vibration résultante est égale au module de la somme géométrique des vecteurs qui représentent les vibrations à additionner. Le vecteur OM (Fig. 5) représente la vibration résultante des trois vibrations représentées par les trois vecteurs OM_1, OM_2, OM_3. Son amplitude est égale à la longueur OM et sa phase Φ est l'angle de OM avec Ox.

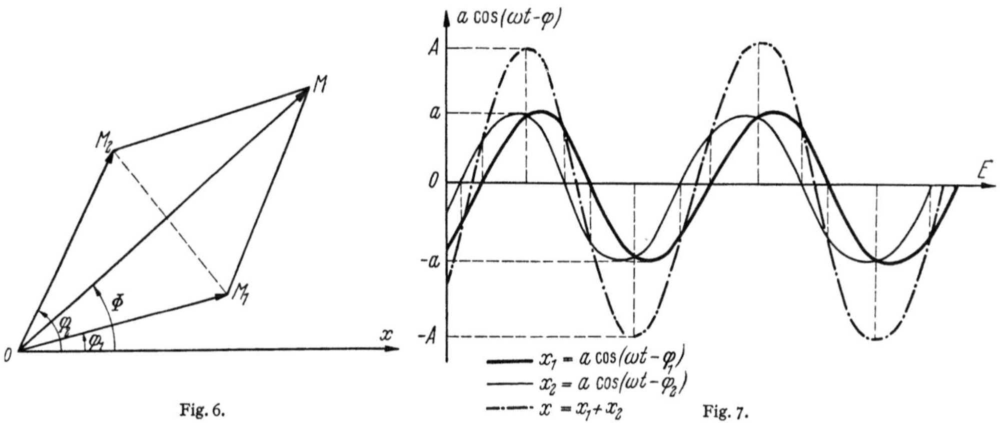

Fig. 6.   Fig. 7.

6. Composition de deux vibrations parallèles de même amplitude. Lorsque les deux vibrations composantes ont même amplitude les formules (5.2) et (5.3) deviennent

$$A^2 = 4a^2\cos^2\left(\frac{\varphi_1-\varphi_2}{2}\right) \tag{6.1}$$

et

$$\tan\Phi = \frac{\sin\varphi_1 + \sin\varphi_2}{\cos\varphi_1 + \cos\varphi_2} = \frac{\sin\dfrac{\varphi_1+\varphi_2}{2}\cos\dfrac{\varphi_1-\varphi_2}{2}}{\cos\dfrac{\varphi_1+\varphi_2}{2}\cos\dfrac{\varphi_1-\varphi_2}{2}} = \tan\frac{\varphi_1+\varphi_2}{2}$$

d'où

$$\Phi = \frac{\varphi_1+\varphi_2}{2}. \tag{6.2}$$

On pourra retrouver ces formules par la construction de Fresnel: le parallélogramme construit sur les deux vecteurs devient un losange et la résultante OM (Fig. 6) est la bissectrice des composantes. Sa phase Φ est la moyenne des phases des vibrations composantes.

La Fig. 7 montre le mécanisme d'addition des deux vibrations composantes: on a porté les élongations $a\cos(\omega t - \varphi)$ en ordonnée et le temps t en abscisse.

7. Composition de deux vibrations rectangulaires. Soit à composer les deux vibrations sinusoïdales de même période

$$x = a\cos(\omega t - \varphi_1), \quad y = b\cos(\omega t - \varphi_2).$$

Les deux vecteurs OM_1 et OM_2 qui les représentent sont dirigés suivant deux directions rectangulaires Ox et Oy (Fig. 8). Posons $\varphi = \varphi_2 - \varphi_1$ en choisissant convenablement l'origine des temps, on peut écrire les deux vibrations sous la forme

$$x = a \cos \omega t, \quad y = b \cos (\omega t - \varphi). \quad (7.1)$$

Développons et éliminons t

$$\frac{y}{b} = \cos \omega t \cos \varphi + \sin \omega t \sin \varphi$$

$$= \frac{x}{a} \cos \varphi + \sin \omega t \sin \varphi$$

d'où

$$\sin \omega t \sin \varphi = \frac{y}{b} - \frac{x}{a} \cos \varphi$$

et

$$\cos \omega t \sin \varphi = \frac{x}{a} \sin \varphi;$$

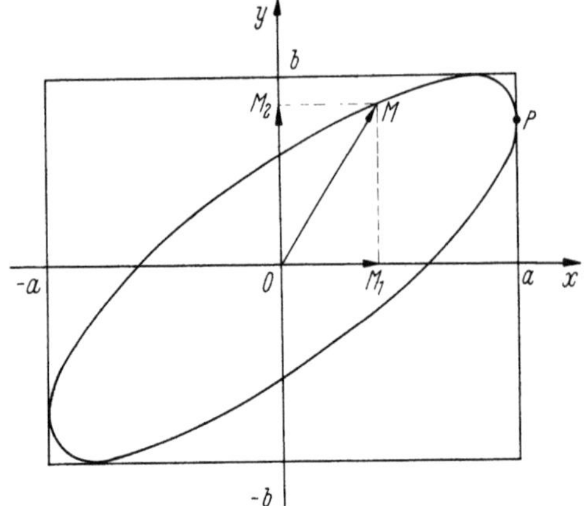
Fig. 8.

élevons au carré et ajoutons membre à membre, il vient:

$$\frac{x^2}{a^2} + \frac{y^2}{b^2} - \frac{2xy}{ab} \cos \varphi = \sin^2 \varphi. \quad (7.2)$$

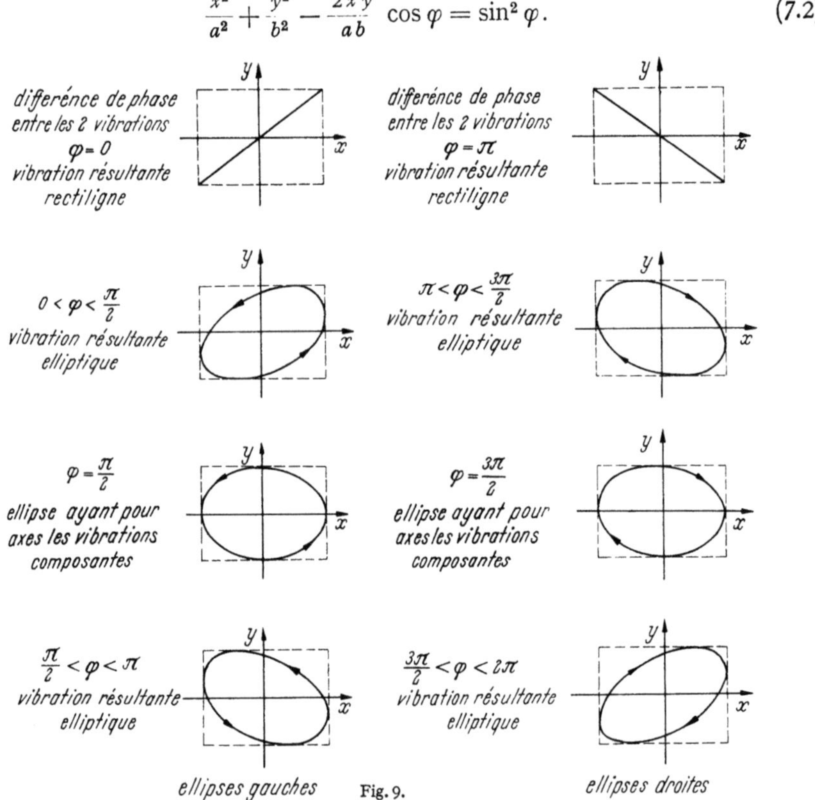

Fig. 9.

C'est l'équation de la trajectoire de l'extrémité M du vecteur OM. L'équation (7.2) représente une ellipse, la vibration résultante est une vibration elliptique. A l'instant $t = 0$, on a $x = a$ et le point M se trouve en P d'ordonnée

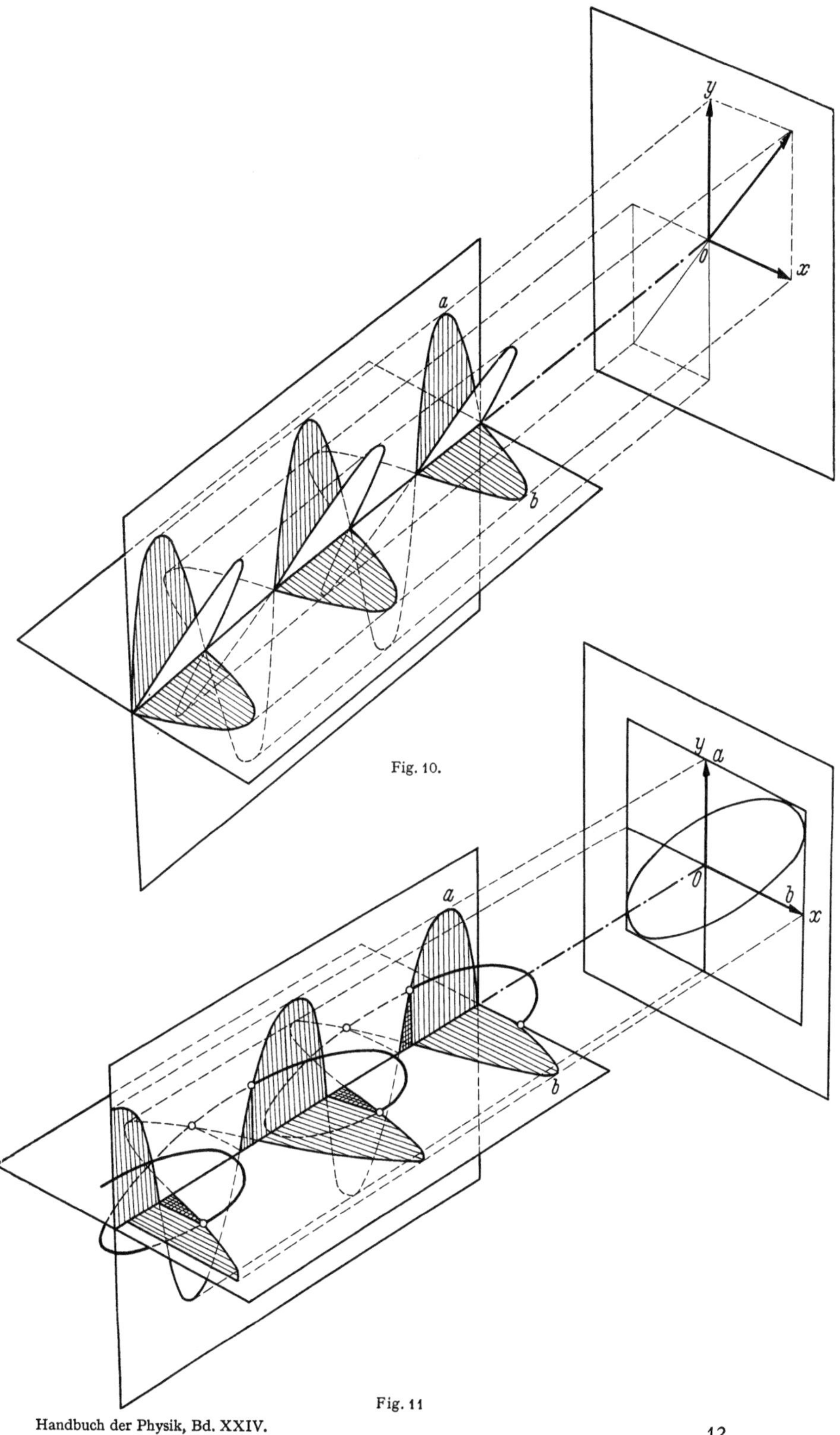

Fig. 10.

Fig. 11

$y = b\cos\varphi$. Si t augmente, le point M se déplace dans le sens de la flèche pour φ ocmpris entre 0 et π. En effet, pour ces valeurs de φ, $dy/dt = b\,\omega\sin\varphi$ est positif et y croît avec t. On dira que l'ellipse est une ellipse gauche. Si φ est compris entre π et 2π, OM tourne en sens inverse, nous aurons une ellipse droite.

Suivant les valeurs de φ, la vibration résultante donnée par (7.2) prendra les formes indiquées sur la Fig. 9. La Fig. 10 montre dans l'espace la composition de deux vibrations en phase ($\varphi = 0$) et la Fig. 11, la composition de deux vibration présentant la différence de phase φ quelconque (vibration elliptique). L'énergie E de la vibration résultante est égale à la somme des énergies des vibrations rectangulaires composantes et ne dépend pas de leur différence de phase, on a $E = a^2 + b^2$.

8. Décomposition d'une vibration elliptique en deux composantes rectangulaires. Toute vibration elliptique peut être considérée comme résultant de la composition de deux vibrations d'amplitudes A et B égales aux axes de l'ellipse, dirigées suivant ces axes et présentant entre elles une différence de phase égale à $\pi/2$ (Fig. 12). Si l'ellipse est gauche, on peut la décomposer en deux vibrations rectilignes

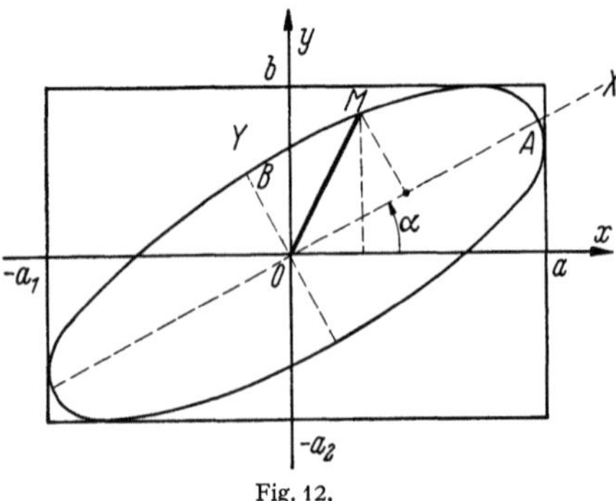
Fig. 12.

$$X = A\cos\omega t, \qquad Y = B\sin\omega t,$$

et si c'est une ellipse droite

$$X = A\cos\omega t, \qquad Y = -B\sin\omega t.$$

On peut également décomposer l'ellipse suivant deux axes quelconques $0x$, $0y$. Considérons un point de l'ellipse, la formule de changement de coordonnées donne pour une ellipse gauche en posant égal à α l'angle entre $0x$ et $0X$

$$x = X\cos\alpha - Y\sin\alpha = A\cos\alpha\cos\omega t - B\sin\alpha\sin\omega t,$$
$$y = X\sin\alpha + Y\cos\alpha = A\sin\alpha\cos\omega t + B\cos\alpha\sin\omega t.$$

La vibration composante x a pour amplitude a d'après ce qui précède

$$a^2 = A^2\cos^2\alpha + B^2\sin^2\alpha. \tag{8.1}$$

De même la vibration y a une amplitude égale à b

$$b^2 = A^2\sin^2\alpha + B^2\cos^2\alpha. \tag{8.2}$$

Ces deux dernières relations montrent que

$$a^2 + b^2 = A^2 + B^2. \tag{8.3}$$

Si φ est la différence de phase entre les deux vibrations rectilignes composantes x et y (7.1), l'aire de l'ellipse est égale à

$$\pi a b \sin\varphi = \pi A B. \tag{8.4}$$

Divisons les deux relations (8.3) et (8.4), on obtient en posant $B/A = \tan\Psi$, $b/a = \tan\vartheta$

$$\sin 2\Psi = \sin 2\vartheta \sin\varphi. \tag{8.5}$$

La relation (8.4) peut s'écrire

$$a^2 b^2 (1 - \cos^2\varphi) = A^2 B^2$$

ou encore

$$a^2 b^2 \cos^2\varphi = a^2 b^2 - A^2 B^2. \tag{8.6}$$

Remplaçons $a^2 b^2$ tiré de (8.1) et (8.2), il vient

$$a\,b\cos\varphi = (A^2 - B^2)\sin\alpha\cos\alpha. \tag{8.7}$$

Divisions (8.4) par (8.7), on obtient

$$\tan 2\Psi = \tan\varphi\sin 2\alpha. \tag{8.8}$$

Si on donne l'ellipse par ses axes A et B, les formules (8.5) et (8.8) permettent le calcul de la différence de phase φ et de l'angle ϑ ($b/a = \tan\vartheta$) de ses composantes suivant $0x$ et $0y$ à partir de l'angle Ψ défini par $B/A = \tan\Psi$.

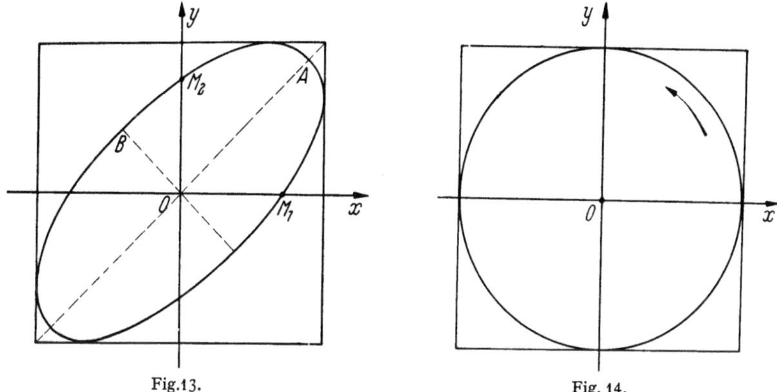

Fig.13.   Fig. 14.

9. Composition de deux vibrations rectangulaires de même amplitude. Posons $a = b$, l'équation (7.2) devient

$$x^2 + y^2 - 2\,x\,y\cos\varphi = a^2\sin^2\varphi, \tag{9.1}$$

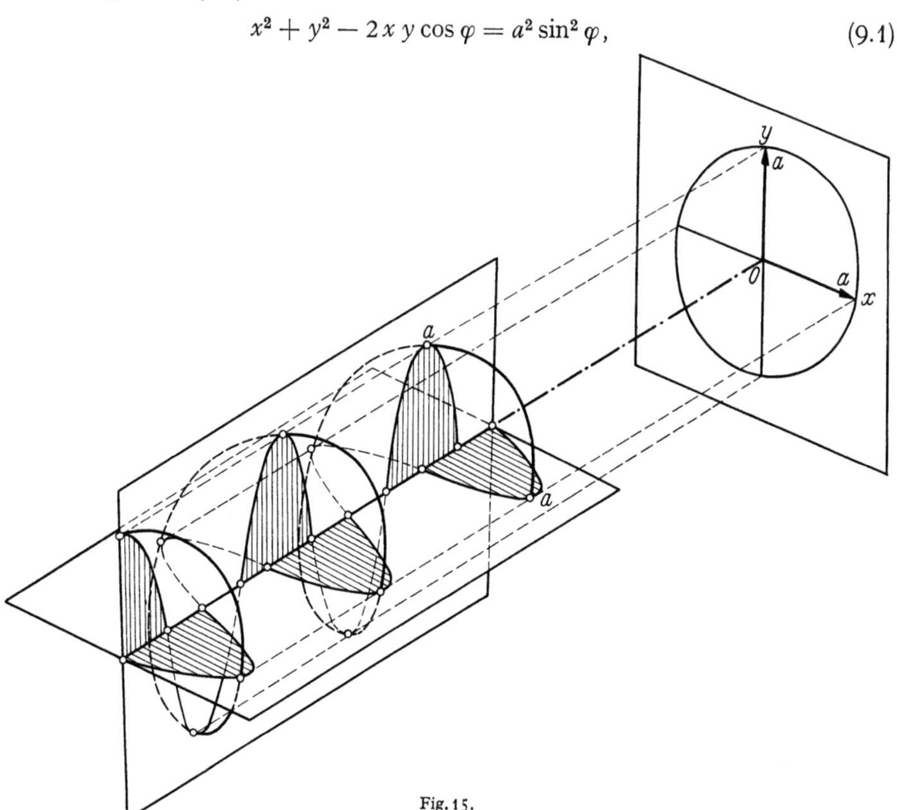

Fig. 15.

c'est une ellipse inscrite dans un carré de côté $2a$ (Fig. 13). D'après le paragraphe précédent, on a $\alpha = \pi/4$ puisque $\tan \vartheta = 1$ et les axes de l'ellipse sont les bissectrices des angles formés par les axes $0x$ et $0y$. On a de plus $\Psi = \varphi/2$ donc le rapport des axes de l'ellipse est

$$\frac{B}{A} = \tan \frac{\varphi}{2}.$$

Dans le cas où $\varphi = (2K+1)\,\pi/2$ l'ellipse devient un cercle, la vibration est une vibration circulaire (Fig. 14).

Comme précédemment, on voit que si $\varphi = \pi/2$, la vibration est une vibration circulaire gauche (sens de la flèche sur la Fig. 14) et si $\varphi = 3\pi/2$, on obtient une vibration circulaire droite.

La Fig. 15 montre dans l'espace la composition de deux vibrations rectangulaires de même amplitude donnant une vibration circulaire.

II. Interférences a deux ondes.

a) Caractères généraux des phénomènes d'interférences.

10. Conditions générales pour que des vibrations lumineuses puissent interférer. On sait que les phénomènes d'interférences résultant de la composition de deux ou plusieurs vibrations lumineuses ne peuvent se produire que dans des conditions bien déterminées qui sont les suivantes:

α) *Les vibrations doivent être cohérentes*, si on place un écran percé de deux petits trous devant une source monochromatique étendue, comme une flamme de sodium par exemple, il n'y a pas interférence. En effet, la théorie classique admet qu'il y a dans l'atome qui émet la vibration monochromatique un oscillateur matériel vibrant avec la période de la vibration. Cet oscillateur est lancé un grand nombre de fois par seconde, à la façon d'un diapason et reprend chaque fois un mouvement de même période que le précédent, mais complètement indépendant car la phase et l'amplitude ont varié brusquement. Si on considère deux sources ou deux points différents d'une même source, ces variations brusques se produisent à des instants différents et n'ont pas de relation entre elles. On dit que ces deux sources ou ces deux points sont des *sources incohérentes* et qu'elles envoient des *vibrations incohérentes*. Pour avoir deux sources cohérentes, c'est-à-dire qui émettent des vibrations cohérentes, il faut que les phases de ces vibrations soient liées.

β) *On ne peut faire interférer que des vibrations ayant la même période.*

γ) *Les vibrations doivent être parallèles.* On sait en effet que la composition de deux vibrations perpendiculaires donne une vibration elliptique.

De ces considérations, il résulte que pour obtenir des phénomènes d'interférences, il est indispensable que les mouvements qui se superposent proviennent d'un même point lumineux. Les rayons partis d'un même point lumineux se superposent après avoir parcouru des chemins différents. Un appareil interférentiel apparaît donc comme un diviseur d'ondes: il divise l'onde incidente en deux ou plusieurs ondes qui, après avoir parcouru des chemins différents, se superposeront en donnant lieu à des phénomènes d'interférences.

Nous supposerons d'abord réalisées les deux conditions permettant d'obtenir les phénomènes les plus simples: la source de lumière est une source ponctuelle et la radiation qu'elle émet est rigoureusement monochromatique.

11. Franges d'interférences, contraste des franges à deux ondes. Soit une source ponctuelle S (Fig. 16) émettant de la lumière monochromatique. Deux rayons tels que SI et SJ parcourent des chemins différents et se rejoignent en S' suivant $I'S'$ et $J'S'$.

Le long de chaque rayon se propage une vibration par exemple $x_1 = a\,e^{j\varphi}$ sur SI et $x_2 = b\,e^{j\varphi'}$ sur SJ. Soient φ_1 la phase de la vibration x_1 en S' et φ_2 la phase de la vibration x_2 en ce même point. Pour avoir la vibration résultante en S', nous aurons à composer les deux vibrations $x_1 = a\,e^{j\varphi_1}$ et $x_2 = b\,e^{j\varphi_2}$. Nous nous limiterons au cas où les deux rayons AI et AJ sont peu inclinés l'un sur l'autre de manière à pouvoir considérer comme parallèles les vibrations à composer.

Dans ces conditions, il est inutile de spécifier l'orientation des vibrations dans l'espace et le principe des interférences s'applique aussi bien en lumière naturelle qu'en lumière polarisée. Les deux vibrations x_1 et x_2 proviennent d'un même point de la source, par conséquent si des perturbations se produisent en ce point, elles affectent identiquement de la même façon l'orientation et la phase des deux vibrations x_1 et x_2. Nous avons donc à chercher la vibration résultante de la composition de deux vibrations parallèles et de même période. Ce problème a été traité à la Sect. 5. Le carré de l'amplitude A

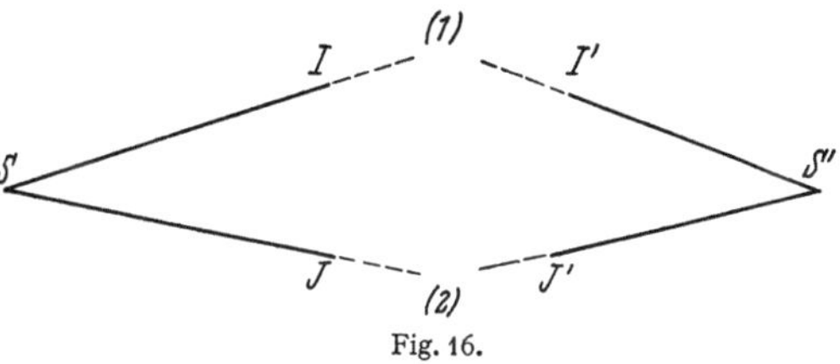

Fig. 16.

de la vibration résultante est donné par l'équation (5.2). Soit ϑ_1 le temps mis par l'une des ondes pour suivre le chemin $SII'S'$, on aura d'aprés (2.1) et (2.2)

$$\vartheta_1 = \frac{\sum n_1 l_1}{V} = \frac{L_1}{V}$$

L_1 étant le chemin optique $SII'S'$. De même pour le temps ϑ_2 mis par l'autre onde pour suivre le trajet $SJJ'S'$, on aura $\vartheta_2 = L_2/V$.

Les deux phases φ_1 et φ_2 de la relation (5.2) s'écrivent ici

$$\varphi_1 = \frac{2\pi L_1}{VT} = \frac{2\pi L_1}{\lambda},$$

$$\varphi_2 = \frac{2\pi L_2}{VT} = \frac{2\pi L_2}{\lambda}.$$

La différence de phase entre les deux vibrations qui arrivent en S' sera, en posant $\delta = L_1 - L_2$

$$\varphi = \varphi_1 - \varphi_2 = 2\pi\frac{\delta}{\lambda}.$$

La différence de chemin optique ou différence de marche δ dépend de la position du point S'. Dans la pratique, on observe les phénomènes dans un certain plan P, alors δ et par suite φ seront des fonctions des coordonnées du point S' dans le plan P. On peut tracer dans ce plan la famille des courbes $\delta = $ constante ou $\varphi = $ constante. Le long de chaque courbe δ et φ étant constants A^2, donné par l'expression (5.2), est constant, et l'éclairement qui lui est proportionnel est également constant. D'une courbe à une autre, l'éclairement variera en fonction de la phase φ. Les courbes correspondant à un éclairement maximum seront données par les valeurs de φ pour lesquelles la condition suivante sera réalisée, K étant un nombre entier

$$\varphi = 2K\pi, \qquad \frac{\delta}{\lambda} = K, \tag{11.1}$$

on aura alors

$$A^2_{\max} = (a + b)^2. \tag{11.2}$$

Dans le plan P (Fig. 17), nous aurons une infinité de courbes correspondant chacune à une valeur entière de K. Ces courbes sont les franges brillantes et le nombre K est le numéro d'ordre de la frange brillante. Le rapport δ/λ est appelé l'ordre d'interférence p.

Les courbes correspondant à un éclairement minimum seront données par les valeurs de φ telles que

$$\varphi = (2K + 1)\,\pi, \qquad \frac{\delta}{\lambda} = K + \frac{1}{2}. \tag{11.3}$$

Ces courbes, en traits ponctués sur la Fig. 17, s'intercalent entre les courbes d'éclairement maximum. Les courbes définies par (11.3) sont les franges sombres. En chaque point d'une frange sombre, on a

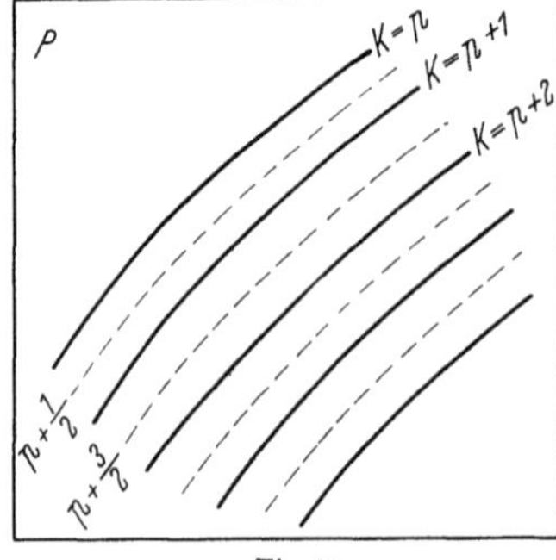

Fig. 17.

$$A^2_{\min} = (a - b)^2. \tag{11.4}$$

Posons $E_1 = A^2_{\max}$ qui est l'éclairement des franges brillantes et $E_2 = A^2_{\min}$ qui est l'éclairement des franges sombres. Nous définirons le contraste des franges par le rapport

$$\gamma = \frac{E_1 - E_2}{E_1}. \tag{11.5}$$

Si $E_2 = 0$, c'est-à-dire si les franges sombres sont parfaitement noires, on a $\gamma = 1$ et le contraste du phénomène est maximum. Si $E_1 = E_2$, $\gamma = 0$, les franges brillantes et les franges sombres produisant le même éclairement, le contraste est nul et le phénomène disparaît. D'après (11.2) et (11.4), on voit que le contraste des franges données par les interférences à deux ondes sera

$$\gamma = \frac{(a + b)^2 - (a - b)^2}{(a + b)^2} = \frac{4ab}{(a + b)^2}, \tag{11.6}$$

ce contraste sera maximum $\gamma = 1$ si $a = b$, c'est-à-dire si les vibrations qui interfèrent ont même amplitude.

b) Etude des phénomènes d'interférences à franges non localisées.

Les franges que l'on observe dans les différents phénomènes d'interférences peuvent se classer en trois catégories:

1. Les franges non localisées, que l'on observe en n'importe quelle région de l'espace où les deux faisceaux sont superposées,

2. les franges d'égale épaisseur ou franges localisées sur les lames d'épaisseur variable,

3. les franges localisées à l'infini, ou franges d'égale inclinaison.

12. Observation des phénomènes. α) *Dispositif des trous d'*Young. Considérons deux trous très petits A_1 et A_2 identiques, percés dans un écran opaque et équidistants de la source lumineuse S (Fig. 18). Si on ne faissait intervenir que l'optique géométrique, nous aurions seulement deux taches lumineuses en A'_1 et A'_2. En fait, chaque petite ouverture *diffracte* la lumière dans un cône d'angle α d'autant plus grand que l'ouverture qui lui donne naissance est plus petite. La diffraction par des ouvertures telles que A_1 et A_2 sera étudiée au chapitre diffraction.

Tout se passe comme si A_1 et A_2 étaient de véritables sources, mais les vibrations qu'elles diffractent sont dues à une source unique et sont par conséquent cohérentes. Les vibrations émises par les deux ouvertures A_1 et A_2 sont cohérentes et en phase si $SA_1 = SA_2$; si $SA_1 \neq SA_2$ les deux ouvertures A_1 et A_2 émettent des vibrations qui ne sont plus en phase mais qui sont toujours cohérentes, c'est-à-dire dont les phases sont liées. Dans les deux cas, les vibrations émises par A_1 et A_2 sont susceptibles d'interférer dans les régions où elles se rencontrent. En plaçant un écran P pas trop près de A_1 et A_2, les deux faisceaux diffractés se recouvrent

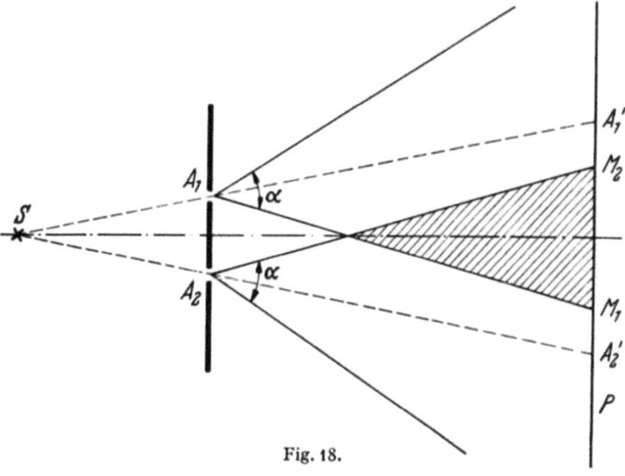

Fig. 18.

dans une région $M_1 M_2$ de cet écran et c'est dans cette partie commune aux deux faisceaux que l'on peut observer les franges d'interférences (Fig. 19).

Fig. 19. Franges d'YOUNG.

$\beta)$ *Bilentille de* BILLET. Une lentille L convergente est sciée en deux suivant un plan méridien et l'on écarte un peu les deux demi-lentilles L_1 et L_2 ainsi obtenues (Fig. 20). On éclaire L_1 et L_2 au moyen de la source S et on obtient les

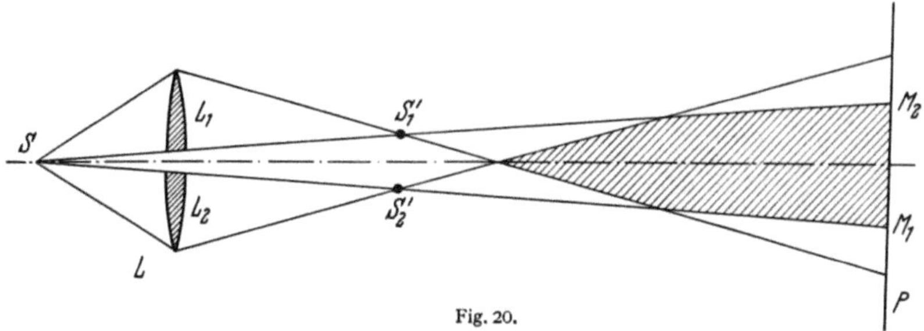

Fig. 20.

deux images S'_1 et S'_2. Les faisceaux issus de S et qui convergent en S'_1 et S'_2 continuent leur trajet, de sorte que tout se passe comme si on avait deux sources de lumière cohérentes S'_1 et S'_2 éclairant l'écran P. Les deux faisceaux se recouvrent dans la région $M_1 M_2$ où l'on peut observer les franges d'interférences.

γ) Miroirs de Fresnel. La lumière émise par la source S tombe sur deux miroirs plans M_1 et M_2 faisant entre eux un petit angle ϑ (Fig. 21). Les deux miroirs donnent de la source S deux images S_1' et S_2' d'autant plus voisines que ϑ est plus petit. Ces deux images S_1' et S_2' sont réfléchies dans les mêmes conditions par les miroirs M_1 et M_2, elles jouent le rôle de deux sources en phase. Les faisceaux provenant de S_1' et S_2' sont limités par les bords des miroirs et ont une partie commune, hachurée sur la Fig. 21. C'est dans cette région qu'il faut placer l'écran P pour observer les franges d'interférences.

Dans ces trois dispositifs et dans les autres systèmes analogues, les franges d'interférences peuvent s'observer pour n'importe quelle position de l'écran

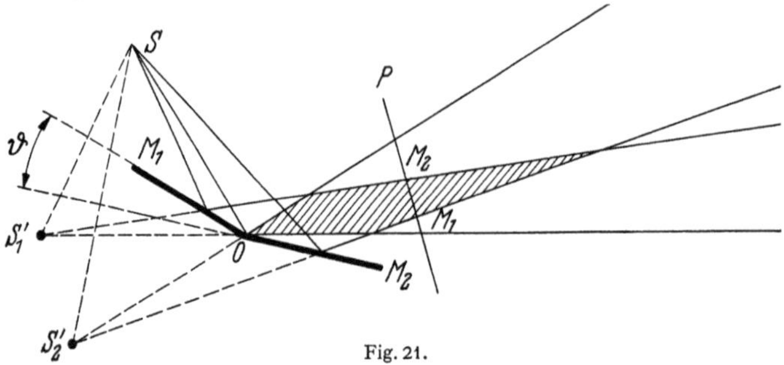

Fig. 21.

pourvu que ce soit dans une région commune aux deux faisceaux. Les franges d'interférences sont ici des franges non localisées.

13. Position des franges. Dans les trois expériences précédentes, on peut admettre que l'écran P est éclairé par deux sources en phase S_1' et S_2'. En un point M de l'écran P (Fig. 22), la différence de phase φ et la différence de marche δ entre les deux vibrations issues de S_1' et S_2', sont

$$\varphi = \frac{2\pi(S_2'M - S_1'M)}{\lambda}, \qquad \delta = S_2'M - S_1'M. \qquad (13.1)$$

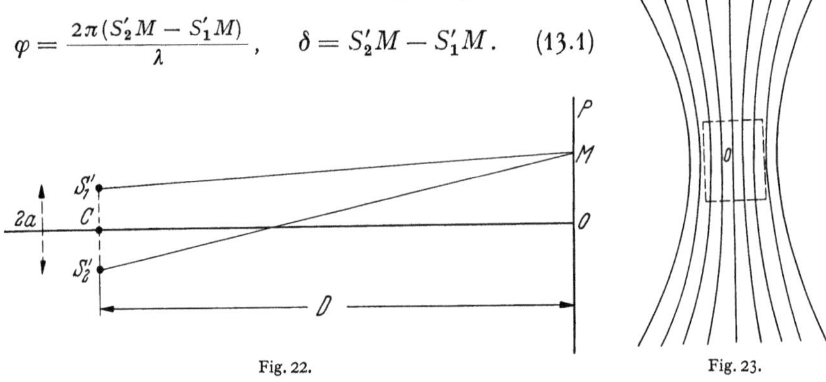

Fig. 22.   Fig. 23.

La phase φ est constante sur les surfaces $S_2'M - S_1'M = $ constante qui sont des hyperboloïdes de révolution admettant S_1' et S_2' pour foyers. Les franges que l'on observe dans le plan P sont donc les intersections par le plan P des hyperboloïdes. Le point M étant toujours loin de S_1' et S_2', on peut confondre les hyperboloïdes avec leurs cônes asymptotes et les franges sont des arcs d'hyperboles intersections des cônes asymptotes avec le plan P. Les franges ont l'aspect de la Fig. 23. La frange centrale ($K = 0$) se trouvera dans le plan perpendiculaire au plan de la Fig. 22 mené par le milieu C de S_1' et S_2', elle est donc rectiligne. En fait, comme on ne peut observer les phénomènes que dans une étendue réduite, pratiquement chaque hyperbole se confond avec sa tangente au sommet et les franges apparaissent rectilignes. Il est donc indiqué d'utiliser comme source S une fente fine parallèle aux franges, chaque point de la fente est incohérent par rapport aux autres, mais tous les systèmes de franges dus à tous les points de la source se décalent parallèlement aux franges et le contraste du phénomène ne change pas.

Calculons δ donné par (13.1) en considérant la Fig. 22.

Posons: $CO = D$, $\quad OM = y$, $\quad S_1' S_2' = 2a$.

On a: $\qquad \delta = S_2'M - S_1'M = \sqrt{D^2 + (y+a)^2} - \sqrt{D^2 + (y-a)^2}$.

Comme nous observons les franges au voisinage du point O et $2a$ étant très petit devant D pour avoir une équidistance des franges suffisantes, on peut extraire les racines par approximation, d'où

$$\delta = \frac{2a\,y}{D}. \tag{13.2}$$

D'après les expressions (11.1) et (11.3), on voit que les franges brillantes seront données par

$$\frac{\delta}{\lambda} = \frac{2a\,y}{\lambda D} = K, \qquad y = K\frac{\lambda D}{2a} \tag{13.3}$$

et les franges sombres par

$$\frac{\delta}{\lambda} = \frac{2a\,y}{\lambda D} = K + \frac{1}{2}, \qquad y = \left(K + \frac{1}{2}\right)\frac{\lambda D}{2a}. \tag{13.4}$$

Les ouvertures S_1' et S_2' étant identiques, elles diffractent le même flux de la même façon et les franges sombres sont parfaitement noires, le phénomène a le contraste maximum.

Les relations (13.3) et (13.4) montrent que les franges sont équidistantes, la distance entre deux franges brillantes ou deux franges noires consécutives étant égale à $\lambda D/2a$. La distance entre une frange noire et la frange brillante suivante est égale à la moitié de la valeur précédente. En O, $\delta = 0$ et on a une frange brillante.

14. Influence de la largeur de la source. Nous avons vu au paragraphe précédent qu'on peut utiliser comme source une fente fine bien parallèle aux franges, donc perpendiculaire au plan de la Fig. 22. Reprenons l'expérience des trous d'YOUNG et plaçons une fente fine en S_0 (Fig. 24). La fente S_0 est sur la médiatrice S_0C de A_1A_2 et la frange centrale brillante se trouve en O sur cette même droite S_0C. Sur la Fig. 24 on a représenté l'éclairement produit par les franges en chaque point M du plan P en prenant pour abscisses les positions de M et pour ordonnées les éclairements correspondants, les ordonnées étant dirigées suivant OC.

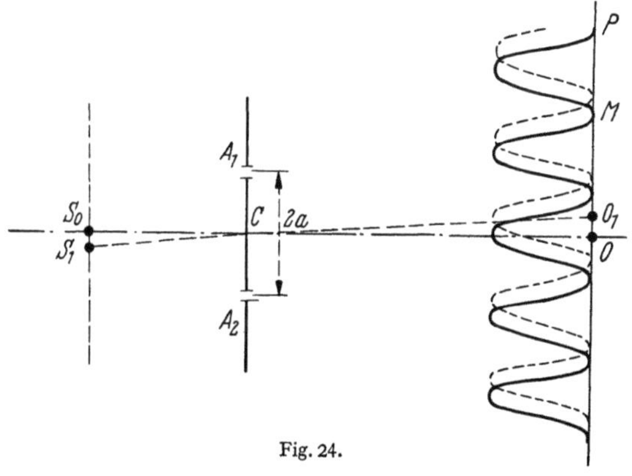

Fig. 24.

En un point M quelconque du plan P, la différence de phase des deux vibrations qui interfèrent est d'après (13.2)

$$\varphi = \frac{4\pi a\,y}{\lambda D}.$$

Les amplitudes interférentes sont égales, posons-les égales à l'unité; l'éclairement au point M où φ est la différence de phase entre les vibrations sera d'après (6.1)

$$E = 4 \cos^2 \frac{\varphi}{2} = 4 \cos^2 \frac{2\pi a y}{\lambda D}. \tag{14.1}$$

Cette expression est représentée par la courbe en trait plein sur la Fig. 24.

Déplaçons la fente fine de S_0 en S_1. La frange centrale vient en O_1 tel que les chemins optiques $S_1 A_1 O_1$ et $S_1 A_2 O_1$ soient égaux. En M, la différence de marche entre les vibrations interférentes est

$$\Delta = (S_1 A_2 + A_2 M) - (S_1 A_1 + A_1 M),$$
$$\Delta = S_1 A_2 - S_1 A_1 + A_2 M - A_1 M = S_1 A_2 - S_1 A_1 + \delta, \tag{14.2}$$

c'est une expression identique à (13.1) à laquelle on a ajouté la constante $S_1 A_2 - S_1 A_1$. On a donc le même phénomène, même équidistance des franges mais le système de franges s'est déplacé en bloc sur l'écran P. Posons $S_0 C = d$ et $S_0 S_1 = x$. La différence de marche en M peut s'écrire d'après (14.2)

$$\Delta = 2a\left(\frac{x}{d} + \frac{y}{D}\right). \tag{14.3}$$

Elle est la même pour tous les points de la fente fine de largeur dx placée en S_1. L'éclairement dE en M dû à cette fente fine sera d'après (6.1)

$$dE = E_0\left(1 + \cos\frac{2\pi\Delta}{\lambda}\right)dx.$$

E_0 étant une constante fixant l'éclairement maximum et Δ ayant la valeur donnée par (14.3). L'éclairement en M dû à une fente de largeur s sera donc

$$E = E_0 \int_{-\frac{s}{2}}^{+\frac{s}{2}} \left[1 + \cos\frac{4\pi a}{\lambda}\left(\frac{x}{d} + \frac{y}{D}\right)\right] dx,$$

$$E = E_0\left[x + \frac{\lambda d}{4\pi a}\sin\frac{4\pi a}{\lambda}\left(\frac{x}{d} + \frac{y}{D}\right)\right]_{-\frac{s}{2}}^{+\frac{s}{2}} = E_0\left(s + \frac{\lambda d}{2\pi a}\cos\frac{4\pi a y}{\lambda D}\sin\frac{2\pi a s}{\lambda d}\right).$$

Posons

$$m = \frac{2\pi a}{\lambda d}, \qquad p = \frac{4\pi a}{\lambda D},$$

l'expression donnant l'éclairement devient

$$E = E_0 s\left(1 + \frac{\sin ms}{ms}\cos p y\right), \tag{14.4}$$

la position des maxima de lumière, c'est à dire des franges brillantes est donnée par

$$\frac{4\pi a y}{\lambda D} = 2K\pi, \qquad y = K\frac{\lambda D}{2a},$$

par comparaison aux relations (13.3) on voit que les franges brillantes se trouvent aux mêmes endroits qu'avec une source rigoureusement fine. De même pour les franges sombres. Calculons le contraste des franges: l'éclairement maximum est égal à

$$E_{\max} = E_0 s\left(1 + \frac{\sin ms}{ms}\right)$$

et l'éclairement minimum

$$E_{\min} = E_0 s\left(1 - \frac{\sin ms}{ms}\right),$$

d'où le contraste

$$\gamma = \frac{E_{\max} - E_{\min}}{E_{\max}} = \frac{2 \sin ms}{ms + \sin ms}. \qquad (14.5)$$

Si s est très petit (fente fine) γ est égal à 1 et décroît si s augmente. Chaque point de la fente tel que S_1 sur la Fig. 24 donne un système de franges décalées par rapport aux autres systèmes. Tous ces systèmes de franges produisent des éclairements qui s'ajoutent puisque les sources qui leur donnent naissance sont incohérentes: le contraste baisse. L'expression (14.5) montre que les franges disparaissent si $ms = \pi$ ou $s = \lambda d/2a$.

La Fig. 25 montre les variations du contraste γ en fonction de s d'après l'expression (14.5). Si la largeur de la source est supérieure à $\lambda d/2a$, les franges réapparaissent avec un contraste inversé et faible, puis disparaissent à nouveau pour $s = \lambda d/a$ etc. ... A chaque nouvelle apparition, le contraste est plus faible. On voit sur la Fig. 25 qu'un élargissement de la source tel que $s \leq \pi/4m$ soit $ms \leq \pi/4$ changera très peu le contraste; le phénomène paraîtra pratiquement aussi net qu'avec un contraste 1.

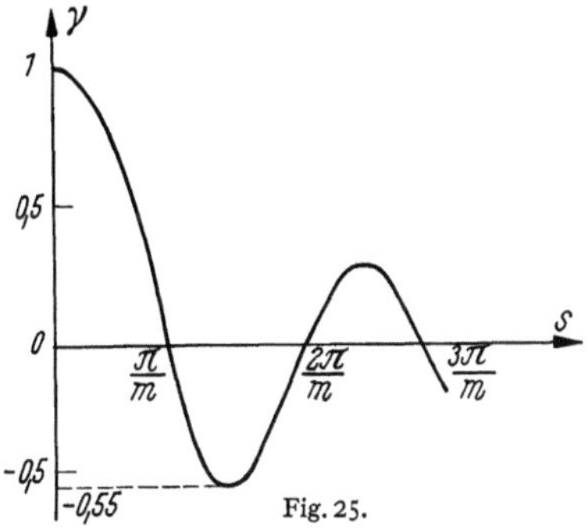

D'après la définition de m, l'angle ϑ sous lequel la fente est vue du point C devra donc être égal au maximum à

$$\vartheta = \frac{s}{d} \leq \frac{\lambda}{8a}. \qquad (14.6)$$

Fig. 25.

15. Influence du chromatisme de la source. Lorsque la source n'est plus monochromatique, le phénomène observé est la somme des phénomènes correspondant aux différentes radiations monochromatiques composant la source. L'intervalle des franges n'étant pas le même pour toutes les radiations, la superposition de tous ces systèmes de franges diminue le contraste du phénomène.

On peut évaluer une tolérance sur le monochromatisme de la source à partir des résultats précédents. Pour que le contraste des franges ne soit pratiquement pas changé, nous avons vu que le diamètre apparent ϑ de la source vu du point C ne devait pas dépasser la valeur donnée par (14.6) et correspondant à $ms \leq \pi/4$.

Si la source est infiniment fine, la différence de marche est donnée par (13.2) et si sa largeur est x, la différence de marche au bord est donnée par (14.3). Lorsqu'on passe d'un bord l'autre d'une source de largeur s, la différence de marche varie donc de la quantité

$$d\Delta = \frac{2as}{d}, \qquad \frac{d\Delta}{\lambda} = \frac{2as}{\lambda d} = \frac{ms}{\pi}$$

pour que le contraste des franges ne change pas, on a été amené à écrire $ms \leq \pi/4$ soit donc une variation de l'ordre d'interférence

$$\frac{d\Delta}{\lambda} \leq \frac{1}{4}. \qquad (15.1)$$

Considérons maintenant une source ponctuelle mais non monochromatique. Dans ce cas, la variation de l'ordre d'interférence en un point où l'on observe le phénomène n'est plus due à une variation de largeur de source, mais à une variation de longueur d'onde. Ecrivons que la variation de l'ordre d'interférence consécutive à une variation de la longueur d'onde de la lumière incidente, supposée assez petite pour que le coefficient de visibilité reste constant, ne dépasse pas $\frac{1}{2}$; on a

$$p = \frac{\delta}{\lambda}$$

d'où

$$dp = -\frac{\delta\, d\lambda}{\lambda^2} = -p\,\frac{d\lambda}{\lambda} \tag{15.2}$$

et d'après (15.1)

$$p\,\frac{d\lambda}{\lambda} \leqq \frac{1}{4}. \tag{15.3}$$

Le rapport $\lambda/d\lambda$ est appelé coefficient de finesse de la radiation utilisée. Si on s'écarte du centre O, l'ordre d'interférence p augmente, et pour que le contraste ne change pas il faudra utiliser une radiation dont le monochromatisme devra satisfaire à (15.3).

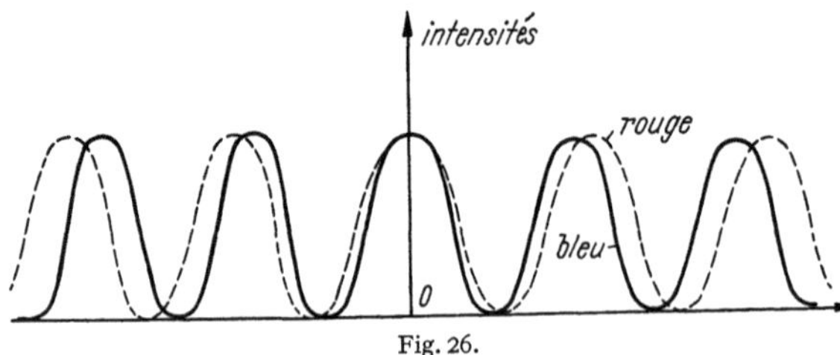

Fig. 26.

16. Phénomènes en lumière blanche. Considérons une source ponctuelle de lumière blanche. A chaque radiation monochromatique correspond un système de franges et tous ces systèmes s'ajoutent dans le plan P d'observation. D'après (13.2) on voit que la différence de marche est nulle en O pour toutes les radiations: on a donc en O une frange brillante de lumière blanche. L'interfrange étant égal à $\lambda D/2a$ il varie avec la longueur d'onde, les franges rouges sont plus larges que les franges bleues (Fig. 26). Pour chaque radiation, l'éclairement est donné à un facteur constant près par (14.1)

$$E = \cos^2\frac{\varphi}{2} = \cos^2\frac{2\pi a y}{\lambda D}.$$

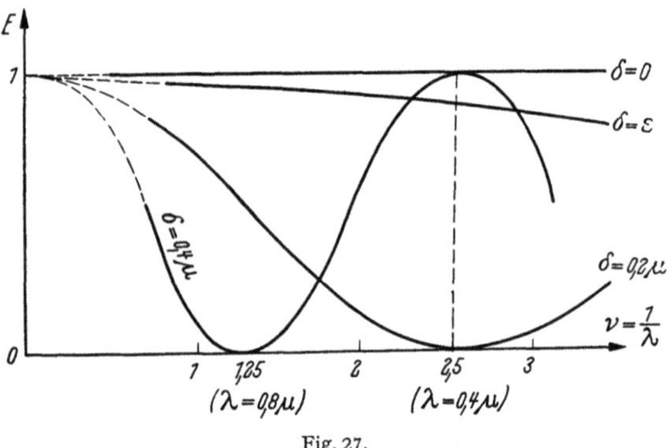

Fig. 27.

Supposons que l'amplitude soit la même pour toutes les radiations et traçons sur un graphique les variations de E en fonction de $\nu = 1/\lambda$ (λ étant exprimé en microns) pour différentes valeurs de δ (Fig. 27). La frange centrale correspond à $\delta = 0$ pour toutes les radiations et la courbe représentative est une droite parallèle à l'axe des abscisses. Si δ est petit, $\delta = \varepsilon$, la courbe de l'éclairement est une sinusoïde très allongée. L'éclairement est plus faible pour les courtes longueurs d'onde que pour les grandes longueurs d'onde. La couleur résultante sera une couleur dans laquelle le violet est plus atténué que le rouge, la teinte obtenue est rougeâtre. Pour $\delta = 0,2\,\mu$ la sinusoïde est tangente à l'axe des fréquences au point $\nu = 2,5$ ($\lambda = 0,4\,\mu$) correspondant au violet. La teinte obtenue est franchement rouge. Pour $\delta = 0,4\,\mu$ la courbe présente un minimum dans le rouge $\nu = 1,25$ ($\lambda = 0,8\,\mu$) et un maximum dans le violet pour $\nu = 2,5$ ($\lambda = 0,4\,\mu$). La teinte résultante est bleue. Pour des valeurs croissantes de δ la sinusoïde se resserre de plus en plus, on a une série de maxima et de minima. Les colorations cessent d'être visibles dès que δ dépasse 3 ou 4 microns. L'oeil perçoit une impression de blanc. Finalement le phénomène présente donc une frange centrale blanche bordée de rouge et encadrée de franges de plus en plus colorées jusqu'au moment où tout est noyé dans un blanc uniforme. On observe donc de part et d'autre de la frange centrale blanche une succession de teintes dont nous parlerons plus loin et qui sont les teintes de NEWTON relatives aux interférences à centre blanc.

17. Spectres cannelés. Lorsque δ augmente, des radiations de plus en plus nombreuses s'éteignent dans le spectre visible. Par exemple, pour $\delta = 2\,\mu$ on a 3 minima nuls entre 0,4 et 0,8 μ (Fig. 28). Observé au spectroscope, le spectre apparaitra sillonné de cannelures noires, 3 dans l'exemple précédent. Il suffirait de placer la fente du spectroscope sur la Fig. 24 en M assez loin de O là où se trouve le blanc uniforme dit blanc d'ordre supérieur. L'ordre d'interférence $p = \delta/\lambda$ varie d'un point à un autre du spectre par suite de la variation de λ.

Quand p est un nombre entier K on observe un maximum et

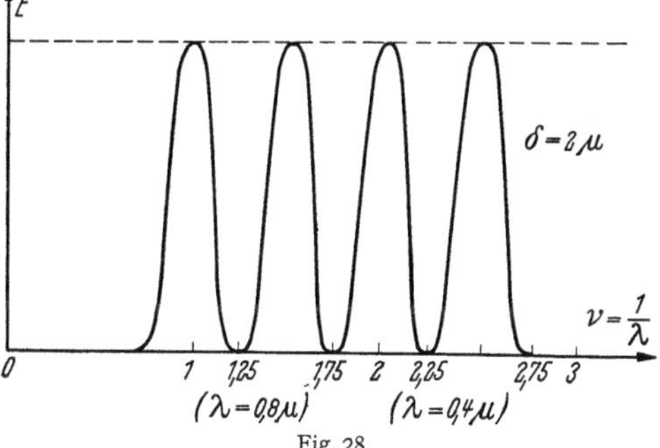

Fig. 28.

quand il est égal à $K + \frac{1}{2}$ on observe un minimum. Les cannelures noires seront donc données par

$$p = \frac{\delta}{\lambda} = K + \frac{1}{2}, \qquad \lambda = \frac{\delta}{K + \frac{1}{2}}. \tag{17.1}$$

Prenons par exemple $\delta = 10\,\mu$ les cannelures brillantes seront données par $\lambda = 10\,\mu/K$

$$K = 12 \quad \lambda = 0,83\,\mu \ \text{(extrême rouge)}$$
$$K = 13 \quad \lambda = 0,77\,\mu$$
$$K = 14 \quad \lambda = 0,71\,\mu$$
$$\cdots\cdots\cdots\cdots\cdots\cdots$$
$$K = 20 \quad \lambda = 0,50\,\mu \ \text{(vert)}$$
$$\cdots\cdots\cdots\cdots\cdots\cdots$$
$$K = 25 \quad \lambda = 0,40\,\mu \ \text{(violet)}.$$

Toutes les franges correspondantes aux valeurs de K comprises entre 13 et 25 figurent dans le spectre visible. Il y en a une douzaine environ. Si le spectre est normal (spectre de réseau), les franges ne sont pas équidistantes, elles sont plus serrées dans le violet que dans le rouge. Avec un prisme les franges sont plus serrées dans le rouge que dans le violet. Les franges sont plus ou moins serrées suivant la différence de marche: soient λ_1 et λ_2 les longueurs d'onde des radiations correspondant aux extrémités du spectre visible. Les ordres d'interférences relatifs à ces longueurs d'onde seront $p_1 = \delta/\lambda_1$ et $p_2 = \delta/\lambda_2$. La différence $\delta\left(\dfrac{1}{\lambda_1} - \dfrac{1}{\lambda_2}\right)$ donne le nombre de franges. Avec $\lambda_1 = 0,4\,\mu$ et $\lambda_2 = 0,8\,\mu$ on a 1,25 δ. Pour $\delta = 1$ mm. il y aura 1250 cannelures visibles dans le spectre. Lorsque δ augmente progressivement, les franges entrent par le violet et sortent par le rouge, mais il en entre plus qu'il n'en sort.

18. Frange achromatique. Dans les phénomènes précédents, nous avons supposé δ indépendant de la longueur d'onde. Plaçons une lame de verre d'épaisseur e et d'indice n devant l'une des deux sources des expériences que nous venons d'étudier (Fig. 29). La différence de marche en M est

$$\Delta = S_2' M - S_1' M - (n-1)\,e = \frac{2a\,y}{D} - (n-1)\,e \tag{18.1}$$

et l'ordre d'interférence

$$p = \frac{2a\,y}{\lambda D} - \frac{(n-1)\,e}{\lambda}.$$

Pour une radiation déterminée, toutes les franges se déplacent d'une longueur y (vers le haut dans le cas de la figure)

$$y = \frac{D(n-1)\,e}{2a}, \tag{18.2}$$

ceci peut servir à mesurer la différence de marche $(n-1)e$ introduite par la lame. Si on mesure le déplacement y des franges, on aura $(n-1)e$ par (18.2). Si $dp/d\lambda$ s'annule pour une longueur d'onde voisine de la longueur d'onde du maximum de sensibilité de l'oeil, l'ordre d'interférence reste constant pour les radiations les plus efficaces et on a pratiquement une frange blanche ou à peine radiations les plus efficaces et on a pratiquement une frange blanche ou à peine teintée qui rappelle la frange brillante centrale blanche des phénomènes précédents. C'est la frange achromatique. Dérivons $p = \Delta/\lambda$ par rapport à λ; on a d'après (18.1)

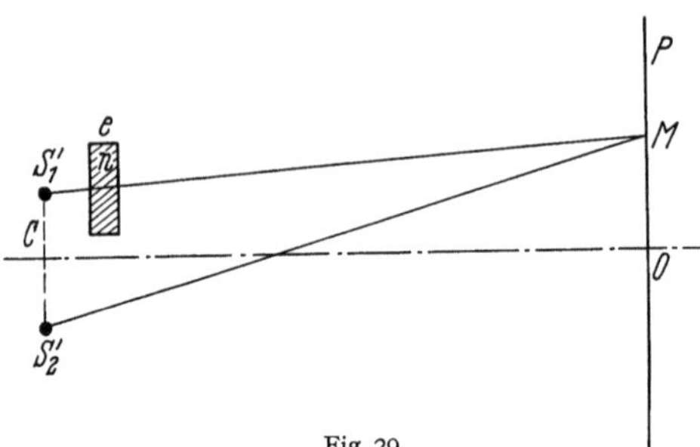

Fig. 29.

$$\frac{dp}{d\lambda} = \frac{\lambda\,\dfrac{d\Delta}{d\lambda} - \Delta}{\lambda^2} = 0$$

d'où

$$p = \frac{\Delta}{\lambda} = \frac{d\Delta}{d\lambda} = -e\,\frac{dn}{d\lambda},$$

l'ordre d'interférence de la frange achromatique sera donc égal à $-e\,\dfrac{dn}{d\lambda}$ et sa position:

$$y = -\frac{\lambda D e}{2a}\left(\frac{dn}{d\lambda} - \frac{n-1}{\lambda}\right),$$

elle sera d'autant plus écartée de O que la lame sera plus épaisse. Pour un verre tel que $n = A + \dfrac{B}{\lambda^2}$ on aura

$$y = \frac{D e}{2a}\left(\frac{3B}{\lambda^2} + A - 1\right).$$

c) Franges d'égale épaisseur des lames transparentes éclairées en lumière parallèle.

19. Franges par réflexion. $\alpha)$ *Incidence normale.* Envoyons normalement un faisceau de lumière monochromatique sur une lame transparente à faces non parfaitement parallèles d'épaisseur e et d'indice n placée dans l'air par exemple (Fig. 30). Chaque rayon incident donne naissance à un rayon réfléchi a en I et à un rayon réfléchi b en J se trouvant situés respectivement sur chaque face de la lame. La différence de marche entre ces deux rayons est $2ne$ mais il faut lui ajouter le chemin supplémentaire $\lambda/2$ dû à la réflexion de l'air sur le verre comme on le verra plus loin à propos des ondes stationnaires. On a donc:

$$\delta = 2ne + \frac{\lambda}{2}. \tag{19.1}$$

Les deux rayons a et b peuvent interférer: le carré de l'amplitude a du rayon réfléchi en I est de l'ordre de $a^2 \approx 0{,}04$ et en J on aura $0{,}04 \times 0{,}96 \approx 0{,}04$ en négligeant l'absorption. Pratiquement les amplitudes des deux rayons qui interfèrent sont égales et si on observe la lame dans la région IJ, on aura un éclairement donné par (6.1)

$$E = \cos^2 \frac{\pi\,\delta}{\lambda} = \cos^2 \frac{\pi}{\lambda}\left(2ne + \frac{\lambda}{2}\right). \qquad (19.2)$$

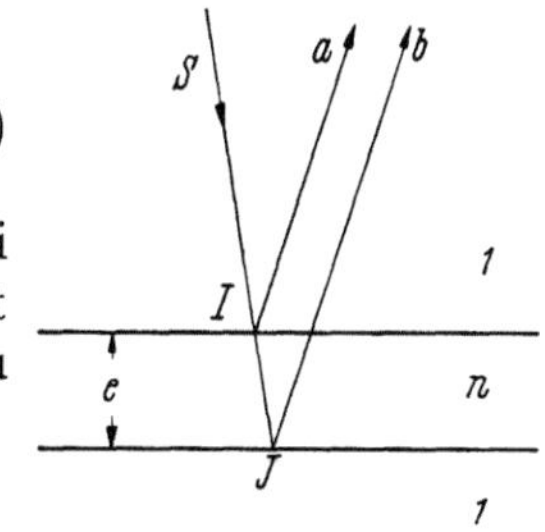

L'indice de réfraction n étant supposé constant ainsi que la longueur d'onde, la différence de marche est fonction de l'épaisseur e de la lame. En un point où l'épaisseur est e, l'ordre d'interférence est

$$p = \frac{\delta}{\lambda} = \frac{2ne}{\lambda} + \frac{1}{2}.$$

Fig. 30.

Les franges tracent donc les courbes d'égale épaisseur de la lame, elles en dessinent la carte topographique (Fig. 31). Les franges brillantes seront données par

$$\frac{\delta}{\lambda} = \frac{2ne}{\lambda} + \frac{1}{2} = K \qquad (19.3)$$

et les franges noires par

$$\frac{\delta}{\lambda} = \frac{2ne}{\lambda} + \frac{1}{2} = K + \frac{1}{2}. \qquad (19.4)$$

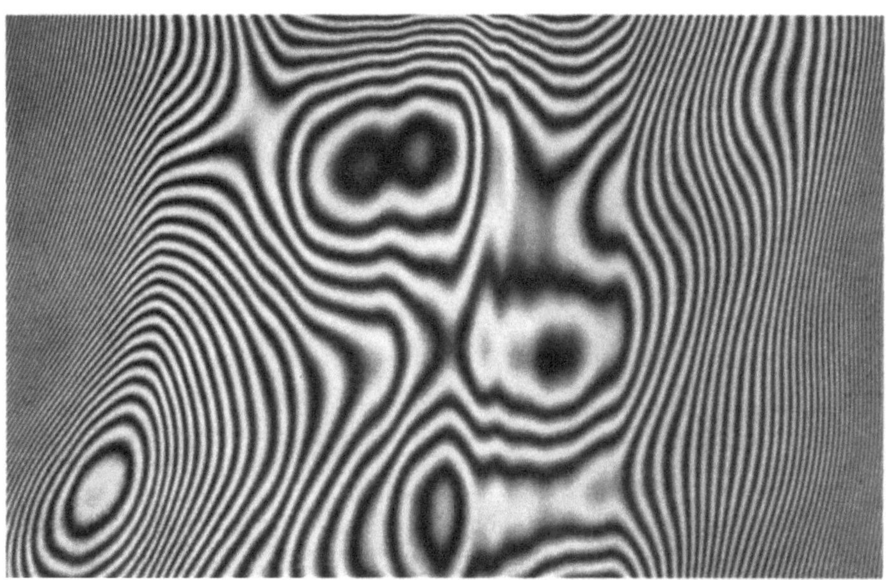

Fig. 31. Franges d'égale épaisseur d'une lame de verre. Franges par réflexion sous l'incidence normale.

Le phénomène a le maximum de contraste puisque les amplitudes interférentes sont égales (11.6). Les franges noires correspondent aux épaisseurs $e = K\dfrac{\lambda}{2n}$ et par conséquent lorsqu'on passe d'une frange à la suivante, l'épaisseur de la lame augmente de $\lambda/2n$. Avec une lame d'air, comprise entre deux lames de verre, éclairée par une radiation de longueur d'onde $0{,}6\,\mu$ on a $\lambda/2n = 0{,}3\,\mu$.

β) *Incidence oblique.* Considérons un faisceau de rayons monochromatiques et parallèles tombant sous une incidence oblique sur une lame d'épaisseur e et d'indice n.

Soit SI un rayon incident faisant avec la normale l'angle d'incidence i (Fig. 32). Il donne naissance à deux rayons: l'un est réfléchi sur la face supérieure de la lame en I, l'autre est réfracté et se réfléchit sur la face inférieure en J, puis rencontre la face supérieure en K et sort après réfraction parallèlement au premier rayon réfléchi en I. Calculons la différence de marche entre ces deux rayons. Ils sont en phase en I et leur différence de phase reste constante à partir de K et H, H étant le pied de la perpendiculaire abaissée de K sur le premier rayon. Entre ces deux limites, le rayon réfléchi en I accomplit le trajet IH dans l'air et le rayon réfracté en I, le trajet IJK dans le milieu d'indice n. La différence de marche est donc $2n\,\overline{IJ} - \overline{IH}$ à laquelle il faut ajouter $\lambda/2$ pour les mêmes raisons qu'au paragraphe précédent

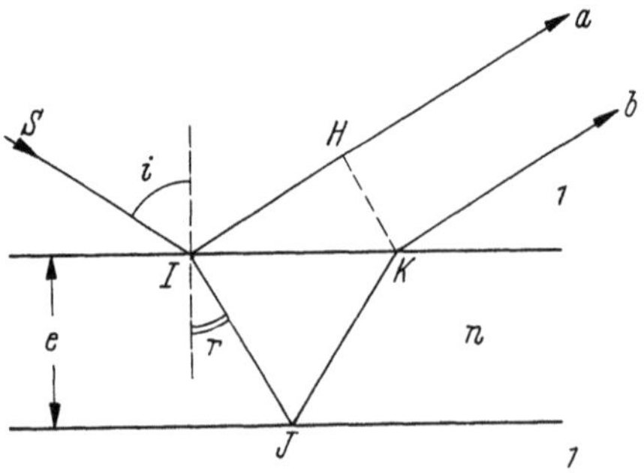

$$\delta = 2n\,IJ - IH + \frac{\lambda}{2}.$$

Or

$$IJ = \frac{e}{\cos r},$$

$$IH = IK \sin i, \qquad IK = 2e \tan r,$$

d'où

$$\left. \begin{aligned} \delta &= \frac{2ne}{\cos r} - 2e \sin i \tan r + \frac{\lambda}{2}, \\ \delta &= 2ne \cos r + \frac{\lambda}{2}. \end{aligned} \right\} \quad (19.5)$$

Fig. 32.

Les amplitudes a et b des rayons interférents sont pratiquement égales et l'éclairement est donné par la formule (19.2). Nous observerons encore des franges qui seront les lieux d'égale épaisseur si l'angle d'incidence i et par suite l'angle de réfraction r sont constants. Les franges brillantes seront données par

$$\frac{\delta}{\lambda} = \frac{2ne \cos r}{\lambda} + \frac{1}{2} = K \tag{19.6}$$

et les franges noires par

$$\frac{\delta}{\lambda} = \frac{2ne \cos r}{\lambda} + \frac{1}{2} = K + \frac{1}{2}. \tag{19.7}$$

Les franges noires correspondent aux épaisseurs $e = \dfrac{K\lambda}{2n \cos r}$.

20. Franges par transmission. α) *Incidence normale.* Considérons maintenant le rayon directement transmis SIa et le rayon deux fois réfléchi $SIJb$ (Fig. 33). Ces deux rayons peuvent interférer. Le rayon deux fois réfléchi $SIJb$ subit deux réflexions de même nature en I et J qui ne lui font subir aucune différence de phase au total. Dans le cas d'une lame d'indice n et d'épaisseur e plongée dans l'air, on a, entre les deux rayons interférents, la différence de marche

$$\delta = 2ne.$$

Les franges brillantes sont données par

$$p = \frac{\delta}{\lambda} = \frac{2ne}{\lambda} = K \tag{20.1}$$

et les franges sombres par

$$p = \frac{\delta}{\lambda} = \frac{2ne}{\lambda} = K + \frac{1}{2}. \tag{20.2}$$

La comparaison des formules (19.3), (19.4) et (20.1), (20.2) montre que les franges seront de même forme que les franges obtenues par réflexion, mais à chaque frange obscure du système réfléchi correspondra une frange brillante du système transmis. Les deux systèmes de franges sont complémentaires.

β) *Incidence oblique.* Si on opère sous une incidence oblique, nous aurons les deux rayons $SIJa$ et $SIJI'Kb$ (Fig. 34) et la différence de marche $2n\,\overline{I'J}-\overline{JH}$

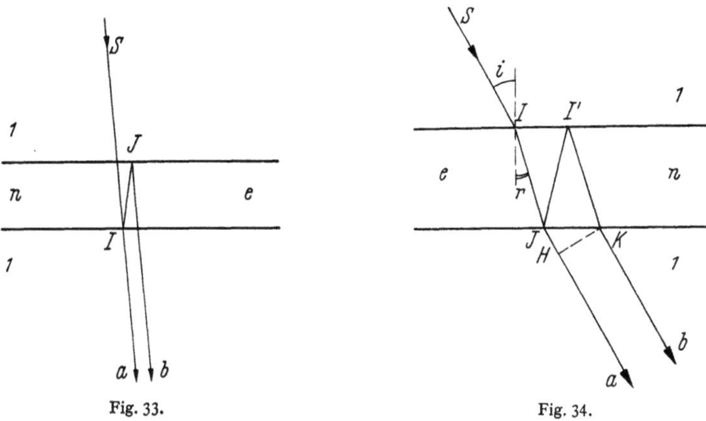

Fig. 33.   Fig. 34.

se calculera comme à la Sect. 19β. Si les deux milieux situés de part et d'autre de la lame sont identiques, on aura

$$\delta = 2n\,e\cos r$$

d'où les franges brillantes

$$p = \frac{\delta}{\lambda} = \frac{2n\,e\cos r}{\lambda} = K$$

et les franges sombres

$$p = \frac{\delta}{\lambda} = \frac{2n\,e\cos r}{\lambda} = K + \frac{1}{2}.$$

Les franges sont complémentaires des franges par réflexion sous la même incidence.

γ) *Contraste des franges par transmission.* Le contraste des franges par transmission n'est plus égal au contraste, pratiquement maximum, des franges par réflexion. En effet, dans le cas de l'incidence normale (Fig. 33) on a pour le carré de l'amplitude du rayon transmis directement SIa

$$a^2 \approx 1 - 2\left(\frac{n-1}{n+1}\right)^2$$

et pour le rayon $SIJa_2$

$$b^2 \approx \left(\frac{n-1}{n+1}\right)^4$$

d'où pour $n = 1{,}5$

$$a^2 \approx 0{,}92 \qquad b^2 \approx 0{,}0016.$$

On aura donc le contraste

$$\gamma = 0{,}15.$$

Le phénomène est visible mais avec un mauvais contraste.

δ) *Remarque.* Nous avons jusqu'ici négligé les rayons plusieurs fois réfléchis car, par suite de la décroissance rapide de leurs amplitudes, ils ne produisent que des perturbations négligeables. Considérons le cas des phénomènes sous l'incidence normale (Fig. 35): les carrés des amplitudes réfléchies et transmises sont données par le tableau suivant pour $n = 1,5$:

Réflexion	$a^2 = R$	$b^2 = R T^2$	$R^3 T^2$
	0,04	0,04	0,00006
Transmission	$a^2 = T^2$	$b^2 = R^2 T^2$	$R^4 T^2$
	0,92	0,0016	0,000002

On constate que le carré de l'amplitude du troisième rayon réfléchi ou transmis est déjà négligeable. Prenons par exemple une lame d'air entre deux lames de verre d'indice n et supposons que les deux faces en verre qui limitent la lame d'air soient bien parallèles. Observons la lame par transmission: en tenant compte de toutes les réflexions, on verra plus loin que la lame est vue avec un éclairement [36]

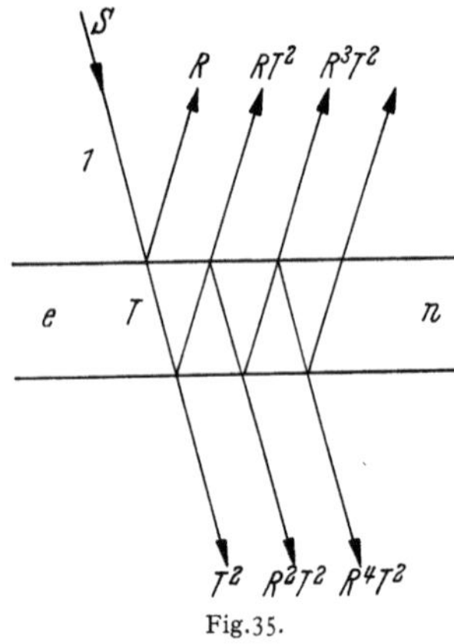

Fig.35.

$$E = \frac{T^2}{(1-R)^2 + 4R \sin^2 \frac{\varphi}{2}}, \qquad \varphi = \frac{4\pi e}{\lambda}.$$

Faisons varier l'épaisseur e de la lame d'air en écartant les deux lames de verre qui la limite, tout en gardant leur parallélisme. Les maxima d'éclairement, c'est à dire les franges brillantes, correspondront à des valeurs de φ données par $\varphi = 2K\pi$ ou $p = \delta/\lambda = K$ et les minima par $p = \delta/\lambda = K + \frac{1}{2}$. Les franges brillantes et sombres correspondent donc aux mêmes valeurs de l'épaisseur que dans les calculs précédents ne faisant intervenir que deux ondes. Si E_1 est l'éclairement maximum, E_2 l'éclairement minimum, le contraste sera donné par

$$\gamma = \frac{4R}{(1-R)^2 + 4R} = 0,148.$$

On obtient une valeur qui diffère de moins d'un centième de la valeur obtenue à la Sect. 20γ. La différence n'est pas visible à l'oeil.

21. Cas particuliers des franges des lames minces: Anneaux de Newton. Une application intéressante au point de vue historique est celle de la production de franges circulaires au moyen d'une lame d'air comprise entre une surface plane et une surface sphérique de grand rayon de courbure. Ce sont les anneaux de Newton (Fig. 37).

Représentons en coupe (Fig. 36) la sphère S de rayon R: c'est, par exemple, la face courbe d'une lentille plan-convexe. Elle est posée sur la surface plane d'une lame de verre P. Entre la lentille et le plan en A se trouve une lame d'air très mince d'épaisseur e_0 qui ne disparaît que s'il y a contact optique entre les deux surfaces.

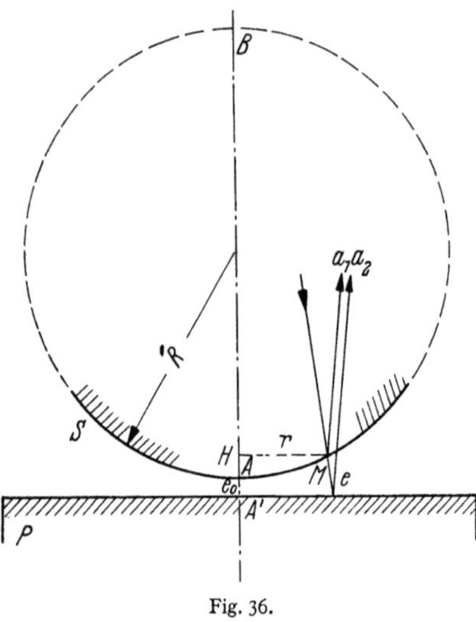

Fig. 36.

En un point M quelconque, l'épaisseur d'air est e et dans le triangle BMA non tracé sur la figure, on aura, en posant $HM = r$ et AH étant petit

$$r^2 = AH(2R - AH) \approx 2R \cdot AH$$

d'où

$$e = AH + e_0 = \frac{r^2}{2R} + e_0.$$

La différence de marche en M sera, en observant les phénomènes par réflexion:

$$\delta = 2e + \frac{\lambda}{2} = \frac{r^2}{R} + 2e_0 + \frac{\lambda}{2}.$$

Il y aura un anneau noir en M si

$$p = \frac{\delta}{\lambda} = \frac{r^2}{R\lambda} +$$

$$+ \frac{2e_0}{\lambda} + \frac{1}{2} = K + \frac{1}{2},$$

c'est à dire si

$$\frac{r^2}{R\lambda} + \frac{2e_0}{\lambda} = K.$$

Le rayon r de cet anneau noir sera donné par

$$r = \sqrt{R\lambda\left(K - \frac{2e_0}{\lambda}\right)}$$

et en posant

$$K_0 = \frac{2e_0}{\lambda},$$

$$r = \sqrt{R(K - K_0)\lambda} \quad (21.1)$$

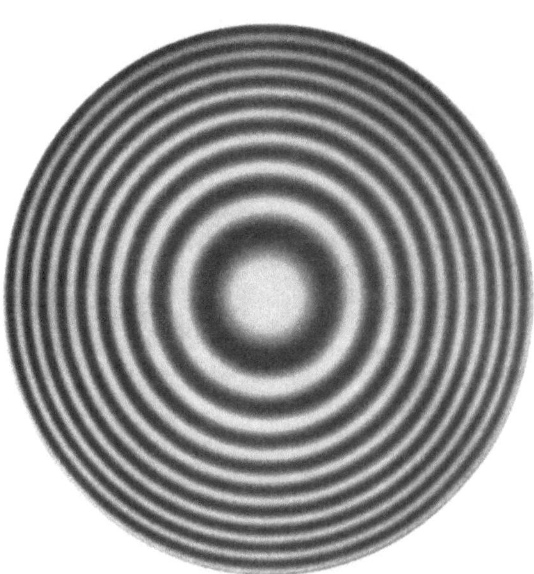

Fig. 37. Anneaux de Newton.

si $K_0 = 0$ (contact optique) $r = \sqrt{RK\lambda}$; les rayons des anneaux noirs varient comme les racines carrées des nombres entiers successifs.

La différence $K - K_0$ est la différence des ordres d'interférence en M et en A: elle représente le nombre de franges vues entre le centre A et le point M considéré.

22. Cas particuliers des franges des lames minces: franges du coin d'air. Considérons les interférences produites par une lame d'air comprise entre deux surfaces de verre planes faisant un petit

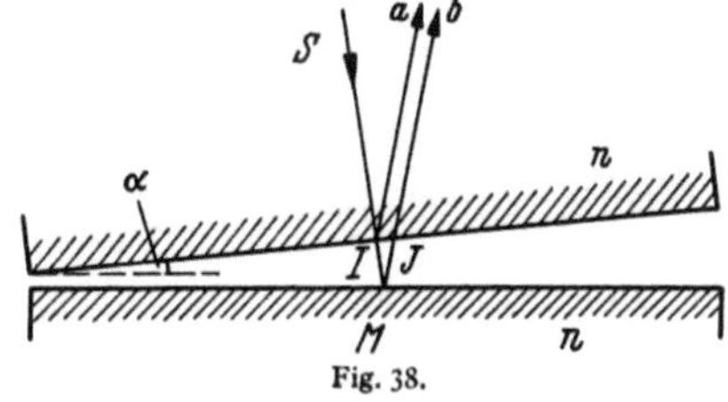

Fig. 38.

angle α (Fig. 38) et observons par réflexion sous incidence normale. La différence de marche entre les deux rayons réfléchis en I et J est égale à

$$\delta = 2e + \frac{\lambda}{2},$$

e étant l'épaisseur d'air au point M où l'on observe le phénomène.

On aura en ce point une frange noire si

$$p = \frac{\delta}{\lambda} = \frac{2e}{\lambda} + \frac{1}{2} = K + \frac{1}{2}$$

ou

$$\frac{2e}{\lambda} = K.$$

Si on considère la frange noire suivante, l'ordre d'interférence est $p' = p + 1$, et l'épaisseur e' telle que

$$\frac{2e'}{\lambda} = K + 1.$$

On verra donc le coin d'air sillonné de franges rectilignes équidistantes et parallèles à l'arête du coin, l'épaisseur du coin d'air augmentant de $e' - e = \lambda/2$ lorsqu'on passe d'une frange à la suivante.

Si on déplace d'un mouvement continu les deux lames l'une par rapport à l'autre, les franges conservent leur forme, mais se meuvent dans le champ d'observation. Lorsque le déplacement atteint une demi-longueur d'onde, chaque frange se sera substituée à sa voisine.

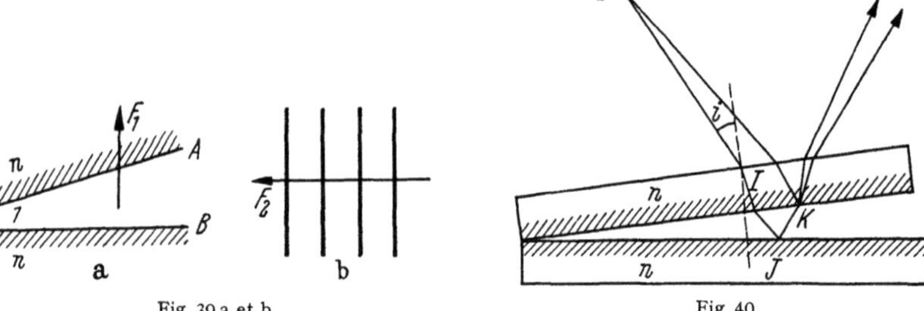

Fig. 39 a et b.   Fig. 40.

Les franges se déplacent vers le lieu d'épaisseur moindre puisque l'ordre des franges va en croissant avec l'épaisseur et que si on éloigne les surfaces, la première frange visible sera d'ordre de plus en plus élevé. Si on éloigne la surface A de la surface B dans le sens de la flèche F_1 (Fig. 39a), les franges se déplaceront dans sens de la flèche F_2 (Fig. 39b). Sous incidence oblique, le retard deviendra (Fig. 40)

$$\delta = 2e \cos i + \frac{\lambda}{2}$$

si l'incidence augmente, la différence de marche diminue et les franges s'écartent les unes des autres.

23. Localisation des franges. Soit S une source ponctuelle et M un point de l'espace où l'on observe les interférences (Fig. 41). $SABM$ et $SA'B'M$ sont

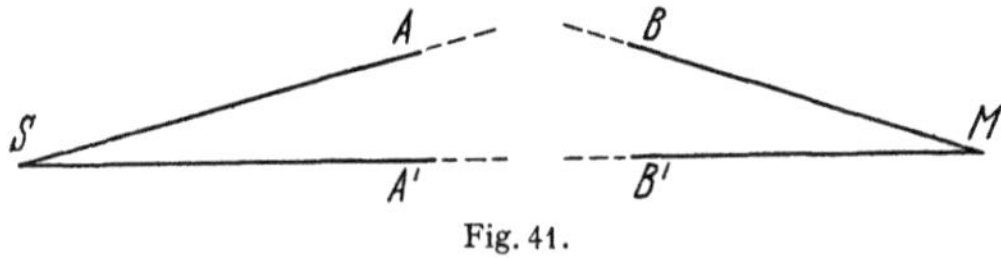

Fig. 41.

deux rayons issus de S et qui interfèrent en M. La différence de marche de ces deux rayons en M est δ et l'éclairement en M est donné par l'expression (5.2). Déplaçons le point M, δ varie mais reste bien déterminé et on peut observer également les interférences dans cette nouvelle position. Si la source est ponctuelle, on obtient donc des franges d'interférences parfaitement nettes, quelle que soit la position où l'on se place pour observer le phénomène, c'est-à-dire quelle que soit la position donnée à l'écran d'observation. Les franges ne sont pas localisées.

Considérons maintenant une source lumineuse étendue S (Fig. 42).

Pour observer des franges en M, il faut que la différence de marche des deux rayons qui partent de S_1 et qui se rencontrent en M soit constante quelle que soit la position du point S_1 à l'intérieur de la source S. Dans ces conditions, chaque point tel que S_1 donnera le même système de franges en M et le phénomène restera parfaitement net. Or le lieu des points S_1 pour lesquels le chemin optique S_1ABM est constant, est un petit élément plan S perpendiculaire à S_1A. Il faut aussi que ce petit élément plan soit le lieu des points pour lesquels le chemin optique $S_1A'B'M$ reste constant. Le rayon S_1A' doit être également normal à cet élément plan S. Il faut donc que les deux rayons S_1A et S_1A' soient confondus car il n'y aura inter-férences que s'ils proviennent du même point de la source S. On pourra observer des franges avec la source étendue S sur un écran placé en M.

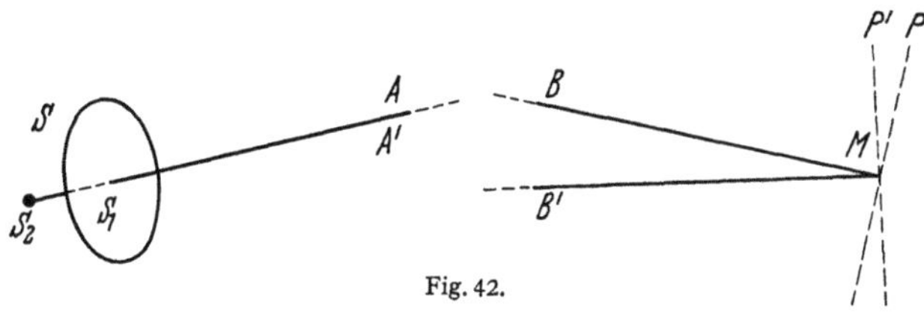

Fig. 42.

Si le point M se déplace le long de BM, le rayon S_1B ne peut plus se confondre avec S_1A et il n'est plus possible d'observer des franges avec la source étendue S. Toutefois, si l'appareil produisant les interférences admet un plan de symétrie au voisinage de la région où l'on observe les franges, on peut avoir des franges avec une source étendue, ces franges étant observables au voisinage d'une surface de localisation. Soit S_1A un rayon du plan de symétrie se dédoublant en deux rayons BM et $B'M$ et P et P', les deux plans d'onde correspondants passant par M. La différence des chemins optiques qui vont de S à P et de S à P' est égale à la différence des chemins optiques S_2ABM et $S_2AB'M$. On peut donc étendre la source dans l'espace environnant, sans craindre une baisse du contraste des franges en M. Le point M où l'on observe les franges est défini par l'intersection avec le rayon BM du rayon $B'M$ provenant du rayon S_1A' confondu avec S_1A. La surface de localisation des franges est le lieu des points de rencontre des rayons interférents provenant d'un même rayon incident.

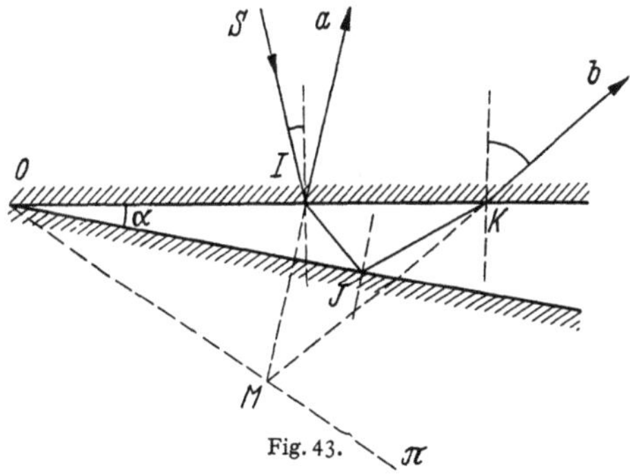

Fig. 43.

Examinons le cas du coin d'air observé par réflexion en supposant la source à l'infini. Le coin d'air est éclairé par un collimateur et soit SI un rayon du faisceau dans un plan de section principale (Fig. 43).

· Le point de localisation M des franges correspondant au rayon SI est obtenu en prolongeant les deux rayons réfléchis Ia et Kb. En répétant la même construction pour différents rayons parallèles à SI, on trouve que le lieu du point M est un plan π passant par l'arête O. Le plan π est le plan de localisation des franges du coin d'air. Si l'incidence est très oblique, le plan π peut faire un angle non négligeable avec la lame, mais si le faisceau incident est normal à la lame, le plan π est à l'intérieur de la lame et on peut même dire confondu avec la lame puisque l'angle α est toujours très petit, de l'ordre de la minute, et e de l'ordre du micron.

d) Franges d'égale inclinaison ou franges localisées à l'infini.

24. Franges par réflexion. Dans les phénomènes que nous avons précédemment étudiés, on faisait tomber un faisceau de rayons parallèles sur une lame d'épaisseur variable. Dans les phénomènes que nous étudions maintenant, nous prendrons des lames dont l'épaisseur e est rigoureusement constante et nous ferons varier l'inclinaison des rayons incidents.

Considérons une source lumineuse S étendue et monochromatique envoyant des rayons diversement inclinés sur une lame à faces parallèles L (Fig. 44). Une lame semi-transparente G inclinée à 45° permet d'envoyer les rayons sous des incidences variées voisines de la normale.

Parmi tous ces rayons, considérons ceux qui tombent sur la lame sous l'incidence i. Entre un rayon qui est réfléchi en I et celui qui est réfléchi en J, on a la différence de marche (19.5)

$$\delta = 2\,n\,e\,\cos r + \frac{\lambda}{2}.$$

Plaçons derrière la lame un objectif O dont l'axe est normal à la lame. Les rayons parallèles à la direction i coupent le plan focal de l'objectif suivant une

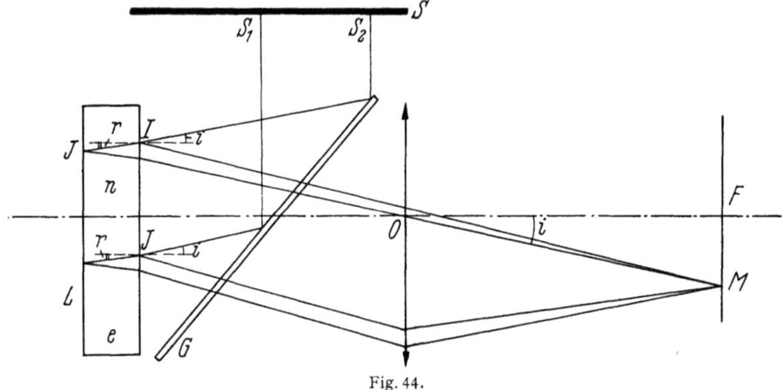

Fig. 44.

circonférence de centre F. Pour tous les points de cette circonférence, l'ordre d'interférence p est constant et égal à

$$p = \frac{\delta}{\lambda} = \frac{2\,n\,e\,\cos r}{\lambda} + \frac{1}{2}.$$

Si $p = K$, nous aurons un anneau brillant et pour $p = K + \frac{1}{2}$ un anneau noir. Ces franges, dites franges d'égale inclinaison, sont des anneaux de centre F. Tous les rayons parallèles à une direction donnée se coupent en un point déterminé du plan focal, les franges sont localisées dans le plan focal de l'objectif L, d'où le nom de franges à l'infini qu'on leur donne également.

Au centre F, l'ordre d'interférence est

$$p_0 = \frac{\delta_0}{\lambda} = \frac{2\,n\,e}{\lambda} + \frac{1}{2}.$$

Si l'on s'éloigne de ce point, l'ordre d'interférence diminue et au point M il sera

$$p = p_0 - K,$$

K représentant le nombre de franges que l'on voit entre F et M, c'est un nombre entier ou fractionnaire.

La relation précédente peut s'écrire

$$\frac{2\,n e \cos r}{\lambda} = \frac{2\,n e}{\lambda} - K$$

et en supposant les angles petits

$$\frac{2\,n e}{\lambda}\left(1 - \frac{r^2}{2}\right) = \frac{2\,n e}{\lambda} - K$$

d'où

$$i = \sqrt{\frac{n\lambda}{e}}\,\sqrt{K}. \qquad (24.1)$$

Si on a une frange noire en F, cette formule donnera le rayon du $K^{\text{ième}}$ anneau noir.

Les rayons des anneaux correspondant au même état d'interférences que le centre varient comme les racines carrées des nombres entiers. Ils se resserrent à mesure qu'on s'écarte du centre. Comme dans les phénomènes par réflexion que nous avons déjà étudiés, le contraste de ces franges est maximum puisque les amplitudes des rayons interférents sont égales. Notons que la constance d'épaisseur de la lame doit être de l'ordre de 1/10 de micron. En effet, si sur une moitié de la lame, l'épaisseur diffère de $\lambda/4n$ de l'épaisseur de l'autre moitié, les deux systèmes d'anneaux produits par les deux points S_1 et S_2 de la source se détruisent (r est supposé petit).

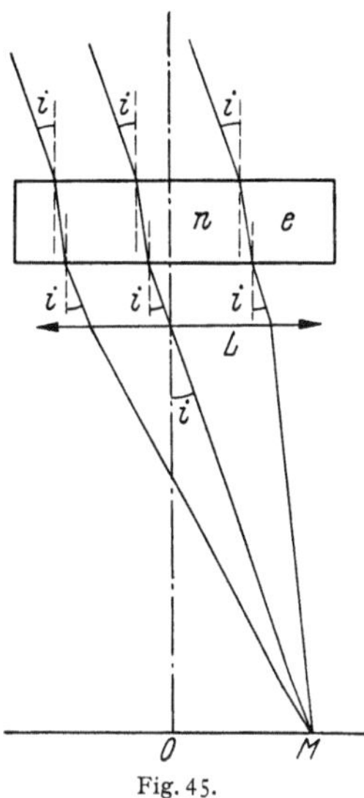

25. Franges par transmission. On peut observer les phénomènes par transmission suivant le schéma de la Fig. 45. La différence de marche est seulement $\delta = 2n\,e\cos r$ et les phénomènes sont complémentaires des phénomènes par réflexion. Les franges d'égale inclinaison par transmission ont un contraste assez faible pour les raisons indiquées précédemment à la Sect. 20γ.

Fig. 45.

e) Influence du diamètre apparent et du monochromatisme de la source sur le contraste des franges.

26. Influence du diamètre apparent de la source sur le contraste des franges d'égale épaisseur. α) *Calcul du contraste.* Reprenons le cas de l'observation des

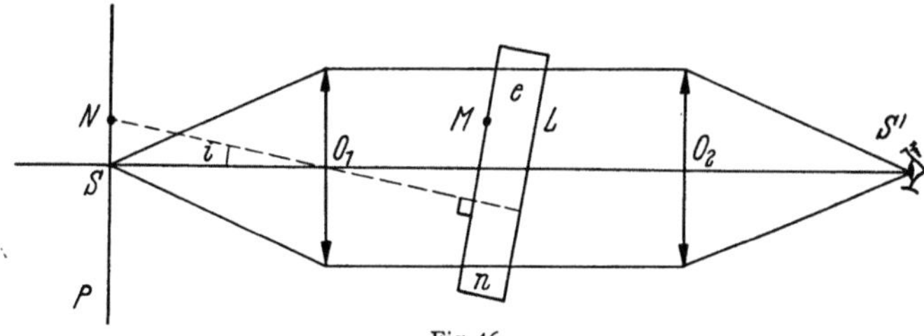

Fig. 46.

franges d'égale épaisseur d'une lame L d'épaisseur e et d'indice n, éclairée par un faisceau de lumière parallèle et monochromatique (Fig. 46). Une source lumineuse S est au foyer d'un objectif O_1, les rayons émergents parallèles traversent la lame L, rencontrent un objectif O_2 et forment en S' une image de S sur la pupille de l'oeil. La lame L est placée au foyer objet de O_2 de façon que l'oeil puisse la voir sans fatigue. On observe ainsi les franges par transmission mais les résultats qui suivent s'appliquent aussi bien au cas des franges par réflexion correspondant au schéma de la Fig. 47. Si la source S est ponctuelle, les rayons lumineux tombent sur la lame sous l'incidence constante i, le point N représentant la trace de la normale à la lame. Nous calculerons le contraste des franges lorsque la source S a la forme d'un carré dont le côté est vu du point O_1 sous l'angle ϑ et en fonction de l'angle i. Si l'angle i n'est pas trop grand, les anneaux à l'infini de la lame L sont des cercles de centre N situés dans le plan P passant par S et de rayons angulaires donnés par (24.1).

Lorsqu'on se déplace de N jusqu'à l'anneau K, la différence de marche varie de

$$\Delta\delta = K\lambda = \frac{i^2 e}{n}.$$

Faisons passer par N et dans le plan P deux axes de coordonnées (Fig. 48) parallèles aux côtés de la source lumineuse dont le centre S se déplacera sur $N\,x$.

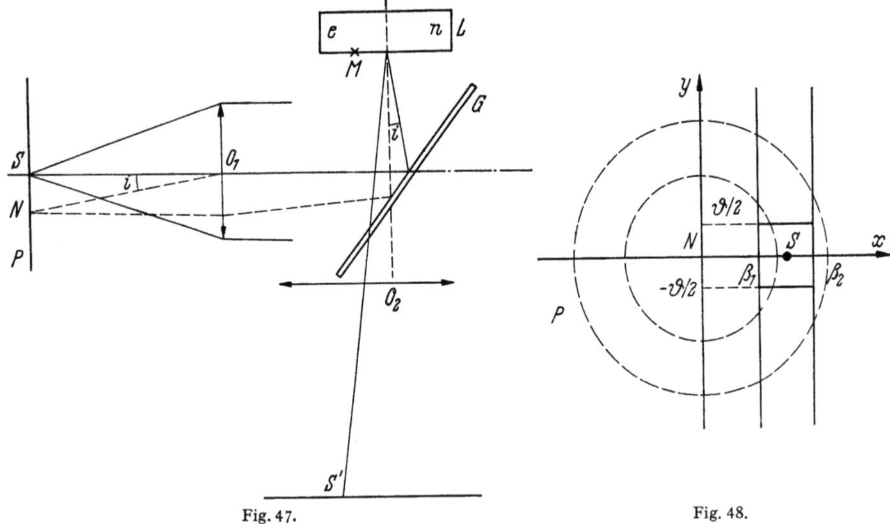

Fig. 47.　　　　　　　　　　　　　　　　Fig. 48.

On peut décomposer i en deux composantes i_x et i_y correspondant aux deux directions $N\,x$ et $N\,y$. Les angles étant petits

$$i^2 = i_x^2 + i_y^2, \qquad \Delta\delta = \frac{e\,(i_x^2 + i_y^2)}{n}.$$

L'éclairement en un point M quelconque de la lame dû à un point quelconque de l'anneau d'ordre K du plan P supposé lumineux, sera

$$a^2 + b^2 + 2a\,b\cos\frac{2\pi}{\lambda}\left[\delta + \frac{e}{n}\,(i_x^2 + i_y^2)\right],$$

δ étant la différence de marche en M pour les rayons émanant du centre N. D'où l'éclairement dû à la source carrée

$$E = \int_{-\frac{\vartheta}{2}}^{+\frac{\vartheta}{2}}\int_{\beta_1}^{\beta_2}\left\{a^2 + b^2 + 2a\,b\cos\frac{2\pi}{\lambda}\left[\delta + \frac{e}{n}\,(i_x^2 + i_y^2)\right]\right\} di_x\,di_y,$$

β_1 et β_2 étant les distances angulaires séparant les cotés de la source parallèles à oy du centre N

$$\beta_2 - \beta_1 = \vartheta, \qquad \frac{\beta_1 + \beta_2}{2} = i.$$

Posons

$$u = \vartheta \sqrt{\frac{e}{n\lambda}}, \qquad v = 2i \sqrt{\frac{e}{n\lambda}},$$

$$F_1(u, v) = G(v + u) \pm G(v - u),$$

$$F_2(u, v) = F(v + u) \pm F(v - u)$$

en prenant le signe $+$ si $v - u$ est négatif et le signe $-$ pour $v - u$ positif, $G(u)$ et $F(u)$ étant les intégrales de FRESNEL

$$G(u) = \int_0^u \cos \frac{\pi}{2} u^2 \, du, \qquad F(u) = \int_0^u \sin \frac{\pi}{2} u^2 \, du$$

et enfin

$$A = F_1(u, v)\, G(u) - F_2(u, v)\, F(u),$$

$$B = F_1(u, v)\, F(u) + F_2(u, v)\, G(u).$$

L'éclairement en M devient

$$E = (a^2 + b^2)\, \vartheta^2 + \frac{n\,a\,b\,\lambda}{e} \sqrt{A^2 + B^2} \cos \left(\frac{2\pi \delta}{\lambda} + \arctan \frac{B}{A} \right).$$

Pour étudier les variations de E, il faut faire varier seulement δ dans l'expression précédente car le rapport e/n qui intervient dans le facteur $n\,a\,b\,\lambda/e$ ainsi que dans les expressions A et B provient de la relation $\varDelta\,\delta = i^2 e/n$ dans laquelle i varie seul, e étant l'épaisseur moyenne de la lame dont les défauts sont toujours petits. Les franges brillantes et les franges sombres sont alors données par

$$\cos \left(\frac{2\pi \delta}{\lambda} + \arctan \frac{B}{A} \right) = \pm 1$$

d'où le contraste

$$\gamma = \frac{2a\,b \sqrt{A^2 + B^2}}{(a^2 + b^2)\, u^2 + a\,b \sqrt{A^2 + B^2}} . \quad (26.1)$$

β) *Franges d'égale épaisseur observées sous l'incidence normale.* Dans le cas des franges d'égale épaisseur observées sous l'incidence normale $i = 0$ et $v = 0$.

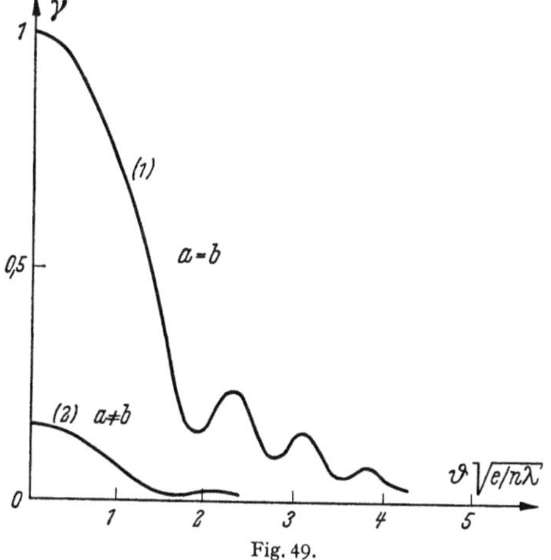

Fig. 49.

La courbe (1) de la Fig. 49 montre les variations du contraste en fonction de $u = \vartheta \sqrt{e/n\lambda}$ lorsque $a = b$. C'est le cas des franges observées par réflexion. Si a est différent de b, les contrastes sont diminués. Ainsi dans le cas des franges par transmission données par une lame de verre placée dans l'air, on a sensiblement $a = 0{,}96$ et $b = 0{,}04$: les résultats sont donnés par la courbe (2). Ces courbes montrent que pour avoir une bonne visibilité des phénomènes, il faudra donner à la source un diamètre apparent satisfaisant à la relation

$$\vartheta \sqrt{\frac{e}{n\lambda}} \leq 1 \qquad\qquad (26.2)$$

d'où les valeurs de ϑ pour quelques lames d'air d'épaisseurs variées avec $\lambda = 0,6\ \mu$

$$e = 0,03\ \text{mm.} \qquad \vartheta \leq 8°$$
$$e = 0,50 \qquad\qquad \vartheta \leq 2°$$
$$e = 10 \qquad\qquad\ \ \vartheta \leq 27'.$$

Pour réaliser expérimentalement cette tolérance il suffit d'examiner les anneaux à l'infini dans l'image S' de la source étendue S. En effet, si on compare les relations (26.2) et (24.1) en faisant $K = 1$ dans (24.1) c'est à dire en supposant que l'ordre d'interférence au centre des anneaux à l'infini est un entier, on voit que la règle (26.2) peut s'écrire

$$\vartheta \leq \alpha$$

donc si en examinant l'image S' on voit plusieurs anneaux à l'infini, on réduira la source S jusqu'au moment où sa largeur sera égale au rayon du premier anneau.

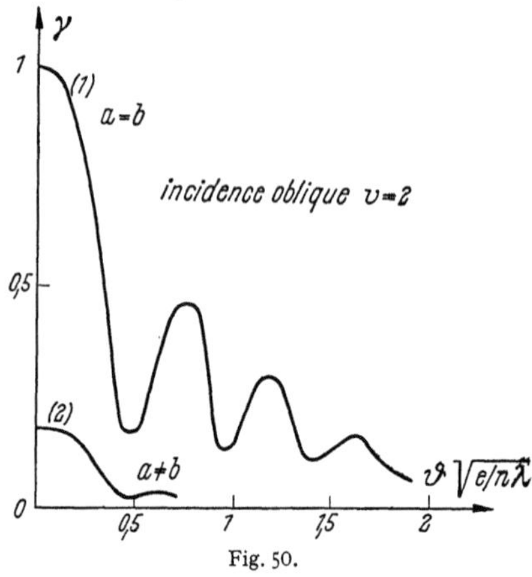

Fig. 50.

γ) *Franges d'égale épaisseur observées sous incidence oblique.* On a maintenant les deux paramètres u et v, l'angle d'incidence étant déterminé par

$$v = 2i\ \sqrt{\frac{e}{n\,\lambda}}\,.$$

La courbe (1) de la Fig. 50 montre les variations du contraste en fonction de $u = \vartheta\sqrt{e/n\lambda}$ lorsque $a = b$ et pour une incidence définie par $v = 2$. La courbe (2) donne les résultats lorsque a est différent de b et que l'on considère comme précédemment le cas des franges par transmission ($a = 0,96$ et $b = 0,04$).

Pour avoir le même contraste que dans le cas de l'incidence normale, il faudra écrire ici

$$u = \vartheta\ \sqrt{\frac{e}{n\,\lambda}} \leq 0,2, \qquad v = 2i\ \sqrt{\frac{e}{n\,\lambda}} = 2\,.$$

On a donc une tolérance 5 fois plus sévère pour $v = 2$ que pour l'incidence normale $v = 0$. On peut observer expérimentalement ce résultat de la façon suivante: prenons une source S dont la largeur satisfait à la relation (26.2) et opérons en incidence normale. En regardant l'image S' on ne voit pratiquement pas d'anneaux puisque la source S' n'en laisse passer qu'une fraction. Déplaçons la source de façon à opérer en incidence oblique; par suite du resserrement des anneaux à l'infini suivant la loi (24.1), on voit apparaître des portions de plus en plus nombreuses d'anneaux dans S' et le contraste des franges d'égale épaisseur diminue rapidement. Il faut donc toujours observer les franges d'égale épaisseur sous l'incidence normale, ce qui permettra d'utiliser des sources de plus grands diamètres et d'avoir des phénomènes plus lumineux.

27. Influence du monochromatisme de la source sur la visibilité des franges. Lorsque la source n'est pas monochromatique, le phénomène observé est la somme des phénomènes correspondant aux différentes radiations monochromatiques composant la source. L'intervalle des franges n'étant pas le même pour toutes les radiations, la superposition de tous ces systèmes de franges diminue le contraste du phénomène.

On peut évaluer la tolérance sur le monochromatisme de la source à partir des résultats précédents. On a vu que pour avoir une bonne visibilité des franges d'égale épaisseur, il fallait que la source ait une largeur ne dépassant pas la valeur ϑ donnée par (26.2). Ceci conduit à admettre que l'ordre d'interférence ne varie pas d'une quantité supérieure à $\frac{1}{4}$ lorsqu'on passe du centre au bord de la source. Considérons maintenant une source ponctuelle mais non monochromatique. Comme à la Sect. 15, on peut dire que la variation de l'ordre d'interférence en un point où l'on observe le phénomène n'est plus dûe à une variation d'inclinaison des rayons incidents, mais à une variation de la longueur d'onde. En appliquant la règle précédente, écrivons que la variation de l'ordre d'interférence consécutive à une variation de la longueur d'onde de la lumière incidente, variation supposée assez faible pour que le coefficient de visibilité de l'oeil, restant constant, ne dépasse pas $\frac{1}{4}$. D'après (15.2) on aura

$$p\,\frac{d\lambda}{\lambda} \leq \frac{1}{4}. \tag{27.1}$$

Dans la comparaison avec la source le contraste ne dépend en effet que de la distance du centre au bord de la source dans le cas présent.

Si on considère de grandes différences de marche, c'est à dire des valeurs de p élevées, pour satisfaire à la relation précédente, il faudra utiliser une radiation pour laquelle le rapport $d\lambda/\lambda$ soit très petit, donc une radiation à grand coefficient de finesse (Sect. 15).

Pour obtenir des raies très fines, il faut employer un gaz sous faible pression à température peu élevée et se servir d'un corps ayant un fort poids atomique: ce sont des sources de luminance relativement faible. Les sources comme les lampes à vapeur de mercure à haute pression donnent des raies intenses mais larges à faible coefficient de finesse. Il n'est pas possible d'observer des franges correspondant à des ordres d'interférence élevés. Ainsi avec une lampe à vapeur de mercure à pression moyenne, un coefficient de finesse pour le raie verte de l'ordre de

$$\frac{\lambda}{d\lambda} = 2500,$$

permettra d'aller jusqu'à un ordre d'interférence donné par

$$p \leq \frac{1}{4}\,\frac{\lambda}{d\lambda} = 625,$$

auquel correspondra une différence de marche $\delta \leq p\lambda$ soit 0,3 mm. Pour la radiation rouge du cadmium (lampe à basse pression), on pourra observer dans les mêmes conditions une différence de marche égale à 15 cm. Notons que l'on peut observer des franges avec des ordres d'interférence plus élevés mais le contraste diminuera.

Ainsi avec la raie rouge du cadmium il sera possible de voir des franges avec une lame de 40 cm. d'épaisseur puisque le contraste sera encore égal à 0,2.

Utilisant l'isotope 84 du krypton isolé par diffusion thermique, Kösters et Engelhard ont construit une lampe qui fonctionne à la température du point triple de l'azote (63° K) sous une pression de 0,03 mm. Des franges d'interférence ont pu être observées avec cette source jusqu'à 80 centimètres de différence de marche représentant un ordre d'interférence de 1 400 000.

Meggers a construit et étudié des lampes utilisant l'isotope 198 du mercure à la pureté 99,9%. La finesse des raies est également très grande: avec la raie 5461, la différence de marche atteinte est de 60 cm., soit un ordre d'interférence égal à 1 000 000.

f) Observation des phénomenes.

28. Franges d'égale épaisseur. Les résultats précédents (26.2) montrent que plus la lame est mince et plus la source utilisée peut être étendue. Soit (Fig. 51) une lame mince éclairée par une source à l'infini et observée à l'oeil nu, et supposons que l'oeil mette au point dans le plan de la lame en M. Puisque l'épaisseur e de la lame est très faible, le plan de localisation des franges correspondant à une direction S_1M est très voisin du plan de la lame. De même pour une autre direction S_2M des rayons provenant d'un autre point S_2 de la source. Donc lorsque l'oeil nu met au point dans le plan de la lame mince, on peut dire que s'il lui était possible de voir les franges avec une source étendue, il les verrait parfaitement nettes et pratiquement dans le plan de la lame. En fait, l'état d'interférence en M varie avec l'inclinaison des rayons: il n'est pas le même pour le rayon S_1M et le rayon S_2M et les phénomènes seront brouillés. Mais la pupille

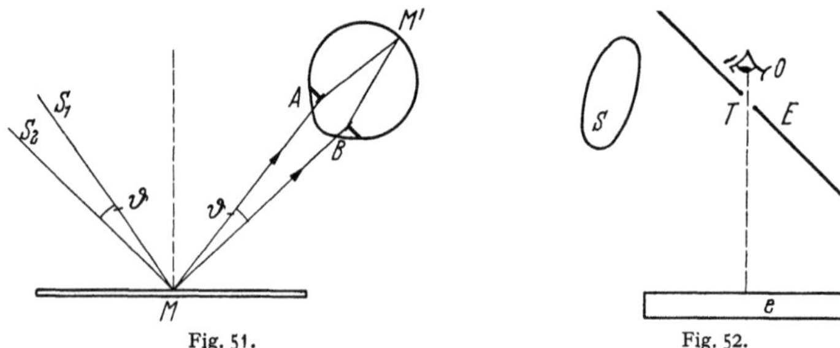

Fig. 51.   Fig. 52.

AB de l'oeil ayant une faible ouverture, la variation d'inclinaison ϑ des rayons lorsqu'on passe d'une extrémité à l'autre du diamètre pupillaire, est petite. Tout se passe comme si l'on utilisait de la source qu'une faible portion vue du point M sous l'angle ϑ. Par exemple, pour une pupille de l'oeil égale à 4 mm. et pour une distance de l'oeil à la lame égale à 25 cm., on aura $\vartheta = 0,016$.

D'après (26.2), on voit qu'il est possible d'observer avec un très bon contraste les franges d'une lame mince d'air d'épaisseur moyenne de l'ordre de 2 mm. en lumière monochromatique.

Pour observer les franges en lumière blanche, l'épaisseur de 2 mm. sera beaucoup trop forte. La tolérance exprimée par la relation (27.1) est sévère: elle indique qu'avec une radiation dont le coefficient de finesse satisfait à (27.1) on a un très bon contraste. Il est possible d'aller plus loin: avec un ensemble continu de radiations dont les longueurs d'onde extrêmes correspondent à une variation de l'ordre d'interférence égale à 1,5, on a encore un contraste très suffisant. La courbe (1) de la Fig. 49 montre que dans ces conditions $\gamma = 0,47$.

Si λ_1 et λ_2 sont les longueurs d'ondes extrêmes et p_1 et p_2 les ondes d'interférence correspondantes, on pourra écrire

$$p_1 - p_2 = \delta\left(\frac{1}{\lambda_1} - \frac{1}{\lambda_2}\right) = \frac{3}{2}$$

avec

$$\delta = 2e + \frac{\lambda}{2},$$

ce qui suppose la lame mince observée par réflexion sous l'incidence normale. En prenant $\lambda_1 = 0,5\ \mu$, $\lambda_2 = 0,7\ \mu$ on obtient $e = 1\ \mu$. Des lames d'air d'épaisseur voisines du micron sont parfaitement visibles en lumière blanche. Avec des lames aussi minces, on peut utiliser le mode d'observation indiqué sur la Fig. 52. Une source S éclaire un écran diffusant blanc E percé d'un petit trou T à travers lequel l'oeil O regarde la surface de la lame mince e sous l'incidence normale. Grâce à l'écran diffusant E, l'oeil voit la lame e éclairée sur toute son étendue et voit les franges dessinant les lignes d'égale épaisseur. Si on observe en lumière monochromatique des franges très fines et très serrées, on utilisera un viseur V (microscope à faible grossissement) de grossissement convenable, au moyen duquel on mettra au point sur la lame mince e (Fig. 53). Une glace G inclinée à 45° permettra d'éclairer la lame e avec la source S et d'observer par réflexion sous incidence normale. Le viseur devra avoir une ouverture convenable de façon que l'inclinaison des rayons admis par le viseur V soit dans la tolérance voulue.

Dans le cas des lames épaisses atteignant quelques centimètres par exemple, il est nécessaire d'opérer en incidence normale avec un faisceau très bien parallèle (Sect. 26). On peut utiliser alors le dispositif représenté sur la Fig. 54. Un trou T de diamètre variable et éclairé par la source lumineuse S (par exemple une lampe à vapeur de mercure) est placé au foyer d'une lentille O_1. Le faisceau de rayons parallèles qui émerge tombe sur une glace G semi-transparente inclinée à 45° puis sur la lame L à étudier. Après réflexion, le faisceau est rendu convergent par la lentille O_2 et on place l'oeil en T' à travers le système. De cette façon l'oeil voit la lame bien éclairée et les franges sont nettes et vues sans accommoder si l'on a pris soin de placer la lame L au foyer antérieur de la lentille O_2. Notons que O_1 peut être une lentille plan-

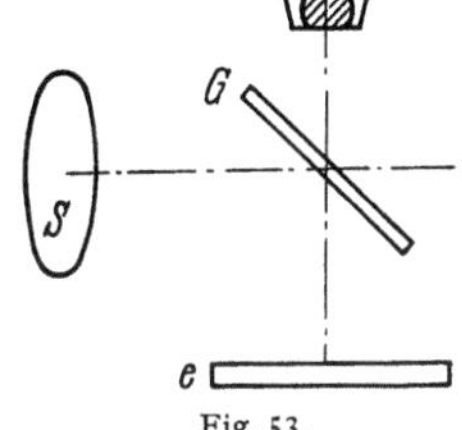

Fig. 53.

convexe dont la convexité est tournée vers la glace G en prenant un diamètre dix fois plus petit environ que sa focale: avec des trous de l'ordre de quelques dixièmes de millimètres et une focale de 1 m., il sera possible d'observer des

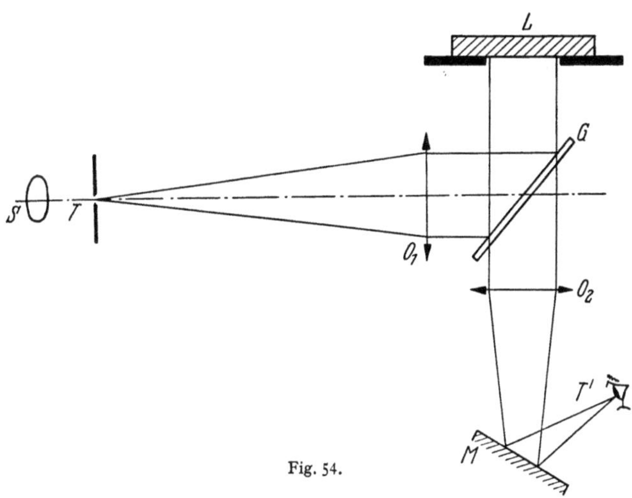

Fig. 54.

lames L de 1 à 2 cm. d'épaisseur si le monochromatisme est suffisant (lampe à vapeur de mercure basse pression et filtre isolant une raie).

29. Franges d'égale inclinaison. Le rayon angulaire des anneaux à l'infini est donné par la relation (24.1). Si l'ordre d'interférence au centre est un nombre entier, le rayon angulaire du premier anneau sera

$$i = \sqrt{\frac{n\lambda}{e}}.$$

Dans le cas d'une lame de verre d'indice 1,5 et en prenant $\lambda = 0,55\,\mu$ le rayon angulaire du premier anneau pour des lames de différentes épaisseurs est donné par le tableau suivant

$e = 0,1$ mm.	1 mm.	10 mm.	20 mm.
$i = 6°$	$2°$	$38'$	$22'$

Le rayon angulaire du premier anneau, pour une lame de 20 mm. d'épaisseur est à peu près égal au diamètre apparent de la lune, donc visible directement à l'oeil nu.

On peut alors observer les franges à l'infini en plaçant l'oeil contre une feuille de papier blanc E (Fig. 55) percée d'un trou de 3 à 4 mm. à travers lequel on regarde la lame. La feuille E est éclairée par exemple par une lampe au mercure S. Par réflexion sur la lame, on voit l'image du trou T comme une petite tache noire entourée des anneaux.

Pour observer les franges à l'infini par transmission, il suffit de placer la lame contre l'oeil et d'observer une surface étendue uniformément éclairée par une lampe à vapeur de mercure. On trouvera les anneaux en orientant convenablement la lame.

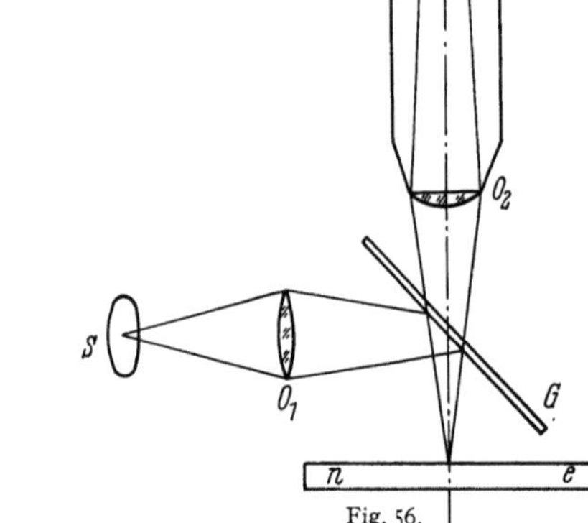

Fig. 55.   Fig. 56.

La difficulté pour observer les anneaux à l'infini est d'avoir une lame à faces bien parallèles. Le dispositif suivant dû à Fabry (Fig. 56) permet d'observer les anneaux même avec une lame grossièrement à faces planes et parallèles. On forme l'image d'une source lumineuse S, lampe à vapeur de mercure par exemple, sur la lame e au moyen d'une lentille O_1, et après réflexion sur une lame semi-transparente G inclinée à 45°. On vise les anneaux à l'infini au moyen d'une petite lunette L réglée pour l'infini et de grossissement $\frac{1}{2}$ environ. Contre l'oculaire est placé un oeilleton percé d'un trou petit T dont l'image se forme sur la lame e. Par ce procédé on utilise une très faible partie de la lame et si celle-ci n'est pas parfaite dans son ensemble, la région utilisée est assez petite pour qu'on puisse la considérer comme une lame pratiquement à faces parallèles dont les variations d'épaisseur sont négligeables.

30. Phénomènes en lumière blanche. Colorimétrie des lames minces. Lorsqu'on observe une lame mince en lumière blanche, à chaque radiation correspond un ordre d'interférence $p = \delta/\lambda$. En un point quelconque de la lame, l'ordre d'interférence varie avec la longueur d'onde. Dans la lumière réfléchie par la lame mince dans la région observée, les radiations pour lesquelles $p = \delta/\lambda = K + \frac{1}{2}$ seront absentes. Si le nombre de ces radiations éteintes n'est pas trop grand, on verra la lame mince colorée. Dans le cas où la lame n'est pas d'épaisseur constante, les couleurs observées ne seront pas les mêmes en tous points de la lame. A chaque épaisseur correspondra une valeur de δ et une teinte déterminée. Que la lame mince soit une lame d'air, ou d'une substance d'indice n, les phénomènes sont pratiquement les mêmes. En effet, tandis que la longueur d'onde λ varie du simple au double d'une extrémité à l'autre du spectre, l'indice n varie tout au plus de 1 à 2%.

Si la lame mince est à faces rigoureusement parallèles, e est constant et la couleur est la même sur toute la lame, on obtient une teinte plate.

Considérons une lame mince d'air observée par réflexion; si l'épaisseur est nulle, on voit d'après (19.2) que $E = 0$ pour toutes les radiations on a donc un noir. Augmentons progressivement l'épaisseur e de la lame en supposant que ses faces restent bien parallèles, c'est-à-dire que la teinte soit uniforme sur toute la lame. Lorsque e très petit, ce sont d'abord les courtes longueurs d'onde qui sont le mieux réfléchies, les premières teintes sont bleues. Pour $e = \lambda/4$ avec $\lambda = 0{,}56\,\mu$ (jaune moyen) on a un maximum de réflexion pour cette même longueur d'onde. Comme elle correspond au maximum de sensibilité de l'oeil, on voit la lame avec une teinte jaune voisine du blanc. Pour $e = \lambda/2$ avec $\lambda = 0{,}56\,\mu$ il y a extinction pour cette longueur d'onde: la lumière réfléchie est composée des extrémités du spectre, on a le pourpe dit «teinte sensible». Le moindre déplacement dans le spectre de la radiation éteinte entraîne un changement de couleur très rapide. On retrouve le même phénomène pour $e = \lambda, 3\lambda/2$, etc. mais à mesure que le retard augmente, les couleurs sont de moins en moins pures. On convient de classer les teintes en différents ordres séparés les uns des autres par les teintes sensibles.

On a des phénomènes semblables par transmission, mais le contraste est mauvais comme il a été vu et les teintes ne sont pas pures. On peut observer des teintes identiques et pures dans l'expérience des trous D'YOUNG à frange centrale brillante. Au centre, le retard est nul pour toutes les radiations et on a maintenant un blanc. Au fur et à mesure que l'on s'écarte de l'axe, c'est à dire que le retard augmente on voit toute une succession de teintes différentes des précédentes.

Les courbes résultant de ces deux types de phénomènes débutant par du noir ou du blanc, ont été cataloguées dans l'échelle des teintes donnée ci après. Dans ce tableau, on représente la différence de marche entre les deux vibrations qui interfèrent par δ' indépendamment des changements de phase éventuels, c'est le retard du à la lame seule. Par exemple, pour une lame mince observée sous l'incidence normale par réflexion

$$\delta = 2ne + \frac{\lambda}{2} = \delta' + \frac{\lambda}{2}.$$

Les définitions de ces couleurs sont assez arbitraires et il est préférable de les représenter au moyen des données colorimétriques. Soit un rayonnement simple de longueur d'onde λ transportant un flux d'énergie Φ_0. Ce flux Φ_0 est divisé par l'appareil interférentiel en deux flux que l'on supposera égaux. Après une différence de marche δ, ils interfèrent et le flux résultant Φ_1 est donné par

$$\Phi_1 = \Phi_0\left(1 - \cos\frac{2\pi\delta'}{\lambda}\right) \quad \text{(échelle 1)}$$

dans le cas des lames minces observées par réflexion ou dans le cas des franges d'Young par exemple

$$\Phi_2 = \Phi_0\left(1 + \cos\frac{2\pi\delta'}{\lambda}\right) \quad \text{(échelle 2)}.$$

Tableau 1.

δ'		$\delta' + \dfrac{\lambda}{2}$	δ'
μ		(1) Noir	(2) Blanc
Premier ordre	0,040	Gris de fer	Blanc
	0,097	Gris lavande	Blanc jaunâtre
	0,158	Bleu gris	Blanc brunâtre
	0,218	Gris plus clair	Brun jaune
	0,234	Blanc verdâtre	Brun
	0,259	Blanc	Rouge clair
	0,267	Blanc jaunâtre	Rouge carmin
	0,275	Jaune paille pâle	Brun rouge sombre
	0,281	Jaune paille	Violet sombre
	0,306	Jaune clair	Indigo
	0,332	Jaune vif	Bleu
	0,430	Jaune brun	Bleu gris
	0,505	Orangé rougeâtre	Vert bleuâtre
	0,536	Rouge chaud	Vert pâle
	0,551	Rouge plus foncé	Vert jaunâtre
Deuxième ordre	0,565	Pourpre	Vert plus clair
	0,575	Violet	Jaune verdâtre
	0,589	Indigo	Jaune d'or
	0,664	Bleu de ciel	Orangé
	0,728	Bleu verdâtre	Orangé brunâtre
	0,747	Vert	Rouge carmin clair
	0,826	Vert plus clair	Pourpe
	0,843	Vert jaunâtre	Pourpre violacé
	0,866	Jaune verdâtre	Violet
	0,910	Jaune pur	Indigo
	0,948	Orangé	Bleu sombre
	0,998	Orangé rougeâtre vif	Bleu verdâtre
	1,101	Rouge violacé foncé	Vert

Si on considère une petite bande spectrale $d\lambda$, il lui correspondra un flux élémentaire

$$d\Phi_0 = E(\lambda)\, d\lambda,$$

$E(\lambda)$ étant la répartition énergétique du spectre de la source lumineuse. Après interférences, on aura

$$d\Phi_1 = E(\lambda)\left(1 - \cos\frac{2\pi\delta'}{\lambda}\right) d\lambda,$$

$$d\Phi_2 = E(\lambda)\left(1 + \cos\frac{2\pi\delta'}{\lambda}\right) d\lambda.$$

La superposition de ces deux types d'interférences redonne la lumière blanche incidente car

$$d\Phi_1 + d\Phi_2 = 2E(\lambda)\, d\lambda = 2\, d\Phi_0.$$

On peut calculer les composantes trichromatiques XYZ par les relations:

$$X_1 = \int \bar{x}\, d\Phi_1, \qquad Y_1 = \int \bar{y}\, d\Phi_1, \qquad Z = \int \bar{z}\, d\Phi_1$$

pour l'échelle (1),

$$X_2 = \int \bar{x}\, d\Phi_2, \qquad Y_2 = \int \bar{y}\, d\Phi_2,$$

$$Z = \int \bar{z}\, d\Phi_2$$

pour l'échelle (2),
$\bar{x}\,\bar{y}\,\bar{z}$ étant des fonctions de la longueur d'onde données dans les tables. Les calculs ont été faits par Y. Le Grand en prenant comme source de lumière blanche l'étalon A de température de couleur 2848° K. Ayant calculé $X_1 Y_1 Z_1$, on peut en déduire aussitôt les coefficients $X_2 Y_2 Z_2$. Les sommes telles que $X_1 + X_2$ sont égales au double des valeurs correspondant à la lumière blanche. Si on a pris par convention le flux lumineux incident émis par A égal à l'unité, on aura

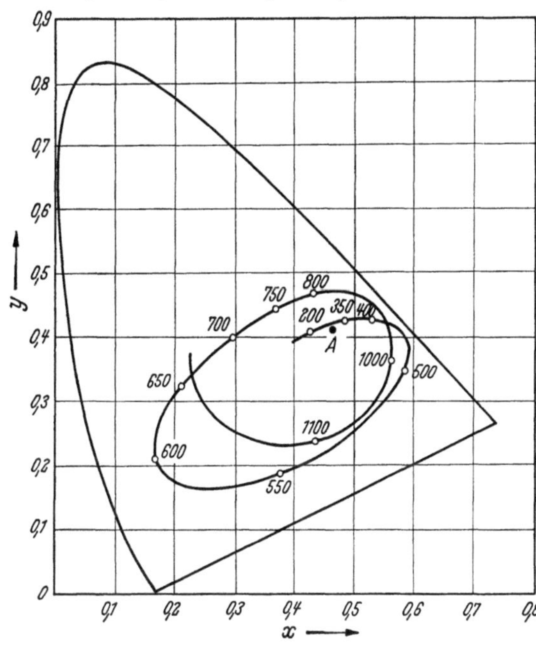

Fig. 57.

$$X_2 = 2,1968 - X_1;$$
$$Y_2 = 2 - Y_1;$$
$$Z_2 = 0,711 - Z_1$$

les coefficients trichromatiques de la source étalon A étant 1,0984, 1 et 0,3555.

On utilise généralement en colorimétrie les coefficients trichromatiques définis par

$$x = \frac{X}{X + Y + Z};$$

$$y = \frac{Y}{X + Y + Z};$$

$$z = \frac{Z}{X + Y + Z}.$$

Les résultats sont donnés par les Fig. 57 et 58 sur lesquelles les courbes représentent le déplacement du point symbolisant la couleur d'interférence lorsque δ' varie. Le triangle des couleurs est représenté sur ces deux figures par un triangle rectangle isocèle et en ne figurant que les coordonnées rectangulaires x et y. On

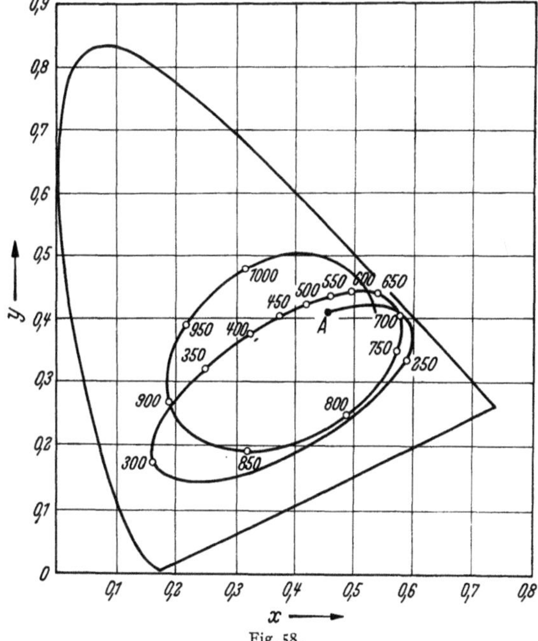

Fig. 58.

remarquera le déplacement rapide du point figuratif dans la région des pourpres. Pour se rendre compte de la précision, on peut représenter les couleurs d'interférences sur un diagramme à chromaticité uniforme tel que celui de Judd. On passe des coordonnées x et y aux coordonnées u et v de ce nouveau système par

$$u = \frac{0,4661\,x + 0,1593\,y}{y - 0,15735\,x + 0,2424}, \qquad v = \frac{0,6581\,y}{y - 0,15735\,x + 0,2424}.$$

On obtient la Fig. 59.

Pour une même variation $\varDelta\,\delta'$ de la différence de marche, la couleur variera d'autant plus rapidement que la longueur correspondant à $\varDelta\,\delta'$ sur la courbe sera plus grande. Ainsi, le premier pourpre teinte sensible P_2 correspondant aux couleurs de l'échelle (2) (couleurs commençant par un blanc) est pratiquement deux fois plus sensible que le premier pourpre teinte sensible P_1 de l'échelle (1) (couleurs commençant par un noir).

Une variation $\varDelta\,\delta' = \pm 10$ Å autour de P_2 correspond à une longueur deux fois plus grande que la même variation autour de P_1.

H. Kubota a étudié l'influence de la température de couleur de la source

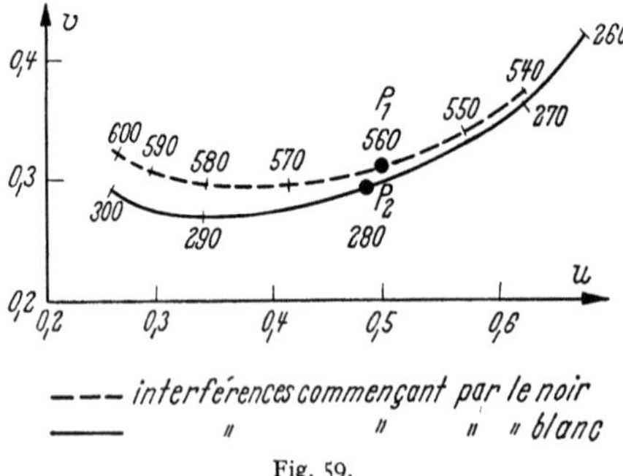

--- *interférences commençant par le noir*
———   "   "   "   *blanc*

Fig. 59.

sur la teinte sensible. Il a montré que la différence de marche δ' donnant la teinte la plus sensible est une fonction linéaire de la température de couleur T. Si la température de couleur augmente, δ' diminue. Ainsi $\delta' = 0,575\ \mu$ donne la teinte la plus sensible pour $T = 2000°$ K et si on utilise la même lame avec le soleil, la sensibilité de l'oeil aux variations de δ' est diminuée 2 fois environ.

31. Spectres cannelés des lames minces. Les calculs de la Sect. 17 sont applicables. La différence de marche δ est produite maintenant par la lame. Pour une lame d'indice n dans l'air $\delta = 2ne + \dfrac{\lambda}{2}$ en observant par réflexion. Pour opérer sous l'incidence normale, on utilise le montage de la Fig. 60. On projette l'image d'une petite région M de la lame e sur la fente d'un spectroscope. La lame mince est éclairée par une source S

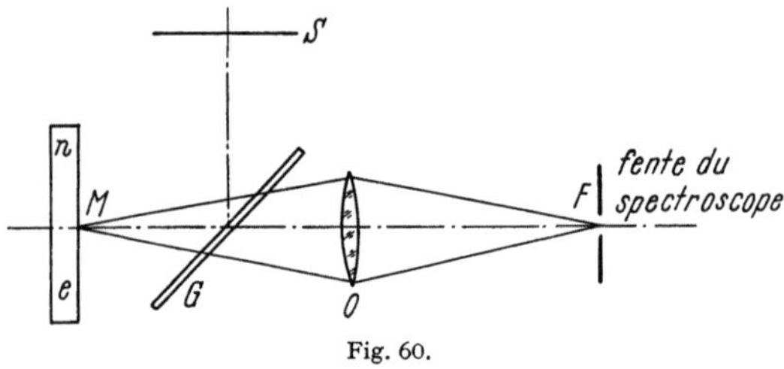

Fig. 60.

après réflexion sur la lame semi-réfléchissante G inclinée à 45°. On peut ainsi étudier la composition spectrale de la lumière réfléchie dans la région M. Si l'épaisseur de la lame est de quelques microns, la lumière réfléchie n'est plus colorée, on a du blanc d'ordre supérieur.

L'examen au spectroscope montre le spectre sillonné de cannelures. Les cannelures noires correspondent aux longueurs d'onde pour lesquelles $p = \delta/\lambda = K + \frac{1}{2}$ et les cannelures brillantes à $p = \delta/\lambda = K$.

Si on connaît δ pour calculer le nombre de cannelures comprises entre λ_1 et λ_2, on a vu [17] qu'il suffit de former les rapports $p_1 = \delta/\lambda_1$ et $p_2 = \delta/\lambda_2$. Il y a autant de cannelure brillantes entre λ_1 et λ_2 que de nombres entiers compris entre p_1 et p_2. L'exemple de la Sect. 17 montre qu'une lame introduisant une différence de marche égale à 1 mm., donnera environ 1250 cannelures dans le spectre ($e = 0,33$ mm. pour $n = 1,5$). Inversement, si on détermine le nombre des cannelures, on pourra mesurer le retard introduit par la lame. Si on a K cannelures brillantes comprises entre les longueurs d'onde λ_1 et λ_2, on aura p étant inconnu

$$p = \frac{\delta}{\lambda_1}, \qquad p + K = \frac{\delta}{\lambda_2}, \qquad \delta = \frac{K \lambda_1 \lambda_2}{\lambda_1 - \lambda_2}. \tag{31.1}$$

Si l'épaisseur est telle que l'on puisse compter une centaine de cannelures, les franges sont assez fines pour avoir une précision de l'ordre de $\dfrac{1}{1000}$.

g) Ondes stationnaires.

32. Incidence normale. $\alpha)$ *Expérience de* WIENER. Envoyons sous l'incidence normale un faisceau de rayons parallèles sur un miroir plan. Les rayons réfléchis

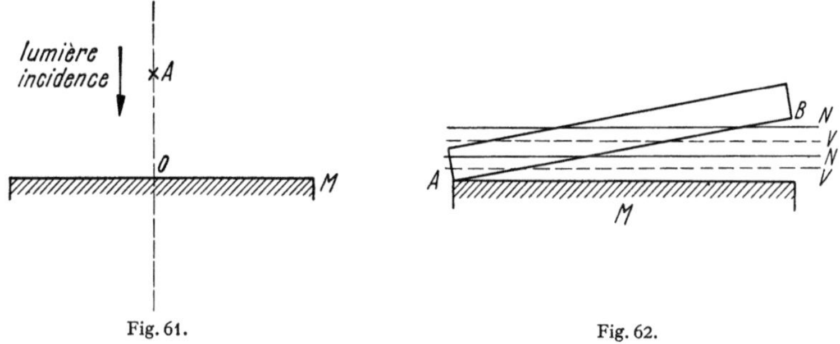

Fig. 61.   Fig. 62.

peuvent interférer avec les rayons incidents, il y a production d'ondes stationnaires.

Soit M le miroir (Fig. 61); en prenant l'origine en O sur le miroir, la vibration incidente et la vibration réfléchie en O peuvent être représentées respectivement par les expressions $x_1 = a \cos \omega t$ et $x_2 = b \cos(\omega t - \varphi)$. En A, ces deux vibrations peuvent s'écrire au même instant si $OA = x$

$$x_1 = a \cos\left(\omega t + \frac{2\pi x}{\lambda}\right), \qquad x_2 = b \cos\left(\omega t - \frac{2\pi x}{\lambda} - \varphi\right).$$

Le carré de l'amplitude résultante sera:

$$A^2 = a^2 + b^2 + 2ab \cos\left(\frac{4\pi x}{\lambda} + \varphi\right),$$

A^2 sera maximum pour

$$\frac{4\pi x}{\lambda} + \varphi = 2K\pi, \qquad x = \left(K - \frac{\varphi}{2\pi}\right)\frac{\lambda}{2}. \tag{32.1}$$

Les plans situés aux distances x du miroir satisfaisant à (32.1) sont les plans ventraux. A^2 sera minimum pour

$$\frac{4\pi x}{\lambda} + \varphi = (2K + 1)\pi, \qquad x = \left(K + \frac{1}{2} - \frac{\varphi}{2\pi}\right)\frac{\lambda}{2}. \tag{32.2}$$

Les plans définis par (32.2) sont les plans nodaux. Tous ces plans sont équidistants et la distance d'un plan nodal à un plan ventral est égale à $\lambda/4$. Il y a maximum de lumière sur les plans ventraux et minimum de lumière sur les plans nodaux. Malgré la très faible distance de l'ordre de $0,1\,\mu$ séparant un plan ventral d'un plan nodal, WIENER a réussi à les mettre en évidence en photographiant le phénomène sur une pellicule de collodion au bromure d'argent ne dépassant pas une épaisseur d'environ $\lambda/30$. La pellicule est déposée sur la face plane AB d'une lame de verre G faisant un très petit angle avec le miroir M (Fig. 62).

La pellicule coupe les plans ventraux et nodaux suivant un système de droites parallèles à l'arête A du coin. On observe effectivement après développement une série de franges parallèles et équidistantes correspondant aux intersections avec les plans ventraux V.

Pour savoir si le plan du miroir correspond à un plan nodal ou à un plan ventral, c'est-à-dire si la réflexion se fait avec ou sans changement de signe, on applique la pellicule sensible, non sur une surface plane mais sur une surface sphérique afin d'obtenir le contact optique. On applique AB sur une lentille en verre O (Fig. 63). La face inférieure de O est noircie pour éviter toute réflexion parasite. La photographie donne un système d'anneaux analogues aux anneaux de Newton et la partie centrale M correspond à un minimum de lumière: il y a donc un noeud au contact de O en M.

Cette même expérience a été répétée par Drude et Nernst en remplaçant la couche photographique AB par une pellicule fluorescente également très mince: les plans nodaux et les plans ventraux sont toujours les mêmes.

Ives et Fry ont mesuré la distribution des éclairements sur AB par un procédé photoélectrique en remplaçant la couche photographique AB par une couche très mince de césium. On mesure les courants photoélectriques obtenus en éclairant suc-

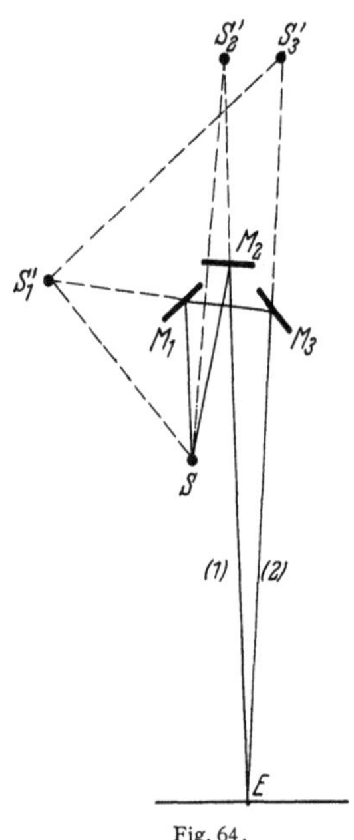

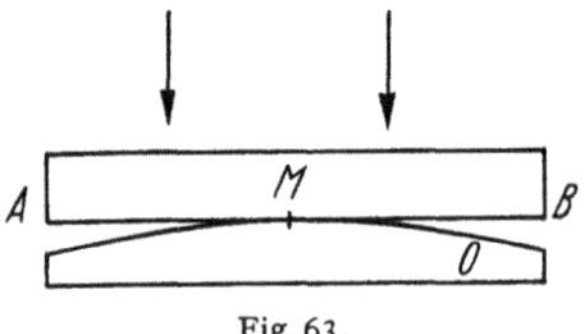

Fig. 63.

Fig. 64.

cessivement les différentes régions de la lame: on trouve que les maxima de courant correspondent aux maxima de la photographie et de la lumière fluorescente.

La réflexion air-verre introduit donc un changement de phase π. Le fait a été également constaté par Fresnel au moyen de l'expérience des trois miroirs (Fig. 64). Une source S de lumière blanche envoie un faisceau (1) de rayons réfléchis par le miroir M_2. La même source envoie un faisceau (2) de rayons réfléchis, d'abord par M_1, puis par M_3. Ces deux faisceaux interfèrent en E où l'on constate que la frange centrale est noire. Tout se passe comme si les deux sources S_2' et S_3' produisant les interférences étaient en opposition, il faut donc admettre que les trois réflexions produisent une différence de phase π.

Que l'on caractérise l'amplitude de la vibration lumineuse par ses effets physiologiques sur l'oeil comme dans l'expérience de Fresnel, par ses effets chimiques comme dans l'expérience de Wiener, par ses effets fluorescents comme dans l'expérience de Drude et Nernst ou par ses effets photoélectriques comme dans l'expérience de Ives et Fry, la réflexion air-verre, ou d'un milieu moins réfringent sur un milieu plus réfringent, introduit toujours un retard d'une demi-longueur d'onde. C'est la comparaison avec la réflexion sur un corps absorbant qui a permis de montrer qu'on a ici un retard et non une avance de $\lambda/2$.

Sur la Fig. 30, le rayon Ia qui se réfléchit de l'air sur le verre, est donc en avance de $\lambda/2$ et le retard $2ne$ du rayon Jb sur le rayon Ia doit être augmenté de $\lambda/2$.

β) *Photographie des couleurs.* Considérons un miroir M recouvert d'une couche sensible de chlorure d'argent assez épaisse mais à grains extrêmement fins (Fig. 65). Eclairons normalement par un faisceau de lumière monochromatique λ sur tous les plans ventraux l'argent est réduit et on obtient une série de lamelles réfléchissantes d'argent réduit, équidistantes de $\lambda/2$. Eclairons en lumière blanche: pour la radiation λ, la lumière réfléchie par tous ces plans est en concordance de phase puisque par réflexion, la différene de marche entre deux plans est deux fois $\lambda/2$. Pour une autre radiation même très voisine de λ, il n'en est plus ainsi s'il y a un grand nombre de lamelles (voir théorie des réseaux), l'amplitude réfléchie est pratiquement

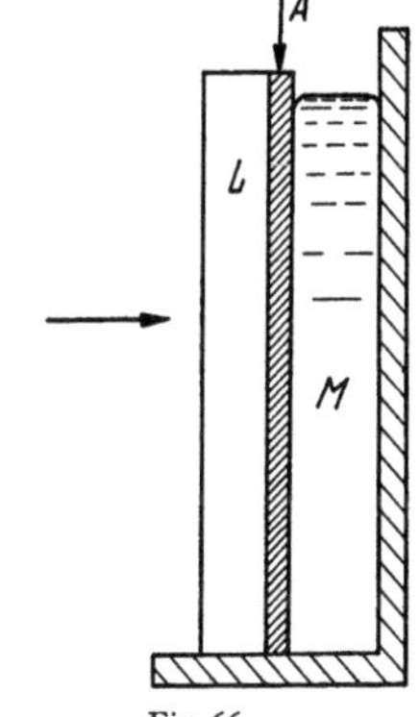

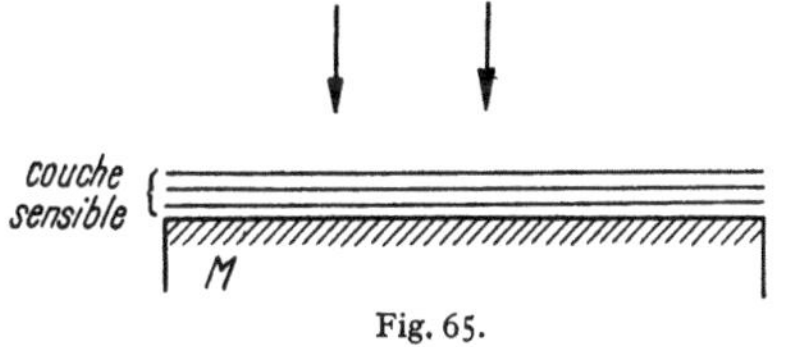

Fig. 65.

Fig. 66.

nulle pour toutes les radiations autres que λ. Donc la lame éclairée en lumière blanche réfléchit de la lumière monochromatique de longueur d'onde λ avec laquelle elle a été primitivement impressionnée. L'application de ce procédé à la photographie des couleurs a été réalisée par LIPPMAN. Si on éclaire en lumière blanche une couche sensible à grains extrêmement fins adossée à un miroir réfléchissant, en chaque point de la lame on obtient tout un système de lamelles réfléchissantes caractéristiques des radiations reçues en ce point. En regardant la plaque par réflexion sous incidence normale en lumière blanche, la lumière réfléchie par chaque point de la plaque reproduit la couleur qui l'a impressionnée.

Les plaques ordinaires ont des grains beaucoup trop gros pour pouvoir être utilisées: il faut employer des plaques spéciales dont le développement et le fixage ont lieu comme pour des plaques ordinaires. La couche sensible A est maintenue entre une lame de verre L et un miroir M constitué par un bain de mercure (Fig. 66).

33. Phénomènes en incidence oblique.

α) *Incidence presque rasante.* Envoyons un faisceau de rayons parallèles sur un miroir M sous une incidence presque ra-

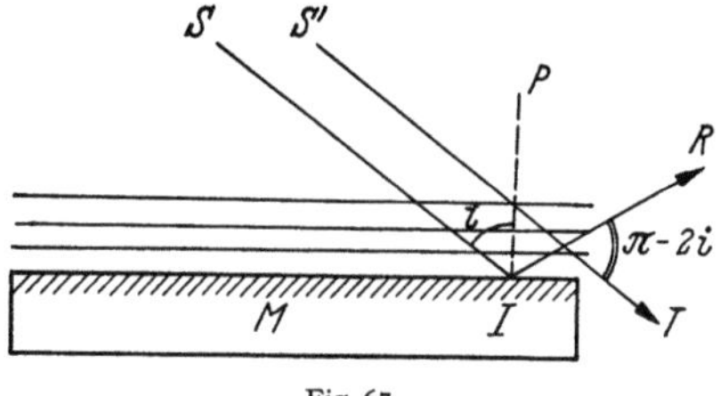

Fig. 67.

sante: un rayon tel que SI donne un réfléchi IR qui peut interférer avec un rayon incident tel que $S'T$ (Fig. 67). On réalise ainsi des ondes stationnaires planes et parallèles au miroir M. Soit i l'angle d'incidence, l'équidistance des plans devient ici $\lambda/2 \cos i$.

Si l'incidence est presque rasante, la distance de plans est très grande et on peut voir dans un plan P normal à M des franges sombres qui sont les intersections des plans nodaux par le plan d'observation P. On constate qu'il y a une frange sombre dans le plan du miroir, ce qui montre encore une fois le retard égal à $\lambda/2$ introduit dans la réflexion air-verre.

β) *Incidence oblique non rasante.* Supposons le miroir M éclairé par un faisceau de rayons parallèles sous l'incidence de 45°: les rayons interférents ne sont plus parallèles. Or, pour pouvoir ramener l'addition vectorielle des vibrations lumineuses à une addition algébrique des élongations, nous nous sommes limité jusqu'ici au cas de vibrations dont les directions de propagation sont sensiblement parallèles. Dans ces conditions, il n'était pas utile de préciser l'orientation des vibrations dans l'espace et le principe des interférences s'appliquait, aussi bien en lumière naturelle qu'en lumière polarisée. Il n'en est plus de même maintenant.

On verra plus tard comment on peut polariser la lumière, c'est-à-dire fixer la direction de la vibration lumineuse dans un seul azimut. Supposons les vibrations incidentes situées dans le plan d'incidence, les vibrations réfléchies deviennent

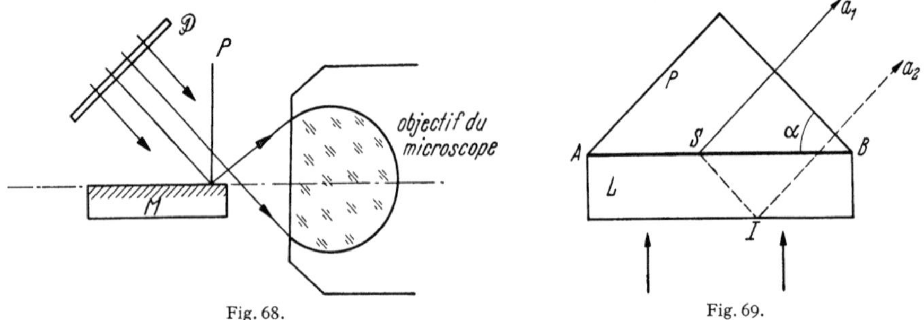

Fig. 68.   Fig. 69.

perpendiculaires aux vibrations incidentes et il n'y a plus d'interférences. Si les vibrations incidentes sont situées dans le plan perpendiculaire au plan d'incidence, leur orientation ne change pas après la réflexion et elles peuvent interférer avec les vibrations incidentes.

On peut observer ces franges de la façon suivante (Fig. 68): le faisceau de rayon est polarisé par un polariseur P de manière que la vibration soit dirigée perpendiculairement au plan d'incidence et on observe le plan P sur lequel on voit les franges avec un microscope dont l'objectif a une ouverture numérique suffisante pour admettre à la fois le faisceau incident et le faisceau réfléchi.

γ) *Remarque.* Nous avons dit que pour pouvoir ramener l'addition vectorielle des vibrations lumineuses à une addition algébrique des élongations, il fallait considérer des rayons interférents peu inclinés l'un sur l'autre au point où ils interfèrent. En fait, cette condition n'est pas suffisante et il faut que les rayons qui interfèrent aient quitté la source sous un petit angle également.

Considérons le dispositif de Selenyi (Fig. 69) dans lequel les rayons qui interfèrent sont partis très inclinés l'un sur l'autre d'un point de la source. Une plaque fluorescente très mince AB est placée entre une lame à faces parallèles mince L et un prisme P en verre d'angle $\alpha = 45°$. Un rayonnement ultraviolet excite la fluorescence visible de la couche AB. Chaque point S de la couche émet des vibrations dans toutes les directions. Les rayons Sa_1 et SIa_2 partent à angle droit du point S qui joue le rôle de source. Les deux rayons Sa_1 et SIa_2 sont parallèles et interfèrent à l'infini. On peut effectivement voir des franges à l'infini sous forme d'anneaux, mais les minima ne sont pas nuls. Plaçons un polariseur entre la lame L et la lunette d'observation: si le polariseur est orienté de façon que la vibration soit perpendiculaire au plan de la Fig. 69, le contraste des franges devient maximum. Les deux vibrations qui interfèrent proviennent d'une même direction de vibration en S sur la source. Si le polariseur est parallèle au plan de figure, les franges disparaissent car les deux vibrations qui interfèrent proviennent de deux vibrations de la sources S perpendiculaires aux rayons Sa_1 et SI. Ces deux vibrations orthogonales ne sont pas cohérentes.

Dans l'expérience du paragraphe précédent, les rayons sont partis d'un même point de la source et très peu inclinés l'un sur l'autre.

Dans l'expérience de Selenyi, les rayons partent très inclinés l'un sur l'autre d'un point de la source et sont sensiblement parallèles au point d'interférence. Les deux expériences sont réciproques.

h) Interferomètres à deux ondes.

Dans certains cas, il est utile d'avoir un appareil interférentiel où les deux faisceaux, avant de se réunir, sont complètement séparés. On peut agir séparément sur chaque faisceau, ce qui rend les mesures plus faciles. Nous allons décrire quelques uns de ces dispositifs.

34. Miroirs de Jamin. L'appareil de Jamin se compose de deux lames de verre épaisses L_1 et L_2 (Fig. 70) à faces parallèles. Leurs faces postérieures $A_1 B_1$ et $A_2 B_2$ sont argentées à fond. Plaçons les parallèlement l'une à l'autre et considérons le rayon SI qui tombe sur la première lame sous une incidence i.

Une partie du rayonnement incident est réfléchie suivant II', la plus grande partie est réfractée et se réfléchit en J, puis sort de la lame en K. Les rayons

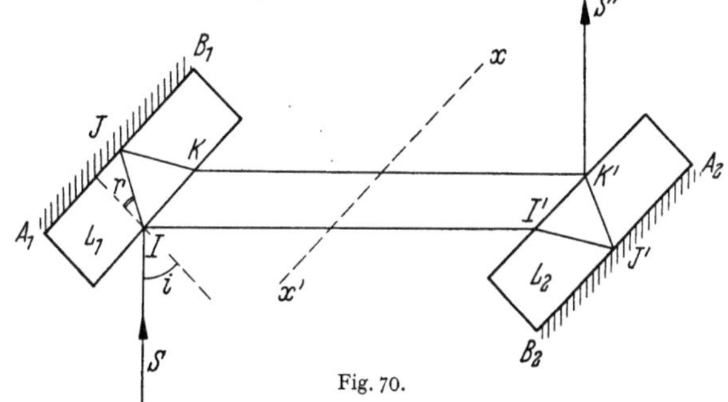

Fig. 70.

II' et KK' sont parallèles d'intensités très différentes et présentent une grande différence de marche. En arrivant sur la lame L_2 ils vont donner divers rayons réfléchis ou réfractés. Nous considérerons seulement le rayon réfléchi $K'S'$ que donne KK' et le rayon $I'J'K'S'$ que donne II'. Ce sont donc bien des interférences à deux ondes que nous observons. Il y a d'autres réflexions possibles mais nous les laisserons de côté.

Les deux rayons $SII'J'K'S'$ et $SIJKK'S'$ qui se superposent ont suivi un chemin symétrique, ont subi les mêmes accidents et possèdent la même amplitude. Si les lames L_1 et L_2 sont rigoureusement parallèles, la différence de marche est nulle et en lumière blanche, on observe une teinte plate. On fait apparaître les franges en faisant faire un petit angle aux deux miroirs. La différence de marche ne dépend évidemment que de la direction des rayons qui interfèrent: on obtient des franges à l'infini. La source peut avoir une position quelconque et l'observation des franges se fera au moyen d'une lunette visant à l'infini. Supposons que la lumière tombe sous un angle i voisin de 45° et que l'arête du toit formé par les deux lames se projette suivant xx'. Les franges à l'infini sont des franges parallèles à xx' dont l'équidistance α est donnée par

$$\alpha = \frac{\lambda}{4\,e\,\vartheta}\,\sqrt{2n^2 - 1}$$

où e et n sont l'épaisseur et l'indice des lames, 2ϑ l'angle de leurs normales.

Pour régler l'appareil, on opère en lumière monochromatique: les franges apparaissent même si le parallélisme des miroirs n'est que grossièrement approché. En agissant sur les vis de réglage des supports des lames, on élargit les franges de façon à avoir presque la teinte plate. On remplace alors la lumière monochromatique par de la lumière blanche. Un très petit mouvement des vis suffit à faire apparaître les franges: on observe une frange centrale blanche entourée de franges colorées. Les miroirs de Jamin ont permis en particulier d'étudier l'indice des gaz

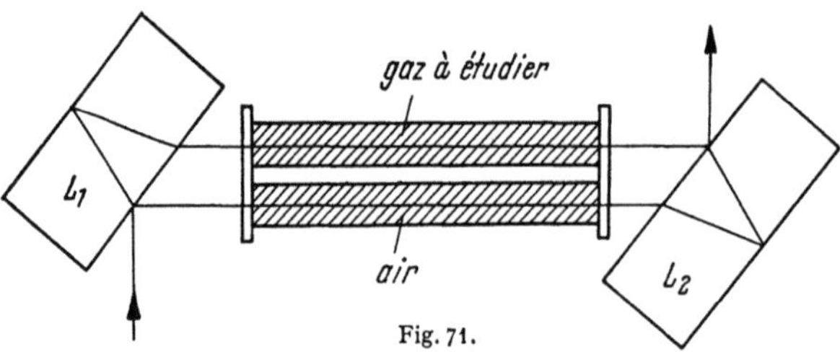

Fig. 71.

et de vérifier la loi $\dfrac{n-1}{d}$ = constante, d étant la masse spécifique du gaz (Fig. 71).

35. Les interféromètres de Michelson et de Mach-Zehnder. α) *Interféromètre de* Michelson. L'appareil imaginé par Michelson, dont il a fait de nombreuses applications, est basé sur le principe suivant: les rayons partis d'une source lumineuse S tombent sur une lame de verre G_1 incliné à 45° (Fig. 72) appelée séparatrice dont la première face A rencontrée par les rayons est semi-réfléchissante.

Le faisceau se sépare en deux parties en A: l'une suit le trajet AB_1, se réfléchit en B_1 sur un miroir plan M_1 et revient sur elle-même, traverse la lame G_1 et pénètre dans un objectif O_2. L'autre partie du faisceau traverse la lame G_1, se réfléchit en B_2 sur le miroir plan M_2, revient en arrière, se réfléchit sur la face A de G_1, traverse à nouveau G_1 en se superposant au premier faisceau. Le premier

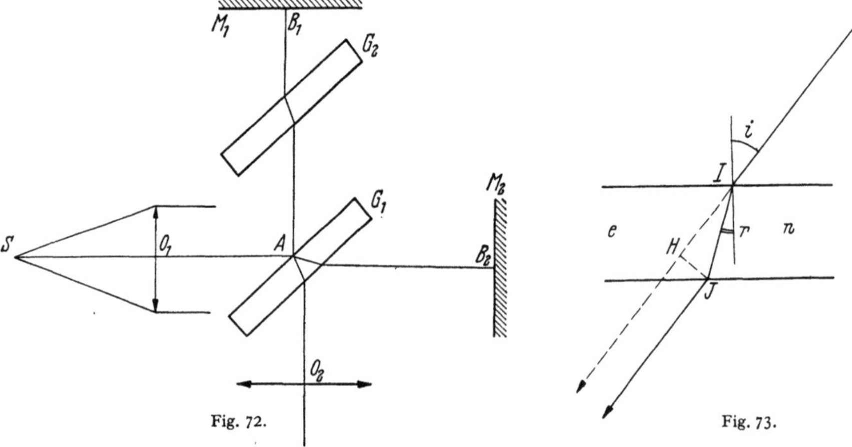

Fig. 72.   Fig. 73.

faisceau ne traverse qu'une fois la séparatrice G_1 alors que le deuxième faisceau la traverse trois fois. Pour rendre les trajets parcourus aussi identiques que possible, on interpose une lame G_2 dite «compensatrice» sur le premier faisceau. Cette lame G_2 doit avoir même épaisseur et être faite du même verre que la lame G_1.

Comme une égalité rigoureuse des épaisseurs est pratiquement impossible à réaliser dans le cas de grandes lames, on peut compenser la très faible différence qui peut exister en inclinant légèrement la compensatrice G_2. Si i est l'angle d'incidence des rayons sur la lame, r l'angle de réfraction, e l'épaisseur de la compensatrice et n son indice, calculons la différence de marche δ entre un rayon qui passe comme s'il n'y avait pas de lame et un rayon qui traverse la lame. On a (Fig. 73)

$$\delta = n\,IJ - IH = \frac{ne}{\cos r} - \frac{e}{\cos r}\cos(i-r),$$

$$\delta = e\,(n\cos r - \cos i).$$

Si on incline la compensatrice d'un petit angle di à partir de cette position. la différence de marche va varier de $d\delta$ et on aura

$$d\delta = e(-n \sin r \, dr + \sin i \, di) = \frac{e \sin(i-r)}{\cos r}\, di.$$

On pourra donc compenser une faible inégalité des épaisseurs de G_1 et G_2 par une petite variation d'inclinaison de la compensatrice.

(1) *Observation des franges d'égale épaisseur.* Par réflexion sur la face semi-réfléchissante A de la séparatrice G_1 (Fig. 74), le miroir M_2 a une image en M_2'; tout se passe comme s'il y avait interférences produites par une lame d'air d'épaisseur e égale à la distance du miroir M_1 à l'image M_2' du miroir M_2. Si l'image du miroir M_2 n'est pas parallèle à M_1, on a alors un coin d'air et on observera les franges localisées sur la lame. Si les différentes pièces optiques constituant l'appareil sont parfaites (lames G_1 et G_2 miroirs M_1 et M_2), les franges sont rectilignes parallèles et équidistantes. Pour les observer, on éclaire l'interféromètre par un collimateur SO_1 et on place la pupille de l'oeil au foyer F de l'objectif O_2. La focale de l'objectif O_2 doit être choisie convenablement pour observer la région de localisation des franges voisine généralement de $M_1 M_2$. Le diamètre du diaphragme placé en S et qui joue le rôle de source lumineuse devra être

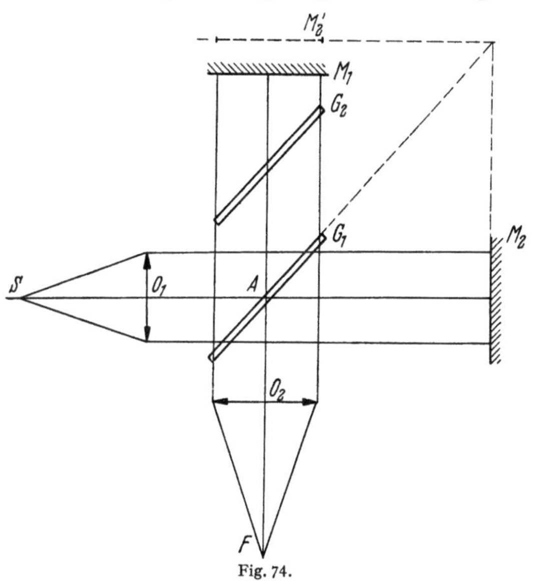

Fig. 74.

d'autant plus petit que la lame $M_1 M_2$ sera plus épaisse d'après ce qui a été vu précédemment. Si la lame d'air comprise entre M_1 et M_2' a une épaisseur très faible, on pourra observer les franges en lumière blanche à condition que les chemins optiques suivis par les deux faisceaux soient rigoureusement identiques. En effet, si les lames G_1 et G_2 n'ont pas le même indice, il pourra y avoir différence de marche nulle pour une radiation, mais pas pour les autres, la différence de marche variera avec la longueur d'onde et on ne pourra observer les franges en lumière blanche.

(2) *Observation des franges d'égale inclinaison.* Si l'image M_2 est rigoureusement parallèle à M_1, on peut observer les anneaux à l'infini de la lame d'air ainsi formée au foyer F de l'objectif O_2. On pourra alors examiner le plan focal F au moyen d'une loupe. L'ensemble de l'objectif O_2 et de la loupe constituera une lunette astronomique dont le grossissement sera choisi en fonction du diamètre angulaire des anneaux observés.

On utilisera en S, non plus un petit trou, mais une surface large, on peut même supprimer le collimateur et le remplacer par une grande surface diffusante éclairée par une source lumineuse monochromatique.

(3) *Réglage de l'interféromètre.* On réalise les opérations suivantes:

1. amener en coïncidence les deux faisceaux à interférer;

2. orienter convenablement les miroirs M_1 et M_2 de façon à travailler en incidence normale.

Ces deux opérations peuvent être conduites simultanément en observant les images de S par autocollimation sur les miroirs M_1 et M_2. En plaçant un écran dans le plan du trou S du collimateur, on voit les deux images de retour correspondant aux deux trajets. On oriente les deux miroirs pour amener ces deux images en coïncidence sur le trou S. Pour parfaire le réglage, on observera les deux images de S dans le plan focal de l'objectif O_2 au moyen d'une loupe. La coïncidence étant bien réalisée, on place l'oeil en F et on voit les franges d'égale épaisseur. On règle ensuite l'orientation d'un des deux miroirs M_1 ou M_2 de façon à élargir au maximum les franges d'égale épaisseur, c'est-à-dire à obtenir la teinte plate.

On réalise ainsi le parallélisme rigoureux de M_2' et M_1 et il est alors possible d'observer les franges d'égale inclinaison;

3. régler la largeur de la source S.

Pour revenir aux franges d'égale épaisseur et les observer avec le meilleur contraste, il faut s'arranger pour que la source S bien soit centrée sur les anneaux à l'infini et qu'elle ne laisse passer qu'une portion de l'anneau central (Sect. 26). Si les anneaux sont étroits, c'est à dire si la lame $M_2' M_1$ est épaisse, il faut donc employer une source S de très petit diamètre. Si l'on veut opérer avec une source large, il faut réduire la distance $M_2' M_1$. Pour cela, on appuie légèrement sur M_2 par exemple en observant les anneaux à l'infini. Si on constate que les anneaux naissent au centre, c'est que l'épaisseur $M_2' M_1$ augmente. On fait alors l'opération inverse, on rapproche M_2 de G_1 et on voit les anneaux disparaître au centre et s'élargir de plus en plus. On continue jusqu'au moment où les anneaux s'élargissent tellement qu'on finit par avoir la teinte plate. L'interféromètre est alors réglé à la différence de marche nulle. On peut supprimer le collimateur SO_1 et observer avec un excellent contraste les franges d'égale épaisseur on dérègle très légèrement l'orientation de M_1 et M_2 en utilisant une source très large et même en lumière blanche.

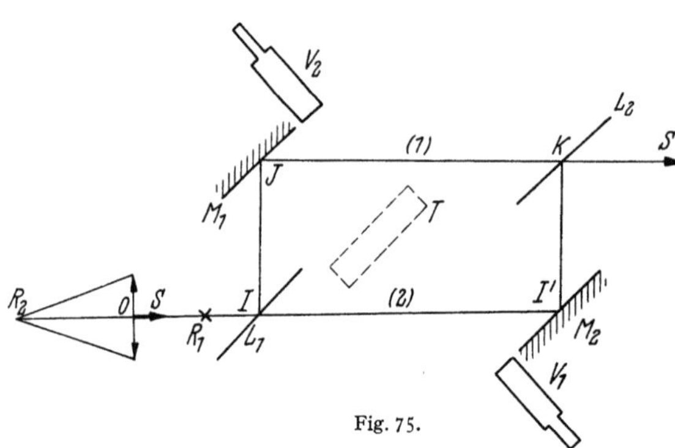

Fig. 75.

β) *Interféromètre de* Mach-Zehnder. Le principe de l'interféromètre de Mach-Zehnder est indiqué par la Fig. 75. Un rayon incident SI tombe sur une lame L_1 semi-réfléchissante et se divise en deux parties. Le rayon réfléchi J tombe sur un miroir M_1 et traverse en K une deuxième lame réfléchissante L_2. Le rayon transmis en I est réfléchi en I' sur le miroir M_2 et se réfléchit en K sur la lame semi-réfléchissante L_2 pour sortir confondu avec le premier rayon.

Pour régler l'interféromètre, il faut rendre $L_1 L_2 M_1 M_2$ exactement parallèles et annuler la différence des deux chemins optiques (1) et (2). Au moyen d'une lunette autocollimatrice V_1 on commence par régler le parallélisme des miroirs M_1 et M_2. Le réglage se fait en observant par autocollimation l'image du repère situé dans le plan focal de la lunette. On oriente le miroir M_1 sur lequel on fait le réglage de façon à faire coïncider l'image du repère avec le repère lui-même. Le miroir M_1 est alors normal à l'axe optique de la lunette. On fait l'autocollimation en même temps sur une lame transparente T placée entre M_1 et M_2. Les miroirs M_1 et T sont parallèles. On transporte la lunette en V_2 et on la règle sur T: l'axe optique de V_2 est parallèle à la normale à M_1 ou T. On fait l'autocollimation sur M_2 en orientant ce miroir. Les deux miroirs M_1 et M_2 sont maintenant parfaitement parallèles.

Le réglage des lames semi-transparentes L_1 et L_2 se fait de la façon suivante: on utilise deux repères R_1 et R_2, l'un situé à distance finie, l'autre au foyer d'un collimateur O, c'est à dire à l'infini. On fait coïncider les deux images du repère R_1 en observant les images au moyen d'un viseur placé en S'. Le réglage s'obtient par de petites rotations de la lame L_2. Le viseur en S' est remplacé par une lunette astronomique et on fait coïncider les deux images du repère R_2 par de petites rotations de la lame L_1. Lorsque les images du repère R_1 à distance finie coïncident ainsi que celles du repère R_2 à distance infinie, les quatre éléments $L_1 L_2 M_1 M_2$ sont parallèles et la différence des chemins optiques voisine de zéro. En lumière monochromatique, on observe les anneaux à l'infini. On déplace alors l'un des éléments parallèlement à lui-même de façon à faire disparaître les anneaux au centre. Il disparaît un anneau chaque fois qu'un des miroirs se déplace d'une longueur égale à $(\lambda/2)\sqrt{2}$. On arrive finalement à obtenir la teinte plate. Si on incline très peu la lame L_1 par exemple, on obtient immédiatement des franges rectilignes localisées au voisinage de l'élément incliné. On peut modifier la localisation en inclinant un deuxième élément, L_2 par exemple.

Pour observer les franges en lumière blanche, il faut se placer au voisinage de la teinte plate, c'est-à-dire de la différence de marche nulle. On éclaire en lumière blanche et en observant la lumière qui a traversé l'interféromètre au moyen d'un spectroscope, on voit un spectre cannelé. Il suffit d'élargir les cannelures pour pouvoir observer les franges en lumière blanche. Naturellement les deux lames L_1 et L_2 doivent être identiques.

III. Interférences à ondes multiples

Dans les divers cas que nous venons d'étudier, nous n'avons considéré que les phénomènes d'interférence produits par deux ondes parties d'un même point de la source et présentant une certaine différence de marche. Nous n'avons pas tenu compte des ondes multiples qui se produisent car, en général la lame a un faible pouvoir réflecteur, et elles étaient négligeables.

Si on donne à la lame un pouvoir réflecteur élevé, on ne peut plus négliger ces ondes multiples et nous allons étudier maintenant les phénomènes qu'elles produisent.

Indiquons tout de suite que les appareils décrits aux Sect. 28 et 29 peuvent être utilisés pour observer les franges à ondes multiples ainsi que des dispositifs analogues utilisant la transmission.

a) Franges par transmission.

36. Franges à l'infini. Calcul de l'éclairement. Considérons une lame d'air limitée par deux faces planes AB et $A'B'$ parallèles et semi-réfléchissantes (Fig. 76). Un rayon incident SMI donne naissance à toute une série de rayons réfléchis J_1K_1, K_1J_2, J_2K_2, etc. ... et il sort de la lame un faisceau de rayons parallèles R_1R_2 ... Tous ces rayons sont rassemblés en un point M du plan focal O de la lentille L. On suppose l'axe optique CO de l'objectif L confondu avec la normale à la lame e. Les deux rayons R_1 et R_2 ont acquis dans la lame d'air d'épaisseur e une différence de marche $\delta = 2e \cos i$. Cette même différence de marche se retrouve entre R_2 et R_3, R_3 et R_4 etc. ... Si δ/λ est égal à un nombre entier K, tous les rayons $R_1 R_2 R_3$ etc. ... sont en phase au point M et il y a certainement maximum de lumière. Dans les phénomènes étudiés aux Sect. 24 et 25, nous n'avons considéré que les deux rayons R_1 et R_2 mais jusqu'ici seul le nombre des rayons qui arrivent en M a changé: on peut dire que les maxima de lumière occupent la même position dans les phénomènes à deux ondes ou à ondes multiples. Les franges brillantes du nouveau phénomène occupent la même position que les franges brillantes des interférences à deux ondes, ce sont des anneaux centrés sur la normale à la lame définie dans le cas de la Fig. 76 par le foyer principal O de l'objectif L. L'ordre d'interférence au centre O des anneaux est $2e/\lambda$ et le rayon des anneaux brillants est donné par la relation (24.1).

Calculons la structure de ces franges. Soit R le facteur de réflexion des faces AB et $A'B'$ et T leur facteur de transmission. L'onde qui a subi $2p$

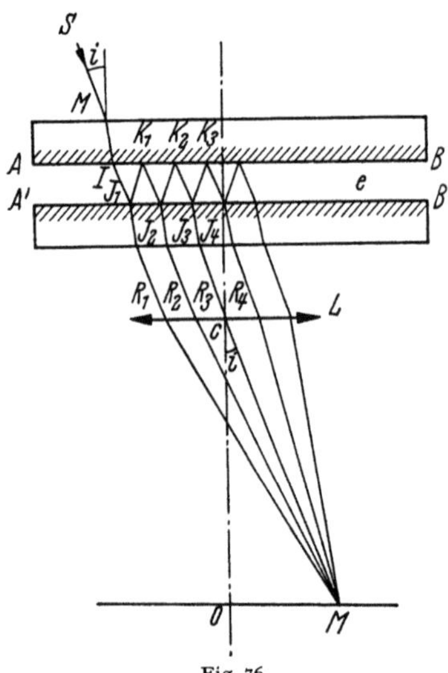

Fig. 76.

réflexions a pour amplitude TR^p, l'amplitude de l'onde incidente étant prise pour unité. On a donc à la sortie de la lame une infinité d'ondes dont les amplitudes sont

$$T \quad TR \quad TR^2 \quad ... \quad TR^p \quad ...$$

et qui présentent avec l'onde directe des différences de marche

$$0 \quad \delta \quad 2\delta \quad ... \quad p\delta \quad ...$$

On obtiendra l'éclairement résultant au point M en composant tous ces mouvements vibratoires par la règle de Fresnel. On aura

$$E = T^2 \sum R^{2p} + 2T^2 \sum R^{p+q} \cos (p - q) \frac{2\pi \delta}{\lambda}.$$

Cette expression contient un terme constant

$$T^2 \sum R^{2p} = \frac{T^2}{1 - R^2}$$

et une série de termes contenant en facteur le cosinus d'un multiple de $\dfrac{2\pi \delta}{\lambda}$. Posons $p - q = m$ le facteur $\cos\left(m \dfrac{2\pi \delta}{\lambda}\right)$ se trouve dans tous ces termes et le coefficient du cosinus est

$$2 T^2 R^m (1 + R^2 + R^4 + \cdots) = \frac{2 T^2 R^m}{1 - R^2}.$$

Finalement l'expression de l'éclairement en M devient

$$E = \frac{2 T^2}{1 - R^2} \left(\frac{1}{2} + R \cos \frac{2\pi \delta}{\lambda} + R^2 \cos \frac{4\pi \delta}{\lambda} + \cdots R^p \cos \frac{2p\pi \delta}{\lambda} + \cdots \right). \quad (36.1)$$

On peut écrire

$$E = \frac{2 T^2}{1 - R^2} \left(\frac{1}{2} + R \cos \varphi + R^2 \cos 2\varphi + \cdots + R^p \cos p\varphi + \cdots \right). \quad (36.2)$$

Calculons la somme de la série entre parenthèse. On peut remarquer que l'expression (36.2) est la partie réelle de la fonction imaginaire S

$$S = \frac{2 T^2}{1 - R^2} \left(\frac{1}{2} + R\, e^{j\varphi} + R^2\, e^{j2\varphi} + \cdots \right),$$

ou encore

$$S = \frac{2\,T^2}{1 - R^2}\left(-\frac{1}{2} + 1 + R\,e^{j\varphi} + R^2\,e^{j\varphi} + \cdots\right).$$

On voit apparaître une progression géométrique de raison $R\,e^{j\varphi}$. La somme de la série est alors donnée par

$$1 + R\,e^{j\varphi} + R^2\,e^{j\varphi} + \cdots = \frac{1}{1 - R\,e^{j\varphi}}\,,$$

d'où

$$S = \frac{2\,T^2}{1 - R^2}\left(\frac{1}{1 - R\,e^{j\varphi}} - \frac{1}{2}\right).$$

En multipliant haut et bas par la quantité conjuguée, on a

$$S = \frac{2\,T^2}{1 - R^2}\left(\frac{1 - R\cos\varphi + j\,R\sin\varphi}{1 - 2\,R\cos\varphi + R^2} - \frac{1}{2}\right),$$

d'où, en prenant la partie réelle

$$E = \frac{2\,T^2}{1 - R^2}\left(\frac{1 - R\cos\varphi}{1 - 2\,R\cos\varphi + R^2} - \frac{1}{2}\right) = \frac{T^2}{1 - 2\,R\cos\varphi + R^2}\,,$$

on a identiquement

$$1 + R^2 - 2\,R\cos\varphi = (1 - R)^2 + 2\,R\,(1 - \cos\varphi) = (1 - R)^2 + 4\,R\sin^2\frac{\varphi}{2}\,,$$

d'où finalement l'éclairement au point M

$$E = \frac{T^2}{(1 - R)^2}\,\frac{1}{1 + \dfrac{4\,R}{(1 - R)^2}\sin^2\dfrac{\varphi}{2}}\,. \tag{36.3}$$

Cette équation peut encore s'écrire

$$E = \frac{E_0}{1 + A\sin^2\dfrac{\varphi}{2}} \tag{36.4}$$

où $E_0 = \dfrac{T^2}{(1 - R)^2}$ est l'éclairement maximum et $A = \dfrac{4\,R}{(1 - R)^2}\cdot$

Lorsque le pouvoir réflecteur est élevé, le terme A est grand. Donnons par exemple à R la valeur 0,8. A est égal à 80. Pour $R = 0{,}9$, on a $A = 360$. Il en

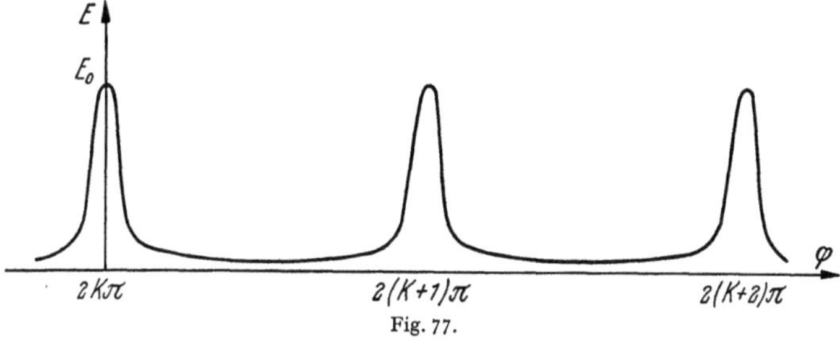

Fig. 77.

résulte que les minima dont la valeur est $\dfrac{E_0}{1 + A}$ sont presque complètement sombres.

De plus, dès que δ/λ diffère un peu d'un nombre entier, le dénominateur de (36.4) devient très grand et l'éclairement tombe presque immédiatement aux valeurs minima. La courbe qui donne la variation de E en fonction de φ (Fig. 77) présente donc des maxima très abrupts. Les franges ont un aspect très caractéristique. Elles apparaissent sous la forme d'anneaux brillants très fins se détachant sur un fond presque noir (Fig. 78a). Plus le pouvoir réflecteur R est grand, plus les franges sont fines. La Fig. 78a montre l'aspect des anneaux à l'infini obtenus

a

b

Fig. 78 a et b. Anneaux à l'infini Fabry-Perot.

avec la raie verte du mercure et un temps de pose faible (lampe basse pression). La Fig. 78b est identique, mais avec un temps de pose plus grand, elle montre la structure fine de la raie verte (voir Sect. 38).

37. Franges d'égale épaisseur. Au lieu d'observer les franges à l'infini d'une lame à faces parallèles, on peut étudier une lame d'épaisseur variable et observer ses franges d'égale épaisseur. Le dispositif est par exemple celui de la Fig. 79. Un collimateur composé d'une source S au foyer d'un objectif O_1 éclaire la lame L.

L'oeil est placé au foyer S' d'un deuxième objectif O_2 qui sert en même temps de loupe pour examiner la surface de L.

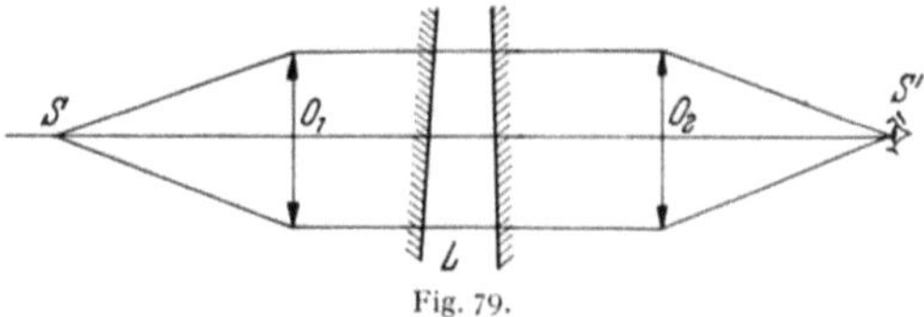

Fig. 79.

Si la lame est épaisse et inclinée, par suite des réflexions multiples les divers rayons réfléchis à partir d'un même rayon incident, s'écartent les uns des autres dans toute la lame et on n'a pas à proprement parler des franges d'égale épaisseur.

En fait, les variations de parallélisme sont toujours très faibles et si on observe en incidence voisine de l'incidence normale, on peut alors parler de franges d'égale épaisseur. En effet, en observant la source S, on voit toute une succession

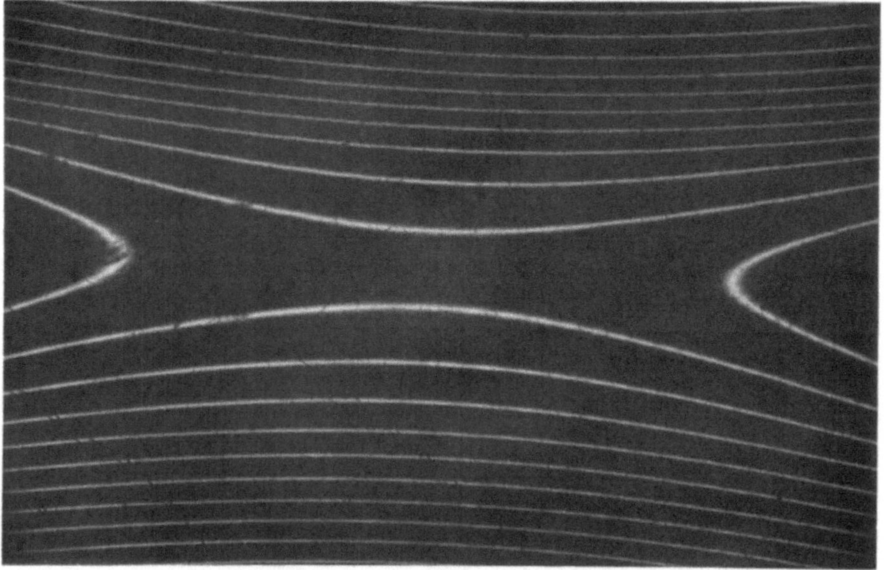

Fig. 80. Franges d'égale épaisseur à ondes multiples par transmission.

d'images s'affaiblissant de plus en plus si on s'écarte de la normale. Ces images sont toujours peu écartées les unes des autres si l'on veut que les franges soient visibles. La Fig. 80, qui peut être comparée à la Fig. 31, montre l'aspect des franges à ondes multiples.

38. Phénomènes produits lorsque la lumière incidente est composée de deux radiations simples. Quand plusieurs radiations simples éclairent en même temps un appareil à interférences à ondes multiples, chaque radiation donne un système de franges brillantes et grâce à la finesse de ces franges, elles peuvent se juxtaposer sans se confondre. Etudions le cas où le faisceau incident contient deux radiations monochromatiques. Soit une radiation de longueur d'onde λ: en un point du plan d'observation où la différence de marche des deux premières ondes interférentes est δ, l'ordre d'interférence sera $\delta/\lambda = p$. Si l'on a une seconde radiation λ', $\lambda < \lambda'$ par exemple, elle donnera également un système de franges brillantes, mais les franges correspondant à λ' seront moins serrées que les franges correspondant à λ.

Marquons les positions des franges des deux systèmes l'un au-dessus de l'autre (Fig. 81) en supposant qu'en O on ait $\delta = 0$ et que les valeurs de δ croissent proportionnellement à la distance au point O. Les deux systèmes de franges d'abord confondus, se séparent peu à peu; puis une frange du système en longueur d'onde λ se trouve intercalée sensiblement au milieu de l'intervalle qui sépare deux franges de longueurs d'onde λ'. Plus loin, il se produit une nouvelle coïncidence des deux systèmes, mais en général cette coïncidence n'est qu'approchée: elle se caractérise par le fait qu'il y a deux franges de longueur d'onde λ comprises entre deux franges consécutives de longueur d'onde λ'. En continuant, la séparation se produit de nouveau, puis une nouvelle coïncidence, etc. . . .

Calculons la période de coïncidence des deux radiations.

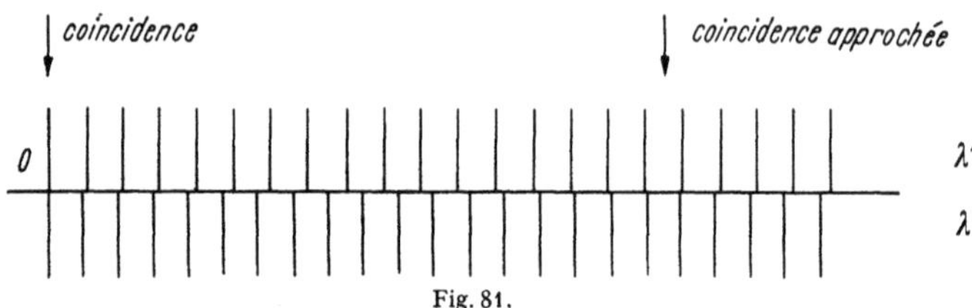

Fig. 81.

Soit δ_1 la différence de marche dans une région où les deux systèmes sont en coïncidence. La différence de marche augmentant (ou diminuant), les deux systèmes de franges se séparent, puis on retombe sur une nouvelle coïncidence lorsque la différence de marche devient δ_2. Entre ces deux coïncidences, l'un des systèmes a une frange de plus que l'autre. Donc en passant d'une coïncidence à la suivante, on pourra écrire

$$\frac{\delta_1}{\lambda} = \frac{\delta_1}{\lambda'} + K,$$

$$\frac{\delta_2}{\lambda} = \frac{\delta_2}{\lambda'} = K + 1.$$

Pour la $n^{\text{ième}}$ coïncidence à partir de la première observée, on aura

$$\frac{\delta}{\lambda} = \frac{\delta}{\lambda'} + K + n.$$

Si $N = K + n$ est le véritable numéro de cette coïncidence comptée depuis la différence de marche nulle, on aura

$$\frac{\delta}{\lambda} = \frac{\delta}{\lambda'} + N, \quad \delta = N \frac{\lambda \lambda'}{\lambda' - \lambda}. \tag{38.1}$$

La quantité $\dfrac{\lambda \lambda'}{\lambda' - \lambda}$ s'appelle la période de coïncidence des deux radiations. Les coïncidences sont périodiques et se reproduisent pour toutes les valeurs de δ qui sont des multiples de la longueur $\dfrac{\lambda \lambda'}{\lambda' - \lambda}$.

La $N^{\text{ième}}$ coïncidence se produit pour la différence de marche donnée par (38.1) et les ordres d'interférences correspondant pour les deux radiations sont

$$p = \frac{\delta}{\lambda} = \frac{N \lambda'}{\lambda' - \lambda}, \quad p' = \frac{\delta}{\lambda'} = \frac{N \lambda}{\lambda' - \lambda}.$$

Si p est un nombre entier K, il y a coïncidence exacte entre la $K^{\text{ième}}$ frange du système λ et la $(K+N)^{\text{ième}}$ frange du système λ'. Dans le cas général, p sera un nombre fractionnaire et l'aspect sera celui de la Fig. 81, la coïncidence ne sera qu'approchée.

Si les radiations deviennent très voisines, la période devient très longue et il arrive un moment où il n'est plus possible de les séparer. La séparation pourra être poussée d'autant plus loin que les franges seront plus fines. Définissons la largeur d'une frange par l'intervalle à l'intérieur duquel l'éclairement reste supérieur à la moitié $E_0/2$ de sa valeur maximum.

On aura d'après (36.4)

$$\frac{E_0}{1 + A \sin^2 \dfrac{\varphi}{2}} = \frac{E_0}{2} \,.$$

Si les franges sont très fines, cette valeur est atteinte pour une valeur de φ petite (Fig. 77) par exemple

$$\varphi = 2\pi\,(K + \varepsilon),$$

d'où

$$\frac{E_0}{1 + A \sin^2 \pi\,(K + \varepsilon)} = \frac{E_0}{2} \,.$$

Soit

$$A \sin^2 \pi\,\varepsilon = 1\,,$$

d'après la définition de $A = \dfrac{4R}{(1-R)^2}$ et puisque ε est petit

$$2\pi\,\varepsilon = \frac{1-R}{\sqrt{R}} \,. \tag{38.2}$$

L'expression (38.2) donne la demi-largeur de la frange en fonction de R.

Soient les deux radiations très voisines λ et $\lambda + d\lambda$, on aura en un point du champ d'interférences

$$\varphi = \frac{2\pi\,\delta}{\lambda}\,, \qquad \frac{d\varphi}{\varphi} = -\frac{d\lambda}{\lambda} \,.$$

Pour que ces deux radiations soient séparées, il faut que la distance séparant le centre d'une frange brillante de longueur d'onde λ du centre de la frange brillante de longueur d'onde $\lambda + d\lambda$ soit égale à $2\pi\varepsilon$. D'où

$$d\varphi = \varphi\,\frac{d\lambda}{\lambda} \geq 2\pi\,\varepsilon\,, \qquad d\lambda \geq \lambda\,\frac{1-R}{2K\pi\sqrt{R}} \,. \tag{38.3}$$

Pour pouvoir séparer deux radiations très voisines, il faut donc que le facteur de réflexion R soit le plus voisin possible de 1. Pour augmenter le facteur de réflexion R, il faut augmenter l'épaisseur des couches métalliques déposées sur les lames. De ce fait, on diminue beaucoup la transparence de l'appareil et on est rapidement limité. On améliore considérablement les résultats en déposant sur les lames des couches minces transparentes. On dépose des couches alternées de grand indice (sulfure de zinc) et de faible indice (fluorure de magnésium); il est alors possible d'arriver à un facteur de réflexion supérieur à 0,9 avec une absorption inférieure à 0,2. La photographie de la Fig. 78b a été obtenue avec ce procédé, les franges sont assez fines pour permettre la séparation des composantes de la raie verte d'une lampe à vapeur de mercure basse pression.

b) Franges par reflexion.

On peut observer aussi les franges à ondes multiples par réflexion (Fig. 82) mais les calculs sont plus compliqués parce que le premier rayon $I a_1$ qui n'a pas traversé la couche métallique AB n'appartient pas à la même série que les autres qui l'ont tous traversée deux fois.

39. Calcul de l'amplitude réfléchie. Dans le cas où les couches semi-transparentes ne possèdent aucune absorption, la figure d'interférence par réflexion est exactement complémentaire de la figure par transmission et il faut tenir compte dans le calcul, de l'absorption des couches et des changements de phase par réflexion qui sont différents de zéro ou de π pour une lame semi-transparente.

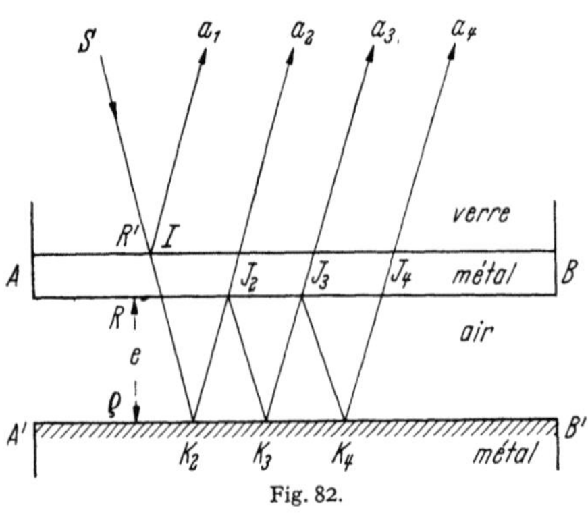

Fig. 82.

Considérons une lame d'air à faces paralléles et d'épaisseur e (Fig. 82) comprise entre une surface métallique réfléchissante $A'B'$ de facteur de réflexion en amplitude ϱ et une lame de verre recouverte d'une couche métallique semi-transparente AB. Le facteur de réflexion air-métal AB en amplitude est r et le facteur de réflexion verre-métal AB en amplitude r'. Soient $\alpha\alpha'\psi$ les changements de phase dus respectivement aux réflexions air-métal AB, verre-métal AB, air-métal $A'B'$ et β le changement de phase dû à la transmission de la couche métallique AB, dont le facteur de transmission en amplitude est égal à t. Les amplitudes des rayons réfléchis sont les suivantes

$$\text{rayon } I a_1 \qquad r' e^{j\alpha'},$$
$$\text{rayon } K_2 a_2 \qquad t^2 \varrho\, e^{j(2\varphi+\psi+2\beta)},$$
$$\text{rayon } K_3 a_3 \qquad t^2 \varrho\, r \varrho\, e^{j(2\varphi+\psi+\alpha+2\varphi+\psi+2\beta)}$$

où

$$\varphi = \frac{2\pi e \cos i}{\lambda}.$$

L'amplitude résultante A dans la direction des rayons réflechis est alors donnée par

$$A = r' e^{j\alpha'} + t^2 \varrho\, e^{j(2\varphi+\psi+2\beta)} \left[1 + r \varrho\, e^{j(\alpha+2\varphi+\psi)} + \cdots\right].$$

D'où

$$A = r' e^{j\alpha'} + \frac{t^2 \varrho\, e^{j(2\varphi+\psi+2\beta)}}{1 - r \varrho\, e^{j(\alpha+2\varphi+\varphi)}}. \tag{39.1}$$

Multiplions la fraction haut et bas par la quantité conjuguée du dénominateur, on a

$$A = r' e^{j\alpha'} + \frac{t^2 \varrho\, e^{j(2\varphi+\varphi+2\beta)} - t^2 r \varrho^2\, e^{j(2\beta-\alpha)}}{1 + r^2 \varrho^2 - 2 r \varrho \cos(\alpha+2\varphi+\psi)}$$

dont la quantité conjuguée A' s'écrit

$$A' = r' e^{-j\alpha'} + \frac{t^2 \varrho\, e^{-j(2\varphi+\psi+2\beta)} - t^2 r \varrho^2\, e^{-j(2\beta-\alpha)}}{1 + r^2 \varrho^2 - 2 r \varrho \cos(\alpha+2\varphi+\psi)}.$$

En faisant le produit, on obtient l'éclairement E. Posons $r^2 = R$, $t^2 = T$, $r'^2 = R'$, T est le facteur de transmission de la couche AB, RR' les facteurs de réflexion air-métal AB, verre-métal AB. On a

$$E = r'^2 + \cfrac{T^2\varrho^2 + T^2 R\varrho^4 - 2T^2 r\varrho^3 \cos(2\varphi + \psi + \alpha)}{[1 + R\varrho^2 - 2r\varrho\cos(\alpha + 2\varphi + \psi)]^2} +$$
$$+ \cfrac{2r'T\varrho\cos(2\varphi + \psi + 2\beta - \alpha') - 2r'Tr\varrho^2\cos(2\beta - \alpha - \alpha')}{1 + R\varrho^2 - 2r\varrho\cos(\alpha + 2\varphi + \psi)} . \tag{39.2}$$

On peut obtenir plus simplement l'amplitude résultante par la méthode graphique suivante dûe à P. Cotton et Ch. Dufour.

40. Détermination graphique de l'amplitude réfléchie. L'expression (39.1) peut s'écrire

$$A = r'\,e^{j\alpha'} - \frac{t^2}{r}\,e^{j(2\beta - \alpha)} + \frac{t^2}{r}\,\frac{e^{j(2\beta - \alpha)}}{1 - r\varrho\,e^{j(2\varphi + \psi + \alpha)}} ,$$

en posant

$$\varphi' = 2\varphi + \psi + \alpha,$$

on a

$$A = r'\,e^{j(\alpha + \alpha' - 2\beta)} + \frac{t^2}{r}\,e^{j\pi} + \frac{t^2}{r}\,\frac{1}{1 - r\varrho\,e^{j\varphi'}} ,$$

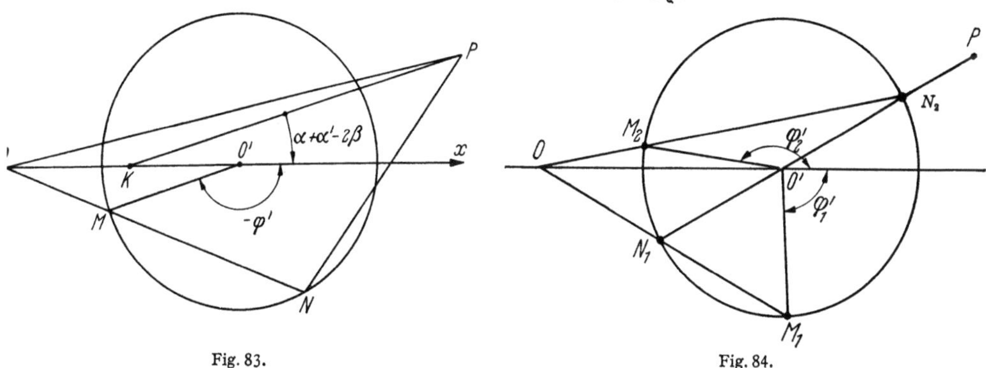

Fig. 83.　　　　　　　　　　　　　　　　　　　　Fig. 84.

Le carré du module de cette expression donne l'éclairement

$$E = \left| \underbrace{r'\,e^{j(\alpha + \alpha' - 2\beta)} + \frac{t^2}{r}\,e^{j\pi}}_{\text{vecteur } PO} + \underbrace{\frac{t^2}{r}\,\frac{1}{1 - r\varrho\,e^{j\varphi'}}}_{\text{vecteur } ON} \right|^2 .$$

Si les faces de la lame restent parallèles, mais si leur écartement e varie en observant sous l'incidence normale, la seule variable est φ'. L'éclairement E n'est fonction que de φ' est les variations de e entraînent des variations de E que nous allons étudier.

La Fig. 83 indique la construction graphique de E.

Le vecteur fixe $\overrightarrow{PO}$ est obtenu en portant $OK = t^2/r$ sur l'axe Ox et en menant ensuite $\overrightarrow{KP}$ de module $KP = r'$ et d'argument $\alpha + \alpha' - 2\beta$. On porte ensuite $OO' = \dfrac{t^2}{r}\,\dfrac{1}{1 - r^2\varrho^2}$ et on trace le cercle de rayon $O'M = r\varrho\overline{OO'}$.

Considérons le point M tel que $\widehat{Ox\cdot O'M} = -\varphi'$ si on trace OM qui coupe le cercle en N, la longueur PN donne l'amplitude cherchée, d'où l'éclairement E.

L'application de la formule (39.2) et plus simplement de la construction graphique précédente, montre que la répartition des éclairements en fonction de φ' n'est plus symétrique de chaque côté d'un maximum.

En effet: la position de PN correspondant à un maximum est PN_1, passant par O' à laquelle correspond la phase φ_1' (Fig. 84).

15*

La position de PN donnant un minimum est PN_2 passant également par O' et à laquelle correspond la phase φ_2'.

Les position des maxima correspondent par conséquent à φ_1', $\varphi_1'+2\pi$, $\varphi_1'+4\pi$... et les positions des minima à φ_2', $\varphi_2'+2\pi$, $\varphi_2'+4\pi$ ... Pour que les minima soient équidistants des maxima, il faudrait que $\varphi_2'=\varphi_1'+\pi$, $\varphi_2'+2\pi=\varphi_1'+3\pi$ ... et la Fig. 84 montre qu'il ne peut en être ainsi. Les phénomènes ont le même aspect que par transmission, mais la fonction $E=f(\varphi')$ n'est plus symétrique par rapport aux maxima ou aux minima.

c) Franges de superposition.

41. Franges d'égale épaisseur. Il n'est possible d'observer les franges en lumière blanche d'une lame mince que si la différence de marche ne dépasse pas quelques longueurs d'onde. On peut les observer avec des lames épaisses en utilisant deux lames, de manière que la différence de marche prise dans la première lame soit compensée dans la seconde.

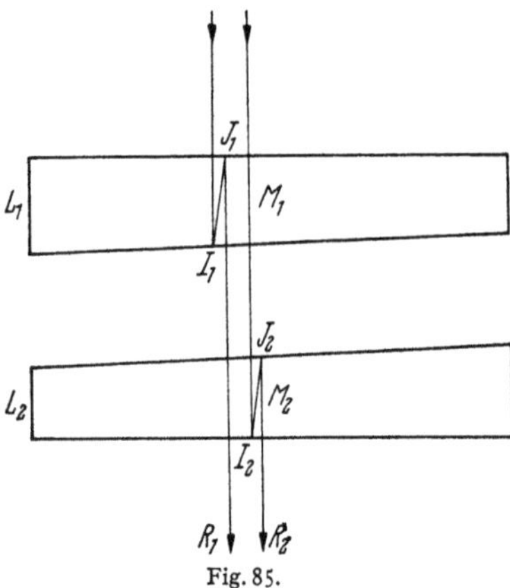

Fig. 85.

Soient deux lames L_1 et L_2, dont les faces sont semi-réfléchissantes (Fig. 85): considérons le rayon R_1 qui se réfléchit en I_1 et J_1 sur les faces semi-réfléchissantes de la lame L_1 et traverse L_2. Considérons de même le rayon R_2 qui traverse L_1, se réfléchit en I_2 et J_2 sur les faces de L_2. Si l'épaisseur de la lame L_1 en M_1 est e_1 et e_2 celle de la lame L_2 en M_2, la différence de marche des deux rayons R_1 et R_2 sera, en prenant deux lames d'air par exemple

$$\delta = 2(e_1 - e_2).$$

$\delta=0$ pour $e_1=e_2$ et si on éclaire les lames en lumière blanche, on voit dans la région correspondante une frange blanche. Cette frange dessine le lieu des points tels que $e_1=e_2$. Elle est bordée de franges présentant des colorations symétriques, les lignes d'égale coloration dessinant les lignes d'égale différence d'épaisseur des deux lames.

Il faut remarquer que les franges en lumière blanche n'apparaissent pas seulement lorsque les deux épaisseurs sont égales, mais encore toutes les fois que les deux épaisseurs sont dans un rapport simple. En effet, si R_1 subit $2m_1$ réflexions dans L_1 et $2m_2$ réflexions dans L_2 et si d'autre part, R_2 subit $2p_1$ réflexions dans la lame L_1 et $2p_2$ réflexions dans la lame L_2, la différence de marche entre R_1 et R_2 sera

$$\delta = (p_1 - m_1)\, \Delta_1 + (p_2 - m_2)\, \Delta_2.$$

Δ_1 et Δ_2 étant les différences de marche correspondent à chacune des lames. La différence de marche δ est nulle quelle que soit la longueur d'onde si

$$\frac{\Delta_2}{\Delta_1} = \frac{p_1 - m_1}{m_2 - p_2}. \tag{41.1}$$

Si Δ_2/Δ_1 est un rapport commensurable, il existe toujours des nombres entiers $p_1\, m_1\, p_2\, m_2$ satisfaisant à l'équation (41.1). Les rayons considérés donnent alors un système de franges en lumière blanche dont la frange centrale dessine le lieu des points pour lesquels le rapport Δ_2/Δ_1 a une valeur simple constante. Au fur et à mesure que Δ_2/Δ_1 devient moins simple, une partie de plus en plus importante de la lumière blanche ne peut interférer et les franges deviennent de plus en plus pâles.

Calculons la distribution des éclairements dans ces franges en supposant que l'épaisseur de chaque lame est assez grande pour que, prise isolément, elle ne donne pas de frange en lumière blanche.

Considérons la première lame L_1: pour une longueur d'onde λ, l'éclairement dans la région M_1 est donné par une expression analogue à (36.1) dans laquelle on affectera $TR\delta$ de l'indice 1. De même, la deuxième lame L_2 supposée seule a un éclairement dans la région M_2 donné par (36.1) en affectant $TR\delta$ de l'indice 2. Si la lumière traverse successivement les deux lames, l'oeil observant dans une direction normale aux lames de manière que M_1 se projette sur M_2, on voit la région $M_1 M_2$ avec un éclairement donné par

$$E = \frac{4\,T_1^2 T_2^2}{(1-R_1^2)\,(1-R_2^2)} \left(\frac{1}{2} + R_1 \cos \frac{2\pi \Delta_1}{\lambda} + R_1^2 \cos \frac{4\pi \Delta_1}{\lambda} + \cdots \right) \times$$

$$\times \left(\frac{1}{2} + R_2 \cos \frac{2\pi \Delta_2}{\lambda} + R_2^2 \cos \frac{4\pi \Delta_2}{\lambda} + \cdots \right)$$

où Δ_1 et Δ_2 sont les différences de marche définies plus haut. Si on effectue le produit de ces deux séries, on a

$$E = \frac{4\,T_1^2 T_2^2}{(1-R_1^2)\,(1-R_2^2)} \left[\frac{1}{4} + \sum \frac{1}{2} R_1^p \cos \left(\frac{2p\pi \Delta_1}{\lambda} \right) + \sum \frac{1}{2} R_2^p \cos \left(\frac{2p\pi \Delta_2'}{\lambda} \right) + \right.$$

$$\left. + \sum R_1^p R_2^q \cos \left(\frac{2p_1 \pi \Delta_1}{\lambda} \right) \cos \left(\frac{2p_2 \pi \Delta_2}{\lambda} \right) \right]$$

ou encore

$$E = \frac{4\,T_1^2 T_2^2}{(1-R_1^2)\,(1-R_2^2)} \left\{ \frac{1}{4} + \sum \frac{1}{2} R_1^p \cos \left(\frac{2p\pi \Delta_1}{\lambda} \right) + \sum \frac{1}{2} R_2^p \cos \left(\frac{2p\pi \Delta_2}{\lambda} \right) + \right. $$
$$\left. + \sum \left[\frac{1}{2} R_1^p R_2^q \cos 2\pi \left(\frac{p_1 \Delta_1 + p_2 \Delta_2}{\lambda} \right) + \frac{1}{2} R_1^p R_2^q \cos 2\pi \left(\frac{p_1 \Delta_1 - p_2 \Delta_2}{\lambda} \right) \right] \right\} . \tag{41.2}$$

Cette expression donne l'éclairement pour une longueur d'onde. Superposons les éclairements dus à toutes les longueurs d'onde.

On a supposé que chaque lame prise isolément ne donnait pas de franges, donc en faisant la somme des expressions analogues à (41.2) pour toutes les longueurs d'onde, les termes qui contiennent $\cos \left(\frac{2p\pi \Delta_1}{\lambda} \right)$, $\cos \left(\frac{2p\pi \Delta_2}{\lambda} \right)$ et $\cos 2\pi \left(\frac{p_1 \Delta_1 + p_2 \Delta_2}{\lambda} \right)$ disparaissent dans la somme. Ces termes ne font en effet intervenir que les franges de chacune des lames ou d'une lame d'épaisseur encore plus grande. L'éclairement E_B en lumière blanche, pour une source à spectre d'égale énergie, devient

$$E_B = \frac{2\,T_1^2 T_2^2}{(1-R_1^2)\,(1-R_2^2)} \left[\frac{1}{2} + \sum \left(R_1^p R_2^q \sum_{\lambda_1}^{\lambda_2} \cos 2\pi \, \frac{p_1 \Delta_1 - p_2 \Delta_2}{\lambda} \right) \right] . \tag{41.3}$$

Si le rapport Δ_1/Δ_2 n'est pas commensurable, aucune des expressions $p_1 \Delta_1 - p_2 \Delta_2$ n'est nulle. Ces expressions ne sont pas petites si Δ_1/Δ_2 n'est pas voisin d'un rapport commensurable. Le terme entre parenthèses dans l'expression (41.3) est nul et l'éclairement E_B est uniforme. Si au contraire Δ_1/Δ_2 est voisin d'un rapport commensurable m/n où m et n sont deux nombres entiers premiers entre eux, on a alors une série de termes qui ne s'annulent pas.

L'éclairement devient égal à

$$E_B = \frac{2T_1^2 T_2^2}{(1-R_1^2)(1-R_2^2)} \left[\frac{1}{2} + R_1^n R_2^m \sum_{\lambda_1}^{\lambda_2} \cos 2\pi \, \frac{n\varDelta_1 - m\varDelta_2}{\lambda} + \right.$$
$$\left. + (R_1^n R_2^m)^2 \sum_{\lambda_1}^{\lambda_2} \cos 4\pi \, \frac{n\varDelta_1 - m\varDelta_2}{\lambda} + \cdots \right]. \tag{41.4}$$

En comparant (41.4) à (36.1), on voit qu'à un facteur constant près c'est l'éclairement donné par une lame mince unique produisant une différence de marche $n\varDelta_1 - m\varDelta_2$ et dont chaque face aurait un facteur de réflexion égal à $R_1^n R_2^m$.

On aura donc un système de franges visibles en lumière blanche chaque fois que $\varDelta_1/\varDelta_2$ sera voisin d'un rapport commensurable m/n; la frange centrale dessinant le lieu des points tels que $\varDelta_1/\varDelta_2 = m/n$.

Supposons que les facteurs de réflexion R_1 et R_2 soient égaux: la lame fictive a alors un facteur de réflexion égal à R^{m+n} pour chacune de ses faces. Si $m+n$ augmente, R^{m+n} diminue puisque R est toujours inférieur à l'unité et par conséquent les colorations sont d'autant moins vives que $m+n$ est plus grand.

Si le facteur de réflexion n'est pas très élevé, le contraste des franges d'un système diminue rapidement au delà du premier.

Le tableau suivant donne le contraste des premiers systèmes de franges en lumière blanche pour deux facteurs de réflexion:

Tableau 2.

$\dfrac{\varDelta_1}{\varDelta_2}$	$m+n$	R^{m+n} $(R=0,75)$	γ	R^{m+n} $(R=0,65)$	γ
1	2	0,56	0,52	0,42	0,35
2 ou $\frac{1}{2}$	3	0,42	0,35	0,27	0,15
3 ou $\frac{1}{3}$	4	0,32	0,24	0,18	0,09

avec:

$$\gamma = \left(\frac{2R^{m+n}}{1+R^{m+n}} \right)^2.$$

42. Franges à l'infini. On peut observer au moyen d'une lunette visant à l'infini des franges de superposition entre deux lames dont les faces sont rigoureusement parallèles.

Considérons par exemple deux lames d'air L_1 et L_2, d'épaisseurs e_1 et e_2 très peu différentes (Fig. 86). Les deux lames sont limitées par des surfaces planes semi-réfléchissantes et parfaitement parallèles. Soit ω l'angle du dièdre formé par les deux lames L_1 et L_2 et prenons pour plan de figure le plan normal à l'arête du dièdre. Calculons la différence de marche dans la direction ϑ par rapport à l'axe xx' également incliné sur les deux lames L_1 et L_2. Cet axe coïncide avec l'axe optique d'un objectif O et on observe les phénomènes dans le plan focal F'.

On a

$$\delta = 2e_1 \cos r_1 - 2e_2 \cos r_2$$

où r_1 et r_2 sont les angles de réfraction dans les deux lames L_1 et L_2. Les angles r_1 et r_2 étant supposés petits ainsi que la différence $e_1 - e_2$, on aura

$$\delta = 2e_1 \left(1 - \frac{r_1^2}{2}\right) - 2e_2 \left(1 - \frac{r_2^2}{2}\right) = 2(e_1 - e_2) + e_1(r_2^2 - r_1^2),$$

d'où

$$\delta = 2(e_1 - e_2) + 2\omega \vartheta e_1. \tag{42.1}$$

Les épaisseurs e_1 et e_2 étant constantes, δ varie donc proportionnellement à ϑ et les franges à l'infini sont ici des droites parallèles à l'arête du dièdre formé par les lames. On aura une frange brillante dans la direction ϑ si

$$p = \frac{\delta}{\lambda} = \frac{2(e_1 - e_2)}{\lambda} + \frac{2\omega\vartheta e_1}{\lambda} = K.$$

La frange brillante suivante se trouvera dans la direction ϑ' donnée par

$$\frac{2(e_1 - e_2)}{\lambda} + \frac{2\omega\vartheta' e_1}{\lambda} = K + 1.$$

La distance angulaire de deux franges consécutives sera

$$\vartheta' - \vartheta = \frac{\lambda}{2e_1\omega}. \qquad (42.2)$$

Les franges sont équidistantes et la frange centrale correspond à

$$2(e_1 - e_2) + 2\omega\vartheta e_1 = 0,$$

c'est à dire

$$\vartheta = \frac{e_2 - e_1}{e_1\omega}.$$

Si l'on fait varier l'une des épaisseurs, les franges se déplacent parallèlement à elles-mêmes. Si on diminue l'angle ω, les franges s'élargissent et on obtient la teinte plate pour $\omega = 0$; cette teinte est le blanc si les épaisseurs sont égales. Les phénomènes sont les mêmes si l'épaisseur e_2 est voisine de $e_1/2$, de $e_1/3$, etc. ... mais les franges deviennent de moins en moins visibles à mesure que le rapport e_2/e_1 devient moins simple.

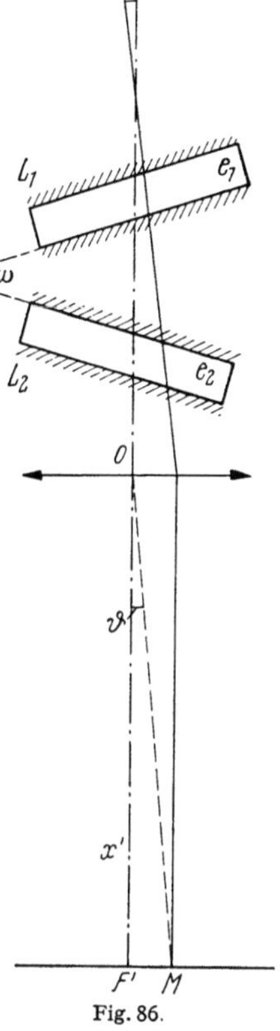

Fig. 86.

d) Appareils interférentiels à ondes multiples et à franges de superposition.

Comme nous l'avons déjà dit, les appareils décrits aux Sect. 28 et 29 peuvent être utilisés pour l'observation des phénomènes à ondes multiples.

Alors que dans le cas des franges à deux ondes, on observe de préférence par réfléxion, on peut ici faire les observations par transmission qui s'interprêtent théoriquement plus facilement.

Dans ce qui suit, nons décrivons des appareils destinés surtout aux mesures utilisant les franges à ondes multiples.

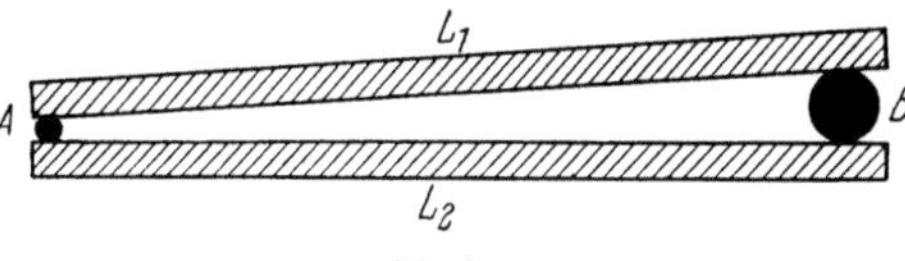

Fig. 87.

43. Lames étalons en lumière parallèle. Une lame étalon est constituée par deux lames planes de verre L_1 et L_2 dont les faces en regard sont argentées et maintenues à une petite distance l'une de l'autre en plaçant aux extrémités deux cales d'épaisseurs légèrement différentes A et B (Fig. 87).

La lame d'air comprise entre les deux faces semi-réfléchissantes a ainsi la forme d'un prisme de très petit angle (coin d'air).

Les franges d'égale épaisseur obtenues sont des droites parallèles entre elles et équidistantes, que l'on observe sous forme de fines lignes brillantes. Sur l'une des faces semi-réfléchissantes, on a tracé dans un sens perpendiculaire aux franges une division: on peut donc repérer les franges et si l'on a déterminé l'ordre d'interférence de l'une d'elles, on connaîtra l'ordre d'interférence en tous les points de la lame par une simple interpolation.

On peut opérer de la façon suivante: observons avec une lame d'air d'épaisseur e constante limitée également par des faces semi-réfléchissantes. On règle e

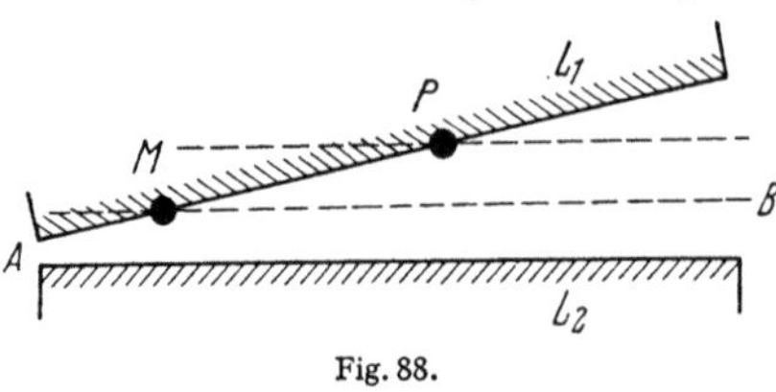
Fig. 88.

pour voir la frange brillante blanche d'égalité des épaisseurs en M (Fig. 88). On ne peut se tromper sur ce système, les autres étant moins constrastés. Supposons que le deuxième système soit en P: l'épaisseur de la lame étalon en ce point est double de son épaisseur en M car on sait que la cale B est plus épaisse que A.

On observe alors la lame étalon seule en lumière monochromatique et on compte le nombre des franges entre M et P, ce qui donne la différence des épaisseurs en M et P. Puisque l'épaiseur en P est deux fois plus grande qu'en M, cette différence représente l'épaisseur de la lame étalon en M, d'où l'étalonnage de la lame.

On verra plus loin comment une telle lame peut jouer le rôle de compensateur et permettre de déterminer finalement un ordre d'interférence compris entre les ordres d'interférence correspondant aux deux extrémités.

44. L'interféromètre Fabry-Pérot (Fig. 89) se compose de deux lames épaisses en verre dont la distance est réglable. Les faces en regard $A_1 B_1$ et $A_2 B_2$ sont

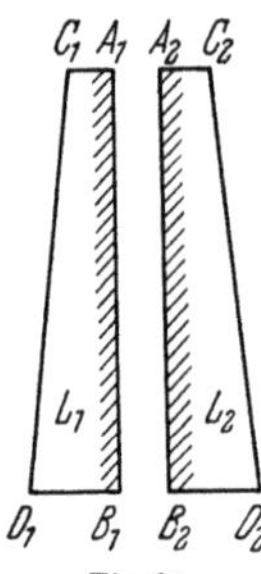
Fig. 89.

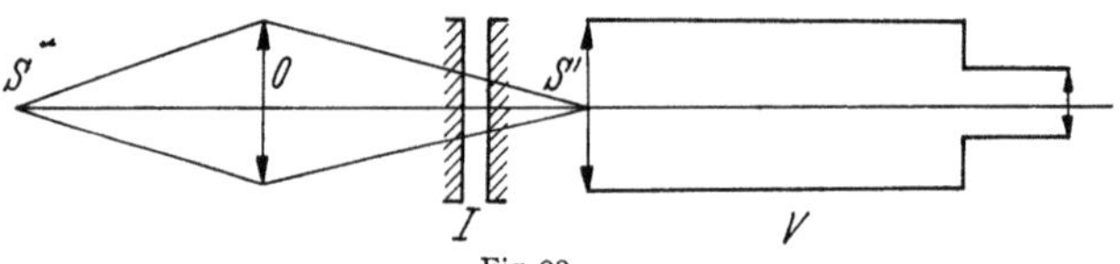
Fig. 90.

semi-réfléchissantes et les faces $C_1 D_1$, $C_2 D_2$ font entre elles un petit angle de façon que les anneaux qu'elles produisent soient rejetés hors du champ. On observe alors des franges d'égale inclinaison produites par la lame d'air comprise entre $A_1 B_1$ et $A_2 B_2$ au moyen d'une lunette V visant à l'infini (Fig. 90). Si on

dispose d'une source S monochromatique de petite dimension, il est commode de projeter son image S' près de la pupille d'entrée de la lunette au moyen d'une

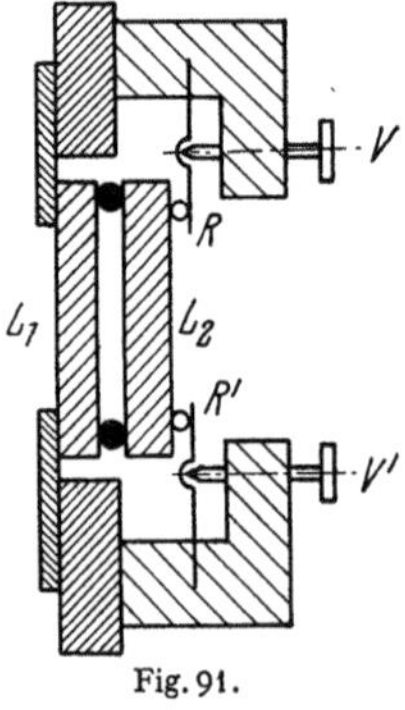
Fig. 91.

lentille O. L'interféromètre I est placé devant l'objectif de la lunette. Dans certains cas, il est intéressant de maintenir les deux faces $A_1 B_1$, $A_2 B_2$ à distance aussi invariable que possible, par l'emploi de trois cales de même épaisseur; on réalise ainsi un étalon interférentiel. On utilise généralement des cales en invar, leur dilatation est très faible et l'augmentation de l'épaisseur de l'étalon due à 'lélévation de température se trouve sensiblement compensée par la diminution correspondante de l'indice de l'air. On ne peut jamais réaliser trois cales rigoureusement identiques pour avoir le parallélisme exigé. On règle alors le parallélisme en serrant les lames L_1 et L_2 contre les cales au moyen des ressorts R et R' réglés par les vis V et V' (il y a en réalité trois ressorts et trois vis à 120°)

(Fig. 91). Grâce aux couches multiples que l'on dépose sur $A_1 B_1$ et $A_2 B_2$, on peut avoir de hauts facteurs de réflexion sans forte absorption et par conséquent, des appareils très lumineux donnant des franges d'une grande finesse.

45. Dispositif de Lummer et Gehrcke. Dans l'appareil de Lummer et Gehrcke, l'augmentation du facteur de réflexion sur les faces de la lame est obtenue en opérant sous une incidence voisine de la réflexion totale. Un rayon incident SI pénètre dans une lame L à faces planes et parallèles au moyen d'un prisme rectangle-isocèle P (Fig. 92). On peut ainsi faire pénétrer le rayon SI de manière qu'il fasse avec la normale en I un angle r un peu inférieur à l'angle de réflexion

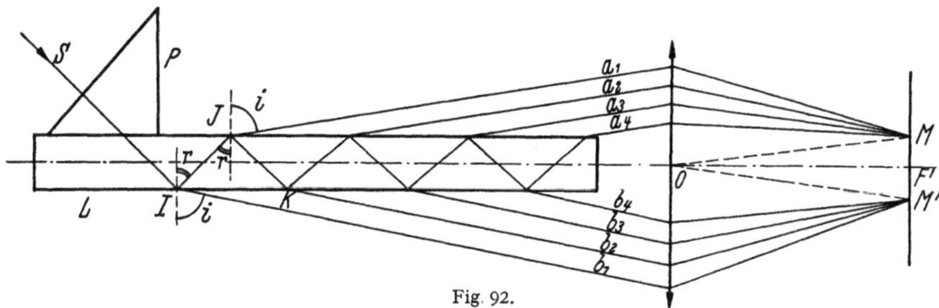

Fig. 92.

totale. Une petite fraction de la lumière émerge sous forme du rayon b_1 et le reste se réfléchit suivant IJ. De même en J, une petite fraction émerge suivant a_1 et l'autre est réfléchie suivant JK. Finalement on obtient toute une série de rayons a_1, a_2, a_3 ... et b_1, b_2, b_3 ... Les amplitudes décroissent suivant une progression géométrique et la différence de phase entre deux rayons successifs est constante. Les rayons parallèles correspondant à un rayon incident sont rassemblés dans le plan focal F' d'un objectif O en deux points M et M' symétriques par rapport à F'. On observe des franges brillantes fines sur fond noir; elles sont à peu près rectilignes et normales au plan de la Fig. 92. Comme la longueur de la lame L est limitée, le nombre des rayons interférents est limité et une formule analogue à (36.3) n'est plus applicable, les raies sont moins fines.

Si on éclaire le dispositif avec deux radiations très voisines, on verra deux groupes de raies, l'un en M, l'autre en M' et on peut ainsi analyser les composantes d'une raie.

46. Dispositif pour l'observation des franges de superposition. *α) Observation des franges d'égale épaisseur.* On projette l'image de la lame L_1 sur la lame L_2 au moyen d'un système afocal composé des deux lentilles O_1 et O_2 (Fig. 93).

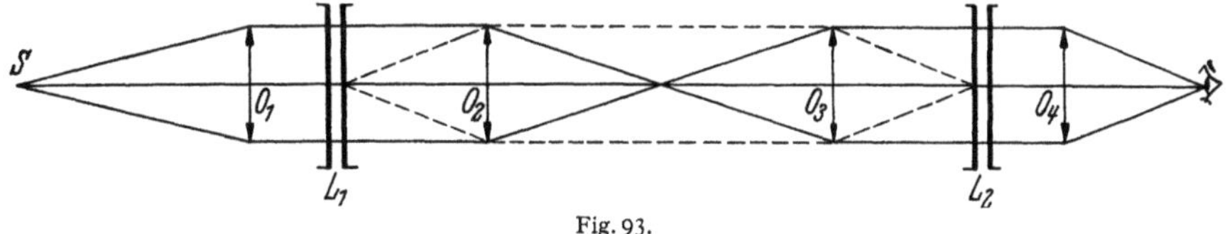

Fig. 93.

Les deux lames L_1 et L_2 sont éclairées en lumière parallèle au moyen du collimateur SO_1. La lame L_2, qui peut être une lame étalon, est examinée au moyen d'un oculaire O_4. En faisant glisser la lame étalon L_2 dans une direction perpendiculaire à son arête, on voit apparaître les franges dans une région où les épaisseurs de L_1 et L_2 sont égales.

β) Observation des franges à l'infini. On utilise une lame d'épaisseur invariable (étalon interférentiel) et un interféromètre Fabry-Pérot. Les deux appareils sont placés l'un à la suite de l'autre, l'étalon interférentiel étant placé sur un support qui permet de l'orienter. Un faisceau de lumière blanche traverse les deux lames et on reçoit la lumière émergente dans une lunette visant à l'infini.

Pour orienter les deux lames l'une par rapport à l'autre, soit pour les rendre parallèles, soit pour qu'elles fassent un petit angle, on observe les images par réflexions multiples.

47. Interféromètre auto-compensé Arnulf-Cagnet. L'interféromètre à lames semi-réfléchissantes A_1B_1 et A_2B_2 est en L (Fig. 94). Il est traversé dans sa moitié supérieure par un faisceau de rayons parallèles provenant du collimateur SO_1. Après traversée de l'objectif O_2, le faisceau est réfléchi symétriquement par un miroir M_1 placé au foyer de O_2 et traverse à nouveau l'objectif O_2. Le faisceau de rayons parallèles émergents passe dans la partie inférieure de l'interféromètre L, se réfléchit sur le miroir plan M_2, traverse l'objectif O_3 qui concentre les rayons dans l'image S'' de S et S' sur la pupille de l'oeil. L'interféromètre L étant placé au foyer objet de l'objectif O_3, celui-ci peut servir de loupe pour

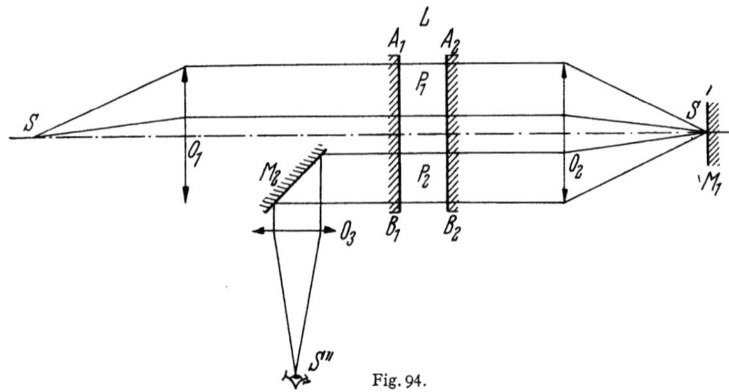

Fig. 94.

examiner L. Le faisceau de rayons parallèles traverse le même interféromètre L dans les régions P_1 et P_2 et par conséquent, si A_1B_1 et A_2B_2 sont bien parallèles, on a un éclairement maximum uniforme. On a la frange brillante blanche dûe aux franges de superposition des deux parties P_1 et P_2 jouant le rôle de deux lames. Le dispositif produit un système de franges de superposition très simplement grâce à une auto-compensation. Si les deux faces A_1B_1 et A_2B_2 font entre elles un petit angle, on voit les franges de superposition sous forme de franges d'égale épaisseur.

IV. Application des interférences.

a) Application des interférences à l'étude des surfaces.

48. Etude de la planéité d'une surface. α) *Défauts inférieurs à une frange.* Pour étudier la planéité d'une surface, on pose cette surface sur un plan étalon et on étudie les franges d'égale épaisseur de la lame d'air comprise entre les deux surfaces. Si la planéité de la surface étudiée est telle que les défauts soient inférieurs à une frange, on place alors une cale mince entre la surface et le plan étalon: des franges apparaissent et elles sont d'autant plus serrées que la cale est plus épaisse. Si la surface étudiée est parfaite, les franges sont rectilignes et équidistantes, sinon elles sont courbes er leurs distances varient.

Soit, Fig. 95, l'aspect des franges ainsi obtenues. Supposons la cale placée en E et soit A_1B_1 la ligne de plus grande pente du plan étudié par rapport au plan étalon: étudions la forme du plan suivant le diamètre A_1B_1. Pour cela, on mesure les abscisses des points d'intersection des franges avec A_1B_1 et on les reporte sur un graphique où l'on a porté en ordonnées des intervalles égaux à $\lambda/2$, sachant qu'en passant d'une frange à la suivante, l'épaisseur varie de $\lambda/2$.

En joignant par une droite les 2 franges extrêmes (droite D), on a les écarts par rapport au plan parfait. Il y a une petite indétermination aux extrémités du diamètre A_1B_1, car on ne connaît pas l'ordre d'interférence en ces 2 points. On peut alors, soit extrapoler, soit rapporter tout aux franges extrêmes. On pourra étudier la surface par rapport à un diamètre quelconque de la même façon.

Les mesures faites suivant le diamètre A_2B_2 fournissent directement l'écart avec le plan étalon, mais le nombre de points devient petit si la surface n'est pas trop mauvaise, aussi il est préférable, pour étudier le diamètre A_2B_2, de placer la cale, soit en A_2, soit en B_2. On utilisera pour cette étude l'un quelconque des dispositifs décrits pour l'étude des franges d'égale épaisseur. On opérera, de préférence, par réflexion, le contraste des franges étant supérieur à celui que l'on obtient par transmission. Ayant affaire à une lame très mince, on pourra utiliser une source de grand diamètre apparent sans

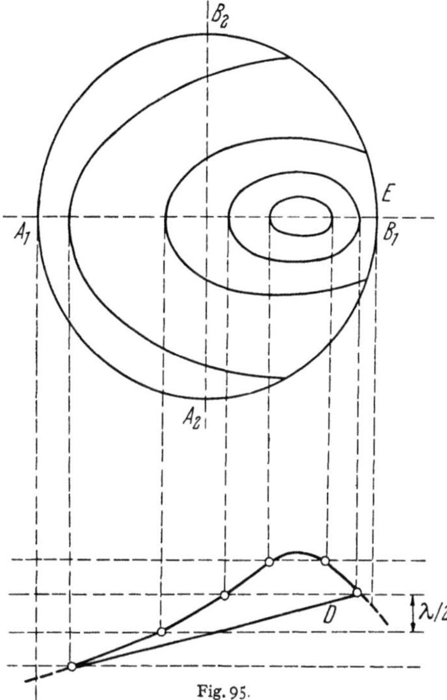

Fig. 95.

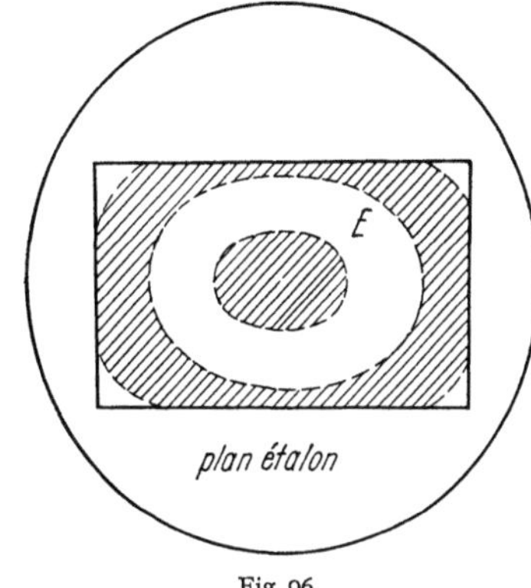

Fig. 96.

craindre une diminution notable de la visibilité. En opérant avec soin, on peut mesurer ainsi des défauts de planéité allant jusqu'à $\lambda/20$.

β) Surface plane dont les défauts sont supérieurs à une frange. Lorsque les défauts d'une surface plane sont supérieurs à une frange, on peut se contenter d'un examen rapide à l'oeil nu.

1. Lumière monochromatique. Supposons que l'aspect des franges soit celui de la Fig. 96. On peut dire que la surface à étudier E est, soit concave, soit convexe, la flèche de la section normale passant par le milieu de la surface à étudier représente le défaut.

La flèche est ici de l'ordre de $\lambda/2$ la surface est à une frange.

Pour savoir si la surface est concave ou convexe, il suffit d'exercer une légère pression avec les doigts sur la surface à étudier.

Les franges s'étalent dans les régions où l'inclinaison relative des deux surfaces diminue; elles s'étalent toujours en se déplaçant de la région où ces surfaces sont en contact vers les régions où l'épaisseur augmente. Si, en appuyant sur les bords de la surface, on voit les franges de la figure s'écarter du centre vers le bord, la surface à étudier est convexe. Si, en appuyant sur le milieu de la surface on voit les franges se rapprocher du centre, la surface est concave.

2. Lumière blanche. Si la lame d'air comprise entre la surface à étudier et le plan étalon est très mince, on peut observer les franges en lumière blanche. L'ordre de succession des couleurs donne immédiatement le sens du défaut, en se rappelant que l'épaisseur de la lame d'air augmente dans le sens où se succèdent les couleurs violettes bleues, jaunes, rouges, c'est-à-dire dans le sens où les longueurs d'onde augmentent.

49. Etude du parallélisme d'une lame. L'étude du parallélisme d'une lame à faces parallèles peut se faire de deux façons, soit par les franges d'égale épaisseur, soit par les franges d'égale inclinaison.

α) *Etude du parallélisme par les franges d'égale épaisseur.* On étudie les franges d'égale épaisseur produites entre les faces de la lame, par les dispositifs décrits aux Sect. 28 et 29, on opérant de préférence par réflexion. Si les faces de la lame sont

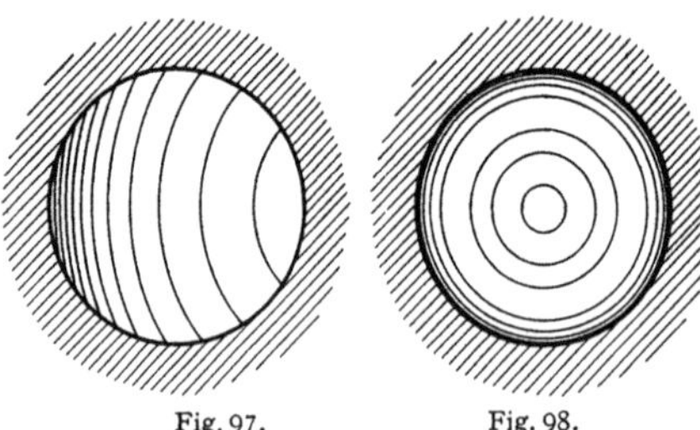

Fig. 97.   Fig. 98.

rigoureusement parallèles, on aura la teinte plate, sinon le nombre de franges observées indiquera le défaut de parallélisme. En supposant que les franges soient à peu près rectilignes, parallèles et équidistantes, les faces de la lame feront entre elles un petit angle ϑ égal à

$$\vartheta = \frac{\lambda}{2nl}$$

l étant l'équidistance des franges observées en incidence normale et n l'indice de la lame. En incidence oblique, on aura

$$\vartheta = \frac{\lambda}{2nl\cos r},$$

r étant l'angle de réfraction dans la lame. Dans l'étude des défauts du parallélisme des lames, il faudra prendre soin d'opérer en incidence normale et d'utiliser une source à faible diamètre apparent. Ces deux conditions doivent être réalisées d'autant mieux que la lame à étudier sera plus épaisse.

Le réglage de la perpendicularité des faisceaux sur la lame à étudier se fera en examinant les anneaux à l'infini produits par la lame. On prend une source à grand diamètre apparent et on examine l'image de la source fournie par l'appareil. Si on utilise le montage représenté par la Fig. 54, il suffira d'observer l'image S'. Les anneaux à l'infini pourront avoir, par exemple, l'aspect de la Fig. 97.

On agira alors sur l'orientation de la lame à étudier par rapport au faisceau lumineux, de façon à centrer les anneaux à l'infini sur l'image S' de la source lumineuse S (Fig. 98). On réduira ensuite convenablement le diamètre apparent de la source pour observer commodément les franges d'égale épaisseur de la lame à étudier.

β) *Etude du parallélisme par les franges d'égale inclinaison.* Cette étude se fait au moyen du dispositif décrit précédemment à propos de l'observation des anneaux à l'infini (Fig. 56, Sect. 29).

L'oculaire de la lunette étant muni d'un fil de réticule fixe, on déplace la lame dans son propre plan. Si la variation d'épaisseur de la lame fait remplacer, devant le réticule, un anneau par le suivant, la différence de marche a varié de λ et l'épaisseur de $\lambda/2n$, c'est à dire environ 0,18 μ.

Lorsque les défauts de la lame sont très faibles, on ne voit plus qu'une fraction d'anneau se déplacer devant le réticule. On procède alors de la façon suivante: soit p l'ordre d'interférence inconnu correspondant au premier anneau; l'ordre d'interférence au centre de la lame sera $p + \varepsilon$, ε étant une fraction comprise entre 0 et 1 et on aura

$$p + \varepsilon = \frac{2ne}{\lambda}.$$

Déplaçons la lame à étudier dans son propre plan, l'épaisseur devient e' et l'ordre d'interférence au centre

$$p + \varepsilon' = \frac{2ne'}{\lambda}.$$

La variation d'épaisseure de la lame est alors

$$e' - e = \frac{\lambda(\varepsilon' - \varepsilon)}{2n}.$$

Le contrôle du parallélisme de la lame revient donc à calculer la variation de l'ordre d'interférence au centre des anneaux.

Considérons l'anneau numéro N à partir du centre: l'ordre d'interférence correspondant à cet anneau sera $p - (N - 1)$ et son rayon angulaire sera

$$\alpha_N = \sqrt{\frac{n\lambda}{e}} \sqrt{(p + \varepsilon) - (p - N + 1)}.$$

Le rayon angulaire α_{N+1} de l'anneau suivant sera

$$\alpha_{N+1} = \sqrt{\frac{n\lambda}{e}} \sqrt{(p + \varepsilon) - (p - N)}$$

d'où

$$\frac{n\lambda}{e} = \alpha_{N+1}^2 - \alpha_N^2.$$

Or on a

$$\alpha_N^2 = \frac{n\lambda}{e}(\varepsilon + N - 1)$$

d'où

$$\alpha_N^2 = (\alpha_{N+1}^2 - \alpha_N^2)\left[\varepsilon + (N - 1)\right]$$

et enfin

$$\varepsilon = \frac{\alpha_N^2}{\alpha_{N+1}^2 - \alpha_N^2} + 1 - N.$$

On fera le calcul de ε pour plusieurs anneaux et on prendra la moyenne des valeurs trouvées.

Le calcul de ε' s'effectuera de la même façon pour la deuxième position de la lame, d'où la variation d'épaisseur cherchée.

Il faut remarquer que cette dernière méthode ne peut s'appliquer à des lames trop épaisses pour lesquelles les anneaux à l'infini sont angulairement très petits. Pour avoir une précision suffisante, il serait alors nécessaire d'employer une lunette à fort grossissement, ce qui est incompatible avec le dispositif de la Fig. 56.

Ce procédé sera malgré tout applicable à des lames épaisses de grandes dimensions si la variation des défauts est assez petite devant les dimensions de la lame pour qu'il soit possible d'employer une lunette à fort grossissement, donc d'utiliser pour la mesure de ε une région de la lame égale à la pupille d'entrée de la lunette.

50. Mesure interférentielle des rayons de courbure des surfaces sphériques. $\alpha)$ *Principe de la méthode.* On pose la surface à étudier sur une surface de référence de rayon de courbure peu différent, mais de signe contraire et on étudie les franges de la lame d'air comprise entre les deux surfaces.

Connaissant le rayon de courbure de la surface de référence et le nombre des franges, on en déduit le rayon de courbure inconnu.

Soient (Fig. 99) les 2 sphères de centres O_1 et O_2 et de rayons $R_1 = \overline{O_1 M_1}$, $R_2 = \overline{O_2 M_2}$. On suppose que les 2 sphères ne sont pas tout à fait au contact en S_1 et S_2, il reste une très mince couche d'air d'épaisseur

$$e = \overline{S_1 S_2}.$$

Si on abaisse de M_1 et M_2 les deux perpendiculaires $M_1 H_1$ et $M_2 H_2$ (cette dernière non représentée), on a

$$\overline{H_1 S_1} = \frac{x^2}{2 R_1}, \quad \overline{H_2 S_2} = \frac{x^2}{2 R_2},$$

x étant la distance à l'axe des points M_1 et M_2. D'où

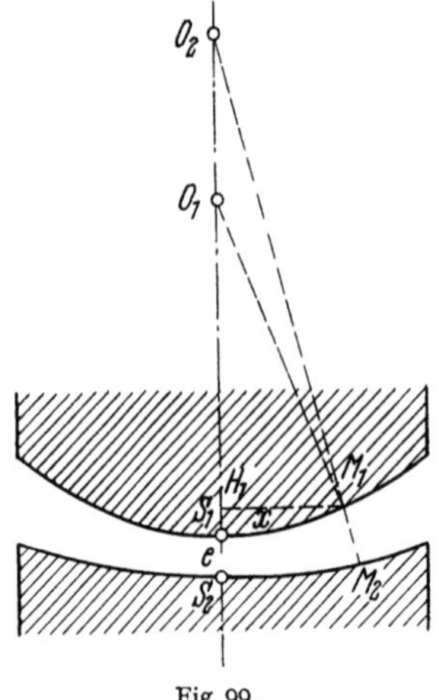

$$\overline{M_1 M_2} = \overline{H_1 S_1} + \overline{S_1 S_2} + \overline{S_2 H_2},$$

$$\overline{M_1 M_2} = e + \frac{x^2}{2}\left(\frac{1}{R_1} - \frac{1}{R_2}\right).$$

On sait que l'épaisseur d'air $M_1 M_2$ augmente de $\lambda/2$ quand on passe d'un anneau au suivant. Donc, si K désigne le nombre de franges (K peut être entier ou fractionnaire) comprises entre M_1 et M_2, on aura

$$e + \frac{x^2}{2}\left(\frac{1}{R_1} - \frac{1}{R_2}\right) = K \frac{\lambda}{2}.$$

Posons

$$e = K_0 \frac{\lambda}{2}.$$

D'où

$$\frac{1}{R_1} = \frac{1}{R_2} + \frac{(K - K_0)\,\lambda}{x^2}.$$

Fig. 99.

$K - K_0$ représente le nombre de franges et de fractions de franges comprises entre le centre et le bord de la surface à étudier.

Si l'on connaît R_2, on pourra ainsi mesurer R_1. Il faut donc déterminer, pour l'anneau de rayon égal à x, le nombre de franges $K - K_0$.

Si p est l'ordre d'interférence inconnu correspondant au premier anneau l'ordre d'interférence au centre sera

$$K_0 = p - \varepsilon$$

et l'ordre d'interférence K correspondant au $N^{\text{ième}}$ anneau observé à partir du centre, sera

$$K = p + N - 1.$$

D'où

$$K - K_0 = N - 1 + \varepsilon.$$

On peut écrire

$$\frac{1}{R_1} - \frac{1}{R_2} = \frac{(N - 1 + \varepsilon)\,\lambda}{x^2}$$

ou encore

$$x^2 = \frac{\lambda}{\dfrac{1}{R_1} - \dfrac{1}{R_2}} N + \frac{\lambda(\varepsilon - 1)}{\dfrac{1}{R_1} - \dfrac{1}{R_2}}.$$

On mesure au comparateur les rayons des premiers anneaux $N = 1$, $N = 2$, $N = 3$, $N = 4$, etc. et on porte en abscisses les numéros de ces anneaux, en ordonnées les carrés des rayons (Fig. 100). D'après l'expression ci-dessus, la courbe représentative doit être une droite: on trace donc la droite D qui représente la moyenne des pointés expérimentaux et dont le coefficient angulaire

$$\frac{\lambda}{\dfrac{1}{R_1} - \dfrac{1}{R_2}} = \tan \alpha$$

donne le rayon de courbure cherché.

On voit que la valeur N_0 de N au centre des anneaux ($x = 0$) est donnée par

$$N_0 - 1 + \varepsilon = 0$$

d'où

$$\varepsilon = 1 - N_0.$$

La distance séparant l'intersection de la droite D avec l'axe des abscisses du point d'abscisse $N = 1$ représente donc la partie fractionnaire de l'ordre d'interférence au centre des anneaux.

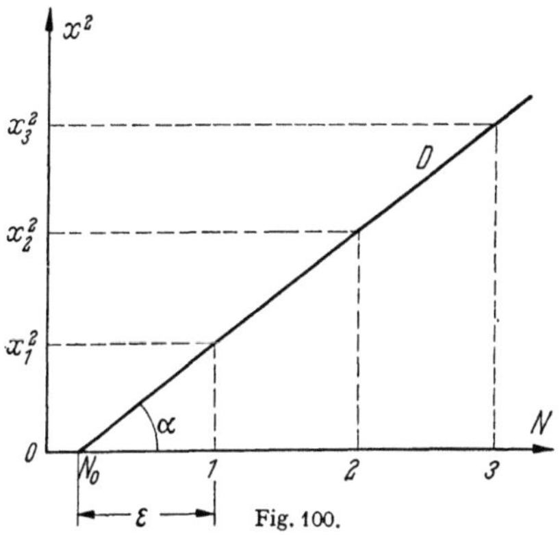

β) *Précision de la méthode.* La précision de la méthode est limitée par les deux causes d'erreurs suivantes:

1. On fait une erreur dN sur l'ordre d'interférence correspondant à l'anneau mesuré parce qu'on ne pointe pas exactement le milieu de la frange.

On attribue ainsi à l'anneau considéré le numéro N, alors que le numéro exact est $N + dN$.

2. Le comparateur donne une précision dx sur la mesure des longueurs x.

La relation

$$\frac{1}{R_1} = \frac{1}{R_2} + \frac{(K - K_0)\,\lambda}{x^2}$$

peut s'écrire, en considérant le $(N + 1)^{\text{ième}}$ anneau

$$\frac{1}{R_1} = \frac{1}{R_2} + \frac{N\lambda}{x^2},$$

puisque

$$K - K_0 \approx N,$$

en différentiant

$$\frac{dR_1}{R_1^2} = \frac{\lambda\,dN}{x^2} - \frac{2N\lambda\,dx}{x^3},$$

$$\frac{dR_1}{R_1} = \frac{\lambda R_1\,dN}{x^2} - \frac{2R_1 N\lambda}{x^2}\,\frac{dx}{x}.$$

D'après la relation initiale, on a

$$1 - \frac{R_1}{R_2} = \frac{R_1 N\lambda}{x^2}$$

d'où

$$\frac{dR_1}{R_1} = \frac{\lambda R_1\,dN}{x^2} - 2\left(1 - \frac{R_1}{R_2}\right)\frac{dx}{x}$$

et, en prenant la somme des erreurs relatives

$$\frac{dR_1}{R_1} = \frac{\lambda R_1\,dN}{x^2} + 2\left(1 - \frac{R_1}{R_2}\right)\frac{dx}{x}$$

on a donc avantage à considérer des anneaux de grands diamètres et par conséquent d'utiliser une surface de référence dont le rayon soit voisin de celui de la surface à étudier. Dans ce dernier cas, on peut écrire

$$\frac{dR_1}{R_1} = \frac{\lambda R_1}{x^2}\, dN.$$

Posons

$$\Omega = \frac{2x}{R_1}$$

on aura

$$dR_1 = \frac{4\lambda}{\Omega^2}\, dN.$$

Exemple: Prenons le cas suivant: $\Omega = \tfrac{1}{5}$, $\lambda = 0{,}6\,\mu$. On a $dR = 0{,}06\, dN$ mm., si $dN = \tfrac{1}{2}$ (une demi-frange), alors $dR_1 = 0{,}03$ mm.

Remarque: La formule

$$\frac{1}{R_1} = \frac{1}{R_2} + \frac{(K-K_0)\,\lambda}{x^2}$$

est valable tant que le rapport Ω ne dépasse pas des valeurs de l'ordre de $\tfrac{1}{5}$.

γ) *Application à la construction des calibres de courbure.* Pour contrôler une surface en cours de travail, on étudie les franges produites entre cette surface et une surface de référence ayant le rayon voulu et dont la qualité est parfaite.

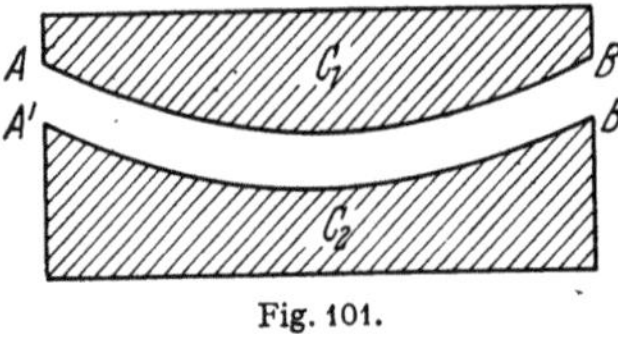

Fig. 101.

Ce procédé s'applique à l'obtention de la surface de référence elle-même, qui est associée à une autre surface de même rayon et de signe contraire: l'ensemble constitue un calibre interférentiel.

On polit les 2 surfaces AB et $A'B'$ du calibre (Fig. 101) et on constate, en posant les surfaces l'une contre l'autre, l'existence de quelques franges; on mesure le rayon de courbure de C_1 et, en comptant les franges, on en déduit facilement le rayon de courbure de C_2 par la formule

$$\frac{1}{R_1} = \frac{1}{R_2} + \frac{(K-K_0)\,\lambda}{x^2}.$$

On calcule alors le nombre de franges N qui devraient exister entre C_1 et C_2 pour que C_2 ait le rayon cherché par le relation trouvée précédemment

$$N = \frac{\Omega^2}{4\lambda}\, dR.$$

on retouche la surface $A'B'$ de C_2 en gardant C_1 comme référence, jusqu'à ce que le nombre de franges calculé soit observé entre C_1 et C_2. La surface $A'B'$ a le rayon de courbure désiré.

51. Traitement des surfaces. α) *Principe.* Lorsqu'un faisceau de lumière traverse un instrument d'optique, une partie se réfléchit sur chaque surface et retourne dans la direction de la source, on peut la considérer comme perdue pour la formation de l'image. Mais ce qui est plus grave, c'est qu'une fraction de cette lumière se réfléchit à nouveau et revient vers l'image en formant un halo parasite. La Fig. 102 montre schématiquement les trajets lumineux que nous venons de considérer. La source S envoie un rayon SI qui se réfléchit en I suivant le rayon (1). Le rayon (1) peut être considéré comme perdu s'il n'y a pas d'autres surfaces entre S et l'objectif O. En J une partie JS' continue et concoure à la

formation de l'image normale S'. Mais une partie se réfléchit également en J suivant JK. En K, une fraction sort suivant le rayon (2) qui est perdu, une autre fraction suit le trajet KS'' et concoure à la formation d'une véritable image en S''; c'est une image parasite qui éclaire uniformément le plan d'observation E et y produit le halo parasite MM'. Naturellement ce halo est très faible mais lorsque l'instrument a beaucoup de surfaces, les pertes par réflexion finissent par diminuer l'éclairement dans l'image S' (l'objet est un objet étendu dont S est l'un de ses points) et augmente l'importance du halo para-

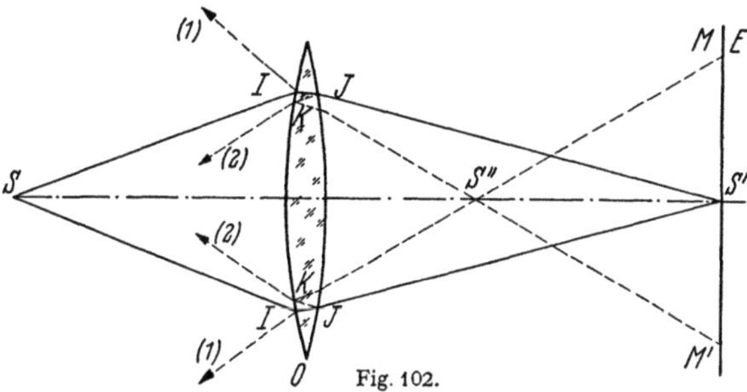
Fig. 102.

site MM': les pertes par réflexion sur les surfaces réduisent l'éclairement de l'image et augmentent l'éclairement du halo parasite.

Par exemple, dans un périscope ou un cystoscope, il peut y avoir jusqu'à 30 surfaces, et dans ce cas, les $\frac{4}{5}$ de la lumière incidente sont perdus par la seule réflexion sur les surfaces. On conçoit dans ces conditions l'intérêt qu'il y a à diminuer la réflexion, on dit «traiter» les surfaces.

Le principe du traitement des surfaces est le suivant:

Soit S la surface de verre d'indice N à traiter (Fig. 103). Déposons sur elle une couche mince d'épaisseur e et d'in-

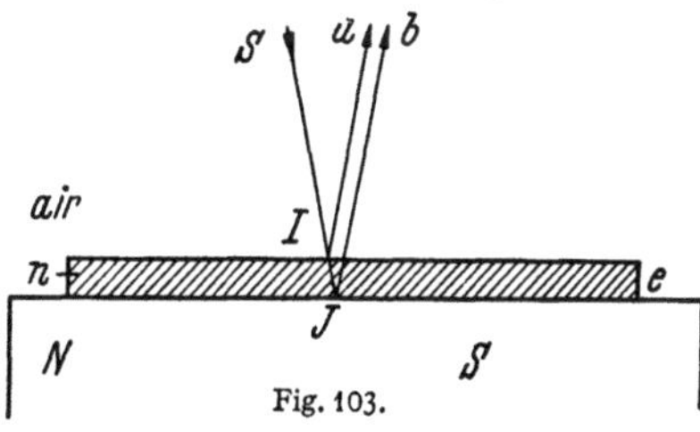
Fig. 103.

dice n. Nous avons vu que la lumière réfléchie est dûe aux interférences des rayons tels que Ia et Ib. L'amplitude réfléchie A est donnée par

$$A^2 = a^2 + b^2 + 2ab\cos\varphi$$

où les amplitudes a et b ont les valeurs suivantes données par les formules de FRESNEL (voir: polarisation par réflexion) pour l'incidence normale

$$a = \frac{1-n}{1+n}, \qquad b = \frac{n-N}{n+N}. \tag{51.1}$$

φ est la différence de phase entre les rayons a et b.

Supposons $n < N$, on a

$$\varphi = \frac{2\pi\delta}{\lambda} = \frac{2\pi}{\lambda}(2ne).$$

Pour la longueur d'onde λ, il y aura minimum de lumière réfléchie si

$$\frac{2\pi\delta}{\lambda} = (2K+1)\pi$$

d'où

$$ne = (2K+1)\frac{\lambda}{4}. \tag{51.2}$$

Le facteur de réflexion de la lame de verre N seule, sans la couche n, serait

$$R = \left(\frac{1-N}{1+N}\right)^2.$$

Après traitement, c'est à dire dépôt de la couche n, il devient d'après (51.2)

$$R' = \frac{2(N - n^2)}{(N + n)(n + 1)}.$$

Si on calcule R' pour différentes valeurs de N et pour une couche d'indice $n = 1{,}34$, on a les résultats suivants:

N verre à traiter	1,40	1,50	1,60	1,70
R facteur de réflexion avant traitement	0,028	0,040	0,053	0,067
R' facteur de réflexion après traitement	0,015	0,008	0,003	0,001

Le facteur de réflexion est peu diminué pour les indices faibles alors que la diminution est importante pour les verres à indice élevé.

Pour avoir extinction complète de la réflexion en lumière monochromatique de longueur d'onde λ, il faudrait d'abord réaliser la condition $a = b$. D'après (51.1), on trouve:

$$n = \sqrt{N}.$$

Pour le verre d'optique dont les indices s'échelonnent entre 1,5 et 1,8, la racine carrée varie entre 1,22 et 1,34. Comme il n'existe pas, pour réaliser les couches,

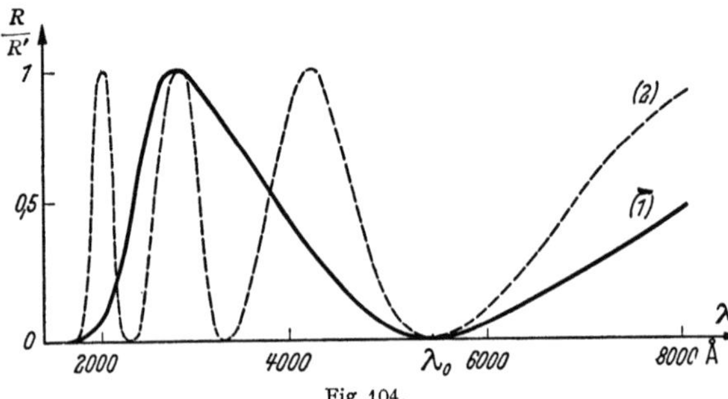

Fig. 104.

de substances utilisables d'indice inférieur à 1,34, la condition $a = b$ ne peut donc pratiquement pas être réalisée.

β) *Lumière blanche.* Lorsqu'on observe en lumière blanche une lame de verre traitée pour la longueur d'onde λ, cette radiation est presque absente dans la lumière réfléchie dont la composition spectrale est modifiée. Comme on ne peut réaliser la diminution de la réflexion de la même façon pour toutes les longueurs d'onde du spectre, on a avantage à choisir comme longueur d'onde λ_0 du minimum, celle pour laquelle le récepteur a le maximum de sensibilité. Par exemple, pour l'oeil, il sera convenable de prendre λ_0 voisin de $0{,}55\,\mu$. La formule (51.2) montre que l'on peut prendre $ne = \lambda/4$, $3\lambda/4$, etc.

La Fig. 104 donne pour $n = \sqrt{N}$ la variation du rapport R'/R en fonction de la longueur d'onde pour deux lames d'épaisseur $ne = \lambda/4$ (courbe 1) et $ne = 3\lambda/4$ (courbe 2). Dans le cas d'une couche d'épaisseur $3\lambda/4$, on voit que la lumière est rétablie rapidement dès que λ s'écarte un peu de λ_0. Pour la couche la plus mince $\lambda/4$, l'augmentation de la réflexion croît beaucoup moins vite lorsque l'on s'écarte de λ_0. Il y a évidemment intérêt à déposer des couches aussi minces que possible. On prend presque exclusivement des couches en $\lambda/4$, c'est-à-dire des couches dont l'épaisseur est de l'ordre du dixième de micron. On peut employer également plusieurs couches superposées et réduire encore les pertes par réflexion.

b) Application des interférences à la mesure des aberrations des systèmes optiques.

Plaçons au foyer image d'un objectif corrigé pour l'infini une source lumineuse ponctuelle. Les ondes émergentes seront planes si l'objectif est bien corrigé des aberrations. Pour étudier la forme d'une surface d'onde, on peut la faire interférer evec une onde de référence. Dans l'exemple précédent, on choisira une onde de référence plane. Les interférences entre l'onde fournie par l'objectif et l'onde plane de référence doivent donner une teinte plate. Si l'on observe des franges, c'est que la surface d'onde donnée par l'objectif n'est pas plane. Ces franges dessinent les lignes d'égale épaisseur entre les 2 ondes et permettent de détermine la forme de la surface d'onde donnée par l'objectif, donc de connaître ses aberrations.

52. Méthode de Twyman-Green. La méthode de Twyman-Green est basée sur l'utilisation de l'interféromètre de Michelson modifié suivant la Fig. 105.

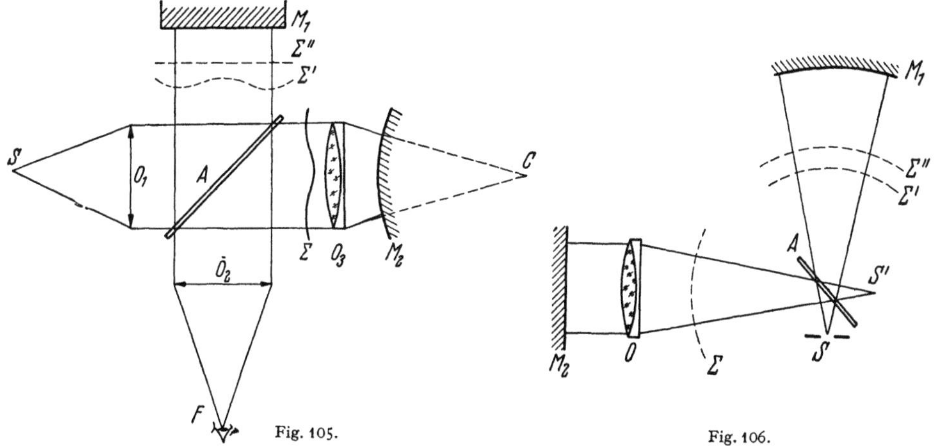

Fig. 105.   Fig. 106.

Le miroir plan M_2 de la Fig. 72 est remplacé par le système à étudier et une surface de référence convenable. Prenons le cas d'un objectif photographique O_3: on place derrière O_3 un miroir convexe M_2 dont le centre de courbure C coïncide sensiblement avec le foyer de l'objectif à étudier. L'objectif O_3 ayant généralement des aberrations, l'onde émergente Σ est déformée. Tout se passe comme si l'oeil placé en F observait les franges d'égale épaisseur entre l'onde plane de référence Σ'' fournie par le miroir plan M_1 et l'onde à étudier Σ'. Si l'objectif O_3 est parfait, Σ est une onde plane et on a une teinte plate entre Σ' et Σ'' l'onde Σ' étant symétrique de Σ par rapport à A. En modifiant un peu l'orientation de M_1, on peut observer des franges rectilignes et équidistantes. On remarque que l'objectif O_3 à étudier est traversé deux fois par les rayons lumineux; les déformations de l'onde Σ sont donc deux fois plus grandes que les déformations à mesurer. La Fig. 106 représente un montage analogue imaginé par Martin, Watt et Weinstein.

Une source lumineuse ponctuelle S éclaire l'objectif O à étudier au moyen de la séparatrice A. Elle éclaire également un miroir concave M_1 parfaitement sphérique dont le centre de courbure coïncide avec S. L'objectif O est placé de façon que son foyer coïncide avec S et on place un miroir d'autocollimation M_2. Si l'objectif O est parfait, la surface d'onde émergente Σ est une sphère de centre S'. Elle interfère avec la surface d'onde sphérique Σ'' donnée par M_1 et on a une teinte plate. Tout se passe comme si on observait les franges d'égale épaisseur entre Σ'' et Σ' symétrique de Σ par rapport à A.

Si l'objectif O présente des aberrations, la surface d'onde Σ est déformée et ses interférences avec Σ'' font apparaître des franges que l'on pourra observer, l'oeil placé en S'.

Le dispositif est intéressant car on peut, soit observer la surface d'onde Σ comme il vient d'être dit, soit examiner directement la tache image S' dont la structure rèvéle également la qualité de l'objectif O, comme nous le verrons plus loin (diffraction).

53. Interprétation des observations dans la méthode de Twyman-Green. Considérons le cas général où l'on étudie un objectif sur l'axe ou en dehors de l'axe, par la méthode de Twyman-Green.

Dans ce dispositif, l'onde plane de référence est constituée par le miroir M_1 convenablement orienté et l'onde de référence sphérique par le miroir convexe étalon M_2.

L'objectif n'étant pas parfait, l'onde Σ' est déformée d'une façon quelconque et l'oeil aperçoit la surface du système étudié sillonnée de franges, elles-mêmes

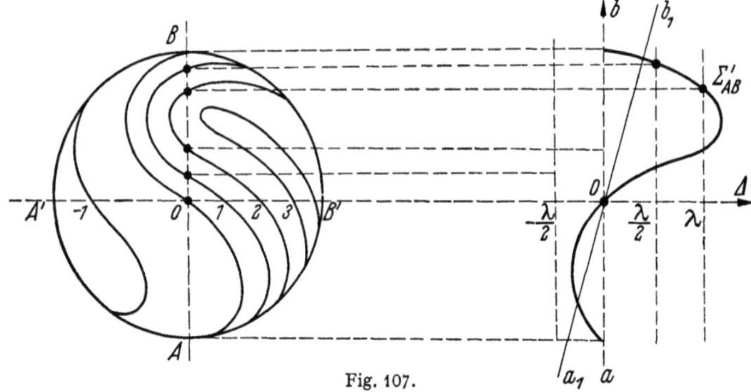

Fig. 107.

quelconques. Le problème qui se pose est le suivant: de l'aspect et de la position des franges, déduire:

a) les écarts Δ de la surface d'onde Σ' par rapport à un plan parallèle au miroir M_1 orienté convenablement,

b) la modification des écarts précédents lorsque l'on fait varier la position du centre du miroir M_2 le long de la caustique de l'instrument à étudier (défaut de mise au point),

c) les retouches à faire subir au système pour le rendre parfait.

α) *Etude des écarts Δ de la surface d'onde Σ' par rapport à un plan de référence parallèle au miroir M_1.* La forme de la surface d'onde Σ' ne dépend évidemment que des aberrations de l'objectif à étudier, mais les franges d'égale épaisseur de la lame d'air comprise entre M_1 et Σ' (ou entre Σ' et Σ'') dépendent, non seulement de la forme de Σ' mais de son orientation par rapport à M_1, c'est-à-dire du choix du plan de référence M_1. Comme les écarts Δ de la surface d'onde par rapport au plan M_1 sont déduits de la forme des franges, on conçoit qu'il est très important de choisir convenablement le plan de référence M_1. Connaissant la forme et la position des franges sur la surface de l'objectif pour une orientation déterminée du plan M_1, nous allons prévoir les différents aspects de la figure observée si nous modifions la position du plan M_1. Parmi toutes ces positions, nous choisirons évidemment celle qui correspond aux écarts Δ les plus faibles puisque c'est cette figure que l'on aurait dû obtenir si le réglage du miroir M_1 avait été correct. Il suffit donc de savoir comment varie l'écart Δ si l'on change les éléments de référence. Pour étudier la forme de l'onde Σ' et de ses écarts Δ par rapport à un plan de référence parallèle au miroir M_1, nous allons déterminer les sections de Σ' par divers plans diamétraux en procédant de la même façon que pour la mesure des défauts de planéité d'un plan par les franges d'égale épaisseur (Sect. 48). Supposons que nous observons sur la surface de l'objectif les franges représentées par la Fig. 107, le numérotage des franges étant supposé être celui de cette figure. Etudions la surface suivant deux diamètres perpendiculaires quelconques AB et $A'B'$. Si on porte en abscisses les écarts Δ et en ordonnées les intersections des franges par le diamètre AB, on obtient immé-

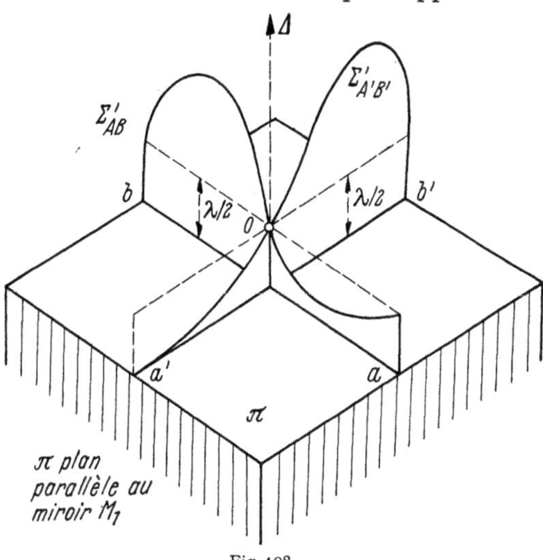

Fig. 108.

diatement le profil Σ'_{AB} des intersections de la surface d'onde Σ' par le plan diamétral AB (Fig. 107). De même, pour $A'B'$ (Fig. 109 et 110).

Les écarts Δ ainsi mesurés correspondent à une orientation déterminée du plan M_1 par rapport à Σ', orientation qui est représentée par l'axe des ordonnées sur les Fig. 109 et 110.

Calculons graphiquement les nouveaux écarts Δ si l'on modifie l'orientation du plan de référence M_1, et pour cela représentons les figures dans l'espace (Fig. 108) où ab et $a'b'$ sont les intersections des plans diamétraux AB et $A'B'$ avec le miroir M_1. Faisons tourner le plan de référence M_1 d'un angle α autour de l'axe $a'b'$ contenu dans ce plan et perpen-

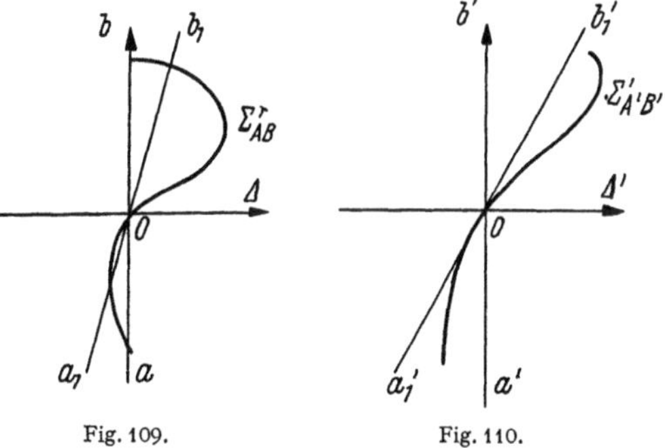

Fig. 109.   Fig. 110.

diculaire au diamètre examiné, rien n'est changé pour les écarts du profil $\Sigma'_{A'B'}$ par rapport à $a'b'$, et les nouveaux écarts Δ_1 de Σ'_{AB} par rapport à $a_1 b_1$ (nouvelle position de ab, Fig. 109) sont donnés par

$$\Delta_1 = \Delta - \alpha H.$$

H étant la distance à l'axe du point considéré. Si on fait subir ensuite au miroir M_1 une translation $OO' = \beta$ (Fig. 111), on obtient finalement les nouveaux écarts Δ_2

$$\Delta_2 = \Delta - (\alpha H + \beta).$$

La droite $\alpha H + \beta$ s'appelle la correctrice du profil Σ'_{AB} de l'onde Σ'. On obtient ainsi les écarts Δ_2 rapportés à la nouvelle position $a_2 b_2$ de l'axe ab (Fig. 113). En opérant exactement de la même façon pour le profil $\Sigma'_{A'B'}$ c'est-à-dire en faisant tourner M_1 d'un angle α' autour de ab, on a successivement les écarts Δ_1' puis Δ_2' (Fig. 112 et 114)

$$\Delta_2' = \Delta' - (\alpha' H + \beta).$$

Comme nous étudions pour l'instant les écarts de la surface d'onde par rapport à un plan parallèle au miroir M_1, il est évident que si le choix des paramètres α et α' est absolument arbîtraire, le paramètre β doit être le même dans les deux cas puisqu'il est nécessaire que la nouvelle origine O' soit commune aux deux profils $\Sigma'_{A_2 B_2}$ et $\Sigma'_{A_2' B_2'}$ (Fig. 113 et 114) pour qu'ils soient rapportés à un même plan de référence. Les paramètres $\alpha \alpha' \beta$ seront alors choisis de façon que les nouveaux écarts Δ_2 et Δ_2' soient les plus faibles possibles et pour cela il suffira de tracer sur les figures deux droites $a_2 b_2$, $a_2' b_2'$ qui épousent au mieux les profils

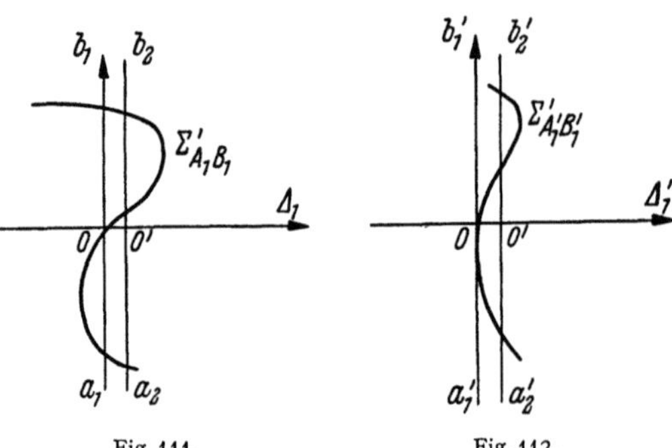

Fig. 111.   Fig. 112.

$\Sigma'_{A_1 B_1}$ et $\Sigma'_{A_1' B_1'}$ dont les intersections O' avec l'axe des abscisses soient situées à la même distance de l'origine O des figures initiales. On prendra ensuite la différence d'abscisses entre le profil tracé $\Sigma'_{A_1 B_1}$ et la droite $\alpha H + \beta$.

En faisant les mêmes opérations pour $\Sigma'_{A_1' B_1'}$, on obtiendra les nouveaux profils $\Sigma'_{A_2 B_2}$ et $\Sigma'_{A_2' B_2'}$ des Fig. 113 et 114.

Si l'on veut étudier des sections de la surface d'onde Σ' par des plans diamétraux autres que AB et $A'B'$, le choix des paramètres α et β pour ces nouvelles

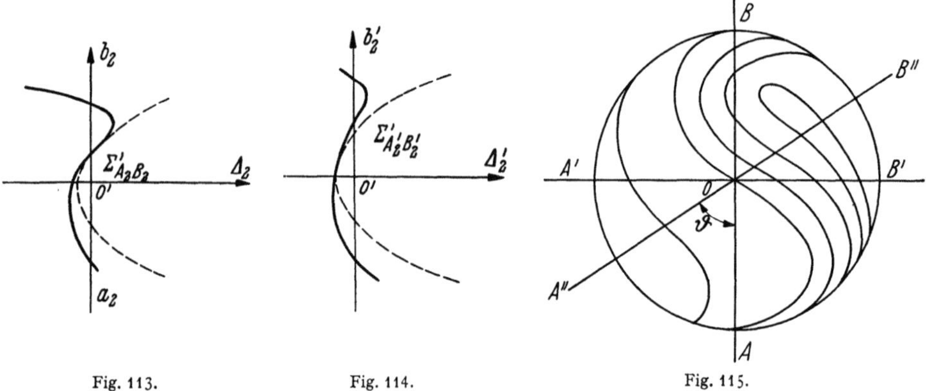

Fig. 113.   Fig. 114.   Fig. 115.

sections n'est plus arbîtraire, la position du plan de référence étant complètement définie par les opérations précédentes relatives aux diamètres AB et $A'B'$.

En effet, si l'on étudie la section de l'onde Σ' par un plan diamétral $A''B''$ faisant l'angle ϑ avec le plan AB (Fig. 115), le nouveau profil obtenu $\Sigma'_{A''B''}$ aura pour axe des ordonnées l'intersection $a''b''$ du plan $A''B''$ avec le plan M_1. Or, si l'on fait tourner ce plan d'un angle α autour de l'intersection $a'b'$ du plan $A'B'$ avec le plan M_1, tout se passe comme si l'on avait fait tourner l'axe des ordonnées $a''b''$ du nouveau profil $\Sigma'_{A''B''}$ d'un angle α_1 donné par $\alpha_1 = \alpha \cos \vartheta$ en supposant que les angles α et α_1 soient petits. Donc, pour le profil $\Sigma'_{A''B''}$ les paramètres définissant la correctrice, seront α_1 et β. En procédant de même pour quelques plans diamétraux dont le nombre sera jugé suffisant par l'observateur, on aura ainsi un ensemble de profils définissant bien la surface d'onde. Le choix des paramètres de départ α et β sera fixé par l'aspect de tous les profils $\Sigma'_{A_2 B_2} \Sigma'_{A'_2 B'_2} \Sigma'_{A''_2 B''_2} \ldots$ que l'on aura étudiés. Comme l'aboutissement de cette étude est la travail de retouche à faire subir au système, il sera naturel de choisir ces paramètres α et β de façon que pour l'ensemble des profils obtenus, les écarts Δ soient les plus faibles possible. Nous avons supposé connus les numéros des franges observées, ce qui n'est jamais réalisé dans la pratique. On procède alors de la façon suivante: on attribue à chaque frange un numéro arbitraire, mais logique par rapport aux franges voisines; on en déduit les profils Σ'_{AB}. De ces profils résulte la connaissance des aspects possibles de la figure observée lorsque l'on modifie la position du plan de référence M_1 grâce aux correctrices.

En appuyant doucement avec le doigt sur le miroir M_1, on regarde si les modifications que l'on observe, correspondent bien, tout au moins qualitativement, aux variations prévues.

β) Modification des écarts précédents lorsque l'on fait varier la position du centre du miroir M_2 le long de la caustique de l'instrument à étudier (défaut de mise au point). Supposons que l'observateur étudiant un objectif ne puisse arriver, par suite de mauvais réglages, à la teinte plate. On ne doit pas en conclure que l'instrument est défectueux car les franges observées peuvent n'être dues qu'à un simple défaut de mise au point, c'est-à-dire à une non-coïncidence du centre du miroir M_2 et du foyer de l'objectif.

Il faut donc étudier les variations du profil de la surface d'onde lorsque l'on déplace le miroir M_2, le miroir M_1 restant fixe. Cette étude se fait de la même façon que précédemment, c'est-à-dire graphiquement.

Le mouvement du miroir M_2 peut se décomposer en un déplacement de son centre le long de l'axe et un déplacement latéral de ce même point. Le déplacement latéral du centre du miroir M_2 produit des effets identiques à ceux que nous venons d'étudier (correctrice linéaire). Il n'y a donc pas lieu d'en tenir compte à nouveau, la correctrice linéaire précédente rendant compte, à la fois des modifications d'orientation du plan de référence M_1 et du déplacement latéral du centre du miroir M_2. Nous n'avons plus qu'à nous préoccuper d'un déplacement du centre du miroir M_2 le long de l'axe.

Si le centre du miroir M_2 ne coïncide pas avec le foyer de l'objectif O_3 supposé parfait, l'onde Σ' est sphérique et les franges correspondant aux profils tels que $\Sigma'_{A_2 B_2} \Sigma'_{A'_2 B'_2} \ldots$ relatifs a la meilleure orientation possible du miroir plan M_1, sont des cercles de centre O'. En effet, les franges que l'on obtient ainsi sont les franges d'égale épaisseur d'une lame d'air comprise entre l'onde sphérique Σ' et un plan: ce sont des anneaux de Newton. On doit donc pouvoir tracer une parabole qui épouse exactement tous les profils $\Sigma'_{A_2 B_2} \Sigma'_{A'_2 B'_2}$ etc. Si R est le rayon de courbure de l'onde émergente, η et ν deux constantes dont le choix sera précisé plus loin, l'équation de cette parabole, dite correctrice parabolique, pourra s'écrire

$$\Delta_3 = \frac{\eta}{2R^2} H^2 + \nu.$$

Quelle que soit la forme de l'onde étudiée, il sera prudent de corriger les profils $\Sigma'_{A_2 B_2} \Sigma'_{A'_2 B'_2}$ par rapport à cette parabole, de la même façon que l'on a corrigé les profils $\Sigma'_{AB} \Sigma'_{A'B'}$ par rapport aux droites $a_2 b_2$ et $a'_2 b'_2$, afin d'éliminer tout défaut de mise au point et d'obtenir une surface d'onde de référence qui soit centrée au mieux par rapport à l'onde émergente examinée. Le travail de retouche pourra alors être entrepris dans les meilleures conditions. On peut choisir la correctrice parabolique de la façon suivante: on trace sur un papier calque toute une série de paraboles passant par l'origine et correspondant à différentes valeurs de η positives et négatives. On superpose le calque aux profils $\Sigma'_{A_2 B_2}$ $\Sigma'_{A'_2 B'_2}$ en faisant coïncider les ordonnées. On le déplace d'une longueur ν le long des abscisses en cherchant quelle est la parabole qui épouse au mieux tous les profils tracés. En prenant enfin les différences d'abscisses entre chaque profil et la correctrice parabolique choisie, on obtient les profils $\Sigma'_{A_3 B_3}$ $\Sigma'_{A'_3 B'_3}$ (Fig. 116 et 117). C'est par rapport à ces profils définitifs, que le travail de retouche va être entrepris. Comme on l'a vu précédemment, le choix des paramètres α et α' est absolument arbitraire. Il en est de même du paramètre η qui est choisi d'après les considérations précédentes. Par contre, les paramètres β et ν sont astreints à la condition que sur les profils $\Sigma'_{A_3 B_3}$ $\Sigma'_{A'_3 B'_3}$, les abscisses,

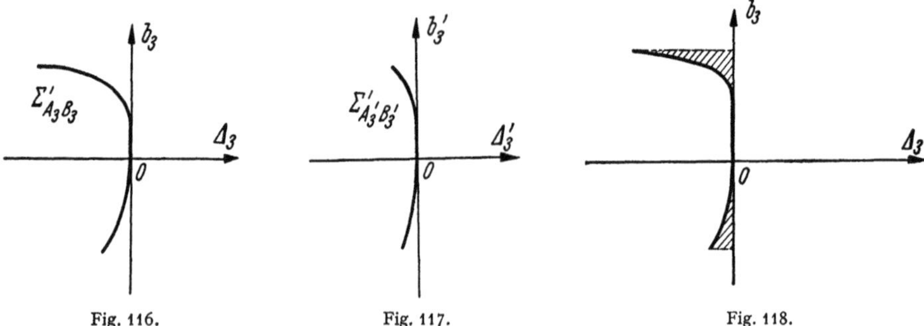

Fig. 116.

Fig. 117.

Fig. 118.

à l'origine, aient la même valeur puisque ces abscisses concernent un point commun à tous les profils. Si le système à étudier présente de l'astigmatisme, on choisira comme plans diamétraux initiaux AB et $A'B'$, les deux plans principaux d'astigmatisme.

γ) *Retouches à faire subir au système.* Le numérotage des franges nous a indiqué les points où l'onde est en avance, et ceux où l'onde est en retard par rapport aux éléments de référence. Considérons simplement les profils $\Sigma'_{A_3 B_3}$ et $\Sigma'_{A'_3 B'_3}$ où nous supposons que les Δ positifs indiquent une avance de l'onde étudiée par rapport aux éléments de référence. Deux cas sont à considérer:

    1. le système à étudier est un système réfringent;

    2. le système étudié est une surface réfléchissante.

1. Cas d'un système réfringent. Il y a surépaisseur de matière aux points où Δ est négatif, par conséquent sur le profil $\Sigma'_{A_3 B_3}$, on fera un transport de l'axe des ordonnées du côté droit de façon que toute la matière à enlever apparaisse d'un seul côté (région hachurée sur la Fig. 118).

La matière à enlever sera donnée par

$$e = \frac{\Delta}{n-1},$$

n étant l'indice de la surface sur laquelle on fera les retouches car si le système comporte plusieurs surfaces, on peut toujours attribuer le défaut total Δ à l'une quelconque d'entre elles.

2. Cas d'une surface réfléchissante. Il y a surépaisseur de matière là où Δ est positif; il faut donc faire un transport de l'axe des ordonnées du côté gauche. La matière à enlever sera donnée par

$$e = \frac{\Delta}{2}.$$

54. Méthodes de Michelson et de Cotton. Dans la méthode de Michelson, on place devant l'objectif O à étudier 2 fentes F_1 et F_2 dont les hauteurs sont petites devant le diamètre de O (Fig. 119). La fente F_2 est fixe et la distance entre les fentes F_1 et F_2 peut varier. Pour une position déterminée de la fente F_1, on observe dans le plan F' conjugué de la source, un système de franges d'interférences d'Young modulé par le phénomène de diffraction de chaque fente (voir diffraction). Si l'objectif O est parfait, la différence de marche $\delta = F_1 F' - F_2 F'$ est nulle en F' centre de l'onde sphérique Σ, quelle que soit la position de F_1. La frange centrale du système de franges est fixe et se trouve en F'. Si la surface d'onde Σ n'est pas sphérique, c'est-à-dire si l'objectif O est aberrant, la différence de marche $\delta = F_1 F' - F_2 F'$ n'est plus nulle en F', la frange centrale se

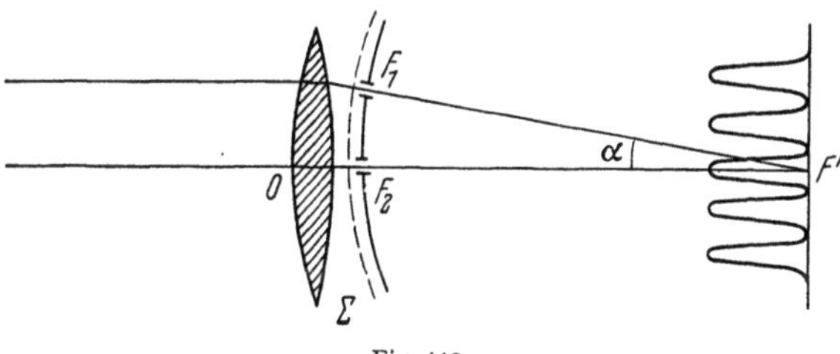

Fig. 119.

déplace d'une distance $y = \delta/\alpha$ (Sect. 13). De la mesure de ce déplacement, on peut donc en déduire la forme de la surface d'onde et, par suite, les aberrations de l'objectif.

La méthode est d'application délicate car, lorsque F_1 s'écarte de F_2, la distance λ/α entre les franges devient très petite et il est difficile de suivre le déplacement de la frange centrale.

Dans la méthode de Cotton, les 2 fentes F_1 et F_2 sont à une distance fixe l'une de l'autre, distance qui est petite par rapport au diamètre de l'objectif. Ces 2 fentes produisent en F' un système de franges dont on pointe la frange centrale. En déplaçant les fentes, on observe la frange centrale, si elle reste fixe, l'objectif O est parfait, si elle se déplace, l'objectif O est aberrant. Notons que la méthode de Cotton n'utilise les interférences que pour faciliter les pointés, ce n'est pas une méthode interférentielle. La position de la frange centrale définit la trace du rayon lumineux qui passerait au milieu de la distance des 2 fentes. La méthode donne par conséquent l'aberration angulaire et non pas directement la différence de marche δ. Il faut faire une intégration pour passer de l'aberration angulaire donnée par la méthode de Cotton au profil de la surface d'onde.

c) Application des interférences à l'observation des objets transparents et réfléchissants.

On a souvent à étudier et à mesurer des objets transparents qui ne diffèrent du milieu qui les entoure que par des variations d'indice. Par exemple, les défauts d'homogénéité du verre, des matières plastiques, des liquides etc. De même, lorsqu'on considère des objets réfléchissants, comme les miroirs, la surface de certains cristaux etc.

Le problème qui se pose est par conséquent de rendre visibles des objets caractérisés seulement par des variations de phase. Les méthodes interférentielles décrites précédemment s'adaptent parfaitement à ce type de problème.

55. Emploi des interférences à ondes multiples. Considérons deux lames L_1 et L_2 dont les faces en regard $A_1 B_1$ et $A_2 B_2$ sont semi-réfléchissantes et bien parallèles (Fig. 120).

Soient e la distance des deux faces $A_1 B_1$ et $A_2 B_2$ et n l'indice du milieu situé entre les deux lames. Si λ est la longueur d'onde de la lumière incidente, la différence de phase φ sous l'incidence normale entre le rayon R_2 et le rayon R_1 est égale à

$$\varphi = \frac{4\pi n e}{\lambda},$$

c'est la différence de phase entre un rayon ayant subi un nombre quelconque de réflexions et le rayon suivant. Soient T le facteur de transmission supposé commun aux deux surfaces semi-réfléchissantes, R leur facteur de réflexion.

L'éclairement E_1 avec lequel on voit les lames est d'après (36.3)

$$E_1 = \left(\frac{T}{1-R}\right)^2 \frac{1}{1 + A \sin^2 \dfrac{\varphi}{2}} = f(\varphi)$$

avec

$$A = \frac{4R}{(1-R)^2}.$$

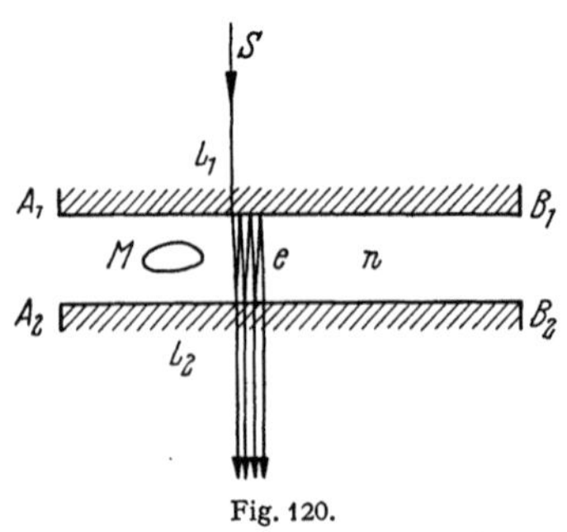

Fig. 120.

Si l'on fait varier e tout en gardant le parallélisme des faces $A_1 B_1$ et $A_2 B_2$, l'éclairement varie en suivant la loi $E_1 = f(\varphi)$.

Si, dans une région M, se trouve un objet transparent d'épaisseur e' et d'indice n' très peu différent de n, la phase en ce point est modifiée. Elle devient

$$\varphi_0 = \frac{4\pi}{\lambda} [n e + (n' - n) e'].$$

Nous supposons que la largeur de l'objet M est suffisamment grande pour que la diffraction n'intervienne pas.

Posons

$$\varphi' = \frac{4\pi (n' - n) e'}{\lambda},$$

on aura

$$\varphi_0 = \varphi + \varphi'.$$

Dans tout ce qui suit, nous admettons que φ' est très petit pour négliger les puissances supérieures à 1. Par conséquent, à l'endroit où se trouve M, l'éclairement devient égal à

$$E_2 = \left(\frac{T}{1-R}\right)^2 \frac{1}{1 + A \sin^2 \left(\dfrac{\varphi + \varphi'}{2}\right)}.$$

Il est différent de E_1, c'est à dire de celui du milieu qui entoure l'objet M. Celui-ci devient visible. La méthode permet donc de transformer les variations de phase en variations d'éclairement.

C'est la méthode interférentielle par teinte plate préconisée par Tolansky. Précisons ces résultats. Puisque φ' est petit, on peut écrire

$$E_2 = E_1 + dE_1,$$

c'est à dire au facteur constant près $(T/1-R)^2$

$$E_2 = \frac{1}{1 + A \sin^2 \dfrac{\varphi}{2}} - \frac{A \sin \varphi}{2\left(1 + A \sin^2 \dfrac{\varphi}{2}\right)^2} \, \varphi'.$$

Posons

$$a = \frac{1}{1 + A \sin^2 \dfrac{\varphi}{2}}, \qquad b = - \frac{A \sin \varphi}{2\left(1 + A \sin^2 \dfrac{\varphi}{2}\right)^2},$$

on aura

$$E_2 = a + b\,\varphi'. \tag{55.1}$$

Les variations d'éclairement sont donc proportionnelles aux variations de phase. Le contraste d'un objet correspondant à la différence de phase φ est donné par

$$\gamma_1 = \frac{E_1 - E_2}{E_1} = \frac{A \sin \varphi}{2\left(1 + A \sin^2 \dfrac{\varphi}{2}\right)} \, \varphi'.$$

Notons que

$$E_1 - E_2 = - \, dE_1$$

et

$$d\varphi = \varphi_0 - \varphi = \varphi'$$

d'ou si $E' = \dfrac{dE_1}{d\varphi}$

$$\gamma_1 = - \, \frac{dE_1}{E_1} = - \, \frac{E'\varphi'}{E_1}.$$

Pour que le contraste soit maximum, il faut donc que E'/E_1 soit maximum, c'est-à-dire que

$$E_1 E'' - E'^2 = 0$$

et, par conséquent, le maximum de visibilité du phénomène ne correspond pas au point d'inflexion de la courbe $E_1 = f(\varphi)$.

Calculons la valeur maximum de ce contraste. Les valeurs de φ et de A doivent satisfaire à l'équation précédente, c'est à dire à

$$\cos \varphi \left(1 + A \sin^2 \frac{\varphi}{2}\right) - \frac{A}{2} \sin^2 \varphi = 0$$

et, si φ est voisin de $2K\pi$,

$$\varphi = \frac{2}{\sqrt{A+2}} + 2K\pi.$$

En reportant cette valeur dans l'expression de γ_1, on trouve

$$\gamma_1 = \frac{A \sqrt{A+2}}{2(A+1)} \, \varphi' \approx \sqrt{\frac{A}{4}} \, \varphi'. \tag{55.2}$$

56. Méthode des franges de superposition. On peut obtenir un résultat analogue par la méthode des franges de superposition telle qu'elle a été préconisée par Sir Thomas Merton. On associe alors aux lames de la Fig. 120 un deuxième système identique, d'épaisseur e'. Si e'/e est égal à un rapport commensurable α/β (α et β étant deux nombres entiers premiers entre eux) on a une certaine teinte. En partant de $e = e'$ pour laquelle on a du blanc, on fait apparaître toutes les couleurs des lames minces en faisant croître $e - e'$.

On peut admettre que l'éclairement suit une loi analogue à celle d'une lame mince unique dont le pouvoir réflecteur serait $R^{\alpha+\beta}$ avec $A' = \dfrac{4 R^{\alpha+\beta}}{(1 - R^{\alpha+\beta})^2}$ et produisant une différence de marche $2n(\beta e' - \alpha e)$. On peut alors faire un calcul identique à celui vient d'être fait dans le cas d'une lame unique.

On trouve un contraste

$$\gamma_2 = \sqrt{\frac{A'}{4}}\,\varphi' \qquad (56.1)$$

avec, pour A' la valeur la plus favorable

$$A' = \frac{4R^2}{(1-R^2)^2}\,.$$

57. Comparaison de ces deux méthodes au contraste de phase. Formons le rapport γ_1/γ_2 d'après (55.2) et (56.1), on a

$$\frac{\gamma_1}{\gamma_2} = \frac{1+R}{\sqrt{R}} \approx 2\,.$$

La sensibilité obtenue par les franges de superposition est donc deux fois moins bonne que par les franges de Fabry-Pérot d'une seule lame.

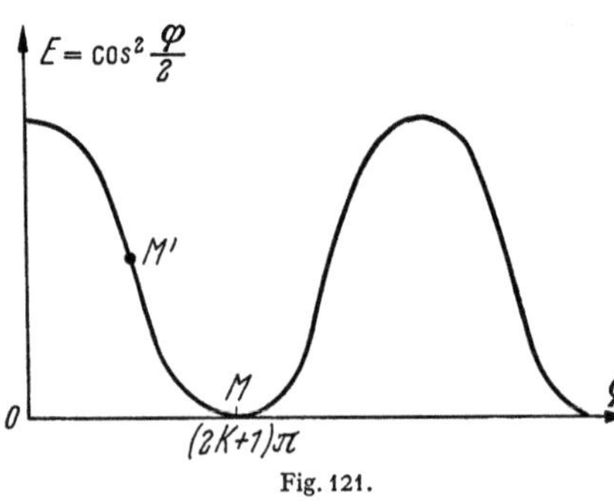

Fig. 121.

Comparons maintenant cette dernière méthode au contraste de phase dont nous parlerons au chapitre sur la diffraction. En admettant qu'il soit possible d'observer le même objet par contraste de phase, on peut comparer les valeurs du contraste obtenues par les deux méthodes. Si d est la densité optique de la lame de phase, on verra que dans le cas du contraste de phase, le contraste est donné par

$$\gamma = 10^{d/2}\,\varphi'$$

car le déphasage φ' correspond à une double traversée de l'objet dans la méthode interférentielle et à une seule traversée dans la méthode du contraste de phase.

Les deux méthodes donneront la même sensibilité si

$$\sqrt{\frac{A}{4}} = 10^{d/2},$$

d'où

$$d = \mathrm{Log}\,\frac{A}{4}\,.$$

En prenant $A = 2500$, ce qui correspond à un facteur de réflexion des lames égal à 0,96, on obtient

$$d = 2,8\,.$$

Il est très facile d'obtenir des lames de phase de densité 2, mais l'obtention de facteurs de réflexion de l'ordre de 0,96 ne peut se faire qu'au moyen de couches multiples si l'on ne veut pas perdre trop de lumière. Il sera donc plus difficile d'obtenir par le procédé interférentiel une sensibilité aussi grande que par les lames de phase de forte densité optique dépassant largement 3.

Toutefois, dans le cas où les franges sont des franges à deux ondes et non à ondes multiples, comme par exemple dans l'interféromètre de Michelson, on obtient, tout au mois théoriquement, le maximum de sensibilité. Ceci est évident sur la Fig. 121 qui représente les variations d'intensité des franges ordinaires. Si l'on règle l'épaisseur, c'est-à-dire la différence de phase de telle sorte que le point représentatif se trouve en M, le champ est noir et un objet correspondant à la phase φ' se détache en clair sur fond noir; on a nécessairement un contraste ègal à 1. On a une image analogue à une image strioscopique. On peut alors effectuer la comparaison de ces différentes méthodes comme le montre le tableau ci-dessous.

Tableau 3.

Méthode	Eclairement de l'image proportionnel à:	Analogie
1. Une lame FABRY-PÉROT . . .	φ'	
2. Franges de superposition (2 lames FABRY-PÉROT) . . .	φ'	contraste de phase
3. Interférences ordinaires . . .	φ'^2	strioscopie

58. Influence du diamètre apparent de la source. Dans tout ce qui précède, le diamètre apparent de la source était supposé négligeable. En fait, il n'en est pas ainsi dans la pratique et la source a toujours une étendue finie. L'étude de

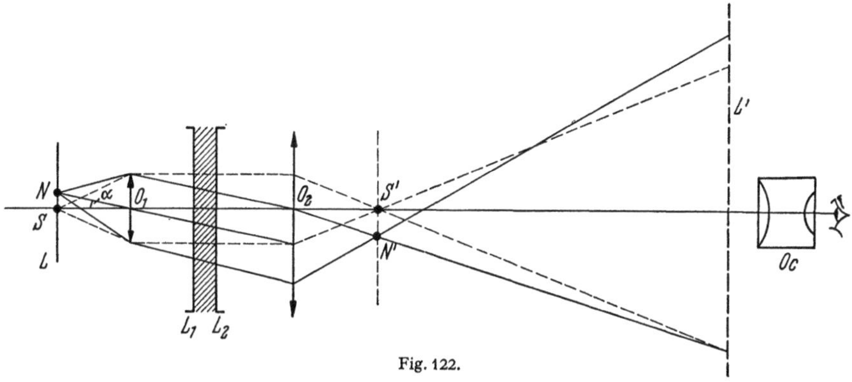

Fig. 122.

l'influence du diamètre apparent de la source sur la visibilité du phénomène est importante, car elle permettra de préciser les conditions optima pour l'observation des objets de phase.

Les lames semi-réfléchissantes étant observées par transmission, nous réalisons le montage schématique de la Fig. 122.

La source lumineuse monochromatique est en S dans le plan focal d'une lentille O_1. Les faisceaux de lumière parallèle traversent les lames $L_1 L_2$ et tombent sur l'objectif O_2. L'objectif O_2 donne en L' une image des lames $L_1 L_2$, image observée au moyen de l'oculaire O_c. L'ensemble de O_2 et de O_c constitue un viseur permettant d'observer les objets de phase situés entre les lames L_1 et L_2. L'image S' de la source se forme dans le plan focal de l'objectif O_2 dont la focale est toujours petite vis-à-vis de la distance qui le sépare de l'oculaire. Dans ces conditions, l'image de la source se forme sur la pupille de l'oeil et le champ est uniformément éclairé.

En supposant les lames $L_1 L_2$ normales à l'axe $O_1 O_2$, la différence de phase pour une source ponctuelle placée en N est

$$\Phi = \frac{4\pi n e \cos r}{\lambda}.$$

L'éclairement obtenu est différent suivant que les lames sont éclairées par la source N ou par la source S. Si on utilise une source étendue, l'éclairement résultant est la somme des éclairements dûs à chaque source élémentaire composant la source étendue. Ici, lorsque l'on fait varier l'écartement des lames $L_1 L_2$, le passage par les maxima brillants se fait d'une façon beaucoup moins brusque.

Le passage entre un maximum d'éclairement et un minimum se fait d'une façon plus lente. Une faible variation de phase n'entraîne plus qu'une faible variation d'éclairement, la sensibilité de la méthode diminue. Nous allons calculer le diamètre maximum à donner à la source pour que la baisse de contraste ne soit pas trop importante. La source est une source circulaire de centre S et de rayon angulaire ϑ vu du point O_1.

Comme les angles ne sont pas trop grands, on peut écrire

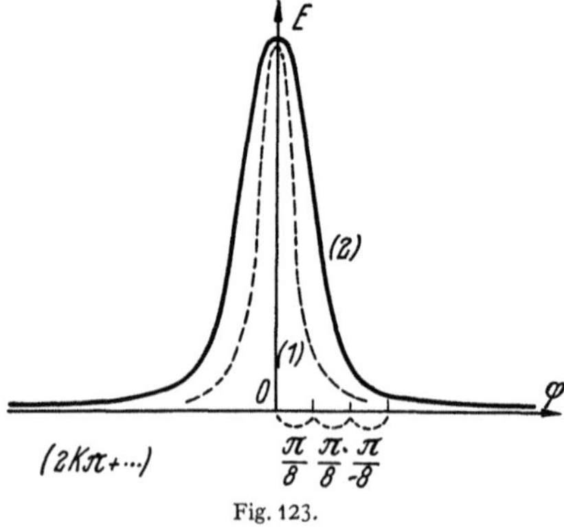

Fig. 123.

$$\Phi = \frac{4\pi n e \cos r}{\lambda} \approx \frac{4\pi n e}{\lambda}\left(1 - \frac{r^2}{2}\right).$$

Or, à l'angle de réfraction r dans le milieu n, correspond l'angle d'incidence α tel que

$$\alpha = n r$$

d'où en posant $\varphi = 4\pi n e/\lambda$

$$\Phi = \varphi - \frac{2\pi e \alpha^2}{n \lambda}.$$

Pour tous les points d'une petite couronne de rayon $\overline{NS} = \overline{SO_1} \cdot \alpha$ et de centre S, la différence de phase est constante et égale à la valeur précédente, d'où l'éclairement dû à cette petite couronne et à un facteur constant près

$$E = \frac{2\pi \alpha\, d\alpha}{1 + A \sin^2\left(\dfrac{\varphi}{2} - \dfrac{\pi \alpha^2 e}{n \lambda}\right)}.$$

Enfin, l'éclairement dû à la somme des couronnes élémentaires composant la source sera

$$E = \int_0^{\vartheta} \frac{2\pi \alpha\, d\alpha}{1 + A \sin^2\left(\dfrac{\varphi}{2} - \dfrac{\pi \alpha^2 e}{n \lambda}\right)}.$$

On peut écrire

$$E = -\frac{n \lambda}{e} \int_{\alpha=0}^{\alpha=\vartheta} \frac{d\left(\dfrac{\varphi}{2} - \dfrac{\pi \alpha^2 e}{n \lambda}\right)}{1 + A \sin^2\left(\dfrac{\varphi}{2} - \dfrac{\pi \alpha^2 e}{n \lambda}\right)}.$$

On obtient, en posant

$$p' = \frac{\pi \vartheta^2 e}{n \lambda},$$

$$E = \frac{n \lambda}{e \sqrt{A+1}} \left[\arctan\left(\sqrt{A+1}\tan\frac{\varphi}{2}\right) - \arctan\left\{\sqrt{A+1}\tan\left(\frac{\varphi}{2} - p'\right)\right\}\right].$$

Pour $A = 150$ ($R = 0{,}85$), la courbe (2) de la Fig. 123 donne la loi de variation de l'intensité au voisinage d'un maximum pour $p' = \pi/8$.

La courbe 1 représente la loi de variation lorsque la source est infiniment fine. Les maxima ne se produisent pas au même endroit et les intensités ne sont par les mêmes, mais nous avons rassemblé les deux courbes pour comparer leur structure du point de vue des formes. On constate un élargissement qui se traduira nécessairement par une baisse de contraste.

La Fig. 124 montre des courbes analogues pour $A = 2500$ avec $p' = \pi/8$ (courbe 2), $p' = \pi/20$ (courbe 3), $p' = \pi/30$ (courbe 4). La courbe 1 représente le cas de la source infiniment fine.

Calculons pour ces différents cas, le contraste optimum qu'il est possible d'obtenir. On a, si $a = \sqrt{A+1}$,

$$\gamma = \frac{E'}{E}\,\varphi' = \frac{\dfrac{a}{2}\Big/\cos^2\dfrac{\varphi}{2}}{1 + a^2\tan^2\dfrac{\varphi}{2}} - \dfrac{\dfrac{a}{2}\Big/\cos^2\left(\dfrac{\varphi}{2}-p'\right)}{1 + a^2\tan^2\left(\dfrac{\varphi}{2}-p'\right)} \Big/ \left[\arctan\left(a\tan\dfrac{\varphi}{2}\right) - \arctan\left[a\tan\left(\dfrac{\varphi}{2}-p'\right)\right]\right]\,\varphi'.$$

Pour une valeur déterminée de $a = \sqrt{A+1}$ il est possible de tracer la courbe γ en fonction de φ et chercher le maximum de γ pour une valeur quelconque, mais fixée, de φ'.

Prenons par exemple $A = 150$ et $p' = \pi/8$, on trouve comme contraste maximum

$$\gamma = 5\varphi'$$

pour

$$\varphi = 2K\pi + \frac{9\pi}{32},$$

c'est-à-dire

$$n\,e = K\frac{\lambda}{2} + \frac{9\lambda}{128}.$$

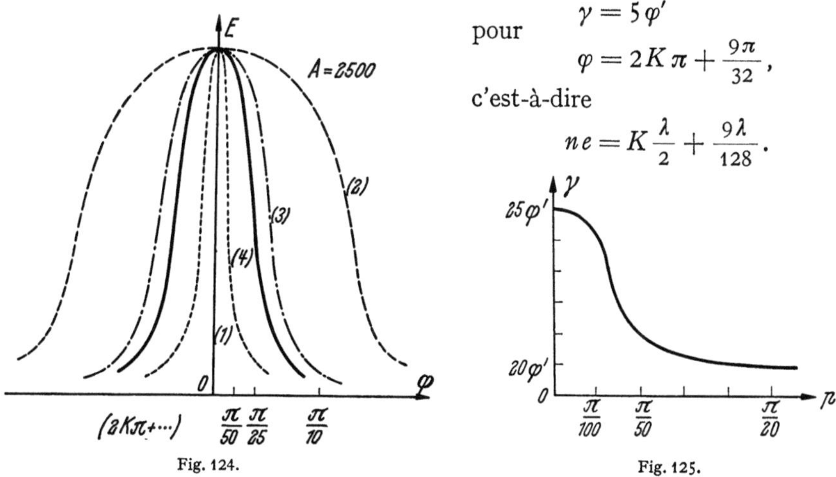

Fig. 124.   Fig. 125.

La source infiniment fine donnant $\gamma = 6\varphi'$, la baisse de contraste est faible mais cette valeur de A ne donne que très peu de sensibilité.

Le tableau suivant et la courbe de la Fig. 125 donnent, pour $A = 2500$, les contrastes maxima et les valeurs correspondantes de φ et ne pour différentes ouvertures de la source.

Tableau 4.

	Source infiniment fine	$p' = \dfrac{\pi}{100}$	$p' = \dfrac{\pi}{50}$	$p' = \dfrac{\pi}{30}$	$p' = \dfrac{\pi}{20}$	$p' = \dfrac{\pi}{8}$
γ	$25\varphi'$	$24\varphi'$	$21\varphi'$	$20,5\varphi'$	$20\varphi'$	$16\varphi'$
$\varphi - 2K\pi$	$\dfrac{3\pi}{100}$	$\dfrac{\pi}{36}$	$\dfrac{\pi}{20}$	$\dfrac{13\pi}{180}$	$\dfrac{19\pi}{180}$	$\dfrac{21\pi}{180}$
$ne - K\dfrac{\lambda}{2}$	$\dfrac{3\lambda}{800}$	$\dfrac{\lambda}{144}$	$\dfrac{\lambda}{80}$	$\dfrac{13\lambda}{720}$	$\dfrac{19\lambda}{720}$	$\dfrac{21\lambda}{320}$

On voit que la baisse de contraste est assez importante pour des valeurs de p' dépassant $\pi/100$.

Si l'on veut utiliser au mieux la finesse donnée par des lames à haut facteur de réflexion, le diamètre apparent de la source devra satisfaire à l'équation

$$\frac{\pi \vartheta^2 e}{n \lambda} \leq \frac{\pi}{100},$$

c'est à dire

$$\vartheta \leq \sqrt{\frac{n \lambda}{100 e}}.$$

D'où les valeurs de ϑ pour quelques épaisseurs e en prenant $n = 1,5$ et $\lambda = 0,6\,\mu$

$$e = 0,01 \text{ mm.} \quad 2\vartheta \leq 1° \, 43',$$

$$e = 0,10 \text{ mm.} \quad 2\vartheta \leq \quad 32',$$

$$e = 1 \text{ mm.} \quad 2\vartheta \leq \quad 10'.$$

On aura donc généralement des sources petites.

Dans le cas des franges de superposition, le calcul se conduira de la même façon puisque l'on peut admettre que l'on a affaire à une seule lame Fabry-Pérot de pouvoir réflecteur $R^{\alpha+\beta}$.

Pour les franges ordinaires à deux ondes, la Fig. 121 montre que si la source n'est pas très fine, le contraste diminuera rapidement. En se plaçant en un point M' correspondant au maximum de contraste, on obtient

$$\gamma_3 = \frac{2 \sin p' \sqrt{1 - \dfrac{\sin^2 p'}{4 p'^2}}}{4 p' - \dfrac{\sin^2 p'}{p'}} \, \varphi'.$$

Si p reste petit, on a alors

$$\gamma_3 = \frac{\varphi'}{\sqrt{3}},$$

ce qui donne les valeurs suivantes pour le contraste

Tableau 5.

	Source infiniment fine	$p' = \dfrac{\pi}{100}$	$p' = \dfrac{\pi}{50}$
Une seule lame Fabry-Pérot $A = 2500$	$25\,\varphi'$	$24\,\varphi'$	$21\,\varphi'$
Franges ordinaires à 2 ondes	1	$\dfrac{\varphi'}{\sqrt{3}}$	$\dfrac{\varphi'}{\sqrt{3}}$

59. Influence du chromatisme de la source. Dans tout ce qui précède, nous avons supposé que les radiations émises par la source étaient rigoureusement monochromatiques. En réalité, il n'en est jamais ainsi et les différentes radiations monochromatiques émises par la source produisent des effets analogues aux différents éléments d'une source étendue monochromatique. Nous admettrons maintenant que la source est une source ponctuelle émettant des radiations comprises dans un petit domaine spectral $d\lambda$. Nous venons de voir qu'une variation de phase $2p' = \pi/50$ due à une variation d'inclinaison des rayons pouvait être tolérable. On peut considérer maintenant que la variation de phase est produite, non plus par une inclinaison des rayons, mais bien par un défaut de monochromatisme de la source.

Admettre une variation de phase de $\pi/50$, c'est admettre une variation dp de l'ordre d'interférence p égale à

$$dp \leq \frac{1}{100},$$

d'après (27.1), on écrira

$$p\frac{d\lambda}{\lambda} \leq \frac{1}{100},$$

soit

$$p \leq \frac{1}{100}\frac{\lambda}{d\lambda}.$$

Le coefficient de finesse $\lambda/d\lambda$ de la radiation utilisée doit donc satisfaire à la relation

$$\frac{\lambda}{d\lambda} \geq \frac{200\,n\,e}{\lambda},$$

en prenant $n = 1{,}5$, on a les valeurs suivantes pour $\lambda = 0{,}6\,\mu$:

$$
\begin{aligned}
e &= 0{,}01 \text{ mm.} \quad \lambda/d\lambda \geq 5\,000,\\
e &= 0{,}10 \text{ mm.} \quad \lambda/d\lambda \geq 50\,000,\\
e &= 1 \text{ mm.} \quad \lambda/d\lambda \geq 500\,000.
\end{aligned}
$$

d) Application des interférences à la mesure des longueurs.

L'application des interférences à la mesure des longueurs fait connaître la longueur cherchée par un certain nombre de longueurs d'onde d'une radiation connue. On passe ensuite de ce nombre à la valeur en unités ordinaires (mètre, centimètre) avec toute la précision voulue. Les longueurs que l'on a à mesurer sont de deux espèces: les longueurs à bouts et les longueurs à traits. Les longueurs à bouts définies par la distance de deux surfaces se présentent naturellement dans la mesure des dimensions des objets. Ces déterminations sont assez courantes dans l'industrie, par exemple la vérification des jauges ou calibres. Les longueurs à traits définies par la distance de deux traits fins tracés sur une surface polie sont employés dans la construction des étalons fondamentaux de longueur. La précision avec laquelle la longueur peut être définie dépend de la finesse et de la régularité des traits.

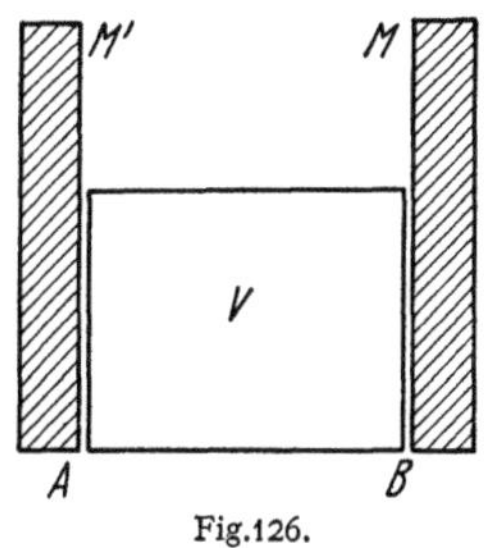
Fig.126.

60. Mesure interférentielle des longueurs à bouts. Soit V le corps solide dont l'épaisseur $A\,B$ constitue la longueur à bouts à mesurer (Fig. 126). Les faces A et B étant très bien polies, on applique la surface B contre un miroir plan M et la face A contre un miroir plan M'. Les faces M et M' étant semi-réfléchissantes, on règle leur parallélisme au moyen des images par réflexion multiples. On parfait le réglage en observant les anneaux à l'infini de la lame d'air comprise entre M et M' et si, en utilisant différentes portions de cette lame les anneaux à l'infini restent immobiles, le réglage est correct. La mesure de l'épaisseur du corps solide entre ses faces A et B se ramène alors à la détermination de l'ordre d'interférence au centre des anneaux à l'infini de la lame d'air comprise entre M et M'.

Naturellement, il faudra tenir compte des lames d'air comprises entre M et B d'une part et M' et A d'autre part. L'épaisseur de ces lames sera mesurée au moyen de la lame étalon (Sect. 43) par l'observation des franges de superposition. Soit $p + \varepsilon$ l'ordre d'interférence au centre des anneaux à l'infini de la lame d'air comprise entre M et M'; aux épaisseurs près des lames d'air AM' et BM' qui sont connues, on a

$$p + \varepsilon = \frac{2e}{\lambda}.$$

L'ordre d'interférence correspondant au premier anneau à partir du centre étant p, l'ordre d'interférence correspondant au $N^{\text{ième}}$ anneau sera $p - (N - 1)$ et on aura, si α_N est son rayon angulaire

$$p - (N - 1) = (p + \varepsilon)\cos\alpha_N$$

d'où, comme α_N est petit

$$\varepsilon = p\,\frac{\alpha_N^2}{2} - (N - 1).$$

Dans cette expression figure l'ordre d'interférence p du premier anneau, c'est-à-dir ela partie entière de l'ordre d'interférence au centre dont nous allons indiquer différents procédés de mesure.

61. Mesure de la partie entière de l'ordre d'interférence au centre. Pour déterminer l'ordre d'interférence correspondant à une frange donnée, on connaît plusieurs méthodes. Nous indiquerons les trois méthodes suivantes:

α) méthode des excédents fractionnaires,

β) méthode des coïncidences,

γ) méthode des franges de superposition.

α) *Méthode des excédents fractionnaires.* On éclaire la lame d'air comprise entre les plans M et M' avec une radiation monochromatique de longueur d'onde λ. L'ordre d'interférence au centre des anneaux à l'infini est

$$p + \varepsilon = \frac{2e}{\lambda},$$

d'où

$$e = (p + \varepsilon)\,\frac{\lambda}{2}.$$

Si on éclaire le système avec une deuxième radiation monochromatique, de longueur d'onde λ', on aura

$$e = (p' + \varepsilon')\,\frac{\lambda'}{2}.$$

En utilisant ainsi plusieurs radiations λ, λ', λ'' etc. ..., on pourra écrire les égalités suivantes

$$p' + \varepsilon' = (p + \varepsilon)\,\frac{\lambda}{\lambda'},$$

$$p'' + \varepsilon'' = (p + \varepsilon)\,\frac{\lambda}{\lambda''}$$

$$\cdot \quad \cdot \quad \cdot \quad \cdot \quad \cdot \quad \cdot \quad \cdot \quad \cdot$$

On peut déterminer $\varepsilon\ \varepsilon'\ \varepsilon''$ ... en mesurant les diamètres de quelques anneaux de chaque système; d'autre part, on connaît toujours une valeur très approchée de la grandeur à mesurer par des moyens mécaniques (cales Johannsen par exemple), on a donc déjà la valeur de p avec une bonne approximation. On peut enfin calculer $(p + \varepsilon)\,\frac{\lambda}{\lambda'}$, $(p + \varepsilon)\,\frac{\lambda}{\lambda''}$, etc. ..., c'est-à-dire $p' + \varepsilon'$, $p'' + \varepsilon''$, ... On compare ces valeurs ainsi calculées aux valeurs expérimentales et on voit si les parties fractionnaires ainsi calculées sont identiques aux parties fractionnaires expérimentales.

Considérons l'exemple suivant: on utilise les raies du cadmium dont les longueurs sont en angströms:

raie rouge   6438,4696
raie verte   5085,8220
raie bleue   4799,9088
raie violette   4678,1500.

La mesure du diamètre des anneaux à l'infini correspondant à la lame étudiée, donne les valeurs suivantes:

raie rouge $\varepsilon = 0,82$
raie verte $\varepsilon' = 0,00$
raie bleue $\varepsilon'' = 0,79$
raie violette $\varepsilon''' = 0,93$.

Par des moyens mécaniques, on sait que l'ordre d'interférence au centre des anneaux rouges est voisin de $p = 31\,050$. On dresse alors le tableau suivant: on donne à p différentes valeurs voisines de $31\,050$ et on calcule les quantités

$$p' + \varepsilon' = (p + \varepsilon)\frac{\lambda}{\lambda'} \qquad \text{(raie verte)},$$

$$p'' + \varepsilon'' = (p + \varepsilon)\frac{\lambda}{\lambda''} \qquad \text{(raie bleue)},$$

$$p''' + \varepsilon''' = (p + \varepsilon)\frac{\lambda}{\lambda'''} \qquad \text{(raie violette)}.$$

Tableau 6.

Raie rouge $p + \varepsilon$	Raie verte $(p + \varepsilon)\,\lambda/\lambda'$	Raie bleue $(p + \varepsilon)\,\lambda/\lambda''$	Raie violette $(p + \varepsilon)\,\lambda/\lambda'''$
31 048,82	39 306,70	41 648,06	42 732,04
49,82	07,97	49,40	33,41
50,82	09,23	50,74	34,79
51,82	10,50	52,08	36,17
52,82	11,76	53,42	37.54
53,82	**13,03**	**54,77**	**38,92**
54,82	14,30	56,11	40,30
55,82	15,56	57,45	41,67
56,82	16,83	58,79	43,05
57,82	18,09	60,13	44,42
58,82	19,36	61,47	45,80
59,82	20,63	52,81	47,18

Parmi toutes ces valeurs calculées, on voit que la valeur de $p + \varepsilon = 31\,053,82$ donne pour les autres raies des ordres d'interférence dont les parties fractionnaires correspondent bien aux valeurs expérimentales. Ce sont ces valeurs qu'il faut choisir. Cette méthode s'applique de la même façon pour des interférences à deux ondes ou à ondes multiples, pour les franges d'égale épaisseur ou pour les anneaux à l'infini.

β) Méthode des coïncidences. Cette méthode s'applique seulement au cas des franges à ondes multiples. On éclaire la lame à étudier simultanément au moyen de deux radiations de longueurs d'onde λ et λ'. La période des coïncidences est $\dfrac{\lambda\lambda'}{\lambda' - \lambda}$ (voir: Sect. 38) et comme on connaît toujours l'épaisseur à mesurer, donc δ, avec une bonne approximation on compare cette valeur aux valeurs obtenues par le calcul

$$\delta_1 = \frac{\lambda\lambda'}{\lambda'-\lambda}, \qquad \delta_2 = 2\,\frac{\lambda\lambda'}{\lambda'-\lambda}, \qquad \delta_3 = 3\,\frac{\lambda\lambda'}{\lambda'-\lambda}, \text{ etc.}$$

La valeur de δ qu el'on doit choisir est celle qui se rapproche le plus de la valeur connue d'avance par des moyens mécaniques. Cette méthode, très simple, n'exige aucune mesure, le simple examen de l'aspect des franges suffit. Ajoutons qu'elle s'applique très bien aux franges d'égale épaisseur, au coin d'air par exemple, pourvu que l'angle ne soit pas trop petit afin d'avoir assez de franges pour observer les coïncidences. Elle s'applique moins bien aux anneaux à l'infini lorsque l'on considère une lame d'épaisseur invariable, car on ne peut voir qu'un petit nombre d'anneaux et si on s'éloigne du centre, ils deviennent rapidement trop serrés pour être observables.

γ) *Méthode des franges de superposition.* Si on emploie la lame étalon (Sect. 43), on étudie les franges de superposition entre la lame à étudier et la lame étalon.

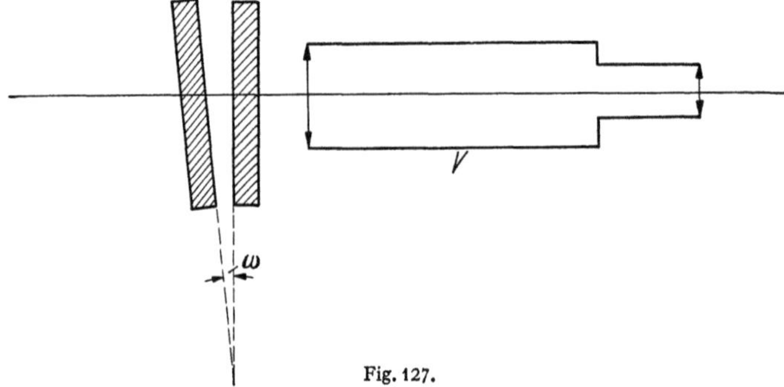

Fig. 127.

L'observation de la frange centrale blanche permettra de connaître immédiatement le rapport des épaisseurs. Connaissant déjà avec une bonne approximation l'épaisseur à mesurer, il sera facile de trouver ce rapport sans ambiguïté et d'avoir ainsi l'épaisseur cherchée avec une grande précision.

Pour faire ces mesures, il est nécessaire d'étalonner au préalable la lame étalon, comme il a été vu à la Sect. 43.

On peut aussi employer l'interféromètre Fabry-Pérot en lui donnant une épaisseur très voisine de l'épaisseur à mesurer par des moyens mécaniques (cales Johannsen par exemple). On dispose les 2 lames perpendiculairement à l'axe optique d'une lunette V réglée sur l'infini (Fig. 127) et on éclaire en lumière blanche.

En inclinant la lame dont l'épaisseur est la plus grande, on voit apparaître les franges en lumière blanche pour une certaine inclinaison. On amène la frange centrale blanche ($\delta = 0$) sur le réticule. On a, pour cette frange, d'après la Sect. 42

$$r_1 = 0, \qquad r_2 = \omega$$

or

$$\delta = 2e_1 \cos r_1 - 2e_2 \cos r_2,$$

d'où

$$\delta = 2e_1 - 2e_2 \cos \omega$$

et

$$e_1 = e_2 \cos \omega.$$

62. Mesure de l'épaisseur d'une lame mince par la méthode des franges d'égale inclinaison. Lorsque l'épaisseur de la lame à étudier n'est pas trop grande, on peut obtenir directement l'ordre d'interférence au centre, partie entière et partie fractionnaire, en mesurant simplement les diamètres angulaires d'anneaux éloignés du centre.

Soient p l'ordre d'interférence correspondant au premier anneau et $p + \varepsilon$ l'ordre d'interférence au centre. Si on considère le $N^{\text{ième}}$ anneau, l'ordre d'interférence qui lui correspond est $p - (N-1)$.

Mesurons les diamètres angulaires α et β de ces deux anneaux, on a

$$p = (p + \varepsilon) \cos \frac{\alpha}{2}, \quad p - (N-1) = (p + \varepsilon) \cos \frac{\beta}{2}.$$

De ce système de deux équations à deux inconnues, on peut tirer p et ε. On aura

$$p = \frac{\cos \dfrac{\alpha}{2}}{\cos \dfrac{\alpha}{2} - \cos \dfrac{\beta}{2}} \, (N-1).$$

Fabry a employé cette méthode pour mesurer l'épaisseur d'un étalon de 2,5 mm. environ. Voici les valeurs trouvées

1. diamètre du premier anneau $\alpha = 0°\ 57'\ 28''$
2. diamètre du 21ème anneau $\beta = 7°\ 37'\ 48''$
3. diamètre du 89ème anneau $\beta = 15°\ 55'\ 8''$
4. diamètre du 313ème anneau $\beta = 29°\ 59'\ 44''$.

Les mesures n° 2, 3, 4, associées successivement à la mesure n° 1, donnent les résults suivants

$$p = 9169{,}8$$
$$p = 9167{,}5$$
$$p = 9168{,}3$$

La valeur exacte obtenue par la méthode des coïncidences est $p = 9169$.

Pour éviter de compter 89 ou 313 franges, on peut procéder ainsi: on mesure le diamètre du 21ème anneau par exemple, ce qui donne une valeur approchée de p, soit p_A. On peut écrire

$$p_A = \frac{\cos \dfrac{\alpha}{2}}{\cos \dfrac{\alpha}{2} - \cos \dfrac{\beta}{2}} \, (N-1).$$

On pointe ensuite des franges quelconques éloignées du centre et l'équation précédente donne pour $N-1$ une valeur assez approchée pour que ce nombre entier soit déterminé sans ambiguïté. La valeur entière de $N-1$ ainsi trouvée, permet d'obtenir dans l'équation précédente une nouvelle valeur de p plus approchée.

63. Mesure des longueurs à traits. Le trait définissant l'extrémité d'une longueur ne peut être localisé que par un pointé transversal effectué au microscope. Il faut donc pouvoir relier la mesure interférentielle à la mesure réalisé par pointés.

Fabry et Pérot ont réalisé un étalon interférentiel à traits au moyen de l'interféromètre à lames argentées. Pour cela, on polit et on argente la tranche supérieure des glaces de verre; sur chacune d'elles, on trace parallèllement à la face argentée des traits très fins presque au bord de la face. Un des couples de traits pris sur les deux faces constitue ainsi une «longueur à trait». Cette longueur peut être comparée à la longueur à étudier, qui n'en diffère que d'une très petite quantité, au moyen du comparateur. La somme des distances des traits aux faces étant mesurée une fois pour toutes, en faisant varier la distance des deux glaces, on obtient un étalon à traits de longueur connue.

Voici comment Fabry et Pérot ont mesuré les distances qui sont comprises entre les surfaces et les traits tracés sur les tranches des glaces.

Au moyen des deux glaces de l'étalon, on construit successivement deux étalons tels que la distance des traits de l'un soit égale au double de la distance des traits de l'autre.

Soient A le nombre de demi-longueurs d'onde contenues dans la somme des distances comprises entre les surfaces et les traits tracés sur la tranche des glaces, p_1 et p_2 les ondes d'interférence de ces deux étalons. Les distances des traits des deux étalons seront

$$(p_1 + A)\,\frac{\lambda}{2}\,, \qquad (p_2 + A)\,\frac{\lambda}{2}$$

et

$$2(p_1 + A)\,\frac{\lambda}{2} = (p_2 + A)\,\frac{\lambda}{2}\,,$$

d'où

$$A = p_2 - 2p_1.$$

Pour réaliser la condition indiquée plus haut, on procède de la façon suivante: considérons une règle auxiliaire sur laquelle on a tracé 3 traits séparés par des distances x_1 et x_2 très voisines de la longueur du plus petit étalon. Au moyen du comparateur, confrontons celle-ci à la longueur y_1, on aura

$$(p_1 + A)\,\frac{\lambda}{2} = x_1 + \varepsilon_1,$$

ε_1 étant très petit, de l'ordre de quelques microns. Exprimons x_1 et ε_1 en longueurs d'onde en posant

$$x_1 = N_1\,\frac{\lambda}{2}\,, \qquad e_1 = n_1\,\frac{\lambda}{2}\,,$$

on aura

$$p_1 + A = N_1 + n_1.$$

Comparons la longueur de ce même étalon à la longueur x_2, on obtient de même

$$p_1 + A = N_2 + n_2.$$

Comparons maintenant l'étalon de longueur double à la distance $x_1 + x_2$ des deux traits extrêmes de la règle, on aura

$$p_2 + A = N_1 + N_2 + n_3.$$

n_3 étant une petite quantité du même ordre que n_1 et n_2. Les trois équations précédentes permettent d'éliminer N_1 et N_2; on obtient

$$A = p_2 - 2p_1 + n_1 + n_2 - n_3.$$

p_1 et p_2 sont connus; n_1, n_2, n_3 sont obtenus en divisant les petites longueurs correspondantes $\varepsilon_1\,\varepsilon_2\,\varepsilon_3$ (mesurées au comparateur) par une valeur approchée de la demi-longueur d'onde.

64. Mesure du mètre international en longueurs d'onde, méthode de Michelson. La première détermination expérimentale de la longueur du mètre en longueurs d'onde a été faite par Michelson en 1892 au Bureau international des Poids et Mesures.

Pour mesurer le mètre international en longueur d'onde, Michelson a constitué une série de 9 étalons de longueur, formés par des surfaces réfléchissantes (Fig. 128). La longueur l d'un étalon est celle qui sépare les deux miroirs superposés M_1 et M_2. Les longueurs des étalons sont:

étalons N°	1	2	3	4	5	6	7	8	9
longueurs l en mm.	0,39	0,7	1,5	3,1	6,2	12,5	25	50	100

Les opérations à réaliser sont les suivantes:

α) mesure directe de l'étalon 1 en longueurs d'onde,

β) mesure de la longueur de l'étalon 9 par comparaisons successives des étalons,

γ) comparaison de l'étalon 9 au mètre international.

α) *Mesure directe de l'étalon 1.* On place l'étalon 1 constitué par les 2 surfaces M_1 et M_2 sur l'interféromètre de MICHELSON (Fig. 129). On incline très légèrement le miroir M_3 de façon que son image M_3' par rapport à la séparatrice G fasse un petit angle avec M_1 et M_2. Lorsque M_1 coupe M_3', on voit les franges en lumière

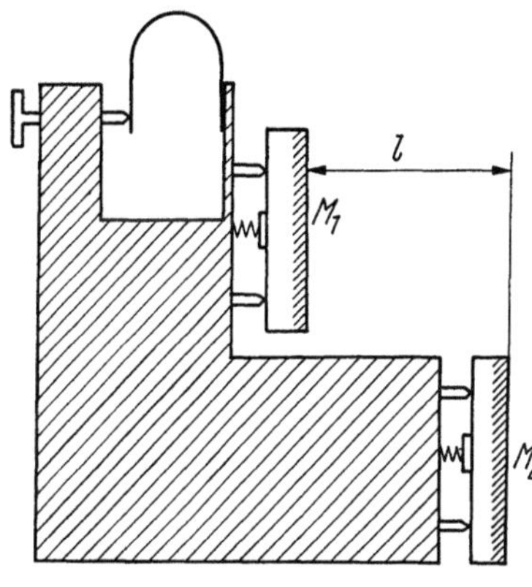

Fig. 128.

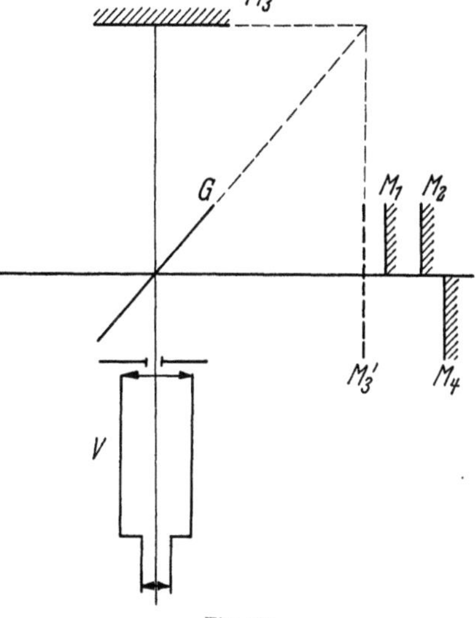

Fig. 129.

blanche puisque la différence de marche est très faible. On repère la frange noire au moyen d'un quadrillage tracé sur M_3. Déplaçons M_3 parallèlement à lui-même jusqu'au moment où M_3' coupe M_2, la frange noire occupant la même position. Le miroir M_3 s'est déplacé d'une longueur égale exactement à la longueur l de l'étalon. Pour compter le nombre de franges qui ont défilé, on observe les anneaux à l'infini de la lame d'air comprise entre M_3' et un miroir auxiliaire M_4 rigoureusement parallèle à M_3'. On peut observer en même temps les anneaux à l'infini et le quadrillage, c'est à dire les franges localisées au voisinage de M_3' en diaphragmant fortement le viseur V utilisé pour viser le quadrillage. L'aspect du champ d'observation est celui de la Fig. 130: il est divisé en 3 plages correspondant aux 3 miroirs $M_1 M_2 M_3$. Dans le cas de la Fig. 130, le miroir M_3 est disposé pour faire apparaître les franges d'égale épaisseur en lumière blanche sur le miroir M_4. On éclaire alors en lumière

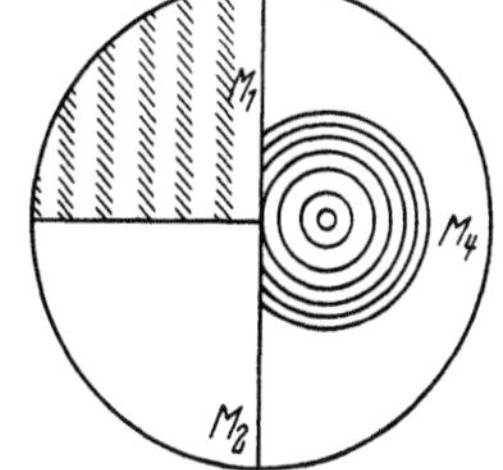

Fig. 130.

monochromatique avec la radiation dont on veut comparer la longueur d'onde à la longueur l de l'étalon 1. On compte le nombre d'anneaux qui défilent jusqu' à ce que le rétablissement de la lumière blanche montre les franges d'égale épaisseur sur le miroir M_2, la frange centrale noire occupant la même position sur le quadrillage qu'au moment du pointé sur M_1. Le nombre des anneaux qui défilent est très grand c'est un nombre entier plus une partie fractionnaire. On vérifie la partie entière par la méthode des excédents fractionnaires en étudiant les anneaux donnés par plusieurs radiations.

β) *Comparaison des étalons.* Pour comparer l'étalon 1 (miroirs $M_1 M_2$) à l'étalon 2 (miroirs $M_1' M_2'$), on les dispose côte à côte (Fig. 131). L'étalon 1 est porté par un chariot mobile et ses faces sont parallèles aux faces de l'étalon 2. On oriente M_3 de façon que son image M_3' fasse un petit angle avec les 4 miroirs $M_1 M_2$ $M_1' M_2'$. En déplaçant M_3 et l'étalon 1 parallèlement à eux-mêmes, on peut faire couper simultanément M_1 et M_1' par M_3' et voir les franges en lumière blanche.

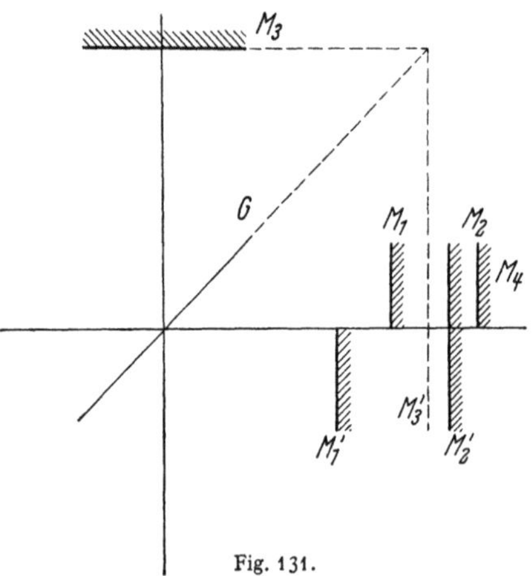

Fig. 131.

On fait coïncider la frange noire visible sur M_1 et M_1' avec un repère du quadrillage. On déplace ensuite M_3 parallèlement à lui-même de façon que M_3' coupe M_2' d'où l'apparition des franges en lumière blanche. La position de M_3' est réglée exactement pour que la frange noire soit sur le repère. On déplace maintenant l'étalon 1 parallèlement à lui-même pour que M_1 prenne la place de M_2. Il en est ainsi lorsque la frange noire apparaît sur le repère. A son tour, le miroir M_3 est déplacé pour que son image M_3' coupe M_2 toujours dans les mêmes conditions. Le plan M_3' s'est donc déplacé d'une longueur exactement égale au double de la longueur l de l'étalon 1. Comme l'étalon 2 n'est pas tout à fait égal au double de l'étalon 1, M_2 n'étant pas exactement dans le même plan que M_2', il reste à mesurer la petite différence de longueur entre ces 2 étalons. On observe alors les anneaux à l'infini de la lame $M_3' M_4$ en enlevant l'étalon 1, le miroir M_4 étant rigoureusement parallèle à M_3'. Il suffit de compter les anneaux qui défilent lorsque M_3' se déplace de la position précédente à la position qui fait apparaître les franges en lumière blanche avec M_2'. Au moment où la frange noire apparaît en M_2' sur le repère, il y a contact entre M_3' et M_2'. La lame $M_3' M_4$ a diminué un peu d'épaisseur et le nombre des anneaux qui ont disparu au centre, donne la différence entre l'étalon 2 et le double de l'étalon 1. La longueur de l'étalon 2 est donc connue en fonction de la longueur d'onde utilisée pour observer les anneaux à l'infini. De proche en proche, on finit par arriver à l'étalon 9, dont on obtient ainsi la longueur.

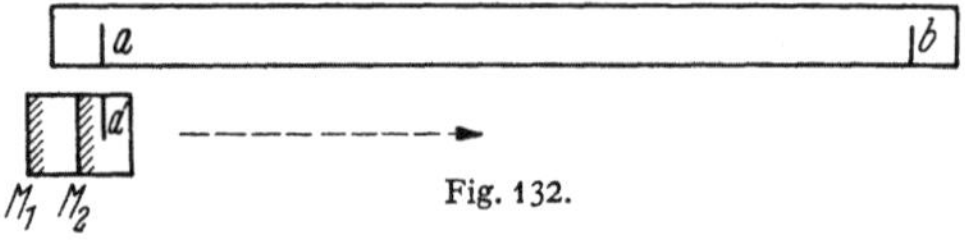

Fig. 132.

γ) *Comparaison de l'étalon 9 au mètre international.* On a tracé un trait très fin a' sur la tranche de l'étalon 9 (Fig. 132) et on amène ce trait sur le prolongement du trait a qui délimite la longueur du mètre étalon. Cette opération se fait au moyen de microscopes (comparateur). Par le procédé des contacts optiques décrits ci-dessus (β), on déplace l'étalon 9 de 10 fois sa longueur. Le trait a' arrive alors presqu'en face du second trait b du mètre. La petite distance qui les sépare est mesurée au comparateur.

δ) *Résultats des mesures.* La radiation choisie par Michelson a été la raie rouge du cadmium et il a défini l'étalon de longueur d'onde comme la longueur d'onde dans l'air sec à 15° C et sous la pression normale de la radiation rouge émise par un tube à vapeur de cadmium.

Le nombre de franges correspondant à la longueur du mètre étalon est déterminé avec une erreur qui ne dépasse pas 2 franges, soit une longueur d'onde, puisqu'un déplacement d'une frange correspond à une longueur égale à $\lambda/2$.

Michelson a trouvé pour la longueur du mètre étalon exprimé en nombre de longueurs d'onde de la raie rouge du cadmium, la valeur N

$$N = 1\,553\,165,3 \quad \text{avec} \quad \lambda = 6438,472\ \text{Å}$$

avec une erreur relative sur la valeur de la longueur d'onde inférieure à 1 milionième.

65. Mesure du mètre international en longueurs d'onde: méthode de Benoît, Fabry et Pérot. En 1913, Benoît, Fabry et Pérot ont repris la mesure du mètre étalon par des procédés entièrement différents, basés sur les phénomènes d'interférence à ondes multiples produits par les lames argentées.

Le principe de la méthode est de construire un étalon optique à traits et de le comparer à la longueur d'une règle étalon (au moyen du comparateur) exactement connue elle-même en fonction du prototype international.

L'étalon optique est constitué, comme nous l'avons vu précédement, par 2 miroirs plans semi-réfléchissants, bien parallèles. Sur la tranche supérieure de

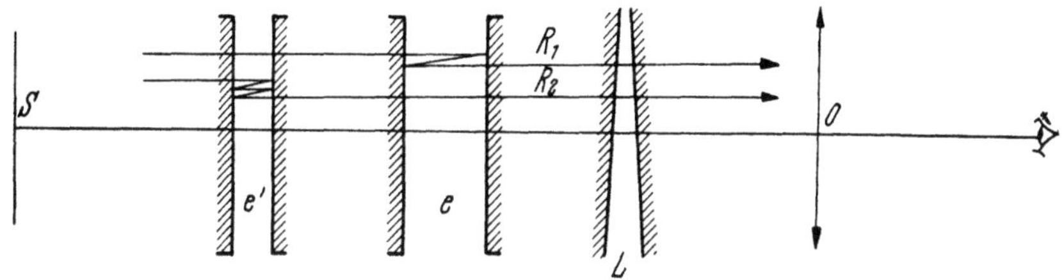

Fig. 133.

ces glaces, on a tracé un trait fin parallèle à la face et c'est la distance de ces deux traits, distance très voisine du mètre (99,92 cm.), qui constitue l'étalon optique.

Les mesures peuvent se diviser en trois parties:

1. la mesure, en fonction du mètre, de la longueur comprise entre les deux traits de l'étalon optique,

2. la mesure en longueurs d'onde de la distance des deux faces des glaces,

3. la mesure du nombre A de longueurs d'onde contenues dans la somme des distances des traits aux surfaces.

La première opération se fait comme toute mesure de longueur à traits sur le comparateur et par pointés transversaux. Quant à la détermination de A, nous avons vu de quelle façon il fallait procéder.

La deuxième opération ne peut se faire directement; en effet, on ne connaît pas de radiation suffisamment fine pour permettre d'atteindre une différence de marche voisine du mètre.

Benoît, Fabry et Pérot ont été obligés de passer par un certain nombre d'intermédiaires en réalisant quatre étalons optiques analogues à celui de 1 mètre et de longueurs

$$6,25\ \text{cm.}; \quad 12,49\ \text{cm.}; \quad 24,98\ \text{cm.}; \quad 49,96\ \text{cm.}$$

L'étalon de 6,25 cm. permet d'observer facilement les anneaux à l'infini de la radiation rouge du cadmium utilisée pour ces mesures.

On peut donc mesurer rigoureusement son épaisseur par l'une des méthodes indiquées précédemment. On compare alors l'étalon de 6,25 cm. à l'étalon de 12,5 cm. au moyen des franges de superposition. Pour cela, on dispose les deux étalons de façon que les faces soient parallèles (Fig. 133) et on éclaire l'ensemble par une source de lumière blanche S.

Considérons:

1. un rayon lumineux R_1 qui traverse directement l'étalon $e' = 6,25$ cm. et subit deux réflexions dans l'étalon $e = 12,5$ cm.,

2. un rayon R_2 qui se réfléchit quatre fois dans e' et passe directement à travers e. La différence de marche de ces deux rayons est

$$\delta = 2e - 4e'$$

malgré la petite différence qui existe entre l'étalon e et l'épaisseur $2e'$, elle est encore trop grande pour permettre d'observer les franges en lumière blanche. On place alors à la suite des deux lames e et e' une lame étalon L jouant le rôle de compensateur. Déplaçons la lame L dans son plan; pour une position donnée, on obtient un système de franges dont la frange blanche dessine le lieu des points où l'épaisseur e_1 de la lame L est égale à la différence des épaisseurs des deux étalons, c'est-à-dire

$$2e - 4e' = e_1.$$

On obtient ainsi la longueur exacte de l'étalon de 12,5 cm. On passe ensuite à l'étalon de 24,98 cm., puis à celui de 49,96 cm. et, enfin, à l'étalon de 1 mètre par les mêmes mesures.

Benoît, Fabry et Pérot ont trouvé comme longueur du mètre exprimée en longueurs d'onde de la radiation rouge du cadmium

$$N = 1\,553\,164,13.$$

66. Application des interférences au repérage d'une direction. α*) Utilisation des franges non localisées.* A.C.S. Van Heel a donné une méthode très précise

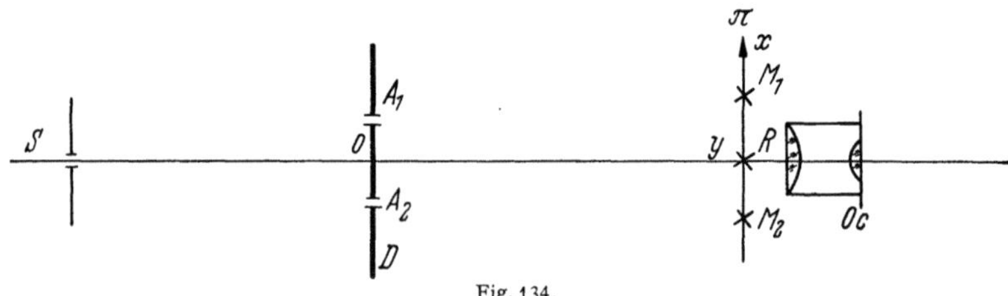

Fig. 134.

dont le principe est le suivant (Fig. 134): une source lumineuse ponctuelle de lumière blanche S et un point O constituent une direction dont on veut repérer la trace sur un certain plan d'observation π.

Considérons deux axes de coordonnées dans le plan d'observation π, l'un Rx dans le plan de figure, l'autre Ry perpendiculaire au plan de figure. Supposons d'abord que l'on ne veuille déterminer que l'abscisse x de la trace. Plaçons en O un écran D percé de 2 fentes fines A_1 et A_2 perpendiculaires au plan de figure. La source S est elle-même une fente fine perpendiculaire au plan de figure. La direction est matérialisée par S et le point O situé à égale distance de A_1 et A_2. Sur le plan π on observe des franges d'Young perpendiculaires au plan de figure, dont la frange centrale blanche contient la trace de la direction SO. On a donc l'abscisse de la trace, mais pas son ordonnée. Il s'agit de déterminer avec la plus grande précision possible le centre de la frange blanche. On sait que lorsqu'on s'écarte de la frange centrale, des colorations apparaissent: ce sont les teintes de l'échelle de Newton. Lorsque la différence de marche est telle que les vibrations sont en opposition pour le jaune, cette couleur est éteinte et on a une teinte pourpre, dite teinte sensible. Dès que l'on s'écarte un peu de cette position, la couleur change rapidement par suite de la grande sensibilité de l'oeil pour le jaune. On peut donc localiser cette teinte sur l'écran π et repérer l'abscisse de M_1 avec une extrême précision. On observe de même la teinte pourpre symétrique par rapport au centre du phénomène. Ayant déterminé la position des 2 teintes pourpres, on en déduit le milieu de la distance qui les sépare, c'est à dire le centre du phénomène. Pour déterminer l'ordonnée de la direction, il suffit de tourner les fentes A_1 et A_2 de 90° et de recommencer la mesure. On peut repérer la direction SO par une seule mesure en remplaçant les deux fentes par une petite ouverture circulaire, figurant le point O, entourée d'une fine ouverture annulaire. Les franges deviennent alors des anneaux colorés concentriques que l'on peut repérer avec l'oculaire dont le réticule est constitué par des anneaux noirs très fins. Le repérage de la direction SO, soit par les fentes en 2 opérations, soit par le trou et l'anneau en une seule opération, est obtenu avec une précision de l'ordre de $0{,}''1$.

La méthode peut être appliquée au repérage de l'orientation d'un miroir en fixant sur lui l'écran D percé des 2 fentes. On peut l'appliquer à d'autres problèmes comme la mesure des angles, la mesure des aberrations, etc. ...

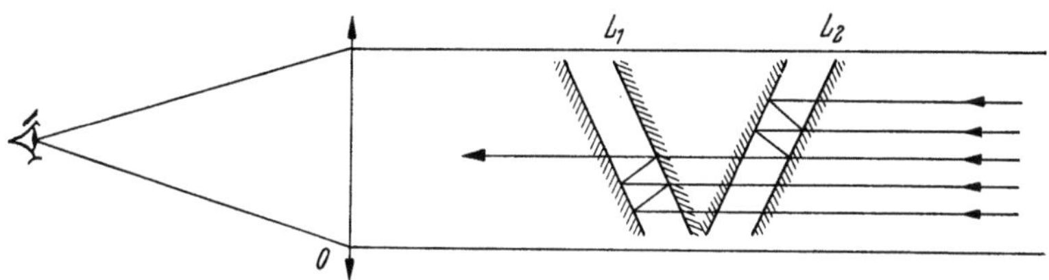

Fig. 135.

β) *Utilisation des franges de superposition.* Les franges d'égale épaisseur d'une lame d'air ou de verre ont été souvent utilisées pour repérer une direction. Ces mesures nécessitent l'emploi d'une source de lumière très monochromatique. A. Arnulf et M. Cagnet opèrent en lumière blanche au moyen des franges de superposition et avec le dispositif suivant (Fig. 135): deux lames à faces presque parallèles semi-métallisées L_1 et L_2 de même épaisseur moyenne, font entre elles un angle α et sont éclairées par un faisceau de rayons parallèles faisant un angle ϑ avec le plan bissecteur des lames L_1 et L_2. Le faisceau de rayons parallèles constitue la direction à repérer. L'objectif O forme l'image de la source ponctuelle située à l'infini sur la pupille de l'oeil qui observe L_1 et L_2 au moyen de O. Les lames n'étant pas rigoureusement à faces parallèles, on observe sur les lames un système de franges de superposition qui sont des franges d'égale différence de marche. Ces franges ont été observées pour la première fois par Brewster sur des lames de verre non semi-réfléchissantes. Les franges sont alors beaucoup moins fines que dans l'interféromètre à lames semi-réfléchissantes où tout changement $d\vartheta$ d'orientation du faisceau incident se traduit par un déplacement visible des franges, même si $d\vartheta$ est très petit.

Pour pouvoir repérer la direction avec une précision $d\vartheta$, il faut que l'on puisse déterminer la position d'une frange avec une précision correspondant à une variation de l'ordre d'interférence dp et on a pour la frange centrale ($\vartheta = 0$)

$$d\vartheta = \frac{\lambda}{2e} \frac{\sqrt{n^2 - \sin^2 \frac{\alpha}{2}}}{\sin \alpha} dp,$$

n étant l'indice des lames L_1 et L_2.

La plus grande précision est obtenue par α voisin de 90°. En admettant une variation $dp = 1{,}3 \cdot 10^{-3}$ correspondant à un pointé dont la précision serait égale au 1/25e de la largeur de la frange et pour des lames ayant un facteur de réflexion $R = 0{,}95$, les précisions calculées sont les suivantes:

	e mm.				
	1	5	10	30	50
$d\vartheta$ en secondes d'arc . .	0,055	0,010	0,005	0,002	0,001

Notons que la largeur des lames doit être assez grande pour recueillir un nombre suffisant de faisceaux interférents, sans quoi les franges deviennent moins fines et la précision diminue. Le calcul et l'expérience montrent que 25 faisceaux suffisent en général.

B. Diffraction

Introduction sur les phénomènes à l'infini et à distance finie.

Soit O un objectif donnant en S' une image de la source lumineuse ponctuelle S (Fig. 136). La surface utile de l'objectif O est limitée par un diaphragme D.

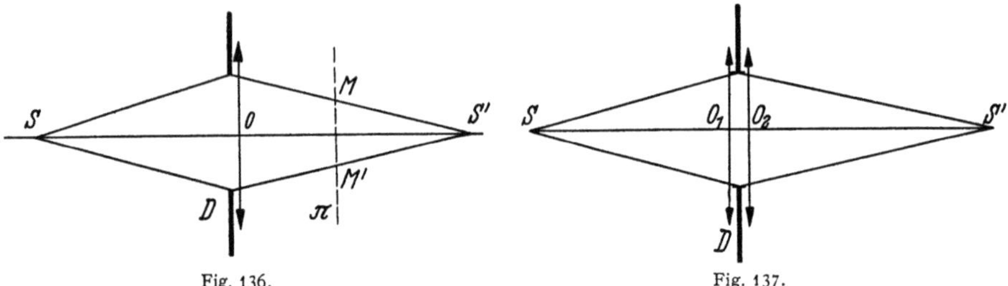

Fig. 136.   Fig. 137.

L'étude complète de la structure de l'image S' ne peut se faire en se basant uniquement sur des considérations d'optique géométrique. Il est nécessaire de faire intervenir la diffraction comme l'a fait l'astronome Airy qui, le premier, a calculé la distribution des éclairements à l'intérieur de l'image S'. Dans le cas de la Fig. 136, on dit que l'on étudie les phénomènes de diffraction à l'infini ou phénomènes de Fraunhofer. On peut en effet remplacer la lentille O par 2 lentilles O_1 et O_2 telles que S et S' restent conjugués (Fig. 137); O_1 a une focale égale à la distance SO_1, et O_2 une focale égale à la distance $O_2 S'$. Tout se passe pour la lentille O_2 comme si elle était éclairée par une source à l'infini, c'est à dire par une onde plane: les phénomènes sont identiques dans les deux cas, en supposant les lentilles minces.

On peut aussi étudier la structure de l'image dans un plan très voisin de S', c'est à dire en présence d'un défaut de mise au point. Tant que le défaut de mise au point est petit, les phénomènes de diffraction étudiés sont encore de la classe des phénomènes de Fraunhofer, mais si on les observe dans un plan π très éloigné de S', il n'en est plus de même. Pratiquement les phénomènes s'observent seulement vers les bords M et M' du faisceau, c'est-à-dire vers la limite de l'ombre géométrique de l'écran D. Puisque l'on est loin de l'image S', il revient au même d'observer les phénomènes quand l'image S' est virtuelle (Fig.138) et même sans lentille (Fig. 139): on dit qu'on observe les phénomènes de diffraction à distance finie ou phénomènes de Fresnel.

Les phénomènes de cette classe sont observés suivant le schéma de la Fig. 139: l'écran D dont on veut étudier la diffraction à distance finie est interposé entre la source lumineuse S et l'écran d'observation π. Dans l'étude des phénomènes de FRAUNHOFER, nous commencerons d'abord par étudier le cas où le système optique produisant l'image, par exemple l'objectif O sur la Fig. 136, est rigoureusement parfait. Nous étudierons ensuite l'influence des défauts du système optique sur la structure des phénomènes de diffraction au voisinage de l'image.

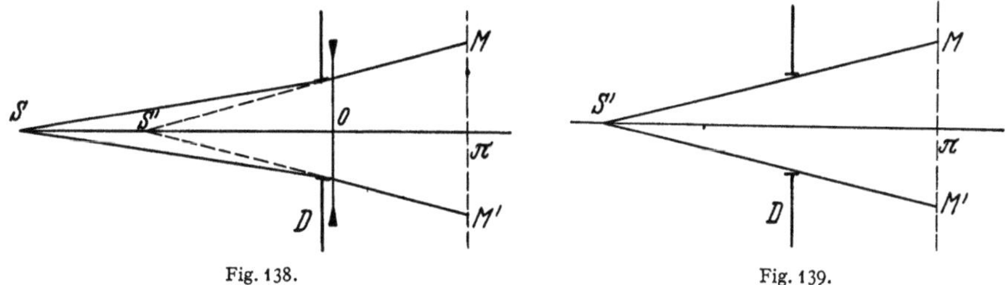

Fig. 138.   Fig. 139.

I. Diffraction de FRAUNHOFER ou diffraction à l'infini.

a) Image d'une source ponctuelle par un système optique parfait.

67. Calcul de l'amplitude dans le cas général. Le calcul général des phénomènes de diffraction est basé sur le principe d'HUYGENS-FRESNEL que nous rappelons brièvement.

Considérons une source S qui émet des ondes et soit Ω la surface d'onde au temps t (Fig. 140). Dans l'hypothèse d'HUYGENS, tout se passe comme si chaque point M de Ω émettait a son tour une onde sphérique secondaire qui se propage et dont le rayon est $v\vartheta$ à l'époque $t+\vartheta$, v désignant la vitesse de la lumière. A l'époque $t+\vartheta$ l'onde résultante Ω' est l'enveloppe des ondes sphériques de rayon $v\vartheta$. C'est dans un milieu homogène et isotrope une sphère de rayon $v(t+\vartheta)$. HUYGENS a donc ainsi mis en évidence le mécanisme de propagation du phénomene entre les divers points de l'espace. L'hypothèse d'HUYGENS a été complétée par FRESNEL qui a admis qu'un point quelconque M de Ω peut être considéré comme une source dont l'amplitude et la phase sont précisément l'amplitude et la phase de la vibration produite en M par la source S.

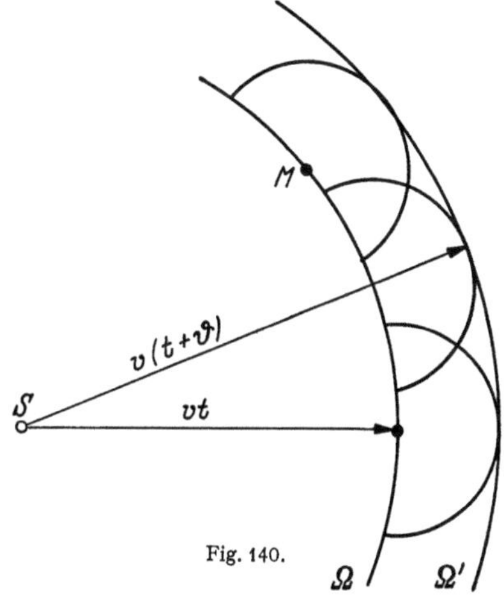

Fig. 140.

Ainsi énoncé, le principe de HUYGENS-FRESNEL donne non seulement une amplitude dans la direction de propagation, mais aussi une amplitude dans le sens inverse de la propagation, amplitude produite par une onde de retour Ω'' (Fig. 141) qui est l'autre surface enveloppe des ondes secondaires. Une étude mathématique des phénomènes a permis de justifier le principe d'HUYGENS-FRESNEL, et d'éviter la présence de l'onde de retour Ω'' contraire à l'expérience. La difficulté a été levée pour les ondes scalaires telles que les ondes acoustiques, par KIRCHHOFF. Le problème des ondes électromagnétiques qui sont des ondes vectorielles, est plus difficile, et différentes formules ont été proposées, qui rendent compte des phénomènes. Nous nous bornerons, dans ce qui suit, à l'application du principe d'Huygens-Fresnel sous sa forme simplifiée indiquée précédemment.

Soit (Fig. 142) une surface d'onde Σ qui s'écarte peu de la forme sphérique: nous voulons calculer l'amplitude au voisinage d'un point C centre d'une sphère de référence S et de pôle O, dont la forme est voisine de la surface d'onde Σ. La surface d'onde Σ est produite

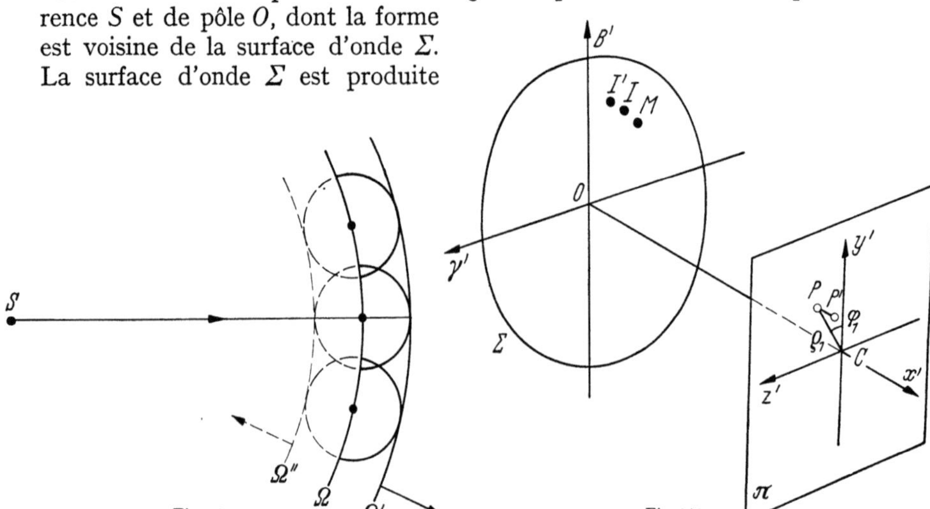

Fig. 141. Fig. 142.

par un objectif L (Fig. 143) et on cherche à déterminer l'amplitude au voisinage de centre C de la sphère moyenne S. Calculons l'amplitude au point P' de coordonnées cylindriques ε, $\varrho_1 \cos\varphi_1$, $\varrho_1 \sin\varphi_1$: ε étant la distance $P'P$ de P' au plan de front π passant par C et $\overline{CP} = \varrho_1$ avec l'angle $y'CP = \varphi_1$ (Fig. 142). Traçons une sphère S' de centre P' et qui passe par le pôle O de la sphère S (Fig. 143). Considérons

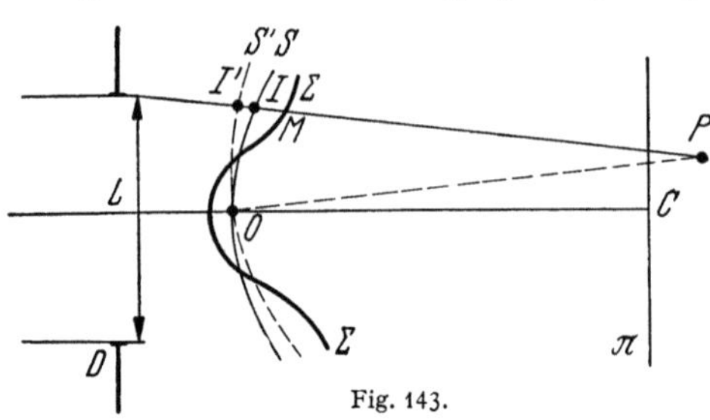

Fig. 143.

un rayon lumineux MP' qui traverse la sphère S en I, la sphère S' en I' et la surface d'onde Σ' en M. On a

$$OP' - OC = I'P' - IC. \quad (67.1)$$

Les coordonnées sphériques de I par rapport à C sont $R = IC$, α' et φ, et les cosinus directeurs de IC sont $\cos\alpha'$, $\sin\alpha'\cos\varphi$ et $\sin\alpha'\sin\varphi$ (Fig. 144). La différence $I'P' - IC$ peut être obtenue en calculant la projection de $I'P'$ sur IC et en lui retranchant IC puisque P' est voisin de C. Le second membre de (67.1) est donc égal à la projection de II' et de CP' sur IC. L'inclinaison de II' sur IC est faible, donc la projection de II' sur IC peut être prise égale à II'; reste à calculer la projection de CP' sur IC. Les coordonnées de P' et les cosinus directeurs de IC étant connus, on a aussitôt la projection de CP':

coordonnées de P' ε, $\varrho_1 \cos\varphi_1$, $\varrho_1 \sin\varphi_1$,

cosinus directeurs de IC $\cos\alpha'$, $\sin\alpha'\cos\varphi$, $\sin\alpha'\sin\varphi$,

projection de CP' sur $IC = \varepsilon\cos\alpha' + \varrho_1\sin\alpha'\cos(\varphi - \varphi_1)$ et (67.1) peut s'écrire

$$\varepsilon = II' + \varepsilon\cos\alpha' + \varrho_1\sin\alpha'\cos(\varphi - \varphi_1). \quad (67.2)$$

Appelons $\varDelta = IM$ l'écart en M entre la surface d'onde $\varSigma$ et la sphère de référence S et $\varDelta' = I'M$ l'écart entre $\varSigma$ et la sphère S' de centre P': d'après (67.2), on a

$$\varDelta' = \varDelta + \varepsilon(1 - \cos\alpha') - \varrho_1 \sin\alpha' \cos(\varphi - \varphi_1). \tag{67.3}$$

Soit $e^{jK\overline{MP'}}$ le mouvement vibratoire envoyé en P' par le point M de la surface d'onde $\varSigma$. Prenons comme origine des phases, la phase en O. Le mouvement vibratoire en P' dû au point M est alors $e^{jK(MP'-OP')}$

$$e^{jK(MP'-OP')} = e^{jK(MP'-I'P')} = e^{jK\varDelta'}.$$

Le mouvement en P' dû à tous les points d'une petite surface ds entourant M est le même et peut s'écrire $e^{jK\varDelta'}\,ds$ d'où le mouvement vibratoire en P' dû à l'onde $\varSigma$ et à un facteur constant près

$$U_{P'} = \iint_{\varSigma} e^{jK\varDelta'}\,ds.$$

Si on exprime sous une forme rigoureuse le principe de HUYGENS, on trouve que l'expression précédente doit être multipliée par le facteur $j/\lambda R$. Le terme $1/R$ est dû à la décroissance en $1/R$ de l'amplitude des ondes secondaires et le facteur j permet de trouver en P' la phase correcte. D'où l'expression de l'amplitude en P'

$$U_{P'} = \frac{j}{\lambda R} \iint_{\varSigma} e^{jK\varDelta'}\,ds. \tag{67.4}$$

Il suffit de remplacer $\varDelta'$ par sa valeur tirée de (67.3) pour avoir la vibration résultante en P'.

Dans ce qui suit, nous considérons des instruments d'ouverture modérée et on peut écrire avec une approximation suffisante l'expression (67.3) sous la forme

$$\varDelta' = \varDelta + \varepsilon\,\frac{\alpha'^2}{2} - \varrho_1 \alpha' \cos(\varphi - \varphi_1). \tag{67.5}$$

Posons

$$\alpha' \cos\varphi = \beta', \qquad \alpha' \sin\varphi = \gamma',$$

et comme

$$\varrho_1 \cos\varphi_1 = y', \qquad \varrho_1 \sin\varphi_1 = z'$$

on aura

$$\varDelta' = \varDelta + \varepsilon\,\frac{\alpha'^2}{2} - (\beta' y' + \gamma' z'). \tag{67.6}$$

Si l'objectif L est parfait, $\varSigma$ est une sphère et on peut la confondre avec la sphère S de centre C, c'est-à-dire faire $\varDelta = 0$. Le point C est l'image géométrique de la source lumineuse. Le terme $\varepsilon\,\dfrac{\alpha'^2}{2}$ représente la différence de marche dûe à un défaut de mise au point $\varepsilon = PP'$. Posons $\varepsilon = 0$, on étudie alors l'amplitude au point P du plan de front π passant par le centre de l'onde sphérique $\varSigma$. On a, puisque $ds = R^2\,d\beta'\,d\gamma'$

$$U_P = \frac{jR}{\lambda} \iint_{\varSigma} e^{-jK(\beta' y' + \gamma' z')}\,d\beta'\,d\gamma' \tag{67.7}$$

c'est l'amplitude en un point de l'image fournie par un instrument parfait rigoureusement mis au point, l'image géométrique de la source étant en C.

68. Répartition des éclairements dans l'image d'un point en présence d'une ouverture circulaire. Supposons la surface d'onde limitée par un contour circulaire de rayon α_0 (Fig. 144). On peut changer l'origine des azimuts et écrire, d'après (67.5) et (67.6)

$$\beta' \, y' + \gamma' \, z' = \alpha' \, \varrho_1 \cos \varphi$$

d'où l'amplitude en P

$$U_P = j \, \frac{R}{\lambda} \int\limits_0^{\alpha'_0} \int\limits_0^{2\pi} e^{-j K \alpha' \varrho_1 \cos \varphi} \, \alpha' \, d\alpha' \, d\varphi. \tag{68.1}$$

Comme

$$\int\limits_0^{2\pi} e^{-j K \alpha' \varrho_1 \cos \varphi} \, d\varphi = 2\pi \, J_0(K \alpha' \varrho_1),$$

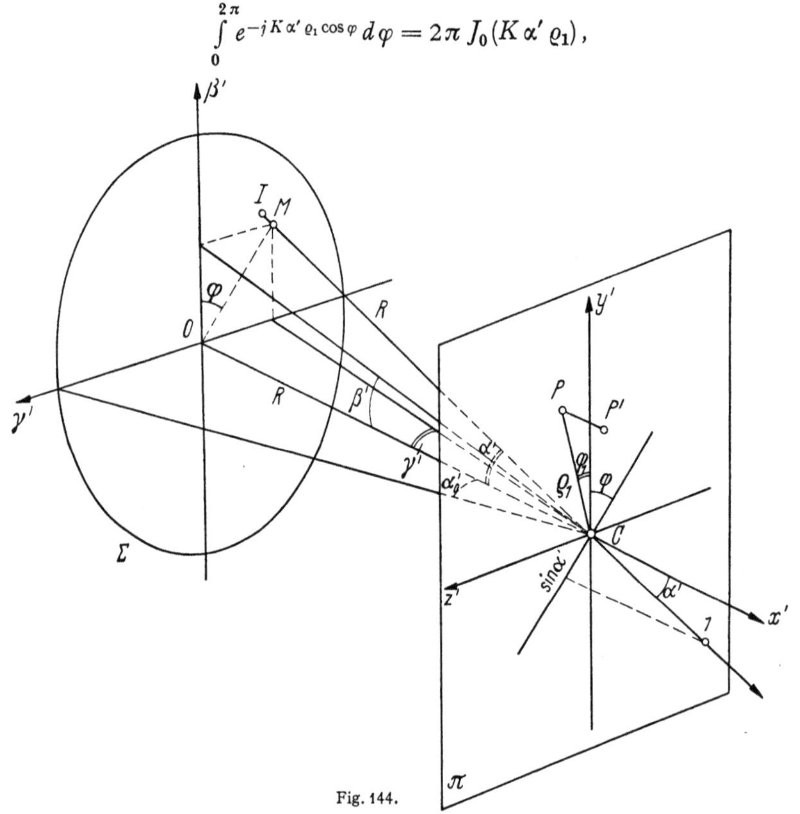

Fig. 144.

J_0 étant la fonction de Bessel d'ordre zéro,

$$U_P = j \, \frac{2\pi R}{\lambda} \int\limits_0^{\alpha'_0} J_0(K \alpha' \varrho_1) \, \alpha' \, d\alpha' = j \, \frac{2\pi R}{\lambda K^2 \varrho_1^2} \int\limits_0^{K \alpha'_0 \varrho_1} J_0(K \alpha' \varrho_1) \, (K \alpha' \varrho_1) \, d(K \alpha' \varrho_1),$$

$$U_P = j \, \frac{\pi R \alpha_0'^2}{\lambda} \, \frac{2 J_1(K \alpha'_0 \varrho_1)}{K \alpha'_0 \varrho_1}. \tag{68.2}$$

car

$$\int\limits_0^x x J_0(x) \, dx = x J_1(x).$$

Si D est le diamètre de l'objectif, on peut poser

$$Z = K \alpha'_0 \varrho_1 = \frac{\pi D}{\lambda R} \varrho_1 = \frac{\pi D}{\lambda R} \sqrt{y'^2 + z'^2} \tag{68.3}$$

d'où la répartition des éclairements dans le plan π

$$E = \frac{(\pi \alpha_0'^2)^2 R^2}{\lambda^2} \left[\frac{2 J_1(Z)}{Z} \right]^2. \tag{68.4}$$

L'éclairement en un point de l'image de diffraction est donc proportionnel au carré de la surface libre de l'objectif. La Fig. 145 représente les variations de l'éclairement dans l'image d'une source lumineuse ponctuelle donnée par un instrument parfait et rigoureusement mis au point. C'est la courbe représentant la fonction $\left[\frac{2 J_1(Z)}{Z} \right]^2$ dont les valeurs, ainsi que celles de l'amplitude $\frac{2 J_1(Z)}{Z}$ sont données dans le tableau 7 page 274.

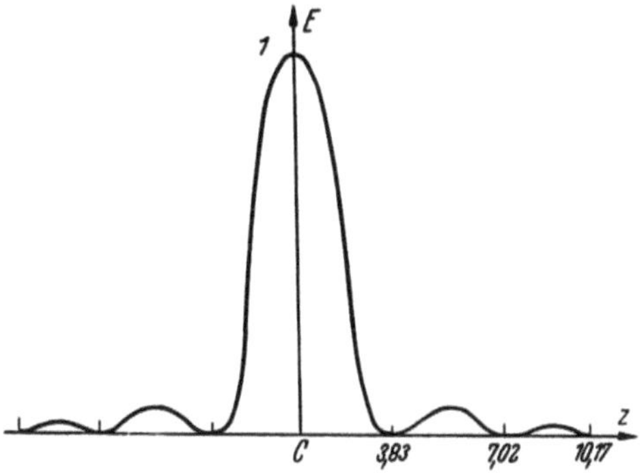

Dans l'expression (68.4), le facteur λ^2 intervient pour maintenir constant le flux, quelle que soit la longueur d'onde. Si on fait tourner la courbe de la Fig. 145 autour de l'axe des ordonnées, on obtient un volume, le volume de diffraction, qui est proportionnel au flux reçu par l'objectif. Si on trace la courbe $E = f(\varrho_1)$, les diamètres des anneaux augmentent avec la longueur d'onde et pour que l'énergie soit constante, quelle que soit λ, il faut diviser les ordonnées de chaque courbe par λ^2. Si on trace la courbe $E = f(Z)$ il n'y a pas lieu de faire intervenir le facteur λ^2, le volume de diffraction étant fonction du paramètre Z seul.

Dans un instrument parfait limité par un contour circulaire, la figure de diffraction ou disque d'AIRY se compose d'une tache centrale très lumineuse entourée d'anneaux alternativement brillants et obscurs. Les maxima des anneaux brillants sont beaucoup moins intenses que le maximum central, et décroissent rapidement (Fig. 146). Les anneaux noirs correspondent aux racines de $J_1(Z) = 0$ c'est à dire

Fig. 145.

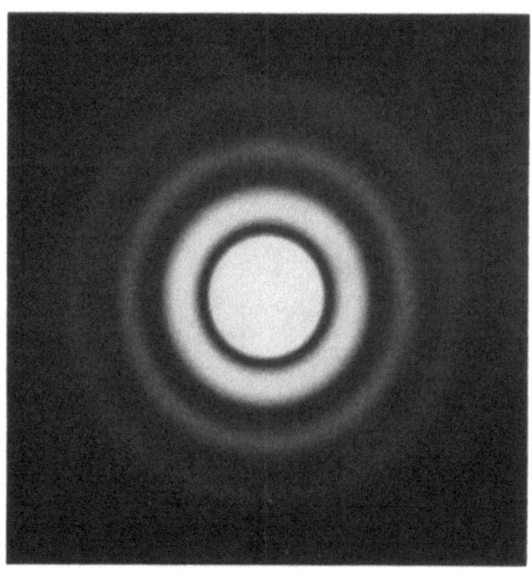

Fig. 146. Figure de diffraction par une ouverture circulaire. Disque d'AIRY.

$$Z = 3{,}83 \quad 7{,}02 \quad 10{,}17 \quad 13{,}32 \ldots .$$

Les maxima correspondent aux valeurs suivantes

$$Z = 5{,}14 \quad 8{,}46 \quad 11{,}62 \ldots$$

avec les éclairements correspondants

$$E = 0{,}0175 \quad 0{,}0042 \quad 0.0016.$$

Le premier minimum est donné par $Z = 3{,}83$.

D'après la définition de Z, on voit que le rayon linéaire du premier anneau noir sera égal à

$$\varrho_1 = \frac{Z}{K\alpha_0'} = \frac{3{,}83\,\lambda}{3{,}14 \times 2\alpha_0'} = \frac{1{,}22\,\lambda}{2\alpha_0'}, \tag{68.5}$$

le rayon angulaire étant égal à

$$\vartheta = \frac{\varrho_1}{R} = \frac{1{,}22\,\lambda}{D}. \tag{68.6}$$

Tableau 7.

Z	$\dfrac{2J_1(Z)}{Z}$	$\left[\dfrac{2J_1(Z)}{Z}\right]^2$	Z	$\dfrac{2J_1(Z)}{Z}$	$\left[\dfrac{2J_1(Z)}{Z}\right]^2$
0,0	$+1{,}0000$	1,0000	6,2	$-0{,}0751$	0,0056
0,2	$+0{,}9950$	0,9900	6,4	$-0{,}0567$	0,0032
0,4	$+0{,}9801$	0,9607	6,6	$-0{,}0379$	0,0014
0,6	$+0{,}9557$	0,9154	6,8	$-0{,}0192$	0,0004
0,8	$+0{,}9221$	0,8503	7,0	$-0{,}0013$	0,0000
1,0	$+0{,}8801$	0,7746	7,2	$+0{,}0151$	0,0003
1,2	$+0{,}8305$	0,6897	7,4	$+0{,}0296$	0,0009
1,4	$+0{,}7742$	0,5994	7,6	$+0{,}0419$	0,0018
1,6	$+0{,}7124$	0,5075	7,8	$+0{,}0516$	0,0027
1,8	$+0{,}6461$	0,4175	8,0	$+0{,}0587$	0,0034
2,0	$+0{,}5767$	0,3326	8,2	$+0{,}0629$	0,0040
2,2	$+0{,}5043$	0,2555	8,4	$+0{,}0645$	0,0042
2,4	$+0{,}4335$	0,1879	8,6	$+0{,}0634$	0,0040
2,6	$+0{,}3622$	0,1312	8,8	$+0{,}0600$	0,0036
2,8	$+0{,}2926$	0,0856	9,0	$+0{,}0545$	0,0030
3,0	$+0{,}2260$	0,0511	9,2	$+0{,}0473$	0,0022
3,2	$+0{,}1633$	0,0267	9,4	$+0{,}0386$	0,0015
3,4	$+0{,}1054$	0,0111	9,6	$+0{,}0291$	0,0008
3,6	$+0{,}0530$	0,0028	9,8	$+0{,}0189$	0,0004
3,8	$+0{,}0067$	0,0000	10,0	$+0{,}0087$	0,0001
4,0	$-0{,}0330$	0,0011	10,2	$-0{,}0013$	0,0000
4,2	$-0{,}0660$	0,0044	10,4	$-0{,}0107$	0,0001
4,4	$-0{,}0922$	0,0085	10,6	$-0{,}0191$	0,0004
4,6	$-0{,}1115$	0,0124	10,8	$-0{,}0263$	0,0007
4,8	$-0{,}1244$	0,0155	11,0	$-0{,}0321$	0,0010
5,0	$-0{,}1310$	0,0172	11,2	$-0{,}0345$	0,0012
5,2	$-0{,}1320$	0,0174	11,4	$-0{,}0379$	0,0014
5,4	$-0{,}1279$	0,0164	11,6	$-0{,}0440$	0,0016
5,6	$-0{,}1194$	0,0153	11,8	$-0{,}0394$	0,0015
5,8	$-0{,}1073$	0,0115	12,0	$-0{,}0372$	0,0014
6,0	$-0{,}0922$	0,0085			

Si l'on considère des instruments de même ouverture α_0' mais de dimensions linéaires quelconques, le disque central et les anneaux de diffraction sont les mêmes pour tous ces instruments. Le tableau suivant donne quelques valeurs numériques en microns du rayon du 1er anneau noir:

$2\alpha_0'$	$\tfrac{1}{3}$	$\tfrac{1}{6}$	$\tfrac{1}{10}$	$\tfrac{1}{20}$
$\lambda = 0{,}556\,\mu$	2,0	4,1	6,8	13,6
$\lambda = 0{,}435\,\mu$	1,6	3,2	5,3	10,6

On peut avoir à calculer E pour des grandes valeurs de Z. Il est alors possible d'utiliser le développement asymptotique suivant, lorsque Z est beaucoup plus grand que 1

$$J_1(Z) = \frac{\sin Z - \cos Z}{\sqrt{\pi Z}} \qquad z \gg 1$$

et à un facteur constant près

$$E(Z) = 4\left(\frac{\sin Z - \cos Z}{Z\sqrt{\pi Z}}\right)^2. \tag{68.7}$$

Les minima nuls sont donnés par $\tan Z = 1$ et l'on voit que lorsque l'on est assez loin du centre de la figure de diffraction $(Z > 15)$, la distance de deux minima nuls consécutifs est pratiquement constante et égale à π.

La formule (68.5) permet de calculer le pouvoir séparateur d'une lunette lorsqu'on observe deux étoiles. La distance angulaire des deux étoiles est ϑ. Ces deux images sont constituées par des taches de diffraction dont on vient d'étudier la structure. Pour que les deux étoiles soient vues séparées, il faut que les deux taches de diffraction n'empiètent pas trop l'une sur l'autre. On admet qu'il en est ainsi lorsque le maximum de l'une des taches correspond au premier minimum de l'autre (Fig. 147). L'expression (68.6) donne par conséquent le pouvoir séparateur de l'objectif O de la lunette

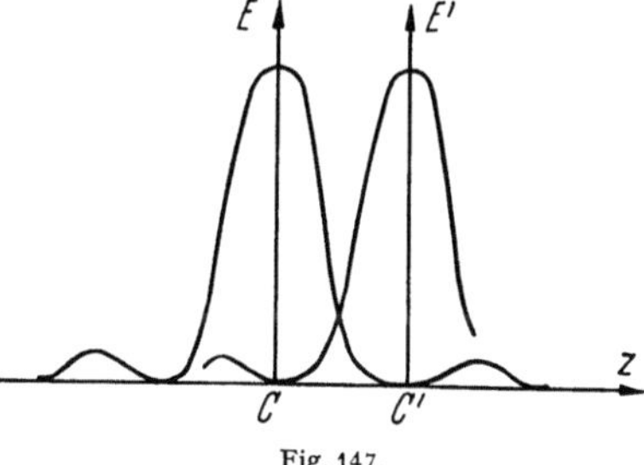

Fig. 147.

$$\vartheta = \frac{1{,}22\,\lambda}{D}. \tag{68.8}$$

Pour dédoubler deux étoiles dont la distance angulaire est $\vartheta = 1''$, il faut un objectif de diamètre D égal à

$$D = 1{,}22 \cdot 6 \cdot 10^{-4} \cdot 206\,000 = 150 \text{ mm.}$$

pour $\lambda = 6 \cdot 10^{-4}$ mm.

69. Distribution de l'énergie dans la figure de diffraction par une ouverture circulaire. Le phénomène de diffraction donné par (68.4) étant de révolution, considérons une petite couronne élémentaire de surface $2\pi\,\varrho_1\,d\varrho_1$: le flux lumineux réparti sur cet élément d'aire sera à une constante près

$$2\pi E(K\alpha_0'\varrho_1)\,\varrho_1\,d\varrho_1.$$

Le flux situé à l'intérieur du cercle de rayon P_1 sera donné par:

$$\mathcal{F}_i = 2\pi \int\limits_0^{\varrho_1} E(K\alpha_0'\varrho_1)\,\varrho_1\,d\varrho_1 = 2\pi\alpha_0'^2 R^2 \int\limits_0^{z} \frac{J_1^2(Z)}{Z}\,dZ.$$

La dernière intégrale donne le flux situé à l'intérieur du cercle de rayon Z: le flux est proportionnel à la surface de l'objectif. On aura

$$\mathcal{F}_i = \pi R^2 \alpha_0'^2 \left[1 - J_0^2(Z) - J_1^2(Z)\right]$$

en utilisant les relations élémentaires

$$J_1'(Z) = -\frac{J_1(Z)}{Z} + J_0(Z)$$

$$J_1(Z) = -J_0'(Z).$$

En prenant pour unité, le flux total $(Z=\infty)$, l'expression donnant le flux situé à l'intérieur du cercle Z sera

$$\mathcal{F}_i = 1 - J_0^2(Z) - J_1^2(Z) \tag{69.1}$$

et le flux extérieur à ce cercle

$$\mathcal{F}_e = J_0^2(Z) + J_1^2(Z). \tag{69.2}$$

Ces expressions donnent les flux quel que soit λ: le volume de diffraction est exprimé en fonction de Z, il ne dépend pas de λ mais pour un cercle de rayon ϱ déterminé, le paramètre Z correspondant, dépend de λ.

Le tableau ci-dessous donne $\mathcal{F}_i$ et $\mathcal{F}_e$ pour quelques valeurs de Z

Z	3,832	7,015	10,175	13,324
$\mathcal{F}_i$	0,839	0,910	0,938	0,953
$\mathcal{F}_e$	0,161	0,090	0,062	0,047

Si l'on considère la lumière située hors de la tache centrale de rayon $Z=3,83$ comme de la lumière parasite, on voit que près de 20% du flux total est ainsi perdu, d'où l'intérêt de l'apodisation de la figure de diffraction, comme on le verra plus loin.

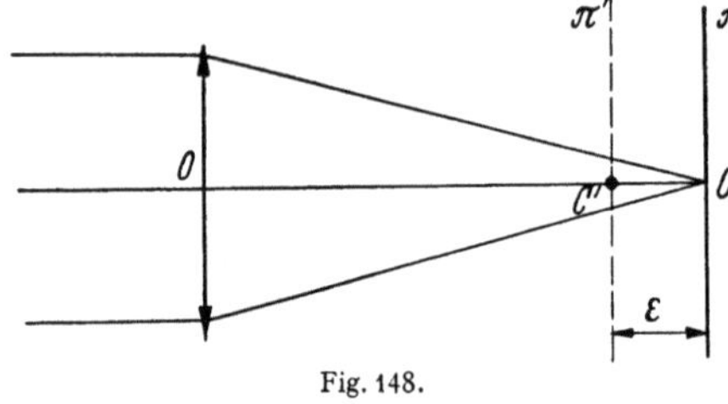
Fig. 148.

70. Image en présence d'un défaut de mise au point (ouverture circulaire). On veut calculer la structure de l'image dans un plan de front π' situé à une petite distance ε du plan π passant par l'image parfaite (Fig. 148). L'amplitude en un point P' du plan π' est donnée par l'expression (67.4) où Δ' a la valeur suivante tirée de (67.5)

$$\Delta' = \varepsilon\,\frac{\alpha'^2}{2} - \varrho_1\alpha'\cos\varphi.$$

On a donc

$$U_{P'} = \frac{jR}{\lambda} \int_0^{\alpha_0'}\int_0^{2\pi} e^{jK\varepsilon\frac{\alpha'^2}{2}}\, e^{-jK\alpha'\varrho_1\cos\varphi}\,\alpha'\,d\alpha'\,d\varphi \tag{70.1}$$

ou

$$U_{P'} = j\,\frac{2\pi R}{K^2\lambda\varrho_1^2}\int_0^{K\alpha_0'\varrho_1} e^{jK\varepsilon\frac{\alpha'^2}{2}}\, J_0(K\alpha'\varrho_1)\,(K\alpha'\varrho_1)\,d(K\alpha'\varrho_1). \tag{70.2}$$

Posons

$$\left(\frac{\alpha'}{\alpha_0'}\right)^2 = t \quad\text{et}\quad \Phi = \frac{K\varepsilon\alpha_0'^2}{2},$$

Φ représente la différence de phase en C' entre les rayons diffractés par le bord et le centre de l'onde. On aura

$$U_{P'} = j\,\frac{\pi R\alpha_0'^2}{\lambda}\int_0^1 e^{j\Phi t}\, J_0(K\alpha'\varrho_1)\,dt. \tag{70.3}$$

Les valeurs de l'éclairement E à une constante près pourront être calculées numériquement au moyen des intégrales

$$E_{P'} = \left[\int_0^1 J_0\left(Z\sqrt{t}\right) \cos\left(\Phi t\right) dt \right]^2 + \left[\int_0^1 J_0\left(Z\sqrt{t}\right) \sin\left(\Phi t\right) dt \right]^2 . \qquad (70.4)$$

Le tableau suivant, extrait des travaux de CONRADY, BUXTON et MARTIN, donne la répartition des éclairements dans l'image d'un point pour différents défauts de mise au point.

Tableau 8.

z	Φ							
	0	$\pi/2$	π	$3\pi/2$	2π	3π	4π	8π
0,000	1,0000	0,8102	0,4051	0,0900	0,0000	0,0453	0,0000	0,0005
0,698	0,8845	0,7175	0,3590	0,0802	0,0004	0,0097	0,0001	0,0005
1,396	0,6008	0,4890	0,2498	0,0617	0,0047	0,0277	0,0012	0,0007
2,094	0,2953	0,2477	0,1437	0,0542	0,0174	0,0155	0,0043	0,0015
2,793	0,0871	0,0884	0,0853	0,0654	0,0353	0,0065	0,0090	0,0028
3,491	0,0066	0,0302	0,0721	0,0760	0,0475	0,0052	0,0120	0,0038
4,189	0,0042	0,0283	0,0708	0,0794	0,0474	0,0048	0,0119	0,0037
4,887	0,0164	0,0315	0,0574	0,0433	0,0393	0,0067	0,0095	0,0028
5,585	0,0145	0,0212	0,0336	0,0393	0,0330	0,0104	0,0067	0,0017
6,283	0,0046	0,0076	0,0153	0,0269	0,0330	0,0156	0,0059	0,0010
6,981	0,0000	0,0023	0,0100	0,0240	0,0349	0,0143	0,0047	0,0014
7,679	0,0021	0,0041	0,0112	0,0209	0,0320	0,0138	0,0047	0,0008
8,378	0,0041	0,0055	0,0102	0,0180	0,0238	0,0143	0,0050	0,0012
9,076	0,0027	0,0031	0,0058	0,0236	0,0144	0,0146	0,0061	0,0011
9,774	0,0004	0,0005	0,0021	0,0057	0,0124	0,0179	0,0052	0,0016
10,472	0,0002		0,0018	0,0031	0,0121	0,0162	0,0051	0,0010
11,170	0,0014		0,0028	0,0055	0,0090	0,0064	0,0062	0,0011
11,868	0,0015		0,0025	0,0044	0,0077	0,0128	0,0073	0,0010
12,566	0,0006		0,0011	0,0023	0,0045	0,0114	0,0095	0,0014
13,264	0,0000		0,0004	0,0012	0,0033	0,0113	0,0108	0,0020
13,963	0,0003		0,0007	0,0021	0,0035	0,0106	0,0112	0,0024
14,661	0,0008		0,0011	0,0020	0,0033	0,0120	0,0097	0,0026
15,358	0,0006		0,0007	0,0015	0,0023	0,0050	0,0088	0,0027
16,057	0,0005		0,0002	0,0007	0,0013	0,0030	0,0082	0,0028
16,755	0,0000		0,0001	0,0003	0,0009	0,0016	0,0074	0,0030
17,453	0,0003		0,0005	0,0006	0,0021	0,0032	0,0068	0,0035

Admettons avec Lord RAYLEIGH qu'un défaut de mise au point donnant au bord de l'onde une différence de phase $\Phi \leq \dfrac{\pi}{2}$ soit tolérable, c'est à dire que l'image correspondante ne diffère pratiquement pas de l'image parfaite. On aura

$$\Phi = \frac{K\,\varepsilon\,\alpha_0'^2}{2} \leq \frac{\pi}{2}$$

d'où

$$\varepsilon \leq \frac{2\lambda}{(2\alpha_0')^2} . \qquad (70.5)$$

C'est le défaut de mise au point que l'on peut tolérer à condition que l'oeil placé derrière l'instrument n'introduise pas de défauts propres, c'est à dire qu'il puisse être considéré comme parfait. La formule (70.5) donne par conséquent la précision d'un pointé longitudinal par un instrument d'ouverture $2\alpha_0'$. La formule (70.4) se prête bien aux calculs numériques par intégration. On peut également calculer l'intégrale (70.2) par développement en série en utilisant les fonctions de Bessel. On a

$$\int\limits_0^Z Z^{2m+1}\, J_0(Z)\, dZ = Z^{2m+1} \sum_0^m (-2)^\nu\, \nu! \binom{m}{\nu} \frac{J_{\nu+1}(Z)}{Z^\nu},$$

$$\nu! \binom{m}{\nu} = m\,(m-1)\,(m-2)\cdots(m-\nu+1)$$

d'où

$$\left.\begin{aligned}
U_{P'} = -j\,\frac{\pi R\,\alpha_0'^2}{\lambda}\Bigg[&\sum_{\nu=0}^{\infty}\sum_{\mu=1}^{2\nu+1} (-1)^{\nu+\mu}\,\frac{2^\mu\,\Phi^{2\nu}}{(2\nu+1-\mu)!}\,\frac{J_\mu(Z)}{Z^\mu} + \\
&+ j\,\Phi\sum_{\nu=0}^{\infty}\sum_{\mu=1}^{2\nu+2} (-1)^{\nu+\mu}\,\frac{2^\mu\,\Phi^{2\nu}}{(2\nu+2-\mu)!}\,\frac{J_\mu(Z)}{Z^\mu}\Bigg].
\end{aligned}\right\} \tag{70.6}$$

Posons

$$U_1 = -\sum_{\mu=0}^{\infty} (-1)^\mu \left(\frac{2\,\Phi}{Z}\right)^{2\mu+1} J_{2\mu+1}(Z),$$

$$U_2 = \sum_{\mu=0}^{\infty} (-1)^\mu \left(\frac{2\,\Phi}{Z}\right)^{2\mu+2} J_{2\mu+2}(Z)$$

avec

$$C = U_1 \cos\Phi - U_2 \sin\Phi,$$

$$S = U_1 \sin\Phi + U_2 \cos\Phi$$

on obtient l'éclairement $E_{P'}$ au point P'

$$\left.\begin{aligned}
E_{P'} &= \frac{\pi^2 R^2}{\lambda^2}\,\frac{\alpha_0'^4}{\Phi^2}\,(C^2+S^2) = \frac{\pi^2 R^2}{\lambda^2}\,\frac{\alpha_0'^4}{\Phi^2}\,(U_1^2+U_2^2), \\
E_{P'} &= \frac{R^2}{\varepsilon^2}\,(U_1^2+U_2^2).
\end{aligned}\right\} \tag{70.7}$$

Si on fait $Z=0$ dans l'une des expressions (70.7), on obtient la variation de l'éclairement le long de l'axe optique de l'objectif.

En utilisant les relations

$$(U_1)_{Z=0} = -\sum_0^{\infty} (-1)^\mu\,\frac{\Phi^{2\mu+1}}{(2\mu+1)!} = -\sin\Phi,$$

$$(U_2)_{Z=0} = \sum_0^{\infty} (-1)^\mu\,\frac{\Phi^{2\mu+2}}{(2\mu+2)!} = 1 - \cos\Phi$$

on a

$$(E_{P'})_{Z=0} = \frac{\pi^2 R^2 \alpha_0'^4}{\lambda^2}\left(\frac{\sin\dfrac{\Phi}{2}}{\dfrac{\Phi}{2}}\right)^2. \tag{70.8}$$

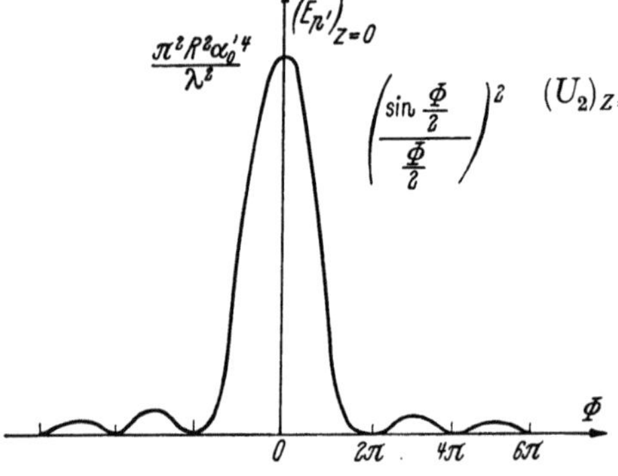

Fig. 149.

La courbe représentative est donnée par la Fig. 149. On a donc des figures de diffraction à centre noir chaque fois que la différence de phase Φ est égale à un multiple de 2π. La Fig. 149 montre que deux figures de diffraction symétriques en Φ par rapport au plan passant par l'image parfaite, sont identiques. Ce fait est souvent utilisé pour juger de la qualité d'un instrument. Il suffit de changer rapidement la mise au point pour passer d'un côté à l'autre de l'image de meilleure mise au point: une petite différence de structure s'observe facilement et permet de déceler la présence de faibles aberrations.

71. Distribution de l'énergie dans la figure de diffraction en présence d'un défaut de mise au point (ouverture circulaire). Dans le cas ou il y a défaut de mise au point, l'éclairement en un point de l'image est proportionnel à $|U_{P'}|^2$, $U_{P'}$ étant donné par l'expression (70.1). Le flux contenu dans une couronne élémentaire de surface $2\pi\,\varrho_1\,d\varrho_1$ sera $2\pi\,|U_{P'}|^2\,\varrho_1\,d\varrho_1$ et le flux à l'intérieur du cercle de rayon ϱ_1

$$\mathcal{F}_i = \frac{2\pi R^2}{\lambda^2} \int_0^{\varrho_1} \left[\int_0^{\alpha_0'} \int_0^{2\pi} e^{jK\varepsilon\frac{\alpha'^2}{2}} e^{-jK\alpha'\varrho_1\cos\varphi}\,\alpha'\,d\alpha'\,d\varphi \right]^2 \varrho_1\,d\varrho_1. \tag{71.1}$$

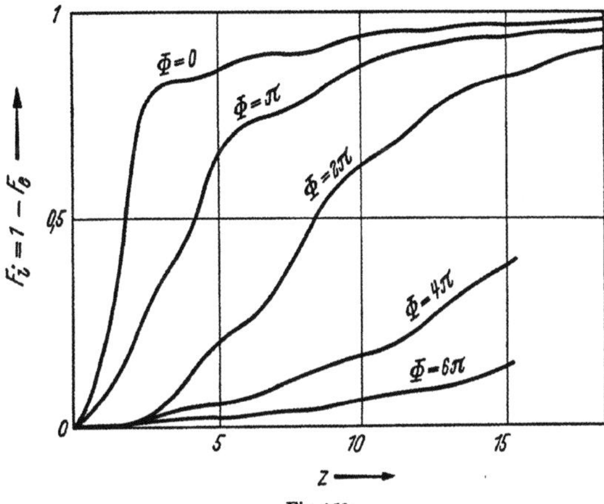

Fig. 150.

Cette intégrale a été calculée par E. WOLF en partant des formules de LOMMEL. Posons

$$Y_n = \sum_{s=0}^{\infty} (-1)^s\,(n+2s)\left(\frac{Z}{2\Phi}\right)^{n+2s} J_{n+2s}(Z),$$

$$P_{2n}(Z) = \sum_{s=0}^{2n} (-1)^s\,[J_s(Z)\,J_{2n-s}(Z) + J_{s+1}(Z)\,J_{2n+1-s}(Z)].$$

Si $|Z/2\Phi| \leq 1$ le flux extérieur au cercle de rayon Z est donné par

$$\left. \begin{aligned} \mathcal{F}_e &= 1 - \left(\frac{Z}{2\Phi}\right)^2\left[1 + \sum_{n=0}^{\infty} \frac{(-1)^n}{2n+1}\left(\frac{Z}{2\Phi}\right)^{2n} P_{2n}(Z)\right] + \\ &+ \frac{4}{2\Phi}\left[Y_1\cos\frac{1}{2}\left(2\Phi + \frac{Z^2}{2\Phi}\right) + Y_2\sin\frac{1}{2}\left(2\Phi + \frac{Z^2}{2\Phi}\right)\right]. \end{aligned} \right\} \tag{71.2}$$

Si $|Z/2\Phi| \geq 1$ on a

$$\mathcal{F}_e = \sum_{n=0}^{\infty} \frac{(-1)^n}{2n+1}\left(\frac{2\Phi}{Z}\right)^{2n} P_{2n}(Z). \tag{71.3}$$

En faisant $\Phi = 0$ l'équation (71.3) se réduit à (69.2) et quand $2\Phi = \pm Z$, les deux équations (71.2) et (71.3) se réduisent a

$$\mathcal{F}_e = J_0(Z)\cos Z + J_1(Z)\sin Z; \tag{71.4}$$

c'est le flux situé en dehors du cercle formé par l'ombre géométrique du bord du diaphragme limitant l'objectif sur le plan d'observation.

Les courbes de la Fig. 150 donnent les résultats numériques obtenus par E. Wolf. La Fig. 151 donne les courbes d'égal éclairement courbes isophotes, au voisinage du foyer d'après Zernike et Nijboer. Les deux lignes droites

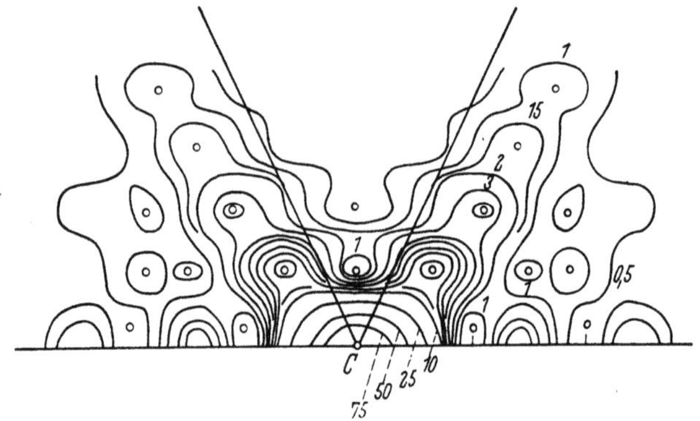

Fig. 151.

indiquent les deux rayons lumineux les plus écartés de l'axe. Le plan de la figure représente un plan méridien et correspond à des valeurs de Z et de Φ comprises entre les valeurs suivantes

$$0 < Z < 15, \quad -35 < 2\Phi < +35.$$

72. Image d'un point en présence d'une ouverture annulaire. Supposons la surface d'onde Σ limitée par deux diaphragmes circulaires de rayons angulaires α_1' et α_2' laissant une surface libre annulaire $\alpha_2' - \alpha_1'$ (Fig. 152).

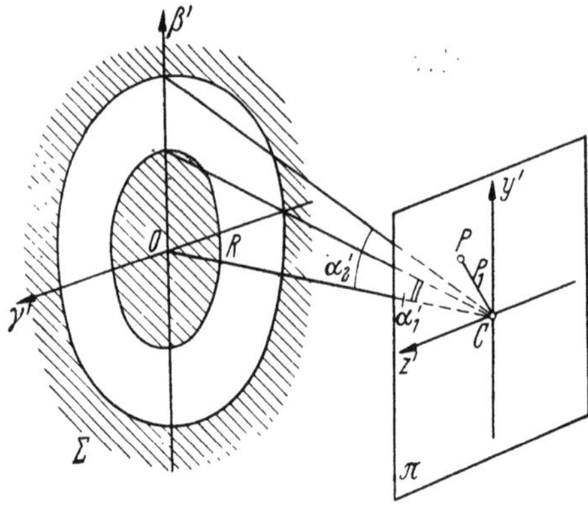

Fig. 152.

D'après la Sect. 68, l'amplitude U_P en P peut s'écrire

$$U_P = j\,\frac{2\pi R}{\lambda} \int_{\alpha_1'}^{\alpha_2'} J_0(K\alpha'\varrho_1)\,\alpha'\,d\alpha' = j\,\frac{2\pi R}{\lambda}\left[\int_0^{\alpha_2'} J_0(K\alpha'\varrho_1)\,\alpha'\,d\alpha' - \int_0^{\alpha_1'} J_0(K\alpha'\varrho_1)\,\alpha'\,d\alpha'\right],$$

$$U_P = \frac{j\pi R}{\lambda}\left[\alpha_2'^2 \frac{2J_1(Z_2)}{Z_2} - \alpha_1'^2 \frac{2J_1(Z_1)}{Z_1}\right] \tag{72.1}$$

où

$$Z_2 = K\alpha_2'\varrho_1, \quad Z_1 = K\alpha_1'\varrho_1$$

et la répartition des éclairements dans l'image

$$E = \frac{\pi^2 R^2}{\lambda^2} \left\{ \alpha_2'^{\,4} \left[\frac{2 J_1(Z_2)}{Z_2} \right]^2 + \alpha_1'^{\,4} \left[\frac{2 J_1(Z_1)}{Z_1} \right]^2 - 2\alpha_1'^{\,2} \alpha_2'^{\,2} \frac{2 J_1(Z_2)}{Z_2} \frac{2 J_1(Z_1)}{Z_1} \right\}. \tag{72.2}$$

L'image de diffraction est représentée sur la Fig. 153 par la courbe en trait plein. La courbe en trait ponctué représente l'image parfaite. Dans le cas d'une ouverture annulaire, le maximum central de la figure de diffraction est réduit et l'éclairement dans les anneaux augmente. La tache centrale s'amincit et les minima restent nuls: ils sont donnés d'après (72.1) par

$$\alpha_2' J_1(K \alpha_2' \varrho_1) = \alpha_1' J_1(K \alpha_1' \varrho_1).$$

Par exemple pour $\alpha_1'/\alpha_2' = 0{,}5$, le rayon du premier anneau noir correspond à

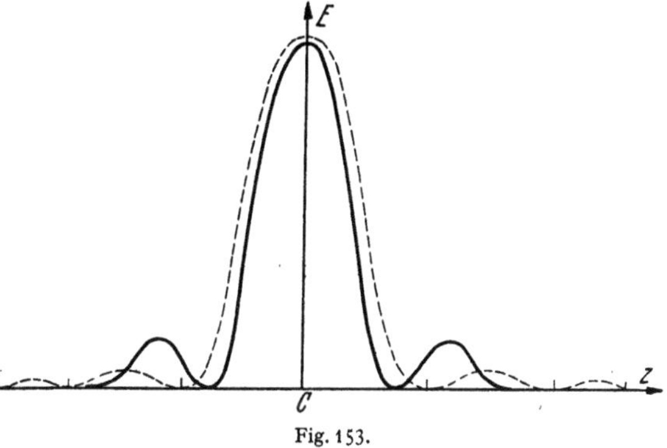

Fig. 153.

$K\alpha_2'\varrho_1 = 3{,}144$ et le rayon du deuxième anneau noir à $K\alpha_2'\varrho_1 = 7{,}183$. Sans l'obturation, on avait $K\alpha_2'\varrho_1 = 3{,}832$ pour le premier anneau noir, le diamètre de la tache centrale est donc réduit de 18% environ.

La relation (72.2) donne l'éclairement du maximum central en faisant $\varrho_1 = 0$ c'est à dire $Z_1 = Z_2 = 0$. On a

$$E_0 = \frac{\pi^2 R^2}{\lambda^2} \cdot (\alpha_2'^{\,4} + \alpha_1'^{\,4} - 2\alpha_1'^{\,2} \alpha_2'^{\,2}) = \frac{\pi^2 R^2}{\lambda^2} (\alpha_2'^{\,2} - \alpha_1'^{\,2})^2. \tag{72.3}$$

Dans le cas d'un objectif ordinaire de rayon angulaire α_2', on aurait

$$E = \frac{\pi^2 R^2}{\lambda^2} \alpha_2'^{\,4}. \tag{72.4}$$

Prenons cet éclairement pour unité, l'éclairement du maximum central pour une ouverture annulaire sera, d'après (72.3):

$$E_0 = \left(1 - \frac{\alpha_1'^{\,2}}{\alpha_2'^{\,2}} \right)^2. \tag{72.5}$$

73. Exemple. Un exemple des ouvertures annulaires nous est fourni par les objectifs de microscope à miroirs. Le schéma de principe de ces objectifs est donné par

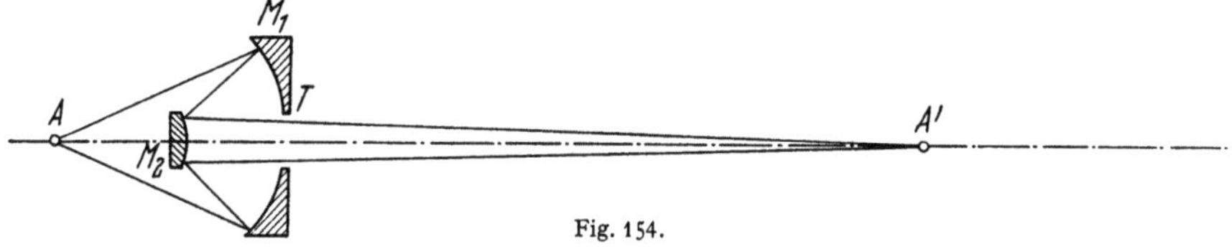

Fig. 154.

la Fig. 154. Les rayons émis par un point de l'objet A sont réfléchis par le miroir concave M_1 puis par un petit miroir convexe M_2. L'ensemble des deux miroirs donne de l'objet A une image A' agrandie, les rayons lumineux passant à travers l'ouverture T pratiquée dans le miroir M_1. On a un faisceau lumineux annulaire dont les limites peuvent être définies dans l'espace image par les angles α_1' et α_2'. L'image A' d'un point A est donc du type de la Fig. 153. Pour que la qualité des images ne soit pas affectée, il faut que la baisse du maximum central ne soit pas trop grande. Toute perte d'énergie par la tache centrale se traduit par un renforcement des anneaux de diffraction, c'est à dire par un étalement de l'image, donc par une diminution de la netteté.

Admettons comme tolérable une baisse du maximum central égal à 0,10. D'après (72.5), on aura

$$\left(1 - \frac{\alpha_1'^2}{\alpha_2'^2}\right)^2 \geq 0,90$$

d'où

$$\frac{\alpha_1'}{\alpha_2'} \leq 0,22$$

ce qui permettra de déterminer les dimensions du petit miroir M_1 par rapport au miroir M_2.

74. Image en lumière blanche (ouverture circulaire). Considérons une source ponctuelle S de lumière blanche à spectre d'égale énergie (Fig. 155). L'image correspondant au plan de mise au point est le disque d'Airy pour une certaine

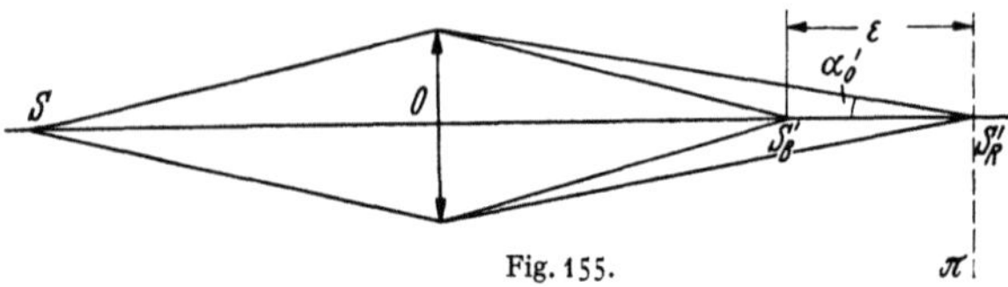

Fig. 155.

longueur d'onde. Par suite du chromatisme de l'objectif O, les images relatives aux autres radiations monochromatiques ne sont pas au point. Soient S_B' et S_R' les foyers de deux radiations quelconques B et R. Le plan π de mise au point passe par exemple par le foyer S_R' et l'image correspondante est le disque d'Airy. Par contre, le foyer S_B' se trouvant à une distance ε du plan π n'est pas au point. La structure de l'image correspondant à la radiation B peut être calculée au moyen des considérations de la Sect. 70. La différence de phase Φ en F_R entre les rayons diffractés par le bord et le centre de l'onde et pour la radiation B est

$$\Phi = \frac{K_B \alpha_0'^2 \varepsilon}{2}.$$

L'éclairement en un point quelconque du plan π pour la radiation B non au point, sera donné par (70.4) ou (70.7). Les images correspondant aux différentes radiations étant en incohérence de phase, l'éclairement en un point du plan π sera égal à la somme, en ce point, des éclairements dûs aux différentes radiations.

Dans bien des cas, il est inutile de calculer un grand nombre de figures de diffraction, on peut se contenter des résultats donnés par la table précédente (voir: Sect. 70). On découpe le spectre en un nombre convenable de bandes assez petites pour que l'on puisse supposer Φ constant dans chaque bande. A chacune des valeurs de Φ correspond une figure de diffraction et on en fait la somme. Pour chaque figure, on porte évidemment le paramètre ϱ_1 en abscisse, donc il faut diviser les ordonnées données par la table, c'est à dire les éclairements, par λ^2 comme il a été vu à la Sect. 68. Si l'image doit être observée par l'oeil supposé optiquement parfait, il faudra tenir compte du coefficient de visibilité des radiations en multipliant chaque ordonnée par le coefficient de visibilité relative à cette radiation d'après la courbe de Gibson et Tyndall.

75. Cas d'un grand nombre d'ouvertures identiques de même orientation et irrégulièrement distribuées. Supposons l'objectif O (Fig. 156) recouvert par un écran D percé d'un grand nombre d'ouvertures identiques de même orientation et irrégulièrement distribuées. Soit M_1 un point d'une de ces petites ouvertures a; posons

$$\delta_1 = M_1 P - OP$$

pour un point quelconque N_1 de cette même ouverture, la différence de marche par rapport à M_1 s'écrit

$$\delta_1' = N_1 P - M_1 P.$$

L'amplitude en P due à l'ouverture a considérée est alors représentée par

$$U_P = j \frac{R}{\lambda} \iint_a e^{j K (\delta_1 + \delta_1')} d\beta'\, d\gamma'. \quad (75.1)$$

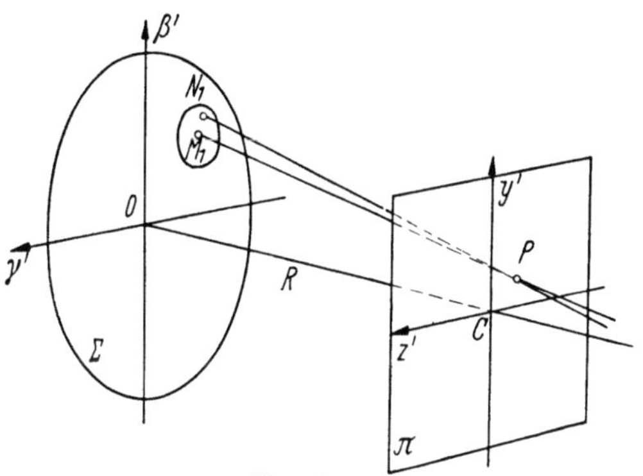

Fig. 156.

En rapportant les différences de marche $\delta_1\, \delta_2 \ldots$ à des points correspondants $M_1 M_2 \ldots$ dans chaque ouverture, l'amplitude dûe à la totalité des ouvertures sera

$$U_P = j \frac{R}{\lambda} \left(e^{j K \delta_1} + e^{j K \delta_2} + \cdots\right) \iint_a e^{j K \delta_1'} d\beta'\, d\gamma'$$

$$= j \frac{R}{\lambda} \left(\Sigma \cos K \delta_1 + j\Sigma \sin K \delta_1\right) \iint_a e^{j K \delta_1'} d\beta'\, d\gamma'$$

d'où l'éclairement

$$E_P = \frac{R^2}{\lambda^2} \left[(\Sigma \cos K \delta_1)^2 + (\Sigma \sin K \delta_1)^2\right] \left[\iint_a e^{j K \delta_1'} d\beta'\, d\gamma'\right]^2. \quad (75.2)$$

On peut écrire

$$(\Sigma \cos K \delta)^2 + (\Sigma \sin K \delta)^2 = \Sigma (\cos^2 K \delta + \sin^2 K \delta) + 2\Sigma \left[\cos K (\delta_i - \delta_j)\right].$$

Chacun des termes de la première somme du second membre étant égal à 1, s'il y a n ouvertures, cette somme est égale à n. Comme il y a un grand nombre d'ouvertures réparties au hasard, les termes de la seconde somme prennent autant de valeurs positives que de valeurs négatives et leur somme est pratiquement nulle.

En appelant E_0 l'éclairement en P dû à une seule ouverture, on aura

$$E_P = n E_0.$$

Le phénomène est donc le même qu'avec une seule ouverture, mais n fois plus lumineux. Si on a n ouvertures circulaires égales et de rayon a, on aura

$$E_P = \frac{n R^2}{\lambda^2} (\pi \alpha_0'^2)^2 \left[\frac{2 J_1(Z)}{Z}\right]^2. \quad (75.3)$$

Considérons l'écran complémentaire du précédent, c'est-à-dire un écran dans lequel les vides sont remplacés par des pleins et vice-versa. La surface libre de l'objectif recouvert de ce nouvel écran est $D - \Sigma a$ et l'amplitude en P devient en posant $\delta = \delta_1' + \delta_1$

$$U_P' = \frac{jR}{\lambda} \iint_{D-\Sigma a} e^{jK\delta} \, d\beta' \, d\gamma'. \tag{75.4}$$

Si le point P où l'on calcule l'amplitude se trouve loin du phénomène correspondant à l'ouverture de l'écran D supposé entièrement libre (objectif O du diamètre D découvert), on a pratiquement dans cette région

$$\frac{jR}{\lambda} \iint_{D} e^{jK\delta} \, d\beta \, d\gamma' \approx 0. \tag{75.5}$$

L'amplitude en P due à l'écran considéré au début de ce paragraphe étant

$$U_P = \frac{jR}{\lambda} \iint_{\Sigma a} e^{jK\delta} \, d\beta' \, d\gamma' \tag{75.6}$$

les relations (75.4), (75.5) et (75.6) montrent que

$$U_P' + U_P = 0$$

et les éclairements

$$|U_P'|^2 = |U_P|^2,$$

d'où le théorème de Babinet: les figures de diffraction produites par deux écrans complémentaires sont identiques.

On doit remarquer que ce théorème n'est valable que dans les régions où l'éclairement dû à toute l'ouverture est pratiquement nul.

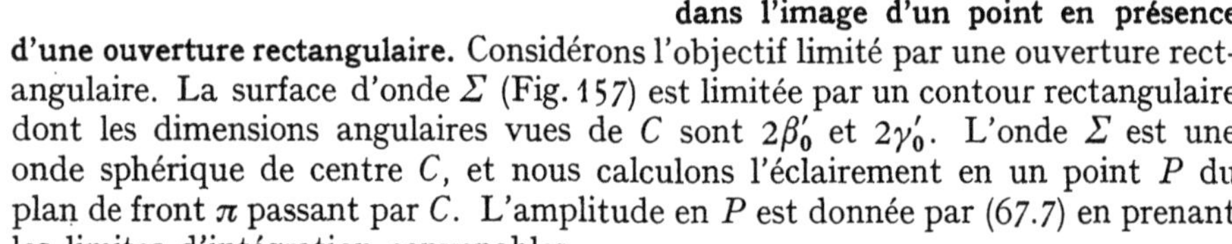

Fig. 157.

76. Répartition des éclairements dans l'image d'un point en présence d'une ouverture rectangulaire.

Considérons l'objectif limité par une ouverture rectangulaire. La surface d'onde Σ (Fig. 157) est limitée par un contour rectangulaire dont les dimensions angulaires vues de C sont $2\beta_0'$ et $2\gamma_0'$. L'onde Σ est une onde sphérique de centre C, et nous calculons l'éclairement en un point P du plan de front π passant par C. L'amplitude en P est donnée par (67.7) en prenant les limites d'intégration convenables.

On a

$$U_P = \frac{jR}{\lambda} \int_{-\gamma_0'}^{+\gamma_0'} \int_{-\beta_0'}^{+\beta_0'} e^{-jK(\beta' y + \gamma' z)} \, d\beta' \, d\gamma' \tag{76.1}$$

d'où

$$U_P = \frac{jR}{\lambda} \left(\frac{e^{-jK\gamma' z'}}{-jKz'} \right)_{-\gamma_0'}^{+\gamma_0'} \left(\frac{e^{-jK\beta' y'}}{-jKy'} \right)_{-\beta_0'}^{+\beta_0'} = \frac{4jR}{\lambda} \beta_0' \gamma_0' \frac{\sin K\gamma_0' z'}{K\gamma_0' z'} \frac{\sin K\beta_0' y'}{K\beta_0' y'}$$

et l'éclairement

$$E_P = \frac{16R^2 \beta_0'^2 \gamma_0'^2}{\lambda^2} \left(\frac{\sin K\beta_0' y'}{K\beta_0' y'} \right)^2 \left(\frac{\sin K\gamma_0' z'}{K\gamma_0' z'} \right)^2. \tag{76.2}$$

Il est égal au produit de deux facteurs qui dépendent uniquement: le premier de la coordonnée y', le second, de la coordonnée z'. On observe dans le plan π

deux séries de franges noires entièrement obscures formant une sorte de quadrillage (Fig. 158). Les franges parallèles à cy' sont données par

$$K\gamma_0' z' = p\pi, \quad z' = p\,\frac{\lambda}{2\gamma_0'} \quad (p \neq 0) \tag{76.3}$$

et les franges parallèles à Cz' par

$$K\beta_0' y' = p\pi, \quad y' = p\,\frac{\lambda}{2\beta_0'} \quad (p \neq 0). \tag{76.4}$$

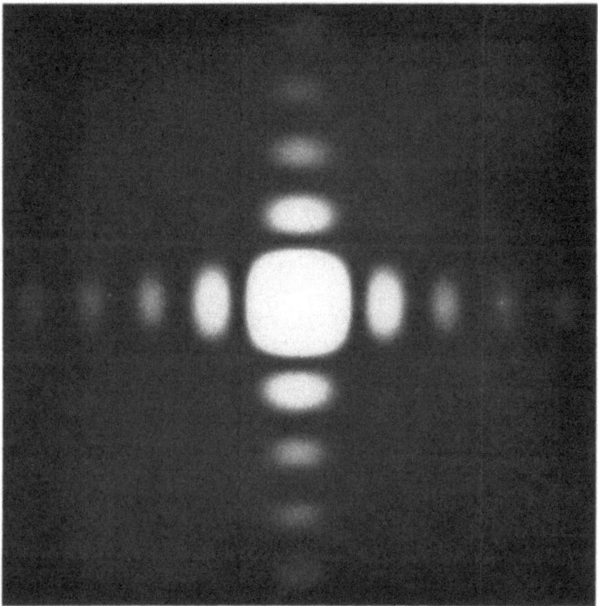

Fig. 158. Figure de diffraction par une ouverture carrée.

La répartition des éclairements suivant une parallèle à Cy' ou Cz' est donnée par les variations de la fonction $\left(\dfrac{\sin x}{x}\right)^2$ en prenant x égal à $K\beta_0' y'$ ou $K\gamma_0' z'$.

Tableau 9.

x	x		$x' = x + \pi$		$x' = x + 2\pi$		$x' = x + 3\pi$	
	$\dfrac{\sin x}{x}$	$\left(\dfrac{\sin x}{x}\right)^2$	$\dfrac{\sin x'}{x'}$	$\left(\dfrac{\sin x'}{x'}\right)^2$	$\dfrac{\sin x'}{x'}$	$\left(\dfrac{\sin x'}{x'}\right)^2$	$\dfrac{\sin x'}{x'}$	$\left(\dfrac{\sin x'}{x'}\right)^2$
0	1	1	0	0	0	0	0	0
$\pi/12$	0,9889	0,9774	$-0,0760$	0,0058	0,0396	0,0016	$-0,0267$	0,0007
$2\pi/12$	9546	9119	1364	0186	0735	0054	0503	0025
$3\pi/12$	9003	8105	1801	0324	1001	0100	0692	0048
$4\pi/12$	8270	6839	2067	0427	1183	0140	0828	0069
$5\pi/12$	7379	5445	2170	0471	1274	0162	0900	0081
$6\pi/12$	6366	4053	2122	0450	1274	0162	0909	0083
$7\pi/12$	5271	2778	1942	0377	1191	0142	0858	0074
$8\pi/12$	4135	1710	1654	0274	1035	0107	0752	0057
$9\pi/12$	3001	0901	1286	0165	0819	0067	0600	0036
$10\pi/12$	1910	0365	0868	0075	0507	0026	0416	0017
$11\pi/12$	0899	0081	0430	0018	0282	0008	0210	0004

Le long de cz' par exemple, on a comme distribution des éclairements, la courbe de la Fig. 159. Tous les minima sont équidistants et donnés par (76.3).

Supposons que l'une des dimensions de l'ouverture rectangulaire devienne petite devant l'autre: γ_0' est par exemple très petit par rapport à β_0'. L'ouverture

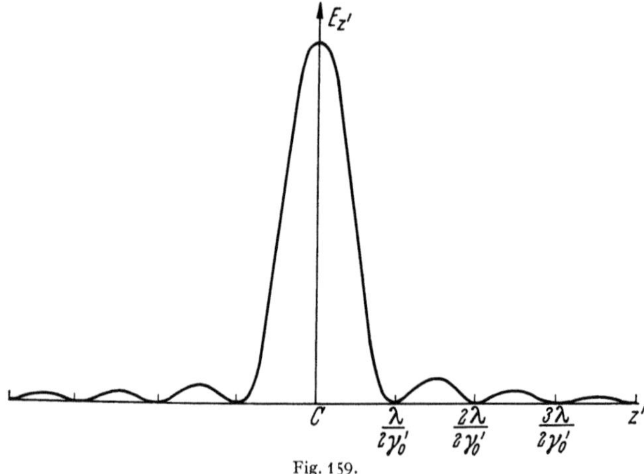

Fig. 159.

se réduit à une fente dirigée suivant $O\beta'$. Les franges parallèles à cz' sont extrêmement serrées par rapport aux franges parallèles à cy': la diffraction suivant cy' est négligeable devant la diffraction qui étale la lumière le long de cz'. Pratiquement la lumière s'accumule le long de cz' suivant la loi donnée par $(y'=0)$

$$E_{z'} = \frac{16 R^2 \gamma_0'^2 \beta_0'^2}{\lambda^2} \left(\frac{\sin K \gamma_0' z'}{K \gamma_0' z'} \right)^2. \tag{76.5}$$

C'est la distribution des éclairements dans la figure de diffraction donnée par un efente fine. La structure de cette image est représentée par la Fig. 159.

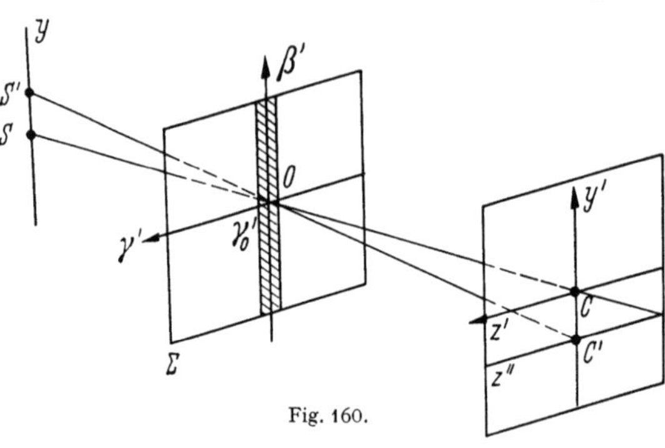

Fig. 160.

Dans ce qui précède, on a toujours considéré une source ponctuelle: la source S a pour image géométrique C et si la surface d'onde Σ est limitée par une fente fine de largeur γ_0' (Fig. 160), la lumière diffractée s'étend le long de cz'. Soit S' une autre source ponctuelle située sur la parallèle sy à la fente fine qui est dirigée suivant $O\beta'$. Son image géométrique C' se trouve sur Cy' et la lumière diffractée s'étend le long de Cz'' parallèle à Cz'. Si tous les points de Sy sont lumineux, leurs images situées sur Cy' produisent des phénomènes de diffraction qui se juxtaposent. Les phénomènes sont donc parfaitement nets si on remplace la source ponctuelle S par une fente fine lumineuse (Fig. 161). Ceci est également vrai, même si la fente source n'est pas parallèle à la fente placée contre l'objectif: les franges sont parallèles à la fente source et le contraste du phénomène n'est pas changé.

Si le diaphragme placé contre l'objectif est un rectangle, les phénomènes sont modifiés, il faut faire l'intégration pour tous les points de la source. Prenons comme source une fente fine dirigée parallèlement à l'un des côtés de l'ouverture rectangulaire, $O\beta'$ par exemple (Fig. 157), et soit $2y'_0$ sa hauteur. L'éclairement $E_{z'}$ le long de Cz' s'écrira

$$E_{z'} = \frac{16R^2\beta_0'^2\gamma_0'^2}{\lambda^2}\left(\frac{\sin K\gamma_0'z'}{K\gamma_0'z'}\right)^2 \int\limits_{-y'_0}^{+y'_0}\left(\frac{\sin K\beta_0'y'}{K\beta_0'y'}\right)^2 d\,\frac{y'}{R}\,, \tag{76.6}$$

$$E_{z'} = -\frac{8R\beta_0'\gamma_0'^2}{\pi\lambda}\left(\frac{\sin K\gamma_0'z'}{K\gamma_0'z'}\right)^2 \int\limits_{-K\beta_0'y'_0}^{+K\beta_0'y'_0}\sin^2 K\beta_0'y'\,d\left(\frac{1}{K\beta_0'y'}\right)$$

et en intégrant par parties

$$E_{z'} = \frac{16R\beta_0'\gamma_0'^2}{\pi\lambda}\left(\frac{\sin K\gamma_0'z'}{K\gamma_0'z'}\right)^2\left[\,\mathrm{Si}\,(2K\beta_0'y'_0) - \frac{\sin^2 K\beta_0'y'_0}{K\beta_0'y'_0}\right] \tag{76.7}$$

Fig. 161. Figure de diffraction par une fente fine.

où Si désigne la fonction sinus-intégral. Comme y'_0 est toujours très grand par rapport à la longueur d'onde, pratiquement $\mathrm{Si}\,(2K\beta_0'y'_0)\approx\pi/2$ et le deuxième terme entre crochets est nul, d'où

$$E_{z'} = \frac{8R\beta_0'\gamma_0'^2}{\lambda}\left(\frac{\sin K\gamma_0'z'}{K\gamma_0'z'}\right)^2. \tag{76.8}$$

A un facteur constant près, la figure de diffraction le long de cz' a la même structure que si l'on avait utilisé une source ponctuelle. La courbe de la Fig. 159 représente encore la distribution des éclairements le long de cz' lorsqu'on place contre l'objectif une ouverture rectangulaire $4\beta_0'y'_0$ et que la source est une fente fine parallèle à $O\beta_0'$. Le calcul montre facilement, que cette même courbe représente également la distribution des éclairements le long d'une parallèle à cz' si l'on ne s'approche pas trop près des extrémités de la fente.

Ces résultats permettent de calculer le pouvoir séparateur obtenu par une ouverture rectangulaire. Considérons comme sources, deux fentes fines incohérentes parallèles à $O\beta_0'$. Chacune de ces fentes a pour image dans le plan π une figure de diffraction donnée par (76.8). Comme pour l'ouverture circulaire, on peut admettre que les deux images seront vues séparées, si le maximum de l'une correspond au premier minimum nul de l'autre.

On aura donc

$$z' = \frac{\lambda}{2\gamma_0'}\,. \tag{76.9}$$

Si $D = 2R\gamma_0'$ est la largeur de l'ouverture rectangulaire suivant une parallèle à CZ', le pouvoir séparateur angulaire $\vartheta = z'/R$ s'écrira

$$\vartheta = \frac{\lambda}{D}\,. \tag{76.10}$$

L'exemple d'un instrument limité par une ouverture rectangulaire nous est fourni par le spectroscope à prisme (Fig. 162) où les faisceaux sont limités par les faces d'entrée ou de sortie des prismes qui ont la forme de rectangles. Considérons deux radiations très voisines λ et $\lambda + d\lambda$: chacune donne naissance à une image de diffraction dont la structure vient d'être étudiée. Les centres de ces figures de diffraction sont en F_1' et F_2' par exemple, et sont vues du centre optique

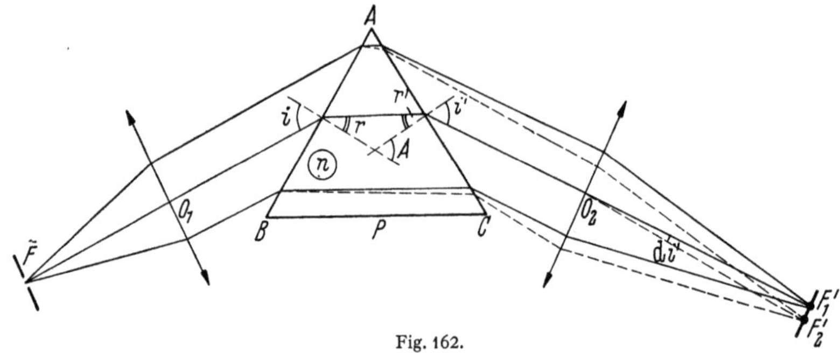

Fig. 162.

de l'objectif O_2 sous l'angle di'. Si $|di'| \geq \vartheta = \lambda/D$, on peut admettre que les deux raies spectrales correspondantes sont vues séparées. Les formules du prisme permettent de calculer di'

$$\left. \begin{aligned} \sin i &= n \sin r, \\ \sin i' &= n \sin r', \\ r + r' &= A. \end{aligned} \right\}$$

Différentions

$$\left. \begin{aligned} n \cos r \, dr + \sin r \, dn &= 0, \\ \cos i' \, di' = n \cos r' \, dr' &+ \sin r' \, dn, \\ dr + dr' &= 0. \end{aligned} \right\}$$

De la première et de la troisième équation, on tire dr' et en remplaçant dans la deuxième, on a

$$\cos i' \, di' = \frac{\sin A}{\cos r} \, dn.$$

Si $e = BC$ est la base du prisme et D la largeur du faisceau émergent, on a

$$\frac{e}{\sin A} = \frac{AC}{\sin\left(\dfrac{\pi}{2} - r\right)} = \frac{D}{\cos i' \cos r}$$

d'où

$$di' = \frac{e}{D} \, dn.$$

Il faut donc que l'on ait

$$\frac{e\,|dn|}{D} \geq \frac{\lambda}{D}, \qquad \frac{\lambda}{|d\lambda|} \leq e \left| \frac{dn}{d\lambda} \right|.$$

Le quotient $R = \lambda/d\lambda$ s'appelle le pouvoir de résolution du spectroscope, il est proportionnel à l'épaisseur e de verre traversé. Cherchons par exemple quelle doit être l'épaisseur e pour séparer les deux raies D_1 et D_2 du sodium. On a

$$\lambda = 5893 \,\text{Å}, \quad d\lambda = 6 \,\text{Å}, \quad R = \frac{5893}{6} \approx 1000.$$

Prenons la formule

$$n = A + \frac{B}{\lambda^2}$$

on a

$$\left|\frac{dn}{d\lambda}\right| = \frac{2B}{\lambda^3}$$

pour les flints

$$B \simeq 10^{-10}\,\text{cm.}^2$$

d'où

$$\left|\frac{dn}{d\lambda}\right| = \frac{2\cdot10^{-10}}{205\cdot10^{-15}} \approx 1000$$

et

$$e = \frac{R}{\left|\dfrac{dn}{d\lambda}\right|} = 1\,\text{cm.}$$

Pour augmenter e c'est-à-dire la résolution, on augmente le nombre des prismes. Avec des spectroscopes à prismes, on peut atteindre $R = 100000$, chiffre que l'on dépasse avec les réseaux. Notons que dans les spectrographes à grande ouverture, où l'objectif O_2 peut atteindre et même dépasser $f/1$, les effets de la diffraction sont moins importants que les résidus d'aberration qui limitent généralement le pouvoir de résolution. Enfin, le grain du récepteur, oeil ou plaque photographique, intervient également et dans beaucoup de cas, c'est lui qui fixe la limite de séparation.

77. Diffraction par deux fentes. α) *Cas d'une source ponctuelle.* Considérons la surface d'onde Σ recouverte d'un écran percé de deux fentes identiques angulairement caractérisées par leur largeur $2\gamma_0'$, leur hauteur $2\beta_0'$ leur intervalle $AA' = \eta$ et parallèles à $O\beta'$ (Fig. 163). La source S est une source ponctuelle située sur l'axe OC dont l'image en C est une figure de diffraction que l'on veut étudier. D'après la relation (67.7), l'amplitude en un point P du plan π passant par le centre C de l'onde Σ est donnée par

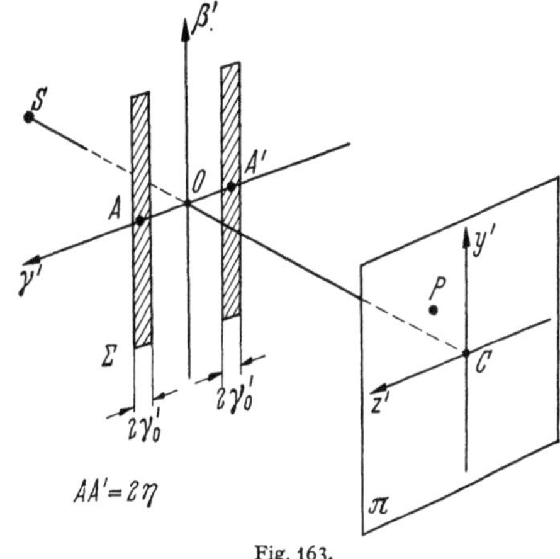

Fig. 163.

$$U_P = \frac{jR}{\lambda}\left[\int_{-\beta_0'}^{+\beta_0'} e^{-iK\beta'y'}\,d\beta' \int_{-\left(\frac{\eta}{2}+\gamma_0'\right)}^{-\left(\frac{\eta}{2}-\gamma_0'\right)} e^{-iK\gamma'z'}\,d\gamma' + \int_{-\beta_0}^{+\beta_0} e^{-iK\beta'y'}\,d\beta' \int_{\frac{\eta}{2}-\gamma_0'}^{\frac{\eta}{2}+\gamma_0'} e^{-iK\gamma'z'}\,d\gamma'\right]. \quad (77.1)$$

On peut supposer comme précédemment, que la hauteur $2\beta_0'$ des fentes est grande par rapport à leur largeur $2\gamma_0'$. Dans ces conditions, les intégrales en β' n'interviennent pas: le flux lumineux est pratiquement réparti le long de cz' et l'amplitude en un point quelconque de Cz' sera:

$$U_{z'} = \frac{2jR\beta_0'}{\lambda}\left[\int_{-\left(\frac{\eta}{2}+\gamma_0'\right)}^{-\left(\frac{\eta}{2}-\gamma_0'\right)} e^{-iK\gamma'z'}\,d\gamma' + \int_{\frac{\eta}{2}-\gamma_0'}^{\frac{\eta}{2}+\gamma_0'} e^{-iK\gamma'z'}\,d\gamma'\right] \quad (77.2)$$

d'où

$$U_{z'} = \frac{8jR\gamma_0'\beta_0'}{\lambda}\,\frac{\sin K z'\gamma_0'}{K z'\gamma_0'}\cos\frac{K z'\eta}{2} \quad (77.3)$$

et la répartition des éclairements

$$E_{z'} = \frac{64\,R^2\,\gamma_0'^2\,\beta_0'^2}{\lambda^2}\left(\frac{\sin K z'\gamma_0'}{K z'\gamma_0'}\right)^2 \cos^2\frac{K z'\eta}{2}. \qquad (77.4)$$

L'éclairement est le produit de deux facteurs: le premier $\left(\dfrac{\sin K z'\gamma_0'}{K z'\gamma_0'}\right)^2$ représente la figure de diffraction par l'une ou l'autre des deux fentes. Les variations de ce facteur sont données par la courbe en trait ponctué sur la Fig. 164. Le second facteur $\cos^2\dfrac{K z'\eta}{2}$ est un facteur sinusoïdal variant entre 0 et 1. Il présente un maximum pour $z'=0$, c'est à dire qu'il y a en C un maximum de lumière. On

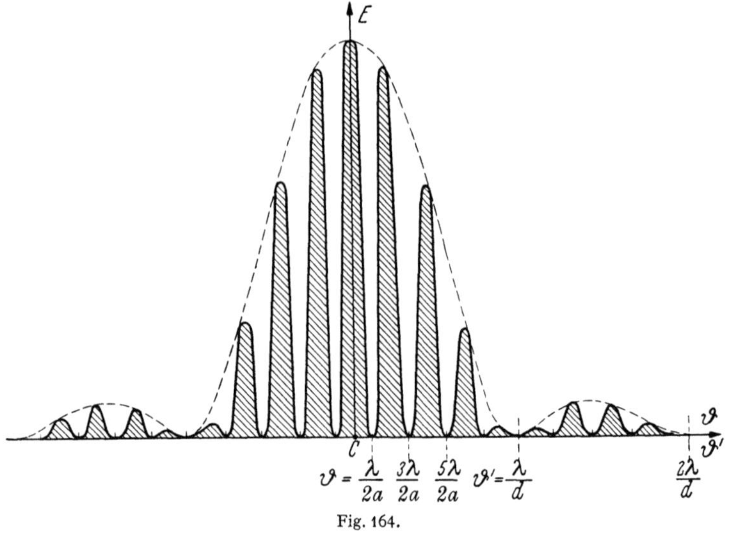

Fig. 164.

ne change rien encore aux phénomènes si on remplace la source ponctuelle S par une fente fine parallèle à $O\beta'$. Les minima nuls correspondent à:

$$K z'\eta = (2p+1)\,\pi$$

ou angulairement en posant $\vartheta = z'/R$ et $a = R\eta$

$$\left.\begin{aligned}\vartheta &= (2p+1)\frac{\lambda}{2R\eta} = (2p+1)\frac{\lambda}{2a}\,,\\[2mm]\vartheta &= \frac{\lambda}{2a}\ \frac{3\lambda}{2a}\ \frac{5\lambda}{2a}\cdots.\end{aligned}\right\} \qquad (77.5)$$

Ce sont des franges d'interférences équidistantes et parallèles à $O\beta'$, l'intervalle entre deux franges noires consécutives étant égal à λ/a. Le premier minimum nul de la figure de diffraction est donné par $z'=\lambda/2\gamma_0'$ ou angulairement par

$$\vartheta' = \frac{\lambda}{d}$$

en posant $d = 2R\gamma_0'$. Comme d est plus petit que a, l'angle ϑ' est supérieur à ϑ et les franges d'interférences sont plus serrées que les franges de diffraction. Elles sont représentées sur la Fig. 164 par la courbe en trait plein. On a des franges d'interférences d'Young modulées par le phénomène de diffraction dû à l'une ou l'autre des fentes. La Fig. 165 a montre la figure de diffraction de l'une des fentes: elle est représentée par le facteur $\left(\dfrac{\sin K z'\gamma_0'}{K z'\gamma_0'}\right)^2$ de l'expression

(77.4). La Fig. 165 b montre l'ensemble du phénomène tel quil est donné par (77.4). On retrouve l'aspect général de la Fig. 165 a mais avec les franges d'interférences qui apparaissent dans les franges de diffraction successives.

a

b

Fig. 165 a et b. a Figure de diffraction par une fente fine. b Figure de diffraction par deux fentes fines. Les franges d'interférences sont modulées par le phénomène de diffraction d'une fente.

β) Cas de deux sources ponctuelles incohérentes. Considérons deux sources ponctuelles incohérentes S et S' (Fig. 166) décalées, non plus dans un sens parallèle à $O\beta'$ comme à la Sect. 65, mais perpendiculairement à la direction des deux fentes. Chacune des deux sources donne dans le plan π passant par le centre C de l'onde Σ une figure de diffraction composée de franges d'interférences. Les centres de ces deux figures de diffraction sont en C et C' et elles sont identiques ; l'écartement des franges d'interférences est donné par (77.5). Supposons que l'écartement angulaire α des deux sources soit tel que les franges brillantes de S coïncident avec les franges noires

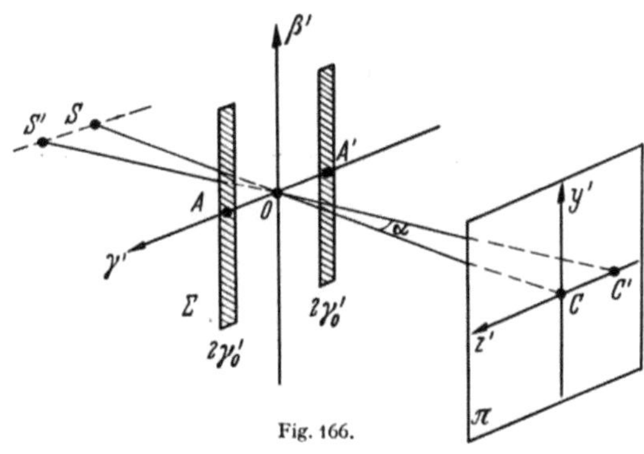

de S': le phénomène disparaît, il n'y a plus de franges d'interférences visibles. La première disparition a lieu lorsque $\alpha = \lambda/2a$, la deuxième lorsque $\alpha = 3\lambda/2a$ etc. La condition de disparition des franges est donc

$$\alpha = (2p + 1)\frac{\lambda}{2a}.\tag{77.6}$$

L'expérience peut être réalisée avec un objectif astronomique en prenant comme sources S et S' les composantes d'une étoile double. On fait varrier maintenant l'écartement a des deux fentes placées contre l'objectif. Les fentes

A et A' sont d'abord très rapprochées et on voit des franges nettes. En les écartant, on voit les franges d'interférences disparaître, réapparaître, etc. On peut donc mesurer la distance angulaire α des composantes d'une étoile double en mesurant l'écartement a des fentes A et A' au moment de la première disparition par exemple. La distance angulaire α des deux étoiles est donnée par (77.6) en faisant $p = 1$.

On a pu mesurer ainsi des composantes dont la distance angulaire était de l'ordre de quelques centièmes de seconde d'arc avec une précision du millième. L'application de la même expérience au cas d'une source étendue sera traitée au chapitre C de cet article.

$\gamma)$ *Remarque.* Reprenons l'expression (77.2) que l'on peut écrire

$$U_{z'} = \frac{4j\,R\,\gamma_0'\,\beta_0'}{\lambda}\left[\frac{\sin K\,z'\,\gamma_0'}{K\,z'\,\gamma_0'}\,e^{\frac{jK z'\eta}{2}} + \frac{\sin K\,z'\,\gamma_0'}{K\,z'\,\gamma_0'}\,e^{-\frac{jK z'\eta}{2}}\right]. \tag{77.7}$$

Les termes entre parenthèses représentent les amplitudes en un point de cz' dûes à chaque fente. L'amplitude dûe à une fente peut s'écrire

$$\frac{\sin K\,z'\,\gamma_0'}{K\,z'\,\gamma_0'}\,e^{\frac{jK z'\eta}{2}}.$$

On voit que si l'on déplace la fente sur l'onde Σ dont un suppose la courbure faible, η varie et par conséquent la phase $\dfrac{K\,z'\,\eta}{2}$ de la vibration résultante en un point de $C z'$ est modifiée. Par contre le facteur $\dfrac{\sin K\,z'\,\gamma_0'}{K\,z'\,\gamma_0'}$ reste constant et la distribution des éclairements ne change pas. De même si l'on fait varier l'orientation de la fente, la figure de diffraction tourne autour de C en même temps que la fente, mais sa structure ne change pas. Quelle que soit la position de la fente dans son plan, la figure de diffraction reste inchangée. Ceci est d'ailleurs un fait général, lorsqu'on déplace le diaphragme percé d'une ouverture quelconque dans son plan, la répartition des éclairements reste la même, seules les phases sont modifiées.

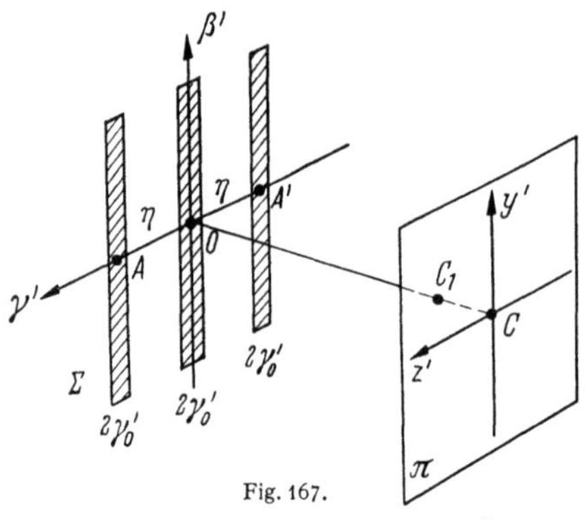

Fig. 167.

78. Diffraction par plusieurs fentes.

Plaçons trois fentes identiques devant l'objectif O (Fig. 167): leurs largeurs sont égales à $2\gamma_0'$ et leurs intervalles sont $OA = OA' = \eta$. D'après la remarque du paragraphe précédent, l'amplitude en un point de cz' sera

$$U_{z'} = \frac{4j\,R\,\gamma_0'\,\beta_0'}{\lambda}\left[\frac{\sin K\,z'\,\gamma_0'}{K\,z'\,\gamma_0'}\,e^{jK z'\eta} + \frac{\sin K\,z'\,\gamma_0'}{K\,z'\,\gamma_0'} + \frac{\sin K\,z'\,\gamma_0'}{K\,z'\,\gamma_0'}\,e^{-jK z'\eta}\right] \tag{78.1}$$

d'où

$$U_{z'} = \frac{4j\,R\,\gamma_0'\,\beta_0'}{\lambda}\,\frac{\sin K z'\,\gamma_0'}{K\,z'\,\gamma_0'}\,(1 + 2\cos K z'\,\eta)$$

et la répartition des éclairements

$$E_{z'} = \frac{16\,R^2\,\gamma_0'^2\,\beta_0'^2}{\lambda^2}\left(\frac{\sin K z'\,\gamma_0'}{K z'\,\gamma_0'}\right)^2 (1 + 2\cos K z'\,\eta)^2$$

que l'on peut encore écrire

$$E_{z'} = \frac{16\,R^2\gamma_0'^2\,\beta_0'^2}{\lambda^2}\left(\frac{\sin K z'\gamma_0'}{K z'\gamma_0'}\right)^2\left(\frac{\sin 3\,\dfrac{K z'\eta}{2}}{\sin\dfrac{K z'\eta}{2}}\right)^2.\tag{78.2}$$

la structure de la figure de diffraction est donnée par la Fig. 168. Le premier facteur correspond aux franges de diffraction de l'une des trois fentes (courbe

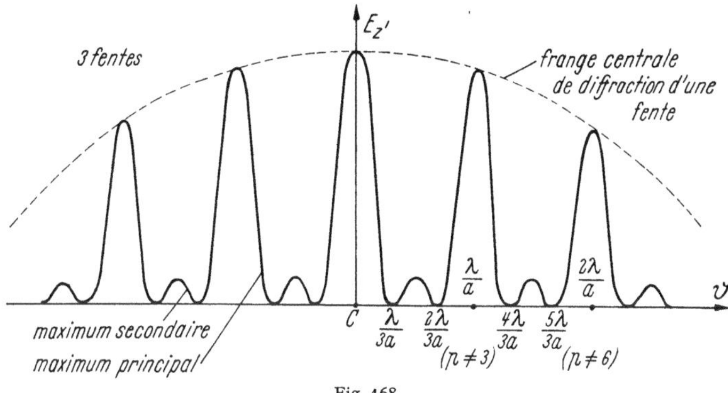

Fig. 168.

en traits ponctués) et le second facteur, à des franges d'interférences (courbe en trait plain).

Etudions ce deuxième facteur: il passe par des maxima si

$$K z'\eta = 2 p\,\pi, \qquad \vartheta = \frac{z'}{R} = p\,\frac{\lambda}{a}\tag{78.3}$$

et on a

$$\left(\frac{\sin 3\,\dfrac{K z'\eta}{2}}{\sin\dfrac{K z'\eta}{2}}\right)_{K z'\eta = 2 p\pi} = 3.$$

Il passe par des minima nuls si

$$\begin{aligned}\sin 3\,\frac{K z'\eta}{2} &= 0 \\ \sin\frac{K z'\eta}{2} &\neq 0\end{aligned}\left\{\begin{aligned}&K z'\eta = \frac{2 p\,\pi}{3},\quad \vartheta = \frac{z'}{R} = p\,\frac{\lambda}{3a}\\ &p \neq 0,\quad p \neq \text{d'un multiple de 3}\end{aligned}\right\}.\tag{78.4}$$

Enfin, en dérivant ce facteur, on obtient les maxima donnés par

$$3\tan\frac{K z'\eta}{2} = \tan 3\,\frac{K z'\eta}{2}.\tag{78.5}$$

Les maxima donnés par (78.3) sont appelés maxima principaux. Les maxima donnés par (78.5) sont peu marqués, ce sont les maxima secondaires. D'après ce qui précède, on voit que l'on a deux minima nuls et un maximum secondaire entre deux maxima principaux.

Sur la Fig. 168 on a représenté seulement une portion de la frange centrale de diffraction de l'une des fentes. On retrouve ce que nous avons déjà vu précédemment: la figure de diffraction due à l'une des fentes module les maxima principaux d'interférences.

La Fig. 169 montre l'aspect des franges. Le phénomène de diffraction produit par trois fentes a des propriétés intéressantes qui ont été utilisées par Zernike, de la façon suivante: au lieu d'observer le phénomène dans le plan π passant par le centre C de l'onde Σ, plaçons-nous un peu en avant ou un peu en arrière, et soit C_1 (Fig. 167) l'intersection avec l'axe de la nouvelle position π' du plan d'observation. La structure du phénomène de diffraction change rapidement

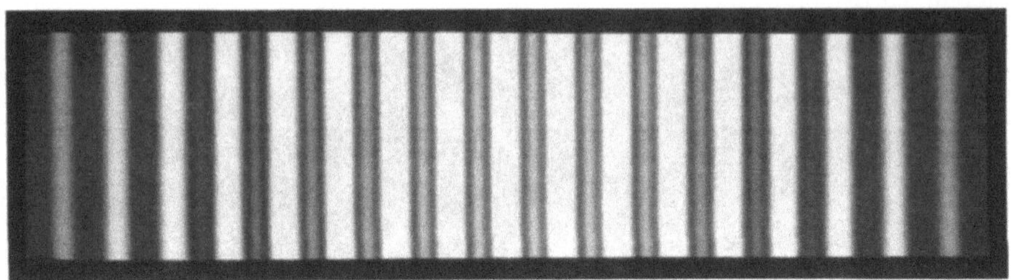

Fig. 169. Figure de diffraction par trois fentes.

d'aspect. Lorsque le plan d'observation vient à une distance de π telle qu'en C_1 la différence de marche $A'C_1 - OC_1 = AC_1 - OC_1 = \pm \dfrac{\lambda}{4}$ le phénomène de diffraction prend l'aspect particulier de la Fig. 170.

Dans sa partie centrale, le phénomène de diffraction montre des franges serrées et d'intensités à peu près égales. Si l'on déplace très peu le plan de mise au point, l'aspect des franges change aussitôt. Le pointé de ces plans est extrêmement sensible, c'est à dire qu'un très faible changement de mise au point, par conséquent de la différence de marche, modifie beaucoup l'aspect des franges. Au lieu de changer la mise au point, on peut placer sur la fente centrale une lame transparente introduisant une différence de marche égale à $\lambda/4$. On observera le même phénomène. Si, au moyen d'un compensateur, on fait varier un peu la différence de marche introduite par la lame, l'aspect des franges change aussitôt. En plaçant une lame d'épaisseur telle que la différence de marche qu'elle introduise soit voisine de $\lambda/4$, on peut avec le compensateur, déterminer exactement la différence de marche $\lambda/4$, c'est à dire mesurer la différence de marche produite par la lame avec une très grande précision ($\lambda/1000$).

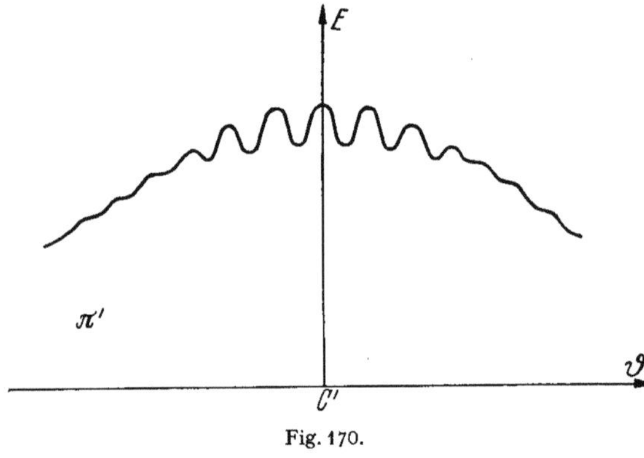

Fig. 170.

Considérons maintenant le cas de n fentes placées devant l'objectif O. L'amplitude en un point de cz' sera avec les mêmes notations si n est impair:

$$U_{z'} = \frac{4\,j\,R\,\gamma_0'\,\beta_0'}{\lambda}\,\frac{\sin K z'\gamma_0'}{K z'\gamma_0'}\Big[1 + e^{-jKz'\eta} + e^{-2jKz'\eta} + \cdots$$
$$+ e^{-\left(\frac{n-1}{2}\right)jKz'\eta} + e^{jKz'\eta} + e^{2jKz'\eta} + \cdots + e^{\left(\frac{n-1}{2}\right)jKz'\eta}\Big].$$

En faisant la somme des deux progressions géométriques

$$1 + e^{-jKz'\eta} + e^{-2jKz'\eta} + \cdots + e^{-\left(\frac{n-1}{2}\right)jKz'\eta}$$

et

$$e^{jKz'\eta}\left(1 + e^{jKz'\eta} + \cdots + e^{\frac{n-3}{2}jKz'\eta}\right)$$

on obtient

$$U_{z'} = \frac{4j\,R\,\gamma_0'\,\beta_0'}{\lambda}\,\frac{\sin K z'\gamma_0'}{K z'\gamma_0'}\,\frac{\sin n\,\dfrac{K z'\eta}{2}}{\sin\dfrac{K z'\eta}{2}} \tag{78.6}$$

et la répartition des éclairements

$$E_{z'} = \frac{16\,R^2\,\gamma_0'^2\,\beta_0'^2}{\lambda^2}\left(\frac{\sin K z'\gamma_0'}{K z'\gamma_0'}\right)^2\left(\frac{\sin n\,\dfrac{K z'\eta}{2}}{\sin\dfrac{K z'\eta}{2}}\right)^2. \tag{78.7}$$

On trouverait les mêmes formules pour n pair.

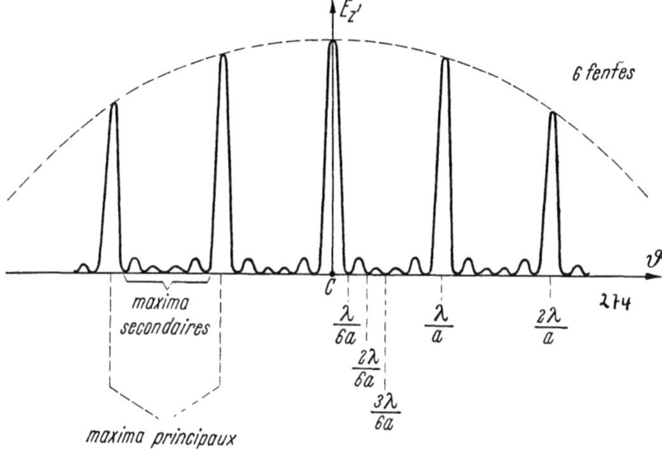

Fig. 171.

Les Fig. 171 et 172 ,correspondant respectivement à $n = 6$ et $n = 10$, peuvent être comparées à la Fig. 168 et montrent l'évolution du phénomène de diffraction

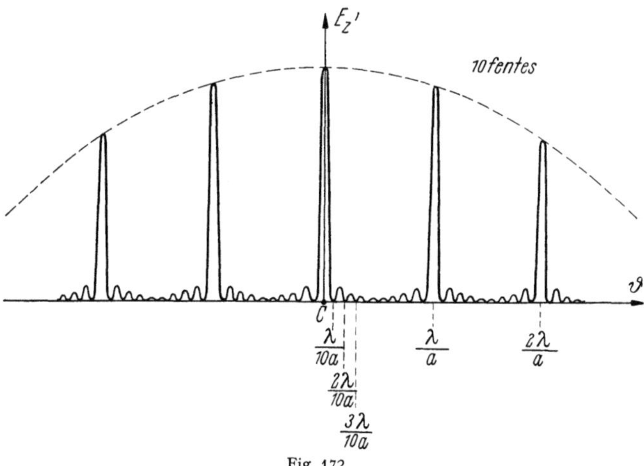

Fig. 172.

lorsque le nombre des fentes augmente. Les Fig. 173 a et b peuvent être également comparées à la Fig. 169. On voit que les maxima principaux ont une position fixe et qu'ils deviennent de plus en plus fins, le nombre des maxima secondaires entre deux maxima principaux augmentant de plus en plus. Lorsque n devient très grand, on est en présence du phénomène de diffraction produit par un réseau que nous étudions maintenant.

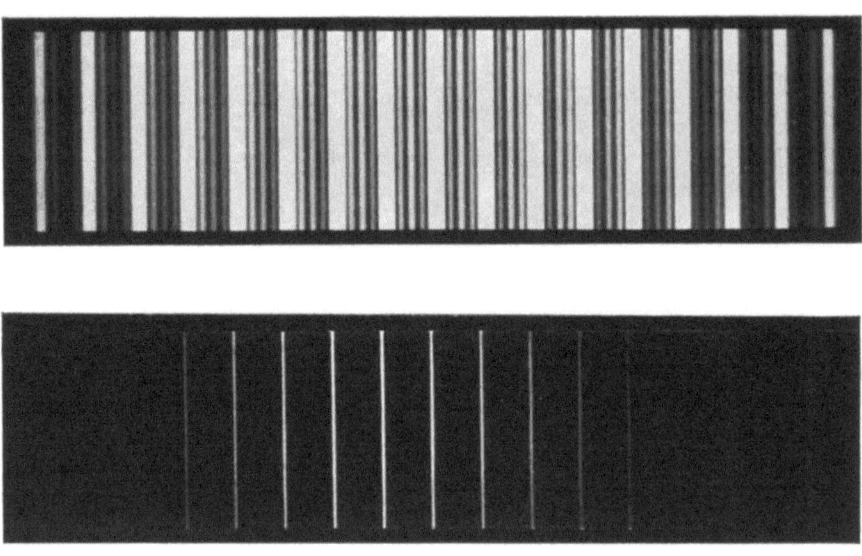

b

Fig. 173a et b. a Figure de diffraction par 5 fentes fines. b Figure de diffraction par 40 fentes.

79. Réseau plan éclairé en lumière monochromatique. On appelle réseau plan un écran percé d'un grand nombre de fentes fines parallèles, situées dans un même plan, égales et équidistantes. La distance qui sépare deux points homologues de deux fentes voisines, est la période du réseau. Les premiers réseaux ont été construits par Fraunhofer au début du XIX° siècle; ils étaient formés par des fils métalliques tendus sur deux vis identiques et parallèles. La période du réseau était celle du pas de la vis et ne pouvait guère être inférieure au $\frac{1}{10}$ de millimètre.

Fraunhofer a obtenu ensuite des réseaux à traits plus serrés en traçant à l'aide d'un diamant, des traits équidistants sur une lame de verre. Rowland, puis Michelson, ont construit des réseaux à grand nombre de traits par ce procédé. Généralement les réseaux employés en spectrographie contiennent rarement plus de 500 traits au millimètre et la longueur de la surface striée est de l'ordre de 10 cm. Tous les traits, soit 50000 dans l'exemple précédent, doivent être identiques c'est à dire que la vis de la machine à diviser utilisée doit être parfaitement étudiée et ses irrégularités parfaitement corrigées pour les 50000 traits, d'où la difficulté de la construction des réseaux.

Les réseaux ainsi obtenus sont des *réseaux par transmission*. On utilise plus souvent les *réseaux par réflexion* en spectrographie. Pour obtenir ces réseaux, on trace sur une surface métallique plane des traits réguliers qui se comportent comme s'ils n'étaient pas réfléchissants. On peut tracer le réseau sur une couche métallique mince opaque déposée par évaporation dans le vide sur une lame de verre bien plane. Dans les réseaux par réflexion, le support en verre n'est pas traversé par la lumière et ses défauts d'homogénéité n'interviennent donc pas, ce qui est un avantage sur les réseaux par transmission. D'autre part, les réseaux par réflexion peuvent être utilisés dans l'ultraviolet ou l'infrarouge sans difficulté, alors qu'un support en verre ne pourrait plus convenir dans ces régions.

On peut obtenir la réplique d'un réseau original en versant sur le réseau une solution de collodion dans l'éther. Après évaporation de l'éther, on détache du réseau une pellicule transparente qui en reproduit les traits. Une réplique transparente peut être transformée en réplique utilisable par réflexion en déposant une mince couche métallique par évaporation dans le vide.

Soit G un réseau par réflexion éclairé par un collimateur SO_1 (Fig. 174). La lumière diffractée par le réseau est recueillie par l'objectif O_2 et on observe le phénomène de diffraction dans son plan focal π. Considérons une petite portion du réseau (Fig. 175) et soient deux rayons incidents S_1I_1, S_1I_2 qui tombent en deux points correspondants en I_1 et I_2 sur

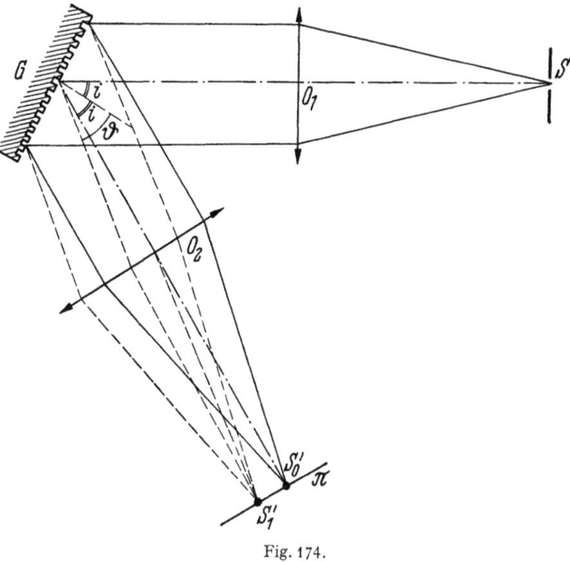

Fig. 174.

deux parties réfléchissantes voisines et sous l'incidence i. Les régions réfléchissantes séparées les unes des autres par les traits considérés comme non réfléchissants agissent comme les fentes transparentes d'un réseau par transmission. Considérons les rayons diffractés I_1S_1' et I_2S_1' dans la direction ϑ et menons par I_3 point correspondant d'un autre trait, les ondes planes correspondantes Σ et Σ'. Du point correspondant I_4 du trait suivant, abaissons les perpendiculaires I_4H et I_4H' sur Σ et Σ'. La différence de marche δ entre est rayons diffractés en I_1 et I_2 les

$$\delta = I_4H' - I_4H$$

si $a = I_1I_2 = I_2I_3 = I_3I_4 = \cdots$ est la période ou pas du réseau, on aura

$$\delta = a(\sin\vartheta - \sin i). \quad (79.1)$$

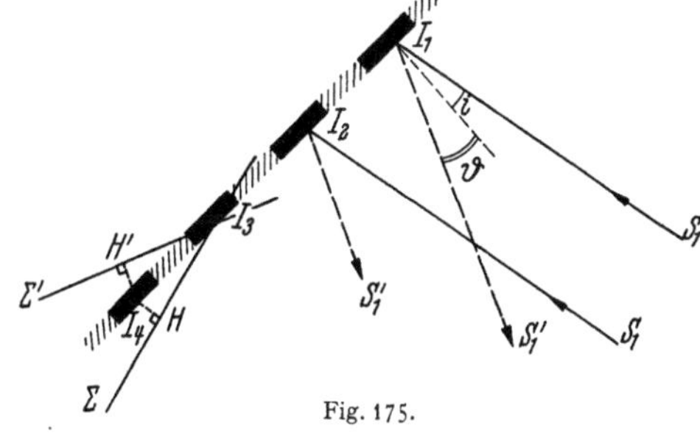

Fig. 175.

Tous les rayons diffractés par les bandes réfléchissantes sont en concordance de phase dans la direction ϑ si

$$\delta = a(\sin\vartheta - \sin i) = p\lambda. \quad (79.2)$$

Il y a donc maximum de lumière dans toutes les directions ϑ telles que

$$\sin\vartheta = \sin i + p\frac{\lambda}{a}$$

et si le faisceau indicent est normal $(i=0)$

$$\sin\vartheta = p\frac{\lambda}{a}. \quad (79.3)$$

Ce sont les maxima principaux étudiés au paragraphe précédent où l'on avait supposé ϑ petit. Si δ n'est pas égal à un nombre entier de longueurs d'onde, il n'y a pratiquement plus de lumière diffractée, même si la différence $\delta - p\lambda$ est très petite. En effet, si $\delta = p\lambda + \dfrac{\lambda}{2n}$ la vibration diffractée par la $(n+1)^{\text{ième}}$ bande réfléchissante présente une différence de marche égale à $np\lambda + \dfrac{\lambda}{2}$ avec la première: elles se détruisent. Les vibrations diffractées par les bandes réfléchissantes se détruisent deux à deux et il n'y a pratiquement pas de lumière entre les maxima principaux. Si la source est une fente fine parallèle aux bandes réfléchissantes du réseau, on voit que la figure de diffraction se compose d'une série de raies lumineuses étroites séparées par de larges intervalles obscurs. Ces fines raies lumineuses ne sont autres que les maxima principaux. Les résultats de la Sect. 78 permettent une étude plus complète du phénomène. Si le réseau comporte n bandes réfléchissantes distantes d'intervalles égaux à $a = R\eta$ et de largeur $d = 2R\gamma_0$, l'éclairement dans la direction $\vartheta = \dfrac{z'}{R}$ est donnée par (78.7), soit à un facteur constant près

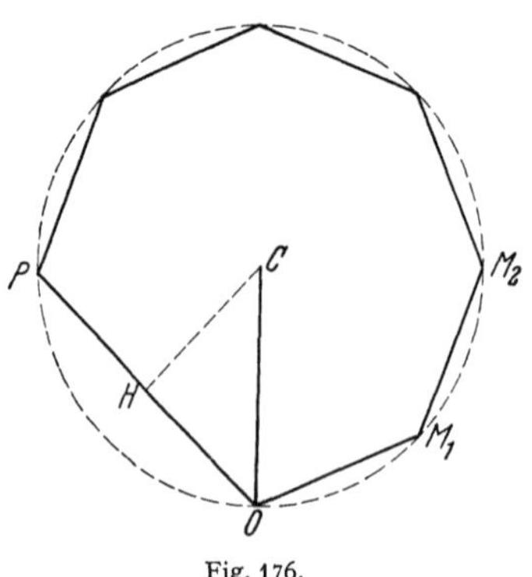

Fig. 176.

$$E_{z'} = \left(\frac{\sin \dfrac{K\vartheta d}{2}}{\dfrac{K\vartheta d}{2}} \right)^2 \left(\frac{\sin n \dfrac{Ka\vartheta}{2}}{\sin \dfrac{Ka\vartheta}{2}} \right)^2. \qquad (79.4)$$

Dans cette formule γ_0' et η représentent les angles sous lesquels on voit de O_2 la largeur de chaque fente et l'intervalle qui les sépare. Le premier facteur $\left(\dfrac{\sin \dfrac{K\vartheta d}{2}}{\dfrac{K\vartheta d}{2}} \right)^2$ n'a pas d'intérêt dans le cas d'un réseau; il représente la répartition de la lumière entre les maxima principaux et on sait que l'intensité des maxima principaux dépend essentiellement de la forme des sillons qui constituent les traits. Les intensités relatives des maxima principaux varient beaucoup d'un réseau à un autre. Nous avons donné la formule (79.4) seulement pour montrer l'évolution des phénomènes en augmentant la nombre des fentes ou des bandes réfléchissantes. Les formules (78.7) et (79.4) sont valables dans le cas d'un nombre n de fentes ou de bandes réfléchissantes très régulières, mais dans l'étude des réseaux il suffira de garder le deuxième facteur de (79.4)

$$E_{z'} = \frac{E_0}{n^2} \left(\frac{\sin n \dfrac{Ka\vartheta}{2}}{\sin \dfrac{Ka\vartheta}{2}} \right)^2 \qquad (79.5)$$

où E_0 représente l'éclairement des maxima principaux. La formule (79.5) peut d'ailleurs se retrouver facilement par des considérations géométriques sur la Fig. 176. Le vecteur OP représentant la vibration résultante est la somme de n vecteurs de même amplitude, dont les phases croissent, en progression arithmétique. On a

$$\overline{OP} = 2\overline{OH} = 2\overline{OC} \sin \frac{\widehat{OCP}}{2} = 2\overline{OC} \sin n \frac{Ka\vartheta}{2}$$

or

$$\overline{OM_1} = 2\,\overline{OC}\,\sin\frac{Ka\vartheta}{2}$$

d'où

$$\frac{\overline{OP}}{\overline{OM_1}} = \frac{\sin n\,\dfrac{Ka\vartheta}{2}}{\sin\dfrac{Ka\vartheta}{2}}\,.$$

Les maxima principaux correspondent à

$$\frac{Ka\vartheta}{2} = p\,\pi, \qquad \vartheta = \frac{z'}{R} = p\,\frac{\lambda}{a}\,, \tag{79.6}$$

alors

$$\left(\frac{\sin n\,\dfrac{Ka\vartheta}{2}}{\sin\dfrac{Ka\vartheta}{2}}\right)^2 = n^2$$

et

$$E_{z'} = E_0.$$

Ce sont les résultats déjà obtenus dans le cas $n=3$ à la Sect. 78.

De même les valeurs de $\dfrac{Ka\vartheta}{2}$ qui annulent la dérivée de (79.5) correspondent aux maxima secondaires

$$n\tan\frac{Ka\vartheta}{2} = \tan n\,\frac{Ka\vartheta}{2}\,.$$

Les minima nuls sont donnés par $\sin n\,\dfrac{Ka\vartheta}{2} = 0$ avec $\sin\dfrac{Ka\vartheta}{2} \neq 0$, ou

$$\frac{Ka\vartheta}{2} = p\,\frac{\pi}{n}\,, \qquad \vartheta = p\,\frac{\lambda}{na} \tag{79.7}$$

p étant un entier différent de zéro et différent d'un multiple de n.

Les relations (79.6) montrent que pour $\dfrac{Ka\vartheta}{2} = p\pi$ et $\dfrac{Ka\vartheta}{2} = (p+1)\,\pi$ on a 2 maxima principaux consécutifs entre lesquels on aura $n-1$ minima nuls et $n-2$ maxima secondaires. Les maxima secondaires dont l'intensité est assez grande pour qu'ils soient perceptibles sont ceux qui sont voisins des maxima principaux, c'est-à-dire ceux pour lesquels $\dfrac{Ka\vartheta}{2}$ est voisin de $p\pi$. Posons

$$\frac{Ka\vartheta}{2} = p\pi + \frac{Ka\vartheta'}{2}\,;$$

pour $\vartheta' = 0$ on est au maximum principal d'ordre p et si l'on veut étudier les maxima secondaires au voisinage de $\vartheta' = 0$, $\dfrac{Ka\vartheta'}{2}$ reste petit et l'expression (79.5) peut s'écrire

$$E_{z'} = \frac{E_0}{n^2}\,\frac{\sin^2 n\,\dfrac{Ka\vartheta'}{2}}{\left(\dfrac{Ka\vartheta'}{2}\right)^2} = E_0\left(\frac{\sin n\,\dfrac{Ka\vartheta'}{2}}{n\,\dfrac{Ka\vartheta'}{2}}\right)^2\,. \tag{79.8}$$

Au voisinage d'un maximum principal, la lumière diffractée par un réseau est identique à la lumière diffractée par une fente de largeur na égale à la largeur du réseau.

Il suffit de remplacer $na = 2\gamma'_0 R$ et $\vartheta' = z'/R$ dans (79.8) pour retrouver (76.5).

Si l'angle ϑ n'est pas petit, ϑ' reste toujours lui-même très petit et on écrira

$$\frac{K a \sin(\vartheta + \vartheta')}{2} = \frac{Ka}{2}(\sin\vartheta + \vartheta'\cos\vartheta).$$

On pose

$$\frac{K a \sin\vartheta}{2} = p\pi + \frac{K a \cos\vartheta}{2}\cdot\vartheta'$$

d'où

$$E_{z'} = E_0 \left(\frac{\sin n \dfrac{K a \cos\vartheta}{2}\vartheta'}{n \dfrac{K a \cos\vartheta}{2}\vartheta'}\right)^2. \tag{79.9}$$

La lumière diffractée par le réseau au voisinage d'un maximum principal est identique à la figure de diffraction donnée par une fente de largeur $na\cos\vartheta$.

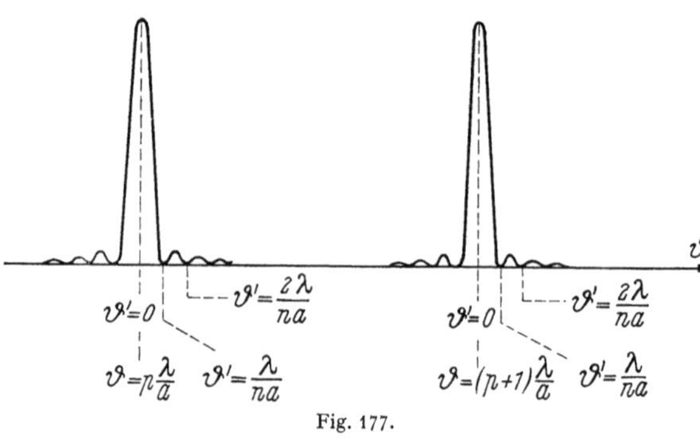

Fig. 177.

La grandeur $na\cos\vartheta$ représente la largeur de faisceau lumineux pénétrant dans l'objectif dans la direction ϑ correspondant au maximum principal considéré. Chaque maximum principal apparaît comme une raie fine lumineuse sur un fond pratiquement noir. Les Fig. 171 et 172 prennent l'aspect de la Fig. 177 dans le cas du réseau.

80. Réseau plan éclairé en lumière non monochromatique. Dans le paragraphe précédent, on a supposé le réseau éclairé en lumière monochromatique. Supposons maintenant que la source émette de la lumière blanche: à chaque longueur d'onde correspond un système de maxima principaux et de maxima secondaires dont on vient d'étudier la structure. Les maxima principaux sont donnés par l'équation (79.6)

$$\vartheta = p\,\frac{\lambda}{a}$$

et pour $p = 0$ (en incidence normale $\vartheta = 0$) il y a superposition de tous les maxima principaux pour toutes les longueurs d'onde: on a une image centrale blanche (Fig. 178).

Considérons par exemple deux radiations bleu et rouge de longueurs d'onde λ_B et λ_R. Le premier maximum principal en λ_B se trouve plus près de l'image centrale que le premier maximum principal en λ_R, il y a séparation des radiations et formation d'un spectre. On obtient ainsi deux spectres du premier ordre pour $p = 1$ symétriques par rapport à l'image centrale (incidence normale), de même deux spectres du 2ème ordre pour $p = 2$ etc. En reprenant la formule (79.3) avec la valeur numérique $a = 2\mu$, on a

p	$\sin\vartheta$ pour $\lambda = 0,4\,\mu$	ϑ	$\sin\vartheta$ pour $\lambda = 0,7\,\mu$	ϑ
0	0	0	0	0
1	0,20	11°,5	0,35	20°,5
2	0,40	23°,6	0,70	44°,5
3	0,60	36°,8	non observable	

On ne peut voir tout au plus que trois spectres de part et d'autre de l'image centrale, la partie rouge du troisième spectre n'étant pas visible.

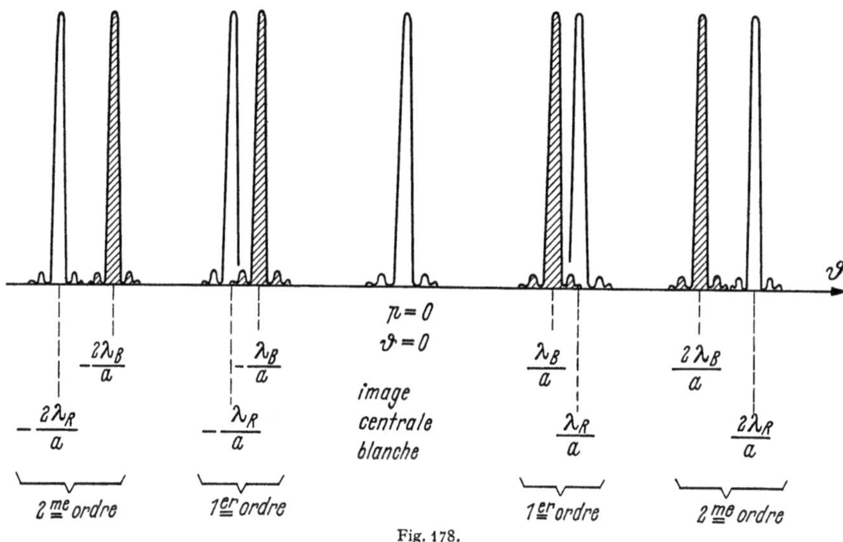

Fig. 178.

Ces spectres empiètent d'ailleurs les uns sur les autres comme le montre le tableau précédent où l'on voit que le troisième spectre empiète sur le second. Les spectres successifs sont disposés comme l'indique la Fig. 179. Pour supprimer

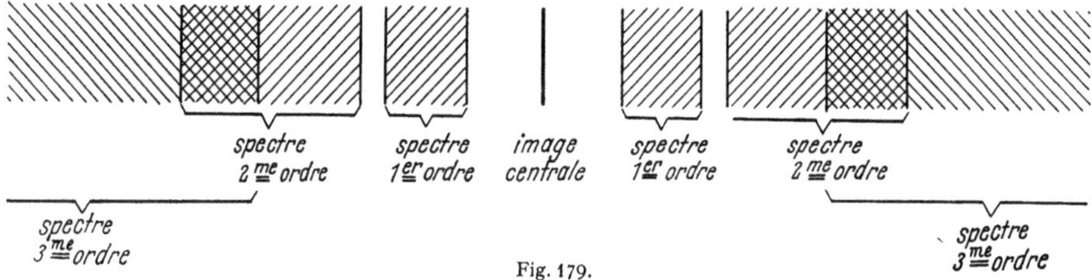

Fig. 179.

les empiètements gênants, on utilise des écrans colorés convenables. Considérons un réseau à traits peu serrés pour que plusieurs spectres se trouvent dans le plan focal d'un objectif O (Fig. 180). Le réseau est par exemple un réseau par trans-

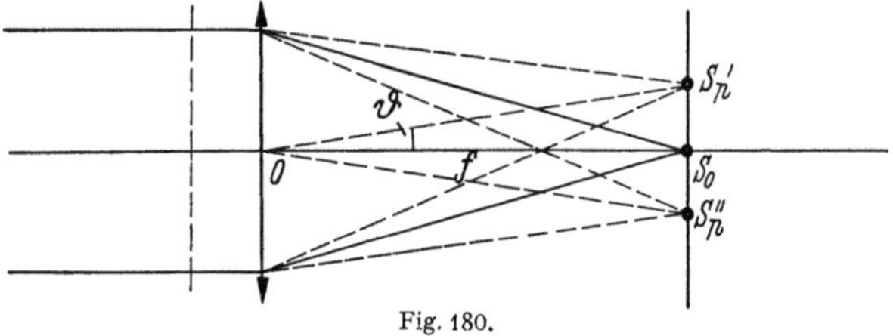

Fig. 180.

mission travaillant sous l'incidence normale. La focale de l'objectif est f et si ϑ est petit, la distance $(S_0 S'_p)_\lambda$ de l'image d'ordre p de la raie λ à l'image centrale S_0 est

$$(S_0 S'_p)_\lambda = (S_0 S''_p)_\lambda = f \cdot \vartheta = p\,\frac{f}{a}\,\lambda.$$

Les distances $S_0 S'_p$ sont proportionnelles aux longueurs d'onde. Pour une autre longueur d'onde λ', on aurait

$$(S_0 S'_p)_{\lambda'} = p\,\frac{f}{a}\,\lambda', \quad \frac{(S_0 S'_p)_{\lambda'}}{(S_0 S'_p)_\lambda} = \frac{\lambda'}{\lambda}.$$

La mesure du rapport des déviations correspondant dans le même spectre à deux radiations donne le rapport de leurs longueurs d'onde. Alors que dans le cas d'un spectre fourni par un prisme, le rapport des déviations dépend de la loi de dispersion de la matière constituant le prisme, le rapport des déviations ne dépend que des longueurs d'onde dans le cas d'un réseau. Lorsque la distance de deux raies est proportionnelle à la différence de leurs longueurs d'onde, le spectre est appelé spectre normal.

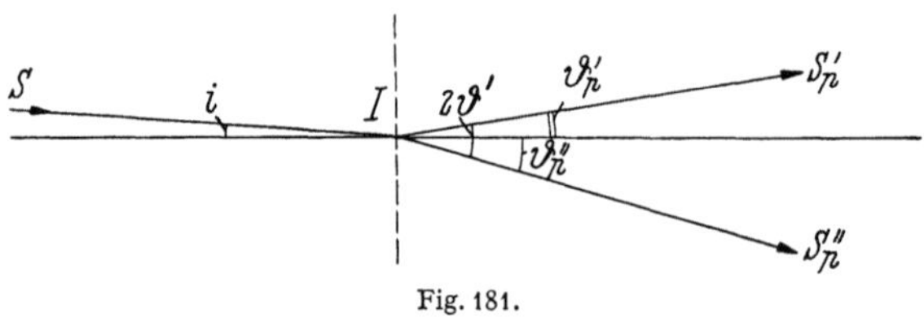

Fig. 181.

La formule (79.2) montre que la position des maxima principaux dépend uniquement de l'écartement des traits du réseau et de la longueur d'onde considérée. On peut donc, avec un réseau, mesurer les valeurs absolues des longueurs d'onde des différentes raies spectrales. Il suffit de messurer ϑ et i et d'appliquer la formule (79.2) connaissant le pas du réseau. Pour augmenter la précision, on opère le plus près possible de l'incidence normale $i=0$ et on mesure l'angle $2\vartheta'$

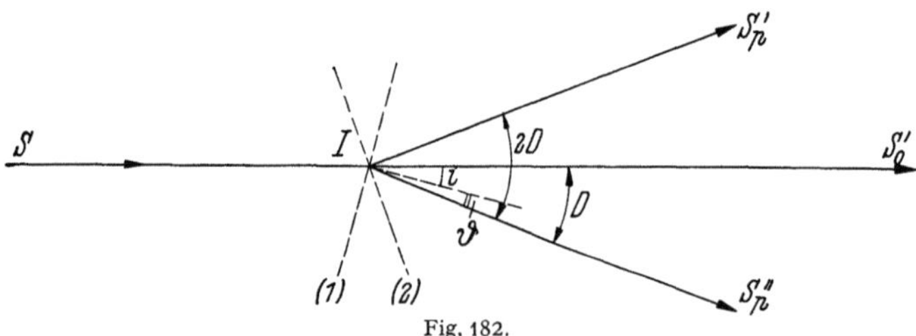

Fig. 182.

des spectres d'ordre p situés de part et d'autre de la normale (Fig. 181). Soient ϑ'_p et ϑ''_p les directions correspondant aux deux spectres d'ordre p, on a

$$\left.\begin{aligned} \sin\vartheta'_p + \sin i &= p\,\frac{\lambda}{a}\\[1ex] \sin\vartheta''_p - \sin i &= p\,\frac{\lambda}{a} \end{aligned}\right\} \sin\vartheta'_p + \sin\vartheta''_p = 2p\,\frac{\lambda}{a}$$

d'où $\vartheta'_p - \vartheta''_p$ étant très petit

$$\sin\vartheta'_p + \sin\vartheta''_p \approx 2\sin\vartheta', \quad \sin\vartheta' = p\,\frac{\lambda}{a}. \tag{80.1}$$

La mesure de ϑ' évite le repérage de la normale et donne λ si on connaît a.

La mesure d'une longueur d'onde peut se faire également en utilisant le procédé du minimum de déviation.

En tournant le réseau, on trouve une position (1) pour laquelle il y a déviation minimum D pour une raie donnée (Fig. 182). On cherche la position (2) du réseau, symétrique de (1) par rapport à $S S'_0$, qui donne également un minimum de déviation. On mesure l'angle $2D$ formé par ces deux directions et on en déduit λ.

En effet, on a

$$D = i + \vartheta, \qquad \frac{dD}{di} = 1 + \frac{d\vartheta}{di}, \qquad \cos\vartheta\,\frac{d\vartheta}{di} + \cos i = 0.$$

Le minimum de déviation est donné par

$$\frac{dD}{di} = 0, \qquad \frac{d\vartheta}{di} = -1, \qquad \cos\vartheta = \cos i.$$

On a $i = \vartheta$, ces deux angles étant situés de part et d'autre de la normale au réseau, d'où, en utilisant la relation $\sin\vartheta + \sin i = p\,\dfrac{\lambda}{a}$ et $D = i + \vartheta = 2\vartheta$

$$2\sin\frac{D}{2} = p\,\frac{\lambda}{a}. \tag{80.2}$$

De la mesure de $2D$ on déduit donc λ. La précision des mesures dépend de la dispersion du réseau: différentions la relation (79.2)

$$\cos\vartheta\,d\vartheta = p\,\frac{d\lambda}{a}$$

d'où la dispersion caractérisée par la variation $d\vartheta$ de ϑ correspondant à une variation $d\lambda$ de λ

$$d\vartheta = \frac{p}{a\cos\vartheta}\,d\lambda. \tag{80.3}$$

La dispersion est d'autant plus grande que l'ordre du spectre est plus élevé.

Considérons le réseau R placé devant l'objectif O (Fig. 183).

L'angle $d\vartheta$ correspondant à la précision de pointé d'une raie est déterminé par les règles des pointés transversaux. La figure de diffraction donnée par l'objectif O devant lequel se trouve le réseau R est la même que si le réseau est enlevé, d'après (79.8). Si D_0 est le diamètre de l'objectif O, on admet que la précision d'un pointé transversal est de l'ordre du dixième du diamètre de la tache de diffraction, soit

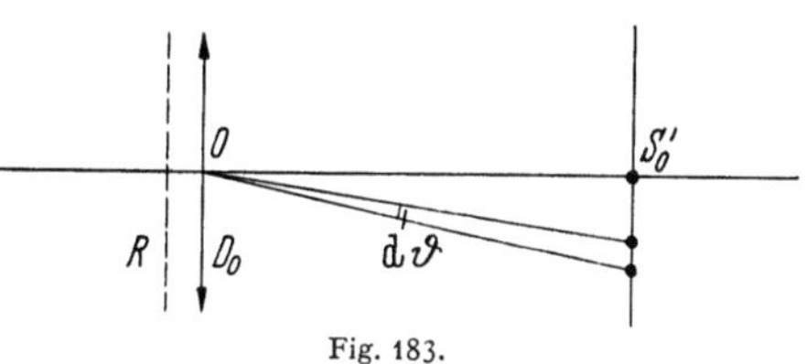

Fig. 183.

$$d\vartheta \geq \frac{\lambda}{5\,D_0}.$$

On aura donc avec (80.3)

$$\frac{p}{a\cos\vartheta}\,d\lambda \geq \frac{\lambda}{5\,D_0}, \qquad d\lambda \geq \frac{\lambda\,a\cos\vartheta}{5p\,D_0};$$

pour $a = 2\mu$ (500 traits au millimètre), $p = 2$ et $D_0 = 5$ cm., la précision sera $d\lambda = \frac{1}{50}$ Å pour $\lambda = 0{,}5\,\mu$.

Il est intéressant également de connaître la plus petite distance de deux raies spectrales que le réseau puisse séparer, c'est-à-dire le pouvoir séparateur du réseau.

On a vu que la structure d'une raie spectrale est identique à la figure de diffraction donnée par une fente de largeur $na\cos\vartheta$ (79.9). On admettra comme précédemment que deux radiations sont séparées lorsque le maximum de l'une des figures de diffraction correspond au premier minimum nul de l'autre. L'écart angulaire ϑ' séparant un maximum principal du premier minimum nul est donné par (79.9)

$$n\,\frac{K\,a\cos\vartheta}{2}\,\vartheta' = \pi, \qquad \vartheta' = \frac{\lambda}{n\,a\cos\vartheta}.$$

Or la distance angulaire $d\vartheta$ séparant les maxima principaux de deux longueurs d'onde λ et $\lambda + d\lambda$ est donnée par (80.3)

$$d\vartheta = \frac{p}{a\cos\vartheta}\,d\lambda.$$

En écrivant que $d\vartheta \geq \vartheta'$ on obtient

$$\frac{\lambda}{d\lambda} \leq n\,p. \tag{80.4}$$

L'expression (80.4) donne le pouvoir séparateur du réseau.

Prenons par exemple $n = 150\,000$ et opérons dans le deuxième ordre ($p = 2$). Aux environs de $\lambda = 6000$ Å, on aura $d\lambda = 0{,}02$ Å.

81. Echelon de Michelson. L'échelon de Michelson est constitué par N lames de verre de même épaisseur e et de même indice n (Fig. 184). Chaque lame est en retrait sur la précédente d'une longueur $A_1 B_1 = A_2 B_2 = A_2 B_3 = \cdots = d$ Un faisceau de rayons parallèles tombe sur l'échelon qui diffracte la lumière par les espaces $A_1 B_1\ A_2 B_2 \ldots$ Ces

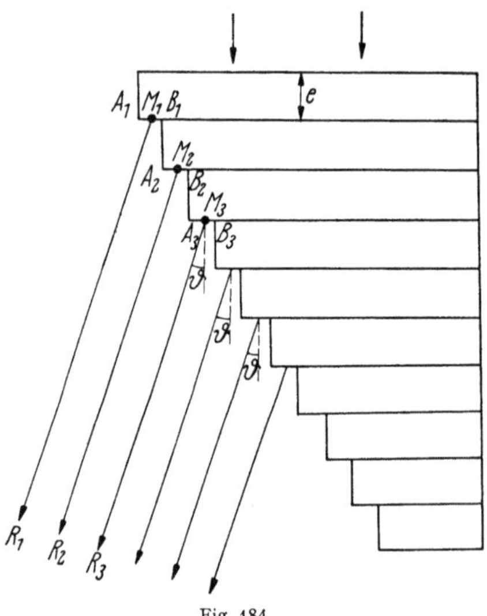

Fig. 184.

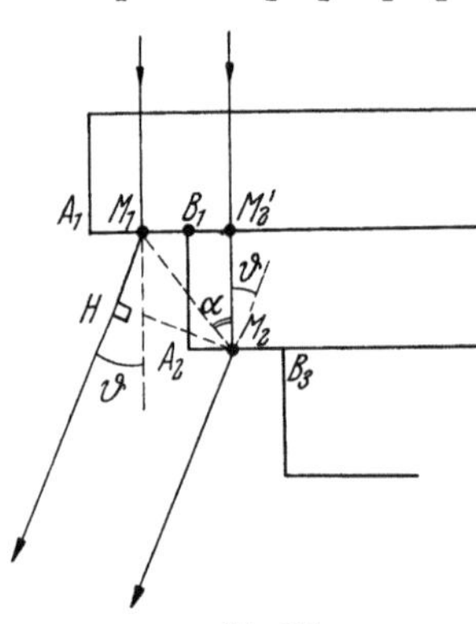

Fig. 185.

espaces agissent comme des fentes et la différence de marche entre deux rayons diffractés dans la direction ϑ tels que $M_1 R_1$ et $M_2 R_2$ est (Fig. 185)

$$\delta = n \cdot \overline{M_2' M_2} - \overline{M_1 H} = n\,e - \overline{M_1 H}$$

or

$$\overline{M_1 H} = M_1 M_2 \cos(\alpha + \vartheta) = \frac{e}{\cos\alpha}\cos(\alpha + \vartheta) = e\,(1 - \vartheta\tan\alpha)$$

car ϑ est toujours petit, donc

$$\delta = (n - 1)\,e + \vartheta e\tan\alpha = (n - 1)\,e + \vartheta d.$$

On aura le spectre d'ordre p dans la direction ϑ si

$$(n - 1)\,e + \vartheta d = p\,\lambda;$$

la dispersion est donnée par

$$\frac{d\vartheta}{d\lambda} = \frac{e}{d}\left[\frac{n-1}{\lambda} - \frac{dn}{d\lambda}\right] = \frac{A\,e}{\lambda\,d}$$

où A est un coefficient numérique compris entre 0,5 et 1 pour les verres ordinaires. On pourra vérifier que la dispersion de la lumière dans l'échelon est du même ordre de grandeur que dans les réseaux. Mais les images données par l'échelon sont extrêment rapprochées et pour éviter l'empiètement des spectres il faut employer une lumière incidente déjà bien monochromatique: l'échelon de Michelson est un analyseur de raies Il peut décomposer une raie déjà isolée par un spectroscope.

Pour $e = 2$ cm. et $n = 1,5$ on a $\delta \approx 1$ cm. et pour $\lambda = 0,5\,\mu$ l'ordre d'interférence est $p = 20000$. Si le nombre des lames est $n = 30$, nombre qu'il est difficile de dépasser par suite des pertes de lumière, le pouvoir de résolution $\dfrac{\lambda}{d\lambda} = n\,p$ $= 600000$. On obtient donc un pouvoir de résolution supérieur à celui des meilleurs réseaux mais inférieurs à celui des appareils interférentiels.

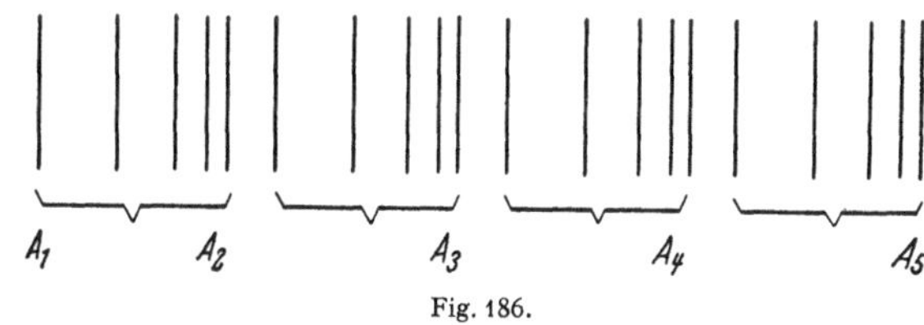

Fig. 186.

82. Erreurs périodiques dans le tracé des réseaux. On sait que dans le tracé mécanique des réseaux, le nombre de traits correspondant à un tour de la vis mère est toujours très grand. Si la vis mère est légèrement excentrée, les traits ne sont plus équidistants et le défaut se reproduit identiquement à chaque tour.

Ces phénomènes sont représentés schématiquement sur la Fig. 186 en supposant que l'on effectue cing tracés par tour de vis. A_1A_2, A_2A_3, A_3A_4 ... représentent les groupes de traits tracés pendant un tour de la vis mère. En réalité, le pas de la vis mère est de l'ordre du millimètre et dans les réseaux employés ordinairement en spectroscopie, on a environ 500 traits par mm., c'est-à-dire par tour de vis. Une erreur périodique du type précédent affecte la forme de l'onde diffractée. Considérons un réseau par réflexion (Fig. 187).

Si le réseau est parfait, les milieux $O_1O_2O_3$ ... des traits réfléchissants sont équidistants. Ils envoient des amplitudes vibratoires déphasées d'un multiple de 2π dans la direction ϑ correspondant au spectre d'ordre p. On a

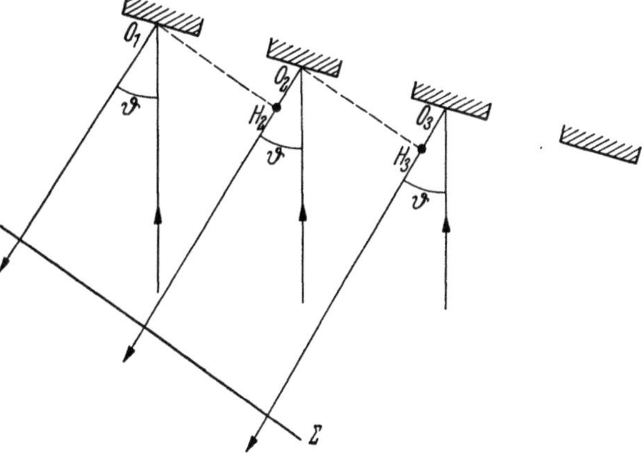

Fig. 187.

$$O_2H_2 = O_3H_3 = \cdots = p\,\lambda. \tag{82.1}$$

Tout se passe comme si les mouvements vibratoires étaient en phase et la surface d'onde Σ diffractée à une certaine distance dans la direction ϑ est une surface d'onde plane. L'onde Σ est l'enveloppe des ondelettes diffractées par les différents traits du réseau et dont les rayons sont tels que la condition (82.1) soit satisfaite (Fig. 188).

Si maintenant le réseau est entaché d'une erreur de tracé, il n'en est plus ainsi et les milieux des traits réfléchissants ne sont plus équidistants. Ils sont déplacés par rapport à leurs positions théoriques, ce qui affecte la position des ondelettes diffractées et par conséquent, la forme de la surface enveloppe qui n'est plus plane.

Pour relier la déformation de la surface d'onde au déplacement ε du trait du réseau, il suffit de remarquer que si l'on déplace un trait pour l'amener sur le trait voisin, la déformation δ est égale à $p\lambda$. On a donc

$$\delta = \frac{p\lambda}{a}\,\varepsilon. \qquad (82.2)$$

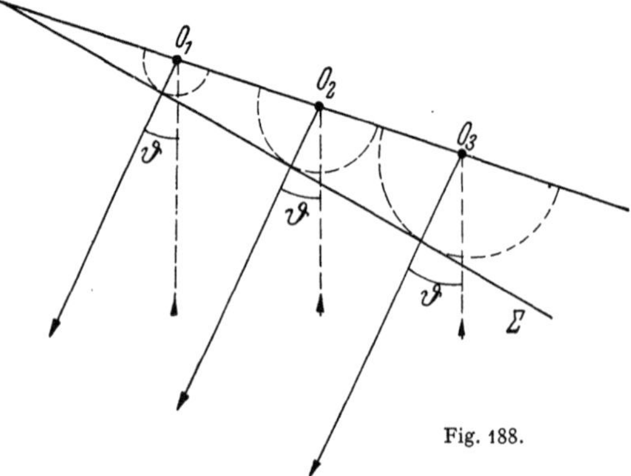

Fig. 188.

Calculons le phénomène de diffraction que l'on observera dans le plan focal d'un objectif suivant le montage de la Fig. 174 par exemple. Le déplacement ε des traits du réseau étant périodique, la déformation δ de la surface d'onde l'est également. Supposons les traits du réseau parallèles à l'axe $O\beta'$ de la Fig. 167, le plan du réseau passant par O. On peut représenter la déformation δ par une série de Fourier de la forme

$$\delta = \frac{p\lambda}{a}\sum_{q=1}^{q=\infty}\left[\varepsilon_q\cos\frac{2\pi q\gamma'}{m} + \varepsilon_q'\sin\frac{2\pi q\gamma'}{m}\right] \qquad (82.3)$$

où m représente la période angulaire de la déformation, la période de l'erreur de position dans le plan focal de l'objectif de focale f étant mf. Si l'erreur de position des traits est sinusoïdale, on a

$$\delta = \frac{p\lambda}{a}\,\varepsilon_0\sin\frac{2\pi\gamma'}{m}. \qquad (82.4)$$

L'amplitude vibratoire $U_{z'}$ le long de Cz' est donnée par (67.4); d'où, à un facteur constant près

$$U_{z'} = \int_{-\gamma_0'}^{+\gamma_0'} e^{j\frac{2p\pi}{a}\varepsilon_0\sin\frac{2\pi\gamma'}{m}}\,e^{jK\gamma'z'}\,d\gamma'$$

$2\gamma_0'$ étant la largeur striée du réseau.

Supposons ε_0 petit et posons

$$e^{j\frac{2p\pi}{a}\varepsilon_0\sin\frac{2\pi\gamma'}{m}} \approx 1 + \frac{2j\,p\,\pi}{a}\varepsilon_0\sin\frac{2\pi\gamma'}{m}$$

d'où

$$U_z = \int_{-\gamma_0'}^{+\gamma_0'} e^{jK\gamma'z'}\,d\gamma' + \frac{2j\,p\,\pi}{a}\varepsilon_0\int_{-\gamma_0'}^{+\gamma_0'}\sin\frac{2\pi\gamma'}{m}\,e^{jK\gamma'z'}\,d\gamma'$$

et

$$U_{z'} = 2\gamma_0' \frac{\sin K\gamma_0' z'}{K\gamma_0' z'} + \left. + \frac{2p\,\gamma_0'\,\pi\,\varepsilon_0}{a}\left[\frac{\sin\left(\frac{2\pi\gamma_0'}{m} + K\gamma_0' z'\right)}{\frac{2\pi\gamma_0'}{m} + K\gamma_0' z'} - \frac{\sin\left(\frac{2\pi\gamma_0'}{m} - K\gamma_0' z'\right)}{\frac{2\pi\gamma_0'}{m} - K\gamma_0' z'}\right].\right\} \quad (82.5)$$

Le premier terme du second membre représente la figure de diffraction du réseau parfait. On a représenté ses variations au voisinage de l'origine sur la Fig. 189 (courbe 1). Les deux autres termes sont analogues au premier (courbes 2 et 3), mais ils sont décalés en abscisse de la quantité $z' = \pm \lambda/m$ et leur amplitude est plus petite. Le rapport des maxima qu'ils présentent au maximum du premier terme est égal à $\pm p\,\pi\,\varepsilon_0/a$. Pratiquement les trois termes n'ont de valeur appréciable qu'au voisinage de leurs maxima respectifs et sont séparés les uns des autres. En définitive, l'image principale est accompagnée de deux images parasites symétriques, que l'on a coutume d'appeler images fantômes (ghosts) et qui sont bien entendu préjudiciables à la qualité du réseau.

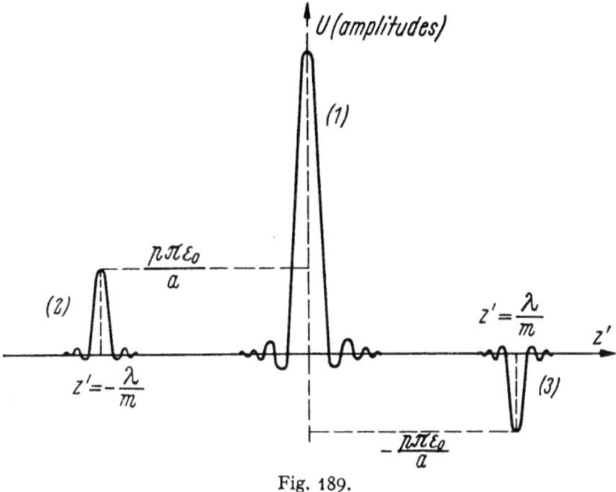

Fig. 189.

L'intensité des ghosts est égale, si l'on prend comme unité celle du maximum de l'image principale, à $p^2\pi^2\varepsilon_0^2/a^2$ c'est à dire qu'elle est proportionnelle au carré de l'ordre p du spectre, ainsi qu'au carré du module ε_0 de l'erreur de mise en place. Si l'on veut par exemple que cette intensité relative soit inférieure à 10^{-3}, on aura (en supposant $p = 1$) $\varepsilon_0/a \approx 10^{-2}$ ce qui signifie que les traits du réseau doivent être mis en place avec une excellente précision.

Si l'erreur de tracé a la forme donnée par l'expression (82.3), on aura un résultat analogue pour les différentes composantes sinusoïdales du développement en série de FOURIER, d'où l'apparition d'un nombre plus ou moins grand de ghosts de chaque côté de l'image principale, et symétriques par rapport à cette image.

83. Réseaux concaves. Les réseaux plans doivent être employés en combinaison avec un collimateur et une lentille. On peut supprimer les lentilles au moyen des réseaux concaves obtenus par tracé de traits parallèles et équidistants sur un miroir concave. Soit R un réseau concave (Fig. 190) et plaçons en F une source lumineuse. Nous aurons en S' l'image réfléchie de la fente et de part et d'autre, des images diffractées. La relation (79.2) est encore applicable au cas des réseaux concaves: soit C le centre de la sphère sur laquelle est tracé le réseau et supposons les traits perpendiculaires au plan de la Fig. 190. Pour qu'on obtienne en S'_p (Fig. 191) un maximum principal d'ordre p, il faut que le trajet $FM + MS'_p$ correspondant à la lumière diffractée par un point M du réseau, augmente de $p\lambda$ lorsqu'on passe d'un trait au suivant. On peut assimiler le réseau R au voisinage de M à un petit réseau plan. Pour calculer la différence de marche δ en S_p' entre les vibrations diffractées par deux traits consécutifs de ce petit réseau, on peut assimiler les portions d'ondes sphériques de centre F et S_p' à des ondes planes normales aux rayons FM et MS_p'. Si i est l'angle d'incidence et ϑ l'angle de diffraction, on a

$$\delta = a\,(\sin\vartheta - \sin i).$$

Il faut que δ soit égal à $p\lambda$ quelle que soit la position de M sur le réseau. Considérons le cas particulier où cette condition est réalisée en disposant les appareils de façon que i et ϑ soient constants. Le lieu de M pour que l'angle $i = FMC = FTC =$ constante est la circonférence Γ passant par F T et C. Plaçons la fente F de façon que la circonférence Γ soit tangente en T à R. Comme l'ouverture du réseau concave est toujours faible, R et Γ sont pratiquement confondus et i

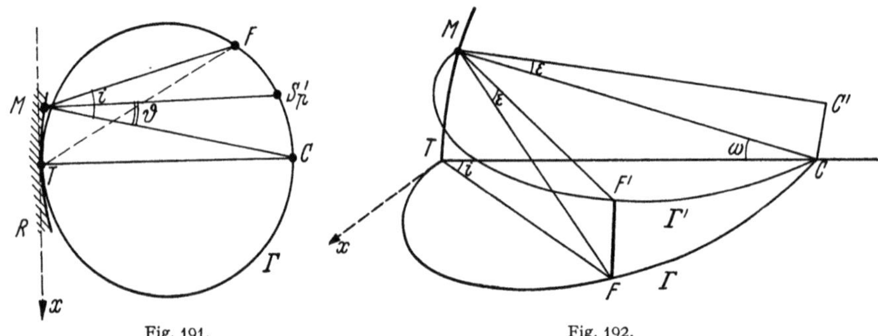

Fig. 191.   Fig. 192.

reste constant pour tous les points M du réseau. En répétant le même raisonnement pour l'angle ϑ, on voit que la fente lumineuse F donne une série de maxima principaux situés sur la circonférence Γ. La position de ces maxima est définie par les angles i et ϑ reliés par la relation (79.2).

En fait, ce raisonnement n'est valable que pour les éléments du réseau situés dans le plan de symétrie horizontal de l'appareil qui a été pris pour plan de figure dans le cas de la Fig. 191. Pour un point M situé en dehors de ce plan, les traits ne sont plus perpendiculaires au plan de figure et FM n'est plus perpendiculaire aux traits du réseau. Plaçons-nous dans le cas du montage de Rowland décrit plus loin: le maximum principal que l'on étudie est en C, centre de courbure du réseau (Fig. 192). Le plan de section principale correspondant au point M est le plan $MF'C$.

Comme précédemment, on assimilera le réseau en M à un petit élément de réseau plan et la source ponctuelle est en F qui se projette en F' sur le plan $MF'C$. Comme ω est toujours petit, l'angle $F'MC \approx FTC = i$. Le rayon diffracté MC' par le petit réseau plan en M est dans le plan TMC et l'angle $FMF' = CMC' = \varepsilon$, ces deux angles étant de part et d'autre du plan $MF'C$. En posant $l = TM$, on trouve facilement que $CC' = l\sin i\tan i$. Toute la lumière issue de F est répartie sur une droite verticale de hauteur $2CC'$. Le réseau ne donne pas une image mais une droite focale, le réseau est astigmate. Pour $i = 45°$ avec $2l = 40$mm. on a $2CC' = 28$ mm. En prolongeant le rayon MC', on verrait qu'il s'appuie sur une deuxième focale horizontale. L'astigmatisme du réseau concave n'est pas gênant car on emploie toujours comme source une fente F parallèle aux traits. Les focales telles que CC' données par les différents points de la source coïncident, et l'image garde sa finesse.

Dans le montage de ROWLAND, le réseau est utilisé de façon à ce que la plaque P soit normale à l'axe du miroir constituant le réseau (Fig. 193). Pour que les spectres successifs viennent se projeter sur P, il faudrait déplacer la fente F, ce qui ne serait pas pratique. ROWLAND remédie à cet inconvénient de la façon suivante:

L'appareil se compose de deux poutres rectangulaires (Fig. 194) le long desquelles peuvent se déplacer deux glissières. Ces deux glissières sont réunies l'une à l'autre par une poutre de longueur égale au rayon de courbure du réseau. L'une porte le réseau R, l'autre la plaque photographique P. La fente est fixée et placée au sommet du triangle rectangle. Le réseau, le milieu de la plaque et la fente sont situés sur la circonférence d'un cercle dont le diamètre est égal au rayon de courbure du réseau. Pour passer d'une région à l'autre, il suffit de faire glisser la poutre PR. Notons que dans le montage de ROWLAND, $\vartheta = 0$ et le spectre obtenu est un spectre normal.

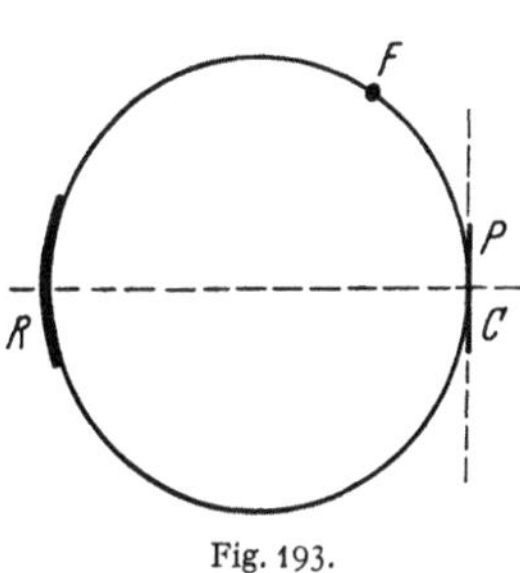

Fig. 193.

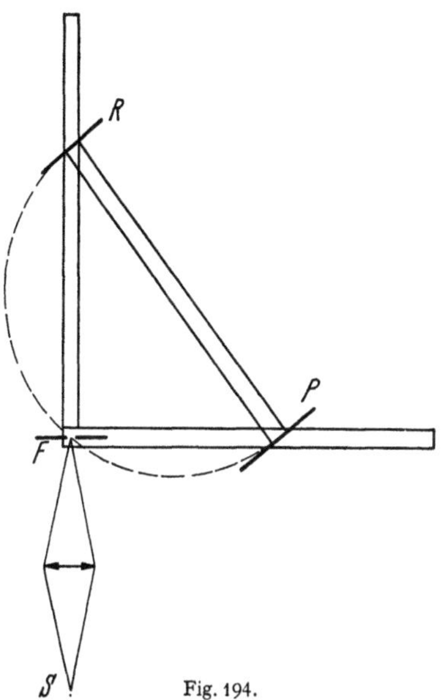

Fig. 194.

84. Applications des réseaux. Réseaux particuliers. α) *Détermination des étalons de longueur d'onde.* A cause des imperfections inévitables des réseaux, la constante d'un réseau n'est pas connue avec une précision suffisante. Aussi les étalons primaires de longueur d'onde constitués par quelques raies du cadmium sont comparés, par les méthodes interférentielles, à la raie rouge du cadmium, elle-même déterminée directement en fonction du mètre étalon. Le nombre des étalons primaires n'est pas suffisant et on a choisi un spectre donnant un très grand nombre de raies bien réparties: le spectre du fer.

Une centaine de raies du fer ont été comparées aux raies du cadmium par les méthodes interférentielles et on a obtenu ainsi des étalons secondaires répartis dans tout le spectre et constituant le système international des repères spectroscopiques. C'est ensuite, par interpolation entre ces repères qu'on a déterminé les raies du fer comprises entre les étalons secondaires. Ces dernières mesures ont été faites au moyen de réseaux qui sont les appareils dispersifs les plus commodes pour effectuer une interpolation à cause de la loi simple de leur dispersion. On dispose ainsi d'un grand nombre d'étalons tertiaires dont les longueurs d'onde sont données dans les tables.

β) *Mesures courantes dans les laboratoires.* Les spectrographes à réseaux sont employés couramment dans les laboratoires d'analyse spectrale où l'on a besoin de mesurer une longueur d'onde avec la précision de quelques centièmes d'angströms.

On photographie sur la même plaque, comme avec un spectrographe à prisme, le spectre du fer et le spectre à étudier, l'un au-dessous de l'autre, en découvrant chaque fois une moitié de la fente. Par interpolation, on obtient la longueur d'onde cherchée.

La plus grande luminosité des spectrographes à prismes les rendait autrefois indispensables à l'étude des sources à faible luminosité. On sait maintenant concentrer la lumière diffractée par un réseau dans un petit nombre de spectres en donnant au tracé des traits un profil convenable. Enfin, si l'on cherche un grand pouvoir de résolution, il sera nécessaire d'employer un réseau ou même des dispositifs interférentiels à ondes multiples.

γ) *Application des réseaux en dehors du spectre visible.* La relation (79.2) montre que les maxima principaux seront observables si $|\sin\vartheta| < 1$, c'est à dire

$$\left| \frac{p\,\lambda}{a} + \sin i \right| < 1$$

d'où

$$\lambda < \frac{2a}{|p|}.$$

Donc, pour donner une image de diffraction, le réseau doit être éclairé avec une radiation de longueur d'onde inférieure à $2a$. Pour un réseau de 500 traits au millimètre, on doit avoir $\lambda < 4\mu$. Pour les longueurs d'onde plus grande, il faut écarter les traits du réseau.

Du côté des courtes longueurs d'onde, ϑ diminue avec λ et les spectres se rapprochent de l'image centrale. A partir d'une certaine valeur de λ, la diffraction n'est plus observable. On peut alors opérer sous une incidence très grande et observer des spectres correspondant à des longueurs d'onde beaucoup plus petites. La formule (80.3) montre qu'en incidence normale, la dispersion est

$$\left(\frac{d\vartheta}{d\lambda} \right)_{i=0} = \frac{p}{a},$$

en incidence rasante

$$\sin\vartheta = -\frac{p\,\lambda}{a} + 1$$

et λ étant petit

$$\sin^2\vartheta = 1 - \frac{2p\,\lambda}{a}$$

d'où

$$\cos\vartheta = \sqrt{\frac{2p\,\lambda}{a}},$$

en remplaçant dans l'expression (80.3)

$$\left(\frac{d\vartheta}{d\lambda} \right)_{i=\frac{\pi}{2}} = \sqrt{\frac{p}{2a\,\lambda}}$$

pour $a = 1\,\mu$ et $p = 1$ on voit que la dispersion est 3 fois plus grande en incidence rasante pour $\lambda = 0,05\,\mu$ et 7 fois plus grande pour $\lambda = 0,01\,\mu$. Tout se passe pour $\lambda = 0,05\,\mu$ avec le réseau à 500 traits éclairé en incidence rasante comme si on opérait en incidence normale avec un réseau à 2240 traits au millimètre.

On a pu étudier par le procédé de l'incidence rasante des radiations s'étendant jusqu'au domaine des rayons X mous.

δ) *Réseaux diffractant dans des directions privilégiées.* La lumière diffractée par un réseau se répartit dans un certain nombre de spectres et par conséquent un spectrographe à prisme donnant un seul spectre est beaucoup plus lumineux qu'un spectrographe à réseau ordinaire. Toutefois, en donnant aux traits du réseau un profil convenable, on peut faire disparaître certains spectres et concentrer la lumière dans les autres.

Considérons un réseau R_1 (Fig. 195) composé de traits réfléchissants $A_1 B_1$ $A_2 B_2$ ... séparés par des intervalles non réfléchissants $B_1 A_2$ $B_2 A_3$ etc. ... C'est un réseau par réflexion du type de ceux que nous avons considéré précédemment.

Si le faisceau incident tombe sur R_1 sous l'incidence i, on a l'image centrale dans la direction S_0' de réflexion régulière et un spectre d'ordre p dans la direction S_p' par exemple. Imaginons un réseau R_2 dont les bandes réfléchissantes ont la forme de dents de scie. L'orientation des bandes est telle que pour le faisceau incident, la direction S_p' correspond à la réflexion régulière. Dans la direction S_p' la différence de marche entre deux rayons consécutifs est

$$\delta = 2a \sin \alpha \cos \vartheta$$

et on aura dans la direction ϑ un maximum principal d'ordre p pour la longuer d'onde λ, telle que

$$\lambda = \frac{2a \sin \alpha \cos \vartheta}{p}$$

du fait que la diffraction dans la direction ϑ coïncide avec la réflexion régulière, l'image correspondante sera particulièrement intense.

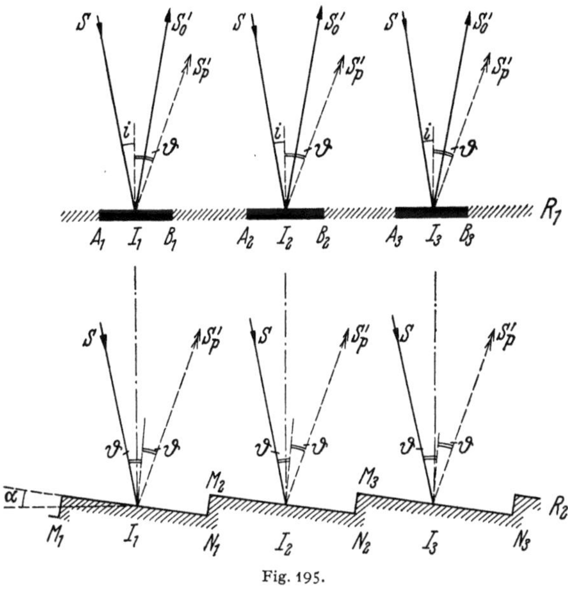

Fig. 195.

On peut donc concentrer l'énergie dans une direction particulière en donnant une orientation convenable aux bandes réfléchissantes du réseau.

WOOD a réussi à tracer des réseaux dont les bandes réfléchissantes avaient une orientation déterminée en employant comme outil traceur l'angle naturel d'un cristal de carbure de silicium. Il a employé ensuite le diamant et obtenu pour une longueur d'onde d'environ $30\,\mu$, une concentration de presque toute l'énergie dans un ou deux ordres d'un seul côté de l'image centrale.

Ces réseaux intermédiaires entre les réseaux ordinaires et l'échelon de MICHELSON sont appelés «réseaux échelettes».

Le même physicien a pu également tracer des réseaux échelettes pour le spectre visible sur des lames de verre recouvertes d'une couche d'aluminium évaporée dans le vide. WOOD a utilisé des pointes de diamant d'orientation et de formes déterminées et obtenu des réseaux à 600 traits par millimètre, dans lesquels 80 pour cent de la lumière verte du mercure $(0{,}5461\,\mu)$ est concentrée dans le premier ordre.

$\varepsilon)$ *Application a l'étude des aberrations.* L'image centrale fournie par un réseau est en phase avec les différents spectres et on peut les faire interférer. C'est le procédé qui a été utilisé par VASCO RONCHI en employant des réseaux à petit nombre de traits au millimètre. L'objectif O dont on veut étudier les aberrations, forme une image S' de la source au voisinage de laquelle on place le réseau R

(Fig. 196). En observant O à travers R on voit plusieurs images de l'objectif, une image centrale S'_0 et plusieurs images latérales $S'_1 \, S''_1 \, S'_2 \, S''_2 \ldots$ qui correspondent aux divers spectres. Choisissons un réseau R de pas convenable pour que ces images empiètent les unes sur les autres (Fig. 197). Elles interfèrent en donnant dans les parties communes des franges rectilignes et parallèles, si l'onde issue de l'objectif O est sphérique, c'est à dire si l'objectif est parfait. Dans le cas où les franges ne sont pas rectilignes, la surface d'onde n'est pas sphérique et de la déformation des franges on peut déduire la forme de la surface d'onde.

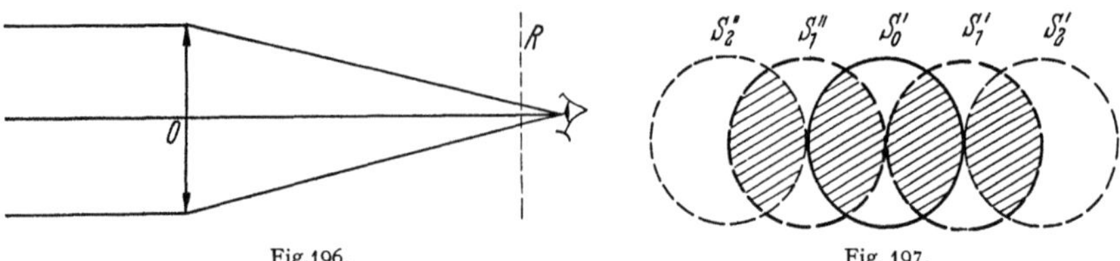

Fig.196.   Fig. 197.

85. Diffraction en présence d'un fond cohérent. On a vu précédemment (Sect. 76) que la figure de diffraction d'une fente était donnée en amplitude et à un facteur constant près, par l'expression

$$U_{z'} = 2\gamma'_0 \, \frac{\sin K \gamma'_0 z'}{K \gamma'_0 z'}. \tag{85.1}$$

L'oeil n'est pas sensible aux variations de phase et la figure de diffraction observée est $E_{z'} = |U_{z'}|^2$.

Zernike a montré qu'il est possible d'observer les variations de phase en utilisant un fond cohérent.

L'expérience est la suivante (Fig. 198): une fente F éclaire un objectif O contre lequel est placé un diaphragme D percé d'une fente de largeur angulaire

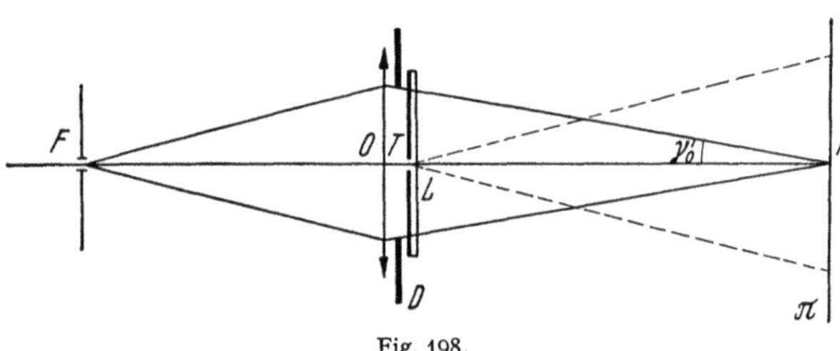

Fig. 198.

$2\gamma'_0$. Sur la Fig. 198, on a donné une grande largeur à la fente seulement pour la clarté du dessin. L'objectif O donne en F' une image de la fente F: l'image F' est une figure de diffraction caractéristique de l'ouverture du diaphragme placé contre l'objectif O. L'image F' est représentée, en amplitude, par l'expression (85.1). Plaçons contre D une lame de verre L recouverte d'une couche mince semi-transparente, sauf dans la région T. La partie découverte T a la forme d'une fente très fine de largeur angulaire $2\gamma''_0 \ll 2\gamma'_0$. La fente T donne dans le plan π un phénomène de diffraction dont la structure est donnée par une expression analogue à (85.1)

$$U_{z'} = 2\gamma''_0 \, \frac{\sin K \gamma''_0 z'}{K \gamma''_0 z'}. \tag{85.2}$$

Comme γ''_0 est très petit devant γ'_0, le phénomène de diffraction produit par γ'_0 en F' est très serré et pratiquement tout entier dans la partie centrale du phénomène de diffraction produit par γ''_0. Tout se passe comme si on superposait à la figure de diffraction donnée par $2\gamma'_0$ (85.1) un fond uniforme dû à la partie centrale de la tache centrale de diffraction produite par $2\gamma''_0$ (85.2).

Ce fond uniforme est en cohérence de phase avec la figure de diffraction en F' due à $2\gamma_0'$ (85.1) et interfère avec elle. Finalement, on observe dans le plan π' les interférences entre la figure de diffraction donnée par (85.1) avec le «fond cohérent» produit par la fente fine T. Pratiquement dans la région F', γ_0'' étant très petit, l'expression (85.2) se réduit à

$$U_{z'} = 2\gamma_0'' \frac{\sin K \gamma_0'' z'}{K \gamma_0'' z'} \approx 2\gamma_0''.$$

Supposons que la couche semi-transparente déposée sur L n'introduise pas de déphasage mais seulement une absorption caractérisée par le coefficient m. La figure de diffraction que l'on observera dans le plan π sera:

$$U_{z'} = 2\gamma_0' m \frac{\sin K \gamma_0' z'}{K \gamma_0' z'} + 2\gamma_0''. \tag{85.3}$$

Réglons l'absorption de la couche semi-réfléchissante de façon que l'amplitude du fond cohérent soit voisine en valeur absolue de celle du premier minimum de

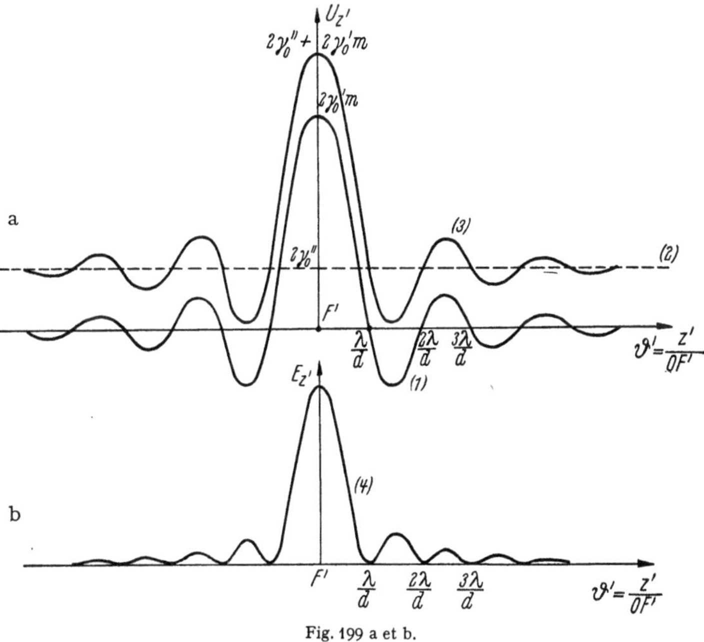

Fig. 199 a et b.

la figure de diffraction étudiée. Les maxima impairs de $|U_{z'}|^2$ correspondent à des minima d'amplitude qui sont en opposition avec le fond cohérent et sont détruits. Les maxima pairs correspondant à des maxima d'amplitude, en phase avec le fond cohérent sont renforcés. C'est ce qu'il est possible de voir sur la Fig. 199 a où l'on a porté en ordonnées les amplitudes $U_{z'}$ et en abscisses $\vartheta' = z'/\overline{OF'}$ avec $d = 2\gamma_0' \cdot \overline{OF'}$, d étant la largeur de la fente placée contre l'objectif O. La courbe (1) représente la figure de diffraction en amplitude due à la fente d (85.1), la droite en traits ponctués (2) représente le fond cohérent. C'est la partie centrale du phénomène de diffraction très étalé produit par la fente fine T. La figure de diffraction résultante en amplitude est donnée par la courbe (3) somme des courbes (1) et (2). La figure de diffraction observée est donnée par la courbe (3) dont toutes les ordonnées seraient élevées au carré. La courbe (3) étant toute entière audessus de l'axe des abscisses, la forme de la figure de diffraction observée reste analogue à celle de la courbe (3) et c'est pourquoi nous ne l'avons pas représentée sur la Fig. 199a. Dans ce qui suit, nous admettrons que la courbe (3) représente donc la distribution des éclairements dans la figure de diffraction résultante. Sur la Fig. 199b, la courbe (4) donne la répartition des éclairements dans la figure de diffraction donnée par la fente $2\gamma'_0$ (85.1) sans fond cohérent. La comparaison des courbes (3) et (4) montre que la figure de diffraction avec fond cohérent est beaucoup plus étalée que la figure de diffraction sans fond cohérent. Elle paraît à l'oeil tout à fait semblable à la figure de diffraction sans fond cohérent et deux fois plus large puisque les maxima impairs de (4) sont remplacés par des minima. D'autre part, la décroissance des éclairements est beaucoup plus lente sur la courbe (3) que sur la courbe (4): on peut percevoir un beaucoup plus grand nombre de franges sur la figure de diffraction avec fond

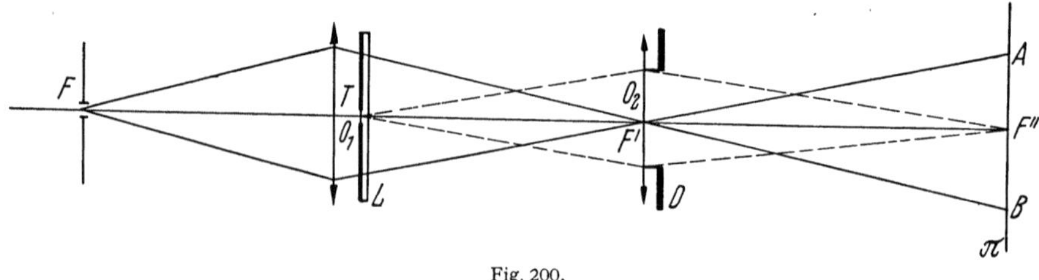

Fig. 200.

cohérent que sans fond cohérent. Si, par exemple, l'éclairement dû au fond cohérent est égal à 5% de celui du maximum central de (4) les éclairements du 10e et 11e maximum sans fond cohérent, sont de l'ordre de 0,1% de celui du maximum central. Avec fond cohérent, on obtient un maximum et un minimum dont les éclairements sont 6,5 et 3,8%.

La Fig. 200 montre un deuxième procédé de réalisation d'un fond cohérent. La fente F éclaire un objectif O_1 qui donne en F' une image de F sur un deuxième objectif O_2. Contre l'objectif O_1 on place la lame de verre L semi-transparente percée de la fente fine T. Celle-ci diffracte dans un angle très grand et on peut dire que l'objectif O_2 est pratiquement couvert par de la lumière diffractée dont l'amplitude et la phase sont constantes. Contre l'objectif O_2 est placé le diaphragme D percé de la fente dont on veut étudier l'influence sur la structure du phénomène de diffraction en F''. Dans ce montage, le fond cohérent est produit par le faisceau de lumière directe qui s'étale en $A\,B$ sur le plan π et le phénomène de diffraction est produit par la lumière diffractée par T et recueillie par l'objectif O_2 limité par la fente à étudier. Si l'on veut produire un fond cohérent d'intensité faible par rapport au phénomène de diffraction à étudier, la première méthode est plus avantageuse. La seconde méthode conviendra mieux pour obtenir des fonds cohérents plus intenses que le maximum de la figure de diffraction à étudier. Naturellement, la méthode du fond cohérent que nous venons d'exposer peut s'appliquer à des figures de diffraction absolument quelconques aussi bien dans la classe des phénomènes de Fraunhofer que dans la classe des phénomènes de Fresnel à distance finie.

Dans ce qui précède, nous avons supposé le fond cohérent en phase avec le maximum central de la figure de diffraction à étudier. Il est possible de faire varier la phase du fond cohérent en déposant sur L des couches minces métalliques semi-transparentes donnant une absorption et un déphasage déterminés.

86. Apodisation d'une figure de diffraction. La possibilité de séparer dans un instrument d'optique les images de deux sources ponctuelles rapprochées est limitée par l'ouverture de l'instrument. Si les deux sources ont la même intensité, le pouvoir séparateur ne dépend que de la largeur de la tache centrale de la figure de diffraction (68.8), l'éclairement dû aux anneaux étant négligeable. Mais si l'une des sources est beaucoup plus brillante que l'autre, on ne peut négliger les anneaux de diffraction et ce sont eux, bien souvent, qui déterminent un pouvoir séparateur très différent de la valeur conventionelle. On a vu d'autre part que près de 20% de l'énergie se trouve hors de la tache centrale de diffraction (voir Sect. 69): cette énergie est non seulement perdue pour la formation d'une image, mais doit être considérée comme nuisible par le voile parasite qu'elle produit.

Il est donc intéressant de réduire le plus possible l'importance des anneaux de diffraction, ce qui s'appelle «apodiser» la tache de diffraction, et d'arriver ainsi à améliorer la vision d'une source faible au voisinage d'une source intense et certainement de la même façon, la vision des objets à faible contraste.

Dans les phénomènes que nous avons étudiés précédemment, l'amplitude sur la surface d'onde Σ était constante et s'annulait brusquement au bord de l'écran limitant Σ. Cette discontinuité joue un rôle essentiel dans l'étalement de la lumière diffractée. Si on la remplace par une décroissance continue, on peut réduire l'importance des anneaux, mais on augmente la largeur de la tache centrale. Le problème de l'apodisation consiste donc à chercher des répartitions d'amplitude, réalisables au moyen d'écrans dégradés, ou par d'autres moyens, qui produisent la meilleure atténuation des anneaux de diffraction tout en élargissant le moins possible la tache centrale.

Considérons le problème à une seule dimension: le phénomène de diffraction le long de l'axe Cz' dû à une ouverture rectangulaire de largeur $2\gamma_0'$ petite devant sa hauteur $2\beta_0'$, est donné à un facteur constant près, par la Sect. 76

$$U_{z'} = \int_{-\gamma_0'}^{+\gamma_0'} e^{-jK\gamma'z'}\, d\gamma' \tag{86.1}$$

si la surface d'onde a partout la même amplitude.

Supposons maintenant que l'on dépose sur la surface de l'objectif une couche absorbante telle que l'amplitude sur la surface d'onde Σ suive la loi $\cos\pi\dfrac{\gamma'}{2\gamma_0'}$. L'amplitude $U_{z'}$ devient (Sect. 91)

$$U_{z'} = \int_{-\gamma_0'}^{+\gamma_0'} \cos\frac{\pi\gamma'}{2\gamma_0'} e^{-jK\gamma'z'}\, d\gamma' = 2\gamma_0' \; \frac{2\cos K\gamma_0'z'}{\pi\left(1 - 4\dfrac{K^2\gamma_0'^2 z'^2}{\pi^2}\right)}$$

et l'éclairement

$$(E_{z'})_A = 4\gamma_0'^2 \cdot \frac{4}{\pi^2}\left(\frac{\cos K\gamma_0'z'}{1 - 4\dfrac{K^2\gamma_0'^2 z'^2}{\pi^2}}\right)^2. \tag{86.2}$$

Sans apodisation, on aurait obtenu d'après (86.1)

$$E_{z'} = 4\gamma_0'^2\left(\frac{\sin K\gamma_0'z'}{K\gamma_0'z'}\right)^2. \tag{86.3}$$

Au centre, $z'=0$, on a $(E_{z'})_A = 4\gamma_0'^2 \cdot \dfrac{4}{\pi^2}$ et $E_{z'} = 4\gamma_0'^2$.

En comparant les deux équations (86.2) et (86.3), on voit que pour la figure de diffraction non apodisée (86.3), l'éclairement des maxima ($\sin K\gamma_0'z' = \pm 1$)

tend vers $\dfrac{1}{K^2\gamma_0'^2 z'^2}$ quand z' augmente, alors que pour la figure de diffraction apodisée (86.2), il tend vers $\dfrac{\pi^4}{16\,K^4\gamma_0'^4 z'^4}$, ce qui produit une diminution consi-

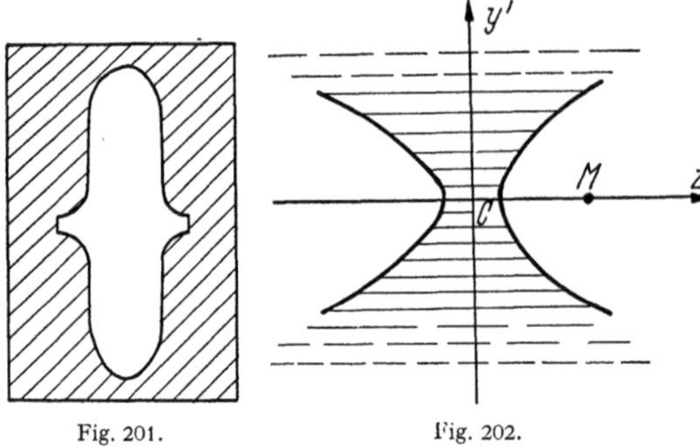

Fig. 201. Fig. 202.

dérable de la lumière diffractée loin de centre, diminution de l'ordre de 100 vers la dixième frange.

Par contre, la largeur de la tache centrale augmente: le premier minimum d'éclairement a lieu pour $\vartheta = \dfrac{z'}{R} = \dfrac{3\lambda}{2D}$ avec $D = 2R\gamma_0'$ alors qu'il a lieu pour $\vartheta = \dfrac{\lambda}{D}$ pour l'ouverture uniforme.

L'apodisation précédente se traduit également par une diminution de l'éclairement au centre de la tache de diffraction égale à $4/\pi^2$ soit 40% environ.

On peut aussi obtenir une apodisation de la tach de diffraction en employant un objectif de transparence uniforme, mais limité par un contour de forme convenable. Nous signalerons l'écran en «chapeau de gendarme» de Jacquinot (Fig. 201) qui, placé devant un objectif, donne une figure de diffraction schématiquement représentée par la Fig. 202. Toute la lumière diffractée est pratiquement

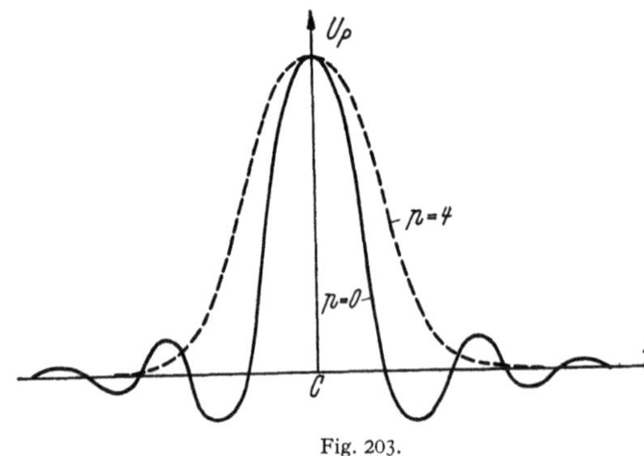

Fig. 203.

concentrée dans la région hachurée et si on se place en un point M situé à environ 6 ou 7 fois le diamétre de la tache centrale, l'éclairement est seulement égal à 10^{-7} en prenant comme unité l'éclairement au centre. La forme de cet écran est définie par l'équation

$$\beta' = 0{,}493 + 0{,}486\cos\pi\gamma' + 0{,}020\cos 2\pi\gamma'.$$

On peut également réaliser une apodisation dans le cas des ouvertures circulaires: prenons par exemple une variation d'absorption suivant la loi $e^{-p\alpha'^2}$ D'après (68.1), l'amplitude U_P deviendra

$$U_P = \frac{jR}{\lambda}\int\limits_0^{\alpha_0'}\int\limits_0^{2\pi} e^{-p\alpha'^2}\,e^{-jK\alpha'\varrho_1\cos\varphi}\,\alpha'\,d\alpha'\,d\varphi.$$

Les résultats du calcul sont donnés par la Fig. 203 d'après Lansraux. Sur cette figure on a porté l'amplitude U en ordonnées et le paramètre $Z = K\alpha_0'\varrho_1$ en abscisse, en ramenant les amplitudes au centre à la même valeur. Pour $p=0$ on a le disque d'Airy en amplitude. En donnant au paramètre p la valeur $p=4$, on a une apodisation assez marquée: pour $Z=15$ on a $U_P = +0{,}0017$, alors que dans le disque d'Airy on a $U_P = +0{,}0274$.

On retrouve ici l'évolution générale du phénomène, la diminution de l'éclairement des anneaux de diffraction se traduit par un élargissement de la tache centrale. On peut penser à faire l'opération inverse de la précédente, c'est-à-dire affiner la tache centrale de la figure de diffraction, au risque d'augmenter l'éclairement des anneaux. Il serait possible ainsi d'améliorer le pouvoir séparateur sur des sources ponctuelles au prix d'une perte de contraste sur les objets étendus. On peut obtenir une solution du problème en prenant un écran complémentaire de l'écran qui a servi à apodiser le phénomène de diffraction. D'autres procédés sont possibles: par exemple, si l'on introduit un déphasage de π sur la lumière qui traverse une portion centrale de l'ouverture dont la surface est une fraction m de l'aire totale, la distance entre les 2 premiers minima nuls de la figure de diffraction décroît comme $\sqrt{1-2m}$ quand m tend vers $1/2$. On peut donc obtenir une tache centrale aussi petite que l'on veut, mais il faut noter que l'éclairement au centre étant proportionnel à $(1-2m)^2$ tend rapidement vers zéro. D'autre part, l'énergie passe dans les anneaux de diffraction qui deviennent si importants que toute observation devient impossible. On peut démontrer que ce résultat est général: l'amélioration du pouvoir séparateur s'accompagne d'une perte de lumière et d'une perte de contraste qui limitent le gain à des valeurs de l'ordre de 25%.

b) Image d'une source ponctuelle par un système optique imparfait.

87. Formule générale. Dans tout ce qui précède, nous avons étudié la structure des phénomènes de diffraction dans le cas d'une onde Σ sphérique fournie par un objectif parfait. Dans ce cas, la figure de diffraction ne dépend que de la forme et des dimensions du diaphragme limitant la surface utile de l'objectif. Il est intéressant de voir comment évolue la tache de diffraction lorsque la surface d'onde Σ cesse d'être sphérique, c'est-à-dire lorsque l'objectif utilisé pour former l'image de la source possède des aber-

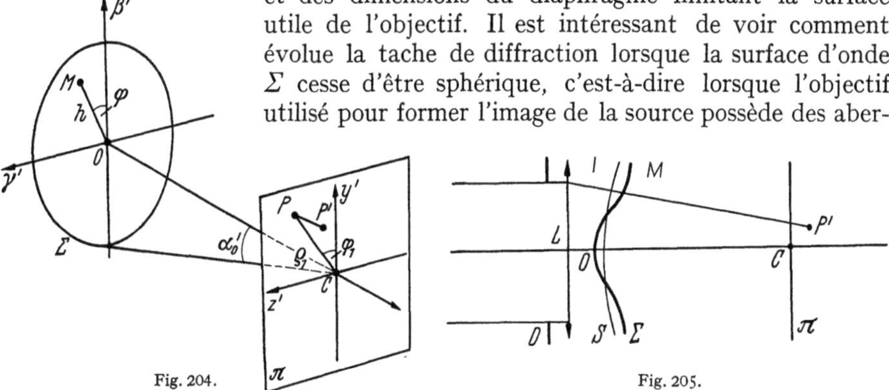

Fig. 204.   Fig. 205.

rations. Le problème a été étudié pour la première fois par Lord RAYLEIGH dans le cas de l'aberration sphérique. Il a montré que si la surface d'onde Σ donnée par l'instrument, peut être enfermée entre deux sphères distantes d'un quart d'onde, l'image est encore très voisine de celle fournie par un objectif parfait; c'est la règle célèbre de Lord RAYLEIGH.

L'amplitude en un point P' (Fig. 204) est donnée par la formule (67.4)

$$U_{P'} = \frac{i}{\lambda R} \iint\limits_{\Sigma} e^{jK\Delta'}\, ds \qquad (87.1)$$

avec

$$\Delta' = \Delta - (\beta' y' + \gamma' z').$$

Δ représente l'écart IM (Fig. 205) entre la surface d'onde Σ déformée par les aberrations de l'objectif L et la sphère S de centre C. A la différence de l'équation (67.6), nous avons inclus le défaut de mise au point ($\varepsilon = PP'$) représenté par $\frac{\varepsilon \alpha'^2}{2}$ dans le terme Δ.

Nous ne considérerons dans ce chapitre que le cas des ouvertures circulaires et il sera commode d'utiliser les coordonnées polaires h et φ, h variant de 0 à 1 du centre au bord de l'ouverture. On écrira

$$h = \frac{1}{\alpha_0'} (\beta'^2 + \gamma'^2)^{\frac{1}{2}}. \tag{87.2}$$

Dans ces conditions, l'écart Δ pourra se développer sous la forme suivante que l'on peut établir en optique géométrique

$$\Delta = d\,h^2 + s_1\,h^4 + s_2\,h^6 + (l\,h + c_1\,h^3 + c_2\,h^5)\cos\varphi + a\,h^2\cos 2\varphi. \tag{87.3}$$

Dans cette expression, les coefficients ont les significations suivantes:

d représente le défaut de mise au point ou la courbure de champ, c'est à dire l'écart maximum au bord de la pupille ($h=1$) entre la sphère de référence et la surface d'onde dans le cas où elle est sphérique;

s_1 est le coefficient d'aberration sphérique du 3ème ordre. C'est l'écart maximum (au bord de l'ouverture) entre la surface d'onde déformée Σ et la sphère de référence qui lui est ici surosculatrice;

s_2 est dans les mêmes conditions, le coefficient d'aberration sphérique du 5ème ordre;

l est un paramètre introduit pour pouvoir éventuellement tenir compte d'un basculement de la sphère de référence;

c_1 est le coefficient de coma du 3ème ordre, c'est-à-dire l'écart marginal maximum entre la surface d'onde entachée de coma et la sphère de référence centrée au foyer paraxial;

c_2 est, dans les mêmes conditions, le coefficient de coma du 5ème ordre;

a est le coefficient d'astigmatisme, c'est à dire l'écart marginal maximum entre la surface d'onde astigmatique et la sphère de référence centrée au milieu de la distance qui sépare les deux focales.

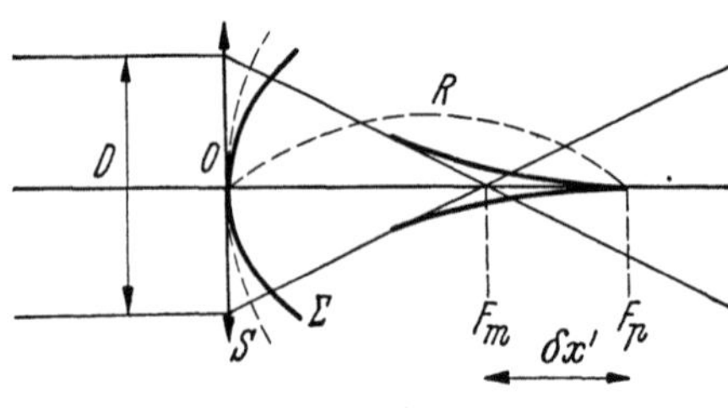

Fig. 206.

88. Image d'un point en présence d'aberration sphérique. En combinant l'aberration sphérique du 3ème ordre avec un défaut de mise au point, on peut écrire

$$\Delta = d\,h^2 + s_1\,h^4.$$

La théorie des aberrations de 3ème ordre montre que le coefficient s_1 est donné par

$$s_1 = \frac{D^2\,\delta x'}{16\,R^2},$$

$\delta x'$ étant l'aberration sphérique longitudinale, c'est à dire la distance séparant le foyer marginal F_m du foyer paraxial F_p (Fig. 206), ce dernier étant pris pour origine avec

$$\Omega = \frac{D}{R}, \qquad \Psi = K\frac{\Omega^2}{8}\,\varepsilon, \qquad \Psi' = K\frac{\Omega^2}{16}\,\delta x', \qquad t = h^2,$$

$$d = \frac{\Omega^2}{8}\,\varepsilon, \qquad Z = K\alpha_0'\varrho_1, \qquad K = \frac{2\pi}{\lambda},$$

on peut écrire

$$K\Delta = \Psi t + \Psi' t^2$$

d'où l'amplitude en P'

$$U_{P'} = \frac{i}{\lambda R} \iint\limits_{\Sigma} e^{jK\Delta'}\,ds = j\,\frac{\pi R\,\alpha_0'^2}{\lambda} \int\limits_0^1 e^{jK\Delta}\,J_0(Z\sqrt{t})\,dt$$

et l'éclairement à und facteur constant près

$$E_{P'}=\left[\int_0^1 J_0\!\left(Z\sqrt{t}\right)\cos\left(K\Delta\right)dt\right]^2+\left[\int_0^1 J_0\!\left(Z\sqrt{t}\right)\sin\left(K\Delta\right)dt\right]^2.$$

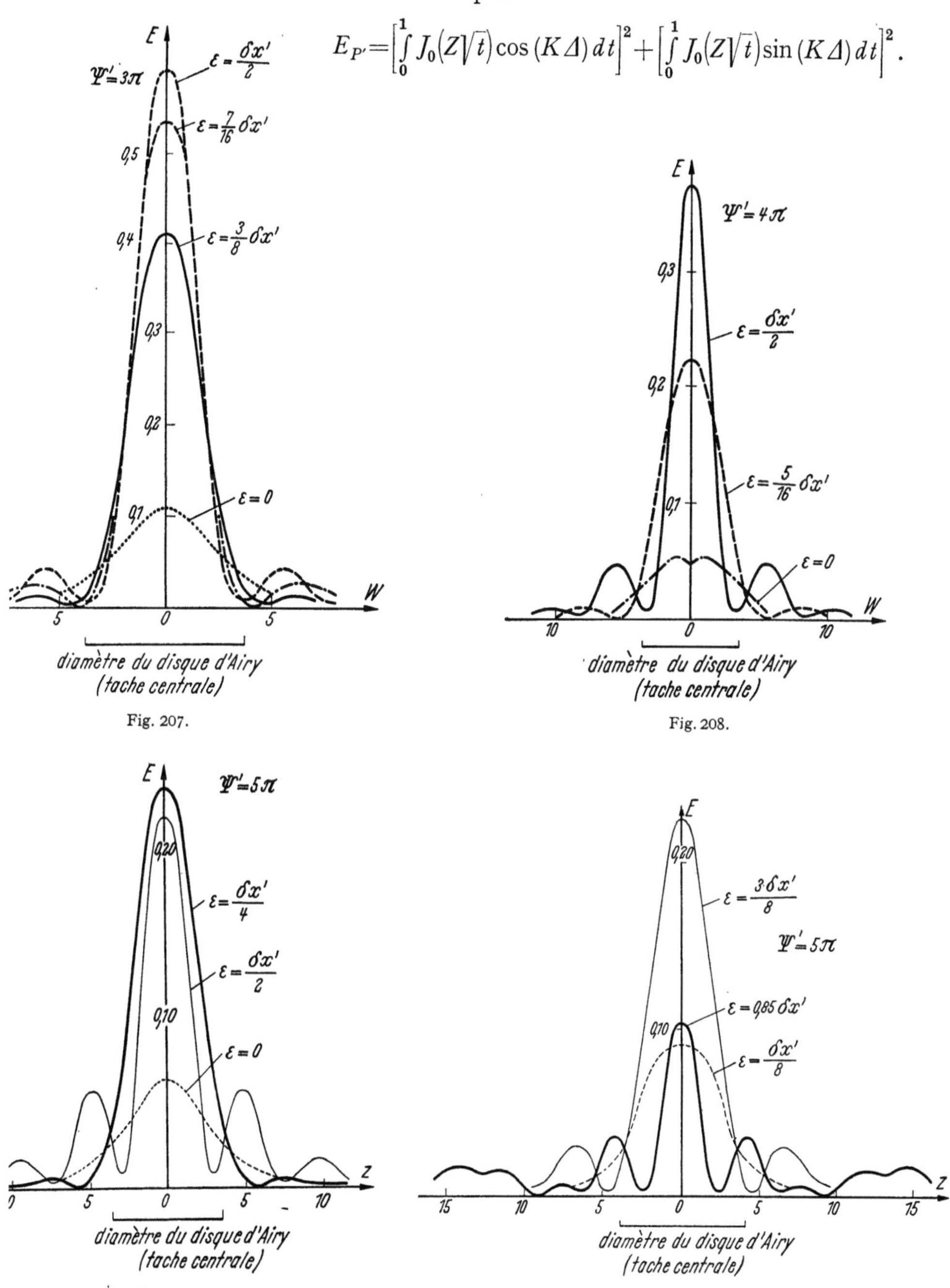

Fig. 207.

Fig. 208.

Fig. 209.

Fig. 210.

De nombreux physiciens ont calculé numériquement ces intégrales. Le tableau ci-après donne les éclairements calculés par CONRADY, BUXTON et MARTIN dans quelques figures de diffraction. Les Fig. 207, 208, 209, 210, correspondent à des aberrations supérieures à celles qui ont été considérées dans le tableau.

Tableau 10. *Eclairements dans les figures de diffraction en présence d'aberration spherique du 3ème ordre et avec divers défauts de mise au point. D'après les travaux de* Conrady, Buxton *et* Martin.

Ψ'/K	$\lambda/4$	$\lambda/4$	$\lambda/4$	λ	λ	λ
Ψ/K	0	$\lambda/4$	$\lambda/2$	λ	$3\lambda/4$	$5\lambda/4$
z	Eclairements					
0,000	0,8001	0,9861	0,8001	0,8001	0,6306	0,6600
0,698	7108	8726	7052	7077	5650	5746
1,396	4913	5740	4748	4822	4243	3362
2,094	2556	2917	2341	2351	2370	1768
2,793	0949	0958	0806	0683	0950	0522
3,491	0295	0066	0312	0062	0237	0289
4,189	0211	0046	0361	0100	0042	0570
4,887	0228	0171	0410	0265	0069	0761
5,585	0158	0256	0276	0290	0343	0625
6,283	0066	0119	0098	0197	0151	0311
6,981	0038	0021	0016	0097	0150	0070
7,679	0056	0023	0024	0044	0119	0002
8,378	0062	0040	0046	0023	0072	0034
9,076	0035	0032	0031	0018	0035	0058
9,774	0012	0005	0006	0025	0024	0040
10,472		0004		0032	0036	0009
11,170	0019	0015		0035	0042	0003
11,868	0019	0015		0027	0030	0013
12,566	0008	0006		0009	0016	0012
13,264	0002	0001		0005	0014	0005
13,963	0009	0004		0012	0018	0000
14,661	0012	0008		0013	0019	0000
15,358	0008	0006		0005	0012	0005
16,057	0002	0001		0003	0005	0004
16,755	0000	0000		0003	0010	0001
17,453	0002	0003		0006	0012	0000

La comparaison des Fig. 207, 208, 209 et 210 montre que l'apparition et l'augmentation de l'aberration se traduit par une diminution de l'éclairement au centre de la tache de diffraction et un renforcement des anneaux de diffraction entourant la tache centrale. C'est d'ailleurs un fait général: toute aberration se traduit par un étalement de la figure de diffraction et une diminution de l'éclairement au centre.

Considérons par exemple l'aberration $\Psi' = 5\pi$ (Fig. 209 et 210): lorsqu'on passe du foyer paraxial au foyer marginal, c'est-à-dire lorsque ε augmente, on voit que la tache centrale devient plus fine, mais des anneaux de diffraction apparaissent de plus en plus nombreux. En calculant pour une même aberration un grand nombre de figures de diffraction, on peut tracer le réseau des courbes d'égal éclairement dans un plan passant par l'axe optique de l'objectif. Les Fig. 211, 212 et 213 montrent les résultats pour des aberrations $\Psi' = 3,8\pi, 8\pi, 20\pi$.

Considérons une aberration Ψ' inférieure à 2π et admettons que l'image la meilleure est celle dont la structure se rapproche le plus de celle du disque d'Airy. Considérons l'expression

$$K\Delta = \Psi t + \Psi' t^2$$

et portons en ordonnées $K\Delta$ et t en abscisses.

Pour différentes valeurs de ε on obtient les trois courbes de la Fig. 214 situées le long de la caustique. Pour $\varepsilon < \dfrac{\delta x'}{2}$ la différence de phase maximum est égale

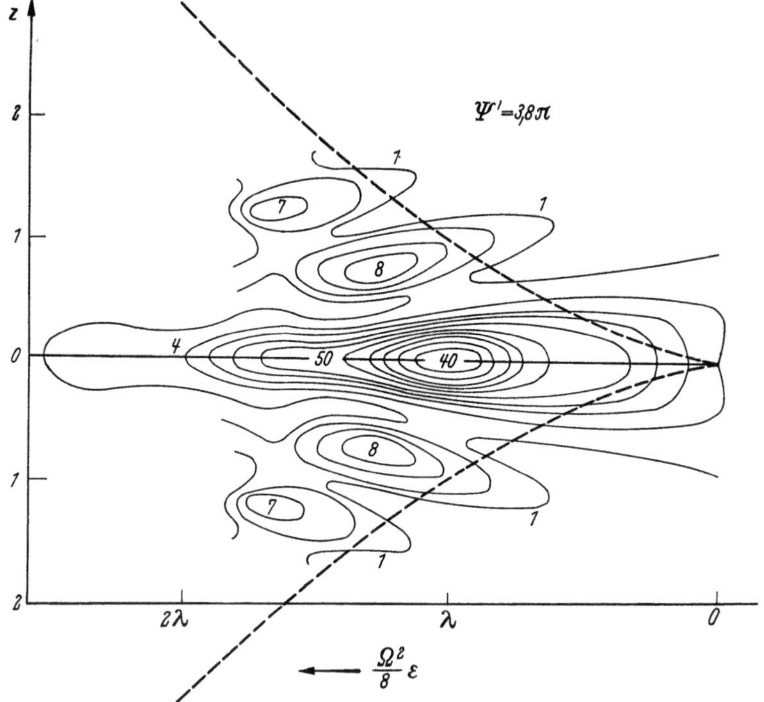

Fig. 211. Courbes isophotes d'aberation sphérique d'après J. Picht.

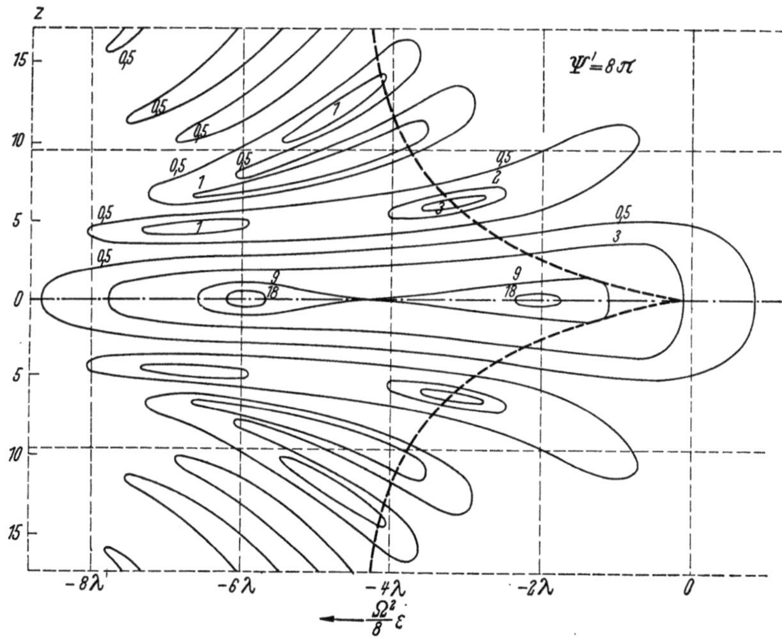

Fig. 212. Courbes isophotes d'aberation sphérique d'après A. Maréchal.

à l'ordonnée du minimum de la fonction $K\varDelta=f(t)$ augmentée de l'ordonnée correspondant à $t=1$. Pour $\varepsilon>\dfrac{\delta x'}{2}$ la différence de phase maximum est seulement

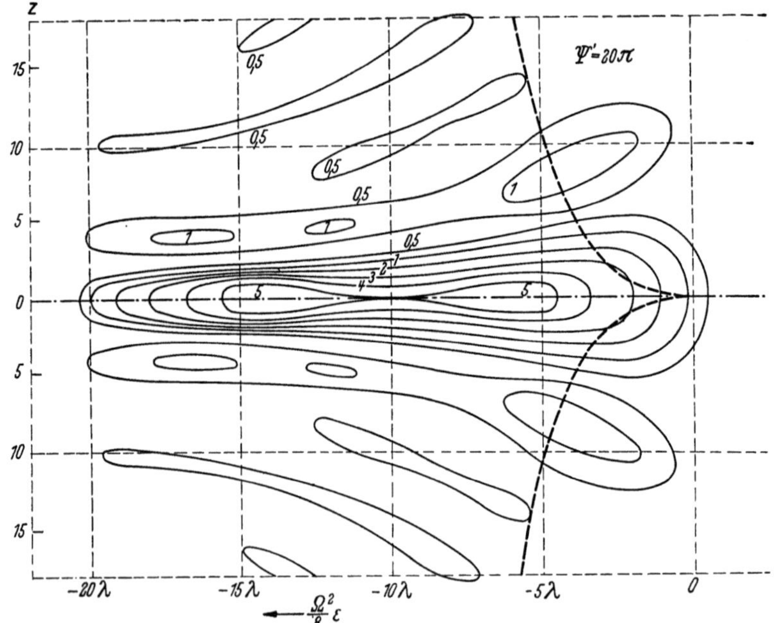

Fig. 213. Courbes isophotes d'aberration sphérique d'après A. Maréchal.

égale à l'ordonnée du minimum de la fonction précédente. On trouve facilement
que la différence de phase maximum a la plus petite valeur possible pour:

$$\varepsilon = \frac{\delta x'}{2}.$$

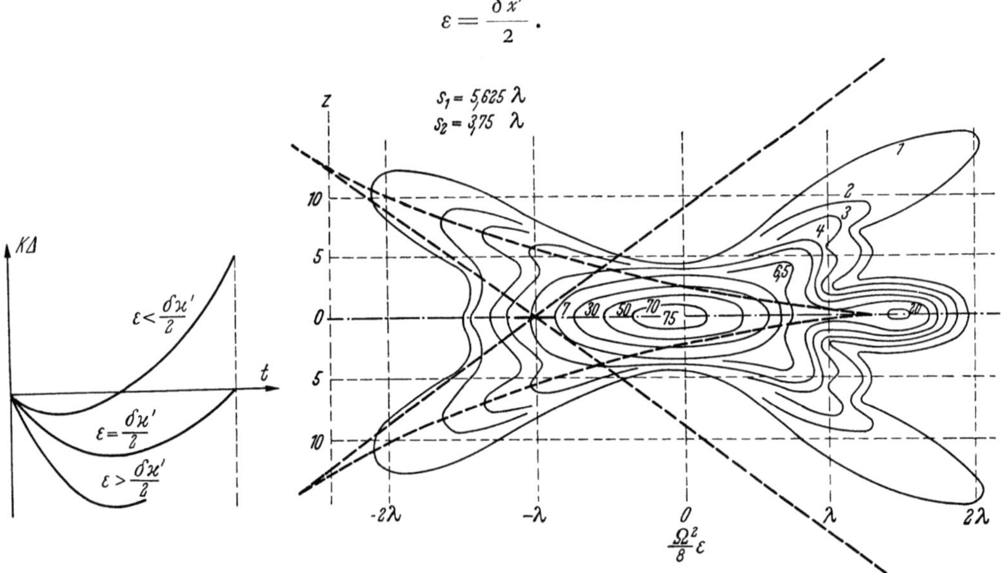

Fig. 214.

Fig. 215. Courbes isophotes d'aberration sphérique d'après A. Maréchal.

C'est cette image dont la structure se rapproche le plus du disque d'Airy. Le
plan de meilleure mise au point en présence d'aberration sphérique du 3ème
ordre se trouve donc à la moitié de l'aberration longitudinale.

　　L'évolution de la tache de diffraction avec aberration sphérique et défaut
de mise au point permet de détecter de faibles résidus d'aberration. Par exemple,

on veut examiner la qualité d'un objectif de microscope: il suffit d'observer la tache de diffraction de part et d'autre de la meilleure mise au point. On donne au mouvement lent du microscope de petits déplacements rapides pour observer les images en avant et en arrière du meilleur plan de mise au point. Si l'aspect des taches de diffraction est identique de part et d'autre de la meilleure image, l'objectif est parfait. De faibles valeurs de l'aberration sphérique font apparaître immédiatement des différences auxquelles l'oeil est très sensible. Un observateur exercé peut facilement détecter la présence d'une aberration sphérique de l'ordre de $\lambda/4$.

Pour calculer la figure de diffraction en présence de l'aberration sphérique du 5ème ordre, on prend le terme suivant dans l'expression (87.3)

$$\Delta = d\,h^2 + s_1\,h^4 + s_2\,h^6.$$

La Fig. 215 montre les courbes isophotes pour

$$s_2 = 3,75\,\lambda$$

$$s_1 = 5,625\,\lambda.$$

La concentration de l'énergie est évidemment bien meilleure.

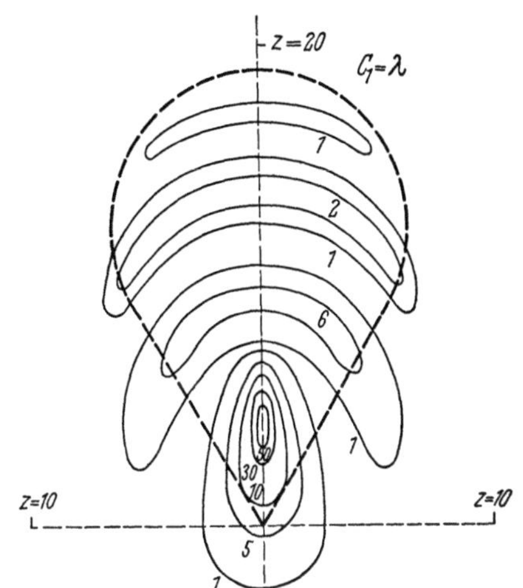

Fig. 216. Courbes isophotes de coma d'après A. Maréchal.

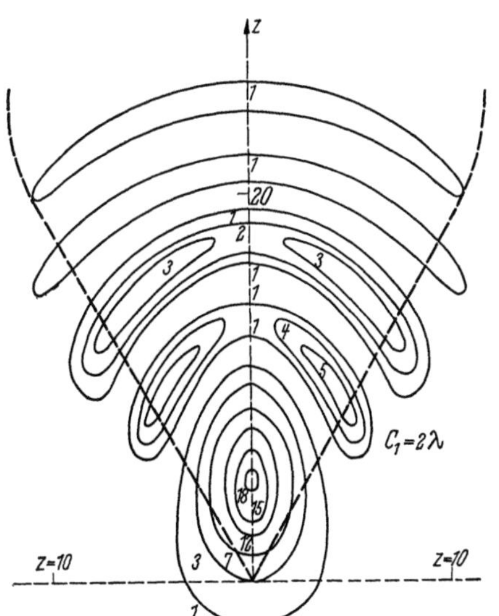

Fig. 217. Courbes isophotes de coma d'après A. Maréchal.

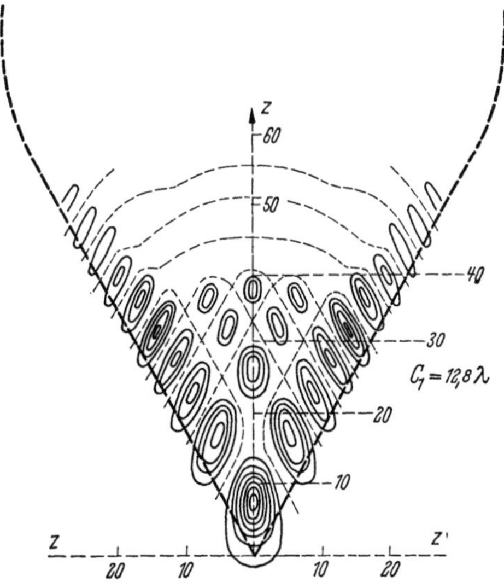

Fig. 218. Courbes isophotes de coma d'après R. Kingslake.

89. Image d'un point en présence de coma d'astigmatisme et de la combinaison des différentes aberrations. Les calculs s'effectuent en combinant les expressions (87.3) et (87.1). Les Fig. 216, 217 et 218 se rapportent au cas de la coma du 3ème ordre

$$\Delta = c_1\,h^3 \cos \varphi.$$

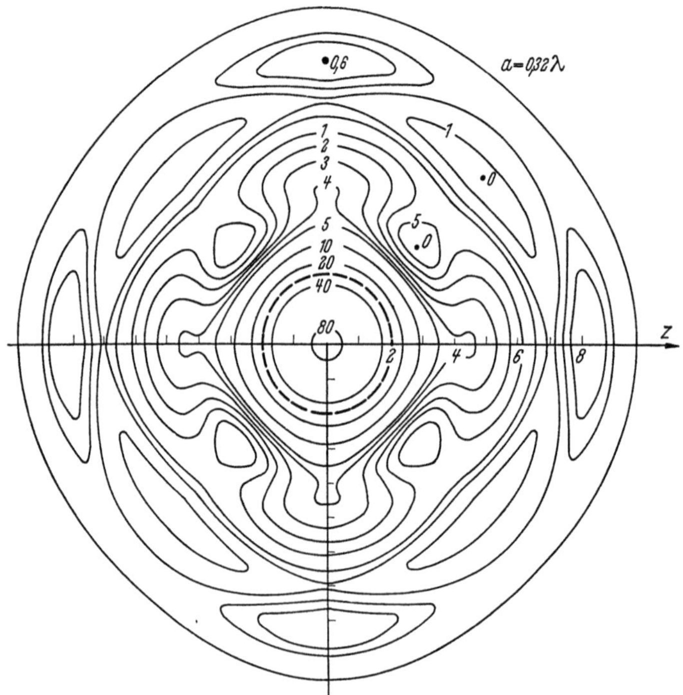

Fig. 219. Courbes isophotes d'astigmatisme d'après B. R. A. Nijboer (cercle de moindre diffusion).

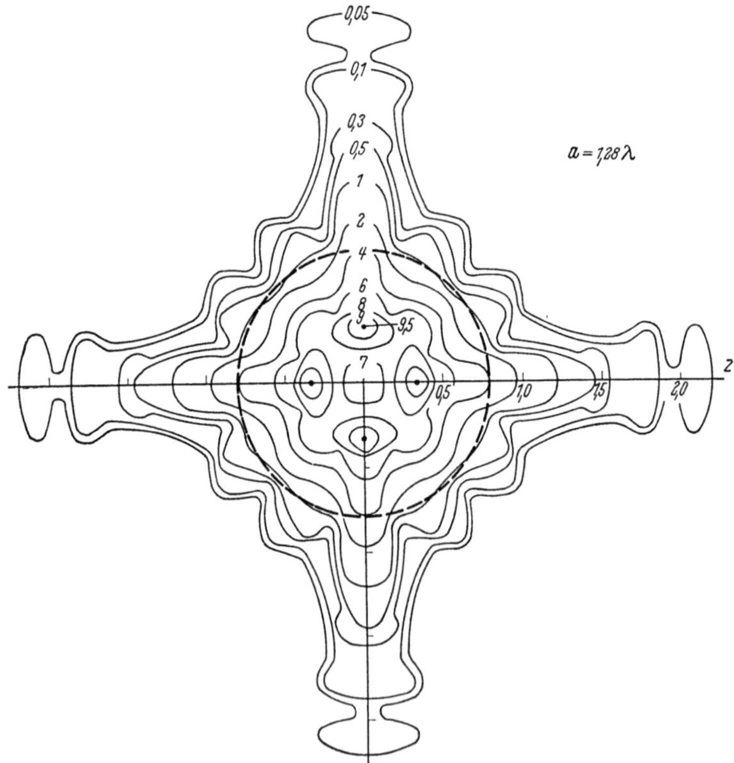

Fig. 220. Courbes isophotes d'astigmatisme d'après K. Nienhuis.

Les Fig. 219, 220 et 221 donnent l'aspect de la figure de diffraction en pré-
nce d'astigmatisme.

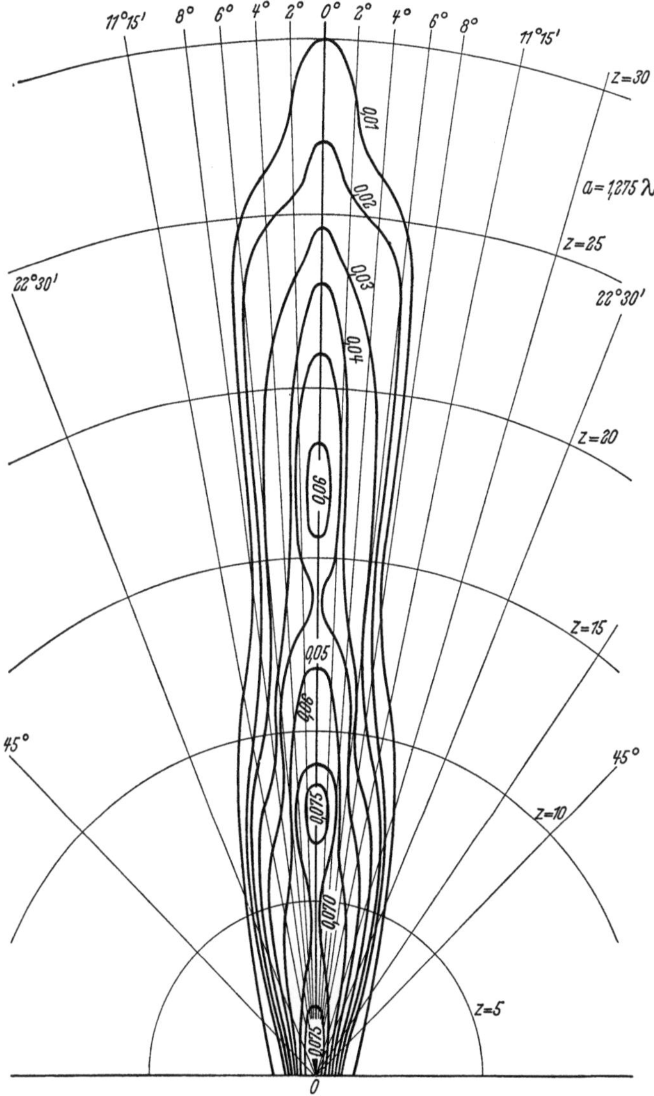

Fig. 221. Isophotes d'astigmatisme au voisinage d'une focale d'après A. MARÉCHAL.

Notons que la courbure de champ n'introduit pas autre chose qu'un défaut
de mise au point.

On peut voir enfin sur la Fig. 222 une remarquable tache de diffraction
calculée par A. MARÉCHAL, et montrant la combinaison de l'aberration sphérique
de la coma et de l'astigmatisme pour $s_1 = 2\lambda$ $c_1 = \lambda$ et $a = \lambda$. Pour ses calculs,
A. MARÉCHAL a étudié et réalisé un intégrateur mécanique spécialement conçu
pour le calcul des figures de diffraction.

L'ensemble des figures de diffraction en présence des aberrations que nous
venons de considérer, montre que l'évolution de l'image n'est pas la même suivant
l'aberration étudiée. On peut constater que si l'aberration augmente de plus en

plus, la tache de diffraction en présence d'astigmatisme, se rapproche très rapidement de l'aspect géométrique de l'aberration. L'évolution de la coma est moins rapide et celle de l'aberration sphérique, encore moins que la coma. Il faut une aberration sphérique considérable pour retrouver l'aspect de la caustique géométrique.

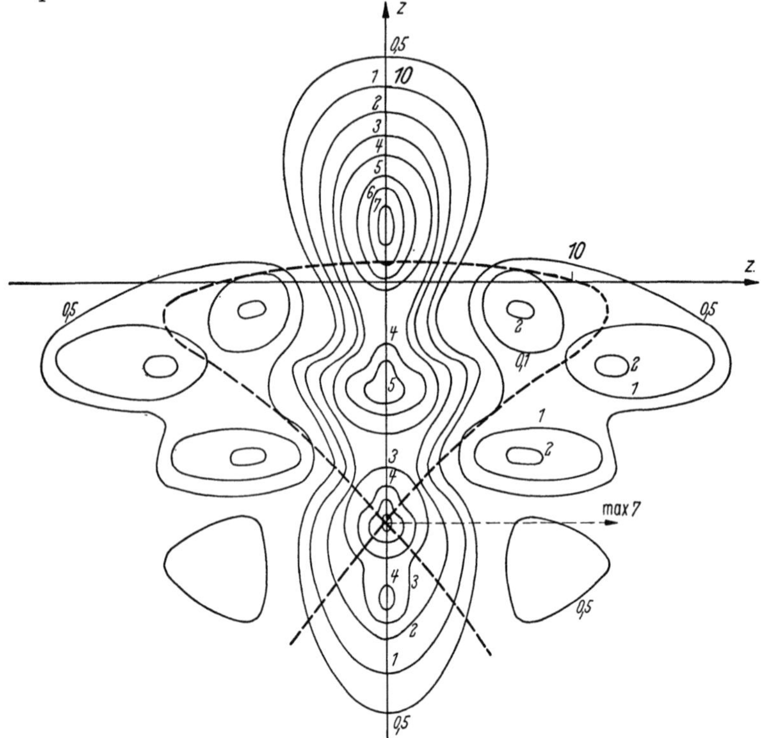

Fig. 222. Isophotes d'une combinaison d'aberration sphérique de coma et d'astigmatisme d'après A. Maréchal.

90. Onde moyenne sphérique possédant des accidents locaux.

On rencontre constamment en optique des ondes dont la forme moyenne est sphérique et qui possèdent des accidents locaux en plus ou moins grand nombre. Ces accidents peuvent être produits par les défauts d'homogénéité tels que les «fils» ou même par les défauts de poli qui se traduisent par un grand nombre de très petites oscillations de la surface d'onde autour de sa position moyenne. De tels défauts perturbent la surface d'onde et modifient la structure de la tache de diffraction.

α) *Phénomène de diffraction produit par un accident de forme rectangulaire.* C'est par exemple le cas d'un défaut d'homogénéité appelé «fil» qui produit dans la surface d'onde un accident constitué par une rapide variation de phase dans un sens perpendiculaire à la longueur du fil que l'on suppose rectiligne. On supposera le fil parallèle à l'axe $O\beta'$ et passant par le milieu O de l'onde Σ (Fig. 223), l'indice variant dans une direction perpendiculaire à $O\beta'$ suivant $O\gamma'$. Cette variation d'indice se traduit par une variation de phase particulière à chaque cas étudié. La largeur du fil étant égale à $l = 2\gamma_0' R$ on admettra qu'entre les abscisses $-\dfrac{l}{2}$ et $+\dfrac{l}{2}$, la phase φ est donnée par la fonction

$$\frac{\varphi}{2}\left(1 + \cos\frac{\pi\gamma'}{\gamma_0'}\right).$$

Pour tous les autres points de l'onde Σ, la phase est constamment égale à zéro (Fig. 224).

L'amplitude en P peut s'écrire d'après (67.4) ou (87.1)

$$U_P = \frac{jR}{\lambda} \iint\limits_{\Sigma} e^{j\frac{\varphi}{2}\left(1+\cos\frac{\pi\gamma'}{\gamma_0'}\right)} e^{-jK(\beta'y'+\gamma'z')}\, d\beta'\, d\gamma',$$

$$U_P = \frac{jR}{\lambda} \iint\limits_{\Sigma} e^{-jK(\beta'y'+\gamma'z')}\, d\beta'\, d\gamma' - \frac{jR}{\lambda} \iint\limits_{l} e^{-jK(\beta'y'+\gamma'z')}\, d\beta'\, d\gamma' +$$

$$+ \frac{jR}{\lambda} \iint e^{j\frac{\varphi}{2}\left(1+\cos\frac{\pi\gamma'}{\gamma_0'}\right)} e^{-jK(\beta'y'+\gamma'z')}\, d\beta'\, d\gamma',$$

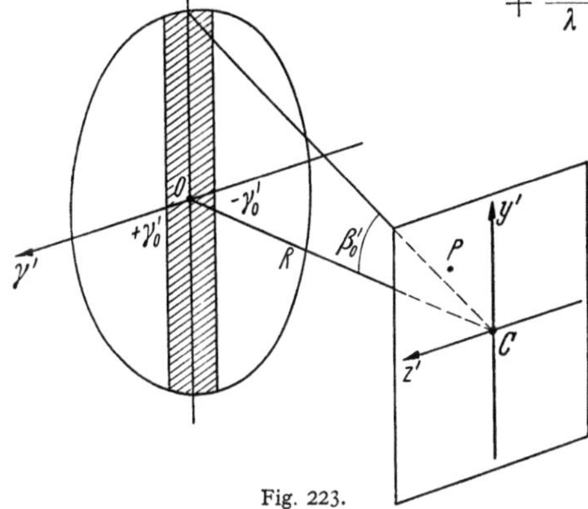

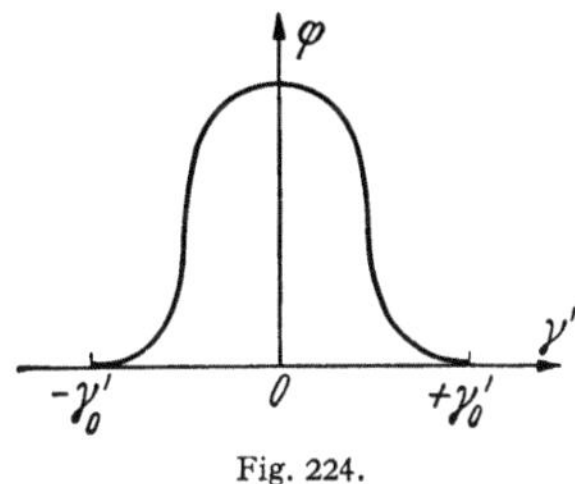

les 2 dernières intégrales étant étendues à la surface du fil l seulement. En remarquant que le fil peut être

Fig. 223. Fig. 224.

assimilé à un rectangle très allongé, on peut intégrer par rapport à β' et écrire la 3ème intégrale sous la forme

$$\frac{2j\beta_0'R}{\lambda}\, \frac{\sin K\beta_0'y'}{K\beta_0'y'} \int\limits_{-\gamma_0'}^{+\gamma_0'} e^{j\frac{\varphi}{2}\left(1+\cos\frac{\pi\gamma'}{\gamma_0'}\right)} e^{-jK\gamma'z'}\, d\gamma', \tag{90.1}$$

$2\beta_0'$ étant le diamètre angulaire de l'objectif, c'est-à-dire la hauteur angulaire du fil

Posons

$$\frac{\pi\gamma'}{\gamma_0'} = \vartheta \quad \text{et} \quad \frac{2\gamma_0'z'}{\lambda} = m$$

l'expression (90.1) devient:

$$\frac{2j\beta_0'\gamma_0'R}{\pi\lambda}\, e^{j\frac{\varphi}{2}}\, \frac{\sin K\beta_0'y'}{K\beta_0'y'} \left[\int\limits_{-\pi}^{+\pi} e^{j\frac{\varphi}{2}\cos\vartheta} \cos m\vartheta\, d\vartheta - j \int\limits_{-\pi}^{+\pi} e^{j\frac{\varphi}{2}\cos\vartheta} \sin m\vartheta\, d\vartheta \right]. \tag{90.2}$$

La deuxième intégrale est nulle et la première peut se calculer au moyen des fonctions de Bessel

$$e^{j\frac{\varphi}{2}\cos\vartheta} = J_0\left(\frac{\varphi}{2}\right) + 2\sum_{n=1}^{\infty} j\, J_n\left(\frac{\varphi}{2}\right) \cos n\vartheta$$

et l'expression (90.2) est égale à

$$\frac{2j\beta_0'\gamma_0'R}{\pi\lambda}\, e^{j\frac{\varphi}{2}}\, \frac{\sin K\beta_0'y'}{K\beta_0'y'} \left[\frac{2\sin m\pi}{m}\, J_0\left(\frac{\varphi}{2}\right) + 4\sin m\pi \sum_{n=1}^{\infty}{}' (-1)^n j^n\, \frac{1}{m^2-n^2}\, J_n\left(\frac{\varphi}{2}\right) \right]$$

le signe Σ' indiquant que l'on donne à n toutes les valeurs 1, 2, 3, 4, … sauf la valeur $n=m$ si m est entier. L'amplitude en P devient

$$U_P = \frac{jR}{\lambda} \iint\limits_{\Sigma} e^{-jK(\beta' y' + \gamma' z')}\, d\beta'\, d\gamma' - \frac{4jR\beta_0'\gamma_0'}{\lambda} \frac{\sin K\gamma_0' z'}{K\gamma_0' z'} \frac{\sin K\beta_0' y'}{K\beta_0' y'} \times$$

$$\times \left[1 - e^{j\frac{\varphi}{2}} J_0\left(\frac{\varphi}{2}\right) - 2m^2 e^{j\frac{\varphi}{2}} \sum_{n=1}^{\infty}{}' j^n (-1)^n \frac{1}{m^2 - n^2} J_n\left(\frac{\varphi}{2}\right) \right]$$

si le fil a une largeur l petite devant le diamètre D de l'objectif, et en étudiant ce qui se passe au voisinage de la tache de diffraction dûe à l'ouverture entière de l'onde Σ et sur l'axe Cz' on est pratiquement dans la partie stationnaire du phénomène de diffraction correspondant au fil et on peut poser

$$\frac{\sin K\gamma_0' z'}{K\gamma_0' z'} \frac{\sin K\beta_0' y'}{K\beta_0' y'} \approx 1 .$$

D'autre part, le fil a en général une surface petite par rapport à celle de l'objectif et on pourra négliger le terme $2m^2 e^{j\frac{\varphi}{2}} \sum\limits_{n=1}^{\infty}{}'$ … L'objectif étant limité par un contour circulaire de diamètre D, on aura, à un facteur constant près, l'éclairement le long de Cz' par l'expression

$$E_{z'} = \left[\frac{2J_1(Z)}{Z}\right]^2 + \frac{2{,}546}{p} \frac{2J_1(Z)}{Z} \left[\cos\frac{\varphi}{2} J_0\left(\frac{\varphi}{2}\right) - 1 \right] + \left.\begin{matrix} \\ \\ \\ \end{matrix}\right\}$$
$$+ \frac{1{,}621}{p^2} \left[1 - 2\cos\frac{\varphi}{2} J_0\left(\frac{\varphi}{2}\right) + J_0^2\left(\frac{\varphi}{2}\right) \right] \qquad (90.3)$$

avec, d'après (68.3) et (68.4)

$$Z = K\beta_0' \varrho_1 = K\alpha_0' \varrho_1$$

et en posant

$$\frac{l}{D} = \frac{1}{p} .$$

Dans le cas où $\dfrac{\varphi}{2} = 1$ et $\dfrac{l}{D} = \dfrac{1}{100}$ on a les résultats suivants:

$Z=0$	$E_{z'}=0{,}9668$	$E_{z'}$ (tache d'Airy) $=1{,}0000$
3,83	0,0004	0,0000
5,14	0,0223	0,0174
7,02	0,0004	0,0000
8,42	0,0024	0,0042
10,17	0,0004	0,0000
11,62	0,0033	0,0017

Les modifications de la tache de diffraction sont caractérisées comme dans le cas des aberrations par une diminution de l'éclairement du maximum central et un renforcement des anneaux de diffraction avec disparition des minima nuls (Fig. 225). La relation (90.3) montre que si l'on s'éloigne de la tache centrale suivant Cz' on a une ligne fine lumineuse d'éclairement constant donné par le dernier terme de (90.3). Cette ligne fine lumineuse produite par le fil et dirigée dans une direction perpendiculaire au fil, peut s'apercevoir à l'oeil en observant la tache de diffraction dans une région où les deux premiers termes du second membre de (90.3) deviennent négligeables, c'est-à-dire en s'égloignant de la tache centrale.

Les variations d'homogénéité des verres d'un objectif produisent donc des effets semblables à ceux des aberrations, une partie de l'énergie située dans la tache centrale se répand dans les anneaux et il en résulte une baisse de qualité des images d'objets à faibles contrastes.

β) Phénomène de diffraction produit par un grand nombre d'accidents. Toutes les surfaces optiques présentent des défauts de poli qui se traduisent par un grand nombre de petites oscillations de la surface d'onde autour de sa forme moyenne sphérique. Les défauts dont les dimensions latérales sont supérieures à 0,1 mm. environ correspondent à des variations de phase très petites devant la longueur d'onde. On peut supposer que les irrégularités de poli entraînent des déformations de la surface d'onde qui prend l'aspect de la Fig. 226.

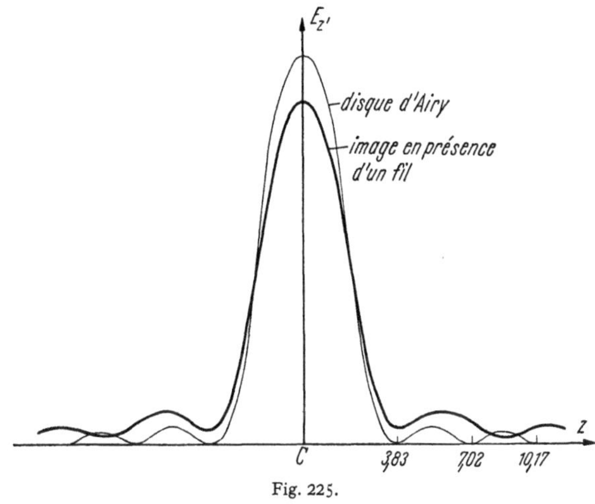

Fig. 225.

Faisons les hypothèses suivantes:

1. La surface d'onde Σ est comprise entre deux sphères S et S' de centre C. L'écart Δ entre ces deux sphères correspond à une différence de phase φ supposée suffisamment petite pour négliger les puissances de φ supérieures à la deuxième.

2. On prend comme origine des phases, la sphère S ou la sphère S', de telle sorte que les irrégularités de la surface d'onde Σ peuvent être considérées, soit comme des bosses, soit comme des creux. Si on choisit par exemple la sphère S comme origine, les irrégularités seront

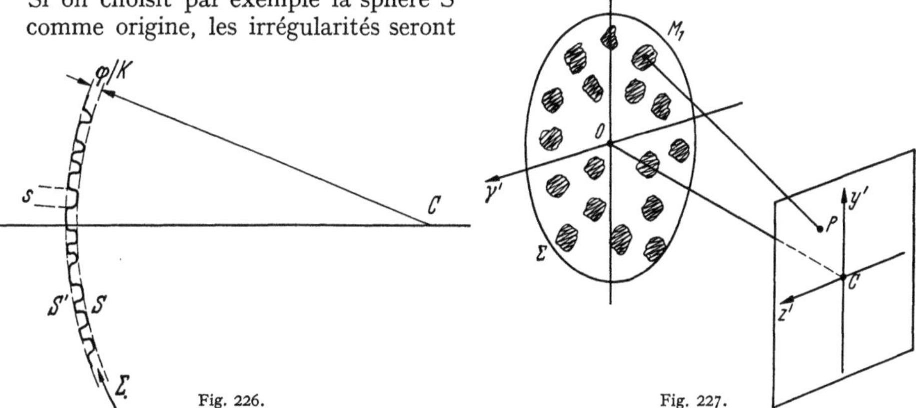

Fig. 226.   Fig. 227.

assimilées à des petites bosses que l'on supposera toutes de même forme, de même orientation et de même surface s, mais distribuées d'une façon irrégulière.

3. On admet que le nombre n des irrégularités est grand. L'objectif a un diamètre D très supérieur aux défauts.

Soit un petit défaut autour d'un point quelconque M_1 de la surface d'onde (Fig. 227) si φ est la phase de l'onde au point M_1, phase due seulement aux irrégularité de poli, l'amplitude au point P peut s'écrire, à un facteur constant près

$$\iint_{\Sigma} e^{j\varphi}\, e^{jK(\beta'y'+\gamma'z')}\, d\beta'\, d\gamma' = \iint_{\Sigma} e^{jK(\beta'y'+\gamma'z')}\, d\beta'\, d\gamma' + \iint_{\Sigma}\left(j\varphi - \frac{\varphi^2}{2}\right) e^{jK(\beta'y'+\gamma'z')}\, d\beta'\, d\gamma'.$$

La première intégrale représente la figure de diffraction dûe à l'onde entière supposée rigoureusement sphérique. La seconde intégrale représentant la lumière diffractée par les irrégularités de la surface d'onde est nulle en dehors de ces irrégularités.

La différence de marche Δ_1 pour le point M_1 est $\Delta_1 = M_1 P - O P$ et pour un point quelconque N_1 du défaut considéré, la différence de marche par rapport à M_1 est $\Delta_1' = N_1 P - M_1 P$.

L'épaisseur des défauts étant petite par rapport à leur largeur et en étudiant ce qui se passe dans une région voisine du phénomène de diffraction correspondant à l'onde Σ entière, l'amplitude en P produite par le défaut considéré est donnée approximativement par l'expression suivante

$$\iint_{s}\left(j\varphi - \frac{\varphi^2}{2}\right) e^{jK(\Delta_1+\Delta_1')}\, d\beta'\, d\gamma' \approx s\,\varphi\left(j - \frac{\varphi}{2}\right) e^{jK\Delta_1}.$$

L'amplitude en P due à la totalité des irrégularités de la surface d'onde pourra s'écrire sous la forme

$$s\,\varphi\left(j - \frac{\varphi}{2}\right)[e^{jK\Delta_1} + e^{jK\Delta_2} + \cdots] = s\,\varphi\left(j - \frac{\varphi}{2}\right)[\Sigma \cos K\Delta + j\,\Sigma \sin K\Delta].$$

Les défauts étant nombreux et distribuées d'une façon quelconque, on a, d'après la Sect. 75

$$(\Sigma \cos K\Delta)^2 + (\Sigma \sin K\Delta)^2 \approx n.$$

Posons

$$\frac{4\,\varepsilon}{\pi\, D^2} = m.$$

En groupant les termes réels et les termes imaginaires, on obtient l'èclairement en P

$$E_P = \left[\frac{2J_1(Z)}{Z}\right]^2 - 4\varphi\, m\, \frac{J_1(Z)}{Z}\, \Sigma \sin K\Delta - 2m\,\varphi^2\, \frac{J_1(Z)}{Z}\, \Sigma \cos K\Delta + n\, m^2\, \varphi^2.$$

Considérons le flux produit par le deuxième terme dans une petite couronne élémentaire du plan image. L'intégration de ce terme ne porte que sur $\Sigma \sin K\Delta$, car la distribution représentée par le terme $J_1(Z)/Z$ est de révolution, donc constante pour tous les points de la petite couronne envisagée. Or, la valeur moyenne de $\Sigma \sin K\Delta$ pour tous les points de la couronne est nulle. L'énergie dûe à ce terme et située dans une petite couronne élémentaire, est donc nulle. Comme Δ varie lorsqu'on se déplace le long de la zône élémentaire, le deuxième terme prend des valeurs positives et négatives, et par conséquent, la tache de diffraction n'est plus de révolution.

Le troisième terme devient très faible dès que l'on s'écarte de l'axe, car les termes de la somme $\Sigma \cos K\Delta$ prennent tous une série de valeurs comprises entre -1 et $+1$, le facteur $\dfrac{J_1(Z)}{Z}$ diminuant également lorsque Z augmente.

Le quatrième terme n'est autre que le maximum du phénomène de diffraction de l'ensemble des irrégularités. On peut donc considérer que l'onde étudiée produit un phénomène de diffraction qui résulte de la superposition de deux phénomènes:

1. Une tache de diffraction analogue au disque d'Airy. L'énergie répartie sur une zone élémentaire est la même que dans la figure d'Airy, mais elle ne l'est plus uniformément: elle se localise par endroits et les anneaux apparaîtront disloqués. Ce phénomène est bien marqué dans les objectifs astronomiques qui présentent du «mamelonnage».

2. Un halo pratiquement uniforme, toujours faible et dû à la présence des irrégularités.

c) Influence de la diffraction sur les images d'objets étendus formées par un système optique parfait.

Dans les paragraphes précédents, nous n'avons envisagé que des sources lumineuses ponctuelles et nous avons étudié les phénomènes de diffraction à l'état pur. Il est intéressant pour beaucoup d'applications pratiques, de considérer des objets étendus et d'étudier l'influence des phénomènes de diffraction sur la structure de leurs images. Le problème semble d'abord assez simple: on pense à ajouter les éclairements produits par les divers éléments de l'objet dans le plan de l'image pour obtenir la répartition globale des éclairements. Ceci n'est justifié que si les divers éléments de l'objet émettent des vibrations incohérents entre elles, c'est à dire ne présentent aucune liaison permanente de phase. Il en est ainsi pour la plupart des instruments d'optique, instruments d'observation ou de photographie d'objets éloignés, spectrographes, etc. ... Ce n'est plus vrai dans le cas du microscope: l'objet est éclairé à l'aide d'une source auxiliaire et les vibrations émises par les différents points de l'objet ne sont pas indépendantes. A l'exception de la Sect. 93, nous ne considérerons que le cas d'un système optique parfait limité par une ouverture circulaire.

91. Transformation de Fourier. Nous allons d'abord rappeler brièvement quelques définitions et propriétés des séries et intégrales de Fourier en vue des applications qui suivront.

α) Représentation d'une fonction périodique par une série de Fourier. Toute fonction périodique $f(x)$ peut être représentée par une somme de fonctions sinusoïdales de périodes $p\ p/2,\ p/3 \ldots$

$$\left.\begin{aligned}
f(x) = a_0 &+ a_1 \cos \frac{2\pi x}{p} + b_1 \sin \frac{2\pi x}{p} \\
&+ a_2 \cos 2\, \frac{2\pi x}{p} + b_2 \sin 2\, \frac{2\pi x}{p} + \ldots \\
&+ a_n \cos n\, \frac{2\pi x}{p} + b_n \sin n\, \frac{2\pi x}{p} + \cdots .
\end{aligned}\right\} \qquad (91.1)$$

Connaissant une fonction périodique $f(x)$, si l'on veut la développer en série de Fourier suivant l'expression (91.1), il faut calculer les coefficients $a_n,\ b_n$. Pour cela, on intègre dans une période le produit

$$f(x) \cos \frac{2\pi n x}{p} \quad \text{ou} \quad f(x) \sin \frac{2\pi n x}{p} \quad \text{et on obtient}$$

$$a_n = \frac{2}{p} \int f(x) \cos \frac{2\pi n x}{p}\, dx, \quad b_n = \frac{2}{p} \int f(x) \sin \frac{2\pi n x}{p}\, dx. \qquad (91.2)$$

Représentons par une série de Fourier la fonction dite «fonction créneaux» représentée par la Fig. 228. C'est une fonction qui prend alternativement la valeur $+1$ ou -1, et dont la période est égale à p.

En intégrant dans l'intervalle $-p/2$, $+p/2$ on trouvera

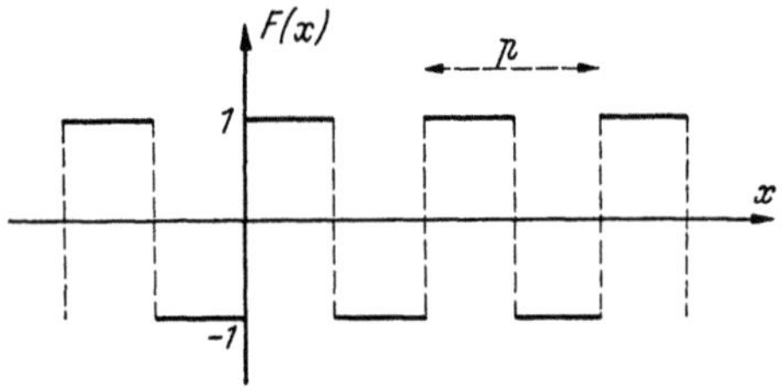

$$a_n = 0, \quad b_n = \frac{4}{\pi n};$$

la «fonction créneaux» pourra donc être représentée par la série

$$f(x) = \frac{4}{\pi}\left(\sin\frac{2\pi x}{p} + \frac{1}{3}\sin 3\frac{2\pi x}{p} + \right.$$

$$\left. + \frac{1}{5}\sin 5\frac{2\pi x}{p} + \cdots\right).$$

Fig. 228.

$\beta)$ *Représentation d'une fonction quelconque par une intégrale de* Fourier; *transformation de* Fourier. Lorsque la fonction n'admet plus de périodicité, on peut faire tendre p vers 'infini, l'intervalle fondamental s'étend alors indéfiniment et on pourra représenter une onction non périodique.

En posant $\nu = n/p$ on écrira:

$$\left.\begin{aligned} a(\nu) &= \int_{-\infty}^{+\infty} f(x)\cos 2\pi\nu x\, dx, \quad b(\nu) = \int_{-\infty}^{+\infty} f(x)\sin 2\pi\nu x\, dx, \\ f(x) &= 2\int_{0}^{\infty}[a(\nu)\cos 2\pi\nu x + b(\nu)\sin 2\pi\nu x]\, d\nu. \end{aligned}\right\} \tag{91.3}$$

Les amplitudes $a(\nu)$ et $b(\nu)$ fixent l'importance de telle ou telle fréquence (ν est une fré quence si la variable x est le temps) dans la représentation de $f(x)$. Par analogie avec l problème de la décomposition d'une lumière complexe, en composantes monochromatiques on dit souvent que l'ensemble $a(\nu)$, $b(\nu)$ représente le «spectre» de la fonction $f(x)$.

Dans le calcul des phénomènes de diffraction, l'emploi des notations complexes est plus pratique et on peut passer des fonctions trigonométriques aux fonctions complexes en remplaçant dans les formules précédentes, les lignes trigonométriques par leurs valeurs tirées des formules d'Euler. On écrira

$$g(\nu) = \int_{-\infty}^{+\infty} f(x)\, e^{j2\pi\nu x}\, dx = a(\nu) + j\, b(\nu),$$

$$g(-\nu) = \int_{-\infty}^{+\infty} f(x)\, e^{-j2\pi\nu x}\, dx = a(\nu) - j\, b(\nu).$$

L'équation (91.3) deviendra

$$f(x) = \int_{0}^{\infty}[g(\nu)\, e^{-j\pi\nu x} + g(-\nu)\, e^{j\pi\nu x}]\, d\nu$$

et, en convenant de faire varier la fréquence ν de $-\infty$ à $+\infty$, on pourra écrire:

$$g(\nu) = \int_{-\infty}^{+\infty} f(x)\, e^{j2\pi\nu x}\, dx, \quad f(x) = \int_{-\infty}^{+\infty} g(\nu)\, e^{-j2\pi\nu x}\, d\nu. \tag{91.4}$$

Ces deux équations définissent la transformation de Fourier, elles jouent des rôles symétriques, l'une des fonctions étant le spectre de l'autre.

Ces résultats peuvent se généraliser au cas des fonctions de deux variables x et y et la transformation de Fourier s'écrit

$$\left.\begin{aligned} g(\mu,\nu) &= \iint f(x,y)\, e^{j2\pi(\mu x+\nu y)}\, dx\, dy \\ f(x,y) &= \iint g(\mu,\nu)\, e^{-j2\pi(\mu x+\nu y)}\, d\mu\, d\nu. \end{aligned}\right\} \tag{91.5}$$

En posant $x = y'/\lambda$, $\mu = \beta'$, $y = z'/\lambda$, $\nu = \gamma'$ et en comparant ces équations à l'équation 67.7) on constate que l'équation (67.7) s'identifie à la transformation de Fourier. On pourra écrire

$$g(\beta', \gamma') = - \frac{j}{\lambda R} \iint f(y', z') \, e^{jK(\beta'y' + \gamma'z')} \, dy' \, dz', \qquad (91.6\text{a})$$

$$f(y', z') = \frac{jR}{\lambda} \iint g(\beta', \gamma') e^{-jK(\beta'y' + \gamma'z')} \, d\beta' \, d\gamma'. \qquad (91.6\text{b})$$

La transformation de FOURIER peut donc s'interpréter comme suit: étant donnée une certaine répartition d'amplitude complexe $g(\beta', \gamma')$ sur la surface d'onde Σ donnée par l'objectif O (Fig. 229), l'équation (91.6b) donne la structure du phénomène de diffraction en C, c'est à dire de l'image de la source ponctuelle. Inversement, si on connaît la structure de phénomène de diffraction en C donnée par $f(y', z')$, on peut connaître la structure de la surface d'onde Σ qui lui a donnée naissance au moyen de l'équation (91.6a).

Dans le chapitre a, nous avons supposé l'objectif O parfaitement transparent (sauf pour l'apodisation) et dénué d'aberration, on avait donc $g(\beta', \gamma') = $ constante.

Dans le cas d'un objectif parfait mais en présence d'un défaut de mise au point $CC' = \varepsilon$ petit (Sect. 70), on a vu qu'il fallait considérer une variation de phase $\dfrac{K \varepsilon \alpha'^2}{2}$ proportionnelle au carré de la distance d'un point M de Σ au centre de l'onde. Tout se passe pour le calcul du phénomène de diffraction en C' comme si on faisait le calcul en C, mais en déposant sur l'objectif un dépôt transparent d'épaisseur variable du centre au bord et produisant une variation de phase $\dfrac{K \varepsilon \alpha'^2}{2}$. Dans le cas d'un défaut de mise au point, le calcul du phénomène de diffraction se fait avec l'équation (91.6b) dans laquelle $g(\beta', \gamma') = e^{j\frac{K \varepsilon \alpha'^2}{2}}$. De même pour les aberrations, et finalement on peut dire que le phénomène de diffraction produit par une source ponctuelle, est donné dans le cas général par (91.6b) où $g(\beta', \gamma')$ prend la forme suivante d'après (87.3)

$$g(\beta', \gamma') = e^{jK(d h^2 + s_1 h^4 + \cdots)}.$$

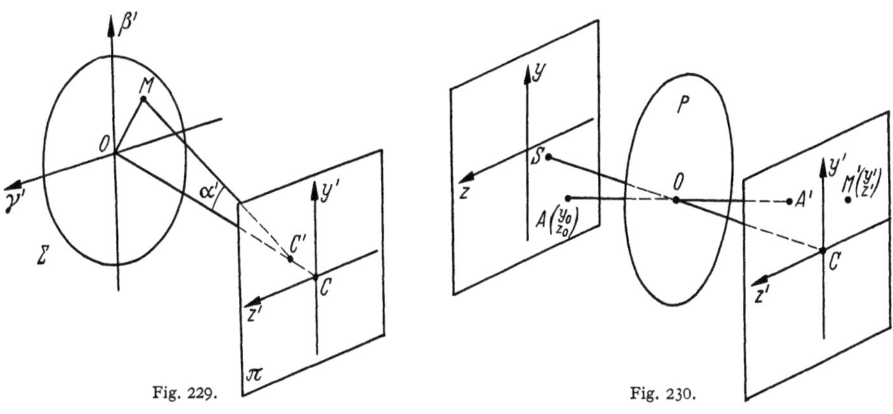

Fig. 229.   Fig. 230.

92. Objets étendus en éclairage incohérent et cohérent. $\alpha)$ *Eclairage incohérent.* Soit un instrument d'optique P destiné à donner d'un objet situé dans un plan ySz une image dans le plan $y'Cz'$ (Fig. 230). Considérons un point objet A de coordonnées y, z et son image idéale A' de coordonnées $\beta y, \beta z$ où β est le grandissement transversal de l'instrument.

Comme nous ne considérons maintenant que des instruments parfaits, l'image du point A est en réalité une figure de diffraction donnée par exemple par (68.4) et dont le centre est en A'.

La figure de diffraction de centre C est donnée par $E(y',z')$ qui n'est autre que l'expression (68.4) et l'image de A sera une figure de diffraction de centre A' produisant en un point $M'(y'z')$ l'éclairement $E(y'-\beta y, z'-\beta z)$. Pour simplifier, nous supposerons dans ce qui suit $\beta = 1$, ce qui ne diminue en rien la généralité des calculs. Supposons que la répartition des luminances dans l'objet soit représentée par une fonction $O(y, z)$.

La répartition des luminances dans l'image géométrique sera représentée par la même fonction ($\beta = 1$). La distribution des éclairements dans l'image en tenant compte de la diffraction s'obtiendra en faisant la somme des éclairements produits par les images des divers points de l'objet. Si $I(y',z')$ est l'éclairement en un point $M'(y',z')$ quelconque de plan image, on aura

$$I(y',z') = \iint O(y,z) \, E(y'-y, z'-z) \, dy \, dz. \tag{92.1}$$

Soient $\omega(\mu'\nu')$ et $\Omega(\mu'\nu')$ les transformées de Fourier des fonctions $O(y, z)$ et $E(y'z')$

$$O(y, z) = \iint \omega(\mu',\nu') \, e^{-jK(\mu'y+\nu'z)} \, d\mu' \, d\nu',$$

$$E(y',z') = \iint \Omega(\mu',\nu') \, e^{-jK(\mu'y'+\nu'z')} \, d\mu' \, d\nu'.$$

D'après les propriétés de la transformation de Fourier (91.5), on peut écrire à une constante près

$$\omega(\mu',\nu') = \iint O(y,z) \, e^{-jK(\mu'y+\nu'z)} \, dy \, dz,$$

$$\Omega(\mu',\nu') = \iint E(y',z') \, e^{jK(\mu'y'+\nu'z')} \, dy' \, dz'.$$

La relation (92.1) devient

$$I(y',z') = \iint O(y,z) \left[\iint \Omega(\mu',\nu') \, e^{-jK[\mu'(y'-y)+\nu'(z'-z)]} d\mu' \, d\nu'\right] dy \, dz.$$

On peut renverser l'ordre des intégrations et écrire

$$I(y',z') = \iint \Omega(\mu',\nu') \, e^{-jK(\mu'y'+\nu'z')} \left[\iint O(y,z) \, e^{jK(\mu'y+\nu'z)} \, dy \, dz\right] d\mu' \, d\nu'$$

d'où, à une constante près

$$I(y',z') = \iint \Omega(\mu',\nu') \, \omega(\mu',\nu') \, e^{-jK(\mu'y'+\nu'z')} \, d\mu' \, d\nu'. \tag{92.2}$$

L'éclairement en un point $y'\,z'$ de l'image est donnée par la fonction $I(y',z')$ qui est la transformée de Fourier du produit $\Omega(\mu',\nu') \, \omega(\mu',\nu')$. $\Omega(\mu',\nu')$ représente la transformée de Fourier de l'image d'un point isolé, $\omega(\mu',\nu')$, la transformée de Fourier de l'objet. Le résultat donnée par l'équation (92.2) peut encore s'énoncer: la transformée de Fourier de l'image d'un objet étendu incohérent est égale au produit de la transformée de Fourier de l'objet par la transformée de Fourier de l'image d'un point isolé.

$\beta)$ *Eclairage cohérent.* L'objet A n'est plus lumineux par lui-même, il est éclairé par une source unique de petites dimensions par exemple à l'infini (Fig. 231), de sorte que les phases des vibrations émises par les divers points de l'objet sont parfaitement liées.

Sur la Fig. 231, on a supposé avoir placé un objectif O_1 contre le plan objet $y\,Sz$, de sorte que l'image de la source à l'infini se trouve en O_2 au foyer de O_1. En O_2, on place un deuxième objectif qui donne du plan objet $y\,Sz$ une image en $y'\,Cz'$. Au lieu que l'objet $y\,Sz$ et l'image $y'\,Cz'$ soient conjugués par rapport à un seul objectif, comme sur la Fig. 230, la conjugaison s'effectue maintenant au moyen des deux objectifs O_1 et O_2 et on suppose toujours $\beta = 1$.

C'est le cas du microscope lorsqu'on diaphragme fortement le condenseur: la petite ouverture du diaphragme joue le rôle de la source unique qui sert à éclairer la préparation. Si $U(y' - y, z' - z)$ est l'amplitude en un point $M'(y' z')$ quelconque produite par la tache de diffraction de centre $A'(y z)$ image géométrique du point source A, et si l'objet est caractérisé par une répartition d'amplitude complexe $P(y, z)$ la répartition des amplitudes dans l'image sera

$$C(y' z') = \iint P(y, z)\, U(y' - y, z' - z)\, dy\, dz. \tag{92.3}$$

Si l'objet est parfaitement transparent et caractérisé seulement par des variations de phase (objet de phase), la fonction $P(y, z)$ sera purement imaginaire.

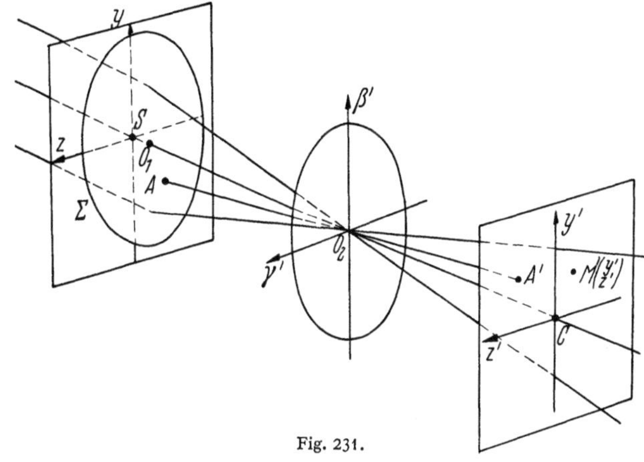

Fig. 231.

Pour un objet n'introduisant que des variations d'absorption (objet d'amplitude), la fonction $P(y, z)$ sera réelle.

On peut aussi calculer la structure de l'image en opérant en deux temps: on calcule d'abord la structure $F(\beta', \gamma')$ du phénomène de diffraction en O_2 produit par la surface d'onde Σ sur laquelle la distribution des amplitudes complexes est donnée par $P(y, z)$. On aura, d'après (91.6), en tenant compte des nouvelles notations

$$F(\beta', \gamma') = - \frac{i}{\lambda R} \iint P(y, z)\, e^{jK(\beta'y + \gamma'z)}\, dy\, dz. \tag{92.4}$$

Puis on applique la transformation de FOURIER pour passer du phénomène de diffraction situé en O_2, à la surface d'onde Σ qui lui a donnée naissance, ou, ce qui revient au même, à son image. On a

$$P(y, z) = \frac{iR}{\lambda} \iint F(\beta', \gamma')\, e^{-jK(\beta'y + \gamma'z)}\, d\beta'\, d\gamma'. \tag{92.5}$$

Dans cette intégration, on suppose que le phénomène de diffraction $F(\beta' \gamma')$ est utilisé, c'est-à-dire que l'ouverture de l'objectif O_2 est infinie. Alors le mouvement vibratoire est le même en A ou A' et la transformation de FOURIER redonne comme amplitude dans l'image, l'amplitude $P(y z)$ de l'objet. Si maintenant, l'ouverture de l'objectif O_2 est finie, ce qui est évidemment le cas pratique, dans l'intégration (92.5) tout le phénomène de diffraction $F(\beta'\gamma')$ n'est plus utilisé, et l'intégrale (92.5) ne redonne plus une amplitude dans l'image égale à celle de l'objet. Comme nous considérons un grandissement du système égal à 1, nous remplacerons y et z par y' et z' et nous aurons l'amplitude $C(y'z')$ dans l'image

$$C(y', z') = \frac{jR}{\lambda} \iint F(\beta', \gamma') e^{-jK(\beta'y' + \gamma'z')} \, d\beta' \, d\gamma'. \tag{92.6}$$

Nous résumons les notations qui seront employées par la suite:

$U(y', z')$ — structure de l'image d'un point isolé de l'objet (amplitude)

$E(y', z')$ — structure de l'image d'un point isolé de l'objet (éclairement)

éclairage incohérent $\begin{cases} O(y, z) \\ I(y', z') \end{cases}$ — répartition des luminances dans l'objet / répartition des éclairements dans l'image

éclairage cohérent $\begin{cases} P(y, z) \\ C(y', z') \\ F(\beta', \gamma') \end{cases}$ — répartition de l'amplitude complexe dans l'objet / répartition de l'amplitude complexe dans l'image / structure du phénomène de diffraction dans le plan de l'image de la source ponctuelle (amplitude).

93. Image d'un disque lumineux incohérent lorsque l'objectif est recouvert d'un diaphragme percé de deux fentes. Considérons un objectif O recouvert d'un

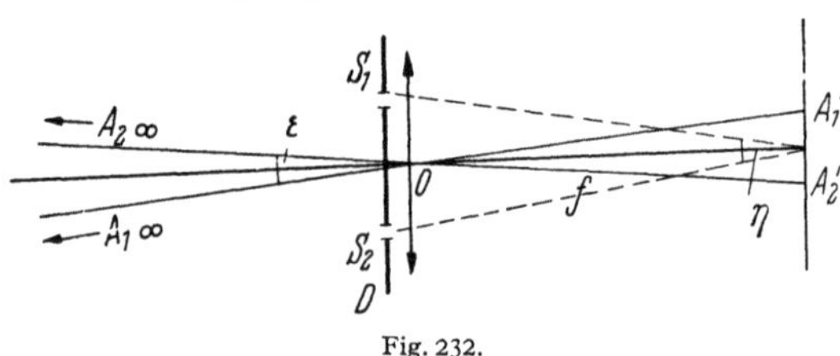

Fig. 232.

diaphragme D percé de deux fentes S_1 et S_2 identiques et parallèles (Fig. 232) distantes angulairement de 2η. L'objet est un petit disque lumineux $A_1 A_2$ à l'infini vu sous l'angle ε et dont l'image se forme en $A_1' A_2'$ dans le plan focal de l'objectif O. Dans le cas où ε est plus petit que la tache de diffraction de l'objectif O, tout se passe comme si l'objet $A_1 A_2$ était une source ponctuelle et on a vu (Sect. 77) que l'image dans le plan focal de O est un système de franges d'Young données par (77.4)

$$E_{z'} = \frac{64 R^2 \gamma_0'^2 \beta_0'^2}{\lambda^2} \left(\frac{\sin K z' \gamma_0'}{K z' \gamma_0'} \right)^2 \cos^2 \frac{K z' \eta}{2}. \tag{93.1}$$

Si la source n'a plus un diamètre apparent ε très petit, il faut faire l'intégration: chaque point de la source donne un système de franges et tous ces systèmes incohérents s'ajoutent. Le contraste des franges diminue. On peut calculer la répartition des éclairements dans l'image $A_1' A_2'$ en utilisant le résultat de la Sect. 92. La transformée de Fourier de l'image du disque est donnée par le produit de la transformée de Fourier de l'objet (ou de son image géométrique) et de la transformée de Fourier de l'image d'un point isolé.

Cherchons d'abord la transformée de Fourier de l'image géométrique de l'objet. L'objet a pour image *géométrique* dans le plan de l'objectif, un cercle de diamètre $f\varepsilon$ et la transformée de Fourier est (Sect. 92)

$$\omega(\mu', \nu') = \iint O(y, z) e^{jK(\mu'y + \nu'z)} \, dy \, dz.$$

Puisque l'image géométrique est un cercle de luminance uniforme $O(y, z) =$ constante, et d'après (67.7) et (68.4), on aura, à un facteur constant près

$$\omega(\mu', \nu') = \iint\limits_{f\varepsilon} e^{jK(\mu'y+\nu'z)}\,dy\,dz = \frac{2\,J_1\left(\dfrac{K\varrho_1\,\varepsilon}{2}\right)}{\dfrac{K\varrho_1\,\varepsilon}{2}}. \tag{93.2}$$

Tout se passe comme si on avait dans un plan yz un diaphragme circulaire de diamètre $f\varepsilon$ donnant dans le plan $\mu'\nu'$ le phénomène de diffraction (93.2).

Par rapport à la Sect. 68, on a maintenant (f focale de l'objectif)

$$\varrho_1 = f\sqrt{\mu'^2 + \nu'^2} \quad \text{et} \quad Z = \frac{K\varepsilon\varrho_1}{2}.$$

Cherchons la transformée de FOURIER de l'image d'un point isolé.

Pour cela, nous ne considérons qu'un seul point de l'objet et on sait que l'image est constituée par un système de franges d'YOUNG.

En reprenant l'expression (93.1) à un facteur constant près, et en négligeant la modulation par la diffraction, ce système de franges est donné par

$$E_{z'} = 2\cos^2\frac{Kz'\eta}{2} = 1 + \cos Kz'\eta = 1 + \frac{1}{2}\left(e^{jKz'\eta} + e^{-jKz'\eta}\right) \tag{93.3}$$

et la transformée de FOURIER de $E_{z'}$ sera (Sect. 92)

$$\Omega(\mu', \nu') = \iint E_{z'}\, e^{jK(\mu'y'+\nu'z')}\,dy'\,dz'. \tag{93.4}$$

Cherchons la transformée en donnant à $E_{z'}$ successivement les trois valeurs correspondant aux trois termes de (93.3).

Tout se passe pour calculer la transformée de FOURIER du premier terme égal à l'unité, comme si on avait dans un plan $y'z'$ un diaphragme circulaire de diamètre D, infiniment grand, puisque l'on a négligé le facteur $\dfrac{\sin Kz'\gamma'_0}{Kz'\gamma'_0}$ de (93.1). La transformée de FOURIER d'une ouverture circulaire infiniment grande, c'est à dire le phénomène de diffraction produit par une telle ouverture, se réduit à un point de coordonnées $\nu' = 0$.

Nous prendrons l'intensité de ce point égal à 1.

De même pour le terme $\dfrac{1}{2}\left(e^{jKz'\eta} + e^{-jKz'\eta}\right)$.

Tout se passe comme si on avait dans un plan $y'z'$ un diaphragme infiniment grand sur lequel la distribution des amplitudes suivrait la loi

$$\frac{1}{2}\left(e^{jKz'\eta} + e^{-jKz'\eta}\right) = \cos Kz'\eta.$$

La transformée de FOURIER de cette expression, c'est à dire son phénomène de diffraction dans le plan $\mu'\nu'$ est constitué par deux images ponctuelles d'intensité égale à $\frac{1}{2}$ par rapport à l'image précédente et d'abscisses $\nu' = -\eta$, $\nu' = +\eta$. En effet, dans un plan $\mu'\nu'$ deux sources distantes de 2η donneront bien dan. le plan $y'z'$ un système de franges d'YOUNG représenté en amplitude par $\cos Kz'\eta$. La transformée de FOURIER $\Omega(\mu'\nu')$ de E_z sera finalement donnée dans le plan $\mu'\nu'$ par trois images ponctuelles d'abscisses $\nu' = -\eta$ (intensité $\frac{1}{2}$), $\nu' = 0$ (intensité 1) $\nu' = +\eta$ (intensité $\frac{1}{2}$).

La transformée de FOURIER de l'image finale est égale au produit des transformées que nous venons de trouver, c'est à dire au produit de (93.2) par les trois valeurs précédentes.

Pour $v' = 0$,

$$\frac{2J_1\left(\dfrac{K\varrho_1\varepsilon}{2}\right)}{\dfrac{K\varrho_1\varepsilon}{2}} = 1 \quad \text{d'où} \quad \Omega(\mu'v') = 1.$$

Pour $v' = -\eta \; (\varrho_1 = fv')$

$$\frac{2J_1\left(\dfrac{K\varrho_1\varepsilon}{2}\right)}{\dfrac{K\varrho_1\varepsilon}{2}} = \frac{2J_1\left(\dfrac{K\varepsilon\eta f}{2}\right)}{\dfrac{K\varepsilon\eta f}{2}} \quad \text{d'où} \quad \Omega(\mu'v') = \frac{1}{2}\frac{2J_1\left(\dfrac{K\varepsilon\eta f}{2}\right)}{\dfrac{K\varepsilon\eta f}{2}}$$

et pour $v' = \eta$

$$\Omega(\mu',v') = \frac{1}{2}\frac{2J_1\left(\dfrac{K\varepsilon\eta f}{2}\right)}{\dfrac{K\varepsilon\eta f}{2}}.$$

La transformée de Fourier de $E_{z'}$ se compose donc de trois images ponctuelles, l'image centrale ayant pour intensité 1 et les deux autres pour intensité $\dfrac{2J_1\left(\dfrac{K\varepsilon\eta f}{2}\right)}{\dfrac{K\varepsilon\eta f}{2}}$. L'image $E_{z'}$ ayant pour transformée de Fourier trois sources lumineuses que l'on doit considérer comme cohérentes, $E_{z'}$ a pour structure le phénomène de diffraction produit par trois sources.

On remonte donc de ces trois sources à l'image $E_{z'}$ par la transformation de Fourier:

$$E_{z'} = 1 + \frac{J_1\left(\dfrac{K\varepsilon\eta f}{2}\right)}{\dfrac{K\varepsilon\eta f}{2}}\, e^{-jK\eta z} + \frac{J_1\left(\dfrac{K\varepsilon\eta f}{2}\right)}{\dfrac{K\varepsilon\eta f}{2}}\, e^{jK\eta z}$$

d'où

$$E_{z'} = 1 + \frac{2J_1\left(\dfrac{K\varepsilon\eta f}{2}\right)}{\dfrac{K\varepsilon\eta f}{2}}\cos K\eta z'.$$

L'image d'un disque circulaire de diamètre apparent ε est donc formé par un système de franges dont le contraste n'est plus égal à 1. Prenons comme contraste des franges, le contraste défini précédemment (11.5), on aura ici

$$(E_{z'})_{\max} = 1 + \frac{2J_1\left(\dfrac{K\varepsilon\eta f}{2}\right)}{\dfrac{K\varepsilon\eta f}{2}}, \qquad (E_{z'})_{\min} = 1 - \frac{2J_1\left(\dfrac{K\varepsilon\eta f}{2}\right)}{\dfrac{K\varepsilon\eta f}{2}}$$

d'où

$$\gamma = \frac{4J_1\left(\dfrac{K\varepsilon\eta f}{2}\right)}{\dfrac{K\varepsilon\eta f}{2} + 2J_1\left(\dfrac{K\varepsilon\eta f}{2}\right)}.$$

Les franges disparaîtront si $J_1\left(\dfrac{K\varepsilon\eta f}{2}\right) = 0$.

Pour un objet de diamètre apparent donnée ε, le contraste des franges est fonction de l'écartement η des fentes placées contre l'objectif. Il y a disparition et réapparition des franges au fur et à mesure que les fentes s'écartent. La première disparition a lieu lorsque $\dfrac{K\varepsilon\eta f}{2} = 3{,}83$, d'où

$$\varepsilon = \frac{1{,}22\,\lambda}{\eta f}. \tag{93.5}$$

On peut déduire de la mesure de l'écartement η des fentes au moment de la première disparition, le diamètre angulaire de la source lumineuse. En utilisant un objectif astronomique, on peut mesurer des diamètres apparents d'objets célestes d'autant plus petits que l'objectif aura un diamètre plus grand. Le principe de la méthode a été indiqué par FIZEAU en 1868. Essayée d'abord sans succès par STÉPHAN en 1874, elle a permis à MICHELSON de mesurer les diamètres apparents d'un certain nombre d'étoiles.

Dans les cas les plus favorables, les diamètres apparents stellaires sont encore très petits, $\varepsilon = 0''{,}047$ pour Bételgeuse et $\varepsilon = 0''{,}024$ pour Arcturus. D'après la formule (93.5), on voit que pour mesurer un diamètre apparent de $0''{,}05$ il faudrait un diamètre au moins égal à

$$D = \eta f \approx 300 \text{ cm.}$$

Pour éviter cette difficulté, MICHELSON a employé le dispositif indiqué sur la Fig. 233. Devant l'objectif O se trouve un système de quatre miroirs $M_1 M_1' M_2 M_2'$ et les deux fentes S_1 et S_2 sont placées devant les miroirs M_1 et M_2. Tout se passe comme si on avait un objectif de diamètre égal à la distance des fentes S_1 et S_2, distance qui pouvait varier de 3 à 6 mètres dans l'appareil réalisé par MICHELSON.

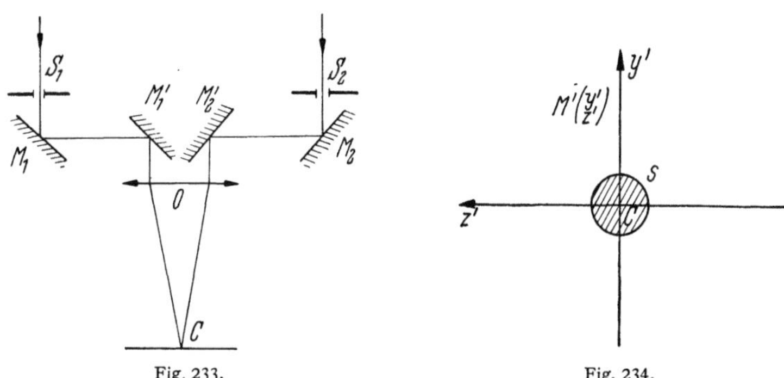

Fig. 233.   Fig. 234.

94. Image d'un petit disque noir sur fond blanc en éclairage incohérent. Nous considérons comme objet, un disque noir dont les dimensions sont faibles par rapport à la tache de diffraction donnée par l'instrument. L'image géométrique de l'objet est un petit cercle de surface s centré sur l'origine C (Fig. 234). La fonction $O(y, z)$ est partout égale à l'unité, sauf à l'intérieur du petit cercle s où elle est nulle.

L'éclairement en M' est égal à l'éclairement que l'on aurait si tout le plan $y'Cz'$ était éclairé, moins l'éclairement dû à la portion s de ce plan.

Puisque s est petit par rapport à la tache de diffraction, pratiquement la lumière qu'enverraient en M' les points de s, s'ils étaient lumineux, serait égale à la lumière envoyée par C. La lumière en M' due à une tache de diffraction de centre C est $E(y', z')$ et la lumière qu'enverraient tous les points du cercle $s \times E(y', z')$. L'éclairement en M' peut donc s'écrire, d'après (92.1)

$$I(y'\,z') = \iint\limits_{-\infty}^{+\infty} E(y' - y, z' - z)\, dy\, dz - s \cdot E(y', z'). \tag{94.1}$$

Or l'intégrale représente le volume total de la tache de diffraction. C'est l'expression $\mathcal{J}_i$ donnée à la Sect. 69 en y faisant $Z = \infty$. On a donc

$$\iint\limits_{-\infty}^{+\infty} E(y' - y, z' - z)\, dy\, dz = \pi R^2 \alpha_0'^2$$

d'où, d'après (68.4)

$$I(Z) = \pi R^2 \alpha_0'^2 - s\, \frac{R^2}{\lambda^2}\, (\pi \alpha_0'^2)^2 \left[\frac{2 J_1(Z)}{Z}\right]^2$$

et au facteur $\pi R^2 \alpha_0'^2$ près

$$I(Z) = 1 - \frac{\pi s \alpha_0'^2}{\lambda^2} \left[\frac{2 J_1(Z)}{Z} \right]^2, \qquad (94.2)$$

ce résultat ne s'appliquant qu'à un disque de diamètre petit par rapport à la tache de diffraction. On a comme répartition des éclairements $I(Z)$ dans l'image, une courbe du type de celle de la Fig. 235. La valeur maximum du facteur entre-crochets du second membre de (94.2) étant égale à 1, le contraste de l'image du petit disque sera

$$\gamma = \frac{\pi s \alpha_0'^2}{\lambda^2}. \qquad (94.3)$$

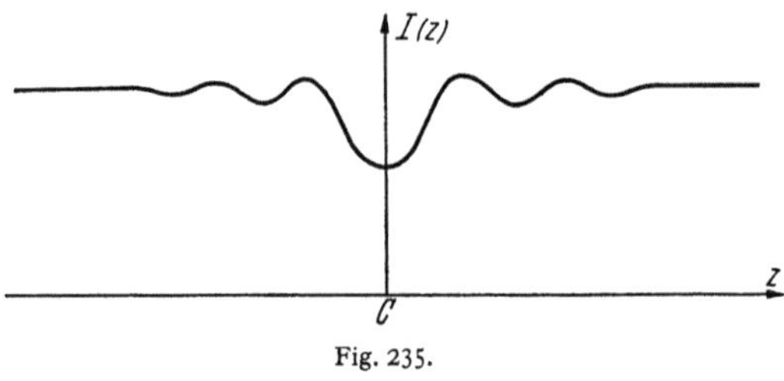

Fig. 235.

95. Cas d'un disque noir de diamètre quelconque sur fond blanc en éclairage incohérent. Lorsque le disque noir a un diamètre quelconque, les formules (94.2) et (94.3) ne peuvent plus s'appliquer. Limitons-nous au calcul de l'éclairement au centre de l'image du disque. Comme au paragraphe précédent, on peut dire que l'éclairement au centre C (Fig. 234) est égal à l'éclairement que l'on aurait si tout le plan $y'Cz'$ était lumineux, moins l'éclairement dû au disque s.

On aura donc en C $(y' = z' = 0)$

$$I(0, 0) = \int\limits_{-\infty}^{+\infty}\!\!\! E(y, z)\, dy\, dz - \iint\limits_{s} E(y, z)\, dy\, dz. \qquad (95.1)$$

La deuxième intégrale a été calculée à la Sect. 69, elle représente le flux intérieur à un cercle de surface s. On aura donc

$$I(0, 0) = \pi R^2 \alpha_0'^2 - \pi R^2 \alpha_0'^2 \left[1 - J_0^2(Z) - J_1^2(Z) \right] \qquad (95.2)$$

avec

$$Z = K \alpha_0' \varrho_1, \qquad s = \pi \varrho_1^2$$

et, en prenant le volume total de la figure de diffraction pour unité (95.2) devient

$$I(0, 0) = J_0^2(Z) + J_1^2(Z). \qquad (95.3)$$

Loin de l'image géométrique du disque, l'éclairement est pratiquement donné par la première intégrale de (95.1), soit $I_{\max} = 1$. Au centre du disque, l'éclairement est donné par (95.3), donc le contraste γ de l'image d'un disque noir de diamètre quelconque, est donné par

$$\gamma = 1 - \left[J_0^2(Z) + J_1^2(Z) \right], \qquad (95.4)$$

si Z devient très petit

$$J_0(Z) \approx 1 - \frac{Z^2}{4}, \qquad J_1(Z) \approx \frac{Z}{2} + \cdots$$

d'où

$$\gamma = \frac{Z^2}{4} = \frac{\pi s \alpha_0'^2}{\lambda^2}$$

et on retrouve l'expression (94.3).

Le cas d'un disque blanc de diamètre quelconque sur fond noir, se résout immédiatement d'après ce qui précède.

En se limitant au calcul de l'éclairement au centre de l'image du disque, on a

$$I(0, 0) = \iint\limits_{s} E(y, z)\, dy\, dz = \pi R^2 \alpha_0'^2 \left[1 - J_0^2(Z) - J_1^2(Z) \right].$$

96. Image d'une ligne noire fine sur fond blanc en éclairage incohérent. Supposons une ligne noire fine de largeur ε en coïncidence avec l'axe Sy (Fig. 230). La fonction $O(y, z)$ est partout égale à l'unité, sauf dans le petit intervalle $-\dfrac{\varepsilon}{2} < z < \dfrac{\varepsilon}{2}$ où elle est nulle.

On a, comme dans le cas du petit disque noir

$$I(y', z') = \iint\limits_{-\infty}^{+\infty} E(y' - y, z' - z)\, dy\, dz - \varepsilon \int\limits_{-\infty}^{+\infty} E(y' - y, z')\, dy'$$

où

$$I(y', z') = \pi R^2 \alpha_0'^2 - \varepsilon \int\limits_{-\infty}^{+\infty} E(Y, z')\, dY \tag{96.1}$$

en posant $Y = y' - y$. En effet (Fig. 236), la ligne étant très étroite pour tous ses points z est très petit. Comme dans le cas du petit disque noir, la formule (96.1) n'est valable que pour une ligne de largeur petite devant le diamètre de la tache de diffraction. On voit que l'intégrale de (96.1) représente la surface de section du solide de diffraction par le plan de coordonnée z'. La courbe donnant la répartition des éclairements le long de cz' ou d'une parallèle à cz', est donnée par la variation de la surface de section du solide de diffraction.

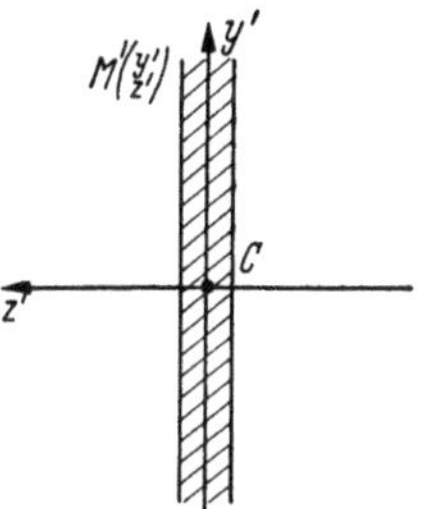

Fig. 236.

Calculons l'intégrale de l'expression (96.1).

On sait que $U(y', z')$ représente la tache de diffraction (en amplitude) image d'un point isolé de l'objet. C'est le phénomène de diffraction produit par la lumière diffractée issue d'un point de l'objet et concentrée dans l'image par l'objectif O_2.

Appliquons la transformation de FOURIER; $U(y', z')$ est le phénomène de diffraction produit par l'objectif O_2 couvert de lumière diffractée et soit $G(\beta', \gamma')$ la distribution des phases et des amplitudes sur l'objectif O_2. La fonction $G(\beta', \gamma')$ est dûe à la lumière diffractée par un point de l'objet, elle est constante sur O_2 et nulle à l'extérieur. La fonction $G(\beta', \gamma')$ ne doit pas être confondue avec la fonction $F(\beta', \gamma')$ qui représente le phénomène de diffraction sur O_2 dû à la distribution des phases et des amplitudes $P(y, z)$ sur l'objet.

Reprenons l'équation (91.6b). On peut l'écrire en remplaçant $f(y', z')$ par $U(y', z')$ et $g(\beta', \gamma')$ par $G(\beta', \gamma')$

$$U(y', z') = \frac{iR}{\lambda} \iint G(\beta', \gamma')\, e^{-iK(\beta' y' + \gamma' z')}\, d\beta'\, d\gamma' \tag{96.2}$$

et en posant $G(\beta', \gamma') = 1$

$$U(y', z') = \frac{iR}{\lambda} \int e^{-iK\beta' y'}\, d\beta' \int e^{-iK\gamma' z'}\, d\gamma'.$$

$U(y', z')$ est donc la transformée de FOURIER à une variable de la fonction

$$L(y', \gamma') = \int e^{-iK\beta' y'}\, d\beta'. \tag{96.3}$$

Imaginons une fente confondue avec $o\gamma'$ et sur laquelle l'amplitude est définie par la fonction $L(y', \gamma')$. Elle donne le long de Cy' un phénomène de diffraction défini par

$$U(y', z') = \frac{iR}{\lambda} \int L(y', \gamma')\, e^{-iK\gamma' z'}\, d\gamma'.$$

Toute l'énergie émise par cette fente doit être répartie dans le phénomène de diffraction, on pourra donc écrire

$$\int |U(y', z')|^2 \, dy' = \frac{R^2}{\lambda} \int |L(y' \, \gamma')|^2 \, d\gamma' \tag{96.4}$$

qui exprime le principe de la conservation de l'énergie.

Or, on a

$$\int E(Yz') \, dY = \int U(y' z')|^2 \, dy'$$

d'où $(\alpha'^2 = \beta'^2 + \gamma'^2)$

$$\int E(Yz') \, dY = \frac{R^2}{\lambda} \int |L(y' \, \gamma')|^2 \, d\gamma' = \frac{R^2}{\lambda} \int_{-\alpha'_0}^{+\alpha'_0} \left| \int_{-\sqrt{\alpha'^2_0 - \gamma'^2}}^{+\sqrt{\alpha'^2_0 - \gamma'^2}} e^{-jK\beta'y'} \, d\beta' \right|^2 \, d\gamma'.$$

$$\int E(Yz') \, dY = \frac{4R^2}{K^2 y'^2 \lambda} \int_{-\alpha'_0}^{+\alpha'_0} \sin^2 K \, y' \, \sqrt{\alpha'^2_0 - \gamma'^2} \, d\gamma'. \tag{96.5}$$

Cette intégrale peut s'obtenir au moyen de la fonction de Struve du premier ordre $H_1(2Z)$. A un facteur constant près, l'intégrale (96.5) s'écrit en faisant le changement de variable $\gamma' = \alpha'_0 \sin \vartheta$

$$\alpha'_0 \int_0^{\pi/2} [1 - \cos(2Z \cos \vartheta)] \cos \vartheta \, d\vartheta = \frac{\pi \alpha'_0}{2} H_1(2Z)$$

d'où

$$\int E(Yz') \, dY = \frac{2\pi R^2 \alpha'^3_0}{\lambda} \frac{H_1(2Z)}{Z^2}$$

et, d'après (96.1), la répartition des éclairements le long de $C z'$, au facteur constant près $\pi R^2 \alpha'^2_0$

$$I(Z) = 1 - \frac{2\varepsilon \alpha'_0}{\lambda} \frac{H_1(2Z)}{Z^2} . \tag{96.6}$$

Cette relation donne la distribution des éclairements dans l'image d'une ligne fine noire sur fond blanc. Au centre de l'image de la ligne $Z = 0$ on a $\dfrac{H_1(2Z)}{Z^2} = \dfrac{8}{3\pi}$ d'où

$$I(0) = 1 - \frac{16 \, \varepsilon \, \alpha'_0}{3\pi \, \lambda} .$$

Loin de la ligne, l'éclairement est égal à 1, d'où le contraste

$$\gamma = \frac{16 \, \varepsilon \, \alpha'_0}{3\pi \, \lambda} . \tag{96.7}$$

Tableau 11.

Z	$H_1(2Z)$	Z	$H_1(2Z)$
0	0,0000	5	0,8918
1	0,6468	6	0,5839
2	1,0697	7	0,4732
3	0,4782	8	0,8171
4	0,4881		

L'expression (96.6) permet de calculer la répartition des éclairements dans l'image d'une ligne fine lumineuse sur fond noir

$$I(Z) = \frac{3\pi}{8} \frac{H_1(2Z)}{Z^2} . \tag{96.8}$$

97. Image d'un objet périodique en éclairage incohérent. Considérons une mire de Foucault où la luminance suit la loi représentée par la Fig. 237, les parties claires étant égales aux parties noires. D'après (91.2), et dans le cas général où le contraste est quelconque on peut la représenter par la série

$$O(z) = 1 + \frac{4\,C_0}{\pi}\left(\sin\frac{2\pi z}{p} + \frac{1}{3}\sin 3\,\frac{2\pi z}{p} + \cdots\right) \tag{97.1}$$

où p est le pas de la mire et C_0 une constante qui fixe le contraste de la mire-objet. Pour $C_0 = 1$, $\gamma = 1$ et pour $C_0 = 0$, $\gamma = 0$.

La mire de Foucault peut être représentée comme une somme de mires sinusoïdales de pas p, $p/3$, $p/5$, ….

Si l'on est au voisinage du pouvoir séparateur sur la mire précédente, c'est que la mire sinusoïdale fondamentale de pas p est déja à la limite de visibilité. Dans ces conditions, on est certain que les mires de pas $p/3$, $p/5$ …

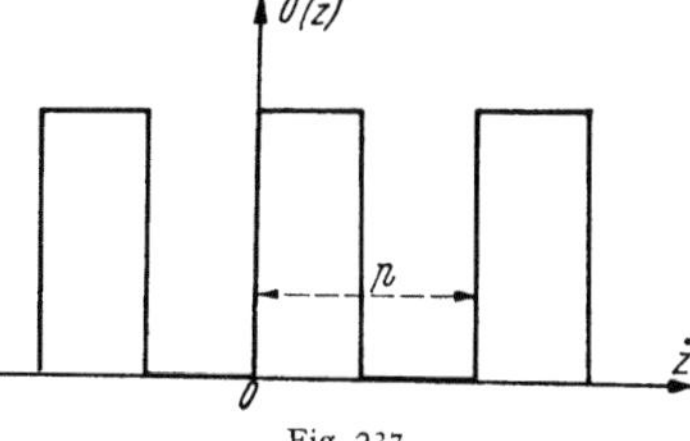

Fig. 237.

encore plus serrées, ne seront pas vues. Au voisinage de la limite de séparation, il suffit de considérer le terme fondamental seulement, et d'étudier l'image de la mire

$$O(z) = 1 + \frac{4\,C_0}{\pi}\sin\frac{2\pi z}{p} \tag{97.2}$$

ou encore

$$O(z) = 1 + \frac{2\,C_0}{j\,\pi}\left(e^{j\,\frac{2\pi z}{p}} - e^{-j\,\frac{2\pi z}{p}}\right).$$

La transformée de Fourier de cette expression est

$$\omega(v') = 1 + \frac{2\,C_0}{j\,\pi}\left(e^{j\,\frac{2\pi z}{p}}\,e^{j\,K\,v'\,z} - e^{-j\,\frac{2\pi z}{p}}\,e^{j\,K\,v'\,z}\right)$$

ou

$$\omega(v') = 1 + \frac{2\,C_0}{j\,\pi}\left[e^{j\,K\left(\frac{\lambda}{p}+v'\right)z} - e^{-j\,K\left(\frac{\lambda}{p}-v'\right)z}\right].$$

La transformée $\omega(v')$ a donc les 3 valeurs $-\dfrac{2j\,C_0}{\pi}$, 1 et $\dfrac{2j\,C_0}{\pi}$ correspondant à $v' = -\dfrac{\lambda}{p}$, $v' = 0$ et $v' = +\dfrac{\lambda}{p}$.

Cherchons maintenant la transformée de l'image d'un point isolé et appliquons la relation (92.2).

Reprenons l'éq. (96. 2) et sa réciproque

$$U(y',z') = \frac{j\,R}{\lambda}\iint G(\beta',\gamma')\,e^{-j\,K(\beta'y'+\gamma'z')}\,d\beta'\,d\gamma',$$

$$G(\beta',\gamma') = -\frac{j}{\lambda R}\iint U(y',z')\,e^{j\,K(\beta'y'+\gamma'z')}\,d y'\,d z'.$$

On peut montrer que si deux fonctions $U(y'z')$ et $G(\beta'\gamma')$ satisfont aux deux équations précédentes, il existe entre elles la relation

$$\iint |U(y',z')|^2\,e^{j\,K(\mu'y'+v'z')}\,d y'\,d z' = R^2\iint G(\beta',\gamma')\,G(\beta'+\mu',\gamma'+v')\,d\beta'\,d\gamma'. \tag{97.3}$$

Notons que si l'on fait $\mu' = v' = 0$ on obtient l'équation

$$\iint U(y',z')|^2\,d y'\,d z' = R^2\iint |G(\beta',\gamma')|^2\,d\beta'\,d\gamma'$$

qui exprime la conservation de l'énergie.

Faisons $v' = \lambda/p$ et $\mu' = 0$ dans (97.3)

$$\iint |U(y',z')|^2 \, e^{jK\frac{\lambda}{p}z'} \, dy' \, dz' = R^2 \iint G(\beta',\gamma') \, G\left(\beta', \gamma' + \frac{\lambda}{p}\right) d\beta' \, d\gamma'.$$

Cette équation donne la transformée de l'image du point en intensité pour $v' = \lambda/p$. Nous ne sommes en effet, intéressés que par les valeurs de la transformée de l'image du point pour les valeurs $v' = 0$, $v' = \pm \dfrac{\lambda}{p}$. La fonction $G(\beta' \gamma')$, transformée de l'image du point en amplitude, est représentée par un cercle de rayon α_0'. De même $G\left(\beta', \gamma' + \dfrac{\lambda}{p}\right)$ est représentée par un cercle de rayon α_0' mais décalé de λ/p sur le précédent (Fig. 238).

La partie commune à ces deux cercles donne la valeur de l'intégrale précédente. On a ainsi les valeurs de la transformée de l'image du point en intensité pour

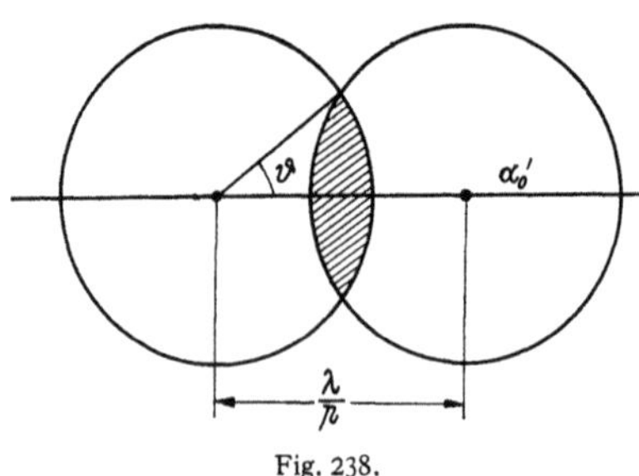

Fig. 238.

$v' = 0$, $v' = \pm \dfrac{\lambda}{p}$, valeurs qu'il faut multiplier par $-\dfrac{2j\,C_0}{\pi}$, 1 et $+\dfrac{2j\,C_0}{\pi}$ pour obtenir la transformée de l'image de la mire.

Sur la Fig. 238 on voit que $\cos\vartheta = \dfrac{\lambda}{2p\,\alpha_0'}$ et la partie commune aux deux cercles a pour surface

$$S = 2\alpha_0'^2(\vartheta - \sin\vartheta\cos\vartheta)$$

si $\mu' = v' = 0$, la partie commune aux deux cercles donne pour la transformée de l'image du point la valeur $\pi R^2 \alpha_0'^2$. Prenons cette valeur pour unité, S devient

$$S' = \frac{2}{\pi}(\vartheta - \sin\vartheta\cos\vartheta).$$

Au voisinage du pouvoir séparateur, ϑ est petit et on peut écrire

$$S' \approx \frac{4}{3\pi}\left[2(1-\cos\vartheta)\right]^{\frac{3}{2}}.$$

Admettons que l'image de la mire disparaît lorsque le contraste est égal au minimum de contraste perceptible par l'oeil, soit $\gamma = 0{,}02$.

La transformée de l'image de la mire sera donnée par les trois valeurs

$$-\frac{2j\,C_0}{\pi}\,S'_{-\frac{\lambda}{p}}, \qquad S'_0, \qquad +\frac{2j\,C_0}{\pi}\,S'_{\frac{\lambda}{p}}$$

d'où la structure de l'image

$$I(z') = S'_0 + \frac{4C_0}{\pi}\,S'_{\frac{\lambda}{p}} \sin\frac{2\pi z'}{p} = 1 + \frac{4C_0}{\pi}\,S'_{\frac{\lambda}{p}} \sin\frac{2\pi z'}{p}$$

d'où le contraste de l'image

$$\gamma = \frac{8C_0}{\pi}\,S'_{\frac{\lambda}{p}} = \frac{32\,C_0}{3\pi^2}\left[2(1-\cos\vartheta)\right]^{\frac{3}{2}} = 0{,}02.$$

Pour une mire de contraste 1 $C_0 = 1$ et

$$\frac{32}{3\pi^2}\left[2(1-\cos\vartheta)\right]^{\frac{3}{2}} = 0{,}02$$

d'où

$$1 - \cos\vartheta = 0{,}035\,.$$

et

$$\cos\vartheta = \frac{\lambda}{2p\,\alpha_0'} = 1 - 0{,}035$$

Le pas de la mire p correspondant au pouvoir séparateur sera donné par

$$p = \frac{1{,}03\,\lambda}{2\alpha_0'}\,.$$

98. Image d'un petit disque noir sur fond blanc en éclairage cohérent. Nous considérons un petit disque noir de surface s, le champ qui l'entoure étant maintenant éclairé par une source auxiliaire.

C'est, par exemple, un petit disque de centre S sur la Fig. 231. Dans ce cas, les vibrations qui attaignent les différents points entourant le disque et qui constituent l'objet lumineux, présentent entre elles une différence de phase constante: elles sont cohérentes. Il devient alors nécessaire d'additionner les amplitudes provenant des divers éléments de l'objet. La relation (92.3) peut s'écrire, en faisant le même raisonnement qu' à la Sect. 94

$$C(y', z') = \int\!\!\int\limits_{-\infty}^{+\infty} U(y' - y, z' - z)\,dy\,dz - s \cdot U(y', z'), \qquad (98.1)$$

la fonction $P(y, z)$ étant égale à l'unité, sauf à l'intérieur de s où elle est nulle.

La première intégrale représente le volume total du solide de diffraction en amplitude. On le calcule facilement en remplaçant dans la Sect. 69 la fonction E donnée par l'expression (68.4) par la fonction d'amplitude U donnée par (68.2). On trouve la valeur $jR\lambda$ d'où $C(y', z')$

$$C(y', z') = jR\lambda - s\,U(y', z')$$

et, d'après (68.2)

$$C(y', z') = jR\lambda - j\,\frac{\pi R\alpha_0'^2}{\lambda}\,s\,\frac{2J_1(Z)}{Z}$$

et au facteur $jR\lambda$ près

$$C(Z) = 1 - \frac{\pi s\alpha_0'^2}{\lambda^2}\,\frac{2J_1(Z)}{Z}\,. \qquad (98.2)$$

Cette expression n'est valable bien entendu que si le diamètre du disque noir est petit devant celui de la tache de diffraction. La répartition des éclairements s'obtiendra en élevant (98.2) au carré, et la structure de l'image est analogue à celle de la Fig. 235.

La valeur maximum de $\dfrac{2J_1(Z)}{Z}$ étant 1, le contraste maximum de l'image d'un petit disque noir en éclairage cohérent sera

$$\gamma = \frac{2\pi s\alpha_0'^2}{\lambda^2}\,.$$

Le contraste est donc deux fois meilleur en éclairage cohérent qu'en éclairage incohérent.

99. Image d'une ligne noire fine sur fond blanc en éclairage cohérent. Nous considérons une ligne fine en coïncidence avec oy (Fig. 231). En raisonnant de la même façon qu' à la Sect. 96, on peut écrire

$$C(y', z') = \int\!\!\int\limits_{-\infty}^{+\infty} U(y' - y, z' - z)\,dy\,dz - \varepsilon \int U(y' - y, z')\,dy$$

et, d'après la Sect. 98

$$C(y', z') = jR\lambda - \varepsilon \int U(y' - y, z')\,dy.$$

Posons $Y = y' - y$. L'éclairement le long de Cz' est

$$C(z') = jR\lambda - \varepsilon \int U(Y, z')\,dY. \tag{99.1}$$

Prenons la relation réciproque de l'expression (96.2) et faisons $\beta' = 0$. On a

$$G(o, \gamma') = -\frac{j}{\lambda R} \iint U(Y, z')\, e^{jK\gamma' z'}\,dY\,dz'.$$

Posons

$$U'(z') = \int U(Y, z')\,dY$$

d'où

$$G(o, \gamma') = -\frac{j}{\lambda R} \int U'(z')\, e^{jK\gamma' z'}\,dz'.$$

Réciproquement, l'ouverture de O_2 étant circulaire ($\gamma_0' = \alpha_0'$) et $U'(z')$ étant la transformée de FOURIER à une variable de $G(o\gamma')$

$$U'(z') = jR \int_{-\alpha_0'}^{+\alpha_0'} G(o, \gamma')\, e^{-jK\gamma' z'}\,d\gamma'.$$

On peut poser $G(o, \gamma') = 1$ d'où

$$U'(z') = 2jR\alpha_0' \frac{\sin K\alpha_0' z'}{K\alpha_0' z'}$$

et l'expression (99.1) devient

$$C(z') = jR\lambda - 2j\varepsilon R\alpha_0' \frac{\sin K\alpha_0' z'}{K\alpha_0' z'}$$

et au facteur $j\lambda R$ près

$$C(Z) = 1 - \frac{2\varepsilon\alpha_0'}{\lambda} \frac{\sin Z}{Z}. \tag{99.2}$$

La courbe de répartition des intensités s'obtiendra en élevant au carré l'expression (99.2).

La valeur maximum de $C(Z)$ étant

$$C(o) = 1 - \frac{2\varepsilon\alpha_0'}{\lambda},$$

en élevant au carré, le contraste sera

$$\gamma = \frac{4\varepsilon\alpha_0'}{\lambda}. \tag{99.3}$$

Si on compare (99.3) et (96.7), on voit que le contraste de la ligne noire est 2,35 fois meilleur en éclairage cohérent qu'en éclairage incohérent. Quant à l'expression donnant les éclairements dans l'image de la ligne fine lumineuse sur fond noir, elle sera donnée par (99.2) en supprimant le premier terme égal à l'unité.

100. Cas particulier des objets transparents: contraste de phase. On rencontre souvent des objets qui ne sont pas visibles, non pas parce qu'ils sont trop petits, mais seulement parce qu'ils sont transparents et ne présentent pratiquement pas de contraste avec le champ qui les entoure. Ces objets, caractérisés par des variations d'indice ou d'épaisseur, sont appelés «objets de phase». Considérons par exemple une préparation contenant une bactérie A transparente, d'épaisseur e et d'indice n (Fig. 239). Si n' est l'indice du milieu dans lequel la bactérie A est immergée, la différence entre les chemins optiques parcourus par le rayon 1 qui traverse la bactérie et le rayon qui passe à côté, est $\Delta = (n - n')\,e$ et la différence de phase $2\pi\Delta/\lambda$.

Les objets réfléchissants constituent également des objets de phase. Soit A un tel objet (Fig. 240): les irrégularités de poli modifient les chemins optiques parcourus par les rayons 1 et 2 et réfléchis par la surface A. Par rapport à une

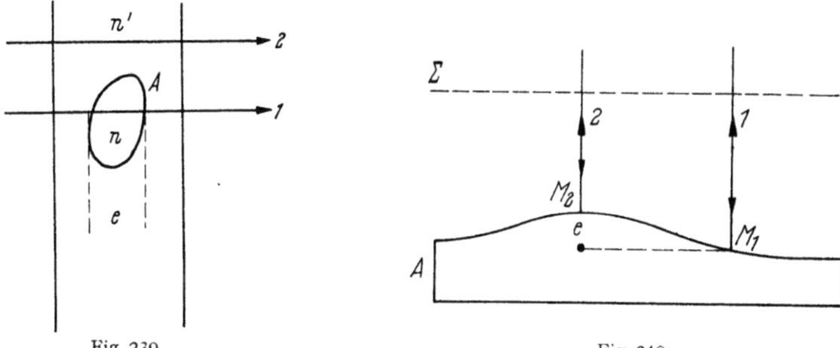

Fig. 239.
Fig. 240.

surface plane de référence Σ parallèle à la surface moyenne de A, la différence des chemins optiques est égale à $2e$, e étant la différence des épaisseurs de l'objet A dans les régions M_1 et M_2.

Des objets de ce type caractérisés seulement par des variations du chemin optique et non par des variations d'amplitude, ne sont pas visibles par les procédés ordinaires d'observation.

La méthode du contraste de phase imaginée par le physicien hollandais ZERNIKE permet de transformer les variations de phase de l'objet en variations correspondantes d'amplitude dans l'image. Les objets de phase deviennent ainsi visibles.

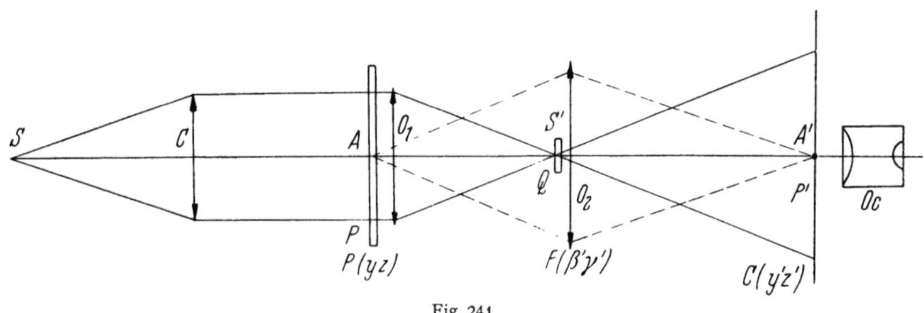

Fig. 241.

Considérons un objet transparent P (Fig. 241) éclairé en faisceau parallèle au moyen d'une source S placée au foyer d'un objectif collimateur C.

Après traversée de l'objet P et d'un objectif O_1, le faisceau éclairant, d'abord concentré en S' image de la source lumineuse S, traverse ensuite l'objectif O_2, puis s'étale sur le plan P' image de P donnée par l'objectif O_2. Nous appelerons ce faisceau, le faisceau de «lumière directe». Un détail déphasant A de l'objet diffracte la lumière qui est recueillie et concentrée par l'objectif O_2 dans l'image A' de A. L'image A' observée au moyen de l'oculaire Oc résulte des interférences de la lumière directe qui produit un fond cohérent avec la lumière diffractée.

Représentons les phénomènes sur le diagramme de Fresnel (Fig. 242). Soit O un point origine. A partir de ce point, portons un vecteur OM dont la longueur est proportionnelle à l'amplitude au point considéré de l'objet. Avec un axe origine Ox il fait un angle φ représentant la phase au même point. L'objet étant transparent, l'extrémité M se trouve sur un cercle de centre O et de rayon pris pour unité. Seule la phase φ varie d'un point à un autre de l'objet. Si toute la lumière diffractée est recueillie par l'objectif O_2, le diagramme de la Fig. 242 représente également les amplitudes et les phases dans l'image. Décomposons le vecteur OM en 2 vecteurs OH et HM. En supposant φ petit, le point H est pratiquement confondu avec A. Le mouvement vibratoire en un point de l'image résulte alors de la composition de 2 vibrations en quadrature: $OH \approx OA$ et AM. L'amplitude OA est l'amplitude dans les régions de l'objet où il n'y a pas de détail déphasant ($\varphi = 0$): c'est l'amplitude que l'on aurait s'il n'y avait pas d'objet, c'est donc l'amplitude de la lumière directe. L'amplitude $HM \approx AM$ est l'amplitude diffractée par un détail déphasant φ de l'objet. L'éclairement E dans l'image $E = \overline{OM}^2 = \overline{OH}^2 + \overline{HM}^2$ est évidemment le même dans l'image du détail ou dans le reste du champ, et l'objet n'est pas visible.

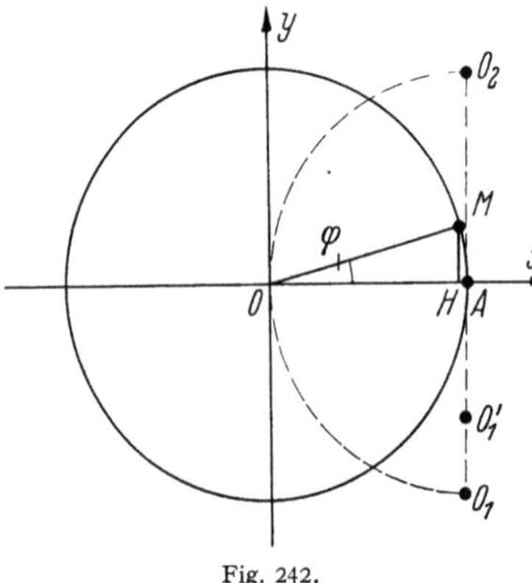

Fig. 242.

Plaçons en S' (Fig. 241) une petite lame transparente très étroite Q. Toute la lumière directe passe à travers Q, mais comme le faisceau de lumière diffractée est au contraire très étalé dans cette région, on peut admettre que pratiquement la lame Q n'agit pas sur la lumière diffractée. Donnons à la lame Q une épaisseur optique telle que les vibrations de lumière directe qui la traverse soient retardées d'un quart d'onde par rapport aux vibrations de la lumière diffractée qui passent à côté de Q. Sur le diagramme de Fresnel, tout se passe comme si l'origine venait en O_2. Dans ces conditions, l'amplitude dans l'image du détail déphasant φ devient $O_2 A - AM = 1 - \varphi$ et l'éclairement $E_1 = 1 - 2\varphi$. A côté de l'image de ce détail dans une région où $\varphi = 0$ l'éclairement est $E_2 = 1$. L'image du détail apparaît donc avec un contraste:

$$\gamma = 2\varphi \qquad (100.1)$$

A une avance de phase φ ($\varphi > 0$) c'est-à-dire à une diminution du chemin optique correspond une diminution d'éclairement de l'image du détail, le contraste de phase est négativ.

Au lieu de retarder les vibrations directes par rapport aux vibrations diffractées, plaçons en S' une lame Q qui avance les vibrations directes d'un quart d'onde. Il suffit de prendre une lame Q plus épaisse qui retarde les vibrations directes de $3\lambda/4$, ce qui équivaut à une avance de $\lambda/4$ par rapport aux vibrations diffractées. Dans ce cas, l'origine vient en O_1 et on a $E_1 = 1 + 2\varphi$. Le contraste est encore égal à 2φ, mais à une avance de phase φ ($\varphi > 0$) correspond une augmentation d'éclairement de l'image du détail, le contraste de phase est positiv.

La lame Q permettant de modifier la phase des vibrations directes par rapport aux vibrations diffractées, s'appelle la lame de phase.

En admettant que l'oeil ne puisse percevoir deux plages dont le contraste soit inférieur à 0,02, la différence de phase limite que l'on pourra mettre en évidence sera donnée par $2\varphi = 0,02$, soit une différence de marche de 10 Å environ. En fait, il faut un contraste de l'ordre de 0,1 pour voir confortablement de petits détails, ce qui limite pratiquement la différence de marche minimum observable à 50 Å.

Il est possible d'augmenter beaucoup la sensibilité de la méthode en rendant la lame de phase absorbante. En effet, caractérisons l'absorption de la lame de phase par le nombre N par lequel elle divise l'intensité de la lumière directe incidente. C'est une lame de phase de densité optique d donnée par

$$d = \log_{10} N.$$

Fig. 243. Lentille bien polie observée en contrasts de phase.

Sur le diagramme de FRESNEL tout se passe comme si l'origine venait en O_1' tel que

$$\left(\frac{A\,O_1}{A\,O_1'}\right)^2 = N.$$

L'éclairement dans l'image du détail déphasant devient

$$E_1 = \left(\frac{1}{\sqrt{N}} + \varphi\right)^2 = \frac{1}{N}\left(1 + \varphi\sqrt{N}\right)^2 = \frac{1}{N}\left(1 + 2\varphi\sqrt{N}\right)$$

et dans le reste du champ $(\varphi = 0)$ $E_2 = 1/N$. Le contraste de l'image devient

$$\gamma = 2\varphi\sqrt{N}. \tag{100.2}$$

Par rapport à la lame de phase transparente, le contraste de l'image est multiplié par $\sqrt{N}$.

En principe si $N = 2500$ $(d = 3{,}4)$, on pourra observer des différences de marche de l'ordre de 1 angström avec un contraste égal à 0,1. La méthode du contraste de phase permet donc d'obtenir une sensibilité considérable, à condition toutefois que le système optique soit lui-même de qualité suffisante et en particulier bien dépourvu de lumière parasite.

La Fig. 243 montre une lentille bien polie vue au contraste de phase. On a réalisé le montage de la Fig. 241 en supprimant l'objectif de collimateur C. La lentille à étudier est en O_1 et sert elle-même à former l'image de S sur la lame de phase Q. Il n'y a donc aucun système optique entre la source S et la lame de phase Q autre que la lentille à étudier et on est ainsi dans les meilleures conditions au point de vue lumière parasite. La lame de phase utilisée est formée par une couche mince d'aluminium de densité optique voisine de 4, déposée sur une lame de verre par évaporation dans le vide. Les inégalités de poli correspondent ici à des variations du chemin optique de l'ordre de 20 Å en moyenne et se traduisent par de petites vagues plus ou moins foncées. On peut également voir des inhomogénéités dans la masse du verre; ce sont ici des fils facilement reconnaissables.

101. Calcul de l'amplitude dans l'image en contraste de phase. La source S (Fig. 241) est supposée très petite, nous sommes donc en éclairage cohérent.

Soit $P_R(y,z)$ la partie réelle de $P(y,z)$ et $\varphi(y,z)$ sa partie imaginaire. Si nous considérons un objet de phase, $P_R(y\,z)$ est constant et la répartition des amplitudes dans le plan passant par S' est donnée par la transformée de Fourier de $P(y,z)$ (92.4)

$$F(\beta',\gamma') = -\frac{j}{\lambda R} \iint (\cos\varphi + j\sin\varphi)\, e^{jK(\beta'y+\gamma'z)}\,dy\,dz. \qquad (101.1)$$

Choisissons comme origine des phases, la phase moyenne définie par la condition

$$\iint \sin\varphi\,dy\,dz = 0$$

et soit m la valeur moyenne de $\cos\varphi$ à l'intérieur de l'objectif O_1. L'expression (101.1) peut s'écrire

$$\left.\begin{aligned} F(\beta',\gamma') = -\frac{j}{\lambda R}\Big[& m\iint_{O_2} e^{jK(\beta'y+\gamma'z)}\,dy\,dz + \\ & + \iint_{O_2} [(\cos\varphi - m) + j\sin\varphi]\, e^{jK(\beta'y+\gamma'z)}\,dy\,dz\Big]. \end{aligned}\right\} \qquad (101.2)$$

Le premier terme représente la figure de diffraction de l'onde parfaite limitée par le contour de l'objectif O_1 située tout contre l'objet. La source S étant très petite si l'objectif O_1 est limité par un contour circulaire, le premier terme représente la distribution en amplitude dans la tache d'Airy. La présence du deuxième terme est due à la lumière diffractée par les variations de phase caractérisant l'objet A. Au point S', $\beta'y+\gamma'z=0$ et on a

$$\iint(\cos\varphi - m)\,dy\,dz = 0.$$

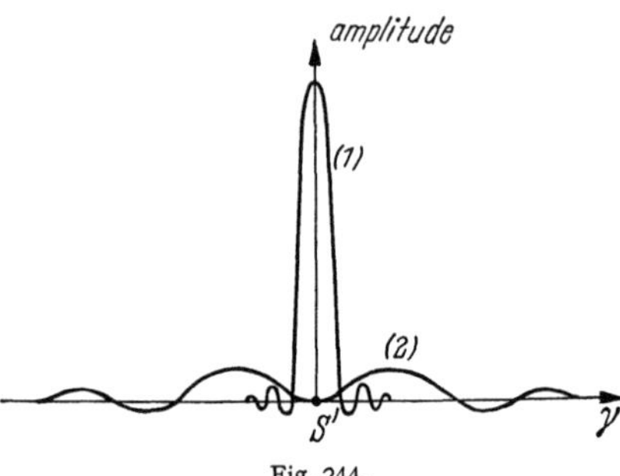

Fig. 244.

Le dernier terme de l'équation (101.2) est donc nul en ce point. On peut dire que la lumière diffractée par les variations de phase de l'objet, nulle en S', s'étale autour de la tache centrale du phénomène de diffraction dû à l'onde parfaite. Elle s'étale d'autant plus loin que les irrégularités de l'objet qui lui donnent naissance ont des dimensions plus petites. La Fig. 244 représente schématiquement ce résultat. La courbe (1) donne la distribution des amplitudes correspondant au premier terme de (101.2); c'est, par exemple, la tache d'Airy. La courbe (2) donne la distribution des amplitudes dûes au second terme de (101.2). La courbe (1) représente la lumière directe et la courbe (2), la lumière diffractée dans le plan passant par S'. On voit donc qu'en plaçant une lame de phase étroite en S', elle n'agit pratiquement pas sur la lumière diffractée, ou en tout cas, elle ne modifie que des termes de basse fréquence dont la disparition n'entraîne que des modifications négligeables dans l'image A'.

Pour passer enfin de la figure de diffraction en S' à l'image A', nous avons

$$C(y', z') = \frac{jR}{\lambda} \iint_{O_3} F(\beta', \gamma')\, e^{-jK(\beta'y + \gamma'z)} d\beta'\, d\gamma'. \tag{101.3}$$

Si les détails de l'objet A ne sont pas trop petits, toute la lumière qu'ils diffractent est recueillie par O_3, et on a évidemment

$$C(y', z') = P(y, z).$$

Dans le cas où φ est petit, on peut écrire

$$C(y', z') = e^{j\varphi(yz)} \approx 1 + j\,\varphi.$$

Si on place une lame de phase en S' qui retarde le lumière directe, l'amplitude $C(y', z')$ dans l'image deviendra égale à $1 - \varphi$ et l'éclairement $1 - 2\varphi$. On retrouve les résultats obtenus directement par le diagramme de Fresnel. Les équations (101.1) (101.2) et (101.3) permettront le calcul complet des phénomènes et de montrer en particulier que les franges de diffraction qui entourent toujours les images en contraste de phase sont dues à l'absorption de la lumière diffractée par la lame de phase.

102. Cas des objets périodiques en éclairage cohérent: réseau d'amplitude. Considérons un objet périodique en éclairage cohérent.

Reprenons la relation (92.3) qui exprime l'amplitude dans l'image d'un objet cohérentet désignons par $F'(\beta', \gamma')$ la transformée de Fourier de $C(y', z')$. On aura, en laissant de côté dans tout ce qui va suivre les facteurs constants

$$F(\beta', \gamma') = \iint P(y, z)\, e^{-jK(\beta'y + \gamma'z)}\, dy\, dz,$$

$$F'(\beta', \gamma') = \iint C(y', z')\, e^{-jK(\beta'y' + \gamma'z')}\, dy'\, dz'.$$

En remplaçant $C(y', z')$ par sa valeur tirée de l'équation (92.3) et en posant $Y = y' - y$, $Z = z' - z$

$$F'(\beta', \gamma') = \iint P(y, z)\, e^{-jK(\beta'y + \gamma'z)}\left[\iint U(Y, Z)\, e^{-jK(\beta'Y + \gamma'Z)}\, dY\, dZ\right] dy\, dz.$$

L'intégrale entre crochets est la transformée de Fourier de l'image du point en amplitude (Sect. 97). On peut donc écrire à un facteur constant près

$$F'(\beta', \gamma') = F(\beta', \gamma')\, G(\lambda\beta', \lambda\gamma').$$

L'instrument étant parfait, $G(\lambda\beta', \lambda\gamma')$ est constant à l'intérieur du cercle de rayon α'_0.

Considérons d'abord le cas d'un objet périodique d'amplitude: la mire sinusoïdale, dont les traits dont parallèles à y, y (Fig. 231), est de la forme

$$P(y, z) = \sin\frac{2\pi z}{p}$$

que l'on peut écrire

$$P(z) = \frac{1}{2j}\left[e^{j\frac{2\pi z}{p}} - e^{-j\frac{2\pi z}{p}}\right].$$

$F(\beta',\gamma')$ sera donc égal à $\pm\dfrac{1}{2j}$ pour $\gamma'=\pm\dfrac{1}{p}$ et $\beta'=0$. $F'(\beta',\gamma')$ présentera les deux composantes suivantes

$$F'(0,\gamma') = \frac{-1}{2j}\, G\left(0,\ -\frac{\lambda}{p}\right) \qquad \left(\gamma'=-\frac{1}{p}\right),$$

$$F'(0,\gamma') = \frac{1}{2j}\, G\left(0,\ \frac{\lambda}{p}\right) \qquad \left(\gamma'=+\frac{1}{p}\right)$$

et l'image de la mire sera représentée par l'expression

$$C(z') = \frac{1}{2j}\, G\left(0,\ -\frac{\lambda}{p}\right) e^{-j\frac{2\pi z'}{p}} + \frac{1}{2j}\, G\left(0,\ \frac{\lambda}{p}\right) e^{j\frac{2\pi z'}{p}}.$$

La fonction G est, soit constante, soit nulle. Si elle est constante, on peut la prendre égale à l'unité. On aura

$$C(z') = \frac{1}{2j}\left[-e^{-j\frac{2\pi z'}{p}} + e^{j\frac{2\pi z'}{p}}\right] \tag{102.1}$$

d'où

$$C(z') = \sin\frac{2\pi z'}{p}.$$

L'image est parfaitement transmise par l'instrument. On peut interpréter physiquement ce résultat de la façon suivante (Fig. 245): le réseau sinusoïdal A placé contre l'objectif O_1 a une structure déterminée par $P(z)$. La transformée de

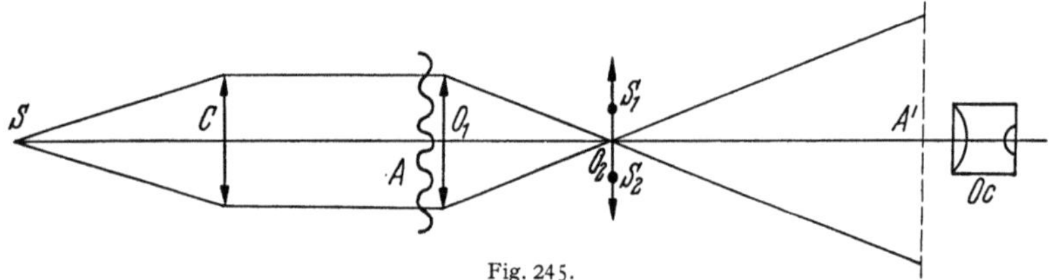

Fig. 245.

Fourier de $P(z)$ est constituée par les deux spectres S_1 et S_2 situés dans le plan O_2 conjugué de la source S. Ces deux spectres sont caractérisés par la fonction $F(\beta',\gamma')$. Si ces deux spectres ne sont pas trop écartés, ils sont recueillis par l'ojectif O_2 et la fonction $F'(\beta',\gamma')$ étant identique à $F(\beta',\gamma')$ il en est de même de la fonction $C(z')$ qui est identique à $P(z)$. L'image A' reproduit fidèlement le réseau sinusoïdal objet A: tout se passe comme si l'on avait en A' des franges d'Young produites par les deux sources synchrones S_1 et S_2.

Si les deux spectres S_1 et S_2 se trouvent en dehors de l'objectif O_2, la fonction G est nulle et il n'y a pas d'image en A'.

Considérons maintenant, non plus un réseau sinusoïdal, mais un réseau ordinaire formé d'intervalles clairs séparés par des intervalles opaques de même largeur. La fonction $P(z)$ donnant la distribution en amplitude d'un tel réseau peut être représentée par la courbe de la Fig. 246.

C'est la «fonction créneaux» étudiée au Sect. 91, et à laquelle on a ajouté un terme constant.

On peut écrire

$$P(z) = 1 + \frac{4}{\pi}\left(\sin\frac{2\pi z}{p} + \frac{1}{3}\sin 3\,\frac{2\pi z}{p} + \cdots\right).$$

Les différents termes du développement en série de FOURIER précédent, donnent des mires sinusoïdales et à chacune d'elles correspondent deux spectres analogues à S_1 et S_2 et d'autant plus écartés l'un de l'autre que le pas de la mire considéré est plus petit. La relation (102.1) devient alors

$$C(z') = 1 + \frac{2}{j\pi}\left[-G\left(0 - \frac{\lambda}{p}\right)e^{-j\frac{2\pi z'}{p}} + G\left(0\frac{\lambda}{p}\right)e^{j\frac{2\pi z'}{p}} - \right.$$
$$\left. - \frac{1}{3}G\left(0 - \frac{3\lambda}{p}\right)e^{-j3\frac{2\pi z'}{p}} + \frac{1}{3}G\left(0\frac{3\lambda}{p}\right)e^{j3\frac{2\pi z'}{p}} + \cdots\right]. \quad (102.2)$$

Cette expression permet de calculer immédiatement la structure de l'image du réseau représenté par la «fonction créneaux» de la Fig. 246.

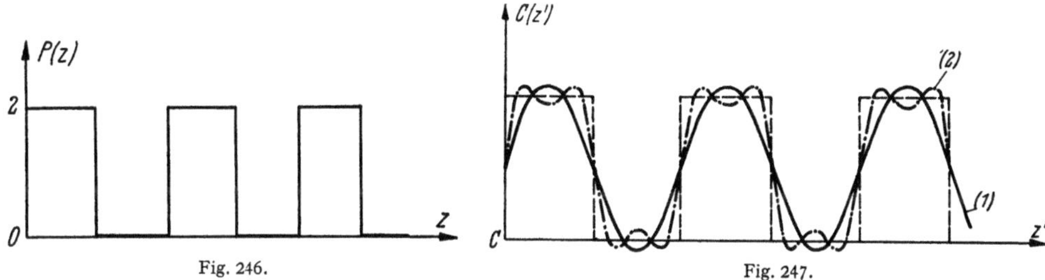

Fig. 246.   Fig. 247.

Le premier terme égal à l'unité représente l'image directe de la source en O_2. Si aucun des spectres des différentes mires sinusoïdales composantes ne sont admis par l'objectif O_2, c'est-à-dire si la fonction G est nulle pour toutes les mires, seule l'image directe est transmise et le plan image A' est uniformément éclairé. Il n'y a pas d'image. Si la fonction G n'est pas nulle pour $\pm\frac{\lambda}{p}$, seule l'image directe et la mire sinusoïdale fondamentale sont transmises. L'image du réseau est une image sinusoïdale représentée par la courbe 1 sur la Fig. 247. Si la fonction G n'est également pas nulle pour $\pm\frac{3\lambda}{p}$, l'image du réseau est donnée par

$$C(z') = 1 + \frac{2}{j\pi}\left[e^{j\frac{2\pi z'}{p}} - e^{-j\frac{2\pi z'}{p}} + \frac{1}{3}e^{j3\frac{2\pi z'}{p}} - \frac{1}{3}e^{-j3\frac{2\pi z'}{p}}\right].$$

Elle est représentée par la courbe (2).

Au fur et à mesure que l'objectif O_2 admet un nombre de plus en plus grand de spectres, l'image s'améliore et représente de mieux en mieux le réseau initial. En particulier, les discontinuités d'amplitude dans l'objet seront d'autant plus fidèlement reproduites que l'objectif O_2 amettra des termes de fréquences plus élevées.

L'expression (102.2) montre que, pour avoir une image sinusoïdale du réseau, il suffit que l'objectif O_2 laisse passer au moins l'image directe et l'un des deux spectres correspondant à $\pm\frac{\lambda}{p}$.

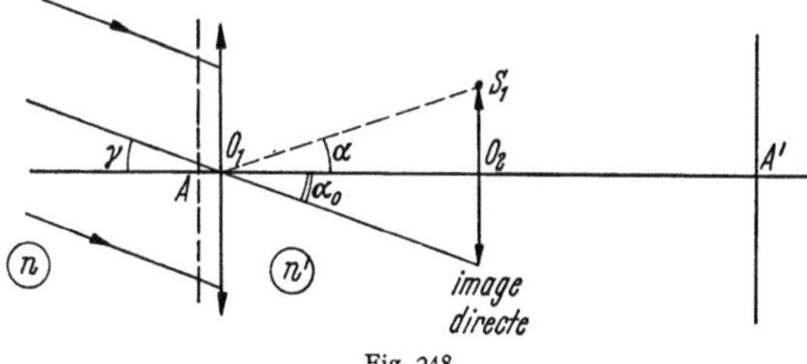

Fig. 248.

La disposition optimum est obtenue lorsque les rayons correspondant à l'image directe et à l'un de ces deux spectres, sont symétriques par rapport à l'axe et couvrent juste l'objectif O_2 (Fig. 248).

Soit γ l'angle des rayons incidents avec la normale au réseau A, n l'indice du milieu situé avant le réseau, n' l'indice du milieu situé entre le réseau et l'objectif. La disposition précédente est réalisée si

$$n \sin \gamma + n' \sin u = \frac{\lambda}{p}$$

avec $u = \alpha_0$. On a alors

$$n \sin \gamma = n' \sin \alpha_0 = n' \sin u$$

d'où

$$p = \frac{\lambda}{2n \sin u}.$$

On peut donc recueillir la mire fondamentale avec un objectif O_2 d'ouverture deux fois plus petite en éclairage oblique convenable qu'en éclairage normal.

103. Réseau de phase: observation en contraste de phase. Nous étudions maintenant un réseau constitué seulement par des variations de phase. L'amplitude est constante sur toute la surface du réseau et la phase suit une loi analogue à celle que nous avons considérée pour l'amplitude dans le cas du réseau d'amplitude. Les «traits» du réseau de phase qui ne sont ici que des variations d'indice ou d'épaisseur, sont parallèles à l'axe $O\,y$ (Fig. 231) et nous étudions les phénomènes en fonction du seul paramètre z. La fonction $P(z)$ est purement imaginaire

$$P(z) = e^{j\,\varphi(z)}$$

et $\varphi(z)$ peut être représenté par la «fonction créneaux» du paragraphe précédent

$$\varphi(z) = \varphi_0 \left(\sin \frac{2\pi z}{p} + \frac{1}{3} \sin 3 \cdot \frac{3\pi z}{p} + \cdots \right) \tag{103.1}$$

où φ_0 est une constante qui fixe les valeurs maximum et minimum de φ. Si l'on est au voisinage du pouvoir séparateur seul le réseau de phase sinusoïdal est à considérer

$$\varphi(z) = \varphi_0 \sin \frac{2\pi z}{p}.$$

Si φ_0 est petit, on peut négliger toutes les puissances de φ_0 supérieures à 1

$$P(z) = 1 + j\,\varphi_0 \sin \frac{2\pi z}{p}$$

que l'on peut écrire

$$P(z) = 1 + \frac{\varphi_0}{2} \left(e^{j\frac{2\pi z}{p}} - e^{-j\frac{2\pi z}{p}} \right).$$

$F(\beta', \gamma')$ est égal à $\pm \frac{\varphi_0}{2}$ pour $\gamma' = \pm \frac{1}{p}$ et à 1 pour $\gamma' = 0$ dans les 3 cas $\beta' = 0$.

On aura donc les composantes suivantes

$$F'(0\gamma') = -\frac{\varphi_0}{2} G\left(0 - \frac{\lambda}{p}\right), \quad \gamma' = -\frac{1}{p},$$

$$F'(0\gamma') = +\frac{\varphi_0}{2} G\left(0 \frac{\lambda}{p}\right), \qquad \gamma' = \frac{1}{p},$$

$$F'(0,\,0) = G(0,\,0), \qquad\qquad \gamma' = 0$$

et l'image du réseau de phase sinusoïdal sera donnée par

$$C(z') = G(0,\,0) - \frac{\varphi_0}{2} G\left(0 - \frac{\lambda}{p}\right) e^{-j\frac{2\pi z'}{p}} + \frac{\varphi_0}{2} G\left(0 \frac{\lambda}{p}\right) e^{j\frac{2\pi z'}{p}}.$$

Si les spectres de ce réseau sinusoïdal sont admis par l'objectif O_2, la fonction G est constante et l'on prend sa valeur pour unité, d'où

$$C(z') = 1 + j\,\varphi_0 \sin\frac{2\pi z'}{p}. \tag{103.2}$$

L'éclairement dans l'image sera donc constant puisque nous négligeons φ_0^2. A la différence du réseau d'amplitude, nous voyons qu'ici l'image $\varphi_0 \sin\dfrac{2\pi z'}{p}$ du réseau de phase est, non plus en phase avec l'image directe représentée par le terme égal à l'unité, mais en quadrature avec lui. Si nous plaçons sur l'image directe de la source une lame de phase, celle-ci n'agit pas sur les deux spectres et fait disparaître le facteur j: elle remet en phase ces images. En contraste de phase, l'image devient

$$C(z') = 1 + \varphi_0 \sin\frac{2\pi z'}{p}.$$

Les variations de phase sont reproduites sous forme de variations d'amplitude. Le contraste de phase permet de transformer un réseau de phase en réseau d'amplitude. Nous pouvons alors opérer sur les différents termes du developpement (103.1) exactement comme dans le cas du réseau d'amplitude.

Inversement, si nous considérons un réseau d'amplitude pour lequel les minima sont peu différents des maxima, son image disparaîtra en contraste de phase. En effet, considérons un réseau d'amplitude tel que dans l'expression (102.2) on ne prenne que le terme égal à l'unité et les deux premières exponentielles. C'est un réseau donnant l'image directe et les deux premiers spectres. En admettant que l'objectif O_2 sur la Fig. 245 laisse passer ces images, l'image du réseau est donnée par l'expression

$$C(z') = 1 + \frac{4}{\pi} \sin\frac{2\pi z'}{p}.$$

Dans le cas d'un réseau d'amplitude à faible contraste dont les minima sont peu différents des maxima, on écrira

$$C(z') = 1 + a_0 \sin\frac{2\pi z'}{p}$$

où a_0 est un coefficient constant qui fixe le contraste du réseau. Si a_0 est petit, les régions absorbantes le sont très peu et leur transparence est très voisine de celle des régions parfaitement transparentes. Si $a_0 = 4/\pi$ nous sommes dans le cas du réseau d'amplitude à contraste 1 étudié précédemment.

Considérons un réseau d'amplitude à faible contraste (a_0 petit) et observons son image en contraste de phase en plaçant la lame de phase sur l'image directe sans modifier les deux spectres admis.

L'image peut s'écrire

$$C(z') = 1 + j\,a_0 \sin\frac{2\pi z'}{p}. \tag{103.3}$$

Comparons cette expression à l'expression (103.2): tout se passe comme si le réseau d'amplitude à faible contraste était transformé en réseau de phase. Par contraste de phase, un réseau d'amplitude à faible contraste est transformé en réseau de phase et son image disparaît.

Ces résultats sont valables quel que soit le nombre des spectres admis par la pupille puisque tous ces spectres sont en phase avec l'image directe dans le réseau d'amplitude ou en quadrature avec elle dans le cas du réseau de phase.

104. Résumè des résultats. Les résultats des Sect. 86, 94, 96, 97, 98, 99, 102 peuvent se résumer de la façon suivante: admettons qu'une image disparaît pour l'oeil lorsque le contraste de l'image atteint la valeur $\gamma = 0{,}02$. Cette valeur est effectivement celle que l'on observe pour les lignes et les mires périodiques, elle est un peu favorable pour les disques.

Rappelons les notations antérieures: dans le cas du microscope, nous désignerons l'ouverture numérique dans l'espace-objet par le terme $n \sin u$, n étant l'indice du milieu objet (Fig. 249).

Pour la lunette astronomique, nous désignerons par D le diamètre de l'objectif, et par f sa focale (Fig. 250). Dans les deux cas, α_0' est l'ouverture dans l'espace image, ϑ l'angle sous lequel on voit une longueur ϱ_1 dans le plan image.

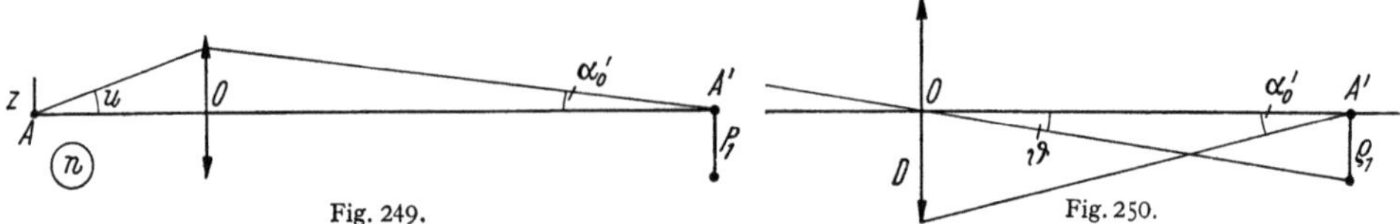

Fig. 249.   Fig. 250.

Dans le cas de la ligne fine noire sur fond blanc en éclairage incohérent, on a trouvé comme contraste de l'image $\dfrac{16\,\varepsilon\,\alpha_0'}{3\pi\lambda}$. Ecrivons que ce contraste est égal à 0,02, on aura la limite de perception de la ligne

$$\frac{16\,\varepsilon\,\alpha_0'}{3\pi\lambda} = 0{,}02$$

pour la lunette astronomique $\vartheta = \dfrac{\varepsilon}{f}$, ϑ étant l'angle sous lequel on voit de O l'image géométrique $\varepsilon = \varrho_1$ de la ligne au moment de la disparition de l'image. D'où

$$\vartheta = \frac{0{,}02\,\lambda}{D}.$$

Pour le microscope, si l'objectif O satisfait à la condition des sinus

$$n\,z \sin u = \varrho_1\,\alpha_0'$$

z étant la largeur de la ligne elle-même dans le plan-objet au moment de la disparation de l'image. On aura

$$z = \frac{0{,}02\,\lambda}{2n \sin u}.$$

En opérant de la même façon pour les différents tests étudiés, on obtient le tableau suivant:

Tableau 12.

	Éclairage cohérent	Éclairage incohérent	
	Microscope	Microscope	Lunette astronomique
Limite de séparation d'une mire de Foucault	$z = \dfrac{\lambda}{2n \sin u}$	$z = \dfrac{1{,}03\,\lambda}{2n \sin u}$	$\vartheta = \dfrac{1{,}03\,\lambda}{D}$
Séparation de 2 points brillants sur fond noir	$z = \dfrac{1{,}63\,\lambda}{2n \sin u}$	$z = \dfrac{1{,}22\,\lambda}{2n \sin u}$	$\vartheta = \dfrac{1{,}22\,\lambda}{D}$
Limite de perception d'un disque noir sur fond blanc[1]	$z = \dfrac{0{,}06\,\lambda}{2n \sin u}$	$z = \dfrac{0{,}09\,\lambda}{2n \sin u}$	$\vartheta = \dfrac{0{,}09\,\lambda}{D}$
Limite de perception d'une ligne fine noire sur fond blanc	$z = \dfrac{0{,}01\,\lambda}{2n \sin u}$	$z = \dfrac{0{,}02\,\lambda}{2n \sin u}$	$\vartheta = \dfrac{0{,}02\,\lambda}{D}$

[1] z = rayon du disque.

Nous avons donné dans ce tableau la limite de séparation de deux points brillants sur fond noir dans le cas de l'éclairage cohérent. On déduira ce résultat sans difficulté de la Sect. 68. Notons enfin qu'il sera possible de recalculer les coefficients numériques qui apparaissent dans ce tableau en fonction du minimum de contraste perceptible du récepteur.

II. Diffraction de FRESNEL ou diffraction à distance finie.

a) Formules générales.

105. Expression de l'amplitude. Soit S une source lumineuse ponctuelle et P nn point où l'on veut calculer l'amplitude (Fig. 251). Les surfaces d'onde émises par S sont des sphères de centre S telles que Σ. Désignons par R' le rayon de Σ: l'amplitude vibratoire produite par S sur cett sphère est proportionnelle à $\dfrac{1}{R'}\, e^{-jKR'}$. Un élément dS de Σ autour d'un point M produit en P situé à la distance r le mouvement vibratoire $\dfrac{A}{rR'}\, e^{-jK(r+R')}\, dS$.

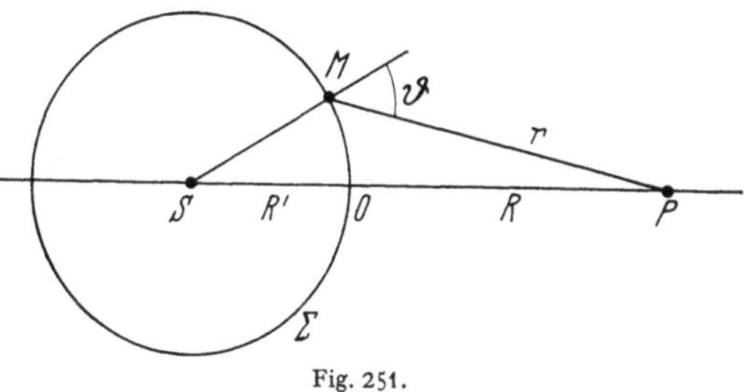

Fig. 251.

Comme il a été dit (Sect. 67), une étude mathématique correcte du principe D'HUYGENS conduit à écrire l'amplitude vibratoire en P sous la forme

$$U_P = \frac{j}{\lambda R'} \iint \frac{A}{r}\, e^{-jK(R'+r)}\, dS. \tag{105.1}$$

A est un coefficient qui dépend de l'angle ϑ de diffraction, il diminue lorsque ϑ augmente.

Posons $R' + r = R + R' + \varDelta$, c'est-à-dire $\varDelta = r - R$. $\varDelta$ représente la différence de marche en P entre les vibrations émises par les points M et O de la surface d'onde Σ.

L'amplitude U_P peut encore s'écrire

$$U_P = \frac{j}{\lambda R'} \iint \frac{A}{r}\, e^{-jK\varDelta}\, dS, \tag{105.2}$$

car on peut sans inconvénient supposer que $R + R'$ est un multiple entier de λ et poser $e^{-jK(R+R')} = 1$. Le calcul de l'amplitude diffractée lorsqu'on veut tenir compte des variations de A peut se faire d'une façon simple par la méthode des zones de FRESNEL.

106. Zones de FRESNEL. Considérons une surface d'onde Σ émise par S (Fig. 252): tous les phénomènes de propagation plus éloignés de S que Σ peuvent être considérés comme engendrés par l'onde Σ. Recouvrons l'onde Σ d'un écran opaque percé d'une petite ouverture T en dehors du voisinage du point O, inter-

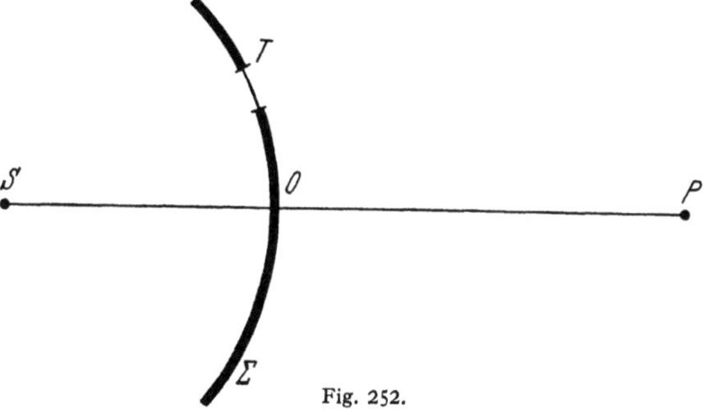

Fig. 252.

section de Σ et de SP. Les mouvements vibratoires émis par T arrivent en P en discordance et se détruisent deux à deux. En effet: traçons du point P comme centre des sphères distantes de $\lambda/2$ et qui coupent l'ouverture T (Fig. 253). Ces sphères découpent dans T des tranches contigues et circulaires telles qu'à chaque point de l'une correspond un point de la tranche suivante qui envoie en P un

mouvement en retard d'une demi-longueur d'onde, donc en opposition. Le mouvement vibratoire envoyé par une tranche est détruite par celui de la tranche voisine et l'ouverture T n'envoie pas de lumière en P. Il n'en est plus de même si l'ouverture T se trouve au voisinage du point O.

Considérons la Fig. 254 où l'on a divisé l'onde Σ par des sphères de centre P et de rayons

Fig. 253.

$$PO, \quad PO + \frac{\lambda}{2}, \quad PO + \lambda, \quad PO + \frac{3\lambda}{2}, \dots$$

Calculons la surface dS de ces zones (Fig. 255). On a

$$dS = 2\pi MH \cdot MM' = 2\pi R'^2 \sin\alpha\, d\alpha.$$

Or, dans le triangle SMP on a

$$r^2 = R'^2 + (R + R')^2 - 2R'(R + R')\cos\alpha. \tag{106.1}$$

Différentions

$$r\, dr = R'(R + R')\sin\alpha\, d\alpha \tag{106.2}$$

d'où

$$dS = \frac{2\pi r R'}{R + R'}\, dr = \frac{\pi r R' \lambda}{R + R'}.$$

Comme r est grand par rapport à λ, il varie peu lorsqu'on passe d'une zone à l'autre et les mouvements vibratoires produits en P par deux zones consécutives

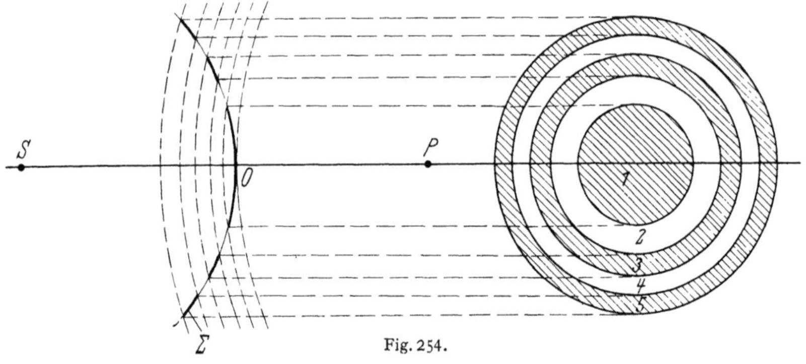

Fig. 254.

se détruisent. En réalité, par suite de la décroissance du coefficient A (105.1), leur résultante n'est pas tout à fait nulle et le mouvement vibratoire total se présente sous la forme d'une série à termes alternativement positifs et négatifs, décroissant lentement en valeur absolue.

Calculons les rayons des zones successives lorsque α est assez petit pour pouvoir poser $\cos\alpha \approx 1 - \frac{\alpha^2}{2}$. On aura d'après (106.1)

$$(R + \Delta)^2 = R^2 + R'(R + R')\,\alpha^2.$$

Négligeons Δ^2 devant $2R\Delta$, il vient

$$\Delta = \frac{R'(R + R')}{2R}\,\alpha^2. \tag{106.3}$$

Posons $\varrho = R'\alpha$, le rayon ϱ_n de la $n^{\text{ième}}$ zone sera donné par $\Delta = n\dfrac{\lambda}{2}$, c'est-à-dire

$$\varrho_n = \sqrt{\frac{R R' \lambda}{R + R'}}\,\sqrt{n}.$$

Les rayons des différentes zones varient comme les racines des nombres entiers successifs et suivent donc la même loi que les rayons des anneaux de NEWTON. Prenons par exemple $R = R' = 5$ mètres. Le rayon de la première zone est

$$\varrho_1 = \sqrt{\frac{R R' \lambda}{R + R'}} = 1{,}22\ \text{mm}.$$

pour $\lambda = 0{,}6\ \mu$. Le rayon de la deuxième zone est $1{,}22\,\sqrt{2} = 1{,}72$ mm., celui dela troisième zone $1{,}22\,\sqrt{3} = 2{,}11$ mm. Les rayons des zones ont des dimensions

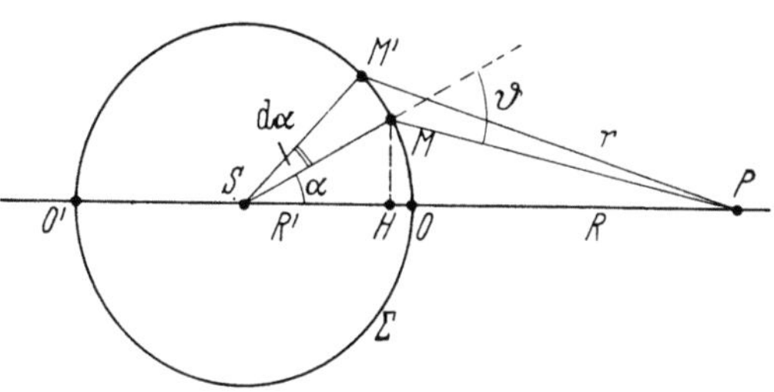

Fig. 255.

qui rendent les expériences possibles. Calculons la vibration résultante en P en supposant que ϑ et par conséquent A, sont constants pour tous les points d'une même zone. L'élément d'aire dS est pris égal à celui d'une petite zone $2\pi R'^2 \sin \alpha\, d\alpha$ de la sphère Σ. D'après (106.2), on aura

$$dS = 2\pi R'^2 \sin \alpha\, d\alpha = \frac{2\pi R' r}{R + R'}\, dr.$$

En reprenant l'équation (105.1) avec $e^{-jKR'} = 1$, l'amplitude en P, due la $n^{\text{ième}}$ zone s'écrira

$$(U_P)_n = \frac{2\pi j A}{\lambda(R + R')} \int\limits_{R+(n-1)\frac{\lambda}{2}}^{R+n\frac{\lambda}{2}} e^{-jKr}\, dr. \tag{106.4}$$

D'où

$$(U_P)_n = \frac{2\pi j A}{\lambda(R + R')} \left[\frac{e^{-jKr}}{-jK}\right]_{R+(n-1)\frac{\lambda}{2}}^{R+n\frac{\lambda}{2}} = (-1)^{n-1}\frac{2A\, e^{-jKR}}{R + R'}.$$

Les vibrations envoyées par les différentes zones en P ont toutes la même phase et leurs amplitudes alternativement positives et négatives, vont en décroissant régulièrement à cause du coefficient A. L'amplitude résultante en P dûe à toutes les zones, sera de la forme

$$U_P = U\, e^{-jKR}$$

avec

$$U = u_1 - u_2 + u_3 - u_4 + \cdots \pm u_n \ldots,$$

le terme général de cette série alternée étant égal en valeur obsolue à

$$u_n = \frac{2A}{R + R'}.$$

Supposons qu'un terme quelconque soit plus petit que la moyenne arithmétique des deux termes qui l'encadrent. Si la sphère contient un nombre pair de zones par exemple, on a

$$U = \frac{u_1}{2} + \left(\frac{u_1}{2} - u_2 + \frac{u_3}{2}\right) + \cdots + \frac{u_{n-1}}{2} - u_n$$

avec

$$U \geq \frac{u_1}{2} + \frac{u_{n-1}}{2} - u_n,$$

ou

$$U = u_1 - \frac{u_2}{2} - \left(\frac{u_2}{2} - u_3 + \frac{u_4}{2}\right) + \cdots - \frac{u_n}{2}$$

avec

$$U \leq u_1 - \frac{u_2}{2} - \frac{u_n}{2}.$$

Puisque les termes varient lentement, on peut écrire $u_1 = u_2 \ldots u_n = u_{n-1}$. Les deux inégalités se réduisent à

$$U = \frac{u_1}{2} - \frac{u_n}{2}.$$

Dans le cas où l'onde Σ entière est découverte, c'est à dire s'il n'y a pas d'écran entre S et P, $u_n = 0$ en admettant que A soit nul lorsque M vient en O' diamétralement opposé à O. On a alors

$$U = \frac{u_1}{2} = \frac{A}{R + R'}. \tag{106.5}$$

L'amplitude produite en P par l'onde entière est égale à la moitié de l'amplitude produite par la zone centrale.

107. Intégrales de Fresnel. Désignons par s l'arc OM correspondant à l'angle α, c'est à dire posons $s = R'_\alpha$: l'expression (106.3) devient

$$\Delta = \frac{R + R'}{2RR'} s^2. \tag{107.1}$$

Plaçons contre la surface d'onde Σ un écran opaque percé d'une ouverture T petite par rapport aux distances de l'écran aux points S et P. D'après (105.2) l'amplitude au point P pourra s'écrire avec une approximation suffisante et à un facteur constant près

$$U_P = \iint e^{-jK\Delta}\, dS \tag{107.2}$$

où Δ a la valeur donnée par l'expression (107.1). Posons

$$\frac{R + R'}{RR'\lambda} s^2 = \frac{v^2}{2},$$

$$v = \sqrt{\frac{2(R + R')}{RR'\lambda}} \cdot s = ms. \tag{107.3}$$

Si on rapporte les différents points de l'ouverture T à un système d'axes rectangulaires Oyz, on aura

$$\Delta = \frac{\lambda m^2 s^2}{4} = \frac{\lambda m^2 (y^2 + z^2)}{4}.$$

L'intégrale (107.2) s'écrira alors

$$U_P = \iint e^{-j\frac{\pi m^2}{2}(y^2 + z^2)}\, dy\, dz.$$

Si l'ouverture est rectangulaire, on choisira pour axes $Oy\,Oz$ des directions parallèles à ses bords, ce qui permettra d'écrire

$$U_P = \int e^{-j\frac{\pi m^2 y^2}{2}}\,dy \int e^{-j\frac{\pi m^2 z^2}{2}}\,dz \qquad (107.4)$$

et le calcul numérique de l'éclairement en P reviendra au calcul de l'une ou l'autre de ces intégrales

$$(U_P)_y = \int e^{-j\frac{\pi m^2 y^2}{2}}\,dy = \int \cos\frac{\pi m^2 y^2}{2}\,dy - j \int \sin\frac{\pi m^2 y^2}{2}\,dy.$$

En posant

$$P = \int \cos\frac{\pi m^2 y^2}{2}\,dy, \qquad Q = \int \sin\frac{\pi m^2 y^2}{2}\,dy$$

l'éclairement dû à une intégrale de ce type serait

$$E_P = P^2 + Q^2. \qquad (107.5)$$

D'après (107.3) les intégrales P et Q peuvent prendre la forme suivante

$$P = \frac{1}{m} \int \cos\frac{\pi v^2}{2}\,dv, \qquad Q = \frac{1}{m} \int \sin\frac{\pi v^2}{2}\,dv$$

et on appelle intégrales de FRESNEL les deux intégrales définies

$$\left.\begin{array}{l} \xi = \displaystyle\int_0^v \cos\frac{\pi v^2}{2}\,dv, \\[2em] \eta = \displaystyle\int_0^v \sin\frac{\pi v^2}{2}\,dv. \end{array}\right\} \qquad (107.6)$$

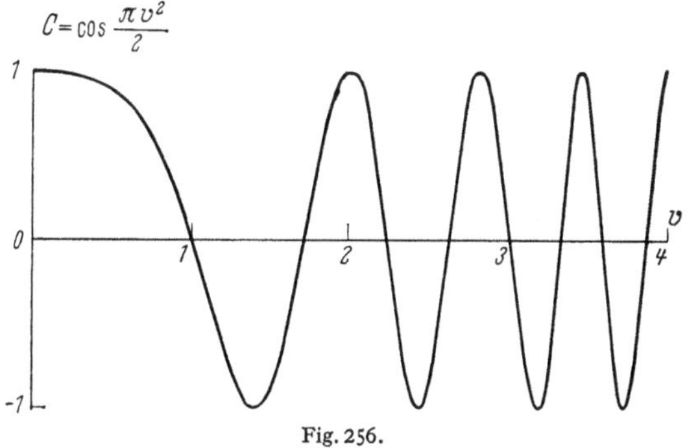

Fig. 256.

Ce sont ces intégrales qui interviennent dans le calcul numérique des phénomènes comme on le verra plus loin. Le tableau suivant donne quelques valeurs numériques des intégrales de FRESNEL.

Etudions les variations de la fonction ξ lorsque v varie de 0 à $+\infty$ et pour cela traçons la courbe $c = \cos\dfrac{\pi v^2}{2}$ (Fig. 256) qui limite l'aire mesurée par l'intégrale ξ. Les ordonnées de la courbe C oscillent entre $+1$ et -1 et s'annulent

Tableau 13.

v	ξ	η	v	ξ	η	v	ξ	η	v	ξ	η
0	0,0000	0,0000	1,3	0,6386	0,6863	2,6	0,3889	0,5500	3,9	0,4223	0,4752
0,1	099	0005	1,4	5431	7135	2,7	3926	4529	4,0	4984	4205
0,2	1999	0042	1,5	4453	6975	2,8	4675	3915	4,1	5737	4758
0,3	2994	0141	1,6	3655	6386	2,9	5624	4102	4,2	5417	5632
0,4	3975	0334	1,7	3238	5492	3,0	6057	4963	4.3	4494	5540
0,5	4923	0647	1,8	3363	4509	3,1	5616	5818	4,4	4383	4623
0,6	5811	1105	1,9	3945	3734	3,2	4663	5933	4,5	5258	4342
0,7	6597	1721	2,0	4883	3434	3,3	4057	5193	4,6	5672	5162
0,8	7230	2493	2,1	5814	3743	3,4	4385	4297	4,7	4914	5669
0,9	7648	3398	2,2	6362	4556	3,5	5326	4153	4,8	4338	4968
1,0	7799	4383	2,3	6268	5525	2,6	5880	4923	4,9	5002	4352
1,1	7648	5365	2,4	5550	6197	3,7	5419	5750	5,0	5636	4992
1,2	7154	6234	2,5	4574	6192	3,8	4481	5656	∞	0,5000	0,5000

pour

$$\frac{\pi v^2}{2} = (2K+1)\frac{\pi}{2}, \qquad v = \sqrt{2K+1} = 1; \quad 1{,}73; \quad 2{,}24 \ldots.$$

Les maxima et minima de l'intégrale ξ correspondent à ces valeurs de v qui annulent sa dérivée. On voit que les aires des boucles de la courbe C sont alternativement positives et négatives et vont en décroissant. Les oscillations de la

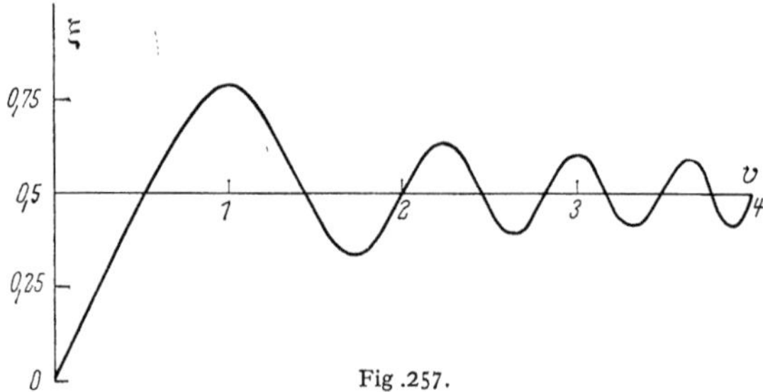

Fig. 257.

fonction ξ iront également en décroissant. La courbe représentant les variations de la fonction ξ oscille autour d'une droite d'ordonnée 0,5 dont elle se rapproche indéfiniment (Fig. 257). On représentera de même la fonction $s = \sin\dfrac{\pi v^2}{2}$ et

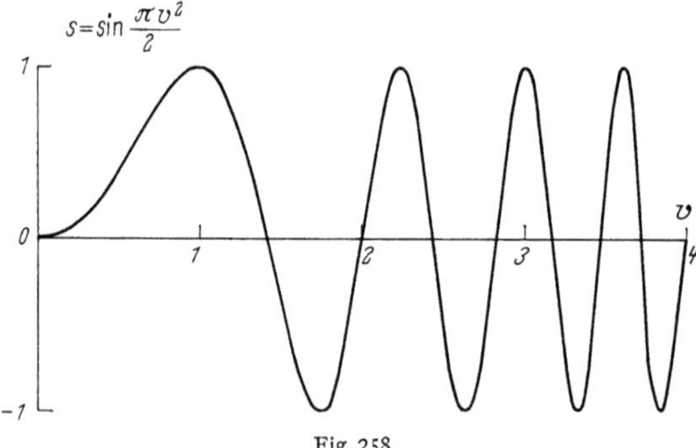

Fig. 258.

la fonction η (Fig. 258 et 259). Les maxima et minima de η correspondent à $v = \sqrt{2K}$ et la courbe $\eta = f(v)$ se rapproche asymptotiquement d'une droite d'ordonnée 0,5. On a enfin $\xi(-v) = -\xi(v)$ et $\eta(-v) = -\eta(v)$.

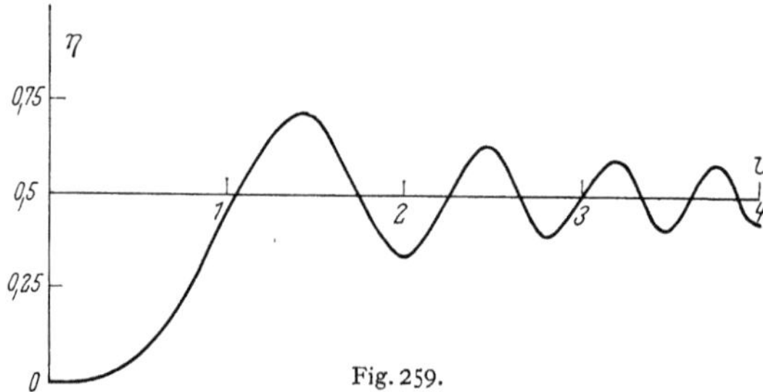

Fig. 259.

108. Spirale de Cornu. On peut représenter simultanément les intégrales de Fresnel par une courbe et déterminer la vibration résultante par une construction géométrique. Portons en abscisses les valeurs de ξ et en ordonnées les valeurs de η. A chaque valeur de v correspond une valeur de ξ et de η. La courbe ainsi obtenue porte le nom de spirale de CORNU. Considérons un élément de cette courbe: à une variation dv de v correspondent des variations $d\xi$ et $d\eta$ de ξ et de η données par

$$d\xi = \cos\frac{\pi v^2}{2}\,dv, \quad d\eta = \sin\frac{\pi v^2}{2}\,dv.$$

L'élément dl de la courbe a pour valeur

$$dl = \sqrt{d\xi^2 + d\eta^2} = \sqrt{\cos^2\frac{\pi v^2}{2} + \sin^2\frac{\pi v^2}{2}}\,dv = dv.$$

Le paramètre v représente donc la longueur de la courbe et en comptant les longueurs à partir de l'origine des coordonnées, on a $l = v$.

A des valeurs égales de Δv correspondent des longueurs équidistantes prises le long de la courbe. Le coefficient angulaire de la tangente en un point de la courbe est donné par

$$\tan\vartheta = \frac{d\eta}{d\xi} = \tan\frac{\pi v^2}{2}, \quad \vartheta = \frac{\pi v^2}{2}.$$

L'angle ϑ de la tangente à la courbe avec l'axe des abscisses a pour valeur $\dfrac{\pi l^2}{2}$ et la courbe s'incline proportionnellement au carré de la longueur de l'arc. Calculons le rayon de courbure ϱ

$$\varrho = \frac{dl}{d\vartheta} = \frac{dv}{\pi v\,dv} = \frac{1}{\pi v} = \frac{1}{\pi l}.$$

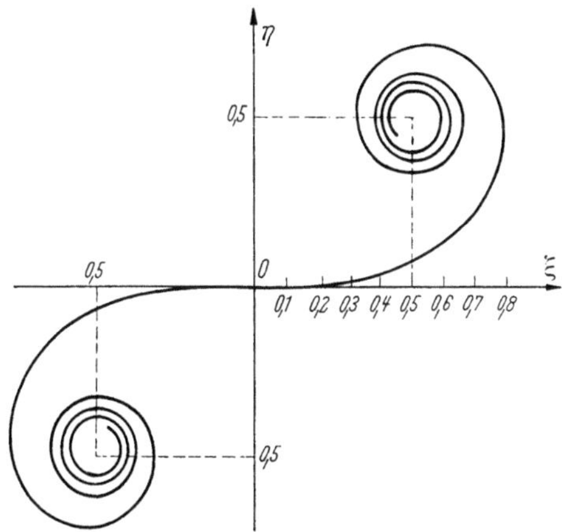

Le rayon de courbure varie en raison inverse de la longueur de l'arc. D'après ces données il est facile de voir quelle est l'allure de la courbe. Pour $v = 0$, ξ, η et ϑ sont nuls, le rayon de courbure est infini. La courbe part de l'origine. En ce point, sa tangente est horizontale et elle présente un point d'inflexion. Pour des valeurs croissantes de v elle s'enroule en spirale autour d'un point asymptotique dont les coordonnées sont $\xi = \eta = \frac{1}{2}$. La tangente à la courbe devient verticale pour $v = 1$, horizontale pour $v = \sqrt{2}$, verticale pour $v = \sqrt{3}$, horizontale pour $v = 2$ et ainsi de suite. Pour des valeurs négatives de v la courbe est symétrique par rapport à l'origine de celle que nous venons d'étudier. La Fig. 260 représente la spirale de CORNU au moyen de laquelle nous étudierons quelques phénomènes de diffraction.

b) Diffraction par les écrans circulaires.

109. Ouverture circulaire: diffraction suivant son axe. Plaçons entre la source S et le point P où l'on veut calculer l'éclairement, un écran E percé d'une ouverture circulaire de rayon s (Fig. 261). L'écran E qui limite l'onde Σ est placé de façon que l'ouverture soit normale à SP et que son centre soit sur cette même droite.

Considérons une ouverture assez petite pour que l'on puisse considérer A (105.1) comme constant. En reprenant l'intégrale (106.4), on aure avec $\Delta = r - R$

$$U_P = \frac{2\pi j A}{\lambda(R + R')} \int_0^\Delta e^{-jK\Delta}\,d\Delta = \frac{2\pi j A}{\lambda(R + R')}\,\frac{e^{-jK\Delta} - 1}{-jK}$$

que l'on peut écrire

$$U_P = \frac{2\pi j A}{\lambda(R+R')}\; \frac{e^{-j\frac{\pi\Delta}{\lambda}}\cdot e^{-j\frac{\pi\Delta}{\lambda}} - e^{-j\frac{\pi\Delta}{\lambda}}\cdot e^{j\frac{\pi\Delta}{\lambda}}}{-jK} = \frac{2A}{R+R'}\sin\frac{\pi\Delta}{\lambda}\, e^{-j\pi\left(\frac{\Delta}{\lambda}-\frac{1}{2}\right)}$$

et d'après (106.5)

$$U_P = 2U\sin\frac{\pi\Delta}{\lambda}\, e^{-j\pi\left(\frac{\Delta}{\lambda}-\frac{1}{2}\right)}. \tag{109.1}$$

Donnons à l'ouverture circulaire un rayon ϱ égal au rayon de la première zone. On a alors $\Delta = \lambda/2$, d'où l'éclairement en P

$$E_P = 4\,|U|^2.$$

L'éclairement en P est quatre fois plus grand que lorsque l'écran n'existe pas. La relation (109.1) permet de calculer l'éclairement le long de l'axe SP. Si

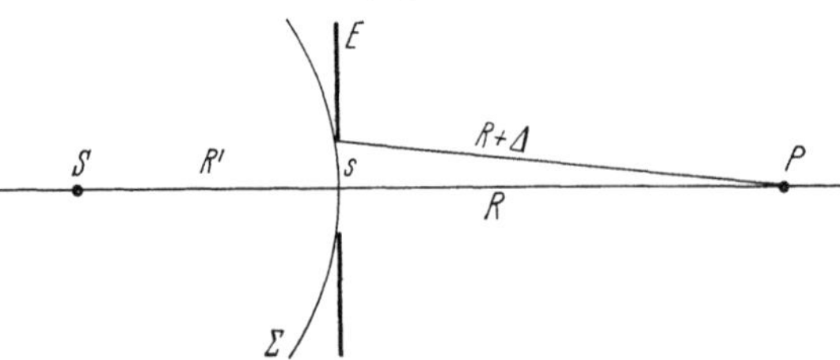

on déplace le point P sur l'axe de l'ouverture circulaire, R varie et on peut calculer dans chaque cas le retard Δ

$$\Delta = \frac{R+R'}{2RR'}\,s^2$$

d'où l'éclairement au point considéré.

Lorsque R varie, l'éclairement le long de l'axe passe alternativement par des maxima et des minima nuls de lumière. On a pour les maxima

$$\frac{R+R'}{RR'\lambda}\,s^2 = 2p+1\,, \qquad \frac{1}{R}+\frac{1}{R'} = (2p+1)\frac{\lambda}{s^2}$$

et pour les minima nuls

$$\frac{R+R'}{RR'\lambda}\,s^2 = 2p\,, \qquad \frac{1}{R}+\frac{1}{R'} = 2p\,\frac{\lambda}{s^2}.$$

Si $s = 2$ mm. avec $R' = 2$ métres et $\lambda = 0{,}5$ μ, les positions des minima seront données par

$$R = \frac{2000}{0{,}5\,p - 1}\ \text{mm}.$$

d'où

$p =$	2	3	4	5	6
$R =$	∞	4 m.	2 m.	133,4 cm.	100 cm.

110. Écran circulaire: diffraction suivant son axe. Plaçons un écran circulaire opaque E entre la source S et le point P où l'on veut calculer l'éclairement (Fig. 262).

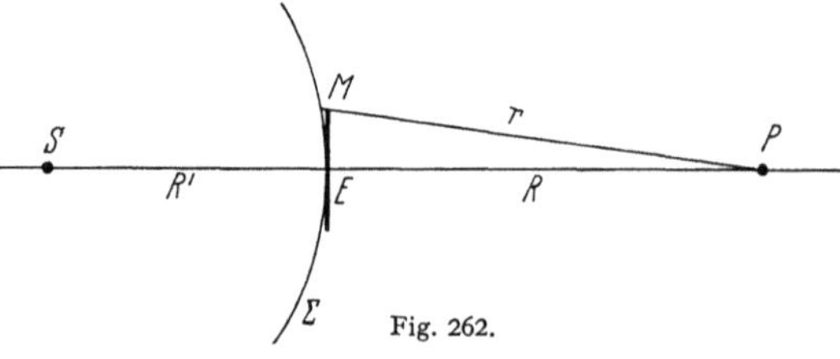

Fig. 262.

Soit M un point du bord de l'écran et $r = MP$. Décomposons la partie libre de l'onde Σ en zones limitées par les points dont les distances à P sont

$$r+\frac{\lambda}{2}, \qquad r+2\frac{\lambda}{2}, \dots$$

D'après la Sect. 106, l'amplitude totale en P est égale à la moitié de l'amplitude produite par la première des zones précédentes comprise entre r et $r+\dfrac{\lambda}{2}$. Si l'écran E est assez petit, cette zone est voisine de la zone centrale de l'onde Σ et produit le même éclairement. Or l'éclairement produit par l'onde Σ entièrement découverte, est égal à l'éclairement produit par la moitié de la zone centrale. Donc, lorsque l'écran E est petit, l'éclairement en P est le même que s'il n'y avait pas d'écran.

On peut remarquer que si l'éclairement en P avec l'écran circulaire E est $|U|^2$ avec un trou circulaire de même diamètre, il est $4\,|U|^2\sin^2\dfrac{\pi\Delta}{\lambda}$ et par conséquent l'éclairement $|U|^2$ produit par l'onde entière n'est pas égal à la somme des éclairements produits par les deux parties qui la constituent. C'est d'ailleurs un fait général que l'on rencontre dans tous les phénomènes d'interférences. Comme le point P est en fait le centre d'une figure de diffraction qui s'étale dans un plan perpendiculaire à l'axe, le principe de la conservation de l'énergie est respecté dans l'ensemble de la tache de diffraction et non au point particulier P. Les interférences modifient la structure de la tache, c'est à dire la répartition de la lumière diffractée. Le flux total reçu par le plan d'observation lorsqu'on utilise l'onde entière, est égal à la somme des flux reçus dans le cas des deux écrans complémentaires.

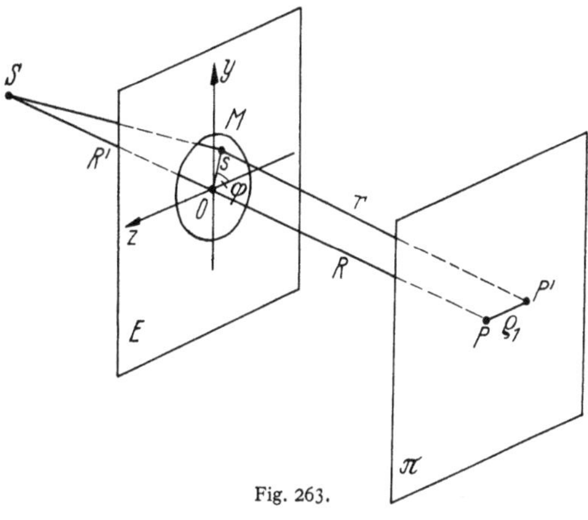

Fig. 263.

111. Ouverture circulaire: diffraction en dehors de l'axe. Soit S la source lumineuse et E l'écran percé d'une ouverture circulaire (Fig. 263). Joignons S au point O, centre de l'ouverture, SO étant supposé normal à E, rencontre le plan d'observation en P. On veut calculer l'éclairement en un point P' voisin du point P. On considère une ouverture circulaire assez petite pour que le coefficient A (105.1) puisse être considéré comme constant. Avec une approximation suffisante, on peut écrire

$$U_{P'} = \frac{jA}{\lambda RR'}\iint e^{-jK\Delta}\,s\,ds\,d\varphi\,,$$

φ représentant l'azimut du point M. La différence de marche Δ n'est plus la même que pour le point P. Si le point P' est à une distance ϱ_1 du point P, à l'expression (107.1) doit être ajouté le retard $-\dfrac{s\varrho_1}{R}\cos\varphi$ et l'amplitude en P' prend la forme

$$U_{P'} = \frac{jA}{\lambda RR'}\int_0^s\int_0^{2\pi} e^{-jK\left[\left(\frac{1}{R}+\frac{1}{R'}\right)\frac{s^2}{2}-\frac{s\varrho_1}{R}\cos\varphi\right]}\,s\,ds\,d\varphi$$

d'où

$$U_{P'} = \frac{2\pi jA}{\lambda RR'}\int_0^s e^{-jK\left(\frac{1}{R}+\frac{1}{R'}\right)\frac{s^2}{2}}\,J_0\left(Ks\,\frac{\varrho_1}{R}\right)s\,ds\,.$$

Posons

$$K\left(\frac{1}{R}+\frac{1}{R'}\right)=p\,,\qquad K\,\frac{s}{R}=q\,,$$

il vient

$$U_{P'} = \frac{2\pi j A}{\lambda R R'} \int_0^s e^{-j\frac{p}{2}s^2} J_0(q\,\varrho_1)\, s\, ds,$$

intégrale qui a été calculée par Lommel.

En posant $w = qs$ et $W = ps^2$, Lommel intègre par partie et trouve des séries qui se calculent à l'aide des fonctions de Bessel.

On obtient

$$\mathcal{A} = \int_0^s e^{-j\frac{p\,s^2}{2}} J_0(q\,\varrho_1)\, s\, ds = \frac{s^2}{j\,W}\left\{ e^{j\frac{w^2}{2W}} - e^{-j\frac{W}{2}}\left[V_0(w,W) + j V_1(w,W)\right]\right\}$$

pour les points P' situés à l'intérieur de l'ombre géométrique. Pour les points P' situés à l'extérieur, l'intégrale précédente devient égale à

$$\frac{s^2}{j\,W}\, e^{-j\frac{W}{2}}\left[-U_2(w,W) + j U_1(w,W)\right].$$

Les fonctions $U_1 U_2 V_0 V_1$ sont définies par le séries

$$U_n = \sum_{p=0}^{\infty} (-1)^p \left(\frac{W}{w}\right)^{n+2p} J_{n+2p}(w),$$

$$V_n = \sum_{p=0}^{\infty} (-1)^p \left(\frac{W}{w}\right)^{n+2p} J_{n+2p}(w).$$

Indiquons que Zernike et Nijboer ont donné une formule qui permet de calculer l'expression conjuguée de l'intégrale $\mathcal{A}$

$$\mathcal{A}^* = e^{j\frac{W}{4}} (2\pi)^{\frac{1}{2}} W^{-\frac{1}{2}} \sum_{n=0}^{\infty} (-j)^n (2n+1) J_{n+\frac{1}{2}}\left(\frac{W}{4}\right) \frac{J_{2n+1}(w)}{w}.$$

C'est ce développement qu'ils ont utilisé pour calculer les courbes isophotes de la Fig. 151. Lommel a calculé les phénomènes pour les valeurs du paramètre $W = \pi,\ 2\pi,\ \ldots\ 7\pi$ et pour quelques valeurs de $q\varrho_1$. Le tableau suivant donne les valeurs de l'éclairement E (unités relatives).

Tableau 14.

$q\varrho_1$	$W = \pi$	$W = 2\pi$	$W = 3\pi$	$W = 4\pi$	$W = 5\pi$	$W = 6\pi$	$W = 7\pi$
0	8105	4053	901	0	324	450	165
1	6286	3158	720	14	256	351	132
2	2772	1558	544	152	156	169	92
3	623	780	703	398	130	64	105
4	269	719	816	487	131	45	118
5	306	539	578	379	154	72	91
6	121	213	305	323	242	124	56
7	18	100	246	350	301	149	43
8	51	114	217	287	250	141	58
9	37	62	112	165	187	159	100
10	4	17	52	121	185	192	128
11	13	26	59	115	166	170	119
12	16	23	41	70	104	123	113

Les positions et les valeurs des maxima et minima correspondent aux valeurs donnés au tableau suivant.

Tableau 15.

	$W = \pi$		$W = 2\pi$		$W = 3\pi$		$W = 5\pi$	
	$q\,\varrho_1$	E	$q\,\varrho_1$	E	$q\,\varrho_1$	E	$q\,\varrho^1$	E
Minimum . . .	3,831	263	3,598	720	1,997	544	3,031	130
Maximum . . .	4,715	320	3,832	721	3,831	822	3,626	131
Minimum . . .	7,016	18	7,016	99	7,016	246	3,832	131
Maximum . . .	8,306	55	7,888	114	7,088	246	7,016	301
Minimum . . .	10,173	3	10,173	16	10,173	51	9,441	181
Maximum . . .	11,578	19	11,414	9	11,036	60	10,173	185

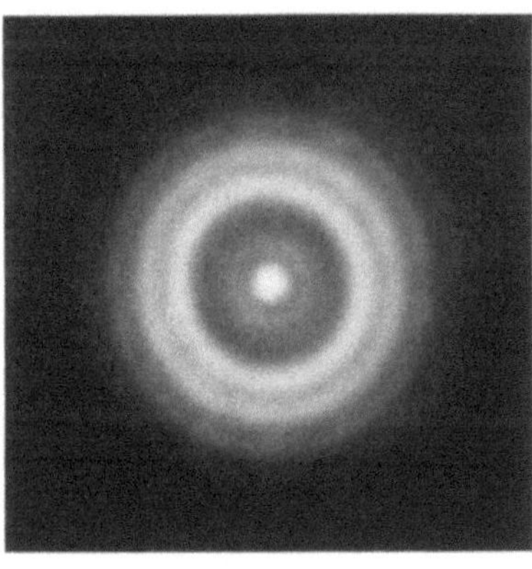

La structure des phénomènes de diffraction est d'une grande complication comme le montrent les Figs. 264a et b.

a   b

Fig. 264a et b. Figures de diffraction de LOMMEL. a Diamètre de l'ouverture circulaire 11 mm. Distance à la source 13 mètres. Distance à la plaque photographique 8 mètres. b Diamètre de l'ouverture circulaire 5 mm. Distance à la source 7 mètres. Distance à la plaque photographique 14 mètres.

112. Réseaux zonés (SORET). Reprenons les Fig. 254 et 255 relatives aux zones de FRESNEL. Toutes les zones impaires 1, 3, 5 ... envoient en P des vibrations qui sont en phase. Dem ême, les zones paires, mais elles produisent en P des vibrations en opposition de phase avec les vibrations envoyées par les zones impaires.

Recouvrons l'onde Σ par un écran ne laissant passer que les zones impaires ou seulement les zones paires: on augmente considérablement l'éclairement en P puisque tous les mouvements vibratoires qui parviennent en ce point sont en phase. Un écran masquant, soit les zones paires, soit les zones impaires, constitue un réseau zoné.

Pour réaliser de tels réseaux, on peut tracer sur une feuille de papier blanc une série de circonférences dont les rayons varient comme les racines carrées des nombres entiers successifs. On noircit de deux en deux les intervalles compris entre ces cercles, puis on réduit par photographie la figure obtenue. Suivant que le centre est blanc ou noir, on constitue un réseau positif ou négatif (Fig. 265a et b).

Plaçons sur l'axe d'un réseau zoné Z positif (par exemple) une source lumineuse ponctuelle S émettant de la lumière monochromatique de longueur d'onde λ (Fig. 266). Considérons un point P_1 de l'axe et supposons que le réseau Z soit

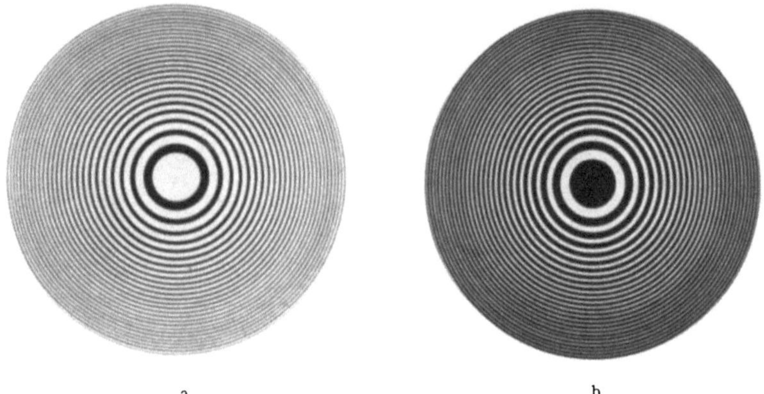

Fig. 265a et b. a Réseau zoné positif. b Réseau zoné négatif.

à des distances R' et R de S et de P, telles que le rayon du premier cercle du réseau Z soit égal au rayon de la première zone (Sect. 106)

$$\varrho_1^2 = \frac{R R' \lambda}{R + R'}.$$

Dans ces conditions, toutes les vibrations envoyées par le réseau arrivent en phase en P_1 et il y a accumulation de lumière en ce point. Si on se déplace sur l'axe en s'écartant du point P_1, il n'en est plus ainsi. En un point quelconque, les vibrations émises par le réseau présentent toutes les phases possibles et l'amplitude résultante est pratiquement nulle. Par contre, chaque fois que l'on se trouvera à une distance du réseau telle que le rayon du premier cercle soit égal au rayon ϱ_{2n+1} d'une zone impaire

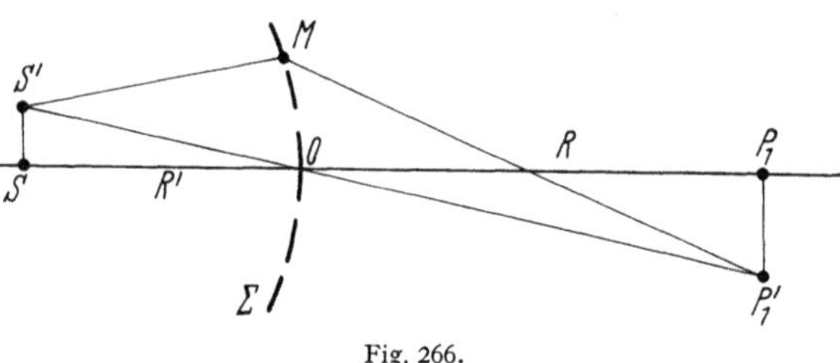

Fig. 266.

quelconque, il y aura à nouveau accumulation de lumière. On a donc sur l'axe une série de points $P_1 P_3 P_5 \ldots$ pour lesquels il y a accumulation de lumière; ce sont de véritables images de la source ponctuelle S et le réseau zoné Z se comporte comme une lentille à foyers multiples.

Le rayon d'une zone impaire est donné par

$$\varrho_{2n+1}^2 = \frac{R R' \lambda}{R + R'} (2n + 1).$$

Si ϱ est le rayon du premier cercle du réseau zoné, on aura une image si

$$\varrho^2 = \frac{R R' \lambda}{R + R'} (2n + 1)$$

qeu l'on peut écrire:

$$\frac{1}{R} + \frac{1}{R'} = \frac{(2n+1)\,\lambda}{\varrho^2}\,.$$

Le réseau zoné se comporte comme une lentille de longueur focale

$$f = \frac{1}{(2n+1)}\cdot\frac{\varrho^2}{\lambda}\,.$$

La répartition des éclairements le long de l'axe du réseau zoné a été calculé par A. Boivin. Si N est le nombre des zones découvertes, on a, à un facteur constant près

$$E = \left(\frac{\varrho^2}{W}\right)^2 \left(\frac{\sin\dfrac{NW}{2}}{\cos\dfrac{W}{4}}\right)^2 .$$

La courbe de la Fig. 267 représente les variations du facteur entre-crochets. On voit que deux maxima principaux consécutifs sont séparés par une série de

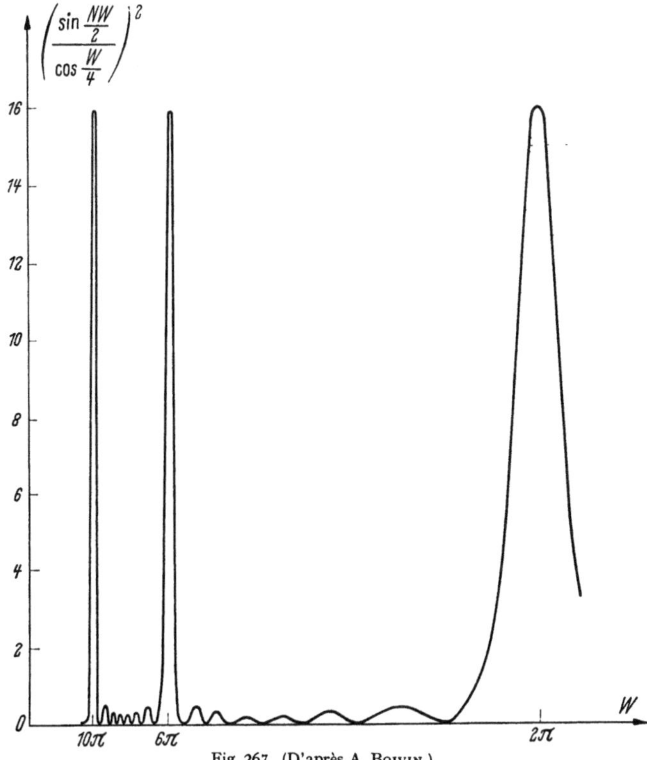

Fig. 267. (D'après A. Boivin.)

maxima secondaires faibles. L'expression précédente montre que par suite de facteur de proportionnalité, les éclairements des maxima principaux vont rapidement en diminuant. Ceci tient à ce que si toutes les zones concourent à la formation de la première image, la seconde n'est produite que par une zone impaire sur trois, la troisième une zone impaire sur cinq, etc.

Considérons un point lumineux S' voisin de S (Fig. 266).

Le réseau en donne une image P_1' voisine de P_1. En effet, les variations du chemin optique $S'MP_1'$ lorsqu'on change un peu l'inclinaison du réseau, sont très faibles par rapport à la différence des chemins optiques $S'M + MP_1'$ et $S'P_1'$. Par conséquent, le réseau zoné peut donner d'un objet SS' une petite image P_1P_1' comme une lentille.

Notons qu'en lumière blanche, le réseau zoné se comporte comme une lentille d'aberration chromatique considérable, puisque la distance focale varie en raison inverse de la longueur d'onde.

c) Diffraction par les écrans rectangulaires.

113. Diffraction par le bord d'un écran rectiligne indéfini. Soit S la source ponctuelle, E un écran opaque et π le plan d'observation (Fig. 268). On veut calculer l'éclairement au voisinage du bord N de l'ombre géométrique de l'écran E. Prenons un point P défini par sa distance l au point N. Soit s la longueur de l'arc OM, en assimilant SOM à un triangle, on a

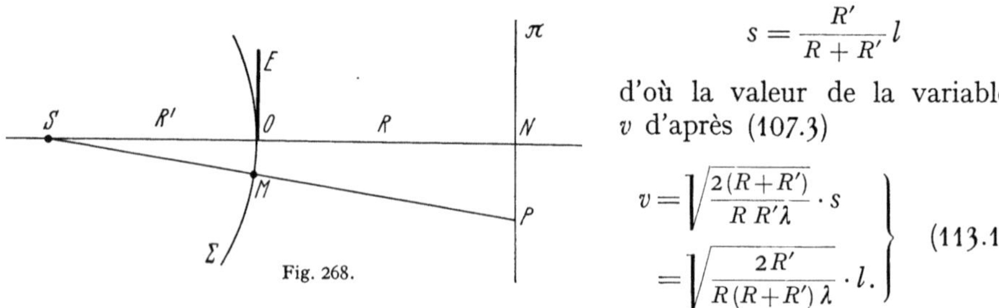

Fig. 268.

$$s = \frac{R'}{R + R'} l$$

d'où la valeur de la variable v d'après (107.3)

$$v = \sqrt{\frac{2(R+R')}{R R' \lambda} \cdot s} \left. \atop = \sqrt{\frac{2R'}{R(R+R')\lambda} \cdot l.} \right\} \quad (113.1)$$

En dehors de l'ombre géométrique, l'amplitude U_P en P sera donnée par l'expression (Sect. 107)

$$U_P = \int\limits_{-z}^{+\infty} \int\limits_{-\infty}^{+\infty} e^{-j\frac{\pi m^2}{2}(y^2+z^2)} \, dy \, dz$$

en prenant l'axe des y parallèle au bord de l'écran E, l'axe des z dirigé vers le bas de la Fig. 268 et l'origine en M, pôle de l'onde Σ. Cette écriture est justifiée par le fait que les intégrales de Fresnel atteignent rapidement leurs valeurs limites ξ_∞ et η_∞ dès que v prend des valeurs notables et que par conséquent les parties de l'onde Σ un peu éloignées de O ne donnent que des termes négligables dans le calcul.

A un facteur constant près, l'amplitude devient

$$U_P = \int\limits_{-v}^{+\infty} e^{-j\frac{\pi v^2}{2}} \, dv$$

et l'éclairement

$$E_P = \left[\int\limits_{-v}^{+\infty} \cos\frac{\pi v^2}{2} \, dv \right]^2 + \left[\int\limits_{-v}^{+\infty} \sin\frac{\pi v^2}{2} \, dv \right]^2 = \left[\frac{1}{2} + \xi(v) \right]^2 + \left[\frac{1}{2} + \eta(v) \right]^2, \quad (113.2)$$

Employons la spirale de Cornu. Considérons les deux points P_1 et P_2 de coordonnées

$$P_1: \qquad \xi_1 = \int\limits_0^{+\infty} \cos\frac{\pi v^2}{2} \, dv, \qquad \eta_1 = \int\limits_0^{+\infty} \sin\frac{\pi v^2}{2} \, dv,$$

$$P_2: \qquad \xi_2 = \int\limits_0^{-v} \cos\frac{\pi v^2}{2} \, dv, \qquad \eta_2 = \int\limits_0^{-v} \sin\frac{\pi v^2}{2} \, dv.$$

Le point P_1 est confondu avec le point asymptotique J (Fig. 269). Le point P_2 se trouve sur la portion de la spirale qui se trouve en dessous de l'axe des ξ. L'expression (113.2) montre que l'éclairement en P est représenté par $\overline{P_2 J}^2$.

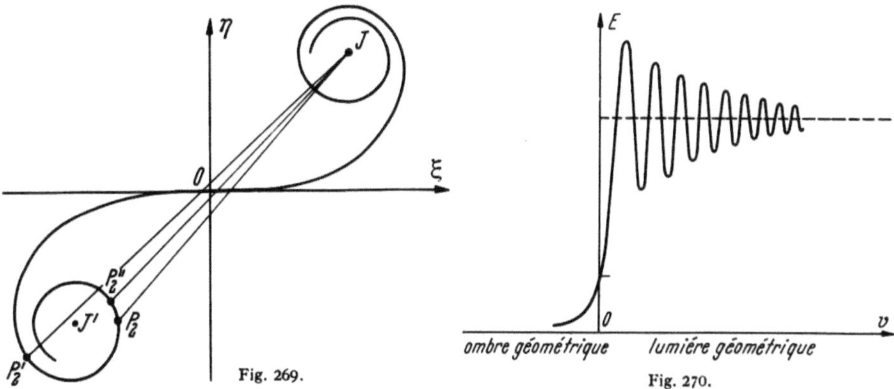

Fig. 269.

Fig. 270.

A chaque point P de l'écran π correspond une valeur différente de ν et par conséquent du point P_2 sur la spirale. En N, l'éclairement est représenté par $\overline{OJ}^2$. Ecartons-nous du point N vers la région éclairée. Le point P_2 se déplace sur la spirale et on arrive en un point P_2' pour lequel $P_2 J$ est maximum. On arrive

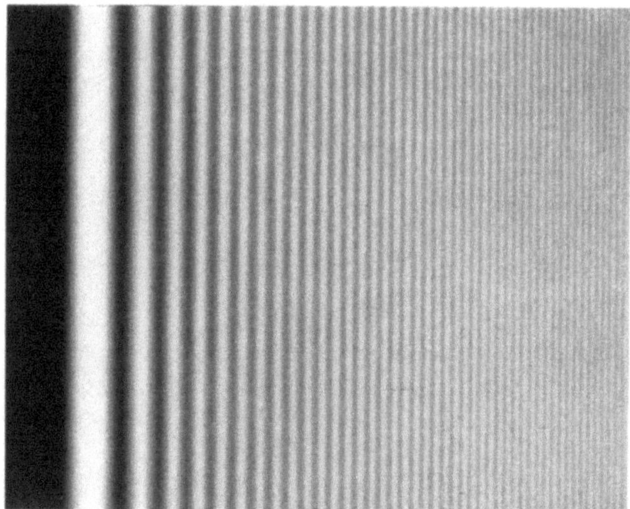

Fig. 271. Franges de diffraction dans l'ombre d'une écran.

ensuite en P_2'' pour lequell $P_2 J$ est minimum. Le point P_2 continuant à se déplacer sur la spirale de gauche, on aura ainsi une série de maxima et de minima successifs de plus en plus voisins comme intensité et comme distance. Le point P_2 tend vers J', l'éclairement devient constant et a pour limite $\overline{JJ'}^2$. Si on se place dans l'ombre géométrique, le point représentatif vient en P_2 sur la spirale de droite. En s'éloignant du point N, le segment JP_2 diminue régulièrement: il n'y a pas de franges dans l'ombre géométrique. L'allure générale du phénomène est représenté par la courbe de la Fig. 270. La Fig. 271 montre l'aspect des franges.

Ces phénomènes ne peuvent s'observer que si la source est bien ponctuelle. Avec une source étendue chaque point donne une figure de diffraction décalée par rapport aux figures de diffraction dûes aux autres points et l'ensemble est noyé dans une pénombre.

Reprenons la relation (113.1)

$$l = v \sqrt{\frac{R(R + R')\,\lambda}{2R'}}$$

et posons

$$\sqrt{\frac{R(R+R')\,\lambda}{2R'}} = h,$$

on a

$$l = v\,h.$$

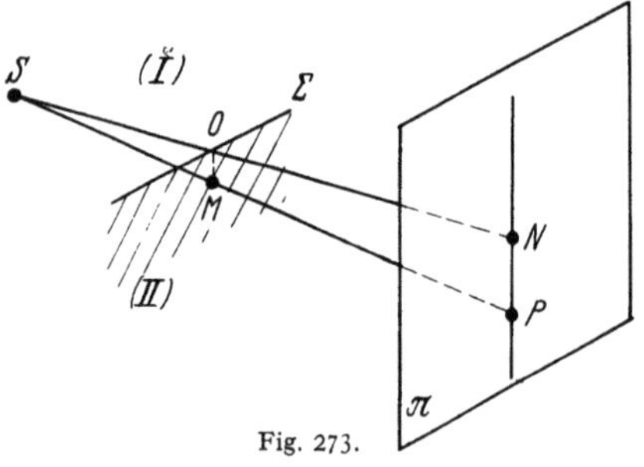

Fig. 272.

Table des maxima et minima.

	v	E
	0	0,5
Maximum 1	1,2172	2,7407
Minimum 1	1,8725	1,5562
Maximum 2	2,3445	2,3985
Minimum 2	2,7390	1,6864
Maximum 3	3,0820	2,2913
Minimum 3	3,3913	1,7437
Maximum 4	3,6741	2,2521
Minimum 4	3,9371	1,7780
Maximum 5	4,1832	2,2207
Minimum 5	4,4159	1,8012
Maximum 6	4,6367	2.1987
Minimum 6	4,8473	1,8185

Tous les phénomènes intéressants correspondent à des valeurs de v très faibles. Si h est petit l est aussi petit. Pour une source à l'infini, on a $h = \sqrt{\dfrac{R\lambda}{2}}$ et avec $D = 1$ m., $\lambda = 0,5\,\mu$ on trouve $h = 0,5$ mm. C'est dans une zone de quelques millimètres de largeur que tous les phénomènes de diffraction seront visibles.

Si ϑ est le diamètre apparent de la source (Fig. 272) on a une zone de pénombre $h' = R\vartheta$. La condition de visibilité des phénomènes est que h' soit petit devant h

$$R\vartheta \ll \sqrt{\frac{R\lambda}{2}}$$

ou

$$\vartheta \ll \sqrt{\frac{\lambda}{2R}}.$$

Avec $R = 1$ m., $\lambda = 0,5\,\mu$ on a $\sqrt{\dfrac{\lambda}{2R}} = 0,5 \cdot 10^{-3}$ radian, soit environ $1',5$. Le soleil ayant un diamètre apparent de $30'$, on ne peut donc pas observer ces phénomènes sur le bord de l'ombre des objets éclairés par le soleil, sauf pour de grandes valeurs de R.

114. Diffraction par deux demi-plans en opposition (Kastler). Supprimons l'écran E et remplaçons-le par une lame transparente à bord rectiligne d'épaisseur optique égale à $\lambda/2$. Pratiquement le dispositif pourra être réalisé en plaçant côte à côte deux lames de verre prismatiques d'angles très faibles. En faisant glisser l'une des lames par rapport à l'autre, on pourra trouver une région assez étendue dans laquelle la différence des chemins optiques sera égale à $\lambda/2$. On peut aussi utiliser un dispositif à polarisation.

Fig. 273.

Le plan de l'onde Σ (Fig. 273) est donc divisé en 2 régions I et II et soit P_1 un point figuratif sur la spirale qui correspond au point M (Fig. 274). Le point

M est dans l'ombre géométrique par rapport à la portion II de l'onde Σ. Supposons d'abord que les vibrations des 2 portions I et II soient en concordance de phase: les vibrations de la portion I produiront en P la resultante $J'P_1$ et les vibrations de la portion II la résultante JP_1. L'onde entière produira en P la vibration résultante constante $\overrightarrow{J'P_1} + \overrightarrow{P_1J} = \overrightarrow{J'J}$ indépendante de P et P_1. Supposons maintenant que les vibrations qui ont traversé la portion II soient en opposition avec celles qui ont traversé la portion I. La résultante observée en P sera $\overrightarrow{J'P_1} + \overrightarrow{JP_1}$. Soit P_1' le point symétrique de P_1 par rapport à l'origine O, on a $\overrightarrow{JP_1} = \overrightarrow{P_1'J'}$ et la résultante en P pourra s'écrire

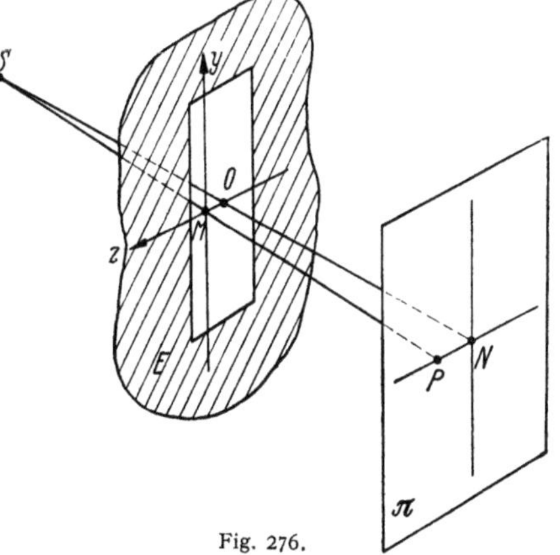

Fig. 274.

$$\overrightarrow{P_1'J'} + \overrightarrow{J'P_1} = \overrightarrow{P_1'P_1} = 2\overrightarrow{OP_1}.$$

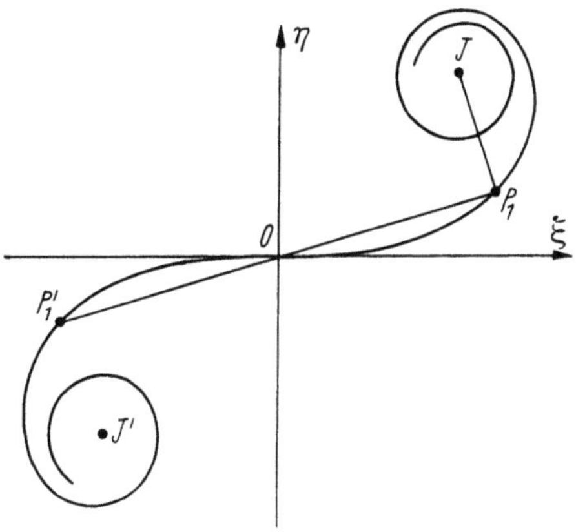

L'amplitude de la vibration lumineuse en P est proportionnelle à la longueur du vecteur $\overrightarrow{OP_1}$ qui relie l'origine O au point figuratif P_1 de la spirale. On obtient donc un système de franges symétriques par rapport à la projection sur le plan π de la ligne de séparation des 2 portions I et II. La frange centrale très large est complètement noire en son milieu. L'allure du phénomène est représentée par la Fig. 275.

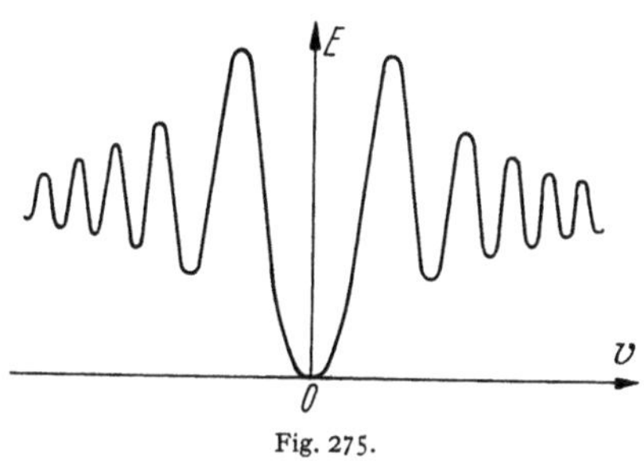

Fig. 275.

Fig. 276.

115. Diffraction par une ouverture rectangulaire. Considérons une ouverture rectangulaire dont les bords sont parallèles aux axes $Mx\,My$. L'amplitude en P (Fig. 276) est donnée par

$$U_P = \int\limits_{y_1}^{y_2}\int\limits_{z_1}^{z_2} e^{-j\frac{\pi m^2}{2}(y^2+z^2)}\,dy\,dz = \int\limits_{y_1}^{y_2} e^{-j\frac{\pi m^2 y^2}{2}}\,dy \int\limits_{z_1}^{z_2} e^{-j\frac{\pi m^2 z^2}{2}}\,dz = f(y_1,y_2)\,f(z_1,z_2).$$

L'éclairement en P est égal au produit de 2 facteurs dont l'un ne dépend que de la position des côtés de l'ouverture parallèles à Mz et l'autre, que de la position des côtés parallèles à My. Si l'ouverture est une fente étroite et parallèle à My, les limites y_1 et y_2 sont grandes en valeur absolue et de signes contraires, le facteur $f(y_1 y_2)$ a une valeur constante indépendante de y_1 et y_2. L'éclairement en P devient alors dans le cas d'une fente étroite

$$U_P = f(z_1\,z_2) = \int\limits_{z_1}^{z_2} e^{-j\frac{\pi m^2 z^2}{2}}\,dz.$$

D'après la Sect. 107, on aura l'éclairement à un facteur constant près

$$E_P = [\xi(v_2) - \xi(v_1)]^2 + [\eta(v_2) - \eta(v_1)]^2,$$

v_1 et v_2 étant les valeurs du paramètre v correspondant à z_1 et z_2. Etudions les phénomènes au moyen de la courbe de Cornu. On a (Fig. 277)

$$v_1 = \sqrt{\frac{2(R+R')}{R R' \lambda}}\,s_1, \qquad v_2 = \sqrt{\frac{2(R+R')}{R R' \lambda}}\,s_2,$$

e étant la largeur de la fente

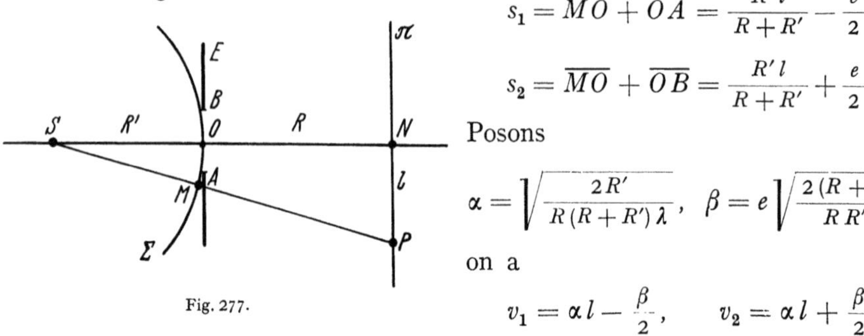

Fig. 277.

$$s_1 = \overline{MO} + \overline{OA} = \frac{R'l}{R+R'} - \frac{e}{2},$$

$$s_2 = \overline{MO} + \overline{OB} = \frac{R'l}{R+R'} + \frac{e}{2}.$$

Posons

$$\alpha = \sqrt{\frac{2R'}{R(R+R')\lambda}}, \qquad \beta = e\sqrt{\frac{2(R+R')}{R R' \lambda}},$$

on a

$$v_1 = \alpha l - \frac{\beta}{2}, \qquad v_2 = \alpha l + \frac{\beta}{2}.$$

La différence $v_2 - v_1 = \beta$ est constante. Pour obtenir sur la spirale de Cornu l'éclairement en P, on prend sur la courbe un point P_1 tel que la longueur de l'arc OP_1 soit égale à αl et on porte de part et d'autre deux arcs $P_1 P_2$, $P_1 P_3$ égaux à $\beta/2$ (Fig. 278). L'éclairement est représenté par le carré de la droite

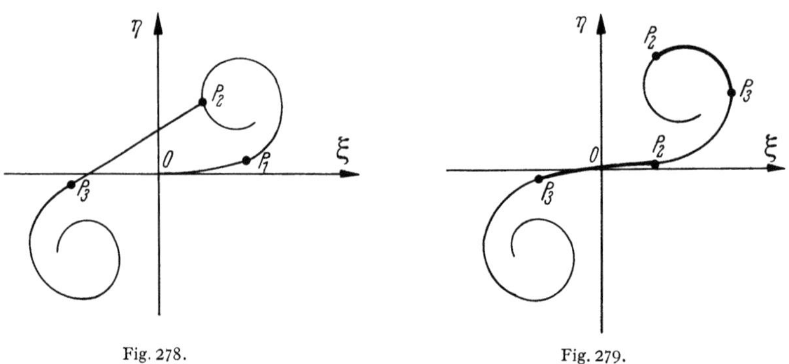

Fig. 278.　　　　　　　　　　　　Fig. 279.

$P_2 P_3$. Pour étudier les variations de l'éclairement en fonction de la position du point P, il suffit d'étudier les variations de la longueur de la corde $P_2 P_3$ qui sous-tend un arc constant égal à β.

Partons du point N (Fig. 277) pour lequel P_1 se trouve en O (Fig. 279). Si la fente est très étroite, l'arc $P_2 P_3$ de longueur constante en glissant le long de la courbe sous-tend une corde $P_2 P_3$ dont la longueur ne varie pratiquement pas. On a autour de N une zone d'éclairement uniforme d'autant plus large que la fente est plus étroite. Si on continue à faire glisser l'arc $P_2 P_3$, la courbure augmente et la longueur de la corde $P_2 P_3$ diminue: l'éclairement décroît régulièrement. Lorsque l'arc $P_2 P_3$ s'engage sur une spire, la corde $P_2 P_3$ passe par des maxima et des minima: on obtient dans l'ombre géométrique une série de franges obscures et de franges brillantes, parallèles à la fente, les franges obscures étant presque complètement noires. Notons que l'on peut remplacer la source ponctuelle par une fente fine parallèle à la fente diffringente. Les franges obscures correspondent à une position de $P_2 P_3$ telle que la longueur de la corde soit égale à celle d'une spire (Fig. 280). La $K^{\text{ième}}$ frange obscure correspond à une valeur de v telle que l'arc $P_2 P_3$ ait une longueur égale à K spires. Si on assimile ces spires à des circonférences de rayon égal au rayon de courbure $\varrho = \dfrac{1}{\pi v}$ de la spirale au point P_1 milieu de l'arc $P_2 P_3$, la longueur de K spires sera $2 K \pi \varrho = \dfrac{2K}{v}$. La longueur de l'arc $P_2 P_3$ étant égale à β, les franges obscures seront données par

$$\frac{2K}{v} = \beta = \frac{2K}{\alpha l}$$

d'où

$$l = \frac{2K}{\alpha \beta} = K \frac{\lambda R}{e}.$$

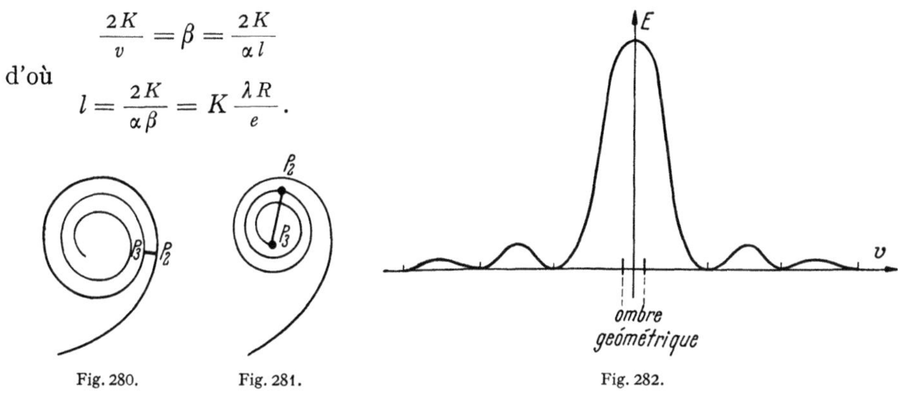

Fig. 280.   Fig. 281.   Fig. 282.

Les franges obscures sont donc équidistantes. Les franges brillantes correspondent aux valeurs de v pour lesquelles l'arc $P_2 P_3$ est égal à $2K+1$ demi-spires ou $2K+1$ demi-circonférences de rayon ϱ (Fig. 281). On a alors

$$\frac{2K+1}{v} = \beta = \frac{2K+1}{\alpha l}$$

d'où

$$l = \frac{2K+1}{\alpha \beta} = \left(K + \frac{1}{2} \right) \frac{\lambda R}{e}.$$

La longueur correspondante $P_2 P_3$ est égale à 2ϱ et l'éclairement des franges brillantes est donné par

$$E = 4 \varrho^2 = \frac{4}{\pi^2 v^2} = \frac{4}{\pi^2 \alpha^2 l^2} = \frac{4 \beta^2}{\pi^2 (2K+1)^2}.$$

L'éclairement des franges brillantes varie en raison inverse du carré de leur distance au centre N. L'allure du phénomène est représentée par la Fig. 282.

116. Diffraction par un fil. Plaçons un écran opaque étroit E (fil) entre la source S et le plan π d'observation (Fig. 283). Soit v_0 la valeur du paramètre v qui correspond au bord de l'écran. En reprenant l'équation (113.2) on a ici

$$E_P = \left[\int_{-\infty}^{-(v_0+v)} \cos \frac{\pi v^2}{2} \, dv + \int_{v_0-v}^{+\infty} \cos \frac{\pi v^2}{2} \, dv \right]^2 + \left[\int_{-\infty}^{-(v_0+v)} \sin \frac{\pi v^2}{2} \, dv + \int_{v_0-v}^{+\infty} \sin \frac{\pi v^2}{2} \, dv \right]^2,$$

$$E_P = [1 - \xi(v_0 - v) - \xi(v_0 + v)]^2 + [1 - \eta(v_0 - v) - \eta(v_0 + v)]^2.$$

Employons la spirale de Cornu. L'arc de la spirale correspondant à la partie de l'onde Σ arrêtée par le fil est un arc $P_1 P_2$ de longueur constante (Fig. 284). La partie découverte de l'onde Σ située en dessous du fil produit la vibration $\overrightarrow{J'P_1}$ et la partie supérieure, la vibration $\overrightarrow{P_2 J}$. La vibration résultante est

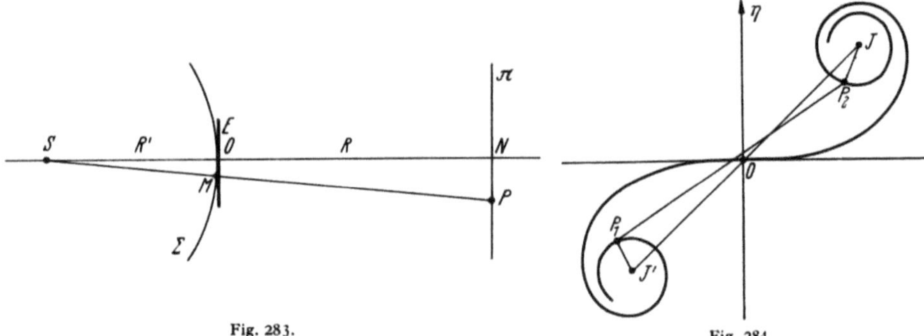

Fig. 283.   Fig. 284.

$\overrightarrow{J'P_1} + \overrightarrow{JP_2}$. On a $\overrightarrow{J'P_1} + \overrightarrow{P_1 P_2} + \overrightarrow{P_2 J} = \overrightarrow{J'J}$ donc la vibration résultante est aussi représentée par $\overrightarrow{J'P_1} + \overrightarrow{P_2 J} = \overrightarrow{J'J} - \overrightarrow{P_1 P_2}$. Ici également l'éclairement en P produit par l'onde entière n'est par égal à la somme des éclairements produits par le fil et la fente de même largeur. Le principe de la conservation de l'énergie est respecté dans l'ensemble du phénomène, mais les énergies sont réparties de façons

Fig. 285. Franges de diffraction dans l'ombre d'une fil.

différentes. L'aspect de la figure de diffraction est généralement très compliqué, comme le montre la Fig. 285.

Etudions les franges situées dans l'ombre géométrique d'un fil un peu large de façon que les points P_1 et P_2 correspondant au point N (centre de la figure de diffraction) soient sur les spires. Représentons ces spires par des cercles de rayons $\varrho = \dfrac{1}{\pi v} = \dfrac{2}{\pi \beta}$, les arcs $O P_1$ et $O P_2$ étant égaux à $\dfrac{\beta}{2}$ (Fig. 286). P_1 et P_2 sont symétriques par rapport à O, $\overrightarrow{J'P_1}$ et $\overrightarrow{P_2 J}$ sont parallèles de même sens et de longueurs égales à ϱ. Ecartons-nous du point N, c'est-à-dire déplaçons P_1 et P_2 sur la spirale: P_1 et P_2 se déplacent de la même longueur dans le même sens. Les vecteurs $\overrightarrow{J'P_1}$ et $\overrightarrow{P_2J}$ varient peu et tournent d'angles égaux en sens inverses.

Lorsque ces deux vecteurs on tourné de $(2K+1)\dfrac{\pi}{2}$ ils sont opposés et l'éclairement est nul. Lorsqu'ils ont tourné de $K\pi$ ils sont parallèles et l'éclairement est maximum. En N au centre du phénomène, il y a donc un maximum et on observe dans l'ombre géométrique une série de franges brillantes et sombres approximativement équidistantes. Lorsque, à partir de sa position initiale correspondant à N, l'arc P_1P_2 s'est allongé de $K\pi\varrho$ on a la $K^{\text{ième}}$ frange brillante. D'après la Sect. 115, on a

$$v = K\,\pi\,\varrho = \frac{2K}{\beta}$$

d'où

$$l = \frac{v}{\alpha} = \frac{2K}{\alpha\beta} = K\,\frac{\lambda R}{e}\,.$$

Si on avait placé 2 sources synchrones sur les bords de l'écran E, on aurait eu un système de franges d'équidistance $\dfrac{R\lambda}{e}$ à frange centrale brillante. Les franges dans l'ombre du fil ont donc la même position que si elles étaient dues

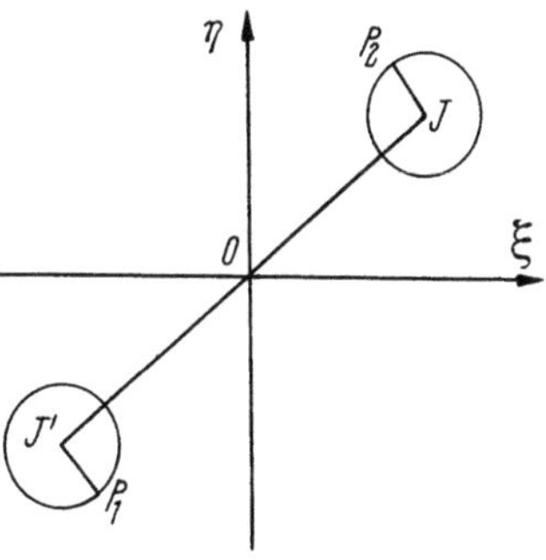

Fig. 286.

à l'interférence des vibrations provenant de deux sources synchrones placées sur les bords de l'écran constituant le fil. Ce résultat a été vérifié par FRESNEL au moyen de l'écran représenté par la Fig. 287. Dans la partie haute du champ d'observation, on voit les franges dans l'ombre de l'écran E, dans la partie basse les franges dues à deux fentes fines F et F'. On constate effectivement que ces deux systèmes de franges sont dans le prolongement l'un de l'autre dans la partie correspondant à l'ombre géométrique de E.

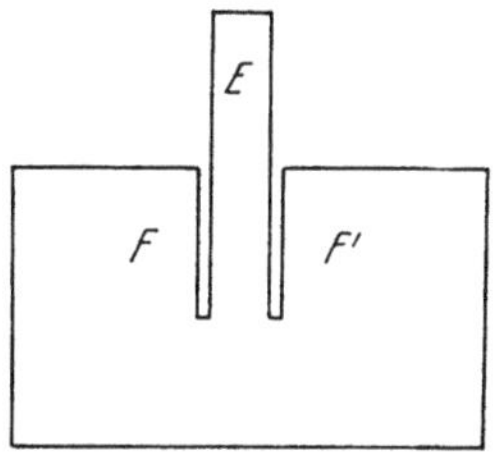

Fig. 287.

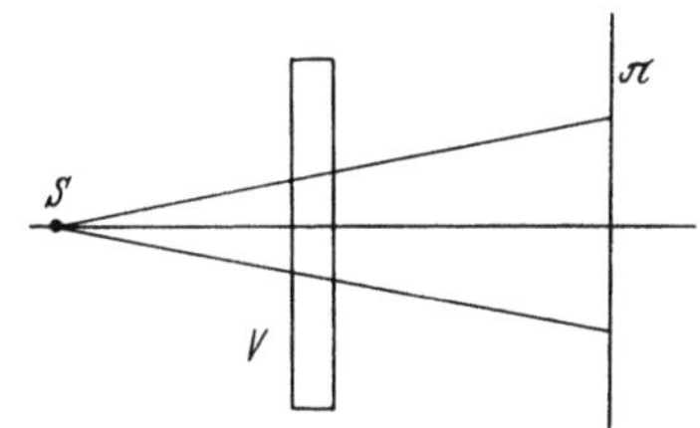

Fig. 288.

117. Diffraction par les objets déphasants. On peut appliquer les résultats précédents à la diffraction par les objets introduisant seulement des variations de phase et non plus des variations d'amplitude. Un exemple de ce type d'objet nous est fourni par les inhomogénéités que l'on rencontre dans les verres d'optique, même de bonne qualité.

Si on éclaire une lame de verre à faces parallèles au moyen d'une source ponctuelle, les variations d'indice produites par les défauts d'homogénéité de la masse produisent des figures de diffraction qui signalent la présence des défauts. On a là une méthode extrêmement simple pour juger de la qualité d'un verre, méthode qui est d'ailleurs couramment employée dans l'industrie.

Soit S la source pontuelle, V la lame de verre contenant des variations d'indice et π le plan d'observation (Fig. 288).

Considérons une variation d'indice ayant la forme d'un fil situé dans la masse et représentée par la loi de la Fig. 289 suivant une section perpendiculaire au fil. Si l'écran π s'éloigne, ces 2 groupes de franges interfèrent et donnent une figure de diffraction plus ou moins compliquée. Sur la Fig. 290 le fil est supposé dans la région C hachurée, son indice est n et sa largeur l. L'indice de l'échantillon est n'. Calculons l'éclairement au point P. M' étant un point quelconque, posons

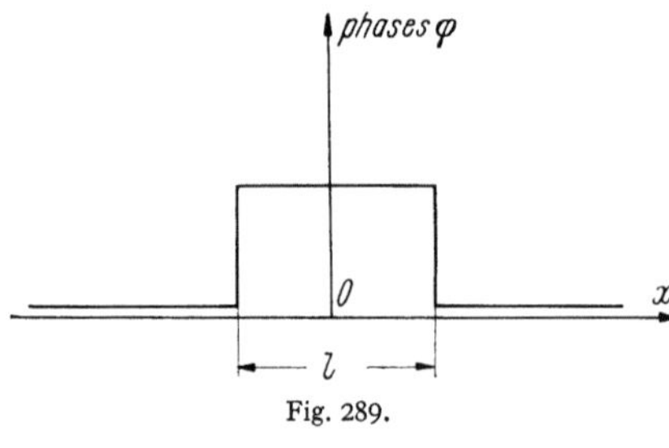

Fig. 289.

$$\varDelta' = R' + r - (R + R') + (n' - n)\, e,$$

e étant l'épaisseur du fil. Si $\widehat{OM'} = s$, on a, d'après (107.1)

$$\varDelta = R' + r - (R + R') = \frac{R + R'}{2\,R\,R'}\, s^2.$$

Posons

$$\varphi = \frac{2\pi}{\lambda}\,(n' - n)\, e = K\,(n' - n)\, e.$$

En étudiant les phénomènes à 1 paramètre, on aura l'amplitude au point P par l'intégrale

$$U_P = \int e^{-jK\varDelta} \cdot e^{-j\varphi}\, dz.$$

Posons $\widehat{MP} = s_1$, $\widehat{MQ} = s_2$. P et Q correspondent aux bords du fil

$$v_1 = \sqrt{\frac{2\,(R + R')}{R\,R'\,\lambda}} \cdot s_1, \qquad v_2 = \sqrt{\frac{2\,(R + R')}{R\,R\,\lambda'}} \cdot s_2.$$

L'éclairement en U à l'intérieur de l'ombre géométrique est

$$E_P = 1 + (1 - \cos\varphi)\,[(\xi_1 - \xi_2)^2 + (\eta_1 - \eta_2)^2 + (\xi_1 - \xi_2) + (\eta_1 - \eta_2)] + \left. \atop + \sin\varphi\,[(\xi_1 - \xi_2) - (\eta_1 - \eta_2)]. \right\} \quad (117.1)$$

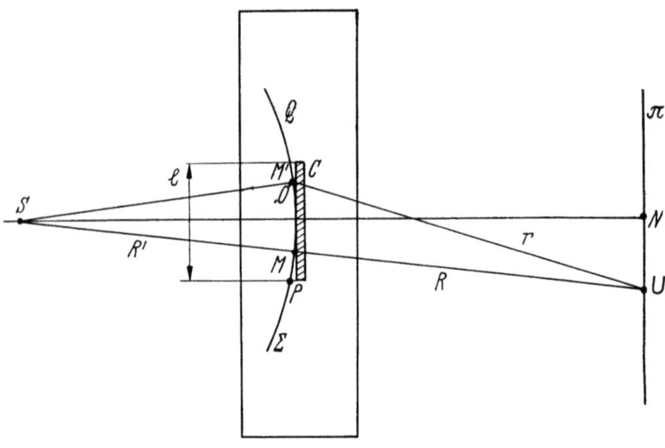

Fig. 290.

ξ_1, ξ_2, η_1, η_2 étant les intégrales de Fresnel correspondant aux valeurs v_1 et v_2 du paramètre v. En dehors de l'ombre géométrique, les limites d'intégration changent et on a

$$E_P = 1 + (1 - \cos\varphi)\,[(\xi_1 - \xi_2)^2 + (\eta_1 - \eta_2)^2 + (\xi_1 - \xi_2) + (\eta_1 - \eta_2)] - \left. \atop - \sin\varphi\,[(\xi_1 - \xi_2) - (\eta_1 - \eta_2)]. \right\} \quad (117.2)$$

Supposons φ petit, on peut écrire (117.1) et (117.2) sous la forme:

$$E_P = 1 \pm \varphi \left[(\xi_1 - \xi_2) - (\eta_1 - \eta_2)\right]$$

en convenant de prendre le signe $+$ pour un point à l'intérieur de l'ombre géométrique et le signe $-$ pour un point situé en dehors de l'ombre géométrique. Au point N, centre de la figure de diffraction $\xi_1 = -\xi_2$, $\eta_1 = -\eta_2$ et on a

$$E_N = 1 - 2\varphi(\xi_2 - \eta_2).$$

Si l'on s'écarte assez loin de l'ombre géométrique, l'éclairement est constant et égal à E_0. Le contraste est alors donné par

$$\gamma = \frac{E_0 - E_N}{E_0} = 2\varphi(\xi_2 - \eta_2),$$

or c'est au point N que l'on a l'éclairement maximum ou minimum (φ petit) car si on considère l'éclairement en un point P quelconque, on peut l'écrire

$$E_P = 1 \pm \varphi \left[(\xi_1 - \eta_1) - (\xi_2 - \eta_2)\right].$$

L'expression entre crochets est la plus grande possible en valeur absolue si $\xi_1 - \eta_1$ et $\xi_2 - \eta_2$ sont de signes contraires. La valeur maximum est alors donnée par

$$\left|\xi_1 - \eta_1\right| = \left|\xi_2 - \eta_2\right| = 0{,}5$$

c'est à dire qu'on est au centre du phénomène.

La meilleure valeur du contraste est donc donnée par l'expression

$$\gamma = 2\varphi(\xi_2 - \eta_2).$$

Comme la valeur maximum de $\xi_2 - \eta_2$ est de l'ordre de 0,5, le contraste maximum est $\gamma = \varphi$. On sait d'autre part que la méthode du contraste de phase avec lame transparente permet d'obtenir $\gamma = 2\varphi$.

En fait, la région centrale est toujours bordée de deux franges importantes. Si on considère le contraste de la partie centrale avec ces franges, on trouve qu'il est environ deux fois plus grand. Dans ces conditions, on peut dire que les méthodes, ombre portée, contraste de phase avec lame transparente, ont la même sensibilité.

Or, dans la pratique courante, il faut un contraste de 0,1 au moins pour que les défauts soient bien visibles; au-dessous de cette valeur les phénomènes sont difficilement observables.

En effet, la granulation de l'écran peut diminuer le contraste si le grain du diffusant est au-dessous de la limite de résolution de l'oeil dans les conditions de l'expérience. D'autre part, si on observe les ombres directement avec un oculaire, les phénomènes entoptiques sont très gênants et il en résulte, ici encore, une baisse apparente de contraste.

Par conséquent la limite de la méthode correspond à une différence de phase $\varphi = 0{,}05$ ou à une différence de marche $\lambda/120$.

La méthode permettra donc l'observation de fils en dessous de la limite tolérable en pratique.

L'expression $v = \sqrt{\dfrac{2(R + R')}{R R' \lambda} \dfrac{l}{2}}$ montre que si R et R' sont très grands, v devient petit, le contraste est faible. Si R et R' deviennent très petits, v devient très grand et le contraste tend vers zéro. Dans ce dernier cas, les franges de chaque discontinuité sont trop séparées pour interférer. L'éclairement au centre est sensiblement le même que celui du fond et le contraste entre ces deux régions très faible. D'autre part le contraste des franges de chaque discontinuité est deux fois plus faible que le contraste des franges du phénomène d'ensemble. Il existe

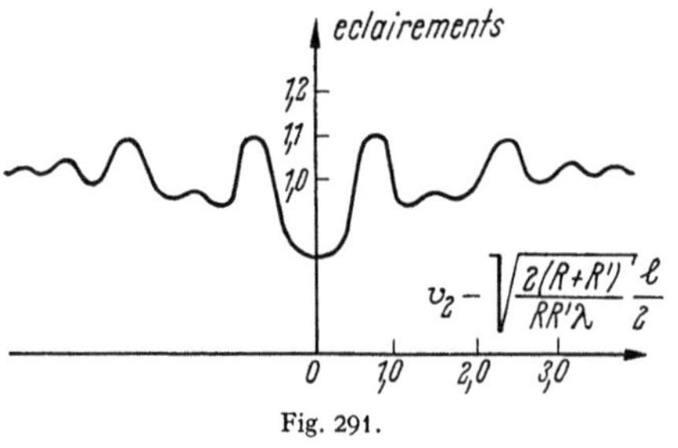

Fig. 291.

donc une position favorable de l'échantillon par rapport à la source et à l'écran, permettant d'obtenir une sensibilité maximum.

La courbe de la Fig. 291 représente la figure de diffraction (éclairements) en fonction de $v_2 - \sqrt{\dfrac{2(R+R')}{R R'\lambda}} \cdot \dfrac{l}{2}$ dans le cas où

$$\frac{R+R'}{R R'} = 6 \cdot 10^{-4} \quad (R \text{ et } R' \text{ en mm.})$$

avec $\varphi = 0{,}2$, $\lambda = 0{,}6\,\mu$ et $v_2 = 2333\,(NP/R) + 0{,}7\,l$ et en choisissant les conditions les plus favorables

$$\xi_2 - \eta_2 = 0{,}5 , \quad v = 0{,}7 .$$

C. Polarisation.

I. Polarisation par réflexion.

a) Expériences fondamentales.

118. Etude de la lumière réfléchie par deux glaces: appareil de Nörremberg. La polarisation de la lumière réfléchie a été observée pour la première fois par Malus en 1810. L'expérience consiste à étudier la lumière après réflexion sur deux

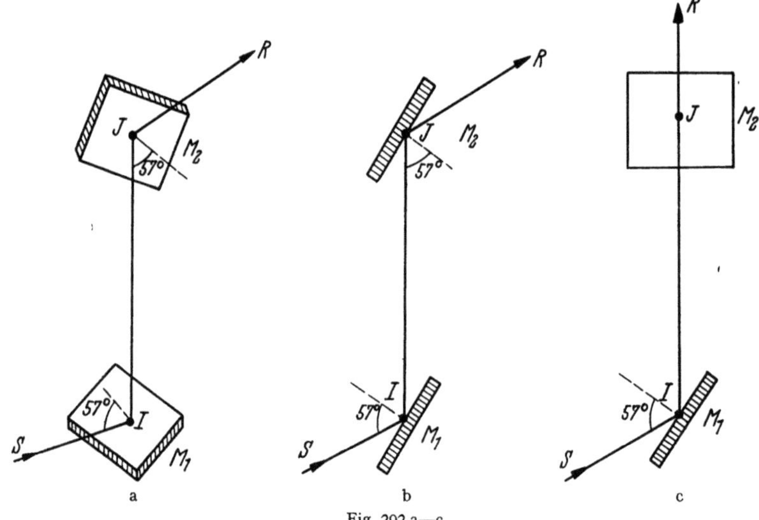

Fig. 292 a—c.

miroirs dont les plans d'incidence peuvent occuper différentes positions l'un par rapport à l'autre. Soit M_1 un miroir plan en verre dont une face est noircie pour éviter la réflexion sur la face postérieure après traversée du verre (Fig. 292a). Envoyons sur ce miroir un faisceau de rayons parallèles dont SI est l'un d'eux. Orientons le miroir M_1 pour que la réflexion en I ait lieu sous l'incidence de 57°. Le rayon réfléchi IJ est reçu en J également sous l'incidence de 57° par un deuxième miroir plan en verre M_2 semblable à M_1. Soit π_1 le plan d'incidence SIJ sur le miroir M_1 et π_2 le plan d'incidence IJR sur le miroir M_2. Faisons tourner le miroir M_2 autour du rayon IJ tout en maintenant l'angle d'incidence égal à 57°

et observons l'intensité lumineuse du rayon réfléchi JR en le recevant sur un écran blanc. Si les deux plans d'incidence π_1 et π_2 sont parallèles (Fig. 292b), on constate que la tache lumineuse reçue sur l'écran blanc est très lumineuse. Tournons le miroir M_2, l'intensité de la tache lumineuse diminue et devient nulle lorsque les deux plans π_1 et π_2 sont perpendiculaires (Fig. 292c). Il n'y a donc plus de lumière réfléchie. En continuant à tourner M_2, la tache lumineuse réapparaît, augmente d'intensité pour passer à nouveau par un maximum lorsque π_1 et π_2 redeviennent parallèles, c'est-à-dire pour la position à 180° de la position initiale. On a donc deux maxima de lumière et deux minima nuls à chaque rotation de 360° du miroir M_2 autour de IJ. L'expérience peut être répétée au moyen de l'appareil de Nörremberg (Fig. 293). Le miroir M_1, constitué par une glace transparente, peut tourner autour d'un axe horizontal $x_1 x_1'$ de façon à régler l'incidence à 57° pour le rayon incident SI. Le rayon réfléchi II' subit la réflexion normale en I' sur un miroir plan métallique M_3 et revient sur luimême. Il traverse la glace M_1 et se réfléchit en J sur le miroir M_2 en verre noir. Le miroir M_2 peut tourner autour d'un axe horizontal $x_2 x_2'$ pour régler l'incidence à 57° et autour d'un axe vertical IJ, de façon à régler à volonté l'angle des plans d'incidence de M_1 et M_2. Le rayon SI provenant d'une source telle qu'une lampe à vapeur de mercure, une lampe au sodium, un arc électrique, etc. ...,

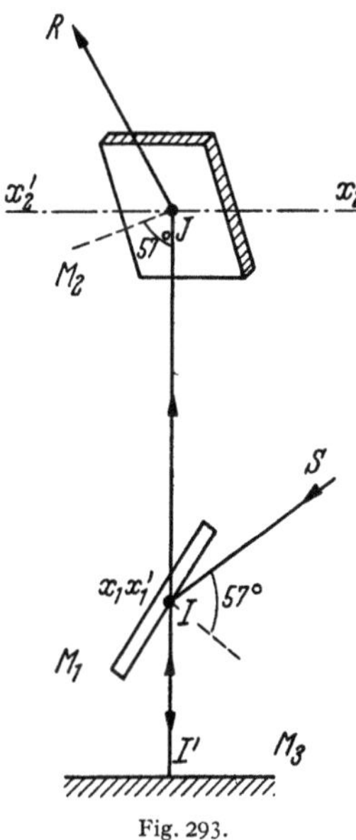

Fig. 293.

donne toujours un rayon réfléchi II' quelle que soit l'orientation du miroir M_1 par rapport à SI le rayon SI est un rayon de lumière naturelle. Par contre, si l'incidence est égale à 57° sur M_1, le rayon réfléchi $II'J$ donne lieu aux phénomènes que nous venons de décrire. En tournant le miroir M_2 autour de l'axe vertical $I'J$, l'intensité du rayon réfléchi JR varie. Si les plans π_1 et π_2 sont parallèles, l'intensité du rayon JR est maximum, si les plans π_1 et π_2 sont perpendiculaires, l'intensité du rayon JR devient nulle. Le rayon $II'J$ qui a subi sur M_1 la réflexion sous l'angle de 57° est un rayon de lumière polarisée. Supposons qu'il soit possible de faire tourner le miroir autour du rayon SI tout en gardant l'incidence égale à 57°: on constaterait que l'intensité du rayon II' resterait constante. Le faisceau de lumière naturelle, qui se propage parallèlement à SI, possède la symétrie de révolution autour de l'axe SI. Quant au rayon $II'J$, ses propriétés sont très différentes, le faisceau de lumière polarisée qui se propage parallèlement à $II'J$ ne possède plus la symétrie de révolution autour de l'axe $I'J$ puisque la rotation du miroir M_2 autour de $I'J$ et sous l'incidence constante de 57° donne un rayon réfléchi JR d'intensité variable. Le miroir M_1 qui produit la polarisation est un polariseur et le miroir M_2 qui permet de constater la polarisation du rayon $II'J$ est un analyseur. Le plan d'incidence SII' qui est un plan de symétrie de la vibration lumineuse, est appelé plan de polarisation du rayon $II'J$.

119. Loi de Brewster. Si dans l'expérience précédente, nous modifions l'inclinaison de la glace M_1 pour avoir une incidence différente de 57°, on constate que l'intensité du rayon JR varie encore lorsque l'on tourne M_2 autour de la verticale, mais les minima ne sont plus nuls. On dit que le rayon $II'J$ est partiellement polarisé. Pour que le rayon $II'J$ soit complètement polarisé et ne donne aucune lumière réfléchie suivant JR lorsque les plans π_1 et π_2 sont perpendiculaires, il faut que l'angle d'incidence i sur la glace M_1 ait une valeur bien déterminée. L'angle d'incidence i appelé angle de Brewster, dépend de l'indice n de la matière constituant le miroir transparent M_1. Il est donné par la loi de Brewster

$$\tan i = n. \tag{119.1}$$

Soit SI un rayon incident sur la glace plane M_1 d'indice n (Fig. 294). On a un rayon réfléchi IR et un rayon réfracté dans la glace IT. Si l'angle d'incidence i est l'angle de Brewster, il satisfait à la relation (119.1). L'angle r étant l'angle de réfraction du rayon réfracté IT, on a $\sin i = n \sin r$.

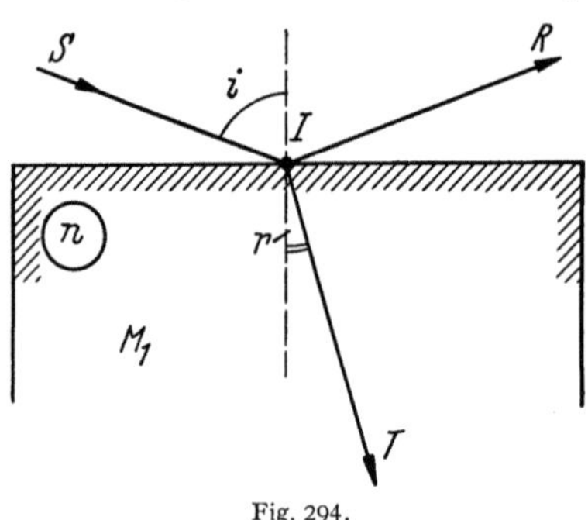

Fig. 294.

En comparant ces deux relations, on obtient

$$\sin r = \cos i$$

d'où

$$r = \frac{\pi}{2} - i.$$

A l'incidence Brewsterienne, le rayon réfléchi IR est normal au rayon réfracté IT.

Notons que l'angle de Brewster dépend de l'indice n et varie par conséquent avec la longueur d'onde de la source employée, mais comme les variations de n sont très faibles, pratiquement la polarisation est complète pour toutes les radiations du spectre visible, et l'appareil de Nörremberg peut être employé en lumière blanche.

120. Loi de Malus. Soit α l'angle des plans d'incidence ou plans de polarisation des deux miroirs M_1 et M_2. Malus a montré que l'intensité I du rayon JR (Fig. 293) est proportionelle au carré du cosinus de l'angle α. A étant une constante, on a

$$I = A^2 \cos^2 \alpha. \tag{120.1}$$

Cette loi peut être vérifiée en disposant à la suite du miroir M_2 un photomètre permettant de mesurer l'intensité du rayon réfléchi JR.

b) Interprétation des expériences précédentes.

121. Hypothèse des vibrations transversales. Les phénomènes d'interférences et de diffraction peuvent s'interpréter par la théorie ondulatoire de la lumière dont l'hypothèse fondamentale est que chaque lumière monochromatique est constituée par une vibration du type $y = a \cos(\omega t - \varphi)$ où y est l'élongation de la vibration, a son amplitude, ω sa pulsation, φ sa phase et t le temps. Quant à la grandeur variable y s'il est inutile d'en préciser la nature pour expliquer les phénomènes d'interférences et de diffraction, il en sera autrement en présence des phénomènes de polarisation. La lumière polarisée montre l'existence de rayons qui n'ont pas la symétrie de révolution autour de la direction de propagation, la vibration lumineuse ne peut donc pas être une vibration longitudinale, car la réflexion d'une telle vibration sur un miroir ne dépendrait pas de l'azimut du plan d'incidence. Les vibrations lumineuses sont des vibrations transversales: dans un milieu isotrope, la vibration lumineuse est une grandeur vectorielle perpendiculaire au rayon lumineux et située par conséquent dans le plan d'onde. Dans le cas de la lumière polarisée, la vibration lumineuse est représentée par un vecteur de direction fixe dans le plan d'onde et dont la grandeur varie suivant une loi sinusoïdale si la lumière est monochromatique.

122. La vibration lumineuse et la théorie électromagnétique. Fresnel a été le premier à expliquer l'ensemble des phénomènes dans la théorie mécanique qu'il a établie au début du XIX° siècle.

Dans sa théorie, Fresnel assimile les vibrations lumineuses aux vibrations élastiques transversales des solides. Puisque la lumière se propage dans le vide, il faut donc admettre que l'espace est rempli d'un milieu doué de propriétés particulières «l'éther» dont les vibrations constituent les vibrations lumineuses. En réalité, cette hypothèse ne nous apprend rien sur la nature des vecteurs qui interviennent dans l'étude des vibrations lumineuses et aboutit à de nombreuses contradictions. Aussi la théorie mécanique de Fresnel a-t-elle été remplacée par la théorie électromagnétique de Maxwell. Pour Fresnel, la fonction $y = a \cos(\omega t - \varphi)$ représentait l'élongation des particules d'éther, elle représente dans la théorie électromagnétique le champ électrique dans l'espace où se propage la lumière, mais rien n'est changé aux calculs par lesquels nous rendons compte des phénomènes.

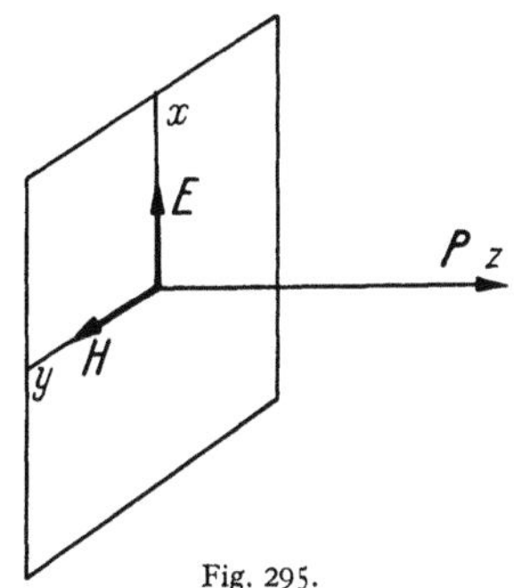

Fig. 295.

A son tour, la théorie de Maxwell s'est trouvée impuissante devant les problèmes posés par les questions d'interactions entre la lumière et la matière. Il a fallu revenir à une théorie qui rappelle celle de l'émission et introduire la notion de photon. Les théories quantiques et la mécanique ondulatoire ont montré la voie dans la recherche d'une théorie réalisant la synthèse de l'électromagnétisme et de la théorie des photons. C'est ainsi que Dirac a réussi à unifier les deux théories en une synthèse capable d'expliquer les phénomènes. L'exposé des questions qui vont suivre pourrait se fonder sur cette théorie unifiée, mais avec un appareil mathématique complexe, elle n'apporte pas de changement dans lese conclusions déduites de la théorie électromagnétique dans le domaine de l'optique que nous considérons ici. C'est pourquoi nous choisirons le mode d'exposé classique en adoptant la théorie électromagnétique. Dans la théorie de Maxwell, la lumière est dûe à la propagation simultanée d'un champ électrique et d'un champ magnétique. L'étude de la polarisation par réflexion de l'air sur le verre montre que la vibration lumineuse doit être assimilée au vecteur champ électrique que nous appellerons vecteur de Fresnel. Si OP est un rayon lumineux se propageant dans un milieu isotrope (Fig. 295), le champ électrique E et le champ magnétique H sont situés dans le plan d'onde et perpendiculaires l'un à l'autre.

Dans le cas d'un rayon OP de lumière polarisée, la direction du vecteur E dans le plan d'onde est fixe et le plan de polarisation est le plan POH perpendiculaire au vecteur champ électrique. Nous définirons dans ce qui suit l'état de polarisation d'un faisceau lumineux par la direction ox de la vibration lumineuse (champ électrique) ou par la direction du plan de vibration xoz perpendiculaire au plan de polarisation. Ainsi, un faisceau lumineux polarisée sous l'incidence brewstérienne fournit une vibration lumineuse perpendiculaire au plan d'incidence qui est le plan de polarisation et par conséquent, parallèle à la surface du miroir.

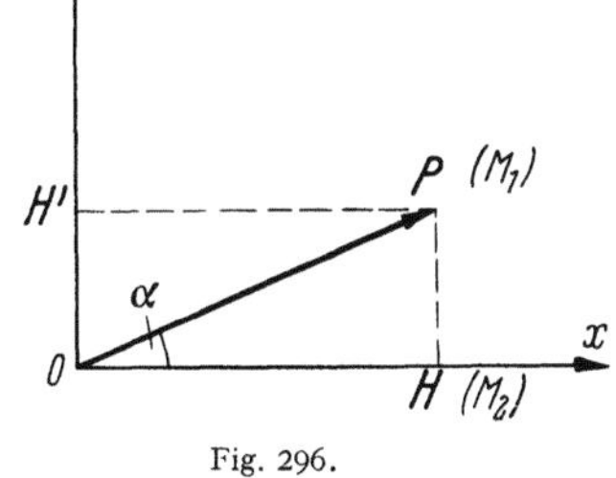

Fig. 296.

123. Interprétation de la loi de Malus. Soit OP la direction de la vibration lumineuse fournie par le miroir M_1 dans l'appareil de Nörremberg (Fig. 296). Le plan d'incidence sur le miroir M_1 est donc perpendiculaire à OP. Représentons par Ox et Oy les directions respectivement perpendiculaires et parallèles au plan d'incidence π_2 sur le deuxième miroir M_2. On peut considérer la vibration OP comme la résultante des 2 vibrations OH et OH' projections de OP sur OX

et OY. Tout se passe comme si on avait pour le miroir M_2 deux vibrations incidentes dirigées suivant OH et OH'. Or la vibration OH' étant parallèle au plan d'incidence π_2 sur M_2 est arrêtée par celui-ci (Sect. 118 et 120). Par contre, la composante OH, perpendiculaire à π_2, est totalement réfléchie. Si $y = A \cos \omega t$ est la vibration incidente OP, la vibration OH sera représentée par $y = A \cos \alpha \cos \omega t$ d'amplitude $A \cos \alpha$. D'où l'intensité de la vibration émergente

$$I = A^2 \cos^2 \alpha.$$

L'hypothèse des vibrations transversales perment donc d'expliquer d'une façon très simple la loi de Malus.

c) Propagation des ondes électromagnétiques dans un milieu isotrope transparent.

124. Equations fondamentales de MAXWELL. Comme nous l'avons déjà dit (Sect. 122) dans la théorie électromagnétique de Maxwell, la lumière est dûe à la propagation simultanée d'un champ électrique et d'un champ magnétique. Cette théorie s'appuie essentiellement sur les deux équations fondamentales de l'électromagnétisme: la première exprime le travail du champ magnétique en fonction du flux du courant électrique (théorème d'Ampère) et la seconde, le travail du champ électrique en fonction de la variation de flux magnétique (loi de l'induction). Ces lois avaient été établies dans le cas des courants et des champs de grandeur fixe ou lentement variable, et Maxwell les généralisa en admettant qu'elles restaient valables pour des grandeurs très rapidement variables.

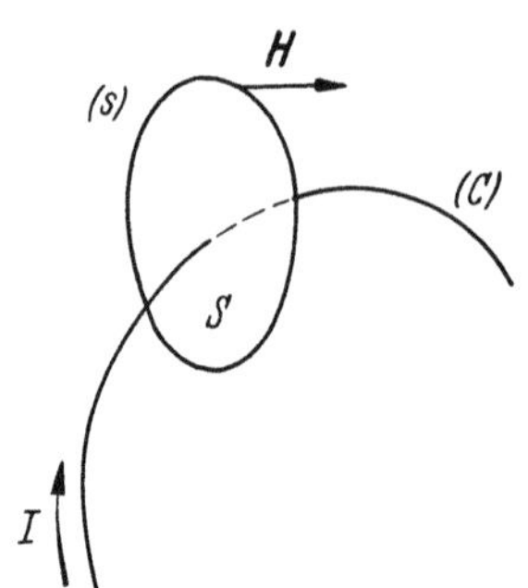

Fig. 297.

Soit C un circuit parcouru un courant électrique d'intensité I en unités électrostatiques (Fig. 297). Ce courant crée dans l'espace environnant un champ magnétique H. D'après le théorème d'Ampère, le travail du vecteur H le long d'une courbe fermée s à l'intérieur de laquelle passe le courant I est

$$\int\limits_{(s)} \boldsymbol{H} \cdot \boldsymbol{ds} = 4\pi \frac{I}{C} \qquad (124.1)$$

où c exprime le rapport des unités électromagnétiques et électrostatiques de quantité d'électricité.

Dans le cas, non plus d'un courant mais d'un système de courants, le théorème d'Ampère est encore valable si dans l'expression (124.1) le courant I représente le courant total qui traverse le contour fermé.

Soient $\boldsymbol{j}$ le vecteur densité de courant et dS un élément d'aire. Le courant total I qu traverse la surface S est

$$I = \iint \boldsymbol{j} \cdot \boldsymbol{dS}.$$

C'est le flux du vecteur $\boldsymbol{j}$ et le théorème d'Ampère s'écrit

$$\int\limits_{(s)} \boldsymbol{H} \cdot \boldsymbol{ds} = \frac{4\pi}{c} \iint\limits_{S} \boldsymbol{j} \cdot \boldsymbol{dS}. \qquad (124.2)$$

Or, d'après le théorème de Stokes

$$\int\limits_{(s)} \boldsymbol{H} \cdot \boldsymbol{ds} = \iint\limits_{S} \mathrm{rot}\,\boldsymbol{H} \cdot \boldsymbol{dS} = \frac{4\pi}{c} \iint\limits_{S} \boldsymbol{j} \cdot \boldsymbol{dS}.$$

Le vecteur $\dfrac{4\pi}{c} \boldsymbol{j}$ est le rotationnel du vecteur $\boldsymbol{H}$ et la relation (124.2) s'écrit également

$$\frac{4\pi \boldsymbol{j}}{c} = \mathrm{rot}\,\boldsymbol{H}. \qquad (124.3)$$

En projetant cette relation sur trois axes rectangulaires $Ox\,Oy\,Oz$, on obtient

$$\frac{4\pi j_x}{c} = \frac{\partial H_z}{\partial y} - \frac{\partial H_y}{\partial z}, \quad \frac{4\pi j_y}{c} = \frac{\partial H_x}{\partial z} - \frac{\partial H_z}{\partial x}, \quad \frac{4\pi j_z}{c} = \frac{\partial H_y}{\partial x} - \frac{\partial H_x}{\partial y}. \qquad (124.4)$$

Avant de relier $\boldsymbol{j}$ au champ électrique $\boldsymbol{E}$ et d'obtenir ainsi le premier groupe des équations de MAXWELL, utilisons la loi fondamentale de l'induction pour établir un deuxième groupe d'équations.

D'après cette loi, si V est la force électromotrice d'induction dans un circuit (s) fermé, Φ le flux d'induction qui traverse à l'instant t la surface S limitée par le circuit (s), on a:

$$V = -\frac{1}{c}\frac{d\Phi}{dt} \qquad (124.5)$$

où V est exprimé en unités électrostatiques et Φ en unités électromagnétiques. La force électromotrice d'induction V est égale au travail $\int_{(s)} \boldsymbol{E} \cdot \boldsymbol{ds}$ du champ électrique E le long du circuit s. Le flux Φ est relié au vecteur induction magnétique B par la relation

$$\Phi = \iint_{S} \boldsymbol{B} \cdot \boldsymbol{dS}. \qquad (124.6)$$

En combinant les expressions (124.5) et (124.6), on a

$$\int_{(s)} \boldsymbol{E} \cdot \boldsymbol{ds} = -\frac{1}{c}\iint_{S} \frac{\partial \boldsymbol{B}}{\partial t} \cdot \boldsymbol{dS} \qquad (124.7)$$

qui s'écrite également

$$\frac{1}{c}\frac{\partial \boldsymbol{B}}{\partial t} = -\operatorname{rot} \cdot \boldsymbol{E} \qquad (124.8)$$

d'où les expressions correspondant aux 3 axes Ox, Oy, Oz.

$$\frac{1}{c}\frac{\partial B_x}{\partial t} = -\frac{\partial E_z}{\partial y} + \frac{\partial E_y}{\partial z}, \quad \frac{1}{c}\frac{\partial B_y}{\partial t} = -\frac{\partial E_x}{\partial z} + \frac{\partial E_z}{\partial x}, \quad \frac{1}{c}\frac{\partial B_z}{\partial t} = -\frac{\partial E_y}{\partial x} + \frac{\partial E_z}{\partial y}. \quad (124.9)$$

Les deux relations (124.3) et (124.8) expriment les lois fondamentales de l'électromagnétisme, elles sont indépendantes de la nature du milieu. Il faut établir maintenant deux autres relations entre les vecteurs $\boldsymbol{E}, \boldsymbol{H}, \boldsymbol{J}, \boldsymbol{B}$ pour arriver à l'équation de propagation. Ces deux nouvelles relations font intervenir la nature du milieu et nous considérerons maintenant un milieu diélectrique isotrope.

Relions $\boldsymbol{j}$ au champ électrique $\boldsymbol{E}$. Comme il ne peut pas exister de courant dans un diélectrique, l'idée essentielle de MAXWELL est d'admettre qu'au point de vue électromagnétique, un circuit ouvert peut être considéré comme fermé par un courant fictif appelé courant de déplacement.

Considérons par exemple un condensateur dont les armatures ont le surface S et soit l'intensité du courant qui parcourt à l'instant t le conducteur extérieur. Si le condensateu reçoit la charge dq pendant le temps dt, on aura

$$i = \frac{dq}{dt}.$$

On considère que le courant i est fermé dans le diélectrique par un courant fictif, le couran- de déplacement, dont la densité de courant est

$$j = \frac{1}{S} \cdot \frac{dq}{dt} = \frac{d\sigma}{dt}$$

où $d\sigma$ est la charge reçue par unité de surface dans le temps dt. Or, d'après le théorème de COULOMB, la charge σ est reliée au champ électrique E par la relation $4\pi q = \varepsilon E$ où ε est la constante diélectrique du milieu. Le courant de déplacement aura donc pour densité

$$j = \frac{\varepsilon}{4\pi}\frac{\partial \boldsymbol{E}}{\partial t}. \qquad (124.10)$$

Quant au champ magnétique $\boldsymbol{H}$ et à l'induction magnétique $\boldsymbol{B}$, ils satisfont à la relation:

$$\boldsymbol{B} = \mu\,\boldsymbol{H} \qquad (124.11)$$

où μ est la perméabilité. La relation (124.10) jointe à la relation (124.3) permet d'éliminer $\boldsymbol{j}$ et de relier directement $\boldsymbol{E}$ à $\boldsymbol{H}$. La relation (124.11) jointe à la relation (124.8) permet d'éliminer $\boldsymbol{B}$ et de relier également $\boldsymbol{E}$ à $\boldsymbol{H}$. On obtient ainsi les deux équations de Maxwell

$$\frac{\varepsilon}{c}\,\frac{\partial \boldsymbol{E}}{\partial t} = \operatorname{rot}\boldsymbol{H}, \qquad \frac{\mu}{c}\,\frac{\partial \boldsymbol{H}}{\partial t} = -\operatorname{rot}\boldsymbol{E} \tag{124.12}$$

ou encore par projection sur les axes, les deux groupes d'équations suivantes

$$\left.\begin{aligned}
\frac{\varepsilon}{c}\,\frac{\partial E_x}{\partial t} &= \frac{\partial H_z}{\partial y} - \frac{\partial H_y}{\partial z}, & -\frac{\mu}{c}\,\frac{\partial H_x}{\partial t} &= \frac{\partial E_z}{\partial y} - \frac{\partial E_y}{\partial z}, \\[2mm]
\frac{\varepsilon}{c}\,\frac{\partial E_y}{\partial t} &= \frac{\partial H_x}{\partial z} - \frac{\partial H_z}{\partial x}, & -\frac{\mu}{c}\,\frac{\partial H_y}{\partial t} &= \frac{\partial E_x}{\partial z} - \frac{\partial E_z}{\partial x}, \\[2mm]
\frac{\varepsilon}{c}\,\frac{\partial E_z}{\partial t} &= \frac{\partial H_y}{\partial x} - \frac{\partial H_x}{\partial y}, & -\frac{\mu}{c}\,\frac{\partial H_z}{\partial t} &= \frac{\partial E_y}{\partial x} - \frac{\partial E_x}{\partial y}.
\end{aligned}\right\} \tag{124.13}$$

Ces formules supposent que le trièdre $Oxyz$ est un trièdre à gauche défini par la règle du tire-bouchon de Maxwell, où la règle du bonhomme d'Ampère: un observateur placé sur Oz et regardant Ox voit Oy à sa gauche.

125. Propagation du champ électrique. Considérons la deuxième eq. (124.12) et prenons le rotationnel des deux membres

$$\frac{\mu}{c}\operatorname{rot}\frac{\partial \boldsymbol{H}}{\partial t} = \frac{\mu}{c}\,\frac{\partial}{\partial t}\operatorname{rot}\boldsymbol{H} = -\operatorname{rot}\operatorname{rot}\boldsymbol{E},$$

En utilisant la première équation (124.12), on a

$$\operatorname{rot}\operatorname{rot}\boldsymbol{E} = -\frac{\varepsilon\mu}{c^2}\,\frac{\partial^2 \boldsymbol{E}}{\partial t^2}, \tag{125.1}$$

où l'identité vectorielle

$$\operatorname{rot}\operatorname{rot}\boldsymbol{E} = \operatorname{grad}(\operatorname{div}\boldsymbol{E}) - \Delta\boldsymbol{E} \tag{125.2}$$

peut être appliquée au premier membre.

D'après l'équation de Poisson

$$\operatorname{div}\boldsymbol{E} = 4\pi\varrho$$

ϱ étant la densité de charge qui est nulle dans l'espace libre, on a

$$\operatorname{div}\boldsymbol{E} = 0 \tag{125.3}$$

et l'équation (125.1) se réduit à

$$\frac{\varepsilon\mu}{c^2}\,\frac{\partial^2 \boldsymbol{E}}{\partial t^2} = \Delta\boldsymbol{E}.$$

En posant

$$v^2 = \frac{c^2}{\varepsilon\mu} \tag{125.4}$$

on obtient l'équation vectorielle de propagation du champ électrique

$$\frac{\partial^2 \boldsymbol{E}}{\partial t^2} = v^2\,\Delta\boldsymbol{E}. \tag{125.5}$$

Dans cette équation aux dérivées partielles, la constante v représente la vitesse de propagation de l'onde électromagnétique. Dans le cas du vide $\varepsilon = 1$, $\mu = 1$, et on a $v = c$. La théorie de Maxwell montre que la vitesse de propagation d'une onde électromagnétique dans le vide est égale au rapport des unités électromagnétique et électrostatique de quantité d'électricité, soit environ $3\cdot 10^{10}$ cm./sec. L'expérience confirme ce résultat en donnant pour v et c des valeurs dont la différence est inférieure à $1/10000$.

Dans le cas d'un diélectrique, on a, d'après l'équation (125.4)

$$v = \frac{c}{\sqrt{\varepsilon}} \tag{125.6}$$

car, pour la plupart des corps, la perméabilité μ est pratiquement égale à l'unité. Comme d'autre part, le rapport de la vitesse c dans le vide à la vitesse v dans un diélectrique ε, égal à l'indice de réfraction n du milieu considéré, on aura donc

$$\varepsilon = n^2. \tag{125.7}$$

Les équations de MAXWELL telles qu'elles ont été employées, ne permettent donc pas de prévoir la dispersion des milieux matériels transparents, c'est-à-dire la variation de n en fonction des différentes radiations du spectre. Pour expliquer le phénomène de la dispersion, les équations de MAXWELL doivent être complétées par des termes qui font intervenir les périodes correspondant aux longueurs d'onde des bandes d'absorption du milieu considéré. La dispersion et l'absorption sont en effet deux phénomènes liés l'un à l'autre, le premier étant sous la dépendance étroite du second.

Dans le cas d'une vibration monochromatique de période T, l'équation suivante remplace l'équation (124.10)

$$\boldsymbol{j} = \frac{\varepsilon'}{4\pi} \frac{\partial \boldsymbol{E}}{\partial t}. \tag{125.8}$$

Dans cette équation, ε' est maintenant une constante complexe qui dépend de la période T, c'est la constante diélectrique complexe du milieu. Si n est l'indice complexe, v l'indice de réfraction, x l'indice d'extinction, on a

$$n^2 = (v - j\,x)^2 = \varepsilon'.$$

Lorsque la longueur d'onde augmente indéfiniment, ε' tend vers ε et v tend vers $\sqrt{\varepsilon}$. La relation de MAXWELL $v^2 = \varepsilon$ est donc une relation limite d'autant mieux vérifiée que la longueur d'onde est plus grande. Remarquons toutefois que la relation (125.7) est bien vérifiée dans le cas de substances à faible dispersion en prenant comme indice n, l'indice pour le jaune moyen du spectre.

126. Propagation d'une onde plane. Considérons une onde plane parallèle au plan xOy et se propageant dans la direction Oz. La phase de la vibration correspondante est donc constante en tous points d'un plan d'équation $z = \text{con-}$ stante, et ne dépend que de z. Pour cette onde, le champ électrique satisfait à l'équation de propagation (125.5) dont les solutions sinusoïdales pour les trois composantes E_x, E_y, E_z du champ électrique sont de la forme

$$E_x = a\,e^{j\left(\omega t - \frac{2\pi z}{\lambda}\right)}, \qquad E_y = b\,e^{j\left(\omega t - \frac{2\pi z}{\lambda} - \varphi\right)}, \qquad E_z = c\,e^{j\left(\omega t - \frac{2\pi z}{\lambda} - \psi\right)} \tag{126.1}$$

où les amplitudes a, b, c et les différences de phase φ et ψ sont des constantes. Puisque E_x et E_y ne dépendent pas de x et de y, l'équation (125.3) se réduit à

$$\frac{\partial E_z}{\partial z} = -\frac{2\pi j\,c}{\lambda}\,e^{j\left(\omega t - \frac{2\pi z}{\lambda} - \psi\right)} = 0.$$

Cette équation ne peut être satisfaite à chaque instant que si l'on a $c = 0$, c'est-à-dire

$$E_z = 0.$$

Le vecteur électrique est par conséquent situé dans le plan d'onde et la théorie électromagnétique montre ainsi que les vibrations lumineuses sont des vibrations transversales.

La vibration la plus générale se propageant dans la direction Oz est donc une vibration elliptique située dans le plan d'onde et dont les équations sont

$$E_x = a\, e^{j\left(\omega t - \frac{2\pi z}{\lambda}\right)}, \qquad E_y = b\, e^{j\left(\omega t - \frac{2\pi z}{\lambda} - \varphi\right)}, \qquad E_z = 0. \tag{126.2}$$

127. Propagation d'une onde plane polarisée rectilignement. Si nous fixons la direction du champ électrique dans le plan d'onde, nous avons une onde polarisée rectilignement.

Posons par exemple $E_y = 0$; une onde plane polarisée rectilignement aura donc pour équations

$$E_x = a\, e^{j\left(\omega t - \frac{2\pi z}{\lambda}\right)}, \qquad E_y = 0, \qquad E_z = 0. \tag{127.1}$$

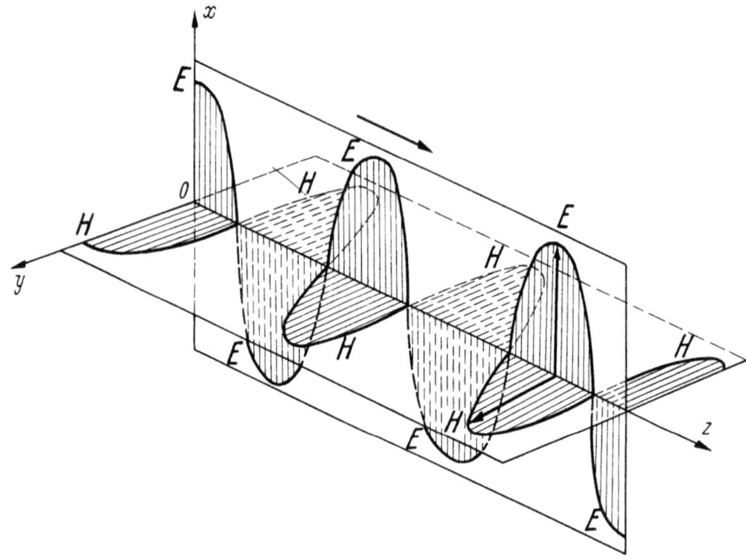

Fig. 298.

Cherchons la direction du vecteur champ magnétique $\boldsymbol{H}$.

Puisque E_x ne dépend que de z, les équations (127.1) jointes aux équations (124.13) donnent

$$\frac{\partial H_x}{\partial t} = 0, \qquad \frac{\partial H_z}{\partial t} = 0$$

et

$$\frac{\partial H_y}{\partial t} = -\frac{c}{\mu}\frac{\partial E_x}{\partial z} = \frac{2\pi j\, c}{\lambda\mu}\, a\, e^{j\left(\omega t - \frac{2\pi z}{\lambda}\right)}$$

d'où, en intégrant

$$H_x = 0, \qquad H_y = \frac{2\pi c}{\lambda\mu\omega}\, a\, e^{j\left(\omega t - \frac{2\pi z}{\lambda}\right)} = \sqrt{\frac{\varepsilon}{\mu}}\, a\, e^{j\left(\omega t - \frac{2\pi z}{\lambda}\right)}, \qquad H_z = 0. \tag{127.2}$$

Le vecteur champ magnétique $\boldsymbol{H}$ est dirigé suivant Oy dans le plan d'onde; il est perpendiculaire au vecteur champ électrique et a même phase. Les vecteurs $\boldsymbol{E}$, $\boldsymbol{H}$ et la direction de propagation Oz forment un trièdre à gauche (Fig. 295).

La Fig. 298 donne la représentation dans l'espace des deux vecteurs E et H à un instant donné: ce sont deux sinusoïdes situées respectivement dans les plans xOy, yOz et en phase.

L'étude de la polarisation par réflexion de l'air sur le verre, et l'étude d'autres phénomènes, comme par exemple celui des ondes stationnaires, montrent que la vibration lumineuse doit être assimilée au vecteur champ électrique E, ou vecteur de Fresnel. Dans le cas de la lumière polarisée, le plan de vibration est le plan xOz perpendiculaire au plan de polarisation yOz contenant le vecteur champ magnétique H.

128. Vecteur de Poynting et énergie transportée. Le diélectrique constituant le milieu dans lequel se propage l'onde électromagnétique est soumis à un champ électrique E et à un champ magnétique H. De la part du champ électrique E, l'unité de volume contient l'énergie électrostatique $\dfrac{\varepsilon E^2}{8\pi}$ et de la part du champ magnétique H, l'énergie $\dfrac{\mu H^2}{8\pi}$. L'énergie totale par unité de volume (cm.³) sera

$$\frac{\varepsilon E^2}{8\pi} + \frac{\mu H^2}{8\pi}.$$

Mais les équations (127.1) et (127.2) montrent que:

$$H = \sqrt{\frac{\varepsilon}{\mu}}\, E.$$

L'énergie totale sera donc $\dfrac{\varepsilon E^2}{4\pi}$. Calculons de flux d'énergie P_z par centimètre carré, c'est-à-dire l'énergie qui traverse une surface de 1 cm.² normale à la direction de propagation, en une seconde. Par suite de la propagation de l'onde plane, celle-ci remplit v cm.³ de plus en une seconde et le flux P_z sera égal à

$$P_z = \frac{v\,\varepsilon}{4\pi} E^2 = \frac{c}{4\pi} \sqrt{\frac{\varepsilon}{\mu}}\, E^2 = \frac{c}{4\pi} E H.$$

Le vecteur P normal au plan des vecteurs rectangulaires E et H représente au facteur $\dfrac{c}{4\pi}$ près, leur produit vectoriel

$$\boldsymbol{P} = \frac{c}{4\pi}\, \boldsymbol{E} \times \boldsymbol{H}. \tag{128.1}$$

Le vecteur P ainsi introduit est le vecteur de Poynting. Bien qu'il ait été obtenu ici dans le cas d'une onde plane, on démontre que son expression donnée par la relation (128.1) est générale. Elle est encore valable dans le cas d'un diélectrique anisotrope, quelle que soit la forme de la surface d'onde. Les lignes de flux du vecteur de Poynting représentent les trajectoires de l'énergie, c'est-à-dire les rayons lumineux qui dans tous les cas sont perpendiculaires aux plans définis par les vecteurs E et H.

L'énergie transportée $\dfrac{v\varepsilon E^2}{4\pi}$ peut s'écrire en prenant la valeur moyenne $\dfrac{a^2}{2}$ de E^2

$$\frac{v\,\varepsilon\, a^2}{8\pi} = \frac{c\,\varepsilon\, a^2}{8\pi n} \approx \frac{c\, n}{8\pi} a^2$$

car $n = c/v$ est pratiquement égal à $\sqrt{\varepsilon}$. On voit donc que pour un milieu donné l'énergie transportée, c'est-à-dire l'intensité du faisceau lumineux est proportionnelle au carré de l'amplitude a.

d) Application de la théorie électromagnétique à l'étude de la lumière réfléchie.

129. Equations de passage. Nous allons maintenant chercher comment varient les vecteurs champ électrique et champ magnétique lorsque l'on franchit la surface de séparation de deux diélectriques. L'étude des conditions aux limites en électricité montre que la composante tangentielle du champ magnétique et la composante normale de l'induction magnétique varient d'une façon continue lorsqu'on traverse la surface de séparation des deux milieux. De mêmel il y a continuité de la composante tangentielle du champ électrique et de la composante normale de l'induction électrostatique.

Soient (Fig. 299) deux milieux diélectriques de constantes ε et ε' séparés par la surface plane xOy. Si $\boldsymbol{E}$ et $\boldsymbol{H}$ sont les champs électriques et magnétiques

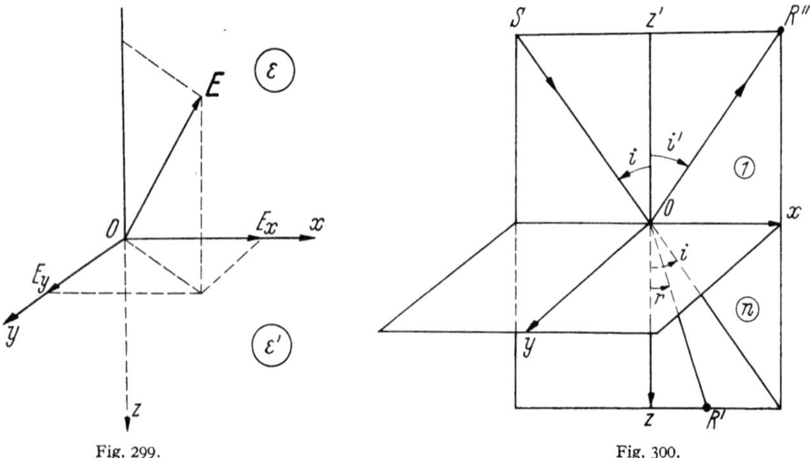

Fig. 299.   Fig. 300.

dans le milieu ε, $\boldsymbol{E}'$ et $\boldsymbol{H}'$ les champs électriques et magnétiques dans le milieu ε', on pourra écrire pour les composantes tangentielles

$$E_x = E'_x, \qquad H_x = H'_x \left.\right\}$$
$$E_y = E'_y, \qquad H_y = H'_y \left.\right\} \tag{129.1}$$

et pour les composantes normales de l'induction électrostatique et de l'induction magnétique

$$\varepsilon E_z = \varepsilon' E'_z, \qquad \mu H_z = \mu' H'_z. \tag{129.2}$$

Dans le deuxième milieu ε', le champ est uniquement le champ de l'onde transmise, tandis que dans le premier milieu ε, le champ est dû à la superposition des champs de l'onde incidente et de l'onde réfléchie. Les équations de passage s'obtiendront donc en écrivant qu'à la surface de séparation ($z=0$) le champ transmis est égal à la somme des champs incidents et réfléchi. Si $\boldsymbol{E}''$ et $\boldsymbol{H}''$ désignent les champs électriques et magnétiques réfléchis, on aura

$$(z=0) \quad E_x + E''_x = E'_x, \qquad H_x + H''_x = H'_x \left.\right\}$$
$$E_y + E''_y = E'_y, \qquad H_y + H''_y = H'_y. \left.\right\} \tag{129.3}$$

Ce sont les équations de passage qui vont nous permettre d'étudier l'amplitude et l'intensité de la lumière réfléchie.

Soient, (Fig. 300), xOy la surface plane séparant les deux milieux, et xOz le plan d'incidence. Pour simplifier, nous prendrons l'indice du premier milieu égal à l'unité et celui du second milieu égal à n. Nous considérons une onde plane polarisée se propageant dans la direction SO. SO est le rayon incident, OR', le rayon transmis, et OR'' le rayon réfléchi. Les angles d'incidence et de réfraction i et r seront comptés positivement en allant de Oz vers Ox. La vibration transportée par le rayon incident SO peut être décomposée en 2 vibrations rectangulaires composantes, l'une située dans le plan d'incidence xOz, l'autre dans le plan perpendiculaire. Nous étudierons successivement ces deux composantes.

130. Amplitude de la vibration réfléchie: vibration dans le plan d'incidence. La vibration incidente est caractérisée par le champ électrique $\boldsymbol{E}$ située dans le plan xOz (Fig. 301) et dont la direction fait l'angle i avec Ox. L'amplitude du champ électrique étant égale à a, nous avons pour l'onde incidente, φ étant la phase

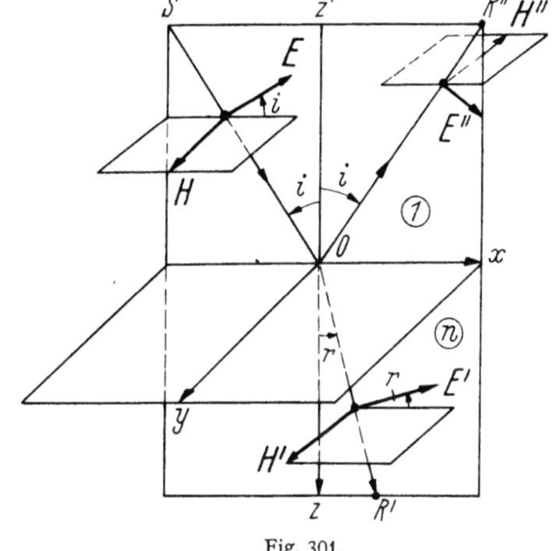

Fig. 301.

$$\left.\begin{aligned}
E_x &= a\cos i\, e^{j(\omega t - \varphi)},\\
E_y &= 0,\\
H_x &= 0,\\
H_y &= a\, e^{j(\omega t - \varphi)}.
\end{aligned}\right\} \quad (130.1)$$

Pour l'onde transmise, le champ électrique fait l'angle r avec Ox et a une certaine amplitude a'. Les équations (127.1) et (127.2) montrent que pratiquement, l'amplitude du champ magnétique est égale à na', on aura donc

$$\left.\begin{aligned}
E'_x &= a'\cos r\, e^{j(\omega t - \varphi')}, \qquad E'_y = 0\\
H'_x &= 0, \qquad\qquad\qquad H'_y = n\, a'\, e^{j(\omega t - \varphi')}.
\end{aligned}\right\} \quad (130.2)$$

Enfin, pour l'onde réfléchie, les amplitudes du champ électrique et du champ magnétique étant égales à a'', on aura

$$\left.\begin{aligned}
E''_x &= a''\cos i\, e^{j(\omega t - \varphi'')}, \qquad E''_y = 0\\
H''_x &= 0, \qquad\qquad\qquad H''_y = -\, a''\, e^{j(\omega t - \varphi'')}
\end{aligned}\right\} \quad (130.3)$$

car la direction du champ $\boldsymbol{H}$ est telle que le trièdre $\boldsymbol{E}''$, $\boldsymbol{H}''$, OR'', doit être direct.

En admettant que les trois vibrations aient des phases égales ou opposées en O (Fig. 301), les équations de passage (129.3) donnent

$$\left.\begin{aligned}
(a + a'')\cos i &= a'\cos r,\\
a - a'' &= n\, a'
\end{aligned}\right\} \quad (130.4)$$

d'où les formules de FRESNEL

$$a' = a\,\frac{2\cos i\sin r}{\sin(i+r)\cos(i-r)} \qquad a'' = -\, a\,\frac{\tan(i-r)}{\tan(i+r)} \qquad (130.5)$$

et l'intensité de la lumière réfléchie

$$I_{\parallel} = \left(\frac{a''}{a}\right)^2 = \frac{\tan^2(i-r)}{\tan^2(i+r)}. \qquad (130.6)$$

Si i et r sont très petits, on peut confondre les sinus et les tangentes avec les angles, on a alors

$$I_0 = \left(\frac{n-1}{n+1}\right)^2 \qquad (130.7)$$

qui n'est autre que le facteur de réflexion du milieu d'indice n sous l'incidence normale. Pour $n = 1{,}5$ (verre) on a $I_0 = 0{,}04$. La courbe (1) de la Fig. 302 représente les variations de l'intensité lumineuse réfléchie (130.6) en fonction de

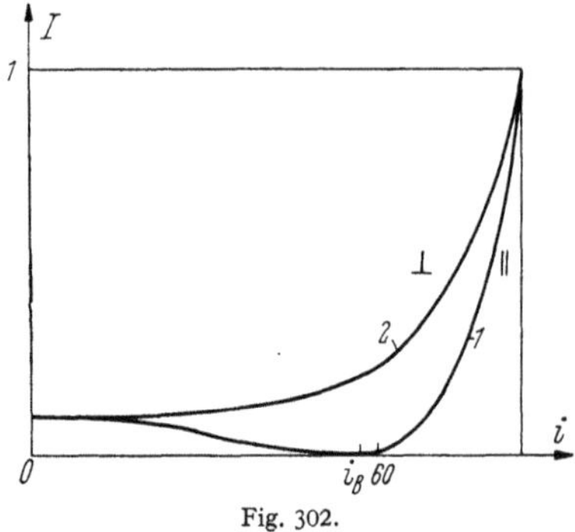

Fig. 302.

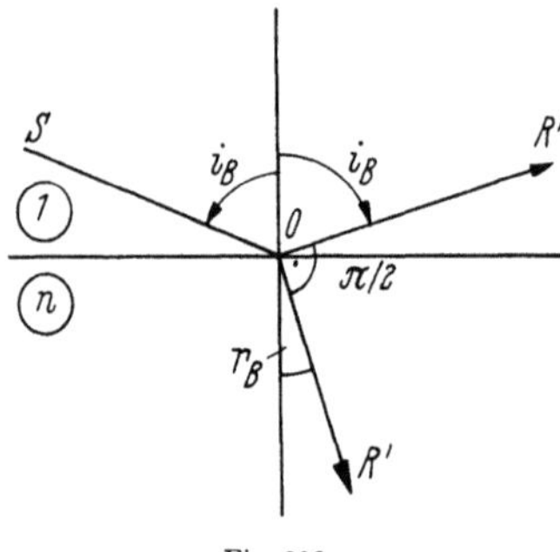

Fig. 303.

l'angle d'incidence i. Lorsque l'angle d'incidence augmente, la lumière réfléchie commence par décroître, s'annule pour une certaine incidence i_B, puis croît jusqu'a la valeur 1 pour l'incidence rasante. L'incidence i_B, telle que la lumière réfléchie soit nulle pour la vibration située dans le plan d'incidence, est l'incidence Brewsterienne. L'angle d'incidence i_B est donné par

$$i_B + r_B = \frac{\pi}{2}. \qquad (130.8)$$

Or

$$\sin i_B = n \sin r_B = n \cos i_B.$$

D'où

$$\tan i_B = n \qquad (130.9)$$

qui est l'expression de la loi de Brewster.

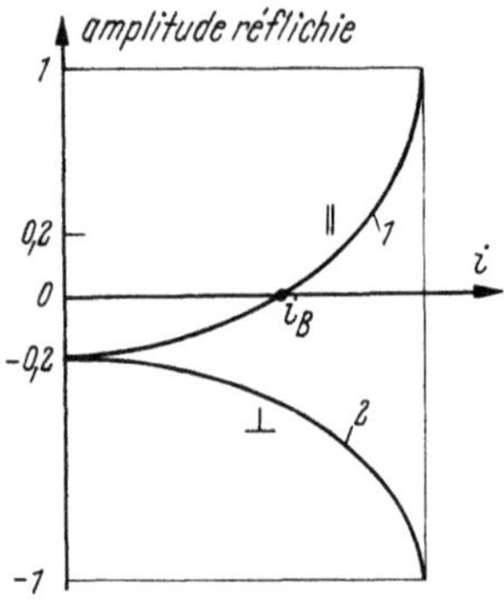

Fig. 304.

L'équation (130.8) montre qu'à l'incidence Brewsterienne, la rayon transmis OR' (Fig. 303) est perpendiculaire au rayon réfléchi OR''.

Le fait que la vibration dans le plan d'incidence soit éteinte par réflexion montre une nouvelle fois que c'est au vecteur champ électrique qu'il faut assimiler la vibration lumineuse ou vecteur de Fresnel.

Sur la Fig. 304, la courbe 1 représente les variations de l'amplitude a''/a (130.5) en fonction de l'incidence. On constate que l'amplitude change de signe lorsqu'on

131. Amplitude de la vibration réfléchie: vibration perpendiculaire au plan d'incidence. La vibration incidente, caractérisée par le champ électrique $\boldsymbol{E}$ est maintenant perpendiculaire au plan xoz, c'est-à-dire parallèle à Oy (Fig. 305). passe de l'incidence normale à l'incidence rasante. Il y a donc un changement brusque de la phase égal à π lorsque le second milieu est plus réfringent ($n > 1$). Le champ $\boldsymbol{H}$ fait l'angle i avec ox et le champ $\boldsymbol{H}'$ l'angle r avec Ox. Soit b l'amplitude du champ électrique. On aura pour l'onde incidente, φ étant la phase

$$\left.\begin{array}{ll} E_x = 0, & E_y = b\,e^{j(\omega t - \varphi)}, \\ H_x = -b\cos i\,e^{j(\omega t - \varphi)}, & H_y = 0. \end{array}\right\} \qquad (131.1)$$

Si b' est l'amplitude du champ électrique transmis, on aura pour l'onde transmise

$$E'_x = 0, \qquad\qquad E'_y = b'\, e^{j(\omega t - \varphi')}, \left.\right\}$$
$$H'_x = -\, n\, b' \cos r \; e^{j(\omega t - \varphi')}, \qquad H'_y = 0. \qquad\qquad (131.2)$$

Enfin, pour l'onde réfléchie, si b'' est l'amplitude du champ électrique réfléchi

$$E''_x = 0, \qquad\qquad E''_y = b''\, e^{j(\omega t - \varphi'')}, \left.\right\}$$
$$H''_x = b'' \cos i \; e^{j(\omega t - \varphi'')}, \qquad H''_y = 0. \qquad\qquad (131.3)$$

Les équations de passage (129.3) donnent

$$(b - b'') \cos i = n\, b' \cos r, \left.\right\}$$
$$b + b'' = b'. \qquad\qquad (131.4)$$

D'où les formules de Fresnel:

$$b' = b\, \frac{2 \cos i \sin r}{\sin (i + r)}, \qquad b'' = -\, b\, \frac{\sin (i - r)}{\sin (i + r)} \qquad (131.5)$$

et l'intensité de la lumière réfléchie

$$I_\perp = \left(\frac{b''}{b}\right)^2 = \frac{\sin^2 (i - r)}{\sin^2 (i + r)} \, . \qquad (131.6)$$

On obtient évidemment la même intensité que pour la vibration parallèle au plan d'incidence lorsque i et r sont très petits (130.7). La courbe (2) de la Fig. 302 représente les variations de l'intensité lumineuse réfléchie (131.6) en fonction de l'angle d'incidence i. L'intensité croît constamment de $\left(\dfrac{n-1}{n+1}\right)^2$ à 1.

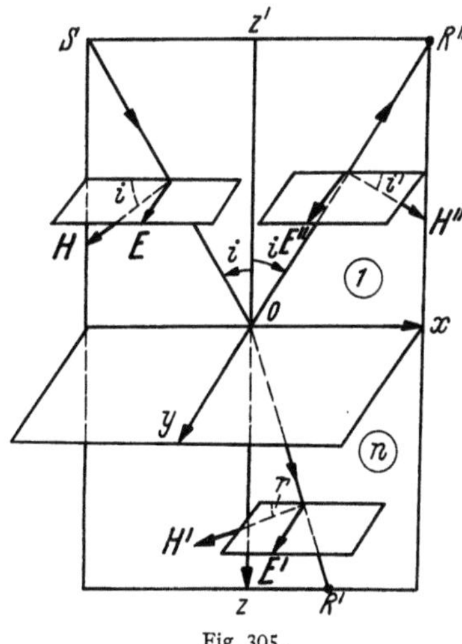

La courbe (2) de la Fig. 304 donne la variation de l'amplitude réfléchie (131.5) en fonction de i: le rapport b''/b conserve toujours le même signe.

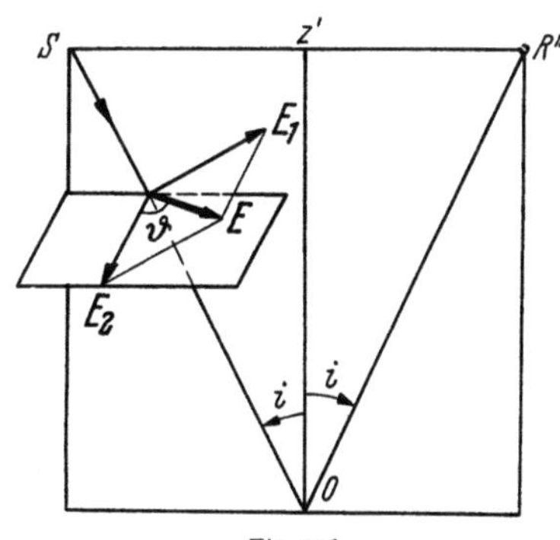

Fig. 305.

Fig. 306.

132. Cas d'une vibration faisant l'angle ϑ avec la normale au plan d'incidence. Lorsque la vibration incidente $\boldsymbol{E}$ est orientée d'une façon quelconque par rapport au plan d'incidence, on peut la décomposer en deux vibrations rectangulaires cohérentes $\boldsymbol{E_1}$ et $\boldsymbol{E_2}$: $\boldsymbol{E_1}$ dans le plan d'incidence et $\boldsymbol{E_2}$ perpendiculaire au plan d'incidence (Fig. 306). Soient ϑ l'angle de la vibration incidente $\boldsymbol{E}$ avec la normale au plan d'incidence, et A l'amplitude de $\boldsymbol{E}$.

Les amplitudes de $\boldsymbol{E}_1$ et $\boldsymbol{E}_2$ seront

$$\boldsymbol{E}_1:\; a = A \sin\vartheta, \qquad \boldsymbol{E}_2:\; b = A \cos\vartheta$$

d'où

$$\frac{a''}{A} = -\frac{\tan(i-r)}{\tan(i+r)}\sin\vartheta, \qquad \frac{b''}{A} = -\frac{\sin(i-r)}{\sin(i+r)}\cos\vartheta.$$

La vibration réfléchie, résultante de deux vibrations rectilignes rectangulaires cohérentes et de même phase, est une vibration rectiligne d'amplitude A'' donnée par

$$\frac{a''^2 + b''^2}{A^2} = \frac{A''^2}{A^2} = \frac{\tan^2(i-r)}{\tan^2(i+r)}\sin^2\vartheta + \frac{\sin^2(i-r)}{\sin^2(i+r)}\cos^2\vartheta. \tag{132.1}$$

Elle fait avec la normale au plan d'incidence l'angle ϑ'' tel que

$$\tan\vartheta'' = \frac{a''}{b''} = \frac{\cos(i+r)}{\cos(i-r)}\tan\vartheta. \tag{132.2}$$

133. Lumière incidente naturelle: polariseurs par réflexion. Les phénomènes d'interférences montrent que la lumière naturelle est formée par des vibrations transversales périodiques. Un rayon de lumière naturelle présente une symétrie complète par rapport à lui-même, mais cette symétrie apparente ne peut être expliquée par aucune forme permanente de vibration. On est amené à admettre que dans le plan d'onde, la vibration est complètement irrégulière si on la prend pendant un temps suffisant pour qu'elle produise un effet sensible. On peut la considérer comme elliptique pendant un certain nombre de périodes, mais avec de continuelles variations d'orientation et de forme.

On peut décomposer la vibration en deux vibrations rectilignes rectangulaires de même intensité moyenne, mais complètement indépendantes au point de vue des phases.

Puisque $a = b$, les formules (130.5) et (131.5) donnent l'intensité de la lumière réfléchie

$$\frac{a''^2 + b''^2}{A^2} = \frac{\tan^2(i-r)}{\tan^2(i+r)} + \frac{\sin^2(i-r)}{\sin^2(i+r)}. \tag{133.1}$$

Si on forme le rapport a''^2/b''^2

$$\frac{a''^2}{b''^2} = \frac{\cos^2(i+r)}{\cos^2(i-r)}. \tag{133.2}$$

On voit que, sauf pour $i=0$ et $i=\pi/2$ on a toujours $a''^2 < b''^2$. Dans la réflexion de la lumière naturelle, il y a prépondérance de la vibration perpendiculaire au plan d'incidence: la lumière réfléchie est partiellement polarisée dans le plan d'incidence.

A l'incidence Brewsterienne $a''=0$ et la formule (133.1) montre que la lumière réfléchie fournit uniquement des vibrations b'' perpendiculaires au plan d'incidence. Si un faisceau de lumière naturelle se réfléchit sous l'incidence Brewsterienne, la lumière réfléchie est totalement polarisée et la vecteur champ électrique qui lui correspond, est perpendiculaire au plan d'incidence. L'équation (133.1) montre que le facteur de réflexion, pour la vibration perpendiculaire au plan d'incidence, est

$$\frac{b''^2}{A^2} = \frac{\sin^2(i_B - r_B)}{\sin^2(i_B + r_B)}.$$

D'après (130.8) et (130.9), on obtient dans le cas d'une lame d'indice n

$$\frac{b''^2}{A^2} = \cos^2 2i = \left(\frac{n^2-1}{n^2+1}\right)^2. \tag{133.3}$$

La réflexion BREWSTERienne de la lumière naturelle sur du verre fournit des polariseurs peu lumineux. En effet, avec les verres dont on dispose, le facteur de réflexion donné par (133.3) est de l'ordre de 0,20 à 0,26 environ. Après réflexion sous cette incidence, on ne retrouve donc que 10 à 13 pour cent de l'énergie incidente.

On peut augmenter beaucoup la luminosité des polariseurs à réflexion par l'utilisation des verres recouverts d'une couche mince transparente, comme l'ont fait F. ABELÈS et H. SCHRÖDER. Soit e l'épaisseur de la couche mince et n_1 son indice (Fig. 307). Elle est déposée sur un verre d'indice n_2. L'angle d'incidence est i et dans la couche mince, l'angle de réfraction est r. Pour la vibration située dans le plan d'incidence, l'amplitude réfléchie à la surface de séparation air-couche, est

$$r_1 = -\frac{\tan(i-r)}{\tan(i+r)}$$

et l'amplitude réfléchie à la surface de séparation couche-verre, est

$$r_2 = -\frac{\tan(r-r')}{\tan(r+r')} \cdot$$

Posons

$$\varphi = \frac{4\pi n_1 e \cos r}{\lambda} \cdot$$

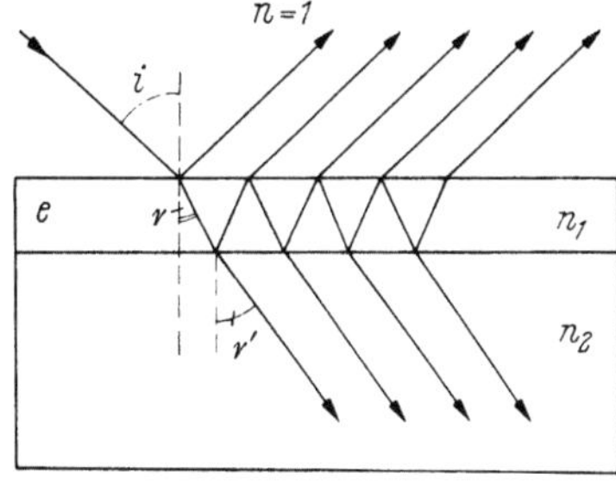

Fig. 307.

Un calcul analogue à celui de la Sect. 36 peut être fait dans le cas de réflexions multiples sur des surfaces de facteurs de réflexion r_1^2 et r_2^2 différents.

L'amplitude totale réfléchie A par l'ensemble de la couche mince et du support en verre, est alors donnée par l'expression

$$A = \frac{r_1 e^{j\varphi} + r_2}{e^{j\varphi} + r_1 r_2} \cdot$$

Puisque l'on donne à r_1 et r_2 les valeurs précédentes, on a l'amplitude réfléchie pour la vibration située dans le plan d'incidence. Si on annule la lumière réfléchie correspondant à cette vibration, après réflexion sur le système, nous aurons seulement une vibration perpendiculaire au plan d'incidence et par conséquent, un polariseur par réflexion.

Pour qu'il en soit ainsi, il faut annuler le carré du module de $r_1 e^\varphi + r_2$ ce qui entraîne les deux conditions suivantes

$$\varphi = (2K+1)\pi, \qquad r_1 = r_2.$$

La première condition fixe l'épaisseur de la couche qui doit être telle que l'on ait

$$e = \frac{(2K+1)\lambda}{4 n_1 \cos r},$$

la seconde s'écrit en explicitant r_1 et r_2

$$a \sin^4 i - b \sin^2 i + c = 0$$

avec

$$a = n_1^8 - n_2^4,$$
$$b = n_1^2 n_2^2 (n_1^6 - 2n_2^2),$$
$$c = n_1^4 n_2^2 (n_1^4 - n_2^2).$$

On a donc une incidence Brewsterienne pour l'incidence i, et le facteur de réflexion pour la vibration perpendiculaire au plan d'incidence est dans ces conditions

$$R_\perp = \left(\frac{n_1^2 \cos^2 r - n_2 \cos i \cos r'}{n_1^2 \cos^2 r + n_2 \cos i \cos r'} \right)^2 .$$

Pour avoir un polariseur plus lumineux que le polariseur par réflexion vitreuse ordinaire, il faut que l'expression précédente donne une valeur plus élevée que l'expression (133.3). Ceci a lieu dès que $n_1 > n_2$, c'est-à-dire dès que l'on utilise une couche plus réfringente que le support. La Fig. 308 montre les variations de $R_\perp$ avec n_1 pour deux indices du support. Par exemple, si on utilise une couche mince d'oxyde de titane d'indice $n_1 = 2,5$, on aura $R_\perp = 0,80$, soit un polariseur réfléchissant 40 pour 100 du flux incident et travaillant sous une incidence égale à 74°.

Un inconvénient des couches minces est leur réflexion sélective. Considérons le cas précédent d'une couche mince d'oxyde de titane: prenons une épaisseur e telle que $\varphi = \pi$ pour $\lambda = 0,550\,\mu$ et $i = 74° \, 25' \, 30''$. Pour cette longueur d'onde, le facteur de réflexion pour la vibration parallèle au plan d'incidence est nul. Il est égal à 0,03 pour $\lambda = 0,4\,\mu$ et à 0,02 pour $\lambda = 0,8\,\mu$. On voit que des polariseurs de ce type seraient malgré tout utilisables assez bien dans tout le spectre visible.

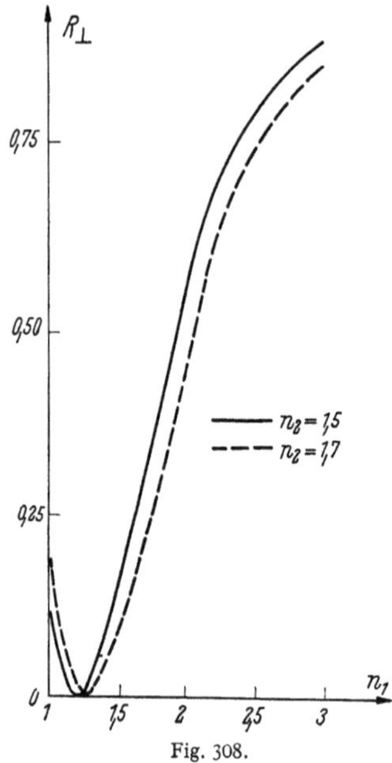

Fig. 308.

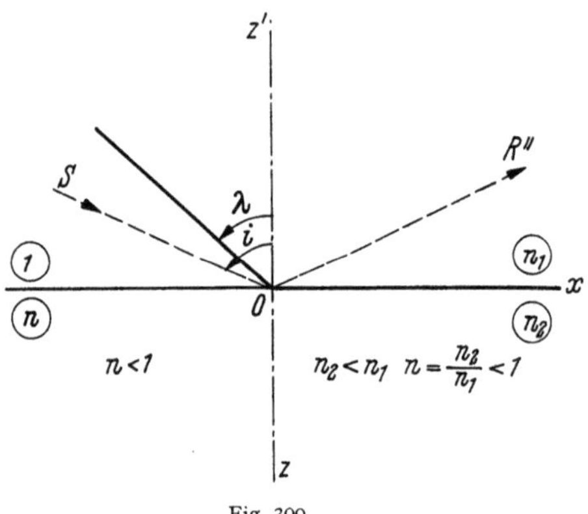

Fig. 309.

134. Réflexion totale. Les formules précédentes sont parfaitement valables lorsqu'il y a réflexion d'un milieu plus réfringent sur un milieu moins réfringent. L'indice n, indice relatif du deuxième milieu par rapport au premier, est alors inférieur à l'unité. Lorsqu'on arrive à la réflexion totale (Fig. 309), l'angle d'incidence λ est tel que $\sin \lambda = n$ et si l'angle d'incidence dépasse cette valeur ($\sin i > n$), la loi de Descartes $\sin r = \dfrac{\sin i}{n}$ donne pour $\sin r$ une valeur supérieure à l'unité et à laquelle ne correspond aucun rayon réfracté.

Cherchons ce que deviennent les formules de Fresnel dans ce cas.

Ces formules s'expriment en fonction de $\sin i$, $\cos i$, $\sin r$, $\cos r$. Or, pour $i > \lambda$, $\sin i$ et $\cos i$ ont des valeurs bien déterminées et quoiqu'il n'y ait pas de rayon réfracté, on peut considérer que $\sin r = \dfrac{\sin i}{n}$ a une valeur réelle, supérieure à l'unité.

Quant à $\cos r$, il est imaginaire puisque l'on a

$$\cos r = \sqrt{1 - \sin^2 r} = \sqrt{1 - \frac{\sin^2 i}{n^2}}.$$

En portant la valeur réelle de $\sin r$ et la valeur imaginaire de $\cos r$ dans les formules de FRESNEL, on obtient pour les amplitudes réfléchies a'' et b'' des valeurs complexes. Posons

$$\left.\begin{aligned} \sin i &= \alpha, & \sin r &= \beta, \\ \cos i &= \alpha', & \cos r &= \pm j\, \frac{\sqrt{\sin^2 i - n^2}}{n} = j\beta'. \end{aligned}\right\} \tag{134.1}$$

on aura

$$\frac{a''}{a} = - \frac{\alpha'\beta - j\alpha\beta'}{\alpha'\beta + j\alpha\beta'} \cdot \frac{\alpha\beta - j\alpha'\beta'}{\alpha\beta + j\alpha'\beta'}; \qquad \frac{b''}{b} = \frac{\alpha'\beta - j\alpha\beta'}{\alpha'\beta + j\alpha\beta'}. \tag{134.2}$$

a''/a et b''/b sont égaux à des quotients de quantités conjuguées, leurs modules sont donc égaux à 1 et on a

$$\left(\frac{a''}{a}\right)^2 = \left(\frac{b''}{b}\right)^2 = 1.$$

Les amplitudes réfléchies sont égales aux amplitudes incidentes, ce qui montre bien que l'énergie lumineuse incidente est totalement réfléchie.

Soient $\varphi_{a''}$ et $\varphi_{b''}$ les phases des vibrations réfléchies a'' et b'', c'est-à-dire les arguments de ces quantités complexes. La différence de phase $\varphi_{a''} - \varphi_{b''}$ entre les deux vibrations réfléchies a'' et b'' est l'argument de la quantité complexe a''/b''

$$\frac{a''}{b''} = \frac{\cos(i+r)}{\cos(i-r)}. \tag{134.3}$$

Soit ϑ l'argument du numérateur et ϑ' celui du dénominateur, on a

$$\tan\vartheta = -\frac{\alpha'\beta'}{\alpha\beta}, \qquad \tan\vartheta' = \frac{\alpha'\beta'}{\alpha\beta} = -\tan\vartheta$$

d'où

$$\vartheta' = \pi - \vartheta$$

et l'argument φ du rapport a''/b'' sera

$$\varphi = (\pi - \vartheta') - \vartheta' = \pi - 2\vartheta'.$$

Calculons l'argument ϑ' du dénominateur. On a

$$\cos(i-r) = \sin i \sin r + \cos i \cos r = \frac{\sin^2 i}{n} \pm j \cdot \frac{\cos i \sqrt{\sin^2 i - n^2}}{n}.$$

L'étude des ondes évanescentes montre que l'on doit prendre le signe moins. on aura donc

$$\tan\frac{\varphi}{2} = \cot\vartheta' = -\frac{\sin^2 i}{\cos i \sqrt{\sin^2 i - n^2}}. \tag{134.4}$$

Pour $\sin i = n$ (incidence limite) et pour $\cos i = 0$ (incidence rasante) on a $\tan \varphi/2 = \infty$, c'est-à-dire $\varphi = \pi$. Il est plus naturel d'avoir la même phase pour les deux vibrations a et a'' lorsque $i = \lambda$ et $i = \pi/2$. On considère alors, non pas les différences de phase φ, mais les différences de phase $\varphi' = \varphi - \pi$.

Pour les autres valeurs de i comprises entre ces deux limites, φ varie et passe par un maximum que l'on peut calculer en dérivant (134.4).

En annulant la dérivée de l'expression (134.4) on obtient

$$\sin^2 i = \frac{2n^2}{1 + n^2} \qquad (134.5)$$

d'où la différence de phase maximum φ_m entre les deux vibrations réfléchies

$$\tan \frac{\varphi_m}{2} = - \frac{2n}{1 - n^2}. \qquad (134.6)$$

Par conséquent, si on considère un faisceau de lumière polarisée rectilignement, correspondant à deux vibrations a et b en phase, après réflexion totale d'un milieu plus réfringent sur un milieu moins réfringent, les deux vibrations a'' et b'' réfléchies totalement présentent entre elles une différence de phase maximum donnée par l'expression (134.6).

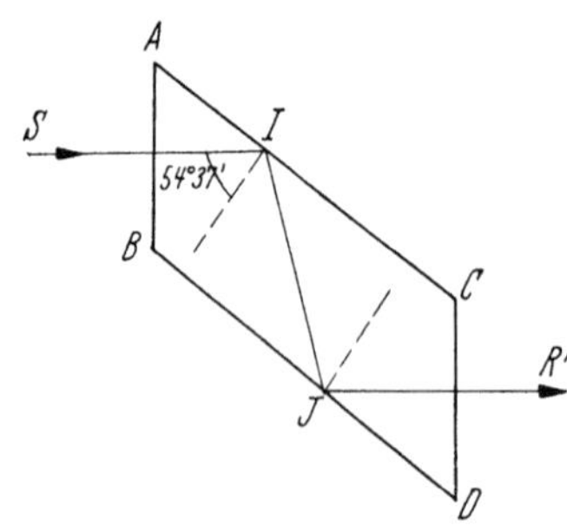

Fig. 310.

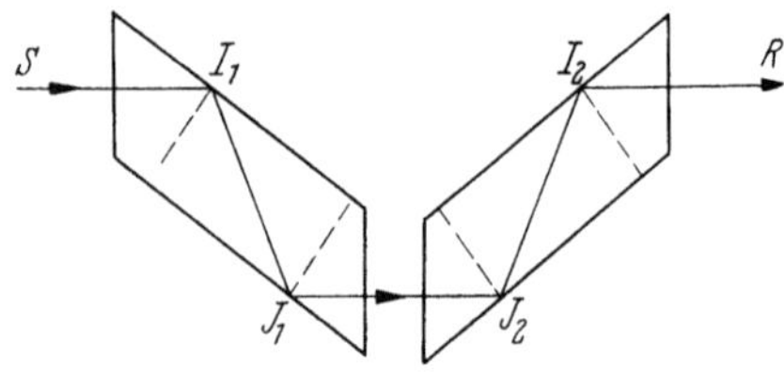

Fig. 311.

Pour un verre d'indice 1,51, on doit donner à n dans la formule (134.5) la valeur

$$n = \frac{n_2}{n_1} = \frac{1}{1,51} = 0,66.$$

On trouve $i = 51° 20'$ et

$$\varphi_m = \pi + 45° 36'; \qquad \varphi'_m = \varphi_m - \pi = 45° 36'.$$

On peut choisir un angle d'incidence tel que le déphasage φ'_m soit exactement égal à 45°. La formule (134.4) donne alors les deux valeurs $i = 48° 37'$ et $i = 54° 37'$ pour $n = 1,51$. Si la vibration incidente est orientée à 45° du plan d'incidence, on doit avoir après deux réflexions internes, un déphasage égal à $\pi/2$ entre les deux vibrations réfléchies. La vibration émergente est donc une vibration circulaire. L'expérience a été réalisée par Fresnel au moyen d'un parallélépipède de verre $ABCD$ (Fig. 310). On a l'angle $BAC = CDB = 54° 37'$ et le rayon incidente SI est normal à la face d'entrée AB. Après les deux réflexions en I et J sous l'incidence de $54° 37'$, le rayon émergent transporte effectivement des vibrations circulaires si les vibrations incidentes sont des vibrations rectilignes à 45° du plan d'incidence. Le parallélépipède de Fresnel est une lame quart d'onde qui peut être employée en lumière blanche car l'indice n du verre varie peu avec la longueur d'onde. L'association de deux parallélépipèdes (Fig. 311) permet de réaliser une lame demi-onde en redonnant au rayon émergent la direction du rayon incident. Malheureusement, par suite des dimensions que l'on doit donner au parallélépipède de Fresnel pour qu'il soit utilisable, l'épaisseur de verre traversé est importante et il est difficile d'éviter la trempe et la biréfringence.

135. Ondes évanescentes. Nous avons vu que le champ électrique, solution de l'équation de propagation, avait dans le second milieu, la forme suivante

$$E' = a' e^{j(\omega t - \varphi)}. \qquad (135.1)$$

En prenant comme origine des phases la phase au point O (Fig. 312), la phase φ en un point M' quelconque $M'(xz)$ du rayon réfracté OR' sera

$$\varphi = \frac{2\pi}{\lambda}\, n\, (x \sin r + z \cos r),$$

n étant l'indice relatif des 2 milieux. L'expression (135.1) devient

$$E' = a'\, e^{j\omega\left(t - n\frac{x\sin r + z\cos r}{v}\right)}.$$

v/n étant la vitesse de propagation suivant OR' dans le second milieu. Dans le cas où il y a réflexion totale, l'amplitude $\boldsymbol{a}'$ et cos r sont imaginaires. L'amplitude a' est de la forme $a' = a'_0 e^{-i\varphi}$ et cos r est donné par (134.1) où l'on prend le signe moins. Le champ électrique prend alors la forme

$$E' = a'_0\, e^{-\frac{2\pi n\beta'}{\lambda}z}\, e^{-i\varphi}\, e^{j\omega\left(t - \frac{x\sin i}{v}\right)}. \qquad (135.2)$$

La partie réelle $Re\, E'$ de cette expression représente les vibrations dans le second milieu. On a

$$\left.\begin{aligned} Re\, E' = a'_0\, e^{-\frac{2\pi n\beta'}{\lambda}z}\, \cos\, \times \\ \times \left[\omega\left(t - \frac{x\sin i}{v}\right) - \varphi\right]. \end{aligned}\right\} \qquad (135.3)$$

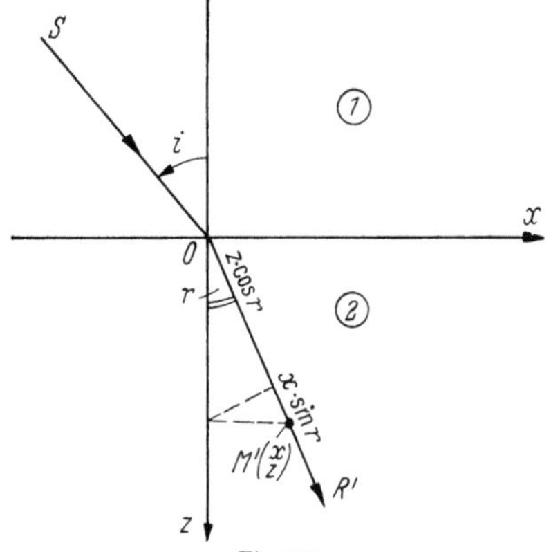

Fig. 312.

Il existe donc dans le second milieu une onde qui se propage parallèlement à la surface de séparation des deux milieux et son amplitude diminue dans une direction perpendiculaire à ce plan, c'est-à-dire suivant Oz. Le facteur $e^{-j\frac{2\pi n\beta'}{\lambda}}$ montre que l'amplitude a déjà une valeur très petite à une distance de quelques longueurs d'onde du plan xOy. Une telle onde est appelée «onde évanescente».

Notons que le choix du signe moins est conforme à l'expérience car un signe plus aurait donné une amplitude croissant rapidement avec z et un tel résultat serait absolument contraire aux observations.

Les ondes évanescentes ont une existence réelle et l'expérience montre qu'il y a réflexion totale d'un milieu plus réfringent sur un milieu moins réfringent, à la condition que le second milieu n'ait pas une épaisseur trop faible. Si le second milieu est constitué par une couche très mince, la lumière s'échappe. Ce phénomène, produit par l'existence des ondes évanescentes, a été utilisé par A. F. Turner dans son filtre interférentiel dit à «reflexion totale frustrée».

136. Polarisation de la lumière transmise. $\alpha)$ *Lumiére incidente naturelle.* La lumière transmise sera encore formée de deux composantes rectangulaires incohérentes et on aura, en faisant $a = b$ dans les deux équations (130.5) et (131.5)

$$\frac{b'^2}{a'^2} = \cos^2(i - r).$$

Observons le phénomène avec une lame de verre à faces parallèles (Fig. 313): entre O et O' le rapport des deux composantes incohérentes transmises est égal au rapport précédent. Après O', on a encore la même chose, mais il y a permutation des angles i et r. Le rapport des deux composantes transmises suivant

$O'R'$ est donc

$$\frac{b'^{\,2}_{R'}}{a'^{\,2}_{R'}} = \cos^4 (i - r).$$

Il y a prépondérance de la vibration a' située dans le plan d'incidence, la lumière est maintenant partiellement polarisée dans un plan perpendiculaire au plan d'incidence.

La polarisation par transmission est maximum lorsque $a'_{R'}$ correspond à l'incidence Brewsterienne, car pour cette incidence $a'' = 0$ et il n'y a pas de composante dans le plan d'incidence qui soit réfléchie. D'après (130.8) et (130.9), on a

$$\left(\frac{b'^{\,2}_{R'}}{a'^{\,2}_{R'}}\right)_B = \left(\frac{2n}{1 + n^2}\right)^4.$$

Pour $n = 1,5$, ce rapport est égal à 0,73. On peut augmenter la proportion de lumière polarisée en utilisant une série de lames à faces parallèles (pile de glaces).

Avec p lames sous l'incidence Brewsterienne, on a

$$\left(\frac{2n}{1 + n^2}\right)^{4p},$$

soit 0,04 pour $p = 10$; la lumière n'est pas encore totalement polarisée, d'autant plus que calcul ne tient pas compte d'une partie notable qui est dépolarisée par les réflexions multiples ou les défauts des lames.

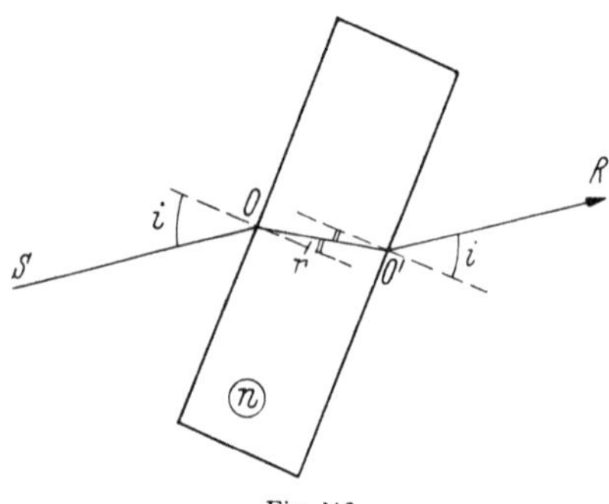

Fig. 313.

β) *Lumière incidente polarisée rectilignement*. Soit ϑ l'angle de la vibration incidente avec la normale an plan d'incidence. Les deux vibrations a et b sont maintenant cohérentes, de même phase et $a/b = \tan \vartheta$. Après traversée de la lame à faces parallèles, on a

$$\frac{b'_{R'}}{a'_{R'}} = \frac{b}{a} \cos^2 (i - r) = \cot \vartheta \cos^2 (i - r).$$

La vibration transmise est une vibration rectiligne qui a tourné d'un angle ϑ' donné par

$$\cot \vartheta' = \cot \vartheta \cos^2 (i - r).$$

La vibration transmise se rapproche du plan d'incidence.

II. Polarisation par double réfraction.

a) Expériences sur la double réfraction.

137. Rayon ordinaire et rayon extraordinaire. Le phénomène de la double réfraction, qui peut s'observer avec la plupart des corps cristallisées transparentes, est particulièrement marqué pour le carbonate de calcium $CaCO^3$ ou spath d'Islande. C'est un cristal du système ternaire qui se présente sous forme de rhomboèdres. Les faces des cristaux de spath ne sont pas nécessairement des losanges, mais peuvent être des parallélogrammes à côtés inégaux (Fig. 314) dont les angles des faces qui se réunissent en A sont toujours égaux à $101°\,53'$. La direction AA'' qui fait en A des angles égaux avec les plans des trois faces qui s'y réunissent, est l'axe ternaire ou axe optique du spath. Une rotation de $2\pi/3$ autour d'une parallèle à AA'' ramène les faces parallèlement à leurs positions primitives. Le plan contenant l'axe AA'' et la normale AN à la face d'entrée, est le plan de section principale du cristal.

Considérons un faisceau étroit de rayons lumineux parallèles tombant normalement sur la face d'entrée $ABCD$ du cristal de spath (Fig. 315).

On constate que le faisceau lumineux se dédouble à l'entrée du spath. Le premier faisceau de lumière suit les lois de DESCARTES et n'est pas dévié, c'est le faisceau ordinaire. Le deuxième faisceau n'obéit pas à ces lois, c'est le faisceau extraordinaire qui est dévié.

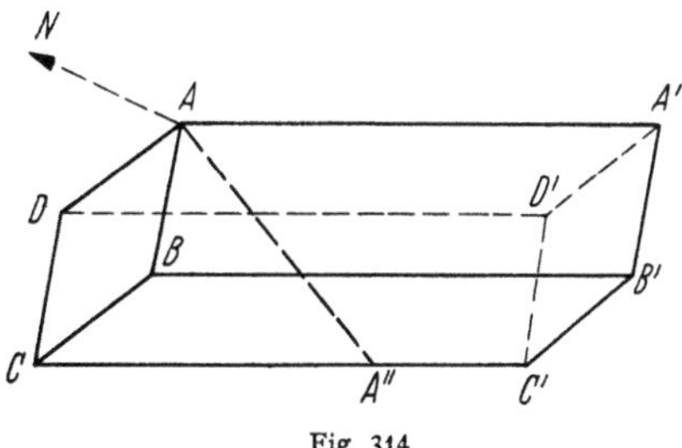

Fig. 314.

A tout rayon incident correspond un rayon ordinaire et un rayon extraordinaire qui se trouve dans le plan de section principale. En faisant tourner le path autour d'un axe normal aux faces AC et $A'C'$, le rayon ordinaire est fixe et le rayon extraordinaire tourne autour du rayon ordinaire.

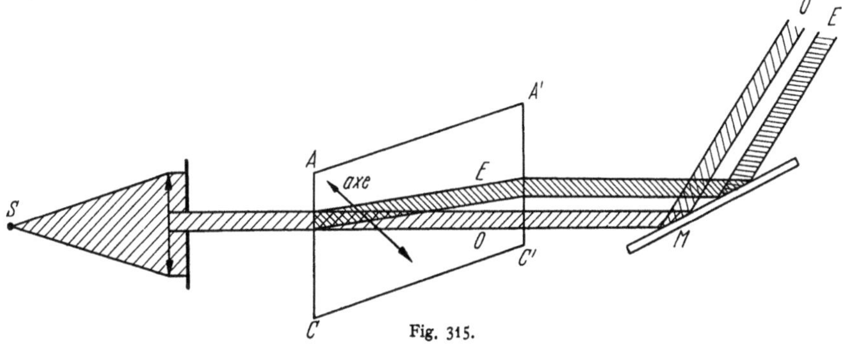

Fig. 315.

Plaçons-nous comme l'indique la Fig. 315, et recevons les deux faisceaux sur un miroir M (réflexion vitreuse) sous l'incidence BREWSTERienne. On constate qu'après réflexion sur M, le faisceau extraordinaire est éteint. Les vibrations du rayon extraordinaire sont situées dans le plan de la figure (Fig. 316). Tournons

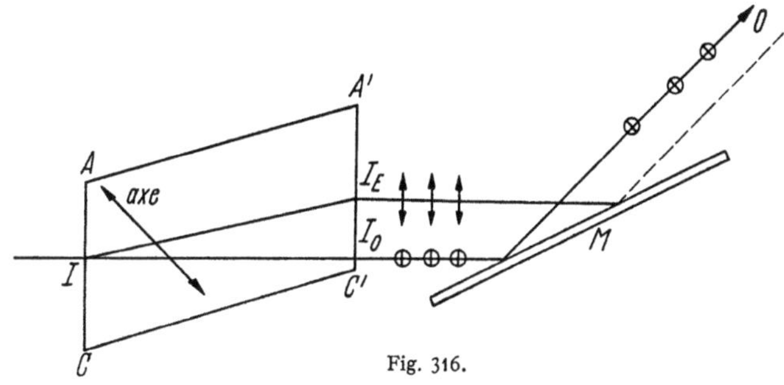

Fig. 316.

de 90° le miroir M, ce sont maintenant les rayons ordinaires qui sont éteints: les vibrations du rayon ordinaire sont perpendiculaires au plan de la Fig. 315. Le rayon ordinaire est polarisé dans le plan de section principale et vibre dans le plan perpendiculaire au plan de section principale. Le rayon extraordinaire est polarisé dans le plan perpendiculaire au plan de section principale, et vibre dans le plan de section principale. Le rayon ordinaire et le rayon extraordinaire, transmis par un prisme biréfringent, sont polarisés à angle droit.

La loi de Malus (Sect. 120) est bien vérifiée: le plan de polarisation correspondant à M est le plan d'incidence, le rayon E est éteint lorsque le plan perpendiculaire au plan de section principale (plan de polarisation du rayon extraordinaire) est perpendiculaire au plan d'incidence sur M. Lorsque ces deux plans font entre eux l'angle α, l'intensité transmise est donnée par (120.1). De même pour le rayon ordinaire.

138. Prisme de Nicol. Un prisme biréfringent tel que le prisme de spath précédent, donne deux faisceaux, l'un ordinaire, l'autre extraordinaire, qui

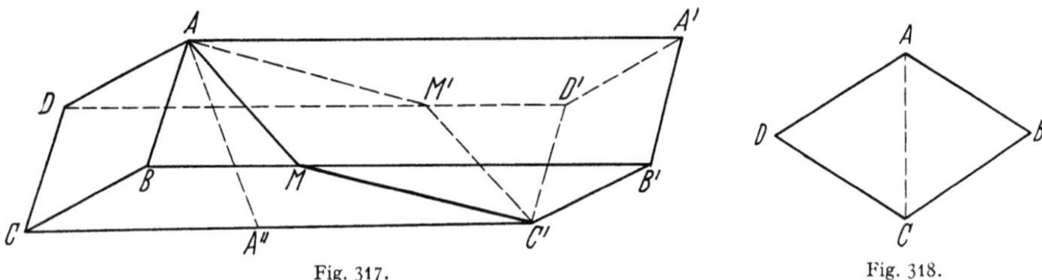

Fig. 317.   Fig. 318.

empiètent plus ou moins l'un sur l'autre si le faisceau est un peu large. Il est commode de supprimer l'un des faisceaux et de conserver au faisceau utilisé sa direction primitive.

Une solution a ètè fournie en 1864 par Nicol.

Ces prismes appelés «nicols» ne laissent passer que le rayon extraordinaire. Ils sont obtenus de la façon suivante: on part d'une lame épaisse de spath limitée par les plans de clivage et dont les arêtes AB et AD sont égales (Fig. 317), l'arête AA' ayant une longueur égale à trois fois AB environ. L'axe AA'' est dans

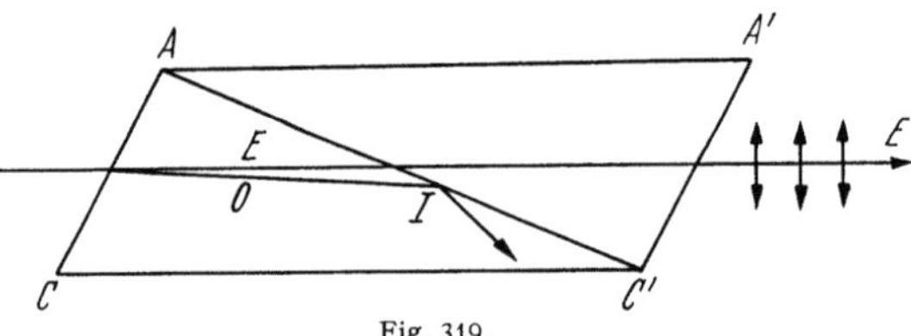

Fig. 319.

le plan $ACA'C'$ qui est le plan de section principale du nicol. Si on regarde de face le nicol, on a l'aspect de la Fig. 318: le plan $ACA'C'$, qui est le plan de section principale, contient la petite diagonale AC.

On scie le prisme suivant le plan $AMM'C'$ perpendiculaire au plan $ACA'C'$ et on recolle les deux moitiés au baume de Canada. On verra plus loin que le spath possède pour le rayon ordinaire un indice $n_0 = 1{,}658$ et pour le rayon extraordinaire l'indice $n_e = 1{,}486$. L'indice du baume de Canada étant voisin de 1,55, il peut y avoir réflexion totale du rayon ordinaire sur le plan $AMM'C'$. L'inclinaison de ce plan a été choisie pour qu'il en soit ainsi et on a la Fig. 319. Le rayon ordinaire subit la réflexion totale en I, seul le rayon extraordinaire est transmis, ses vibrations étant dirigées dans le plan de section principale, c'est-à-dire parallèlement à la petite diagonale.

Nous donnerons plus loin quelques détails complémentaires concernant les nicols ainsi que d'autres types de prismes polariseurs.

b) Propagation d'une onde plane dans un milieu anisotrope transparent.

139. Equations fondamentales. Certaines propriétés d'un milieu peuvent être représentées par un nombre algébrique, comme par exemple la masse spécifique, la température, etc. ... Ce sont les proppriétés scalaires du milieu. D'autres

propriétés, les propriétés vectorielles, comme la dilatation, l'aimantation, etc. ...
sont représentées par un vecteur. Dans un milieu isotrope et homogène, les propriétés vectorielles sont les mêmes dans toutes les directions; elles dépendent au
contraire de la direction si le milieu est anisotrope. Les cristaux comme le spath,
dont nous venons de parler, sont des milieux anisotropes.

Les deux équations (124.3) et (124.8) qui sont indépendantes de la nature du
milieu, peuvent s'appliquer à un milieu anisotrope. Des quatre équations (124.3),
(124.8), (124.10) et (124.11)

$$\left. \begin{array}{ll} \dfrac{4\pi\boldsymbol{j}}{c} = \mathrm{rot}\,\boldsymbol{H}, & \dfrac{1}{c}\dfrac{\partial \boldsymbol{B}}{\partial t} = -\,\mathrm{rot}\,\boldsymbol{E}, \\[3mm] \boldsymbol{B} = \mu\,\boldsymbol{H}, & \boldsymbol{j} = \dfrac{\varepsilon}{4\pi}\dfrac{\partial \boldsymbol{E}}{\partial t} \end{array} \right\} \tag{139.1}$$

seule l'équation reliant $\boldsymbol{j}$ à $\boldsymbol{E}$ sera modifiée dans un milieu anisotrope. Pour
obtenir cette équation dans un milieu anisotrope, on utilise le vecteur induction
électrique $\boldsymbol{D}$ défini en électrostatique. Il est lié au courant de déplacement $\boldsymbol{j}$
dans un diélectrique par la relation

$$\boldsymbol{j} = \frac{1}{4\pi}\frac{\partial \boldsymbol{D}}{\partial t} \tag{139.2}$$

valable aussi bien pour un diélectrique isotrope que pour un diélectrique anisotrope.

Les trois premières éq. (139.1) et (139.2) donnent en prenant toujours $\mu = 1$,
c'est-àdire $\boldsymbol{B} = \boldsymbol{H}$,

$$\frac{1}{c}\frac{\partial \boldsymbol{D}}{\partial t} = \mathrm{rot}\,\boldsymbol{H}, \qquad \frac{1}{c}\frac{\partial \boldsymbol{H}}{\partial t} = -\,\mathrm{rot}\,\boldsymbol{E}. \tag{139.3}$$

Il faut relier maintenant $\boldsymbol{D}$ au vecteur champ électrique $\boldsymbol{E}$.

Dans le cas d'un diélectrique isotrope, ces deux vecteurs ont même direction
et on a $\boldsymbol{D} = \varepsilon\boldsymbol{E}$ où ε est la constante diélectrique du milieu isotrope.

Dans le cas d'un milieu anisotrope, $\boldsymbol{D}$ et $\boldsymbol{E}$ n'ont plus la même direction. On
montre qu'il existe dans un milieu diélectrique anisotrope trois axes de symétrie
électrique par rapport auxquels les relations entre $\boldsymbol{D}$ et $\boldsymbol{E}$ prennent une forme
simple.

Prenons comme axes de coordonnées rectangulaires $O\,x\,y\,z$ les trois axes de
symétrie électrique du milieu anisotrope.

Soient $D_x D_y D_z E_x E_y E_z$ les projections des vecteurs $\boldsymbol{D}$ et $\boldsymbol{E}$ sur ces axes, on a

$$D_x = \varepsilon_1 E_x, \qquad D_y = \varepsilon_2 E_y, \qquad D_z = \varepsilon_3 E_z. \tag{139.4}$$

$\varepsilon_1\,\varepsilon_2\,\varepsilon_3$ sont les constantes diélectriques principales. Les relations (139.3) et
(139.4) résolvent le problème de la propagation dans le milieu anisotrope.

Comme dans le cas de la propagation dans un milieu isotrope, ces équations
ne tiennent pas compte de la dispersion et on lève la difficulté de la même façon
(Sect. 125).

On a vu d'autre par (Sect. 126) qu'un vecteur dont la divergence est nulle
(125.3) est situé dans le plan d'onde. Par conséquent les vecteurs $\partial \boldsymbol{D}/\partial t$ et $\partial \boldsymbol{H}/\partial t$
égaux à des rotationnels sont des vecteurs de divergence nulle: ils sont situés
dans le plan d'onde. D'après la Sect. 127, on voit que les deux vecteurs $\boldsymbol{D}$ et $\boldsymbol{H}$
sont perpendiculaires, en phase et que le trièdre qu'ils forment avec la normale
$O N$ à l'onde dirigée dans le sens de propagation, est un trièdre à gauche défini
par la règle d'Ampère (Fig. 320). On montrerait de même que le vecteur $\boldsymbol{E}$

est normal au champ magnétique H et en phase avec lui, mais n'étant pas confondu avec D, le vecteur E n'est plus dans le plan d'onde. Il est dans le plan DON puisque E est perpendiculaire à H et sa projection sur le plan d'onde est dirigée dans le même sens que D.

D'après la Sect. 128, la direction de propagation de l'énergie, c'est à dire la direction du rayon lumineux est la direction OP normale aux deux vecteurs E et H. Alors que dans un milieu isitrope, OP serait confondu avec ON et E avec D ($\alpha = 0$) il n'en est plus de même dans un milieu anisotrope.

Dans les milieux isotropes, on est conduit à représenter la vibration lumineuse par le vecteur électrique E car il n'y a pas lieu de distinguer dans ce cas entre les vecteurs E et D. Pour représenter à la fois les phénomènes dans les milieux anisotropes et isotropes, il est avantageux d'assimiler la vibration lumineuse au vecteur induction électrique D. On peut interpréter ainsi les phénomènes observés dans les milieux isotropes où D et E sont proportionnels et dans les milieux anisotropes, ce qui permettra de conserver la propriété fondamentale de transversalité des ondes.

140. Equation de propagation du champ électrique. Considérons une onde plane monochromatique et polarisée rectilignement. Soit A l'amplitude de la vibration, c'est-à-dire l'amplitude du vecteur induction électrique dont les cosinus directeurs sont a_x, a_y, a_z, les axes de coordonnées étant les axes de symétrie électrique. De même B étant l'amplitude du champ électrique dont les cosinus directeurs sont b_x, b_y, b_z on peut écrire

$$D_x = A\,a_x\,e^{j\,(\omega t - \varphi)}, \quad D_y = A\,a_y\,e^{j\,(\omega t - \varphi)}, \quad D_z = A\,a_z\,e^{j\,(\omega t - \varphi)}, \left.\begin{matrix} \\ \\ \end{matrix}\right\} \quad (140.1)$$
$$E_x = B\,b_x\,e^{j\,(\omega t - \varphi)}, \quad E_y = B\,b_y\,e^{j\,(\omega t - \varphi)}, \quad E_z = B\,b_z\,e^{j\,(\omega t - \varphi)},$$

E et D étant en phase (Sect. 139). Par analogie avec les milieux isotropes, on peut poser

$$\varepsilon_1 = n_1^2, \quad \varepsilon_2 = n_2^2, \quad \varepsilon_3 = n_3^2,$$

n_1, n_2, n_3 étant les indices principaux du milieu anisotrope. Les éq. (139.4) s'écrivent

$$D_x = n_1^2 E_x, \quad D_y = n_2^2 E_y, \quad D_z = n_3^2 E_z. \quad (140.2)$$

Le comparaison des éq. (124.12) et (125.2) avec (139.3) permet d'écrire

$$\frac{1}{c^2}\,\frac{\partial^2 D}{\partial t^2} = \Delta E - \operatorname{grad}(\operatorname{div} E).$$

Dans le cas des milieux isotropes, on avait $\operatorname{div} E = 0$; il n'en est plus ainsi maintenant. D'après (140.2) on aura pour les trois composantes E_x, E_y, E_z:

$$\frac{n_1^2}{c^2}\,\frac{\partial^2 E_x}{\partial t^2} = \Delta E_x - \frac{\partial}{\partial x}(\operatorname{div} E), \left.\begin{matrix} \\ \\ \\ \\ \\ \end{matrix}\right.$$
$$\frac{n_2^2}{c^2}\,\frac{\partial^2 E_y}{\partial t^2} = \Delta E_y - \frac{\partial}{\partial y}(\operatorname{div} E), \left.\begin{matrix} \\ \end{matrix}\right\} \quad (140.3)$$
$$\frac{n_3^2}{c^2}\,\frac{\partial^2 E_z}{\partial t^2} = \Delta E_z - \frac{\partial}{\partial z}(\operatorname{div} E).$$

Ces équations constituent les équations de propagation du champ électrique dans un milieu anisotrope.

Remplaçons E_x, E_y, E_z par leurs valeurs tirées des équations (140.1). Soient α, β, γ les cosinus directeurs de ON (Fig. 320) et v la vitesse de propagation mesurée suivant la normale à l'onde. Posons $n = c/v$ indice du milieu anisotrope pour l'onde étudiée. D'après la Sect. 135, on pourra écrire

$$\omega t - \varphi = \omega\left(t - n\,\frac{\alpha\,x + \beta\,y + \gamma\,z}{c}\right)$$

d'où

$$\frac{\partial^2 E_x}{\partial t^2} = -\,B\,b_x\,\omega^2\,e^{j\omega\left(t - n\,\frac{\alpha\,x+\beta\,y+\gamma\,z}{c}\right)}.$$

On calcule de même $\varDelta E_x$ et $\dfrac{\partial}{\partial x}\,(\operatorname{div}\boldsymbol{E})$

$$\varDelta E_x = -\,B\,b_x\,\omega^2\,\frac{n^2}{c^2}\,(\alpha^2 + \beta^2 + \gamma^2)\,e^{j\omega\left(t - n\,\frac{\alpha\,x+\beta\,y+\gamma\,z}{c}\right)},$$

$$\frac{\partial}{\partial x}\,(\operatorname{div}\boldsymbol{E}) = -\,B\,\omega^2\,\frac{n^2}{c^2}\,\alpha\,(b_x\,\alpha + b_y\,\beta + b_z\,\gamma)\,e^{j\omega\left(t - n\,\frac{\alpha\,x+\beta\,y+\gamma\,z}{c}\right)}$$

en remplaçant dans la première des équations (140.3) et avec la relation $\alpha^2 + \beta^2 + \gamma^2 = 1$ on a

$$n_1^2\,b_x = n^2\,b_x - n^2\,\alpha\,(b_x\,\alpha + b_y\,\beta + b_z\,\gamma),$$

ainsi que les deux autres équations

$$n_2^2\,b_y = n^2\,b_y - n^2\,\beta\,(b_x\,\alpha + b_y\,\beta + b_z\,\gamma),$$
$$n_3^2\,b_z = n^2\,b_z - n^2\,\gamma\,(b_x\,\alpha + b_y\,\beta + b_z\,\gamma),$$

d'où

$$\left.\frac{(n^2 - n_1^2)\,b_x}{\alpha} = \frac{(n^2 - n_2^2)\,b_y}{\beta} = \frac{(n^2 - n_3^2)\,b_z}{\gamma}.\right\}\quad(140.4)$$

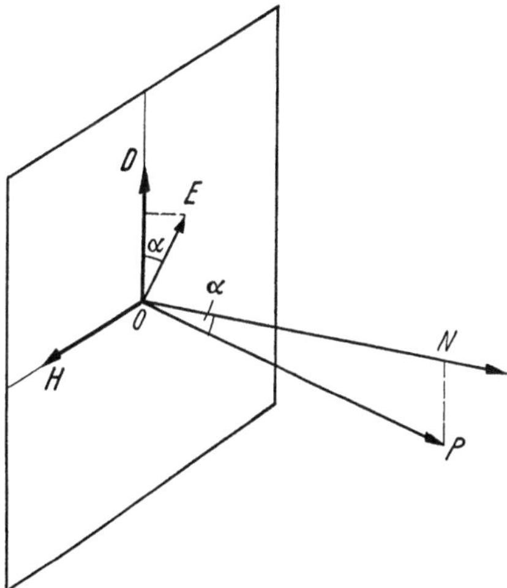

Fig. 320.

Or, d'après les équations (140.2), on a entre les cosinus directeurs des vecteurs $\boldsymbol{D}$ et $\boldsymbol{E}$ les relations

$$\frac{n_1^2\,b_x}{a_x} = \frac{n_2^2\,b_y}{a_y} = \frac{n_3^2\,b_z}{a_z}.\quad(140.5)$$

L'expression (140.4) devient

$$\frac{(n^2 - n_1^2)\,a_x}{n_1^2\,\alpha} = \frac{(n^2 - n_2^2)\,a_y}{n_2^2\,\beta} = \frac{(n^2 - n_3^2)\,a_z}{n_3^2\,\gamma}\quad(140.6)$$

que l'on peut écrire

$$\frac{\alpha\,a_x}{\dfrac{n_1^2\,\alpha^2}{n^2 - n_1^2}} = \frac{\beta\,a_y}{\dfrac{n_2^2\,\beta^2}{n^2 - n_2^2}} = \frac{\gamma\,a_z}{\dfrac{n_3^2\,\gamma^2}{n^2 - n_3^2}} = \frac{\alpha\,a_x + \beta\,a_y + \gamma\,a_z}{\dfrac{n_1^2\,\alpha^2}{n^2 - n_1^2} + \dfrac{n_2^2\,\beta^2}{n^2 - n_2^2} + \dfrac{n_3^2\,\gamma^2}{n^2 - n_3^2}}.$$

Le vecteur $\boldsymbol{D}$ étant perpendiculaire au vecteur ON, on a $\alpha a_x + \beta a_y + \gamma a_z = 0$ et de même

$$\frac{n_1^2\,\alpha^2}{n^2 - n_1^2} + \frac{n_2^2\,\beta^2}{n^2 - n_2^2} + \frac{n_3^2\,\gamma^2}{n^2 - n_3^2} = 0.$$

Cette équation permet de calculer l'indice n si on connait les cosinus directeurs α, β, γ de la normale ON au plan d'onde. On peut la mettre sous la forme

$$\left.\begin{aligned}n_1^2\,\alpha^2\,(n^2 - n_2^2)\,(n^2 - n_3^2) + n_2^2\,\beta^2\,(n^2 - n_3^2)\,(n^2 - n_1^2) + \\ + n_3^2\,\gamma^2\,(n^2 - n_1^2)\,(n^2 - n_2^2) = 0.\end{aligned}\right\}\quad(140.7)$$

C'est une équation bicarrée en n qui peut avoir 4 racines réelles. Nous ne retiendrons que les 2 racines positives n' et n'' : si on prend $n_1 > n_2 > n_3$ on aura $n_1 > n' > n_2$ et $n_2 > n'' > n_3$.

141. Surface des indices et surface d'onde. C'est la surface qui représente géométriquement les variations de l'indice en fonction de l'orientation du plan d'onde.

A partir d'une origine O, on porte sur une direction parallèle à ON deux longueurs ON' et ON'' égales aux deux indices n' et n''. Le lieu des points N'

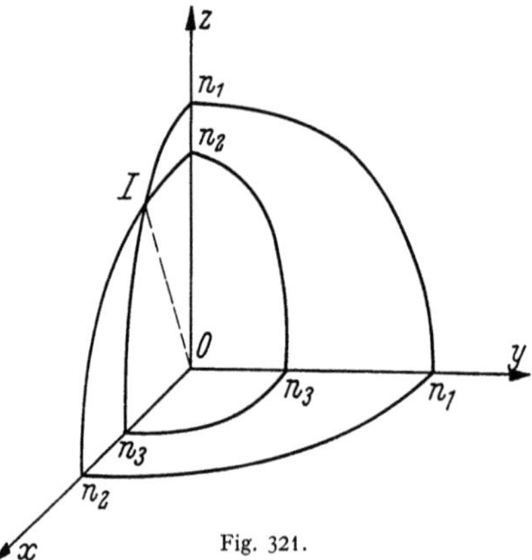

Fig. 321.

et N'' est la surface des indices. Les coordonnées x, y, z de N' ou N'' sont $x = \alpha n$, $y = \beta n$, $z = \gamma n$: reportons ces valeurs dans l'équation (140.7), on obtient l'équation de la surface des indices

$$\left. \begin{array}{l} (n_1^2 x^2 + n_2^2 y^2 + n_3^2 z^2)(x^2 + y^2 + z^2) - \\ -n_1^2 x^2 (n_2^2 + n_3^2) - n_2^2 y^2 (n_3^2 + n_1^2) - \\ -n_3^2 z^2 (n_1^2 + n_2^2) + n_1^2 n_2^2 n_3^2 = 0. \end{array} \right\} \quad (141.1)$$

Les intersections de la surface des indices par le plan $z = 0$ donnent les deux courbes

$$x^2 + y^2 = n_3^2, \qquad \frac{x^2}{n_2^2} + \frac{y^2}{n_1^2} = 1,$$

de même pour les intersections par les deux autres plans de coordonnées. Les intersections de la surface des indices par les plans de coordonnées se composent donc d'un cercle et d'une ellipse (Fig. 321). On voit que dans le plan xOz le cercle et l'ellipse se coupent en 4 points deux à deux symétriques par rapport à l'origine (Fig. 322).

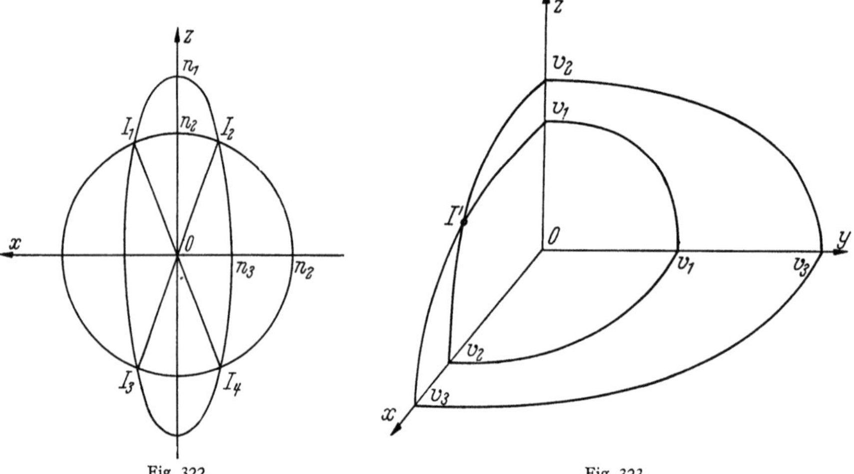

Fig. 322.   Fig. 323.

Suivant les deux directions OI_1 et OI_2, les deux indices n' et n'' sont égaux et une vibration quelconque se propage sans altération suivant ces directions. Les deux directions OI_1 et OI_2 s'appellent les axes optiques du milieu anisotrope. Les quatre points I sont des ombilics de la surface.

Connaissant les variations de n en fonction de la direction ON (surface des indices), on peut passer à la surface d'onde en étudiant les variations de $v = c/n$. La surface d'onde se déduit de la surface des indices par une transformation par polaires réciproques par rapport à une sphère de centre O. Elle a une forme analogue à celle de la surface des indices, comme le montre la Fig. 323. Les quatre points I' sont des ombilics de la surface d'onde.

142. Ellipsoïde des indices. La surface des indices donne les variations de l'indice en fonction de la direction de la normale ON à l'onde. Cherchons maintenant les variations de l'indice en fonction de la direction de la vibration.

On peut écrire l'équation (140.6) sous la forme

$$\frac{\left(\dfrac{n^2}{n_1^2} - 1\right) a_x^2}{\alpha\, a_x} = \frac{\left(\dfrac{n^2}{n_2^2} - 1\right) a_y^2}{\beta\, a_y} = \frac{\left(\dfrac{n^2}{n_3^2} - 1\right) a_z^2}{\gamma\, a_z} = \frac{n^2 \left(\dfrac{a_x^2}{n_1^2} + \dfrac{a_y^2}{n_2^2} + \dfrac{a_z^2}{n_3^2}\right) - 1}{\alpha\, a_x + \beta\, a_y + \gamma\, a_z}.$$

On a $\alpha a_x + \beta a_y + \gamma a_z = 0$ et le numérateur de la dernière fraction est également nul

$$\frac{a_x^2}{n_1^2} + \frac{a_y^2}{n_2^2} + \frac{a_z^2}{n_3^2} = \frac{1}{n^2}. \tag{142.1}$$

A partir d'une origine O, menons sur une droite parallèle à la vibration a_x, a_y, a_z une longueur égale à n. On obtient un point M de coordonnées $\alpha = n a_x$, $y = n a_y$, $z = n a_z$ et l'équation (142.1) devient

$$\frac{x^2}{n_1^2} + \frac{y^2}{n_2^2} + \frac{z^2}{n_3^2} = 1. \tag{142.2}$$

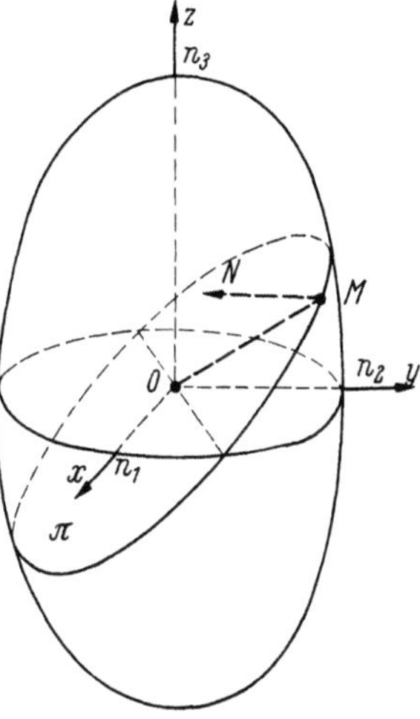

Fig. 324.

Le lieu du point M est un ellipsoïde appelé «ellipsoïde des indices». Les paramètres directeurs de la normale MN (Fig. 324) à l'ellipsoïde sont x/n_1^2, y/n_2^2, z/n_3^2, c'est-à-dire a_x/n_1^2, a_y/n_2^2, a_z/n_3^2. D'après (140.5) et (139.4) MN est donc la direction du champ électrique correspondant à l'onde considérée, et OM qui donne la direction de la vibration, doit être la projection de la normale MN sur le plan d'onde. Connaissant l'ellipsoïde des indices et la vibration OM, qui est la direction de l'induction électrique, il suffit, pour avoir le plan d'onde correspondant à la vibration OM, de mener le plan passant par OM perpendiculairement au plan OMN. Le plan d'onde coupe l'ellipsoïde suivant l'ellipse π.

143. Vibrations privilégiées. D'après les résultats précédents à un plan d'onde donné, peuvent correspondre deux indices différents n' et n''. A l'indice n' correspond la vibration de cosinus directeurs a_x', a_y', a_z' donnée par (140.6)

$$\frac{(n'^2 - n_1^2)\, a_x'}{\alpha\, n_1^2} = \frac{(n'^2 - n_2^2)\, a_y'}{\beta\, n_2^2} = \frac{(n'^2 - n_3^2)\, a_z'}{\gamma\, n_3^2}$$

et à l'indice n'', la vibration a_x'', a_y'', a_z''

$$\frac{(n''^2 - n_1^2)\, a_x''}{\alpha\, n_1^2} = \frac{(n''^2 - n_2^2)\, a_y''}{\beta\, n_2^2} = \frac{(n''^2 - n_3^2)\, a_z''}{\gamma\, n_3^2}.$$

Si on calcule $a_x' a_x'' + a_y' a_y'' + a_z' a_z''$ on trouve que cette quantité est nulle, par conséquent les deux vibrations correspondant aux deux indices n' et n'' sont perpendiculaires. On peut donc dire que pour une orientation déterminée du plan d'onde, il existe dans ce plan deux vibrations rectangulaires qui peuvent se propager sans altération dans le milieu anisotrope. A ces deux vibrations privilégiées correspondent les deux indices n' et n'' et les vitesses de propagation $v' = c/n'$ et $v'' = c/n''$.

On peut vérifier expérimentalement l'existence de ces deux directions privilégiées en plaçant une lame cristalline à faces parallèles L entre deux nicols dont les sections principales sont croisées, c'est-à-dire entre deux nicols arrêtant la lumière sans l'interposition de L (Fig. 325). L'orientation des faces de L par rapport aux axes cristallographiques étant quelconques, on constate que l'extinction ne subsiste pas lorsqu'on interpose la lame L. Faisons tourner L autour d'un axe parallèle aux rayons lumineux: on trouve qu'il existe deux positions rectangulaires pour lesquelles l'extinction est rétablie. Suivant ces deux directions, la vibration rectiligne incidente est transmise sans altération, ce sont les deux vibrations privilégiées correspondant au plan d'onde parallèle aux faces de la lame L.

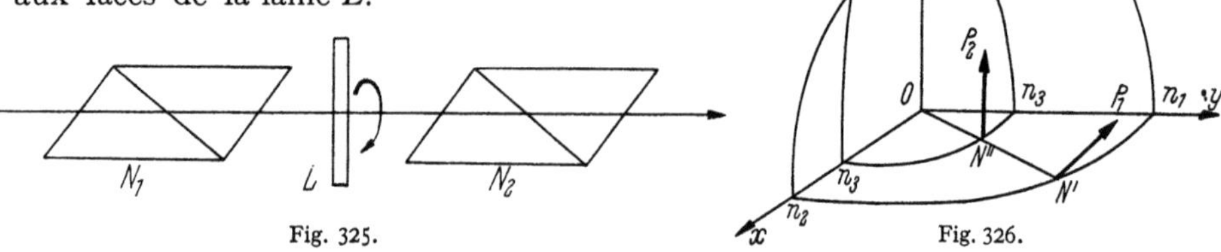

Fig. 325.   Fig. 326.

Si le plan d'inde est normal à un axe optique, les deux vibrations privilégiées se propagent alors avec la même vitesse puisque $n' = n''$: il n'y a plus à proprement parler de vibrations privilégiées. Pour cette orientation du plan d'onde, une vibration quelconque se propage sans altération dans le milieu anisotrope.

Supposons la vibration parallèle à l'un des axes de coordonnées oz par exemple (Fig. 326) $a_x = a_y = 0$ et comme le plan d'onde contient la vibration il est parallèle

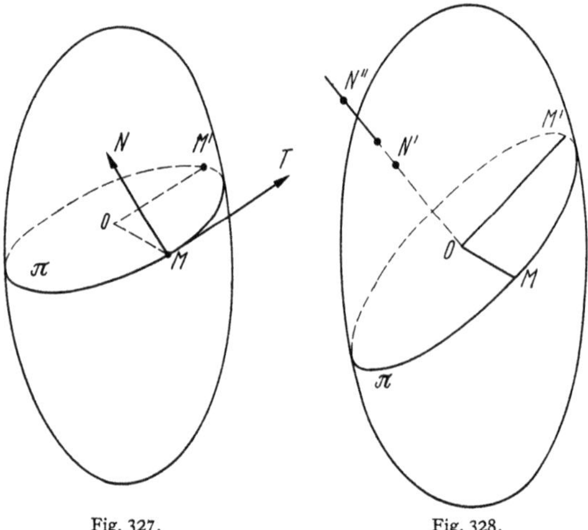

à oz: l'une des vibrations privilégiées P_2 est parallèle à oz l'indice correspondant étant n_3. On a le point N'' sur la surface des indices. L'autre vibration privilégiée P_1 perpendiculaire à P_2 se trouve dans le plan xoy. C'est l'intersection du plan d'onde et du plan xoy; elle correspond au point N' de la surface des indices.

On peut chercher également les positions des deux directions privilégiées correspondant à un plan d'onde déterminé par rapport à l'ellipsoïde des indices. Considérons

Fig. 327.   Fig. 328.

la Fig. 324: nous avons vu comment on peut construire le plan d'onde π correspondant à une direction OM de vibration. OM est la direction de l'induction électrique, MN celle du champ électrique. L'intersection MT du plan d'onde π et du plan tangent en M à l'ellipsoïde (Fig. 327), est perpendiculaire à l'induction électrique MN, c'est la direction du champ magnétique. MT également tangente à l'ellipse d'intersection du plan d'onde π et de l'ellipsoïde est normale au rayon vecteur OM. OM est donc un des axes de l'ellipse, l'autre étant la seconde direction OM' du plan π. Les deux vibrations privilégiées que peut propager un plan d'onde π sont les axes de l'ellipse d'intersection de l'ellipsoïde des indices par le plan d'onde. Les indices correspondants sont égaux aux demi-longueurs de ces axes. Ces résultats permettent de construire facilement la surface des indices à partir de l'ellipsoïde des indices (Fig. 328): on coupe l'ellipsoïde des indices par le plan d'onde π et on détermine sur l'ellipse d'intersection les deux demi-axes OM et OM'. A partir de O sur la normale au plan d'onde, on porte deux longueurs $ON'=OM, ON''=OM'$, le lieu des points N' et N'' est la surface des indices. L'ellipsoïde des indices possède deux sections cycliques dont les normales donnent la direction des axes optiques.

144. Milieux uniaxes. Considérons un milieu anisotrope pour lequel deux des trois indices principaux sont égaux, par exemple $n_2 = n_3$. D'après (142.2) l'ellipsoïde des indices devient

$$\frac{x^2}{n_1^2} + \frac{y^2 + z^2}{n_2^2} = 1.$$

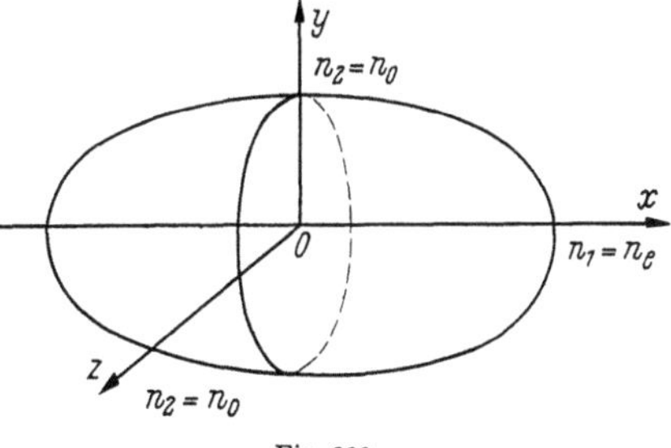

Fig. 329.

L'ellipsoïde des indices coupe le plan yoz suivant une circonférence, il est de révolution autour de ox (Fig. 329). Le milieu présente la symétrie de révolution autour de cet axe: ox est l'axe optique du milieu qui est appelé «milieu uniaxe». Une onde plane perpendiculaire à l'axe optique se propage dans le milieu aniso-trope comme dans un milieu isotrope. L'indice correspondant $n_2 = n_3$ est appelé l'indice ordinaire n_0 et l'indice n_1 correspondant aux vibrations parallèles à l'axe optique est appelé l'indice extraordinaire principal n_e. L'équation de l'ellipsoïde des indices pour un milieu uniaxe s'écrira

$$\frac{x^2}{n_e^2} + \frac{y^2 + z^2}{n_0^2} = 1.\tag{144.1}$$

Soit un plan d'onde π quelconque qui coupe l'ellipsoïde des indices suivant une ellipse d'axe OM_0 et OM_E (Fig. 330). Les axes OM_0 et OM_E sont les directions des vibrations privilégiées correspondant au plan d'onde π d'après ce que l'on

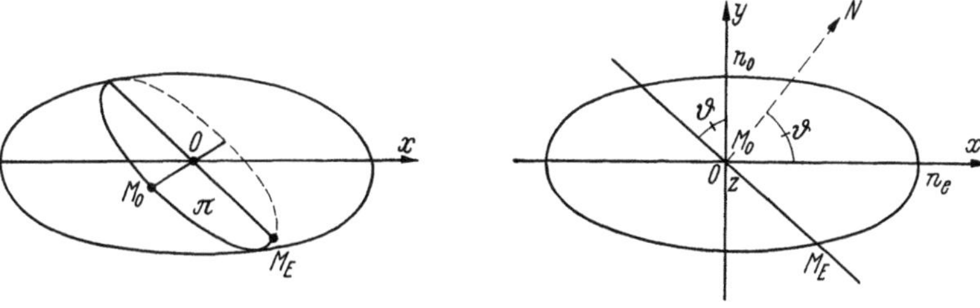

Fig. 330.　　　　　　　　　　　　　　　　Fig. 331.

a vu. La vibration OM_0 perpendiculaire à l'axe optique est la vibration ordinaire, elle se propage comme dans un milieu isotrope. La vibration OM_E est la vibration extraordinaire et il lui correspond l'indice extraordinaire n compris entre l'indice ordinaire n_0 et l'indice extraordinaire principal n_e. La vibration OM_E est la projection de l'axe optique sur le plan d'onde.

Les phénomènes étant de révolution autour de ox on peut prendre OM_0 comme axe oz et prendre pour plan de figure, le plan xoM_E perpendiculaire à OM_0 (Fig. 331). OM_E est le rayon vecteur de l'ellipse d'intersection de l'ellipsoïde des indices par le plan de figure. L'équation de cette ellipse est

$$\frac{x^2}{n_e^2} + \frac{y^2}{n_0^2} = 1.\tag{144.2}$$

Les coordonnées de M_E sont $x = n \sin \vartheta$ et $y = n \cos \vartheta$. Portons ces valeurs dans (144.2), il vient

$$n^2 \left(\frac{\sin^2 \vartheta}{n_e^2} + \frac{\cos^2 \vartheta}{n_0^2} \right) = 1 \,, \tag{144.3}$$

relation qui donne l'indice extraordinaire n en fonction de ϑ.

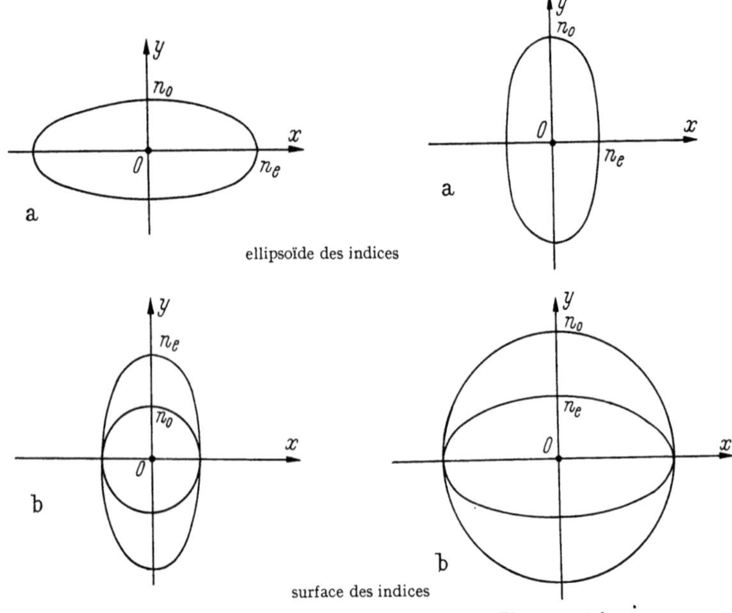

ellipsoïde des indices

surface des indices

Fig. 332 a et b.   Fig. 333 a et b.

Construisons la surface des indices à partir de l'ellipsoïde des indices comme il a été dit (Sect. 143): elle se compose de deux nappes, une sphère et un ellipsoïde de révolution, qui sont tangents. Si $n_0 < n_e$ on a la Fig. 332, le milieu uniaxe

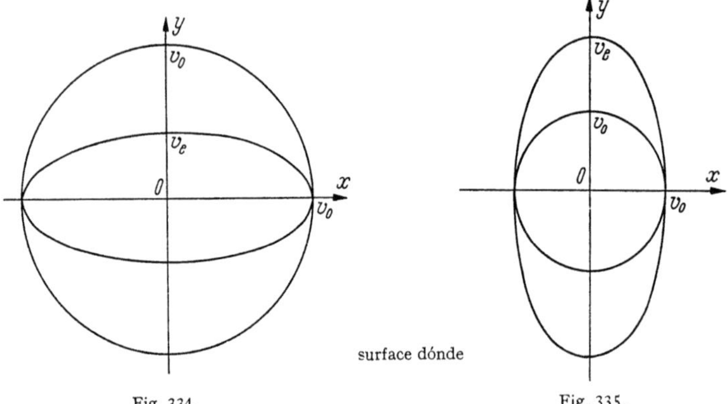

surface dónde

Fig. 334.   Fig. 335.

est positif: c'est le cas du quartz. Si $n_0 > n_e$ le milieu est négatif, c'est par exemple le cas du spath (Fig. 333). Ces résultats peuvent se retrouver à partier de l'équation (140.7) en faisant $n_2 = n_3 = n_0$ et $n_1 = n_2$.

La surface d'onde se déduit de la surface des indices par une transformation par polaires réciproques, elle a une forme analogue à celle de la surface des indices. On a la Fig. 334 pour un milieu positif et la Fig. 335 pour un milieu négatif. La

Fig. 336 montre la position des plans d'onde π_0 et π_e correspondant à un rayon OA_0. Représentons la section E de l'ellipsoïde des indices par le plan $x\,o\,y$. L'angle ε du rayon OA_0 et de la normale à l'onde extraordinaire est égal à l'angle de la direction OM de la vibration et de la normale MN_E à l'ellipsoïde des indices. Si V est l'angle de la tangente à l'ellipse en M avec la direction OM, on a

$$\tan \varepsilon = \cot V = \frac{1}{n}\,\frac{d\,n}{d\,\varepsilon} = -\frac{1}{2}\,n^2\,\frac{d\left(\dfrac{1}{n^2}\right)}{d\,\varepsilon}$$

et d'après (144.3)

$$\tan \varepsilon = \frac{n^2}{2}\left(\frac{1}{n_0^2} - \frac{1}{n_e^2}\right) - \sin 2\vartheta. \qquad (144.4)$$

Les variations de n étant faibles, on peut admettre que la valeur maximum de ε correspond à $\sin 2\vartheta = 1$. La valeur de n correspondante donnée par (144.3) sera $\dfrac{1}{n^2} = \dfrac{1}{n_0^2} + \dfrac{1}{n_e^2}$ d'où la valeur maximum de ε

$$\tan \varepsilon_m = \frac{n_e^2 - n_0^2}{n_e^2 + n_0^2} \approx \frac{n_e - n_0}{n'} \qquad (144.5)$$

n' étant une valeur moyenne de l'indice. Pour le quartz, $n_e - n_0 = 9 \cdot 10^{-3}$ et $n' = 1{,}54$ d'où $\varepsilon_m \approx 17'$.

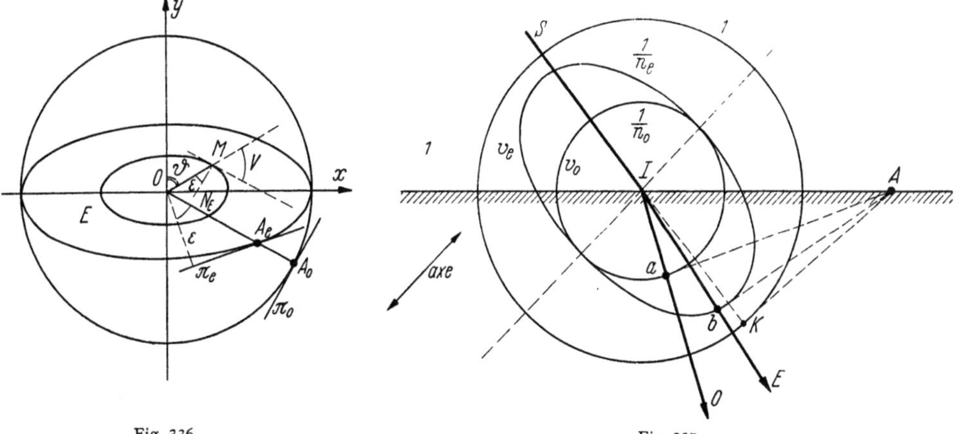

Fig. 336.   Fig. 337.

145. Rayons réfractés dans un milieu uniaxe. On peut appliquer la construction d'Huygens, mais on aboutit à un problème de géométrie dans l'espace car les deux rayons réfractés ne sont pas en général dans un même plan. La construction se simplifie lorsque le plan d'incidence est un plan de symétrie du cristal; dans ce cas, les deux rayons sont dans un même plan. Ceci a lieu lorsque le plan d'incidence contient l'axe du cristal ou lorsque ce plan est perpendiculaire à l'axe du cristal.

Dans le premier cas, nous avons la Fig. 337. On prolonge le rayon incident SI qui coupe la surface d'onde relative au premier milieu en K. La tangente en K à cette surface d'onde coupe le plan de séparation des deux milieux suivant A. Par A, on mène les deux tangentes $A\,a$ et $A\,b$ aux surfaces d'ondes ordinaire et extraordinaire. On obtient ainsi les deux rayons réfractés IO, ordinaire, et IE extraordinaire.

La Fig. 338 correspond au deuxième cas lorsque l'axe du cristal est perpendiculaire au plan d'incidence. La construction est alors exactement semblable à celle employée pour la simple réfraction. On obtient deux rayons réfractés qui suivent les lois de Descartes avec l'indice ordinaire n_0 et l'indice extraordinaire n_e. Les Figs. 337 et 338 se rapportent au cas d'un cristal négatif. Appliquons les constructions précédentes à une lame de spath de clivage (Fig. 339) recevant de la lumière naturelle normalement à ses faces. Prenons comme plan de figure, un plan de section principale. Le point A est ici rejeté à l'infini et les points a et b sont donnés en menant les tangentes aux surfaces d'onde parallèlement aux faces de la lame de spath. On obtient les deux rayons réfractés ordinaire IJ_0 et extraordinaire IJ_E. Le rayon ordinaire n'est pas dévié par la lame à faces parallèles

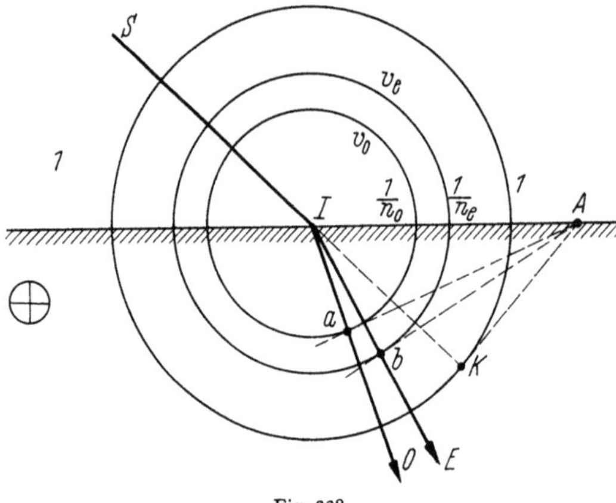

Fig. 338.

constituée par la lame de spath. Le rayon extraordinaire sort parallèlement au rayon ordinaire. D'après ce qui précède, la direction de la vibration extraordinaire est la projection du rayon Ib sur le plan d'onde π_E, elle est dans le plan de figure, c'est-à-dire dans le plan de section principale. La vibration ordinaire est alors perpendiculaire au plan de section principale.

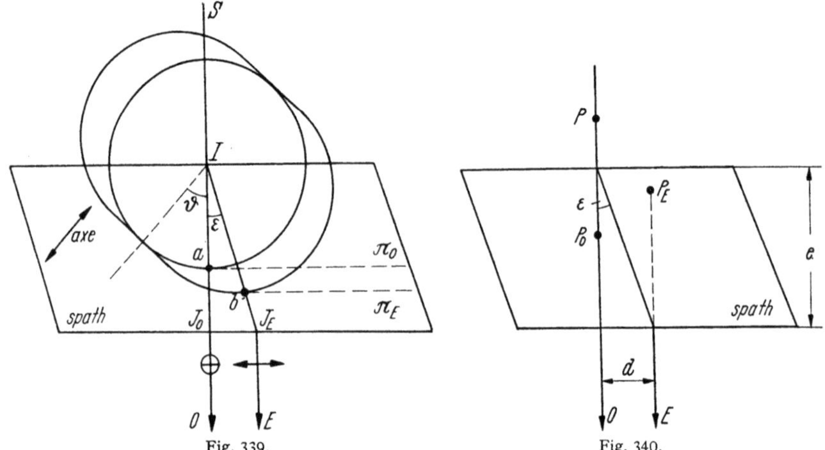

Fig. 339.   Fig. 340.

La Fig. 339 montre que si l'on observe un objet P situé à distance finie à travers la lame de spath, on verra deux images P_0 et P_E (Fig. 340) séparées par la distance d. Par suite de la différence des indices ordinaire et extraordinaire, l'image ordinaire P_0 est un peu plus rapprochée de l'obervateur que l'image extraordinaire $(n_0 > n_e)$. On aura $d = e \tan \varepsilon$ et comme pour un cristal naturel de spath, on a $\vartheta \approx 45°$ l'angle ε est voisin de sa valeur maximum ε_m d'où

$$d = e \tan \varepsilon_m = \frac{n_e - n_0}{n'}\, e \tag{145.1}$$

soit $d = 0,106\,e$ pour le spath.

146. Rayons réfractés dans un milieu bi-axe. *α) Construction des rayons réfractés dans un cas particulier.* Nous avons vu aux Sect. 141 et 142 la structure de la surface des indices de la surface d'onde et de l'ellipsoïde des indices pour un milieu possédant deux axes optiques, c'est-à-dire un milieu bi-axe.

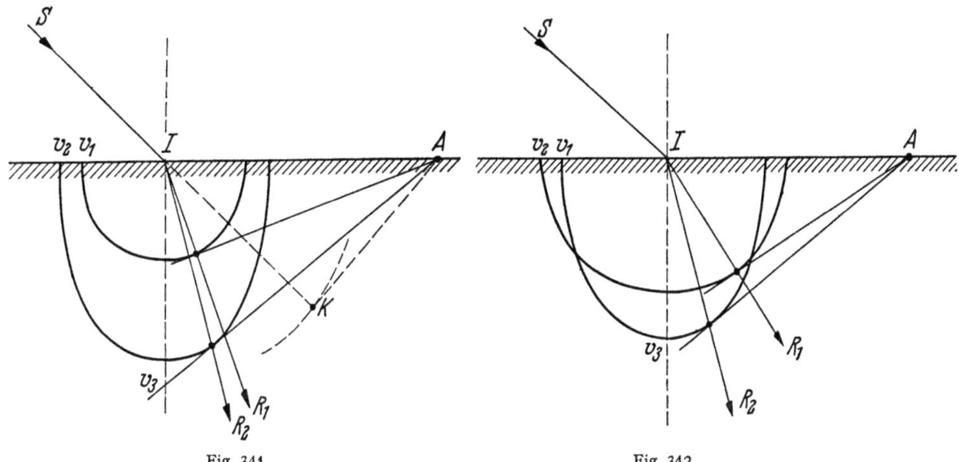

Fig. 341.   Fig. 342.

L'ensemble des résultats précédents montrent que les rayons réfractés ne sont pas généralement situés dans le plan d'incidence et que les lois de DESCARTES ne sont pas applicables aux rayons lumineux. On peut les appliquer aux normales aux ondes qui ne coïncident pas avec les rayons lumineux dans un milieu anisotrope.

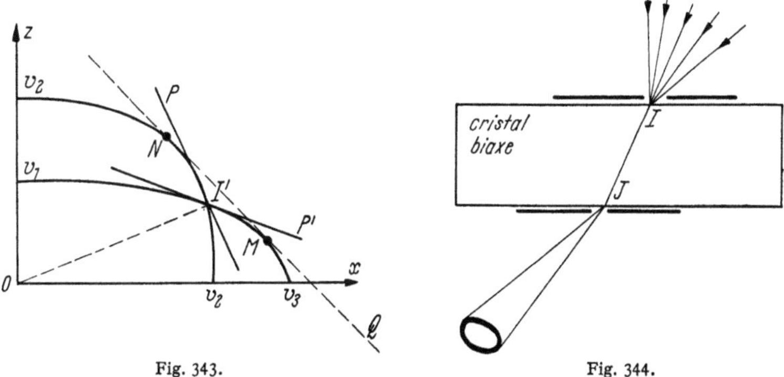

Fig. 343.   Fig. 344.

Le plan d'incidence contient les deux rayons réfractés lorsqu'il est confondu avec un plan de symétrie de la surface d'onde.

Supposons d'abord le plan d'incidence confondu avec le plan *yoz* de la Fig. 323. La construction D'HUYGENS donne la Fig. 341; le rayon OR_1 correspondant à la tangente à la section circulaire est polarisé dans le plan d'incidence. Le rayon OR_2 correspondant à la section elliptique est polarisé dans le plan perpendiculaire au plan d'incidence. La Fig. 342 correspond où la plan d'incidence est confondu avec le plan *xoz*.

β) Réfraction conique externe. HAMILTON a prévu théoriquement que dans certains cas un rayon à l'intérieur d'un cristal biaxe peut donner naissance à un cône de rayons émergents, étant donné que l'on peut mener une infinité de plans tangents à la surface d'onde aux quatre ombilics. De même à un rayon incident peut correspondre un cône de rayons à l'intérieur du cristal. Ces phénomènes constituent dans le premier cas le phénomène de la réfraction conique externe et dans le second, le phénomène de la réfraction conique interne. Leur observation constitue une vérification expérimentale de la théorie de la propagation de la lumière dans un milieu anisotrope.

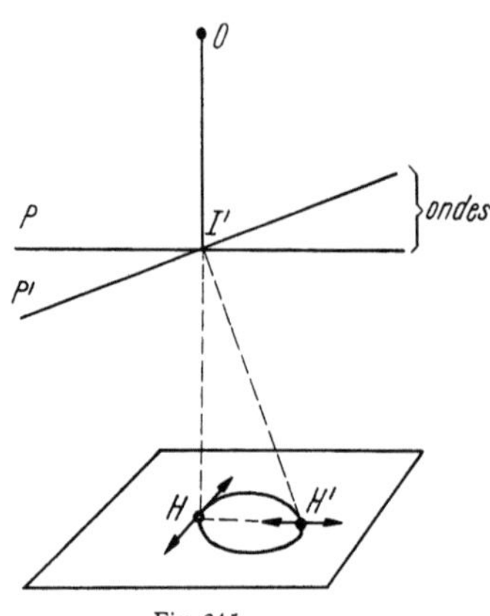

Fig. 345.

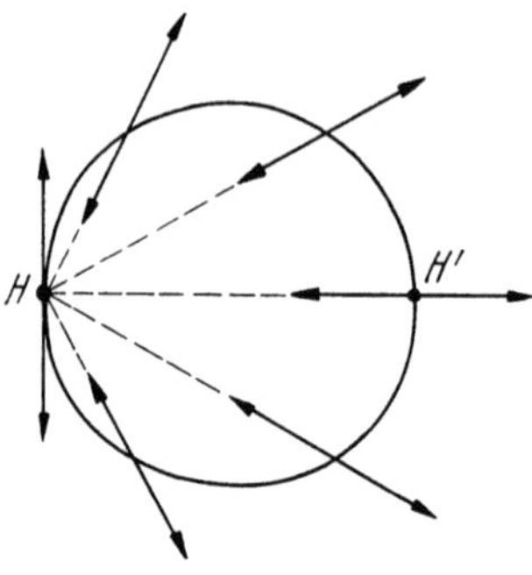

Fig. 346.

Considérons d'abord la réfraction conique externe: la Fig. 343 représente la section de la surface d'onde par le plan xoz. Si un rayon suit la direction OI' à l'intérieur du cristal, il lui correspond une infinité d'ondes planes. A chaque onde, correspond dans le milieu isotrope extérieur une direction de rayon bien définie. Leur ensemble forme un cône creux de lumière. Sur la Fig. 344 le rayon IJ est limité par deux petites ouvertures à l'entrée et à la sortie de la lame cristal-

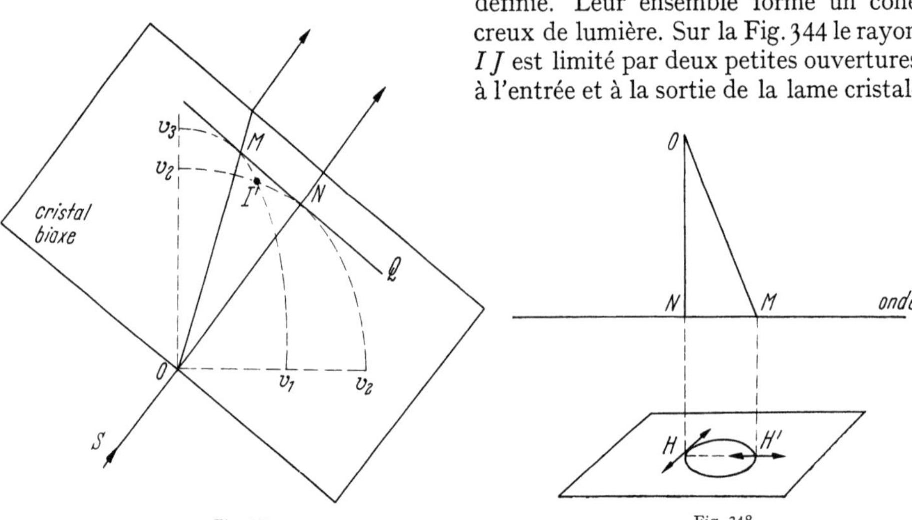

Fig. 347.

Fig. 348.

line, il est parallèle à la direction OI' de la Fig. 343. Pour faire l'expérience, on envoie en I un faisceau de lumière convergent assez ouvert pour qu'il contienne des rayons parallèles à toutes les directions du cône qu'on veut obtenir. A la sortie, on a un cône de rayons normaux aux ondes planes de réfraction conique. Les Fig. 345 et 346 montrent l'orientation de la vibration transmise par chacun d'eux, orientation que l'on peut vérifier avec un nicol tournant. Sur la Fig. 345, OI' représente la même direction que sur la Fig. 343 ou IJ sur la Fig. 344.

γ) *Réfraction conique interne.* Considérons la Fig. 343 et menons le plan Q parallèle à oy, tangent à la surface d'onde. On montre que ce plan touche la surface d'onde suivant un cercle MN. Ce plan est parallèle à l'une des deux sections cycliques de l'ellipsoïde des indices et ON est un axe optique. A chaque point de la circonférence MN correspond une direction de vibration et une direction de propagation. Sur la Fig. 347, on considère un cristal dont la section représentée par le plan de figure est parallèle au plan xoz de la Fig. 323. Les faces du cristal sont supposées parallèles au plan Q défini sur la Fig. 343. On fait tomber sur le cristal une onde incidente parallèle aux faces, de façon à avoir à l'intérieur du cristal des ondes parallèles au plan Q. A un rayon incident SO correspond un cône creux de rayons qui s'appuient sur le cercle MN. A la sortie, ce cône se transforme en un cylindre creux. Pour observer le phénomène, on limite le faisceau incident par deux petites ouvertures (Fig. 348), en examinant O à travers le cristal au moyen d'une loupe, on voit un anneau lumineux. La Fig. 347 est un cas particulier, mais le phénomène s'observe aussi bien si le plan Q n'est pas parallèle aux faces du cristal. On voit alors normalement au faisceau émergent une ellipse lumineuse. La Fig. 348 montre l'orientation des vibrations.

147. Réflexion totale. $\alpha)$ *Milieu uniaxe.* D'après la Sect. 145, on voit qu'il ne peut y avoir d'ondes réfractées que si le point A (Fig. 337) est extérieur aux deux nappes de la surface d'onde relatives au milieu uniaxe. Considérons un milieu isotrope d'indice N supérieur aux indices du cristal; pour une certaine inclinaison des rayons incidents, le point A ne sera plus extérieur aux deux nappes de la surface d'onde

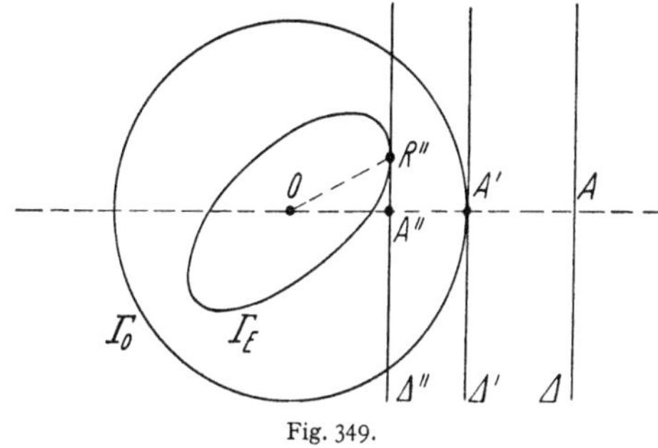

Fig. 349.

et il y aura réflexion totale. Au lieu de représenter les phénomènes comme sur la Fig. 337, représentons-les en prenant pour plan de figure le plan de la surface de séparation des deux milieux (Fig. 349). Nous considérons un rayon qui tombe du milieu isotrope d'indice N sur un cristal uniaxe en O. Γ_0 et Γ_E sont les sections de la surface d'onde par la surface de séparation. Pour une certaine incidence, le plan tangent à la surface d'onde correspondant au milieu isotrope, coupe le plan de figure suivant Δ.

Faisons varier l'angle d'incidence sans changer le plan d'incidence qui est perpendiculaire au plan de figure suivant OA. Le point A et la droite Δ se rapprochent de O. Lorsque Δ occupe la position Δ', il y a réflexion totale pour le rayon ordinaire dans le cas de figure (cristal positif). Quand Δ vient en Δ'' il y a réflexion totale pour le rayon extraordinaire. Si i' et i'' sont les angles d'incidence correspondant aux deux positions Δ' et Δ'' on a

$$\sin i' = \frac{c}{N \cdot \overline{OA}}\ , \quad \sin i'' = \frac{c}{N \cdot \overline{OA''}}$$

et l'angle limite i'' varie avec l'orientation de la surface d'onde extraordinaire. On peut alors calculer l'indice extraordinaire n

$$\frac{\sin i''}{\sin r''} = \frac{n}{N}\ ,$$

r'' étant l'angle de réfraction correspondant à i'' et qui n'est pas en général égal à 90°. L'angle r'' est égal à 90° lorsque OR'' est maximum ou minimum, c'est-à-dire pour les indices principaux n_0 et n_e.

On peut utiliser ces résultats pour mesurer les indices principaux n_0 et n_e d'un cristal au moyen du réfractomètre de Pulfrich (Fig. 350). Le prisme de référence est ici un cylindre d'indice N sur lequel est posé le cristal dont on veut mesurer les indices. En éclairant au moyen d'un faisceau convergent, on voit dans le plan focal de l'objectif L de la lunette, non plus une limite comme dans

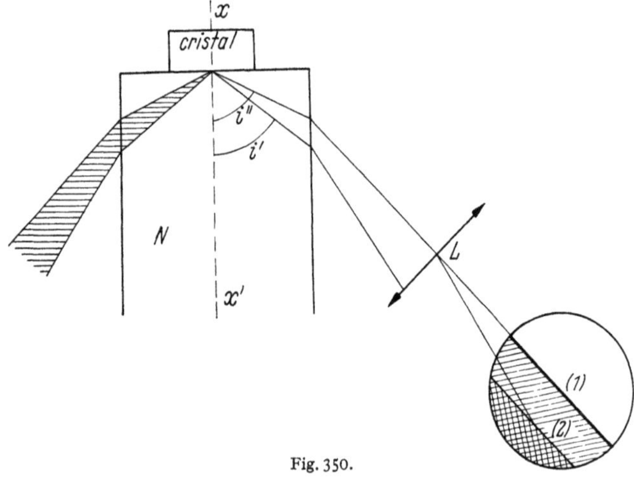

Fig. 350.

la méthode ordinaire appliquée à un échantillon isotrope, mais deux limites nettes (1) et (2) correspondant aux deux indices. En faisant tourner l'ensemble cylindre-cristal autour de l'axe xx', on voit l'une des limites rester fixe et l'autre se déplacer. La limite fixe correspond à l'indice n_0 et l'indice extraordinaire principal n_e est celui qui diffère le plus de l'indice n_0. On l'obtient quand les deux limites sont le plus écartées possible et pour cette position $n_e = N \sin i''$.

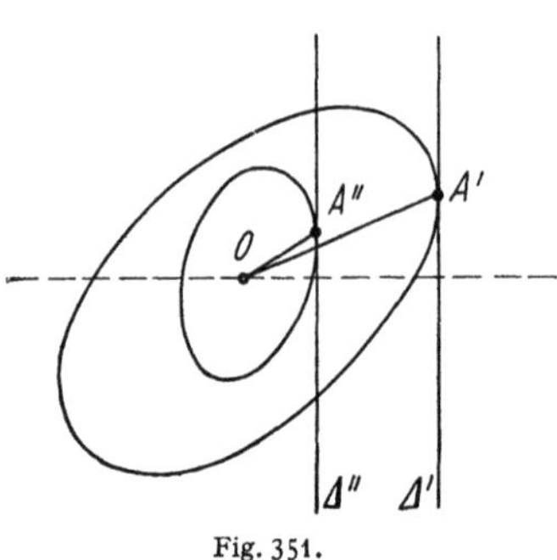

Fig. 351.

β) *Milieu biaxe.* On raisonne de la même façon qu'au paragraphe précédent. Les sections de la surface d'onde par le plan de séparation des deux milieux se composent de deux courbes fermées Γ et Γ' (Fig. 351) qui ne se rencontrent pas en supposant que le plan de séparation ne contienne pas l'axe OI' de la Fig. 323.

On a donc deux angles d'incidence i' et i'' correspondant aux deux positions Δ' et Δ'' et produisant la réflexion totale des deux rayons réfractés. Comme dans le cas précédent, on peut mesurer au réfractomètre de Pulfrich les angles i' et i''. En faisant tourner l'ensemble cylindre-cristal, on obtient maintenant deux valeurs maxima et deux valeurs minima. Trois de ces angles limites i_m correspondent aux trois indices principaux par les relations $n_m = N \sin i_m$. Le plus grand correspond au plus grand indice principal n_1, le plus petit, au plus petit indice principal n_3. Il y a indétermination pour l'indice moyen n_2 qui peut correspondre au minimum de OA' ou au maximum de OA''. Pour savoir quelle valeur choisir, il est nécessaire de tailler une autre face dans le cristal.

148. Différents types de cristaux. On classe les cristaux dans différents systèmes suivant les éléments de symétrie qu'ils possèdent et qui sont des éléments de symétrie de l'ellipsoïde des indices. Un plan de symétrie du cristal est nécessairement un plan de symétrie optique. Un axe de symétrie d'ordre 2 est un axe de symétrie optique, un axe d'ordre $p > 2$ est un axe tel que des rotations de $2\pi/p$, $4\pi/p$, ... amènent une direction quelconque à coïncider avec une autre direction ayant les mêmes propriétés. C'est une direction qui ramène l'ellipsoïde des indices à coïncider avec lui-même, c'est un axe de révolution de cet ellipsoïde.

Les sept systèmes cristallins se classent en trois groupes:

a) Les cristaux cubiques qui possèdent plusieurs axes d'ordre supérieur à 2; l'ellipsoide des indices possédant plusieurs axes de révolution est une sphère et le cristal se comporte comme un corps isotrope. Exemples: sel gemme (NaCl), fluorine (CaF_2), chlorate de sodium ($NaClO_3$), etc.

b) Les cristaux sénaires, quaternaires et ternaires, qui possèdent un axe principal d'ordre 2. L'axe principal est l'axe de révolution de l'ellipsoïde des indices, c'est un axe optique. Ces cristaux sont les cristaux uniaxes.

c) Les cristaux terninaires, binaires et asymétriques, qui ne possèdent pas d'axe d'ordre supérieur à 2. L'ellipsoïde des indices a ses axes inégaux et possède deux sections cycliques dont les normales donnent la direction des deux axes optiques; ce sont les cristaux biaxes.

Nous donnons ci-après les indices de quelques cristaux de ces différents types.

Cristaux uniaxes positifs.

Quartz	SiO_2	$n_0 = 1{,}5442$	$n_e = 1{,}5533$
Calomel	Hg_2Cl_2	$n_0 = 1{,}9732$	$n_e = 2{,}6559$
Zircon	SiO_2ZrO_2	$n_0 = 1{,}92$	$n_e = 1{,}97$

Cristaux uniaxes négatifs.

Spath	$CaCO_3$	$n_0 = 1{,}6584$	$n_e = 1{,}4865$
Azotate de Sodium	$NaNO_3$	$n_0 = 1{,}5874$	$n_e = 1{,}3361$
Emeraude	$Cl_3Al_2Si_6O_{18}$	$n_0 = 1{,}582$	$n_e = 1{,}576$
Apatite	$Ca_5P_3O_{12}(ClF)$	$n_0 = 1{,}639$	$n_e = 1{,}635$

La différence $n_e - n_0$ ou «biréfringence» de cristal est généralement faible, sauf pour le spath où elle atteint la valeur 0,17.

Cristaux biaxes.

Aragonite . .	$CaCO_3$	$n_1 = 1{,}6859$	$n_2 = 1{,}6816$	$n_3 = 1{,}5301$
Topaze . . .	$Al_2Si(OF_2)_5$	$n_1 = 1{,}6211$	$n_2 = 1{,}6138$	$n_3 = 1{,}6116$
Mica	$(KNa)_4H_8Al_{12}Si_{12}O_{48}$	$n_1 = 1{,}5977$	$n_2 = 1{,}5936$	$n_3 = 1{,}5601$
Gypse . . .	$CaSO_4, 2H_2O$	$n_1 = 1{,}5296$	$n_2 = 1{,}5226$	$n_3 = 1{,}5205$

Suivant que la bissectrice de l'angle aigu des axes est, soit l'axe principal de grand indice, soit l'axe principal de petit indice on dit que l'on a affaire à un cristal biaxe positif (gypse) ou négatif (aragonite, mica).

Dans le premier cas, si les axes se rapprochent de plus en plus, on a à la limite, l'ellipsoïde d'un uniaxe positif, et dans le second cas, l'ellipsoïde d'un uniaxe négatif (Fig. 352 et 353).

On notera que les axes optiques ne sont pas des axes de symétrie, ils sont situés dans un plan de symétrie de l'ellipsoïde des indices, mais leur angle dépend des indices principaux. En effet, reprenons la Fig. 322, et soit $2V$ l'angle aigu des axes. Le point I_1 est sur le cercle $x = n_2 \sin V$, $z = n_2 \cos V$ et également sur l'ellipse $\dfrac{x^2}{n_3^2} + \dfrac{z^2}{n_1^2} = 1$. On a donc

$$\frac{n_2^2 \sin^2 V}{n_3^2} + \frac{n_2^2 \cos^2 V}{n_1^2} = 1 = \sin^2 V + \cos^2 V$$

d'où

$$\frac{n_2^2}{n_3^2} \tan^2 V + \frac{n_2^2}{n_1^2} = \tan^2 V + 1$$

et

$$\tan^2 V = \frac{n_1^2 - n_2^2}{n_2^2 - n_3^2} \cdot \frac{n_3^2}{n_1^2} \cdot \frac{n_2^2}{n_2^2} = \frac{\dfrac{1}{n_2^2} - \dfrac{1}{n_1^2}}{\dfrac{1}{n_3^2} - \dfrac{1}{n_2^2}} \approx \frac{n_1 - n_2}{n_2 - n_3}.$$

Comme les indices dépendent de la longueur d'onde avec une loi qui n'est pas la même pour les trois, la position des axes varie avec la longueur d'onde c'est le phénomène de la dispersion des axes.

Dans tout ce qui précède, nous avons considéré des cristaux transparents, or il existe des cristaux absorbants dans lesquels les phénomènes sont modifiés.

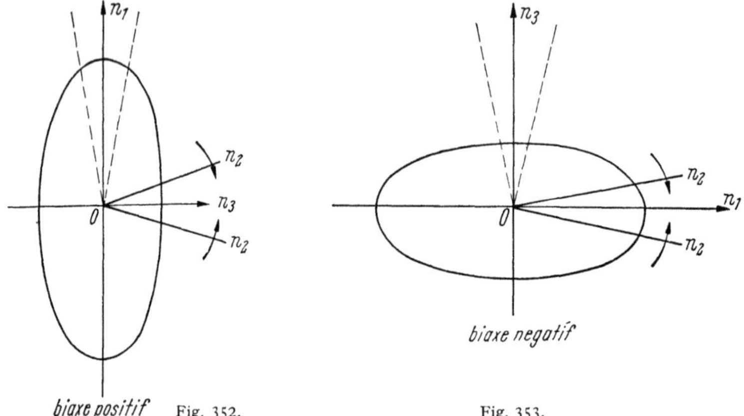

Fig. 352.   Fig. 353.

On montre que dans un cristal peu absorbant, les lois de la propagation sont les mêmes que dans un cristal transparent au point de vue indices et vitesses de propagation, et que l'indice d'extinction dépend de la direction de la vibration. Les deux directions privilégiées d'un plan d'onde ne correspondent donc pas au même indice d'extinction et sont inégalement affaiblies.

Si on examine avec un nicol un tel cristal taillé sous forme de lame à faces parallèles, on constate effectivement que l'intensité dépend de celle des vibrations privilégiées que laisse passer le nicol. En lumière blanche, les deux directions privilégiées ayant des indices d'extinction qui varient avec la longueur d'onde, les faisceaux transmis correspondants ont des couleurs différentes qui dépendent de l'orientation de la lame par rapport aux axes de symétrie du cristal: le cristal est dit «pléochroïque».

Dans le cas d'un cristal uniaxe, pour un plan d'onde quelconque, on observe une teinte ordinaire et une teinte extraordinaire; le cristal est «dichroïque». On peut observer ainsi avec du quarts enfumé une teinte ordinaire rouge pâle et une teinte extraordinaire jaune terne.

Certains échantillons de tourmaline colorés en vert, sont fortement dichroïques, ils absorbent beaucoup la vibration ordinaire et peu la vibration extraordinaire.

III. Étude de la lumière transmise par une lame cristalline éclairée en faisceau parallèle.

a) Lumière elliptique produite à la traversée d'une lame, mesure de la différence de marche.

149. Différence de marche produite par une lame cristalline. Considérons une lame cristalline à faces planes et parallèles d'épaisseur e et faisons tomber sur cette lame une onde plane Ω parallèle aux faces de la lame. L'onde Ω peut propager sans altération l'une ou l'autre des deux vibrations privilégiées rectangulaires P_1 et P_2 (Sect. 143). Si une vibration se propage suivant P_1, elle accomplit le chemin optique $\dfrac{c}{v'}\, e = n'e$ en traversant la lame. Si elle se propage suivant P_2, elle accomplit le chemin optique $\dfrac{e}{v''}\, e = n''e$. Pour les vibrations qui sont parallèles à P_1 ou P_2, la lame cristalline se comporte comme une lame isotrope. Nous appellerons P_1 et P_2 les lignes neutres de la lame ou axes de la lame et les plans que définissent ces directions avec la normale à la lame, les sections principales de la lame.

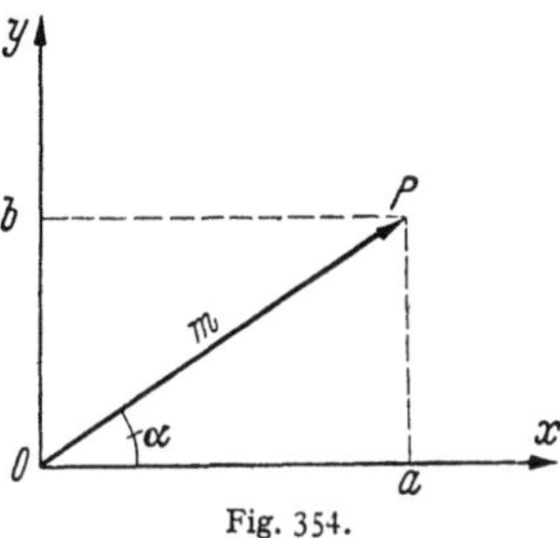

Fig. 354.

Pour étudier les phénomènes, nous prendrons comme axes de coordonnées xOy deux axes rectangulaires dans un plan parallèle à la lame, Oy confondu avec la direction privilégiée P_2 correspondant à n'' et Ox confondu avec P_1 correspondant à n' avec $n'' > n'$. L'axe Ox de la lame sera l'axe «rapide» et Oy l'axe «lent».

α) *Vibration incidente rectiligne.* Soit OP la vibration rectiligne incidente (Fig. 354). Cette vibration peut être décomposée avant d'entrer dans la lame en deux vibrations dirigées suivant les axes de la lame cristalline et présentant une différence de marche nulle

$$x = m \cos \alpha \cos \omega t = a \cos \omega t,$$

$$y = m \sin \alpha \cos \omega t = b \cos \omega t.$$

La lame cristalline étant supposée transparente, à la sortie les amplitudes sont les mêmes, mais une différence de marche $\delta = (n'' - n')\, e$ ou une différence de phase $\varphi = \dfrac{2\pi}{\lambda}\,(n'' - n')\, e$ s'est introduite entre les deux composantes

$$x = a \cos \omega t,$$

$$y = b \cos (\omega t - \varphi).$$

La vibration résultante n'est plus rectiligne. Le point de coordonnées x et y décrit une ellipse dont on trouve l'équation en éliminant le temps

$$\frac{x^2}{a^2} + \frac{y^2}{b^2} - \frac{2xy}{ab} \cos \varphi = \sin^2 \varphi,$$

c'est l'équation d'une ellipse inscrite dans un rectangle de côtés $2a$ et $2b$. La Fig. 355 montre la forme de l'ellipse en fonction de φ. Pour $\varphi = 0$ l'ellipse se réduit à son grand axe. Pour $\varphi > 0$ et petit, l'ellipse est aplatie. Pour $\varphi = \pi/2$ l'équation de l'ellipse devient $\dfrac{x^2}{a^2} + \dfrac{y^2}{b^2} = 1$ l'ellipse est rapportée à ses axes.

Pour $\varphi > \pi/2$ l'ellipse s'aplatit de nouveau, mais dans l'autre sens. Pour $\varphi = \pi$ l'ellipse se réduit à l'autre diagonale du rectangle. Pour $\varphi = 3\pi/2$ elle est de nouveau rapportée à ses axes.

Pour $\varphi = \pi/2$, $\delta = \lambda/4$ les deux vibrations présentent une différence de marche égale à $\frac{1}{4}$ de longueur d'onde. La lame qui a l'épaisseur correspondante est dite lame quart d'onde. De même, pour $\varphi = (2K+1)\dfrac{\pi}{2}$ ou $\delta = \dfrac{\lambda}{4} + K\dfrac{\lambda}{2}$.

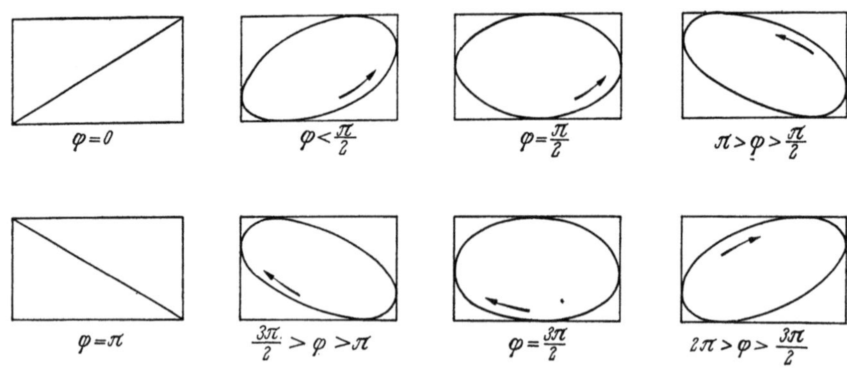

Fig. 355.

La lame qui a une épaisseur telle qu'elle introduise une différence de phase égale à π, $\delta = \lambda/2$ ou encore $\varphi = (2K+1)\pi$ avec $\delta = \dfrac{\lambda}{2} + K\lambda$ est une lame demi-onde. La vibration émergente est encore rectiligne et symétrique de la vibration incidente par rapport aux axes de la lame (Sect. 152). Pour $\varphi = 2\pi$ ou $\delta = \lambda$ ou encore $\varphi = 2K\pi$ avec $\delta = K\lambda$, la lame est dite lame onde. La vibration transmise est identique à la vibration incidente.

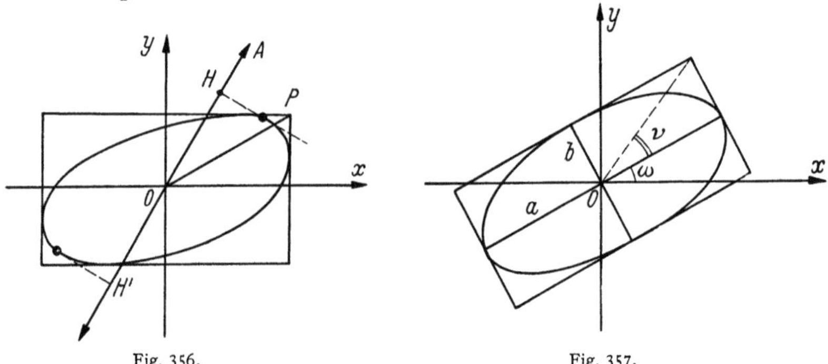

Fig. 356.   Fig. 357.

Si la vibration incidente OP est orientée à 45° des lignes neutres de la lame $a = b$ et les rectangles de la Fig. 355 deviennent des carrées. Les axes de l'ellipse sont les diagonales du carré et l'un d'eux coïncide avec la direction de la vibration incidente.

Reprenons l'une des vibrations elliptiques de la Fig. 355 et reçevons-la sur un nicol analyseur (Fig. 356). L'amplitude de la vibration transmise est représentée par la projection OH de l'ellipse sur la direction OA suivant laquelle le nicol laisse passer les vibrations. On voit que si on tourne le nicol analyseur, il n'y aura pas extinction complète. La lumière transmise passera par des maxima et des minima non nuls.

Il n'y aura extinction que si la vibration incidente est dirigée suivant l'un des axes ox ou oy de la lame. On peut ainsi déterminer expérimentalement la direction des lignes neutres d'une lame cristalline.

β) Représentation d'une vibration elliptique sur la sphére de POINCARÉ. Soient $oxoy$ deux axes rectangulaires tracés dans un plan parallèle aux faces de la lame (Fig. 357). La vibration elliptique est caractérisée par l'angle ω que fait l'un de ses axes avec ox, par l'angle v, tan $v = b/a$, de ce même axe avec la diagonale du rectangle et par son sens de parcours dextrorsum (sens de oy vers ox) ou sinistrorsum (sens de ox vers oy).

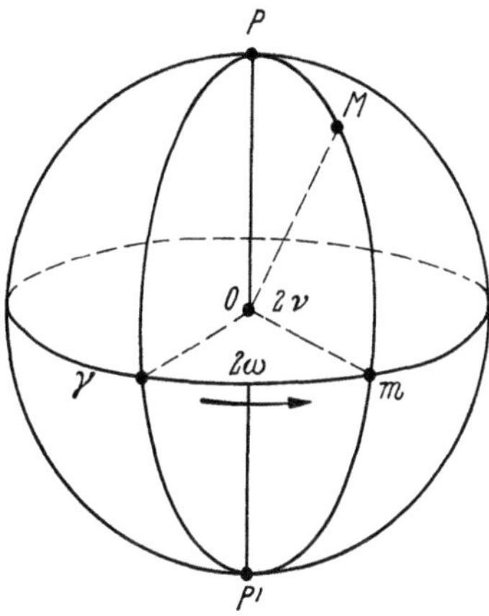
Fig. 358.

Cette ellipse est représentée sur une sphère de rayon 1 (Fig. 358) par un point M de longitude 2ω et de latitude $2v$. L'hémisphère boréal correspond aux ellipses sinistrorsum et l'hémisphère austral, aux ellipses dextrorsum. Le pôle boréal P ($b = a$ car $2v = 90°$) correspond à une vibration circulaire sinistrorsum, le pôle austral P' à une vibration circulaire dextrorsum. Les points d'un même parallèle correspondent à des ellipses de même forme, de même sens de parcours et d'orientations différentes. Les points de l'équateur correspondent à des vibrations rectilignes. Le point γ origine des longitudes correspond à la vibration ox, le point diamétralement opposé à γ correspond à la vibration oy. Ainsi la vibration elliptique représentée

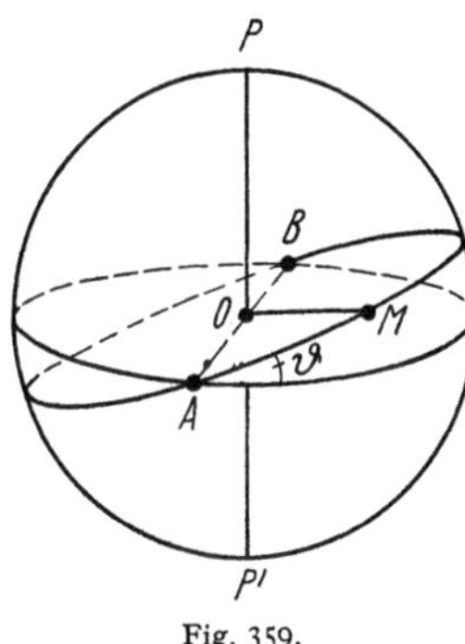
Fig. 359.

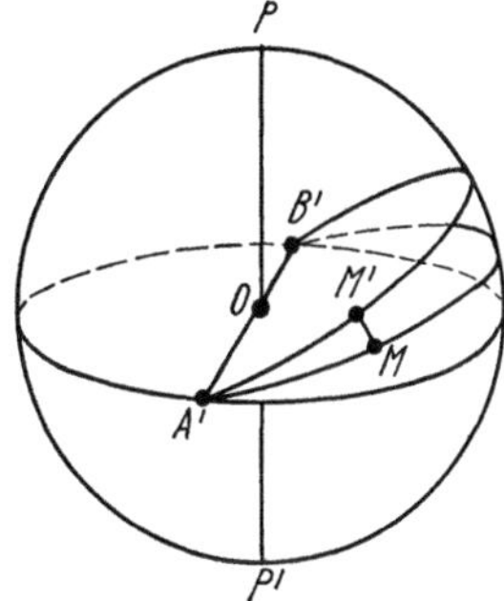
Fig. 360.

par M (Fig. 359) peut être considérée comme la résultante des deux vibrations rectilignes rectangulaires A et B, le plan MAB coupant la sphère suivant un grand cercle. Ecrivons la vibration elliptique sous la forme

$$x = \cos \alpha \cos \omega t,$$

$$y = \sin \alpha \cos (\omega t - \varphi).$$

Un calcul simple de trigonométrie sphérique montrera que $\vartheta = \varphi$ et l'arc $AM = 2\alpha$. Supposons que la vibration précédente tombe sur une lame cristalline introduisant la différence de phase φ' entre les deux directions privilégiées. Soient A', B' les vibrations qui peuvent se propager suivant les lignes neutres (Fig. 360) et M la vibration elliptique. L'angle ϑ mesure la différence de phase φ entre les deux composantes A' et B' avant l'entrée dans la lame (A' vibration en retard sur B'). Après traversée de la lame, il faut augmenter le retard de phase φ de φ';

on fait donc tourner le point M d'un angle φ' autour de $A'B'$ dans le sens inverse des aiguilles d'une montre (on regarde de A' vers B') et on obtient le point M'. Le point M' représente la vibration elliptique à la sortie de la lame.

150. Mesure de la différence de marche produite par une lame cristalline au moyen du compensateur de Babinet. La mesure de la différence de marche produite par une lame cristalline peut se faire en compensant sa biréfringence par une biréfringence de signe opposé obtenue au moyen d'un compensateur. Dans le compensateur de Babinet, on peut produire

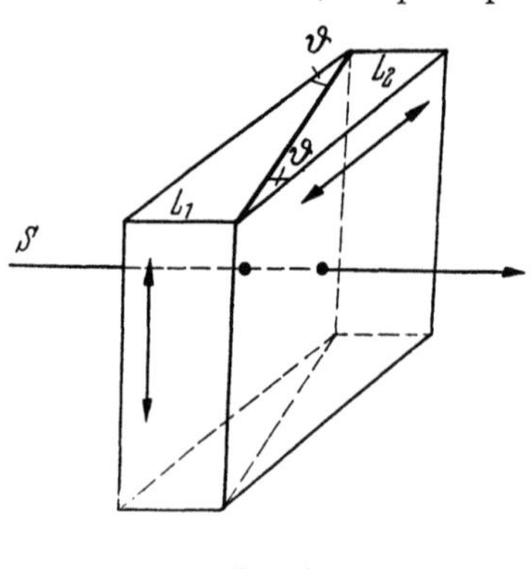

Fig. 361.

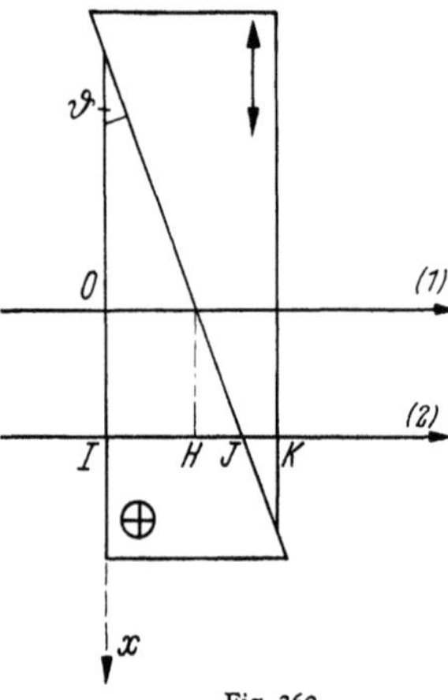

Fig. 362.

une différence de marche variable et mesurable de la façon suivante: une lame à faces parallèles formée de deux coins de quartz L_1 et L_2 de même angle ϑ très petit et d'axes optiques croisés, est traversée par de la lumière monochromatique dans une direction normale aux axes optiques (Fig. 361). On peut déplacer l'un des coins par rapport à l'autre dans une direction perpendiculaire à l'arête des coins au moyen d'une vis micrométrique. La Fig. 362 represente la compensateur coupé par un plan perpendiculaire à l'axe optique du coin L_1.

Soit le rayon (1) qui traverse les coins au point O où ils ont même épaisseur et considérons un rayon (2) qui les traverse à une distance x de o. Comme les deux lames ont leurs axes croisés, leurs actions sont de signes contraires. Si $e = IJ$, $e' = JK$ on a $e - e' = 2\overline{HJ} = 2x \tan \vartheta$ et la différence de marche totale est

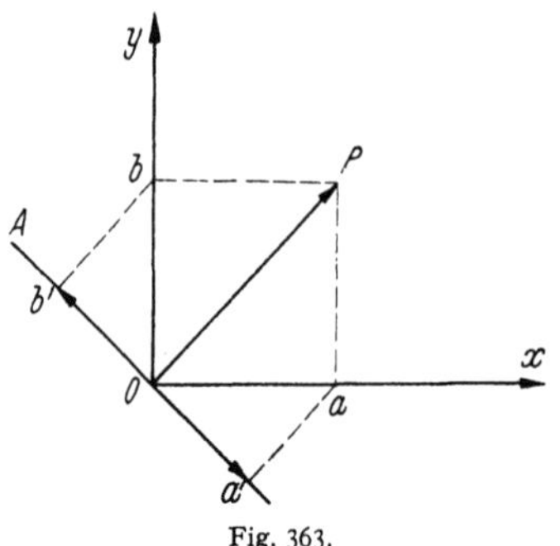

Fig. 363.

$$\delta = (n_e - n_0)(e - e') = 2(n_e - n_0)\, x \tan \vartheta$$

elle est constante sur des parallèles aux arêtes des coins. Envoyons une vibration polarisée rectilignement OP à 45° des axes ox et oy parallèles aux axes optiques des coins (Fig. 363). On peut la décomposer avant l'entrée dans la lame en deux composantes Oa et Ob dirigées suivant ox et oy. Ces deux vibrations sont transmises par les éléments du compensateur mais leur différence de phase est changée. A la sortie, on aura deux vibrations analogues à a et b présentant une certaine différence de phase. Recevons-les sur un nicol analyseur qui laisse passer les vibrations suivant la direction OA perpendiculaire à OP (polariseur et analyseur croisés). Les projections Oa' et Ob' sont des vibrations parallèles qui peuvent interférer suivant les règles classiques établies au chapitre sur les interférences.

Elles interfèrent ici dans les meilleurs conditions puisque leurs amplitudes sont égales, la vibration incidente étant à 45° des axes ox, oy. Les franges auront donc le contraste I, c'est-à-dire que les franges sombres seront parfaitement noires.

Nous venons de décrire dans un cas particulier le phénomène des interférences en lumière polarisée dont nous parlerons plus en détails au chapitre IV (Sect. 165). Si la lame introduit entre a et b la différence de marche $\delta = K\lambda$ les vibrations a' et b' sont en opposition, il n'y a pas de lumière transmise. Pour $\delta = (2K+1)\frac{\lambda}{2}$ les projections a' et b' sont en phase, il y a maximum de lumière. On voit sur le compensateur une série de franges d'interférences équidistantes et parallèles aux arêtes des coins. On aura une frange noire si

$$2(n_e - n_0)\, x \tan\vartheta = K\lambda,$$

c'est-à-dire

$$x = \frac{K\lambda}{2(n_e - n_0)\tan\vartheta}$$

et l'interfrange sera

$$i = \frac{\lambda}{2(n_e - n_0)\tan\vartheta}. \tag{150.1}$$

Si le compensateur est éclairé en lumière blanche, on observe des franges colorées, la frange centrale correspondant à $e = e'$ ($\delta = 0$) étant noire. Comme $n_e - n_0$ varie peu dans le spectre, les teintes sont sensiblement celles de l'échelle de NEWTON à centre noir (Sect. 166).

La formule (150.1) montre que l'interfrange ne dépend pas de la position du point O sur les lames. Si on déplace l'une des lames avec la vis micrométrique,

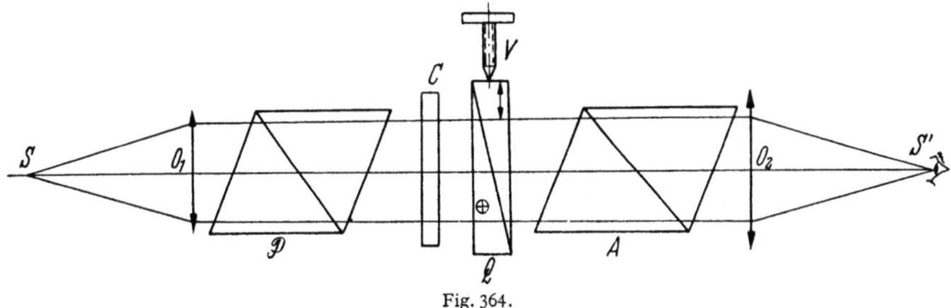

Fig. 364.

le point O se déplace par rapport aux lames et avec lui tout le système de franges, qui ne change pas d'aspect. C'est cette propriété qui est utilisée pour mesurer une différence de marche. Le compensateur Q (Fig. 364) est placé en faisceau parallèle au moyen du collimateur SO_1. Un nicol $\mathcal{P}$ polariseur donne une vibration à 45° des axes du compensateur et l'analyseur A placé après Q est croisé avec $\mathcal{P}$. Pour faire ce réglage, on commence par enlever Q et on règle $\mathcal{P}$ et A à l'extinction. On interpose Q et en le faisant tourner dans son plan, on rétablit l'exteinction puis on le tourne à 45°. L'oeil placé au foyer S' d'une loupe O_2 met au point sur le compensateur afin de voir les franges bien nettes. Au moyen de la vis micrométrique V, on règle la position des franges pour qu'une frange noire soit encadrée par les fils d'un repère placé contre Q par exemple. La lame cristalline C à étudier est placée entre $\mathcal{P}$ et A avant ou après Q et est orientée de façon que ses lignes neutres coïncident avec celles du compensateur Q. On constate que les franges sont déplacées. Au moyen de la vis V étalonnée au préalable, on ramène la frange noire entre les repères. On a ainsi la partie fractionnaire $\varepsilon\lambda$ ($0 < \varepsilon < 1$) de la différence de marche $\delta = \varepsilon\lambda + K\lambda$. Pour avoir la partie entière K, on peut examiner la teinte de la lame au Nörremberg (voir polarisation chromatique en lumière parallèle Sect. 166) si elle est assez mince. On peut aussi observer le compensateur en lumière blanche et voir où se trouve la frange noire. Ceci est possible car généralement la dispersion de la biréfringence varie peu d'une substance à l'autre. La précision des mesures est de l'ordre du millième de longueur d'onde sur la différence de marche, à condition que la qualité des surfaces du compensateur soit suffisante.

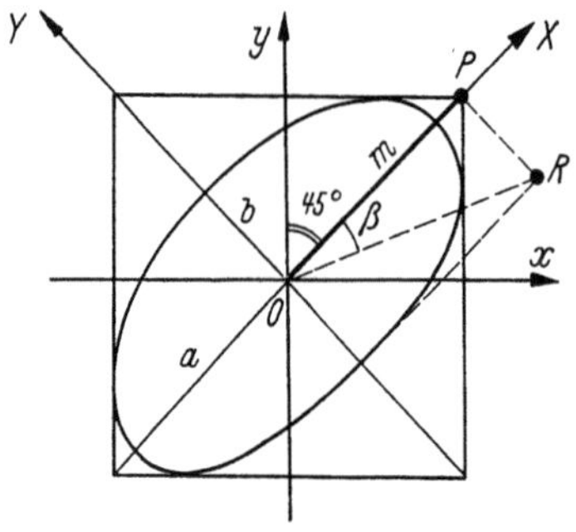

151. Mesure de la différence de marche produite par une lame cristalline par la méthode du quart d'onde. On a vu que lorsqu'une vibration rectiligne tombe sur une lame cristalline à 45° de ses axes, la vibration émergente est une vibration elliptique dont un des axes coïncide avec la direction de la vibration incidente (Sect. 149). Calculons le rapport des axes: l'équation de l'ellipse est (Fig. 365)

Fig. 365.

$$x = \tfrac{1}{2}\sqrt{2}\, m \cos \omega t, \quad y = \tfrac{1}{2}\sqrt{2}\, m \cos (\omega t - \varphi).$$

Faisons un changement de coordonnées en prenant les nouveaux axes OX, OY. On a

$$X = m \cos \frac{\varphi}{2} \cos \left(\omega t - \frac{\varphi}{2}\right),$$

$$Y = m \sin \frac{\varphi}{2} \sin \left(\omega t - \frac{\varphi}{2}\right),$$

d'où le rapport des axes

$$\frac{b}{a} = \tan \frac{\varphi}{2} = \tan \beta$$

et

$$\varphi = 2\beta.$$

C'est le résultat que nous avons déjà obtenu à la Sect. 9.

Recevons la vibration elliptique précédente sur une lame quart d'onde dont une ligne neutre coïncide avec OP. Supposons par exemple l'axe lent dirigé suivant OY. On peut écrire la vibration elliptique sous la forme (vibration gauche)

$$X = a \cos \omega t, \quad Y = b \sin \omega t.$$

La vibration transmise par le quart d'onde sera

$$X = a \cos \omega t,$$

$$Y = b \sin \left(\omega t - \frac{\pi}{2}\right) = - b \cos \omega t.$$

C'est une vibration rectiligne dirigée suivant OR et faisant avec OP un angle β compris entre O et $-\dfrac{\pi}{2}$. Pour une vibration droite, on aurait un angle β compris entre O et $+\dfrac{\pi}{2}$.

D'où la méthode de mesure d'une différence de marche: on oriente la lame cristalline à étudier de façon que ses lignes neutres soient à 45° de la vibration incidente OP, fournie par le polariseur. La vibration émergente est une vibration elliptique dont un des axes coïncide avec OP. On reçoit cette vibration elliptique sur une lame quart d'onde dont une ligne neutre OX coïncide avec la direction OP. Après traversée du quart d'onde, on a une vibration rectiligne dirigée suivant OR et faisant l'angle $\beta = \varphi/2$ avec OP. L'analyseur et le polariseur étant primitivement à l'extinction, celle-ci n'est pas conservée. Pour rétablir l'extinction, il faut tourner l'analyseur d'un angle égal à β. On obtient ainsi φ, c'est-à-dire la différence de marche produite par la lame cristalline. Le schéma du montage est indiqué sur la Fig. 366. $\mathcal{P}$ et $\mathcal{A}$ sont à l'extinction. On interpose la lame cristalline C en l'orientant de façon à ne pas détruire l'extinction; de même pour la lame quart d'onde Q. Cette dernière a une ligne neutre parallèle à OP comme d'ailleurs C. Tournons la lame C de 45°: nous sommes dans les conditions précédentes: pour rétablir

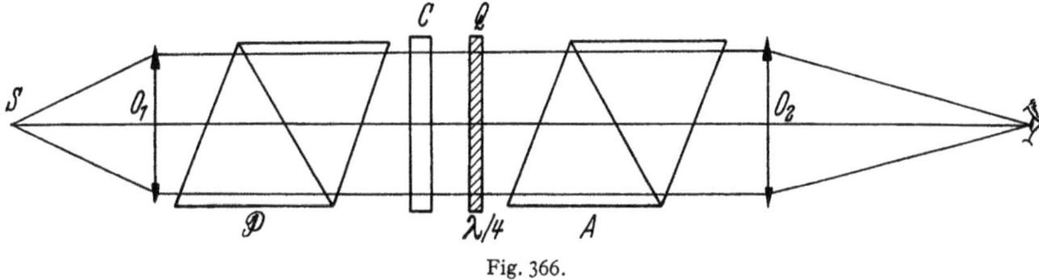

Fig. 366.

l'extinction, il faut tourner $\mathcal{A}$ d'un angle $\beta = \varphi/2$. Si l'axe rapide du quart d'onde coïncide avec OP et si $|\beta| < \pi/2$ la vibration elliptique est de sens contraire à la rotation. Comme précédemment on aura la différence de phase à $2K\pi$ près. Pour lever l'indétermination, on procèdera comme au paragraphe précédent.

Notons qu'on peut améliorer beaucoup la précision des mesures en utilisant un analyseur à pénombre à la place de $\mathcal{A}$ (Sect. 152). On peut mesurer β avec une précision de l'ordre 18″, soit une précision de l'ordre de 1/36000 de la longueur d'onde sur la différence de marche. Bien entendu, la qualité du montage doit être en rapport avec la précision cherchée. La précision est supérieure à celle obtenue avec le compensateur de BABINET en particulier grâce à l'emploi de l'analyseur à pénombre.

b) Analyse d'une vibration lumineuse.

152. Analyseur à pénombre à lame demi-onde. Pointé d'une vibration rectiligne.
Dans l'exemple du paragraphe précédent, la mesure de β a été effectuée avec un simple nicol analyseur en cherchant l'orientation correspondant à l'extinction. Par suite de la lumière parasite, la recherche d'un minimum n'est pas une opération très précise et on préfère remplacer l'appréciation d'un minimum par l'appréciation de l'égalité d'éclairement de deux plages séparées par une ligne fine. C'est ce qui est réalisé dans l'analyseur à pénombre. Il est composé d'une lame demi-onde Q solidaire d'un nicol $\mathcal{A}$, la lame demi-onde ne recouvrant que la moitié de la face d'entrée du nicol $\mathcal{A}$ (Fig. 367).

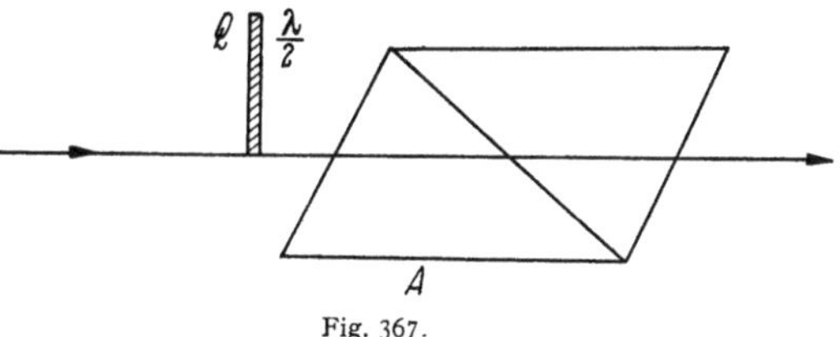

Fig. 367.

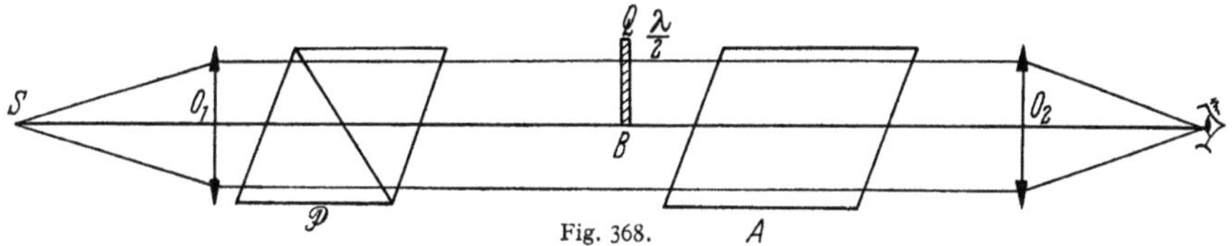

Fig. 368.

Considérons d'abord l'action d'une lame demi-onde, d'axes ox, oy (Fig. 368) sur une vibration rectiligne OP. Avant l'entrée dans la lame, on peut décomposer

OP en deux vibrations dirigées suivant ox et oy

$$x = m \cos\alpha \cos\omega t, \quad y = m \sin\alpha \cos\omega t.$$

Après traversée de la lame demi-onde, on a

$$x = m \cos\alpha \cos\omega t,$$

$$y = m \sin\alpha \cos(\omega t - \pi) = -m \sin\alpha \cos\omega t.$$

La vibration transmise est encore rectiligne mais sa direction OP' est symétrique de OP par rapport aux lignes neutres de la lame demi-onde.

Si la vibration incidente est elliptique

$$x = a \cos\omega t, \quad y = b \cos(\omega t - \varphi),$$

après traversée de la lame demi-onde

$$x = a \cos\omega t,$$

$$y = b \cos(\omega t - \varphi - \pi) = -b \cos(\omega t - \varphi).$$

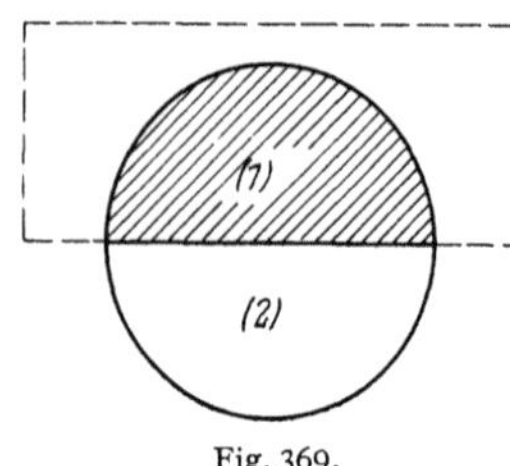

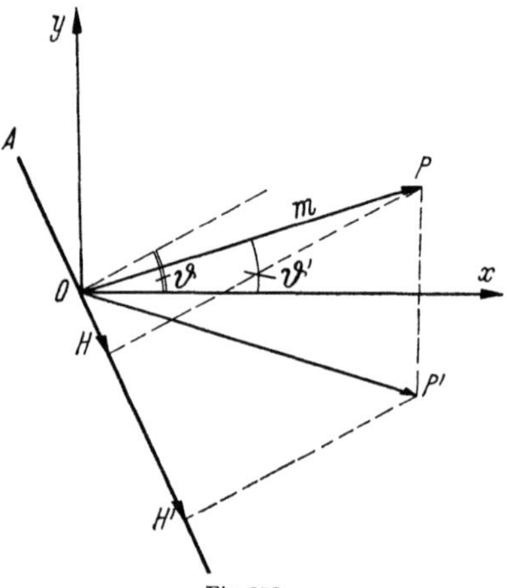

Fig. 369.

Fig. 370.

La vibration transmise est une vibration elliptique symétrique de la vibration incidente par rapport aux lignes neutres de la lame demi-onde et est parcourue en sens inverse.

Considérons alors le montage représenté sur la Fig. 368. C'est le même dispositif que celui de la Fig. 366, mais on remplace la lame quart d'onde par une lame demi-onde couvrant la moitié du champ, on supprime C et on fait la mise au point avec la loupe O_2 sur le bord B de la lame demi-onde. Le champ est donc divisé en deux régions, la région (1) couverte par la lame demi-onde, et la région (2) éclairée par les rayons qui n'ont pas traversé la lame (Fig. 369). On veut pointer l'azimut de la vibration rectiligne fournie par le polariseur $\mathcal{P}$.

Soient ox et oy (Fig. 370) les lignes neutres de la lame demi-onde et OA la direction des vibrations que laisse passer le nicol A. On orientera le nicol analyseur de façon que la perpendiculaire à oA fasse un angle ϑ petit avec ox. La vibration rectiligne à pointer est la vibration OP d'amplitude m.

Dans la région (2) (Fig. 369), la vibration est OP, dans la région (1), on a la vibration OP' symétrique de OP par rapport à ox.

Dans la région (2) on a un éclairement E_2 proportionnel à $\overline{OH}^2$

$$E_2 = m^2 \sin^2(\vartheta - \vartheta').$$

Dans la région (1) on a un éclairement E_1 proportionnel à $\overline{OH'}^2$

$$E_1 = m^2 \sin^2(\vartheta + \vartheta').$$

Les deux plages (1) et (2) ont même éclairement si $\vartheta' = 0$, c'est-à-dire si ox est dirigé suivant OP (Fig. 371). Tournons légèrement Q et A par exemple dans le sens des aiguilles d'une montre sur la Fig. 371. La projection OH diminue tandis

que OH' augmente, et inversement si on tourne l'analyseur à pénombre dans l'autre
sens. Le pointé est donc précis puisque les éclairements des 2 plages (1) et (2)
varient en sens inverse dès que l'on s'écarte de la position donnant l'égalité.
A l'égalité des éclairements, la vibration incidente est la direction de la ligne neutre
de la lame demi-onde qui fait un petit angle avec la perpendiculaire à $o\mathcal{A}$. L'éclairement des deux plages est alors égal à

$$E_1 = E_2 = m^2 \sin^2 \vartheta.$$

Il est faible puisque ϑ est petit, d'où le nom d'analyseur à pénombre. L'angle
2ϑ s'appelle l'angle de pénombre. On peut noter que l'on a encore $E_1 = E_2$ si
OP est confondu avec oy, mais alors les éclairements sont voisins de leurs valeurs
maxima et leurs variations relatives sont petites si on tourne l'analyseur à pénombre. Le pointé n'est pas précis. Dans le mesure de la différence de marche par la
méthode du quart d'onde, on aura avantage à remplacer l'analyseur $\mathcal{A}$ par un
analyseur à pénombre. Le pointé de la vibration rectiligne transmise par la lame
quart d'onde s'effectuera alors avec le maximum de précision. La précision d'un
pointé avec un analyseur à pénombre est d'environ $\vartheta/100$ soit dans les meilleures
conditions $18''$ pour $\vartheta = 30'$.

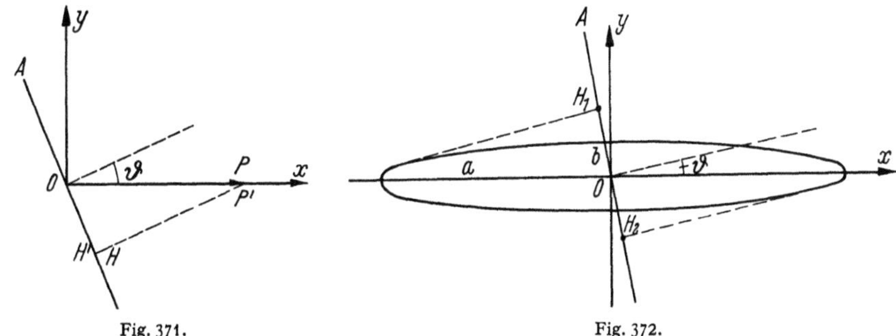

Fig. 371.   Fig. 372.

153. Pointé d'une vibration elliptique allongée. Au lieu d'une vibration rectiligne, considérons une vibration elliptique allongée (Fig. 372). Si l'un des axes
de l'ellipse coïncide avec ox, il y a égalité des éclairements des deux plages. Dans
le cas où c'est le grand axe qui coïncide avec ox, une petite rotation de l'analyseur
à pénombre fait diminuer l'éclairement de l'une des plages et augmenter celle de
l'autre. On est dans les mêmes conditions qu'au paragraphe précédent et il est
posible de pointer le grand axe d'une vibration elliptique allongée. Pour avoir
la même précision, il faut que b/a soit suffisamment petit. En posant $b/a = \tan\beta$,
il faut que $\beta < \vartheta$ par exemple $\beta = \vartheta/2$.

154. Analyse d'une vibration elliptique d'axes connus. Pour que la vibration
elliptique soit déterminée, il faut connaître l'orientation de l'ellipse et le rapport
de ses axes. Nous nous plaçons d'abord dans le cas où l'orientation des axes est
connue et où il suffit de déterminer le rapport des axes.

La méthode de DE SÉNARMONT résulte de la Sect. 151. La vibration elliptique
à analyser est reçue sur une lame quart d'onde et on fait coïncider les axes de
l'ellipse avec les lignes neutres de la lame quart d'onde. La vibration transmise
est une vibration rectiligne que l'on pointe avec un analyseur à pénombre. Ayant
au préalable repéré l'orientation des axes du quart d'onde, on mesure ainsi l'angle
β qui caractérise l'ellipticité de la vibration ($\tan \beta = b/a$).

On repère l'orientation des axes du quart d'onde avant d'interposer le système qui produit la vibration elliptique à étudier: le polariseur et l'analyseur

à pénombre étant en place, on oriente ce dernier pour avoir l'égalité des éclairements. On interpose ensuite le quart d'onde et on l'oriente pour que l'égalité soit maintenue; l'une des lignes neutres du quart d'onde est alors parallèle à la vibration pointée par l'analyseur à pénombre. On place entre le polariseur et le quart d'onde le système produisant la vibration elliptique à étudier. La direction de ses axes étant connue, on les dispose parallèlement aux lignes neutres du quart d'onde. Pour pointer la vibration rectiligne rétablie, il faut tourner l'analyseur à pénombre d'un angle β dont la tangente est égale au rapport des axes de la vibration elliptique. L'axe oy du quart d'onde étant parallèle à l'axe b de l'ellipse, l'axe ox parallèle à l'axe a de l'ellipse, β est l'angle dont il faut tourner l'analyseur à pénombre à partir de la direction oc pointée primitivement par lui. Le sens de l'ellipse est le sens dans lequel il faut tourner l'analyseur à pénombre pour l'amener de la vibration rectiligne rétablie sur ox par une rotation inférieure à 90°.

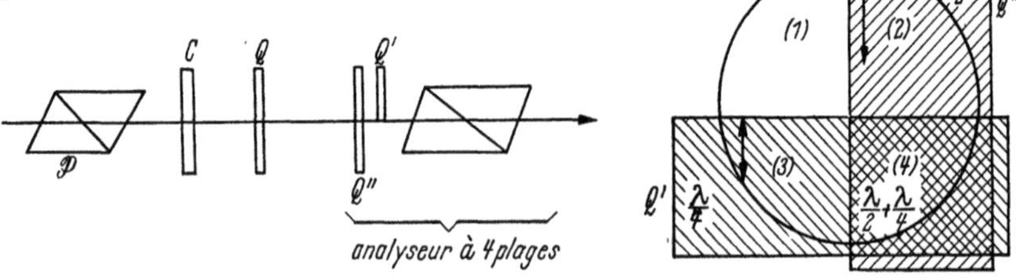

Fig. 373.   Fig. 374.

155. Analyse d'une vibration elliptique d'orientation quelconque. Lorsque les axes de l'ellipse produite par la lame cristalline C ne sont pas connus, on cherche par tâtonnements, à orienter les lignes neutres d'un quart d'onde Q suivant les axes de la vibration elliptique (Fig. 373). Le réglage est obtenu lorsque la vibration rétablie est rectiligne. Il faut donc d'abord être sûr que la vibration rétablie est bien rectiligne. Pour cette opération, l'analyseur à pénombre n'est pas précis, étant donné que l'on ne voit pas de différence entre une vibration elliptique allongée et une vibration rectiligne (Sect. 153). On utilise alors un analyseur à quatre plages formé par une lame quart d'onde Q' et une lame demi-onde Q'' dont les lignes neutres sont parallèles (Fig. 374). La lame Q'' et le nicol forment un analyseur à pénombre ordinaire. Les régions (1) et (2) ont même éclairement si la vibration elliptique a son grand axe en coïncidence avec l'une des lignes neutres commune à Q' et Q''. D'après la Sect. 153, c'est la ligne neutre qui fait un petit angle avec la perpendiculaire à la direction de vibration du nicol placé derrière les lames. Nous appellerons N cette ligne neutre.

Par rapport à la plage (3), la vibration elliptique incidente a alors ses axes parallèles à ceux du quart d'onde et après traversée de Q', on a une vibration rectiligne faisant l'angle β avec N et les deux régions (3) et (4). Si la vibration incidente est rectiligne et orientée d'une façon quelconque, Q' en donne une vibration elliptique dont un des axes est dirigé suivant N. Les régions (3) et (4) ont même éclairement, mais pas les plages (1) et (2), sauf si la vibration rectiligne incidente est parallèle N. Pour que les quatre plages de l'analyseur aient le même éclairement, il faut que la vibration incidente soit rectiligne et dirigée suivant la ligne neutre N commune à Q' et Q''.

Lorsque les quatre plages ont même éclairement, on repère l'orientation du quart d'onde Q, ce qui donne l'orientation des axes de l'ellipse par rapport à des directions connues. L'orientation de l'ensemble $Q' Q''$ A donne l'angle β dont la tangente est égale au rapport des axes.

c) Polariseurs et prismes biréfringents.

156. Prisme de Nicol. Nous avons déjà donné quelques indications sur ces prismes à la Sect. 138. La Fig. 375 donne les principales caractéristiques d'un nicol limité par ses faces de clivage. Tant que les rayons lumineux se n'écartent pas trop du rayon moyen SI, les rayons ordinaires subissent la réflexion totale sur AC et seuls les rayons extraordinaires sont transmis. Mais si les rayons s'inclinent trop dans le sens $S'I$, il arrivera un moment où les rayons ordinaires ne subiront plus la réflexion totale et traversont le prisme. La partie du champ correspondante ne sera plus polarisée.

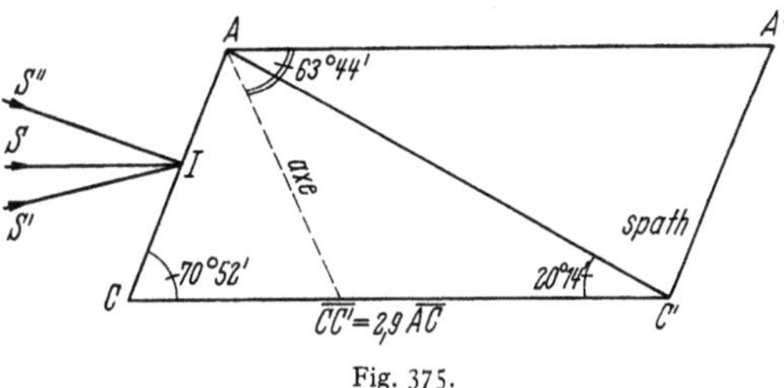

Fig. 375.

On constate que le champ polarisé qui ne transmet que les vibrations extraordinaires est voisin de 30° environ. Les valeurs

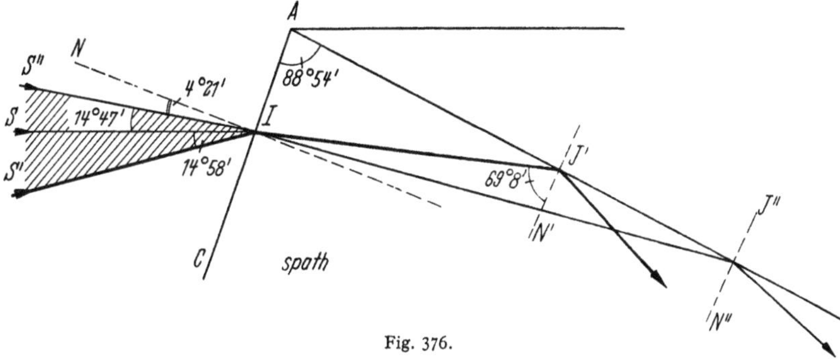

Fig. 376.

numériques sont données sur la Fig. 376; le champ est représenté par l'angle $S'IS''$. La Fig. 376 correspond à un nicol dont les deux éléments sont collés au baume de Canada; l'angle limite de réflexion totale spath-baume est pour le rayon ordinaire

$$\sin \widehat{IJ'N'} = \frac{n_{\text{baume}}}{n_0} = \frac{1{,}549}{1{,}658}\,,$$

$$\widehat{IJ'N'} = 69°\ 8'.$$

Considérons deux rayons de lumière, l'un parallèle à SI c'est-à-dire AA' ou CC' (Fig. 375) et imaginons l'autre parallèle à l'axe optique à l'intérieur du nicol. Recevons ces deux rayons dans le plan focal d'un objectif. La trace dans ce plan du rayon (Fig. 377) parallèle à AA' est H, la trace du deuxième est J_1. Soit M la trace d'un rayon quelconque. On peut admettre en première approxima-

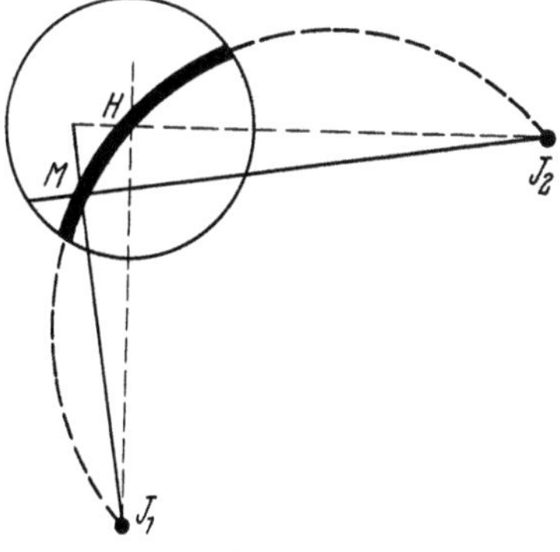

Fig. 377.

tion que la vibration transmise par le nicol et correspondant au rayon M quelconque est situé dans le plan J_1M. Plaçons à la suite de ce nicol un deuxième nicol orienté à l'extinction. Pour le premier nicol, la trace du plan de section principale est HJ_1, pour le deuxième nicol elle sera HJ_2 perpendiculaire à HJ_1. Il y aura bien extinction en H puisqu'en ce point, les deux sections principales sont perpendiculaires. Pour un point M quelconque du champ d'observation, il n'en sera plus de même sauf si l'angle $J_1 M J_2$ est droit. Il en sera ainsi pour tous les points M d'une circonférence de diamètre $J_1 J_2$. Donc seuls les points du champ situés sur cette circonférence seront éteints. On observe une frange noire à 45° des sections principales des nicols, c'est la frange de Lippich. Lorsque deux nicols sont croisés, le champ n'est pas uniformément noir sauf s'ils sont éclairés en faisceau parfaitement parallèle.

En polarimétrie, l'ouverture des faisceaux est de l'ordre du degré et la frange de Lippich est visible. Dans l'analyseur à pénombre, les deux moitiés du champ n'ont pas un éclairement uniforme et il en résulte des erreurs qui peuvent atteindre plusieurs minutes. Les appareils de précision sont munis de polariseurs à champ normal dont nous parlerons plus loin.

157. Prisme de Foucault. Au lieu de recoller les deux parties du rhomboèdre de spath ACC' et $AC'A'$ avec du baume (Fig. 375), laissons seulement une lame d'air mince. On diminue l'inclinaison de la face AC qui n'est plus que de 50° (Fig. 378) et le rapport AA'/AC est égal à 0,9. Le cristal est moins long et moins coûteux, mais le champ n'est que de 8°.

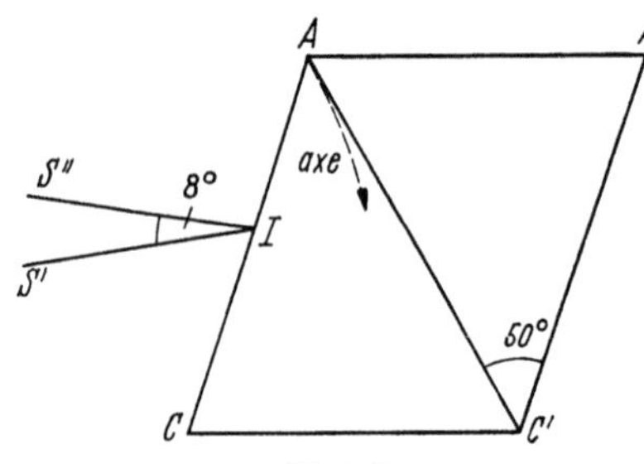

Fig. 378.

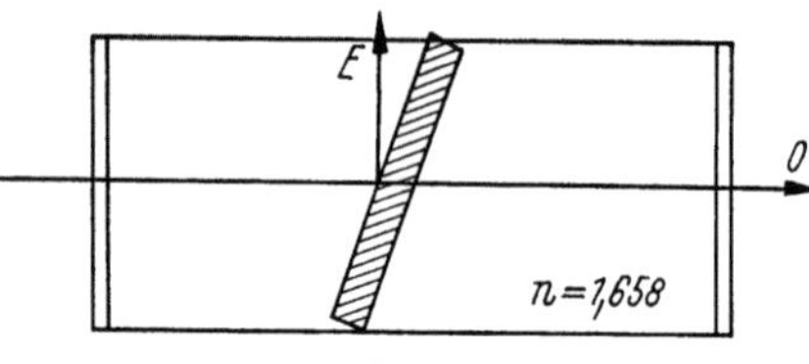

Fig. 379.

158. Polariseur de Brace. Il est constitué par une lame mince de spath taillée parallèlement à l'axe et immergée dans une cuve contenant un liquide d'indice élevé comme de la naphthaline monobromée ($n = 1{,}658$). Le rayon extraordinaire subit la réflexion totale sur la première face et c'est ici le rayon ordinaire qui est transmis. Le champ de ce polariseur est égal à 36° (Fig. 379).

159. Polariseur de Glazebrook. Reprenons la Fig. 377: si les points J_1 et J_2 s'éloignent à l'infini, tous les angles $J_1 M J_2$ sont égaux à 90° pour tous les

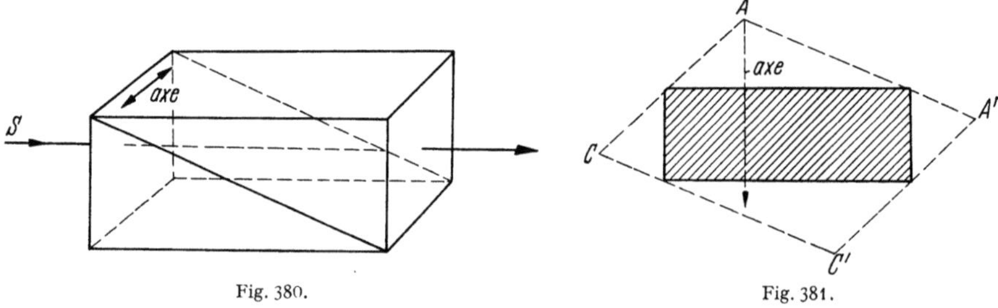

Fig. 380.

Fig. 381.

points M du champ. Il n'y a plus de frange de Lippich, le champ tout entier est éteint. Pour réaliser un polariseur de ce type, il faut donc que les faces d'entrée et de sortie soient taillées parallèlement à l'axe optique. On obtient un polariseur à champ normal tel que le prisme de Glazebrook par exemple (Fig. 380). C'est un prisme qui n'utilise pas les faces de clivage et qui entraîne beaucoup de perte de matière dans sa taille (Fig. 381).

Si le prisme est collé au baume de Canada d'indice $n = 1,55$, on a $\alpha = 19° 36'$ et $A A'/A C = 2,81$ avec un champ symétrique $i' + i'' = 8°$ (Fig. 382). S'il est collé à l'huile de lin d'indice $n = 1,485$ on a $\alpha = 25°$ et $A A'/A C = 2,15$ pour le même champ. Pour l'ultra-violet on colle les prismes de Glazebrook à la glycérine d'indice $n = 1,474$. On a alors $\alpha = 17° 8'$, et $A A'/A C = 3,2$ et $i' + i'' = 32° 6'$.

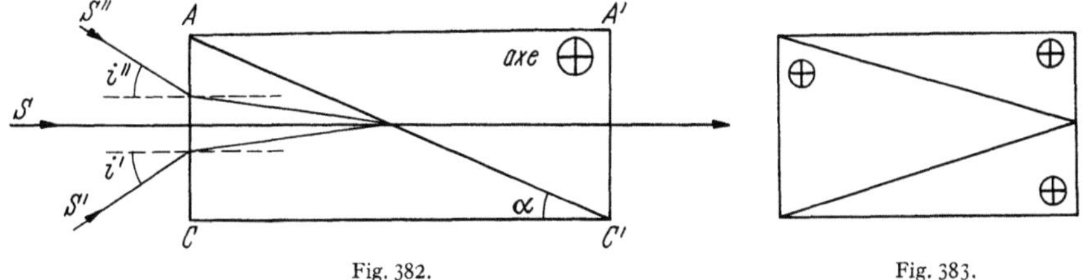

Fig. 382.   Fig. 383.

160. Polariseurs de Glan et d'Ahrens. Le prisme de Glan est un prisme analogue au Glazebrook, mais comme dans le prisme de Foucault, les deux éléments ne sont pas collés, une mince couche d'air les sépare. Pour $\alpha = 50° 15'$ et $A A'/A C = 0,85$ on a $i' + i'' = 8° 6'$.

Comme le prisme de Foucault, le prisme de Glan est très court à un champ faible et présente des pertes de lumière par réflexions multiples dans la couche d'air.

Le prisme d'Ahrens (Fig. 383) a deux plans de coupe, il est deux fois plus court que le Glazebrook à ouverture égale.

161. Polaroïdes. Nous avons cité Sect. 148 l'exemple des cristaux de tourmaline qui absorbent beaucoup plus le rayon ordinaire que le rayon extraordinaire. Pour une épaisseur convenable, on peut arrêter complètement le rayon ordinaire et constituer ainsi un véritable polariseur. Le défaut des cristaux de tourmaline est d'affaiblir aussi un peu le rayon extraordinaire et de colorer fortement la lumière.

Des films polarisant la lumière par absorption de l'une des vibrations ont été réalisés par la Polaroïd Corporation aux U.S.A. Ces films appelés «polaroïdes» et mis au point par E. H. Land contiennent des petits cristaux fortement dichroïques inclus dans une feuille de matière plastique (polaroïdes type H). Ces cristaux sont orientés en étirant le film. On obtient ainsi un polariseur qui transmet 80 pour cent de lumière polarisée dans un plan et moins de I pour cent dans le plan perpendiculaire. Deux polaroïdes à l'extinction se comportent comme une lame de densité optique 4 et ne laissent passer que le dix millième de la lumière incidente. Les résultats sont moins bons à l'extrémité bleu du spectre (1 millième de lumière transmise) et deux polaroïdes croisés donnent une lumière bleu lorsqu'on les observe avec une source intense.

Les polaroïdes offrent de nombreuses possibilités grâce à leurs dimensions qui peuvent être considérables et à leur facilité d'emploi. B. Lyot les a utilisés dans certains de ses filtres monochromatiques polarisants (Sect. 174).

162. Prismes biréfringents de Wollaston et de Rochon. On peut réaliser des prismes biréfringents qui séparent les faisceaux ordinaires et extraordinaires en associant un prisme de spath à un prisme de verre par exemple (Fig. 384). Le verre ayant un pouvoir dispersif différent de celui du spath, on peut annuler la dispersion, c'est-à-dire achromatiser l'ensemble sans annuler la déviation. L'achromatisme n'est évidemment parfaitement réalisé que pour l'une des deux images. Malheureusement le verre possède souvent de la biréfringence accidentelle et il est avantageux d'utiliser des prismes biréfringents entièrement en quartz ou en spath, ce dernier étant choisi pour les fortes séparations angulaires. Le prisme de Wollaston est représenté sur la Fig. 385. Les faces d'entrée et de sortie sont parallèles aux axes optiques qui sont croisés entre eux. La Fig. 386

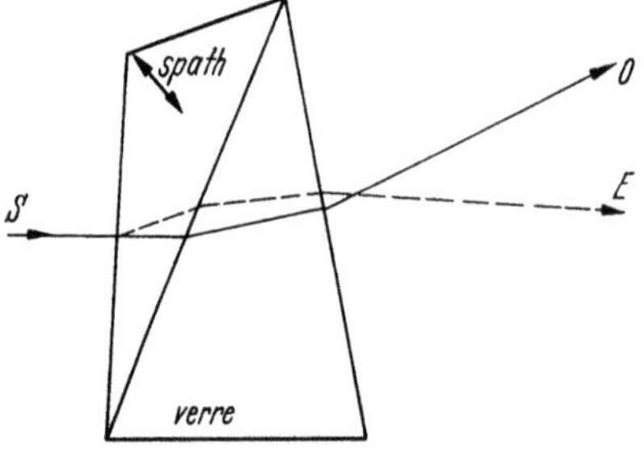

Fig. 384.

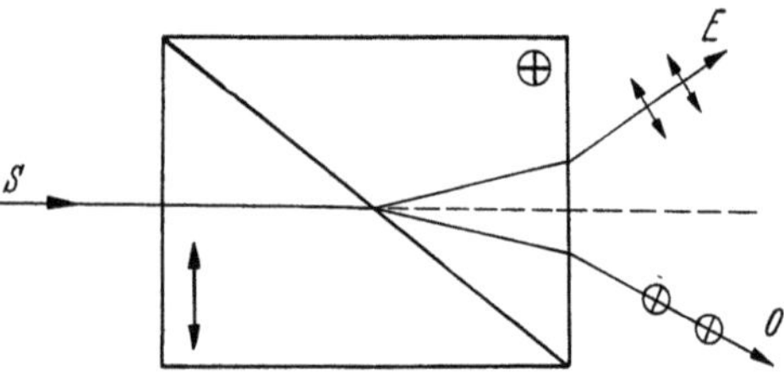

Fig. 385.

montre la marche des rayons dans le cas de deux prismes en quartz. La surface d'onde ordinaire relative au premier prisme P_1 est une sphère dont la trace sur le plan de figure est la circonférence 3. La surface d'onde extraordinaire relative à P_1 est un ellipsoïde dont la trace est l'ellipse 2 tangente en Z et Z' à la circonférence 3. La direction ZZ' donne l'axe optique du prisme P_1. Dans le prisme P_2 la surface d'onde ordinaire a pour trace la circonférence 3 et la surface d'onde extraordinaire aura pour trace la circonférence 1 puisque l'axe optique de ce prisme est perpendiculaire au plan de figure. Un rayon incident SI normal au prisme n'est

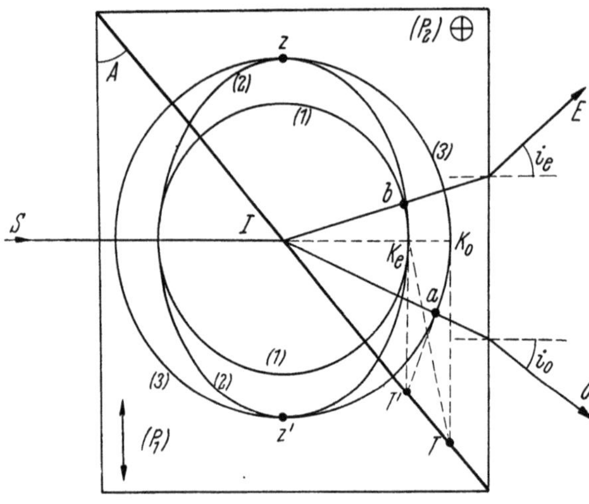

Fig. 386.

pas dédoublé en pénétrant dans P_1, il transporte des vibrations ordinaires dirigées perpendiculairement au plan de section principale (plan de figure) et des vibrations extraordinaires dans le plan de figure.

Comme les sections principales de P_1 et P_2 sont croisées, les vibrations ordinaires dans P_1 deviendront extraordinaires dans P_2 et les vibrations extraordinaires dans P_1 deviendront ordinaires dans P_2. On prolonge

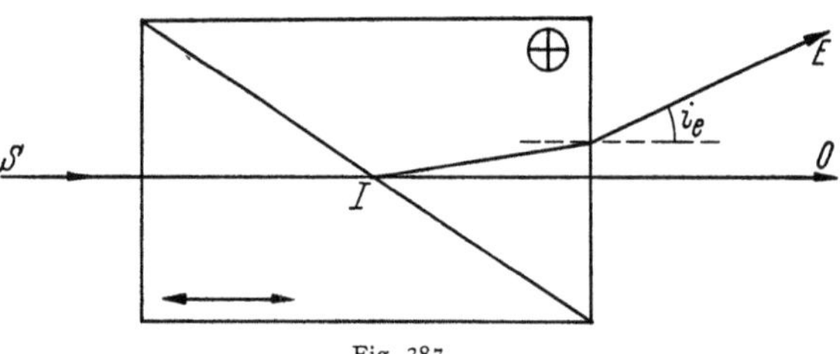

Fig. 387.

le rayon incident SI jusqu'en K_0, la tangente à la surface d'onde ordinaire 3 donne le point T. Par T nous menons la tangente Tb à la surface d'onde extraordinaire de P_2 d'où le rayon extraordinaire Ib. De même pour le rayon ordinaire Ia correspondant à des vibrations extraordinaires dans P_1.

La séparation angulaire est

$$i_0 + i_e = 2(n_c - n_0)\,\mathrm{tg}\,A\,.$$

Le prisme de ROCHON est constitué de la même façon, mais les rayons entrent par la face inférieure sur les Fig. 385 et 386· La Fig. 387 montre la marche schématique des rayons dans ce prisme. Il produit une séparation angulaire i_e deux fois plus faible

$$i_e = (n_e - n_0)\tan A\,.$$

IV. Interférences en lumière polarisée.

a) Interférences en lumière parallèle monochromatique.

163. Expérience de FRESNEL et ARAGO. Reprenons l'expérience des trous d'YOUNG et plaçons devant les deux fentes A_1 et A_2 deux tourmalines ou deux polaroïdes croisés P_1 et P_2 (Fig. 388). On observe un plan π quelconque avec un oculaire O. En un point M du plan π on a à composer deux vibrations rectangulaires: la vibration résultante est une vibration elliptique et l'éclairement

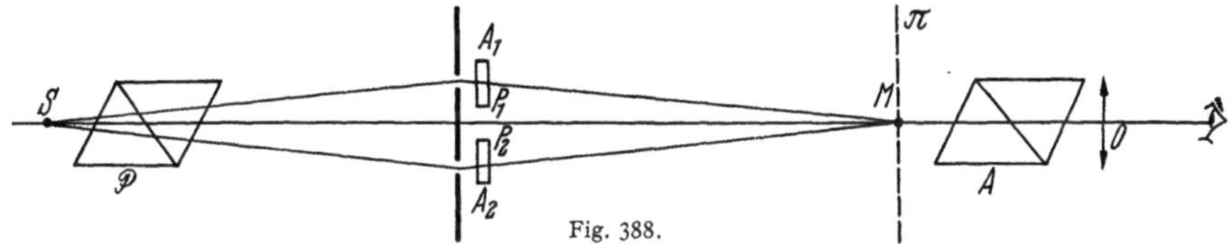

Fig. 388.

dans le plan π est constant. On constate en effet que si les polaroïdes sont croisés, il n'y a pas de franges d'interférences. Les franges apparaissent si les polaroïdes sont parallèles.

L'expérience réalisée par FRESNEL et ARAGO avec des tourmalines, montre bien que les vibrations lumineuses sont des vibrations transversales.

Plaçons alors un analyseur A devant l'oeil et examinons le plan π. Les vibrations émises par P_1 et P_2, que nous supposons croisés, ont pour composantes parallèles celles qui se projettent suivant la direction des vibrations admises par l'analyseur. Si celui-ci est à 45° des directions des polaroïdes, les deux composantes parallèles susceptibles d'interférer sont égales. Nous sommes donc les conditions les plus favorables et pourtant il n'apparaît pas de franges. La raison en est que les deux composantes émises par P_1 et P_2 proviennent d'une source de lumière naturelle et sont incohérentes (Sect. 133).

Pour rendre les deux composantes émises par P_1 et P_2 cohérentes, il suffit d'interposer un polariseur $\mathcal{P}$ entre la source S et les fentes A_1 et A_2. Les deux vibrations émises par P_1 et P_2 proviennent maintenant d'une vibration polarisée d'orientation déterminée et sont cohérentes. Si le polariseur est à 45° des vibrations émises par P_1 et P_2 comme l'analyseur, les composantes qui interfèrent ont même amplitude et les franges ont le maximum de contraste. Si le polariseur donne une vibration parallèle à P_1 ou P_2, les franges disparaissent car le polariseur étant croisé avec P_2 dans le premier cas, et P_1 dans le second cas, tout se passe comme si l'on avait obturé 'lune ou l'autre des fentes.

164. Calcul de l'éclairement et du contraste des franges. Soit OP la direction de vibration fournie par le polariseur P (Fig. 389). Les axes ox et oy sont parallèles aux directions des vibrations émises par les polaroïdes P_1 et P_2. L'analyseur A laisse passer les vibrations dirigées suivant oA. La vibration incidente dirigée suivant OP peut être représentée par $A\cos\omega t$ où A est son amplitude. Les vibrations que transmettent les polaroïdes sont $OH = A\cos\alpha\cos\omega t$ et

$OH' = A \sin\alpha \cos\omega t$. En un point M du plan d'observation, la différence de phase φ entre les deux vibrations émises par P_1 et P_2 est donnée par (13.2). On peut les écrire

$$OH = A \cos\alpha \cos\omega t, \qquad OH' = A \sin\alpha \cos(\omega t - \varphi).$$

Posons

$$a = A \cos\alpha, \qquad\qquad b = A \sin\alpha,$$

d'où

$$OH = a \cos\omega t, \qquad\qquad OH' = b \cos(\omega t - \varphi). \qquad (164.1)$$

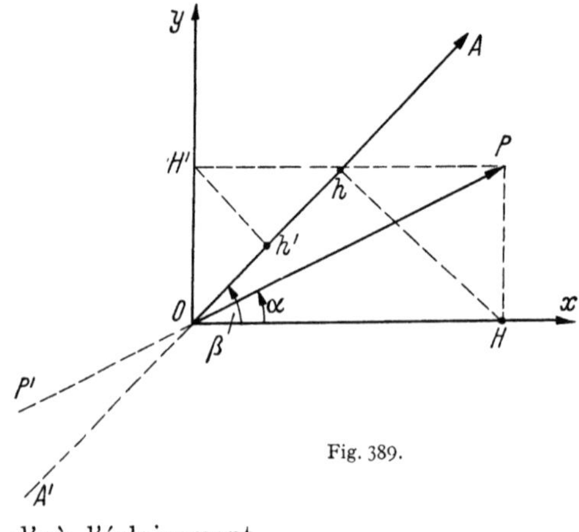

Fig. 389.

L'analyseur ne laisse passer que les vibrations dirigées suivant oA c'est-à-dire Oh et Oh' projections de OH et OH'. Les deux vibrations parallèles qui vont interférer seront donc

$$\left.\begin{aligned} Oh &= a \cos\beta \cos\omega t, \\ Oh' &= b \sin\beta \cos(\omega t - \varphi). \end{aligned}\right\} \quad (164.2)$$

La vibration résultante est alors

$$Oh + Oh' = (a \cos\beta + b \sin\beta \cos\varphi) \times$$
$$\times \cos\omega t + b \sin\beta \sin\varphi \sin\omega t$$

d'où l'éclairement

$$E = (a \cos\beta + b \sin\beta \cos\varphi)^2 + b^2 \sin^2\beta \sin^2\varphi,$$

$$E = a^2 \cos^2\beta + b^2 \sin^2\beta + 2ab \cos\beta \sin\beta \cos\varphi.$$

On peut écrire

$$E = (a \cos\beta + b \sin\beta)^2 - 2ab \sin 2\beta \sin^2\frac{\varphi}{2}$$

ou encore

$$E = (a \cos\beta - b \sin\beta)^2 + 2ab \sin 2\beta \cos^2\frac{\varphi}{2}.$$

Remplaçons a et b par leurs valeurs, on obtient finalement les deux expressions de l'éclairement $(A=1)$

$$E = \cos^2(\alpha - \beta) - \sin 2\alpha \sin 2\beta \sin^2\frac{\varphi}{2}, \qquad (164.3)$$

$$E = \cos^2(\alpha + \beta) + \sin 2\alpha \sin 2\beta \cos^2\frac{\varphi}{2}. \qquad (164.4)$$

Si on change α en $\alpha + \pi$ ou β en $\beta + \pi$, ces deux formules ne changent pas, puisqu'on ne peut distinguer entre OA et OA', ou bien entre OP et OP'. Il suffit donc d'étudier ce qui se passe pour $0 < \alpha < \pi$ et $0 < \beta < \pi$.

Supposons d'abord OP et OA dans le même quadrant, le premier ou le deuxième; $\sin 2\alpha \sin 2\beta$ est positif et il y a maximum de lumière (164.4) si

$$\cos^2\frac{\varphi}{2} = 1, \qquad \varphi = 2K\pi, \qquad \delta = K\lambda; \qquad (164.5)$$

on a

$$E_{\max} = \cos^2(\alpha + \beta) + \sin 2\alpha \sin 2\beta.$$

Il y aura minimum de lumière si

$$\cos^2\frac{\varphi}{2} = 0, \qquad \varphi = (2K+1)\pi, \qquad \delta = (2K+1)\frac{\lambda}{2} \qquad (164.6)$$

et
$$E_{\min} = \cos^2(\alpha + \beta)$$

d'où le contraste

$$\gamma = \frac{E_{\max} - E_{\min}}{E_{\max}} = \frac{\sin 2\alpha \sin 2\beta}{\cos^2(\alpha + \beta) + \sin 2\alpha \sin 2\beta}. \qquad (164.7)$$

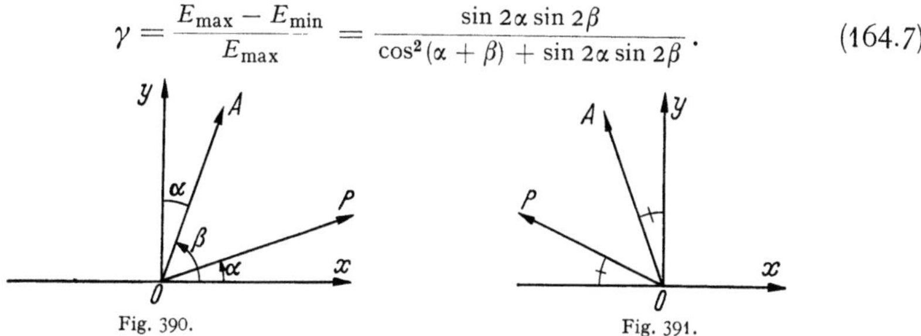

Fig. 390.   Fig. 391.

Il est maximum et égal à 1 si $\cos^2(\alpha + \beta) = 0$, c'est-à-dire $\alpha + \beta = \dfrac{\pi}{2}$ ou $\dfrac{3\pi}{2}$. On a les dispositions des Fig. 390 et 391. Pour ces positions, l'éclairement des franges brillantes est

$$E_{\max} = \sin 2\alpha \sin 2\beta.$$

On aura le plus de lumière possible si

$$|\sin 2\alpha| = |\sin 2\beta| = 1,$$

c'est-à-dire

$$\alpha = \beta = \frac{\pi}{4} \quad \text{ou} \quad \alpha = \beta = \frac{3\pi}{4}.$$

Les nicols doivent être parallèles et à 45° des vibrations émises par P_1 et P_2.

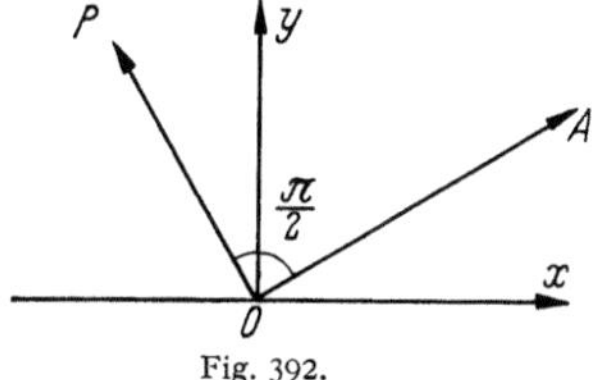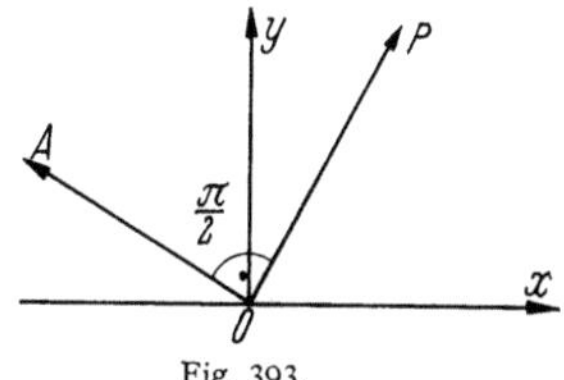

Fig. 392.   Fig. 393.

Supposons maintenant OP et PA dans deux quandrants différents. Le produit $\sin 2\alpha \sin 2\beta$ est négatif. Il y aura maximum de lumière (164.3) si

$$\sin^2 \frac{\varphi}{2} = 1, \qquad \varphi = (2K + 1)\pi, \qquad \delta = (2K + 1)\frac{\lambda}{2}. \qquad (164.8)$$

On a

$$E_{\max} = \cos^2(\alpha - \beta) - \sin 2\alpha \sin 2\beta.$$

Les minima de lumière seront donnés par

$$\sin^2 \frac{\varphi}{2} = 0, \qquad \varphi = 2K\pi, \qquad \delta = K\lambda \qquad (164.9)$$

et

$$E_{\min} = \cos^2(\alpha - \beta)$$

d'où le contraste des franges $\sin 2\alpha \sin 2\beta$ étant négatif

$$\gamma = \frac{|\sin 2\alpha \sin 2\beta|}{\cos^2(\alpha - \beta) + |\sin 2\alpha \sin 2\beta|}.$$

Le contraste est maximum ($\gamma = 1$) si $\cos^2(\alpha - \beta) = 0$, c'est-à-dire $\alpha - \beta = \pm\dfrac{\pi}{2}$. On a les dispositions des Fig. 392 et 393. L'éclairement des franges brillantes sera

$$E_{\max} = |\sin 2\alpha \sin 2\beta|.$$

28*

On aura encore le plus de lumière possible si

$$|\sin 2\alpha| = |\sin 2\beta| = 1,$$

c'est-à-dire

$$\alpha = \frac{3\pi}{4}, \quad \beta = \frac{\pi}{4} \quad \text{ou} \quad \alpha = \frac{\pi}{4}, \quad \beta = \frac{3\pi}{4}.$$

Les nicols doivent être croisés et à 45° des vibrations émises par P_1 et P_2.

Notons d'après les relations (164.5) que les franges occupent la même position qu'en lumière naturelle lorsque les nicols sont parallèles. Les formules (164.8)

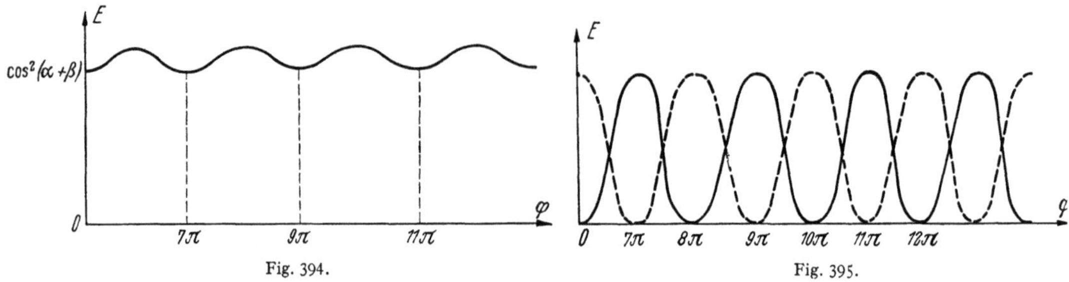

Fig. 394.　　　　　　　　　　　　　　　　Fig. 395.

montrent au contraire qu'entre nicols croisés, on a un système de franges occupant la position complémentaire des franges en lumière naturelle.

Les Fig. 394 et 395 traduisent ces résultats.

La Fig. 394 se rapporte au cas où OA et OP sont dans le même quadrant: le contraste des franges est inférieure à 1 si α et β ont des valeurs quelconques.

La Fig. 395 concerne le cas des nicols parallèles (courbe en traits ponctués) ou croisés (courbe en trait plein). Les expressions (164.3) et (164.4) donnant l'éclairement, se réduisent à

$$E = \cos^2 \frac{\varphi}{2} \quad \text{(nicols parallèles)}, \tag{164.10}$$

$$E = \sin^2 \frac{\varphi}{2} \quad \text{(nicols croisés)}. \tag{164.11}$$

165. Interférences produites par une lame cristalline. Considérons une lame cristalline C éclairée en faisceau parallèle (Fig. 396). La lumière incidente est

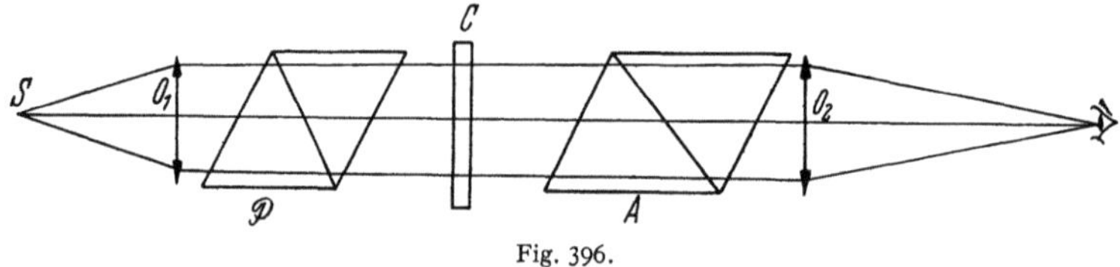

Fig. 396.

polarisée par un nicol $\mathcal{P}$. La vibration incidente est OP et les lignes neutres de la lame C sont dirigées suivant ox et oy (Fig. 397). Avant l'entrée dans la lame, on peut décomposer la vibration OP en deux composantes dirigées suivant ox et oy; avec les notations précédentes, on aura

$$OH = A \cos\alpha \cos\omega t, \quad OH' = A \sin\alpha \cos\omega t,$$

$$OH = a \cos\omega t, \quad OH' = b \cos\omega t.$$

On sait que lame transmet sans altération les deux vibrations OH et OH' puisque celles-ci sont dirigées suivant ses lignes neutres (Sect. 149).

La direction OH correspond à un indice n' et la direction OH' à l'indice n''. A la sortie de la lame, ces deux vibrations auront une différence de phase dûe à ce que les indices ou les vitesses de propagation suivant les deux axes de la lame, ne sont pas les mêmes. A la sortie de la lame, si e est son épaisseur, on aura

$$OH = a \cos\left(\omega t - \frac{2\pi n' e}{\lambda}\right),$$

$$OH' = b \cos\left(\omega t - \frac{2\pi n'' e}{\lambda}\right).$$

En posant

$$\varphi = \frac{2\pi (n'' - n') e}{\lambda},$$

on pourra écrire

$$OH = a \cos \omega t, \quad OH' = b \cos (\omega t - \varphi).$$

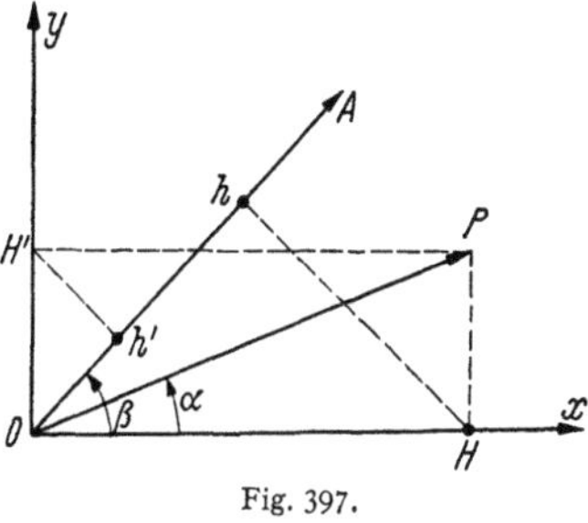

Fig. 397.

Le calcul est identique à celui du paragraphe précédent. Si oA est la direction des vibrations que laisse passer l'analyseur, on aura l'éclairement par les formules (164.3) et (164.4)

$$E = \cos^2 (\alpha - \beta) - \sin 2\alpha \sin 2\beta \sin^2 \frac{\varphi}{2}, \tag{165.1}$$

$$E = \cos^2 (\alpha + \beta) + \sin 2\alpha \sin 2\beta \cos^2 \frac{\varphi}{2}. \tag{165.2}$$

Au lieu que la différence de phase φ soit produite par la différence des chemins optiques accomplis par les vibrations émises par A_1 et A_2 en M (Fig. 388), elle est produite ici par la biréfringence de la lame. Si la phase φ n'est pas constante en tous les points de la lame, l'éclairement varie. L'oeil observant avec la loupe O_2 voit des franges d'interférences sur la lame. La phase φ varie si l'épaisseur ou la biréfringence de la lame ne sont pas constantes. Dans le compensateur de Babinet, on fait varier la phase φ au moyen de deux prismes d'axes croisés, et on obtient un système de franges d'interférences parallèles et équidistantes.

La discussion est la même que celle du paragraphe précédent. Les franges ont le contraste maximum égal à 1 et l'éclairement des franges brillantes a la plus grande valeur possible si les nicols sont croisés ou parallèles, leurs sections principales étant à 45° des axes de la lame cristalline.

L'éclairement en un point de la lame est

$$E_\perp = \sin^2 \frac{\varphi}{2} \quad \text{(nicols croisés)}, \tag{165.3}$$

$$E_{\parallel} = \cos^2 \frac{\varphi}{2} \quad \text{(nicols parallèles)}. \tag{165.4}$$

b) Interférences en lumière blanche.

166. Polarisation chromatique en lumière parallèle. Considérons une lame cristalline mince observée entre nicols et éclairons en lumière blanche. Les expressions (165.1) et (165.2) montrent que si la longueur d'onde varie, φ et par conséquent E varient également. L'intensité lumineuse transmise par la lame varie avec la longueur d'onde, des colorations apparaissent. Si la lame est suffisamment mince, les colorations peuvent être aussi vives que celles données par les lames isotropes. Ces phénomènes portent le nom de phénomènes de polarisation chromatique.

A chaque longueur d'onde correspond une valeur de φ et entre nicols croisés, l'éclairement est donné par (165.3). Pour la même longueur d'onde, l'éclairement entre nicols parallèles est donné par (165.4) avec la même valeur de φ. Si par un moyen quelconque, nous superposons ces deux éclairements, on a

$$E_\perp + E_\| = \sin^2\frac{\varphi}{2} + \cos^2\frac{\varphi}{2} = 1 .$$

L'éclairement résultant est constant; quelle que soit la longueur d'onde, on reconstitue la lumière blanche incidente. Les deux teintes obtenues entre nicols parallèles et croisés sont complémentaires. Si les nicols ne sont ni parallèles ni croisés, les expressions (165.1) et (165.2) montrent qu'il s'ajoute un terme de la forme $\cos^2(\alpha - \beta)$ ou $\cos^2(\alpha + \beta)$.

Ces termes ne font pas intervenir la longueur d'onde, on les appelle «termes incolores» qui produisent seulement un voile blanc parasite. Il en est de même du coefficient $\sin 2\alpha \sin 2\beta$ de sorte que, quelle que soient les orientations de la lame ou des nicols, on ne peut obtenir que deux teintes complémentaires l'une de l'autre et plus ou moins lavées de blanc. Les teintes ont le maximum de pureté et de vivacité si les nicols sont croisés ou parallèles et à 45° des lignes neutres de la lame cristalline.

On peut les observer soit au Nörremberg (Fig. 398), soit avec le montage de la Fig. 396 où l'on peut également utiliser des polaroïdes.

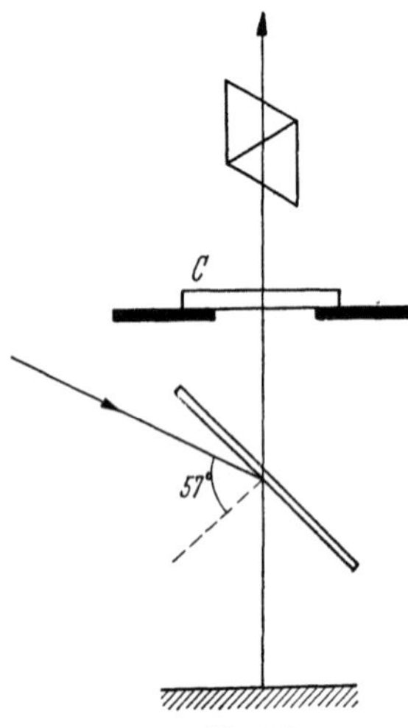

Fig. 398.

L'étude des teintes par polarisation peut se faire somme à la Sect. 30. On peut considérer sans grande erreur que $n'' - n'$ est constant dans tout le spectre et par conséquent, les teintes par polarisation entre nicols croisés ou entre nicols parallèles sont très sensiblement les mêmes que celles données par les interférences à centre noir et à centre blanc. Ce sont les échelles données à la Sect. 30.

L'échelle des teintes pourrait s'observer facilement au moyen d'un compensateur de Babinet d'angle suffisamment faible donnant un grand étalement des franges. De chaque côté de la frange noire centrale, se trouvent deux échelles de teintes identiques et symétriques.

On peut ainsi déterminer la différence de marche produite par une lame cristalline en examinant sa teinte au Nörremberg. En opérant entre polariseurs parallèles et croisés, on a deux teintes complémentaires qui permettent de choisir celle donnant la meilleure détermination.

On a avantage à se placer au voisinage d'une teinte sensible correspondant à l'extinction du jaune moyen du spectre. Il est possible de réaliser par exemple une lame teinte sensible en quartz pour lequel $n_e - n_0 = 9 \cdot 10^{-3}$. Il faut tailler une lame de quartz parallèle à l'axe d'épaisseur e

$$e = \frac{\lambda}{n_e - n_0} = \frac{0{,}6\,\mu}{9 \cdot 10^{-3}} = 66\,\mu .$$

Elle fournit alors le pourpre teinte sensible du premier ordre entre nicols croisés.

Superposons cette lame teinte sensible à une lame trop mince pour donner des teintes par elle-même. Immédiatement, le pourpre vire au rouge ou au bleu et on peut déterminer la différence de marche introduite par la lame très mince avec une grande précision. Si la lame à étudier est assez épaisse pour donner les teintes délavées des ordres supérieurs, on peut lui superposer une lame connue d'épaisseur voisine orientée de façon que les différences de marche se retranchent. En choisissant convenablement la lame étalonnée, on peut obtenir une teinte du premier ordre donnant une bonne détermination.

Il y a d'avantage quand cela est possible, à utiliser le pourpre teinte sensible, entre nicols parallèles. On a vu en effet qu'au voisinage de cette teinte, la détermination d'une différence de marche est deux fois plus sensible qu'entre nicols croisés (Sect. 30). Hiroshi Kubota a montré qu'il est possible d'obtenir des teintes encore plus sensibles: on a vu qu'entre nicols parallèles, l'intensité transmise par une lame cristalline est proportionnelle à $\cos^2 \frac{\varphi}{2}$; ajoutons une différence de phase supplémentaire ψ indépendante de la longueur d'onde. L'éclairement devient

$$E = \cos^2 \tfrac{1}{2}(\varphi + \psi).$$

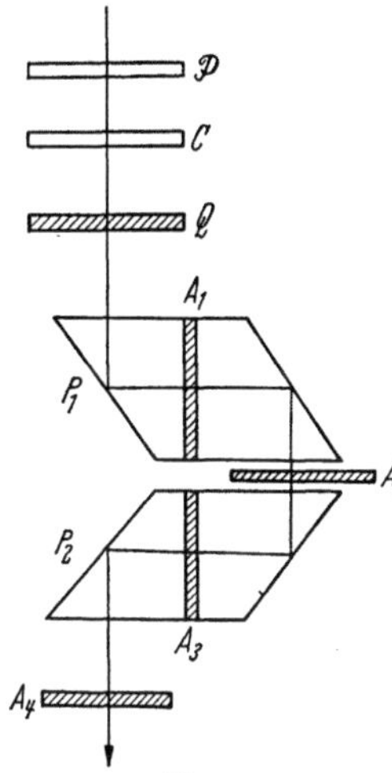

La couleur de l'étalon C vu à travers ce système a été calculée par Hiroshi Kubota. Ce physicien a trouvé que la variation du chemin optique $\Delta[(n' - n'')e]$ qu'il faut produire pour avoir un changement ΔC de la chromaticité est proportionnel à $\left(1 - \dfrac{\psi}{\pi}\right) \Delta c$ pour ψ donné. La sensibilité chromatique $\dfrac{\Delta c}{\Delta[(n'' - n')e]}$ est alors proportionnelle à $1 - \dfrac{\psi}{\pi}$. La sensibilité est d'autant plus grande que ψ est plus voisin de π. Théoriquement elle pourrait être aussi grande que l'on veut, mais plus la sensibilité théorique augmente et plus la luminance devient faible. La différentiation des couleurs devient de plus en plus difficile.

Fig. 399.

Le dispositif utilisé par Hiroshi Kubota pour produire une différence de marche achromatique est indiqué sur la Fig. 399. C'est un ensemble de deux prismes du type Fresnel produisant la différence de marche à la réflexion totale. $\mathscr{D}$ est le polariseur, C la lame à étudier, Q la lame teinte sensible, P_1 et P_2 les deux prismes. Si δ est le retard correspondant à une réflexion, l'analyseur étant en A_1, on aura les retards 2δ, 3δ ou 4δ suivant que l'analyseur est placé en A_2 A_3 ou A_4.

167. Spectres cannelés. Lorsque la lame cristalline à étudier devient plus épaisse, les radiations affaiblies ou éteintes sont réparties dans tout le spectre comme pour une lame isotrope et le champ d'observation est incolore: on a du blanc d'ordre supérieur qu'on peut analyser au spectroscope. L'observation se fait suivant le schéma de la Fig. 400. Un collimateur SO_1 éclaire la lame cristalline C avant laquelle on a placé le polariseur $\mathscr{D}$. La lumière transmise traverse l'analyseur A le prisme V et l'objectif O_2 donne en S' une image de la fente S dans le plan π où l'on observe le spectre. L'ensemble $SO_1 VO_2$ constitue un spectroscope ou un spectrographe et on a simplement interposé la lame C et les deux nicols entre l'objectif du collimateur et le prisme V. Si α et β sont les angles des directions de vibrations que laissent passer le polariseur $\mathscr{D}$ et l'analyseur A avec la ligne neutre ox de la lame cristalline C (Fig. 397), l'éclairement en π est donné par les formules (165.1) et (165.2).

Nous supposerons que la dispersion de double réfraction, c'est-à-dire la variation de la différence $n' - n''$ en fonction de la longueur d'onde, est négligeable. Dans le cas des cristaux biaxes, les directions privilégiées varient d'une radiation à l'autre mais on peut supposer avec une approximation suffisante, que les directions privilégiées sont les mêmes pour les radiations les plus lumineuses du spectre. Pour un cristal uniaxe, cette approximation ne se pose pas car les directions des vibrations privilégiées sont rigoureusement les mêmes pour toutes les couleurs. Si les nicols sont parallèles, l'éclairement est

$$E = \cos^2 \frac{\varphi}{2}. \tag{164.10}$$

Si les nicols sont croisés, il est

$$E = \sin^2 \frac{\varphi}{2}. \tag{164.11}$$

Dans les deux cas, la différence de phase φ varie inversement proportionnellement à la longueur d'onde. Lorsque les nicols sont parallèles, l'éclairement est nul pour

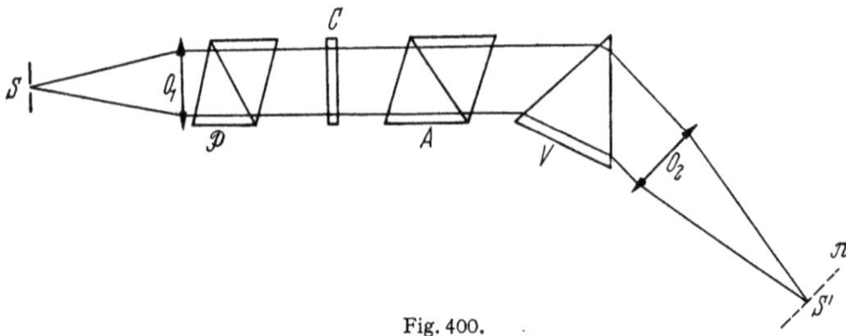

Fig. 400.

$\varphi = (2K + 1)\pi$. Les longueurs d'onde données par $(n'' - n')e = (2K + 1)\frac{\lambda}{2}$ sont éteintes dans le spectre et on observe un système de cannelures noires séparées par des cannelures brillantes correspondant à $(n'' - n')e = K\lambda$. De même, entre nicols croisés on observera un système de cannelures noires données par $(n'' - n')e = K\lambda$ et de cannelures brillantes correspondant à $(n'' - n')e = (2K + 1)\frac{\lambda}{2}$. Lorsque l'on passe de la position nicols parallèles à la position nicols croisés, les deux systèmes de cannelures s'intervertissent. Les résultats des Sect. 164 et 165 montrent bien que le spectre cannelé a le maximum de contraste (cannelures sombres entièrement noires) et que les cannelures brillantes ont le maximum de luminosité si les nicols sont parallèles ou croisés, les axes de la lame cristalline étant à 45° des sections principales des nicols.

Les formules (165.1) et (165.2) montrent également que les cannelures ont toujours la même position, quelles que soient les valeurs des angles α et β. Il y a seulement interversion des deux systèmes si on change le signe du produit $\sin 2\alpha \sin 2\beta$. Lorsqu'on tourne le nicol A, les cannelures s'estompent et se renforcent sur plcae.

Le calcul de la position des cannelures en fonction de la différence de marche $(n'' - n')\,e$ s'effectue de la même façon que pour les lames isotropes (Sect. 31). Si les nicols sont croisés, on a le même calcul que pour une lame isotrope observée par réflexion en remplaçant $2n\,e$ par $(n'' - n')\,e$. Si les ordres d'interférences correspondant aux longueurs d'ondes λ_1 et λ_2 sont $p_1 = \dfrac{(n'' - n')\,e}{\lambda_1}$ et $p_2 = \dfrac{(n'' - n')\,e}{\lambda_2}$ il y aura entre ces deux longueurs d'onde autant de cannelures noires entre nicols croisés que de nombres entiers compris entre p_1 et p_2. Aux valeurs entières de $2ne/\lambda_1$ et de $2ne/\lambda_2$ correspondent également des cannelures noires pour une lame isotrope observée par réflexion car au Sect. 31 nous avons posé $p = \dfrac{\delta}{\lambda}$ $= K + \dfrac{1}{2}$ pour les cannelures noires mais la lame étant observée par réflexion $\delta = 2ne + \dfrac{\lambda}{2}$ et on a bien $\dfrac{2ne}{\lambda} = K$ pour ces cannelures.

Par exemple, avec une lame de quartz taillée parallèlement à l'axe et d'épaisseur $e = 5$ mm., on a pour les radiations extrêmes du spectre 0,4 et 0,8 μ

$$p_1 = \frac{0{,}009 \times e}{0{,}4} = 112{,}2; \quad p_2 = \frac{0{,}009 \times e}{0{,}8} = 56{,}3.$$

Entre nicols croisés, les cannelures noires correspondent aux valeurs entières de $p = 56, 57, 58 \ldots 112$, soit en tout 56 cannelures noires. On peut remarquer la différence qui existe entre $n'' - n'$ pour une lame anisotrope et $2n$ pour une lame isotrope. Le rapport $\dfrac{2n}{n'' - n'}$ est de l'ordre de 300 pour le quartz. On observe donc les mêmes phénomènes pour les lames isotropes et les lames anisotropes, mais avec des épaisseurs beaucoup plus grandes dans le cas des lames anisotropes. Si le spectre est un spectre normal (spectre de réseau), les franges sont plus écartées dans le rouge que dans le violet. Avec un prisme, les radiations rouges sont beaucoup plus tassées et les cannelures peuvent être plus étroites que dans le violet.

Si l'épaisseur de la lame cristalline diminue le nombre de cannelures observées dans le spectre diminue. Lorsque son épaisseur devient trop faible, l'observation d'un spectre cannelé n'est plus possible. Toutefois, on peut réaliser une expérience intéressante dans ce cas: considérons une lame cristalline trop mince pour donner un spectre cannelé et semi-métallisons ses faces. Observons le spectre de la lumière transmise par la lame: on constate la présence de fines cannelures. Ces cannelures proviennent des interférences entre les ondes qui se sont réfléchies sur les deux faces et celles qui ont traversé directement la lame. Mais les différences de marches $2n' e$ et $2n'' e$ ne sont pas égales et chaque cannelure est en réalité un doublet. Soit p l'ordre d'interférence et $\delta = p\lambda$ la différence de marche correspondant à un doublet. Lorsqu'on passe d'une composante à l'autre de ce doublet, la différence de marche varie de $d\delta$

$$d\delta = p\,\frac{d\lambda}{\lambda},$$

$d\lambda$ étant l'intervalle spectral entre les deux composantes.

On a

$$n'' - n' = n\,\frac{d\lambda}{\lambda}.$$

L'intervalle $d\lambda$ ne dépend pas de l'épaisseur et peut donc servir à la mesure de la biréfringence de la lame.

168. Polarisation chromatique en lumière convergente. Calcul de la différence de marche pour un cristal uniaxe. Considérons une lame cristalline C à faces parallèles placée entre un polariseur $\mathscr{P}$ et un analyseur $\mathscr{A}$, deux polaroïdes par exemple (Fig. 401). A un rayon incident SI correspondent deux rayons transmis JT et KT' parallèles. Ces deux rayons se coupent en un point M du plan focal π d'un objectif O. Ils interfèrent en M. En éclairant le dispositif par un faisceau contenant des rayons de diverses inclinaisons, on observe dans le plan focal π une figure d'interférence. Les franges obtenues sont les franges à l'infini de la lame cristalline C. L'expérience est analogue à celle du Sect. 24 concernant les lames

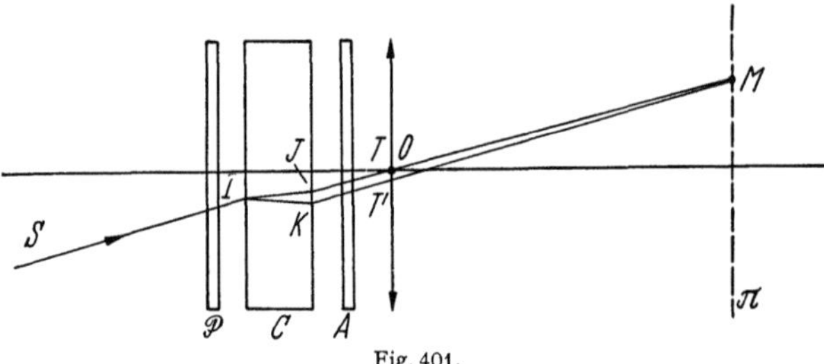

Fig. 401.

isotropes, mais ici les rayons qui interfèrent sont produits par la biréfringence de la lame cristalline. On dit qu'on observe les interférences de polarisation en lumière convergente.

Soit SI un rayon incident toubant sur une lame cristalline à faces parallèles d'épaisseur e (Fig. 402). Au rayon incident SI correspond une onde plane Σ qui, après traversée de la lame, devient l'onde Σ'. L'onde Σ est par exemple l'onde

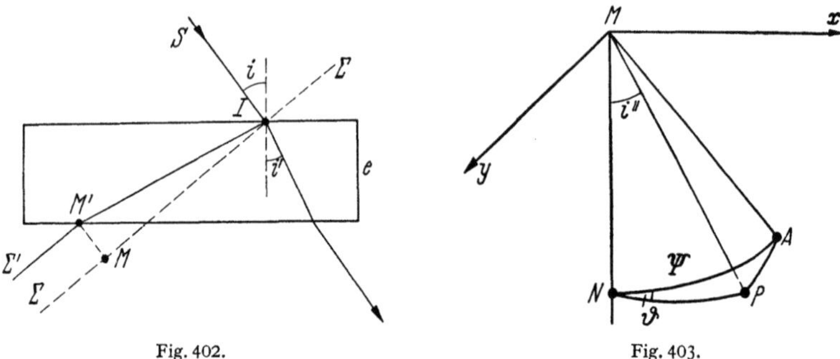

Fig. 402.   Fig. 403.

dont les vibrations sont parallèles à l'une des deux directions privilégiées dans le cristal. Le retard de l'onde émergente Σ' sur la position Σ qu'elle occuperait si la lame n'existait pas, est

$$\Delta' = M M' = e \left(n' \cos i' - \cos i \right),$$

de même pour la deuxième onde Σ'' non représentée sur la figure et correspondant à l'autre direction privilégiée

$$\Delta'' = e \left(n'' \cos i'' - \cos i \right).$$

La différence de marche Δ entre ces deux ondes est

$$\Delta = e \left(n'' \cos i'' - n' \cos i' \right); \tag{168.1}$$

c'est la différence de marche en M (Fig. 401).

Calculons Δ dans le cas d'un cristal uniaxe.

Soit xMy la surface de la lame (Fig. 403), MA l'axe du cristal MN la normale au plan xMy. Ψ l'angle de l'axe MA avec la normale à la lame, ϑ l'angle de la section principale NMA avec le plan d'incidence NMx, MP la normale à l'onde extraordinaire, ϑ' l'angle de MP et MA. Le triangle sphérique ANP donne

$$\cos \vartheta' = \cos \Psi \cos i'' + \sin \Psi \sin i'' \cos \vartheta.$$

On peut alors calculer $n'' \cos i''$ et on trouve

$$n'' \cos i'' = \frac{(a^2 - b^2) \sin 2\,\Psi \cos \vartheta \sin i}{2c^2} + \left. \vphantom{\frac{1}{1}} \right\}$$
$$+ \frac{1}{c} \sqrt{1 - a^2 \sin^2 \vartheta \sin^2 i - \frac{a^2 b^2}{c^2} \cos^2 \vartheta \sin^2 i} \quad \left. \vphantom{\frac{1}{1}} \right\} \quad (168.2)$$

ou

$$a = \frac{1}{n_e}, \qquad b = \frac{1}{n_0}$$

et

$$c^2 = a^2 \sin^2 \Psi + b^2 \cos^2 \Psi,$$

On a d'autre part

$$(n' \cos i')^2 = \frac{1 - b^2 \sin^2 i}{b^2} = \frac{\cos^2 i'}{b^2}. \qquad (168.3)$$

La différence de marche Δ est donnée en remplaçant (168.2) et (168.3) dans l'expression (168.1).

169. Cristal uniaxe perpendiculaire à l'axe en lumière convergente. Considérons le cas d'un cristal perpendiculaire à l'axe. On a $\Psi = 0$ d'où

$$\Delta = \frac{e}{b} \left[\sqrt{1 - a^2 \sin^2 i} - \sqrt{1 - b^2 \sin^2 i} \right]$$

et pour les angles petits

$$\Delta = \frac{n_0^2 - n_e^2}{2 n_0 n_e} i^2 e. \qquad (169.1)$$

Les phénomènes dans le plan focal de l'objectif O sont alors déterminés par (169.1) et les relations (165.1) et (165.2)

$$E = \cos^2 (\alpha - \beta) - \sin 2\alpha \sin 2\beta \sin^2 \frac{\varphi}{2}, \quad \left. \vphantom{\frac{\varphi}{2}} \right\}$$
ou
$$E = \cos^2 (\alpha + \beta) + \sin 2\alpha \sin 2\beta \cos^2 \frac{\varphi}{2} \quad \left. \vphantom{\frac{\varphi}{2}} \right\} \qquad (169.2)$$

et

$$\varphi = \frac{2\pi \Delta}{\lambda}, \qquad \Delta = \frac{n_e^2 - n_0^2}{2 n_0 n_e} i^2 e.$$

Ces équations montrent que les lignes d'égal retard ou «lignes isochromatiques» d'un cristal perpendiculaire à l'axe sont des cercles concentriques de rayons $R = f i$ (f focale de l'objectif O).

On pouvait prévoir ce résultat étant donnée qu'une lame perpendiculaire à l'axe présente la symétrie de révolution autour de la direction de l'axe optique. La différence de marche est par conséquent la même pour les rayons qui font un même angle avec l'axe, et les lignes isochromatiques sont donc des circonférences ayant pour centre la trace de l'axe optique.

Les expressions donnant l'éclairement montrent qu'il y aura des points où n'apparaîtra aucune coloration. Ce sont les points pour lesquels $\sin 2\alpha = 0$ ou $\sin 2\beta = 0$.

On aura donc les courbes $\alpha = 0$ et $\alpha = \pi/2$ ou $\beta = 0$ et $\beta = \pi/2$. Les phénomènes présentent toujours le maximum de visibilité lorsque les nicols sont croisés ou parallèles; dans le cas des nicols croisés ($\beta - \alpha = \pi/2$) on a deux lignes noires parallèles aux sections principales des nicols. Entre nicols parallèles ($\beta - \alpha = 0$) on a deux lignes blanches correspondant à l'éclairement maximum. Ces lignes, lieux des points où n'apparaît aucune coloration, sont appelées «lignes neutres» de la lame.

En lumière blanche, on observera dans le plan focal de l'objectif O une série de franges colorées circulaires et une croix noire (nicols croisés) ou une croix blanche (nicols parallèles).

La formule (169.1) montre que les anneaux (lignes isochromatiques) suivent la loi de Newton et sont inversement proportionnels à la racine carrée de l'épaisseur. Entre nicols croisés, les teintes sont celles des anneaux de Newton à centre noir. Elles sont pures quel que soit le point du champ observé et leur éclairement diminue au fur et à mesure que l'on s'approche de la croix noire. Entre nicols parallèles, les teintes sont celles des anneaux de Newton à centre blanc. Elles sont pures à 45° des sections principales des nicols et sont de plus en plus noyées de blanc lorsqu'on s'approche de la croix blanche. L'ensemble de ces résultats peut se déduire également de l'étude de la surface des indices ou de la surface d'onde, mais les formules (168.2) et (168.3) sont générales et permettent d'étudier un cas quelconque.

170. Cristal uniaxe parallèle à l'axe en lumière convergente. Considérons une lame uniaxe taillée parallèlement à l'axe. On a $\Psi = \pi/2$ et d'après (168.2) et (168.3) on a

$$(n'' \cos i'')^2 = \frac{1 - (a^2 \sin^2 \vartheta + b^2 \cos^2 \vartheta) \sin^2 i}{a^2},$$

$$\varDelta = e \left[\frac{\sqrt{1 - (a^2 \sin^2 \vartheta + b^2 \cos^2 \vartheta) \sin^2 i}}{a} - \frac{\sqrt{1 - b^2 \sin^2 i}}{b} \right].$$

Pour les petites valeurs de i

$$\varDelta = e \frac{a - b}{a b} \left[1 + \frac{b (a \sin^2 \vartheta - b \cos^2 \vartheta)}{2} i^2 \right]. \tag{170.1}$$

En posant

$$x = f i \cos \vartheta, \qquad y = f i \sin \vartheta \tag{170.2}$$

on aura

$$\varDelta = e \frac{a - b}{a b} \left[1 + \frac{b}{2 f^2} (a y^2 - b x^2) \right].$$

Les lignes isochromatiques d'une lame uniaxe taillée parallèlement à l'axe sont des hyperboles dont les asymptotes sont données par

$$y = \pm \sqrt{\frac{b}{a}} \, x = \pm \sqrt{\frac{n_e}{n_0}} \, x.$$

On sait que la direction de vibration extraordinaire de l'onde extraordinaire est la projection sur son plan de l'axe optique (Sect. 144). Comme on n'observe jamais dans un champ très étendu, on peut considérer que la direction privilégiée extraordinaire reste pratiquement parallèle à l'axe optique. Les angles α et β gardent des valeurs constantes dans tout le champ et on n'observe pas de lignes neutres.

Le relation (170.1) donne en faisant $\vartheta = 0$

$$\varDelta = e \frac{a - b}{a b} \left(1 - \frac{b^2}{2} i^2 \right)$$

ou encore en posant $\varDelta_0 = (n_e - n_0) e$

$$\varDelta_0 - \varDelta = \frac{\varDelta_0 i^2}{2 n_0^2}.$$

Prenons une lame de quartz de 20 mm. d'épaisseur

$$\Delta_0 = 9 \cdot 10^{-3}\, e = 180\,\mu$$

et

$$i = \sqrt{\frac{2\,n_0^2\,(\Delta_0 - \Delta)}{180}}\;.$$

Pour la première frange noire correspondant à la longueur d'onde λ on aura entre nicols croisés

$$i = \sqrt{\frac{2\,n_0^2}{180}\,\lambda}$$

en supposant que Δ_0 soit égal à un nombre entier de longueurs d'onde. Prenons $\lambda = 0{,}6\,\mu$ et $n_0 = 1{,}54$ (quartz). On a $i = 0{,}16$ soit 9° et les lignes isochromatiques ont des dimensions convenables pour l'observation. Or $(n_0 - n_e)\, e = 180\,\mu$, donc les différences de marche correspondent à des ordres d'interférence élevés, on a du blanc d'ordre supérieur et les franges ne sont visibles qu'en lumière monochromatique. Si on diminue l'épaisseur pour avoir un Δ_0 donnant une coloration visible, alors la variation $\Delta_0 - \Delta$ est tellement faible que la coloration est pratiquement constante dans tout le champ. Si on prend deux lames identiques et croisées, le retard pris par l'un des vibrations sur l'autre dans la première lame est compensé par une avance dans la deuxième lame. Les deux vibrations sortent du système avec une différence de marche nulle et le phénomène est observable en lumière blanche.

171. Lame uniaxe inclinée par rapport à l'axe en lumière convergente. Considérons le cas d'une lame taillée à 45° de l'axe optique. Les formules (168.2) et (168.3) donnent en faisant $\Psi = 45°$ et en utilisant les relations (170.2) avec $\xi = x/f$, $\eta = y/f$

$$\Delta = e\left[\sqrt{\frac{2}{a^2 + b^2}} - \frac{1}{b} - \frac{b^2 - a^2}{a^2 + b^2}\,\xi + \cdots\right]. \qquad (171.1)$$

Plaçons derrière cette lame une deuxième lame identique et tournée de 90°. Il suffit de remplacer dans l'expression (171.1) ξ par $-\eta$ pour avoir la différence de marche Δ' dûe à la deuxième lame. La différence de marche δ produite par les deux lames est égale à la différence entre Δ et Δ'

$$\delta = e\,\frac{b^2 - a^2}{a^2 + b^2}\,(\xi + \eta)\,. \qquad (171.2)$$

Les lignes isochromatiques sont les droites $\xi + \eta = $ constante. On a un système de franges visibles en lumière blanche, la frange centrale est une frange noire entre nicols croisés et blanche entre nicols parallèles. De chaque côté et symétriquement, on observe des franges colorées donnant l'echelle des teintes. L'ensemble des deux lames précédentes constitue le polariscope de SAVART. Il donne pratiquement à l'infini les mêmes franges que celles données par le compensateur de BABINET dans le plan des lames. En fait, ceci n'est vrai qu'en première approximation, car si les franges du compensateur de BABINET sont rigoureusement des droites parallèles, il n'en est pas de même dans le cas du polariscope de SAVART. L'expression (171.2) est le premier terme d'un développement en série: les franges rectilignes sont des fragments de courbes du second degré vues à une grande distance de leurs sommets.

172. Lame biaxe perpendiculaire à la bissectrice de l'angle aigu des axes en lumière convergente. Les résultats des Sect. 142 et 143 permettent d'établir facilement ce qui va suivre.

Considérons un plan d'onde π quelconque qui coupe l'ellipsoïde des indices suivant l'ellipse représentée sur la Fig. 404.

Les directions privilégiées correspondant au plan d'onde π sont les axes OM et OM' de l'ellipse π intersection du plan d'onde et de l'ellipsoïde des indices. Les intersections C et C' des sections cycliques de l'ellipsoïde des indices avec le plan d'onde π sont symétriques par rapport à OM et OM'. Soient OI_1 et OI_2 les projections des deux axes optiques sur le plan π. OI_1 est perpendiculaire à OC et OI_2 à OC', donc les vibrations privilégiées d'un plan d'onde π sont les deux bissectrices des projections sur ce plan des deux axes optiques.

Cette remarque va nous permettre d'étudier la figure d'interférence à l'infini d'un biaxe. Soient J_1 et J_2 les images, dans le plan focal de l'objectif, des points à l'infini sur les axes optiques (Fig. 405). OP et OA représentent les directions

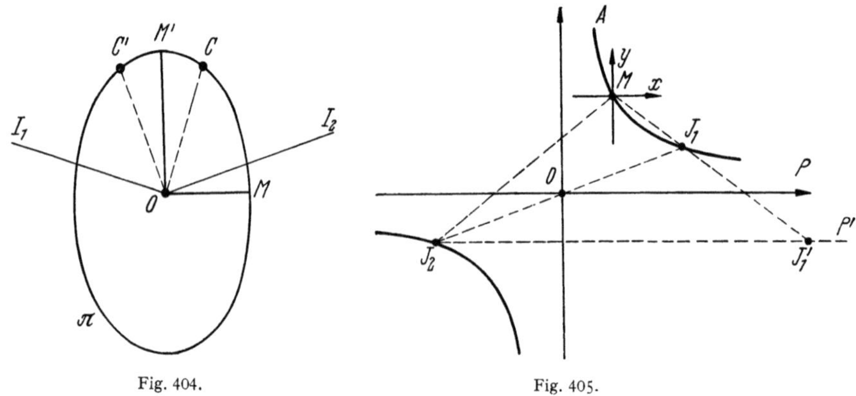

Fig. 404.
Fig. 405.

des vibrations que laissent passer le polariseur et l'analyseur (ils sont croisés sur la figure). Pour trouver les lignes neutres dans le plan focal (lignes noires), il faut chercher le lieu des points M tels que les bissectrices de l'angle $J_1 M J_2$ soient parallèles à OP et OA. En effet, d'après ce qui précède, les bissectrices Mx et My de l'angle $J_1 M J_2$ sont les directions des vibrations privilégiées correspondant au rayon dont le point M est la trace. Si ces directions sont parallèles à OP et OA elles sont éteintes et le point M est bien sur une ligne noire. On a

$$\tan \widehat{M J_2 J_1'} + \tan \widehat{M J_1' P'} = 0$$

d'où x, y étant les coordonnées de M, a, b, celles de J_1 et $-a, -b$, celles de J_2

$$\frac{y-b}{x-a} + \frac{y+b}{x+a} = 0, \qquad y = \frac{ab}{x}.$$

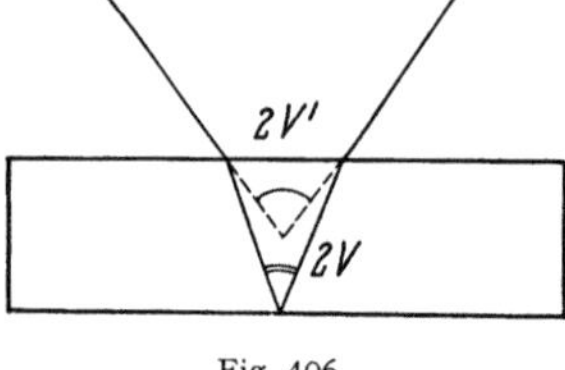

Fig. 406.

Les lignes noires sont les deux branches d'une hyperbole équilatère qui passe par J_1 et J_2. Si les lignes neutres de la lame sont parallèles aux sections principales des nicols, on obtient une croix noire. Lorsque la lame tourne, les lignes noires se déforment.

Pour construire les lignes isochromatiques, on peut remarquer que les points J_1 et J_2 correspondent à des directions où la différence de marche entre les deux ondes est nulle. Les courbes d'égal retard, c'est-à-dire les lignes isochromatiques, seront donc des courbes fermées entourant J_1 et J_2 si le retard est faible. Si la différence de marche augmente, les lignes isochromatiques qui entourent les points J_1 et J_2 s'écartent et finissent par se rejoindre. La ligne isochromatique pour laquelle il en est ainsi, a l'aspect d'une lemniscate. Pour des retards plus grands, les lignes isochromatiques forment des courbes fermées qui entourent les courbes précédentes.

En mesurant la distance $J_1 J_2$, on en déduit l'angle $2V$ des axes optiques: on a $\tan 2V' = \dfrac{J_1 J_2}{f}$ car on mesure l'angle $2V'$ dans l'air (f est la focale de l'objectif) (Fig. 406) et $\sin V = \dfrac{\sin V'}{n}$, n étant l'indice moyen.

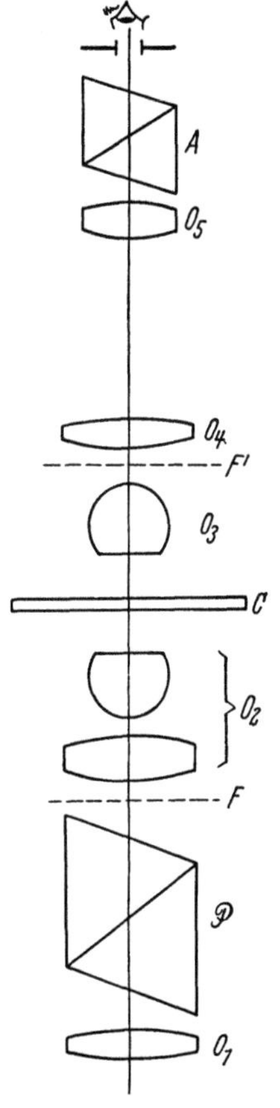

Fig. 407.

173. Observation des phénomènes de polarisation en lumière convergente. On observe ces phénomènes au moyen du microscope polarisant (Fig. 407). Le condenseur est un système afocal composé d'une lentille O_1 qui forme en F l'image d'une source très éloignée. Le plan F est le plan focal de l'autre partie O_2 du condenseur. La lame cristalline C est donc traversée par une infinité de faisceaux parallèles provenant des différents points de F. Comme O_2 est un système fortement convergent, ces faisceaux ont toutes les inclinaisons possibles. L'objectif O_3 également fortement convergent donne du plan F une image F' qui coïncide avec son plan focal. C'est dans le plan F' que se trouve la figure d'interférence à étudier. Une lentille de champ O_4 et un oculaire O_5 qui permet de viser F', complètent l'appareil avec le polariseur $\mathcal{P}$

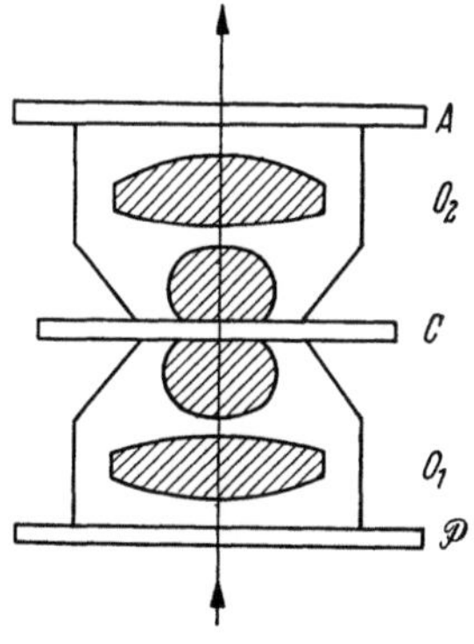

Fig. 408.

et l'analyseur $\mathcal{A}$. Le microscope est muni des graduations nécéssaires pour mesurer les orientations de la lame cristalline et de l'analyseur.

Pour un examen rapide, on peut se servir de deux tourmalines montées sur une pince ou encore de deux polaroïdes.

Lorsqu'il faut un grand champ, le dispositif de la Fig. 408 est commode. Ce sont deux condenseurs de microscopes renversés O_1 et O_2 entre lesquels on a interposé la lame cristalline C. Un polaroïde en $\mathcal{P}$ et un autre en $\mathcal{A}$ permettent une observation facile des phénomènes. Les Figs. 409 à 414, montrent les photographies de quelques phénomènes d'interférences. La Fig. 409 représente les interférences d'une lame de spath perpendiculaire à l'axe entre nicols croisés, et la Fig. 410 concerne une lame de quartz perpendiculaire à l'axe. Nous avons donné cette dernière photographie seulement à titre de comparaison. La croix noire disparaît au centre du champ par suite de la polarisation rotatoire du quartz.

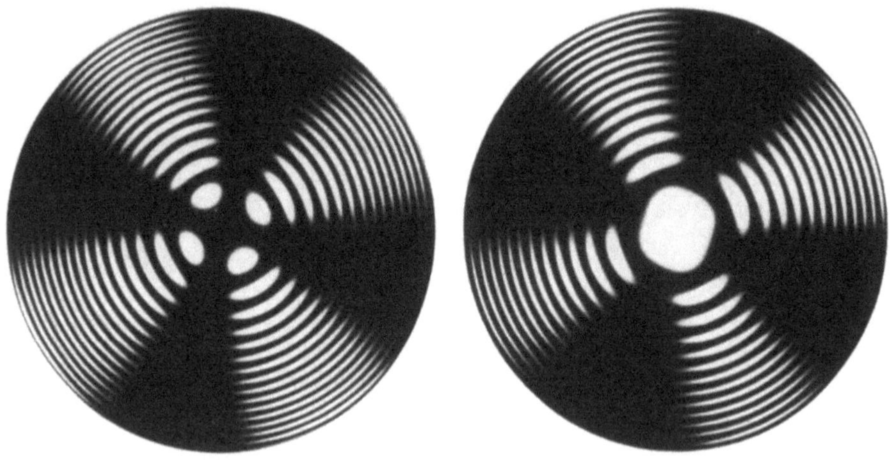

Fig. 409. Spath perpendiculaire à l'axe en lumière convergente (Nicols croisés).

Fig. 410. Quartz perpendiculaire à l'axe en lumière convergente.

Les Fig. 411 et 412 montrent les franges d'une lame de quartz parallèle à l'axe entre nicols croisés et nicols parallèles. La Fig. 413 est donnée par un polariscope de Savart: c'est seulement dans la partie centrale que les franges sont à peu

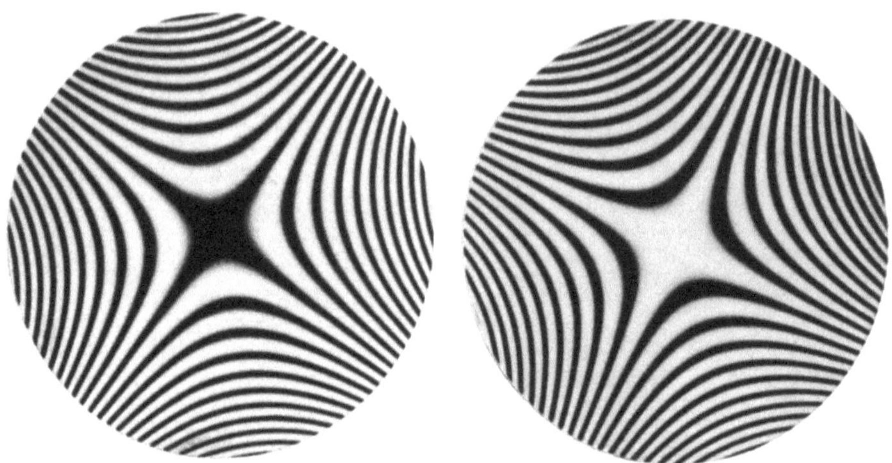

Fig. 411. Quartz parallèle à l'axe en lumière convergente (Nicols croisés).

Fig. 412. Quartz parallèle à l'axe en lumière convergente (Nicols parallèles).

près rectilignes et parallèles. La Fig. 414 représente les lignes neutres et les lignes isochromatiques d'un biaxe (topaze du Brésil) taillé perpendiculairement à la bissectrice de l'angle aigu des axes, les nicols étant croisés.

174. Filtre monochromatique polarisant de B. Lyot. Le filtre de Lyot est formé d'une suite de lames cristallines $Q_1\,Q_2\,Q_3\ldots$ du quartz par exemple, taillées parallèlement à l'axe entre lesquelles sont intercalés des polariseurs $P_1\,P_2\,P_3\ldots$

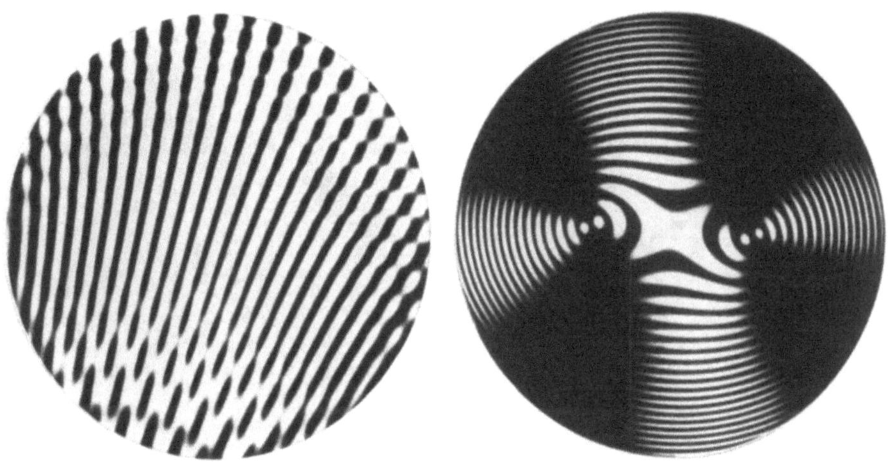

Fig. 413. Polariscope de Savart en lumière convergente.

Fig. 414. Biaxe en lumière convergente (topaze du Brésil).

(Fig. 415). Chaque lame a une épaisseur égale à la moitié de celle qui la suit et tous les axes sont parallèles. Les sections principales des polariseurs $P_1\,P_2\ldots$ sont parallèles et à 45° des axes des lames.

Considérons la première lame Q_1: son épaisseur est telle qu'elle donne au spectroscope un spectre comportant deux cannelures par exemple (Fig. 416a).

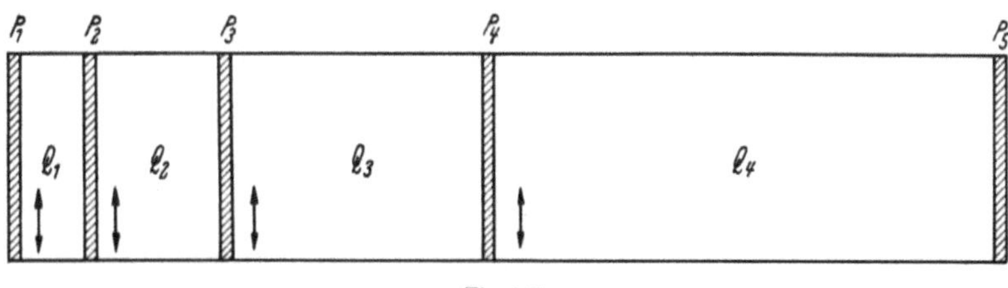

Fig. 415.

La lame Q_2 observée seule entre nicols donne des cannelures deux fois plus serrées puisque son épaisseur est deux fois plus grande (Fig. 416b). De même pour les lames Q_3 et Q_4 qui, prises isolément, donnent les cannelures représentées par les Fig. 416c et d.

Rassemblons les lames entre polariseurs pour constituer le filtre. La lumière transmise par l'ensemble est représentée par une courbe dont les ordonnées sont égales au produit des ordonnées des courbes a, b, c et d. On obtient pratiquement deux maxima fins, là où il y a coïncidence des cannelures brillantes pour toutes les lames.

Le calcul peut se faire comme dans le cas d'un réseau à nombre de traits limité (Sect. 79).

Soient n le nombre des lames, e l'épaisseur de la lame la plus mince et φ la différence de phase

$$\varphi = \frac{2\pi}{\lambda}\,(n_0 - n_e)\,e.$$

Prenons l'amplitude qui sort du premier polariseur égale à l'unité. Après la première lame et le deuxième polariseur, on a deux vibrations parallèles d'amplitude $\frac{1}{2}$ et présentant une différence de phase φ. A la sortie de la deuxième lame et du troisième polariseur, on a 4 vibrations parallèles d'amplitude $\frac{1}{4}$ et de phases $0, \varphi, 2\varphi, 3\varphi$. Finalement, à la sortie du filtre composé de n lames, on

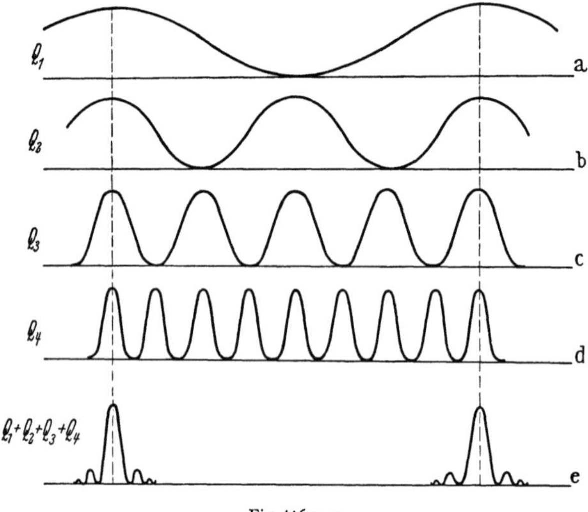

Fig. 416 a—e.

a 2^n vibrations parallèles d'amplitude égale à $1/2^n$ et de phases $0, \varphi, 2\varphi \ldots (2^n-1)\,\varphi$. L'amplitude résultante est égale à celle que donnerait un réseau de 2^n traits

$$U = \frac{\sin 2^n \dfrac{\pi\,(n_0 - n_e)\,e}{\lambda}}{2^n \sin \dfrac{\pi\,(n_0 - n_e)\,e}{\lambda}}. \tag{174.1}$$

La répartition des éclairements donnée par U^2 est comparable à la répartition représentée par la formule (79.5). On a des maxima principaux $U^2 = 1$ pour les valeurs de λ qui rendent la fraction $\dfrac{(n_0 - n_e)\,e}{\lambda}$ entière. Entre ces maxima principaux, se trouvent une série de maxima secondaires très faibles. Le spectre de la lumière transmise par le filtre se compose d'un petit nombre de bandes brillantes et étroites (les maxima principaux) séparées par de larges intervalles pratiquement obscurs. On peut calculer la largeur de chacune des bandes en tenant compte de la courbe de dispersion du cristal.

En posant $\mu = n_0 - n_e$, on trouve

$$\Delta\lambda = \frac{\lambda^2}{2^n e \mu}\,\frac{1}{1 - \dfrac{\lambda}{\mu}\dfrac{\partial\mu}{\partial\lambda}}, \tag{174.2}$$

e étant toujours l'épaisseur de la lame la plus mince.

Pour le quartz $\partial\mu/\partial\lambda = 0{,}923$ pour la raie C, et 0,90 pour la raie F. L'écartement des bandes brillantes est égal à $2^n \varDelta\lambda$. La formule (174.2) permet de calculer un filtre qui isole une radiation de longueur d'onde donnée au centre d'une bande dont la largeur est égale à une valeur donnée $\varDelta\lambda$. Pour faire varier la longueur d'onde du filtre, on peut modifier sa température: la biréfringence varie également et les bandes brillantes se déplacent dans le spectre. On peut ainsi amener l'une des bandes transmises en coïncidence exacte avec la radiation que l'on veut isoler. Pour un filtre en quartz par exemple, entre 10° et 60°, on a un déplacement de 32 Å dans le rouge et de 22 Å dans le bleu. Pour obtenir des déplacements dans un domaine spectral étendu, Lyot a réalisé un filtre avec des lames d'épaisseurs variables. Une faible variation d'épaisseur suffit d'ailleurs pour explorer tout le spectre.

On sait qu'une lame de quartz taillée parallèlement à l'axe et observée en lumière convergente montre des hyperboles. Dans la région centrale, la phase φ est pratiquement constante, mais si on regarde obliquement à travers le filtre, les hyperboles apparaissent; le champ de l'appareil est limité à un carré de 2° environ inscrit dans les hyperboles.

Pour augmenter le champ, Lyot a utilisé les trois artifices suivants:

1. On remplace une lame cristalline d'épaisseur e par deux lames identiques d'épaisseur $e/2$ et croisées en intercalant entre elles une lame demi-onde à 45° des axes des lames. On évite ainsi que les biréfringences se retranchent. Les lignes isochromatiques sont maintenant des cercles très larges et le champ est 26 fois plus grand que celui d'une lame simple pour le quartz.

Par suite de la lame demi-onde, le dispositif est applicable dans un domaine restreint spectral.

2. On remplace une lame cristalline par deux lames taillées dans des cristaux de signes contraires tels que le quartz et le spath, leurs axes de même signe étant parallèles et les axes optiques croisés. Le champ est 9 fois plus grand que celui d'une lame simple en quartz, mais ce système est 2 fois moins épais que le système précédent.

3. Une troisième combinaison consiste à superposer une lame taillée dans un premier cristal et deux lames d'axes croisés, taillées dans un deuxième cristal de signe contraire. Par exemple, une lame de quartz et deux lames de spath taillées parallèlement à l'axe optique.

Ce système est aussi épais que la première combinaison, mais il est utilisable dans un domaine spectral beaucoup plus étendu et donne un champ encore plus grand.

Le premier filtre commencé par Lyot en 1933 comportait 10 polariseurs et 9 lames cristallines de 25 mm. de côté, les deux lames les plus épaisses étant formées de deux cristaux différents. L'appareil avait été étudié pour isoler une bande de 1 Å dans le vert (5303 Å), les autres étant situées au moins à 500 Å. Par manque de spath nécessaire pour les polariseurs, la construction a été interrompue en 1934.

Quelques années plus tard, les polaroïdes font leur apparition et Lyot construit un deuxième filtre simplifié (1938) où les polariseurs sont constitués par des polaroïdes.

Ce nouveau filtre comporte 6 lames de quartz de 36 mm. de côté et permet d'isoler 6 radiations. Les bandes qu'il transmet ont une largeur moyenne de 2 Å dans le vert et de 3 Å dans le rouge. Sa transparence est de 10% dans le vert et 25% dans le rouge. Les longueurs d'onde sont réglées et stabilisées par un thermostat dans lequel se trouve le filtre.

29*

En 1939, Lyot revient à la construction d'un filtre à polariseurs en spath. Il comporte 7 polariseurs en spath et 6 lames en quartz. Sa transparence est égale à 40% dans tout le spectre et il a une largeur de bande de 3 Å dans le rouge.

Ces divers filtres ont permis à B. Lyot d'obtenir des photographies remarquables de la couronne solaire en dehors des éclipses.

Des filtres de ce type sont construits maintenant dans l'industrie. Un monochromateur utilisant les phénomènes de polarisation rotatoire (quartz perpendiculaire à l'axe) a été construit en 1952 par C. S. Hulburt et J. L. Rosenfeld.

c) Interféromètres à polarisation applicables à l'étude des objets isotropes transparents.

Il est possible de réaliser des interféromètres à polarisation utilisables en lumière blanche et d'observer commodément les variations de phase des objets isotropes transparents. Le principe de ces appareils est le suivant: l'onde déformée par les variations de phase de l'objet traverse un système biréfringent et se trouve dédoublée. Au moyen d'un polariseur et d'un analyseur, on peut faire interférer ces deux ondes et déceler ainsi les détails d'un objet transparent déphasant. On peut considérer deux types d'interféromètres: les interféromètres à dédoublement latéral dans lesquels le dédoublement des images se fait dans un plan perpendiculaire à l'axe de l'ensemble de l'appareil, et les interféromètres à dédoublement longitudinal dans lesquels le dédoublement a lieu dans le sens de l'axe, les deux images n'étant plus au point en même temps.

175. Interféromètres à dédoublement lateral. Le schéma de ces appareils est représenté sur la Fig. 417.

Une onde plane Σ polarisée rectilignement par un polariseur $\mathscr{P}$ traverse un objet transparent M constitué dans le cas de la figure par une petite variation d'épaisseur A dans une lame de verre à faces parallèles. Comme dans le cas du contraste

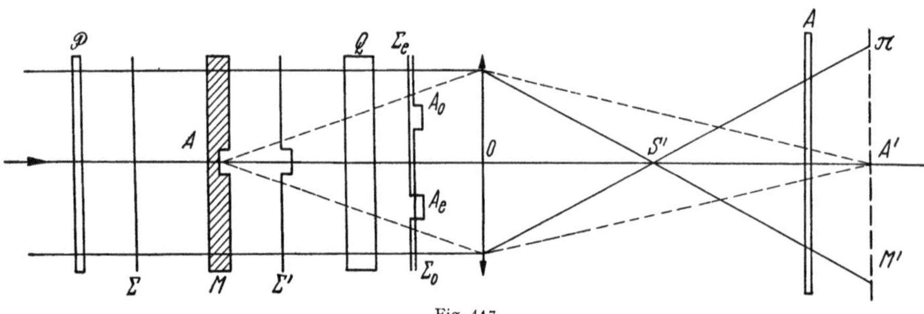

Fig. 417.

de phase, l'objet M peut être remplacé par un objet quelconque transparent. Après traversée de l'objet M, l'onde Σ est déformée par les variations de phase, on a l'onde Σ'. Un système biréfringent Q dédouble l'onde Σ' en deux ondes Σ_0 et Σ_e de façon que le détail A donne deux images A_0 et A_e qui ne se recouvrent pas. L'objectif O donne une image M' de l'objet M sur un plan π que l'on peut observer à l'aide d'un oculaire ou dans lequel on peut placer une plaque photographique. Les deux images A_0 et A_e sont au point dans le plan π et peuvent interférer grâce à l'analyseur A. Nous appellerons ces deux images dans le plan π les images A_0' et A_e' et nous les représentons sur la Fig. 418. Dans les régions (1), (2) et (3), les deux ondes Σ_0 et Σ_e sont à la même distance, l'état d'interférence est le même. Supposons par exemple que ces deux ondes soient distantes de $\lambda/2$

pour le jaune moyen du spectre et opérons entre polariseur et analyseur parallèles. En (1), (2) et (3) on a le pourpre teinte sensible du premier ordre. Dans les régions A_0' et A_e', il n'en est plus de même: en A_e' la distance des deux ondes est $\dfrac{\lambda}{2}$ + l'épaisseur optique de l'objet A; en A_e', elle est $\dfrac{\lambda}{2}$ — l'épaisseur optique de cet objet. La teinte vire en A_0' et A_e' et l'objet devient observable. Les teintes en A_0' et A_e' sont symétriques de la teinte sensible dans l'échelle des teintes. En orientant convenablement le système biré-fringent Q, on peut modifier à volonté la différence de marche entre les deux ondes Σ_0 et Σ_e. Si on opère entre nicols croisés, on peut se placer au voisinage du noir correspondant à une

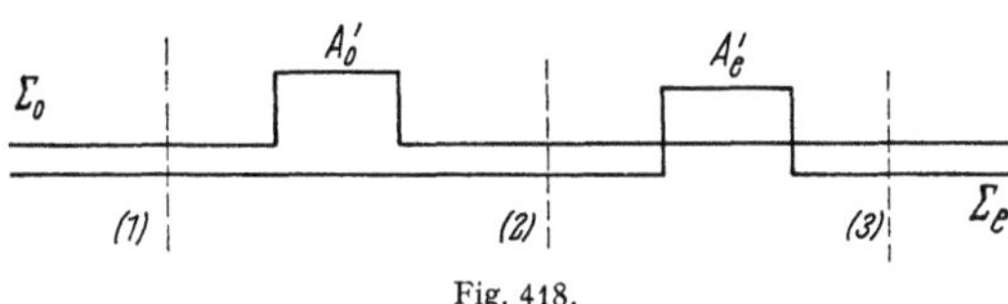

Fig. 418.

différence de marche très faible, la sensibilité du dispositif est excellente. Au lieu de placer le système biréfringent Q en faisceau parallèle, on peut le placer au voisinage de l'image S' de la source donnée par l'objectif O sans rien changer au reste de l'appareil et sans que le principe en soit modifié. Le procédé ne s'applique évidemment que si le dédoublement est plus grand que l'objet, sans cela il y aurait recouvrement.

Nous décrirons ci-après quelques-uns de ces appareils:

176. Interféromètre de Jamin (1868). Le système biréfringent est formé de deux lames de spath identiques Q_1 et Q_2 taillées à 45° de l'axe environ et entre lesquelles on a placé une lame demi-onde L (Fig. 419). Les axes de Q_1 et Q_2 sont parallèles et l'objet M à étudier est interposé entre Q_1 et Q_2. A l'entrée dans le première lame Q_1, le rayon se dédouble en un rayon ordinaire O et un rayon extraordinaire E. La lame demi-onde a ses axes à 45° des axes de Q_1 et Q_2: elle intervertit la direction de vibration des deux rayons et le rayon ordi-naire devient extraordinaire dans la deuxième lame Q_2 et le rayon extraordinaire

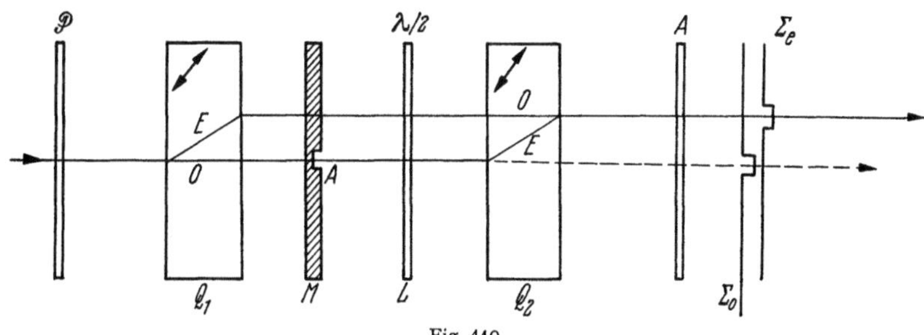

Fig. 419.

devient ordinaire. Les deux rayons sont confondus à la sortie de Q_2 et présentent une différence de marche nulle. Une faible inclinaison de Q_1 ou Q_2 permet de faire varier la différence de marche. L'objet M étant entre Q_1 et Q_2 est dédoublé: on a une onde Σ_0 correspondant aux rayons non déviés par Q_2 (rayons ordinaires pour cette lame) donc aux rayons extraordinaires dans la première lame Q_1. La deuxième onde Σ_e est donnée par les rayons qui sont extraordinaires dans la deuxième lame, elle est décalée par rapport à Σ_0. Un analyseur A permet de faire interférer les deux ondes Σ_0 et Σ_e et on a ainsi une image dont les phases sont rendues visibles par des couleurs ou des changements d'éclairement. Un calcul analogue à ceux de la Sect. 171 montrera facilement que les franges à l'infini de l'ensemble Q_1LQ_2 sont des hyperboles. On peut donc éclairer le dispositif par un colimateur dans le plan focal duquel on placera un diaphragme iris ne laissant provenir la lumière que de la partie centrale des hyperboles. L'appareil a été appliqué par Lebedeff à l'observation au microscope. La préparation est intercalée entre Q_1 et Q_2 avec la lame demi-onde et l'ensemble est placé sur la platine du microscope entre le condenseur et l'objectif.

177. Interféromètre de Françon (1951). Rassemblons les deux lames Q_1 et Q_2 identiques en tournant Q_2 de 90° par rapport à Q_1 et supprimons la lame demi-onde (Fig. 420). Le rayon ordinaire dans la première lame devient extraordinaire

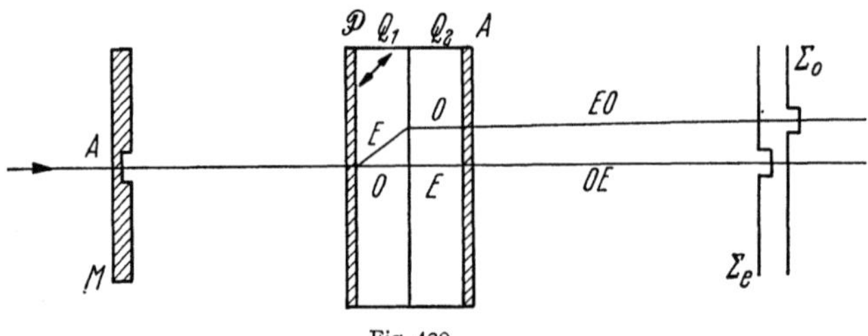

Fig. 420.

dans la deuxième et le rayon extraordinaire dans la première lame devient ordinaire dans le deuxième. La compensation se fait directement sans lame demi-onde, on peut donc utiliser tout le spectre. Le polariseur n'a pas besoin d'être avant l'objet, on peut le placer juste avant les lames Q_1 et Q_2 qui ne constituent d'ailleurs pas autre chose qu'un polariscope de Savart. L'objet M est donc éclairé en lumière blanche naturelle et non pas forcément en lumière polarisée. Les lignes isochromatiques du polariscope de Savart étant des franges rectilignes et parallèles (Sect. 171), la source sera une fente placée au foyer du collimateur. Une inclinaison du polariscope permet de régler la différence de marche à volonté. Le fait de ne plus être obligé de placer l'objet dans le dispositif même permet l'emploi facile de ce dispositif en microscopie. L'ensemble est placé dans l'oculaire du microscope qui peut être, soit un microscope ordinaire, soit un microscope métallographique.

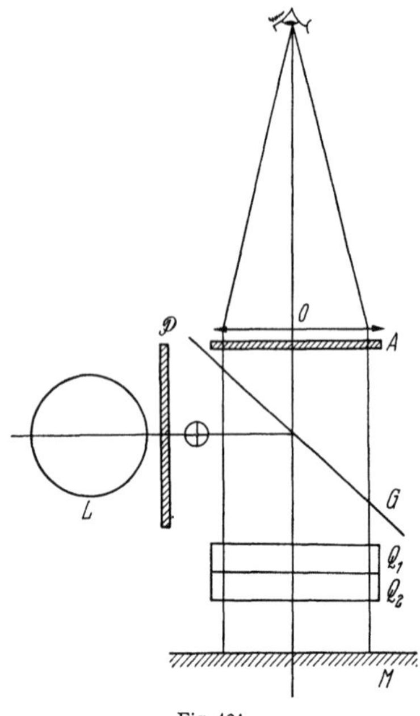

Fig. 421.

Signalons l'emploi de ce même dispositif d'une façon un peu différente dans le cas d'un objet réfléchissant (Fig. 421): la source $\mathcal{L}$ est une lampe opale, le polariseur est en $\mathcal{D}$ et la lumière est renvoyée par le verre semi-réfléchissant G incliné à 45°. Les rayons lumineux traversent le polariscope de Savart $Q_1 Q_2$, se réfléchissent sur l'objet à étudier M, traversent l'analyseur A croisé avec $\mathcal{D}$, puis une loupe O au foyer de laquelle se trouve l'oeil. On a pris la précaution d'orienter $\mathcal{D}$ de façon que la vibration soit dirigée perpendiculairement au plan de figure. Le principe de l'observation est le même que précédemment, mais le polariscope traversé 2 fois donne des lignes isochromatiques hyperboliques très larges qui recouvrent la pupille de l'oeil. La diaphragmation par la pupille suffit à limiter l'inclinaison des faisceaux, toute collimation étant inutile. Notons en passant la lumière parasite réfléchie par la face supérieure de Q_1 est arrêtée par A qui est croisé avec P. De même, la lumière réfléchie par la face inférieure de Q_2, car cette lumière donne naissance sur la pupille de l'oeil à des hyperboles à centre noir. La lumière parasite du dispositif se trouve éliminée.

Au lieu de chercher un grand dédoublement des images, on peut au contraire prendre un dédoublement petit devant l'objet comme cela a été fait par plusieurs auteurs. La methode ne décèle plus la phase elle-même mais ses variations, on a une méthode différentielle. L'objet est vu par ses «pentes». A la différence des appareils existants jusque là, nous avons proposé de prendre un dédoublement, non pas petit devant l'objet, mais petit devant tous les objets observés, c'est-à-dire de l'ordre de grandeur du pouvoir séparateur de l'oeil. Ceci est bien justifié en microscopie par exemple, où les variations de phase sont généralement très brusques et où par conséquent la sensibilité de la méthode est suffisante. On peut réaliser alors un interféromètre de dimensions fixées une fois pour toutes et applicable à n'importante quelle dimension d'objet.

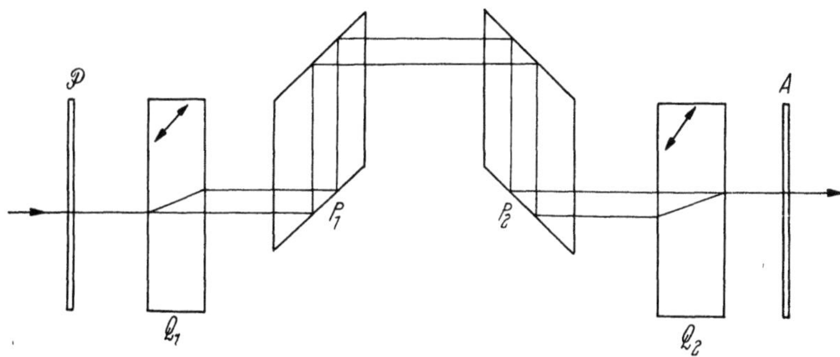

Fig. 422.

178. Interféromètre de Fleischmann (1951). Le grand inconvénient de l'interféromètre de Jamin est l'emploi de la lame demi-onde qui limite le domaine spectral utilisable. Pour toutes les longueurs d'onde pour lesquelles la lame n'est pas demi-onde, il n'y a plus compensation des deux lames Q'_e et Q_2 et de la lumière parasite apparaît. Cette difficulté a été levée d'une façon ingénieuse par Fleischmann qui remplace la lame demi-onde ordinaire de l'interféromètre de Jamin par deux parallélépipèdes de Fresnel P_1 et P_2 (Fig. 422). Chaque parallélépipède agit comme une lame quart d'onde (Sect. 134) et l'ensemble comme une lame demi-onde achromatique. D'autre part, comme pour l'interféromètre de Jamin, les lignes isochromatiques sont des hyperboles très écartées et il est possible d'employer une source large.

179. Interféromètre de Lindberg (1951). Dans les interféromètres à dédoublement latéral, nous avons dit qu'il était possible de placer le système biréfringent Q (Fig. 417) soit en faisceau parallèle, soit au voisinage de l'image S'. Les interféromètres précédents utilisent des lames cristallines à faces parallèles placées en faisceau parallèle. Dans le dispositif de Lindberg, les deux combinaisons sont réalisées en même temps. Deux lames biréfringentes Q_1 et Q_2 sont placées en faisceau parallèle (Fig. 423); un objectif O_1 forme une image de la source sur un polariscope de Savart Q_3 à la suite duquel se trouve l'analyseur A. Le polariseur P est placé avant les lames Q_1 et Q_2. La lame Q_1 est taillée en forme de prisme de petit angle et l'ensemble $Q_1 Q_2$ se comporte comme un compensateur de Babinet.

Si on examinait ce système seul, il donnerait des franges localisées sur la lame Q_1. Le polariscope de Savart donne des franges à l'infini (lignes isochromatiques) qui sont très sensiblement des droites parallèles et équidistantes. Elles se trouvent, sous forme virtuelle, dans le plan focal de l'objectif O_1. Pour réaliser la compensation des deux systèmes, le coin Q_1 est placé dans le plan focal de O_1. De cette manière, les franges à l'infini virtuelles de Savart se peignent sur le coin Q_1. En donnant à ce dernier un angle convenable, on peut avoir même écartement pour les 2 systèmes de franges. Si, de plus, les orientations de $Q_1 Q_2$ et de Q_3 sont convenables, ces deux systèmes s'annulent et on a une teinte plate à l'infini, tout

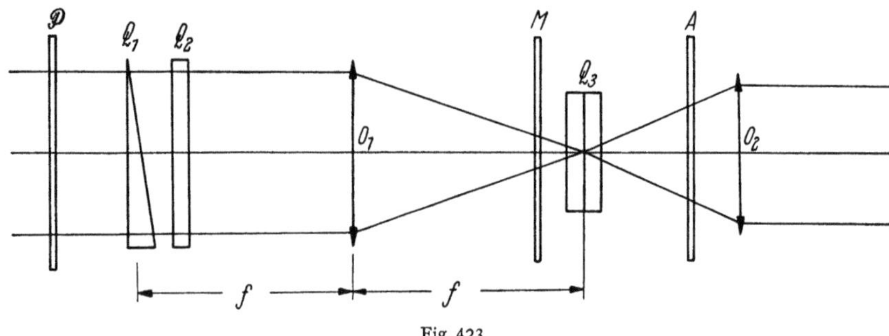

Fig. 423.

au moins dans une région assez étendue. En fait, les lignes isochromatiques de l'ensemble $Q_1 Q_2 Q_3$ sont des hyperboles et ces hyperboles sont deux fois plus larges que si on avait employé deux polariscopes de Savart au lieu de l'ensemble précédent.

L'objet M à examiner est placé avant le polariscope de Savart et on peut l'observer au moyen de la loupe O_2 par exemple. La compensation joue pour l'ensemble $Q_1 Q_2 Q_3$ et on peut employer une source large, mais elle ne joue pas naturellement pour l'objet M lui-même qui est dédoublé par le Savart Q_3, toujours suivant le même principe. Les lames $Q_1 Q_2$ et Q_3 sont taillées dans la même matière, du quartz par exemple, pour avoir la même dispersion.

180. Interféromètre de Smith (1947). Le dispositif de Smith est basé sur l'emploi des prismes biréfringents du type Wollaston. La Fig. 424 donne le

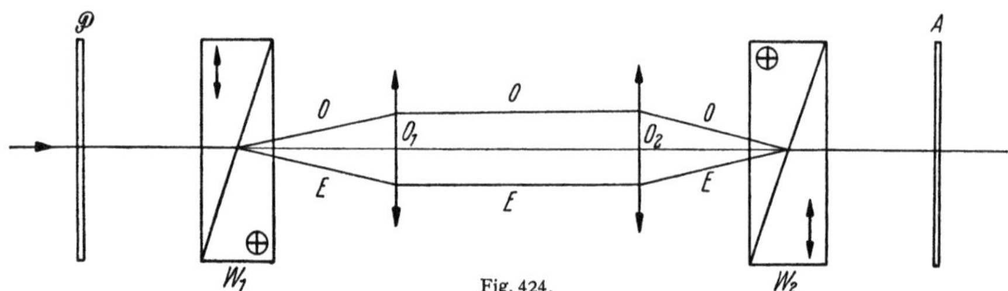

Fig. 424.

schéma de l'appareil. Un prisme de Wollaston W_1 est placé au foyer d'un objectif O_1 et un rayon incident dédoublé en deux rayons, l'un ordinaire O, l'autre extraordinaire E, donne, après traversée de l'objectif O_1 deux rayons parallèles. Ces deux rayons tombent sur un deuxième objectif O_2 identique à O_1 qui les fait converger sur un deuxième prisme de Wollaston W_2 identique à W_1 et orienté de manière que les deux rayons se recombinent pour ne donner finalement qu'un rayon émergent. L'objet à examiner est placé entre les deux objectifs O_1 et O_2 et il est dédoublé par le prisme W_2. Les deux ondes correspondantes interfèrent après traversée de l'analyseur A et les phénomènes observés sont les mêmes que précédemment.

Le système biréfringent étant sur l'image de la source, ce sont maintenants les lignes isochromatiques des prismes qui sont vues dans le plan de l'image. Ce sont les hyperboles des cristaux taillés parallèlement à l'axe optique. Bien entendu, les deux moitiés de chaque prisme se compensant, les hyperboles sont visibles en lumière blanche. Aussi, pour que le champ d'observation soit uniforme, il est nécessaire que l'inclinaison des rayons qui pénètrent dans les prismes ne soit pas trop grande, sinon les hyperboles apparaitront. Cette condition est réalisée en microscopie où l'ouverture des faisceaux dans l'espace image est faible. Notons qu'ici les franges de chaque WOLLASTON sont localisées, comme dans le BABINET, dans chaque prisme; ces franges sont rigoureusement rectilignes et parallèles, de sorte que la compensation se fait pour toute la surface des prismes et on peut employer une source qui couvre les prismes.

Naturellement, il n'est pas nécessaire que les objectifs O_1 et O_2 soient identiques, ainsi que les WOLLASTON W_1 et W_2. Pour deux objectifs O_1 et O_2 de focales déterminées, on calculera facilement l'angle des WOLLASTON pour que la compensation soit réalisée. On peut voir les choses de la façon suivante: les franges localisées sur le WOLLASTON W_1 sont conjuguées de celles qui sont localisées sur le WOLLASTON W_2, il y aure compensation si ces deux système de franges se projettent exactement l'un sur l'autre. Le rapport des focales de O_1 et O_2 doit être dans le rapport des équidistances des deux systèmes de franges.

Comme nous venons de l'indiquer le prisme de WOLLASTON ne peut être employé qu'avec des faisceaux très peu convergents. En combinant ce prisme avec les dispositifs utilisés par LYOT pour augmenter le champ (Sect. 174) on peut réaliser un nouveau type de prisme conservant toutes les propriétés du WOLLASTON et utilisable avec des faisceaux lumineux très convergents.

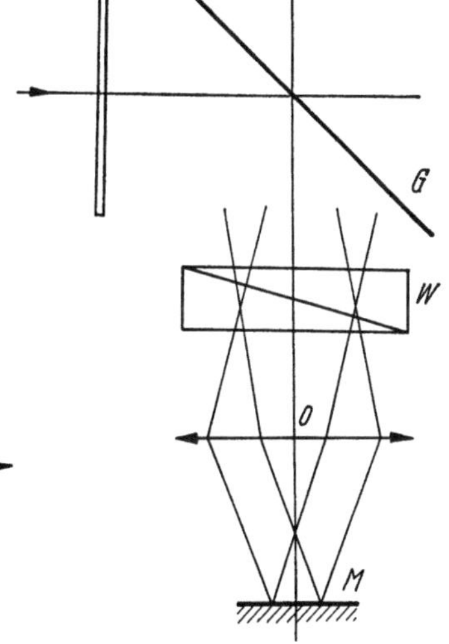

Fig. 425.   Fig. 426.

181. Interféromètre de NOMARSKI (1952). Lorsqu'on veut appliquer le dispositif de SMITH au microscope, on est limité aux faibles grossissements à moins de construire des objectifs spéciaux. En effet, si O_1 est le condenseur et O_2 l'objectif (Fig. 424), les 2 prismes W_1 et W_2 doivent être conjugués à travers O_1 et O_2. Le prisme W_2 est donc pratiquement au foyer de l'objectif du microscope: or le plan focal n'est pas accessible dans les objectifs puissants et le système n'est donc applicable qu'aux systèmes faibles.

NOMARSKI a modifié le prisme W_2 de façon à pouvoir réaliser la conjugaison, même pour les objectifs forts. Le prisme W_2 a maintenant l'aspect de la Fig. 425.

Le deuxième élément est taillé incliné par rapport à l'axe, ce qui a pour effet de sortir le plan de localisation des franges en π. Il est alors possible de faire coïncider ce plan avec le plan focal d'un objectif de microscope, même puissant.

Ce même prisme ou le prisme de Wollaston peut être employé pour l'observation par réflexion (Fig. 426). Il est placé de façon que le plan de localisation des franges soit confondu avec le plan focal de l'objectif O du microscope. Dans ces conditions, après réflexion sur l'objet M, les rayons reviennent symétriquement par rapport à l'axe: si, d'un côté, le rayon ordinaire avait pris une avance sur le rayon extraordinaire, il prend un retard égal de l'autre côté et il y a compensation. Le système est équivalent à l'ensemble des deux prismes de Wollaston de la Fig. 424. Les avantages du prisme de Wollaston utilise en double traversée ont été montrées pour la premiere fois par Lenouvel.

182. Interféromètre à dédoublement longitudinal: dispositif de Philpot (1951).

Tous les interféromètres précédents produisent les interférences par dédoublement latéral.

Dans le dispositif que nous allons décrire (Fig. 427), le dédoublement s'effectue le long de l'axe. L'objet transparent M placé entre deux objectifs O_1 et O_2 est éclairé par la source lumineuse S. Les rayons traversent un système biréfringent Q_1 formé d'une lentille divergente de quartz associée à une lentille convergente de la même matière. La face extérieure de l'élément convergent est parallèle à la face extérieure de l'élément divergent. L'un des deux éléments est taillé parallèlement à l'axe et l'autre perpendiculaire à l'axe. Après traversée de Q_1, la lumière incidente polarisée par le polariseur $\mathcal{P}$ se décompose en deux faisceaux, un faisceau de rayons ordinaires et un faisceau de rayons extraordinaires. Pour les rayons ordinaires, Q_1 est sans puissance et pour les rayons extraordinaires, il se comporte comme une lentille de quelques dioptries. L'objectif O_1 donne par conséquent deux images de S, l'une ordinaire O sur l'objet M, et l'autreextraordinaire E un peu au-dessous. Les rayons pénètrent ensuite dans l'objectif O_2 identique à O_1, puis dans un système Q_2

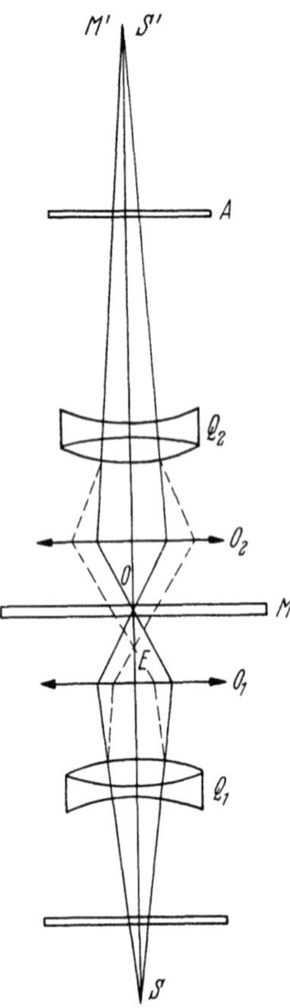

Fig. 427.

également identique à Q_1, mais croisé avec lui. Le système Q_2 recombine les rayons ordinaires et les rayons extraordinaires, assurant ainsi la compensation. L'image M' est observée avec un oculaire ou une loupe. Cette image résulte des interférences, grâce à l'analyseur A, de l'image ordinaire au point, et de l'image extraordinaire non au point. L'image extraordinaire qui est au point avec l'image ordinaire correspond au point E situé au-dessous de M. L'image extraordinaire non au point qui interfère en M' avec l'image ordinaire produit pratiquement un fond cohérent uniforme si le défaut de mise au point de l'image extraordinaire est suffisant. En utilisant un compensateur, on peut faire varier le déphasage à volonté. L'appareil donne directement les phases, comme dans le procédé à grand dédoublement latéral, mais sans être gêné par ce dédoublement. Par contre, les composantes des doublets Q_1 et Q_2, qui sont taillées perpendiculaires à l'axe, introduisent de la dispersion rotatoire et il est nécessaire d'employer une source de lumière monochromatique.

Bibliographie Sommaire.

1. Ouvrages généraux.

BORN, M.: Optik. Berlin: Springer 1933.

BRUHAT, G.: Cours d'Optique, 4° édit. revue par A. KASTLER. Paris: Masson & Cie. 1954.

DITCHBURN, R. W.: Light. London: Blackie & Son 1954.

DRUDE, P.: The theory of optics. 1902.

FABRY, C.: Introduction générale à la photométrie. Rev. Opt. **1927**.

KRONIG, R.: Textbook of physics. London: Pergamon Press 1954.

POHL, R. W.: Optik. Berlin: Springer 1947.

2. Diffraction.

AIRY, G. B.: Trans. Cambridge Phil. Soc. **5**, 283 (1835).

ARNULF, A., O. DUPUY et F. FLAMAT: Rev. Opt. **32**, 529 (1953).

BEREK, M.: Z. Physik **40**, 421 (1926).

BOUGHON, P., B. DOSSIER et P. JACQUINOT: C. R. Acad. Sci. Paris **223**, 661 (1946).

BUXTON, A.: Mon. Not. Roy. Astronom. Soc. **81**, 547 (1921); 1923, Ibid., **83**, 475 (1923).

CHRETIEN, H.: Cours de calcul des combinaisons optiques. Rev. Opt. **1938**.

CITTERT, P. H. VAN: Physica, Haag **1**, 201 (1934).

CONRADY, A. E.: Dictionary of Applied physics, vol. IV, p. 222. 1923.

CONRADY, A. E.: Mon. Not. Roy. Astronom. Soc. **79**, 575 (1919).

CORPUT, J. G. VAN DER: Proc. Kon. Akad. Wetensch. Amst. **51**, 6 (1948).

COUDER, A.: C. R. Acad. Sci. Paris **219**, 677 (1944).

DUFFIEUX, P. M.: L'intégrale de Fourier et ses applications à l'optique (Besancon, Faculté des Sciences). 1946.

DURAND, E.: La théorie des images optiques, p. 127. Paris: Éditions de la Revue d'Optique 1949.

EPSTEIN, L.: J. Opt. Soc. Amer. **39**, 226 (1949).

FRANÇON, M.: Cah. Phys. **1944**, Nr. 26, 15. — Rev. Opt. **26**, 254 (1947a); **26**, 369 (1947b); **27**, 157 (1948a); **27**, 595, 761 (1948b). — C. R. Acad. Sci. Paris **299**, 293 (1949). — Thèse R. O. 1948.

HÖNL, H.: Eine strenge Formulierung des klassischen Beugungsproblems. Z. Physik **131**, 290 (1952).

HOPKINS, H. H.: Proc. Phys. Soc. Lond. **55**, 116 (1943); **56**, 48 (1944). — Proc. Phys. Soc. Lond. B **62**, 22 (1949).

HOPKINS, H. H.: The concept of partial coherence in optics. Proc. Roy. Soc. Lond., Ser. A **208**, 263 (1951).

HOPKINS, H. H., and P. M. BARHAM: Proc. Phys. Soc. Lond. B **63**, 737 (1950).

JACQUINOT, P., P. BOUGHON et B. DOSSIER: La théorie des images optiques, p. 183. Paris: Editionsde la Revue d'Optique 1949.

KAMPEN, N. G. VAN: Physica, Haag **14**, 575 (1949).

KASTLER, A.: Rev. Opt. **29**, 307 (1950).

KATHAWATE, Y. V.: Proc. Ind. Acad. Sci. A **21**, 177 (1945).

KINGSLAKE, R.: Proc. Phys. Soc. Lond. **61**, 147 (1948).

KÖNIG, A.: Handbuch der Experimentalphysik, Bd. 20, Part II, p. 141. 1929.

KORRINGA, J., B. NIJBOER, A. et A. MARECHAL: Document S.O. 48—8, Commission Internationale d'Optique. Paris: Institut d'Optique.

LANSRAUX, G.: Rev. Opt. **26**, 23 (1947).

LANSRAUX, G.: La théorie des images optiques, p. 194. Paris: Editions de la Revue d'Optique 1949.

LAPICQUE, C.: Thèse, Rev. Opt. **1938**.

LINFOOT, E. H.: Recent advances in optics. Oxford: Clarendon Press 1955.

LINFOOT, E. H., and E. WOLF: Phys. Soc. B **66**, 145 (1953).

LINFOOT, E. H., and E. WOLF: M.N.R.A.S. **112**, 452 (1952).

LOMMEL, E.: Abh. Bayer. Akad. **15**, Abth. 2, 229 (1885); 15, Abth. 3, 529 (1886).

LUNEBERG, R. K.: Mathematical theory of optics. Providence, R. I.: Brown University 1944.

LYOT, B., et M. FRANÇON: Etude des défauts d'homogénéité de grands disques de verre. Rev. Opt. **29**, 499 (1950).

MARÉCHAL, A.: Rev. Opt. **1953**.

MARÉCHAL, A.: Thèse. Rev. Opt. **1948**.

MARÉCHAL, A.: C. R. Acad. Sci. Paris **218**, 345 (1944a); **218**, 395 (1944b). — Cah. Phys. **1944**c. Nr. 26, I. — Rev. Opt. **26**, 257 (1947a); **27**, 73 (1948). — J. Opt. Soc. Amer. **37**, 403 (1947b).

Martin, L. C.: Mon. Not. Roy. Astronom. Soc. **82**, 310 (1922).
Nienhuis, K.: Thesis, University of Groningen 1948.
Nienhuis, K., and B. T. A. Nijboer: Physica, Haag **14**, 590 (1949).
Nijboer, B. R. A.: Thesis, University of Groningen 1942. — Physica, Haag **10**, 679 (1943);
    13, 605 (1947).
Osterberg, H., and J. E. Wilkins: J. Opt. Soc. Amer. **39**, 553 (1949).
Osterberg, H., and F. C. Wissler: J. Opt. Soc. Amer. **39**, 558 (1949).
Picht, J.: Optische Abbildung, 1931.
Picht, J.: Ann. Phys., Lpz. **77**, 685 (1925); **80**, 491 (1926). (Also 1931, Optische Abbildung.
    Braunschweig: Vieweg & Sohn.)
Ramachandran, G. N.: Proc. Ind. Acad. Sci. A **21**, 165 (1945).
Rayleigh: Phil. Mag. **8**, 403 (1879). (Also scientific papers **1**, 428.)
Richter, R.: Z. Instrumentenkde. **45**, 1 (1925).
Slevogt, H.: Optik **4**, 349 (1949).
Sommerfeld: Vorlesungen über theoretische Physik, Bd. IV Optik. Wiesbaden: Dieterich-
    sche Verlagsbuchhandlung 1950.
Steward, G. C.: Phil. Trans. Roy. Soc. Lond., Ser. A **225**, 131 (1925). — Trans. Cambridge
    Phil. Soc. **23**, 235 (1926).
Struve, H.: Mém. Acad. Sci., St-Pétersbourg (7), **34**, No 5, I (1886).
Toraldo di Francia: Onde electromagnetiche. Bologne: Zanichelli 1954.
Wagner, D.: Rev. Opt. **29**, 419 (1950).
Wolf, E.: Proc. Roy. Soc. Lond., Ser. A **204**, 533 (1951).
Wolf, E.: Phys. Soc. **14**, 95 (1951).
Zernike, F.: Physica, Haag **5**, 785 (1938).
Zernike, F.: Physica, Haag **1**, 689 (1934). — Proc. Phys. Soc. Lond. **61**, 158 (1948).
Zernike, F., et B. R. A. Nijboer: Contribution à La théorie des images optiques, p. 227.
    Paris: Editions de la Revue d'Optique 1949.

3. Interférences.

Abelès, F.: J. de Phys. **11**, 403 (1950).
Arnulf, A., et M. Cagnet: C. R. Acad. Sci. Paris **230**, 2014 (1950).
Dufour, C.: Rev. Opt. **31**, 1 (1952).
Dufour, C.: Thèse. Paris: Masson et Cie. 1950.
Fabry, C.: Les applications des interférences lumineuses. Rev. Opt. **1923**.
Fabry, C.: Jubilé scientifique. Paris: Gauthier-Villars 1938.
Faust, R. C.: Proc. Phys. Soc. Lond. **65**, 48 (1952).
Heel, A. C. S. van: J. Opt. Soc. Amer. **40**, 809 (1950).
Jacquinot, P., et C. Dufour: J. de Phys. **11**, 427, 431 (1950).
Michelson, A. A.: Phil. Mag. **30**, 1 (1890).
Svensson, H.: Acta chem. scand., Copenh. **5**, 1301 (1951).
Svensson, H.: Acta chem. scand., Copenh. **5**, 1410 (1951).
Svensson, H., u. K. Odengrui: Acta chem. scand., Copenh. **6**, 720 (1952).

4. Polarisation.

Fleischmann, R.: Interferenzverfahren zur Messung der absoluten Phasen bei der Unter-
    suchung absorbierender Medien. Z. Physik **129**, 275 (1951).
Françon, M.: Rev. Opt. **30**, 221 (1951).
Hiroshi Kubota: J. Opt. Soc. Amer. **42**, 144 (1952).
Hiroshi Kubota and Teruji Ose: On the interference color of chromatic polarization.
    J. Applied Phys. **24**, 63 (1955).
Hiroshi Kubota, Tetsuya Ara et Hiroyoshi Saito: J. Opt. Soc. Amer. **41**, 537 (1951).
Lindberg, A.: Verwendung einer Savartschen Doppelplatte zur Herstellung eines phasen-
    gleichen Gesichtsfeldes. Z. Physik **131**, 231 (1952).
Lohmann, A.: Ein neus Dualitätsprinzip in der Optik. Optik **10**, 477 (1954).
Lyot, B.: Thèse, imprimerie Tessier, 1929.
Lyot, B.: Annales d'Astrophysique. Rev. Opt. **7**, 31 (1944).
Nomarski, G.: J. de Phys. **16**, 9 (1955).

Optik dünner Schichten.

Von

H. WOLTER.

Mit 41 Figuren.

I. Allgemeine Übersicht über die Optik dünner Schichten und die Stoffauswahl in dem vorliegenden Artikel.

1. Die Optik dünner Schichten ist heute in stürmischer Entwicklung. Technische Anwendungen in erster Linie sind der Anlaß.

Mit Systemen dünner Schichten werden Interferenzfilter gebaut, die durch Kombination miteinander nahezu jede denkbare und wünschenswerte Filterung des Spektrums in absehbarer Zeit erwarten lassen. Besonders bemerkenswert ist ihre fast absorptionsfreie Form, in der sie einen Teil des Spektrums durchlassen, den Rest reflektieren.

Optische Instrumente wie Fernrohre, Ferngläser, Photooptiken und Mikroskope werden mit dünnen Schichten durch Beseitigung störender Reflexionen lichtstärker und nahezu streulichtfrei gemacht — „vergütet". Nach dem gleichen Prinzip aufgebaute Schichtsysteme heben für schrägen Einfall die Reflexion einer der beiden Polarisationen auf und werden so zu Polarisatoren mit neuen Eigenschaften. Gewisse Ausführungsformen reflektieren die eine und lassen die andere Polarisation hindurch, jede für sich nahezu rein und fast ohne Absorption.

Metallspiegel werden durch dünne dielektrische „Verstärkerfolien" zu korrosionsbeständigen und wischfesten Spiegeln von vorher unerreicht hoher Reflexion.

Dünne dielektrische Schichten übernehmen die Rolle der alten metallischen Strahlenteiler für Interferometer und dergleichen — nun fast ohne zu absorbieren.

Zugleich mit dem Umfang technischer Anwendung dünner Schichten steigt das Interesse an Strukturproblemen und der Frage nach den optischen Konstanten dünner Schichten und nach Methoden, diese zu messen.

Von der im Zuge dieser Entwicklung entstandenen umfangreichen Originalliteratur kann in diesen Handbuchartikel nur eine kleine Auswahl eingehen, wenn die Tradition dieses Handbuches hinsichtlich Gründlichkeit der Einzeldarstellung und Ausführlichkeit in der Behandlung der physikalischen Grundlagen gewahrt werden soll. Historische Fragen werden daher sehr knapp, Prioritätsfragen nicht behandelt; einen Ersatz hierzu mag die Bibliographie am Ende des Artikels bieten, die deshalb — statt der meist zu allgemeinen Titel — den Inhalt kennzeichnende Schlagworte zu fast jeder Originalarbeit beifügt. Vor allem aber sei bezüglich historischer Fragestellungen auf das Buch H. MAYER, „Physik dünner Schichten" Teil I, Stuttgart 1950, hingewiesen, in dem die bis zu seinem Erscheinen bekannten Veröffentlichungen einen bemerkenswert vollständigen Niederschlag gefunden haben, und das zugleich die Optik dünner Schichten in einen weiteren Rahmen fügt. Das Strukturproblem dünner Metallschichten wird in einem anderen Artikel dieses Handbuches gesondert behandelt.

Der vorliegende Artikel bringt zunächst die Lösung der physikalischen Grundaufgabe, die Berechnung der reflektierten und durchgelassenen Intensität für ein System beliebig vieler Schichten. Die Auflösung des sich ergebenden linearen Gleichungssystems geschieht durch Rekursion, der selbst ein physikalischer Gehalt zukommt. Das lineare Gleichungssystem zu einer Matrixgleichung zusammenzufassen — was trivialerweise möglich und in der Literatur in einfachen Fällen auch diskutiert ist — scheint nicht sehr nutzbringend. Auch die Sprechweise der Leitungstheorie wird nicht eingeführt; vorgezogen wird die wellenoptische Behandlung, da diese allgemeiner ist und die Leitungstheorie umgekehrt ihr eingeordnet werden könnte. Die Analogie zwischen den Größen $\hat{g}$ und Leitwerten einerseits und den Größen $\hat{g}$ und Widerständen andererseits (S. 466) kann freilich gelegentlich als Richtschnur dienen, um aus der Vierpoltheorie geläufige Erkenntnisse auf weniger bearbeitete optische Gebiete zu übertragen. Insbesondere ist die Analogie zur Leitungstheorie für senkrechten Einfall der Wellen vollständig. In dem vorliegenden Artikel wird aber bewußt fast durchweg auf beliebigen Einfall verallgemeinert, da die Entwicklung der praktischen Methoden heute gleichfalls diesen Weg geht.

Aus der Lösung der Grundaufgabe ergeben sich am unmittelbarsten die physikalischen Prinzipien der Reflexminderung. Aus diesem Grunde und wegen ihrer bis heute größten praktischen Bedeutung werden sie als erste unter den Prinzipien der Anwendungen dünner Schichten behandelt. Dem ihnen gewidmeten Abschnitt III ist die Untersuchung zur Reflexverstärkung angehängt und unmittelbar eingefügt die Betrachtung der Polarisation mit Hilfe dünner Schichten. Die Polarisation ergibt sich ohnehin als Spezialfall der Reflexfreiheit für schrägen Einfall.

Die Wellenlängenabhängigkeit der Effekte ist in den genannten Abschnitten nur beiläufig betrachtet; sie wird dagegen Hauptgegenstand in dem Abschnitt V „Interferenzfilter". Der letzte Abschnitt des Artikels, VI, behandelt die Methoden zur Messung der optischen Konstanten dünner Schichten, ein Gebiet, das zur Zeit wieder stark im Fluß ist; ihm wurden daher teilweise noch unveröffentlichte Verfahren mit eingefügt. Auch in dem Abschnitt V über Interferenzfilter wurden noch unveröffentlichte Resultate solcher Arbeiten benutzt, die zur Abrundung des Bildes — z.B. bei der Koppelung der Filter — für diesen Handbuchartikel eigens veranlaßt worden waren.

II. Die Grundlagen und die Grundaufgabe der Optik dünner Schichten.

a) Fortpflanzung einer ebenen elektromagnetischen Welle im homogenen, isotropen Medium.

2. In dieser Ziffer sei ein — mit isotroper Materie homogen erfüllter oder auch leerer — Bereich des Raumes betrachtet. Es seien dort

die Dielektrizitätskonstante ε,

die magnetische Permeabilität μ

und die elektrische Leitfähigkeit σ

räumlich und zeitlich konstant. Freie Ladungen seien nicht vorhanden; also sei

die Ladungsdichte $\varrho = 0$.

Dort auftretende elektrische und magnetische Feldstärken $\mathfrak{E}$ bzw. $\mathfrak{H}$ genügen den MAXWELLschen Gleichungen

$$\operatorname{rot} \mathfrak{H} = \frac{4\pi\sigma}{c}\,\mathfrak{E} + \frac{\varepsilon}{c}\,\frac{\partial \mathfrak{E}}{\partial t}\,, \tag{2.1}$$

$$\operatorname{rot} \mathfrak{E} = \qquad -\frac{\mu}{c}\,\frac{\partial \mathfrak{H}}{\partial t}\,, \tag{2.2}$$

$$\operatorname{div} \mathfrak{E} = 0\,, \tag{2.3}$$

$$\operatorname{div} \mathfrak{H} = 0 \tag{2.4}$$

und der aus ihnen folgenden Wellengleichung

$$\Delta \mathfrak{E} = \frac{4\pi\sigma\mu}{c^2}\,\frac{\partial \mathfrak{E}}{\partial t} + \frac{\varepsilon\mu}{c^2}\,\frac{\partial^2 \mathfrak{E}}{\partial t^2}\,, \tag{2.5}$$

bzw. der entsprechenden Gleichung für $\mathfrak{H}$. t bezeichnet die Zeit.

In dem Medium breite sich eine ebene elektromagnetische Welle in Richtung des Vektors $\mathfrak{f}$ aus. Die Feldstärken seien angesetzt als Realteile der Größen

$$\mathfrak{E}(\mathfrak{r},\,t) = \mathfrak{E}_0 \cdot e^{i\omega\left\{t-\frac{\mathfrak{n}}{c}(\mathfrak{r}\cdot\mathfrak{f})\right\}}\,, \tag{2.6}$$

$$\mathfrak{H}(\mathfrak{r},\,t) = \mathfrak{H}_0 \cdot e^{i\omega\left\{t-\frac{\mathfrak{n}}{c}(\mathfrak{r}\cdot\mathfrak{f})\right\}} \tag{2.7}$$

mit einer noch zu bestimmenden Konstanten $\mathfrak{n}$. Darin bezeichnet e die Basis der natürlichen Logarithmen, i die imaginäre Einheit, c die Vakuumlichtgeschwindigkeit, $\mathfrak{r}$ den Ortsvektor des betrachteten Raumpunktes, $\omega = 2\pi\nu = 2\pi\,\dfrac{c}{\lambda}$ die Kreisfrequenz der Welle von der Vakuumwellenlänge λ und $\mathfrak{f}$ einen Einheitsvektor, den „Fortpflanzungsvektor", als dessen Komponenten $\mathfrak{f}_x$, $\mathfrak{f}_y$, $\mathfrak{f}_z$ reelle oder auch komplexe Zahlen zugelassen seien. Bei absorbierenden Stoffen kommt man oft ohne komplexe Fortpflanzungsvektoren nicht aus. Ihre physikalische Bedeutung ist von der Totalreflexion und aus der Metalloptik bekannt und wird auch hier in Ziff. 12 an einem praktischen Falle erläutert werden.

Einsetzen des Ansatzes (2.6) in die Wellengleichung (2.5) oder von (2.7) in die (2.5) entsprechende Wellengleichung für $\mathfrak{H}$ gibt

$$\mathfrak{n}^2 = \varepsilon\mu - i\,\frac{4\pi\sigma\mu}{\omega}\,, \tag{2.8}$$

also eine Bestimmung der Konstanten $\mathfrak{n}$ des Ansatzes (2.6), (2.7) aus den Materialkonstanten und der Frequenz. Im Leiter ist $\mathfrak{n}$ komplex und heißt „komplexer Brechungsindex", im folgenden schlechthin „Brechungsindex". Wir bezeichnen wie üblich

$$\mathfrak{n} = n - i\,k \tag{2.9}$$

mit rellen Zahlen n (genannt „Brechungsindexrealteil") und k (genannt „Absorptionskoeffizient").

Einsetzen von (2.9) in (2.8) gibt

$$n^2 - k^2 = \varepsilon \cdot \mu\,, \tag{2.10}$$

$$n \cdot k = \frac{\sigma \cdot \mu}{\nu} \tag{2.11}$$

(„DRUDEsches Gesetz"). Auflösen nach n und k liefert

$$n = \sqrt{\mu} \cdot \sqrt{\sqrt{\left(\frac{\varepsilon}{2}\right)^2 + \left(\frac{\sigma}{\nu}\right)^2} + \frac{\varepsilon}{2}}\,, \tag{2.12}$$

$$k = \sqrt{\mu} \cdot \sqrt{\sqrt{\left(\frac{\varepsilon}{2}\right)^2 + \left(\frac{\sigma}{\nu}\right)^2} - \frac{\varepsilon}{2}}\,. \tag{2.13}$$

Die Bedeutung dieser Größen z.B. für eine Welle mit reellem Fortpflanzungsvektor $\mathfrak{f}$ erhellt aus (2.6) nach Einsetzen von $n - ik$ für $\mathfrak{n}$:

$$\mathfrak{E}(\mathfrak{r},\,t) = \mathfrak{E}_0 \cdot e^{-\frac{k\omega}{c}(\mathfrak{r}\cdot\mathfrak{f})} \cdot e^{i\omega\left\{t-\frac{n}{c}(\mathfrak{r}\cdot\mathfrak{f})\right\}}\,, \tag{2.14}$$

$$\mathfrak{Re}\,(\mathfrak{E}_x) = |\mathfrak{E}_{0x}| \cdot e^{-\frac{k\omega}{c}(\mathfrak{r}\cdot\mathfrak{f})} \cdot \cos\left(\omega\left\{t-\frac{n}{c}(\mathfrak{r}\cdot\mathfrak{f})\right\} + \delta_x\right)\,. \tag{2.15}$$

Hierin bedeutet δ_x eine Phasenkonstante entsprechend $\mathfrak{E}_{0x} = |\mathfrak{E}_{0x}| \cdot e^{i\delta_x}$. Analoges gilt für die y- und die z-Komponente. c/n ist offenbar die Phasengeschwindigkeit. Die Amplitude fällt längs der durch $\mathfrak{f}$ gegebenen Fortpflanzungsrichtung exponentiell ab und ist auf den e-ten Teil gesunken nach Durchlaufen der Strecke

$$F = \frac{c}{k\,\omega} = \frac{\lambda}{2\pi\,k}\,, \tag{2.16}$$

die als „Eindringtiefe" des Materials bezeichnet wird.

Die Intensität der Welle ist gegeben durch das Zeitmittel des Poynting-Vektors

$$\mathfrak{S} = \frac{c}{4\pi} \cdot \left[\mathfrak{Re}\,(\mathfrak{E}) \times \mathfrak{Re}\,(\mathfrak{H})\right]. \tag{2.17}$$

Für reellen Fortpflanzungsvektor $\mathfrak{f}$ und linear polarisierte Welle erhält man nach Einsetzen von (2.15) bzw. der entsprechenden Gleichung für $\mathfrak{Re}\,(\mathfrak{H})$ und unter Berücksichtigung der Maxwellschen Gl. (2.1) und (2.2) die Intensität (Leistung je cm² Querschnitt):

$$J = |\overline{\mathfrak{S}}| = \frac{c}{8\pi}\,\frac{n}{\mu} \cdot |\mathfrak{E}|^2 = \frac{c}{8\pi}\,\frac{\mu \cdot n}{n^2 + k^2} \cdot |\mathfrak{H}|^2. \tag{2.18}$$

Dabei ist benutzt, daß aus den Maxwellschen Gleichungen z.B. für eine Fortpflanzung in x-Richtung

$$\mu \cdot \mathfrak{H}_y = -\,\mathfrak{n} \cdot \mathfrak{E}_z, \tag{2.19}$$

$$\mu \cdot \mathfrak{H}_z = \mathfrak{n} \cdot \mathfrak{E}_y \tag{2.20}$$

folgt und allgemein eine solche Kopplung zwischen den elektrischen und magnetischen Feldstärken in zwei zueinander und zum Fortpflanzungsvektor $\mathfrak{f}$ senkrechten Ebenen gilt.

Außer diesen hier zusammengestellten Ergebnissen der elektromagnetischen Lichttheorie werden in diesem Artikel nur noch die aus dem Energiesatz folgenden elektromagnetischen Grenzbedingungen gebraucht, die besagen, daß sich die Tangentialkomponenten der elektrischen und magnetischen Feldstärke bei Übergang von einem Medium in ein anderes nicht sprunghaft ändern.

b) Das Gleichungssystem für ebene elektromagnetische Wellen an einem geschichteten Medium.

3. Ansätze für die Wellen. Eine elektromagnetische Welle treffe auf ein System planparalleler Schichten homogener isotroper Medien. Gefragt wird nach der von dem Schichtsystem reflektierten und der durchgelassenen Welle, und zwar sowohl nach ihren Intensitäten als auch nach ihren Phasen. Deshalb werden zunächst die komplexen elektrischen und magnetischen Feldstärken der reflektierten und der durchgelassenen Welle nach Amplitude und Phase berechnet; die auf das Schichtsystem auftreffende Welle und die Daten der Medien gelten als in jeder Weise bekannt. Die hierdurch umrissene Aufgabe heiße die Grundaufgabe dieses Abschnitts. Ist sie gelöst, so lassen sich andere Aufgaben, z.B. Fragen nach den Brechungsindices vorgegebener Schichten, durch Umkehrung der Fragestellung lösen.

Fig. 1 dient zur Festlegung der Bezeichnungen. x, y, z ist ein cartesisches Rechtssystem. m Ebenen mit den Gleichungen

$$z = z_i \quad \text{für} \quad i = 1, \ldots, m \tag{3.1}$$

seien die Grenzflächen zwischen den $(m+1)$ Medien, die nach Fig. 1 von 0 bis m numeriert sind. Das 1. bis $(m-1)$-te Medium haben die Schichtdicken

$$d_l = z_{l+1} - z_l \quad \text{für} \quad l = 1, \dots, m-1. \tag{3.2}$$

Das 0-te und das m-te Medium seien nur durch die Ebene $z = z_1$ bzw. $z = z_m$ begrenzt.

Im m-ten Medium falle eine Welle ein ($\mathfrak{A}_m$ in Fig. 1!). In jedem Medium sei eine „anlaufende" Welle

$$\mathfrak{A}_l(\mathfrak{r}) = \mathfrak{A}_l^\circ \cdot e^{-i\omega \frac{n_l}{c}(x \sin \varphi_l - z \cos \varphi_l)} \tag{3.3}$$

und eine „reflektierte" Welle

$$\mathfrak{R}_l(\mathfrak{r}) = \mathfrak{R}_l^\circ \cdot e^{-i\omega \frac{n_l}{c}(x \sin \varphi_l + z \cos \varphi_l)} \tag{3.4}$$

für $l = 0, \dots m$ zugelassen.

Im letzten, dem 0-ten Medium verschwinde die reflektierte Welle $\mathfrak{R}_0^\circ = 0$, und die einfallende $\mathfrak{A}_m^\circ$ wird als bekannt angesehen. Der Zeitfaktor $e^{i\omega t}$ sei mit in den Größen $\mathfrak{A}_l^\circ$, $\mathfrak{R}_l^\circ$ untergebracht.

Über die Fortpflanzungsvektoren

$$\mathfrak{f}_l = (\sin \varphi_l; 0; -\cos \varphi_l) \tag{3.5}$$

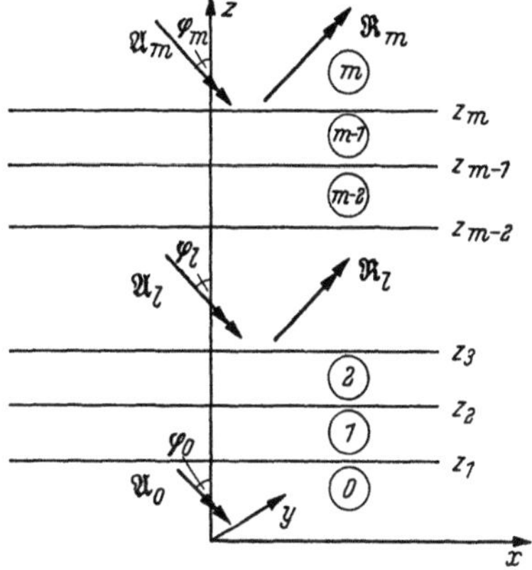

Fig. 1. Bezeichnungen bei einem Schichtsystem mit m Grenzflächen.

ist nur vorausgesetzt, daß ihre y-Komponente verschwindet. Der Ansatz der x und der z-Komponente als sin und cos eines Winkels bedeutet dann für $\mathfrak{f}_l$, das ohnehin der Bedingung $\mathfrak{f}_{lx}^2 + \mathfrak{f}_{ly}^2 = 1$ genügen soll, keine Einschränkung der Allgemeinheit, da wir φ_l rein formal unter anderem auch komplex zulassen, nicht eine anschauliche Bedeutung dieses Winkels verlangen und dementsprechend keine geometrischen Hilfen etwa aus der Fig. 1 ziehen.

In den Fortpflanzungsvektor der reflektierten Wellen

$$\mathfrak{f}_l' = (\sin \varphi_l; 0; \cos \varphi_l) \tag{3.6}$$

ist bereits das „Reflexionsgesetz" hineingesteckt worden, um ein späteres Zurückgreifen auf die Formeln zu erleichtern. Wird $\mathfrak{f}_l'$ mit einem beliebigen Winkel φ_l' angesetzt, so zeigt sich später doch, daß nur bei $\varphi_l' = \varphi_l$ die Grenzbedingungen zu befriedigen sind.

Jede der Gln. (3.3) und (3.4) repräsentiert sechs Gleichungen. $\mathfrak{A}_l$ steht dabei für die drei elektrischen Feldstärkekomponenten $\hat{\mathfrak{A}}_{lx}, \hat{\mathfrak{A}}_{ly}, \hat{\mathfrak{A}}_{lz}$ und die magnetischen Feldstärkekomponenten $\check{\mathfrak{A}}_{lx}, \check{\mathfrak{A}}_{ly}, \check{\mathfrak{A}}_{lz}$. Entsprechend steht $\mathfrak{A}_l^\circ$ für $\hat{\mathfrak{A}}_{lx}^\circ, \hat{\mathfrak{A}}_{ly}^\circ$ usw. Entsprechendes gilt für die $\mathfrak{R}$-Wellen.

4. Die Kopplung der Feldkomponenten. Die mannigfaltigen Polarisationsmöglichkeiten der einfallenden Welle $\mathfrak{A}_m$ kann man durch Addition aus zwei linear polarisierten Wellen entstanden denken. Wegen der Linearität der MAXWELLschen Gleichungen und der Grenzbedingungen, denen allein die Wellen — außer dem Ansatz (3.3) bis (3.6) — noch unterworfen werden, können wir uns begnügen, zwei Lösungen der Aufgabe zu suchen:

1. Lösung: Transversal-E-Wellen mit

$$\hat{\mathfrak{A}}_{lx} = \hat{\mathfrak{A}}_{lz} = \hat{\mathfrak{R}}_{lx} = \hat{\mathfrak{R}}_{lz} = 0, \tag{4.1}$$

bei denen die elektrische Feldstärke überall senkrecht auf der Einfallsebene, der x, z-Ebene, steht,;

2. *Lösung:* Transversal-H-Wellen, mit

$$\hat{\mathfrak{A}}_{lx} = \hat{\mathfrak{A}}_{lz} = \hat{\mathfrak{R}}_{lx} = \hat{\mathfrak{R}}_{lz} = 0, \tag{4.2}$$

bei denen die magnetische Feldstärke überall senkrecht auf der Einfallsebene, der x, z-Ebene, steht. Die allgemeine Lösung kann man dann durch Addition solcher Lösungen erhalten.

Für jede TE-Welle ist ausreichend charakteristisch ihre $\mathfrak{E}_y$-Komponente, da $\mathfrak{E}_x$, $\mathfrak{E}_z$ verschwinden und sich die $\mathfrak{H}$-Komponenten leicht aus $\mathfrak{E}_y$ berechnen lassen nach der Maxwellschen Gleichung; z. B.

$$-\frac{\mu}{c} \cdot \frac{\partial \mathfrak{H}}{\partial t} = -i\omega \frac{\mu}{c} \cdot \mathfrak{H} = \operatorname{rot} \mathfrak{E},$$

$$i\omega \frac{\mu}{c} \mathfrak{H}_x = \frac{\partial \mathfrak{E}_y}{\partial z}. \tag{4.3}$$

Für $\mathfrak{A}$-Wellen bzw. für $\mathfrak{R}$-Wellen folgt durch Differenzieren nach z an den Gln. (3.3) bzw. (3.4)

$$i\omega \frac{\mu_l}{c} \hat{\mathfrak{A}}_{lx} = i\omega \frac{\mathfrak{n}_l}{c} \cos \varphi_l \cdot \hat{\mathfrak{A}}_{ly},$$

$$i\omega \frac{\mu_l}{c} \hat{\mathfrak{R}}_{lx} = -i\omega \frac{\mathfrak{n}_l}{c} \cos \varphi_l \cdot \hat{\mathfrak{R}}_{ly},$$

$$\hat{\mathfrak{A}}_{lx} = \hat{g}_l \hat{\mathfrak{A}}_{ly} \quad \text{für } l = 0, \ldots, m \text{ bei } T\text{-}E\text{-Wellen} \tag{4.4}$$

$$\hat{\mathfrak{R}}_{lx} = -\hat{g}_l \hat{\mathfrak{R}}_{ly} \tag{4.5}$$

mit

$$\hat{g}_l = \frac{\mathfrak{n}_l \cdot \cos \varphi_l}{\mu_l}. \tag{4.6}$$

Für TH-Wellen ist entsprechend nach (2.1)

$$\mu \cdot \operatorname{rot} \mathfrak{H} = \frac{4\pi\sigma\mu}{c} \mathfrak{E} + i\omega \frac{\varepsilon\mu}{c} \mathfrak{E} = \frac{i\omega}{c} \mathfrak{E}\left(\varepsilon \cdot \mu - i \frac{4\pi\sigma\mu}{\omega}\right)$$

$$= i \frac{\omega}{c} \mathfrak{n}^2 \cdot \mathfrak{E},$$

$$i \frac{\omega}{c} \frac{\mathfrak{n}^2}{\mu} \mathfrak{E}_x = -\frac{\partial \mathfrak{H}_y}{\partial z}. \tag{4.7}$$

Für $\mathfrak{A}$-Wellen und $\mathfrak{R}$-Wellen folgt durch Differenzieren nach z an den Gln. (3.3) bzw. (3.4)

$$i \frac{\omega}{c} \cdot \frac{\mathfrak{n}_l^2}{\mu_l} \cdot \hat{\mathfrak{A}}_{lx} = -\hat{\mathfrak{A}}_{ly} \cdot i\omega \frac{\mathfrak{n}_l}{c} \cdot \cos \varphi_l,$$

$$i \frac{\omega}{c} \cdot \frac{\mathfrak{n}_l^2}{\mu_l} \cdot \hat{\mathfrak{R}}_{lx} = +\hat{\mathfrak{R}}_{ly} \cdot i\omega \cdot \frac{\mathfrak{n}_l}{c} \cdot \cos \varphi_l.$$

$$\hat{\mathfrak{A}}_{lx} = -\hat{g}_l \hat{\mathfrak{A}}_{ly} \tag{4.8}$$

$$\text{für } l = 0, \ldots, m \text{ bei } TH\text{-Wellen}$$

$$\hat{\mathfrak{R}}_{lx} = \hat{g}_l \hat{\mathfrak{R}}_{ly} \tag{4.9}$$

mit der Abkürzung

$$\hat{g}_l = \frac{\mu_l \cdot \cos \varphi_l}{\mathfrak{n}_l}. \tag{4.10}$$

5. Die Grenzbedingungen. Somit können die Grenzbedingungen für beide Tangentialkomponenten (y- und x-Komponente) angesetzt werden. Sie lauten für TE-Wellen an der Fläche $z = z_l$, gültig für $l = 1, \ldots, m$ für die elektrischen y-Komponenten

$$\hat{\mathfrak{A}}_{ly}(x; y; z_l) + \hat{\mathfrak{R}}_{ly}(x; y; z_l) = \hat{\mathfrak{A}}_{l-1,y}(x; y; z_l) + \hat{\mathfrak{R}}_{l-1,y}(x; y; z_l), \qquad (5.1)$$

für die $\mathfrak{H}$-Komponenten

$$\hat{\mathfrak{A}}_{lx}(x; y; z_l) + \hat{\mathfrak{R}}_{lx}(x; y; z_l) = \hat{\mathfrak{A}}_{l-1,x}(x; y; z_l) + \hat{\mathfrak{R}}_{l-1,x}(x; y; z_l). \qquad (5.2)$$

Setzt man in die Ansätze (3.3) und (3.4) ein, so zeigt sich, daß eine Befriedigung der Grenzbedingungen (5.1) für alle x nur möglich ist, wenn $\mathfrak{n}_l \sin \varphi_l = \mathfrak{n}_{l-1} \sin \varphi_{l-1}$ für $l = 1, \ldots, m$ d.h.

$$\mathfrak{n}_0 \sin \varphi_0 = \mathfrak{n}_1 \sin \varphi_1 = \cdots = \mathfrak{n}_m \sin \varphi_m \qquad (5.3)$$

(SNELLIUSsches Brechungsgesetz)[1]. Hätte man die $\mathfrak{R}$-Wellen mit einem von φ_l unabhängigen Winkel $\varphi_l{}'$ angesetzt, so hätte sich hier auch das Reflexionsgesetz $\varphi_l = \varphi_l{}'$ ergeben. Da alle $\mathfrak{n}_l$ und φ_m als gegeben angesehen werden, sind alle φ_l bekannt.

Mit der Abkürzung

$$\varrho_l = i \omega \frac{\mathfrak{n}_l}{c} \cos \varphi_l \qquad (5.4)$$

nehmen (5.1) und (5.2) die Form an: Grenzbedingungen für TE-Wellen (mit $l = 1, 2, \ldots, m$)

$$\hat{\mathfrak{A}}^{\circ}_{ly} \cdot e^{\varrho_l z_l} + \hat{\mathfrak{R}}^{\circ}_{ly} e^{-\varrho_l z_l} = \hat{\mathfrak{A}}^{\circ}_{l-1,y} e^{\varrho_{l-1} z_l} + \hat{\mathfrak{R}}^{\circ}_{l-1,y} e^{-\varrho_{l-1} z_l}, \qquad (5.5)$$

$$\hat{g}_l \hat{\mathfrak{A}}^{\circ}_{ly} e^{\varrho_l z_l} - \hat{g}_l \hat{\mathfrak{R}}^{\circ}_{ly} \cdot e^{-\varrho_l z_l} = \hat{g}_{l-1} \hat{\mathfrak{A}}^{\circ}_{l-1,y} e^{\varrho_{l-1} z_l} - \hat{g}_{l-1} \hat{\mathfrak{R}}^{\circ}_{l-1,y} e^{-\varrho_{l-1} z_l}. \qquad (5.6)$$

Für TH-Wellen gelten die Grenzbedingungen (5.1) und (5.2) wenn man $\frown$ und $\frown$ vertauscht. Da für TH und TE-Wellen die Ansätze (3.3) und (3.4) gleich lauten und die Gln. (4.8), (4.9) an die Stelle von (4.4), (4.5) rücken, gehen die Grenzbedingungen für TH-Wellen aus denen für TE-Wellen (5.5), (5.6) hervor, wenn alle $\frown$-Zeichen durch $\frown$ ersetzt werden.

Wir lassen daher in (5.5) und (5.6) die $\frown$-Zeichen fort und haben in

$$\mathfrak{A}^{\circ}_{ly} e^{\varrho_l z_l} + \mathfrak{R}^{\circ}_{ly} e^{-\varrho_l z_l} = \mathfrak{A}^{\circ}_{l-1,y} e^{\varrho_{l-1} z_l} + \mathfrak{R}^{\circ}_{l-1,y} e^{-\varrho_{l-1} z_l}, \qquad (5.7)$$

$$g_l \mathfrak{A}^{\circ}_{ly} e^{\varrho_l z_l} - g_l \mathfrak{R}^{\circ}_{ly} e^{-\varrho_l z_l} = g_{l-1} \mathfrak{A}^{\circ}_{l-1,y} e^{\varrho_{l-1} z_l} - g_{l-1} \mathfrak{R}^{\circ}_{l-1,y} e^{-\varrho_{l-1} z_l} \qquad (5.8)$$

die Grenzbedingungen für beide Wellentypen. Man kann durch Anbringen des $\frown$ bzw. $\frown$ im Resultat auf den jeweils gewünschten Wellentyp spezialisieren. Das lineare Gleichungssystem (5.7), (5.8) umfaßt $2m$ Gleichungen und enthält die $(2m+2)$ Größen $\mathfrak{A}^{\circ}_{ly}$, $\mathfrak{R}^{\circ}_{ly}$ (für $l = 0, \ldots, m$). Von diesen ist aber $\mathfrak{A}^{\circ}_{m}$ gegeben und $\mathfrak{R}^{\circ}_0 = 0$ vorausgesetzt. Nach den restlichen $2m$ Unbekannten kann man sofort nach der bekannten Determinantenmethode auflösen.

---

[1] Daraus folgt, daß z.B. bei reellem $\mathfrak{n}_0$, komplexem $\mathfrak{n}_1$ und reellem $\varphi_0 \neq 0$ jedenfalls φ_1 komplex sein muß. Die physikalische Bedeutung solcher „inhomogener" Wellen erhellt aus Gl. (2.6). Abschnitt III a bringt hierzu ein Beispiel.

6. Spezialfall einer Grenzfläche. Zum Beispiel für $m=1$ ist das Gleichungssystem (5.7), (5.8)

$$\mathfrak{A}^\circ_{1y}\, e^{\varrho_1 z_1} + \mathfrak{R}^\circ_{1y}\, e^{-\varrho_1 z_1} = \mathfrak{A}^\circ_{0y}\, e^{\varrho_0 z_1} + \mathfrak{R}^\circ_{0y}\, e^{-\varrho_0 z_1}, \tag{6.1}$$

$$g_1\, \mathfrak{A}^\circ_{1y}\, e^{\varrho_1 z_1} - g_1\, \mathfrak{R}^\circ_{1y}\, e^{-\varrho_1 z_1} = g_0 \cdot \mathfrak{A}^\circ_{0y}\, e^{\varrho_0 z_1} - g_0\, \mathfrak{R}^\circ_{0y}\, e^{-\varrho_0 z_1}. \tag{6.2}$$

Da $\mathfrak{R}^\circ_{0y} = 0$ und $\mathfrak{A}^\circ_{1y}$ gegeben, folgt durch Addition bzw. Subtraktion nach Multiplikation von (6.1) mit g_1 bzw. g_0

$$\frac{\mathfrak{A}_{0y}(0;0;z_1)}{\mathfrak{A}_{1y}(0;0;z_1)} = \frac{\mathfrak{A}^\circ_{0y}\, e^{\varrho_0 z_1}}{\mathfrak{A}^\circ_{1y} \cdot e^{\varrho_1 z_1}} = \mathfrak{d}^\circ\, \frac{e^{\varrho_0 z_1}}{e^{\varrho_1 z_1}} = \frac{2g_1}{g_1 + g_0}, \tag{6.3}$$

$$\frac{\mathfrak{R}_{0y}(0;0;z_1)}{\mathfrak{A}_{1y}(0;0;z_1)} = \frac{\mathfrak{R}^\circ_{1y}\, e^{-\varrho_1 z_1}}{\mathfrak{A}^\circ_{1y}\, e^{\varrho_1 z_1}} = \mathfrak{r}^\circ\, \frac{e^{-\varrho_1 z_1}}{e^{\varrho_1 z_1}} = \frac{g_1 - g_0}{g_1 + g_0}. \tag{6.4}$$

Der komplexe Durchlässigkeitskoeffizient $\mathfrak{d}^\circ$ und der komplexe Reflexionskoeffizient $\mathfrak{r}^\circ$ seien durch diese Gleichungen definiert. Das sind die Fresnelschen Formeln für eine Grenzfläche.

Einfach ist auch noch die Auflösung der Gln. (5.7) (5.8) für $m=2$ und $m=3$. Doch geschieht das übersichtlicher mit einer Rekursionsformel, die um der Vielfachschichten willen im nächsten Abschnitt ohnehin behandelt werden muß.

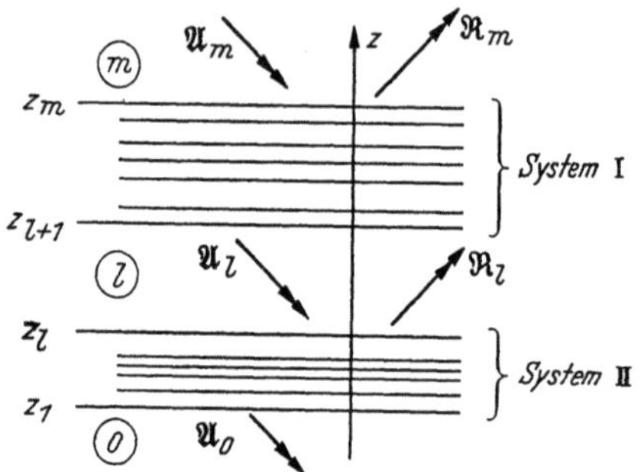

Fig. 2. Zwei hintereinandergesetzte Schichtsysteme.

c) Zwei hintereinandergesetzte Schichtsysteme.

7. Für das erste Schichtsystem I (siehe Fig. 2) für sich allein sei die Grundaufgabe, reflektierte und durchgelassene Amplitude aus der einfallenden zu berechnen, gelöst.

Wegen der Homogenität der Gln. (5.7), (5.8) liegt die Lösung dann in einer Form vor

$$\frac{\mathfrak{A}^\circ (\text{Ausfall})}{\mathfrak{A}^\circ (\text{Einfall})} = \mathfrak{d}^\circ_I \; ; \quad \frac{\mathfrak{R}^\circ (\text{Ausfall})}{\mathfrak{A}^\circ (\text{Einfall})} = \mathfrak{r}^\circ_I$$

Auch für von unten auf das System I auffallende Wellen sei das Problem gelöst und habe zu einer „Durchlässigkeit" $\mathfrak{d}'^\circ_I$ und einen Reflexionsfaktor $\mathfrak{r}'^\circ_I$ geführt. Ebenso sei die Lösung für das zweite Schichtsystem II mit einem Durchlässigkeitskoeffizienten $\mathfrak{d}^\circ_{II}$ und einem Reflexionskoeffizienten $\mathfrak{r}^\circ_{II}$ gegeben. Die ∘ an den Größen $\mathfrak{d}^\circ$, $\mathfrak{r}^\circ$ soll andeuten, daß alle Wellen nach dem Ansatz (3.3), (3.4) auf einen gemeinsamen Nullpunkt bezogen sein sollen und die $\mathfrak{d}^\circ$, $\mathfrak{r}^\circ$ Quotienten von Größen $\mathfrak{A}^\circ$ bzw. $\mathfrak{R}^\circ$ sind.

Nach Hintereinandersetzen der Schichtsysteme I und II in der Art nach Fig. 2 muß dann wegen der Addierbarkeit von Lösungen der linearen Maxwellschen Gleichungen und der ebenfalls linearen Grenzbedingungen sein

$$\mathfrak{R}^\circ_m = \mathfrak{A}^\circ_m \cdot \mathfrak{r}^\circ_I + \mathfrak{R}^\circ_l \cdot \mathfrak{d}'^\circ_I, \tag{7.1}$$

$$\mathfrak{A}^\circ_l = \mathfrak{A}^\circ_m \cdot \mathfrak{d}^\circ_I + \mathfrak{R}^\circ_l \cdot \mathfrak{r}'^\circ_I, \tag{7.2}$$

$$\mathfrak{R}^\circ_l = \mathfrak{A}^\circ_l \cdot \mathfrak{r}^\circ_{II}, \tag{7.3}$$

$$\mathfrak{A}^\circ_0 = \mathfrak{A}^\circ_l \cdot \mathfrak{d}^\circ_{II}. \tag{7.4}$$

Aus (7.2) und (7.3) folgt

$$\mathfrak{A}_l^\circ = \mathfrak{A}_m^\circ \, \mathfrak{d}_{\mathrm{I}}^\circ + \mathfrak{r}_{\mathrm{I}}^{\circ\prime} \mathfrak{r}_{\mathrm{II}}^\circ \cdot \mathfrak{A}_l^\circ;$$

d.h.

$$\mathfrak{A}_l^\circ = \mathfrak{A}_m^\circ \, \frac{\mathfrak{d}_{\mathrm{I}}^\circ}{1 - \mathfrak{r}_{\mathrm{I}}^{\circ\prime} \mathfrak{r}_{\mathrm{II}}^\circ}, \tag{7.5}$$

$$\mathfrak{R}_l^\circ = \mathfrak{A}_m^\circ \, \frac{\mathfrak{d}_{\mathrm{I}}^\circ \mathfrak{r}_{\mathrm{II}}^\circ}{1 - \mathfrak{r}_{\mathrm{I}}^{\circ\prime} \mathfrak{r}_{\mathrm{II}}^\circ}. \tag{7.6}$$

Einsetzen in (7.1) gibt

$$\frac{\mathfrak{R}_m^\circ}{\mathfrak{A}_m^\circ} = \mathfrak{r}_{\mathrm{I}}^\circ + \mathfrak{d}_{\mathrm{I}}^{\circ\prime} \, \frac{\mathfrak{d}_{\mathrm{I}}^\circ \mathfrak{r}_{\mathrm{II}}^\circ}{1 - \mathfrak{r}_{\mathrm{I}}^{\circ\prime} \mathfrak{r}_{\mathrm{II}}^\circ},$$

$$\mathfrak{r}_{\mathrm{I+II}}^\circ = \frac{\mathfrak{R}_m^\circ}{\mathfrak{A}_m^\circ} = \frac{\mathfrak{r}_{\mathrm{I}}^\circ + \mathfrak{r}_{\mathrm{II}}^\circ \left(\mathfrak{d}_{\mathrm{I}}^\circ \mathfrak{d}_{\mathrm{I}}^{\circ\prime} - \mathfrak{r}_{\mathrm{I}}^\circ \mathfrak{r}_{\mathrm{I}}^{\circ\prime}\right)}{1 - \mathfrak{r}_{\mathrm{I}}^{\circ\prime} \mathfrak{r}_{\mathrm{II}}^\circ} \dagger. \tag{7.7}$$

Einsetzen von (7.5) in (7.4) gibt

$$\mathfrak{d}_{\mathrm{I+II}}^\circ = \frac{\mathfrak{A}_0^\circ}{\mathfrak{A}_m^\circ} = \frac{\mathfrak{d}_{\mathrm{I}}^\circ \mathfrak{d}_{\mathrm{II}}^\circ}{1 - \mathfrak{r}_{\mathrm{I}}^{\circ\prime} \mathfrak{r}_{\mathrm{II}}^\circ} \dagger. \tag{7.8}$$

(7.7) und (7.8) geben die Lösung der Grundaufgabe für das aus den Systemen I und II aufgebaute System, wenn die Grundaufgabe für beide Systeme I und II für sich bereits gelöst ist. I und II können beide bereits komplizierte Systeme sein; davon wird bei der Betrachtung über Interferenzfilter in Abschnitt V Gebrauch gemacht werden. Man kann aber auch z.B. als I eine einfache Grenzfläche wählen, für die nach den FRESNELschen Formeln (6.3), (6.4) alles bekannt ist; dann erhält man eine Rekursionsformel, die aus der Lösung für $m-1$ Grenzflächen die Lösung für m Grenzflächen liefert. Diese Rekursionsformel ist der Gegenstand des folgenden Abschnitts. Allgemein gelten auch die Gln. (30.6) auf S. 501.

d) Lösung der Grundaufgabe durch Rekursion.

8. Hinzufügung einer weiteren Grenzfläche zu einem Schichtsystem. Als Schichtsystem I im Sinne des letzten Abschnitts wählen wir speziell eine Grenzfläche zwischen m-tem und $(m\text{-}1)$-tem Medium, wie Fig. 3 das andeutet. Für das System II mit $m-1$ Grenzflächen sei die Grundaufgabe durch Berechnung der Größen

$$\mathfrak{r}_{\mathrm{II}}^\circ = \mathfrak{r}_{m-1}^\circ; \qquad \mathfrak{d}_{\mathrm{II}}^\circ = \mathfrak{d}_{m-1}^\circ \tag{8.1}$$

bereits gelöst. Für das System I ist sie nach den sinngemäß anzuwendenden (schreibe m statt 1) Gln. (6.3) und (6.4) gelöst für die Größen:

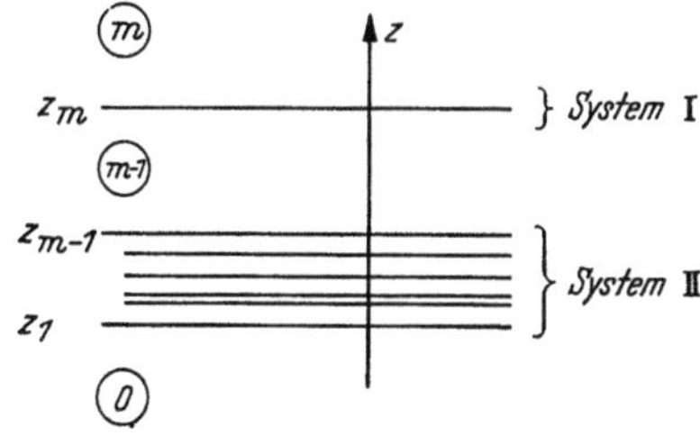

Fig. 3. Zur Ableitung der Rekursionsformeln.

$$\mathfrak{d}_{\mathrm{I}}^\circ = \frac{2 g_m}{g_m + g_{m-1}} \cdot \frac{e^{\varrho_m \, z_m}}{e^{\varrho_{m-1} z_m}}; \qquad \mathfrak{r}_{\mathrm{I}}^\circ = \frac{g_m - g_{m-1}}{g_m + g_{m-1}} \cdot \frac{e^{\varrho_m \, z_m}}{e^{-\varrho_m \, z_m}}. \tag{8.2}$$

$\dagger$ Für Spezialfälle zuerst abgeleitet von G. B. AIRY: Phil. Mag. (3) **2**, 20 (1833) und Pogg. Ann. **41**, 512 (1837). Die Klammer in (7.7) hat nicht allgemeine einfache Werte, wie in der Literatur gelegentlich durch Verallgemeinerung aus dem Fall, daß I nur eine Grenzfläche enthält, unterstellt wird.

Die entsprechenden gestrichenen Größen für Einfall von unten (Fig. 3) auf das System I findet man offenbar durch die Vorschrift: Vertausche die Mediennummern m und $m-1$, belasse jedoch die Grenzflächenkoordinate z_m unverändert, und ersetze jedes ϱ durch $-\varrho$ mit Rücksicht auf die Vertauschung der Rollen von $\mathfrak{A}$-Wellen mit $\mathfrak{R}$-Wellen vgl. (3.3), (3.4) und (5.4). Daher ist

$$\mathfrak{d}_{\mathrm{I}}^{\circ\,\prime} = \frac{2g_{m-1}}{g_{m-1}+g_m} \cdot \frac{e^{-\varrho_{m-1}\cdot z_m}}{e^{-\varrho_m z_m}}, \tag{8.3}$$

$$\mathfrak{r}_{\mathrm{I}}^{\circ\,\prime} = \frac{g_{m-1}-g_m}{g_{m-1}+g_m} \cdot \frac{e^{-\varrho_{m-1} z_m}}{e^{\varrho_{m-1} z_m}}. \tag{8.4}$$

Durch Einsetzen in (7.7) und (7.8) unter Berücksichtigung der Beziehung

$$\mathfrak{d}_{\mathrm{I}}^{\circ}\,\mathfrak{d}_{\mathrm{I}}^{\circ\,\prime} - \mathfrak{r}_{\mathrm{I}}^{\circ}\,\mathfrak{r}_{\mathrm{I}}^{\circ\,\prime} = \frac{4g_m g_{m-1}}{(g_m+g_{m-1})^2} e^{2\varrho_m z_m}\cdot e^{-2\varrho_{m-1} z_m} + \frac{(g_m-g_{m-1})^2}{(g_m+g_{m-1})^2}\cdot e^{2\varrho_m z_m}\cdot e^{-2\varrho_{m-1} z_m},$$

also

$$\mathfrak{d}_{\mathrm{I}}^{\circ}\,\mathfrak{d}_{\mathrm{I}}^{\circ\,\prime} - \mathfrak{r}_{\mathrm{I}}^{\circ}\,\mathfrak{r}_{\mathrm{I}}^{\circ\,\prime} = e^{2(\varrho_m-\varrho_{m-1})z_m}, \tag{8.5}$$

folgt

$$\mathfrak{r}_m^{\circ} = \frac{\mathfrak{R}_m^{\circ}}{\mathfrak{A}_m^{\circ}} = \frac{\dfrac{g_m-g_{m-1}}{g_m+g_{m-1}}\cdot e^{2\varrho_m z_m} + \mathfrak{r}_{m-1}^{\circ}\,e^{2(\varrho_m-\varrho_{m-1})z_m}}{1 + \dfrac{g_m-g_{m-1}}{g_m+g_{m-1}}\cdot \mathfrak{r}_{m-1}^{\circ}\cdot e^{-2\varrho_{m-1} z_m}}, \tag{8.6}$$

$$\mathfrak{d}_m^{\circ} = \frac{\mathfrak{A}_0^{\circ}}{\mathfrak{A}_m^{\circ}} = \frac{\dfrac{2g_m}{g_m+g_{m-1}}\cdot e^{(\varrho_m-\varrho_{m-1})z_m}\cdot \mathfrak{d}_{m-1}^{\circ}}{1 + \dfrac{g_m-g_{m-1}}{g_m+g_{m-1}}\cdot \mathfrak{r}_{m-1}^{\circ}\cdot e^{-2\varrho_{m-1} z_m}}, \tag{8.7}$$

d.h.

$$\left.\begin{aligned}
\mathfrak{d}_m^{\circ}\cdot e^{\varrho_0 z_1-\varrho_m z_m} &= \frac{\mathfrak{A}_0^{\circ}\cdot e^{\varrho_0 z_1}}{\mathfrak{A}_m^{\circ}\cdot e^{\varrho_m z_m}} = \frac{\mathfrak{A}_0(0;0;z_1)}{\mathfrak{A}_m(0;0;z_m)} \\[2ex]
&= \frac{2g_m\cdot \mathfrak{d}_{m-1}^{\circ}\cdot e^{\varrho_0 z_1-\varrho_{m-1} z_m}}{(g_m+g_{m-1}) + (g_m-g_{m-1})\,\mathfrak{r}_{m-1}^{\circ}\,e^{-2\varrho_{m-1} z_m}}
\end{aligned}\right\} \tag{8.8}$$

$$\left.\begin{aligned}
\mathfrak{r}_m^{\circ}\cdot e^{-2\varrho_m z_m} &= \frac{\mathfrak{R}_m^{\circ}\cdot e^{-\varrho_m z_m}}{\mathfrak{A}_m^{\circ}\cdot e^{\varrho_m z_m}} = \frac{\mathfrak{R}_m(0;0;z_m)}{\mathfrak{A}_m(0;0;z_m)}, \\[2ex]
&= \frac{(g_m-g_{m-1}) + (g_m+g_{m-1})\,\mathfrak{r}_{m-1}^{\circ}\,e^{-2\varrho_{m-1} z_m}}{(g_m+g_{m-1}) + (g_m-g_{m-1})\,\mathfrak{r}_{m-1}^{\circ}\,e^{-2\varrho_{m-1} z_m}}.
\end{aligned}\right\} \tag{8.9}$$

Das sind die Rekursionsformeln, die das Problem für m Grenzflächen auf das für $(m-1)$ Grenzflächen und so schließlich auf das mit einer Grenzfläche, das mit Gl. (6.3) und (6.4) gelöst war, zurückführen.

9. Die Rekursionsformeln in koordinatenfreier Formulierung. Aus den Rekursionsformeln verschwinden noch die Koordinaten z_i der Grenzflächen, und es treten statt dessen die vom Koordinatensystem unabhängigen Schichtdicken d_i auf, wenn wir statt der auf den Koordinatenursprung bezogenen Wellen $\mathfrak{A}_i^{\circ}$, $\mathfrak{R}_i^{\circ}$ die an den Endflächen herrschende Wellenamplitude durchweg benutzen und entsprechende Durchlässigkeits- und Reflexionsfaktoren

$$\mathfrak{d}_m = \mathfrak{d}_m^\circ \cdot e^{\varrho_0 z_1 - \varrho_m z_m} = \frac{\mathfrak{A}_0(0;0;z_1)}{\mathfrak{A}_m(0;0;z_m)}, \tag{9.1}$$

$$\mathfrak{d}_{m-1} = \mathfrak{d}_{m-1}^\circ e^{\varrho_0 z_1 - \varrho_{m-1} z_{m-1}} = \frac{\mathfrak{A}_0(0;0;z_1)}{\mathfrak{A}_{m-1}(0;0;z_{m-1})} \tag{9.2}$$

usw. und

$$\mathfrak{r}_m = \mathfrak{r}_m^\circ e^{-2\varrho_m z_m} = \frac{\mathfrak{R}_m(0;0;z_m)}{\mathfrak{A}_m(0;0;z_m)}, \tag{9.3}$$

$$\mathfrak{r}_{m-1} = \mathfrak{r}_{m-1}^\circ e^{-2\varrho_{m-1} z_{m-1}} = \frac{\mathfrak{R}_{m-1}(0;0;z_{m-1})}{\mathfrak{A}_{m-1}(0;0;z_{m-1})} \tag{9.4}$$

usw. verwenden.

Damit gehen (8.8) und (8.9) über in:

$$\mathfrak{r}_m = \frac{(g_m - g_{m-1})\, e^{\varrho_{m-1} d_{m-1}} + (g_m + g_{m-1})\, \mathfrak{r}_{m-1}\, e^{-\varrho_{m-1} d_{m-1}}}{(g_m + g_{m-1})\, e^{\varrho_{m-1} d_{m-1}} + (g_m - g_{m-1})\, \mathfrak{r}_{m-1}\, e^{-\varrho_{m-1} d_{m-1}}}, \tag{9.5}$$

$$\mathfrak{d}_m = \frac{2 g_m \cdot \mathfrak{d}_{m-1}}{(g_m + g_{m-1})\, e^{\varrho_{m-1} d_{m-1}} + (g_m - g_{m-1})\, \mathfrak{r}_{m-1}\, e^{-\varrho_{m-1} d_{m-1}}}. \tag{9.6}$$

Diese Rekursionsformel wird schon für $m=1$ sinnvoll, wenn wir für Vorliegen keiner Grenzfläche $\mathfrak{r}_0 = 0$ und $\mathfrak{d}_0 = 1$ festsetzen. Dann folgt nämlich für $m=1$

$$\mathfrak{r}_1 = \frac{g_1 - g_0}{g_1 + g_0}; \quad \mathfrak{d}_1 = \frac{2 g_1}{g_1 + g_0}, \tag{9.7}$$

also die FRESNELschen Formeln (6.3) und (6.4). Für $m=2$, also ein Schichtsystem mit zwei Grenzflächen ist

$$\mathfrak{r}_2 = \frac{(g_2 - g_1)\, e^{\varrho_1 d_1} + (g_2 + g_1)\, e^{-\varrho_1 d_1} \cdot \dfrac{g_1 - g_0}{g_1 + g_0}}{(g_2 + g_1)\, e^{\varrho_1 d_1} + (g_2 - g_1)\, e^{-\varrho_1 d_1} \cdot \dfrac{g_1 - g_0}{g_1 + g_0}}, \tag{9.8}$$

$$\mathfrak{d}_2 = \frac{2 g_2 \cdot \dfrac{2 g_1}{g_1 + g_0}}{(g_2 + g_1)\, e^{\varrho_1 d_1} + (g_2 - g_1)\, e^{-\varrho_1 d_1} \cdot \dfrac{g_1 - g_0}{g_1 + g_0}}. \tag{9.9}$$

Bezeichnet man

$$\mathfrak{d}_l = \frac{2 g_l \cdot 2 g_{l-1} \cdots 2 g_1}{N_l}, \tag{9.10}$$

$$\mathfrak{r}_l = \frac{Z_l}{N_l}, \tag{9.11}$$

so gehen die Rekursionsformeln (9.5), (9.6) über in

$$\mathfrak{d}_m = \frac{2 g_m \cdot 2 g_{m-1} \cdots 2 g_1}{(g_m + g_{m-1})\, e^{\varrho_{m-1} d_{m-1}}\, N_{m-1} + (g_m - g_{m-1})\, e^{-\varrho_{m-1} d_{m-1}}\, Z_{m-1}}, \tag{9.12}$$

$$\mathfrak{r}_m = \frac{(g_m - g_{m-1})\, e^{\varrho_{m-1} d_{m-1}} \cdot N_{m-1} + (g_m + g_{m-1})\, e^{-\varrho_{m-1} d_{m-1}}\, Z_{m-1}}{(g_m + g_{m-1})\, e^{\varrho_{m-1} d_{m-1}} \cdot N_{m-1} + (g_m - g_{m-1})\, e^{-\varrho_{m-1} d_{m-1}}\, Z_{m-1}}, \tag{9.13}$$

und die auf Z_l und N_l bezogenen Rekursionsformeln werden

$$Z_m = (g_m - g_{m-1})\, e^{\varrho_{m-1} d_{m-1}}\, N_{m-1} + (g_m + g_{m-1})\, e^{-\varrho_{m-1} d_{m-1}}\, Z_{m-1}, \tag{9.14}$$

$$N_m = (g_m + g_{m-1})\, e^{\varrho_{m-1} d_{m-1}}\, N_{m-1} + (g_m - g_{m-1})\, e^{-\varrho_{m-1} d_{m-1}}\, Z_{m-1}, \tag{9.15}$$

Tabelle 1.

Zahl der Grenz-flächen m	$\mathfrak{r}_m = \dfrac{Z_m}{N_m}$	$\mathfrak{d}_m = \dfrac{2^m}{N_m} \prod\limits_{\nu=1}^{m} g_\nu$
1	$\mathfrak{r}_1 = \dfrac{g_1 - g_0}{g_1 + g_0}$	$\mathfrak{d}_1 = \dfrac{2g_1}{g_1 + g_0}$
2	$\mathfrak{r}_2 = \dfrac{(g_2 - g_1)(g_1 + g_0)\,e^{\varrho_1 d_1} + (g_2 + g_1)(g_1 - g_0)\,e^{-\varrho_1 d_1}}{(g_2 + g_1)(g_1 + g_0)\,e^{\varrho_1 d_1} + (g_2 - g_1)(g_1 - g_0)\,e^{-\varrho_1 d_1}}$	$\mathfrak{d}_2 = \dfrac{2g_2 \cdot 2g_1}{N_2}$
3	$\mathfrak{r}_3 = \dfrac{(g_3-g_2)\left\{(g_2+g_1)(g_1+g_0)\,e^{\varrho_1 d_1} + (g_2-g_1)(g_1-g_0)\,e^{-\varrho_1 d_1}\right\}e^{\varrho_2 d_2} + (g_3+g_2)\left\{(g_2-g_1)(g_1+g_0)\,e^{\varrho_1 d_1} + (g_2+g_1)(g_1-g_0)\,e^{-\varrho_1 d_1}\right\}e^{-\varrho_2 d_2}}{(g_3+g_2)\left\{(g_2+g_1)(g_1+g_0)\,e^{\varrho_1 d_1} + (g_2-g_1)(g_1-g_0)\,e^{-\varrho_1 d_1}\right\}e^{\varrho_2 d_2} + (g_3-g_2)\left\{(g_2-g_1)(g_1+g_0)\,e^{\varrho_1 d_1} + (g_2+g_1)(g_1-g_0)\,e^{-\varrho_1 d_1}\right\}e^{-\varrho_2 d_2}}$	$\mathfrak{d}_3 = \dfrac{2g_3 \cdot 2g_2 \cdot 2g_1}{N_3}$
4	$$\begin{aligned} Z_4 = (g_4 - g_3)\Big[&(g_3 + g_2)\left\{(g_2 + g_1)(g_1 + g_0)\,e^{\varrho_1 d_1} + (g_2 - g_1)(g_1 - g_0)\,e^{-\varrho_1 d_1}\right\}e^{\varrho_2 d_2} + \\ &+ (g_3 - g_2)\left\{(g_2 - g_1)(g_1 + g_0)\,e^{\varrho_1 d_1} + (g_2 + g_1)(g_1 - g_0)\,e^{-\varrho_1 d_1}\right\}e^{-\varrho_2 d_2}\Big]e^{\varrho_3 d_3} \\ + (g_4 + g_3)\Big(&(g_3 - g_2)\left\{(g_2 + g_1)(g_1 + g_0)\,e^{\varrho_1 d_1} + (g_2 - g_1)(g_1 - g_0)\,e^{-\varrho_1 d_1}\right\}e^{-\varrho_2 d_2} + \\ &+ (g_3 + g_2)\left\{(g_2 - g_1)(g_1 + g_0)\,e^{\varrho_1 d_1} + (g_2 + g_1)(g_1 - g_0)\,e^{-\varrho_1 d_1}\right\}\cdot e^{-\varrho_2 d_2}\Big)e^{-\varrho_3 d_3} \end{aligned}$$ $$N_4 = (g_4 + g_3)\,[\ \ ]\cdot e^{\varrho_3 d_3} + (g_4 - g_3)\,(\ \ )\cdot e^{-\varrho_3 d_3}$$	$\mathfrak{d}_4 = \dfrac{16 g_4 g_3 g_2 g_1}{N_4}$

Der Inhalt der Klammern [ ] und ( ) ist derselbe wie bei Z_4.

die zusammen mit (9.10), (9.11) in einfacher Weise die Reflexionsfaktoren und Durchlässigkeitsfaktoren berechnen lassen. Tabelle 1 zeigt die Ergebnisse für Systeme von 1 bis 4 Grenzflächen.

Zur Anwendung der Tabelle 1. Die Formeln erfassen zugleich Transversal-E-Wellen, bei denen die elektrische Feldstärke senkrecht zur Einfallsebene steht und Transversal-H-Wellen, bei denen die magnetische Feldstärke senkrecht zur Einfallsebene steht. Die g_l haben für beide Wellentypen verschiedene Bedeutung:

$$TE\text{-Wellen:}\quad \text{Setze}\quad g_l = \hat{g}_l = \frac{\mathfrak{n}_l \cos \varphi_l}{\mu_l},$$

$$TH\text{-Wellen:}\quad \text{Setze}\quad g_l = \hat{g}_l = \frac{\mu_l \cos \varphi_l}{\mathfrak{n}_l}$$

($\mathfrak{n}_l =$ Brechungsindex, $\mu_l =$ Permeabilität, $\varphi_l =$ Winkel zwischen „Fortpflanzungsrichtung der Welle" im l-ten Medium und der Flächennormalen).

$\mathfrak{r}$ und $\mathfrak{d}$ bedeuten Quotienten der zur Einfallsebene senkrechten komplexen Feldstärken, bei TE-Wellen der elektrischen und bei TH-Wellen der magnetischen. $\mathfrak{r}$ ist Quotient aus reflektierter und einfallender Feldstärke, beide am gleichen Punkt der von den Wellen erreichten Grenzfläche genommen. $\mathfrak{d}$ ist Quotient aus durchgegangener und einfallender Feldstärke, beide an zwei von den Wellen erreichten Punkten der letzten bzw. der ersten Grenzfläche des Schichtsystems genommen; die Gerade durch beide Punkte steht senkrecht zu den Grenzflächen. Intensitätsverhältnisse für Durchlaß und Reflexion findet man nach Gl. (2.18), wenn einfallende und resultierende Welle homogen sind,

für TE-Wellen nach

$$J_r = \frac{I_{\text{refl}}}{I_{\text{einf}}} = |\hat{\mathfrak{r}}|^2; \quad J_d = \frac{I_{\text{durchg}}}{I_{\text{einf}}} = \frac{n_0 \cdot \mu_m}{n_m \cdot \mu_0} \cdot |\hat{\mathfrak{d}}|^2, \tag{9.16}$$

für TH-Wellen nach

$$J_r = \frac{I_{\text{refl}}}{I_{\text{einf}}} = |\hat{\mathfrak{r}}|^2; \quad J_d = \frac{I_{\text{durchg}}}{I_{\text{einf}}} = \frac{\mu_0 \cdot n_0}{|n_0|^2} \cdot \frac{|\mathfrak{n}_m|^2}{\mu_m \cdot n_m} \cdot |\hat{\mathfrak{d}}|^2. \tag{9.17}$$

III. Reflexionsfreie Schichtsysteme und linearpolarisierend reflektierende Schichtsysteme.

a) Reflexionsfreiheit bei einer Grenzfläche (Brewster-Zenneck-Fall).

10. Allgemeine Lösung. Die Grenzfläche zwischen zwei verschiedenen Medien reflexionsfrei zu machen, ist eine im Zusammenhang mit der angewandten Optik schon von H. D. Taylor[1] 1896 gestellte Aufgabe.

Bevor wir ihre Lösung betrachten, sei in dieser Ziffer die Frage beantwortet, unter welchen Umständen eine einzige Grenzfläche zwischen verschiedenen Medien von sich aus bereits reflexionsfrei ist. Nach Tabelle 1 auf S. 472 ist das der Fall, wenn $g_1 = g_0$ ist. Das bedeutet

$$\text{für } TE\text{-Wellen}\quad \frac{\mathfrak{n}_1}{\mu_1} \cdot \cos \varphi_1 = \frac{\mathfrak{n}_0}{\mu_0} \cos \varphi_0, \tag{10.1}$$

$$\text{für } TH\text{-Wellen}\quad \frac{\mu_1}{\mathfrak{n}_1} \cdot \cos \varphi_1 = \frac{\mu_0}{\mathfrak{n}_0} \cdot \cos \varphi_0. \tag{10.2}$$

---

[1] H. D. Taylor: The adjustment and testing of telescope objectivs. New York: T. Cook 1896. Die ersten Entdeckungen, die sich auf diesen Gegenstand beziehen, machte bereits Fraunhofer. Einen ausführlichen geschichtlichen Überblick gibt H. Mayer: Physik dünner Schichten, Bd. 1, S. 240ff.

Hinzu kommt als Nebenbedingung das Brechungsgesetz (5.3)

$$\mathfrak{n}_1 \cdot \sin \varphi_1 = \mathfrak{n}_0 \cdot \sin \varphi_0. \tag{10.3}$$

$\alpha)$ Für TE-Wellen folgt durch Division von (10.3) durch (10.1)

$$\mu_1 \cdot \tan \varphi_1 = \mu_0 \cdot \tan \varphi_0. \tag{10.4}$$

Ist $\mu_1 = \mu_0 = 1$, wie praktisch in den meisten Fällen, so folgt $\varphi_1 = \varphi_0$ und $\mathfrak{n}_1 = \mathfrak{n}_0$, also keine außertriviale Lösung. Sind μ_1 und μ_0 wenig verschieden voneinander, so weicht die Lösung wenig von der trivialen ab.

$\beta)$ Für TH-Wellen folgt durch Multiplikation von (10.2) mit (10.3)

$$\mu_1 \cdot \sin 2\varphi_1 = \mu_0 \cdot \sin 2\varphi_0. \tag{10.5}$$

Für $\mu_1 = \mu_0 = 1$, den praktisch wichtigsten Fall, hat das außer der trivialen Lösung $\varphi_1 = \varphi_0$ (d.h. $\mathfrak{n}_1 = \mathfrak{n}_0$) die Lösung

$$\varphi_1 = \frac{\pi}{2} - \varphi_0; \tag{10.6}$$

nach (10.2) und (10.3) heißt das

$$\tan \varphi_1 = \cot \varphi_0 = \frac{\mathfrak{n}_0}{\mathfrak{n}_1}. \tag{10.7}$$

11. Beide Medien Nichtleiter (Brewster-Fall). Sind beide Medien Nichtleiter, also $\mathfrak{n}_1$ und $\mathfrak{n}_0$ reell, so spricht man vom „Brewster-Fall", der „Einfallswinkel" φ_1 ist reell und heißt „Brewsterscher Winkel". (10.6) bedeutet das bekannte Senkrechtstehen des reflektierten „Strahls" auf dem durchgehenden Strahl. Für diesen Einfallswinkel

$$\varphi_1 = \arctan \frac{\mathfrak{n}_0}{\mathfrak{n}_1}. \tag{11.1}$$

wird nur die TE-Welle reflektiert; das reflektierte Licht ist also linear polarisiert, wenn natürliches Licht einfällt.

12. Ein Medium leitend (Zenneck-Fall). Ist das Medium ①, aus dem die Welle einfällt, nichtleitend, Medium ⓪ leitend, so sind φ_1 und φ_0 nach Gl. (10.7) komplex. Man spricht von der Zenneck-Welle[1]. Aus den zur x, z-Einfallsebene senkrechten magnetischen Feldstärken, die nach (3.3) lauten

$$\hat{\mathfrak{A}}_{ly}(x; y; z) = \hat{\mathfrak{A}}_{ly}^{\circ} \cdot e^{-i\frac{\omega}{c}\mathfrak{n}_l(x \sin \varphi_l - z \cos \varphi_l)} \tag{12.1}$$

(für $l = 1, 0$), erhält man für die elektrische x-Komponente nach den Maxwellschen Gleichungen oder direkt aus Gl. (4.8) mit (4.10)

$$\hat{\mathfrak{A}}_{1x} = -\frac{\mu_1 \cos \varphi_1}{\mathfrak{n}_1} \hat{\mathfrak{A}}_{1y}, \tag{12.2}$$

$$\hat{\mathfrak{A}}_{0x} = -\frac{\mu_0 \cos \varphi_0}{\mathfrak{n}_0} \hat{\mathfrak{A}}_{0y} = \hat{\mathfrak{A}}_{1x} \tag{12.3}$$

unter Berücksichtigung der Bedingung für Reflexfreiheit (10.2). Da nach der Maxwellschen Gl. (4.7) stets

$$\mu \cdot \operatorname{rot} \mathfrak{H} = i \frac{\omega}{c} \mathfrak{n}^2 \mathfrak{E} \tag{12.4}$$

---

[1] J. Zenneck: Ann. Phys. **23**, 846 (1907).

ist, so folgt für die $\mathfrak{E}_z$-Komponenten

$$\hat{\mathfrak{A}}_{lz} = \frac{\mu_l}{i\,\dfrac{\omega}{c}\,\mathfrak{n}_l^2}\cdot\frac{\partial\hat{\mathfrak{A}}_{ly}}{\partial x} = -\frac{\mu_l}{\mathfrak{n}_l}\sin\varphi_l\,\hat{\mathfrak{A}}_{lv}. \tag{12.5}$$

Aus (10.7), (12.2), (12.3) und (12.5) folgt für das komplexe Amplitudenverhältnis der elektrischen Feldstärkenkomponenten

$$\frac{\hat{\mathfrak{A}}_{1z}}{\hat{\mathfrak{A}}_{1x}} = \tan\varphi_1 = \frac{\mathfrak{n}_0}{\mathfrak{n}_1} = \cot\varphi_0 = \frac{\hat{\mathfrak{A}}_{0x}}{\hat{\mathfrak{A}}_{0z}}. \tag{12.6}$$

Da $\mathfrak{n}_0$ komplex ist, schwingen also $\hat{\mathfrak{A}}_{lz}$ und $\hat{\mathfrak{A}}_{lx}$ nicht in Phase; der Endpunkt der $\mathfrak{E}$-Vektoren beschreibt in beiden Medien einander ähnliche Ellipsen, wie die Fig. 4 sie für einen praktischen Fall mit $\mathfrak{n}_1 = 1$; $\mathfrak{n}_0 = 4 - 3i$ und für Punkte nahe der Grenzfläche zeigt. Das Seitenverhältnis des der Ellipse koordinatenparallel umschriebenen Rechtecks ist bei $\mathfrak{n}_1 = 1$

$$\left|\frac{\hat{\mathfrak{A}}_{1z}}{\hat{\mathfrak{A}}_{1x}}\right| = \left|\frac{\hat{\mathfrak{A}}_{0x}}{\hat{\mathfrak{A}}_{0z}}\right| = |\mathfrak{n}_0| = \sqrt{n_0^2 + k_0^2} \tag{12.7}$$

(im Beispiel gleich 5). Die Phasenverschiebung ist nach (12.6)

$$\text{arc}\,\frac{\hat{\mathfrak{A}}_{1z}}{\hat{\mathfrak{A}}_{1x}} = \text{arc}\,\frac{\hat{\mathfrak{A}}_{0x}}{\hat{\mathfrak{A}}_{0z}} = \text{arc}\,(n_0 - i\,k_0) = -\,\text{arc}\tan\frac{k_0}{n_0} \tag{12.8}$$

(im Beispiel gleich $-\text{arc}\tan 0{,}75 \approx -37°$).

Die ZENNECK-Welle hat unter anderem Bedeutung für die Ausbreitung elektromagnetischer Wellen über leitendem Erdboden. Näheres findet man in J. ZEN-NECKs Originalarbeit[1] und z.B. bei VILBIG[2].

Die Fig. 4 zeigt ein Beispiel für Wellen mit „komplexen Winkeln" gegen die Koordinatenachse, wie sie wegen des Brechungsgesetzes (5.3) bei Beteiligung leitender Medien meist auftreten. Mindestens eine der Feldstärken, die elektrische oder die magnetische, hat dann nicht mehr die Eigenschaft, zu einer Richtung transversal zu liegen. Die von Nichtleitern geläufigsten Vorstellungen der Transversalwellen sind dann zu modifizieren.

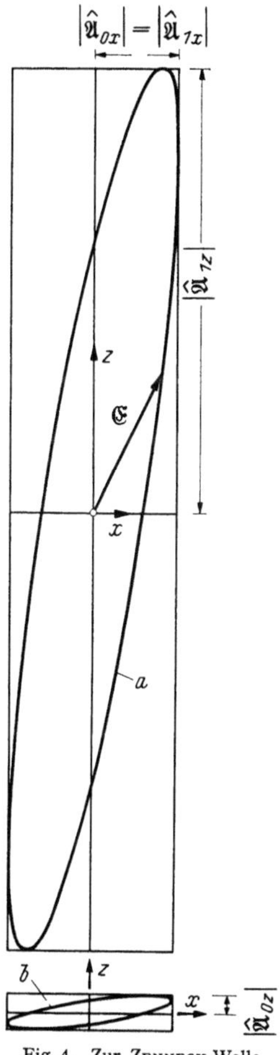

Fig. 4. Zur ZENNECK-Welle. a) Schwingungsellipse der elektrischen Feldstärke im Medium ①, aus dem die Welle einfällt (z.B. Luft); b) dasselbe im Medium ② (z.B. Erdboden).

b) Reflexfreiheit und Polarisation bei zwei Grenzflächen.

13. Die allgemeinen Bedingungen für Reflexfreiheit. Die Grenze zwischen zwei Medien ② und ⓪, z.B. Luft und Glas, kann man reflexionsfrei machen durch Einfügen eines Zwischenmediums ①, dessen optische Konstanten und dessen Dicke passend bemessen werden. Den entscheidenden praktischen Erfolg brachte das Überziehen der Glasoberfläche mit einer dünnen Schicht; diese

[1] J. ZENNECK: Ann. Phys. **23**, 846 (1907).
[2] F. VILBIG: Lehrbuch der Hochfrequenztechnik, Bd. I, 288—293, Leipzig 1942.

Methode, optische Systeme wie Ferngläser oder photographische Objektive zu „vergüten", d.h. von störenden und lichtschwächenden Reflexen zu befreien, wurde von A. SMAKULA in den Werken Carl Zeiss zur technischen Reife entwickelt[1].

Die erforderlichen Daten der Zwischenschicht müssen nach Tabelle 1 so eingerichtet werden, daß $r_2 = 0$ ist, d.h.

$$(g_2 - g_1)(g_1 + g_0) \cdot e^{\varrho_1 d_1} + (g_2 + g_1)(g_1 - g_0) \cdot e^{-\varrho_1 d_1} = 0. \tag{13.1}$$

Diese Bedingung für Reflexfreiheit schreiben wir auch

$$\frac{g_2 - g_1}{g_2 + g_1} \cdot \frac{g_1 + g_0}{g_1 - g_0} = - e^{-2\varrho_1 d_1} ; \tag{13.2}$$

darin bedeutet

$$\varrho_1 = i \cdot \frac{2\pi}{\lambda} \cdot \mathfrak{n}_1 \cos \varphi_1 , \tag{13.3}$$

d_1 die Schichtdicke des Mediums ①, und für die g_i ist zu setzen:

$$\text{bei } TE\text{-Wellen} \quad \hat{g}_i = \frac{\mathfrak{n}_i}{\mu_i} \cos \varphi_i , \tag{13.4}$$

$$\text{bei } TH\text{-Wellen} \quad \hat{g}_i = \frac{\mu_i}{\mathfrak{n}_i} \cos \varphi_i . \tag{13.5}$$

Die Bedingung (13.1) kann auch wie folgt umgeformt werden:

$$\frac{g_2 - g_1}{g_2 + g_1} = \frac{(g_0 - g_1)\, e^{-\varrho_1 d_1}}{(g_0 + g_1)\, e^{+\varrho_1 d_1}} . \tag{13.6}$$

Korrespondierende Subtraktion und Addition gibt

$$\frac{g_2}{g_1} = \frac{g_0 \operatorname{Cos} \varrho_1 d_1 + g_1 \operatorname{Sin} \varrho_1 d_1}{g_0 \operatorname{Sin} \varrho_1 d_1 + g_1 \operatorname{Cos} \varrho_1 d_1} , \tag{13.7}$$

$$\frac{g_2}{g_0} = \frac{1 + \dfrac{g_1}{g_0} \operatorname{Tan} \varrho_1 d_1}{1 + \dfrac{g_0}{g_1} \operatorname{Tan} \varrho_1 d_1} . \tag{13.8}$$

Gl. (13.6) geht in sich über, wenn g_2 mit g_0 und ϱ_1 mit $-\varrho_1$ vertauscht wird; daher folgt aus (13.8) auch die gleichwertige Bedingung

$$\frac{g_0}{g_2} = \frac{1 - \dfrac{g_1}{g_2} \operatorname{Tan} \varrho_1 d_1}{1 - \dfrac{g_2}{g_1} \operatorname{Tan} \varrho_1 d_1} , \tag{13.9}$$

d.h.

$$\operatorname{Tan} \varrho_1 d_1 = \frac{g_1(g_2 - g_0)}{g_1^2 - g_2 g_0} . \tag{13.9a}$$

Wegen der Bedeutung von ϱ_1 nach Gl. (13.3) ist

$$\operatorname{Tan} \varrho_1 d_1 = \frac{\operatorname{Tan}\left(\beta_1 \dfrac{2\pi d_1}{\lambda}\right) + i \tan\left(\alpha_1 \dfrac{2\pi d_1}{\lambda}\right)}{1 + i \operatorname{Tan}\left(\beta_1 \dfrac{2\pi d_1}{\lambda}\right) \tan\left(\alpha_1 \dfrac{2\pi d_1}{\lambda}\right)} , \tag{13.10}$$

---

[1] A. SMAKULA, D.R.P. 685767 vom 1. 11. 1935 der Firma Carl Zeiss, Jena. Eine kurze historische Darstellung gibt A. SMAKULA in Z. Instrumentenkde. **60**, 33 (1940), eine ausführlichere H. MAYER auf S. 240ff. des 1. Bandes seines Buches.

mit reellen Größen α_1 und β_1, die definiert sind durch

$$\alpha_1 - i\beta_1 = \mathfrak{n}_1 \cos\varphi_1 = \sqrt{\mathfrak{n}_1^2 - \mathfrak{n}_2^2 \sin^2\varphi_2}\,, \tag{13.11}$$

d.h. für reelles $\mathfrak{n}_2$ und homogene einfallende Welle (also φ_2 reell) durch

$$\alpha_1^2 - \beta_1^2 = n_1^2 - k_1^2 - n_2^2 \sin^2\varphi_l \quad \text{(abgekürzt als } 2D) \tag{13.12}$$

$$\alpha_1\beta_1 = n_1 k_1 \tag{13.13}$$

oder

$$\alpha_1 = \sqrt{\sqrt{D^2 + (n_1 k_1)^2} + D}\,, \tag{13.14}$$

$$\beta_1 = \sqrt{\sqrt{D^2 + (n_1 k_1)^2} - D}\,. \tag{13.15}$$

Jede der gleichwertigen Bedingungsgleichungen für Reflexfreiheit (13.2), (13.8), (13.9) ist zwei reellen Bedingungsgleichungen äquivalent. Um beide zu befriedigen, stehen sogar bei festgelegtem g_2, g_0, also vorgegebenen Endmedien und vorgegebenem Einfallswinkel noch drei Variable n_1, k_1, d_1, also die Daten der Schicht zur Verfügung. Es gibt daher prinzipiell eine Lösungsmannigfaltigkeit von der Mächtigkeit des Kontinuums. Wenn die Möglichkeiten praktisch auch durch die endliche Anzahl von Stoffen für das Medium ① eingeschränkt sind, so existiert doch auch praktisch eine so große Zahl von Lösungen — insbesondere, wenn man noch den Einfallswinkel φ_2 für Polarisationseffekte variiert — daß hier nur Beispiele aufgeführt werden sollen.

14. Reflexfreiheit und lineare Polarisation, wenn alle Medien Nichtleiter sind[1]. Sind alle Brechungsindices $\mathfrak{n}_2, \mathfrak{n}_1, \mathfrak{n}_0$ reell und ist die einfallende Welle homogen (φ_2 reell), so sind nach SNELLIUS auch die Winkel φ_1, φ_0 reell und daher nach (13.4) und (13.5) auch alle g_i. Die Bedingung (13.2) für Reflexfreiheit enthält dann auf der linken Seite nur reelle Größen; die rechte Seite

$$- e^{-i\frac{4\pi d_1}{\lambda} n_1 \cos\varphi_1} \tag{14.1}$$

muß deshalb ebenfalls reell sein, d.h.

$$\frac{4\pi d_1}{\lambda} n_1 \cos\varphi_1 = M \cdot \pi \quad \text{mit } M \text{ ganz.} \tag{14.2}$$

Für gerades M geht (13.2) über in

$$\frac{g_2 - g_1}{g_2 + g_1} \cdot \frac{g_1 + g_0}{g_1 - g_0} = -1\,,$$

$$\frac{g_2 - g_1}{g_2 + g_1} = \frac{g_0 - g_1}{g_0 + g_1}\,.$$

Korrespondierende Subtraktion und Addition gibt

$$\frac{g_2}{g_1} = \frac{g_0}{g_1}\,; \quad \text{d.h.} \quad g_2 = g_0\,.$$

---

[1] A. SMAKULA: D.R.P. 685767 vom 1. 11. 1935 der Firma Carl Zeiss, Jena. — Z. Instrumentenkde. **60**, 33 (1940). — Glastechn. Ber. **19**, 377 (1941). Dort weitere Literatur. — Phys. Z. **43**, 217 (1942). — L. H. CARTWRIGHT u. A. F. TURNER: Phys. Rev. **55**, 675 (1939). — J. STRONG: J. Opt. Soc. Amer. **26**, 73 (1936). — K. BLODGETT: Phys. Rev. **57**, 921 (1940). — L. HIESINGER: Optik **3**, 485 (1948). — H. SCHRÖDER: Optik **3**, 499 (1948). — H. SCHOPPER: Optik **9**, 498 (1952).

Das ist für $\mu_2 = \mu_0$ eine triviale Lösung, die optische Gleichheit beider Medien (2) und (0) bedeutet, und für $\mu_2 \neq \mu_0$ eine nahezu triviale Lösung, jedenfalls keine Lösung des im Anfang dieses Abschnitts genannten technischen Problems.

Also sei M ungerade und nach (14.2)

$$d_1 = M \cdot \frac{\lambda_1}{4} \cdot \frac{1}{\cos \varphi_1} \; ; \tag{14.3}$$

darin bedeutet $\lambda_1 = \lambda/n_1$ die Wellenlänge im Medium (1). Für das g_1 ist nach (13.2) zu fordern

$$\frac{g_2 - g_1}{g_2 + g_1} = \frac{g_1 - g_0}{g_1 + g_0}, \tag{14.4}$$

Korrespondierende Subtraktion und Addition gibt

$$\frac{g_2}{g_1} = \frac{g_1}{g_0},$$

also

$$g_1 = \sqrt{g_2 g_0} \; . \tag{14.5}$$

Für TE-Wellen heißt das nach Gl. (13.4)

$$\frac{n_1}{\mu_1} \cos \varphi_1 = \sqrt{\frac{n_0 n_2}{\mu_0 \mu_2} \cos \varphi_0 \cos \varphi_2} \, , \tag{14.6}$$

für TH-Wellen nach Gl. (13.5)

$$\frac{\mu_1}{n_1} \cos \varphi_1 = \sqrt{\frac{\mu_0 \mu_2}{n_0 n_2} \cos \varphi_0 \cos \varphi_2} \, . \tag{14.7}$$

Speziell bei senkrechtem Einfall ist $\cos \varphi_l = 1$ für $l = 0; 1; 2$, und dann stimmen beide Bedingungen (14.6) und (14.7) miteinander überein, was physikalisch evident ist. Sie lauten

$$\frac{n_1}{\mu_1} = \sqrt{\frac{n_0}{\mu_0} \cdot \frac{n_2}{\mu_2}} \; ; \tag{14.8}$$

die Schichtdicke sei nach (14.3) ein ungerades Vielfaches der Dicke

$$d_1 = \frac{\lambda_1}{4} \, . \tag{14.9}$$

Im praktisch häufigsten Fall ($\mu_0 = \mu_1 = \mu_2 = 1$) muß der Brechungsindex für Reflexfreiheit bei senkrechtem Einfall gleich dem geometrischen Mittel der beiden anderen Brechungsindices sein, bei Luft und Glas also $n_1 = \sqrt{1,5} = 1,225$. Diese Entspiegelung hat eine große praktische Bedeutung für die Vergütung von Optiken. Vgl. hierzu Tabelle 4 und die ausführliche, auch die technische Herstellung betreffende Literatur[1].

Bei schrägem Einfall einer homogenen Welle (d.h. φ_2 reell) ist auch bei $\mu_0 = \mu_1 = \mu_2 = 1$ die Entspiegelung nicht für beide Polarisationen streng erreichbar; denn die Forderung (14.6) für reflexfreie TE-Wellen

$$n_1 \cos \varphi_1 = \sqrt{n_0 n_2 \cos \varphi_0 \cos \varphi_2} \, , \tag{14.10}$$

nimmt mit dem Brechungsgesetz auch die Form an

$$\tan \varphi_1 = \sqrt{\tan \varphi_0 \cdot \tan \varphi_2} \, . \tag{14.11}$$

---

[1] A. Smakula: D.R.P. 685767 vom 1. 11. 1935 der Frima Carl Zeiss, Jena. — H. Schröder: Z. Naturforsch. **4a**, 515 (1949) und weitere Literatur von S. 477. Tabelle 4 auf S. 488.

Ebenso nimmt die Forderung (14.7) für reflexfreie TH-Wellen

$$\frac{\cos \varphi_1}{n_1} = \sqrt{\frac{\cos \varphi_0}{n_0} \cdot \frac{\cos \varphi_2}{n_2}}, \qquad (14.12)$$

durch Kombination mit dem Brechungsgesetz die Form an

$$\sin 2\varphi_1 = \sqrt{\sin 2\varphi_0 \cdot \sin 2\varphi_2}. \qquad (14.13)$$

Wollte man nun die Gültigkeit von (14.11) und (14.12) zugleich fordern, d.h. die Entspiegelung für TE- und für TH-Wellen, so würde durch Multiplikation von (14.11) und (14.13) bzw. durch Division folgen

$$\sin \varphi_1 = \sqrt{\sin \varphi_0 \cdot \sin \varphi_2}; \quad \cos \varphi_1 = \sqrt{\cos \varphi_0 \cos \varphi_2} \qquad (14.14)$$

und daraus

$$\sin \varphi_0 \sin \varphi_2 = \sin^2 \varphi_1 = 1 - \cos^2 \varphi_1 = 1 - \cos \varphi_0 \cos \varphi_2, \qquad (14.15)$$

$$\cos (\varphi_0 - \varphi_2) = 1; \quad \varphi_0 = \varphi_2. \qquad (14.16)$$

Das Brechungsgesetz liefert daraus $\varphi_0 = \varphi_2 = 0$, wenn $n_0 \neq n_2$ vorausgesetzt wird. Entspiegelung beider Polarisationen gibt es streng daher nur für senkrechten Einfall. Doch ist die Entspiegelung der Optiken, auch wenn sie nur für senkrechten Einfall richtig bemessen ist, für mäßige Aperturen noch gut, da in die Bedingungen (14.10) und (14.12) nur die Cosinusfunktionen der Winkel eingehen.

Für jede einzelne der beiden Polarisationen kann die Entspiegelung erreicht werden.

Für TE-Wellen kann man den dazu erforderlichen Brechungsindex der dünnen Schichten aus (14.10) mit dem Brechungsgesetz wie folgt errechnen:

$$n_1^2 (1 - \sin^2 \varphi_1) = n_0 n_2 \cos \varphi_0 \cos \varphi_2,$$

$$n_1^2 = n_0 n_2 \cos \varphi_0 \cos \varphi_2 + n_0 \sin \varphi_0 n_2 \sin \varphi_2,$$

$$n_1^2 = n_0 n_2 \cos (\varphi_2 - \varphi_0) \qquad (14.17)$$

Stets ist $n_1 < \sqrt{n_0 n_2}$; für Glas und Luft als Medien ist das schwer realisierbar.

Günstiger steht es bei Entspiegelung für TH-Wellen. Für sie folgt aus (14.12) zusammen mit dem Brechungsgesetz

$$n_1^2 (1 - \sin^2 \varphi_1) = n_1^4 \cdot \frac{\cos \varphi_2 \cdot \cos \varphi_0}{n_2 \cdot n_0},$$

$$n_1^2 - n_2 \sin \varphi_2 n_0 \sin \varphi_0 = n_1^4 \frac{\cos \varphi_2 \cos \varphi_0}{n_2 n_0},$$

$$n_1^4 - n_1^2 \frac{n_2 n_0}{\cos \varphi_2 \cos \varphi_0} + n_2^2 n_0^2 \tan \varphi_2 \tan \varphi_0 = 0,$$

$$n_1^2 = \frac{n_0 n_2}{2 \cos \varphi_0 \cos \varphi_2} \left(1 \pm \sqrt{1 - \sin 2\varphi_0 \sin 2\varphi_2}\right). \qquad (14.18)$$

Für beide Vorzeichen kann $n_1 > \sqrt{n_0 n_2}$ werden. Der doppelten Lösung in (14.18) entspricht es, daß auch (14.13) zwei Lösungen für φ_1 hat, die komplementär zueinander sind.

Für das Beispiel $n_0 = 1{,}5$ (Glas); $n_2 = 1$ (Luft) und verschieden gewählte Einfallswinkel φ_2 ist φ_0 aus dem Brechungsgesetz, n_1 aus (14.18) mit dem oberen

Vorzeichen berechnet worden und in Fig. 5 eingetragen. Zugleich ist nach H. Schröder[1] die reflektierte Intensität für die *TE*-Welle angegeben; sie ist nach Tabelle 1 zu erhalten als $|\mathfrak{r}_2|^2$ aus

$$\mathfrak{r}_2 = \frac{g_2\,g_0 - g_1^2}{g_2\,g_0 + g_1^2}. \tag{14.19}$$

Für $g_l = \hat{g}_l = \dfrac{\cos\varphi_l}{n_l}$ wird $\hat{\mathfrak{r}}_2 = 0$.

In Fig. 5 eingetragen ist die Größe

$$|\hat{\mathfrak{r}}_2|^2 = \left| \frac{n_2\cos\varphi_2\,n_0\cos\varphi_0 - n_1^2\cos^2\varphi_1}{n_2\cos\varphi_2\,n_0\cos\varphi_0 + n_1^2\cos^2\varphi_1} \right|^2. \tag{14.20}$$

Der Vorteil der dünn beschichteten Glasplatte als Polarisator liegt ersichtlich darin, daß die Lichtausbeute viel größer sein kann als beim Brewster-Fall des unbeschichteten Glases. Dann muß freilich die dünne Schicht einen hohen Brechungsindex $n_1 > 1,5$ (z.B. TiO_2) haben, und der Einfallswinkel ist groß zu machen (siehe dazu den Abschnitt Reflexverstärkung). Ein Vergleich dieser Einschichtpolarisatoren mit einem Mehrschichtpolarisator wird in Ziff. 21 gegeben.

15. Unterlage und Schicht als absorbierend oder nicht-absorbierend zugelassen; Betrachtung für dicke und dünne Schichten.

α) Die allgemeine Bedingung für Reflexfreiheit (13.8) oder (13.9) ist ohne Voraussetzungen über $\varrho_1 d_1$ nicht sehr übersichtlich. Immer ist aber, wenn das Medium ① absorbiert, für hinreichend große Schichtdicke d_1 erreichbar, daß

$$\mathrm{Tan}\,\varrho_1 d_1 \approx 1 \tag{15.1}$$

ist.

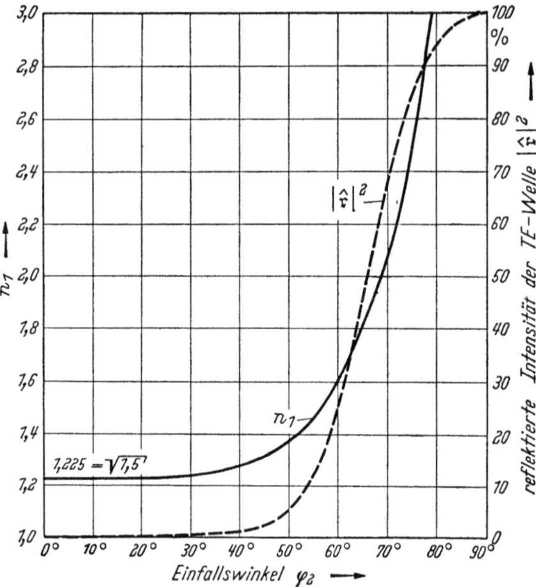

Fig. 5. Brechungsindex n_1 einer auf Glas ($n_0 = 1,5$) aufgebrachten dünnen Schicht, die das reflektierte Licht linear polarisiert bei dem Einfallswinkel φ_2. Medium ② ist Luft. ——— Brechungsindex der Schicht. ——— Reflektierte Intensität der *TE*-Welle. (Die reflektierte Intensität der *TH*-Welle ist 0.)

Fig. 6 zeigt einen typischen Verlauf von $\mathrm{Tan}\,\varrho_1 d_1$ in der Gaussschen Zahlenebene mit dem Wicklungspunkt 1.

Für hinreichend große Schichtdicke d_1 wird daher die Bedingung für Reflexfreiheit (13.8) und die gleichwertige (13.9) befriedigt von der Lösung

$$g_0 = g_1 = g_2.$$

Das führt offenbar auf Lösungen, die in Ziff. 10 bereits in allen wesentlichen Zügen abgehandelt sind. Denken wir nun d_1 kleiner werdend, so gehen aus diesen Lösungen zunächst nur wenig modifizierte hervor, also z.B. eine Lösung, die dem Brewster-Fall zwischen Medien ② und ① ähnelt und durch die Nähe des Mediums ⓪ nur etwas verändert ist.

---

[1] H. Schröder: Optik **3**, 499 (1948).

β) Zu wesentlich neuartigen Lösungen kommt man aber in dem anderen Grenzfall sehr dünner Schichten, für die man die Spirale in Fig. 6 durch die Tangente im Punkte $d_1 = 0$ ersetzen kann. Wir nehmen daher an

$$\text{Tan } \varrho_1 d_1 \approx \varrho_1 d_1, \quad \text{d.h.} \quad |\varrho_1 d_1| \ll 1. \tag{15.2}$$

Die Bedingung für Reflexfreiheit (13.8) nimmt dann die Form an

$$\frac{g_2}{g_0} = \frac{1 + \dfrac{g_1 \varrho_1 d_1}{g_0}}{1 + \dfrac{g_0 \varrho_1 d_1}{g_1}}. \tag{15.3}$$

Wir fragen nach dem $\mathfrak{n}_1$ und d_1, durch die man diese Bedingung befriedigen kann, wenn alle anderen Größen vorgegeben sind. Für TE-Wellen enthält ϱ_1/g_1 die Unbekannte $\mathfrak{n}_1$ nach Gl. (13.3), (13.4) und (13.5) nicht, während sie bei $\varrho_1 g_1$ quadratisch vorkommt, umgekehrt ist es bei TH-Wellen. Man löst deshalb (15.3) auf

$$g_1 \varrho_1 d_1 = g_2 - g_0 + g_0 g_2 d_1 \frac{\varrho_1}{g_1}, \tag{15.4}$$

bzw.

$$\left. \begin{aligned} \frac{\varrho_1}{g_1} d_1 &= \frac{1}{g_2} - \frac{1}{g_0} + \\ &+ \frac{1}{g_0} \cdot \frac{1}{g_2} d_1 \cdot \varrho_1 \cdot g_1. \end{aligned} \right\} \tag{15.5}$$

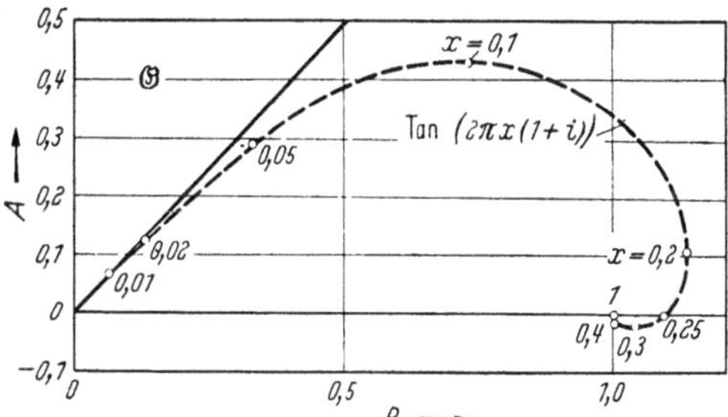

Fig. 6. Tan $\varrho_1 d_1$ in der GAUSSschen Zahlenebene in einem Beispiel, bei dem $n = k$ ist.

Für TE-Wellen mit $g_l = \hat{g}_l$ besagt Gl. (15.4) nach (13.3) und (13.4)

$$i \mathfrak{n}_1^2 \frac{\cos^2 \varphi_1}{\mu_1} \cdot \frac{2\pi d_1}{\lambda} = \mathfrak{n}_2 \cdot \frac{\cos \varphi_2}{\mu_2} - \mathfrak{n}_0 \frac{\cos \varphi_0}{\mu_0} + i \mathfrak{n}_2 \mathfrak{n}_0 \cdot \frac{\cos \varphi_2 \cos \varphi_0}{\mu_2 \mu_0} \mu_1 \frac{2\pi d_1}{\lambda}. \tag{15.6}$$

Für TH-Wellen besagt (15.5) mit $g_l = \hat{g}_l$ nach Gl. (13.3) und (13.5)

$$i \mathfrak{n}_1^2 \frac{1}{\mu_1} \frac{2\pi d_1}{\lambda} = \frac{\mathfrak{n}_2}{\mu_2 \cos \varphi_2} - \frac{\mathfrak{n}_0}{\mu_0 \cos \varphi_0} + i \mathfrak{n}_2 \mathfrak{n}_0 \frac{\mu_1 \cos^2 \varphi_1}{\mu_2 \mu_0 \cos \varphi_2 \cos \varphi_0} \cdot \frac{2\pi d_1}{\lambda}. \tag{15.7}$$

Jede der beiden komplexen Gleichungen hat nichttriviale Lösungen in großer Zahl.

Für senkrechten Einfall z.B. stimmen (15.6) und (15.7) natürlich miteinander überein. Bei $\mu_0 = \mu_1 = \mu_2 = 1$ ist für Reflexfreiheit dann offenbar zu fordern

$$\mathfrak{n}_1^2 = \mathfrak{n}_2 \mathfrak{n}_0 - i (\mathfrak{n}_2 - \mathfrak{n}_0) \frac{\lambda}{2\pi d_1}. \tag{15.8}$$

Setzen wir den Brechungsindex $\mathfrak{n}_2$ des Mediums ②, aus dem die Welle einfällt, als reell voraus und schreiben wir ferner $\mathfrak{n}_l = n_l - i k_l$, so wird unter Benutzung der Beziehung (2.8) aus (15.8)

$$n_1^2 - k_1^2 = \varepsilon_1 = n_2 n_0 + k_0 \frac{\lambda}{2\pi d_1}, \tag{15.9}$$

$$2 n_1 k_1 = 2\lambda \frac{\sigma_1}{c} = n_2 k_0 + (n_2 - n_0) \frac{\lambda}{2\pi d_1}. \tag{15.10}$$

Alle Lösungen haben zwangsläufig $n_1^2 > k_1^2$. Das ist bei dünnen Schichten vieler Metalle im Gegensatz zu dicken Schichten auch im Gebiet des sichtbaren Lichtes der Fall. Bei einer Cu-Unterlage und Luft als Medium des Einfalls ist die Reflexfreiheit mit einer Schicht von $d_1 \approx \lambda/20$ und $n_1 = 3$, $k_1 \approx 1$ erfüllbar. Das sind Werte, wie sie bei dünnen Metallschichten vorkommen.

Noch größere Variationsmöglichkeiten hat man für die Aufgabe, eine Polarisation durch Reflexion auszulöschen[1]. Hier haben jedoch nichtabsorbierende Schichten auf Metall noch bessere Lichtausbeute (s. unten).

Die Gln. (15.9) und (15.10) lassen sich am leichtesten befriedigen, wenn die Unterlage nicht absorbiert ($k_0 = 0$). Diesen Fall behandelt die nächste Ziffer.

Tabelle 2.

Absorptionskoeffizienten k_1, Brechungsindexrealteil n_1 und relative Schichtdicke $\gamma_1 = \dfrac{4\pi d_1}{\lambda}$ für eine Schicht, die zwischen Luft ($n_2 = 1$) und einem Nichtleiter ($n_0 = 1{,}5$) liegt und reflexionsfrei ist für Licht, das vom dichteren Medium her senkrecht einfällt; berechnet nach Gl. (16.3).

k_1	n_1	γ_1
1,0	1,58	0,34
1,5	**1,94**	**0,172**
2,0	2,35	0,106
2,5	2,78	0,072
3,0	**3,24**	**0,0515**
3,5	3,71	0,0384
4,0	3,71	0,0298
4,5	4,67	0,0238
5,0	5,14	0,0195
6,0	6,12	0,0136
7,0	7,10	0,0101

16. Dünne absorbierende Schichten zwischen nichtleitenden Medien. Nur das als sehr dünne Schicht betrachtete Zwischenmedium absorbiert, d.h. n_2, n_0 seien reell, n_1 wird als komplex zugelassen; $|\varrho_1 d_1| \ll 1$. Dann ist die Bedingung für Reflexfreiheit bei senkrechtem Einfall nach Gl. (15.9) und (15.10) wegen $k_0 = 0$

$$n_1^2 - k_1^2 = \varepsilon_1 = n_2 n_0 , \qquad (16.1)$$

$$2 n_1 k_1 = 2\lambda \frac{\sigma_1}{c} = (n_2 - n_0) \frac{\lambda}{2\pi d_1} . \qquad (16.2)$$

Stets muß $n_1^2 > k_1^2$ sein. Da ferner $2 n_1 k_1 > 0$ ist, gibt es hier nur Lösungen, wenn $n_2 > n_0$ ist, also das Licht von der Glasseite her auf eine Schicht auffällt, die sich z.B. zwischen Glas und Luft befindet. In diesem Beispiel ($n_2 = 1{,}5$; $n_0 = 1$) besagen (16.1) und (16.2)

$$n_1 = \sqrt{k_1^2 + 1{,}5} ; \qquad \gamma = \frac{4\pi d_1}{\lambda} = \frac{1}{2 n_1 k_1} . \qquad (16.3)$$

Es existiert dann eine eindimensionale Mannigfaltigkeit von Lösungen, nämlich zu jedem k_1 eine. Tabelle 2 gibt einige Werte.

Die fett gesetzten Daten waren recht genau mit dünnen Goldschichten realisiert bei Beobachtungen von F. Goos[2], die wir in Tabelle 3 wiedergeben.

Tabelle 3.

Von F. Goos beobachtete Reflexfreiheit an dünnen Goldschichten, auf Homosil durch Kathodenzerstäubung niedergeschlagen.

k_1	n_1	γ	Wellenlänge λ	Schichtdicke d_1	Durchgelassene Intensität J_d	Luftseitig reflektierte Intensität J_r	Homosilseitig reflektierte Intensität J_r'
			mμ	mμ	%	%	%
1,6	1,9	0,171	265	3,6	65,3	10,5	0
2,9	3,1	0,0526	578	2,42	67,5	10,0	0

Die Fig. 7 zeigt ferner den von F. Goos beobachteten Verlauf der Intensitäten für eine dünne Silberschicht in Abhängigkeit von der Schichtdicke. Das Minimum bei 9 mμ Schichtdicke zeigt die annähernde Reflexfreiheit.

---

[1] H. Schröder: Optik **3**, 499 (1948). — H. Schopper: Optik **10**, 426 (1953).
[2] F. Goos: Z. Physik **106**, 606 (1937).

Viele dieser dünnen Metallschichten zeigen, wenn wir der Reflexfreiheit nicht genau genügen, immerhin einen ausgeprägten Polarisationseffekt bei passend schrägem Einfall. Eine Untersuchung hierzu stammt von H. Schopper[1] mit dem Ziel, Polarisatoren für das Ultrarot zu gewinnen. Im optischen Gebiet haben die Reflexionspolarisatoren hoher Lichtausbeute, bei denen eine nicht absorbierende Schicht sich auf metallischer Unterlage befindet, größere Bedeutung. Das ist Gegenstand der nächsten Ziffer.

Die Bedingungen (15.9) und (15.10) bzw. (16.1), (16.2) für Reflexbefreiung haben auch im Gebiet der Zentimeter- bis Meterwellen Bedeutung. Die Gl. (16.2) besagt dabei, daß der Flächenwiderstand[2] der Schicht ① die Größe

$$\frac{1}{\sigma_1 d_1} = \frac{4\pi}{c \cdot (n_2 - n_0)}, \quad (16.4)$$

d. h. in praktischen Einheiten den Wert

$$R = \frac{120\pi}{n_2 - n_0}\,\Omega \quad (16.5)$$

haben soll.

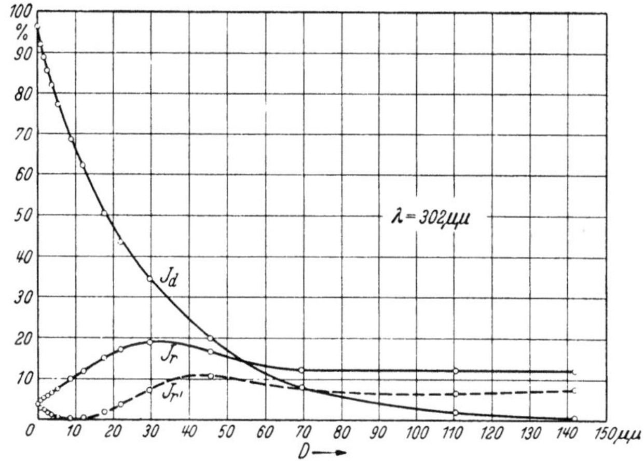

Fig. 7. Intensitäten in Abhängigkeit von der Wägungsschichtdicke D, gemessen von F. Goos an dünnen Silberschichten. Lies $m\mu$ statt $\mu\mu$. [F. Goos: Z. Physik **106**, 606 (1937)].

17. Nur die Unterlage absorbiert. Der Brechungsindex der „Unterlage", n_0, sei komplex; die Brechungsindices der Schicht, n_1 und des Mediums, aus dem die Welle einfällt, n_2, seien rell. Dann ist $\varrho_1 d_1$ rein imaginär und

$$\left.\begin{aligned} \mathrm{Tan}\,\varrho_1 d_1 &= \mathrm{Tan}\left(i\,\frac{2\pi d_1}{\lambda}\,n_1 \cos\varphi_1\right) \\ &= i\,\tan\eta \end{aligned}\right\} \quad (17.1)$$

mit reellem

$$\eta = \frac{2\pi d_1}{\lambda}\,n_1 \cos\varphi_1. \quad (17.2)$$

Die Bedingung (13.9) für Reflexfreiheit besagt dann

$$\frac{g_0 g_2}{g_1^2} = \frac{\dfrac{g_2}{g_1} - i\tan\eta}{\dfrac{g_1}{g_2} - i\tan\eta}, \quad (17.3)$$

d. h.

$$i\left\{1 - \frac{g_0 g_2}{g_1^2}\right\}\tan\eta = \frac{g_2 - g_0}{g_1}. \quad (17.4)$$

g_2, g_1 sind reell; g_0 sei bezeichnet

$$g_0 = v_0 - i\varkappa_0 \quad (17.5)$$

mit reellen v_0, $\varkappa_0$. Gl. (17.4) ist dann

$$i\left\{1 - \frac{g_2}{g_1^2}\,(v_0 - i\varkappa_0)\right\}\tan\eta = \frac{g_2 - v_0}{g_1} + i\,\frac{\varkappa_0}{g_1}. \quad (17.6)$$

---

[1] H. Schopper: Optik **10**, 426 (1953).

[2] „Flächenwiderstand einer dünnen Schicht" heißt der Widerstand eines quadratischen Ausschnittes, wenn extrem gut leitende Elektroden längs gegenüberliegender Seiten als Zuleitung dienen. Er ist von der Größe des Quadrats unabhängig und wird in Ω gemessen.

Es sind also die beiden Bedingungen zu erfüllen

$$\left(1 - \frac{g_2 v_0}{g_1^2}\right)\tan\eta = \frac{\varkappa_0}{g_1}, \qquad (17.7)$$

$$\frac{g_2 \varkappa_0}{g_1^2}\tan\eta = \frac{v_0 - g_2}{g_1}. \qquad (17.8)$$

g_1 hat also der Bedingung

$$g_1^2 = g_2\left\{\frac{\varkappa_0^2}{v_0 - g_2} + v_0\right\} \qquad (17.9)$$

zu genügen und η der Bedingung

$$\tan\eta = \frac{v_0 - g_2}{g_2 \varkappa_0}\cdot g_1. \qquad (17.10)$$

Mit der Annahme $\mu_0 = \mu_1 = \mu_2 = 1 = n_2$ (Einfall aus Luft) ist für eine TE-Welle

$$(v_0 - i\varkappa_0)^2 = g_0^2 = \mathfrak{n}_0^2\cos^2\varphi_0 = \mathfrak{n}_0^2 - \mathfrak{n}_0^2\sin^2\varphi_0 = \mathfrak{n}_0^2 - \sin^2\varphi_2, \qquad (17.11)$$

$$v_0^2 - \varkappa_0^2 = n_0^2 - k_0^2 - \sin^2\varphi_2 \qquad \text{(abgekürzt als } 2G), \qquad (17.12)$$

$$v_0\varkappa_0 = n_0 k_0, \qquad (17.13)$$

$$v_0 = \sqrt{\sqrt{G^2 + (n_0 k_0)^2} + G}, \qquad (17.14)$$

$$\varkappa_0 = \sqrt{\sqrt{G^2 + (n_0 k_0)^2} - G}. \qquad (17.15)$$

Bei vielen Metallen unterscheiden sich v_0 und $\varkappa_0$ nicht sehr von n_0 und k_0 auch bei sehr schrägem Einfall. Für einen Einfallswinkel $\varphi_2 = 80°$ und mit den optischen Konstanten des Rh, nämlich $n_0 = 1,81$ und $k_0 = 5,31$, ist $v_0 = 1,78$ und $\varkappa_0 = 5,39$. Die erste Bedingung für Reflexfreiheit (17.9)

$$g_1^2 = \left(\frac{\varkappa_0^2}{v_0 - \cos\varphi_2} + v_0\right)\cos\varphi_2$$

führt in dem Beispiel zu dem Wert $\hat{g}_1 = 1,86$. Daraus folgt wegen der Definition von $\hat{g}_1$ und dem Brechungsgesetz

$$n_1\cos\varphi_1 = \hat{g}_1 = 1,86; \qquad n_1\sin\varphi_1 = \sin\varphi_2 = 0,985$$

schließlich $\varphi_1 = 27,9°$ und $n_1 = 2,10$.

Die 2. Bedingung für Reflexfreiheit (17.10) liefert dann $\tan\eta = 0,32$; also

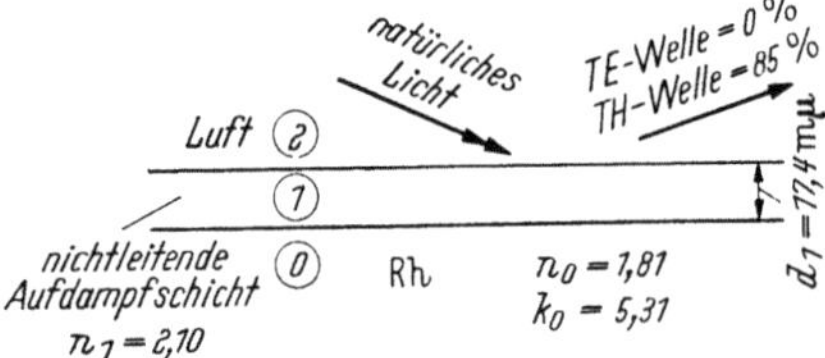

Fig. 8. Beschichteter Metallspiegel
als Reflexionspolarisator hoher Ausbeute.

$$\eta = 17,8° = \frac{d_1}{\lambda}\,n_1\cos\varphi_1\cdot 360°$$

bis auf ein willkürliches ganzes Vielfaches von 180°. Daraus folgt die erforderliche Schichtdicke

$$d_1 = \lambda\left(0,0264 + \frac{M}{2}\,\frac{1}{n_1\cos\varphi_1}\right)$$

mit einer beliebigen ganzen Zahl M. Bei $\lambda = 660\ m\mu$ ist also eine Schicht vom Brechungsindex $n_1 = 2,10$ und einer Dicke $d_1 = 17,4\ m\mu$, aufgebracht auf einem Rh-Spiegel, ein Reflexionspolarisator in der Anordnung nach Fig. 8. Wenn der Einfallswinkel 80° beträgt, wird nur die TH-Welle reflektiert.

Mit der berechneten Schichtdicke und den anderen Daten kann man nach der Formel für $\mathfrak{r}_2$ in Tabelle 1 die reflektierte Intensität $|\mathfrak{r}_2|^2$ berechnen als Bruchteil der einfallenden. Für die *TE*-Welle ergibt sich der Wert 0, für die *TH*-Welle der große Wert 85%. Die mit einer passenden dünnen Nichtleiterschicht bedeckte Metallplatte ist also ein Reflexionspolarisator hoher Lichtausbeute.

Fig. 9a und b zeigen die Abhängigkeit der Reflexion vom Einfallswinkel und von der Wellenlänge für eine KBr-Schicht auf Blei, berechnet und gemessen von H. SCHOPPER[1].

Um den gleichen Rh-Spiegel für senkrechten Einfall reflexfrei zu machen, müßte die dünne Schicht den reichlich hohen Brechungsindex $n_1 = 6{,}06$ und die Dicke $d_1 = \lambda \cdot 0{,}0191$ haben. Da die Reflexfreiheit nur für eine Wellenlänge gilt, haben

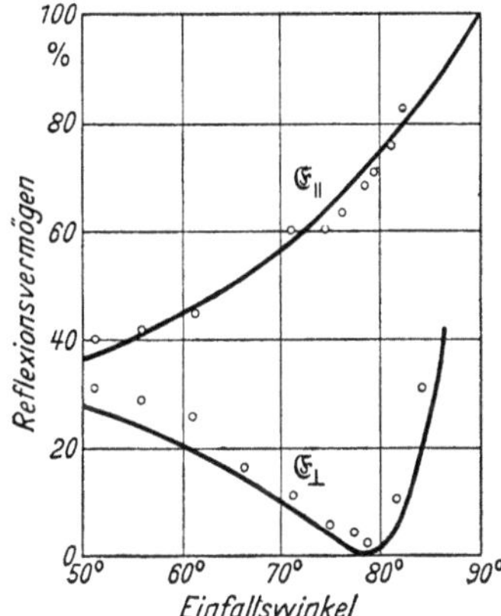

Fig. 9a. Reflexionsvermögen für $\mathfrak{E}_\perp$ und $\mathfrak{E}_\parallel$ in Abhängigkeit vom Einfallswinkel φ_2 für eine KBr-Schicht auf einer Bleiunterlage. ooo Meßpunkte, —— berechnet mit $n_0 = 2$, $k_0 = 3{,}4$, $n_1 = 1{,}56$; $\lambda = 623$ mμ. [Nach H. SCHOPPER: Optik 10, 437 (1953).]

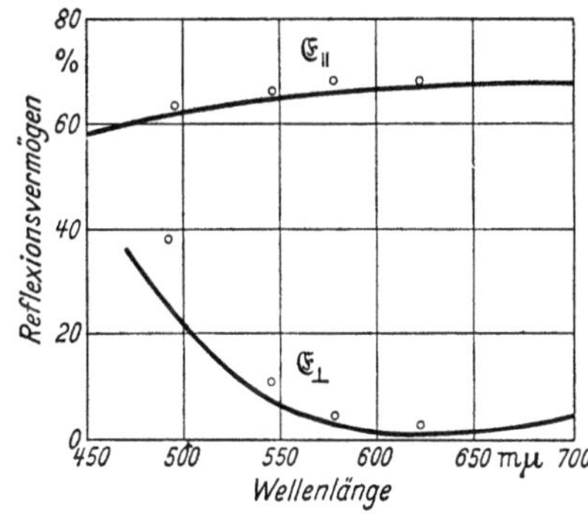

Fig. 9b. Reflexionsvermögen für $\mathfrak{E}_\perp$ und $\mathfrak{E}_\parallel$ in Abhängigkeit von der Wellenlänge für eine KBr-Schicht auf einer Bleiunterlage. ooo Meßpunkte, —— berechnete Werte (optische Konstanten wie bei Fig. 9a). [Nach H. SCHOPPER: Optik 10, 438 (1953).]

Schichten dieser Art auch Eigenschaften, die sie gelegentlich als Reflexions-Interferenzfilter verwendbar erscheinen lassen. Doch sind gewisse Mehrschichtsysteme, wie sie in den nächsten Abschnitten behandelt werden, dafür besser geeignet.

c) Mehrschichtsysteme aus Nichtleitern mit festgelegter Dicke.

18. Rekursionsformeln. Nach Ziff. 14 macht eine nichtleitende Schicht ①die Grenze zwischen zwei nichtleitenden Medien ② und ⓪ reflexfrei, wenn ihre Schichtdicke nach Gl. (14.3) ein ungerades Vielfaches der Dicke

$$d_1 = \frac{\lambda}{4} \cdot \frac{1}{n_1 \cos \varphi_1}$$

ist. Solche Schichten nennen wir „Viertelwellenlängenschichten", $\frac{3}{4}\lambda$-Schichten usw.; und für $\frac{\lambda}{4}$-Mehrfachschichten wollen wir in dieser Ziffer voraussetzen, daß sie aus solchen $\frac{\lambda}{4}$-Schichten, $\frac{3}{4}\lambda$-Schichten usw. nichtleitender Medien aufgebaut sind bis auf das natürlich einseitig unbegrenzte Medium ⓜ, aus dem die homogene Welle unter dem reellen Einfallswinkel φ_m einfällt, und bis auf das ebenfalls einseitig unbegrenzte Endmedium ⓪ (vgl. Fig. 1). m ist die Zahl der Grenzflächen. Dann ist

$$e^{-2\varrho_l d_l} = -1 \quad \text{für} \quad l = 1; \ \ldots m - 1 \tag{18.1}$$

---

[1] H. SCHOPPER: Optik 10, 426 (1953).

und nach den Rekursionsformeln (9.11), (9.13), (9.14) und (9.15) kann man für
die reflektierte Amplitude schreiben

$$r_m = \frac{Z'_m}{N'_m} \tag{18.2}$$

mit

$$Z'_m = (g_m - g_{m-1}) N'_{m-1} + (g_m + g_{m-1}) Z'_{m-1}, \tag{18.3}$$

$$N'_m = (g_m + g_{m-1}) N'_{m-1} + (g_m - g_{m-1}) Z'_{m-1}. \tag{18.4}$$

Dabei unterscheiden sich die Z'_m und N'_m von den früher gebrauchten Z_m und
N_m um den Faktor i.

Für diese speziellen Systeme läßt sich die Rekursion leicht geschlossen ausführen und daraus eine Erweiterung der für reflexfreie Einschichtsysteme geltenden Bedingungen auf Mehrschichtsysteme gewinnen.

19. Geschlossene Ausführung der Rekursion durch vollständige Induktion.
Für das System mit einer Grenzfläche ($m=1$) ist nach Fresnel (Tabelle 1)

$$r_1 = \frac{g_1 - g_0}{g_1 + g_0}; \qquad Z'_1 = g_1 - g_0; \qquad N'_1 = g_1 + g_0. \tag{19.1}$$

Nach den Rekursionsformeln (18.3) und (18.4) folgt für zwei Grenzflächen

$$Z'_2 = (g_2 - g_1)(g_1 + g_0) - (g_2 + g_1)(g_1 - g_0) = 2(g_2 g_0 - g_1^2), \tag{19.2}$$

$$N'_2 = (g_2 + g_1)(g_1 + g_0) - (g_2 - g_1)(g_1 - g_0) = 2(g_2 g_0 + g_1^2) \tag{19.3}$$

und für drei Grenzflächen

$$Z'_3 = (g_3 - g_2) 2(g_2 g_0 + g_1^2) - (g_3 + g_2) 2(g_2 g_0 - g_1^2),$$

$$Z'_3 = 4(g_3 g_1^2 - g_2^2 g_0), \tag{19.4}$$

$$N'_3 = 4(g_3 g_1^2 + g_2^2 g_0). \tag{19.5}$$

Für $m = 4$ folgt entsprechend

$$Z'_4 = 8(g_4 g_2^2 g_0 - g_3^2 g_1^2), \tag{19.6}$$

$$N'_4 = 8(g_4 g_2^2 g_0 + g_3^2 g_1^2). \tag{19.7}$$

Daraus vermuten wir, daß allgemein

$$Z'_m = 2^{m-1}(g_m g_{m-2}^2 \cdots - g_{m-1}^2 g_{m-3}^2 \cdots), \tag{19.8}$$

$$N'_m = 2^{m-1}(g_m g_{m-2}^2 \cdots + g_{m-1}^2 g_{m-3}^2 \cdots) \tag{19.9}$$

für $m \geq 2$ gilt.

Jedes der in (19.8) und (19.9) auftretenden Produkte enthält nur gerade oder
nur ungerade Indices. Das Produkt mit ungeraden Indices endet mit dem Faktor g_1^2; das mit geraden Indices endet mit dem Faktor $g_2^2 g_0$.

Der Beweis wird durch vollständige Induktion geführt; denn aus der Annahme, (19.8) und (19.9) gelten für ein m, folgt mit den umgeschriebenen Gln. (18.3)
und (18.4)

$$Z'_{m+1} = g_{m+1}(N'_m - Z'_m) - g_m(N'_m + Z'_m), \tag{19.10}$$

$$N'_{m+1} = g_{m+1}(N'_m - Z'_m) + g_m(N'_m + Z'_m) \tag{19.11}$$

sofort

$$Z'_{m+1} = 2^{m-1} \cdot 2 \cdot g_{m+1} g_{m-1}^2 g_{m-3}^2 \cdots - 2^{m-1} \cdot 2 \cdot g_m g_m g_{m-2}^2 \cdots, \tag{19.12}$$

$$N'_{m+1} = 2^{m-1} \cdot 2 \cdot g_{m+1} g_{m-1}^2 g_{m-3}^2 \cdots + 2^{m-1} \cdot 2 \cdot g_m g_m g_{m-2}^2 \cdots. \tag{19.13}$$

Das ist aber die Behauptung (19.8), (19.9) für $m+1$ statt m. Also gilt für die reflektierte Amplitude nach (18.2)

$$\mathfrak{r}_m = \frac{g_m\,g_{m-2}^2\cdots - g_{m-1}^2\,g_{m-3}^2\cdots}{g_m\,g_{m-2}^2\cdots + g_{m-1}^2\,g_{m-3}^2\cdots} = \frac{1-v^2}{1+v^2} \tag{19.14}$$

mit der Abkürzung

$$v^2 = \frac{g_{m-1}^2\cdot g_{m-3}^2\cdots}{g_m\cdot g_{m-2}^2\cdots} \tag{19.15}$$

und der Vorschrift, daß bei jedem auftretenden g_0^2 der Exponent 2 durch 1 zu ersetzen ist.

Für die Durchlässigkeit gilt mit derselben Vorschrift

$$\mathfrak{d}_m = i\,\frac{2g_m\,2g_{m-1}\cdots 2g_1}{N_m'}\;, \tag{19.16}$$

$$\mathfrak{d}_m = 2i\,\frac{g_m\,g_{m-1}\cdots g_1}{g_m\,g_{m-2}^2\cdots + g_{m-1}^2\,g_{m-3}^2\cdots}\;, \tag{19.17}$$

daher insbesondere für ungerade m

$$\mathfrak{d}_m = \frac{2i}{1+v^2}\cdot\frac{g_{m-1}\,g_{m-3}\cdots g_4\,g_2}{g_{m-2}\,g_{m-4}\cdots g_3\,g_1}\;, \tag{19.18}$$

und für gerade m

$$\mathfrak{d}_m = \frac{2i}{1+v^2}\cdot\frac{g_{m-1}\,g_{m-3}\cdots g_3\,g_1}{g_{m-2}\,g_{m-4}\cdots g_2\,g_0}\;. \tag{19.19}$$

Für gerade m heiße

$$g_s = \sqrt{g_m}\cdot v\cdot\sqrt{g_0} \tag{19.20}$$

,,Pseudo-g-Wert`` der Mehrfachschicht. Mit ihm gehen die Reflexionsformel (19.14) und die Durchlässigkeitsformel (19.19) über in

$$\mathfrak{r}_m = \frac{g_m\,g_0 - g_s^2}{g_m\,g_0 + g_s^2}\;;\qquad \mathfrak{d}_m = \frac{i\,2g_m\,g_s}{g_m\,g_0 + g_s^2}\;. \tag{19.21}$$

Sie haben die Gestalt, als ob statt der Mehrfachschicht nur *eine* Schicht vom g-Wert g_s zwischen den Medien $\widehat{m}$ und $\widehat{0}$ vorhanden wäre.

20. Die Bedingung für Reflexfreiheit nichtabsorbierender Mehrschichtsysteme aus $\lambda/4$-Schichten. Das Schichtsystem hat nach Gl. (19.14) die Reflexion 0, wenn

$$v = 1;\quad \text{d.h.}\quad g_m\,g_{m-2}^2\cdots = g_{m-1}^2\cdot g_{m-3}^2\cdots \tag{20.1}$$

(auftretendes g_0 hat den Exponenten 1 statt 2). Das ist eine Erweiterung der für zwei Grenzflächen in Ziff. 14 aufgestellten Bedingung (14.5) $g_2 g_0 = g_1^2$. Das allgemeine Gesetz (20.1) wurde zuerst von L. HIESINGER[1] aus den von ihm für $m \leq 3$ bewiesenen Formeln vermutet.

Sind Anfangs- und Endmedium vorgegeben, so legt (20.1) den Brechungsindex der Schicht völlig fest für $m = 2$, also für eine Einzelschicht. Für jede neue hinzukommende Schicht kommt ein neues g_l als verfügbar hinzu, und die Zahl der Freiheitsgrade erhöht sich dabei um 1. Allgemein ist die Anzahl der Freiheitsgrade

$$f = m - 2,$$

---

[1] L. HIESINGER: Optik **3**, 485 (1948). Siehe jedoch auch F. ABELÊS: Rev. Opt. **28**, 11 (1949).

also um 1 kleiner als die Zahl der Schichten. So ist für die Doppelschicht vorgegeben g_3 und g_0; g_1 und g_2 unterliegen der einen Gl. (20.1), die hier

$$g_3 g_1^2 = g_2^2 g_0, \quad \text{d.h.} \quad \frac{g_1}{g_2} = \sqrt{\frac{g_0}{g_3}} \tag{20.2}$$

lautet und also nur das Verhältnis von g_1 und g_2 festlegt. Für gerades m, d.h. 1, 3, 5, ... Schichten läßt sich eine geometrische Folge der g_i mit der Entspiegelungsbedingung (20.1) vereinbaren, da diese dann geschrieben werden kann

$$\frac{g_0}{g_1} \cdot \frac{g_2}{g_1} \cdot \frac{g_2}{g_3} \cdot \frac{g_4}{g_3} \cdots \frac{g_{m-2}}{g_{m-1}} \frac{g_m}{g_{m-1}} = 1 \tag{20.3}$$

und automatisch befriedigt ist, wenn die $(m-1)$ Gleichungen

$$\frac{g_0}{g_1} = \frac{g_1}{g_2} = \frac{g_2}{g_3} = \cdots = \frac{g_{m-2}}{g_{m-1}} = \frac{g_{m-1}}{g_m} \tag{20.4}$$

erfüllt sind. $(m-1)$-Variable $g_1, \ldots g_{m-1}$ stehen aber in der Tat zur Verfügung.

Mit $m = 4$, also drei Schichten zwischen $g_4 = 1$ und $g_0 = 1,5$ kann man also z.B. durch die als geometrische Folge gewählten Werte

$$g_4 = 1; \quad g_3 = 1,106; \quad g_2 = 1,225;$$
$$g_1 = 1,356; \quad g_0 = 1,5$$

entspiegeln.

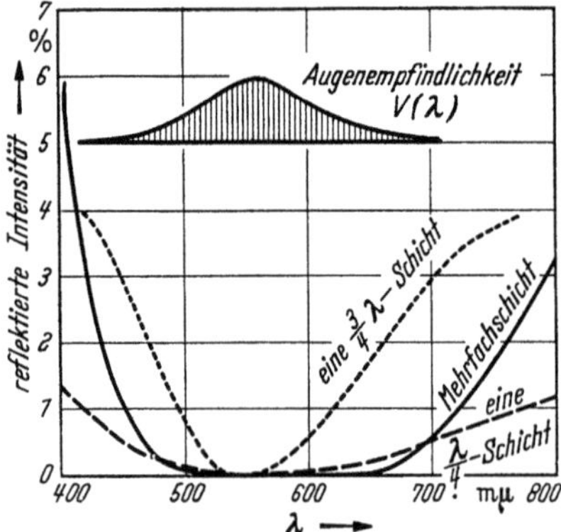

Fig. 10. Reflektierte Intensität als Funktion der Wellenlänge bei reflexmindernden Einfachschichten und einer Mehrfachschicht. [Nach L. Hiesinger: Optik 3, 485—492 (1948).]

Die Fig. 10 zeigt nach L. Hiesinger[1] die Reflexion einer Mehrfachschicht im Vergleich zu einer Viertelwellenlängen-Einfachschicht und zu einer $\tfrac{3}{4}$ Wellenlängen-Einfachschicht. Je mehr Schichten benutzt werden, mit desto höherer Ordnung kann die Reflexionskurve die Wellenlängenachse berühren, desto stärker entfernt sie sich aber schließlich von ihr. Mittelwerte gibt Tabelle 4.

Freilich lassen sich die Schichtdicken und Brechungsindices in der Praxis oft nicht genau nach einem mathematisch gegebenen Plan bemessen; dann tritt die Nullstelle höherer Ordnung zu mehreren benachbarten Nullstellen auseinander. Da zwischen ihnen die Reflexion meist noch nicht zu störenden Beträgen ansteigt, bedeutet das nur eine erwünschte Erweiterung des Wellenbereichs, solange die Fertigungstoleranzen hinreichend gering bleiben[2].

Tabelle 4.

Eigenschaften verschiedener reflexvermindernder Schichten [nach H. A. Tanner und L. B. Lockhardt, German Reflection Reducing Coatings for Glass, J. Opt. Soc. Amer. **36**, 701 (1946)].

Type	Restreflexion für weißes Licht %	Wischfestigkeit
Kryolith, vakuumbedampft	0,93	mäßig
SiO$_2$, Zentrifugalprozeß	0,2 … 0,9	schlecht
3-Schichten-Zentrifugalprozeß	0,6	gut
3-Schichten-Gasreaktionsprozeß . . .	0,2	sehr gut
MgF$_2$ vakuumbedampft	1,4	sehr gut

[1] L. Hiesinger: Optik **3**, 485 (1948).
[2] P. Leurgans: J. Opt. Soc. Amer. **39**, 639 (1949).

21. Hinweise auf weitere Lösungen der Aufgabe, mit Mehrschichtsystemen zu entspiegeln oder zu polarisieren. Fig. 31 c und b auf S. 510 zeigen die Reflexion von Mehrschichtsystemen, „Wechselschichten", die entsprechend der Fig. 31 a aufgebaut sind. Die Abszissenachse ist proportional der Frequenz geteilt. Diese Systeme bilden für gewisse Wellen eine andere Lösung der Entspiegelungsbedingung (20.1); bei ihnen wechseln Schichten hohen und niedrigen Brechungsindexes ab; die Nullstellen der Reflexion in Fig. 31 b treten infolgedessen ziemlich weit (gleichabständig) auseinander. Für die „Vergütung" kommen solche Systeme nicht in Frage, da die Reflexion zwischen den Nullstellen bereits über die Reflexion der unbeschichteten Fläche ansteigt. Deshalb wurde die Behandlung solcher Systeme in den Abschnitt „Interferenzfilter" (S. 500) verwiesen; als solche haben sie große praktische Bedeutung. Da sie in gewissen Wellenbereichen stark reflektieren, spielen sie in dem nächsten Hauptabschnitt über Reflexverstärkung die Hauptrolle.

Für die Entspiegelung in breiten Frequenzbereichen und möglichst auch für eine große Apertur des einfallenden Bündels ist im Gegensatz zu den „Wechselschichten" ein möglichst gleichmäßiger Übergang zwischen den Hauptmedien zu schaffen. Die vollkommenste Lösung dieser Aufgabe bieten die inhomogenen Schichten, die einen Grenzfall der Mehrschichtensysteme von Ziff. 20 bilden und in Ziff. 23 ausführlich behandelt werden.

Auch für die Aufgabe, mit Mehrfachschichten zu polarisieren, ist Gl. (20.1) die maßgebende Bedingung. Eine interessante Lösung gab M. Banning[1], dessen

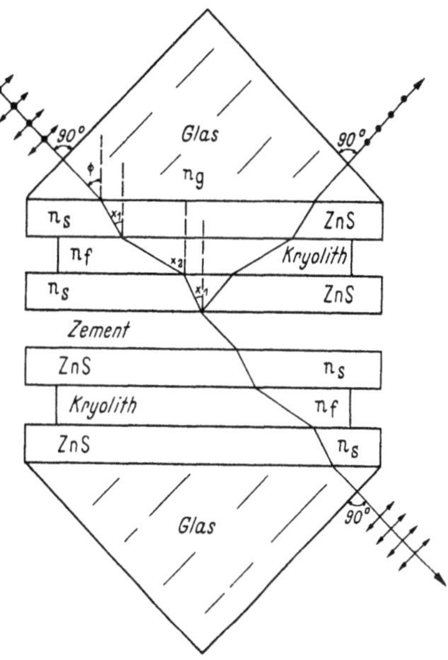

Fig. 11. Mehrschichten-Polarisator nach Banning.

Abbildung hier als Fig. 11 aufgenommen ist. Auf der Hypotenusenfläche eines Glasprismas befindet sich eine Wechselschicht aus hochbrechendem ZnS und niedrigbrechendem Kryolith. Zwei gleichartige Prismen mit beschichteten Hypotenusenflächen sind aufeinandergekittet. Das System reflektiert die *TH*-Welle nicht, da für sie die Entspiegelungsbedingung (20.1) erfüllt wurde. Die *TE*-Welle wird zwar nicht vollständig, aber wegen des Wechsels von hoch- und niedrigbrechenden Substanzen sehr stark reflektiert (vgl. den nächsten Hauptabschnitt IV); deshalb findet sich die eine Polarisation vollständig im durchgehenden, die andere fast vollständig im reflektierten Licht. Der Polarisationsgrad des reflektierten Lichts beträgt 100%, der des durchgehenden 98%.

Mit mehreren Einschichtpolarisatoren (Fig. 5 in Ziff. 14), die nicht planparallel hintereinander geschaltet waren, also einem Satz beiderseitig mit TiO_2 beschichteter Glasplatten, erreichte H. Schröder[2] die entsprechende „Lichtteilung" und sogar noch besseren Polarisationsgrad, z.B. 99,7% bei 6 TiO_2-Schichten.

---

[1] M. Banning: Phys. Rev. **59**, 914, (1947).
[2] H. Schröder: Optik **3**, 499 (1948). — H. Schröder u. R. Schläfer: Z. Naturforsch. **4a**, 576 (1949). H. Schröder: Z. angew. Phys. **3**, 53 (1951).

d) Mehrfachschichten aus Nichtleitern ohne festgelegte Schichtdicken.

22. Die Doppelschicht als Beispiel. Die Festlegung auf $\frac{\lambda}{4}$-, $\frac{3}{4}\lambda$-Schichten usw. führt zwar zu einer besonders einfachen Klasse von Lösungen des Entspiegelungsproblems, aber es gibt Lösungen auch mit anderen Schichtdicken, wenn zwei oder mehr Schichten vorliegen. Für 2 Schichten, also $m=3$, gab K. SCHUSTER[1] einen vollständigen Überblick über die Lösungsmannigfaltigkeit.

Aus Tabelle 1 entnimmt man die Entspiegelungsbedingung für $\mathfrak{r}_3 = 0$ und erhält

$$\left.\begin{aligned}(g_3-g_2)(g_2+g_1)(g_1+g_0)\,e^{\varrho_1 d_1+\varrho_2 d_2}+(g_3-g_2)(g_2-g_1)(g_1-g_0)\,e^{-\varrho_1 d_1+\varrho_2 d_2}+\\ +(g_3+g_2)(g_2-g_1)(g_1+g_0)\,e^{\varrho_1 d_1-\varrho_2 d_2}+(g_3+g_2)(g_2+g_1)(g_1-g_0)\,e^{-\varrho_1 d_1-\varrho_2 d_2}=0\,.\end{aligned}\right\} \quad (22.1)$$

Da die Exponenten rein imaginär sind, hat diese Gleichung die Form

$$A \cdot e^{i(\tau_1+\tau_2)} + B \cdot e^{i(-\tau_1+\tau_2)} + C \cdot e^{i(\tau_1-\tau_2)} + D \cdot e^{-i(\tau_1+\tau_2)} = 0 \qquad (22.2)$$

mit reellen $A, B, C, D, \tau_1, \tau_2$ nach Gl. (22.1) und ist äquivalent den beiden simultanen Bedingungen

$$(A + D)\cos(\tau_1 + \tau_2) + (B + C)\cos(\tau_2 - \tau_1) = 0, \qquad (22.3)$$

$$(A - D)\sin(\tau_1 + \tau_2) + (B - C)\sin(\tau_2 - \tau_1) = 0, \qquad (22.4)$$

d.h.

$$(A + D)\{1 - \tan\tau_1\tan\tau_2\} + (B + C)\{1 + \tan\tau_1\tan\tau_2\} = 0, \qquad (22.5)$$

$$(A - D)\{\tan\tau_1 + \tan\tau_2\} + (B - C)\{\tan\tau_2 - \tan\tau_1\} = 0 \qquad (22.6)$$

oder

$$\tan\tau_1 \cdot \tan\tau_2 = \frac{A + B + C + D}{A - B - C + D}, \qquad (22.7)$$

$$\frac{\tan\tau_1}{\tan\tau_2} = \frac{A + B - C - D}{-A + B - C + D}. \qquad (22.8)$$

Mit den Bedeutungen von $A, B, C, D, \tau_1, \tau_2$, die sich durch Identifizieren von (22.1) mit (22.2) ergeben, folgt

$$\tan^2\tau_1 = \frac{\left(1 - \dfrac{g_3}{g_0}\right)\left(\dfrac{g_1}{g_0} - \dfrac{g_3}{g_1}\right)}{\left(\dfrac{g_2}{g_1} - \dfrac{g_3 g_1}{g_2 g_0}\right)\left(\dfrac{g_3}{g_2} - \dfrac{g_2}{g_0}\right)}, \qquad (22.9)$$

$$\tan^2\tau_2 = \frac{\left(1 - \dfrac{g_3}{g_0}\right)\left(\dfrac{g_3}{g_2} - \dfrac{g_2}{g_0}\right)}{\left(\dfrac{g_2}{g_1} - \dfrac{g_3 g_1}{g_2 g_0}\right)\left(\dfrac{g_1}{g_0} - \dfrac{g_3}{g_1}\right)}. \qquad (22.10)$$

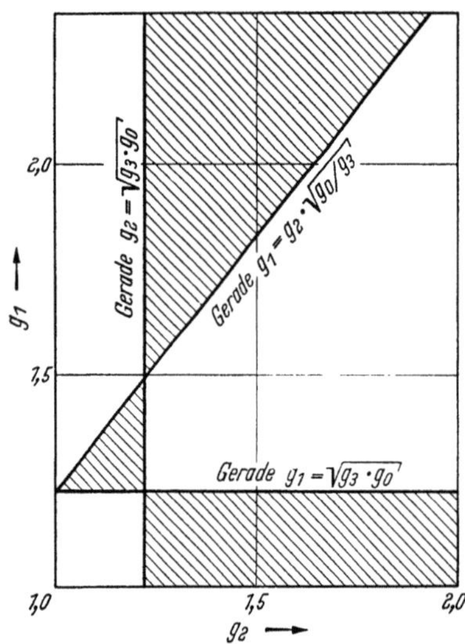

Fig. 12. Zur Entspiegelung mit einer Doppelschicht. [Nach K. SCHUSTER: Ann. Phys. (6) 4, 352 (1949).] Lösbar ist die Aufgabe, mit einer Doppelschicht zu entspiegeln, wenn die Schichtdaten g_1, g_2 den schraffierten Gebieten angehören.

Darin bedeuten nach Gl. (5.4) die Winkel

$$\tau_l = \frac{2\pi d_l}{\lambda}\,\mathfrak{n}_l\cos\varphi_l \quad \text{für } l = 1, 2. \qquad (22.11)$$

Damit $\tan\tau_1$ und $\tan\tau_2$ reell sind, müssen bei vorgegebenen g_3, g_0 die g_1, g_2 den schraffierten Gebieten der Fig. 12 angehören[1].

---

[1] K. SCHUSTER: Ann. Physik (6) **4**, 352 (1949).

23. Inhomogene Schichten[1]. Die Grundaufgabe, d. h. die Berechnung der reflektierten und durchgelassenen Wellenamplitude ist auch für inhomogene Schichten durch die Rekursionsformeln (9.14) und (9.15) des Abschnitts I gelöst, sofern sich die Schichtenkonstanten $\mathfrak{n}$ und μ nur in z-Richtung ändern.

Wir setzen dabei Stetigkeit der Konstanten in der Schicht und stetigen Anschluß an das Medium (m), aus dem die Welle einfällt, und an das Medium (0), in welches die Welle gegebenenfalls hindurchtritt, voraus. Das bedeutet keine wesentliche Einschränkung der Allgemeinheit, da sich nach der Methode des Zusammensetzens endlich viele Unstetigkeitsstellen ohne weiteres anschließend berücksichtigen lassen.

Man nähert die inhomogene Schicht dann durch ein System homogener Schichten $(m), \ldots (0)$ an, wie Fig. 13 das andeutet.

In diesem Teil interessiert vor allem die reflektierte Wellenamplitude; für sie gilt nach Gl. (9.5) die Rekursion

$$\mathfrak{r}_m = \frac{(g_m - g_{m-1})\, e^{\varrho_{m-1} d_{m-1}} + (g_m + g_{m-1})\, e^{-\varrho_{m-1} d_{m-1}}\, \mathfrak{r}_{m-1}}{(g_m + g_{m-1})\, e^{\varrho_{m-1} d_{m-1}} + (g_m - g_{m-1})\, e^{-\varrho_{m-1} d_{m-1}}\, \mathfrak{r}_{m-1}} \, . \tag{23.1}$$

Ist die Reflexion $|\mathfrak{r}_{m-1}| \ll 1$ und $|g_m - g_{m-1}| \ll |g_m + g_{m-1}|$, so wird der zweite Summand im Nenner besonders klein. Wir wollen ihn daher vernachlässigen; physikalisch bedeutet das Vernachlässigung der Zickzackreflexionen innerhalb der inhomogenen Schicht. Für die praktisch wichtigen Fälle der Entspiegelung ist das zulässig. Dann besagt (23.1)

$$\mathfrak{r}_m = \frac{g_m - g_{m-1}}{g_m + g_{m-1}} + \mathfrak{r}_{m-1} \cdot e^{-2\,\varrho_{m-1}\cdot d_{m-1}} \, , \tag{23.2}$$

$$\mathfrak{r}_{m-1} = \frac{g_{m-1} - g_{m-2}}{g_{m-1} + g_{m-2}} + \mathfrak{r}_{m-2} \cdot e^{-2\,\varrho_{m-2} d_{m-2}} \quad \text{usw.} \tag{23.3}$$

Einsetzen gibt für $\mathfrak{r}_m$ eine endliche Reihe

$$\left.\begin{aligned}
\mathfrak{r}_m = q_m &+ q_{m-1} \cdot e^{-2\,\varrho_{m-1} d_{m-1}} + q_{m-2} \cdot e^{-2\,(\varrho_{m-1} d_{m-1} + \varrho_{m-2} d_{m-2})} + \\
&+ q_{m-3} \cdot e^{-2\,(\varrho_{m-1} d_{m-1} + \varrho_{m-2} d_{m-2} + \varrho_{m-3} d_{m-3})} + \\
&+ \cdots + q_1 \cdot e^{-2\,(\varrho_{m-1} d_{m-1} + \cdots + \varrho_1 d_1)};
\end{aligned}\right\} \tag{23.4}$$

dabei ist abgekürzt

$$q_l = \frac{g_l - g_{l-1}}{g_l + g_{l-1}} \, . \tag{23.5}$$

Um der einfachen Rechnung willen setzen wir nun voraus, daß

$$\varrho_l d_l = i\,\delta \quad \text{(konstant für alle } l); \tag{23.6}$$

das ist z. B. für reelle $\mathfrak{n}$ immer durch passende Intervallteilung der ganzen Schichtdicke D (s. Fig. 13) in $(m-1)$-Schichten möglich. Dann ist (23.4)

$$\mathfrak{r}_m = \sum_{l=0}^{m-1} q_{m-l} \cdot e^{-2 i \delta l} \, . \tag{23.7}$$

---

[1] M. SCHLICK: Diss. Berlin 1904. — R. GANS: Ann. Phys. (4) **47**, 709 (1915). — K. FÖRSTERLING: Ann. Phys. (5) **11**, 1 (1931). — G. BAUER: Ann. Phys. (5) **19**, 434 (1934). — H. SCHRÖDER: Ann. Phys. (5) **39**, 55 (1941). — W. GEFFCKEN: Ann. Phys. (5) **40**, 385 (1941). — B. W. KOFINK u. E. MENZER: Ann. Phys. (5) **39**, 388 (1941). — C. D. THOMAS u. C. R. COLWELL: Phys. Rev. **56**, 1214 (1939). — D. KOSSEL: Optik **3**, 266 (1948).

Es ist nach Definition für feine Intervallteilung

$$q_{m-l} = \frac{g_{m-l} - g_{m-l-1}}{g_{m-l} + g_{m-l-1}} \approx \frac{g'(z_{m-l})\, d_{m-l}}{2 \cdot g(z_{m-l})} = \frac{g'(z_{m-l}) \cdot \delta \cdot i}{2 g(z_{m-l})\, \varrho(z_{m-l})}. \tag{23.8}$$

Um (23.7) bequem als geometrische Reihe summieren zu können, setzen wir $q_l = $ const voraus; d.h. wir verlangen, daß die Inhomogenität so beschaffen ist, $\mathfrak{n} = \mathfrak{n}(z)$ also so verläuft, daß

$$i \cdot \frac{g'(z)}{2 \cdot g(z) \cdot \varrho(z)} = K \tag{23.9}$$

ebenfalls konstant ist.

Diese Voraussetzung schränkt die Gültigkeit der Betrachtung wesentlich ein. Für TE-Wellen und $\mu = 1$ bedeutet sie z.B. wegen $\varrho(z) \sim i \cdot g(z)$, daß

$$\frac{g'(z)}{g^2(z)} = C \tag{23.10}$$

konstant ist; d.h.

$$\frac{1}{g(z)} = -\int \frac{g'(z)}{g^2(z)} dz = a(z-b), \tag{23.11}$$

also

$$g(z) = \frac{1}{a(z-b)}; \tag{23.12}$$

a, b sind Konstanten. Für TE-Wellen soll $g(z)$ also einen hyperbolischen Verlauf haben wie in Fig. 13.

Mit (23.9) folgt allgemeiner auch für TH-Wellen

$$\mathfrak{r}_m = K \cdot \delta \cdot \sum_{l=0}^{m-1} e^{-2i\delta l}, \tag{23.13}$$

$$\mathfrak{r}_m = K \cdot \delta \cdot \frac{1 - e^{-2im\delta}}{1 - e^{-2i\delta}}. \tag{23.14}$$

Bei feiner Teilung ist $|\delta| \ll 1$ nach (23.6) und also

$$\mathfrak{r}_m = K \cdot e^{-im\delta} \cdot \frac{e^{im\delta} - e^{-im\delta}}{2i}, \tag{23.15}$$

$$\mathfrak{r}_m = K \cdot e^{-im\delta} \cdot \sin(m\delta). \tag{23.16}$$

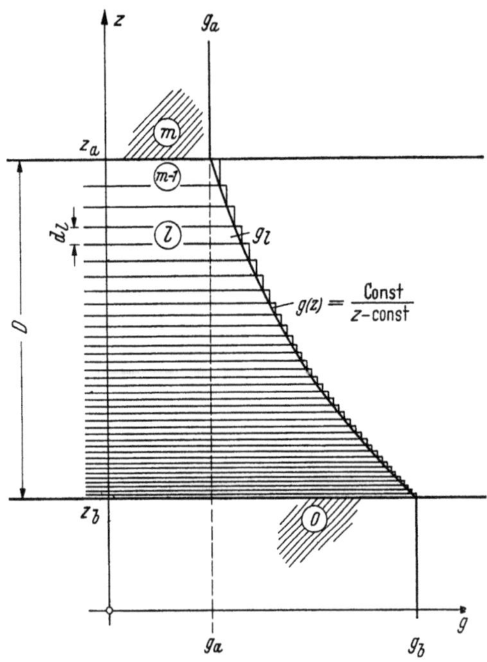

Fig. 13. Inhomogene Schicht mit hyperbolischem g-Verlauf, angenähert durch ein Schichtsystem mit flächengleicher Intervallteilung ($g_l \cdot d_l = $ const).

Nach (23.16) und (23.6) geht offenbar im Limes feiner Teilung

$$m \cdot \delta \to \int_{z_b}^{z_a} \frac{\varrho(z)}{i} dz = \frac{2\pi}{\lambda} \int_{z_b}^{z_a} \mathfrak{n}(z) \cos \varphi(z)\, dz;$$

$$\mathfrak{r}_m = \frac{g'(z)}{g(z)} \cdot \frac{e^{-\frac{2\pi i}{\lambda} \int_{z_b}^{z_a} \mathfrak{n}(z) \cos \varphi(z)\, dz}}{4\pi \mathfrak{n}(z) \cos \varphi(z)} \cdot \lambda \cdot \sin\left(\frac{2\pi}{\lambda} \int_{z_b}^{z_a} \mathfrak{n}(z) \cos \varphi(z)\, dz\right). \tag{23.17}$$

Darin ist $\varphi(z)$ eine Funktion von z, die nach dem Brechungsgesetz

$$\sin \varphi(z) = \frac{1}{\mathfrak{n}(z)} \cdot \mathfrak{n}(z_a) \cdot \sin \varphi(z_a). \tag{23.18}$$

aus dem Brechungsindex $\mathfrak{n}(z_a)$ des Einfallsmediums und dem Einfallswinkel $\varphi(z_a)$ überall entnommen werden kann.

Wegen der Differentialgleichung (23.9) kann man alle in (23.17) auftretenden Größen durch die an den Grenzen z_a, z_b vorhandenen ausdrücken; zunächst folgt aus (23.9) durch Integration

$$\log \frac{g(z_a)}{g(z_b)} = K \cdot \int_{z_b}^{z_a} \mathfrak{n}(z) \cos \varphi(z)\, dz \cdot \frac{4\pi}{\lambda}, \qquad (23.19)$$

oder

$$K = \frac{\lambda}{4\pi s} \cdot \log \frac{g(z_a)}{g(z_b)}; \qquad (23.19')$$

d.h.

$$\mathfrak{r}_m = \frac{1}{2} \cdot e^{-i\frac{2\pi s}{\lambda}} \cdot \log \frac{g(z_a)}{g(z_b)} \cdot \frac{\sin \frac{2\pi s}{\lambda}}{\frac{2\pi s}{\lambda}}, \qquad (23.20)$$

mit der Abkürzung

$$s = \int_{z_b}^{z_a} \mathfrak{n}(z) \cos \varphi(z)\, dz. \qquad (23.21)$$

Für s erhält man sowohl bei TE- als auch TH-Wellen wegen der Beziehung zwischen g und ϱ nach Gl. (23.9) ebenfalls eine Bestimmung aus den Randwerten. Zum Beispiel folgt für TE-Wellen mit $\mu = 1$ nach (23.12)

$$\left.\begin{aligned} g(z) &= \mathfrak{n}(z) \cos \varphi(z) \\ &= \frac{1}{B \cdot z}, \end{aligned}\right\} \qquad (23.22)$$

für passende Wahl des Koordinatenanfangs:

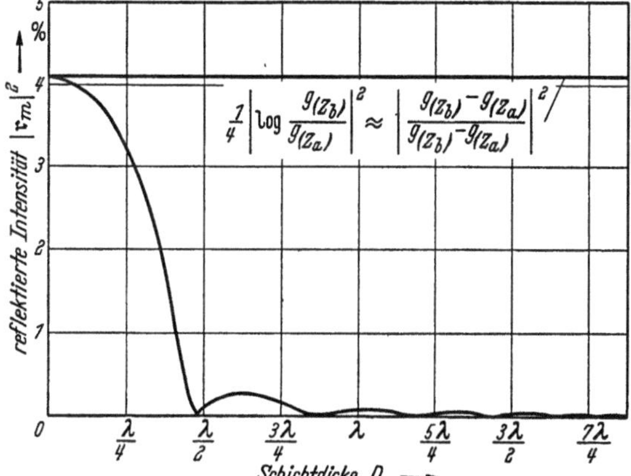

Fig. 14. Senkrecht reflektierte Intensität in Luft an einer Glasoberfläche, die mit einer inhomogenen Schicht der Dicke D (mit hyperbolischem Brechungsindexverlauf) bedeckt ist, nach Gl. (23.17).

$$s = \int_{z_b}^{z_a} g(z)\, dz = \frac{1}{B} \cdot \log \frac{z_a}{z_b} = \frac{1}{B} \cdot \log \frac{g_b}{g_a}, \qquad (23.23)$$

$$z_a = \frac{1}{B \cdot g(z_a)}; \quad z_b = \frac{1}{B \cdot g(z_b)}, \qquad (23.24)$$

$$z_a - z_b = D = \frac{1}{B} \cdot \frac{g(z_b) - g(z_a)}{g(z_a) \cdot g(z_b)}, \qquad (23.25)$$

$$s = D \cdot \frac{g(z_a)\, g(z_b)}{g(z_b) - g(z_a)} \cdot \log \frac{g_b}{g_a}. \qquad (23.26)$$

Die Gln. (23.20) und (23.21) zusammen sind eine Erweiterung auf beliebigen Einfall für die Formel, die D. Kossel[1] als seine Gl. (10) mitteilt.

Aus (23.20) und der danach berechneten Fig. 14 liest man z.B. für reelle $\mathfrak{n}$ ab, daß die reflektierte Intensität bei vorgegebenen $g(z_a)$ und $g(z_b)$ wie $\{\sin(2\pi s/\lambda): 2\pi s/\lambda\}^2$ gegen 0 geht, wenn die optische Dicke der inhomogenen Schicht hinreichend vergrößert, also der Übergang gemildert wird. Das macht die inhomogenen Schichten so vorteilhaft gegenüber den Einfachschichten, insbesondere,

---

[1] D. Kossel: Optik 3, 269 (1948).

weil dieses Verhalten für beide Polarisationen zugleich auch bei großen Aperturen und auch für alle Wellen eines endlichen Wellenlängengebiets gleichmäßig erreichbar ist. Auch bei kleinen geeigneten optischen Dicken kann nach Gl. (23.20) die Reflexion verschwinden, und zwar dann, wenn die optische Dicke s gleich einem ganzen Vielfachen der halben Wellenlänge ist.

Für $s \to 0$ ist die mehrfach in die Herleitung der Formel (23.20) eingehende Voraussetzung eines stetigen $n(z)$ nicht mehr erfüllt. Doch ist nach Gl. (23.20)

$$|\mathfrak{r}_m| = \frac{1}{2}\left|\log \frac{g(z_a)}{g(z_b)}\right| \quad \text{für} \quad s \to 0, \qquad (23.27)$$

nahezu gleich dem richtigen Wert

$$\frac{g(z_a) - g(z_b)}{g(z_a) + g(z_b)},$$

wie Tabelle 5 zeigt.

Tabelle 5.

$\dfrac{g(z_a)}{g(z_b)}$	$\dfrac{1}{2}\log\dfrac{g(z_a)}{g(z_b)}$	$\dfrac{g(z_a)-g(z_b)}{g(z_a)+g(t_b)}$
1	0	0
1,5	0,200	0,203
2,0	0,346	0,333
2,5	0,458	0,429

Eine Genauigkeitssteigerung der Theorie inhomogener Schichten läßt sich erreichen, wenn in Gl. (23.1) für $\mathfrak{r}_{m-1}$ im Nenner ein Näherungswert eingesetzt wird; doch lohnt sich der Aufwand kaum, da man in der Praxis eine inhomogene Schicht ohnehin nicht genau nach einem geforderten Brechungsindexverlauf fertigen kann.

e) Mehrfachschichten mit Leitern und Nichtleitern.

24. Läßt man in Mehrfachschichten auch Leiter zu, so hat die Entspiegelungsbedingung eine sehr große Mannigfaltigkeit von Lösungen. Sie lautet nach Tabelle 1 z. B. für 3 Grenzflächen

$$\mathfrak{r}_3 = 0 = (g_3 - g_2)\{(g_2 + g_1)(g_1 + g_0) + (g_2 - g_1)(g_1 - g_0)e^{-2\varrho_1 d_1}\} + {} \\ + (g_3 + g_2)\{(g_2 - g_1)(g_1 + g_0) + (g_2 + g_1)(g_1 - g_0)e^{-2\varrho_1 d_1}\}e^{-2\varrho_2 d_2}. \Bigg\} \quad (24.1)$$

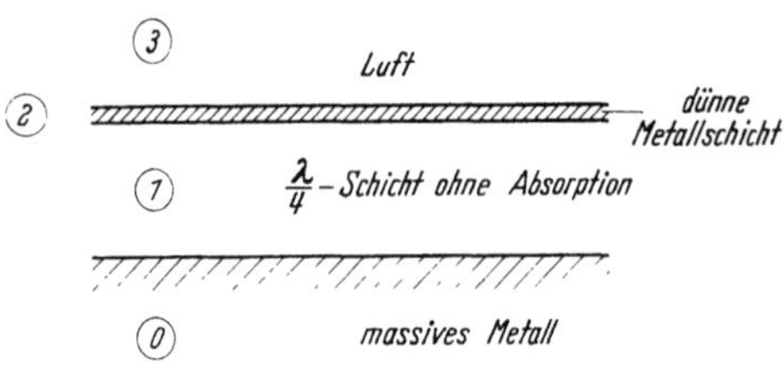

Fig. 15. Entspiegelung massiven Metalls durch λ/4-Nichtleiterschicht und aufgebrachte dünne Metallschicht passenden Flächenwiderstandes.

Hier sei als Beispiel nur der in Fig. 15 skizzierte Fall betrachtet, daß auf einer massiven Metallplatte ⓪ eine nichtleitende Viertelwellenlängenschicht, d. h. eine Schicht solcher Dicke sich befindet, daß

$$e^{-2\varrho_1 d_1} = e^{-i\frac{4\pi d_1}{\lambda}n_1\cos\varphi_1} = e^{-i\pi} = -1.$$

Diese Schicht sei von einer dünnen Metallschicht der Dicke $d_2 \ll \lambda$ bedeckt; dafür ist (24.1) einfacher zu formulieren

$$0 = (g_3 - g_2)\{g_2 g_0 + g_1^2\} + (g_3 + g_2)\{g_2 g_0 - g_1^2\}(1 - 2\varrho_2 d_2), \qquad (24.2)$$

d. h.

$$\varrho_2 d_2 = \frac{g_2(g_3 g_0 - g_1^2)}{(g_3 + g_2)(g_2 g_0 - g_1^2)}. \qquad (24.3)$$

Werden die Konstanten der Schichten und der Einfallswinkel vorgegeben, so kann hieraus nach (24.3) die erforderliche Schichtdicke d_2 berechnet werden. Da (24.3) jedoch zwei reelle Gleichungen repräsentiert, ist außer d_2 auch den Schichtkonstanten noch ein Zwang auferlegt.

Wir betrachten speziell den für lange Wellen interessierenden Fall, daß die Metalle sehr hohe Leitfähigkeit haben und nach (2.8) daher

$$\mathfrak{n}_0^2 \approx -i\frac{2\sigma_0\lambda}{c}; \quad \mathfrak{n}_2^2 \approx -i\frac{2\sigma_2\lambda}{c} \qquad (24.4)$$

gesetzt werden kann mit $|\mathfrak{n}_0| \gg 1$; $|\mathfrak{n}_2| \gg 1$. Dann sind nach dem Brechungsgesetz φ_0 und φ_2 komplexe Winkel von kleinem Betrage und $\cos\varphi_2 \approx \cos\varphi_0 \approx 1$. Für TE-Wellen und $\mu_l = 1$ ist also

$$g_0 \approx \sqrt{\frac{\sigma_0 \lambda}{c}}\,(1 - i); \qquad g_2 \approx \sqrt{\frac{\sigma_2 \lambda}{c}}\,(1 - i) \qquad (24.5)$$

$$\varrho_2 \approx \frac{2\pi}{\lambda}\,\sqrt{\frac{\sigma_2 \lambda}{c}}\,(1 + i). \qquad (24.6)$$

Da $|g_2| \gg 1$, $|g_0| \gg 1$ ist und wir g_1 als von der Größenordnung 1 ansehen, besagt dann (24.3) einfach

$$\varrho_2 d_2 \approx \frac{g_3}{g_2}, \qquad (24.7)$$

also

$$\frac{d_2 \sigma_2}{c} \approx \frac{n_3 \cos\varphi_3}{4\pi}. \qquad (24.8)$$

Dies heißt offenbar, daß der Flächenwiderstand der dünnen Metallschicht die Größe

$$\frac{1}{\sigma_2 d_2} \approx \frac{4\pi}{c \cdot n_3 \cdot \cos\varphi_3}, \qquad (24.9)$$

d.h. in praktischen Einheiten den Wert

$$R = \frac{120\pi}{n_3 \cos\varphi_3}\,\Omega \qquad (24.10)$$

haben soll. Für Luft als Einfallsmedium und senkrechten Einfall soll also der Flächenwiderstand $377\,\Omega$ betragen[1]. Für Entspiegelung von TE-Wellen bei schräg einfallenden Wellen muß R entsprechend (24.10) vergrößert werden.

Für Entspiegelung bei TH-Wellen folgt aus der allgemeinen Bedingung (24.3) wegen $|g_2| \ll 1$; $|g_0| \ll 1$ ebenso

$$\varrho_2 d_2 \approx \frac{g_2}{g_3} = \frac{n_3}{\mathfrak{n}_2} \cdot \frac{1}{\cos\varphi_3}; \qquad (24.11)$$

d.h.

$$\frac{d_2 \sigma_2}{c} \approx \frac{n_3}{4\pi \cos\varphi_3}. \qquad (24.12)$$

Der Flächenwiderstand muß dann

$$R \approx \frac{120\pi}{n_3} \cdot \cos\varphi_3\,\Omega \qquad (24.13)$$

sein. Vollständige Entspiegelung ist also für beide Polarisationen zugleich nur bei senkrechtem Einfall erreichbar.

IV. Steigerung der Reflexion durch dünne Schichten.

a) Nichtleiter.

25. Die Einfachschicht. Wenn eine Welle teils an der Vorder- und teils an der Rückseite einer $\lambda/4$-Schicht reflektiert wird, so treffen beide Wellen interferierend mit einem Gangunterschied von $\lambda/2$ aufeinander. Werden beide Wellen jeweils an einem dichteren Medium — oder beide an einem dünneren Medium — reflektiert, so erleiden beide den gleichen Phasensprung bei ihrer Reflexion, und deshalb ist durch Interferenz eine Reflexminderung evident. Das wurde oben gezeigt und näher erläutert.

---

[1] Siehe Fußnote 1, S. 483.

Wenn aber die dünne Schicht (Medium ①) einen Brechungsindex n_1 größer als die Brechungsindices n_2, n_0 der beiden angrenzenden Medien ② und ⓪ hat, so ist der Phasensprung bei der einen Reflexion um 180° verschieden von dem bei der anderen Reflexion, und man kann Verstärkung der Reflexion durch die $\lambda/4$-Schicht erwarten. Dasselbe gilt, wenn n_1 kleiner als n_0 und n_2 ist.

Die reflektierte Intensität ist nach Tabelle 1 im Falle zweier Grenzflächen

$$|\mathfrak{r}_2|^2 = \left|\frac{g_2 g_0 - g_1^2}{g_2 g_0 + g_1^2}\right|^2 = \left|\frac{1 - \left(\dfrac{g_1}{\sqrt{g_2 g_0}}\right)^2}{1 + \left(\dfrac{g_1}{\sqrt{g_2 g_0}}\right)^2}\right|^2 \tag{25.1}$$

für Schichten mit einer Dicke d_1, die der Bedingung $e^{2\varrho_1 d_1} = e^{\pm i\pi}$ d.h. nach (5.4)

$$\left.\begin{aligned} d_1 &= \frac{\lambda}{4 n_1 \cos \varphi_1} \cdot M \\ [M\ \text{ganz, ungerade}] \end{aligned}\right\} \tag{25.2}$$

gehorcht.

Wenn g_1 das geometrische Mittel von g_2 und g_0 ist, so wirkt die Schicht entspiegelnd; sie erhöht dagegen die Reflexion gegenüber dem Wert ohne Schicht $\left|\dfrac{g_2 - g_0}{g_2 + g_0}\right|^2$ wesentlich, wenn g_1^2 sich weit von $g_2 g_0$ entfernt; das zeigt die Fig. 16.

Ein Spiegel, der 50% bei senkrechtem Einfall reflektiert, entsteht beispielsweise aus einer Glasplatte $(n_0 = 1{,}5)$ in Luft $(n_2 = 1{,}0)$, wenn eine $\lambda/4$ dicke Schicht vom Brechungsindex

$$n_1 = 2{,}4 \cdot \sqrt{1{,}5 \cdot 1} = 2{,}9$$

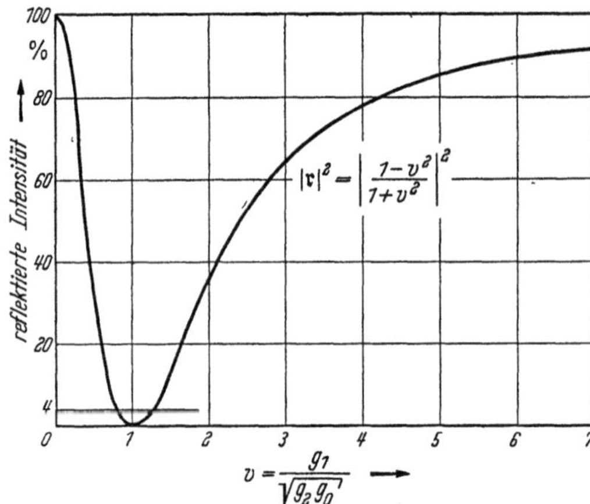

Fig. 16. Reflektierte Intensität bei einer Einzelschicht in Abhängigkeit vom Verhältnis des Schichtbrechungsindexes zum geometrischen Mittel des Brechungsindexes der angrenzenden Medien. Für senkrechten Einfall ist $v = \dfrac{n_1}{\sqrt{n_2 \cdot n_0}}$.

auf die Glasplatte gebracht wird. So hohe Brechungsindices sind in dünnen Metalloxydschichten realisiert.

Einige Werte gibt die Tabelle 6, auszugsweise entnommen einer Arbeit von K. HAMMER[1].

Tabelle 6.

Oxyd von	Reflektierte Intensität	Durchgelassene Intensität	Brechungsindex der Schicht	Farbe in Durchsicht
Antimon	15%	85%	1,85	
Beryllium	7	82	1,6	
Blei	32	67	2,3	
Cer CeO_2)	29	70	2,25	
Nickel	26	69	2,15	grau
Tellur	18	81	1,95	
Titan (TiO_2) . . .	40	59	2,6	
Wismut	36	63	2,45	
Wolfram	20	80	2,00	
Eisen (Fr_2O_3) . . .	49	43	2,95	
Kupfer (Cu_2O) . .	45	49	2,75	braun
Silizium	58	34	3,30	

[1] K. HAMMER: Optik **3**, 496 (1948).

26. Erweiterung auf Mehrfachschichten. Die Fig. 16 gibt zugleich die reflektierte Intensität für eine Mehrfachschicht aus $\lambda/4$-Schichten wieder. Denn nach Gl. (19.14) ist für eine solche Schicht mit m Grenzflächen die reflektierte Intensität

$$|\mathfrak{r}_m|^2 = \left|\frac{1-v^2}{1+v^2}\right|^2,\tag{26.1}$$

wenn nun unter v verstanden wird

$$v = \frac{g_{m-1}\cdot g_{m-3}\cdots}{\sqrt{g_m}\cdot g_{m-2}\cdot g_{m-4}\cdots};\tag{26.2}$$

dabei ist ein im Zähler oder Nenner auftretendes g_0 nicht voll als Faktor sondern statt dessen nur $\sqrt{g_0}$ einzusetzen. Um ein von 1 möglichst verschiedenes v zu erreichen, wird man Schichten mit großen Brechungsindices und solche mit kleinen Brechungsindices abwechseln lassen. Zum Beispiel erhält man bei senkrechtem Einfall ($g_l = n_l$) für fünf Schichten, d.h. $m = 6$

$$v = \frac{n_5\cdot n_3\cdot n_1}{\sqrt{n_6}\cdot n_4\cdot n_2\cdot \sqrt{n_0}}.\tag{26.3}$$

Für Luft ($n_6 = 1$) und Glas ($n_0 = 1{,}5$) erhält man mit $n_4 = n_2 = 1{,}5$ (Zaponlack) und $n_5 = n_3 = n_1 = 2{,}6$ (TiO$_2$)

$$v = \frac{2{,}6^3}{1{,}5^{2,5}} = 6{,}4;$$

das bedeutet nach (26.1) und Fig. 16 eine Reflexion von 92%. Der Pseudo-Brechungsindex, definiert analog Gl. (19.20), für die oben genannte Schicht ist

$$n_s = \frac{2{,}6^3}{1{,}5^2} = 7{,}8.$$

Mehrfachschichten dieser Art können nahezu 100% reflektieren, ohne daß man zu allzu hohen und schwer realisierbaren Brechungsindices Zuflucht zu nehmen braucht. Dafür sorgen die Multiplikationen in (26.3), die schnell zu hohen v-Werten führen[1].

27. Abhängigkeit der Reflexsteigerung von der Wellenlänge. Die Bedingung, daß die Schicht eine $\lambda/4$-Schicht sei, allgemein die Bedingung (25.2), ist nur für diskrete Wellenlängen streng erfüllbar. Daß aber dann auch für Nachbarwellenlängen die Reflexion noch gut bleibt, zeigt eine Berechnung der reflektierten Intensität $|\mathfrak{r}_3|^2$ nach Tabelle 1. Fig. 17 zeigt das Ergebnis. Der sichtbare Wellenbereich ist durch starke senkrechte Linien abgegrenzt. Für Vielfachschichten gibt Abschnitt V (Ziff. 47 und Fig. 31) genauere Auskunft über die Wellenlängenabhängigkeit.

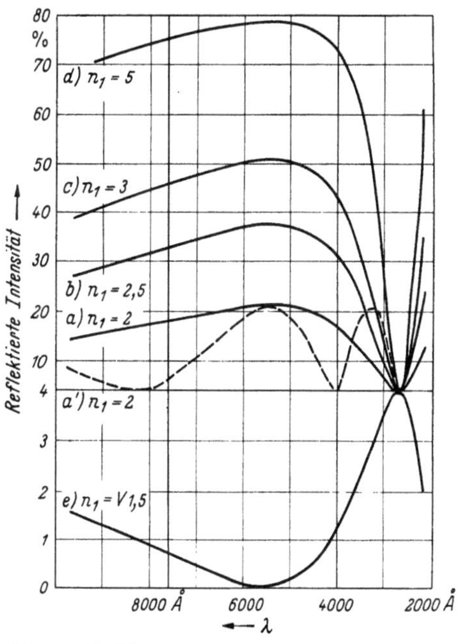

Fig. 17. Die Wellenlängenabhängigkeit der Reflexion von reflexionserhöhenden und reflexionsvermindernden Schichten. Alle Kurven gelten für $n_2 = 1$ und $n_0 = 1{,}5$. Die ausgezogenen Kurven beziehen sich auf eine Dicke $n_1 d_1 = \frac{1}{4}\cdot 540\,\mathrm{m\mu}$, die gestrichelte Kurve auf $n_1 d_1 = \frac{3}{4}\cdot 540\,\mathrm{m\mu}$. [Nach K. HAMMER: Z. techn. Phys. 24, 173 (1943).]

28. Abhängigkeit von der Schichtdicke. Man könnte sich vorstellen, daß auch $\lambda/2$-Schichten gute Reflexion geben, da sich dann auch die vorn und die hinten reflektierten Wellen durch Interferenz verstärken können. Doch muß dann die

---

[1] Vgl. C. H. CARTRIGHT u. A. F. TURNER: Phys. Rev. **55**, 1128 (1939). — H. SCHRÖDER: Z. angew. Phys. **3**, 53 (1951) mit Verzeichnis vieler einschlägiger Originalarbeiten.

Schicht einen zwischen den beiden anderen Brechungsindices liegenden Brechungsindex haben, und das führt zwangsläufig zu schwachen Teilreflexionen.

Nach Tabelle 1 ist für $\lambda/2$-Schichten (d.h. $e^{-2\varrho_1 d_1} = 1$)

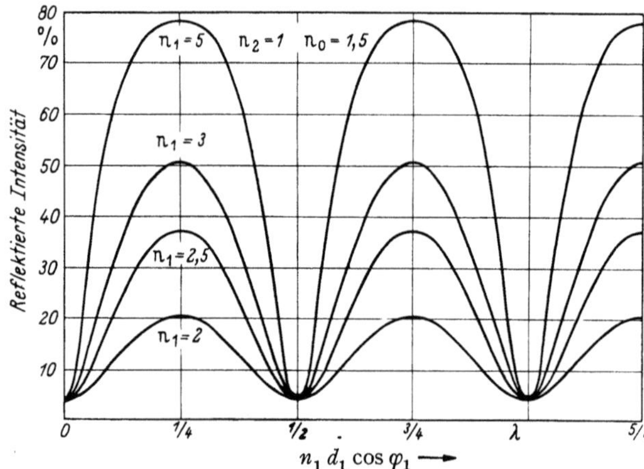

Fig. 18. Die Abhängigkeit des Reflexionsvermögens von der Schichtdicke für absorptionsfreie Schichten mit vier verschiedenen Brechungsindices der Schicht. [Nach K. HAMMER: Z. techn. Physik **24**, 171 (1943).]

$$r_2 = \frac{g_2 g_1 - g_1 g_0}{g_2 g_1 + g_1 g_0} \left.\right\} \quad (28.1)$$
$$= \frac{g_2 - g_0}{g_2 + g_0},$$

d.h. die Reflexion ist dann sogar dieselbe, als ob die Schicht nicht vorhanden wäre, unabhängig von ihrem Brechungsindex. Stellt man also die reflektierte Intensität für senkrechten Einfall als Funktion der optischen Dicke $n_1 d_1$ dar,

so gehen bei $n_1 d_1 = \frac{\lambda}{2} \cdot M$

(M ganz) alle Kurven durch denselben Punkt. Fig. 18 gibt ein solches Bild für eine Schicht auf Glas. $\lambda/4$-Schichten sind für die Reflexsteigerung optimal.

b) Beteiligung von Leitern.

29. Schichten mit Absorption. Die in Tabelle 6 aufgeführten Metalloxyde absorbieren praktisch ein wenig; das hat bei einer $\lambda/4$-Schicht eine geringe Verschlechterung gegenüber der Formel (26.1) zur Folge.

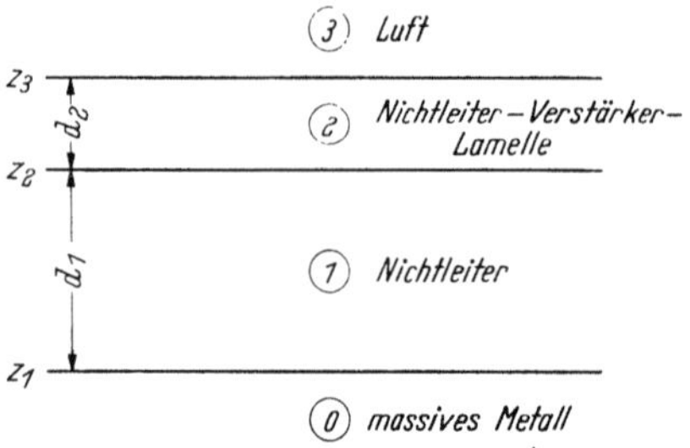

Eine ausführliche experimentelle und theoretische Darstellung gibt K. HAMMER[1].

Bei Mehrfachschichten kann aber auch eine Kombination von Leitern und Nichtleitern zu besonders starken Reflexionen führen. R. MESSNER[2] hat eine Kombination einer Metallplatte mit einer nichtleitenden Verstärkerlamelle angegeben (Fig. 19).

Fig. 19. Reflexionsverstärkung eines massiven Metallspiegels durch nichtleitende Verstärkerlamelle.

Das Schichtsystem besteht aus einer massiven Metallplatte (Medium ⓪) und einer Nichtleiterfolie (Medium ②), die wegen ihres hohen Brechungsindexes selbst stark reflektiert und in einen solchen Abstand von der Metallplatte gelegt wird, daß die von ihr reflektierte Welle sich phasenrichtig zu der vom Metall reflektierten addiert.

Zweckmäßig wählt man die Schicht (2) als $\lambda/4$-Schicht, die besonders stark reflektiert, wenn ihr Brechungsindex größer ist als der ihrer beiderseitigen Umgebung. Ihr Reflexionsfaktor ist dann analog Gl. (25.1) reell

$$r_{\mathrm{I}} = \frac{g_3 g_1 - g_2^2}{g_3 g_1 + g_2^2}, \qquad (29.1)$$

---

[1] K. HAMMER: Z. techn. Phys. **24**, 174 (1943).
[2] R. MESSNER: Optik **2**, 228 (1947).

bezogen auf einen Ort z_3 (Fig. 19). Die Reflexion der Metallplatte ist

$$\mathfrak{r}_{II} = \frac{g_1 - g_0}{g_1 + g_0}\,, \tag{29.2}$$

bezogen auf den Ort z_1.

Nach Gl. (7.7) des Abschnitts II über das Zusammenfügen zweier Schichtsysteme ist die Reflexion des Gesamtsystems

$$\mathfrak{r} = \frac{\mathfrak{r}_I^{\circ} + \mathfrak{r}_{II}^{\circ}\left(\mathfrak{d}_I^{\circ}\,\mathfrak{d}_I^{\prime\circ} - \mathfrak{r}_I^{\circ}\,\mathfrak{r}_I^{\prime\circ}\right)}{1 - \mathfrak{r}_I^{\prime\circ}\,\mathfrak{r}_{II}^{\circ}}\,. \tag{29.3}$$

Die Größen $\mathfrak{r}_I^{\circ}, \mathfrak{r}_{II}^{\circ}, \ldots$ sind aus (29.1) und (29.3) zu berechnen, indem die Wellen auf einen gemeinsamen Nullpunkt bezogen werden. Dann ist entsprechend (8.9)

$$\mathfrak{r}_I^{\circ} = \mathfrak{r}_I\,e^{2\,\varrho_3 z_3}\,; \qquad \mathfrak{r}_{II}^{\circ} = \mathfrak{r}_{II}\,e^{2\,\varrho_1 z_1}\,; \qquad \mathfrak{r}_I^{\prime\circ} = \mathfrak{r}_I\,e^{-2\,\varrho_1 z_1}\,. \tag{29.4}$$

Nach Tabelle 1 und entsprechend Gl. (8.8) ist

$$\mathfrak{d}_I^{\circ} = \pm\,i\,\frac{2\,g_3 g_2}{g_1 g_3 + g_2^2}\cdot e^{\varrho_3 z_3 - \varrho_1 z_2}\,; \qquad \mathfrak{d}_I^{\prime\circ} = \pm\,i\,\frac{2\,g_1 g_2}{g_3 g_1 + g_2^2}\cdot e^{-\varrho_1 z_2 + \varrho_3 z_3}\,. \tag{29.5}$$

Daraus folgt zunächst[1]

$$\begin{aligned}
\mathfrak{d}_I^{\circ}\,\mathfrak{d}_I^{\prime\circ} - \mathfrak{r}_I^{\circ}\,\mathfrak{r}_I^{\prime\circ} &= \left(\frac{-\,4\,g_2^2 g_1 g_3}{(g_1 g_3 + g_2^2)^2} - \frac{(g_1 g_3 - g_2^2)^2}{(g_1 g_3 + g_2^2)^2}\right)\cdot e^{2\,\varrho_3 z_3 - 2\,\varrho_1 z_2} \\
&= -\,e^{2\,\varrho_3 z_3 - 2\,\varrho_1 z_2}\,.
\end{aligned} \tag{29.6}$$

Einsetzen in (29.3) gibt, wenn wir die in $\mathfrak{r}$ ins Verhältnis gesetzten Wellen auch am Orte z_3 nehmen,

$$\mathfrak{r} = \frac{\mathfrak{r}_I - \mathfrak{r}_{II}\,e^{-2\,\varrho_1 d_1}}{1 - \mathfrak{r}_I\,\mathfrak{r}_{II}\,e^{-2\,\varrho_1 d_1}}\,. \tag{29.7}$$

$\mathfrak{r}_I$ ist in unserem Falle reell. Man wird daher d_1 so wählen, daß $\mathfrak{r}_{II}\,e^{-2\,\varrho_1 d_1}$ ebenfalls reell wird und das entgegengesetzte Vorzeichen wie $\mathfrak{r}_I$ annimmt.

Dann wird die reflektierte Intensität

$$|\mathfrak{r}|^2 = \left(\frac{|\mathfrak{r}_I| + |\mathfrak{r}_{II}|}{1 + |\mathfrak{r}_I|\cdot|\mathfrak{r}_{II}|}\right)^2 \tag{29.8}$$

relativ groß. Fig. 20 zeigt mit jeder ausgezogenen Kurve $|\mathfrak{r}|^2$ als Funktion der Verstärkerfolienreflexion $|\mathfrak{r}_I|^2$ für verschiedene Werte der Metallreflexion $|\mathfrak{r}_{II}|^2$.

Die Bedeutung der MESSNERschen Verstärkerfolie erhellt daraus, daß eine solche Folie, die selbst z. B. 20% Intensität reflektiert, die Reflexion eines Metallspiegels z. B. von 56 auf 80%, von 80 auf 92%, von 90 auf 96% und von 96 auf 98% erhöht (Fig. 20!).

Die Wirkung des Verfahrens läßt sich noch steigern, wenn man eine zweite Verstärkerfolie in passendem Abstand vor der ersten anbringt.

Durch andere Wahl des Abstandes d_1 kann die Reflexion auch auf 0 gebracht werden. Dazu muß $-\mathfrak{r}_{II}\,e^{-2\,\varrho_1 d_1}$ die entgegengesetzte Richtung von $\mathfrak{r}_I$ erhalten, ferner muß $|\mathfrak{r}_I| = |\mathfrak{r}_{II}|$ gemacht werden. Bedeutung kann das für Polarisation bei schrägem Einfall oder für „Schwärzung" einer Metallwand durch Vorsetzen einer nichtleitenden Schicht gewinnen. Im ersten Falle ist das Verfahren dem

---

[1] Eine allgemeine Formel analog (29.6), wie sie in der Literatur gelegentlich für beliebige Schichtdicken d_2 unterstellt wird, gilt nicht.

von SCHOPPER[1] behandelten analog (Ziff. 17), im zweiten steht es zu der Anbringung einer dünnen Metallschicht in $\lambda/4$-Abstand vor der Metallwand (Ziff. 24) in Konkurrenz. Auch die in jenen Ziffern beschriebenen Verfahren lassen sich leicht aus Formel (29.7) verstehen.

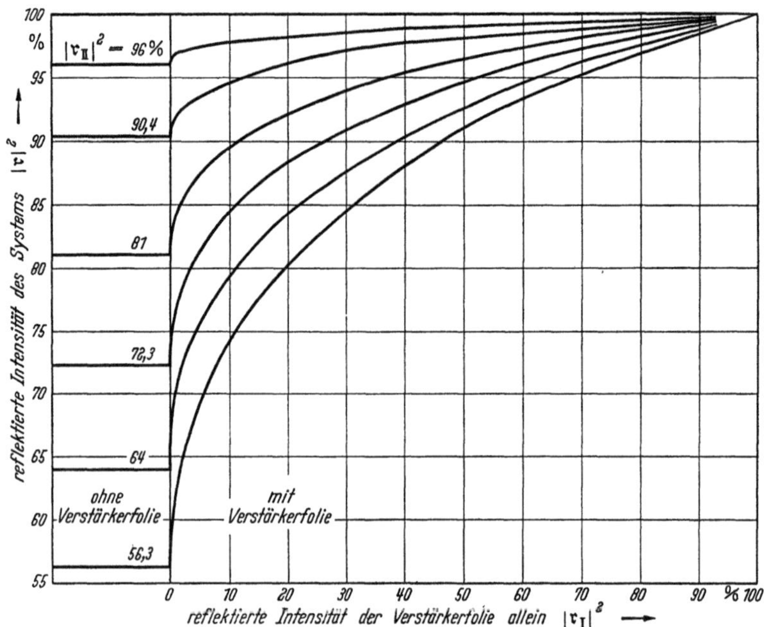

Fig. 20. Wirkung einer MESSNERschen Verstärkerfolie zur Reflexerhöhung bei einem Metallspiegel.

V. Interferenzfilter.

a) Filter mit Metallschichten.

30. Einfachfilter. Bei den bisher in diesem Artikel betrachteten Problemen waren gewisse Eigenschaften wie Reflexfreiheit für eine oder für jede Polarisation, hohe Durchlässigkeit oder Reflexionsverstärkung für möglichst große Wellenbereiche erwünscht.

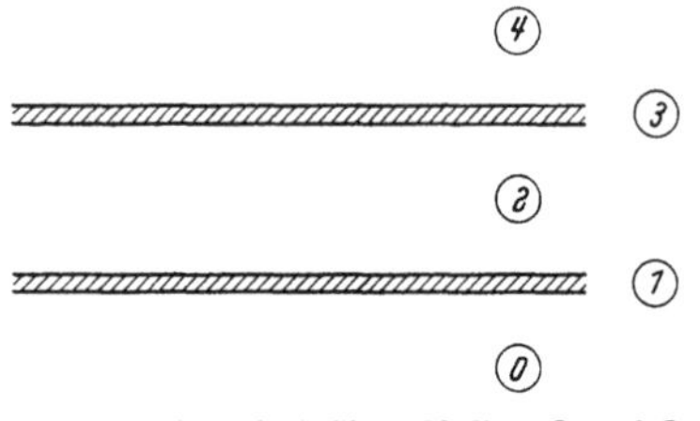

Fig. 21. Einfachschichtfilter. Medium ① und ③ sind dünne Metallschichten, ⓪, ② und ④ sind Nichtleiter.

In der angewandten Optik spielen eine ebenso wichtige Rolle die Schichten und Schichtsysteme, die solche Eigenschaften infolge der Interferenz des Lichtes selektiv besitzen und dann als „Interferenzfilter" bezeichnet werden.

Da alle in den bisherigen Abschnitten diskutierten Schichtsysteme wellenlängenabhängige Eigenschaften haben, können sie als Interferenzfilter im weiteren Sinne bezeichnet werden und zur unterschiedlichen Behandlung verschiedener Wellenlängen benutzt werden. So geben die Fig. 17 und 18 zugleich einen Einblick in die Wirksamkeit eines Reflexions-Interferenzfilters.

Angenehmer für die meisten Anwendungen sind Durchsichtsfilter[2], und sie sollen als Beispiele in diesem Abschnitt daher zunächst betrachtet werden.

---

[1] H. SCHOPPER: Optik **10**, 426 (1953).
[2] W. GEFFCKEN: D.R.P. Nr. 716153, Kl. 42h, Gr. 34,11 vom 8. 12. 1939.

Die wichtigsten Eigenschaften der Interferenzfilter lassen sich bereits an dem in Fig. 21 skizzierten einfachen Schichtsystem übersehen. Es besteht aus einer Schicht ② der Dicke d_2, die sich zwischen zwei Schichten z. B. Metallschichten ① und ③ befindet mit Dicken d_1 und d_3.

Die Lichtwelle falle von ④ her auf das Schichtsystem; für das Amplitudenverhältnis der Durchlässigkeit gilt nach Gl. (7.8) und für die Reflexion nach Gl. (7.7)

$$\mathfrak{d}\circ = \frac{\mathfrak{d}_I^\circ\,\mathfrak{d}_{II}^\circ}{1 - \mathfrak{r}_I^{\prime\circ}\,\mathfrak{r}_{II}^\circ}\,; \qquad \mathfrak{r}\circ = \frac{\mathfrak{r}_I^\circ + \mathfrak{r}_{II}^\circ\big(\mathfrak{d}_I^\circ\,\mathfrak{d}_I^{\prime\circ} - \mathfrak{r}_I^\circ\,\mathfrak{r}_I^{\prime\circ}\big)}{1 - \mathfrak{r}_I^{\prime\circ}\,\mathfrak{r}_{II}^\circ}\,. \tag{30.1}$$

Dabei sind die Reflexionsfaktoren $\mathfrak{r}_I^\circ, \mathfrak{r}_I^{\prime\circ}, \mathfrak{r}_{II}^\circ$ und die Durchlässigkeitsfaktoren $\mathfrak{d}_I^\circ, \mathfrak{d}_I^{\prime\circ}, \mathfrak{d}_{II}^\circ$ auf einen gemeinsamen Nullpunkt $(0, 0, 0)$ bezogen. Da für jede nach unten gerichtete Welle nach Gl. (3.3)

$$\mathfrak{A}_l(0;0;z) = \mathfrak{A}_l(0;0;0) \cdot e^{\varrho_l z} \qquad \text{mit} \quad \varrho_l = 2\pi i\,\frac{n_l \cos \varphi_l}{\lambda} \tag{30.2}$$

und für jede nach oben laufende Welle

$$\mathfrak{R}_l(0;0;z) = \mathfrak{R}_l(0;0;0) \cdot e^{-\varrho_l z} \tag{30.3}$$

gilt, so ist

$$\left.\begin{aligned}
\mathfrak{d}_I &= \mathfrak{d}_I^\circ\, e^{\varrho_2 z_3 - \varrho_4 z_4}; & \mathfrak{d}_{II} &= \mathfrak{d}_{II}^\circ\, e^{\varrho_0 z_1 - \varrho_2 z_2}; & \mathfrak{d} &= \mathfrak{d}\circ\, e^{\varrho_0 z_1 - \varrho_4 z_4}, \\
\mathfrak{r}_I &= \mathfrak{r}_I^\circ \cdot e^{-2\varrho_4 z_4}; & \mathfrak{r}_{II} &= \mathfrak{r}_{II}^\circ\, e^{-2\varrho_2 z_2}; & \mathfrak{r} &= \mathfrak{r}\circ \cdot e^{-2\varrho_4 z_4}.
\end{aligned}\right\} \tag{30.4}$$

Für den Reflexionsfaktor $\mathfrak{r}_I^\prime$ der Schicht I von unten her vertauschen $\mathfrak{A}$- und $\mathfrak{R}$-Wellen ihre Rolle, jedes ϱ ist also durch $-\varrho$ zu ersetzen, ferner tauschen die Medien ④ und ② ihre Rolle und die Grenzen z_3 und z_4. Daher ist aus (30.4) zu schließen

$$\mathfrak{r}_I^\prime = \mathfrak{r}_I^{\prime\circ}\, e^{2\varrho_2 z_3}; \qquad \mathfrak{d}_I^\prime = \mathfrak{d}_I^{\prime\circ}\, e^{-\varrho_4 z_4 + \varrho_2 z_3}. \tag{30.5}$$

Einsetzen in (30.1) gibt

$$\mathfrak{d} = \frac{\mathfrak{d}_I\,\mathfrak{d}_{II} \cdot e^{-\varrho_2 d_2}}{1 - \mathfrak{r}_I^\prime\,\mathfrak{r}_{II}\, e^{-2\varrho_2 d_2}}\,; \qquad \mathfrak{r} = \frac{\mathfrak{r}_I + \mathfrak{r}_{II}\,(\mathfrak{d}_I\,\mathfrak{d}_I^\prime - \mathfrak{r}_I\,\mathfrak{r}_I^\prime)\, e^{-2\varrho_2 d_2}}{1 - \mathfrak{r}_I^\prime\,\mathfrak{r}_{II} \cdot e^{-2\varrho_2 d_2}}\,. \tag{30.6}$$

Das gilt ganz allgemein, wie auch die Schichtsysteme I und II beschaffen sein mögen.

Um einen ungefähren Einblick in die Wellenlängenabhängigkeit der Durchlässigkeit zu bekommen, setzen wir den Fall, daß $|\mathfrak{d}_I| \cdot |\mathfrak{d}_{II}|$ und

$$\mathfrak{r}_I^\prime\,\mathfrak{r}_{II} = q \cdot e^{2\pi i \delta} \tag{30.7}$$

wellenlängenunabhängig seien, wie es z. B. mit Metallschichten ① und ③ für große Wellenlängenbereiche realisierbar ist. Ferner absorbiere das Medium ② nicht. Dann ist

$$\frac{1}{|\mathfrak{d}|} = \frac{1}{|\mathfrak{d}_I||\mathfrak{d}_{II}|} \cdot \left| 1 - q \cdot e^{2\pi i\left(\delta - \frac{2 d_2 n_2 \cos \varphi_2}{\lambda}\right)} \right|. \tag{30.8}$$

Das ist bis auf den wellenlängenunabhängigen Nenner gleich dem Abstand des Punktes 1 in der GAUSSschen Zahlenebene von den Punkten eines Ursprungskreises des Radius q (Fig. 22). Der Kreis wird einmal durchlaufen, wenn $2 d_2 n_2 \cos \varphi_2/\lambda$ um 1 wächst. Bei den Wellenlängen λ_M, für die

$$\frac{2 d_2 n_2 \cos \varphi_2}{\lambda_M} - \delta = M \tag{30.9}$$

(genannt „Ordnung") eine ganze Zahl ist, hat der Abstand ein Minimum, $|\mathfrak{b}|$ ein Maximum. Seine Höhe errechnet man aus

$$|\mathfrak{b}|_{\text{Max}} = \left| \frac{\mathfrak{b}_{\text{I}}\,\mathfrak{b}_{\text{II}}}{1-q} \right| = \frac{|\mathfrak{b}_{\text{I}}| \cdot |\mathfrak{b}_{\text{II}}|}{1 - |\mathfrak{r}'_{\text{I}}| \cdot |\mathfrak{r}_{\text{II}}|}. \tag{30.10}$$

Je stärker die Schichten (1) und (3) reflektieren, je größer also q ist, desto „schneller" wächst in Fig. 22 und 23 der „Abstand" von seinem Minimum $1-q$ auf das $\sqrt{2}$-fache; d.h. desto schmaler ist die „Halbwertsbreite" $2\varDelta$ für die durchgelassene Intensität.

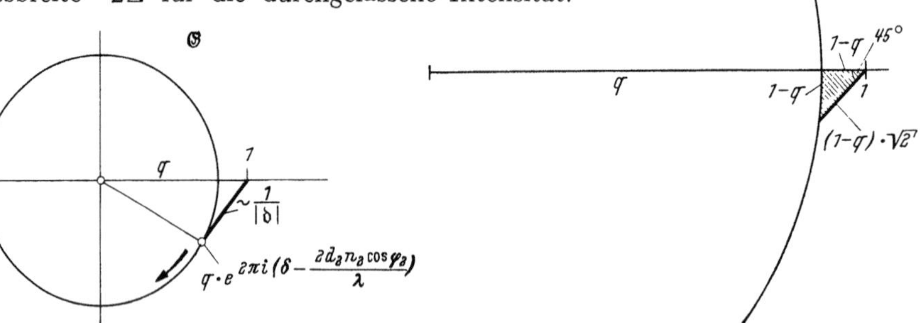

Fig. 22. Reziproke Durchlässigkeitsamplitude in der GAUSS-schen Zahlenebene.

Fig. 23. Zur Halbwertsbreite des Durchlaßbereichs eines Einfach-Interferenzfilters.

Wenn q nur wenig kleiner ist als 1, kann die Halbwertsbreite ungefähr aus dem kleinen schraffierten „Dreieck" in Fig. 23 entnommen werden.

$$\varDelta \left(q \cdot e^{2\pi i \left(\delta - \frac{2\,d_2\,n_2\cos\varphi_2}{\lambda}\right)} \right) \approx 1 - q; \tag{30.11}$$

d.h.

$$|\varDelta\lambda| \cdot \left| \frac{d}{d\lambda} e^{2\pi i \left(\delta - \frac{2\,d_2\,n_2\cos\varphi_2}{\lambda}\right)} \right| \approx \frac{1-q}{q};$$

$$|\varDelta\lambda| \cdot 2\pi \cdot \frac{2\,d_2\,n_2\cos\varphi_2}{\lambda^2} \approx \frac{1-q}{q}.$$

Die relative Halbwertsbreite ist

$$\left| \frac{2\varDelta\lambda}{\lambda} \right| \approx \frac{1-q}{q} \cdot \frac{\lambda}{2\pi\,d_2\,n_2\cos\varphi_2} = \frac{1-q}{q\,(M+\delta)\,\pi}. \tag{30.12}$$

Das Filter erhält also einen schmalen Durchlaßbereich, wenn man die Reflexion der Schichten (1) und (3) hoch wählt; doch ist hier ein Kompromiß nötig, da hohe Reflexion geringe Durchlässigkeit bedingt. Zur quantitativen Diskussion wollen wir annehmen, daß (0), (2), (4) gleiche Medien und die Schichten (1) und (3) miteinander identisch seien. Dann ist nach (30.10) die durchgelassene Intensität im Maximum

$$J_{d\,(\text{Max})} = \frac{J^2_{d\,\text{I}}}{(1 - J_{r\,\text{I}})^2} = \left(\frac{J_{d\,\text{I}}}{J_{d\,\text{I}} + J_{a\,\text{I}}} \right)^2. \tag{30.13}$$

Darin bezeichnet $J_{d\,\text{I}}$, $J_{r\,\text{I}}$, $J_{a\,\text{I}}$ die von einer Schicht (z. B. I oder II) durchgelassene, reflektierte bzw. absorbierte Intensität. $J_{d\,(\text{Max})}$ wäre 1, wenn die Metallschichten nicht absorbierten. Bei dicken Metallschichten überwiegt aber schließlich die absorbierte gegenüber der durchgelassenen Intensität. Wenn beste Lichtausbeute bei dem Filter erwünscht ist, wird man also die Metallschichten (1) und (3) nur etwa so dick wählen, daß $J_{a\,\text{I}}$ kleiner als $J_{d\,\text{I}}$ bleibt. Bei dem in dieser Hinsicht im Sichtbaren günstigen Silber wird man also bis zu einer Reflexion von 94,5% gehen, da dann $J_{d\,\text{I}} = 3\%$ und $J_{a\,\text{I}} = 2,5\%$, also $J_{d\,\text{Max}} = 30\%$ hat.

Die relative Halbwertsbreite nach (30.12) ist dann etwa

$$\frac{2\Delta\lambda}{\lambda} \approx \frac{5,5\%}{(M+\delta)\cdot\pi}. \tag{30.14}$$

Da bei Silber $\delta \approx 0,7$ ist, wie noch unten gezeigt wird, so erhält man die Werte der Tabelle 7.

Tabelle 7.

Ordnung M	0	1	2	3	4	5
Relative Halbwertsbreite	2,3%	1%	0,6%	0,5%	0,4%	0,3%

Wenn hoher Intensitätsverlust um viele Zehnerpotenzen in Kauf genommen werden kann, so ist mit Silberschichten im Sichtbaren eine Reflexion von 96% und daher noch $\frac{2}{3}$ der in Tabelle 7 angegebenen Halbwertsbreite erreichbar. Doch

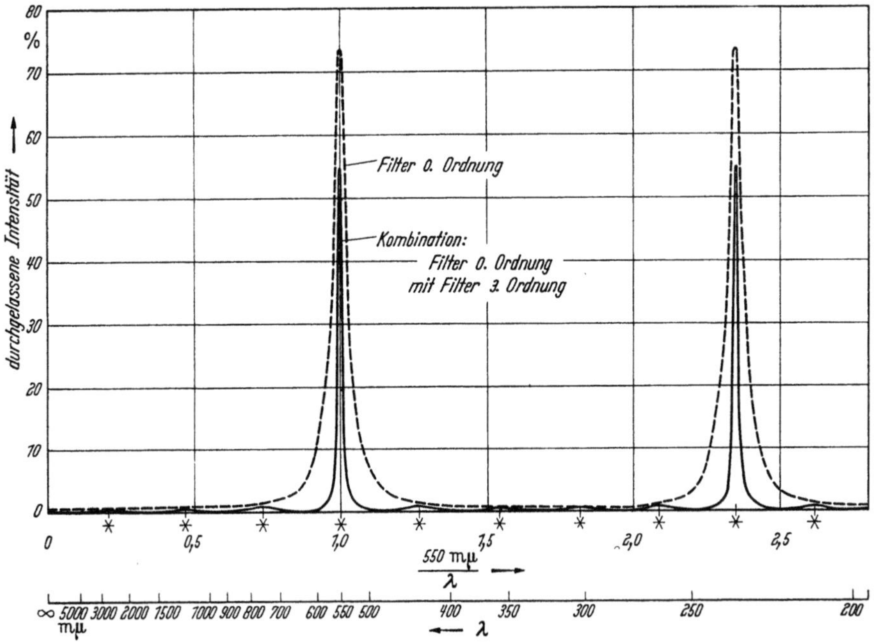

Fig. 24. Durchgelassene Intensität für ein Interferenzfilter nullter Ordnung und für eine Kombination aus dem Filter nullter Ordnung und einem Filter dritter Ordnung.

ist dann die Benutzung höherer Ordnungen vorteilhafter; die dann zugleich durchgelassenen Nachbarordnungen können durch Farbfilter oder ein Interferenzfilter niedriger Ordnung abgefiltert werden. Die Wirkung einer solchen Filterkombination zeigt Fig. 24. Da es sich um ein lichtstarkes Filter handeln sollte, wurde jede der Versilberungen nur auf 86% Reflexion gebracht bei 12% Durchlässigkeit. Die Reflexionsphase ist bei solchen relativ dicken Schichten praktisch gleich der einer unendlich dicken Schicht, für die bei annähernd senkrechtem Einfall gilt

$$\mathfrak{r}_{\mathrm{II}} = \frac{g_2 - g_1}{g_2 + g_1} = \frac{n_2 - n_1 + i\,k_1}{n_2 + n_1 - i\,k_1} = \frac{(n_2^2 - n_1^2 - k_1^2) + 2i\,k_1 n_2}{(n_2 + n_1)^2 + k_1^2}, \tag{30.15}$$

$$\operatorname{arc}\mathfrak{r}_{\mathrm{II}} = \operatorname{arc\,tan} \frac{2 n_2 k_1}{n_2^2 - n_1^2 - k_1^2}. \tag{30.16}$$

Mit den von Krautkrämer[1] gemessenen Werten an Ag für $k_1 = 3{,}5$; $n_1 = 0{,}1$ und $n_2 = 1{,}5$ erhält man arc $\mathfrak{r}_{II} = 2\pi \cdot 0{,}379$. Da $\mathfrak{r}'_I = \mathfrak{r}_I$ ist, folgt $\delta = 0{,}758$.

Um in nullter Ordnung die Wellenlänge $\lambda_0 = 550$ mμ als Durchlaßwelle zu erhalten, ist nach Gl. (30.9) eine Dicke erforderlich

$$d_2 = \frac{0{,}758 \cdot 550 \,\mathrm{m\mu}}{2 \cdot n_2 \cdot \cos \varphi_2} = 150 \,\mathrm{m\mu}$$

bei einem Winkel $\varphi_2 \approx 22°$. Man wählt d_2 zweckmäßig etwas größer, als dem Winkel $\varphi_2 = 0$ entsprechen würde ($d_2 = 139$ mμ), um durch Schwenken des Filters die Durchlaßwelle noch etwas zu kürzeren und längeren Wellen schieben zu können. Fig. 24 zeigt die nach (30.8) mit diesen Daten berechnete Durchlässigkeit $|\mathfrak{d}|^2$ in der punktierten Kurve.

31. Filterkombination ohne Kopplung. Ein wesentlich selektiveres Filter erhält man, wenn man die Ordnung $M = 3$ wählt und also die Dicke

$$d_2 = \frac{(3 + 0{,}758) \cdot 500 \,\mathrm{m\mu}}{2 \cdot n_2 \cdot \cos \varphi_2} = 700 \,\mathrm{m\mu}$$

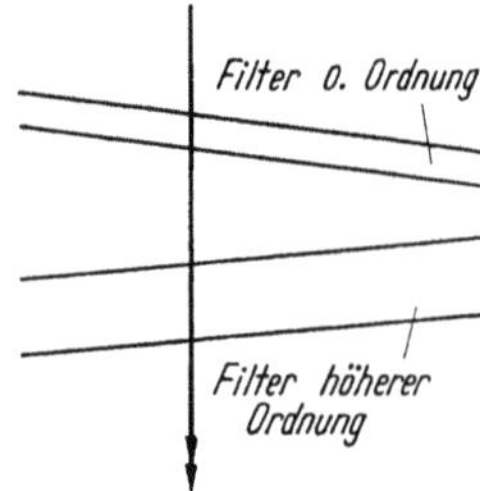

Fig. 25. Kombination zweier Einfach-Interferenzfilter ohne Kopplung.

bei einem Winkel $\varphi_2 \approx 10°$ macht. Doch erhält man dann (Fig. 24) weitere Maxima an allen mit * versehenen Stellen. Sie werden zu unbedeutenden Werten herabgedrückt, indem dieses Filter mit dem Filter nullter Ordnung kombiniert wird. Das Resultat zeigt die ausgezogene Kurve in Fig. 24. Die experimentelle Anordnung ist in Fig. 25 skizziert.

Die zwischen den Filtern reflektierten Strahlen werden durch die Winkelstellung der Filter unter 32° gegeneinander herausgeworfen. Diese lichtstarke Filterkombination läßt bei $\lambda = 550$ mμ 55% des auftreffenden Lichtes hindurch, hat eine relative Halbwertsbreite von 4% und hat sonst nach Rot und Ultrarot zu unbegrenzt weniger als 1% Durchlässigkeit, ebenso für Wellen bis zu 250 mμ herab. Bei $\lambda = 235$ mμ ist das Filter wieder stark durchlässig, da zufällig die erste Ordnung des dünnen und die achte Ordnung des dickeren Filters nahe zusammenfallen. Ein geringes Schwenken des einen Filters führt dann zu einem vollständigen Zusammenfall der Maxima und macht die Kombination zu einem Filter für 235 mμ. Wegen der stärkeren Veränderlichkeit der optischen Konstanten des Silbers im Ultraviolett gibt Fig. 24 die Kurve in diesem Bereich jedoch nur qualitativ wieder.

Bei anderer Bemessung der Filterdicke kann man freilich auch erreichen, daß die nullte Ordnung des dünnen Filters von dem dickeren Filter nahezu ausgelöscht wird und der Sperrbereich des Filters sich auch weiter ins Ultraviolett erstreckt.

Die kleine Apertur eines solchen Filters wird dadurch ausgeglichen, daß es mit sehr großer Fläche hergestellt werden kann. Da nur das Produkt aus Apertur und Filterdurchmesser abbildungsinvariant ist, kann man ein Wellenbündel der kleinen Apertur und der großen Fläche eines Interferenzfilters meist anpassen.

Beachtet man die Schiebbarkeit der Wellenlänge durch Filterschwenken, so zeigt dieses Beispiel die Vielseitigkeit einer solchen Filterkombination, die dadurch auch einem Mehrschichtenfilter aus fest aufeinander aufgebrachten abwechselnden Metall- und Nichtleiterschichten überlegen ist.

---

[1] J. Krautkrämer: Ann. Phys. (5) **32**, 537 (1938).

32. Mehrschichtenfilter mit Kopplung. Als Beispiel eines Mehrschichtenfilters sei das Doppelfilter nach Fig. 26 betrachtet.

Die geradzahligen Medien seien einander gleiche Nichtleiter (z. B. $n_2 = n_4 = n_6 = 1,5$); die Schichten ①, ③, ⑤ seien unter sich identische Metallschichten der Durchlässigkeitsamplitude $\mathfrak{d}_I$ und der Reflexionsamplitude $\mathfrak{r}_I = \mathfrak{r}_I'$; nun ist die Durchlässigkeitsamplitude des Filters nach den Gln. (30.6)

$$|\mathfrak{d}| = \frac{|\mathfrak{d}_I| \cdot |\mathfrak{d}_{II}|}{|1 - \mathfrak{r}_I \mathfrak{r}_{II} \cdot e^{-2\varrho_4 d_4}|}, \tag{32.1}$$

mit

$$|\mathfrak{d}_{II}| = \frac{|\mathfrak{d}_I|^2}{|1 - \mathfrak{r}_I^2 e^{-2\varrho_2 d_2}|} \tag{32.2}$$

und

$$\mathfrak{r}_{II} = \mathfrak{r}_I \frac{1 + (\mathfrak{d}_I^2 - \mathfrak{r}_I^2) \cdot e^{-2\varrho_2 d_2}}{1 - \mathfrak{r}_I^2 e^{-2\varrho_2 d_2}}. \tag{32.3}$$

Sind die Metallschichten nur wenig durchlässig, kann also $\mathfrak{d}_I^2$ neben $\mathfrak{r}_I^2$ vernachlässigt werden, so ist $\mathfrak{r}_{II} = \mathfrak{r}_I$ und daher

$$|\mathfrak{d}| \approx \frac{|\mathfrak{d}_I|^3}{|(1 - \mathfrak{r}_I^2 e^{-2\varrho_2 d_2})(1 - \mathfrak{r}_I^2 e^{-2\varrho_4 d_4})|}. \tag{32.4}$$

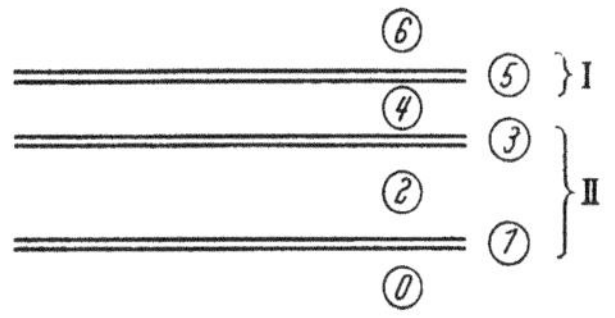

Fig. 26. Mehrschichten-Interferenzfilter mit Kopplung.

Die Durchlässigkeit hat also denselben relativen Verlauf, als wenn zwei Einfachfilter der Dicken d_4 bzw. d_2 kopplungsfrei kombiniert werden. Dazu vergleiche man das Beispiel der vorhergehenden Ziffer. Doch hat das Filter nach Fig. 26 den Vorteil, daß eine Silberschicht eingespart und dadurch in (32.4) rechts nur $|\mathfrak{d}_I|^3$ statt des bei der kopplungsfreien Kombination unvermeidlichen $|\mathfrak{d}_I|^4$ auftritt, also Licht gespart wird.

Die Voraussetzung $|\mathfrak{d}_I|^2 \ll |\mathfrak{r}_I|^2$ bedeutet also eine Entkopplung beider Filterhälften, und die Größe $\mathfrak{d}_I^2$ in Gl. (32.3) erfaßt die in Strenge je nach Dicke der Versilberung ③ mehr oder weniger enge Kopplung beider Filterplatten aneinander. Mit Berücksichtigung dieser Kopplung erhält man aus (32.1), (32.2) und (32.3) streng

$$|\mathfrak{d}| = \frac{|\mathfrak{d}_I|^3}{|(1 - \mathfrak{r}_I^2 \cdot e^{-2\varrho_2 d_2})(1 - \mathfrak{r}_I^2 e^{-2\varrho_4 d_4}) - \mathfrak{d}_I^2 \mathfrak{r}_I^2 e^{-2\varrho_2 d_2 - 2\varrho_4 d_4}|}. \tag{32.5}$$

Fig. 27 zeigt an drei Beispielen in der Gaussschen Zahlenebene einen Teil der vom ersten Summanden im Nenner der Gl. (32.5) durchlaufenen Kurve. Der zweite Summand des Nenners durchläuft einen kleinen Kreis um den Nullpunkt. Der Nenner als Abstand solcher Punkte hat bei kleinen $|\mathfrak{d}_I|^2$ fast dieselben Werte wie bei Entkopplung, die durch Schrumpfung des Kreises auf sein Zentrum dargestellt würde. Nur die Abstände, die von nahe am Nullpunkt gelegenen Punkten der den ersten Summanden darstellenden Kurve ausgehen, sind relativ stark verändert. Die resultierende Durchlässigkeit ist in Fig. 27 rechts für jedes der Beispiele gegeben[1].

Mehrfachfilter zeigen Kopplungseffekte vor allem an den Maxima, ähnlich wie gewisse Bandfilter der Hochfrequenztechnik, die auch mathematisch analog behandelt werden können[2].

---

[1] Für die Ausführung der numerischen Rechnung dankt Verfasser Herrn cand. rer. nat. U. Gradmann. Eine Veröffentlichung erscheint in den Optica Acta.

[2] Vgl. auch W. Geffcken: Z. angew. Phys. **6**, 249 (1954). — Ch. Dufour: J. Phys. Radium **11** (1950).

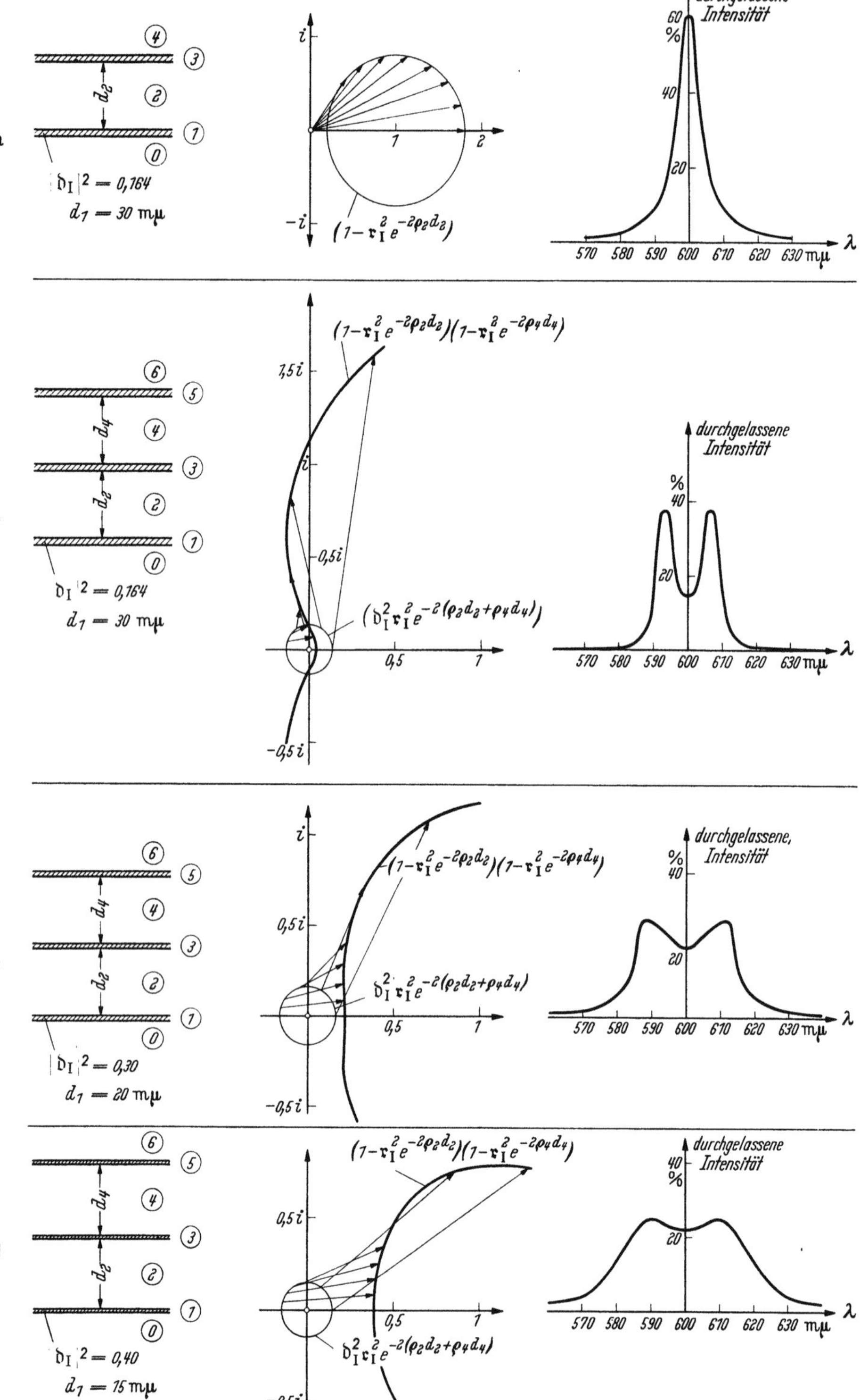

Fig. 27 a—d. Zur Kopplung von Interferenzfiltern. a Einfachfilter $|\mathfrak{d}_I|^2 = 0{,}164$. b Kopplung zweier Einfachfilter durch eine Schicht mit $|\mathfrak{d}_I|^2 = 0{,}164$. c Kopplung zweier Einfachfilter durch eine Schicht mit $|\mathfrak{d}_I|^2 = 0{,}30$. d Kopplung zweier Einfachfilter durch eine Schicht mit $|\mathfrak{d}_I|^2 = 0{,}40$.

Bei einem Vergleich ist zu beachten, daß jedes der Mehrfachfilter nach Fig. 27 unter sich gleiche Metallschichten als Medien ①, ③ usw. enthält und dadurch Kopplung und Resonanzschärfe bei dem Übergang vom Fall b zu c und d zugleich variieren.

Ein Mehrfachschichtenfilter mit vier dünnen Silberschichten ①, ③, ⑤, ⑦ und drei nichtleitenden Zwischenmedien ②, ④, ⑥ mit einem Aufbau nach Fig. 28 links hat einen Durchlaßbereich von 25 mμ in Bandfilterform nach Fig. 28 rechts.

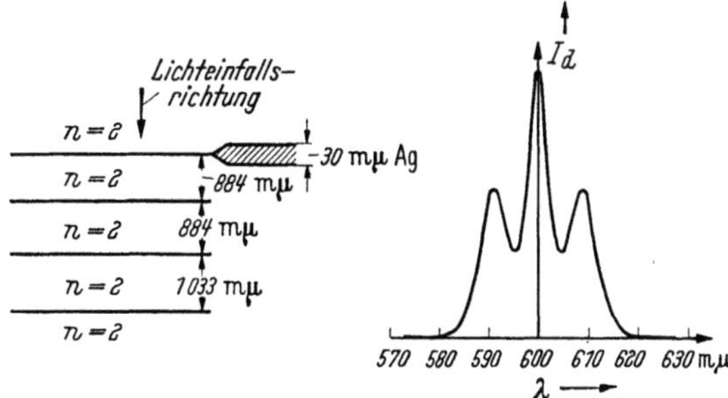

Fig. 28. Interferenzfilter aus drei Nichtleiterschichten und vier Leiterschichten mit Kopplung. Links: Aufbau, rechts: durchgelassene Intensität als Funktion der Wellenlänge.

33. Ultrarotsperre; Ultrarotpaß. Das Filter nullter Ordnung nach Fig. 24 sperrt das gesamte Ultrarot, läßt aber im Sichtbaren um $\lambda = 550$ mμ gut durch. Als Wärmeschutzfilter für Projektoren z.B. interessieren solche Filter, doch soll der Durchlässigkeitsbereich für diesen Zweck breiter sein. Dazu braucht nur die Reflexion der Schichten ① und ③ in Fig. 23 verringert zu werden, z.B. auf je 60%; das Ergebnis zeigt qualitativ die Fig. 29 mit der ausgezogenen Kurve.

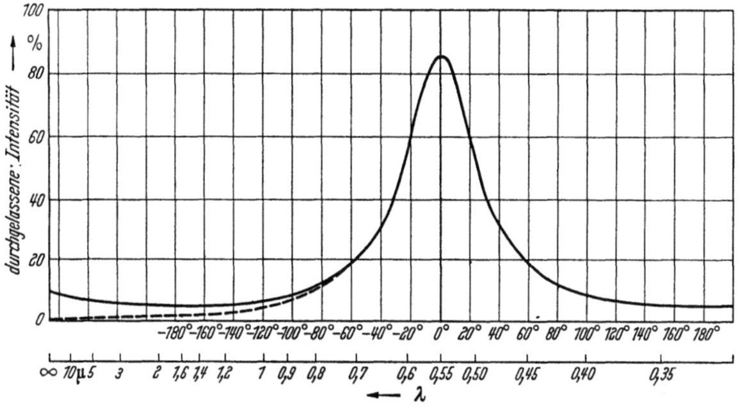

Fig. 29. Durchgelassene Intensität bei einem GEFFCKEN-Filter geringer Selektivität.

Da jedoch Silber im Ultrarot besser als im Sichtbaren reflektiert, ist dieses Filter als Ultrarotsperre praktisch noch wirksamer (gestrichelte Kurve).

Wählt man die Schicht ② dicker, so daß das in Fig. 29 bei $\lambda = 550$ mμ liegende Maximum ins nahe Ultrarot rückt und der zwischen dem Maximum nullter Ordnung und dem Maximum erster Ordnung liegende Sperrbereich das Sichtbare überdeckt, dann erhält man einen Ultrarotpaß, der das Ultrarot durchläßt, aber das sichtbare Licht gut reflektiert. Solche Filterbeläge auf Kugelflächen sind als Hohlspiegel für Projektorbeleuchtungen geeignet, da sie ebenso wie die Ultrarotsperre eine Filmüberhitzung hintanhalten.

Doch gibt es neuerdings für diese beiden Aufgaben bessere Lösungen mit Hilfe nichtabsorbierender Mehrschichtenfilter; sie sind der Gegenstand des nächsten Abschnitts.

b) Filter ohne Absorption.

34. Vorzüge absorptionsfreier Filter[1]. Interferenzfilter, die allein aus absorptionsfreiem Material aufgebaut sind, interessieren vor allem aus zwei Gründen.

Erstens kann man hoffen, aus sehr vielen passend bemessenen Schichten mit abwechselnd hohem und niedrigem Brechungsindex die stark reflektierenden Metallschichten der Filter nach Fig. 21 oder Fig. 26 zu ersetzen und zu lichtstärkeren Interferenzfiltern schmalsten Durchlaßbereiches zu kommen.

Zweitens ist für Filter mit breitem Durchlaßgebiet, wie sie in Ziff. 33 diskutiert wurden, nur eine geringe Reflexion erforderlich, wie man sie auch ohne Metallbelag an den Grenzen verschiedener Nichtleiter hat. Als Ultrarotpaß oder Ultrarotsperre z.B. sind solche absorptionsfreien Interferenzfilter wegen ihrer geringen Erwärmung besonders erwünscht.

35. Das dünne nichtabsorbierende Blättchen als Interferenzfilter in Durchsicht und Reflexion. Die dünne nichtabsorbierende Schicht ① mit der Dicke d_1 liege zwischen nichtabsorbierenden Medien ② und ⓪. Nach Tabelle 1 sind die Amplitudendurchlässigkeit und die Reflexion

$$|\mathfrak{d}| = \frac{\dfrac{4g_1 g_2}{(g_2 + g_1)(g_1 + g_0)}}{\left| 1 - \dfrac{g_2 - g_1}{g_2 + g_1} \cdot \dfrac{g_0 - g_1}{g_0 + g_1} \cdot e^{-2\pi i \, 2 d_1 n_1 \cos \varphi_1/\lambda} \right|} , \tag{35.1}$$

$$|\mathfrak{r}| = \left| \frac{\dfrac{g_2 - g_1}{g_2 + g_1} + \dfrac{g_1 - g_0}{g_1 + g_0} \cdot e^{-2\pi i \, 2 d_1 n_1 \cos \varphi_1/\lambda}}{1 - \dfrac{g_2 - g_1}{g_2 + g_1} \cdot \dfrac{g_0 - g_1}{g_0 + g_1} \cdot e^{-2\pi i \, 2 d_1 n_1 \cos \varphi_1/\lambda}} \right| . \tag{35.2}$$

Daraus erhält man die Intensitäten

$$J_d = \frac{n_0}{n_2} |\mathfrak{d}|^2 , \tag{35.3}$$

$$J_r = |\mathfrak{r}|^2 . \tag{35.4}$$

Der Nenner ist von derselben Gestalt wie in Gl. (30.6). Doch ist hier die Größe

$$q = \frac{g_2 - g_1}{g_2 + g_1} \cdot \frac{g_0 - g_1}{g_0 + g_1} \tag{35.5}$$

auf jeden Fall reell, und zwar negativ, wenn g_1 zwischen g_0 und g_2 liegt, sonst positiv.

1. Fall: g_1 zwischen g_0 und g_2; $q < 0$.

Der zweite Summand im Nenner durchläuft in der GAUSSschen Zahlenebene einen Kreis wie in Fig. 22. Der Nenner in den Gln. (35.1) und (35.2) ist der Abstand des jeweiligen Kreispunktes vom Punkte 1. $|\mathfrak{d}|$ hat Maxima für maximale Annäherung des Kreispunktes an 1, also für $e^{-2\pi i \, 2 d_1 n_1 \cos \varphi_1/\lambda} = -1$ d.h. für

$$\frac{2 d_1 n_1 \cos \varphi_1}{\lambda} = \frac{1}{2} + M \qquad [M \text{ ganze Zahl, ,,Ordnung''}]. \tag{35.6}$$

Ein Ergebnis zeigt Fig. 30a (obere Kurve). Da $|q| \ll 1$ ist, wenn g_1 zwischen g_0 und g_2 liegt, hat man nur schwache Selektivität.

[1] K. HAMMER: Optik **5**, 365 (1949). — K. HAMMER: Z. techn. Phys. **24**, 169 (1943). — H. SCHRÖDER: Z. angew. Phys. **3**, 53 (1951). — K. BLODGETT: Phys. Rev. **55**, 391 (1939).

2. Fall: g_1 *nicht zwischen* g_0 *und* g_2; $q > 0$.

$|\mathfrak{d}|$ hat Maxima für

$$\frac{2 d_1 n_1 \cos \varphi_1}{\lambda} = M. \qquad [M \text{ ganze Zahl, ,,Ordnung''}]. \qquad (35.7)$$

Für $g_0 = 1{,}5$; $g_1 = 2{,}6$; $g_2 = 1{,}0$, wie das bei TiO_2 auf Glas bei senkrechtem Einfall realisiert ist, erhält man einen Verlauf der durchgelassenen Intensität nach Fig. 30a (untere Kurve).

Da sich durchgelassene und reflektierte Intensität zu 1 ergänzen, kann die reflektierte Intensität zugleich entnommen werden. Für $g_0 = g_2 = 1{,}0$; $g_1 = 2{,}6$ erhält man einen Verlauf nach Fig. 30b.

Die Reflexion geht dann auf 0, wenn die Schichtdicke ein ganzes Vielfaches der halben Wellenlänge im Material ist. Eine $\lambda/2$-Schicht zwischen gleichartigen Medien wirkt auf die reflektierte und durchgehende Intensität so, als ob sie nicht vorhanden wäre.

Solche Schichten sind als Reflexionsfilter verwendbar, wenn es darauf ankommt, einen mäßig schmalen Wellenbereich aus dem Spektrum fernzuhalten.

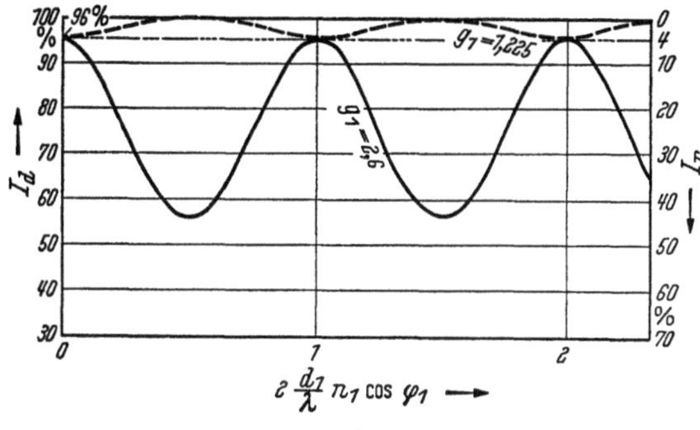

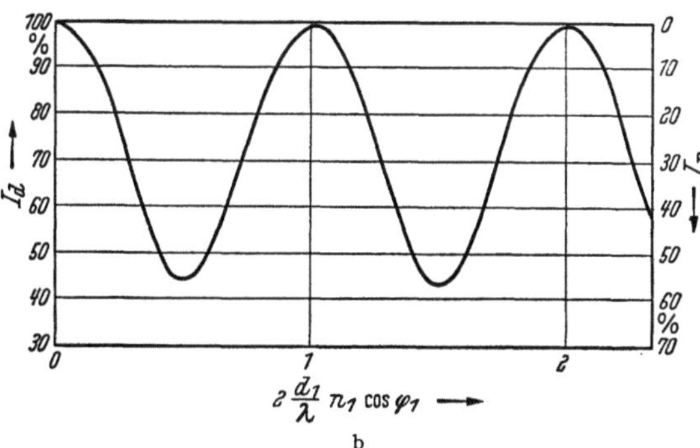

36. Nichtabsorbierende Mehrfachschichten[1].

In Durchsicht hat die ,,freie'' dünne nichtabsorbierende Schicht

Fig. 30a u. b. a Durchlässigkeit und Reflexion einer dünnen Schicht mit $g_1 = 1{,}225$ (- - -) bzw. $g_1 = 2{,}6$ (———) zwischen Medien mit $g_0 = 1{,}5$ und $g_2 = 1{,}0$. b Durchlässigkeit und Reflexion einer dünnen Schicht mit $g_1 = 2{,}6$ zwischen Medien mit $g_0 = 1{,}0 = g_2$.

nach Fig. 30b die höchste denkbare Lichtausbeute 100% für die ,,Sollwelle'' λ_0. Um die Selektivität zu steigern, baut man ein Schichtsystem, das nach Fig. 31a aufgebaut ist und für die Sollwelle aus vielen $\lambda_0/4$-Schichten mit abwechselnd hohem und niedrigem Brechungsindex besteht. Für die Sollwelle hat das System sehr starke Reflexion, wie in Ziff. 26 gezeigt wurde, und zwar ist die reflektierte Intensität nach Gl. (26.1) für das $m = 2s$ Grenzflächen enthaltende Schichtsystem

$$|\mathfrak{r}|^2 = \left(\frac{1-v^2}{1+v^2}\right)^2 = \left(\frac{g_1^{2s}-g_0^{2s}}{g_1^{2s}+g_0^{2s}}\right)^2 \quad \text{mit} \quad v = \left(\frac{g_0}{g_1}\right)^s. \qquad (36.1)$$

Für die genau dazwischenliegende Welle, für die jede der Schichten eine $\lambda/2$-Schicht ist, ist das Schichtsystem gleichsam nicht vorhanden, wie oben bereits mehrfach diskutiert; für diese Wellen ist also das System reflexionsfrei, sofern

<hr>

[1] C. H. CARTWRIGHT u. A. F. TURNER: Phys. Rev. **55**, 1128 (1939). — H. SCHRÖDER: Z. angew. Phys. **3**, 53 (1951) mit Angabe vieler Originalarbeiten.

beide Endmedien ebenfalls gleich sind. Diese Wellen werden völlig durchgelassen. Fig. 31 b zeigt den Verlauf für vier Grenzflächen, und zwar für die Werte $g_0 = g_2 = g_4 = 1,5$; $g_1 = g_3 = 2,6$, Fig. 31 b das Entsprechende für acht Grenzflächen.

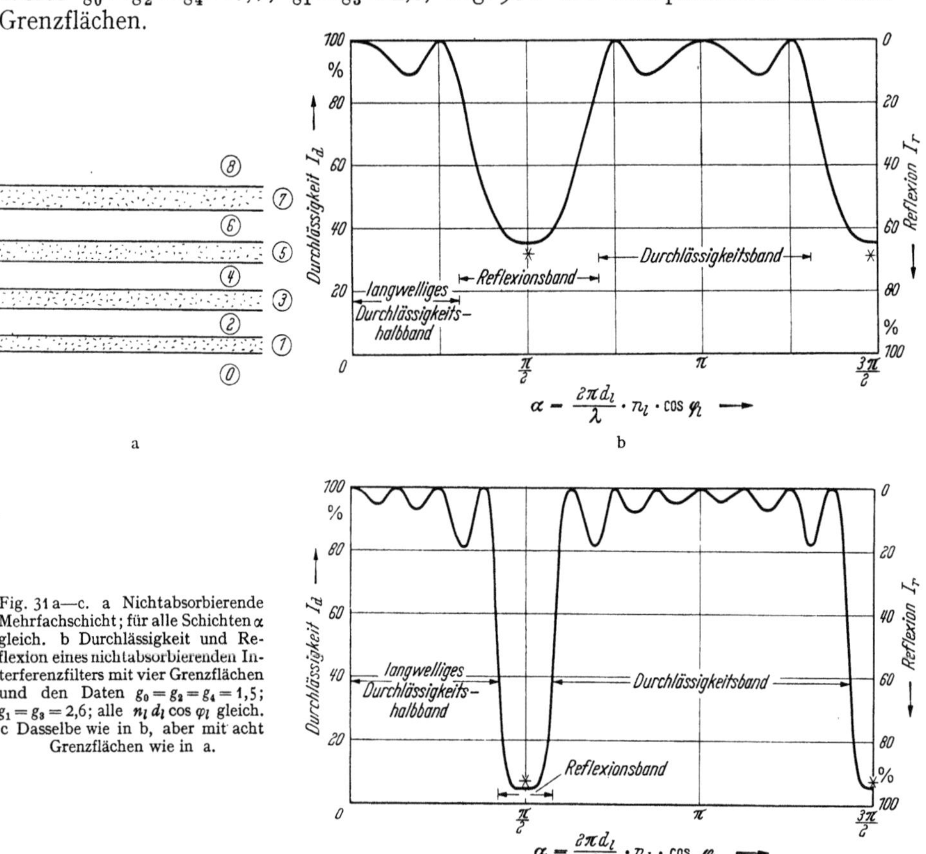

Fig. 31 a—c. a Nichtabsorbierende Mehrfachschicht; für alle Schichten α gleich. b Durchlässigkeit und Reflexion eines nichtabsorbierenden Interferenzfilters mit vier Grenzflächen und den Daten $g_0 = g_2 = g_4 = 1,5$; $g_1 = g_3 = 2,6$; alle $n_l\, d_l \cos \varphi_l$ gleich. c Dasselbe wie in b, aber mit acht Grenzflächen wie in a.

37. Rekursionsformeln für nichtabsorbierende Mehrfachschichten. Die Berechnung der Durchlässigkeit und Reflexion bei solchen Mehrschichtfiltern geschieht mit den Rekursionsformeln (9.14), (9.15). Sie lassen sich in unserem Falle der $\lambda_0/4$-Schichten aus nur zwei sich abwechselnden Stoffen vereinfachen; denn es ist unter unseren Voraussetzungen

$$e^{2\tau i \frac{n_l \cos \varphi_l}{\lambda} d_l} = e^{\varrho_l d_l} = e^{i\alpha} \tag{37.1}$$

unabhängig vom Index.

Wir bezeichnen:

$$g_l - g_{l-1} = -(g_{l-1} - g_{l-2}) = D \qquad \text{für gerade } l \tag{37.2}$$

$$g_l + g_{l+1} = S \qquad \text{für jedes } l. \tag{37.3}$$

Aus Tabelle 1 auf S. 472 folgt für zwei Grenzflächen

$$Z_2 = -e^{i\alpha} D\, S\, (1 - e^{-2i\alpha}), \tag{37.4}$$

$$N_2 = e^{i\alpha} (S^2 - D^2 \cdot e^{-2i\alpha}). \tag{37.5}$$

Wenn m gerade ist, so heißen die Rekursionsformeln (9.14) und (9.15) nun

$$Z_m = - D \cdot e^{i\alpha} N_{m-1} + S \cdot e^{-i\alpha} \cdot Z_{m-1}, \tag{37.6}$$

$$N_m = S \cdot e^{i\alpha} N_{m-1} - D \cdot e^{-i\alpha} Z_{m-1} \tag{37.7}$$

und nach nochmaliger Anwendung der Rekursionsformeln

$$Z_m = S \cdot D \left(1 - e^{2i\alpha}\right) N_{m-2} + \left(S^2 e^{-2i\alpha} - D^2\right) Z_{m-2}, \tag{37.8}$$

$$N_m = \left(S^2 e^{2i\alpha} - D^2\right) N_{m-2} + S D \left(1 - e^{-2i\alpha}\right) Z_{m-2}. \tag{37.9}$$

Zum Beispiel ist für die der Fig. 31 b zugrunde gelegte Anordnung mit vier Grenzflächen

$$e^{-i\alpha} N_4 = \left(S^2 - D^2 e^{-2i\alpha}\right)^2 e^{2i\alpha} - D^2 S^2 \left(1 - e^{-2i\alpha}\right)^2, \tag{37.10}$$

$$e^{-i\alpha} Z_4 = S D \left\{ \left(1 - e^{2i\alpha}\right) \left(S^2 - D^2 e^{-2i\alpha}\right) - \left(1 - e^{-2i\alpha}\right) \left(S^2 e^{-2i\alpha} - D^2\right) \right\}. \tag{37.11}$$

Es ist

$$J_r = |\mathfrak{r}|^2 = \left| \frac{Z_m}{N_m} \right|^2; \quad J_d = 1 - J_r. \tag{37.12}$$

Die Auswertung geschieht zweckmäßig graphisch in der GAUSSschen Zahlenebene[1]. Geschlossene Formeln gibt W. GEFFCKEN[2].

Wie Fig. 31a u. b zeigt, ist die Wirkung dieser Interferenzfilter aus mehreren nichtabsorbierenden $\lambda/4$-Schichten eine völlig andere als die der „Monochromatfilter" nach GEFFCKEN, die in Ziff. 30 behandelt wurden. Man hat es hier mit stark gekoppelten dünnen Plättchen zu tun und erhält dementsprechend eine Art Bandfilterkurve. Die Tiefe des Einschnitts bei $2\alpha = 180°$ kann durch Erhöhung der Schichtanzahl beliebig verstärkt werden. Tabelle 8 gibt einige Werte. Bei passender Wahl der Schichtdicken kann ein Schichtsystem nach Fig. 31a u. b als Ultrarotpaß oder auch als Sperre für das Ultrarot ausgebildet werden[3]. Es hat auch als Reflexionsfilter Bedeutung, das z.B. das sichtbare Licht, nicht aber das Ultrarot reflektiert. Fig 32a gibt gemessene Reflexion und Durchlässigkeit eines „Kaltlichtspiegels" nach SCHRÖDER[4,2]. Fig 32b zeigt die Wirkung eines sog. Blauspiegels, der nur das kurzwellige sichtbare Licht reflektiert, anderes sichtbares Licht durchläßt.

Tabelle 8.

	Zahl der Grenzflächen							
	2	4	6	8	10	12	14	20
Durchlässigkeit am Minimum, also für $\dfrac{2 d_1 n_1 \cos \varphi_1}{\lambda} = \dfrac{1}{2} + M$	75%	36%	13,8%	4,8%	1,6%	0,55%	0,18%	0,007%

Ein solches Schichtsystem nach Fig. 31 ist wegen seiner starken Reflexion an den mit * bezeichneten Stellen der Fig. 31 geeignet als Ersatz für die Metallschichten eines Einfachfilters (Fig. 21). Mit zwei solchen Schichtsystemen, die abwechselnd aus ZnS und Kryolith aufgebaut waren und je sieben Schichten enthielten, baute man fast absorptionsfreie Linienfilter[5]. Zwischen den beiden

[1] Siehe auch N. CABRERA: C. R. Acad. Sci. Paris **234**, 1045, 1146 (1952).
[2] W. GEFFCKEN: Glastechn. Ber. **24**, 148 (1951); D. P. 902 191; Kl. 42h; Gr. 34, 11 (1954).
[3] E. R. BLONT, R. S. CORLEY u. P. L. SNOW: J. Opt. Soc. Amer. **39**, 634 (1949).
[4] H. SCHRÖDER: Z. angew. Phys. **3**, 63 (1951).
[5] H. SCHRÖDER: Naturwiss. **34**, 277 (1947). — H. D. POLSTER: J. Opt. Soc. Amer. **39**, 1054 (1949).

Systemen befand sich eine $\lambda_0/2$-Schicht als wellenlängengebendes Element (der Schicht ② der Fig. 21 entsprechend). Die Durchlässigkeitskurve eines solchen Filters ähnelt der Fig. 31 c; nur erhebt sich an den Stellen * ein schmales

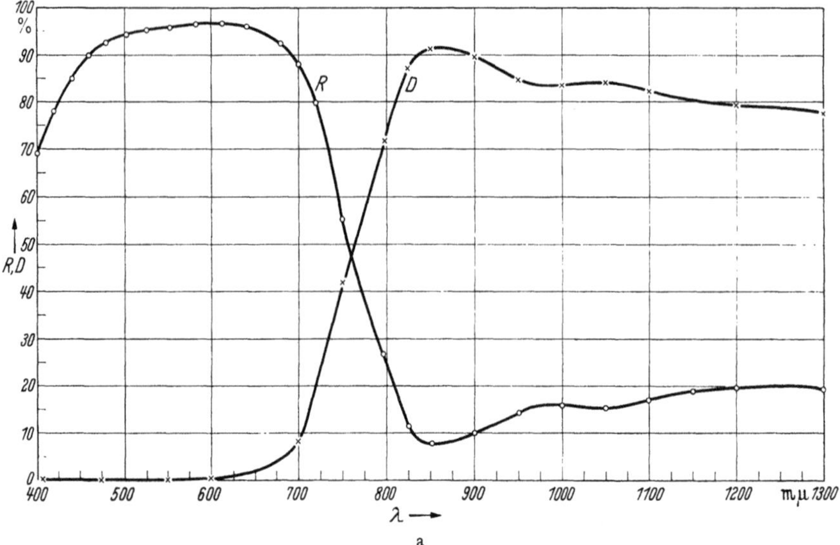

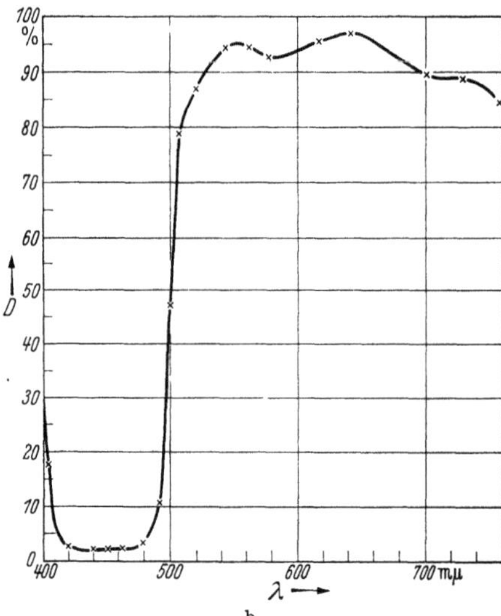

Fig. 32 a u. b. a Intensitätsverlauf bei einem ultrarotdurchlässigen Spiegel für das sichtbare Licht (Kaltlichtspiegel). [Nach H. Schröder: Z. angew. Phys. 3, 63 (1951).] b Intensitätsverlauf bei einem „Blauspiegel". [Nach H. Schröder: Z. angew. Phys. 3, 62 (1951).]

Durchlässigkeitsmaximum. Es lag bei Polsters Filter bei 520 mμ, hatte eine Halbwertsbreite von 6 mμ und eine Maximaldurchlässigkeit von 80%. Der angrenzende Sperrbereich ging bis zu 440 und 620 mμ; dort hatte die Durchlässigkeit wieder 5% und stieg dann weiterhin steil an.

38. Reflexionsfilter. Die in Ziff. 37 betrachteten Interferenzfilter haben auch in Reflexion verwertbare Eigenschaften und sind daher bereits als Reflexionsfilter anzusprechen.

Einen zweiten Typ von Reflexionsfiltern erhält man, wenn man ein Durchsichtsfilter vor eine gut reflektierende Metallplatte legt. Geschieht das in inkohärenter Weise, so erhält man eine wirksame Doppelnutzung des Durchsichtsfilters. Wird das Filter der Metallplatte kohärent aufgelegt, so hat man in Reflexion im wesentlichen dieselbe Wirkung wie bei einem Durchsichtsfilter mit rund doppelter Schichtzahl bei Kopplung beider Teile. Das einfachste Reflexionsfilter dieser Art zeigt Fig. 33a. Die Eigenschaften sind dabei den in Ziff. 37 beschriebenen gleich, sofern der Abstand des Filters von der Platte unter Berücksichtigung des Phaseneinflusses der Metallreflexion zweckmäßig bemessen wird. So läßt sich z.B. durch Aufbringen einer nichtabsorbierenden

Schicht auf eine Metallplatte und anschließendes Aufdampfen einer dünnen Metallschicht ein Ultrarot-Reflexionsfilter herstellen, das dieselben Eigenschaften hat wie die entsprechenden oben beschriebenen Ultrarot-Durchsichtsfilter. (Vgl. Fig. 15 und die Arbeit von HADLEY und DENNISON[1].)

Besonders frei von Absorption sind Mehrfachschichtfilter, die als stark reflektierende Schichten weder Metallschichten noch hochbrechende — und dann meist auch etwas absorbierende — Schichten verwenden sondern statt dessen von der Totalreflexion bei hinreichend schiefem Einfall Gebrauch machen. Freilich ist Totalreflexion selbst nicht erwünscht; aber man macht entweder den Einfallswinkel ein wenig kleiner als den Grenzwinkel der Totalreflexion und erhält dann jede gewünschte reflektierte Amplitude — aber nur für sehr kleine Apertur des Lichtbündels, oder man verwendet die im Totalreflexionsbereich bekannte Reflexbehinderung nach QUINKE, die auftritt, wenn das dünnere Medium in hinreichend dünner Schicht vorliegt[2]. Auch Polarisatoren lassen sich

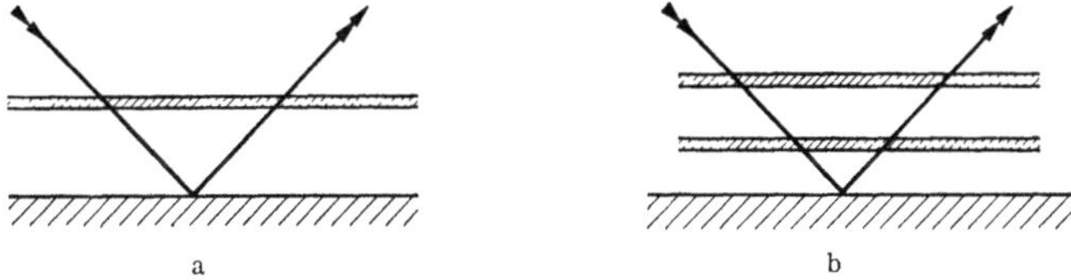

a   b

Fig. 33 a u. b. a Einfaches Reflexionsfilter. b Durchsichtsfilter und Metallspiegel kombiniert als Reflexionsfilter.

nach diesem Prinzip bauen, insbesondere bei Mitbenutzung doppelbrechenden Materials (z.B. Kryolith) für die dünnen Schichten. Die eine Polarisation findet sich dann fast ganz im reflektierten, die andere im durchgegangenen Licht (vgl. Ziff. 21).

VI. Methoden zur Messung der optischen Konstanten und der Dicke dünner Schichten.

a) Allgemeines.

39. Der Brechungsindex und der Absorptionskoeffizient eines in dünner Schicht vorliegenden Mediums interessieren den angewandten Physiker, der mit solchen Schichten Aufgaben der praktischen Optik, wie sie in den vorhergehenden Abschnitten besprochen wurden, lösen will. Diese optischen Konstanten dünner Schichten haben aber bereits lange vor dem Auftauchen praktischer Aufgaben die Physiker beschäftigt, seit KUNDT[3] die optischen Konstanten der Metalle an dünnen, noch durchlässigen Metallprismen zu messen begonnen hat. Die gerade an Metallen beobachteten Anomalien — damit meint man die starken Abweichungen der Konstanten dünner Schichten von denen des massiven Metalls — machten die Frage nach der Struktur dünner Schichten immer wieder aktuell[4].

<hr>

[1] HADLEY u. DENNISON: J. Opt. Soc. Amer. **37**, 451 (1947); **38**, 483 (1948). — E. E. BARR: J. Opt. Soc. Amer. **39**, 634 (1949).
[2] B. H. BILLINGS: J. Opt. Soc. Amer. **39**, 634 (1949). — B. H. BILLINGS u. M. A. PITTMANN: J. Opt. Soc. Amer. **39**, 978 (1949).
[3] A. KUNDT: Wied. Ann. **34**, 469 (1888). — W. VOIGT: Wied. Ann. **24**, 144 (1885). — C. STATESCU: Ann. Phys. **33**, 1032 (1910). — H. FRITZE: Ann. Phys. (4) **47**, 763 (1915). — P. DRUDE: Siehe Bibliographie.
[4] Siehe den Artikel über Metalloptik von J. BOR und J. FRIEDEL in Bd. XXV dieses Handbuches.

Um den Brechungsindex und den Absorptionskoeffizienten einer dünnen Schicht zu messen, ist daher vorwiegend seit der Jahrhundertwende eine sehr große Zahl von Methoden entwickelt worden. Eine vollständige und der historischen Entwicklung Rechnung tragende Darstellung ist im Rahmen dieses Artikels raummäßig ausgeschlossen; sie ist von H. Mayer in seinem Buche „Physik dünner Schichten"[1] gegeben worden, auf das besonders hingewiesen sei.

Dieser Artikel soll nur das Charakteristische der drei Typen solcher Meßmethoden, der Polarisationsmethoden, der Intensitätsmethoden und der Interferenzmethoden, in systematischem Zusammenhange darstellen.

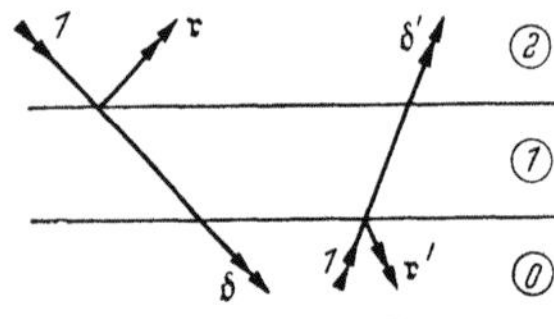

Fig. 34. Dünne Schicht ① zwischen Medien ② und ⓪

Ganz allgemein fragen wir in diesem Artikel nach dem komplexen Brechungsindex n_1 einer dünnen Schicht und nach ihrer Dicke d_1. Die aus ② einfallende Welle sei als $\mathfrak{A}_2$, die reflektierte als $\mathfrak{R}_2$ und die ins Medium ⓪ durchfallende Welle als $\mathfrak{A}_0$ bezeichnet entsprechend den oben benutzten Bezeichnungen dieses Artikels (S. 465).

b) Die Polarisationsmethoden.

40. Das Prinzip der allgemeinen Polarisationsmethode. Drude[2] und seine Schüler hatten mit großem Erfolg die optischen Konstanten massiver Metalle aus der Elliptizität des reflektierten Lichtes gemessen. Kennzeichnet man mit s die Komponenten senkrecht zur Einfallsebene und mit p die parallel zu ihr[3], so nennt man „die Elliptizität des reflektierten Lichtes" das komplexe Verhältnis elektrischer Feldkomponenten[4]

$$\mathfrak{P}_R = \frac{\hat{\mathfrak{R}}_{2p}}{\hat{\mathfrak{R}}_{2s}} = \tan\chi_R \cdot e^{i\delta_R}; \quad \chi_R \text{ und } \partial_R \text{ reell}. \tag{40.1}$$

Das „Azimut" χ_R und die „Phasendifferenz" δ_R sind durch Vermessung der Form und Lage der Schwingungsellipse nach bekannten Verfahren experimentell leicht selbst meßbar.

Försterling[5] hat das Drudesche Verfahren auf dünne Schichten übertragen und wegen der zusätzlichen Unbekannten, der Schichtdicke, auch die „Elliptizität des hindurchgegangenen Lichtes"

$$\mathfrak{P}_D = \frac{\hat{\mathfrak{A}}_{0p}}{\hat{\mathfrak{A}}_{0s}} = \tan\chi_D \cdot e^{i\delta_D}; \quad \chi_D \text{ und } \partial_D \text{ reell} \tag{40.2}$$

und für dünnste Schichten auch Intensitäten herangezogen.

Als Vorläufer der Försterlingschen und der späteren Polarisationsmethoden kann auch die Alkemade-Drudesche[6] Näherungsmethode gelten, bei der man für extrem dünne Schichten am Polarisationswinkel des Schichtträgers die Elliptizität des reflektierten Lichtes mißt und aus ihr die beiden Unbekannten, Dicke und Brechungsindex der Schicht, berechnet.

---

[1] H. Mayer: Physik dünner Schichten, Teil I, S. 206f. Stuttgart 1950.

[2] P. Drude: Wied. Ann. **32**, 613 (1887) (Mass. Metalle); **39**, 492 (1890). — K. Lauch: Ann. Phys. (4) **74**, 55 (1924). — W. Betz: Ann. Phys. (4) **18**, 590 (1905). — C. Statescu: Ann. Phys. **33**, 1032 (1910).

[3] Diese Bezeichnungen werden hier nur für homogene Wellen, die einen reellen Fortpflanzungsvektor haben, verwendet. p- und s-Komponenten sind senkrecht zur Fortpflanzungsrichtung genommen.

[4] ∧ kennzeichnet elektrische Feldstärkenkomponenten, ⌒ magnetische (s. S. 465).

[5] K. Försterling: Göttinger Nachr. **1911**, 449. — N. Galli u. K. Försterling: Göttinger Nachr. **1911**, 58.

[6] R. van Alkemade: Wied. Ann. **20**, 22 (1883); siehe auch Ziff. 63 und die Fußnote 2 von S. 539.

Das einfallende Licht erhält der Einfachheit halber die Elliptizität 1; d.h. man macht

$$\hat{\mathfrak{A}}_{2p} = \hat{\mathfrak{A}}_{2s}.\tag{40.3}$$

Das bedeutet Linearpolarisation des einfallenden Lichtes mit einer Lage der Polarisationsebene unter 45° zur Einfallsebene.

Nimmt man an, daß alle Permeabilitäten $\mu_l = 1$ sind, und kürzt man $\cos\varphi_l = c_l$ ab, so ist[1] nach Tabelle 1 für $m = 2$ Grenzflächen bei Transversal-E-Wellen mit Gl. (4.6)

$$\hat{\mathfrak{r}} = \frac{\hat{\mathfrak{R}}_{2s}}{\hat{\mathfrak{A}}_{2s}} = \frac{(n_2 c_2 - n_1 c_1)(n_1 c_1 + n_0 c_0)\,e^{\varrho_1 d_1} + (n_2 c_2 + n_1 c_1)(n_1 c_1 - n_0 c_0)\,e^{-\varrho_1 d_1}}{(n_2 c_2 + n_1 c_1)(n_1 c_1 + n_0 c_0)\,e^{\varrho_1 d_1} + (n_2 c_2 - n_1 c_1)(n_1 c_1 - n_0 c_0)\,e^{-\varrho_1 d_1}},\tag{40.4}$$

$$\hat{\mathfrak{d}} = \frac{\hat{\mathfrak{A}}_{0s}}{\hat{\mathfrak{A}}_{2s}} = \frac{4 n_1 c_1 n_2 c_2}{(n_2 c_2 + n_1 c_1)(n_1 c_1 + n_0 c_0)\,e^{\varrho_1 d_1} + (n_2 c_2 - n_1 c_1)(n_1 c_1 - n_0 c_0)\,e^{-\varrho_1 d_1}}\tag{40.5}$$

und für Transversal-H-Wellen nach den Gln. (4.10), (2.19), (2.20) und Tabelle 1

$$\begin{aligned}\frac{\hat{\mathfrak{R}}_{2p}}{\hat{\mathfrak{A}}_{2p}} &= \frac{\dfrac{1}{n_2}\hat{\mathfrak{R}}_{2s}}{\dfrac{1}{n_2}\hat{\mathfrak{A}}_{2s}} = \hat{\mathfrak{r}} \\[2mm]
&= \frac{\left(\dfrac{c_2}{n_2} - \dfrac{c_1}{n_1}\right)\left(\dfrac{c_1}{n_1} + \dfrac{c_0}{n_0}\right)e^{\varrho_1 d_1} + \left(\dfrac{c_2}{n_2} + \dfrac{c_1}{n_1}\right)\left(\dfrac{c_1}{n_1} - \dfrac{c_0}{n_0}\right)e^{-\varrho_1 d_1}}{\left(\dfrac{c_2}{n_2} + \dfrac{c_1}{n_1}\right)\left(\dfrac{c_1}{n_1} + \dfrac{c_0}{n_0}\right)e^{\varrho_1 d_1} + \left(\dfrac{c_2}{n_2} - \dfrac{c_1}{n_1}\right)\left(\dfrac{c_1}{n_1} - \dfrac{c_0}{n_0}\right)e^{-\varrho_1 d_1}},\end{aligned}\tag{40.6}$$

$$\begin{aligned}\frac{\hat{\mathfrak{A}}_{0p}}{\hat{\mathfrak{A}}_{2p}} &= \frac{\dfrac{1}{n_0}\hat{\mathfrak{A}}_{0s}}{\dfrac{1}{n_2}\cdot\hat{\mathfrak{A}}_{2s}} = \frac{n_2}{n_0}\cdot\hat{\mathfrak{d}} \\[2mm]
&= \frac{\dfrac{n_2}{n_0}\cdot 4\cdot\dfrac{c_1}{n_1}\cdot\dfrac{c_2}{n_2}}{\left(\dfrac{c_2}{n_2} + \dfrac{c_1}{n_1}\right)\left(\dfrac{c_1}{n_1} + \dfrac{c_0}{n_0}\right)e^{\varrho_1 d_1} + \left(\dfrac{c_2}{n_2} - \dfrac{c_1}{n_1}\right)\left(\dfrac{c_1}{n_1} - \dfrac{c_0}{n_0}\right)e^{-\varrho_1 d_1}}.\end{aligned}\tag{40.7}$$

Wenn wir in (40.6) gegenüber den Gln. (2.19), (2.20) ein Minuszeichen unterdrücken, so bedeutet das nur eine Definition über die Richtung, in der wir $\hat{\mathfrak{R}}_{2p}$ als positiv bezeichnen! Division der Gl. (40.7) durch Gl. (40.5) führt zu der Elliptizität des durchgegangenen Lichtes $\mathfrak{P}_D$:

$$n_2 n_1^2 n_0\,\mathfrak{P}_D = \frac{(n_2 c_2 + n_1 c_1)(n_1 c_1 + n_0 c_0)\,e^{\varrho_1 d_1} + (n_2 c_2 - n_1 c_1)(n_1 c_1 - n_0 c_0)\,e^{-\varrho_1 d_1}}{\left(\dfrac{c_2}{n_2} + \dfrac{c_1}{n_1}\right)\left(\dfrac{c_1}{n_1} + \dfrac{c_0}{n_0}\right)e^{\varrho_1 d_1} + \left(\dfrac{c_2}{n_2} - \dfrac{c_1}{n_1}\right)\left(\dfrac{c_1}{n_1} - \dfrac{c_0}{n_0}\right)e^{-\varrho_1 d_1}}.\tag{40.8}$$

Die Elliptizität des reflektierten Lichtes selbst wird kompliziert; einfacher wird wegen des gleichen Nenners in (40.4) und (40.5) bzw. in (40.6) und (40.7) die Kombination

$$\frac{\mathfrak{P}_R}{\mathfrak{P}_D} = n_2 n_1^2 n_0\,\frac{\left(\dfrac{c_2}{n_2} - \dfrac{c_1}{n_1}\right)\left(\dfrac{c_1}{n_1} + \dfrac{c_0}{n_0}\right)e^{\varrho_1 d_1} + \left(\dfrac{c_2}{n_2} + \dfrac{c_1}{n_1}\right)\left(\dfrac{c_1}{n_1} - \dfrac{c_0}{n_0}\right)e^{-\varrho_1 d_1}}{(n_2 c_2 - n_1 c_1)(n_1 c_1 + n_0 c_0)\,e^{\varrho_1 d_1} + (n_2 c_2 + n_1 c_1)(n_1 c_1 - n_0 c_0)\,e^{-\varrho_1 d_1}}.\tag{40.9}$$

---

[1] Es war nach Gl. (5.4) definiert.

Da $\mathfrak{P}_R$ und $\mathfrak{P}_D$ aus Azimut- und Phasendifferenzmessungen nach Gl. (40.1) und (40.2) bekannt sind, lassen sich die Gln. (40.8) und (40.9) auf die Formen bringen

$$u^*(\mathfrak{n}_1) \cdot e^{\varrho_1 d_1} + v^*(\mathfrak{n}_1) \cdot e^{-\varrho_1 d_1} = 0, \tag{40.8a}$$

$$u(\mathfrak{n}_1) e^{\varrho_1 d_1} + v(\mathfrak{n}_1) \cdot e^{-\varrho_1 d_1} = 0 \tag{40.9a}$$

mit Funktionen u, u^*, v, v^*, die außer bekannten Größen als einzige Unbekannte $\mathfrak{n}_1$ algebraisch enthalten. (40.8a) und (40.9a) lassen sich nach $e^{2\varrho_1 d_1}$ auflösen:

$$e^{2\varrho_1 d_1} = -\frac{v^*(\mathfrak{n}_1)}{u^*(\mathfrak{n}_1)}, \tag{40.10}$$

$$e^{2\varrho_1 d_1} = -\frac{v(\mathfrak{n}_1)}{u(\mathfrak{n}_1)}. \tag{40.11}$$

Gleichsetzen gibt

$$v^*(\mathfrak{n}_1) \cdot u(\mathfrak{n}_1) - u^*(\mathfrak{n}_1) \cdot v(\mathfrak{n}_1) = 0, \tag{40.12}$$

eine algebraische Gleichung in $\mathfrak{n}_1$, die nach der einzigen komplexen Unbekannten $\mathfrak{n}_1$ aufgelöst werden kann, wenn auch im allgemeinen nur in komplizierter iterativer Rechnung oder nomogrammetrisch.

Ist $\mathfrak{n}_1$ auf diese Weise gewonnen, so kann die Schichtdicke aus Gl. (40.10), weniger genau aus Gl. (40.11), die für $\mathfrak{n}_2 \approx \mathfrak{n}_0$ ganz versagt, gewonnen werden.

Freilich wirken sich bei dieser Schichtdickenbestimmung die Meßfehler bei dünnsten Schichten ($d_1 < 10$ mμ) so stark aus, daß andere zusätzliche Messungen (z.B. Intensitätsmessungen) notwendig werden.

In dieser Hinsicht günstiger und vor allem einfacher wird die Auswertung, wenn auch die Reflexion an der Schichtrückseite und deren Elliptizität $\mathfrak{P}'_R$ herangezogen wird[1]. Nach Elimination von $e^{2\varrho_1 d_1}$ erhält man statt (40.12) die quadratische Gleichung

$$y^2 - Q \cdot y = 1 \tag{40.13}$$

für die mit der Unbekannten $\mathfrak{n}_1^2$ einfach zusammenhängende Größe

$$y \equiv \left(\frac{1}{\mathfrak{n}_1^2} - \frac{1}{1 + (n_0^2 - 1)\cos^2\varphi_2} \right) \frac{2n_0^2}{n_0^2 - 1}. \tag{40.14}$$

Darin ist

$$Q \equiv 4 \cdot \frac{n_0^2 - 1}{n_0} \left(\frac{1}{n_0^2 - 1} + \cos^2\varphi_2 \right) \cdot \frac{\mathfrak{P}'_R : \mathfrak{P}_D}{1 + \mathfrak{P}'_R : \mathfrak{P}_R}. \tag{40.15}$$

Die Schichtdicke kann dann z.B. nach Försterling aus

$$\tan\frac{\varrho_1 d_1}{i} = i\,\frac{(n_0^2 - 1)\cos\varphi_2 \sqrt{n_0^2 - \sin^2\varphi_2}}{\mathfrak{n}_1^2 - 2 \cdot \sin^2\varphi_2} \cdot \frac{1 + \mathfrak{P}'_R : \mathfrak{P}_R}{1 - \mathfrak{P}'_R : \mathfrak{P}_R} \tag{40.16}$$

berechnet werden.

Die Vereinfachung der mathematischen Arbeit wird durch überschüssige Messungen erkauft.

41. Die spezielle Polarisationsmethode $(n_0 = n_2)$[2]. Besonders einfach wird die sich an die Gln. (40.4) bis (40.7) anschließende Eliminationsarbeit, wenn die Schicht ① auf beiden Seiten von Medien gleichen Brechungsindexes umgeben ist, also $n_0 = n_2$, d.h. nach dem Brechungsgesetz auch $\varphi_0 = \varphi_2$ ist. In der Praxis stellt man diese Gleichheit her, indem man auf die z.B. auf einer Glasplatte sitzende Schicht ein Glasprisma setzt und den „optischen Kontakt" durch einen

---

[1] F. Försterling: Ann. Phys. (5) **30**, 745 (1937). — G. Essers-Rheindorf: Ann. Phys. **28**, 297 (1937).

[2] K. Försterling: Nachr. Ges. Wiss. Göttingen **1911**, 58, 449. — N. Galli: Diss. Göttingen 1911.

Tropfen Benzol, Immersol usw. herstellt[1]. Denn in diesem Falle kann in (40.9) die Exponentialfunktion herausgekürzt werden:

$$\frac{\mathfrak{P}_R}{\mathfrak{P}_D} = \frac{n_2^2 c_1^2 - \mathfrak{n}_1^2 c_2^2}{n_2^2 c_2^2 - \mathfrak{n}_1^2 c_1^2}. \tag{41.1}$$

Die Gleichung enthält als einzige Unbekannte $\mathfrak{n}_1$, freilich auch versteckt in c_1, das jedoch mit Hilfe des Brechungsgesetzes leicht auf $\mathfrak{n}_1$ und den Einfallswinkel φ_2 zurückgeführt werden kann. Die linke Seite ist nach (40.1) und (40.2) aus den Elliptizitätsmessungen bekannt; (41.1) ist daher

$$\frac{\tan \chi_R}{\tan \chi_D} \cdot e^{i(\delta_R - \delta_D)} = \frac{n_0^2}{\mathfrak{n}_1^2} \sin^2 \varphi_2 - \cos^2 \varphi_2, \tag{41.2}$$

d.h.

$$\left.\begin{aligned} &\frac{\tan \chi_R}{\tan \chi_D} \{\cos(\delta_R - \delta_D) + i \sin(\delta_R - \delta_D)\} \\ &\quad = n_0^2 \cdot \frac{n_1^2 - k_1^2}{(n_1^2 + k_1^2)} \sin^2 \varphi_2 - \cos^2 \varphi_2 + i\, n_2^2 \frac{2 n_1 k_1}{(n_1^2 + k)^2} \sin^2 \varphi_2 \end{aligned}\right\} \tag{41.3}$$

oder bei Zerlegung in Real- und Imaginärteil

$$n_0^2 \cdot \frac{n_1^2 - k_1^2}{(n_1^2 + k_1^2)^2} = \cot^2 \varphi_1 + \frac{\tan \chi_R \cdot \cos(\delta_R - \delta_D)}{\tan \chi_D \cdot \sin^2 \varphi_2} \equiv A, \tag{41.4}$$

$$n_0^2 \frac{2 n_1 k_1}{(n_1^2 + k_1^2)^2} = \frac{\tan \chi_R \cdot \sin(\delta_R - \delta_D)}{\tan \chi_D \cdot \sin^2 \varphi_2} \equiv B, \tag{41.5}$$

d.h.[2]

$$n_1^2 = \frac{n_0^2}{2} \cdot W \cdot (1 + A \cdot W); \quad k_1^2 = \frac{n_0^2}{2} W \cdot (1 - A W) \tag{41.6}$$

mit

$$W = \frac{1}{\sqrt{A^2 + B^2}}. \tag{41.7}$$

Die Schichtdicke d_1 kann dann aus Gl. (40.8a) berechnet werden, die für $n_0 = n_2$ ebenso $e^{\varrho_1 d_1}$ enthält wie für $n_0 \neq n_2$.

42. Bewertung der Polarisationsmethoden. Der Mangel der Polarisationsmethoden liegt in ihrer nicht leicht zu durchbrechenden Beschränkung auf das sichtbare Spektralgebiet. Ihr großer Wert besteht vor allem darin, daß in die Messung der optischen Konstanten nicht explizit die zunächst ja unbekannte Schichtdicke eingeht. Wenn bei anderen Verfahren die Schichtdicke aus der gewogenen Schichtmasse oder aus den Herstellungsbedingungen gemessen wird, kann dort doch nur von einer Schätzung der Schichtdicke gesprochen werden; denn man weiß heute sicher, daß die Dichte in dünnen Schichten wesentlich von der des massiven Stoffes abweicht[3]. Hat man aus Polarisationsmessungen die optischen Konstanten auch nur bei einer Wellenlänge im Sichtbaren gemessen, so lassen sich darauf Schichtdickenbestimmungen nicht nur aus Gl. (40.6), sondern genauer noch aus Intensitätsmessungen aufbauen, während die Intensitätsmessungen allein bei sehr dünnen Schichten bestenfalls die Produkte $d_1 \cdot n_1 \cdot k_1$

---

[1] K. Försterling: Nachr. Ges. Wiss. Göttingen **1911**, 58, 449. — N. Galli: Diss. Göttingen 1911.

[2] Diese Ergebnisse weichen bezüglich der Vorzeichen von denen der Försterlingschen Arbeit ab; die dort unrichtige Vorzeichenbestimmung wird bei der hier gegebenen direkteren Herleitung überflüssig.

[3] Vgl. z.B. A. Sommer: Diss. Hamburg 1940. — H. Schopper: Z. Physik **130**, 565 (1951).

und $d_1 \cdot (n_1^2 - k_1^2 - 1)$ bestimmen lassen, wie unten gezeigt wird. Deshalb ist man auch immer wieder auf die im grundsätzlichen sehr alten Polarisationsmethoden zumindest zur Ergänzung anderer Messungen zurückgekehrt.

c) Die Intensitätsmethoden.

43. Strenge Verfahren. Schon 1913 hat Försterling eine Intensitätsmessung als zusätzliche Messung zu den Polarisationsmessungen herangezogen, um außer dem Brechungsindex n_1 auch die Schichtdicke d_1 noch bei sehr dünnen Schichten genau zu erhalten. Hermann, Goldschmidt und Dember und Smakula haben dann aus Intensitätsmessungen unmittelbar Absorptionskonstanten berechnet. Aber die benutzten Näherungsformeln haben einer späteren Kritik nicht standgehalten.

Den entscheidenden Anstoß erhielten die praktisch in allen Spektralbereichen anwendbaren und darin den Polarisationsmethoden überlegenen Intensitätsmethoden 1933 durch H. Murmann, R. Schulze, C. F. Veenemans[1], die mit graphischen Verfahren die strengen Intensitätsgleichungen nach den Unbekannten n_1 und k_1 auflösten, während sie die Schichtdicke aus nichtoptischen Messungen (Wägung oder Aufdampfverteilung) als bekannt ansahen. Das Versagen beider Methoden bei dünnsten Schichten hat dann das Problem der Intensitätsmethoden aufgerollt, nämlich die Frage, in welchen Fällen Intensitätsmessungen überhaupt zur Bestimmung der optischen Konstanten und möglichst auch der Schichtdicke ausreichen.

44. Die strengen Intensitätsformeln für beliebigen Einfall und beliebige Polarisation. Den Bruchteil der durchgelassenen Intensität (genannt J_d) und den Bruchteil der reflektierten Intensität (genannt J_r) erhält man speziell für $\mu_l = 1$ und nichtabsorbierende Medien ② und ⓪ nach Tabelle 1 durch Bilden der Ausdrücke

$$J_r = |\mathfrak{r}_2|^2 \tag{44.1}$$

$$J_d = \frac{n_0}{n_2} \cdot |\mathfrak{d}_2|^2. \tag{44.2}$$

Das gibt streng sowohl für TE- als auch TH-Wellen[2]:

$$J_r = \frac{\xi\,\tau\,e^{-\gamma\,\mathfrak{Im}(\mathfrak{n}_1 c_1)} + \zeta\,\sigma\,e^{\gamma\,\mathfrak{Im}(\mathfrak{n}_1 c_1)} + 2q\cos(\gamma\,\mathfrak{Re}(\mathfrak{n}_1 c_1)) + 2r\sin(\gamma\,\mathfrak{Re}(\mathfrak{n}_1 c_1))}{\zeta\,\tau\,e^{-\gamma\,\mathfrak{Im}(\mathfrak{n}_1 c_1)} + \xi\,\sigma\,e^{\gamma\,\mathfrak{Im}(\mathfrak{n}_1 c_1)} + 2s\cos(\gamma\,\mathfrak{Re}(\mathfrak{n}_1 c_1)) + 2t\sin(\gamma\,\mathfrak{Re}(\mathfrak{n}_1 c_1))}, \tag{44.3}$$

$$J_d \cdot \frac{c_0}{c_2} = \frac{16\,|g_1|^2\,g_2\,g_0}{\zeta\,\tau\,e^{-\gamma\,\mathfrak{Im}(\mathfrak{n}_1 c_1)} + \xi\,\sigma\,e^{\gamma\,\mathfrak{Im}(\mathfrak{n}_1 c_1)} + 2s\cos(\gamma\,\mathfrak{Re}(\mathfrak{n}_1 c_1)) + 2t\sin(\gamma\,\mathfrak{Re}(\mathfrak{n}_1 c_1))}, \tag{44.4}$$

$$J_r' = \frac{\sigma\,\zeta\,e^{-\gamma\,\mathfrak{Im}(\mathfrak{n}_1 c_1)} + \tau\,\xi\,e^{\gamma\,\mathfrak{Im}(\mathfrak{n}_1 c_1)} + 2q\cos(\gamma\,\mathfrak{Re}(\mathfrak{n}_1 c_1)) - 2r\sin(\gamma\,\mathfrak{Re}(\mathfrak{n}_1 c_1))}{\zeta\,\tau\,e^{-\gamma\,\mathfrak{Im}(\mathfrak{n}_1 c_1)} + \xi\,\sigma\,e^{(\mathfrak{Im}\,\gamma\,n_1 c_1)} + 2s\cos(\gamma\,\mathfrak{Re}(\mathfrak{n}_1 c_1)) + 2t\sin(\gamma\,\mathfrak{Re}(\mathfrak{n}_1 c_1))}. \tag{44.5}$$

J_r' bezeichnet den reflektierten Intensitätsbruchteil bei Einfall von ⓪ her; man erhält daher Gl. (44.5) aus Gl. (44.3), indem man g_0 und g_2 vertauscht. Die Durchlässigkeit (44.4) ist gegenüber dieser Vertauschung invariant. Die Gleichungen

[1] H. Murmann: Z. Physik **80**, 161 (1933). — R. Schulze: Physik. Z. **34**, 24 (1933). — C. F. Veenemans: Arch. néerl. **14**, 84 (1933). — Siehe auch P. Y. Harringhhuizen, D. A. Was u. A. M. Kruithof: Physica, Haag **4**, 695 (1937).

[2] In Gl. (44.4) ist J_d definitionsgemäß auf gleiche Auffängerfläche normiert. Meist mißt man direkt $J_d \cdot \cos\varphi_0/\cos\varphi_2$, da der Bündelquerschnitt durch eine Blende im Medium ② festgelegt wird und sich bei Übergang in das Medium ⓪ mit dem Verhältnis der $\cos\varphi$ transformiert.

enthalten folgende Abkürzungen

$$\begin{aligned}
\xi &= (\Re e\, g_1 - g_2)^2 + \Im m^2 g_1; & q &= |g_1|^2 \cdot (g_2^2 + g_0^2) - |g_1|^4 - g_2^2 g_0^2 - 4 g_2 g_0 \Im m^2 g_1; \\
\zeta &= (\Re e\, g_1 + g_2)^2 + \Im m^2 g_1; & s &= |g_1|^2 \cdot (g_2^2 + g_0^2) - |g_1|^4 + g_2^2 g_0^2 + 4 g_2 g_0 \Im m^2 g_1; \\
\sigma &= (\Re e\, g_1 - g_0)^2 + \Im m^2 g_1; & r &= (g_2 - g_0)(|g_1|^2 + g_2 g_0) \cdot 2 \cdot \Im m\, g_1; \\
\tau &= (\Re e\, g_1 + g_0)^2 + \Im m^2 g_1; & t &= (g_2 + g_0)(|g_1|^2 - g_2 g_0) \cdot 2 \cdot \Im m\, g_1; \\
& & \gamma &= \frac{4\pi d_1}{\lambda}.
\end{aligned} \tag{44.6}$$

Ferner ist für TE-Wellen $g_l = \mathfrak{n}_l c_l$, für TH-Wellen $g_l = c_l/\mathfrak{n}_l$; $c_l = \cos\varphi_l$. Speziell für senkrechten Einfall gehen die Gleichungen in die MURMANNschen[1] über, da dann

$$\begin{aligned}
\Im m\,(\mathfrak{n}_1 c_1) &= -k_1; \quad \Re e\,(\mathfrak{n}_1 c_1) = n_1; \quad \xi = (n_1 - n_2)^2 + k_1^2, \ldots \\
q &= (n_1^2 + k_1^2)(n_2^2 + n_0^2) - (n_1^2 + k_1^2)^2 - n_2^2 u_0^2 - 4 n_2 u_0 k_1^2, \ldots \\
r &= -2 k_1 (n_2 - n_0)(n_1^2 + k_1^2 + n_2 n_0), \ldots
\end{aligned} \tag{44.7}$$

45. Strenge Verfahren für senkrechten Einfall (Verfahren von MURMANN, SCHULZE und VEENEMANS). Die Schichtdicke d_1 wird nichtoptisch (aus MURMANNs Verdampfungsbedingungen) bestimmt. Die Gln. (44.3) und (44.4), für senkrechten Einfall spezialisiert, die wir abkürzend schreiben[2]

$$J_r = J_r(n, k, d), \tag{45.1}$$

$$J_d = J_d(n, k, d), \tag{45.2}$$

enthalten dann nur zwei Unbekannte n und k. Nach ihnen löst MURMANN graphisch auf in der für zwei Unbekannte üblichen Weise. In einer n, k-Ebene stellt jede der beiden Gleichungen eine Kurve dar; ihre Schnittpunkte — es gibt stets deren zwei — geben zwei Paare n, k als Lösungen. Welche Lösung die richtige ist, läßt sich für große d leicht aus dem Anschluß an die Konstanten des massiven Metalls erkennen. Für dünnere Schichten ist die Entscheidung zunächst noch aus Stetigkeitsgründen leicht bis zu einer Grenze, die für Ag bei 7 mµ (für $\lambda = 0,5\mu$) liegt und etwa proportional λ zu wachsen scheint. Unter dieser Grenze schneiden sich die Kurven unter sehr spitzen Winkeln[3], ja berühren sich schließlich für dünnste Schichten. So erhält man dann für n, k nur noch eine grobe Schätzung. Beachtet man nämlich, daß die Kurven ja wegen der Meßfehler eine endliche Breite haben sollten, so kommt dort ein ziemlich langgestrecktes Gebiet für die n, k-Werte in Frage. Eigentümlicherweise aber hat dieses schmale Gebiet eine solche Lage, daß die Werte $n \cdot k$ für alle Punkte dieses Gebietes wenig voneinander abweichen, so daß immerhin das Produkt $n \cdot k$ noch bestimmbar bleibt.

Ähnlich scheinen die Dinge bei SCHULZE zu liegen. VEENEMANS mißt nur die Durchlässigkeit seines Sb-Stufenkeils. Dann benutzt er die Formel

$$J_d = J_d(n; k; d),$$

rechnet mit mehreren angenommenen n, k-Wertepaaren J_d aus und findet dann durch Interpolation die n, k-Werte, die mit der gemessenen Intensität J_d ver-

[1] H. MURMANN: Z. Physik **80**, 161 (1933). — H. WOLTER: Z. Physik **105**, 276 (1937); vgl. auch D. HACMAN: Z. Physik **114**, 173 (1939).
[2] Wir lassen bei den Unbekannten n_1, k_1, d_1 die Indices von jetzt an fort.
[3] Das zeigt auch das neueste graphische Verfahren von D. MALÉ (s. Bibliographie), nach dem die drei Unbekannten aus den drei Intensitäten bestimmt werden.

träglich sind. Diese n, k-Werte werden dann in einer n, k-Ebene als Kurve dargestellt. Dasselbe geschieht für eine benachbarte Stufe und liefert eine zweite Kurve

$$J_d^* = J_d(n; k; d^*);$$

beide schneiden sich (freilich unter spitzem Winkel). Diese Werte wären aber nur dann die Konstanten des Antimons, wenn beide Stufen die gleichen optischen Konstanten hätten. Dort, wo n und k dickenabhängig sind — und diese Fälle interessieren am meisten —, liefert die Methode nur Mittelwerte. Hier interessiert aber vor allem, daß auch die Veenemanssche Methode den Effekt zeigt, der bei der Murmannschen diskutiert wurde. Unterhalb der dort genannten Grenze schmiegen sich Veenemans' Kurven sehr eng einer Hyperbel $n \cdot k = \mathrm{const}$ an und liefern keinen rechten Schnittpunkt mehr. Doch ist auch hier das Produkt $n \cdot k$, das durch die genannte Hyperbel festgelegt ist, noch zu gewinnen.

Dieses Auftreten desselben Versagens der Intensitätsmethoden bei dünnsten Schichten kann kein Zufall sein. Trotzdem scheinen die komplizierten Formeln (44.3) und (44.4) unmittelbar keine Erklärung zu geben, zumal die Größen n, k dort gerade in der Kombination $n \cdot k$ nicht auftreten. Die theoretische Begründung und exaktere Fassung dieses Effekts ist Gegenstand der folgenden Ziffer.

Zur Kritik der strengen Verfahren mit nichtoptischer Schichtdickenbestimmung sei aber noch darauf hingewiesen, daß nach den Messungen Sommers[1] sich die Dichte dünnschichtigen Goldes z. B. als sehr viel kleiner (50% und weniger) als die des massiven Metalls herausgestellt hat und sich damit also die nichtoptischen Schichtdickenbestimmungen als unzuverlässig erwiesen haben. Das hat bei den Verfahren Murmanns, Schulzes, Veenemans' u. a. deshalb so weitgehende Folgen, weil ein falsches d auch fälschend auf n und k selbst, und zwar in zunächst nicht übersehbarer Weise wirkt.

Am nächsten liegt der folgende Gedanke, zu einer optischen Schichtdickenbestimmung zu kommen. Die Intensitäten J_r und J_r' sind bei „Unterlageschichten" (das soll heißen $n_0 \neq n_2$) verschieden. Es müßte also möglich sein, die drei Gln. (44.3), (44.4) und (44.5)

$$J_r = J_r(n; k; d); \quad J_d = J_d(n; k; d); \quad J_r' = J_r'(n; k; d)$$

nach den drei Unbekannten n, k, d aufzulösen. Wo prinzipiell die Grenzen eines solchen Verfahrens liegen, wird in den nächsten Ziffern untersucht[2].

46. Genäherte Intensitätsformeln für dünne Unterlageschichten. Um das Versagen der Intensitätsmethoden bei dünnen Schichten zu erklären, entwickelt man[3] die transzendenten Funktionen in den umgeformten Formeln der Tabelle 1

$$\mathfrak{r}_2 = \frac{g_1(g_2 - g_0) \cos h + i(g_2 g_0 - g_1^2) \sin h}{g_1(g_2 + g_0) \cos h + i(g_2 g_0 + g_1^2) \sin h}, \tag{46.1}$$

$$\left. \begin{aligned} \mathfrak{d}_2 &= \frac{2 g_2 g_1}{g_1(g_2 + g_0) \cos h + i(g_2 g_0 + g_1^2) \sin h} \\ &\left(h = \frac{1}{i}\varrho_1 d_1 = \frac{2\pi d_1}{\lambda}\,\mathfrak{n}_1 c_1 = \frac{\gamma}{2} \cdot \mathfrak{n}_1 c_1\right) \end{aligned} \right\} \tag{46.2}$$

---

[1] A. Sommer: Diss. Hamburg 1940. Vgl. auch H. Schopper: Z. Physik **130**, 565 (1951).
[2] Ein solches Verfahren wurde von D. Malé [Ann. Phys., Paris **9**, 10 (1954)] entwickelt. Malé sieht sich aus den hier diskutierten Gründen bei dünnsten Schichten zum Heranziehen von Phasenmessungen gezwungen. Das wird unten näher begründet.
[3] H. Wolter: Z. Physik **105**, 269 (1937) und Staatsexamensarbeit 1933.

in Potenzreihen nach h,

$$\cos h = 1 - \frac{h^2}{2} + - \cdots; \quad \sin h = h - \frac{h^3}{6} + \cdots. \tag{46.3}$$

Bei Vernachlässigung zweiter und höherer Potenzen von h neben 1 folgt

$$\mathfrak{r} = - \frac{(g_0 - g_2) + i\,\dfrac{\gamma}{2}\,\dfrac{n_1 c_1}{g_1}\,(g_1^2 - g_2 g_0)}{(g_2 + g_0) + i\,\dfrac{\gamma}{2}\,\dfrac{n_1 c_1}{g_1}\,(g_1^2 + g_2 g_0)}, \tag{46.4}$$

$$\mathfrak{d} = \frac{2 g_2}{(g_2 + g_0) + i\,\dfrac{\gamma}{2}\,\dfrac{n_1 c_1}{g_1}\,(g_1^2 + g_2 g_0)}, \tag{46.5}$$

$$\mathfrak{r}' = \frac{(g_0 - g_2) - i\,\dfrac{\gamma}{2}\,\dfrac{n_1 c_1}{g_1}\,(g_1^2 - g_2 g_0)}{(g_2 + g_0) + i\,\dfrac{\gamma}{2}\,\dfrac{n_1 c_1}{g_1}\,(g_1^2 + g_2 g_0)}. \tag{46.6}$$

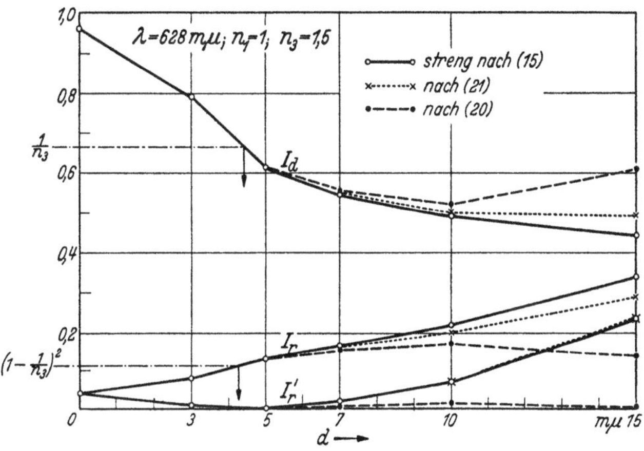

Fig. 35. Vergleich der Näherungen (46.7) bis (46.9) ×···× und (46.10) bis (46.12) •——• mit den strengen Gln. (44.3) bis (44.5) o——o.

Nimmt man Glieder mit h^2 noch mit[1], so erhält man nach (44.1) und (44.2) speziell für senkrechten Einfall die Näherungsgleichungen für die Intensitäten[2] (gültig für $d \leq D_1$):

$$J_r = \frac{(n_0 - n_2 + \gamma n k)^2 + M}{(n_0 + n_2 + \gamma n k)^2 + M}, \tag{46.7}$$

$$J_d = \frac{4 n_2 n_0}{(n_0 + n_2 + \gamma n k)^2 + M}, \tag{46.8}$$

$$J_r' = \frac{(n_0 - n_2 - \gamma n k)^2 + M}{(n_0 + n_2 + \gamma n k)^2 + M}. \tag{46.9}$$

Darin ist abgekürzt

$$M = \frac{\gamma^2}{4}\left\{\left(n^2 - k^2 - \frac{n_0^2 + n_2^2}{2}\right)^2 - \left(\frac{n_0^2 - n_2^2}{2}\right)^2\right\} \approx \frac{\gamma^2}{4}\,(n^2 - k^2)^2.$$

---

[1] Bis zu dritten Potenzen von h geht F. ABELÈS: J. Phys. Radium **14**, 5 (1953).
[2] A. SOMMER: Diss. Hamburg (1940) und H. WOLTER: Z. Physik **105**, 286 (1937).

Da M bei sehr kleinem $\gamma/2 = 2\pi d/\lambda$ nur eine kleine Korrekturgröße gibt, kann man auch mit gröberen Näherungen

$$J_r = \left(\frac{n_0 - n_2 + \gamma n k}{n_0 + n_2 + \gamma n k}\right)^2 , \qquad (46.10)$$

$$J_d = \frac{4 n_2 n_0}{(n_0 + n_2 + \gamma n k)^2} , \qquad (46.11)$$

$$J_r' = \left(\frac{n_0 - n_2 - \gamma n k}{n_0 + n_2 + \gamma n k}\right)^2 \qquad (46.12)$$

(gültig für $d \leq D_2$) für extrem dünne Schichten auskommen. Fig. 35 zeigt an einem Zahlenbeispiel (Konstanten des Silbers) den Vergleich zwischen den streng nach (44.3) bis (44.5) berechneten Intensitäten mit den aus den feineren [(46.7) bis (46.9)] und den gröberen Näherungen [(46.10) bis (46.12)] berechneten Werten. Hiernach scheint die Gültigkeitsgrenze der besseren Formeln bei $D_1 \approx 10$ mμ, die der gröberen bei $D_2 \approx 5$ mμ zu liegen. Doch hängen D_1 und D_2 von der geforderten Genauigkeit, letzten Endes von der erreichbaren Meßgenauigkeit ab.

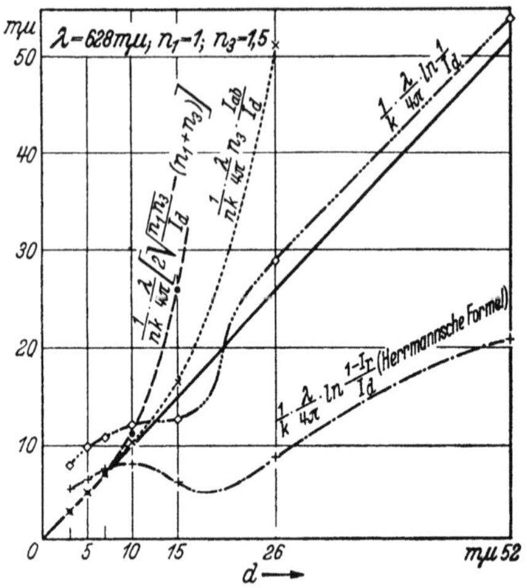

47. Näherungsformeln für $d \cdot n \cdot k$ und $d \cdot (n^2 - k^2)$ bei dünnen Schichten. Im Gegensatz zu den strengen Intensitätsformeln (44.3) bis (44.5) lassen sich die Näherungsgleichungen (46.7) bis (46.9) nach den darin vorkommenden Kombinationen der Unbekannten leicht geschlossen auflösen; durch Elimination von M aus zwei der Gleichungen erhält man[1]

$$\gamma n k = n_0 \frac{1 - J_d - J_r}{J_d} , \qquad (47.1)$$

$$\gamma n k = n_2 \frac{1 - J_d - J_r'}{J_d} , \qquad (47.2)$$

und daraus

$$\gamma n k = \frac{n_0 n_2}{n_0 - n_2} \cdot \frac{J_r - J_r'}{J_d} \qquad (47.3)$$

Fig. 36. Nachprüfung der Gleichungen zur Schichtdickenbestimmung. Es ist $I_{ab} = 1 - I_d - I_r$.

(gültig für $d \leq D_1$). Aus den gröberen Näherungen (46.10), (46.11), (46.12) folgt durch Auflösen

$$\gamma n k = 2 \cdot \sqrt{\frac{n_2 n_0}{J_d}} - (n_0 + n_2) , \qquad (47.4)$$

$$\gamma n k = n_2 \cdot \frac{1 + \sqrt{J_r}}{1 - \sqrt{J_r}} - n_0 \qquad (47.5)$$

(gültig für $d \leq D_2$). Fig. 36 gibt eine Nachprüfung einiger dieser Gleichungen[2] an einem Zahlenbeispiel (Konstanten des Silbers). Sie enthält zugleich eine Nachprüfung der Hermannschen[3] Formel und der Formel von Schulze[4] für dicke Schichten, die unten noch behandelt wird.

---

[1] H. Wolter: Z. Physik **105**, 269 (1937). Verbesserte Gleichungen gibt neuerdings F. Abelès: J. Phys. Radium **14**, 5 (1953).

[2] Die Brauchbarkeit wurde unter anderem auch von A. Sommer (Diss. Hamburg 1940) erwiesen.

[3] H. Goldschmidt u. H. Dember: Z. techn. Phys. **7**, 137 (1926).

[4] R. Schulze: Phys. Z. **34**, 24 (1933).

Durch Auflösen nach M folgt aus (46.7) bis (46.9) die SOMMER-DAVIDsche[1] Formel

$$n^2 - k^2 = \frac{n_0^2 + n_2^2}{2} \pm$$
$$\pm \frac{2}{\gamma} \sqrt{2 n_2 n_0 \frac{J_r + J_r'}{J_d} - (n_0 + n_2)^2 - (\gamma n k)^2 + \frac{\gamma^2}{16} (n_0^2 - n_2^2)} . \tag{47.6}$$

Doch schreibt SOMMER dazu: „Die Formel liefert jedoch $n^2 - k^2$ nicht mit derselben Sicherheit wie die Formeln (47.1) und (47.2) das Produkt $\gamma n k$. Sie ist sehr empfindlich gegen kleine Änderungen von M, die innerhalb der Meßgenauigkeit liegen, weil der Faktor $4/\gamma^2$ bei dünnsten Schichten sehr groß ist. Die Güte der aus der Näherung (47.6) für $n^2 - k^2$ gewonnenen Ergebnisse nimmt also mit abnehmender Schichtdicke ab, weshalb auch der Verlauf der n- und k-Kurven bei Goos im Gebiet dünnster Schichten wenig gesichert erscheint. Da nun bei den hier in dieser Arbeit hauptsächlich benutzten Schichten annähernd $M = 0$ ist, gehen die Meßfehler besonders ungünstig in die Rechnung ein, so daß teilweise $n^2 - k^2$ sogar imaginär wird. Jedenfalls bewirkt die Meßgenauigkeit schon einen so großen Spielraum in $n^2 - k^2$, daß bei einem Vergleich mit den Polarisationsmessungen keine aufschlußreichen Ergebnisse zu erzielen sind."

48. Die „Identitäten" als Kriterien für die Gültigkeitsgrenzen D_1 und D_2 und für die Tragweite der Intensitätsmethoden. Wo die Gln. (46.7), (46.8), (46.9) innerhalb der Meßfehlergrenzen gelten, also für $d \leq D_1$, hängen die drei Intensitäten J_r, J_d, J_r' nur von zwei Kombinationen der Unbekannten, $\gamma n k$ und $2 \sqrt{M} \approx \gamma (n^2 - k^2)$ ab. Diese können aus den Gleichungen daher eliminiert werden. Daraus folgt [z.B. durch Gleichsetzen der rechten Seiten von (47.1) und (47.2)] die Beziehung

$$\frac{n_0}{n_2} = \frac{1 - J_d - J_r'}{1 - J_d - J_r} = \frac{J_{ab}'}{J_{ab}} . \tag{48.1}$$

PARTSCH und HALLWACHS[2] erklärten damit bereits ihre paradox erscheinende Beobachtung, daß der lichtelektrische Effekt größer ist, wenn das Licht von der Quarzseite einfällt, als bei unmittelbarem Auffall auf die Schicht. J_{ab}' und J_{ab} bezeichnet die jeweils absorbierten Intensitäten, und die Gl. (48.1) bedeutet offenbar, daß dünne Schichten n_0/n_2-mal so viel Licht absorbieren, wenn das Licht von der Unterlage einfällt statt von der Luftseite.

Da die durchgehende Intensität in beiden Fällen gleich ist, so geht die größere Absorption ganz auf Kosten der Reflexion. J_r' muß also für dünne Schichten stets kleiner als J_r sein. Auflösen von (48.1) nach J_r gibt die „erste Identität":

$$J_r = \frac{1}{n_0} \{n_2 J_r' + (n_0 - n_2) (1 - J_d)\} . \tag{48.2}$$

Da diese Gleichung die Unbekannten nicht enthält, nennt man sie im Rahmen der algebraischen Eliminations- und Auflöseaufgabe eine „Identität".

Aus den Gln. (46.10), (46.11) und (46.12), die nur eine einzige Kombination der Unbekannten, $\gamma n k$, enthalten, folgt durch Elimination von $\gamma n k$ eine von (48.2) unabhängige Identität

$$J_r = \left(1 - \sqrt{\frac{n_2}{n_0} J_d}\right)^2 , \tag{48.3}$$

---

[1] A. SOMMER: Diss. Hamburg 1940. — F. GOOS: Z. Physik **106**, 606 (1937). — J. KRAUTKRÄMER: Ann. Phys. (5) **32**, 537 (1938).

[2] PARTSCH u. HALLWACHS: Ann. Phys. (4) **41**, 247 (1913).

die sog. „*zweite Identität*". Ihre Bedeutung erhellt aus dem folgenden Satz, der sich aus der absoluten und gleichmäßigen Konvergenz der Reihen (46.3) beweisen läßt:

Zu jedem Meßfehlerbereich $\varepsilon > 0$ gibt es ein D_1 und ein D_2 so, daß für alle Schichtdicken $d \leq D_1$

$$|J_r(\text{Gl. (48.2)}) - J_r(\text{streng})| < \varepsilon$$

und für alle $d \leq D_2$

$$|J_r(\text{Gl. (48.3)}) - J_r(\text{streng})| < \varepsilon$$

ist. Wir müssen dabei nur den Bereich, der den Werten n, k, n_0, n_2 zur Verfügung steht, als beschränkt ansehen.

Im Rahmen des Geltungsbereiches der Identitäten kann man also z.B. J_r aus ihnen ebenso genau berechnen wie messen. Im Bereich beider Identitäten sind daher J_d, J_r, J_r' keine drei unabhängigen Stücke; ihre Messung kommt vielmehr nur einer unabhängigen Messung gleich. Daher liefern die Intensitätsmethoden bei dünnsten Schichten, d.h. für $d \leq \text{Min}\,(D_1; D_2)$ nur noch eine Unbekannte; es ist die Größe $d \cdot n \cdot k$, wie die Gln. (46.10), (46.11) und (46.12) zeigen. Damit ist das Versagen der Intensitätsmethoden bei dünnsten Schichten geklärt. Wo die Grenze liegt, kann mit Hilfe der Identitäten an den gemessenen Intensitäten selbst geprüft werden ohne Kenntnis der gesuchten Größen n, k, d.

Tabelle 9 gibt Überprüfungen der Identitäten an einer Meßreihe von F. Goos für dünne Silberschichten auszugsweise wieder. Die erste Identität ist fast bei allen Schichten erfüllt; die letzte Schicht macht eine Ausnahme infolge ungleichmäßigen Aufbaues. Unterhalb 6 mμ kann mit einer sicheren Messung von $n^2 - k^2$ mittels Intensitäten, wenn diese selbst auf mehrere Promille fehlerhaft gemessen sind, daher nicht gerechnet werden[1].

Tabelle 9.

[Aus H. WOLTER, Z. Physik **105**, 290 (1937)] $\lambda = 578\ m\mu$; $n_2 = 1$; $n_0 = 1,46$.

d aus Wägung $m\mu$	Gemessene Intensitäten			J_r aus 1. Identität	Fehler der 1. Identität	J_r aus 2. Identität	Fehler der 2. Identität
	J_d %	J_r' %	J_r %	%	%	%	%
2	79,0	1,3	7,2	7,5	0,3	7,0	$-0,2$
6	44,3	5,8	21,2	21,5	0,3	20,2	$-1,3$
6	42,0	5,5	21,5	22,0	0,5	21,6	0,1
13	43,2	32,7	40,0	40,3	0,3	20,8	$-19,2$
14	32,9	32,1	42,7	43,1	0,4	27,6	$-15,1$
18	29,6	45,5	52,6	53,3	0,7		
18	31,6	51,0	56,2	55,8	$-0,4$		
19	31,0	54,4	58,8	58,9	0,1		
26	15,0	74,6	77,3	77,6	0,3		
29	12,16	75,3	78,8	79,2	0,4		
29	12,73	78,3	81,2	81,2	0,0		
36	6,77	84,2	85,7	87,0	1,2		
45	3,18	87,2	89,5	90,3	0,8		
33	2,64	48,8	60,3	64,1	3,8		

Natürlich bedeutet die Erfüllung der Identitäten nur dann „Nichtmeßbarkeit", wenn diese Erfüllung auch bei einer gewissen Variation der Schichtdicke, der Wellenlänge und der Metallart erhalten bleibt. Da der Übergang von Silber zu Gold oder einer anderen Wellenlänge im Sichtbaren an der Erfüllung der

---

[1] Diese Grenze kann jedoch zu immer dünneren Schichten verlagert werden, je mehr die Meßgenauigkeit gesteigert wird.

Identitäten nichts Entscheidendes ändert, kann man die Intensitäten in gewissem Rahmen bei dünnsten Schichten nicht mehr als unabhängig ansehen. In zweckmäßiger Weise erfaßt die zweite Identität das Gebiet, in dem $M \approx 0$ ist, ebenfalls. Dort ist sie nach ihrer Herkunft aus den Gln. (46.10), (46.11), (46.12) für $M = 0$ natürlich besonders gut erfüllt und kennzeichnet so das Gebiet, in dem wegen des quadratischen Eingehens von $n^2 - k^2$ in M eine Berechnung von $n^2 - k^2$ aus Intensitäten besonders ungenau ausfallen muß.

Die erste Identität gilt nach der Tabelle 9 nahezu für alle auftretenden Schichtdicken. Der Grund für den großen Gültigkeitsbereich liegt darin, daß die erste Identität näherungsweise auch für stark absorbierende dicke Schichten gilt; denn dann ist nach Tabelle 1 für senkrechten Einfall

$$\frac{1 - J_d - J'}{1 - J_d - J_r} \approx \frac{n_0}{n_2} \cdot \frac{(n + n_2)^2 + k^2}{(n + n_0)^2 + k^2}.$$

Wo der zweite Faktor rechts innerhalb der Meßfehlergrenzen nahezu 1 ist — und das gilt stets für großes k^2 — gilt die erste Identität sogar auch für dicke Schichten. Das ist jedoch z.B. um $\lambda = 265$ mμ bei Silber anders; dort wird die erste Identität nur unterhalb 10 mμ befriedigt. Man kann also ausnahmsweise bei sehr kleiner Absorption (z.B. bei Ag um 265 mμ) und Schichtdicken über 10 mμ aus Intensitätsmessungen die drei Unbekannten n, k, d prinzipiell erhalten[1]. Bei den interessanten dünnen Schichten ist aber stets nur eine Bestimmung von höchstens zwei Kombinationen der Unbekannten und bei allerdünnsten nur noch die Bestimmung des Produktes $d \cdot n \cdot k$ aus Intensitätsmessungen möglich.

Schwach absorbierende Stoffe werden im Teil e) noch getrennt behandelt werden.

Bezüglich weiterer Identitäten (z.B. der „Minimumbedingung") sei nur auf die Originalarbeiten hingewiesen[2]. C. v. FRAGSTEIN[3] gibt eine Verbesserung der Identitäten, führt dabei aber auch Phasen außer den Intensitäten ein[4].

49. Intensitätsmethoden bei schrägem Einfall. Die Gleichungen von Ziff. 46 lassen sich leicht auf TE-Wellen beliebigen Einfalls erweitern. Für TE-Wellen ist bei $\mu_l = 1$ für alle Medien $g_l = \mathfrak{n}_l c_l$; also nach den Gln. (46.4) bis (46.6)

$$\mathfrak{r} = - \frac{g_0 - g_2 + i\dfrac{\gamma}{2}\left(n^2 - k^2 - i\,2n\,k - n_2 n_0 \cos(\varphi_2 - \varphi_0)\right)}{g_0 + g_2 + i\dfrac{\gamma}{2}\left(n^2 - k^2 - i\,2n\,k + n_2 n_0 \cos(\varphi_2 + \varphi_0)\right)}, \qquad (49.1)$$

$$\mathfrak{d} = \frac{2\,g_2}{g_0 + g_2 + i\dfrac{\gamma}{2}\left(n^2 - k^2 - i\,2n\,k + n_2 n_0 \cos(\varphi_2 + \varphi_0)\right)}, \qquad (49.2)$$

$$\mathfrak{r}' = \frac{g_0 - g_2 - i\dfrac{\gamma}{2}\left(n^2 - k^2 - i\,2n\,k - n_2 n_0 \cos(\varphi_2 - \varphi_0)\right)}{g_0 + g_2 + i\dfrac{\gamma}{2}\left(n^2 - k^2 - i\,2n\,k - n_2 n_0 \cos(\varphi_2 + \varphi_0)\right)}. \qquad (49.3)$$

---

[1] Graphische Verfahren hierzu gab D. MALÉ (s. Bibliographie). Diese Verfahren zeigen auch sehr gut die schleifenden Schnitte der Kurven bei dünnsten Schichten; vgl. auch R. PHILIP u. J. TROMPETTE [C. R. Acad. Sci. Paris **241**, 627 (1955)], die nach MALÉs Verfahren arbeiteten.

[2] Siehe H. MAYER: Physik dünner Schichten, S. 216 und F. GOOS: Z. Physik **100**, 95 (1936); **106**, 606 (1937). — H. WOLTER: Z. Physik **105**, 269 (1937).

[3] C. v. FRAGSTEIN: Z. Physik **139**, 163 (1954).

[4] Wie gut die erste Identität (48.1) aber bereits in der Praxis erfüllt ist, zeigten außer den unter Fußnote 1 genannten Arbeiten neuerdings die Arbeiten von D. MALÉ (s. Bibliographie).

Daraus erhält man die Intensitäten[1] nach Tabelle 1

$$J_r = \frac{(n_0 \cos \varphi_0 - n_2 \cos \varphi_2 + \gamma\, n\, k)^2 + M}{(n_0 \cos \varphi_0 + n_2 \cos \varphi_2 + \gamma\, n\, k)^2 + M}\,, \tag{49.4}$$

$$J_d \cdot \frac{\cos \varphi_0}{\cos \varphi_2} = \frac{4\, n_2\, n_0 \cdot \cos \varphi_2 \cdot \cos \varphi_0}{(n_0 \cos \varphi_0 + n_2 \cos \varphi_2 + \gamma\, n\, k)^2 + M}\,. \tag{49.5}$$

$$J_r' = \frac{(n_0 \cos \varphi_0 - n_2 \cos \varphi_2 - \gamma\, n\, k)^2 + M}{(n_0 \cos \varphi_0 + n_2 \cos \varphi_2 + \gamma\, n\, k)^2 + M} \tag{49.6}$$

mit einem $M \approx \dfrac{\gamma^2}{4} \cdot (n^2 - k^2)^2$[†].

Auflösen nach $\gamma\, n\, k$ gibt für TE-Wellen

$$\gamma\, n\, k = n_0 \cos \varphi_0 \cdot \frac{1 - J_r - J_d \cdot \dfrac{\cos \varphi_0}{\cos \varphi_2}}{J_d \cdot \dfrac{\cos \varphi_0}{\cos \varphi_2}}\,, \tag{49.7}$$

$$\gamma\, n\, k = n_2 \cos \varphi_2 \cdot \frac{1 - J_r' - J_d \dfrac{\cos \varphi_0}{\cos \varphi_2}}{J_d \cdot \dfrac{\cos \varphi_0}{\cos \varphi_2}}\,, \tag{49.8}$$

$$\gamma\, n\, k = \frac{n_0\, n_2 \cos \varphi_0 \cos \varphi_2}{n_0 \cos \varphi_0 - n_2 \cos \varphi_2} \cdot \frac{J_r - J_r'}{J_d \dfrac{\cos \varphi_0}{\cos \varphi_2}}\,. \tag{49.9}$$

Entsprechend gelten die Identitäten

$$\frac{n_0 \cos \varphi_0}{n_2 \cos \varphi_2} = \frac{1 - J_r' - J_d \dfrac{\cos \varphi_0}{\cos \varphi_2}}{1 - J_r - J_d \dfrac{\cos \varphi_0}{\cos \varphi_2}} \tag{49.10}$$

und

$$\sqrt{J_r} + \sqrt{\frac{n_2}{n_0}\, J_d} = 1 \tag{49.11}$$

für TE-Wellen.

Die bei Hacman[2] Gl. (6) und (6′) in den Zählern für J_d stehenden Größen $\gamma \cdot \varepsilon_{12}$ sind zu streichen, ebenso die e-Funktionen in den entsprechenden Zählern der Gln. (4) und (4′) bei Hacman und (14) bei Wolter, Z. Phys. **105**, 275 (1937). Statt im Punkte H mit komplexen Koordinaten nimmt man die durchgegangene Welle z. B. am Punkte $(0; 0; -d)$.

Für TH-Wellen ist z. B.

$$\mathfrak{d} = \frac{2g_2}{g_2 + g_0 + i\, \dfrac{\gamma}{2} \left(1 + \dfrac{\mathfrak{n}_1^2}{n_0\, n_2} \cos \varphi_0 \cos \varphi_2 - \dfrac{n_0\, n_2}{\mathfrak{n}_1^2} \sin \varphi_0 \sin \varphi_2\right)}\,.$$

Für $|\mathfrak{n}_1^2| \gg 1$ freilich führt auch das auf Unbekanntenkombinationen $\gamma\, n\, k$, $\gamma\,(n^2 - k^2 \ldots)$, da das letzte Glied im Nenner vernachlässigbar ist; es sei denn,

---

[1] In dem Intensitätsverhältnis J_d sind die Intensitäten stets auf gleiche Auffängerfläche normiert; benutzt man dagegen den gesamten Querschnitt (eines begrenzten Strahlenbündels), der sich bei Übertritt in das neue Medium ändert, so ist $J_d \cdot \cos \varphi_0/\cos \varphi_2$ das gemessene Verhältnis.

[†] In M vorgenommene Vernachlässigungen sind dabei von derselben Größenordnung wie die ohnehin zuvor geschehenen. In γ quadratische Glieder sind nur mitgenommen, soweit sie mit Faktoren behaftet sind, die bei stark absorbierenden Metallen große Werte haben können.

[2] D. Hacman: Z. Physik **14**, 173 (1939).

daß man nahezu streifenden Einfall benutzt. Solche Verfahren sind zur Zeit in Entwicklung, aber noch nicht veröffentlicht.

Sehr zweckmäßig ist die Messung mit TH-Wellen am BREWSTERschen Winkel, da dort die reflektierte Intensität

$$|\hat{\mathfrak{r}}|^2 = |\hat{\mathfrak{r}}'|^2 = \left| \frac{i\,(\hat{g}_2^2 - \hat{g}_1^2) \cdot \dfrac{\gamma}{2}\,\mathfrak{n}_1 c_1}{2\,\hat{g}_2 \hat{g}_1 + i\,(\hat{g}_2^2 + \hat{g}_1^2)\,\dfrac{\gamma}{2}\,\mathfrak{n}_1 c_1} \right|^2$$

nach Art einer Nullmethode sehr empfindlich auch auf dünnste Schichten reagiert.

50. Ergänzungen der dnk-Formeln für dicke Schichten. Die Formeln (49.7) bis (49.9) für dnk geben eine genaue optische Schichtdickenbestimmung, wenn nk aus Polarisationsmessungen bekannt ist. Da sie nur für dünne Schichten gelten, interessiert eine ergänzende Formel für dicke Schichten, die man aus (44.4) erhält, indem man das zweite, dritte und vierte Glied des Nenners neben dem ersten vernachlässigt. Das ist bei dicken Schichten erlaubt, da $-\gamma\,\mathfrak{Jm}\,(\mathfrak{n}_1 c_1)$ dann praktisch eine sehr große positive Zahl ist. Es folgt so aus (44.4)

$$J_d = \frac{16\,g_2\,g_0\,|g_1|^2 \cdot e^{\gamma\,\mathfrak{Jm}\,(n_1 c_1)}}{\{(\mathfrak{Re}\,g_1 + g_2)^2 + \mathfrak{Jm}^2\,g_1\}\,\{(\mathfrak{Re}\,g_1 + g_0)^2 + \mathfrak{Jm}^2\,g_1\}} \,. \tag{50.1}$$

Die Nachprüfung an streng gerechneten Zahlenwerten erweist die Brauchbarkeit der Formel z.B. für Metalle wie Gold und Silber bis herab zu Schichtdicken, bei denen die Näherungen für dünne Schichten bereits einsetzen (vgl. Fig. 36). In der Formel für die Schichtdicke selbst

$$d = - \frac{\lambda}{4\pi\,\mathfrak{Jm}\,(n_1 c_1)} \left(\log \frac{1}{J_d} + \log \frac{16\,g_2\,g_0\,|g_1|^2}{\{(\mathfrak{Re}\,g_1 + g_2)^2 + \mathfrak{Jm}^2\,g_1\}\{(\mathfrak{Re}\,g_1 + g_0)^2 + \mathfrak{Jm}^2 g_1\}} \right) \tag{50.2}$$

kann man den zweiten Summanden in der Klammer oft vernachlässigen, wenn es sich um äußerst dicke Schichten handelt, da $\log \dfrac{1}{J_d}$ dann extrem hohe Werte annimmt.

Die sich aus (50.2) für senkrechten Einfall ergebende Formel[1]

$$k\gamma = \log \frac{1}{J_d} + \log \frac{16\,n_2\,n_0\,(n^2 + k^2)}{\{(n + n_2)^2 + k^2\}\,\{(n + n_0)^2 + k^2\}} \tag{50.3}$$

läßt sich bei den stark absorbierenden Metallen häufig durch die SCHULZEsche Formel[2]

$$k\gamma = \log \frac{1}{J_d}$$

ersetzen. Doch haben (50.2) und (50.3) natürlich den weitaus größeren Geltungsbereich[3].

51. Vorzüge und Nachteile der Intensitätsmethoden. Intensitätsmethoden liefern bei dünnen Schichten ebenso wie andere Methoden sicher falsche Werte für die optischen Konstanten, wenn nichtoptisch bestimmte Schichtdicken in die Rechnung eingesetzt werden.

Der Vorteil der reinen Intensitätsmethoden gegenüber anderen liegt in der experimentellen Durchführbarkeit in nahezu allen Wellengebieten, ihr Nachteil in der Unvollständigkeit der Aussage. Bei den stark absorbierenden Metallen

---

[1] H. WOLTER: Z. Physik **105**, 298 (1937).
[2] R. SCHULZE: Phys. Z. **34**, 24 (1933).
[3] C. v. FRAGSTEIN: Optik **10**, 578 (1953).

kann bei den wichtigen dünnsten Schichten nur noch eine Kombination der Unbekannten, das Produkt $d \cdot n \cdot k$, eindeutig, verläßlich und genau aus Intensitäten bestimmt werden, wenn man von dem noch in Entwicklung befindlichen Verfahren streifend einfallender TH-Wellen absieht. Drei Unbekannte n, k und d sind selbst bei mäßig dicken Schichten genau nur dann zu erhalten, wenn es sich um schwach absorbierende Metalle (Ag im Ultraviolett z. B) oder um Nichtmetalle handelt.

Jedoch hat die Messung der Größe dnk bereits erheblichen Wert. Ihr spektraler Verlauf kann über die Struktur der Schichten Aufschluß geben[1].

Zusammen mit polarisationsoptischen Bestimmungen von n, k bei einer Welle im Sichtbaren führt die Messung von dnk bzw. dk mit Hilfe der Intensitäten sicher zu der heute genauesten optischen Schichtdickenbestimmung nach dem Schema[2]

$$d = \frac{d \cdot n \cdot k \,(\text{aus Intensitäten})}{n \cdot k \,(\text{aus Pol.})}$$

für dünne Schichten und

$$d = \frac{d \cdot k \,(\text{aus Intensitäten})}{k \,(\text{aus Pol.})}$$

bei dicken Schichten.

d) Interferenzmethoden.

52. Zu der experimentellen Methodik. Das bei dünnsten Metallschichten nicht wegzudiskutierende Versagen der Intensitätsmethoden gab nach 1937 den Anstoß, interferometrische Methoden anzuwenden. Bei den Polarisationsmethoden hatte man sich begnügt, von den komplexen Reflexions- und Durchlässigkeitsamplituden

$$\begin{aligned}
\hat{\mathfrak{d}} &= |\hat{\mathfrak{d}}| \cdot e^{i \hat{\psi}_d}; & \hat{\mathfrak{d}} &= |\hat{\mathfrak{d}}| \cdot e^{i \hat{\psi}_d}; \\
\hat{\mathfrak{r}} &= |\hat{\mathfrak{r}}| \cdot e^{i \hat{\psi}_r}; & \hat{\mathfrak{r}} &= |\hat{\mathfrak{r}}| \cdot e^{i \hat{\psi}_r}; \\
\hat{\mathfrak{r}}' &= |\hat{\mathfrak{r}}'| \cdot e^{i \hat{\psi}'_r}; & \hat{\mathfrak{r}}' &= |\hat{\mathfrak{r}}'| \cdot e^{i \hat{\psi}'_r}.
\end{aligned} \right\} \tag{52.1}$$

Verhältnisse wie

$$\mathfrak{P}_R = \frac{\hat{\mathfrak{r}}}{\hat{\mathfrak{r}}} = \frac{|\hat{\mathfrak{r}}|}{|\hat{\mathfrak{r}}|} \cdot e^{i\,(\hat{\psi}_r - \hat{\psi}_r)} = \tan \chi_R \cdot e^{i \delta_R}$$

zu messen, d. h. von den Amplitudenbeträgen nur ihre Verhältnisse und von den Phasen nur ihre Differenzen für verschiedene Polarisationen zu verwenden. Später benutzte man bei den Intensitätsmethoden die absoluten Beträge der Amplituden; und nach ihrem Versagen, das bei dünnsten Metallschichten nicht beseitigt werden konnte, blieb als dritte und letzte Möglichkeit, auch die absoluten Phasen $\hat{\psi}_d$, $\hat{\psi}_d$, $\hat{\psi}_r$, $\hat{\psi}_r$, $\hat{\psi}'_r$, $\hat{\psi}'_r$ heranzuziehen[3].

R. FLEISCHMANN und H. SCHOPPER[1] haben seit 1951 hierzu Meßverfahren erdacht, entwickelt und auf die Messung der optischen Konstanten und der Dicken dünner Schichten angewandt.

[1] H. MAYER: Physik dünner Schichten, Teil I, S. 238. Stuttgart 1950. — H. WOLTER: Z. Physik **115**, 696 (1940). — E. DAVID: Z. Physik **114**, 389 (1939). — H. SCHOPPER: Z. Physik **130**, 565 (1951).

[2] A. SOMMER: Diss. Hamburg 1940.

[3] R. FLEISCHMANN u. H. SCHOPPER: Siehe Bibliographie am Ende dieses Artikels. Siehe auch W. WERNICKE: Über die absoluten Phasenänderungen bei der Reflexion des Lichtes. Pogg. Ann. **159**, 198 (1876) und P. DRUDE: Wied. Ann. **50**, 595 (1893); **51**, 77 (1894). — DRUDE gibt in Wied. Ann. **39**, 483 (1890) bereits eine vollständige Klassifikation der prinzipiell möglichen Methoden und führt auch die ,,absolute Phasenmessung'' dabei auf.

Der Phaseneinfluß beispielsweise einer Glas-Luft-Grenze ist nach den FRESNEL-schen Formeln sowohl in Reflexion als auch in Durchsicht als 0 oder π bzw. 0 bekannt. Um z.B. ψ_d zu messen, braucht man daher eine die zu vermessende Schicht tragende Glasplatte nur teilweise von ihr zu befreien und das Ganze so in irgendein Zweistrahlinterferometer zu setzen, daß ein Wellenbündel die

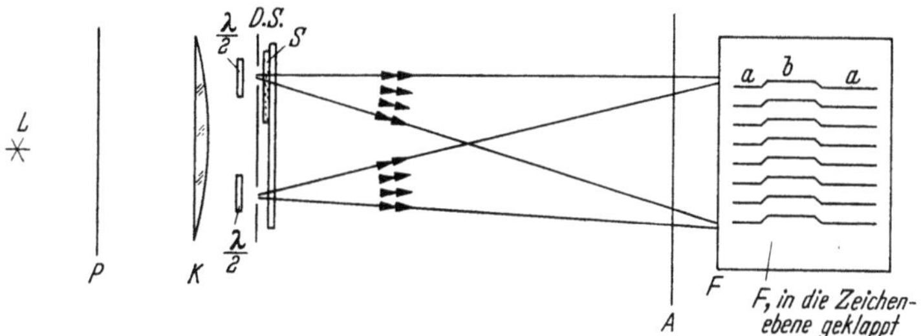

Fig. 37. Doppelspaltinterferometer mit Intensitätsausgleich nach FLEISCHMANN.

Schicht, das andere die reine Glas-Luft-Grenze passiert. Fig. 37 zeigt z.B. ein Doppelspaltinterferometer mit der zu vermessenden Schicht bei S. Auf dem Schirm erscheinen Interferenzstreifen (Minima) a, die verschoben sind gegen die Interferenzstreifen b, die von einem überall von der Schicht befreiten Streifen der Glasplatte (parallel zur Zeichenebene) herrühren. Aber wegen der verschiedenen Intensität beider Wellenbündel sind die Streifen a sehr kontrastarm und zu genauen Messungen ungeeignet, wenn nicht ein Intensitätsausgleich vorgenommen wird. Ihn bewirkt man nach FLEISCH-MANNs Vorschlag durch die beiden um die Bündelachse drehbaren $\lambda/2$-Glimmerplättchen vor den Spalten in Zusammenwirken mit dem Polarisator P und dem Analysator A. Dreht man eine der $\lambda/2$-Platten, so dreht man effektiv die Polarisationsebene und regelt den Intensitätsbruchteil, den der Analysator A aus diesem Bündel auf den Schirm läßt. So ist jedes Intensitätsverhältnis herstellbar.

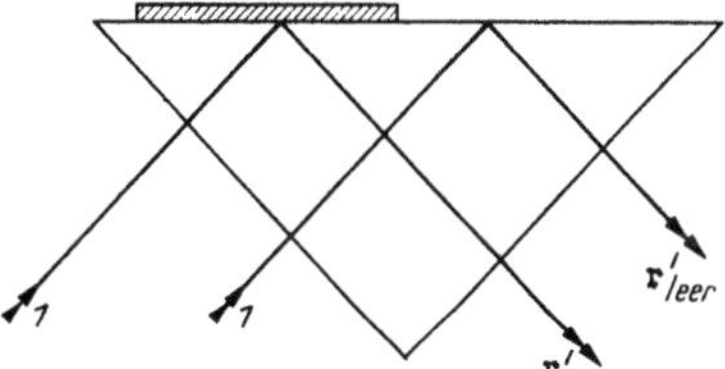

Fig. 38. Einbauteil für Interferenzuntersuchung bei sehr schrägem Einfall ohne Intensitätsausgleich.

Analog kann man ZERNIKEs Dreistrahl-Interferometer oder das JAMINsche oder andere Interferometer z.B. auch das KÖSTERsche[1] mit einem Intensitätsausgleich ausrüsten.

Mit photographischer Methode erreichte H. SCHOPPER[2] eine Genauigkeit bis herab zu 1/250 Streifenbreite, d.h. 1,5° an ψ_d.

Eine Methode, die ohne Intensitätsausgleich auskommt, benutzt Strahlen, die fast streifend, bzw. nahe dem Grenzwinkel der Totalreflexion einfallen (Fig. 38!).

Die Intensitäten der Strahlen sind dann meist von gleicher Größenordnung und die Interferenzstreifen erlauben eine hinreichend genaue Phasenbestimmung und bei quantitativer Photometrie zugleich auch den Intensitätsvergleich; man bestimmt nach Fig. 38 z.B. unmittelbar die Größe

$$\frac{\mathfrak{r}'}{\mathfrak{r}'_{leer}} - 1.$$

Darin bezeichnet $\mathfrak{r}'_{leer}$ die Reflexion an dem von der Schicht befreiten Teil der Hypotenusenfläche [siehe unten Gl. (57.6) und die folgenden].

---

[1] A. C. MÖNCH: Optik **8**, 550 (1951). — R. BAHN u. O. BÖTTGER: Z. Physik **135**, 376 (1953).
[2] Vgl. die unter den Namen FLEISCHMANN oder SCHOPPER aufgeführten Arbeiten in der Bibliographie.

Für die gleichzeitige Phasen- und Amplitudenmessung eignen sich auch Interferometer wie das JAMINsche oder seine Weiterentwicklungen, die eine polarisationsoptische Elliptizitätsmessung als Ausgangspunkt zur Phasen- und Amplitudenberechnung machen[1]. Man vergleiche auch den Artikel Phasenkontrastverfahren in diesem Bande.

Die in dieser Weise gemessene und nach phasenrichtiger Berücksichtigung der Glas-Luft-Grenze gewonnene Phase ist noch nicht ψ_d selbst, da definitionsgemäß $\mathfrak{d}$ das Verhältnis der austretenden Welle an der einen Schichtseite zur einfallenden Welle an der anderen Schichtseite gibt[2]. Die gemessene Phase ist vielmehr (vgl. Fig. 39!)

$$\vartheta_d = \psi_d + \frac{2\pi d}{\lambda} n_2 \cos\varphi_2. \tag{52.2}$$

Die Apparatur nach Fig. 37 läßt sich auf Reflexion umstellen. Da in den Größen $\mathfrak{r}$ und $\mathfrak{r}'$ die beiden Wellen jeweils an der von der einfallenden Welle zunächst getroffenen Seite genommen sind, ist für Einfall von der Trägerseite her

$$\vartheta_r' = \psi_r' \tag{52.3}$$

selbst unmittelbar gemessen, und bei Einfall von der Schichtseite her ergibt die Messung direkt

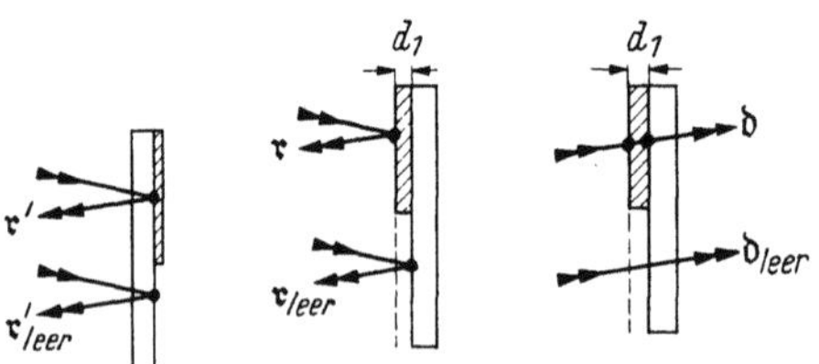

Fig. 39. Zur Phasenkorrektur.

$$\vartheta_r = \psi_r + 2 \cdot \frac{2\pi d}{\lambda} n_2 \cos\varphi_2. \tag{52.4}$$

Der Nullpunkt des Koordinatensystems liege an der Grenze Schicht—Unterlage.

Bezeichnen wir die Reflexions- und Durchlässigkeitskoeffizienten, gebildet als Quotienten der an diesem Nullpunkt genommenen Wellen mit

$$\mathfrak{r}_0 = \mathfrak{r} \cdot e^{2\varrho_2 d} = \mathfrak{r} \cdot e^{2i\frac{2\pi d}{\lambda} n_2 \cos\varphi_2}, \tag{52.5}$$

$$\mathfrak{r}_0' = \mathfrak{r}', \tag{52.6}$$

$$\mathfrak{d}_0 = \mathfrak{d} \cdot e^{\varrho_2 d} = \mathfrak{d} \cdot e^{i\frac{2\pi d}{\lambda} n_2 \cos\varphi_2}, \tag{52.7}$$

wie das bereits oben in Ziff. 9 geschah, dann sind also $\mathfrak{r}_0$, $\mathfrak{r}_0'$, $\mathfrak{d}_0$ als die unmittelbar gemessenen Werte anzusehen, während für $\mathfrak{r}$, $\mathfrak{r}'$ und $\mathfrak{d}$ eine Korrektur nach den Gln. (52.5) bis (52.7) erforderlich ist.

53. Allgemeines Verfahren zur Berechnung der optischen Konstanten aus Phasen- und Betrags- (d. h. Intensitäts)messungen. Aus Phasen- und Intensitätsmessungen mögen zunächst

$$\mathfrak{d} = |\mathfrak{d}| \cdot e^{i\psi_d} = \frac{4g_2 g_1}{(g_2 + g_1)(g_1 + g_0)e^{\varrho_1 d} + (g_2 - g_1)(g_1 - g_0)e^{-\varrho_1 d}}, \tag{53.1}[3]$$

$$\mathfrak{r} = |\mathfrak{r}| \cdot e^{i\psi_r} = \frac{(g_2 - g_1)(g_1 + g_0)e^{\varrho_1 d} + (g_2 + g_1)(g_1 - g_0)e^{-\varrho_1 d}}{(g_2 + g_1)(g_1 + g_0)e^{\varrho_1 d} + (g_2 - g_1)(g_1 - g_0)e^{-\varrho_1 d}}, \tag{53.2}$$

$$\mathfrak{r}' = |\mathfrak{r}'| \cdot e^{i\psi_r'} = \frac{(g_0 - g_1)(g_1 + g_2)e^{\varrho_1 d} + (g_0 + g_1)(g_1 - g_2)e^{-\varrho_1 d}}{(g_2 + g_1)(g_1 + g_0)e^{\varrho_1 d} + (g_2 - g_1)(g_1 - g_0)e^{-\varrho_1 d}} \tag{53.3}$$

---

[1] H. WOLTER: Z. Physik **140**, 57, 565 (1955).
[2] Siehe Ziff. 9.
[3] Läßt man das Licht von der Unterlage aus einfallen, so wird die Durchlässigkeitsamplitude $\mathfrak{d}' = \mathfrak{d} \cdot g_0/g_2$ gleichphasig mit $\mathfrak{d}$. Da die Korrektur ebenfalls dieselbe ist wie bei $\mathfrak{d}$ (wegen $\mathfrak{d}_0' = \mathfrak{d}' \cdot e^{\varrho_2 d}$), bringt die Betrachtung von $\mathfrak{d}'$ nichts Neues.

als bekannt angesehen werden. Zwar sind die Korrekturen in den Gln. (52.5) bis (52.7) nur nach gemessener Schichtdicke bekannt. Aber die Schichtdicke sei zunächst als ungefähr bekannt vorausgesetzt; nachträglich wird ein Iterationsverfahren zu einer strengen optischen Schichtdickenbestimmung führen. Durch Bilden der Kombinationen $\mathfrak{r}/\mathfrak{d}$ und $\mathfrak{r}'/\mathfrak{d}$ bereitet man die Elimination von d vor. Es ist

$$a^* e^{\varrho_1 d} + b^* e^{-\varrho_1 d} = \frac{1}{\mathfrak{d}}, \tag{53.4}$$

$$a \cdot e^{\varrho_1 d} + b\, e^{-\varrho_1 d} = \frac{\mathfrak{r}}{\mathfrak{d}}, \tag{53.5}$$

$$b \cdot e^{\varrho_1 d} + a \cdot e^{-\varrho_1 d} = -\frac{\mathfrak{r}'}{\mathfrak{d}} \tag{53.6}$$

mit den Abkürzungen

$$a^* = \frac{(g_2 + g_1)(g_1 + g_0)}{4 g_2 g_1}, \tag{53.7}$$

$$b^* = \frac{(g_2 - g_1)(g_1 - g_0)}{4 g_2 g_1}, \tag{53.8}$$

$$a = \frac{(g_2 - g_1)(g_1 + g_0)}{4 g_2 g_1}, \tag{53.9}$$

$$b = \frac{(g_2 + g_1)(g_1 - g_0)}{4 g_2 g_1}. \tag{53.10}$$

Paarweise Elimination von $e^{-\varrho_1 d}$ gibt

$$e^{\varrho_1 d} = \left(b\, \frac{1}{\mathfrak{d}} - b^*\, \frac{\mathfrak{r}}{\mathfrak{d}}\right) : (a^* b - a b^*), \tag{53.11}$$

$$e^{\varrho_1 d} = \left(a\, \frac{1}{\mathfrak{d}} + b^*\, \frac{\mathfrak{r}'}{\mathfrak{d}}\right) : (a^* a - b b^*), \tag{53.12}$$

$$e^{\varrho_1 d} = \left(a\, \frac{\mathfrak{r}}{\mathfrak{d}} + b\, \frac{\mathfrak{r}'}{\mathfrak{d}}\right) : (a^2 - b^2). \tag{53.13}$$

Ebenso gibt paarweise Elimination von $e^{+\varrho_1 d}$

$$e^{\varrho_1 d} = -(a^* b - a b^*) : \left(a\, \frac{1}{\mathfrak{d}} - a^*\, \frac{\mathfrak{r}}{\mathfrak{d}}\right), \tag{53.14}$$

$$e^{\varrho_1 d} = -(a^* a - b b^*) : \left(b\, \frac{1}{\mathfrak{d}} + a^*\, \frac{\mathfrak{r}'}{\mathfrak{d}}\right), \tag{53.15}$$

$$e^{\varrho_1 d} = -(a^2 - b^2) : \left(b\, \frac{\mathfrak{r}}{\mathfrak{d}} + a\, \frac{\mathfrak{r}'}{\mathfrak{d}}\right). \tag{53.16}$$

Natürlich sind die sechs Gln. (53.11) bis (53.16) nur in gewissen Tripeln voneinander unabhängig. Durch Gleichsetzen zweier rechter Seiten ist d auf mannigfache Weise eliminierbar. Es gibt entsprechend vielfältige Formen des Verfahrens. Gl. (53.12) und (53.15) ergeben z.B.

$$(a^* a - b b^*)^2 + \left(\frac{a}{\mathfrak{d}} + b^*\, \frac{\mathfrak{r}'}{\mathfrak{d}}\right)\left(\frac{b}{\mathfrak{d}} + a^*\, \frac{\mathfrak{r}'}{\mathfrak{d}}\right) = 0; \tag{53.17}$$

das ist nach Einsetzen von (53.7) bis (53.10)

$$g_1^2 = g_2^2 g_0^2 \frac{(1 - \mathfrak{r}')^2 - \mathfrak{d}^2}{g_2^2 (1 + \mathfrak{r}')^2 - g_0^2 \mathfrak{d}^2}. \tag{53.18}$$

Trennung in Realteil und Imaginärteil gibt z. B. für TE-Wellen

$$n^2 - k^2 - n_2^2 \sin^2 \varphi_2 = \Re\, g_1^2 = g_2^2\, g_0^2\, \frac{u\,u^* + v\,v^*}{u^2 + v^2}, \tag{53.19}$$

$$2nk = -\Im\, g_1^2 = g_2^2\, g_0^2\, \frac{u\,v^* - v\,u^*}{u^2 + v^2} \tag{53.20}$$

mit den Abkürzungen

$$u^* = 1 + |\mathfrak{r}'|^2 \cos 2\psi_r' - 2|\mathfrak{r}'|\cos \psi_r' - |\mathfrak{d}|^2 \cos 2\psi_d, \tag{53.21}$$

$$u = g_2^2(1 + |\mathfrak{r}'|^2 \cos 2\psi_r') + g_2^2\, 2|\mathfrak{r}'|\cos \psi_r' - g_0^2 |\mathfrak{d}|^2 \cos 2\psi_d, \tag{53.22}$$

$$v^* = |\mathfrak{r}'|^2 \sin 2\psi_r' - 2|\mathfrak{r}'|\sin \psi_r' - |\mathfrak{d}|^2 \sin 2\psi_d, \tag{53.23}$$

$$v = g_2^2(|\mathfrak{r}'|^2 \sin 2\psi_r') + g_2^2\, 2|\mathfrak{r}'|\sin \psi_r' - g_0^2 |\mathfrak{d}|^2 \sin 2\psi_d. \tag{53.24}$$

ψ_d, ψ_r, ψ_r' haben die Bedeutung nach Gl. (52.1) bis (52.4).

Zur exakten Schichtdickenbestimmung durch Iteration kann man eine der Gln. (53.11) bis (53.16) oder eine ihrer möglichen Kombinationen verwenden, z. B. das Produkt der Gln. (53.12) und (53.15)

$$e^{2\varrho_1 d} = -\frac{a + b^* \mathfrak{r}'}{b + a^* \mathfrak{r}'}; \tag{53.25}$$

daraus folgt[1]

$$d = \frac{\lambda}{4\pi i\, \mathfrak{n}_1 c_1} \log \frac{(g_1 - g_2)\{g_1(1 + \mathfrak{r}') + g_0(1 - \mathfrak{r}')\}}{(g_1 + g_2)\{g_1(1 + \mathfrak{r}') - g_0(1 - \mathfrak{r}')\}}, \tag{53.26}$$

oder einfacher eine der Gln. (49.7) bis (49.9), falls man sich in deren Gültigkeitsbereich befindet. Wegen

$$\mathfrak{n}_1^2 c_1^2 = \mathfrak{n}_1^2 - n_2^2 \sin^2 \varphi_2; \qquad \log(|w| \cdot e^{i\delta}) = \log|w| + i(\delta + 2\pi M)$$

ist die Auswertung von (53.26) nicht kompliziert. Die willkürliche ganze Zahl M bringt zwar die bekannte Vieldeutigkeit der Schichtdickenbestimmung; doch führt die Messung bei mehreren Wellenlängen zu einer Entscheidung. Die Iteration ist relativ einfach, da nur die Größe ψ_d bei gemessenem ϑ_d sich von Schritt zu Schritt ändert, also in den Gln. (53.21) bis (53.24) jeweils nur der letzte Summand betroffen wird. Schnelle Konvergenz des Iterationsverfahrens kann man in jedem Falle erzwingen, indem man nicht mit dem z. B. aus Gl. (53.26) nach dem m-ten Schritt erhaltenen $_m d$ (Ausg.) unmittelbar in (52.2) und (53.21) bis (53.24) eingeht, sondern vorher eine „konvergenzerzeugende Mittelbildung"

$$_{m+1}d \text{ (Eing.)} = \alpha \cdot {}_m d \text{ (Ausg.)} + \beta \cdot {}_m d \text{ (Eing.)} \tag{53.27}$$

mit zweckmäßig gewählten Konstanten α, β (mit $\alpha + \beta = 1$) vornimmt.

54. Verfahren mit reinen Reflexionsmessungen[2]. Um die unbequeme Umstellung einer justierten Apparatur von Reflexion auf Durchlässigkeit zu vermeiden, kann man bei Unterlageschichten mit Reflexionsmessungen allein arbeiten. Die Gln. (53.2) und (53.3) lassen sich durch Multiplikation mit dem Nenner und Ordnen auf die Form bringen

$$e^{2\varrho_1 d} = \frac{(g_0 - g_1)\{(g_2 + g_1) - (g_2 - g_1)\mathfrak{r}\}}{(g_0 + g_1)\{(g_2 - g_1) - (g_2 + g_1)\mathfrak{r}\}} = \frac{(g_2 - g_1)\{(g_0 + g_1) - (g_0 - g_1)\mathfrak{r}'\}}{(g_2 + g_1)\{(g_0 - g_1) - (g_0 + g_1)\mathfrak{r}'\}}; \tag{54.1}$$

---

[1] (53.26) bringt selbst bei dünnsten Schichten noch hohe Genauigkeit, besonders bei TH-Wellen mit fast streifendem Einfall (s. Ziff. 60).

[2] Siehe Fußnote von S. 535. Über dieses experimentell nach Fig. 38 mit Fresnelschem Biprisma statt Doppelspalt durchgeführte Verfahren erscheint eine Veröffentlichung in Z. Physik **1956**.

daraus folgt

$$\mathfrak{g}_1^2 = \mathfrak{g}_0 \mathfrak{g}_2 \frac{\mathfrak{g}_2(1-\mathfrak{r}) - \mathfrak{g}_0(1-\mathfrak{r}')}{\mathfrak{g}_2(1+\mathfrak{r}') - \mathfrak{g}_0(1+\mathfrak{r})}. \tag{54.2}$$

Zerlegen in Realteil und Imaginärteil liefert

$$\mathfrak{Re}\, \mathfrak{g}_1^2 = \frac{AA' - BB'}{A^2 + B^2} \quad (= n^2 - k^2 - n_2^2 \sin^2 \varphi_2 \quad \text{für } TE\text{-Wellen}), \tag{54.3}$$

$$\mathfrak{Im}\, \mathfrak{g}_1^2 = \frac{BA' + AB'}{A^2 + B^2} \quad (= -2nk \quad \text{für } TE\text{-Wellen}) \tag{54.4}$$

mit den Abkürzungen

$$A = \frac{1}{\mathfrak{g}_2}\,|\mathfrak{r}|\,\sin\psi_r - \frac{1}{\mathfrak{g}_0}\,|\mathfrak{r}'|\,\sin\psi_r', \tag{54.5}$$

$$A' = \mathfrak{g}_2|\mathfrak{r}|\,\sin\psi_r - \mathfrak{g}_0|\mathfrak{r}'|\,\sin\psi_r', \tag{54.6}$$

$$B = \frac{1}{\mathfrak{g}_2}(1 + |\mathfrak{r}|\,\cos\psi_r) - \frac{1}{\mathfrak{g}_0}(1 + |\mathfrak{r}'|\,\cos\psi_r'), \tag{54.7}$$

$$B' = \mathfrak{g}_2(1 - |\mathfrak{r}|\,\cos\psi_r) - \mathfrak{g}_0(1 - |\mathfrak{r}|\,\cos\psi_r'). \tag{54.8}$$

Zur Schichtdickenbestimmung dient eine der Gln. (54.1) bzw. (53.26). Von der Iteration ist in den Gln. (54.5) bis (54.8) nur jeweils der erste Summand betroffen.

55. Strenge Verfahren ohne Iteration für senkrechten Einfall. Die Iteration kann vermieden werden, wenn von zwei Kombinationen der komplexen Amplituden $\mathfrak{r}_0$, $\mathfrak{d}_0$, $\mathfrak{r}_0'$ ausgegangen wird, die keine Korrektur mit $2\pi d/\lambda$ enthalten, also z.B.

$$\mathfrak{r}_0' = \mathfrak{r}'; \qquad \frac{\mathfrak{d}_0^2}{\mathfrak{r}_0} = \frac{\mathfrak{d}^2}{\mathfrak{r}} \qquad \text{oder} \qquad \mathfrak{r}_0' = \mathfrak{r}'; \qquad \frac{\mathfrak{r}_0 \mathfrak{r}_0'}{\mathfrak{d}_0^2} = \frac{\mathfrak{r}\mathfrak{r}'}{\mathfrak{d}^2}.$$

Die zweite Kombination führt nach den Gln. (53.5) und (53.6) zu

$$-\frac{\mathfrak{r}\mathfrak{r}'}{\mathfrak{d}^2} = (a\,e^{\varrho_1 d} + b\,e^{-\varrho_1 d}) \cdot (b \cdot e^{\varrho_1 d} + a\,e^{-\varrho_1 d}), \tag{55.1}$$

$$-\mathfrak{r}' = \frac{b\,e^{\varrho_1 d} + a\,e^{-\varrho_1 d}}{a^*\,e^{\varrho_1 d} + b^*\,e^{-\varrho_1 d}} \tag{55.2}$$

mit a, a^*, b, b^* nach (53.7) bis (53.10). Aus (55.1) folgt

$$-\frac{\mathfrak{r}\mathfrak{r}'}{\mathfrak{d}^2} = a^2 + b^2 + ab\,(e^{2\varrho_1 d} + e^{-2\varrho_1 d}) \tag{55.3}$$

und aus (55.2)

$$e^{2\varrho_1 d} = -\frac{\mathfrak{r}'b^* + a}{\mathfrak{r}'a^* + b}. \tag{55.4}$$

Einsetzen in (55.3) führt zu einer Gleichung für das komplexe $\mathfrak{n}_1^2$ bei senkrechtem Einfall

$$\left.\begin{aligned}
\mathfrak{n}_1^2 &= n^2 - k^2 - 2ink \\
&= n_0 n_2 \frac{n_0|\mathfrak{r}|\,e^{i\vartheta_r}(1 - |\mathfrak{r}'|\,e^{i\vartheta_r'})^2 + J_d\,e^{2i\vartheta_d}[n_2 - n_0(1 - |\mathfrak{r}'|\,e^{i\vartheta_r'})]}{n_2|\mathfrak{r}|\,e^{i\vartheta_r}(1 + |\mathfrak{r}'|\,e^{i\vartheta_r'})^2 - J_d\,e^{2i\vartheta_d}[n_0 - n_2(1 + |\mathfrak{r}'|\cdot e^{i\vartheta_r'})]}.
\end{aligned}\right\} \tag{55.5}[1]$$

Zur Schichtdickenbestimmung dient z.B. (55.4).

---

[1] H. Schopper: Z. Physik **131**, 218 (1951), Gl. (6). Allgemeiner ist

$$\mathfrak{g}_1^2 = \mathfrak{g}_0 \frac{\mathfrak{g}_0^2\mathfrak{r}_0(1-\mathfrak{r}_0')^2 + \mathfrak{g}_2\mathfrak{d}_0^2(\mathfrak{g}_2 - \mathfrak{g}_0(1-\mathfrak{r}_0'))}{\mathfrak{g}_0\mathfrak{r}_0(1+\mathfrak{r}_0')^2 - \mathfrak{d}_0^2(\mathfrak{g}_0 - \mathfrak{g}_2(1+\mathfrak{r}_0'))}. \tag{55.5'}$$

56. Näherungsverfahren für sehr dicke Schichten. Bei sehr dicken Schichten genügt eine Messung von

$$\mathfrak{r}' \approx \frac{g_0 - g_1}{g_0 + g_1} \tag{56.1}$$

zur Bestimmung von g_1. Die Auflösung gibt

$$g_1 \approx g_0 \frac{1 - \mathfrak{r}'}{1 + \mathfrak{r}'} , \tag{56.2}$$

$$g_1 \approx g_0 \frac{(1 - |\mathfrak{r}'|^2) - i\,2\,|\mathfrak{r}'|\sin\vartheta'_r}{1 + |\mathfrak{r}'|^2 + 2\,|\mathfrak{r}'|\cos\vartheta'_r} . \tag{56.3}$$

Für senkrechten Einfall z.B. ist das

$$n \approx n_0 \frac{1 - |\mathfrak{r}'|^2}{1 + |\mathfrak{r}'|^2 + 2\,|\mathfrak{r}'|\cos\vartheta'_r} , \tag{56.4}$$

$$k \approx n_0 \frac{2\,|\mathfrak{r}'|\sin\vartheta'_r}{1 + |\mathfrak{r}'|^2 + 2\,|\mathfrak{r}'|\cos\vartheta'_r} . \tag{56.5}$$

Die Schichtdicke kann aus

$$\frac{g_2 - g_1}{g_2 + g_1} \approx \mathfrak{r} = |\mathfrak{r}|\,e^{i\psi_r} = |\mathfrak{r}| \cdot e^{i\left(\vartheta_r - \frac{4\pi d}{\lambda}\,n_2\cos\varphi_2\right)}$$

entnommen werden. Es ist z.B. für senkrechten Einfall

$$\vartheta_r - \frac{4\pi d}{\lambda}\,n_2 = \arctan \frac{n_2 - n_1}{n_2 + n_1} = \arctan \frac{2n_2 k}{n^2 + k^2 - u_2^2} , \tag{56.6}$$

$$d = \frac{\lambda}{4\pi n_2}\left(\vartheta_r - \arctan \frac{2n_2 k}{n^2 + k^2 - n_2^2}\right) . \tag{56.7}$$

57. Näherung für dünnste Schichten. Ist $\gamma \ll 1$, so kann man die Reihen für $\cos h$ und $\sin h$ in den Gln. (46.1) und (46.2) abbrechen und erhält in erster Näherung

$$\mathfrak{r}^0 = \mathfrak{r} \cdot e^{2\varrho_2 d} \approx \frac{g_1(g_2 - g_0) + i(g_2 g_0 - g_1^2)\dfrac{\gamma}{2}\,n_1 c_1}{g_1(g_2 + g_0) + i(g_2 g_0 + g_1^2)\dfrac{\gamma}{2}\,n_1 c_1}\,(1 + i\gamma\,n_2 c_2), \tag{57.1}$$

$$\mathfrak{d}^0 = \mathfrak{d} \cdot e^{\varrho_2 d} \approx \frac{2g_2 g_1}{g_1(g_2 + g_0) + i(g_2 g_0 + g_1^2)\dfrac{\gamma}{2}\,n_1 c_1}\left(1 + i\,\frac{\gamma}{2}\,n_2 c_2\right) , \tag{57.2}$$

$$\mathfrak{r}'^0 = \mathfrak{r}' \approx \frac{-g_1(g_2 - g_0) + i(g_2 g_0 - g_1^2)\dfrac{\gamma}{2}\,n_1 c_1}{g_1(g_2 + g_0) + i(g_2 g_0 + g_1^2)\dfrac{\gamma}{2}\,n_1 c_1} . \tag{57.3}$$

Kürzen wir ab

$$G = \frac{g_2 g_0 - g_1^2}{g_1(g_2 - g_0)} \;;\quad H = \frac{g_2 g_0 + g_1^2}{g_1(g_2 + g_0)} , \tag{57.4}$$

$$\left.\begin{aligned}
\mathfrak{r}^0_{\text{leer}} &= \frac{g_2 - g_0}{g_2 + g_0} , \\[4pt]
\mathfrak{r}'^0_{\text{leer}} &= \frac{g_0 - g_2}{g_0 + g_2} , \\[4pt]
\mathfrak{d}^0_{\text{leer}} &= \frac{2g_2}{g_2 + g_0} ,
\end{aligned}\right\} \tag{57.5}$$

so nehmen die Gln. (57.1) bis (57.3) in der ersten Näherung die Form an

$$\frac{\mathfrak{r}^{\circ}}{\mathfrak{r}^{\circ}_{\text{leer}}} = 1 + i\,\frac{\gamma}{2}\,(\quad \mathfrak{n}_1 c_1 G - \mathfrak{n}_1 c_1 H + 2 n_2 c_2),\qquad(57.6)$$

$$\frac{\mathfrak{r}^{\circ\prime}}{\mathfrak{r}^{\circ\prime}_{\text{leer}}} = 1 + i\,\frac{\gamma}{2}\,(-\,\mathfrak{n}_1 c_1 G - \mathfrak{n}_1 c_1 H \qquad),\qquad(57.7)$$

$$\frac{\mathfrak{d}^{\circ}}{\mathfrak{d}^{\circ}_{\text{leer}}} = 1 + i\,\frac{\gamma}{2}\,(\qquad -\,\mathfrak{n}_1 c_1 H + \; n_2 c_2).\qquad(57.8)$$

58. Die allgemeine komplexe Identität für dünnste Schichten. Die drei Unbekannten n, k, γ kommen in (57.6), (57.7) und (57.8) in drei Kombinationen $\frac{\gamma}{2}\,\mathfrak{n}_1 c_1 G$, $\frac{\gamma}{2}\,\mathfrak{n}_1 c_1\,H$ und $\frac{\gamma}{2}\,n_2 c_2$ vor; doch besteht trotzdem eine lineare Abhängigkeit; denn die Koeffizientendeterminante dieser drei Kombinationen hat den Wert

$$\begin{vmatrix} i & -i & 2i \\ -i & -i & 0 \\ 0 & -i & i \end{vmatrix} = 0.$$

Eliminiert man die beiden Kombinationen $\gamma\,\mathfrak{n}_1 c_1 G$ und $\gamma\,\mathfrak{n}_1 c_1 H$ (das geht wesentlich nur auf eine Weise), so verschwindet $\gamma n_2 c_2$ zugleich. Zur Elimination von $\gamma\,\mathfrak{n}_1 c_1 G$ müssen (57.6) und (57.7) addiert werden. $\gamma\,\mathfrak{n}_1 c_1 H$ verschwindet nur, wenn man von dem Ergebnis die doppelte Gl. (57.8) abzieht. Daraus resultiert die „allgemeine komplexe Identität" für dünnste Schichten

$$\frac{\mathfrak{r}^{\circ}}{\mathfrak{r}^{\circ}_{\text{leer}}} + \frac{\mathfrak{r}^{\circ\prime}}{\mathfrak{r}^{\circ\prime}_{\text{leer}}} = 2\cdot\frac{\mathfrak{d}^{\circ}}{\mathfrak{d}^{\circ}_{\text{leer}}}\qquad(58.1)$$

oder

$$\left(\frac{\mathfrak{r}^{\circ}}{\mathfrak{r}^{\circ}_{\text{leer}}} - 1\right) + \left(\frac{\mathfrak{r}^{\circ\prime}}{\mathfrak{r}^{\circ\prime}_{\text{leer}}} - 1\right) = 2\cdot\left(\frac{\mathfrak{d}^{\circ}}{\mathfrak{d}^{\circ}_{\text{leer}}} - 1\right)\qquad(58.2)$$

für *TE*- und *TH*-Wellen jeder Art[1].

59. Näherungen für *TE*-Wellen bei dünnsten Schichten. Da allgemein nach der Definition (57.4)

$$-G - H = 2\cdot\frac{g_0}{g_1}\cdot\frac{g_1^2 - g_2^2}{g_2^2 - g_0^2},\qquad(59.1)$$

$$G - H = 2\cdot\frac{g_2}{g_1}\cdot\frac{g_1^2 - g_0^2}{g_0^2 - g_2^2}\qquad(59.2)$$

ist, kann für *TE*-Wellen wegen $g_1 = \mathfrak{n}_1 c_1$ für $\mu_l = 1$ statt der Gln. (57.6), (57.7) und (57.8) einfacher geschrieben werden:

$$\frac{\mathfrak{r}^{\circ}}{\mathfrak{r}^{\circ}_{\text{leer}}} = 1 + i\gamma\,g_2\,\frac{g_1^2 - g_2^2}{g_0^2 - g_2^2} = 1 + i\gamma\,(\mathfrak{n}_1^2 - n_2^2)\,\frac{g_2}{g_0^2 - g_2^2},\qquad(59.3)$$

$$\frac{\mathfrak{r}^{\circ\prime}}{\mathfrak{r}^{\circ\prime}_{\text{leer}}} = 1 - i\gamma\,g_0\,\frac{g_1^2 - g_2^2}{g_0^2 - g_2^2} = 1 - i\gamma\,(\mathfrak{n}_1^2 - n_2^2)\,\frac{g_0}{g_0^2 - g_2^2},\qquad(59.4)$$

$$\frac{\mathfrak{d}^{\circ}}{\mathfrak{d}^{\circ}_{\text{leer}}} = 1 - i\,\frac{\gamma}{2}\,\frac{g_1^2 - g_2^2}{g_2 + g_0} = 1 - i\gamma\,\frac{\mathfrak{n}_1^2 - n_2^2}{2\,(g_2 + g_0)}.\qquad(59.5)$$

---

[1] Die komplexen Identitäten und die anderen unveröffentlichten Teile dieses Kapitels entstammen einer infolge des Kriegsausbruches unvollständig gebliebenen Monographie des Verfassers.

Diese drei Gleichungen für *TE*-Wellen enthalten nur die Kombination $\gamma\,(g_1^2 - g_2^2)$ der Unbekannten; es gilt daher außer der allgemeinen komplexen Identität (58.10) noch die „spezielle komplexe Identität für *TE*-Wellen", die aus (59.3), (59.4) und (59.5) gewonnen werden kann, z.B. durch Elimination von $\gamma\,(g_1^2 - g_2^2)$ aus (59.3) und (59.4):

$$\left(\frac{\mathfrak{r}^\circ}{\mathfrak{r}^\circ_{\text{leer}}} - 1\right) g_0 = -\left(\frac{\mathfrak{r}^{\circ\prime}}{\mathfrak{r}^{\circ\prime}_{\text{leer}}} - 1\right) g_2 \tag{59.6}$$

oder aus (59.4) und (59.5):

$$\left(\frac{\mathfrak{r}^{\circ\prime}}{\mathfrak{r}^{\circ\prime}_{\text{leer}}} - 1\right) = \left(\frac{\mathfrak{d}^\circ}{\mathfrak{d}^\circ_{\text{leer}}} - 1\right) \frac{2g_0}{g_0 - g_2} \tag{59.7}[1]$$

oder aus (59.5) und (59.3):

$$\left(\frac{\mathfrak{r}^\circ}{\mathfrak{r}^\circ_{\text{leer}}} - 1\right) = \left(\frac{\mathfrak{d}^\circ}{\mathfrak{d}^\circ_{\text{leer}}} - 1\right) \frac{2g_2}{g_2 - g_0}. \tag{59.8}$$

Diese drei Gleichungen sind nur zu je zweien unabhängig voneinander, und nur jeweils eine ist unabhängig von der allgemeinen komplexen Identität.

Für *TE*-Wellen bestehen daher zwischen den drei komplexen Größen $\mathfrak{r}^\circ$, $\mathfrak{d}^\circ$ und $\mathfrak{r}^{\circ\prime}$ zwei komplexe Gleichungen, zwischen den sechs reellen Zahlen $|\mathfrak{r}^\circ|$, $|\mathfrak{d}^\circ|$, $|\mathfrak{r}^{\circ\prime}|$, arc $\mathfrak{r}^\circ$, arc $\mathfrak{d}^\circ$ und arc $\mathfrak{r}^{\circ\prime}$ bestehen vier reelle Gleichungen, die z.B. in Realteil und Imaginärteil aufgespaltenen Identitäten. Daher sind nur zwei reelle Kombinationen der Unbekannten γ, n, k berechenbar. Das sind nach (59.3) bis (59.5) offenbar Realteil und Imaginärteil der Größe

$$\begin{aligned}
\gamma\,(g_1^2 - g_2^2) &= \gamma\,\{\mathfrak{n}_1^2\,(1 - \sin^2 \varphi_1) - n_2^2 \cos^2 \varphi_2\} \\
&= \gamma\,(\mathfrak{n}_1^2 - n_2^2) \\
&= \gamma\,(n^2 - k^2 - n_2^2) - i\,\gamma \cdot 2\,n\,k.
\end{aligned} \right\} \tag{59.9}$$

So folgt aus (59.4), (59.3) bzw. (59.5)

$$i\,\gamma\,(n^2 - k^2 - n_2^2) + 2\gamma\,n\,k = \left(1 - \frac{\mathfrak{r}^{\circ\prime}}{\mathfrak{r}^{\circ\prime}_{\text{leer}}}\right) \frac{n_0^2 \cos^2 \varphi_0 - n_2^2 \cos \varphi_2}{n_0 \cos \varphi_0}, \tag{59.10}$$

$$i\,\gamma\,(n^2 - k^2 - n_2^2) + 2\gamma\,n\,k = \left(1 - \frac{\mathfrak{d}^\circ}{\mathfrak{d}^\circ_{\text{leer}}}\right) 2\,(n_0 \cos \varphi_0 - n_2 \cos \varphi_2), \tag{59.11}$$

$$i\,\gamma\,(n^2 - k^2 - n_2^2) + 2\gamma\,n\,k = \left(1 - \frac{\mathfrak{r}^\circ}{\mathfrak{r}^\circ_{\text{leer}}}\right) \frac{n_2^2 \cos^2 \varphi_2 - n_0^2 \cos^2 \varphi_0}{n_2 \cos \varphi_2}. \tag{59.12}$$

Diese Kombinationen der Unbekannten, $\gamma\,(n^2 - k^2 - 1)$ und $\gamma \cdot n \cdot k$, wenn Medium ② Luft ist, sind die einzigen, die mit *TE*-Wellen, also insbesondere bei senkrechtem Einfall bei dünnsten Schichten aus Amplituden- und Phasenmessungen zu entnehmen sind. Nur diese Größen nehmen Einfluß auf das Licht schlechthin (vgl. A. DAVID[2]).

Statt der Rechnung nach den Gln. (59.10) und (59.11) benutzt H. SCHOPPER[3] ein graphisches „Dreigeradenverfahren".

---

[1] Für senkrechten Einfall kann man z.B. (59.7) einfach schreiben $1 + \mathfrak{r}'_0 = \mathfrak{d}_0 \dfrac{n_0}{n_2} = \mathfrak{d}'_0$.

[2] A. DAVID: Z. Physik **115**, 514 (1940) zeigt das für senkrechten Einfall auf andere Weise.

[3] H. SCHOPPER: Z. Physik **130**, 565 (1951) (Senkrechter Einfall).

60. Unabhängige Bestimmung aller Unbekannten durch Hinzunahme von TH-Wellen sehr schrägen Einfalls. Für TH-Wellen, also mit $g_1 = c_1/\mathfrak{n}_1$, besagt z.B. Gl. (57.7) wegen (59.1)

$$\frac{\overset{'}{\mathfrak{r}^\circ}}{\overset{'}{\mathfrak{r}^\circ_{\text{leer}}}} = 1 + i\gamma\, \mathfrak{n}_1 c_1 \frac{g_0}{g_1} \frac{g_1^2 - g_2^2}{g_2^2 - g_0^2} \tag{60.1}$$

oder

$$\frac{\overset{'}{\mathfrak{r}^\circ}}{\overset{'}{\mathfrak{r}^\circ_{\text{leer}}}} = 1 + i\gamma\, \frac{g_0}{g_2^2 - g_0^2}\, (c_1^2 - \mathfrak{n}_1^2 g_2^2). \tag{60.2}$$

Wegen

$$c_1^2 = \cos^2\varphi_1 = 1 - \sin^2\varphi_1 = 1 - \frac{n_2^2 \sin^2\varphi_2}{\mathfrak{n}_1^2}$$

folgt

$$\frac{\overset{'}{\mathfrak{r}^\circ}}{\overset{'}{\mathfrak{r}^\circ_{\text{leer}}}} = 1 + i\gamma\, \frac{g_0}{g_2^2 - g_0^2}\left(1 - \frac{n_2^2 \sin^2\varphi_2}{\mathfrak{n}_1^2} - \mathfrak{n}_1^2 n_2^{-2} \cos^2\varphi_2\right). \tag{60.3}$$

Die Klammer in Gl. (60.3) ist

$$\left.\begin{aligned}
&1 - \frac{n_2^2 \sin^2\varphi_2}{\mathfrak{n}_1^2} - (\mathfrak{n}_1^2 - n_2^2)\frac{\cos^2\varphi_2}{n_2^2} - \cos^2\varphi_2 \\
&\qquad = \sin^2\varphi_2\, \frac{\mathfrak{n}_1^2 - n_2^2}{\mathfrak{n}_1^2} - \cos^2\varphi_2\, \frac{\mathfrak{n}_1^2 - n_2^2}{n_2^2}.
\end{aligned}\right\} \tag{60.4}$$

Daher besagt (60.3)

$$\left[\left(\frac{\overset{'}{\mathfrak{r}^\circ}}{\overset{'}{\mathfrak{r}^\circ_{\text{leer}}}} - 1\right)\frac{g_2^2 - g_0^2}{g_0}\right]_{TH} = i\gamma\,(\mathfrak{n}_1^2 - n_2^2)\left(\frac{\sin^2\varphi_2}{\mathfrak{n}_1^2} - \frac{\cos^2\varphi_2}{n_2^2}\right)_{TH} \tag{60.5}$$

für TH-Wellen. Der Zusatz TH soll darauf hinweisen, daß die aufgeführten Größen $\mathfrak{r}^\circ$, g_2 usw. aus einer Messung mit einer TH-Welle entnommen sind. Andererseits ist aus einer TE-Wellenmessung (z.B. auch für senkrechten Einfall) bekannt nach Gl. (59.4)

$$\left[\left(\frac{\overset{'}{\mathfrak{r}^\circ}}{\overset{'}{\mathfrak{r}^\circ_{\text{leer}}}} - 1\right)\frac{g_2^2 - g_0^2}{g_0}\right]_{TE} = i\gamma\,(\mathfrak{n}_1^2 - n_2^2). \tag{60.6}$$

Da (60.5) außer der Kombination $\gamma\,(\mathfrak{n}_1^2 - n_2^2)$ noch $\mathfrak{n}_1^2$ unmittelbar enthält, kann man $\mathfrak{n}_1^2$ durch Division aus (60.5) und (60.6) gewinnen

$$\mathfrak{n}_1^2 = n^2 - k^2 - i\,2n\,k = \frac{(\sin^2\varphi_2)_{TH}}{\left(\dfrac{\cos^2\varphi_2}{n_2^2}\right)_{TH} + \dfrac{\left[\left(\dfrac{\overset{'}{\mathfrak{r}^\circ}}{\overset{'}{\mathfrak{r}^\circ_{\text{leer}}}} - 1\right)\dfrac{g_2^2 - g_0^2}{g_0}\right]_{TH}}{\left[\left(\dfrac{\overset{'}{\mathfrak{r}^\circ}}{\overset{'}{\mathfrak{r}^\circ_{\text{leer}}}} - 1\right)\dfrac{g_2^2 - g_0^2}{g_0}\right]_{TE}}}. \tag{60.7}$$

Werden TE- und TH-Versuch bei gleichem Ausfallswinkel gemacht, so ist

$$\mathfrak{n}_1^2 = \frac{n_2^2 \sin^2\varphi_2}{\cos^2\varphi_2 - \dfrac{\left[\left(\dfrac{\overset{'}{\mathfrak{r}^\circ}}{\overset{'}{\mathfrak{r}^\circ_{\text{leer}}}} - 1\right)(n_0^2 c_2^2 - n_2^2 c_0^2)\right]_{TH}}{\left[\left(\dfrac{\overset{'}{\mathfrak{r}^\circ}}{\overset{'}{\mathfrak{r}^\circ_{\text{leer}}}} - 1\right)(n_0^2 c_0^2 - n_2^2 c_2^2)\right]_{TE}}}. \tag{60.8}$$

Geht man zu senkrechtem Einfall über, so gehen Zähler und Nenner gegen Null; das ist in Übereinstimmung mit der oben mit Identitäten bewiesenen Tatsache, daß aus Messungen bei senkrechtem Einfall n_1 nicht bei dünnsten Schichten getrennt von der Schichtdicke bestimmbar ist. Wünscht man geringe Auswirkung der Meßfehler, so ist hinreichend schräger Einfall notwendig; Untersuchungen nahe am Grenzwinkel der Totalreflexion bei r' bzw. fast streifendem Einfall bei Benutzung von r sind günstig. Aus n_1^2 und $\gamma(n_1^2 - n_2^2)$ erhält man dann auch γ, d.h. die Schichtdicke.

Ähnlich wie aus r'_0 läßt sich n_1^2 aus $\mathfrak{d}_0$ und r_0 gewinnen oder aus Kombinationen.

Die Gleichungen dieser Ziffer gelten nur für dünnste Schichten, jedenfalls in sehr viel kleinerem Bereich als die Näherungen des Abschnitts über Intensitätsmethoden; sie sollen weniger zur tatsächlichen Berechnung der Größe n_1 und der Schichtdicke d dienen als vielmehr zeigen, daß die Verfahren der Ziff. 53 bis 55 zweckmäßig auf schrägen Einfall angewandt werden, damit sich die Meßfehler auch bei dünnsten Schichten nicht stark auf die Resultate auswirken.

61. Phasenformeln und Phasenidentität für dünnste Schichten. Für TE-Wellen folgt aus den Gln. (59.3), (59.4) und (59.5)[1]

$$\tan \arc \frac{\mathfrak{d}_0}{\mathfrak{d}_0^{\text{leer}}} = \gamma \frac{k^2 - n^2 + n_2^2}{2(g_0 + g_2) - 2\gamma n k}, \tag{61.1}$$

$$\tan \arc \frac{r'_0}{r'_0{}_{\text{leer}}} = \gamma \frac{k^2 - n^2 + n_2^2}{\dfrac{g_0^2 - g_2^2}{g_0} - 2\gamma n k}, \tag{61.2}$$

$$\tan \arc \frac{r_0}{r_0{}_{\text{leer}}} = -\gamma \frac{k^2 - n^2 + n_2^2}{\dfrac{g_0^2 - g_2^2}{g_2} + 2\gamma n k}. \tag{61.3}$$

Man kann danach also aus den reinen Phasen auch die Kombinationen $\gamma(n^2 - k^2 - n_2^2)$ und $\gamma n k$ erhalten. Eliminiert man beide Kombinationen, so erhält man die „Phasenidentität"

$$\frac{g_2}{g_0} = \frac{\tan \arc \dfrac{\mathfrak{d}_0}{\mathfrak{d}_0^{\text{leer}}} - \tan \arc \dfrac{r'_0}{r'_0{}_{\text{leer}}}}{\tan \arc \dfrac{\mathfrak{d}_0}{\mathfrak{d}_0^{\text{leer}}} - \tan \arc \dfrac{r_0}{r_0{}_{\text{leer}}}}, \tag{61.4}$$

die nicht unabhängig von den komplexen Identitäten ist und wie die Gln. (61.1), (61.2), (61.3) auch nur für extrem dünne Schichten gilt. Ihr Gültigkeitsbereich ist jedenfalls wesentlich kleiner als der für die Intensitätsidentitäten der Ziff. 48.

62. Bewertung der absoluten Phasenmethoden. Das Wort „absolute" Phase darf nicht darüber hinwegtäuschen, daß es absolute Phasen im ursprünglichen Sinne des Wortes nicht gibt. Die Phasenverschiebungen sind hier bezogen auf die gleiche Messung an dem von der Schicht befreiten Träger. Das setzt aber voraus, daß der Träger überhaupt von der Schicht befreit werden kann. Die Polarisations- und die Intensitätsmethoden dagegen erfassen auch Schichten, bei denen das nicht möglich ist. In Ziff. 63 behandeln wir ein Beispiel hierzu.

---

[1] Vgl. H. Schopper: Z. Physik **130**, 565 (1951). Verbesserte Gleichungen, die neben in γ linearen Gliedern auch einige quadratische mitnehmen, gab C. v. Fragstein: Z. Physik **139**, 163 (1954). Die entsprechenden Gleichungen mit konsequenter Entwicklung bis zu quadratischen Gliedern einschließlich veröffentlichte F. Abelès: C. R. Acad. Sci. Paris **234**, 2053 (1952).

e) Methoden zur Brechungsindex- und Schichtdickenmessung an nicht absorbierenden Schichten.

63. Polarisationsverfahren. Für absorptionsfreie Schichten sind die meisten der in den Teilen *b*, *c* und *d* beschriebenen Verfahren ohne weiteres anwendbar. Das gilt insbesondere auch für die Polarisationsmethoden der Ziff. 40 und 41. Wenn aber im voraus bereits bekannt ist, daß der Absorptionskoeffizient k der zu vermessenden Schicht Null ist, dann vereinfacht sich die Auswertung; vor allem aber kommt man mit einer geringeren Zahl von Messungen aus, da nur noch nach zwei Unbekannten, n_1 und d_1 gefragt ist. Deshalb genügt bei der Polarisationsmethode z. B. die Elliptizitätsmessung am reflektierten Licht allein.

ALKEMADE[1] und DRUDE[2] haben überdies eine weitere Vereinfachung der Rechnung dadurch erreicht, daß sie als Einfallswinkel den BREWSTERschen Winkel des unbeschichteten Trägers wählten. Dadurch steigt zugleich die Empfindlichkeit des Verfahrens, da man ohne Schicht rein linear polarisiertes Licht erhält und die Elliptizität mit Schicht sich also bei dieser Nullmethode besonders deutlich bemerkbar macht. Es gelang ALKEMADE[1] und DRUDE[2], die geringen Oberflächenschichten, die jede Glasplatte je nach dem Polierverfahren trägt, quantitativ zu erfassen. Qualitativ war die Elliptizität des reflektierten Lichtes am BREWSTER-Winkel, wenn linear polarisiertes Licht einfällt, bereits von BREWSTER[3] selbst, von SEEBECK[4] und AIRY[5] beobachtet und von SEEBECK und LORENZ[6] aus der Hypothese einer oberflächlichen Übergangsschicht gedeutet worden.

Bildet man den Quotienten aus den Gln. (40.9) und (40.8), so erhält man die Polarisation des reflektierten Lichtes $\mathfrak{P}_R = \tan\chi_R \cdot e^{i\,\delta_R}$ [s. Gl. (40.1)] als Funktion der Unbekannten n_1 und d_1. Die Auflösung ist besonders einfach für dünnste Schichten, für die man die Reihenentwicklung in den Gln. (46.1) durchführen und nach zweiten Potenzen von d_1 abbrechen kann. Dann folgt nach DRUDE sowie nach SISSINGH und GROOSMULLER[7]

$$\frac{\tan\chi_R}{\tan\delta_R} = \frac{n_1^4(3n_2^2 + n_0^2) - n_1^2(4n_2^2 n_0^2 + n_2^4 - n_0^4) + n_2^2 n_0^2(n_2^2 - n_0^2)}{(n_2^2 + n_0^2)(n_2^2 - n_1^2)(n_1^2 - n_0^2)}. \tag{63.1}$$

Das ist eine in n_1^2 quadratische Gleichung. Nach ihrer Auflösung kann die Schichtdicke aus

$$d = \frac{\lambda}{\pi}\,\frac{n_0^2 - n_2^2}{\sqrt{n_2^2 + n_0^2}}\,\frac{n_1^2 \cot\chi_R}{(n_0^2 - n_1^2)(n_1^2 - n_2^2)} \tag{63.2}$$

berechnet werden. DRUDE zeigte, daß auch bei inhomogenem Übergang zwischen beiden Medien ② und ⓪ diese Gleichung sinngemäß gültig bleibt in der Form

$$\cot\chi_R = \frac{\pi}{2}\cdot\frac{\sqrt{n_2^2 + n_0^2}}{n_0^2 - n_2^2}\int\frac{(n_0^2 - n^2(z))(n^2(z) - n_2^2)}{n^2(z)}\,dz, \tag{63.3}$$

wenn $n(z)$ den Brechungsindex am Orte mit den Koordinaten x, y, z bezeichnet und nur von der schichtnormalen Koordinate z abhängt[8].

---

[1] R. v. ALKEMADE: Diss. Leiden 1882 und Wied. Ann. **20**, 22 (1883).

[2] P. DRUDE: Siehe Bibliographie, insbesondere Wied. Ann. **36**, 532 (1889) und **43**, 126 (1891).

[3] D. BREWSTER: Phil. Trans. Roy. Soc. Lond. **1815**, 215.

[4] A. SEEBECK: Pogg. Ann. **20**, 27 (1830).

[5] G. B. AIRY: Pogg. Ann. **41**, 512 (1837).

[6] L. LORENZ: Pogg. Ann. **111**, 460 (1860).

[7] R. SISSINGH u. J. TH. GROOSMULLER: Phys. Z. **27**, 518 (1926).

[8] Mit Polarisationsmessungen arbeiten auch K. BLODGETT (1935) (unter Benutzung des Phasenumschlags am Polarisationswinkel) und A. VASICEK [Phys. Rev. **57**, 925 (1940)].

64. Intensitätsmethoden. Außer den im Abschnitt c) besprochenen Methoden kann man auch die Methode von D. KOSSEL[1] zu den Intensitätsmethoden rechnen. Das Verfahren geht auf die bekannten Methoden zur Brechungsindexbestimmung aus dem Grenzwinkel der Totalreflexion zurück.

Auf der Trägerrückseite wird ein leuchtender kleiner Fleck F erzeugt (Fig. 40). Das von der dünnen Schicht reflektierte Licht gibt auf der mattierten Trägerrückseite oder auf einem dort angebrachten Film eine Intensitätsverteilung, die auf den Brechungsindex schließen läßt. Besonders einfach ist der Brechungsindex n_1 der Schicht aus dem äußersten Radius R des von Interferenzerscheinungen ausgefüllten Kreisgebiets zu entnehmen:

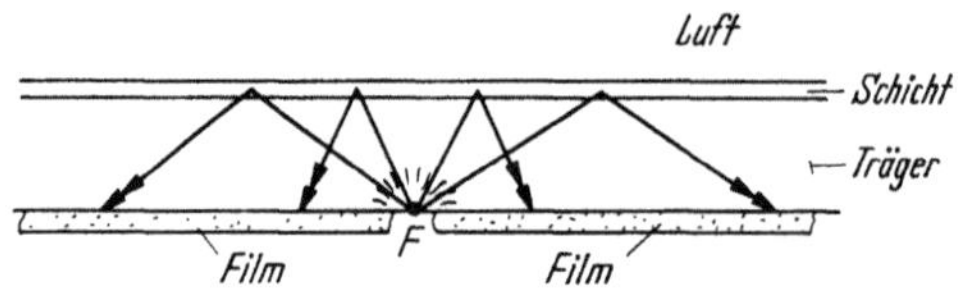

Fig. 40. Zu KOSSELS Methode.

$$n_1 = \frac{n\,(\text{Träger})}{\sqrt{1 + \left(\dfrac{2 \cdot \text{Trägerdicke}}{R}\right)^2}}. \quad (64.1)$$

KOSSEL zeigte so, daß im Vakuum aufgedampfte Kryolithschichten je nach Güte des Vakuums Brechungsindices von 1,05 (bei 10^{-1} Torr) bis 1,274 (bei 10^{-3} Torr) haben.

65. Strenge Reflexionsmethode mit Phasen- und Amplitudenmessung. Nach Tabelle 1 ist die nach der Trägerseite der Schicht reflektierte komplexe Amplitude

$$\mathfrak{r}' = \frac{(g_0 - g_1)(g_1 + g_2) + (g_0 + g_1)(g_1 - g_2)\,e^{-2\varrho_1 d_1}}{(g_0 + g_1)(g_1 + g_2) + (g_0 - g_1)(g_1 - g_2)\,e^{-2\varrho_1 d_1}} \quad (65.1)$$

mit

$$2\varrho_1 d_1 = i\gamma\,n_1 \cos\varphi_1; \quad \gamma = \frac{4\pi d_1}{\lambda}\,.$$

Auflösen nach der e-Funktion liefert

$$e^{i\gamma\,n_1 \cos\varphi_1} = \frac{g_1 - g_2}{g_1 + g_2} \cdot \frac{g_1 + \mathfrak{z}}{g_1 - \mathfrak{z}} \quad \text{mit} \quad \mathfrak{z} = g_0\,\frac{1 - \mathfrak{r}'}{1 + \mathfrak{r}'} = p + iq. \quad (65.2)$$

Da das Medium ① nicht absorbiert, muß die e-Funktion in (65.2) den Betrag 1 haben; daher folgt für g_1 die Gleichung

$$\left| \frac{g_1 - g_2}{g_1 + g_2} \cdot \frac{g_1 + \mathfrak{z}}{g_1 - \mathfrak{z}} \right|^2 = 1,$$

d.h.

$$g_1^2 = g_2 \cdot \frac{|\mathfrak{z}|^2 - p \cdot g_2}{p - g_2}\,. \quad (65.3)$$

Dann kann die Schichtdicke d_1 aus Gl. (65.2) berechnet werden:

$$\frac{4\pi d_1}{\lambda}\,n_1 \cos\varphi_1 = 2\pi M + \mathrm{arc}\left(\frac{g_1 - g_2}{g_1 + g_2} \cdot \frac{g_1 + \mathfrak{z}}{g_1 - \mathfrak{z}}\right) \quad (65.4)$$

(M ganz). Die Mehrdeutigkeit infolge der unbestimmten Ordnung M ist zu beheben, wenn mehrere Messungen bei verschiedenen Einfallswinkeln φ_0 gemacht werden. (65.4) läßt sich umformen zu

$$d_1 = \frac{\lambda}{2 n_1 \cos\varphi_1}\left\{ M \pm \frac{1}{2\pi} \arctan \frac{2q\,g_1}{g_1^2 - |\mathfrak{z}|^2}\right\}; \quad (65.5)$$

[1] D. KOSSEL: Z. Physik **126**, 233 (1949).

darin gilt das Plus-Zeichen für $g_1 > g_2$, sonst das Minus-Zeichen. Eine ähnliche Methode, auf r statt r' aufbauend, ist auch bei komplexem n_0, also metallischem Träger verwendbar[1].

Für dünnste Schichten auf nichtabsorbierendem Träger kann man auch unmittelbar aus Gl. (59.12) entnehmen

$$i\gamma\,(n_1^2 - n_2^2) = \left(1 - \frac{\mathfrak{r}^\circ}{\mathfrak{r}^\circ_{\text{leer}}}\right) \frac{n_2^2 \cos^2 \varphi_2 - n_0^2 \cos^2 \varphi_0}{n_2 \cos \varphi_2}\,. \tag{65.6}$$

66. Näherungsverfahren unter Vernachlässigung der Zickzackreflexionen aus reinen Phasenmessungen am durchgelassenen Licht. Besonders einfach sind die experimentellen Vorrichtungen, wenn nur Phasenmessungen am durchgehenden Licht benötigt werden.

In der Formel nach Tabelle 1 für die komplexe Durchlässigkeitsamplitude

$$\mathfrak{d}_\circ = \mathfrak{d} \cdot e^{\varrho_2 d_1} = \frac{4 g_2 g_1\, e^{\varrho_2 d_1}}{(g_0 + g_1)(g_1 + g_2)\, e^{\varrho_1 d_1} + (g_0 - g_1)(g_1 - g_2)\, \mathfrak{z}^{-\varrho_1 d_1}} \tag{66.1}$$

erfaßt der zweite Summand im Nenner die Zickzackreflexionen. Vernachlässigt man diese, so gilt

$$\mathfrak{d}_\circ = \frac{4 g_2 g_1}{(g_0 + g_1)(g_1 + g_2)} \cdot e^{(\varrho_2 - \varrho_1) d_1}\,, \tag{66.2}$$

d. h.

$$\vartheta = \text{arc } \mathfrak{d}_\circ = \frac{2\pi d_1}{\lambda}\,(n_2 \cos \varphi_2 - n_1 \cos \varphi_1) \tag{66.3}$$

bis auf ein ganzes Vielfaches von 2π, das wir in arc $\mathfrak{d}_\circ$ bzw. ϑ mitgeführt denken. Die beiden Unbekannten n_1 und d_1 erhält man aus den beiden Gleichungen

$$\vartheta_a = \frac{2\pi d_1}{\lambda}\,(n_2 \cos \varphi_{2a} - n_1 \cos \varphi_{1a})\,, \tag{66.4}$$

$$\vartheta_b = \frac{2\pi d_1}{\lambda}\,(n_2 \cos \varphi_{2b} - n_1 \cos \varphi_{1b})\,, \tag{66.5}$$

die eine Messung a bei dem Einfallswinkel φ_{2a} und eine zweite Messung b bei einem anderen Einfallswinkel φ_{2b} erfassen.

Nach (66.3) und dem Brechungsgesetz ist

$$\left(\frac{\lambda \vartheta}{2\pi d_1} - n_2 \cos \varphi_2\right)^2 = n_1^2 \cos^2 \varphi_1 = n_1^2 - n_2^2 \sin^2 \varphi_2\,, \tag{66.6}$$

also

$$\left(\frac{\lambda \vartheta}{2\pi d_1} - n_2 \cos \varphi_2\right)^2 + n_2^2 \sin^2 \varphi_2 \equiv \frac{\lambda \vartheta}{2\pi d_1}\left(\frac{\lambda \vartheta}{2\pi d_1} - 2 n_2 \cos \varphi_2\right) + n_2^2 = n_1^2 \tag{66.7}$$

invariant gegen Änderung des Einfallswinkels. Daher ist

$$\vartheta_a\left(\vartheta_a \frac{\lambda}{2\pi d_1} - 2 n_2 \cos \varphi_{2a}\right) = \vartheta_b\left(\vartheta_b \frac{\lambda}{2\pi d_1} - 2 n_2 \cos \varphi_{2b}\right)\,, \tag{66.8}$$

$$\frac{\lambda}{2\pi d_1}\,(\vartheta_a^2 - \vartheta_b^2) = 2 n_2 (\vartheta_a \cos \varphi_{2a} - \vartheta_b \cos \varphi_{2b})\,, \tag{66.9}$$

$$d_1 = \frac{\lambda}{4\pi n_2} \cdot \frac{\vartheta_a^2 - \vartheta_b^2}{\vartheta_a \cos \varphi_{2a} - \vartheta_b \cos \varphi_{2b}}\,; \tag{66.10}$$

---

[1] Vgl. H. Schopper: Z. Physik **130**, 427 (1951) und die dort angegebene ältere Literatur.

der Brechungsindex folgt dann aus (66.7)

$$n_1^2 = n_2^2 + \frac{\lambda\vartheta}{2\pi d_1}\left(\frac{\lambda\vartheta}{2\pi d_1} - 2n_2\cos\varphi_2\right).$$ (66.11)

67. Abwandlung zu einem eindeutigen Verfahren ohne Wellenwechsel. Für dünne Schichten mit höchstens wenigen Wellenlängen Dicke ist die Mehrdeutigkeit durch einige Messungen bei verschiedenen Einfallswinkeln leicht zu beheben, da die Ergebnisse miteinander übereinstimmen müssen. Bei Schichten von vielen

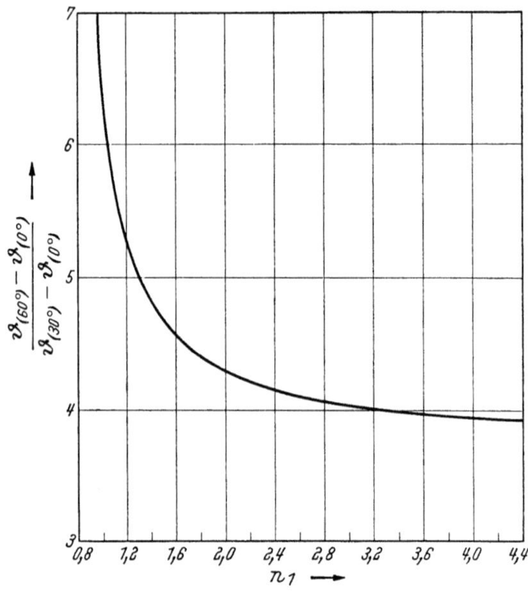

Fig. 41. Verhältnis der Phasenverschiebungen als Funktion des Brechungsindexes. Zur eindeutigen Bestimmung von Brechungsindex und Schichtdicke aus Phasenverschiebungen des durchgehenden Lichtes bei Drehen der Schicht.

Wellenlängen Dicke dagegen verfährt man zweckmäßig systematischer. Die ϑ_a, ϑ_b, ϑ_c usw. sind nur bis auf ein ganzes Vielfaches von 2π meßbar. Aber ihre Differenzen

$$\vartheta_a - \vartheta_b = \frac{2\pi d_1}{\lambda}\{n_2(\cos\varphi_{2a} - \cos\varphi_{2b}) - n_1(\cos\varphi_{1a} - \cos\varphi_{1b})\},$$ (67.1)

$$\vartheta_b - \vartheta_c = \frac{2\pi d_1}{\lambda}\{n_2(\cos\varphi_{2b} - \cos\varphi_{2c}) - n_1(\cos\varphi_{1b} - \cos\varphi_{1c})\}$$ (67.2)

sind vollständig bekannt, wenn die Zahl der durchlaufenden Interferenzstreifen bei Drehen des Objekts vom Einfallswinkel φ_{2a} bis φ_{2b} und φ_{2c} abgezählt wird. Aus (67.1) und (67.2) folgt

$$\begin{aligned}\frac{\lambda}{2\pi d_1} &= \frac{n_1(\cos\varphi_{1b} - \cos\varphi_{1a}) - n_2(\cos\varphi_{2b} - \cos\varphi_{2a})}{\vartheta_a - \vartheta_b}, \\ &= \frac{n_1(\cos\varphi_{1b} - \cos\varphi_{1c}) - n_2(\cos\varphi_{2b} - \cos\varphi_{2c})}{\vartheta_c - \vartheta_b}.\end{aligned}$$ (67.3)

Ist der Brechungsindex n_1 bekannt, so kann aus einer der Gln. (67.1) oder (67.2) sofort die Schichtdicke d_1 berechnet werden. Es genügt dann, bei zwei Einfallswinkeln zu messen. Sind d_1 und n_1 beide unbekannt, so ist zunächst aus der zweiten Gln. (67.3) n_1 zu berechnen. Dazu stellt man zweckmäßig für festes

$n_2 (=1$ bei Luft) und feste Winkel φ_{2a}, φ_{2b}, φ_{2c} die Größe

$$\frac{\vartheta_a - \vartheta_b}{\vartheta_c - \vartheta_b} = \frac{n_1 (\cos \varphi_{1b} - \cos \varphi_{1a}) - n_2 (\cos \varphi_{2b} - \cos \varphi_{2a})}{n_1 (\cos \varphi_{1b} - \cos \varphi_{1c}) - n_2 (\cos \varphi_{2b} - \cos \varphi_{2c})} \tag{67.4}$$

graphisch als Funktion von n_1 dar. Das gemessene $\dfrac{(\vartheta_a - \vartheta_b)}{(\vartheta_c - \vartheta_b)}$ erlaubt dann, aus der Eichkurve (Fig. 41) das n_1 zu entnehmen. Fig. 41 ist für die Einfallswinkel $\varphi_{2b} = 0°$; $\varphi_{2a} = 60°$; $\varphi_{2c} = 30°$ berechnet. Es ist auch apparativ zweckmäßig, feste Einfallswinkel durch Anschläge am Teilkreis zu markieren, damit die Beobachtung der Interferenzstreifen nicht zur Ablesung des Teilkreises unterbrochen zu werden braucht.

68. Abwandlung zu einem strengen Verfahren. Aus der strengen Gl. (66.1) erhält man ohne Vernachlässigung

$$\frac{1}{\mathfrak{d}°} = \frac{(g_0 + g_1)(g_1 + g_2)}{4 g_2 g_1} e^{(\varrho_1 - \varrho_2) d_1} \left\{ 1 + \frac{(g_0 - g_1)(g_1 - g_2)}{(g_0 + g_1)(g_1 + g_2)} e^{-2 \varrho_1 d_1} \right\}, \tag{68.1}$$

also

$$\left. \begin{aligned}
\vartheta = - \text{arc} \, \frac{1}{\mathfrak{d}°} &= \frac{2 \pi d_1}{\lambda} (n_2 \cos \varphi_2 - n_1 \cos \varphi_1) \\
&+ \text{arc tan} \, \frac{\sin \left(\dfrac{4 \pi d_1}{\lambda} n_1 \cos \varphi_1 \right)}{\dfrac{g_0 + g_1}{g_0 - g_1} \cdot \dfrac{g_1 + g_2}{g_1 - g_2} + \cos \left(\dfrac{4 \pi d_1}{\lambda} n_1 \cos \varphi_1 \right)},
\end{aligned} \right\} \tag{68.2}$$

$$\left. \begin{aligned}
\frac{2 \pi d_1}{\lambda} &(n_2 \cos \varphi_2 - n_1 \cos \varphi_1) \\
= \vartheta - &\text{arc tan} \, \frac{\sin \left(\dfrac{4 \pi d_1}{\lambda} n_1 \cos \varphi_1 \right)}{\dfrac{g_0 + g_1}{g_0 - g_1} \cdot \dfrac{g_1 + g_2}{g_1 - g_2} + \cos \left(\dfrac{4 \pi d_1}{\lambda} n_1 \cos \varphi_1 \right)} \equiv \vartheta (\text{korrigiert}).
\end{aligned} \right\} \tag{68.3}$$

Bestimmt man nach dem Verfahren der vorhergehenden Ziffer n_1 und d_1 genähert, so kann man in dem arc tan der Gl. (68.3) diese Werte verwenden und dann mit den so korrigierten Werten ϑ_a (korrigiert) und ϑ_b (korrigiert) erneut in die Gleichungen der vorhergehenden Ziffer eingehen. Mit den so verbesserten n_1, d_1, ist dann der arc tan der Gl. (68.3) genauer zu berechnen usw. Dieses Iterationsverfahren konvergiert sehr schnell.

Ein komplizierteres Iterationsverfahren läßt sich auf der Messung von $\mathfrak{r}$ aufbauen, die vor der Messung von $\mathfrak{r}'$ den Vorteil hat, daß nur eine Grenzfläche der Trägerplatte mitwirkt. Doch wird man dem Verfahren, das mit dem durchgegangenen Licht arbeitet, wegen des ungeknickten Strahlenganges den Vorzug geben. Insbesondere ist hier im Gegensatz zu allen Reflexionsuntersuchungen keine Umjustierung bei Wechsel des Einfallswinkels nötig; es genügt, die planparallele Trägerplatte mit der Schicht im Strahlengang zu schwenken.

69. Interferenzverfahren von Tolansky[1]. Im grundsätzlichen sind die Tolanskyschen Vielstrahl-Interferenzanordnungen den Geffckenschen Interferenzfiltern verwandt, doch wird die scharfe Interferenz der Vielfachreflexion, die nach Gl. (30.9) durch die Größe $d \cdot n / \lambda_M$ bestimmt wird, bei Tolansky nicht zur Kennzeichnung einer Wellenlänge sondern zur Kennzeichnung der Größe $d \cdot n$ benutzt.

[1] S. Tolansky: Multiple beam interferometry of surfaces and films, Oxford: Clarendon Press 1948 und viele Originalarbeiten (siehe Bibliographie). — Phys. Bl. **4**, 472 (1948).

Die Schicht wird von TOLANSKY beidseitig durchlässig, aber kräftig versilbert und dann in durchgehendem monochromatischem Licht beobachtet. Ist die Dicke von Ort zu Ort nicht völlig konstant, so erscheinen in senkrecht hindurchgehendem Licht schmale helle „Linien gleicher Dicke" („multiple-beam FIZEAU fringes") an den Stellen, an denen die Bedingung (30.9)

$$\frac{2nd}{\lambda} = M + \delta \qquad (69.1)$$

erfüllt ist. M ist eine beliebige ganze Zahl, δ ist die durch je einmalige Reflexion an beiden Silberschichten bewirkte Phasenverschiebung [vgl. (30.9) und (30.7)]. Die beidseitig versilberte Schicht wirkt wie ein Monochromatfilter (Fig. 21 und 24). Da jedoch nd nicht völlig konstant ist, stellt sich der Intensitätsverlauf nach Fig. 24 (gestrichelte Kurve) auf der Schicht entsprechend ihrer nd-Ortsveränderlichkeit dar. Zwischen den Kurven muß interpoliert werden. Durch Anwendung mehrerer Spektrallinien kann man die Zahl der Kurven gleicher nd-Werte erhöhen und die Interpolation erleichtern. TOLANSKY geht noch einen ·Schritt weiter, indem er Glühlicht benutzt und die beidseitig versilberte Schicht vor einen Spektrographenspalt setzt oder besser auf ihn abbildet. Das Spektrum wird dann von hellen Linien („fringes of equal chromatic order") durchzogen, die für den jeweils an der abgebildeten Spaltstelle befindlichen Schichtort die Wellenlängen hell anzeigen, für die nach Gl. (30.9)

$$\frac{2nd}{\lambda_M} = M + \delta; \qquad \frac{2nd}{\lambda_{M+1}} = M + 1 + \delta; \ldots \qquad (69.2)$$

ist. δ eliminiert man durch Differenzbildung

$$2nd\left(\frac{1}{\lambda_{M+1}} - \frac{1}{\lambda}\right) = 1. \qquad (69.3)$$

Dieses Verfahren kann auch für Schichten konstanter Dicke durchgeführt werden und liefert das Produkt nd, wenn n mit der Wellenlänge nicht wesentlich veränderlich ist.

Die Dicke einer beliebigen absorbierenden oder nichtabsorbierenden Trägerschicht erhält man nach TOLANSKY so, daß man die Schicht teilweise von dem Träger entfernt (wir nennen diese Stelle unten die „Lücke") und alles mit einer gleichmäßig dicken Silberschicht bedampft. Eine ebene, in etwa gleicher Weise versilberte Glas- oder Glimmerplatte wird aufgelegt. Dann kann nach einem der soeben beschriebenen Verfahren $n_L \cdot d$ für die Luftschicht zwischen beiden Versilberungen als Funktion des Ortes gemessen werden, und der Sprung, den die Dicke der Luftschicht bei Überschreiten der „Lückenkanten" macht, gibt die gesuchte Schichtdicke. Abgesehen von porösen Schichten, die einen Teil des aufgedampften Silbers aufnehmen oder selbst zu anderer Struktur zwingen, als es der reine Träger tut, führt das Verfahren zu Meßfehlergrenzen von wenigen Atomlagen.

70. Hinweise auf weitere Verfahren. TOLANSKYs Methode mit spektraler Beobachtung der Vielstrahl-Interferenzen[1] kann als eine Verfeinerung der alten Methode angesehen werden, die mit Farben dünner Blättchen arbeitete und auch

---

[1] Die Vielstrahl-Interferenzen an sich sind freilich nicht genauer als Zweistrahl-Interferenzen. W. KRUG (s. Bibliographie) zeigte vielmehr, daß Zweistrahl-Interferenzen verläßlichere und genauere Ergebnisse haben, wenn man die steilen Flanken in der Nähe von Nullstellen zur Kennzeichnung ausnutzt. Dieses Ergebnis ist im Einklang mit allgemeineren Erfahrungen über die „Minimumstrahlkennzeichnung" [H. WOLTER, Die Minimumstrahlkennzeichnung als Mittel zur Genauigkeitssteigerung optischer Messungen, Ann. Physik (6) **7**, 354 (1950)].

heute noch oft gebraucht wird[1]. Hierher gehört auch die Methode von G. BAUER[2] mit Reflexionsmessung in Interferenzmaxima und -minima einer Schicht kontinuierlicher Dickenzunahme. Verwandt hiermit sind auch die Interferenzmethoden, die mit anderen Interferometern, z. B. dem KÖSTERschen[3] arbeiten und ebenso wie die TOLANSKYsche Methode auf eine Messung eines Produkts aus Dicke und Brechungsindex hinausführen.

Von ganz anderer Art sind dagegen die Verfahren, die eine Strahlablenkung durch das Objekt benutzen. Sie messen unmittelbar oft nicht das Produkt der Schichtdicke und des Brechungsindexes sondern ihren Quotienten, wie z.B. die Verfahren, die als Kennzeichen des d/n eine Strahlversetzung durch eine Planparallelplatte verwenden. Die Kombination solcher Verfahren mit Interferenzverfahren führt daher zur Kenntnis beider Größen.

Die in diesem Zusammenhange wichtigen Lichtschnittverfahren[4] sind in dem folgenden Artikel im Teil D behandelt. Die Phasenkontrastverfahren, die mit den Schlierenverfahren den Hauptgegenstand jenes Artikels bilden, sind hier bereits mehrfach genannt worden. Sie haben einerseits Bedeutung für die Messungen an dünnen Schichten und machen andererseits von dünnen Schichten selbst wesentlichen Gebrauch.

Literatur.

Die aufgeführten Originalarbeiten sind ohne Titel zitiert; ihr Inhalt ist jeweils durch einige Stichworte umrissen. Bei Büchern ist der volle Titel angegeben.

ABELÈS, F.: C. R. Acad. Sci. Paris **227**, 553 (1949). Determination des indices et des épaisseurs des couches minces.
— Rev. Opt. **28**, 11 (1949). Couches minces simples ou multiples (Travaux théoriques).
— C. R. Acad. Sci. Paris **234**, 198 (1952). Propriétés des couches métalliques très minces.
— C. R. Acad. Sci. Paris **234**, 2053 (1952). Sur les déphasages que subit une onde plane par réflexion ou par transmission à travers une couche métallique très minces.
— Proc. Phys. Soc. Lond. B **65**, 996 (1952). Measurement of optical constants by reflection.
— J. Phys. Radium **14**, 5 (1953). Méthode nouvelle pour déterminer indices et épaisseurs.
— C. R. Acad. Sci. Paris **236**, 1412 (1953). Zwei neue Methoden zur Messung des Brechungsindexes und der Dicke dünner durchsichtiger Schichten.
AIRY, G. B.: Phil. Mag. (3) **2**, 20 (1833) u. Pogg. Ann. **41**, 512 (1837). AIRYsche Formeln.
AITCHISON, P. M.: Nature, Lond. **166**, 522 (1950). Differential phase change at reflection.
ALKEMADE, R. VAN: Wied. Ann. **20**, 22 (1883). Polarisationsmethode zur Feststellung einer dünnen Schicht auf massiver Unterlage.
AMY, M. P.: Rev. Opt. **6**, 305 (1927). Reflexminderung chemisch.
ANDERSON, S., and D. D. KIMPTON: J. Amer. Ceram. Soc. **34**, 141 (1951). Interference films on glass.
ASTOIN, N., et M. B. VODAR: J. Phys. Radium **14**, 424 (1953). Couches minces transparentes dans l'ultraviolet lointain. (SiC- und Al-Schichten.)
AVERY, D. G.: Nature, Lond. **163**, 916 (1949). Interferometric determination of the apparent thickness of thin metallic films.
— Phil. Mag. (7) **41**, 1018 (1950). Optical properties of evaporated layer of silver, copper and tin.
— Proc. Phys. Soc. Lond. B **65**, 425 (1952). An improved method for measurements of optical constants by reflection. (Aus dem Reflexionsverhältnis für beide Polarisationen bei verschiedenen Einfallswinkeln.)
— Proc. Phys. Soc. Lond. B **66**, 134 (1953). Optical constants of lead sulphide, lead selenide and lead telluride in the $0.5—3\,\mu$ region.
BAHN, R., u. O. BÖTTGER: Z. Physik **135**, 376 (1953). Die Brechzahl und die Dicke dünner anodisch erzeugter Aluminiumoxydschichten. [KÖSTERs Interferometer benutzt. S. MÖNCH, Optik **8**, 550 (1951).]

---

[1] G. C. MÖNCH: Optik **9**, 75 (1952).
[2] G. BAUER: Ann. Phys. **19**, 434 (1934).
[3] G. C. MÖNCH: Optik **8**, 550 (1951). — R. BAHN u. O. BÖTTGER: Z. Physik **135**, 376 (1953).
[4] E. MENZEL: Naturwiss. **38**, 332 (1951). — H. RAHBECK u. M. OMAR: Nature, Lond. **169**, 1008 (1952).

BANNING, M.: Phys. Rev. **59**, 914 (1941). Reflectivity of thin Al- and Cu-Films in the ultra-violett. (Interferenzpolarisatoren.)
— J. Opt. Soc. Amer. **37**, 792 (1947). Interferenzpolarisatoren.
BANNON, J., and C. E. CURNOW: Nature, Lond. **161**, 136 (1948). Structure of calcium fluoride film.
BARNES, R. B., and M. CZERNY: Phys. Rev. **38**, 338 (1931). Reflection power of metals in thin layers for the infrared.
BARR, E. E.: J. Opt. Soc. Amer. **39**, 634 (1949). Experimental study of DENNISON-HADLEY infrared reflection filters.
BARELL, H., and J. S. PRESTON: Proc. Phys. Soc. Lond. B **64**, 97 (1951). Die Titanoxyd-schicht als Strahlenteiler.
BAUER, G.: Ann. Phys. (5) **19**, 434 (1934). Optische Absorptionskonstanten von Alkalihalo-genidkristallen im Gebiet ihrer ultravioletten Eigenfrequenzen.
BAUER, J.: Ann. Phys. (5) **20**, 481 (1934). Die Dispersion des Phasensprungs bei der Licht-reflexion an dünnen Metallschichten.
BETZ, W.: Ann. Phys. (4) **18**, 590 (1905). Eine Methode zur Bestimmung der Dicke und optischen Konstanten durchsichtiger Metallschichten.
BILLINGS, B. H., and M. HYMAN: J. Opt. Soc. Amer. **37**, 119 und 395 (1947). The infra-red refractive index and dispersion of evaporated stibnite films.
— J. Opt. Soc. Amer. **39**, 634 (1949). An experimental study of TURNER frustrated total reflection filters in the infra-red.
—, and M. A. PITTMAN: J. Opt. Soc. Amer. **39**, 978 (1949). Frustrated total reflection filter for the infra-red.
BLACKMAN, M.: Phil. Mag. **18**, 262 (1934). The optical properties of thin films.
BLODGETT, K.: J. Amer. Chem. Soc. **57**, 1007 (1935). Erfassung des Polarisationswinkels durch Beobachtung des hier auftretenden Phasenumschlags.
— Phys. Rev. **55**, 391 (1939). Use of interference to extinguish reflection of light from glass.
— Phys. Rev. **57**, 921 (1940). Nonreflecting films.
BLOUT, E. R., R. S. CORLEY and P. L. SNOW: J. opt. Soc. Amer. **39**, 634 (1949). Infra-red transmitting-filters.
BOR, J., and B. G. CHAPMAN: Nature, Lond. **163**, 183 (1949). Simultaneous measurement of the optical constants of metals over a wide wave-length range. (Polarisationsmethode, photographisch.)
BOUSQUET, P.: J. Phys. Radium **13**, 294 (1952). Détermination graphique des coefficients de FRESNEL.
— C. R. Acad. Sci. Paris **237**, 516 (1953). Sur un système de franges se produissant dans une lame transparente an voisinage de l'angle limite de réflexion totale.
— C. R. Acad. Sci. Paris **238**, 1485 (1954). Facteurs de réflexion et de fluorure de calcium.
— C. R. Acad. Sci. Paris **239**, 406 (1954). Lames minces de fluorure de calcium dans la zone de réflexion totale.
— C. R. Acad. Sci. Paris **240**, 2502 (1955). Facteurs de réflexion de lames minces de cryo-lithe.
— C. R. Acad. Sci. Paris **241**, 478 (1955). Étude spectrophotométrique de lames minces transparentes en incidence oblique.
BREIN, R.: Optik **10**, 492 (1953). Interferenzeinrichtungen an Neigungsmessern.
BUCH, S.: Z. wiss. Photogr. **45**, 212 (1951). Über die Farbe reflexionsmindernder Schichten.
BURRIDGE, J. C., H. KUHN and A. PEVY: Proc. Phys. Soc. Lond. B **66**, 963 (1953). Re-flectivity of thin aluminium films and their use in interferometry.
CABALLERO, D. L.: J. Opt. Soc. Amer. **37**, 176 (1947). A theoretical development of exact solution of reflectance of multiple layer optical coatings.
CABRERA, N.: C. R. Acad. Sci. Paris **234**, 1045 (1952). Sur les propriétés optiques des couches multiples alternées.
— C. R. Acad. Sci. Paris **234**, 1146 (1952). Systeme periodisch wechselnder Schichten.
CARIO, G., u. J. H. KALLWEIT: Z. Physik **140**, 47 (1955). Zum Wachstum dünner Schichten.
CARTWRIGHT, C. H., and A. F. TURNER: Phys. Rev. **55**, 595 A (1939). Reducing the reflection from glass by evaporated films.
— Phys. Rev. **55**, 1128 (1939). Multilayer films of high reflecting power.
CAU, M.: C. R. Acad. Sci. Paris **186**, 1293 (1928); **188**, 57; 246 (1929). — Ann. Phys. **11**, 354 (1929). Anisotrope Schichten (in diesem Handbuchartikel nicht behandelt).
CHARBONNIER, W.: Dickemessung an nichtabsorbierenden Schichten. Diss. Hamburg 1930.
CIRKLER, W., u. J. EULER: Phys. Bl. **3**, 105 (1954). Die Vergütung optischer Systeme.
COLEMAN, H. S., A. F. TURNER and O. A. ULBRICH: J. Opt. Soc. Amer. **37**, 521 (1947). (Dicke Mg F_2-Schichten.) Crystal orientation and refractive index of thick evaporated Mg F_2-films.

COLEMAN, H. S., and G. W. CRAWFORD: J. Opt. Soc. Amer. **43**, 326 (1953). In Zweikomponentenschichten beeinflussen die Komponenten gegenseitig ihre Eigenschaften wesentlich.

COLLGER, P. W.: J. Opt. Soc. Amer. **41**, 285 (1951). A method for sharpening Fizeau fringes.

COTTON, P., et P. ROUARD: J. Phys. Radium **11**, 461 (1950). Propriétés optiques des lames minces solides.

CRAWFORD, M. F., W. M. GRAY, A. L. SCHAWLOW and F. M. KELLY: J. Opt. Soc. Amer. **39**, 888 (1949). Reflexion und Durchlässigkeit von Al-Schichten verschiedener Aufdampfgeschwindigkeit.

CZERNY, M.: Z. Physik **65**, 600 (1930). Steinsalz im Ultraroten.

DAVID, E.: Z. Physik **114**, 389 (1939). Deutung der Anomalien der optischen Konstanten dünner Metallschichten.

— Z. Physik **115**, 514 (1940). Lichtstreuung an dünnen Metallschichten.

DÖRNENBURG, G., u. R. FLEISCHMANN: Z. Physik **129**, 300 (1951). Änderung des Phasensprung bei der Reflexion an Silber in Abhängigkeit vom Einfallswinkel.

DREISCH, TH., u. E. RÜTTEN: Z. Physik **60**, 69 (1930). Ultrarote Absorption und Struktur sehr dünner kathodischer Metallschichten.

DRUDE, P.: Wied. Ann. **32**, 584 (1887). Über die Gesetze der Reflexion und Brechung des Lichtes an der Grenze absorbierender Kristalle. (Metallreflexion ab S. 613.)

— Wied. Ann. **36**, 532 (1889). Über Oberflächenschichten I. Erklärung der elliptischen Polarisationen am BREWSTER-Winkel mit Oberflächenschichten (s. ALKEMADE).

— Wied. Ann. **36**, 865 (1889). Über Oberflächenschichten II.

— Wied. Ann. **39**, 481 (1890). Bestimmung der optischen Konstanten der Metalle (Methodendiskussion).

— Wied. Ann. **43**, 126 (1891). Polarisation bei Oberflächenschichten.

— Wied. Ann. **50**, 595 (1893). Monochr. Interferenzverfahren (Fizeau) bei dicken Schichten.

— Wied. Ann. **51**, 77 (1894). [Fortsetzung von **50**, 595 (1893).]

DUFOUR, CH.: C. R. Soc. Franc. Phys. **21**, Beilage zu J. Phys. Radium **12**, Nr. 4 (Apr.) (1951). Sur un procédé de repérage des épaisseurs des couches minces diélectriques.

DÜHMKE, M.: Optik **10**, 501 (1953). Interferenzoptisches Gerät zur Messung dünner Schichten.

EBELING, J.: Z. Physik **58**, 333 (1929). Alkalimetall auf Glas.

EDWARDS, H. W.: Phys. Rev. **38**, 166 (1931). Farben dünner Cu-Filme auf Al oder Ni.

EISNER, E.: Research **4**, 183 (1951). Phase change of light on reflection at a mica-silver interface.

ELLIOT, A., E. J. AMBROSE and R. TEMPLE: J. Opt. Soc. Amer. **38**, 212 (1948). Polarisation ultraroter Strahlung durch Se-Schichten und NaF-TeJ-Mehrfachschichten.

EPSTEIN, L. J.: J. Opt. Soc. Amer. **42**, 806 (1952). The design of optical filters. (Äquivalent-Indizes für Mehrfachfilter).

ESSERS-RHEINDORF, G.: Ann. Phys. **28**, 297 (1937). Dicke und Konstanten von Trägerschichten aus Polarisationsmessungen.

EULER, J.: Z. Physik **137**, 318 (1954). Ultrarotoptische Eigenschaften von Metallen und mittlere freie Weglänge der Leitungselektronen.

FALKENHAGEN, H.: Handbuch der physikalischen Optik (GEHRKE), Bd. 1, S. 795, 1927. Struktur aus Dispersion.

FAUST, R. C.: Phil. Mag. (7) **41**, 1238 (1950). Interferometric study of some optical properties of evaporated silver films.

FICHTER, R.: Helv. phys. Acta **19**, 21 (1946). Ultrarotabsorption von Aluminiumoxydschichten.

FLEISCHMANN, R.: Z. Physik **129**, 275 (1951). Interferenzverfahren zur Messung der absoluten Phasen bei der Untersuchung absorbierender Medien.

—, u. H. SCHOPPER: Z. Physik **129**, 285 (1951). Die Bestimmung der optischen Konstanten und der Schichtdicke absorbierender Schichten mit Hilfe der Messung der absoluten Phasenänderung.

— — Z. Physik **130**, 304 (1951). Verfahren zur genauen Messung absoluter Lichtphasen an nichtabsorbierenden und absorbierenden Schichten.

— — Z. Physik **131**, 225 (1952). Ein photometrisches Präzisionsverfahren zur Messung absoluter Lichtphasen mit Hilfe eines phasengleichen Gesichtsfeldes.

—, u. A. LOHMANN: Z. Physik **137**, 362 (1954). Die Bestimmung einer absoluten Lichtphase durch Intensitätsmessung in der Beugungsfigur eines Gitters.

FOCHS, P. D.: J. Opt. Soc. Amer. **40**, 623 (1950). Dicke und Konstanten interferometrisch gemessen (aus kanneliertem Spektrum).

FÖRSTERLING, K.: Nachr. Ges. Wiss. Göttingen **1911**, 449. Dicke und Konstanten freier Schichten aus Polarisationsmessungen, ergänzt durch Intensitätsmessungen.

— Phys. Z. **14**, 265 (1913). Inhomogene Schichten.

— Phys. Z. **15**, 225 (1914). Inhomogene Schichten.

— Phys. Z. **15**, 940 (1914). Inhomogene Schichten.

— Ann. Phys. (5) **11**, 1 (1931). Über die Ausbreitung des Lichtes in inhomogenen Medien.

Försterling, K.: Ann. Phys. (5) **30**, 745 (1937). Polarisationsmethode zur Messung der Konstanten und Dicke von Metallschichten auf Trägern.

Fragstein, C. v.: Optik **9**, 337 (1951). Reziprozität der Durchlässigkeit einer beliebigen Mehrfachschicht für elektromagnetische Wellen.

— Optik **10**, 578 (1953). Zur optischen Schichtdickenbestimmung „dicker" Metallschichten.

— Z. Physik **139**, 163 (1954). Die Bestimmung von Schichtdicke und optischen Konstanten bei „dünnen" Schichten (Verbesserung der „Indentitäten").

Françon, M.: Optik **10**, 512 (1953). Bernard Lyot und die Beobachtung der Sonnenkorona außerhalb der Finsternisse (Rotationsdispersionsfilter).

Frau, D. C.: Rev. Opt. **31**, 161 (1952). Étude théorique du filtre interférentiel.

Frazer, J. H.: Phys. Rev. **33**, 97 (1929); **34**, 644 (1929). Polarisationsmethode für dünne Oberflächenschichten.

Fritze, H.: Ann. Phys. (4) **47**, 763 (1915). Vergleich der Methoden Försterlings, Drudes, Kundts.

Galli, N.: Das optische Verhalten dünner Metallschichten (Polarisationsmessungen, vgl. Försterling 1911). Diss. Göttingen 1911.

—, u. K. Försterling: Göttinger Nachr. **58** (1911). Konstanten freier Schichten aus Polarisationsmessungen.

Gans, R.: Ann. Phys. (4) **47**, 709 (1915). Inhomogene Schichten.

Gee, A. E., and H. D. Polster: J. Opt. Soc. Amer. **39**, 1044 (1949). Schichtdicke aus spektraler Untersuchung eines Schichtfilters.

Geffcken, W.: Ann. Phys. (5) **40**, 385 (1941). Reflexion an einer inhomogenen Schicht.

— DRP. 716 153/42h, Gr. 34, 11. (Anmelder: Jenaer Glaswerk Schott & Gen.) Interferenzlichtfilter.

— Z. Glaskunde **24**, 143 (1951). Dünne Schichten auf Glas.

— D. P. 848 716; 42h; Gr. 34, 11 (1952). Interferenzfilteranordnung mit nichtmetallischen und abwechselnd aufeinanderfolgenden hoch- und tiefbrechenden Schichten für schrägen Einfall des Lichtes.

— D.P. 899 120; 42h; 21 (1953). Polarisator.

— Z. angew. Phys. **6**, 249 (1954). Das Wellenbandfilter, ein Interferenzfilter mit besonders hoher Leistung.

— D.P. 904 357; 42h; 34, 11 (1954). Lichtfilter aus einer Mehrzahl lichtdurchlässiger nichtmetallischer Schichten, von denen immer je zwei voneinander durch eine ebensolche Schicht, jedoch von anderer Brechungszahl getrennt sind.

— D. P. 913 005; 42h; 8 (1954), D. P. 902 191; 42h; 34,11 (1954) Interferenzlichtfilter.

Goldschmidt, H., u. H. Dember: Z. techn. Phys. **7**, 137 (1926). Genäherte Intensitätsformel für dicke Metallschichten; Absorption von Pt.

Goos, F.: Z. Physik **100**, 95 (1936). Durchlässigkeit und Reflexionsvermögen dünner Silberschichten von Ultrarot bis Ultraviolett.

— Z. Physik **106**, 606 (1937). Die optischen Konstanten dünner Goldschichten aus Durchlässigkeits- und Reflexionsmessungen von Ultrarot bis Ultraviolett.

Grass, J.: Z. Physik **139**, 358 (1954). Zur Anwendung der Drudeschen Theorie der freien Elektronen auf das optische Verhalten der Metalle im Sichtbaren und Ultraviolett.

Hacman, D.: Z. Physik **114**, 170 (1939). Intensitätsformeln für schrägen Einfall bei Metallschichten.

Hadenhorst, H.-G.: Z. angew. Phys. **7**, 487 (1955). Durchlässigkeit inhomogener Schichten.

Hadley, L. N., and D. M. Dennison: J. Opt. Soc. Amer. **37**, 451 (1947). Transmission interference filters.

Hagen, E., u. H. Rubens: Ann. Phys. (4) **8**, 432 (1902). Die Absorption ultravioletter, sichtbarer und ultraroter Strahlen in dünnen Metallschichten.

Hakoila, K. J.: Ann. Univ. Turk A **12**, 99 (1952). Bestimmung des Brechungsindexes und der Dicke von Planparallelplatten mit Hilfe optischer Interferenzen. (Mit Variation des Einfallswinkels.)

Hammer, K.: Z. techn. Phys. **24**, 169 (1943). Teildurchlässiger Spiegel aus nichtmetallischen Stoffen.

— Optik **3**, 495 (1948). Steigerung der Glasreflexion durch Metalloxydschichten.

— Optik **5**, 365 (1949). Absorptionsfreie Mehrfachschichten.

Haringhhuizen, P. J., D. A. Was u. A. M. Kruithof: Physica, **4**, 695 (1937). Strenge Intensitätsmethode, graphische Interpolation.

Harris, L., J. K. Bedsley and A. L. Loeb: J. Opt. Soc. Amer. **40**, 801 (1950). Reflection and transmission of radiation by metal films and the influence of nonabsorbing backings.

— J. Opt. Soc. Amer. **41**, 604 (1951). Reflection and transmission of radiation by metal films and the influence of nonabsorbing backings.

—, and J. K. Bedsley: J. Opt. Soc. Amer. **42**, 634 (1952). The reflectance of thin cellulose nitrate films. (Reflexion als Funktion des Gewichts.)

HASS, G.: Optik **1**, 8 (1946). Dicke Schichten aus Al und Ag, DRUDEs Methode.
— Optik **1**, 134 (1946). Dünne Oxydschichten auf Al nach DRUDEs Polarisationsmethode.
—, and N. W. SCOTT: J. Opt. Soc. Amer. **39**, 279 (1949). Silicon monoxide protected front-surface mirrors.
—, and C. SALZBERG: J. Opt. Soc. Amer. **43**, 326 (1953). [Reflexion und Durchlässigkeit von SiO-Schichten (0,1 bis 10 μ dick) für $0,4\,\mu \leq \lambda \leq 14\,\mu$.]
— H. H. SCHROEDER and A. F. TURNER: J. Opt. Soc. Amer. **43**, 326 (1953). Mirror coatings for low visible and high infrared reflectance.
HEAVENS, O. S.: Proc. Phys. Soc. Lond. B **64**, 419 (1941). Measurement of the thickness of thin films by multiple-beam interferometry.
— Optical properties of thin solid films. London 1955.
HERMANSEN, A.: Nature, Lond. **167**, 104 (1951). A method for production of interference filters for specified wavelengths.
— Nature, Lond. **174**, 218 (1954). Interference filters of greater areas and higher contrast.
HETTNER, G.: Optik **1**, 2 (1946). Durchlässigkeit und Reflexionsvermögen dünner Metallschichten im langwelligen Ultrarot bei beliebigem Einfallswinkel.
HIESINGER, L.: Naturwiss. **34**, 121 (1947). Reflexminderung von Glasoberflächen durch mehrere homogene Interferenzschichten.
— Optik **3**, 485 (1948). Entspiegelung von Glasoberflächen. Theorie der Zweifachschicht und Mehrfachschichten.
HILSCH, R., u. R. W. POHL: Z. Physik **48**, 384 (1928). Über die ersten ultravioletten Eigenfrequenzen einiger einfacher Kristalle.
HOLDEN, J.: J. Opt. Soc. Amer. **41**, 504 (1951). The shape of reflected interference fringes from interferometers coated with thin metal films.
HOLLAND, L.: J. Opt. Soc. Amer. **42**, 686 (1952). The structure of evaporated metal films. (Kritik zu SENNET und SCOTT.)
HÜBNER, W.: Optik **7**, 128 (1950). Zur Anwendung der Vierpoltheorie auf die MAXWELLschen Gleichungen.
ISHIGURO, K.: J. Phys. Soc. Japan **5**, 187 (1950). Doppelspaltinterferometer zur Messung absoluter Phasen.
—, and G. KUWAHAVA: J. Phys. Soc. Japan **6**, 71 (1951). Determination of optical constants of Ag films from the measurements of intensity and phase change. I. u. II.
JACKSON, D. A., K. KUHN and A. H. JARETT: Nature, Lond. **170**, 455 (1952). Silver films and dielectric multiple films in interferometry.
JAGERSBERGER, A.: Z. Physik **89**, 564 (1934). Die spontane Lichtdurchlässigkeitsänderung von dünnen Metallfolien.
JACQUINOT, P., and D. ROUARD: J. Opt. Soc. Amer. **39**, 1048 (1949). Bericht vom Kolloquium über dünne Schichten in Marseille 1949.
JARRETT, A. H.: Z. Astrophys. **34**, 91 (1954). Multilayer coatings of high efficiency and their applications for astrophysical research work.
KAEMPF, F.: Ann. Phys. (4) **16**, 308 (1905). Doppelbrechung in KUNDTschen Spiegeln.
KELLNER, A.: Z. Physik **56**, 215 (1929). Untersuchungen im Spektralgebiet zwischen 20 und 40 μ. Näherungsformel mit Intensitäten und optischen Konstanten einer Schicht.
KIESSIG, H.: Ann. Phys. (5) **10**, 769 (1931). Interferenzen von Röntgenstrahlen an dünnen Schichten.
KINDINGER, M., u. K. KOLLER: Z. Physik **110**, 237 (1938). Die Gaseinsaugung dünner Metallschichten, ein Versuch zur Klärung der Anomalie der optischen Konstanten.
KOBER, H.: Ann. Phys. (6) **11**, 12 (1953). Die optischen Konstanten von Gold und Silber im Ultraviolett. (Polarisationsmethode; Objekt dient mehrmals hintereinander zur Reflexion; s. KRETZMANN 1940.)
KOFINK, B. W., u. E. MENZER: Ann. Phys. (5) **39**, 388 (1941). Inhomogene Schicht.
KOLLMORGEN, F.: Trans. Soc. Illum. Eng. **11**, 220 (1916). Reflexminderung chemisch.
KOSSEL, D.: Optik **3**, 266 (1948). Einfache Darstellung der Reflexion an einer inhomogenen Schicht.
— Z. Physik **126**, 233 (1949). Interferenzen an Schichten mit einer totalreflektierenden Grenze.
KOSSEL, W., u. K. STROHMAIER: Z. Naturforsch. 6a, 504 (1951). Dünne Al_2O_3-Häutchen zur Minimumstrahlkennzeichnung.
KRAUTKRÄMER, J.: Ann. Phys. (5) **32**, 537 (1938). Über optische Konstanten, elektrischen Widerstand und Struktur dünner Metallschichten. (Intensitätsmethoden nach MURMANN, DAVID, WOLTER.)
KRETZMANN, R.: Ann. Phys. (5) **37**, 303 (1940). DRUDEs Polarisationsmethode [Wied. Ann. **32**, 584 (1887)]; jedoch Genauigkeitssteigerung durch Mehrfachreflexion an gleichen Spiegeln.

KRONIG, R., B. S. BLAISSE and J. J. V. D. SANDE: Appl. Sci. Res. B **1**, 63 (1950). Optical impedance and surface coating.

KRUG, W.: Feingerätetechnik, H. 4, 1955. Bessere Wiedergabe der Oberflächenstruktur durch Zweistrahl- als durch Mehrstrahlinterferenzen.

KUHN, H.: Rep. Progr. Phys. **14**, 64 (1951). New techniques in optical interferometry.

KUNDT, A.: Wied. Ann. **34**, 469 (1888). Über die Brechungsexponenten der Metalle (aus Ablenkungen durch dünne Metallprismen).

KUTZELNIGG, A.: Metalloberfläche **5**, B 39 (1953). Interferenzfarben bei elektrolytisch abgeschiedenen dünnen Metallfilmen.

LANDWEHR, R.: Optik **5**, 355 (1949). Zur Messung der Ebenheit von reflektierenden Flächen mittels Interferenzen gleicher Dicke.

LAUCH, K.: Ann. Phys. (4) **74**, 55 (1924). Die optischen Konstanten chemisch reiner, undurchsichtiger durch Kathodenzerstäubung hergestellter Metallschichten. (Polarisation und Intensität am reflektierten Licht.)

LESQUIBE, F., et P. GACOMO: J. Phys. Radium **14**, 4 (1953). Mesure, sons l'incidence normale, de l'indice de couches minces absorbantes. Application au carbone.

LEURGANS, P.: J. Opt. Soc. Amer. **39**, 639 (1949). Reflection films.

— J. Opt. Soc. Amer. **41**, 714 (1951). The impedance concept in thin film optics.

LINDBERG, A.: Z. Physik **131**, 231 (1952). Verwendung einer SAVARTschen Doppelplatte zur Herstellung eines phasengleichen Gesichtsfeldes.

LOCKHARDT, L. B., and P. KLING: J. Opt. Soc. Amer. **37**, 689 (1947). Three-layered reflection-reducing coatings.

LOH, H. Y., and P. CHEV: Phys. Rev. (2) **91**, 221 (1953). Change of optical constants of fresh thin copper and silver films on exposure to air.

LUCY, F. A.: J. Chem. Phys. **16**, 167 (1948). Brechungsindex und Dicke von Oberflächenschichten aus Polarisationsmessungen; Verbesserung des DRUDEschen Verfahrens.

MACEK, O.: Arch. techn. Messen **201**, 235 (1952). Filter-Spektroskopie.

MAHAN, A. L.: J. Opt. Soc. Amer. **42**, 259 (1952). Exact solution for the reflectance of reflection-increasing and reflection reducing films.

MALÉ, D.: C. R. Acad. Sci. Paris **230**, 286 (1950). Calcul des variations des constantes optiques des couches lacunaires.

— C. R. Acad. Sci. Paris **230**, 1349 (1950). Méthode graphique de détermination simultanée des constantes optiques et de l'épaisseur des couches métalliques minces.

— J. Phys. Radium **11**, 332 (1950). Étude graphique des propriétés optiques des lames métalliques minces.

— C. R. Acad. Sci. Paris **235**, 1630 (1952). Sur le choix des mesures à effectuer pour déterminer simultanément les constantes optiques et l'épaisseur des lames minces absorbantes.

— Ann. Phys. **9**, 10 (1954). Sur la détermination simultanée des constantes optiques et de l'épaisseur des lames minces absorbantes.

— Rev. Opt. **34**, 281 (1955). Étude graphique des lames minces.

—, et R. RINALDI: C. R. Acad. Sci. Paris **240**, 2130 (1955). Détermination des incides et de l'epaisseur de couches monomoléculaires stratifiiés d'hémine.

—, et P. ROUARD: J. Phys. Radium (8) **14**, 584 (1953). Sur les déterminations des constantes optiques des métaux massifs faites au moyen des lames épaises.

MAYER, H.: Physik dünner Schichten. Teil I. Herstellung, Dickenmessung, optische Eigenschaften. Band 4, Physik und Technik, herausgegeben von Dr. FRITZ GÖSSLER. Stuttgart: Wissenschaftliche Verlagsgesellschaft 1950.

MENZEL, E.: Naturwiss. **38**, 332 (1951). Mehrfacher mikroskopischer Lichtschnitt mit Minimumstrahlenkennzeichnung.

— Naturwiss. **39**, 398 (1952). Dreispaltverfahren am Mikroskop.

MESSNER, R.: Optik **2**, 228 (1947). Mehrfachschichten.

— Feinwerktechn. **57**, 297 (1953). Elementare graphische Methoden zur Darstellung der Interferenzeigenschaften von ein- und zweikomponentigen absorptionsfreien optischen Interferenzschichten.

— Feinwerktechn. **57**, 396 (1953). Mehrkomponentige absorptionsfreie optische Interferenzschichten.

— Feinwerktechn. **57**, 396 (1953). Interferenzeigenschaften absorbierender dünner Schichten.

MÖNCH, G. C.: Optik **8**, 550 (1951). Messung kleiner Schichtdicken mit Doppelprismeninterferometer.

— Optik **9**, 75 (1952). Interferenzfarben dünner Blättchen und die Bestimmung ihrer Dicken.

— Optik **9**, 97 (1952). Über die Farben dünner Häutchen und ihre Beziehungen zu den Farben dünner doppelbrechender Kristallplatten.

MOHLER, N. M., and J. R. LOOFBOUROW: Amer. J. Phys. **20**, 499, 579 (1952). Optical filters. (Absorptionsfilter, Polarisations-, Interferenz- und CHRISTIANSEN-Filter.)

MOLLOW, E.: Z. Physik **120**, 618 (1943). Zum Nachweis von Interferenzerscheinungen durch Streuteilchen im Interferenzfeld. (Interferenzen gleicher Dicke an LiF-Schicht auf Silber bei streifendem Licht.)

MOOS, T. S.: Proc. Phys. Soc. Lond. B **66**, 141 (1953). Interrelation between optical constants for lead telluride and silicon.

MOSER, H., u. J. WITTMANN: Z. Physik **131**, 48(1951). Ein experimenteller Beitrag zur Minimumstrahlkennzeichnung.

MUCHMORE, R. B.: J. Opt. Soc. Amer. **38**, 20 (1948). Optimum band width for two layer anti-reflection films.

MURMANN, H.: Z. Physik **54**, 741 (1929). Durchlässigkeit dünner Metallschichten für langwellige ultrarote Strahlung.

— Z. Physik **80**, 161 (1933). Die optischen Konstanten durchsichtigen Silbers.

— Z. Physik **101**, 643 (1936). Der spektrale Verlauf der anomalen optischen Konstanten dünnen Silbers.

MURRAY, A. E.: J. Opt. Soc. Amer. **41**, 872 (1951). The effect of antireflection films on color in optical instruments.

NICOLL, F. H.: J. Opt. Soc. Amer. **42**, 241 (1952). Anomalous interference films on glass by chemical treatment.

ODENBACH, F.: Ann. Phys. (5) **38**, 469 (1940). Dicke und Konstanten von Metallschichten nach Polarisationsmethode.

OSTERBERG, H., and N. E. PAGE: J. Opt. Soc. Amer. **42**, 290 (1952). A method of products for analysing multilayers.

—, and W. H. KASHDAN: J. Opt. Soc. Amer. **42**, 291 (1952). Bilayers that simulate monolayers with augmented interfacial reflectivity (Theor.).

PARTSCH, A., u. W. HALLWACHS: Ann. Phys. (4) **41**, 247 (1913). Reflexionsvermögen dünner Metallschichten und Eindringtiefe bei der Lichtelektrizität.

PFUND, A. H.: J. Opt. Soc. Amer. **24**, 99 (1934). Sb_2S_3 zur Reflexionserhöhung.

PHILIP, R.: C. R. Acad. Sci. Paris **241**, 559, 596 (1955). Phasensprung bei Reflexion sichtbaren und ultravioletten Lichts an dünnen Ag-Schichten.

—, et J. TROMPETTE: C. R. Acad. Sci. Paris **241**, 627 (1955). Détermination simultanée des constantes optiques et de l'épaisseur des lames très minces d'argent dans le visible et le proche ultraviolet.

PICK, H.: Z. Physik **126**, 12 (1949). Herstellung spiegelnder Niederschläge durch chemische Reaktionen.

POHLACK, H.: Ann. Phys. (6) **5**, 311 (1950). Vierpoltheorie für Mehrfachschichten.

— Jenaer Jb. **1953**, 241. Abhängigkeit der Intensitäten an Metallschichten von den Außenmedien.

— Ann. Phys. (6) **11**, 383 (1953). Zur Theorie optischer Interferenzschichtsysteme.

POLSTER, H. D.: J. Opt. Soc. Amer. **39**, 1038 (1949). Multilayer filter. (Rekursionsformeln. Effektiver Brechungsindex einer Mehrfachschicht. Filter mit verhinderter Totalreflexion.)

— J. Opt. Soc. Amer. **39**, 1054 (1949). A dielectric interferometer filter.

PRIDE, G. E., and W. H. KASHDAN: J. Opt. Soc. Amer. **42**, 291 (1952). Bilayers that simulate monolayers with augmented interfacial reflectivity (exper).

QUINCKE, G.: Pogg. Ann. **142**, 177 (1871). Optische Experimentaluntersuchungen. (Umfassender Bericht über Interferenzbeobachtungen, unter anderem auch an dünnen Schichten und Metallen; frühere Arbeiten des gleichen Verf. dort aufgeführt.)

RAHBECK, H., and M. OMAR: Nature, Lond. **169**, 1008 (1952). Measurement of the thickness of transparent films with the light-profile microscope.

REICHELT, W.: Zur Bandbreite der Reflexionsauslöschung von Zweifachschichten. (100 Jahre Heräus, Festschrift S. 393. 1951.)

RICHTER, J.: Ann. Phys. (4) **77**, 81 (1925). Phasendifferenzmessungen an dünnen durch Kathodenzerstäubung hergestellten Silber- und Kupferschichten.

RIEDEL, L.: Ann. Phys. **28**, 603 (1937). Versuche zur experimentellen Bestimmung der freien Weglänge der Elektronen in Blei und Cadmium.

RIENITZ, J.: Optik **8**, 561 (1951); **9**, 512 (1952). Winkelspiegelinterferometer.

ROBIN, S.: J. Phys. Radium **14**, 427 (1953). Mesures de pouvoirs réflecteurs des couches métalliques épaisses (rhodium, beryllium, nickel) dans la région de SCHUMANN.

ROBINSON, T. S.: Proc. Phys. Soc. Lond. B **65**, 910 (1952). n, k aus Reflexion über einen Frequenzbereich.

ROOD, J. L.: J. Opt. Soc. Amer. **39**, 884 (1949). Some properties of thin evaporated films on glass.

ROTHEN, A., and M. HANSON: Rev. Sci. Instrum. **19**, 839 (1948); **20**, 66 (1949). DRUDE-Methode experimentell verbessert.

Rouard, P.: Cahiers Phys. **1**, 25 (1941). Sur un phénomène optique se produisant dans les couches métalliques très minces. (Phasen benutzt.)
—, et P. Cotton: C. R. Acad. Sci. Paris **228**, 1706 (1949). Sur les facteurs de réflexion, dans le support transparent sur le métal, de différents métaux en couches très minces.
— Propriétés optiques de lames minces solides. Mémorial des Sciences Physiques, Faszicula LIV, Paris 1952. (Zusammenfassendes Werk.)
— D. Malé et J. Trompette: J. Phys. Radium (8) **14**, 587 (1953). Détermination des facteurs de réflexions, de transmission et d'absorption de lames minces d'or obtenes par évaporation.
Sennet, R. S., and G. D. Scott: J. Opt. Soc. Amer. **42**, 686 (1952). The structure of evaporated metal films.
Shea, D.: Wied. Ann. **47**, 177 (1892). Strengere Form der Kundtschen Prismenmethode.
Simon, J.: J. Opt. Soc. Amer. **41**, 730 (1951). Optical constants of germanium, silicon and pyrite in the infra-red.
Smakula, A.: Phys. Z. **34**, 788 (1933). Absorption der Metalle.
— Z. Physik **86**, 185 (1933). Absorption dünner Cu-, Ag- und Au-Schichten im sichtbaren und ultravioletten Gebiet. (Sehr viele Meßpunkte im Spektrum!)
— Z. Physik **88**, 114 (1934). Über die Lichtabsorption der Metalle.
— DRP. 685767 v. 1. 11. 1935. (Anmelder: Firma Carl Zeiss, Jena.) Reflexminderung durch Aufbringen einer dünnen Schicht.
— Z. Instrumentekde. **60**, 33 (1940). Über die Erhöhung der Lichtstärke optischer Geräte.
— Phys. Z. **43**, 217 (1942). Reflexionsminderung an optischen Gläsern. (Zusammenfassender Bericht.)
Smith, F. D.: J. Opt. Soc. Amer. **42**, 291 (1952). Application of the reversibility principle to a multiple dielectric film.
Sommer, A.: Optische Konstanten dünner Goldschichten aus Intensitäts- und Polarisationsmessungen. Diss. Hamburg 1940.
Suhrmann, R., u. G. Barth: Z. techn. Phys. **15**, 547 (1934). Strukturänderung von Metallschichten.
— Phys. Z. **36**, 843 (1935). Änderung des elektrischen Widerstandes und der Reflexion dünner Metallschichten.
— Z. Physik **103**, 133 (1936). Über die Änderung des elektrischen Widerstandes und des Reflexionsvermögens von bei tiefer Temperatur kondensierten Metallspiegeln.
—, u. H. Schnackenberg: Z. Physik **119**, 287 (1942). Über den Reaktionsablauf beim Übergang reiner Metallschichten aus dem ungeordneten in den geordneten Zustand. (Strukturänderung von Metallschichten; dort weitere Literaturhinweise.)
Schlick, M.: Über die Reflexion des Lichts in einer inhomogenen Schicht, Diss. Berlin 1904.
Schmaltz, G.: Technische Oberflächenkunde. Berlin 1936.
Schopper, H.: Z. Physik **130**, 427 (1951). Die Untersuchung „dicker" Metallschichten und ihrer Oberflächenschichten mit Hilfe der absoluten Phasen.
— Z. Physik **130**, 565 (1951). Die Untersuchung dünner absorbierender Schichten mit Hilfe der absoluten Phase.
— Optik **9**, 498 (1952). Die Erzeugnug von linearpolarisiertem Licht mit Hilfe einer dünnen absorbierenden Schicht.
— Z. Physik **131**, 215 (1942). Die Bestimmung der optischen Konstanten und der Schichtdicke beliebig dicker Schichten mit Hilfe der absoluten Phase.
— Z. Physik **135**, 163 (1953). Zur Deutung der optischen Konstanten der Alkalimetalle.
— Z. Physik **135**, 516 (1953). Ein optisches Kalkspatinterferometer mit wellenlängenunabhängigem Intensitätsausgleich.
— Optik **10**, 426 (1953). Die Erzeugung von linearpolarisiertem Licht durch Reflexion an geschichteten Metallen.
Schröder, H.: Ann. Phys. (5) **39**, 55 (1941). Bemerkung zur Theorie des Lichtdurchgangs durch inhomogene durchsichtige Schichten.
— Naturwiss. **34**, 277 (1947). Neue optische Filter durch absorptionsfreie Interferenzschichten
— Optik **3**, 499 (1948). Die Erzeugung von linear-polarisiertem Licht durch dünne dielektrische Schichten.
— Z. Naturforsch. **4a**, 515 (1949). Die Bildung reflexvermindernder Oberflächenschichten auf Glas durch Lösungen mit p_H-Werten nahe 7.
—, u. R. Schläfer: Z. Naturforsch. **4a**, 576 (1949). Verlustfreie optische Interferenzpolarisatoren.
— Z. angew. Phys. **3**, 53 (1951). Über die Lichtteilungsfunktion dünner Mehrfachschichten und ihre Anwendungen.
Schuch, E.: Ann. Phys. (5) **13**, 297 (1932). Die Durchlässigkeit dünner Platinschichten im Wellenbereich von 0,25 bis 2,5 μ.

Schulz, G.: Ann. Phys. (6) **14**, 177 (1954). Über Interferenzen gleicher Dicke und Längenmessung mit Lichtwellen.

Schulz, H., u. H. Hannemann: Z. Physik **22**, 222 (1924). Metalloberflächen.

— A. T. M., V 44-6 (1938). Messung der Brechungszahl.

Schulz, L. G.: J. Opt. Soc. Amer. **40**, 690 (1950). Fabry-Perot zur Dickenmessung dünner Aufdampfschichten. Siehe Tolansky: J. Opt. Soc. Amer. **41**, 425 (1951).

— J. Opt. Soc. Amer. **40**, 802 (1950). The effect of phase changes in white light interferometry.

— J. Opt. Soc. Amer. **41**, 261 (1951). The effect of phase changes in white light interferometry.

— J. Opt. Soc. Amer. **41**, 871 (1951). Dispersion of evaporated films.

— J. Opt. Soc. Amer. **41**, 1047 (1951). An interferometric method for the determination of the absorption coefficients of metals, with results for silver and aluminium.

Schulze, R.: Phys. Z. **34**, 24 (1933). Optische und lichtelektrische Untersuchungen an dünnen Metallschichten. (Intensitätsmethode.)

Schuster, K.: Ann. Phys. (6) **4**, 352 (1949). Anwendung der Vierpoltheorie auf die Probleme der optischen Reflektionsminderung, Reflexionsverstärkung und der Interferenzfilter.

Statescu, C.: Ann. Phys. (4) **33**, 1032 (1910). Polarisationsmessungen an dünner Metallschicht auf massivem Metall.

Steinbuch, K.: Z. angew. Phys. **1**, 256 (1949). Reflexverstärkung.

Steinheil, A.: Ann. Phys. (5) **19**, 465 (1934). Oxydschichten auf Metallen.

Strohmaier, K.: Z. Naturforsch. **6a**, 508 (1951). Ein einfaches Verfahren zur Herstellung großflächiger durchsichtiger Häute aus Aluminiumoxyd mit einer Dicke zwischen 50 und 300 mμ.

— Z. Physik **135**, 44 (1953). Bemerkungen zur optischen Dickenmessung mit Dreistrahlinterferenzen.

Studer, F. J., and D. A. Cusano: J. Opt. Soc. Amer. **41**, 871 (1951). Titanium dioxide films as selective reflectors of the near infrared.

Tanner, H. A., and L. B. Lockhardt: J. Opt. Soc. Amer. **36**, 701 (1946). German reflection reducing coatings for glass.

Taylor, H. D.: The adjustement and testing of telescope objectivs, York (England), T. Cook (1896), Engl. Pat. 29561. 1904.

Thomas, C. D., and R. C. Colwell: Phys. Rev. **56**, 1214 (1939). Inhomogene Schichten.

Tiller, C., and H. Levinstein: Phys. Rev. (2) **86**, 657 (1952). Evaporated antimony films.

Tolansky, S., and W. L. Wilcock: Proc. Roy. Soc. Lond., Ser. A **191**, 182 (1947). Interference studies of diamond faces. A crossed fringe technique. (Dickenmessungen, dort Hinweis auf weitere Originalarbeiten.)

— Phys. Bl. **4**, 472 (1948). Mehrfachreflex-Interferometrie an Oberflächen und Schichten.

— Multiple Beam Interferometry of Surfaces and Films. Oxford: Clarendon Press 1948.

— J. Opt. Soc. Amer. **41**, 425 (1951). The measurement of thin film thickness by interferometry. [Kritik zu Schulz, J. Opt. Soc. Amer. **40**, 690 (1950).]

— Nature, Lond. **163**, 885 (1949). Kurzer Tagungsbericht über das in Marseille stattgefundene internationale Kolloquium über Eigenschaften dünner Schichten.

— Nature, Lond. **169**, 445 (1952). A high-resolution surfaceprofile microscope.

Tronnier, H., u. H. Wagener: Optik **8**, 165 (1951). Zur Anwendung oberflächenvergüteter Mikroskop-Optik.

Väisälä, Y.: Ann. Univ. Fennicae thoensis, Ser. A 1 **1923**, Nr. 2. Dreispaltverfahren.

Vasiček, A.: Phys. Rev. **57**, 925 (1940). Analyse des reflektierten Lichtes mit Soleil-Babinet.

— J. Opt. Soc. Amer. **37**, 623 (1947). Reflexion einer bedeckten Fläche wird als Reflexion einer unbedeckten Grenzfläche eines hypothetischen Mediums gedeutet.

— Acad. Sci. Moravo-Silesiacae **23**, 355 (1951). Polarisationsverfahren. (Schichtdicke nach Tolansky.)

— Cas. Pest. Mat. **3**, 167 (1953). The study of anodized Al_2O_3 films on aluminium by the polarimetric method.

Veenemans, C. F.: Arch. néerl. Sci. exact. natur., Sér. III A **14**, 84 (1933.) Bestimmung der optischen Konstanten aus Lichtabsorptionsmessungen an dünnen Metallschichten.

Vernier, P.: J. Phys. Radium **14**, 175 (1953). Absorptionskoeffizient dünner PbS-Schichten im Ultraroten.

Vilbig, F.: Lehrbuch der Hochfrequenztechnik, Bd. I, S. 288. Leipzig 1942. (Zenneck-Welle.)

Voigt, W.: Wied. Ann. **24**, 144 (1885). Kundts Prismenmethode nicht anwendbar bei hoher Absorption.

— Wied. Ann. **35**, 76 (1888). Über die Reflexion und Brechung des Lichtes an Schichten absorbierender isotroper Medien.

Watson, J. T., and G. Joos: J. Opt. Soc. Amer. **41**, 876 (1951). A differential surface refractometer for testing for variation of refractive index in large plates of glass.

WEINSTEIN, W.: J. Opt. Soc. Amer. **37**, 576 (1947). The reflectivity and transmission of multiple thin coatings.

WEISKIRCHNER, W.: Z. Naturforsch 6a, 509 (1951). Der Brechungsquotient dünner Al_2O_3-Häutchen.

WERNICKE, W.: Pogg. Ann. **159**, 198 (1876). Absolute Phasenänderungen bei Reflexion.

WIENER, O.: Wied. Ann. **31**, 629 (1887). Phasenänderung bei Reflexion und Dickenbestimmung dünner Blättchen.

WILLIAMSON, D. E.: J. Opt. Soc. Amer. **39**, 613 (1949). Infra-red-interference filter used in calibration.

WOLIN, S.: J. Opt. Soc. Amer. **43**, 373 (1953). Inhomogene Medien.

WOLTER, H.: Z. Physik **105**, 269 (1937). Intensitätsmethode mit Näherungen. Intensitäts-Identitäten.

— Z. Physik **113**, 547 (1939). Resonanzstellen im Absorptionsspektrum dünner Metallschichten.

— Z. Physik **115**, 696 (1940). Erklärung der optischen Anomalien dünner Metallschichten aus Dichteschwankungen.

— Z. Physik **140**, 57, 565 (1955). Phasen- und Amplitudenmessung.

WOLTERSDORFF, W.: Z. Physik **91**, 230 (1934). Über die optischen Konstanten dünner Metallschichten in langwelligem Ultrarot.

WOOD, R. W.: Physikal Optics, S. 371. 1911. Einfluß einer Oberflächenschicht auf die Reflexion.

ZENNECK, J.: Ann. Phys. (4) **23**, 846 (1907). Über die Fortpflanzung ebener elektromagnetischer Wellen längs einer ebenen Leiterfläche und ihre Beziehung zur drahtlosen Telegraphie.

ZERNIKE, F.: J. Opt. Soc. Amer. **40**, 326 (1950). Dreispaltverfahren. (Siehe ferner Literatur zum Artikel „Schlieren-, Phasenkonstrast- und Lichtschnittverfahren".)

ZORLL, U.: Optik **9**, 449 (1952). Die Genauigkeitsgrenze der Dickenmessung von dünnen Schichten mit dem KÖSTERschen Interferenzdoppelprisma.

Schlieren-, Phasenkontrast- und Lichtschnittverfahren.

Von

H. WOLTER.

Mit 54 Figuren.

A. Übersicht.

1. Zur Bezeichnung der Schlieren-, Phasenkontrast-, Lichtschnitt- und Interferenzverfahren. Seit ZERNIKE[1] 1932 die Schlierenverfahren vom Standpunkt der ABBESchen Theorie betrachtet und dabei das Phasenkontrastverfahren gefunden hat, ist ein nahezu erstarrtes Gebiet der Optik in einmaliger Weise in Bewegung geraten. Es befindet sich noch mitten in ihr.

Der Wunsch, über dieses Gebiet hier vollständig berichten zu wollen, wäre von vornherein als unerfüllbar erkennbar. Es kann höchstens angestrebt werden, zu den wichtigsten Teilen des Gebiets *typische* Beispiele zu behandeln.

Leider mußte eine große Zahl von guten Originalarbeiten ungenannt bleiben.

Die Vielfalt der Literatur hat eine Sprachverwirrung mit sich gebracht, die hier einleitend die Umgrenzung der in dem Artikel benutzten Begriffe nötig macht. Aber es ist klar, daß diese Umgrenzung bei dem Fluß der Entwicklung und bei den unvermeidlichen Überschneidungen nicht völlig starr gehandhabt werden kann. Alle hier behandelten Verfahren zeigen „Schlieren" an und werden deshalb von vielen Autoren, vor allem in älteren Arbeiten schlechthin als „Schlierenverfahren" bezeichnet. Alle diese Verfahren machen Phasendifferenzen sichtbar und werden daher gelegentlich „Phasenkontrastverfahren" im weiteren Sinne genannt. Mit dem gleichen Recht kann man sie zu den Interferenzverfahren rechnen, da die entscheidende Wirkung aus Interferenzphänomenen verstanden werden muß.

Im Anschluß an den heute sich mehr und mehr durchsetzenden Sprachgebrauch finden hier enger gewählte Begriffe Verwendung:

Schlierenverfahren heißen in diesem Artikel die optischen Verfahren, bei denen die Richtung, in die eine Objektstelle das Licht lenkt, über die Kennzeichnung entscheidet.

Die Bezeichnung „*Phasenkontrastverfahren*" findet nur dort Verwendung, wo ein Eingriff in die Spektren eine unterschiedliche Phasenbehandlung des Spektrums nullter Ordnung (des direkten Lichtes) gegenüber den Seitenspektren (dem abgebeugten Licht) bewirkt. Phasenkontrastverfahren kennzeichnen eine Objektstelle im Bilde auf Grund der Phasendifferenzen gegenüber der näheren Umgebung.

Als *Dunkelfeldverfahren* können alle jene Abbildungsverfahren gelten, die das Spektrum nullter Ordnung (das direkte, nicht abgebeugte Licht) nicht am Bilde mitwirken lassen; sie können je nach Ausführung ein Grenzfall des Phasenkontrastverfahrens oder ein Schlierenverfahren oder in dem strengen Fall, daß nur genau das Spektrum nullter Ordnung beseitigt wird („strenges Dunkelfeldverfahren"), Schlierenverfahren und Grenzfall des Phasenkontrastverfahrens zugleich sein.

---

[1] F. ZERNIKE: Physica, Haag **1**, 43 (1934). — D.R.P. Nr. 636168/42h/6, 10 vom 26. 11.1932

Lichtschnittverfahren nennt man allgemein die Verfahren, bei denen die Spur beobachtet wird, die eine mit Hilfe des Lichtes gekennzeichnete Ebene auf einer oder mehreren Körperoberflächen bildet. Die Grenze zu den Schlierenverfahren ist fließend; unterscheiden sich die Lichtschnittverfahren z.B. von Wieners Schlierenverfahren doch nur durch die Akkommodation des Auges auf die Objektoberfläche statt auf eine Ebene zwischen ihr und dem Beobachter. Die modernste Form des Lichtschnittverfahrens ist zugleich ein Interferenzverfahren.

Interferenzverfahren erster Art heißen die Abbildungsverfahren, die ohne unmittelbaren Eingriff in die Spektren „gleichmäßiges" Zusatzlicht auf die Bildebene werfen. Sie kennzeichnen eine Objektstelle nach der zu ihr gehörigen Lichtphase. *Interferenzverfahren zweiter Art* benutzen kohärentes Zusatzlicht, das gleichmäßige Intensität, aber nahezu linear ortsabhängige Phase hat. *Interferenzverfahren dritter Art* haben ungleichmäßiges kohärentes Zusatzlicht, z.B. ein zweites, seitlich verschobenes Bild.

Ein Vergleich der wichtigsten Verfahren befindet sich in den Ziff. 64 und 65 am Schluß dieses Artikels.

B. Schlierenverfahren.

I. Geometrisch-optische Betrachtung der Schlierenverfahren.

a) Eindimensionale Kennzeichnung der Ablenkung des Lichtes bei zweidimensionalem Objektfeld.

2. Aufgabe und einfachste Realisierung der abbildenden Schlierenverfahren. „Schliere" nennt man jede Stelle eines Objekts, die das Licht in eine andere als die im Einzelfall definierte „Normalrichtung" schickt. Als Normalrichtung kann z.B. für alle Punkte des Objekts die gleiche Richtung definiert werden. Für jeden Objektpunkt kann aber auch als seine zugehörige Normalrichtung eine andere Richtung definiert sein; z.B. wird man unter Umständen für ein Objektiv oder einen Hohlspiegel alle Normalrichtungen durch einen im Endlichen gelegenen Punkt gehend definieren und die Objektiv- bzw. Spiegelstellen, die das Licht in eine davon abweichende Richtung schicken, als Schlieren bezeichnen. Auch in dem ersten Fall zielen alle Normalrichtungen zu einem festen Punkt. Er liegt im Unendlichen statt im Endlichen; das ist der einzige Unterschied zu dem zweiten Fall. Da man mit abbildenden Mitteln die unendlich ferne Ebene ins Endliche abbilden kann und umgekehrt, lassen beide Beispiele sich aufeinander zurückführen. Deshalb sei im folgenden speziell die Normalrichtung als gleich für alle Objektpunkte definiert angesehen, wenn nicht ausdrücklich anderes bemerkt ist.

„Schlierenverfahren" haben das Ziel, Schlieren aufzuzeigen und die Abweichung der Lichtstrahlrichtung e (Einheitsvektor in Fortpflanzungsrichtung des Lichtes) von der „Normalrichtung" e_0 meßbar zu machen. Für alle Punkte der Objektebene *zugleich* bewirken das die abbildenden Schlierenverfahren, die jeden Objektpunkt in eine Ebene, die „Bildebene", abbilden und dabei dem Licht am Bildort des Objekts ein Kennzeichen für die Ablenkung aufprägen (z.B. eine Amplitude oder eine Farbe).

Das Objekt sei entweder von sich aus nahezu eben, oder aber statt seiner werde eine x, y-Ebene der Fig. 1 als „Objektebene O", dicht hinter ihm festgesetzt; Ersatz für das Objekt sind dann die Punkte dieser Ebene, und untersucht wird das Licht auf die Richtung hin, in der es diese Ebene verläßt. Es tritt nach Fig. 1 in das Objekt ein als Parallelstrahlenbündel der Richtung $e_0 = (0; 0; 1)$, hergestellt mit Hilfe des Kondensors (Fig. 1) aus dem Licht, das von der rückseitig beleuchteten Punktblende P, genannt „Lichtquellenblende", ausgeht.

Die Objektebene (x, y) wird auf die Bildebene (x', y') abgebildet durch das Objektiv.

Die optische Kennzeichnung des Bildpunktes (x', y') einer Objektstelle (x, y) je nach der Größe der Vektorkomponenten

$$\sin \alpha_x = e_x = e_x(x; y); \quad \sin \alpha_y = e_y = e_y(x; y) \tag{2.1}$$

nach Töplers[1] Methode beruht darauf, daß alle Strahlen der Richtung e in dem Raume zwischen Kondensor und Objektiv vom Objektiv gesammelt werden in einem einzigen Punkte

$$\xi = f \cdot \tan \alpha_x; \quad \eta = f \cdot \tan \alpha_y, \tag{2.2}$$

der bildseitigen Objektivbrennebene[2]. Durch Markierung ihrer Punkte ξ, η kann daher eine Markierung der Richtungen geschehen. f ist die Objektivbrennweite.

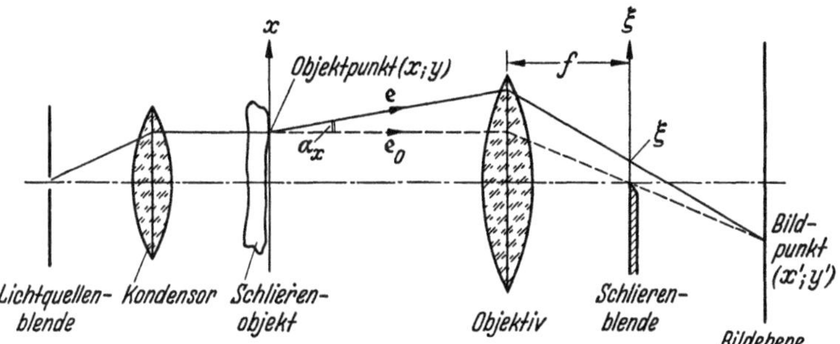

Fig. 1. Abbildendes Schlierenverfahren; Objekt im parallelen Strahlengang.

Wird wie in Fig. 1 in der Objektivbrennebene eine Schneide angebracht, die alle Punkte ξ, η mit $\xi < 0$ bedeckt, so werden auf die Bildebene alle die Punkte $(x; y)$ dunkel abgebildet, die das Licht in Richtungen mit $e_x < 0$ verläßt. Schiebt man die Schneide bis zum Punkte ξ_s, so geht im Bilde die Hell-Dunkelgrenze durch alle Punkte, denen Richtungen mit

$$\alpha_x = \text{arc} \tan \frac{\xi_s}{f} \tag{2.3}$$

zukommen; solche mit größerem α_x sind hell, die mit kleinerem α_x sind dunkel wiedergegeben. Die Fig. 16a zeigt eine solche Schlierenaufnahme einer Glasplatte als Schlierenobjekt. Die Hell-Dunkelgrenze gibt darin eine „Isokampte", d.h. den geometrischen Ort aller Punkte, die gleiche Lichtablenkkomponente α_x haben.

Legt man die Schlierenschneide in der bildseitigen Objektivbrennebene der Fig. 1 parallel zur Zeichenebene statt senkrecht zu ihr, dann kann mit der Apparatur entsprechend die Lichtablenkkomponente e_y oder α_y gemessen werden.

3. Praktische Ausführungsformen des Töplerschen Schlierenverfahrens. Eine Vereinfachung bedeutet das Fortlassen des Kondensors in der Apparatur nach Fig. 1. Dann wird die Schlierenblende in die Bildebene der Lichtquellenblende gesetzt (Fig. 2). Als Normalrichtung dienen dann die von der Lichtquellenblende ausgehenden Radien.

---

[1] A. Töpler: Siehe Bibliographie.
[2] Richtungen durch die eine Brennebene einer Linse und Punkte in der anderen Brennebene sind einander zugeordnet. Hierauf gründet sich ein Dualitätsprinzip der Optik [siehe A. Lohmann: Optik **11**, 478 (1954)].

Das Objektiv muß in dieser Anordnung wesentlich größer sein als der zu untersuchende Teil des Schlierenobjekts, da nur ein solcher Teil des Objekts

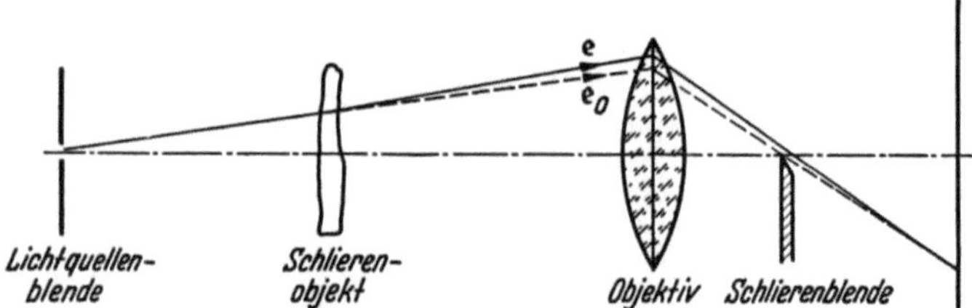

Fig. 2. Abbildendes Schlierenverfahren; Objekt im divergenten Strahlengang.

ausgeleuchtet wird, dessen Normalstrahlen (gestrichelt) und Realstrahlen (ausgezogen) in das Innere der Objektivfassung treffen.

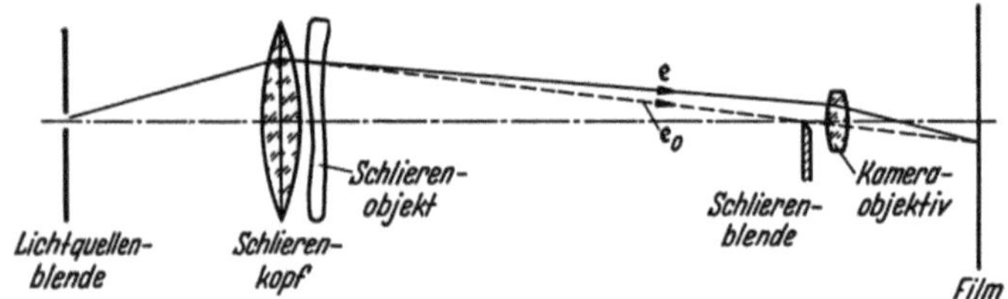

Fig. 3 a. Abbildendes Schlierenverfahren; Objekt im konvergenten Strahlengang.

Eine andere Vereinfachung der Apparatur nach Fig. 1 erhält man, wenn man die dort erforderlichen großen und anspruchsvollen Optiken, Objektiv und Kondensor, zu einer Optik, genannt „Schlierenkopf", zusammenfaßt, wie Fig. 3 a das andeutet.

Das ist die älteste TÖPLERsche Anordnung. Freilich ist dann zur Abbildung des Schlierenobjekts doch wieder ein Objektiv — am zweckmäßigsten dicht hinter der Schlierenblende — einzusetzen. Dieses kann aber ziemlich klein sein. Sein Durchmesser richtet sich nicht nach der Größe des Schlierenobjekts, sondern braucht nur etwas größer als $a \cdot \tan \alpha_m$ zu sein, wenn a der Abstand von der Schliere zur Schlierenblende und α_m die größte zu messende Ablenkung ist.

Die Normalrichtungen sind in diesem Falle konvergent.

Fig. 3 b. Zonenfehler und Schleifspuren am 80 cm-Objektiv des Potsdamer Refraktors, sichtbar gemacht mit der Schneidenmethode (Schneidenkante stand vertikal).

Das Verfahren eignet sich besonders gut, wenn eine sammelnde Optik selbst auf Fehler d.h. „Schlieren" untersucht werden soll. Sie ist dann einfach als Schlierenkopf einzusetzen. Eine solche Aufnahme zeigt Fig. 3 b. Die Optik hat ersichtlich Zonenfehler und Schleifspuren.

Als Schlierenkopf kann auch ein guter Hohlspiegel dienen, und auch Hohlspiegel werden nach Foucaults Beispiel so geprüft, z.B. mit dem eleganten Koinzidenzverfahren für Kugelspiegel (Fig. 4). Im Spiegelmittelpunkt befindet sich eine Lochblende. Sie wird von rechts über den halbdurchlässig versilberten Spiegel HS beleuchtet. Das von „guten" Spiegelstellen reflektierte Licht durchsetzt die Lochblende erneut und erreicht das Objektiv O, das den Spiegel auf die Beobachtungsebene F abbildet. Falsch geneigte Spiegelstellen reflektieren das

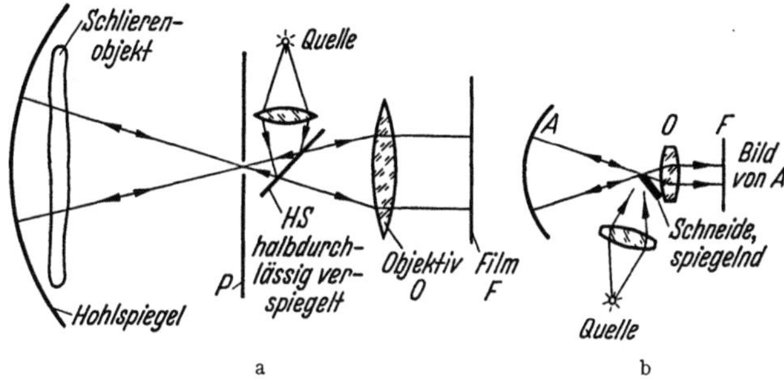

Fig. 4 a u. b. Schlierenverfahren in Autokollimation mit Hohlspiegel.
a Lochblendenverfahren; b Foucaults Schneidenverfahren.

Licht neben die Öffnung von P und erscheinen auf der Beobachtungsebene F dunkel. Die Kreisöffnung in P bewertet nicht α_x oder α_y sondern den Betrag $\sqrt{\tan^2\alpha_x + \tan^2\alpha_y}$. Statt der Lochblende bei P kann auch eine Schneide benutzt werden, die ebenfalls als Lichtquellen- und Schlierenblende zugleich wirkt. Ist sie als reflektierendes Spiegelstückchen ausgebildet, so wird der halbdurchlässige Spiegel erspart (Fig. 4b).

Die Wirkung anderer Blendenformen wird unten noch diskutiert. Für höchst empfindliche Schlierenapparaturen benötigt man einen selbst schlierenfreien

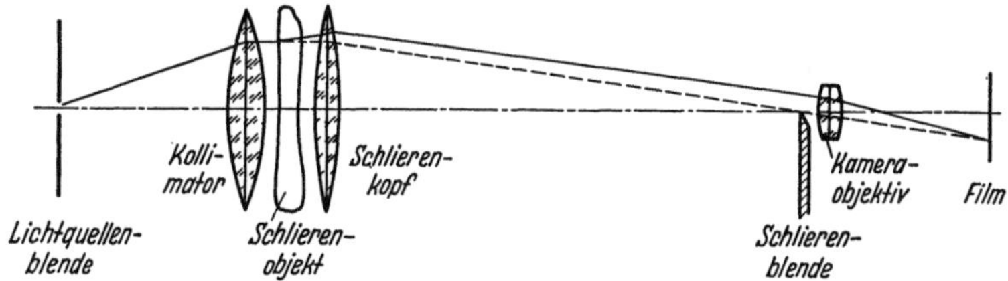

Fig. 5 a. Abbildendes Schlierenverfahren mit Kollimator und Schlierenkopf.

Schlierenkopf großer Brennweite, da ein kleines α_x sich wegen $\xi = f \cdot \tan\alpha_x$ um so mehr auswirkt, je größer f ist. Die Apparatur nach Fig. 1 würde dann sehr lang werden; denn der Abstand von dem Schlierenobjekt bis zur Beobachtungsebene müßte mindestens $4f$ sein. Dann ist die Apparatur nach Fig. 3 günstiger. Falls die Normalrichtungen alle parallel zueinander sein sollen — das schafft oft die am saubersten definierten Verhältnisse — ist der Schlierenkopf wieder aufzuteilen. Die Fig. 5 zeigt eine solche Apparatur, wie sie für empfindliche Schlierenaufnahmen bevorzugt wird.

Kondensor und Schlierenkopf können als Begrenzungen einer „Schlierenkammer" für die Untersuchung in Gasen oder Flüssigkeiten dienen. Fig. 5b und 5c zeigen Aufnahmen mit einer solchen Apparatur. Zahlreiche Anwendungen

Fig. 5b. Töplersches Schneiden-Schlierenbild vom Flug eines Geschosses durch die von zwei Flammen aufsteigende heiße Luft. Es zeigt den Wirbelkanal, die Kopfwelle und Heckwellen, ferner die Brechung der Kopfwelle durch die erwärmte Luft. (Aufnahme O. v. Schmidt. Die Aufnahme wurde dankenswerterweise von W. Lochte-Holtgreven zur Verfügung gestellt.)

und weitere Ausführungsformen abbildender Schlierenverfahren behandelt H. Schardin[1].

4. Verschiedene Formen der Schlierenblende und eindimensionale Farbschlieren-verfahren. $\alpha)$ *Kreisblende.* Setzt man in eine der Apparaturen nach Fig. 1 oder 5 eine Kreisblende vom Radius r, zentriert zur optischen Achse, statt der Schneide, so werden genau die Objektpunkte hell abgebildet, für die

$$\tan^2\alpha_x + \tan^2\alpha_y = \frac{\xi^2 + \eta^2}{f^2} < \frac{r^2}{f^2}$$

ist. Ein Beispiel zeigt Fig. 5d.

---

[1] H. Schardin: Die Schlierenverfahren und ihre Anwendungen. Erg. exakt. Naturwiss. **20**, 302 (1942).

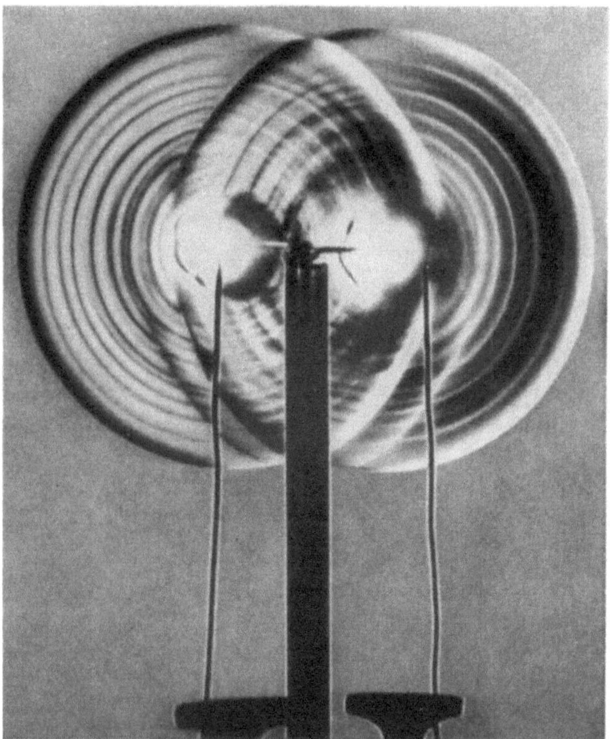

Fig. 5c. Töplersche Schlierenaufnahme zweier Funken-Knallwellen in Luft. Schneide vertikal. Abweichungen von der Kugelform einer Welle im Bereich der anderen (oben Mitte) zeigen die Nichtlinearität der Stoßwellen. In der unteren Hälfte haben die Wellen Vorläufer, die auf höhere Fortpflanzungsgeschwindigkeit im vertikalen Haltestab (unten Mitte) schließen lassen. (Aufnahme aus dem Institut H. Schardin.)

β) *Farb-Ringsystem.* Eine Platte mit einem System von konzentrischen Ringen, die der Reihe nach mit den Farben

Violett für $\qquad \sqrt{\xi^2 + \eta^2} \leq f \cdot 0{,}001$,

Blau für $\quad f \cdot 0{,}001 < \sqrt{\xi^2 + \eta^2} \leq f \cdot 0{,}002$,

Grün für $\quad f \cdot 0{,}002 < \sqrt{\xi^2 + \eta^2} \leq f \cdot 0{,}003$.

Gelb, Orange, Rot, „Weiß", „Schwarz" für weitere Bereiche angefärbt ist, bewährt sich als Schlierenblende besonders, wenn man einen Überblick über die Ablenkungsbeträge ohne Blendenwechsel, also in einem einzigen Bilde haben will.

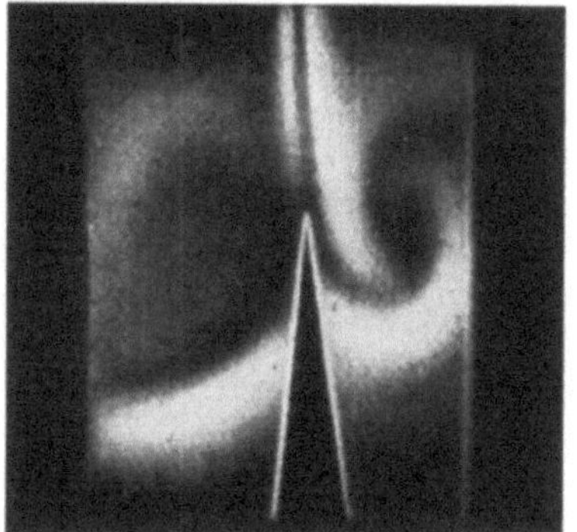

Fig. 5d. Zur Entstehung eines Schneidentones. (Nach H. Schardin.) Rechts Wirbelbildung, sichtbar gemacht durch Anblasen mit Kohlensäure. Abbildendes Schlierenverfahren mit Kreisloch als Lichtquellenblende und komplementärem Kreisscheibchen (Stopper) als Schlierenblende. Die Intensität des Lichtes im Schlierenbild ist proportional dem Ablenkbetrag:

$$J \sim \sqrt{\tan^2 \alpha_x + \tan^2 \alpha_y} \approx |\mathfrak{e} - \mathfrak{e}_0| .$$

γ) *Farbgitter.* In analoger Weise können die Farben bei der Messung von Komponenten α_x oder α_y benutzt werden, wenn statt der Schneide in die Richtungsauswahlebene eine Platte gesetzt wird, die der Reihe nach in rote, gelbe, grüne, blaue Parallelstreifen usw. eingeteilt ist (s. Fußnote S. 560).

δ) Kontinuierliches Farbschlierenverfahren. Eine entsprechende kontinuierliche Kennzeichnung durch Spektralfarben erhält man, wenn man die Dispeision in einem Prisma ausnutzt[1]. Die Fig. 6 zeigt eine solche Anordnung. Sie besteht aus einem üblichen Spektrographen, in dessen parallelen Strahlenteil das Objekt eingesetzt wird. An die Stelle der Kassette tritt ein Spalt als Schlierenblende, und hinter ihm befindet sich die auf das Objekt eingestellte Kamera oder das Auge.

Die Normalrichtung $\alpha_x = 0$ wird durch eine Spektralfarbe, z.B. $\lambda = 550\ \mathrm{m}\mu$, gekennzeichnet, die durch die Lage des als Schlierenblende dienenden Spaltes definiert wird. Jede Objektstelle, die das Licht in eine Richtung mit $\alpha_x > 0$

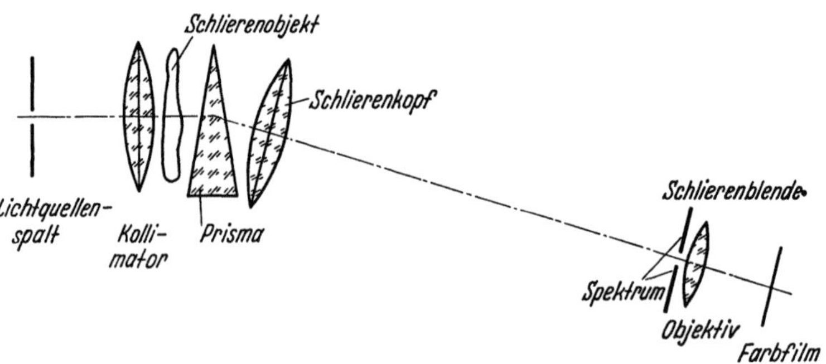

Fig. 6. Farbschlierenapparatur mit Prisma.

schickt, erscheint im Lichte einer längeren Welle (gelb, orange, rot); jede Stelle mit $\alpha_x < 0$ ist durch kurzwelligeres Licht (grün, blau, violett) im Bilde gekennzeichnet.

Die Empfindlichkeit der Apparatur steigt, wenn die Dispersion des Prismas abnimmt. Deshalb gibt bereits ein Kollimator mit Vergrößerungsfarbfehler, der etwas schräg in der Apparatur nach Fig. 1, 2, 3 oder 5 einjustiert wird, ein empfindliches Farbschlierenbild, ohne daß noch ein Prisma eingebaut werden müßte. Statt der spaltförmigen Schlierenblende kann man auch die Schneide verwenden (man erhält dann Mischfarben unterschiedlicher Sättigung) oder besser einen Steg, Draht usw., dessen Breite zweckmäßig etwa einem guten Drittel der Breite des Spektrums selbst entspricht. Man erhält dann gut gesättigte Mischfarben zur Kennzeichnung der Objektorte nach ihrem α_x. Bei dem Ersetzen eines Spaltes durch einen kongruenten Steg gehen die Farben in „komplementäre" Farben über. Farbaufnahmen mit diesem Verfahren veröffentlichte H. Schardin[2]. Fig. 54a auf S. 639 zeigt eine Körnerschliere in Glas, aufgenommen mit einem mikroskopischen Prismenfarbschlierenverfahren, jedoch noch kombiniert mit Phasenkontrast[3].

ε) Gitterblende. Bei der Diskussion verschiedener Schlierenblendenformen sei auch auf die Gitterblende hingewiesen, die an die Stelle der Schneide z.B. in der Apparatur nach Fig. 1 bis 5 treten kann. Helle und dunkle Parallelstreifen wechseln auf ihr ab. Entsprechend wechseln in dem Bild des Objekts helle und dunkle Bereiche ab, die bei monotonem Verlauf der Ablenkung α_x mit dem Objektort leicht den verschiedenen Bereichen von α_x zugeordnet werden können.

ζ) Ringgitter. Ähnlich wirkt eine Schlierenblende aus abwechselnd durchlässigen und undurchlässigen konzentrischen Ringen für die Kennzeichnung von $\sqrt{\xi^2 + \eta^2}$.

---

[1] H. Schardin: Erg. exakt. Naturwiss. **20**, 314 (1942).
[2] H. Schardin: Erg. exakt. Naturwiss. **20**, 302 (1942). Tafel I.
[3] H. Wolter: Ann. Physik (6) **9**, 57 (1951). Das Verfahren ist näher beschrieben in Ziff. 56a.

η) Spalt und Steg. Spalt und Steg sind triviale Formen der Schlierenblende, über die in diesem Teil, der sich auf die geometrisch-optische Betrachtung der Schlierenverfahren beschränkt, nichts weiter zu sagen ist.

5. Die Schneide als Lichtquellenblende und ihre Bedeutung für quantitative Auswertung. Zwischen Lichtquellenblende, die bisher als „Punktblende" vorausgesetzt war, und Schlierenblende herrscht eine gewisse Äquivalenz. So entstehen geometrisch optisch gleichartige Bilder, wenn beide Blenden ausgetauscht und ihre Größen so transformiert werden, wie es dem Abbildungsmaßstab der einen Ebene auf die andere entspricht. Freilich ist die kleinere Öffnung besser vor dem

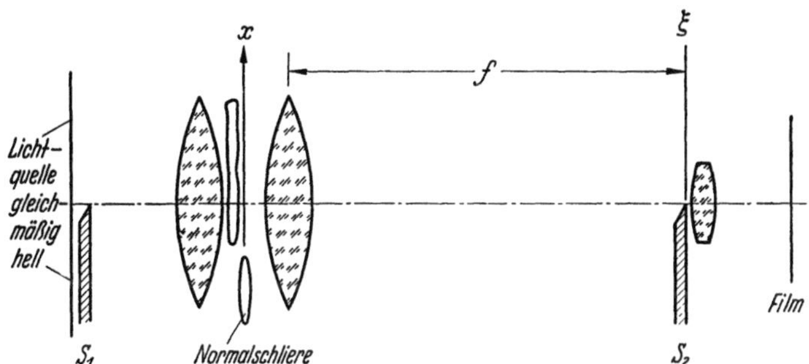

Fig. 7. Zweischneidenverfahren mit Normalschliere zur Ablenkmessung durch Photometrie.

Objekt — von der Lichtquelle her beurteilt — anzubringen, da die kleine Öffnung zwischen Objekt und Film eine auflösungsmäßig verschlechternde Abbildung bewirken würde. Dazu s. Ziff. 17 ff.

Besonders interessant sind solche Blendenpaare, die einander ähnlich sind wie Kreis zum Kreis, Spalt zum Spalt, Schneide zur Schneide und deren Größe so bemessen ist, daß die zweite Blende mit dem Bild der ersten Blende bei Abwesenheit eines Objekts zur Deckung kommt. Töpler selbst benutzte bereits zwei Dreiecke, von denen er jedoch im wesentlichen je eine Seite (aufeinander abgebildet) als entscheidend wirksam erkannte. Im vereinfachten Fall handelt es sich also um zwei Schneiden S_1 und S_2 in der Apparatur nach Fig. 7. Der helle Teil von S_1 ist auf den dunklen von S_2 kongruent abgebildet. Alle Objektstellen mit $\alpha_x \leqq 0$ erscheinen dunkel. Solche mit $\alpha_x > 0$ sind hell abgebildet, und zwar offenbar um so heller, je größer α_x ist; denn zur Abbildung trägt ein Streifen der gleichmäßig ausgeleuchteten Fläche über S_1 bei, dessen Breite proportional zu $\xi \sim \tan \alpha_x$ ist. Deshalb ist die Intensität im Bild

$$J = K \cdot \tan \alpha_x \qquad (5.1)$$

und umgekehrt die Ablenkung

$$\alpha_x = \arctan \frac{J}{K}. \qquad (5.2)$$

Man kann also die Ablenkkomponente α_x durch Photometrieren messen, wenn die Konstante K durch Eichung mit einer bekannten Schliere, genannt „Normalschliere", bestimmt wird. Als Normalschliere" eignet sich ein Prisma sehr kleinen brechenden Winkels. Schardin und seine Mitarbeiter benutzen als „Normalschliere" eine Linse, die z.B. durch eine Brennweitenmessung oder durch Ablenkmessung mit verschieblicher Schneide als genau bekannt gelten kann. Die beobachterseitige Hauptebene der Normalschliere ist in die Objektebene zu legen; dann ist die Ablenkung α_x durch die als Normalschliere dienende Linse der Brennweite f_n eine nahezu lineare Funktion des Abstandes der Koordinate von der Normalschlierenmitte x_0:

$$\alpha_x = \arctan \frac{x_0 - x}{f_n} = \arctan \frac{J}{K}. \qquad (5.3)$$

Vgl. Fig. 8! Da die Linse als Normalschliere unmittelbar im Bilde eine Zuordnung zwischen Intensität und Ablenkung liefert, braucht man zur Photometrie keine weiteren „Stufen" zu machen. Man sucht vielmehr zu einem Objektort, der

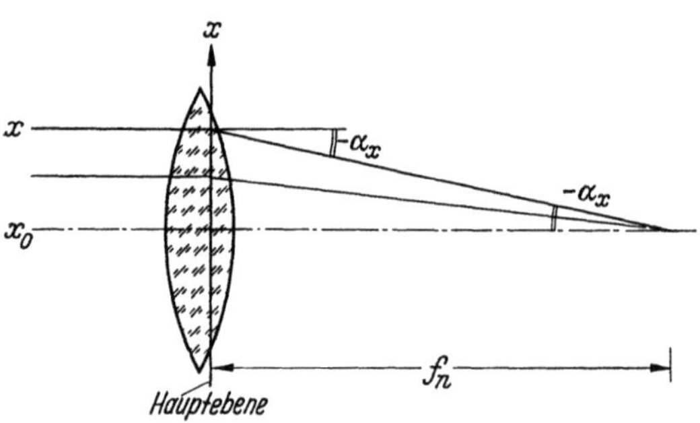

Fig. 8. Linse als Normalschliere.

mit der Intensität J abgebildet ist, den intensitätsgleichen Ort im Bild der Normalschliere x und hat dann nach Gl. (5.3) die Ablenkung des betrachteten Objektorts

$$\alpha_x = \arctan \frac{x_0 - x}{f_n}. \qquad (5.4)$$

Alle Betrachtungen dieser Ziffer galten zunächst für den Fall, daß beide Schneidenkanten aufeinander genau abgebildet wurden. Dadurch wurden alle Stellen mit $\alpha_x \leqq 0$ dunkel wiedergegeben, also von der Messung ausgeschlossen. Diese Grenze des Meßbereichs kann man aber zu negativen Werten von α_x verschieben, indem man die Schneide S_1 (oder S_2) in der Apparatur (Fig. 7) verschiebt. Gl. (5.1) ist dann zu ersetzen durch

$$J \sim \tan \alpha_x + \frac{\xi_s}{f}, \qquad (5.5)$$

wenn sich das Bild der Schneide S_1 um das Stück ξ_s in der Ebene von S_2 nach oben verschiebt. Da hiervon das Objekt und das Vergleichsobjekt, die Normal-

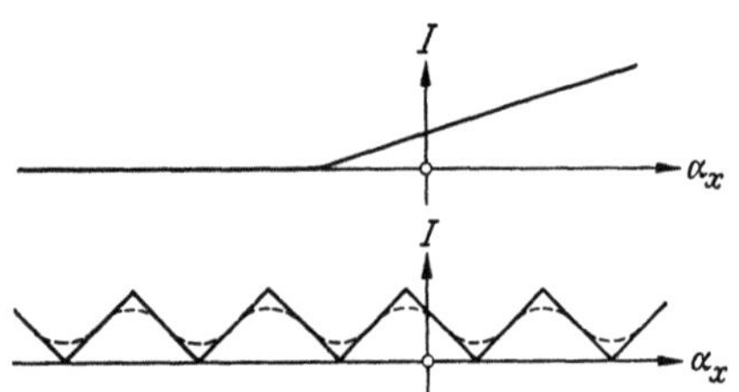

Fig. 9a. Die Intensität als Funktion der Ablenkung bei den Doppelschneidenverfahren (oben) und bei dem Doppelgitterverfahren (unten).

schliere, in völlig gleicher Weise betroffen werden, ändert sich an der Auswerteformel (5.4) nichts; nur ist der Bereich der von 0 verschiedenen Intensitäten erweitert und die Meßbarkeit für α_x ins Negative ausgedehnt worden[1].

6. Allgemeinere abbildende Schlierenverfahren mit komplementären Blendenpaaren. Statt des Schneidenpaares der Schlierenapparatur nach Fig. 7 kann man irgendein Paar komplementärer Blenden verwenden. Komplementär heißen die Blenden, wenn die Öffnungen der einen genau deckend auf die undurchlässigen Stellen der anderen abgebildet werden können. Für sehr kleine Ablenkungen ändert sich nichts an der Auswerteformel (5.4) bei Verwendung einer Vergleichsschliere. Die Empfindlichkeit aber, d.h. hier die Flankensteilheit des Intensitätsanstiegs

$$\frac{\partial J}{\partial \alpha_x} \approx K,$$

vergrößert sich mit der Länge der sich deckenden Kanten (s. Fig. 9a). Nimmt man komplementäre äquidistante Gitter an Stelle von S_1 und S_2 der Apparatur nach Fig. 7, so wächst die Empfindlichkeit also mit der Zahl der Gitterstreifen. Zugleich

---

[1] H. Schardin: Erg. exakt. Naturwiss. **20**, 323 (1942).

nimmt aber der Bereich ab, in dem die Kennzeichnung einer Ablenkkomponente α_x durch die Intensität eindeutig ist; denn Stellen, die gerade so stark ablenken, daß die Gitter um eine Gitterkonstante verschoben aufeinander abgebildet werden, erscheinen genau so hell wie nichtablenkende Stellen (vgl. Fig. 9a). Je mehr Gitterstege aufeinander abzubilden sind, desto größer wird auch die Auswirkung von Gitterfehlern; diese runden die Kurve nach Fig. 9a ab und setzen daher die Empfindlichkeit des Verfahrens gegen kleine Ablenkungen herab.

Bedeutung hat dieses Verfahren dennoch; denn man kann mit seiner Hilfe sehr große Objekte abbildend auf Schlieren auch dann untersuchen, wenn nicht hinreichend große Optiken als Schlierenkopf usw. verfügbar sind. Der Schlierenkopf wird entbehrlich, wenn man nach Fig. 9b vorgeht. Damit freilich das ganze

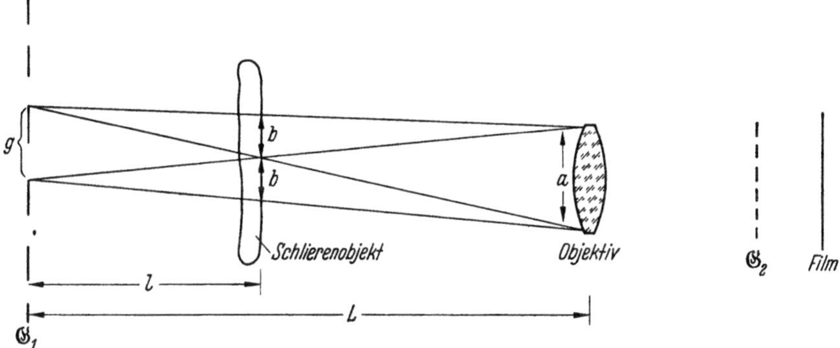

Fig. 9b. Doppelgitterverfahren ohne große Optiken.

Gesichtsfeld ausgeleuchtet ist und nicht die Gitterstruktur auf dem Film erscheint, muß auch das Objektiv eine gewisse Mindestgröße haben.

Diese kann um so kleiner sein, je kleiner die Gitterkonstante ist. Das Verfahren zeigt nämlich bereits dann Schlieren, wenn statt G_1 und G_2 einfache komplementär angeordnete Schneiden verwendet werden; doch ist dann nur ein kleiner Streifen des Gesichtsfeldes ausgeleuchtet, dessen entsprechende Breite im Objekt b sich zur Objektivöffnung a offenbar verhält wie die Abstände l zu L:

$$\frac{b}{a} = \frac{l}{L}. \tag{6.1}$$

Die Gitterkonstante muß nur so klein sein, daß sich die ausgeleuchteten Streifen, die jede Kante gibt, hinreichend überdecken oder, wie Fig. 9 es zeigt, mindestens aneinanderstoßen, d.h.

$$\frac{g}{a} \leq \frac{l}{L-l}. \tag{6.2}$$

Für $L = 2l$ z.B. darf also die Gitterkonstante höchstens gleich a sein. Wegen der kreisrunden Öffnung des Objektivs erfordert eine gleichmäßige Ausleuchtung dann aber ein g, das wesentlich kleiner als a ist[1].

7. Schattenschlierenverfahren. Noch einfacher als bei dem in Ziff. 6 beschriebenen Verfahren kommt man zum Ziel, wenn man das Schlierenobjekt auf einen Schirm nicht streng abbildet, sondern durch Schattenwurf projiziert. Fig. 10a skizziert die einfache Vorrichtung, in der L eine nahezu punktförmige Lichtquelle bedeutet. Bei schlierenfreiem Objekt ist der Schirm gleichmäßig beleuchtet.

---

[1] H. SCHARDIN, Erg. exakt. Naturwiss. **20**, 314 (1942). Das Verfahren wurde zuerst von H. MAECKER aufgebaut.

Schlieren im Objekt stören dieses Bild und werden dadurch erkennbar. Fig. 10b zeigt eine solche Aufnahme von einer nicht schlierenfreien Glasplatte.

Das Verfahren ist den abbildenden unterlegen wegen der Schwierigkeit einer Zuordnung zwischen dem beobachteten Licht und den Objektstellen, von denen es stammt. Aber für schwache Schlieren ist diese Schwierigkeit gering, und bei ihnen bewährt sich auch die große Empfindlichkeit des Verfahrens. Zum Beispiel sind die Oberflächenwellen in einer Wellenwanne ein dankbares Objekt für das Schattenschlierenverfahren. Man erhält äußerst kontrastreiche Bilder, wenn man die Abstände Lichtquelle— Objekt a und Objekt-Schirm b so bemißt, daß

$$\frac{1}{a} + \frac{1}{b} \approx \frac{1}{f}$$

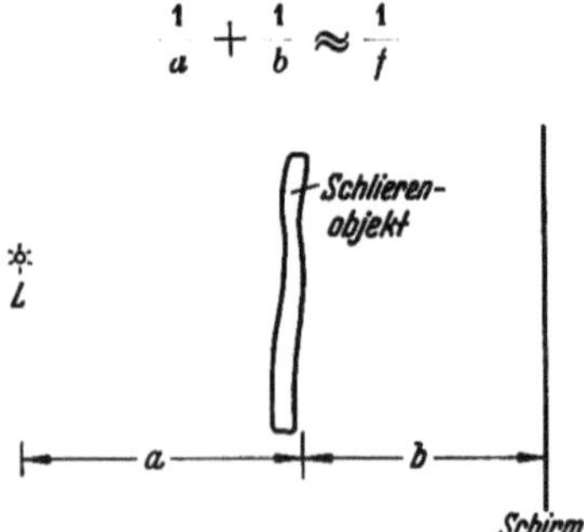

Fig. 10a. Schattenschlierenverfahren.

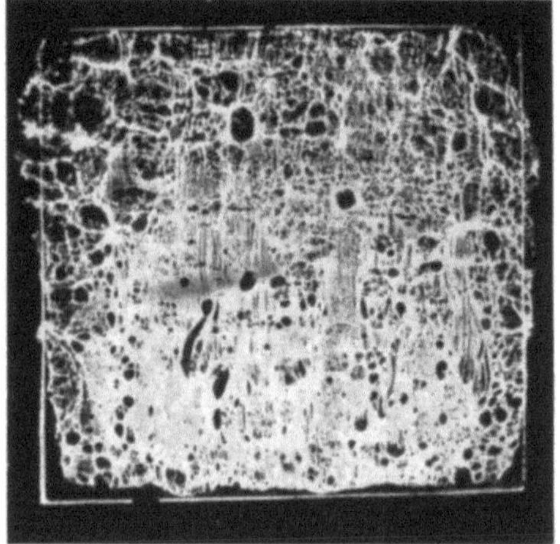

Fig. 10b. Schattenschlierenbild einer Glasplatte.
(Nach H. Schardin.)

ist, wenn f die Brennweite der von den Wellen in ihrem Scheitel gebildeten Zylinderlinsen bezeichnet. Der richtige Abstand wird am einfachsten durch Probieren gefunden.

Filmaufnahmen mit dem Schattenschlierenverfahren werden zweckmäßig nicht durch Filmen des Schirms in einer Apparatur nach Fig. 10a durchgeführt,

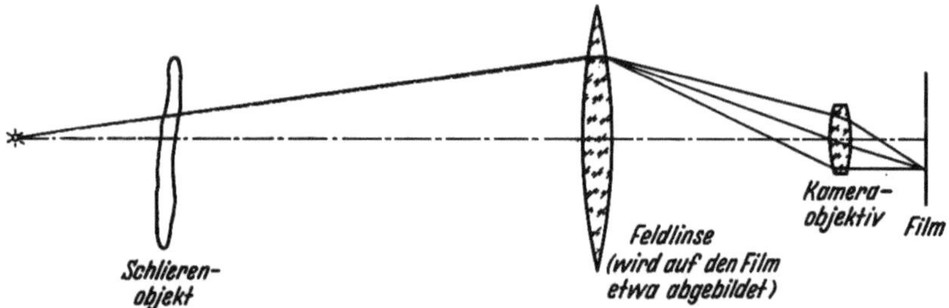

Fig. 10c. Schattenschlierenverfahren mit Feldlinse und Kamera.

da die Bildhelligkeit dazu oft nicht ausreicht. Besser bringt man an die Stelle des Schirms eine große Feldlinse oder einen Hohlspiegel, die das Licht in das Kameraobjektiv werfen (s. Fig. 10c). Die Kamera wird auf die Feldoptik scharf eingestellt, nicht auf das Objekt. Zwar muß nun die Feldoptik sehr groß sein, jedenfalls größer als das Objekt; aber im Gegensatz zu den abbildenden Verfahren mit Schlierenkopf kann die Feldoptik schlecht sein; es kann sich um eine unkorrigierte Einzellinse oder um einen Scheinwerferspiegel handeln. Je schlechter freilich die Feldoptik ist, desto größer muß die Öffnung des Kameraobjektivs sein, damit das ganze Bild der Lichtquelle, so sehr es auch durch die Feldlinse und die Schlieren verzerrt ist, noch in die Öffnung des Kameraobjektivs fällt.

Das Schattenschlierenverfahren eignet sich wegen seiner großen Empfindlichkeit auch zur Prüfung von polierten spiegelnden Oberflächen. Eine sehr kleine Lichtquelle, z. B. durch Abbildung einer Punktlichtlampe mit Hilfe eines umgekehrt benutzten Mikroskopobjektivs realisiert, strahlt über die zu prüfende Spiegelfläche auf einen Beobachtungsschirm. Ungleichmäßige Intensitätsverteilung zeigt die Schlieren des Objekts, d.h. die Polierfehler an (Verfahren von KÖHLER und KRAFT[1]).

b) Zweidimensionale Kennzeichnung der Ablenkung bei zweidimensionalem Objektfeld.

8. Unstetige zweidimensionale Verfahren. Der Mangel des Schattenschlierenverfahrens liegt darin, daß keine unmittelbare Zuordnung zwischen den Ablenkungen und den sie verursachenden Objektstellen erkennbar ist. Er läßt sich dadurch beheben, daß unter Verzicht auf die stetige Abbildung viele Einzelstrahlen von der Lichtquelle ausgeschickt werden, die auf dem lichtempfindlichen Papier, das als Auffangschirm dient, Schwärzungspunkte hervorrufen. Wird das mit und ohne Schlierenobjekt nacheinander bei dem gleichen lichtempfindlichen Papier durchgeführt, so gibt der Abstand je zweier zueinander gehöriger Punkte nach Betrag und Richtung an, wie der von dem betrachteten Strahl getroffene Objektpunkt das Licht ablenkt. Das Verfahren wurde von HEIDENHAIN erdacht[2].

Die Frage, ob das Verfahren zu einem stetigen abgewandelt werden kann, ist Gegenstand der nächsten Ziffer.

9. Zweidimensionale abbildende Farbschlierenverfahren. Die oben beschriebenen abbildenden Schlierenverfahren geben nur eine eindimensionale Kennzeichnung der Ablenkung. Einige kennzeichnen wahlweise die Komponente α_x oder nach Drehung der Blenden um 90° auch α_y; andere geben den Ablenkbetrag $\sqrt{\tan^2\alpha_x + \tan^2\alpha_y}$. Die Ablenkungen, die jede Objektstelle grundsätzlich verursachen kann, bilden eine zweidimensionale Mannigfaltigkeit entsprechend dem zweidimensionalen Kontinuum, in dem der Ablenkvektor $\mathfrak{c}$ nach Festlegung seines Betrages auf 1 noch variieren kann.

Daß die eindimensionale Grauskala nicht zu einer eineindeutigen stetigen Kennzeichnung der zweidimensionalen Ablenkungsmannigfaltigkeit ausreichen kann, ist evident auf Grund eines mathematischen Satzes. Dieser besagt, daß eine zweidimensionale Mannigfaltigkeit nicht eineindeutig und stetig auf eine eindimensionale Mannigfaltigkeit „abgebildet" werden kann. Auch das sog. „Diagonalverfahren" bewirkt lediglich eine unstetige „Abbildung" und ist deshalb für die physikalische Meßtechnik oft unbrauchbar, da wegen der Unstetigkeit beliebig kleine Meßfehler noch begrenzte (nicht mit den Meßfehlern gegen Null gehende) Fehler des Resultats bewirken.

Zur zweidimensionalen stetigen Kennzeichnung der Lichtablenkungen sind daher die Farben unerläßlich. Ein Farbschlierenverfahren[3], das in einem Momentbild beide Dimensionen erfaßt, ist in Fig. 11a skizziert. Die Anordnung entspricht weitgehend der nach Fig. 1; jedoch ist die Punktblende durch eine Kreisöffnung ersetzt, und als Schlierenblende dient eine „Farbsektorplatte", die auf durchsichtiger Unterlage sektorförmige Farbschichten trägt der Reihe nach in den

---

[1] G. SCHMALTZ: Technische Oberflächenkunde, S. 89. Berlin 1936.
[2] H. SCHARDIN: Erg. exakt. Naturwiss. **20**, 362 (1942).
[3] D.P. Nr. 842856, Kl. 42h, Gr. 34,11 vom 14. 4. 1949.

Farben des Farbenkreises (s. Fig. 11b) Purpur, Rot, Orange, Gelb, Grün, Blau, Indigo, Violett.

Auf diese Platte ist die gleichmäßig ausgeleuchtete Kreisscheibe der Lichtquellenblende abgebildet, und zwar für schlierenfreie Gesichtsfeldteile zentrisch,

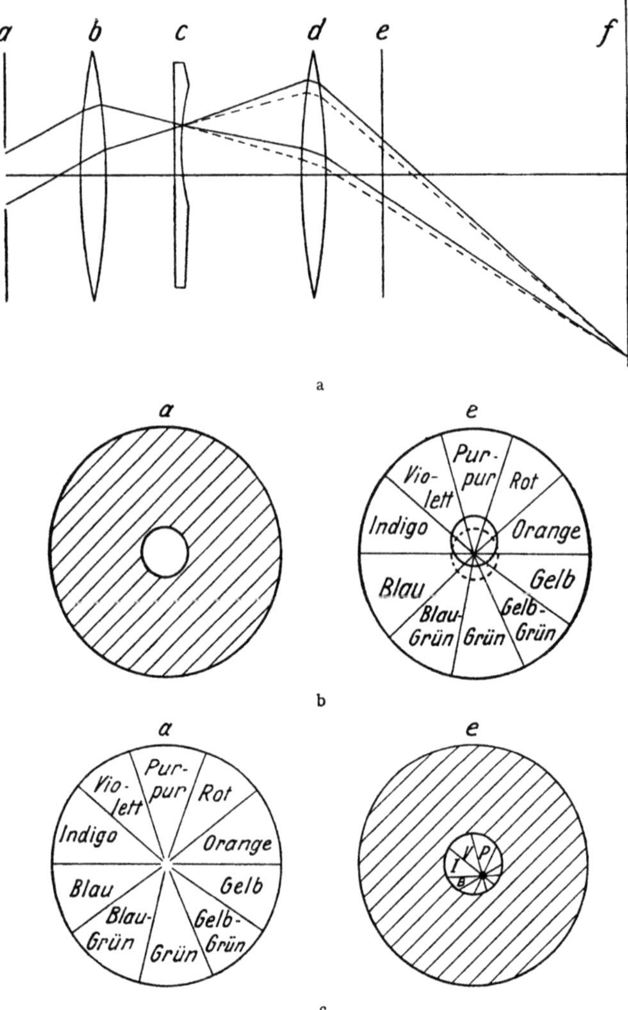

Fig. 11a—c. a Anordnung bei den zweidimensionalen Farbschlierenverfahren. b Blende und Farbsektorscheibe, in die Zeichenebene umgeklappt bei dem Verfahren erster Art. c Blende und Farbsektorscheibe, in die Zeichenebene umgeklappt bei dem Verfahren zweiter Art.

wie der gestrichelte Kreis in Fig. 11b das anzeigt. Schlierenfreie Gesichtsfeldteile erscheinen daher in dem Grau, das als Gemisch gleicher Anteile aller Farben des richtig aufgebauten Farbenkreises entsteht. Lenkt aber eine Objektstelle das Licht nach oben (Fig. 11a und b) ab, so gehört ihr ein nach oben, also zum Purpur auf der Farbsektorplatte verschobener Kreis zu; diese Objektstelle ist daher im Bilde durch eine Purpurfärbung gekennzeichnet, und zwar durch eine um so gesättigtere, je stärker die Ablenkung ist. Der Sättigungsgrad kennzeichnet den Ablenkbetrag,

$$\text{Farbsättigung} \leftrightarrow \sqrt{\tan^2 \alpha_x + \tan^2 \alpha_y},$$

der Farbton (rot, grün usw.) die Ablenkrichtung

$$\text{Farbton} \leftrightarrow \frac{\tan \alpha_y}{\tan \alpha_x}.$$

Eine nach unten ablenkende Objektstelle erscheint im Bilde grün angefärbt, komplementär zum Purpur der nach oben ablenkenden Stellen. Orangegelb zeigt Ablenkung nach rechts, indigoblau Ablenkung nach links usw.

Die Farbfig. 54 am Ende dieses Artikels gibt in Fig. b ein zweidimensionales Farbschlierenbild dieser Art, und zwar eine mikroskopische Momentaufnahme von Kaliumsulfatkristallen, die in ihrer Mutterlauge wachsen. Jedes abbildende Schlierenverfahren läßt sich mühelos auf das Mikroskop übertragen. Die in Fig. 1 und 11a mit Kondensor bezeichnete Optik ist dann durch den Mikroskopkondensor repräsentiert, und das Mikroskopobjektiv übernimmt die Rolle des Objektivs. Die Schlierenblende ist in der beobachterseitigen Objektivbrennebene anzubringen, in der sich bei vielen Mikroobjektiven ohnehin eine Aperturblende befindet.

Das Töplersche Schlierenverfahren hat Töpler selbst auf das Mikroskop übertragen, indem er die „Schrägbeleuchtung" benutzte, bei der das direkte Bild der Lichtquellenblende auf den Rand der Aperturblende fällt. Die Aperturblende wirkt dann als Schlierenschneide.

Wegen der weitgehenden Vertauschbarkeit von Lichtquellenblende und Schlierenblende kann das zweidimensionale Farbschlierenverfahren auch so betrieben werden, daß die Farbsektorplatte vor die ausgedehnte Lichtquelle tritt und eine Blende mit Kreisöffnung in die beobachterseitige Objektivbrennebene gesetzt wird (Verfahren 2. Art nach Fig. 11c). Diese Lösung stellt geringere Anforderungen an die optische Güte der Farbsektorplatte, begrenzt aber durch die Kreisblende zugleich die Objektivapertur, wenn man nicht mit voller Objektivapertur arbeitet und dann geringere Farbsättigung in Kauf nimmt. Die erste Lösung hat auch den Vorteil, daß die Beleuchtungsapertur etwas kleiner als die Beobachtungsapertur ist und deshalb unerwünschtes Streulicht in geringem Maße das Bild fälscht[1].

c) Zweidimensionale Kennzeichnung und Registrierung der Ablenkung bei eindimensionalem Objektfeld.

10. Fig. 12 zeigt einen Teil der Apparatur nach Fig. 5. Doch fehlt das Objektiv; dementsprechend findet keine Abbildung des Objekts auf die Beobachtungsebene statt. Diese liegt vielmehr unmittelbar in der zu den Richtungen dualen Ebene, so daß zwischen den Richtungen α_x, α_y, unter denen die Strahlen das Objekt verlassen und den Punkten ξ, η der Beobachtungsebene die eineindeutige Beziehung besteht

$$\tan \alpha_x = \frac{\xi}{f}; \quad \tan \alpha_y = \frac{\eta}{f}. \tag{10.1}$$

Die in der ξ, η-Ebene entstehende Lichtverteilung gibt also durch ihre Intensität eine Statistik der im Objekt vorkommenden Ablenkungen.

Die oft unübersichtliche Lichtverteilung nimmt die Gestalt einer besser zu deutenden Kurve an, wenn nur eine eindimensionale Mannigfaltigkeit von Objektstellen jeweils zur Zeit untersucht wird, also z. B. das wirksame Objektfeld durch einen Spalt unmittelbar vor dem Objekt begrenzt wird. Die Zuordnung zwischen den Richtungen und den Objektstellen, von denen sie stammen, geschieht z. B.

---

[1] Weitere Einzelheiten und Farbschlierenaufnahmen bringen die Originalarbeiten H. Wolter: Ann. Physik (6) **8**, 1 (1950). — Bild u. Ton **11** u. **12**, 4 (1950).

durch Vorsetzen eines keilförmigen Interferenzfilters vor das Objekt. Jeder Objektstelle kommt damit eine andere Farbe zu, in deren Licht die ihr zugehörige Ablenkstelle erscheint. Statt dessen kann man bei Objekten kontinuierlicher Ablenkänderung auch vor das Objekt ein Gitter mit senkrecht zum Spalt laufenden Stegen setzen; es versieht die Ablenkkurve mit Markierungen zur Identifizierung der Objektorte. Statt eines geraden Spaltes kann gelegentlich auch ein Ringspalt benutzt werden, wenn z.B. ein Ringgebiet des Objekts interessiert.

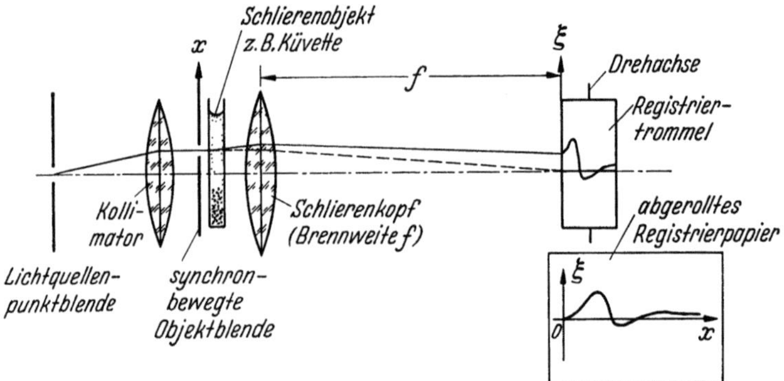

Fig. 12. Ablenkregistrierung mit Registriertrommel und synchron bewegtem Objektspalt.

d) Eindimensionale Kennzeichnung der Ablenkung bei eindimensionalem Objekt.

11. Ablenkregistrierung mit synchron bewegtem Objekt und Film. In vielen Fällen interessiert bei einem Objekt nur die Strukturänderung längs eines Geradenstückes.

Ein Beispiel hierfür ist eine Cuvette, in der Konzentrationsgradienten nur in der x-Richtung vorliegen (Fig. 12). Praktisch wichtig sind solche Fälle bei Diffusionsuntersuchungen und bei der Überprüfung der Sedimentation in der Ultrazentrifuge. Ein anderes Beispiel ist die Luft unter einem Stück Heizrohr, dessen Achse in Längsrichtung zu den Lichtstrahlen liegt, also in Fig. 12 in der optischen Achse liegend gedacht sei.

Im Beispiel der Cuvette treten dann überhaupt nur Ablenkungen α_x auf, es ist $\alpha_y \equiv 0$, und die „Kurve der Ablenkpunkte" in der ξ, η-Ebene ist ein Geradenstück in der ξ-Achse. Bei dem Heizrohr sei durch einen Spalt vor dem Objekt mit Längsausdehnung in der x-Richtung ein radiales Rohrgebiet ausgewählt; dann ist auch hier $\alpha_y \equiv 0$, und die Ablenkkurve in der ξ, η-Ebene liegt ganz in der ξ-Achse. Wir sprechen in beiden Beispielen von einem eindimensionalen Objekt und eindimensionaler Ablenkung.

Die Ablenkung $\xi(x) = f \cdot \tan \alpha_x(x)$ kann dann als Funktion des Objektorts registriert werden, wenn senkrecht zur ξ-Achse eine filmbespannte Registriertrommel läuft, während synchron (z.B. durch Seilzug vermittelt) eine spaltförmige Objektblende in x-Richtung dicht vor der Cuvette oder dem Heizrohr usw. entlanggezogen wird. Handelt es sich um bewegbare Objekte, die eine Erschütterung bei Bewegung auch vertragen (Maschinenglasplatten z.B.), so kann natürlich das Objekt selbst in x-Richtung synchron zur Registriertrommel verschoben werden, während die Objektblende fest bleibt. Dann können die Optiken der Apparatur ziemlich kleine Öffnungen haben.

Erweitert man die Lichtquellenpunktblende zu einem in y-Richtung (zum Beschauer hin in Fig. 12) ausgedehnten Spalt, so wird die Lichtstrecke in der ξ, η-Ebene in η-Richtung zu einem Lichtband ausgezogen. Nun können α_x-Komponenten auch dann registriert werden, wenn α_y nicht verschwindet; vor der Registriertrommel ist dann noch ein Spalt in ξ-Richtung anzubringen. Lichtstärker wird die Apparatur durch Einfügen einer Zylinderlinse, die das in η-Richtung ausgedehnte Lichtband wieder auf die ξ-Achse zusammenzieht, in ξ-Richtung selbst aber keinen wesentlichen Einfluß nimmt.

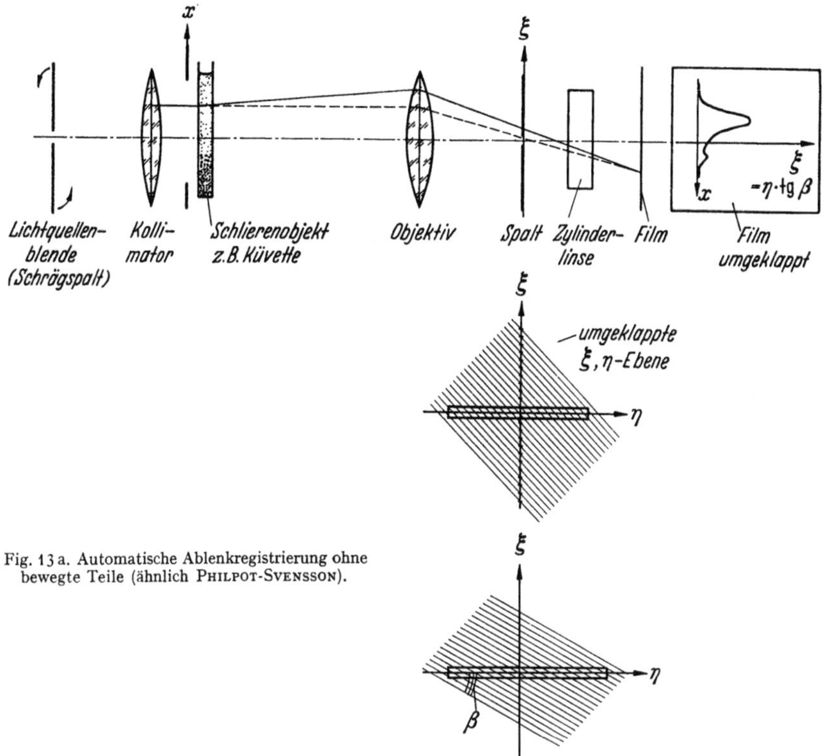

Fig. 13 a. Automatische Ablenkregistrierung ohne bewegte Teile (ähnlich PHILPOT-SVENSSON).

12. Automatische Ablenkregistrierung ohne bewegliche Teile. Dieselbe automatische Darstellung der Ablenkung $\xi = f \cdot \tan \alpha_x(x)$ als Funktion von x, wie die Apparatur nach Fig. 12 sie auf dem Registrierfilm gibt, läßt sich ohne bewegte Teile gewinnen. Die synchron bewegte Objektblende (Fig. 12) wird entfernt; vor dem Objekt befinde sich aber, wie in Ziff. 11 diskutiert, nötigenfalls der feste Objektspalt mit seiner Ausdehnung in x-Richtung.

Die Objektpunkte in der x-Achse und die Ablenkpunkte in der (dazu parallelen) ξ-Achse sollen nun auf denselben Film abgebildet werden. Durch Kombination einer Zylinderlinse und einer gewöhnlichen Linse ist das prinzipiell möglich. Um dabei eine graphische Darstellung der Funktion $\xi(x)$ in Cartesischen Koordinaten zu erhalten, muß freilich die ξ-Strecke noch um 90° gedreht werden. Dazu kann z. B. ein verdrilltes spiegelndes Stahlband benutzt werden, wie H. SCHARDIN es in einem solchen Falle anwendet. Eine sehr elegante Lösung geht im wesentlichen auf SVENSSON zurück. Die Lichtquellenpunktblende wird durch einen Spalt ersetzt, der unter 45° zur x-Achse geneigt ist. Die Ablenkstrecke, die vorher in der ξ-Achse lag (Fig. 13 a) wird dadurch zu einem Lichtband ausgezogen, das nun längs ξ-Achse und η-Achse gleichartigen Verlauf zeigt. In die ξ, η-Ebene wird ein Spalt gelegt mit seiner Ausdehnung in η-Richtung. Damit ist das Ziel - eine senkrecht zur Objektstrecke stehende Ablenkstrecke — erreicht. Die Zylinderlinse bildet nun die horizontale Punktreihe der η-Achse auf den Film ab, während sie die Abbildung der vertikalen Objektpunktreihe durch das Objektiv nicht stört. Freilich zieht die Zylinderlinse jeden Punkt $(x; 0)$ des Objekts zu einer Strecke in y-Richtung auseinander; auf ihr ist jedoch nur der Punkt hell, dessen η im Spalt der ξ, η-Ebene selbst hell ist, da in y-Richtung ja die Punkte des η-Spaltes durch die Zylinderlinsenabbildung getrennt nebeneinander abgebildet sind. Das Ergebnis sind Kurven, wie Fig. 13b und 22 einige zeigen. Die störenden Beugungserscheinungen rühren daher, daß die Objektabbildung durch eine schmale Blende hindurch geschehen muß. Dieses Beugungsproblem, das ganz allgemein bei Schlierenapparaturen auftritt, wird im Abschnitt III betrachtet. Svensson[1] legt den Lichtquellenspalt in y-Richtung und stellt den Schlierenspalt schräg; das ist der beschriebenen Anordnung nahezu äquivalent. Legt man den

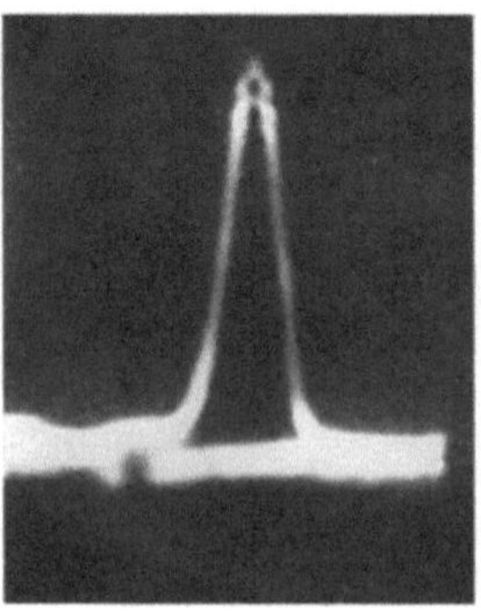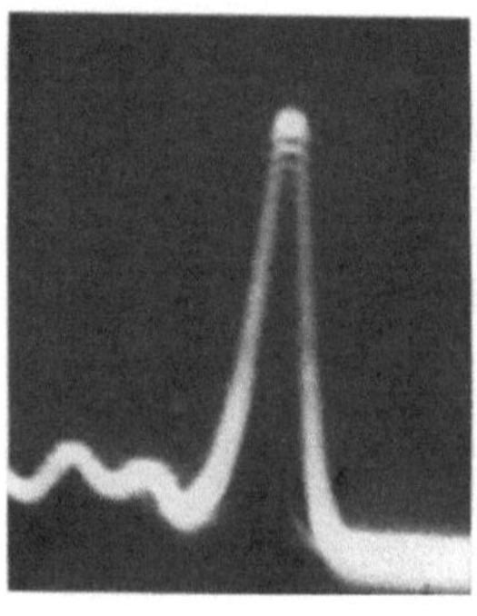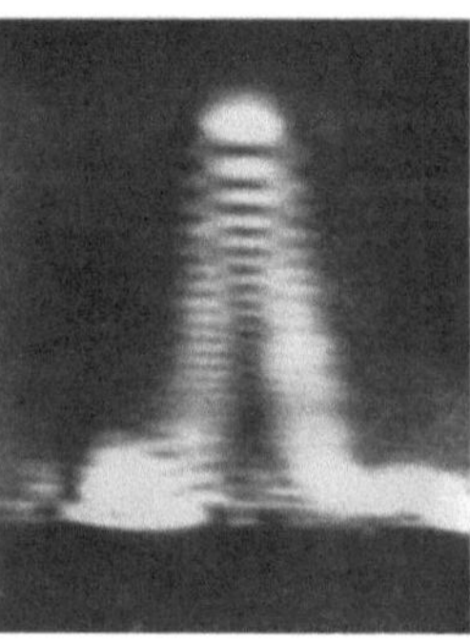

Fig. 13b. Philpot-Svensson-Registrierkurven. (Nach Antweiler und Kayser.) Links normale Spaltbreite, rechts zu schmaler Spalt in der Richtungsauswahlebene. Die Zwiebelform der Spitze in der linken Kurve beruht ebenso auf Beugung wie die Auffiederung der rechten Kurven.

Lichtquellenspalt unter einem Winkel $\beta < 45°$ zur y-Richtung, so wird die Ablenkpunktreihe ξ nicht kongruent in die η-Achse übertragen sondern noch um den Faktor $\cot \beta$ überhöht. Die Empfindlichkeit wird gesteigert, zugleich der Meßbereich verkleinert. Durch Drehen des Lichtquellenspaltes kann der Meßbereich dem Bedarf des jeweils untersuchten Objekts in gewissen Grenzen angepaßt werden.

Statt des Spaltes in der ξ, η-Ebene kann man andere Schlierenblenden, einen Draht z.B., eine Schneide verwenden (s. Teil III). Das Verfahren ist auch in anderer Hinsicht mannigfach variierbar, z.B. infolge Austauschbarkeit der Rollen von Lichtquellenblende und Schlierenblende. In dieser Hinsicht sei auf die Originalliteratur[2] und bezüglich der älteren Verfahren auch auf den mehrfach genannten zusammenfassenden Bericht von H. Schardin[3] hingewiesen.

II. Zum quantitativen Zusammenhang zwischen Lichtablenkungen und ihren physikalischen Ursachen.

a) Schlüsse von der Lichtablenkung auf Form und Brechungsindex der Objekte.

13. Formfehler eines Objekts als Grund der Lichtablenkung. Eine fast senkrecht vom Licht getroffene prismatische Platte vom Brechungsindex n mit sehr

---

[1] H. Svensson: Kolloid-Z. **87**, 181 (1939).

[2] O. Wiener: Wied. Ann. **49**, 105 (1893). — I. Thovert: Ann. de Phys. **2**, 369 (1914). — O. Lamm: Nature, Lond. **132**, 820 (1933). — G. Philpot: Nature, Lond. **141**, 283 (1938). — Svensson: Kolloid-Z. **87**, 181 (1939). — O. Armbruster, W. Kossel u. K. Strohmaier: Z. Naturforsch. **6a**, 510 (1951).

[3] H. Schardin: Erg. exakt. Naturwiss. **20**, 302 (1942), vgl. besonders S. 364.

kleinem brechendem Winkel χ lenkt das Licht bekanntlich um den Winkel

$$\sqrt{\alpha_x^2 + \alpha_y^2} = \alpha = |\mathfrak{e} - \mathfrak{e}_0| = (n-1)\,\chi \tag{13.1}$$

ab. χ ist dabei der Dickengradient dem Betrage nach. Da auch die Richtung der Ablenkung $\mathfrak{e} - \mathfrak{e}_0$ dem Gradienten entspricht, gilt für den Gradienten der Dicke D

$$\operatorname{grad} D = \frac{1}{n-1}\,(\mathfrak{e} - \mathfrak{e}_0); \quad \text{d.h.} \quad \frac{\partial D}{\partial x} = \frac{\alpha_x}{n-1}\,; \quad \frac{\partial D}{\partial y} = \frac{\alpha_y}{n-1}\,. \tag{13.2}$$

Hat man für zwei Punkte $P_1(x_1;\,y_1)$ und $P_2(x_2;\,y_2)$ die Gradienten auf irgendeinem Wege W zwischen P_1 und P_2 gemessen, so erhält man wegen

$$dD = \frac{\partial D}{\partial x}\,dx + \frac{\partial D}{\partial y}\,dy$$

die Dickendifferenz der Platte an beiden Punkten gleich dem Linienintegral

$$\left.\begin{aligned}
D(P_2) - D(P_1) &= \frac{1}{n-1}\int\limits_{(W)} (\mathfrak{e} - \mathfrak{e}_0)\,d\mathfrak{s} \\
&= \frac{1}{n-1}\left\{\int\limits_{(W)}\alpha_x\,dx + \int\limits_{(W)}\alpha_y\,dy\right\}.
\end{aligned}\right\} \tag{13.3}$$

Analog gilt für Reflexion an einer Fläche, die am Orte x, y um das kleine Stück D von der x, y-Ebene abweicht, daß die Lichtablenkung doppelt so groß ist wie die Flächennormalenneigung, also wie der „Dickengradient":

$$\operatorname{grad} D = \frac{\mathfrak{e} - \mathfrak{e}_0}{2}\,; \quad \frac{\partial D}{\partial x} = \frac{\alpha_x}{2}\,; \quad \frac{\partial D}{\partial y} = \frac{\alpha_y}{2}\,. \tag{13.4}$$

Daher folgt die Differenz der Abweichungen an zwei Orten P_1 und P_2

$$\left.\begin{aligned}
D(P_2) - D(P_1) &= \tfrac{1}{2}\int\limits_{(W)} (\mathfrak{e} - \mathfrak{e}_0)\,d\mathfrak{s} \\
&= \tfrac{1}{2}\left\{\int\limits_{(W)}\alpha_x\,dx + \int\limits_{(W)}\alpha_y\,dy\right\}
\end{aligned}\right\} \tag{13.5}$$

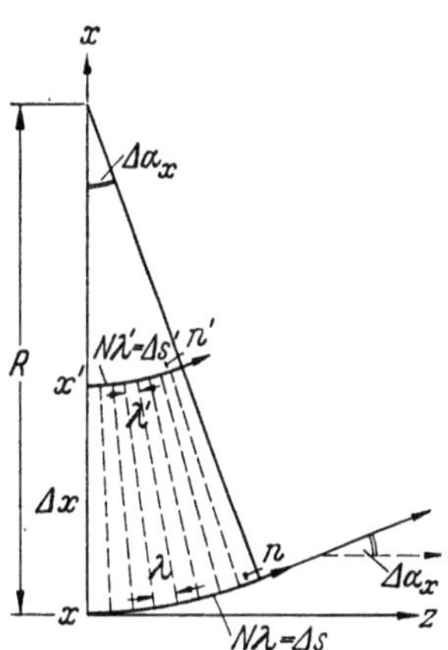

Fig. 14. Zur Ablenkung des Lichtes im inhomogenen Medium.

über einen beliebigen Weg W von P_1 nach P_2. Das läßt sich auf Hohlspiegel übertragen, wenn als Normalrichtung $\mathfrak{e}_0$ die Sollrichtung definiert wird und α_x, α_y gegen sie gemessen werden. D bedeutet dann die Abweichung der Fläche von der Sollform. Entsprechend sind die Formeln (13.1) bis (13.3) für Linsen verwendbar.

14. Der Zusammenhang zwischen der Lichtablenkung und dem relativen Gradienten des Brechungsindexes. Das Objekt habe konstante Dicke, aber kontinuierlich von Ort zu Ort variablen Brechungsindex n.

Analog den Bezeichnungen früherer Abbildungen ist in Fig. 14 die z-Richtung als Strahlrichtung $\mathfrak{e}_0$ gedacht. Alle Ablenkungen

$$\mathfrak{e} - \mathfrak{e}_0 = (\alpha_x;\,\alpha_y;\,0) \tag{14.1}$$

seien klein.

Wir wollen dann für einen betrachteten Objektpunkt P_0 ein Achsensystem x, y, z so legen, daß der Brechungsindex in der Umgebung des Punktes P_0 sich in y-Richtung nicht ändert (d.h. $\operatorname{grad}_y n = 0$). Dann hat der Lichtstrahl aus

Symmetriegründen auch keinen Anlaß zu einer Ablenkung zur positiven oder negativen y-Richtung; es ist also der Beitrag, den die betrachtete kleine Umgebung unseres Objektpunktes zur Ablenkung in y-Richtung liefert,

$$\varDelta \alpha_y = 0 = \operatorname{grad}_y n. \tag{14.2}$$

In x-Richtung kann der eventuell gekrümmte Strahl bei hinreichend kleiner Umgebung des Punktes P_0 jedenfalls als kreisbogenförmig gekrümmt angesehen werden.

Als Strahl bezeichnen wir dabei ein Wellenbündel der Breite $\varDelta x$, dessen Wellennormalen gestrichelt in Fig. 14 angedeutet sind. Die Bogenlängen des oberen Strahlrandes $\varDelta s'$ und des unteren $\varDelta s$ stehen in einfachen Beziehungen zur (gleichen!) Anzahl N der auf sie entfallenden Wellenlängen und den Brechungsindices am oberen und am unteren Strahlrande

$$\left.\begin{aligned}
\varDelta s &= N \cdot \lambda = N \cdot \frac{\lambda_0}{n}, \\
\varDelta s' &= N \cdot \lambda' = N \cdot \frac{\lambda_0}{n'} = \varDelta s \cdot \frac{n}{n'}.
\end{aligned}\right\} \tag{14.3}$$

Andererseits stehen diese Bogenlängen auch zum Krümmungsradius R und dem Ablenkwinkel $\varDelta\alpha_x$ in Beziehung:

$$\left.\begin{aligned}
\varDelta s &= R \cdot \varDelta\alpha_x \\
\varDelta s' &= (R - \varDelta x)\,\varDelta\alpha_x
\end{aligned}\right\} \quad \text{d.h.} \quad \varDelta s - \varDelta s' = \varDelta x\,\varDelta\alpha_x. \tag{14.4}$$

Daher folgt

$$\varDelta\alpha_x = \frac{1}{\varDelta x}\left(1 - \frac{n}{n'}\right)\varDelta s = \frac{\varDelta n}{\varDelta x}\,\frac{1}{n'}\,\varDelta s. \tag{14.5}$$

Für hinreichend kleine Gebiete ist also

$$\varDelta\,(e - e_0)_x = \varDelta e_x = \varDelta\alpha_x = \frac{\partial n}{\partial x}\cdot\frac{1}{n}\cdot\varDelta s, \tag{14.6}$$

$$\varDelta e_x = \frac{1}{n}\cdot\operatorname{grad}_x n\cdot\varDelta s. \tag{14.7}$$

Gl. (14.2) besagt entsprechend trivialerweise

$$\varDelta e_y = 0 = \frac{1}{n}\cdot\operatorname{grad}_y n\cdot\varDelta s. \tag{14.8}$$

Wegen der Transformationseigenschaft dieser Vektoren (Kovarianz von Vektorgleichungen) gilt daher für jedes Koordinatensystem, das aus dem bisher speziell gewählten durch Drehung um die z-Achse hervorgeht,

$$\left.\begin{aligned}
\varDelta\alpha_x &= \varDelta e_x = \frac{1}{n}\cdot\operatorname{grad}_x n\cdot\varDelta s, \\
\varDelta\alpha_y &= \varDelta e_y = \frac{1}{n}\cdot\operatorname{grad}_y n\cdot\varDelta s.
\end{aligned}\right\} \tag{14.9}$$

Da der Lichtweg nahezu in z-Richtung verläuft, folgt durch Integration über die gesamte Objektdicke in erster Näherung:

$$\alpha_x = \int \frac{1}{n}\,\frac{\partial n}{\partial x}\,dz; \qquad \alpha_y = \int \frac{1}{n}\,\frac{\partial n}{\partial y}\,dz. \tag{14.10}$$

15. Integration. Die Integrale der Gl. (14.10) sind trivial auszuführen bei Objekten, die über eine Dicke d in der z-Richtung identische Eigenschaften haben. Dann ist einfach

$$\alpha_x = \frac{1}{n} \cdot \frac{\partial n}{\partial x} \cdot d; \qquad \alpha_y = \frac{1}{n} \cdot \frac{\partial n}{\partial y} \cdot d. \qquad (15.1)$$

Im Beispiel der Sedimentation in einer planparallel begrenzten Cuvette kann durch eine zweite Integration n selbst gewonnen werden, wenn wie in Fig. 13b die Ablenkung $\xi = f \cdot \alpha_{x,\,\text{gem}}$ als Funktion von x gemessen vorliegt. Jedoch muß die Brechung bei dem Cuvettenaustritt berücksichtigt werden; denn durch Brechung wird dort jedes α_x um den Faktor n/n_L bei Übergang in Luft (Brechungsindex n_L) vergrößert. Das heißt in die Gl. (15.1) ist als α_x einzusetzen

$$\alpha_x = \frac{n_L}{n} \cdot \alpha_{x,\,\text{gem}} = \frac{n_L}{n} \cdot \frac{\xi(x)}{f}. \qquad (15.2)$$

Da $\dfrac{\partial n}{\partial y} \equiv 0$ ist, ist $\dfrac{\partial n}{\partial x} = \dfrac{dn}{dx}$ und

$$n_L \cdot \frac{\xi(x)}{f \cdot d} = \frac{dn}{dx}; \qquad (15.3)$$

d. h.

$$n - n_0 = n_L \cdot \frac{1}{f \cdot d} \cdot \int_{x_0}^{x} \xi(x')\, dx'. \qquad (15.4)$$

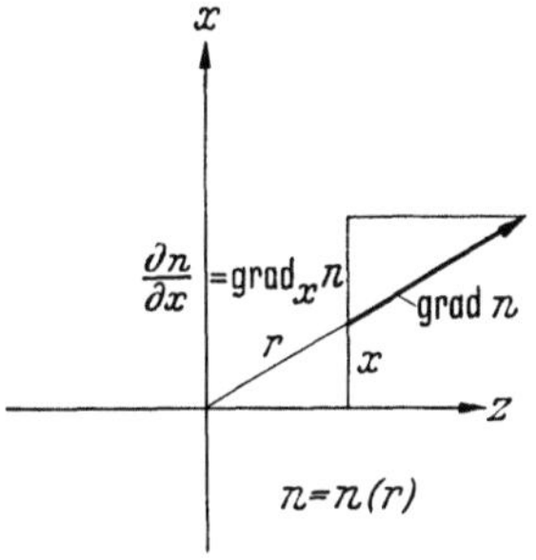

Fig. 15. Zum zylindersymmetrischen Medium.

Der Brechungsindex an irgendeinem Orte x_0 muß bekannt sein (z. B. durch Probeentnahme und Messung nach bekannten Verfahren). Dann kann aus (15.4) für jeden Objektort der Brechungsindex berechnet werden und nach den im folgenden diskutierten Zusammenhängen auch die Konzentration, in anderen Fällen die Dichte, der Druck oder die Temperatur (im wesentlichen mit Hilfe der Dispersionsformel oder aus einem experimentell zuvor ermittelten funktionalen Zusammenhang zwischen den genannten Größen und dem Brechungsindex).

Ähnlich wie Gl. (13.3) folgt hier

$$\frac{d \cdot (n_P - n_0)}{n_L} = \int_{P_0}^{P} \alpha_{x,\,\text{gem}}\, dx + \int_{P_0}^{P} \alpha_{y,\,\text{gem}}\, dy \qquad (15.4')$$

für zweidimensionale Objekte, die in Strahlrichtung keine Brechungsindexänderung im Bereich der konstanten Dicke d haben.

Auch für andere einfach aufgebaute Objekte läßt sich der Brechungsindex als Ortsfunktion gewinnen, z. B. bei Zylindersymmetrie mit einer Zylinderachse in y-Richtung. n sei also eine reine Funktion $n = n(r)$ des Radius $r = \sqrt{x^2 + z^2}$. Dann (s. Fig. 15) ist

$$\frac{\partial n}{\partial x} = \frac{dn}{dr} \cdot \frac{x}{r} \qquad (15.5)$$

und nach Gl. (14.10) für einen Lichtstrahl, der im Abstande x an der Objektachse vorbeizielt, ist

$$\alpha_x = x \cdot \int \frac{1}{n} \frac{dn}{dr} \frac{1}{r}\, dz. \qquad (15.6)$$

Wegen $r^2 = x^2 + z^2$ ist $r\, dr = z\, dz$, also

$$\frac{1}{x}\, \alpha_x(x) = 2 \cdot \int_{x}^{\infty} \frac{d}{dr} (\log n) \frac{dr}{\sqrt{r^2 - x^2}}. \qquad (15.7)$$

Das ist eine Integralgleichung erster Art für die Funktion $\frac{d}{dr}(\log n)$ von r und den Kern $(r^2-x^2)^{-\frac{1}{2}}$. Sie läßt sich durch die Substitution $r^2=1/v$; $x^2=1/u$ auf die Abelsche Form

$$G(u) = \int_0^u \frac{\varphi(v)}{\sqrt{u-v}}\, dv \qquad (15.8)$$

bringen mit den Bedeutungen für G und φ

$$G(u) = \alpha_x\left(\sqrt{\tfrac{1}{u}}\right); \qquad f(r) = \frac{d}{dr}(\log n); \qquad \varphi(v) = \frac{1}{v}f\left(\sqrt{\tfrac{1}{v}}\right). \qquad (15.9)$$

Wenn $\alpha_x(\infty)=0$ ist, so ist die Lösung[1]

$$\varphi(v) = \frac{1}{\pi}\int_0^v \frac{1}{\sqrt{v-u}}\,\frac{d}{du}\,\alpha_x\left(\sqrt{\tfrac{1}{u}}\right)du. \qquad (15.10)$$

Nochmalige Integration liefert

$$n(r) = n(\infty)\cdot\exp\left(-\int_0^{\sqrt{\tfrac{1}{r}}} \frac{\varphi(v)}{2\cdot\sqrt{v}}\,dv\right). \qquad (15.11)$$

Die Integrationen nach (15.10) sind am einfachsten mit dem Planimeter auszuführen; für das reine Bilden der Stammfunktion in (15.11) empfiehlt sich das Auszählen von Quadratmillimetern. Dieses Verfahren lohnt sich vor allem, wenn $\alpha_x(x)$ etwa mit der bei graphischen Verfahren üblichen Genauigkeit bekannt ist. Sonst ist auch H. Schardins Näherungsverfahren sehr zweckmäßig[2].

Statt über v und u zu integrieren, kann man auch die Substitution nach Lösung der Integralgleichung wieder rückgängig machen und findet dann

$$n(r) = n(\infty)\cdot\exp\left(\frac{1}{\pi}\int_r^\infty \frac{1}{\varrho}\int_\varrho^\infty \frac{x}{\sqrt{x^2-\varrho^2}}\,\alpha_x'(x)\,dx\,d\varrho\right); \qquad (15.12)$$

$\alpha_x'(x)$ bezeichnet darin den Differentialquotienten von $\alpha_x(x)$. Doch bewährte sich das Vorgehen nach (15.10) und (15.11) in der Praxis besser, da alle Integrationsgrenzen endlich sind und daher das lästige Abschätzen von „Flächenschwänzchen'' fortfällt.

b) Schlüsse vom Brechungsindex auf Dichte, Konzentration, Druck oder Temperatur.

16. Für Schlüsse vom Brechungsindex eines Stoffes oder Stoffgemisches auf Dichte, Zusammensetzung, Druck, Temperatur oder andere physikalische Daten geben empirisch gewonnene Beziehungen die sicherste Grundlage. In vielen Fällen, vor allem bei Flüssigkeiten und Gasen, folgt die Beziehung der Dispersionsgleichung

$$R = \frac{n^2-1}{n^2+2} = \sum_i C_i\, a_i(\lambda). \qquad (16.1)$$

---

[1] Siehe z.B. W. Magnus u. F. Oberhettinger: Formeln und Sätze für die speziellen Funktionen der mathematischen Physik, S. 141. Berlin 1943.

[2] H. Schardin: Erg. exakt. Naturwiss. **20**, 370 (1942).

Darin heiße R die Refraktion; C_i bedeutet die Zahl der Gramm von der i-ten Komponente im Liter des Gemisches; $a_i(\lambda)$ sind Funktionen allein der Wellenlänge und der Natur der i-ten Komponente. Die $a_i(\lambda)$ sind von der Konzentration, dem Druck, der Temperatur usw. in weiten Grenzen unabhängig.

Für einen Stoff konstanter relativer Zusammensetzung (also z. B. eine Flüssigkeit oder ein Gas), der lediglich als Ganzes einem anderen Druck oder einer anderen Temperatur ausgesetzt wird und dabei seine Dichte ϱ ändert, ist daher die Refraktion proportional der Dichte

$$R \sim \varrho. \tag{16.2}$$

Für Gase mit $n - 1 \ll 1$, die der Gasgleichung idealer Gase

$$p \cdot \frac{1}{\varrho} \sim T \tag{16.3}$$

annähernd genügen, ist

$$R = \frac{(n-1)(n+1)}{n^2 + 2} \approx (n-1)\frac{2}{3} \sim \varrho \sim \frac{p}{T}. \tag{16.4}$$

Hat man die Proportionalitätskonstante durch einen Versuch bestimmt (z. B. nach JAMIN), so kann beispielsweise die Temperatur vermittels der Gleichung

$$T = \text{Konst.} \frac{p}{n-1} \tag{16.5}$$

aus dem Druck und dem Brechungsindex berechnet werden. So hat SPERLING[1] den Temperaturverlauf im Kohlebogen ($p = 1\,\text{Atm} = \text{Konst.}$) im isobaren Fall gemessen.

Im isothermen Fall ist einfach $p \sim n - 1$, und bei den häufigen adiabatischen Zustandsänderungen (Schall u. ä.) ist $p \sim \varrho^\varkappa$ also nach (16.4)

$$p = \text{Konstante} \cdot (n-1)^\varkappa. \tag{16.6}$$

Liegt ein Stoffgemisch vor, bei dem nur die Konzentration C_1 einer Komponente wesentlich variiert, so ist nach (16.1)

$$R \approx C_1 a_1 + b; \quad (b = \text{Konst.}). \tag{16.7}$$

Die Konzentration kann daher aus dem Brechungsindex nach

$$C_1 \approx \frac{1}{a_1}\left(\frac{n^2 - 1}{n^2 + 2} - b\right) \tag{16.8}$$

berechnet werden. So läßt sich also z. B. die Sedimentation in der Ultrazentrifuge quantitativ verfolgen oder auch die Diffusion, wenn der Brechungsindex jeweils durch Schlierenmessungen auf eine der oben besprochenen Arten bestimmt wird.

III. Die Beugungsunschärfe bei Schlierenverfahren und ihre Herabsetzung durch Minimumstrahlkennzeichnung.

a) Beugungserscheinungen bei abbildenden Schlierenverfahren als Meßfehlergrenzen.

17. Die Unschärfebedingung der Optik und ihre Bedeutung für die Grenze der Strahldefinition bei abbildenden Schlierenverfahren. Die Aufgabe der abbildenden Schlierenverfahren, die Lichtablenkung $\alpha_x(x; y)$; $\alpha_y(x; y)$ als Funktion des

---

[1] J. SPERLING: Z. Physik **128**, 269 (1950).

Objektortes x, y zu messen, scheint zunächst nicht mit jeder beliebigen Genauigkeit lösbar zu sein; denn die Unschärfebedingung für Wellen[1]

$$\Delta x \cdot \Delta \alpha_x \approx \lambda \; \Big\} $$
$$\Delta y \cdot \Delta \alpha_y \approx \lambda \; \Big\} \qquad\qquad (17.1)$$

besagt, daß ein Wellenbündel, das eine Richtungsunschärfe $\Delta \alpha_x$ hat, auf seine Herkunft optimal mit der Koordinatenunschärfe Δx untersucht werden kann. Der Gehalt dieser Formulierung ist der bekannte Sachverhalt, den wir mit dem Wort „Beugung" bezeichnen.

Freilich waren die klassischen Beugungsbeziehungen, die als Vorläufer dieser Unschärfebedingungen aufzufassen sind, zunächst nur an Beispielen gewonnen. Nachdem sie aber zur Entstehung der Heisenbergschen Unschärfebedingung beigetragen haben, sind sie dann in der Form (17.1) in die Optik zurückgekehrt mit dem wohlbegründeten weitergehenden Anspruch auf grundsätzliche und allgemeine Gültigkeit. In ihnen sind daher auch bei den Schlierenverfahren die prinzipiellen Meßgenauigkeitsgrenzen für Koordinaten und Winkel aneinander geknüpft — jedoch nur in folgendem Sinn. Der Herkunftsort von Photonen und die Richtung, die sie von dort genommen haben, streuen bei Messung beider Größen stets so, daß $\Delta x \cdot \Delta \alpha_x \gtrsim \lambda$ ist. Dies sagt über die Meßfehlergrenzen nur dann Entsprechendes aus, wenn Ort und Richtung durch Photonenansammlungen, also durch möglichst hohe Konzentration von Licht gekennzeichnet werden.

Kennzeichnet man dagegen Richtungen z.B. durch Flächen, in denen die Lichtintensität Null ist, so gilt für deren Doppelwertsbreite, d.h. für die Streuung der Winkelmessung überhaupt keine prinzipielle untere Grenze, wenn man nur hinreichend viele Photonen zur Ausmittelung der statistischen Streuung zur Verfügung hat, also lange genug belichten kann. Das ist analog zu der Tatsache, daß man den Aufenthaltsort eines Elektrons im Atom für einen bestimmten Zustand nicht beliebig genau angeben kann, wohl aber die Knotenflächen, auf denen mit Sicherheit nie ein Elektron gefunden wird.

Dieses Prinzip der Minimumstrahlkennzeichnung[2] führt gegenüber der Maximumstrahlkennzeichnung zu einem Gewinn g von der Größenordnung 10 bis 100; denn in der Praxis findet auch die Doppelwertsbreite einer „Nullfläche" eine untere Grenze — verursacht durch Streulicht und andere Fehler der benutzten Optiken. Allgemeiner muß man daher die optimale Unschärfe durch

$$\Delta x \cdot \Delta \alpha_x \approx \frac{\lambda}{g} \qquad\qquad (17.2)$$

beschreiben mit $g = 1$ bei Maximumstrahlkennzeichnung und äquivalenten Verfahren (unter anderen Philpot-Svensson, Töpler, Lamm) und z.B. $g \approx 25$ bei der Minimumstrahlkennzeichnung, wie sie unten ab S. 582 beschrieben ist.

Besonders anschaulich ist das Zustandekommen der Unschärfe durch Beugung bei dem Verfahren nach Ziff. 11 und Fig. 12. Der bewegliche Objektspalt bestimmt durch seine Breite Δx die Definitionsgenauigkeit der Koordinate. Zugleich engt er das Lichtbündel in x-Richtung so ein, daß in der ξ, η-Ebene ein durch Beugung in ξ-Richtung verbreiterter Lichtfleck entsteht. Die Halbwertsbreite ist bei dieser Maximumstrahlkennzeichnung durch

$$\frac{\Delta \xi}{f} \approx \frac{\lambda}{\Delta x} \qquad\qquad (17.3)$$

d.h. durch (17.1) gegeben.

---

[1] H. Wolter: Ann. Physik (6) **7**, 341 (1950).
[2] H. Wolter: Habil.-Schr. Kiel, Jan. 1949. — Phys. Bl. **5**, 234 (1949).

Bei dem PHILPOT-SVENSSONschen Verfahren entstehen die automatischen Registrierkurven, z.B. Fig. 13b, wie oben beschrieben, durch Abbildung des eindimensionalen Objekts in der x-Richtung — es hat nur Strukturen in dieser — und durch gleichzeitige seitliche Registrierung der Ablenkung. Dabei wird das Objekt selbst abgebildet durch den schmalen, die Ablenkrichtung α_x kennzeichnenden Spalt hindurch. Er begrenzt um der Richtungskennzeichnung willen zwangsläufig die zur Objektabbildung verfügbare Bündelbreite und erzeugt durch diesen Eingriff die Beugungsunschärfe. Je schmaler der Spalt, desto krasser die Beugung (rechtes Bild). Links ein Bild mit einem Spalt optimaler Breite.

Anders liegen die Dinge, wenn nur die Ablenkkomponente α_y als Funktion von x gemessen werden soll. Dann ist nur ein Richtungsspalt mit Längsausdehnung in x-Richtung erforderlich. Er engt das abbildende Bündel nur in y-Richtung ein, nimmt aber keinen Einfluß auf die Abbildungsschärfe für x-Koordinaten. α_y und x lassen sich also unabhängig voneinander beliebig genau messen.

Der Fehler am Brechungsindexgradienten $\Delta\dfrac{\partial n}{\partial x}$ und der Fehler Δx der Ortszuordnung sind z.B. bei der Cüvette nach den Gln. (17.2), (15.1) und (15.2) bei $n_L \approx 1$ verknüpft durch

$$\Delta x \cdot \Delta \frac{\partial n}{\partial x} \approx \frac{\lambda}{d \cdot g} \tag{17.4}$$

($d = $ Cüvettendicke in Lichtstrahlrichtung; $g = $ Gewinn des Verfahrens). Da nach Gl. (15.4) dann

$$d \cdot \{n(x_e) - n(x_0)\} = \int_{x_0}^{x_e} \alpha_{x,\,\mathrm{gem}}(x)\,dx, \tag{17.5}$$

ist die Auswirkung der Unschärfe auf den Meßfehler am Brechungsindex $n(x_e)$ bei bekanntem festem $n(x_0)$ wie folgt zu berechnen. Durch Variationsbildung an (17.5) folgt

$$\left.\begin{aligned}
d \cdot \Delta n(x_e) &\approx \Delta \int_{x_0}^{x_e} \alpha_{x,\,\mathrm{gem}}(x)\,dx \\
&\approx |x_e - x_0| \cdot \Delta\alpha_x + S(\alpha_x) \cdot \Delta x.
\end{aligned}\right\} \tag{17.6}$$

$S(\alpha_x)$ bedeutet darin die Schwankungssumme der Funktion $\alpha_x(x)$; d.h.

$$S(\alpha_x) = \lim_{\delta - 0} \sum_{\nu=0}^{n} |\alpha_{x,\,\mathrm{gem}}(x_\nu) - \alpha_{x,\,\mathrm{gem}}(x_{\nu-1})| \tag{17.7}$$

bei Intervallteilung $x_0, x_1, x_2, \ldots x_n = x_e$ mit $|x_\nu - x_{\nu-1}| < \delta$. Enthält die Cüvette nur zwei wesentlich verschiedene Gebiete mit monotonem Brechungsindexübergang, so hat $\alpha_x(x)$ nur ein einziges Maximum und die Schwankungssumme ist gleich dem Doppelten dieses Maximums. Dann folgt

$$\left.\begin{aligned}
d \cdot \Delta n(x_e) &\approx |x_e - x_0|\,\Delta\alpha_x + 2 \cdot \alpha_{x,\,\mathrm{max}}\,\Delta x \\
&\approx 2\alpha_{x,\,\mathrm{max}}\,\Delta x + |x_e - x_0|\,\frac{\lambda}{g}\,\frac{1}{\Delta x}.
\end{aligned}\right\} \tag{17.8}$$

Eine Funktion von der Form der rechten Seite hat bekanntlich ein Minimum, wenn Δx so bemessen wird, daß beide Summanden gleich sind. Macht man deshalb optimal

$$\Delta x = \sqrt{\frac{|x_e - x_0|}{2\alpha_{x,\,\mathrm{max}}} \cdot \frac{\lambda}{g}}\;; \qquad \Delta\alpha_x = \sqrt{\frac{2\alpha_{x,\,\mathrm{max}}}{|x_e - x_0|} \cdot \frac{\lambda}{g}}, \tag{17.9}$$

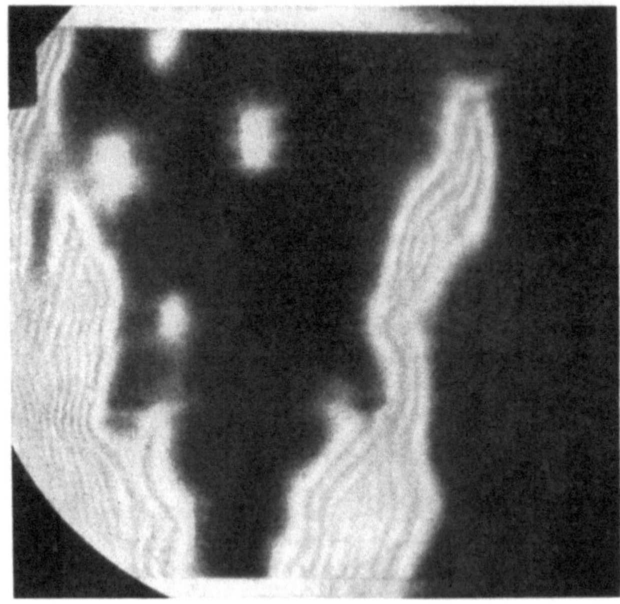

a

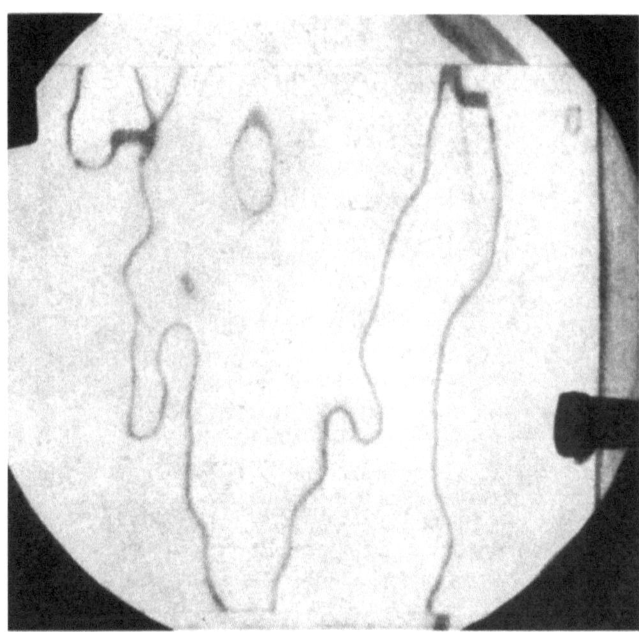

b

Fig. 16a u. b. Schlierenaufnahme einer Glasplatte, a nach TOEPLERs Verfahren, b nach dem Minimumschlierenverfahren bei optimaler Belichtungszeit (Minimum im Graubereich).

so wird

$$d \cdot \Delta n \left(x_e\right) \approx \sqrt{8 \cdot \left| x_e - x_0 \right| \cdot \alpha_{x,\max} \cdot \frac{\lambda}{g}} \,. \tag{17.10}$$

Die Zahl 8 im Radikanden spielt freilich die Rolle eines Vorsichtsfaktors; die Addition in (17.8) setzt den unglücklichsten „Zufall" der Fehlerverteilung voraus, ebenso die Bildung der Schwankungssumme. Wenn man deshalb die 8 im Radikanden fortläßt, wird man „meistens" noch den Tatsachen gerecht. Vorausgesetzt ist aber jedenfalls, daß die prinzipiell mögliche — nur durch Beugung bedingte — Grenze erreicht und die Unschärfe annähernd optimal nach (17.9) auf Koordinate und Winkel verteilt ist.

Die Gl. (17.10) bedeutet, daß bei Maximumstrahlkennzeichnung mit wesentlich größerer Auswirkung der Meßfehler auf den berechneten Brechungsindex gerechnet werden muß, als es nach den verbreiteten Literaturangaben erscheint. Nur dann, wenn man nicht die *Meßfehlergrenzen*, sondern die *Empfindlichkeitsgrenze* meint, darf man $|x_e - x_0|$ mit Δx und α_x mit $\Delta\alpha_x$ in Gl. (17.10) identifizieren. Das gibt die Empfindlichkeitsgrenze[1]

$$d \cdot \Delta n \approx \lambda \cdot \sqrt{\frac{8}{g}}.$$

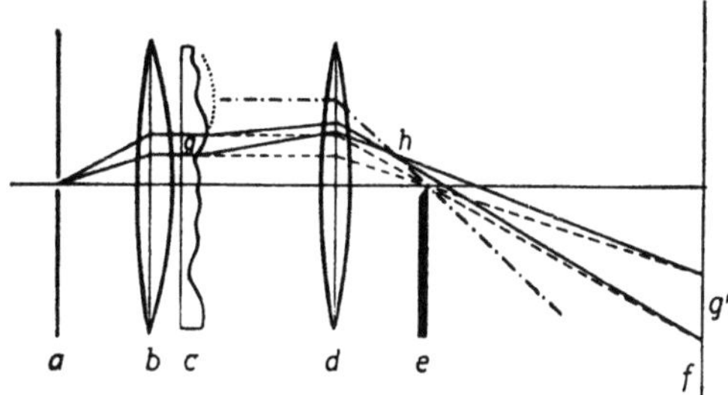
Fig. 17. Zur Wirkungsweise der abbildenden Schlierenverfahren (zweite Näherung), gezeichnet für ein Schneidenverfahren.

18. Deutung der Beugungserscheinungen durch eine zweite Näherung bei der Betrachtung der abbildenden Schlierenverfahren (Beugung an der Schlierenblende). Auch bei den Schlierenverfahren mit zweidimensionaler Objektabbildung ist die Beugung infolge des *Eingriffs*, den die Schlierenblende vornimmt, wirksam. Die Fig. 16a gibt eine Schlierenaufnahme wieder, die mit einer Apparatur nach Fig. 5 gemacht wurde. Objekt war eine leicht schlierige Glasplatte. Die Orte mit der Ablenkung $\alpha_x = 0$ sind als Helldunkelgrenze angezeigt, da das Verfahren als Schlierenblende eine Schneide benutzte. Aber auch diese Grenze ist durch Beugung unscharf, und zwar hat sie die Struktur, die von der FRESNELschen Beugung an der Schirmkante her bekannt ist.

Die Entstehung dieser Beugungserscheinung sei an Fig. 17 erläutert, die eine Apparatur von derselben Art wie Fig. 1 darstellt. Um etwas Bestimmtes vor Augen zu haben, stellen wir uns ein Schlierenobjekt vor, das etwa eine Platte homogenen Glases sei und nur an einer seiner Begrenzungsflächen Abweichungen von der Ebene zeigt.

Die Oberfläche des Schlierenobjekts werde in so kleine Flächenelemente eingeteilt, daß innerhalb

eines jeden die Krümmung in der Zeichenebene praktisch konstant ist. Doch seien die Flächenelemente noch groß gegenüber der Auflösungsgrenze, und die Abbildung der verschiedenen Flächenelemente sei als unabhängig voneinander anzusehen.

Die zweite Voraussetzung hat diese Betrachtung mit der elementarsten gemeinsam; nach dieser wurde jedes Oberflächenelement durch eine Ebene angenähert nach Art einer Näherung erster Ordnung, während die Berücksichtigung der Krümmung die neue Darstellung als Näherung zweiter Ordnung charakterisiert.

___

[1] Vgl. G. HANSEN: Zeiss-Nachr. **3**, 302 (1940). — H. J. ANTWEILER: Mikrochemie **36** 36 (1951). — H. SVENSSON: Kolloid-Z. **87**, 181 (1939); **90**, 141 (1940). — Ark. Kemi. Mineral Geol. A **22**, 10 (1946). — H. LABBARD u. H. STAUB: Helv. chim. Acta **30**, 1954 (1947).

Ist g ein Flächenelement auf dem Schlierenobjekt, so geben die gestrichelten Linien in Fig. 17 die Hauptstrahlenkonstruktion des zu g gehörigen Bildelements g' in der Beobachtungsebene f. Die tatsächlich die Abbildung vermittelnden Strahlen (ausgezogene Linien in Fig. 17) verlaufen im allgemeinen anders; man konstruiert sie, indem man die Oberfläche von g zu einer vollständigen Linse ergänzt (punktiert) und beachtet, daß der strichpunktierte Hauptstrahl ein geometrischer Ort für den resultierenden Brennpunkt h ist. h liegt vor der Ebene e, wenn das Element g sammelt, hinter ihr, wenn g eine „Zerstreuungslinse" ist.

Das Bildelement g' wird immer dann dunkel sein, wenn h unterhalb des von den Hauptstrahlen (gestrichelt) und dem Bildelement g' aufgespannten „Hauptstrahlenkegels" liegen würde. Liegt h oberhalb dieses Hauptstrahlenkegels, so erscheint das Bildelement hell. „Oben" und „unten" sind dabei in dem Sinne gebraucht, der durch die Lage der Schneide in Fig. 17 festgelegt ist.

Nur teilweise erhellt wird das Bildelement dann, wenn der Brennpunkt h in den Hauptstrahlenkegel fällt (dem wir natürlich auch seine rückwärtige Verlängerung, wie die gestrichelten Linien sie andeuten, hinzurechnen). Dann liegt die Schirmkante selbst in dem die Abbildung des Bildelements g vermittelnden Lichtstrahlenbündel, und zwar entweder nach Fig. 18a oder b. Die auf der Beobachtungsebene f dann innerhalb des Bildelements liegende Schattengrenze muß zwangsläufig die bekannte Fresnelsche Beugungserscheinung an der Schirmkante zeigen, wie Fig. 18a und b andeuten.

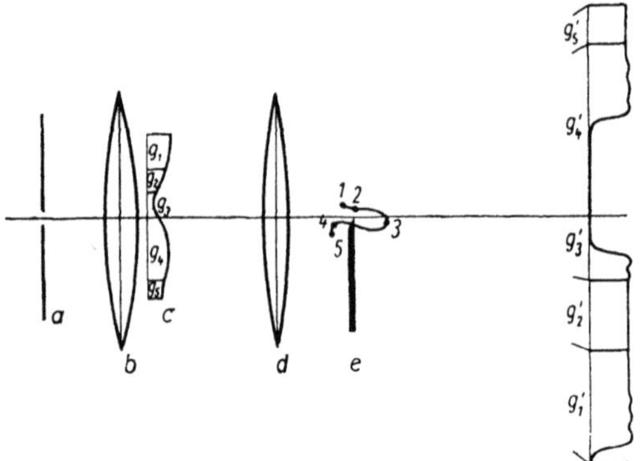

Fig. 19. Brennpunkte 1 bis 5 für ein in fünf Elemente geteiltes Schlierenobjekt und das durch die Lage der Brennpunkte bedingte Schlierenbild (zweite Näherung).

Fig. 19 zeigt für ein in 5 Flächenelemente aufgeteiltes Schlierenobjekt die Brennpunkte h_1 bis h_5 und das durch sie bedingte Schlierenbild. Es entspricht dem experimentellen Befund der Schlierenaufnahme Fig. 16a.

b) Die Minimumstrahldefinition und das abbildende Minimumschlierenverfahren.

19. Die Minimumstrahlkennzeichnung[1]. Wegen des durch Beugung unscharfen Übergangs zwischen Hell und Dunkel können die Kurven gleicher Ablenkung („Isokampten") am Töplerschen Schlierenbild nur durch Photometrie gewonnen werden. Man kann sie in Anlehnung an das Fresnelsche Beugungsbild der Schirmkante definieren als die Kurven, an denen die Helligkeit gleich einem Viertel der „Fernabintensität" ist. Aber die photographische Photometrie ist nur mit Genauigkeiten von einigen Prozenten durchführbar, und das wirkt sich in einer entsprechenden Ungenauigkeit der Isokamptenbestimmung aus.

Die Ursache liegt in der ungünstigen Lichtverteilung im Fresnelschen Schattenbild einer Schirmkante. Aber auch ein Spalt oder ein dünner Draht statt der Schirmkante, auch wenn man ihnen optimale Breite gibt, geben keine schärfere Kennzeichnung. Fig. 20 zeigt in logarithmischem Maßstab die Fresnelsche Lichtverteilung hinter der Schirmkante (a), hinter einem Spalt optimaler

---

[1] D.P. 819728, Kl. 42h, Gr. 34,11 vom 12. 4. 1949.

Breite (b), bei einem engeren (c) und weiteren Spalt (d). Fig. 20e gibt dasselbe bei einem Draht optimaler Dicke.

Eine schärfere Kennzeichnung gibt eine „lineare Nullstelle", die man im Beugungsbild einer 180°-Phasenplatte erhält. Diese wird an die Stelle der Schneide gesetzt und dreht das Licht der einen Halbebene um 180° gegenüber dem Licht der anderen Halbebene. Die Unschärfebedingung besagt nur etwas über die mögliche Energiebündelung und gilt daher nicht für die „Doppelwertsbreite" einer Nullstelle. Realisieren kann man eine solche Phasenplatte z. B. dadurch, daß man eine Planparallelplatte zur Hälfte mit einer geeigneten dünnen Zaponlackschicht oder dergleichen bedeckt[1]. Das FRESNELsche Beugungsbild einer solchen 180°-Phasenplatte hat eine Lichtverteilung nach Fig. 20f. Die Nullstelle hat eine prinzipiell unbegrenzte Schärfe; die praktische Grenze ist durch unvermeidbares Streulicht gegeben. Fig. 21 gibt die entsprechenden experimentellen Aufnahmen. Die Anordnung bei den Aufnahmen entsprach der Fig. 18a, und die gewählten Abstände　Lichtquelle — schattengebendes Mittel — Beobachtungsebene waren in allen Fällen gleich[2]. Der experimentelle Befund stimmt mit der theoretischen Kurve völlig überein. Auch die schwachen Nebenmaxima sind deutlich zu erkennen[3]. Bei der Aufnahme Fig. 21c wurden sie absichtlich überbelichtet, um den Graubereich in das Gebiet hoher Flankensteilheit der Kurve Fig. 20f zu bringen.

Die Minimumstrahlkennzeichnung ist eine Nullmethode; das macht wie bei anderen Nullmethoden (z. B. bei der WHEATSTONEschen Brücke) ein quantitatives Nachweismittel (hier die Photometrie) unnötig. Der Verwendbarkeit eines hochempfindlichen ungeeichten Instruments entspricht hier die Möglichkeit äußerster Überbelichtung, die das Minimum mit großer Schärfe erkennen läßt.

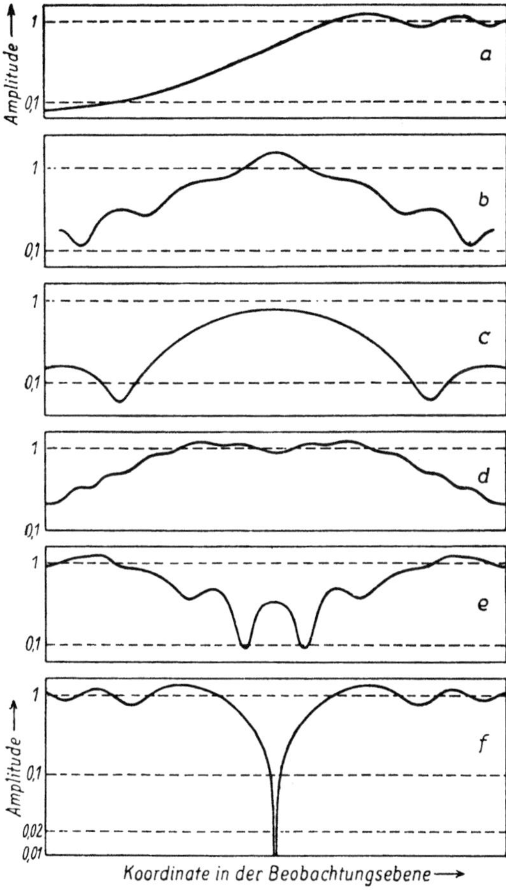

Fig. 20a—f. Theoretischer Amplitudenverlauf auf einer Beobachtungsebene im Lichte einer punktförmigen Lichtquelle. Zwischen Lichtquelle und Beobachtungsebene befindet sich bei a eine undurchlässige Halbebene (Schneide); b ein Spalt „optimaler" Breite; c ein Spalt von 40% der optimalen Breite; d ein Spalt von 167% der optimalen Breite; e ein Draht „optimaler Dicke"; f die 180°-Phasenplatte der Minimumstrahlkennzeichnung. Der logarithmische Maßstab ist dem Problem angemessen, da es stets auf Halbwertsbreiten bzw. Doppelwertbreiten, jedenfalls auf Amplitudenverhältnisse (oder Intensitätsverhältnisse) ankommt.

---

[1] Andere Realisierungen sind genannt bei K. STROHMAIER [Z. Naturforsch. 6a, 508 (1951)] und H. WOLTER [Ann. Physik (6) 7, 341 (1950)], ferner bei H. MOSER u. J. WITTMANN [Z. Physik 131, 48 (1951)].

[2] Bezüglich der Theorie siehe H. WOLTER: Ann. Physik (6) 7, 188 (1950).

[3] Eine besonders schöne Registrierkurve dieser Beugungserscheinung zeigen W. KOSSEL und K. STROHMAIER [Z. Naturforsch. 6a, 504 (1951)].

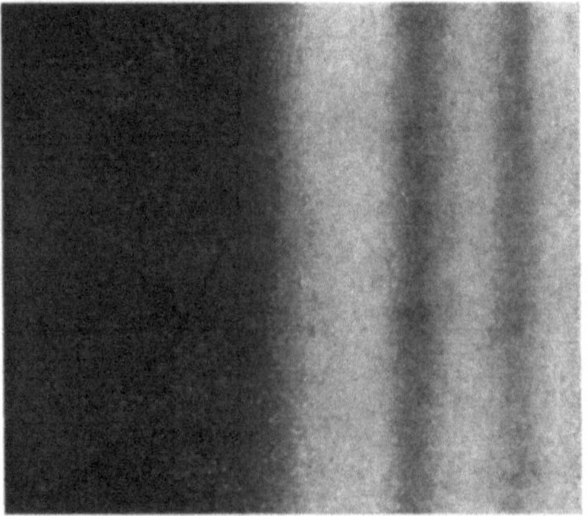

a

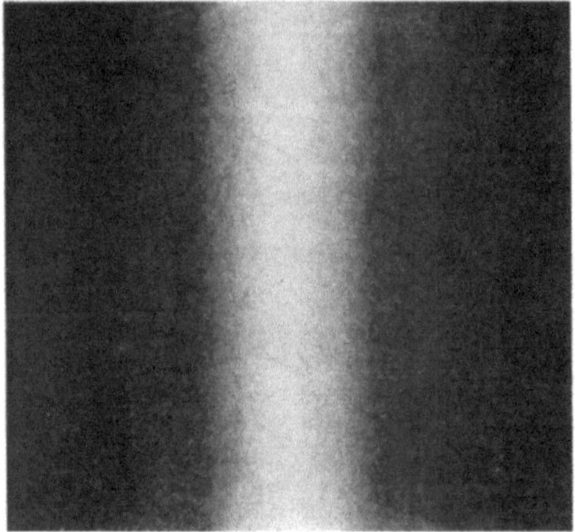

b

c

Fig. 21 a—c. Experimenteller Befund entsprechend der Fig. 20, a für die undurchlässige Halbebene; b für einen Spalt optimaler Breite; c für die 180°-Phasenplatte der Minimumstrahlkennzeichnung. (Alle Aufnahmen jeweils optimal belichtet, so daß die entscheidenden Flanken im Bereich größter Schwärzungssteilheit lagen.)

20. Das abbildende Minimumschlierenverfahren[1]. Setzt man die in Ziff. 19 genannte 180°-Phasenplatte statt der Schneide, des Spaltes, Drahtes usw. als „Schlierenblende" in eine abbildende Schlierenapparatur ein, so muß nun nach dem in Ziff. 19 beschriebenen Befund ein auf Null herabziehendes Minimum hoher Flankensteilheit dort entstehen, wo bei dem Schneidenverfahren die Schattengrenze lag. Die Fig. 16b gibt die mit diesem Verfahren hergestellte Schlierenaufnahme derselben Glasplatte, die in Fig. 16a mit der Schneide als Schlierenblende aufgenommen ist. Beide Aufnahmen entsprechen sich wie die Fig. 20a und b bzw. 21a und c.

Die einfache Theorie erster Näherung der Schlierenverfahren verlagert alle Brennpunkte 1 bis 5 der Fig. 19 in die Ebene e. Nach ihr könnte bei TÖPLERs Verfahren jedes Flächenelement also nur entweder hell oder dunkel abgebildet werden. Die Beugung wird völlig ignoriert, und das Minimumverfahren bliebe dort ganz unverständlich; es müßte alle Flächenelemente gleichmäßig hell abbilden, da die Phasenplatte ja nirgends absorbiert.

Die mit der Schirmkante nur unter Verwendung der Photometrie erreichbare Genauigkeit läßt sich mit dem Minimum ohne jede Photometrie übertreffen; aber es sei auch bemerkt, daß durch die Minimumstrahlkennzeichnung erst eine „Strahldefinition" geschaffen wird und daß eine exakte Auswertung der Ergebnisse ebenfalls nicht geometrisch-optisch geschehen dürfte.

Das gilt auch für die spezielle PHILPOT-SVENSSON-Form des Minimumschlierenverfahrens für automatische Registrierung eindimensionaler Ablenkung bei eindimensionalem Objekt. O. ARMBRUSTER, W. KOSSEL und K. STROHMAIER haben auch hier die 180°-Phasenplatte mit Erfolg eingesetzt.

---

[1] D.P. Nr. 819925, Kl. 42h, Gr. 34,11 vom 12. 4. 1949.

Die Fig. 22 zeigt ihre Registrierkurven im Vergleich zu den mit einem Spalt oder Steg gewonnenen (vgl. auch die mit Spalt hergestellten beugungsverfälschten PHILPOT-SVENSSON-Aufnahmen Fig. 13 b).

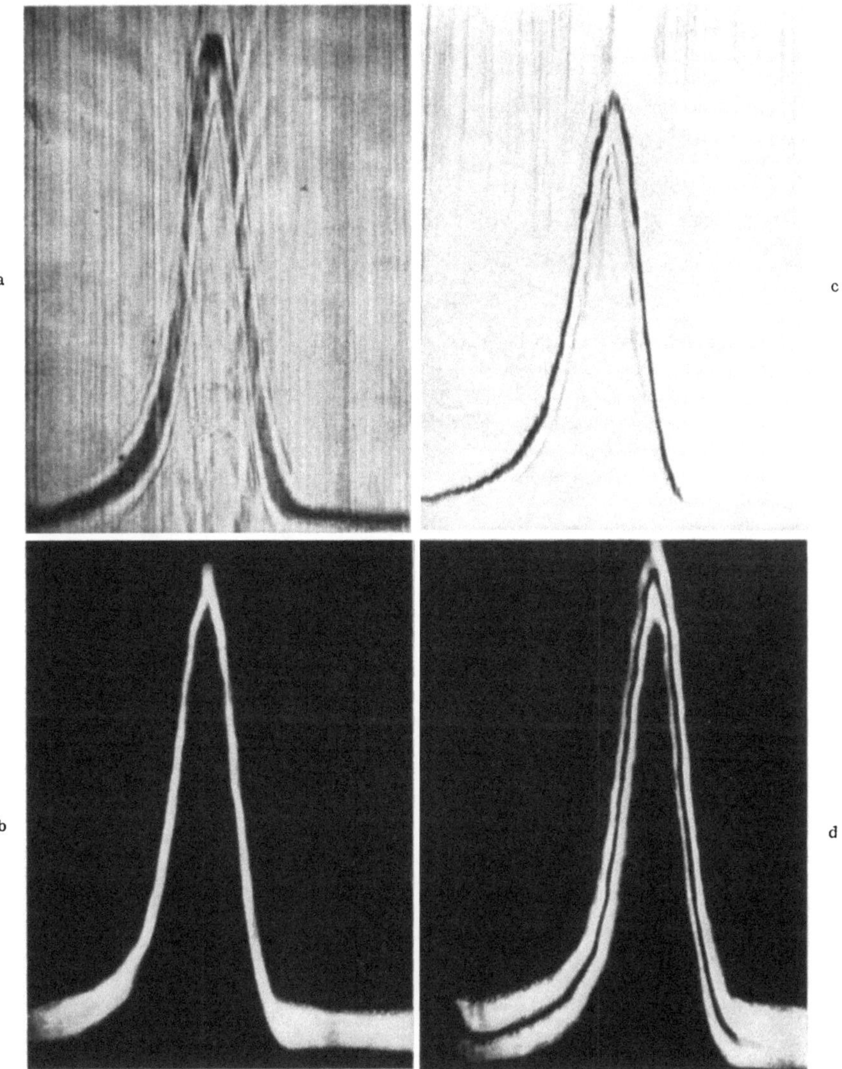

Fig. 22a—d. Vier PHILPOT-SVENSSON-Registrierkurven. [Nach O. ARMBRUSTER, W. KOSSEL und K. STROHMAIER: Z. Naturforsch. 5a, 510 (1951).] Schlierenblende war im Falle a ein Draht; b ein Spalt; c eine 180°-Phasenplatte; d ein Spalt, durch eine 180°-Phasenplatte in zwei gleich große gegenphasige Gebiete aufgeteilt.

c) Beugung und Minimummethode bei Schattenschlierenverfahren mit Ortskennzeichnung.

21. Die Beugung bei dem Schattenverfahren. Ein Objekt O wurde im Schattenwurf von einer als punktförmig geltenden Lichtquelle * aus auf einen Schirm S projiziert (Fig. 23). Um die Herkunft der Strahlen von den Objektpunkten trotz der Ablenkung durch Schlieren erkennbar zu machen, seien die Strahlen „markiert", oder z.B. die interessierenden Objektpunkte seien angefärbt.

Ist eine Objektstelle x auf den Schirm S mit der linearen Unschärfe $\Delta x'$ in x-Richtung abgebildet, so entspricht dem bei geometrisch-optischer Umdeutung im Objekt die Unschärfe Δx nach Fig. 23a:

$$\frac{\Delta x}{\Delta x'} = \frac{a}{a+b}. \tag{21.1}$$

Diese Unschärfe wird bewirkt durch Beugung, die nach der Unschärfebedingung (17.1)

$$\Delta \alpha_x \cdot \Delta x \geqq \lambda \tag{21.2}$$

und Fig. 23 sich so auswirkt, daß

$$\Delta \alpha_x = \frac{\Delta x'}{b}. \tag{21.3}$$

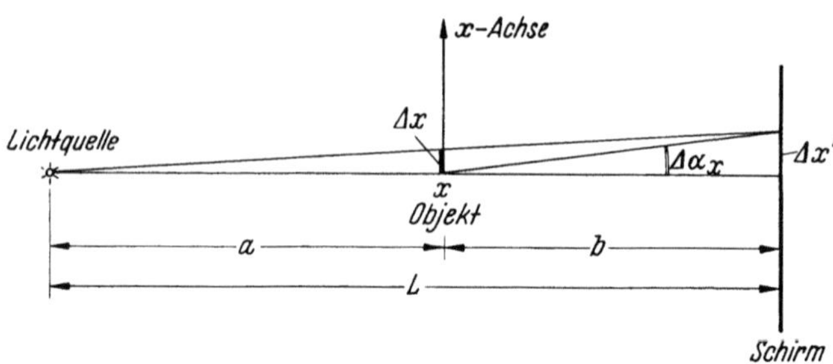

Fig. 23a. Die Beugung bei dem Schattenschlierenverfahren mit Ortsmarkierung.

Eliminiert man je zwei der Größen Δx, $\Delta x'$, $\Delta \alpha_x$, so erhält man

$$(\Delta x')^2 \geqq \lambda \cdot \frac{b}{a}\,(a+b), \tag{21.4}$$

$$(\Delta x)^2 \geqq \lambda \cdot \frac{a \cdot b}{a+b} = \lambda \cdot \frac{1}{\frac{1}{a}+\frac{1}{b}}, \tag{21.5}$$

$$(\Delta \alpha_x)^2 \geqq \lambda \cdot \frac{a+b}{a \cdot b} = \lambda\left(\frac{1}{a}+\frac{1}{b}\right). \tag{21.6}$$

Die beiden ersten Gleichungen sind aus der üblichen Fresnelschen Beugungsrechnung bekannt. Die beiden letzten zeigen, daß man die Koordinatenunschärfe und die Winkelunschärfe jede für sich beliebig wählen kann, wenn die Abstände a und b frei verfügbar sind; doch wird kleine Winkelunschärfe zwangsläufig mit großer Koordinatenunscharfe erkauft.

Ist die verfügbare Gesamtlänge $L = a + b$ begrenzt, z.B. infolge des begrenzten Raumes, so ist $(\Delta \alpha_x)^2$ auch dann nicht beliebig klein zu machen, wenn man auf die Unschärfe Δx keine Rücksicht zu nehmen braucht (z.B. dann, wenn α_x nur wenig ortsabhängig ist); denn nach Ersatz von a und b durch L und a besagt (21.6)

$$(\Delta \alpha_x)^2 \geqq \lambda \cdot \frac{L}{a(L-a)}, \tag{21.7}$$

daß ein Minimum für

$$0 = \frac{\partial}{\partial a}\left(a(L-a)\right) = L - 2a, \quad \text{d.h.} \quad a = \frac{L}{2} = b$$

vorliegt. Dieses Unschärfeminimum hat den Betrag

$$|\Delta\alpha_x| \geq 2\cdot\sqrt{\frac{\lambda}{L}}\,. \tag{21.8}$$

Die zugleich eintretende Koordinatenunschärfe ist

$$|\Delta x| \geq \tfrac{1}{2}\cdot\sqrt{\lambda\cdot L}\,. \tag{21.9}$$

Für ein praktisches Beispiel mit $\lambda = 0{,}5\ \mu$ und $L = 5$ m hat man also $|\Delta\alpha_x| \geq 2\cdot 10^{-3}$ und $|\Delta x| \geq 0{,}8$ mm. Bei $L = 20$ m wäre die Winkelunschärfe $1:1000$ und die Koordinatenunschärfe $1{,}6$ mm.

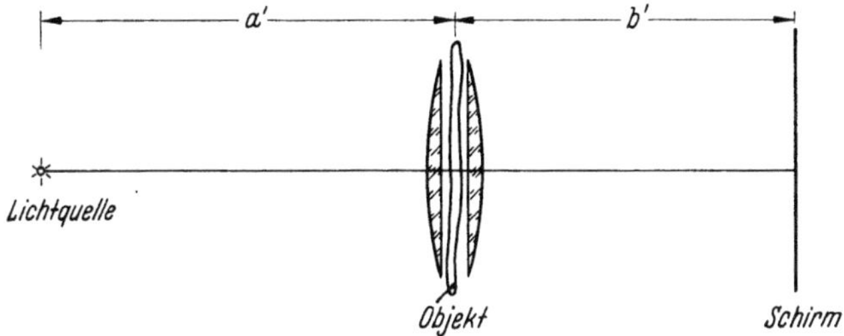

Fig. 23 b. Schattenschlierenverfahren mit zwei Hilfslinsen zur Abstandstransformation.

Man kann volle Freiheit in der Wahl der Länge L trotz begrenzter Untersuchungsräume gewinnen, wenn man die Längen a und b künstlich durch Einsetzen von guten Linsen dicht vor und hinter dem Objekt transformiert (vgl. Fig. 23 b). Sind die Brennweiten f_a und f_b, so ist der tatsächliche Abstand a' bzw. b' nach der Linsenformel zu reduzieren auf den effektiven Abstand a bzw. b:

$$\frac{1}{a} = \frac{1}{a'} - \frac{1}{f_a}; \quad \frac{1}{b} = \frac{1}{b'} - \frac{1}{f_b}, \tag{21.10}$$

d.h. es gilt dann statt (21.5) und (21.6)

$$|\Delta x|^2 \geq \lambda\cdot\frac{1}{\dfrac{1}{a'} + \dfrac{1}{b'} - \dfrac{1}{f_a} - \dfrac{1}{f_b}}, \tag{21.11}$$

$$|\Delta\alpha_x|^2 \geq \lambda\left(\frac{1}{a'} + \frac{1}{b'} - \frac{1}{f_a} - \frac{1}{f_b}\right). \tag{21.12}$$

Die effektiven Abstände sind trotz endlicher Abmessungen des Raumes nun unbegrenzt. Als Spezialfälle sind auch die in den Ziffn. 10 bis 12 behandelten Fälle $a' = f_a$ und $b' = f_b$ in dieser Beobachtung mit erfaßt.

Praktisch wichtig ist der Fall, bei dem das Objekt parallel angestrahlt wird, da dann dicke eindimensionale Objekte (Cuvetten) nur an Orten gleichen Brechungsindexgradienten von den Strahlen durchsetzt werden. Dann ist wegen $a' = f_a$ d.h. $a = \infty$:

$$|\Delta x|^2 \geq \lambda\cdot b, \tag{21.13}$$

$$|\Delta\alpha_x|^2 \geq \frac{\lambda}{b}\,. \tag{21.14}$$

Eine Erhöhung der „Strahlschärfe" bringt auch hier die Minimumstrahlkennzeichnung, die in Ziff. 22 an einem Beispiel betrachtet sei.

22. Die Minimum-Methode bei dem Schattenschlierenverfahren. Die Nullstelle hinter einer 180°-Phasenkante bleibt auf dem ganzen Wege des Lichts erhalten; sie kennzeichnet also eine Ebene im Raum mit prinzipiell unbegrenzter Schärfe. Wird diese „Nullebene" durch ein Schlierenobjekt geschickt, so markiert sie auf einem dahinter angebrachten Schirm die Ablenkung in Kurvenform[1,2].

Bedeutung hat das Verfahren vor allem bei eindimensionalem Objekt und eindimensionalen Ablenkungen. Man schickt z.B. analog zu dem alten Wienerschen Verfahren[3] die „Nullebene" schräg (z.B. unter 45°) durch eine Cuvette

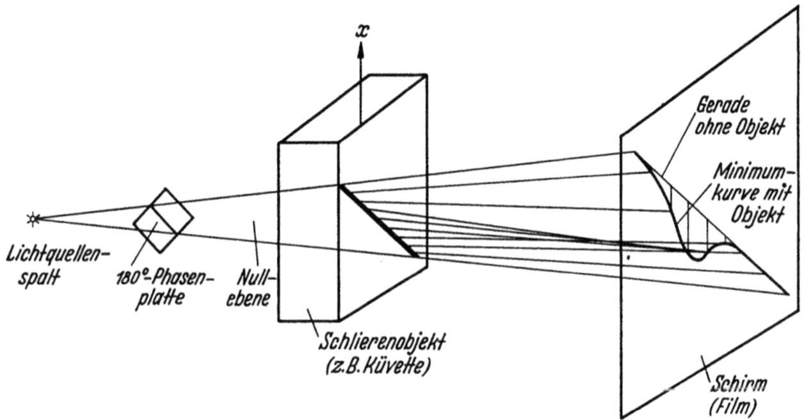

Fig. 24. Minimumstrahlkennzeichnung bei Wienerschem Verfahren (Schattenschlierenverfahren mit Markierung einer Geraden im Objekt).

(Fig. 24). Die dunkle Kurve auf hellem Grund gibt in schrägen Koordinaten direkt die Ablenkungen α_x als Funktion der Höhe x in der Cuvette.

Da die Minimumstrahlkennzeichnung völlig scharfe Minimumkurven auch ohne abbildende Linsen gibt, kann das Verfahren mit Vorteil für alle Zwecke benutzt werden, für die man die Philpot-Svenssonsche Apparatur verwendet.

Statt einer Nullebene kann man ein System von Nullebenen verwenden[2]. An die Stelle der 180°-Phasenplatte in Fig. 24 tritt dann ein Fresnelsches Biprisma. Kreuzt man zwei Biprismen oder zwei Phasenplatten, so erhält man zweidimensionale Verfahren, die auch bei zweidimensionalen Objekten unmittelbar zu der höchsten mit Schlierenverfahren überhaupt erreichbaren Meßgenauigkeit führen. Es ist ferner bei dicken Objekten zweckmäßig, vor das Objekt eine Linse so zu setzen, daß das Objekt parallel durchstrahlt wird. Man macht also $a' = f_a$ im Sinne von Ziff. 21, d.h. $a = \infty$.

Trotz der praktisch erreichbaren hohen Strahlschärfe muß aber hier ebenso wie bei dem abbildenden Minimumschlierenverfahren beachtet werden, daß eine Berechnung der interessierenden Größen aus der Ablenkung neue Auswertungsmethoden (Integralgleichungsmethoden[4] statt Strahlmethoden) voraussetzt, wenn der Genauigkeitszuwachs sich voll auf das Resultat auswirken soll.

---

[1] D.P. 819728, Kl. 42h, Gr. 34,11 vom 12. 4. 1949.
[2] H. Wolter: Ann. Physik (6) **7**, 353 (1950). — Phys. Bl. **5**, 234 (1949).
[3] O. Wiener: Wied. Ann. **49**, 105 (1893).
[4] H. Wolter: Ann. Physik (6) **7**, 191 (1950), zweite Anmerkung zur Theorie des Minimumschlierenverfahrens.

C. Die Beugung am Objekt bei abbildenden Schlierenverfahren und das Phasenkontrastverfahren nach ZERNIKE.

I. Die Beugung am Objekt und ihre Wirkung auf die Schlieren- und Phasenkontrastbilder einfacher Objektformen.

a) Entwicklung des ZERNIKE-Programms aus den Beugungsbetrachtungen der vorhergehenden Abschnitte.

23. Die Fig. 32f (S. 599) zeigt in ihrem untersten Teil die TÖPLERsche Schlieren-aufnahme eines Objekts, das aus einer nahezu schlierenfreien Glasplatte besteht, die drei nichtabsorbierende Zaponlackstreifen exakt konstanter Dicke trägt. Man denkt sich das Objekt zweckmäßig von links beleuchtet und sieht dann die Zaponlackstreifen reliefartig hervortreten. So anschaulich dieses Bild ist, so sehr steht es im Widerspruch zu der geometrisch-optischen Deutung der Schlieren-abbildung. Denn nach ihr sollten die Flächenstücke mit einem konstanten Dicken-gradienten Null auch in konstanter Helligkeit abgebildet werden. Ursache der Abweichung ist wieder die Beugung, und zwar die Beugung am Objekt, die in der Richtungsauswahlebene zu jedem der Zaponlackstreifen seine charakteristischen Beugungsfransen erzeugt, von denen die TÖPLER-Schneide dann die eine Hälfte fortnimmt.

In Ziff. 17 war die Auswirkung der Beugung auf die abbildenden Schlieren-verfahren summarisch aber vollständig nach der Unschärfebedingung geschehen. Einen ersten Einblick in *das Aussehen* der Beugungsbilder, wie es meist bei Objekten mit stetigem Verlauf der Parameter zu beobachten ist, gab Ziff. 18 u. a. an Hand der Fig. 16a für das TÖPLERsche Verfahren. Das Bild wurde dort ge-deutet als ein FRESNELsches Beugungsbild, hervorgerufen durch die *Beugung an der Schlierenblende im Lichte der Brennpunktfläche.*

Durch eine „Brennpunktfläche" kann man aber die Wirkung eines Objekts nur dann beschreiben, wenn es als aus relativ großen Linsenstückchen stetig zusammengesetzt aufgefaßt werden kann. Sind die Linsenstückchen sehr klein — oder sind gar Dickensprünge im Objekt —, so entspricht jedem statt eines Brenn-punktes ein Beugungsscheibchen, und die Brennpunktfläche wird selbst zu einem komplizierten Beugungsgebilde. Dieses zu betrachten ist nun nach HUYGHENS-KIRCHHOFF nicht erforderlich. Es genügt vielmehr, das in irgendeiner Ebene, z.B. der Richtungsauswahlebene (Brennebene des Objekts bzw. Schlierenkopfes) erscheinende Beugungsbild nach Amplitude und Phase zu berechnen, dann die Änderung, die es durch die Schlierenblende erfährt — z.B. teilweise Absorption durch die TÖPLER-Schneide — zu berücksichtigen und schließlich das in dieser Ebene noch vorhandene Licht als Lichtquellensystem anzusehen, dessen Strahlung in der Beobachtungsebene zu dem Schlierenbilde interferiert.

Dieses Programm für die nächsten Ziffern beschreibt das Vorgehen ZERNIKEs[1], der auf diese Weise die ABBEsche Theorie der Bildentstehung auf das TÖPLERsche Schlierenverfahren übertrug und dabei eine verbesserte Schlierenmethode — das Phasenkontrastverfahren — fand.

b) Allgemeiner Zusammenhang zwischen Objektfunktionen und Beugungsfunktionen bei eindimensionalen Objekten.

24. Fig. 25 zeigt eine einfache Schlierenapparatur mit engem linearem Spalt als Lichtquellenblende. Das Objekt sei „eindimensional", habe also in einer zur Spaltrichtung parallelen Koordinatenrichtung keine Änderungen. In der dazu

---

[1] F. ZERNIKE: Physica, Haag **1**, 43. — Roy. Astronom. Soc. Monthly Not. **94**, 377 (1934). — Z. techn. Phys. **16**, 454. — Phys. Z. **36**, 848 (1935).

senkrechten Richtung, der x-Richtung, sei die Wirkung des Objekts auf das Licht durch die komplexe Funktion $O(x)$ beschrieben. Diese „Objektfunktion" gebe die Lichterregung unmittelbar hinter dem Objekt nach Amplitude und Phase. Die Lichterregung, genannt „Bildfunktion", in der Bildebene ist mit der Objektfunktion identisch, wenn wir — mathematisch — Orientierung, Maßstab der x-Koordinate, Intensitätseinheit und Phasennullpunkt im Bilde zweckmäßig wählen — und vor allem physikalisch durch kein Hindernis (Schlierenblende usw.) die Bildübermittlung beeinträchtigen.

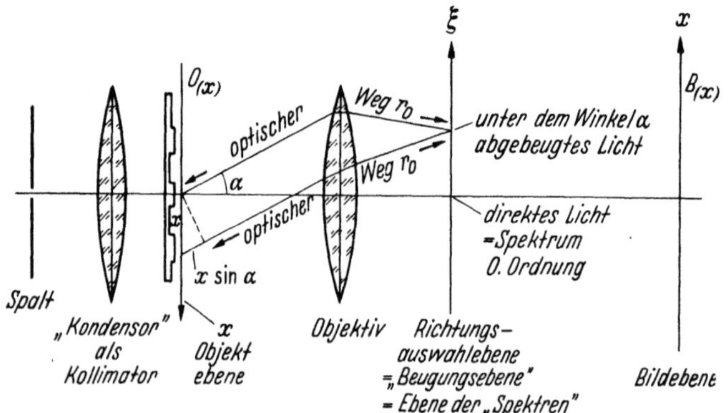

Fig. 25. Schlierenanordnung mit Spektren in der Richtungsauswahlebene.

Zum „Beugungsbild" in der „Richtungsauswahlebene" liefert das Objektstück zwischen x und $x + dx$ den Beitrag

$$dE^*(\xi) = K \cdot dx \cdot O(x) \cdot e^{i\omega t} \cdot e^{-\frac{2\pi i}{\lambda}(r_0 + x \sin \alpha)} \tag{24.1}$$

zu dem Integral

$$E^*(\xi) = K \cdot \int O(x) \cdot e^{-2\pi i x \frac{\sin \alpha}{\lambda}} dx \cdot e^{i\left(\omega t - \frac{2\pi r_0}{\lambda}\right)}. \tag{24.2}$$

Wir ersetzen die Winkelvariable α zweckmäßig durch die „Richtvariable"

$$\gamma = \frac{\sin \alpha}{\lambda} \tag{24.3}$$

und bezeichnen ferner

$$E^*\left(f \cdot \tan(\arcsin \gamma \lambda)\right) \cdot e^{-i\left(\omega t - \frac{2\pi r_0}{\lambda}\right)} = E(\gamma)$$

als die „Beugungsfunktion". Für sie gilt nach (24.2)

$$\frac{1}{K} E(\gamma) = \int_{-\infty}^{\infty} O(x)\, e^{-2\pi i x \gamma} dx. \tag{24.4}$$

Nach Fouriers Integraltheorem ist daher auch aus $E(\gamma)$ die Objektfunktion zu erhalten

$$O(x) = \frac{1}{K} \int_{-\infty}^{\infty} E(\gamma)\, e^{2\pi i x \gamma} d\gamma, \tag{24.5}$$

die mit der Bildfunktion $B(x)$ im Falle der Nichtbehinderung identisch ist. Greifen wir dagegen in die Richtungsauswahlebene ein und machen dabei z.B. durch

Blenden, Phasenschieber oder andere Hilfsmittel aus $E(\gamma)$ die abgewandelte Beugungsfunktion

$$Q(\gamma)\,E(\gamma) \tag{24.6}$$

[dabei heißt $Q(\gamma)$ die „Eingriffsfunktion"], so wird die neue Bildfunktion

$$B(x) = \frac{1}{K} \int\limits_{-\infty}^{\infty} Q(\gamma)\,E(\gamma)\,e^{2\pi i x\gamma}\,d\gamma, \tag{24.7}$$

und schließlich bei Verwendung von (24.4) unter Umbenennung der Integrationsvariablen x in x'

$$B(x) = \int\limits_{-\infty}^{\infty}\int\limits_{-\infty}^{\infty} Q(\gamma)\,O(x')\,e^{2\pi i\gamma(x-x')}\,dx'\,d\gamma. \tag{24.8}$$

Die bei der Schlierenbildberechnung herausfallende Konstante K hat nur Einfluß auf die Bezeichnung der Funktion $E(\gamma)$; sie wird im Einzelfall zweckmäßig so festgesetzt, daß die Funktion $E(\gamma)$ entweder eine besonders anschauliche Bedeutung gewinnt oder möglichst einfach ist. Bei unperiodischen Objekten setzen wir einfach $K = 1$.

c) Betrachtung periodischer Objekte zur Einführung.

25. Zusammenhang zwischen periodischen Objekten und ihrem Beugungsbild (ihren „Spektren")[1]. Wenn das Objekt ein ausgedehntes Gitter ist, so ist die Apparatur nach Fig. 25 von dem Lichtquellenspalt bis zur Richtungsauswahlebene ein Gitterspektrograph. Die Spektren der Ordnungen 0, -1, $+1$, ... entstehen in der Richtungsauswahlebene. Wir setzen monochromatisches Licht voraus; dann hat das Beugungsbild nur diskrete helle Punkte für die Winkel α_m, die mit der Gitterkonstanten g, der ganzzahligen „Ordnung" m und der Wellenlänge λ nach der Gitterbedingung

$$g \cdot \sin\alpha_m = m \cdot \lambda \tag{25.1}$$

zusammenhängen. Die Richtvariable für diese hellen Stellen ist

$$\gamma_m = \frac{\sin\alpha_m}{\lambda} = \frac{m}{g}. \tag{25.2}$$

Aus Gl. (24.4) folgt, daß die „Spektren"

$$E_m \equiv E(\gamma_m) = \frac{1}{g}\int\limits_{0}^{g} O(x)\,e^{-2\pi i\frac{x}{g}m}\,dx \tag{25.3}$$

Koeffizienten der Fourier-Reihe für $O(x)$ sind:

$$O(x) = \sum_{m=-\infty}^{\infty} E_m \cdot e^{2\pi i\frac{x}{g}m} \tag{25.4}$$

ohne Eingriff und

$$B(x) = \sum_{m=-\infty}^{\infty} Q_m E_m\, e^{2\pi i\frac{x}{g}m} \tag{25.5}$$

---

[1] Vgl. außer den Arbeiten Zernikes auch die von E. und C. Menzel und das Buch von A. H. Bennett (s. Bibliographie am Schluß des Artikels, S. 641 ff.).

bei einem Eingriff mittels der „Eingriffsfunktion" Q_m. Dabei ist das Integral (25.3) nicht über das viele Gitterkonstanten breite Objektfeld, sondern der Einfachheit halber nur über eine Gitterkonstante erstreckt worden; die ohnehin frei verfügbare Konstante K wurde zur Kompensation dieser Maßnahme so gewählt, daß in (25.3) vor das Integral schließlich $1/g$ trat. Das ist zweckmäßig; denn nach dieser Normierung der E_m hat ein objektfreies Feld, für das $E_m = 0$ für $|m| \geq 1$, nach Gl. (24.4) ein $E_0 = O(x) \equiv B(x) = $ konst. E_0 gibt dann also selbst die Bildamplitude für objektfreies Gesichtsfeld. Liegt ein Objekt vor, so ist nach (25.3)

$$E_0 = \frac{1}{g} \int_0^g O(x)\, dx, \qquad (25.6)$$

also der Mittelwert der Amplitude.

Die Fourier-Reihe (25.4), (25.5) beschreibt, wie das von den Spektralpunkten ausgehende Licht in der Bildebene miteinander zum Bilde interferiert. Die Normierung ist offenbar so geschehen, daß jede Größe $E_m \cdot e^{2\pi i \frac{x}{g} m}$ gerade das Licht nach der Amplitude und Phase beschreibt, die das m-te Spektrum auf der Bildebene hat.

Die Interferenz der Wellen zum Bilde veranschaulicht Fig. 26 für den Fall eines sog. „Amplitudengitters" aus abwechselnd durchsichtigen und ganz undurchsichtigen Streifen, d.h. mit abwechselnden Werten 1 und 0 der Objektfunktion $O(x)$ in 26a.

Ist das Objekt speziell ein „Phasengitter", besteht es also z.B. aus durchlässigen Zaponlackstreifen auf einer guten Glasplatte, so hat die Objektfunktion überall den Betrag $|O(x)| \equiv 1$.

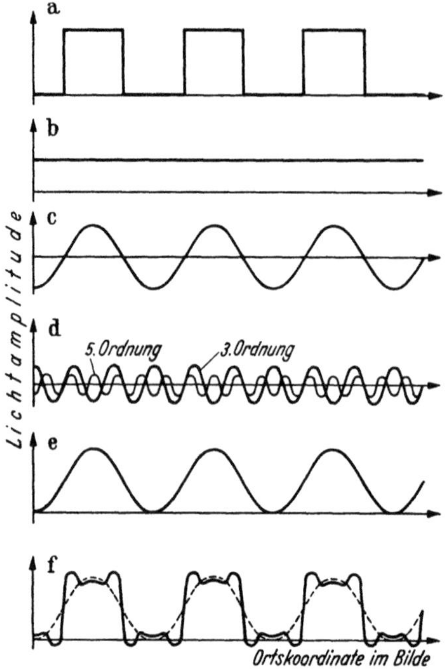

Fig. 26a—f. Bildentstehung (nach Abbe) durch Interferenz der von den Spektren ausgehenden Wellen. Das Spektrum nullter Ordnung (b) allein zeigt nur den Amplitudenmittelwert; das Hinzukommen des Spektrenpaares erster Ordnung (E_1 und E_{-1}) (c) ergibt (e) bereits grob die Struktur des Objekts. Je mehr Spektren berücksichtigt werden, d.h. von dem optischen System durchgelassen werden, desto objektähnlicher wird das Bild. Aus allen Spektren bis zur 5. Ordnung resultiert die ausgezogene Kurve f.

In der Gaußschen Zahlenebene (Fig. 27) haben also die Werte $O(x) = O_1$ hinter den Streifen und die Werte O_2 hinter den Zwischenräumen gleiche Länge 1, aber (je nach Dicke d der Zaponlackstreifen) unterschiedliche Richtung. Der Winkel zwischen beiden vom Nullpunkt zu den Punkten O_1 und O_2 gezogenen Vektoren d.h. der Phasenwinkel, um den die Zaponlackstreifen verzögern, ist bei Vernachlässigung der Zickzackreflexionen

$$\varphi \approx 2\pi \left(\frac{d}{\lambda_E} - \frac{d}{\lambda_Z} \right) = 2\pi \frac{d}{\lambda} (n_E - n_Z). \qquad (25.7)$$

Darin bezeichnet λ_Z und λ_E die Lichtwellenlänge im Zaponlack bzw. im Einschlußmedium (z.B. Luft), λ die Vakuumwellenlänge und n_Z, n_E die Brechungsindizes im Zaponlack bzw. im Einschlußmedium (Luft). Diese einfache Formel (25.7) gilt jedoch nur dann streng, wenn Zickzackreflexionen nicht auftreten; allgemeinere und genauere Formeln wurden im Artikel „Optik dünner Schichten" in diesem Bande (s. S. 530) gegeben.

E_0 ist in Fig. 27a das Spektrum nullter Ordnung, d.h. der Mittelwert (Schwerpunkt) von O_1 und O_2. Bei der Mittelbildung ist O_1 und O_2 jeweils ein „Gewicht" entsprechend der Breite der Zaponlackstreifen bzw. der Zwischenräume zu geben. Stets liegt E_0 auf der Strecke O_1O_2 näher an O_2, wenn die Zwischenräume breiter als die Gitterstreifen sind.

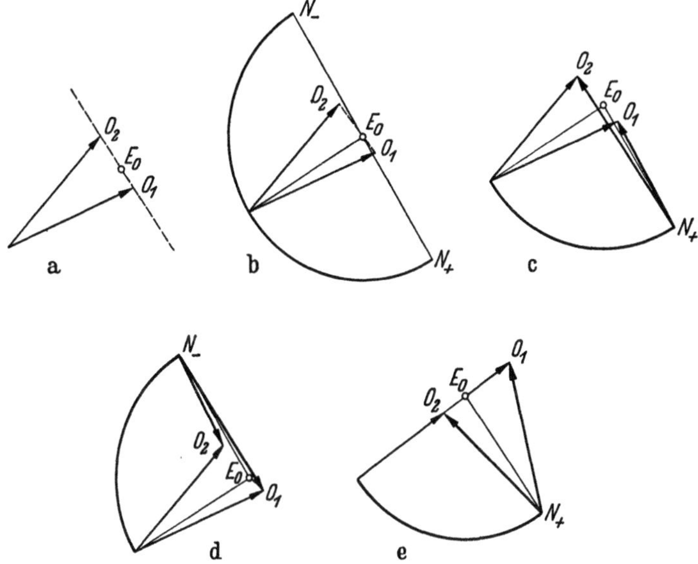

Fig. 27a—e. Komplexe Lichtamplituden in der GAUSSschen Zahlenebene, O_1 für die Gitterstreifen, O_2 für die Zwischenräume. E_0 = Spektrum nullter Ordnung = Mittelwert von O_1 und O_2.

26. Strenges Dunkelfeldverfahren bei Gittern. Entfernt man das Spektrum nullter Ordnung aus dem Licht, indem man es durch einen Draht als Schlierenblende in der Richtungsauswahlebene (hintere Objektivbrennebene in Fig. 25) abdeckt, so geht die neue Bildfunktion nach Gl. (25.5) aus der alten [mit der Objektfunktion $O(x)$ identischen] hervor, indem E_0 von ihr abgezogen wird

$$B_{sD}(x) = O(x) - E_0. \tag{26.1}$$

Der Zusatz sD soll an die ältere Bezeichnung „strenges Dunkelfeldverfahren"[1] erinnern.

Die Subtraktion des E_0 von den beiden Werten O_1 und O_2, deren $O(x)$ bei dem Phasengitter fähig ist, bedeutet in der GAUSSschen Zahlenebene eine Nullpunktsverlegung nach E_0. Die resultierenden Vektoren E_0O_1 und E_0O_2, allein durch das Zusammenwirken der Seitenspektren zustandekommend, sind die komplexen Amplituden im strengen Dunkelfeldbild und unterscheiden sich in ihrer Länge genau dann, wenn Streifen und Zwischenräume verschieden breit sind, E_0 also nicht genau in der Mitte zwischen O_1 und O_2 liegt. Der schmalere Teil wird jeweils heller als der breitere dargestellt. Experimentell wird das belegt durch die strenge

---

[1] H. WOLTER: Ann. Physik (6) **7**, 36 (1950). — Neuerdings wird von der Firma Reichert, Wien, ein Mikroskop herausgebracht, das u. a. nach diesem Verfahren arbeitet (s. S. 608); es heißt dort „Anoptralverfahren" [siehe auch ALVAR WILSKA: Mikroskopie (Wien) **9**, 1 (1954)].

Dunkelfeldaufnahme Fig. 31b unter IV rechts, in der die schmalen Zaponlack-streifen hell auf dunklem Grunde erscheinen. Allgemein kann an dem Dunkel-feldbilde nicht entschieden werden, was Lackstreifen oder Zwischenräume sind. Auch versagt die Methode bei gleicher Breite der Streifen und Zwischenräume; denn dann liegt E_0 in der Mitte der Strecke O_1O_2, und die Bildfunktion nach Fig. 28b hat eine Intensität nach Fig. 28d zur Folge, die Zwischenräume und Lackstreifen gleich hell darstellt. Der experimentelle Befund (Fig. 31b, bei III) bestätigt das. Nur gelegentlich sind einige Zwischenräume dunkler, die versehentlich etwas zu breit geraten waren. Fig. 31b zeigt ganz links die strenge Dunkelfeldaufnahme eines Amplitudengitters; Fig. 31a gibt Hellfeldbilder, d.h. Bilder ohne Schlierenblende, derselben Obejkte jeweils darüber. Ob ein Phasen- oder ein Amplitudengitter

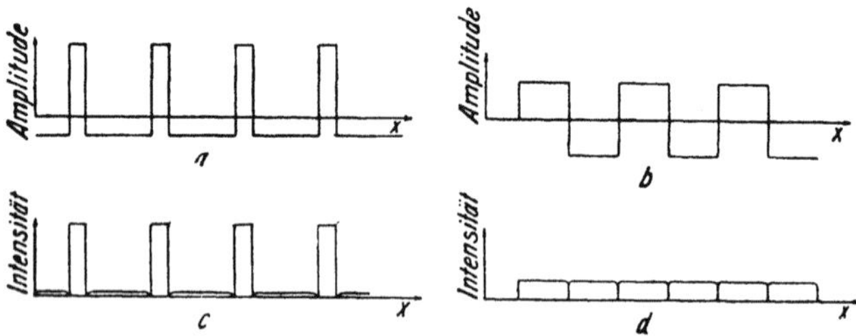

Fig. 28 a—d. Amplituden (oben) und Intensitäten (unten) in strengen Dunkelfeldbildern, links für ein Gitter aus schmalen Streifen und breiten Zwischenräumen, rechts bei gleicher Breite der Streifen und Zwischenräume. Ob es sich um Amplituden-gitter oder Phasengitter handelt, macht bei dem strengen Dunkelfeldverfahren keinen Unterschied.

vorliegt, ist nach Fig. 27a gleichgültig für den Erfolg des strengen Dunkelfeld-verfahrens; denn nur die Länge der Strecke O_1O_2 und die Lage von E_0 auf ihr (d.h. die relative Breite der Streifen und Zwischenräume) haben Einfluß auf die Bildintensität. Schmale dunkle Streifen auf hellem Grund erscheinen im strengen Dunkelfeldbild also hell auf dunklem Grund.

27. Strenges Phasenkontrastverfahren nach Zernike[1]. Die Mängel des strengen Dunkelfeldverfahrens (Versagen bei $|O_1E_0| = |E_0O_2|$, Fehlen einer Entscheidung über das Vorzeichen einer Phasendifferenz) entfallen, wenn der Nullpunkt statt nach E_0 etwa nach einem außerhalb der Strecke O_1O_2 gelegenen Punkt der Ge-raden O_1O_2 verlegt wird; dann bewirken Phasenunterschiede zwischen O_1 und O_2 stets Intensitätsunterschiede im Bilde (vgl. Fig. 27).

Dieses Ziel erreicht Zernike in seinem Phasenkontrastverfahren ungefähr, indem er das Spektrum nullter Ordnung fortnimmt und um 90° phasengedreht wieder zufügt. Der Nullpunkt wird dann nach N_+ oder N_- verlegt. Im ersten Fall spricht man von positivem, im zweiten von negativem Phasenkontrastverfahren. Im ersten Falle (Fig. 27c) werden die Zaponlackstreifen (O_1) dunkler als die Zwischenräume dargestellt. Umgekehrt wirkt „negativer Phasenkontrast" (Fig. 27d).

Realisiert wird die Phasenverschiebung durch eine als „Schlierenblende" dienende Glasplatte, die am Orte des Spektrums nullter Ordnung einen zum Bilde der Lichtquellenblende kongruenten Phasenstreifen trägt, z.B. einen Lackstreifen passender Dicke bei negativem oder eine Rille in einer Lackschicht bei positivem Phasenkontrast. Bilder unsichtbarer Zaponlackgitter, die mit einer solchen Zernike-Schlierenblende aufgenommen wurden, zeigt die Fig. 31c; es sind die

---

[1] F. Zernike: Siehe Bibliographie.

gleichen Gitter, die darüber (b) mit dem strengen Dunkelfeldverfahren und bei a im gewöhnlichen Hellfeld, d. h. ohne Schlierenblende, aufgenommen sind.

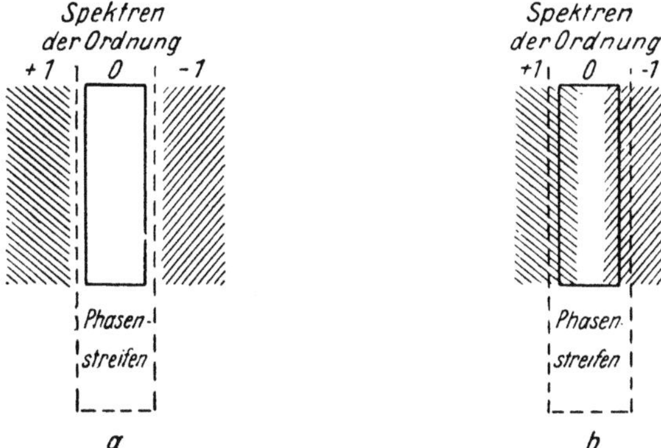

Fig. 29a u. b. Lage des Spektrums zum Phasenstreifen a bei strengem, b bei nicht strengem Phasenkontrastverfahren Spektren erster Ordnung schraffiert.

Das Verfahren arbeitet zuverlässig nur bei nicht beliebig großer Phasendifferenz zwischen O_1 und O_2. Würden O_1 und O_2 sich um 180° in Phase unterscheiden, so würde das Phasenkontrastverfahren keine Amplitudenunterschiede mehr bewirken.

Für dünnste Objekte kann das Verhältnis der Helligkeiten kontrastreicher gestaltet werden, wenn das Spektrum nullter Ordnung nicht nur phasengedreht

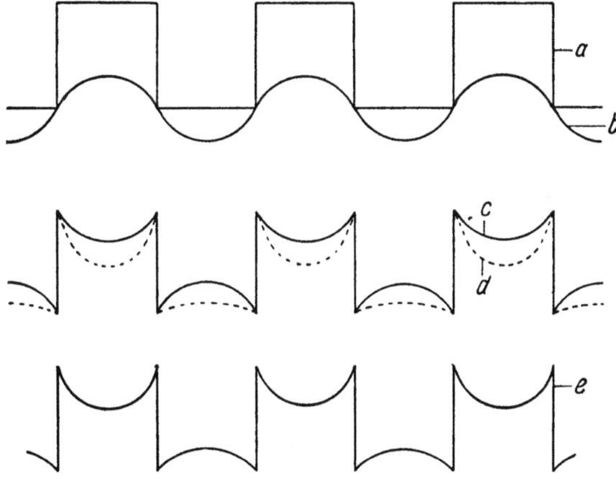

Fig. 30. Kurve a strenges Phasenkontrastbild, Kurve b Interferenzbild der Spektren erster Ordnung im Bilde, Kurve c Phasenkontrastbild-Amplitude nach Entfernung der Spektren erster Ordnung, Kurve d Intensität nach Entfernung der Spektren erster Ordnung, Kurve e Intensität nach phasenverkehrter Zufügung der Spektren erster Ordnung = Bild bei nicht streng durchgeführtem Phasenkontrastverfahren.

sondern auch geschwächt wird, so daß der neue Nullpunkt (Fig. 27c oder d) von N_- oder N_+ näher an einen der Punkte O_1 oder O_2 heranrückt. Bei der Apparatur für die Aufnahmen zu Fig. 31c war davon schon Gebrauch gemacht worden[1].

Je stärker die Absorption des Phasenstreifens gewählt wird, desto empfindlicher werden schwächste Objekte angezeigt, desto kleiner wird aber der φ-Bereich

---

[1] Technische Einzelheiten siehe Ann. Physik (6) **7**, 53 (1950).

I
Amplitudengitter   II
leer   III
Phasengitter mit Stegbreite
= Zwischenraumbreite

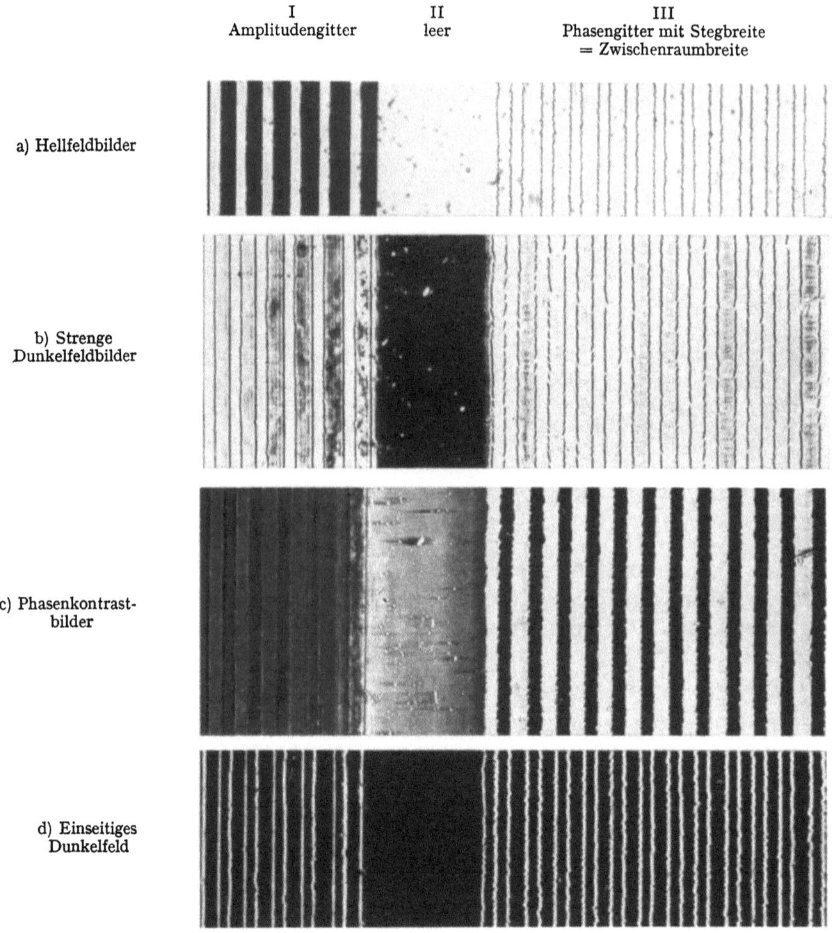

Fig. 31 a—f. Bilder von Amplituden- und Phasengittern. In der Vertikalen bleiben die Objekte,
in der Horizontalen die Verfahren identisch.

für die noch gut und eindeutig dargestellten Objekte. Deshalb wurden Phasenstreifen kontinuierlich veränderlicher Absorption vorgeschlagen[1]. Da mit ihnen zugleich eine Messung der Objektphasenverschiebung möglich ist, werden sie unten bei den quantitativen Methoden diskutiert werden (Ziff. 51).

Unsichtbar wird das *Amplitudengitter* Fig. 31c links (I) mit dem Phasenkontrastverfahren. Wie Fig. 27a und e zeigen, vertauschen bei dem Phasenkontrastverfahren die Amplitudenobjekte und die Phasenobjekte ihre Rolle.

28. Nicht strenges Phasenkontrastverfahren bei Gittern. Das strenge Phasenkontrastverfahren gibt die Phasenobjekte völlig „objekttreu" wieder, wie Fig.31c III zeigt; es setzt jedoch vollständige Trennung des Spektrums nullter Ordnung von den Seitenspektren voraus; denn nur dann kann man das Spektrum nullter

---

[1] Siehe u. a. P. Keck u. A. Brice: Optik **5**, 31 (1949).

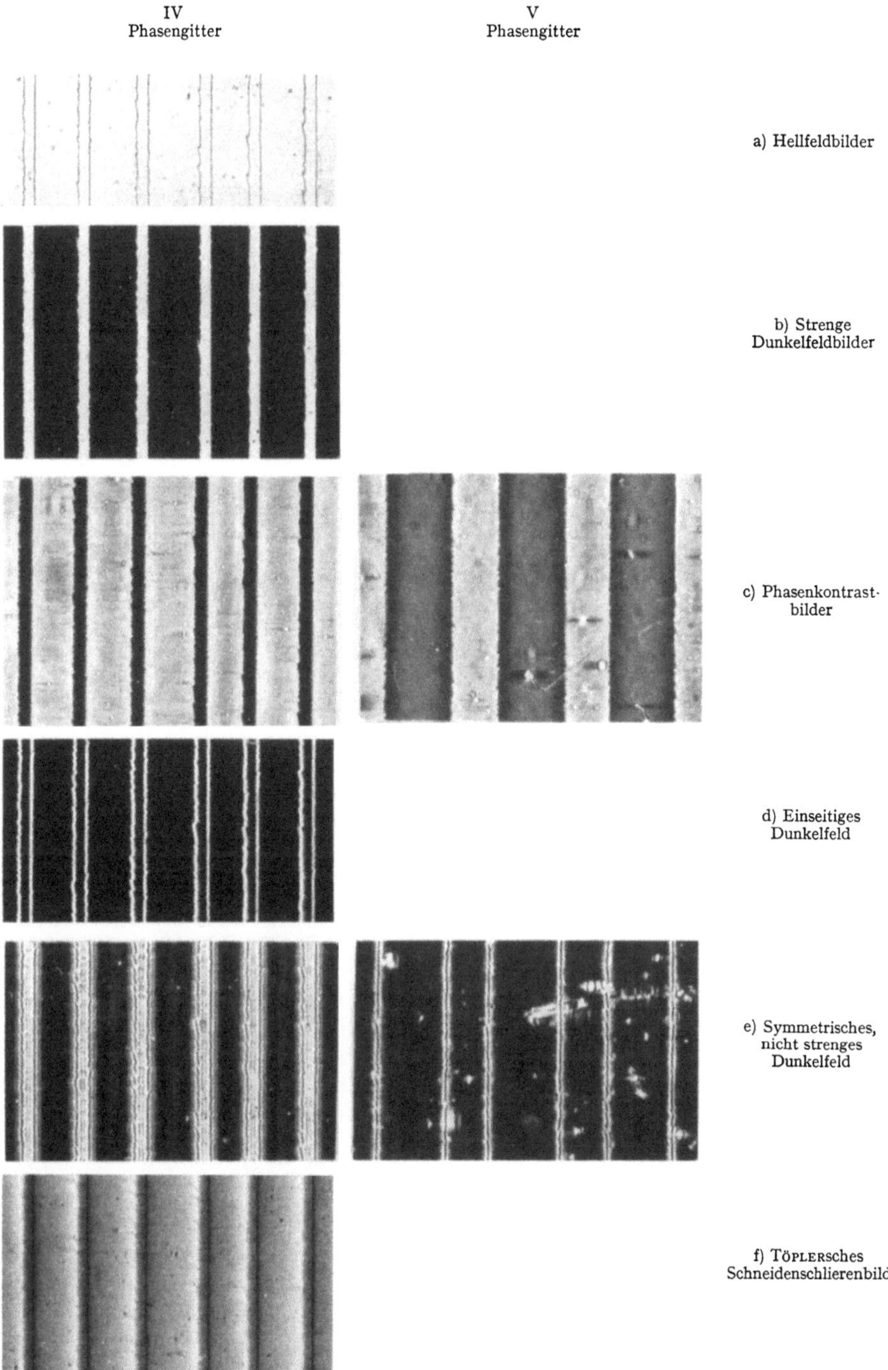

Ordnung phasendrehen, ohne zugleich die Seitenspektren mit zu erfassen. Aber schon um der Bildhelligkeit willen muß man in der Praxis dem Lichtquellenspalt eine gewisse Mindestbreite geben, und die Spektren, die ja bei Monochromasie

Bilder des Lichtquellenspaltes sind, erhalten demgemäß eine endliche Breite wie in Fig. 29. Fig. 29a stellt die noch gerade getrennten Spektren für eine hinreichend kleine Gitterkonstante dar. Der in Fig. 29a gestrichelt umrahmte

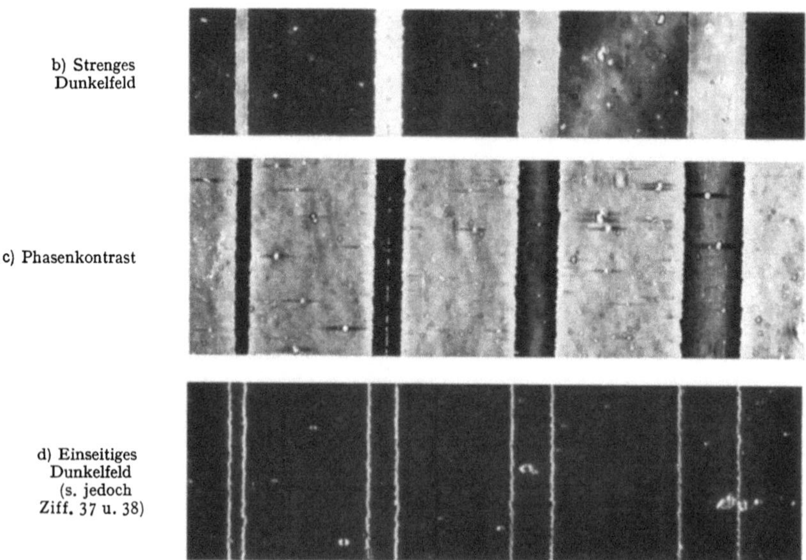

Fig. 32 a—f. Bilder von nichtperiodischen Phasenstreifen. Die Streifenbreite wächst von links nach rechts. In der Vertikalen bleiben die Objekte, in der Horizontalen die Verfahren identisch.

Phasenstreifen kann hier genau das Spektrum nullter Ordnung erfassen, und die Bilder sind dann strenge Phasenkontrastbilder wie in Fig. 31c III.

Gröbere Gitter haben notwendigerweise enger liegende Spektren; bei gleicher Apparatur läßt sich daher das Spektrum nullter Ordnung nicht mehr völlig von den Seitenspektren absondern (Fig. 29b), und die Seitenspektren werden mehr oder weniger von der Phasendrehung mit betroffen. Die Auswirkung zeigt Fig. 31c IV. Die Zwischenräume sind in ihrer Mitte dunkel abschattiert. Sind die Lackstreifen selbst auch breiter (Fig. 31c V), so erscheinen sie in der Mitte aufgehellt. Diese vorgetäuschten Schattierungen, denen im Objekt nichts Wirkliches entspricht, werden verständlich, wenn man bedenkt, daß das Spektrenpaar erster Ordnung nach Gl. (24.2) wie eine cos-Funktion (Kurve b in Fig. 30) zur Amplitude $B(x)$ beiträgt. Beseitigt man die Spektren erster Ordnung, so bleibt nur die Differenz der Kurven a und b, d.h. die Kurve c. Durch Quadrieren erhält man die Intensität (Kurve d). Fügt man die Spektren erster Ordnung um 90° phasengedreht wieder hinzu, so erhält man einen Intensitätsverlauf nach Fig. 30e, der qualitativ den Beobachtungen entspricht. Das phasenmäßig falsch behandelte Spektrenpaar erster Ordnung ist weniger wirksam geworden und fehlt deshalb weitgehend im resultierenden Bilde; das verursacht die Aufhellung in der Mitte der dunklen Bildteile und die Verdunklung in der Mitte der hellen Bildteile.

Die quantitative Untersuchung und die Verallgemeinerung auf andere Abbildungsverfahren wie das Töplersche ist für Gitter ebenfalls leicht möglich;

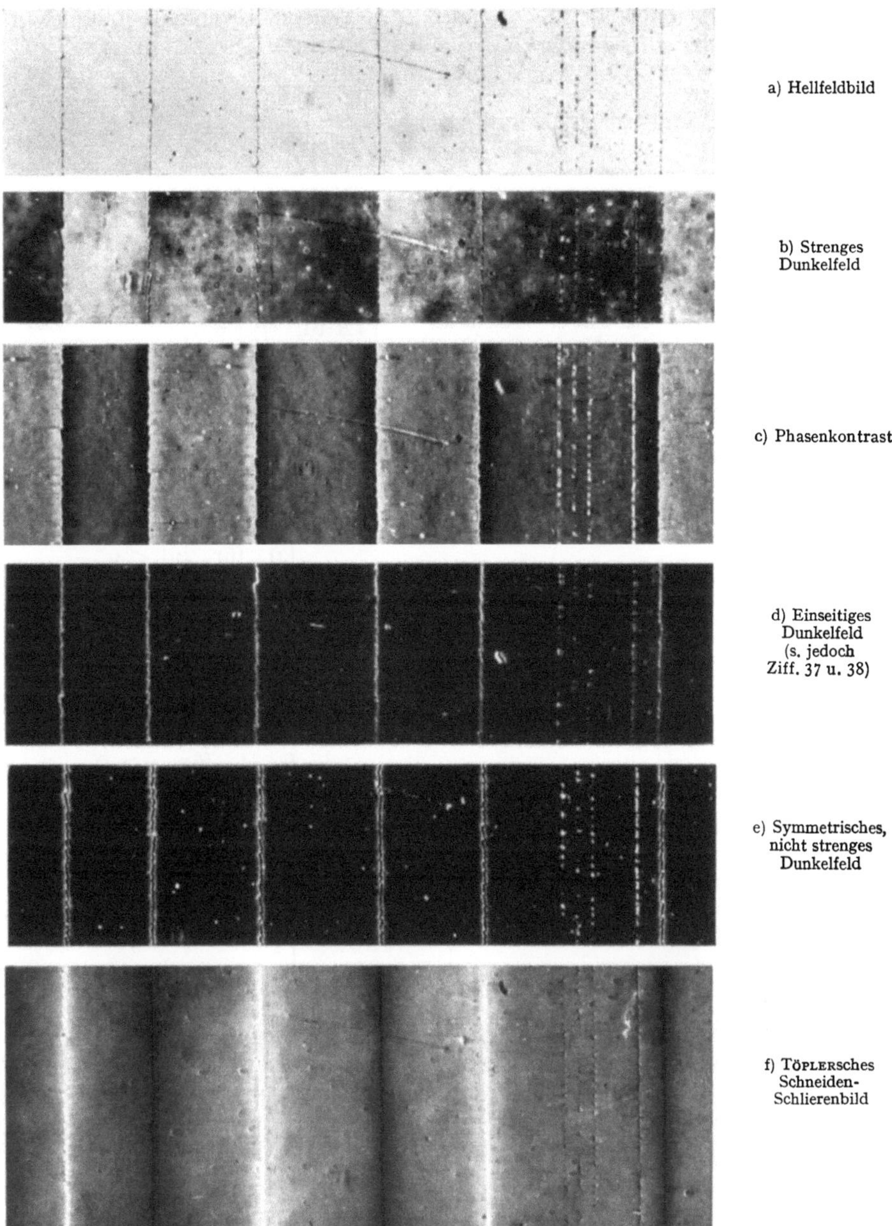

a) Hellfeldbild

b) Strenges
Dunkelfeld

c) Phasenkontrast

d) Einseitiges
Dunkelfeld
(s. jedoch
Ziff. 37 u. 38)

e) Symmetrisches,
nicht strenges
Dunkelfeld

f) Töplersches
Schneiden-
Schlierenbild

doch soll hier darauf nicht eingegangen werden, da dies besser an unperiodischen Objekten wegen ihrer umfassenderen praktischen Bedeutung im nächsten Abschnitt geschieht.

d) Quantitative Betrachtung der Schlieren- und Phasenkontrastbilder bei unperiodischen Objekten allgemein.

29. Die Eingriffsfunktionen $Q(\gamma)$ der Schlieren- und des Phasenkontrastverfahrens. Das Beugungsbild $E(\gamma)$ des Objekts wird durch die in der Beugungsebene angebrachte Schlierenblende umgewandelt in $Q(\gamma) \cdot E(\gamma)$. Die „Eingriffsfunktion" $Q(\gamma)$ ist für einige zu betrachtende Schlierenverfahren:

$$(29.1) \qquad \text{für die Aperturbegrenzung } Q_H(\gamma) = \begin{cases} 1 & \text{für } |\gamma| \leqq \gamma_0 \\ 0 & \text{sonst,} \end{cases}$$

$$(29.2) \qquad \text{strenges Zernike-Verfahren } Q_{sZ}(\gamma) = \begin{cases} iA & \text{für } |\gamma| < \dfrac{100}{G}; \ -1 < A < +1 \\ 1 & \text{sonst,} \end{cases}$$

$$(29.3) \qquad \text{Zernike-Verfahren } Q_Z(\gamma) = \begin{cases} iA & \text{für } |\gamma| < \gamma_i \\ 1 & \text{sonst,} \end{cases}$$

$$(29.4) \qquad \text{strenges Dunkelfeldverfahren } Q_{sD}(\gamma) = \begin{cases} 0 & \text{für } |\gamma| < \dfrac{100}{G} \\ 1 & \text{sonst,} \end{cases}$$

$$(29.5) \qquad \text{Dunkelfeldverfahren (symmetrisch) } Q_D(\gamma) = \begin{cases} 0 & \text{für } |\gamma| < \gamma_i \\ 1 & \text{sonst,} \end{cases}$$

$$(29.6) \qquad \text{einseitiges Dunkelfeldverfahren } Q_{eD}(\gamma) = \begin{cases} 1 & \text{für } \gamma > \gamma_i > 0 \\ 0 & \text{sonst,} \end{cases}$$

$$(29.7) \qquad \text{Minimumschlierenverfahren } Q_M(\gamma) = \begin{cases} 1 & \text{für } \gamma > 0 \\ 0 & \text{für } \gamma = 0 \\ -1 & \text{für } \gamma < 0, \end{cases}$$

$$(29.8) \qquad \text{Töplers Schlierenverfahren } Q_T(\gamma) = \begin{cases} 0 & \text{für } \gamma < 0 \\ \tfrac{1}{2} & \text{für } \gamma = 0 \\ 1 & \text{für } \gamma > 0. \end{cases}$$

Die Aperturbegrenzung (29.1) ist eine zwangsläufige Zugabe jedes realen Abbildungsverfahrens. Bei allen Verfahren sei deshalb der Ausfall aller Spektren mit $|\gamma| > \gamma_0$ stillschweigend zusätzlich vorausgesetzt. Doch interessiert die Wirkung der Aperturbegrenzung, d.h. das begrenzte Auflösungsvermögen, hier nicht immer, und dann sei $\gamma_0 \to \infty$ angenommen.

G sei die Breite des ausgeleuchteten Objektfeldes. Das direkte — auch bei objektfreiem Feld — in der Beugungsebene auftretende Bild des unendlich schmalen Lichtquellenspaltes hat die Breite $\delta\gamma \approx 1/G$. Die Forderung $Q_{sZ}(\gamma) = iA$ für $|\gamma| < 100/G$ soll in (29.2) sicherstellen, daß das gesamte direkte Licht (hier gleichsam das „Spektrum nullter Ordnung") von der 90°-Phasendrehung (Faktor i) und der Schwächung (Faktor A) betroffen wird. G wird als sehr groß gegenüber allen interessierenden Teilen des Objekts angenommen. Daher werden die für das Objekt charakteristischen Beugungsfransen bei $Q_{sZ}(\gamma)$ nicht wesentlich berührt; denn diese Beugungsfransen haben gegenüber der sehr geringen Breite $100/G$ eine sehr große Ausdehnung. Daher wird das Verfahren als strenges Zernike-Verfahren bezeichnet. Für $Q_Z(\gamma)$ dagegen ist dem Phasenstreifen der Zernike-Platte eine beliebige Breite (festgelegt durch $|\gamma| < \gamma_i$) gegeben.

30. Integralformeln für die Bildfunktionen bei beliebigen eindimensionalen Objekten. In dem Doppelintegral für die Bildfunktion nach Gl. (24.8)

$$B(x) = \int\limits_{-\gamma_0}^{\gamma_0} \int\limits_{-\frac{G}{2}}^{+\frac{G}{2}} Q(\gamma) \cdot O(x') \cdot e^{2\pi i \gamma (x - x')} \, dx' \, d\gamma \qquad (30.1)$$

kann für jede Eingriffsfunktion $Q(\gamma)$ aus Ziff. 29 die Integration über γ sofort ausgeführt werden. Man erhält z.B. für einige Bildfunktionen, die wir entsprechend den Eingriffsfunktionen durch Indices kennzeichnen,

$$B_H(x) = \int\limits_{-\frac{G}{2}}^{\frac{G}{2}} O(x') \, \frac{\sin\left(2\pi\gamma_0(x' - x)\right)}{\pi(x' - x)} \, dx' , \qquad (30.2)$$

$$B_T(x) = \int\limits_{-\frac{G}{2}}^{\frac{G}{2}} O(x') \, \frac{1 - e^{-2\pi i (x' - x)\gamma_0}}{2\pi i(x' - x)} \, dx' , \qquad (30.3)$$

$$B_M(x) = \int\limits_{-\frac{G}{2}}^{\frac{G}{2}} O(x') \, \frac{1 - \cos\left(2\pi\gamma_0(x' - x)\right)}{\pi i(x' - x)} \, dx' . \qquad (30.4)$$

Allgemein gilt offenbar

$$B_T(x) = \tfrac{1}{2}\left(B_H(x) + B_M(x)\right) . \qquad (30.5)$$

$$B_Z(x) = B_H(x) - (1 - iA) \int\limits_{-\frac{G}{2}}^{\frac{G}{2}} O(x') \, \frac{\sin\left(2\pi\gamma_i(x - x')\right)}{\pi(x - x')} \, dx' . \qquad (30.6)$$

Das symmetrische Dunkelfeldbild geht daraus hervor, wenn man $A = 0$ setzt:

$$B_D(x) = B_H(x) - \int\limits_{-\frac{G}{2}}^{\frac{G}{2}} O(x') \, \frac{\sin\left(2\pi\gamma_i(x - x')\right)}{\pi(x - x')} \, dx' . \qquad (30.7)$$

Zwischen dem Dunkelfeldbild und dem ZERNIKE-Bild besteht daher allgemein die Beziehung

$$B_Z(x) = (1 - iA) \cdot B_D(x) + iA \cdot B_H(x) . \qquad (30.8)$$

Für großes Auflösungsvermögen wird daraus

$$B_Z(x) = (1 - iA) \cdot B_D(x) + iA \cdot O(x) . \qquad (30.9)$$

Ferner wird dann $(\gamma_0 \to \infty)$

$$B_D(x) = O(x) - \int\limits_{-\frac{G}{2}}^{\frac{G}{2}} O(x') \, \frac{\sin\left(2\pi\gamma_i(x - x')\right)}{\pi(x - x')} \, dx' , \qquad (30.10)$$

$$B_Z(x) = O(x) - (1 - iA) \int\limits_{-\frac{G}{2}}^{\frac{G}{2}} O(x') \, \frac{\sin\left(2\pi\gamma_i(x - x')\right)}{\pi(x - x')} \, dx' . \qquad (30.11)$$

Für „vernünftige" Objektfunktionen kann man in (30.4) im Limes großen Auflösungsvermögens $(\gamma_0 \to \infty)$ die cos-Funktion durch ihren Mittelwert Null ersetzen und erhält

$$B_M(x) = \frac{1}{\pi i} \int\limits_{-\frac{G}{2}}^{\frac{G}{2}} \frac{O(x')}{x' - x}\, dx'. \tag{30.12}$$

Wegen $B_H(x) = O(x)$ für $\gamma_0 \to \infty$ und der Beziehung (30.5) zwischen Töpler-Bild und Minimumschlierenbild gilt für hohes Auflösungsvermögen

$$B_T(x) = \frac{1}{2}\left(O(x) + \frac{1}{\pi i} \int\limits_{-\frac{G}{2}}^{\frac{G}{2}} \frac{O(x')}{x' - x}\, dx'\right). \tag{30.13}$$

Um die einschränkenden Voraussetzungen über die Objektfunktion zu vermeiden, werden wir jedoch bei den Beispielen der nächsten Ziffern unmittelbar von (30.1) ausgehen und zunächst über x' statt über γ integrieren.

e) Schlieren- und Phasenkontrastverfahren bei einem Objektstreifen konstanter Dicke.

31. Die Beugungsfunktion und die Bildfunktionen eines Streifens rechteckigen Querschnitts. Ist das Objekt ein symmetrisch zum Nullpunkt des Koordinatensystems liegender Streifen der Breite b, charakterisiert durch die Objektfunktion

$$O(x) = \begin{cases} a & \text{für } |x| < \dfrac{b}{2} \\[2mm] \dfrac{a+1}{2} & \text{für } |x| = \dfrac{b}{2} \\[2mm] 1 & \text{für } \dfrac{b}{2} < |x| < \dfrac{G}{2}; \quad b \ll G, \end{cases} \tag{31.1}$$

so ist die Beugungsfunktion nach Gl. (24.4) mit $K = 1$

$$\begin{aligned} E(\gamma) &= \int\limits_{-\frac{G}{2}}^{\frac{G}{2}} O(x)\, e^{-2\pi i \gamma x}\, dx \\ &= G \cdot \frac{\sin \pi G \gamma}{\pi G \gamma} + (a - 1)\, b \cdot \frac{\sin \pi b \gamma}{\pi b \gamma}. \end{aligned} \tag{31.2}$$

Daraus berechnet sich die Bildfunktion nach Gl. (24.7) mit $K = 1$

$$B(x) = \int\limits_{-\gamma_0}^{\gamma_0} Q(\gamma)\, E(\gamma)\, e^{2\pi i \gamma x}\, d\gamma. \tag{31.3}$$

Da nun das bei der Integration über γ häufig auftretende Integral

$$\int\limits_{-\gamma_1}^{\gamma_1} G\,\frac{\sin \pi G \gamma}{\pi G \gamma}\, d\gamma = \frac{1}{\pi} \int\limits_{-\pi G \gamma_1}^{\pi G \gamma_1} \frac{\sin u}{u}\, du \to 1 \quad \text{für} \quad G \to \infty \tag{31.4}$$

bei beliebigem γ_1, so erhält man nach kurzer Zwischenrechnung die entsprechend den Eingriffsfunktionen $Q(\gamma)$ nach Ziff. 29 indizierten Bildfunktionen für die verschiedenen Schlierenverfahren und Phasenkontrastverfahren

$$B_H(x) = 1 + (\mathfrak{a} - 1)\,\Pi(u_0;\,v), \tag{31.5}$$

$$B_{sZ}(x) = iA + (\mathfrak{a} - 1)\,\Pi(u_0;\,v) = B_H(x) - 1 + iA, \tag{31.6}$$

$$B_Z(x) = iA + (\mathfrak{a} - 1)\,\Pi(u_0;\,v) + (iA - 1)(\mathfrak{a} - 1)\,\Pi(u_i;\,v), \tag{31.7}$$

$$B_{sD}(x) = (\mathfrak{a} - 1)\,\Pi(u_0;\,v) = B_H(x) - 1, \tag{31.8}$$

$$B_D(x) = (\mathfrak{a} - 1)\{\Pi(u_0;\,v) - \Pi(u_i;\,v)\}, \tag{31.9}$$

$$B_{eD}(x) = \tfrac{1}{2}B_D(x) + i(\mathfrak{a} - 1)\{\Gamma(u_0;\,v) - \Gamma(u_i;\,v)\}, \tag{31.10}$$

$$B_M(x) = i(\mathfrak{a} - 1)\,\Gamma(u_0;\,v), \tag{31.11}$$

$$B_T(x) = \frac{1}{2} + \frac{\mathfrak{a} - 1}{2}\{\Pi(u_0;\,v) + i\,\Gamma(u_0;\,v)\}. \tag{31.12}$$

Darin ist

$$v = \frac{2x}{b};\quad u = \pi b\gamma;\quad u_0 = \pi b\gamma_0;\quad u_i = \pi b\gamma_i; \tag{31.13}$$

$$\frac{2}{\pi}\int_0^u \frac{\sin t}{t}\,e^{ivt}\,dt = \Pi(u;\,v) + i\,\Gamma(u;\,v), \tag{31.14}$$

$$\pi\,\Pi(u;\,v) = \mathrm{Si}\big(u(v+1)\big) - \mathrm{Si}\big(u(v-1)\big);\quad \mathrm{Si}(z) = \int_0^z \frac{\sin t}{t}\,dt, \tag{31.15}$$

$$\pi\,\Gamma(u;\,v) = \widehat{\mathrm{Ci}}\big(u(v+1)\big) - \widehat{\mathrm{Ci}}\big(u(v-1)\big); \tag{31.16}$$

$$\widehat{\mathrm{Ci}}(x) = \int_0^x \frac{1 - \cos t}{t}\,dt = \sum_{\nu=1}^{\infty} (-1)^{\nu-1}\frac{x^{2\nu}}{2\nu\cdot(2\nu)!}. \tag{31.17}$$

$\mathrm{Si}(x)$ ist der Integralsinus, $\widehat{\mathrm{Ci}}(x)$ ist dem Integralcosinus verwandt, ist jedoch überall regulär und daher bequemer für die numerische Rechnung als der Integralcosinus[1].

Fig. 33 zeigt die Funktionen $\Pi(u;\,v)$ und $\Gamma(u;\,v)$ für einige Werte von u. Aus ihnen werden in den nächsten Ziffern die Schlierenbilder abgelesen werden.

32. Hellfeldbild eines Streifenobjekts bei Aperturbegrenzung (begrenztem Auflösungsvermögen der Abbildung). Das Hellfeldbild eines dunklen Streifens mit $\mathfrak{a} = 0$ ist bei Aperturbegrenzung nach Gl. (31.5) durch die Bildamplitude

$$B_H(x) = 1 - \Pi(u_0;\,v) = 1 - \Pi\left(\pi b\gamma_0;\,\frac{2x}{b}\right) \tag{32.1}$$

gegeben. Ein anschauliches Bild dieser Amplitude erhält man, wenn man die obere Hälfte der Fig. 33 auf den Kopf stellt.

Bei einem reinen Phasenobjekt mit $\mathfrak{a} = e^{-i\varphi}$ wirkt sich die Aperturbegrenzung selbst wie ein Schlierenverfahren aus, da die vorwiegend „vom Objektrand ausgehenden" abgebeugten Strahlen großen Winkels α nicht mehr an der Abbildung teilnehmen und die Objektränder daher dunkel erscheinen. Die Fig. 31a und 32a

---

[1] H. Wolter: Ann. Physik (6) **7**, 44 (1950).

zeigen solche Hellfeldaufnahmen mit zwangsläufig begrenzter Apertur des Aufnahmeobjektivs. Quantitativ ist nach Gl. (31.5)

$$B_H(x) = 1 + (e^{-i\varphi} - 1)\,\Pi(u_0; v). \tag{32.2}$$

Das gibt zwar $B_H(x) \equiv 1$ für $u_0 \to \infty$ nach Fig. 33a; aber bei endlichem γ_0, d.h. u_0, ist am Objektrande

$$\Pi(u_0; \pm 1) \approx \tfrac{1}{2}, \tag{32.3}$$

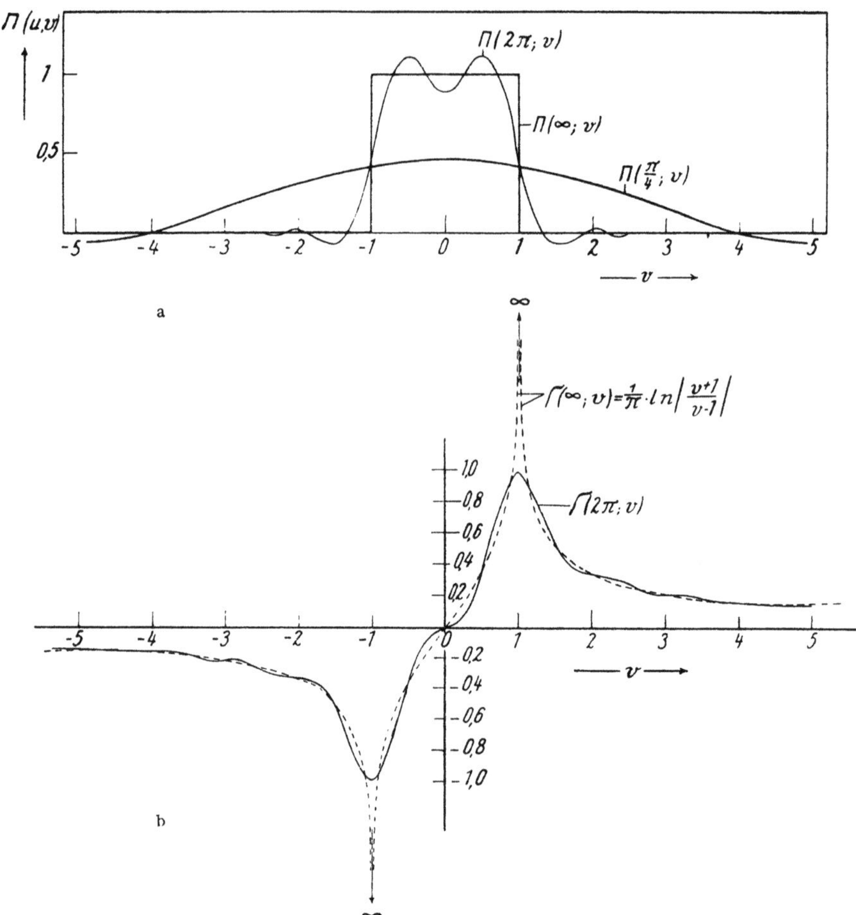

Fig. 33 a u. b. Hilfsfunktionen $\Pi(u; v)$ und $\Gamma(u; v)$ zur Berechnung der Bildamplituden.

sofern das Auflösungsvermögen nicht zu klein ist. Dann ist dort in einem Randbezirk, dessen Breite mit wachsendem Auflösungsvermögen abnimmt:

$$B_H\left(\pm \frac{b}{2}\right) = \frac{1}{2}(1 + e^{-i\varphi}), \tag{32.4}$$

$$\left|B_H\left(\pm \frac{b}{2}\right)\right|^2 = \frac{1}{2}(1 + \cos\varphi). \tag{32.5}$$

Für sehr schwache Objekte ($|\varphi| \ll 1$) ist die „Kontrastierung" gering, für $\varphi = 180°$ ist sie vollständig, da hier die Ränder völlig dunkel wiedergegeben sind. Fig. 34 zeigt mit der punktierten Kurve die Randintensität $V = \left|B_H\left(\pm \frac{b}{2}\right)\right|^2$ als Funktion von φ im Zusammenhang mit der linken Ordinatenskala. Die Bildintensität der objektfreien Gesichtsfeldteile ist 1 gesetzt. $-\log V$ heiße „Konstrast"[1].

Aber nicht die Intensitätsverhältnisse allein entscheiden über die Brauchbarkeit eines Verfahrens sondern die gesamte Intensitätsverteilung im Bilde und deren „Objekttreue". In dieser Hinsicht ist das Phasenkontrastverfahren dem Hellfeldbild noch stärker als bezüglich des Kontrastes überlegen, wie die nächste Ziffer zeigt.

33. Strenges Phasenkontrastverfahren nach Zernike bei vereinzelten Objekten. α) Bei großer Objektivapertur $\pi b \gamma_0 \gg 1$ ist nach Gl. (31.6) das strenge Zernike-Bild des Streifens

$$B_{sZ}(x) = \begin{cases} (\mathfrak{a} - 1) + iA & \text{für} \quad |x| < \dfrac{b}{2} \\ iA & \text{sonst.} \end{cases} \qquad (33.1)$$

Ist $\mathfrak{a} = e^{-i\varphi}$, das Objekt also ein reines Phasenobjekt, das die Phase um den Winkel φ verzögert, so ist

$$|B_{sZ}(x)|^2 = \begin{cases} (1 - \cos\varphi)^2 + (A - \sin\varphi)^2 & \text{für} \quad |x| < \dfrac{b}{2} \\ A^2 & \text{sonst.} \end{cases} \qquad (33.2)$$

Für kleines $\varphi > 0$, also ein „schwaches" Phasenobjekt, das optisch dichter als die Umgebung ist, ist $(1 - \cos\varphi)^2$ klein von zweiter Ordnung. Der dem Objektstreifen zugeordnete Bildteil ist dann dunkler als die Umgebung, wenn $A > 0$ („positiver" Phasenkontrast) ist und heller als die Umgebung, wenn $A < 0$ ist („negativer" Phasenkonstrast). Optimalen Kontrast, d.h. extremalen Wert für das Intensitätsverhältnis

$$V = \frac{\text{Int}\left(\text{für } |x| < \dfrac{b}{2}\right)}{\text{Int (sonst)}} = \frac{(1 - \cos\varphi)^2 + (A - \sin\varphi)^2}{A^2} \qquad (33.3)$$

gibt der Wert A, der den partiellen Differentialquotienten $\partial V/\partial A$ zu Null macht, also

$$A_{\text{opt}} = 2 \cdot \frac{1 - \cos\varphi}{\sin\varphi}. \qquad (33.4)$$

Für kleine φ ist das etwa φ selbst. Für kein A wird der Objektstreifen völlig dunkel abgebildet, da $(1 - \cos\varphi)^2$ mindestens als Restintensität bleibt.

Die Zahl der dunkler als die Umgebung wiedergegebenen Objekte wird um so kleiner, je kleiner A ist. Im Limes des strengen Dunkelfeldverfahrens ($A = 0$) sind alle Objekte heller als die nun völlig dunkle Umgebung. Besonders hell erscheinen bei Phasenkontrastverfahren mit beliebigem $|A| < 1$ Objekte mit Phasenverschiebungen von rund 180°; das Intensitätsmaximum liegt zwischen 135 und 225° je nach dem Vorzeichen und der Größe von A.

Fig. 34 zeigt den Verlauf des Intensitätsverhältnisses $V(\varphi)$ für einige Werte A (ausgezogene Kurven). Für negative A, d.h. negativen Phasenkonstrast, erhält man die entsprechenden Kurven, indem man die Kurven der Fig. 34 um die Gerade $\varphi = 0$ umklappt. $-\log V$ heiße „Kontrast"[1].

Für schwache Objekte ($|\varphi| \ll 1$) wird der Kontrast dem Betrage nach um so größer, je steiler die Kurve vom Punkte $\varphi = 0$; $V = 1$ (Werte der Objektumgebung)

---

[1] Unterschiedliche Definitionen des „Kontrasts" siehe bei K. Michel: Jb. Optik u. Feinmechanik **1955**. — A. H. Bennett: Siehe Bibliographie.

fortstrebt, je kleiner also A ist. Das einfache Hellfeldbild (punktierte Kurve), das die Phasenobjekte nur wegen der Aperturbegrenzung am Rande dunkel von der Umgebung absetzt, gibt die geringste Kontrastempfindlichkeit bei kleinen Phasendifferenzen $|\varphi|$, da die Kurve an $\varphi = 0$ sogar eine horizontale Tangente hat. Das Phasenkontrastverfahren ist stark überlegen, wie auch A bemessen ist. Das strenge Phasenkontrastverfahren mit $A = 0$ hat die höchste Empfindlichkeit am Nullpunkt. Es ist als „strenges Dunkelfeldverfahren" bezeichnet und in Ziff. 35 noch ausführlich behandelt.

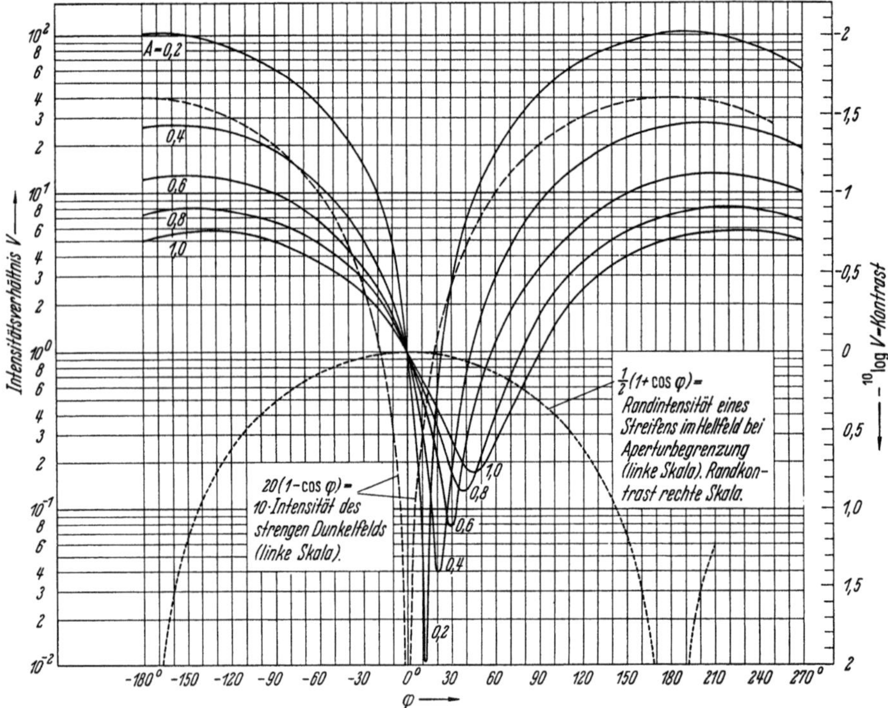

Fig. 34. Intensitätsverhältnis und Kontrast bei strengem positiven Phasenkontrastverfahren mit 90°-Zernike-Phasenstreifen für einige A-Werte (ausgezogen). Die entsprechenden Kurven für negativen Phasenkontrast ($A < 0$) erhält man durch Spiegeln der gezeichneten Kurven an der Geraden $\varphi = 0$. Punktiert und gestrichelt sind Vergleichskurven für Hellfeld mit Aperturbegrenzung bzw. für strenges Dunkelfeld.

Kleines A bedingt außer dem Vorteil hoher Empfindlichkeit aber auch den Nachteil eines kleinen Eindeutigkeitsbereichs für φ. Das Minimum der ausgezogenen Kurven in Fig. 34 liegt nämlich bei

$$\varphi_{\text{Min}} = \arctan A, \tag{33.4a}$$

da die Forderung $\partial V/\partial \varphi = 0$ auf $\tan \varphi = A$ führt.

Alle Objekte mit $V < 1$ (Fig. 34) erscheinen dunkler als die Umgebung. Das sind bei $A = 1$ genau die Objekte mit $0 < \varphi < 90°$ und z.B. bei $A = 0,2$ die Objekte mit $0 < \varphi < 22° 40'$. Ist die Phase der vorkommenden Objekte statistisch verteilt, so werden vom strengen Phasenkontrastverfahren stets mehr Objekte hell als dunkel wiedergegeben, unabhängig davon ob positiver oder negativer Phasenkontrast (A negativ) vorliegt.

β) Für manche Objekte ist es besser, eine andere Phasenverschiebung Φ statt 90° vorzunehmen. Dann ist in der Formel (33.1) iA zu ersetzen durch $A \cdot e^{i\Phi}$.

Die Objektfunktion im Streifen sei ebenfalls allgemein $\mathfrak{a} = a \cdot e^{-i\varphi}$; dann ist nach (33.1)

$$B_{sZ}(x) = \begin{cases} a \cdot e^{-i\varphi} - 1 + A\,e^{i\Phi} & \text{für} \quad |x| < \dfrac{b}{2} \\ A \cdot e^{i\Phi} & \text{sonst,} \end{cases} \tag{33.5}$$

und das Intensitätsverhältnis des Streifens zur Umgebung ist

$$V = \frac{(a\cos\varphi - 1 + A\cos\Phi)^2 + (A\sin\Phi - a\sin\varphi)^2}{A^2}. \tag{33.6}$$

Dann läßt sich stets eine Phasenplatte mit einem geeigneten A und Φ finden, so daß $B_{sZ}(x) = 0$ für $|x| < \dfrac{b}{2}$, also das Objekt ideal dunkel auf hellem Grund erscheint. Die erforderlichen Werte A, Φ ergeben sich aus

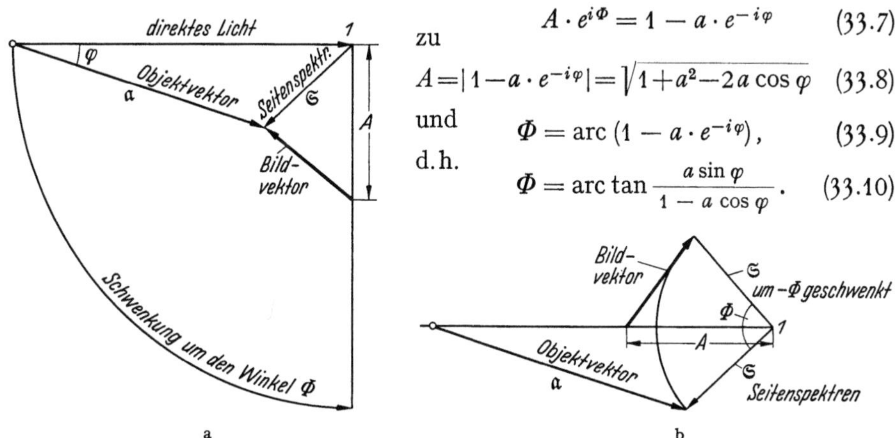

$$A \cdot e^{i\Phi} = 1 - a \cdot e^{-i\varphi} \tag{33.7}$$

zu

$$A = |1 - a \cdot e^{-i\varphi}| = \sqrt{1 + a^2 - 2a\cos\varphi} \tag{33.8}$$

und

$$\Phi = \operatorname{arc}(1 - a \cdot e^{-i\varphi}), \tag{33.9}$$

d.h.

$$\Phi = \arctan \frac{a\sin\varphi}{1 - a\cos\varphi}. \tag{33.10}$$

Fig. 35a u. b. Zwei äquivalente Konstruktionen des Bildvektors bei strengem Phasenkontrastverfahren mit Phasendrehung Φ und Schwächung des Spektrums nullter Ordnung mit dem Faktor A.

Vorrichtungen für veränderliches A und Φ werden in den Ziff. 51 und 52 genannt; dort wird auch die Messung der das Objekt kennzeichnenden Größen a, φ mit Geräten variabler Werte A, Φ beschrieben.

Alle hier über das strenge Phasenkontrastverfahren gemachten Aussagen gelten auch für zweidimensionale Objekte, da sich die Ausgangsformel (33.1) bzw. (33.5) für sie leicht allgemein an Hand der Fig. 35a und b ableiten läßt. In Fig. 35a ist $\mathfrak{a}$ der Lichtvektor am Bildort des Objekts im Hellfeld (also ohne Eingriff), a sein Betrag. Der Vektor der Länge $0 \to 1$ ist der Lichtvektor außerhalb des Objektbildes und zugleich „Spektrum nullter Ordnung", da wir das Objekt als sehr klein zur gesamten Bildfläche annehmen. Der Vektor $\mathfrak{S}$ ist das durch die Seitenspektren (das abgebeugte Licht) hervorgerufene Licht, das am Bildort des Objekts die komplexe Amplitude vom Punkte 1 zum Werte $\mathfrak{a}$ zwingt. Das Phasenkontrastverfahren verlegt den Nullpunkt durch Schwenkung des Spektrums nullter Ordnung um den Winkel Φ und Schwächung auf die Länge A. Für die Bildamplitude am Bildort des Objekts erhält man trigonometrisch die oben angegebenen Werte.

Physikalisch äquivalent ist der Phasendrehung des Spektrums nullter Ordnung die entgegengesetzte Phasendrehung aller Seitenspektren, da Phasen nur relative Bedeutung haben. Dem entspricht Fig. 35b. Natürlich liegt die Bildamplitude hier um den Winkel Φ gedreht gegenüber der in Fig. 35a gezeichneten. Doch betrifft das alle Bildteile und ist daher physikalisch belanglos.

γ) Als Spezialfall des allgemeinsten Phasenkontrastverfahrens ist auch das „Feldabsorptionsverfahren" ERDMANNs (Vortrag Kiel 1952) anzusehen. Dabei wird eine absorbierende Schicht in der Richtungsauswahlebene (Objektivbrennebene) im Bereich des abgebeugten Lichtes angebracht. Diese Absorberschicht hat einen Ausschnitt im Bereich des direkten Lichtes (Spektrums nullter Ordnung). Das ist, wenn wir von der mittleren Bildhelligkeit absehen, äquivalent einer Verstärkung des Spektrums nullter Ordnung allein ($A \gg 1$) unter Festhalten der Seitenspektren. In der GAUSSschen Zahlenebene bedeutet dies eine rückwärtige Verlängerung des Spektrums nullter Ordnung, also eine Verlagerung des Nullpunktes, und zwar in die Nähe des Einheitskreises selbst bei passender Bemessung der Größe A, mit anderen Worten der Absorption. Die Phasenobjekte, deren Pfeilspitzen naturgemäß alle auf dem Einheitskreise liegen, erscheinen nun dunkel, wenn diese Pfeilspitze nahe an dem neuen Nullpunkt liegt, sonst hell. Das Bild eines Gases z. B. ist daher von dunklen Linien durchzogen, die eine ähnliche Beurteilung des Objekts erlauben wie Interferenzlinien. Das Verfahren ist daher gelegentlich ein hinreichender Ersatz für kostspieligere Interferenzverfahren (MACH-ZEHNDER), z. B. bei Strömungsuntersuchungen. Ein ähnliches mikroskopisches Verfahren wird bei der Firma Reichert, Wien, auch als „Anoptral"-Verfahren bezeichnet. S. Fußn. S. 593.

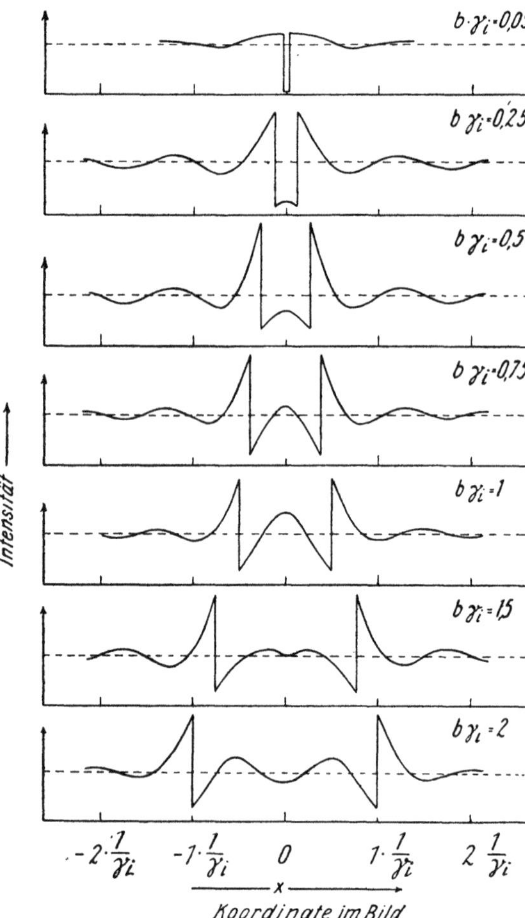

Fig. 36. Theoretischer Intensitätsverlauf im Phasenkontrastbilde verschieden breiter Zaponlackstreifen bei gleicher Apparatur. Nur schmalste Objekte werden unverfälscht wiedergegeben. Vergleiche die experimentellen Ergebnisse Fig. 32 c.

34. Reales Phasenkontrastverfahren. Praktisch kann man schon um der Bildhelligkeit willen die Lichtquellenblende nicht beliebig klein machen. Da nun der ZERNIKE-Streifen mindestens die Breite des Blendenbildes haben muß, wird zugleich mit dem direkten Licht auch ein wesentlicher Teil des charakteristischen Beugungsspektrums der Objekte (besonders der größeren) phasengedreht. Die Bildintensität als Funktion der Koordinate im Bilde zeigt dann ähnliche Abschattierungen, wie sie an Gittern schon in Ziff. 28 behandelt wurden. Aus Gl. (31.7) erhält man quantitativ für reine Phasenobjekte gleicher Dicke mit

$$\mathfrak{a} = 1 \cdot e^{-i\,0,1}$$

und nahezu optimale Lichtschwächung des positiven Phasenkontrastes $A = +0,1$ die Intensitäten im Bilde, die Fig. 36 zeigt. Die Koordinate im Bilde ist darin in

$\dfrac{1}{\gamma_i} = \dfrac{\lambda}{\sin \alpha_i}$ als Einheit angegeben. $\sin \alpha_i$ bedeutet die Apertur des ZERNIKE-Streifens. Ist Z die volle Breite des ZERNIKE-Streifens in cm und f die Objektivbrennweite, so ist

$$\alpha_i = \frac{Z}{2f}; \qquad \frac{1}{\gamma_i} \approx \frac{2\lambda f}{Z}. \tag{34.1}$$

Wie die Kurven[1] zeigen, wird der Objektstreifen noch einigermaßen unverfälscht schwarz auf gleichmäßig hellem Grunde wiedergegeben, solange seine Breite

$$b < 0{,}1 \cdot \frac{2\lambda f}{Z} \tag{34.2}$$

ist; breitere Objekte zeigen helle „Höfe" und Aufhellungen im Innern, denen nichts Wirkliches im Objekt entspricht — sie werden also nicht objekttreu wiedergegeben. Experimentell wird das belegt durch die Phasenkontrastaufnahmen der Fig. 32c; auch hier sind nur die beiden schmalen Streifen ganz links objekttreu abgebildet, während schon der dritte und erst recht die weiter rechts liegenden Streifen Mittenaufhellungen und Höfe haben. Das Photometer zeigt, daß die breiten Objekte nur noch am Rande von ihrer Umgebung kontrastiert sind. In ihrer Mitte ist die Intensität annähernd gleich der in der Mitte der Zwischenräume. Obwohl das infolge einer psychologisch-physiologischen Nebenwirkung nicht glaubhaft scheint, überzeugt man sich davon leicht, wenn man die Fig. 32 so knifft, daß eine Streifenmitte unmittelbar an eine Zwischenraummitte stößt. Die Photometerkurven von U. SCHMIDT[2] und der Vergleich mit seinen theoretischen Untersuchungen bestätigten das vor kurzem von neuem (s. a. Ziff. 45!).

35. Strenges Dunkelfeldverfahren. Bei dem strengen Dunkelfeldverfahren ist genau das direkte Licht in der Richtungsauswahlebene abgedeckt. Der abdeckende Draht z.B. muß dabei so dünn sein, daß von den charakteristischen Spektren des Objekts fast nichts ausgelöscht wird. Das ist freilich nur bei kleinen Objekten möglich, da deren charakteristisches Beugungsspektrum sehr ausgedehnt ist.

Unter diesen Voraussetzungen ist nach Gl. (31.8) das strenge Dunkelfeldbild unmittelbar durch die Fig. 33 oben gegeben bis auf den Faktor $(a-1)$. Das Bild gleicht dem $B_H(x)$, gibt das Objekt — gleichgültig ob Phasen- oder Amplitudenobjekt — so gut objekttreu wieder, wie das Auflösungsvermögen dies zuläßt. Bei hohem Auflösungsvermögen haben wir volle Objekttreue; dann ist die Bildintensität nach Gl. (31.8) und Fig. 33a für reines Phasenobjekt mit $a = 1 \cdot e^{\pm i\varphi}$

$$|B_{sD}(x)|^2 = \begin{cases} (1 - \cos \varphi)^2 + \sin^2 \varphi = 2\,(1 - \cos \varphi) & \text{für} \quad |x| < \dfrac{b}{2} \\ 0 & \text{sonst}. \end{cases} \tag{35.1}$$

Fig. 34 gibt mit der gestrichelten Kurve im Zusammenhang mit der linken Skala das 10fache der Intensität für ein reines Phasenobjekt der Phase φ.

Photometrie im Dunkelfeldbild gibt genaue Phasenmessung für die dünnen Schichten[3].

Für beliebiges Objekt ist nach (31.8)

$$|B_{sD}(x)|^2 = \begin{cases} |a - 1|^2 & \text{für} \quad |x| < \dfrac{b}{2} \\ 0 & \text{sonst}. \end{cases} \tag{35.2}$$

---

[1] Vgl. auch E. MENZEL: Optik **5**, 385 (1949). — H. WOLTER: Naturwiss. **37**, 272. — Ann. Physik (6) **7**, 33, 147 (1950). — K. SCHUSTER: Jenaer Jb. **1951**, 22.
[2] U. SCHMIDT: Ann. Physik (6) **16**, 68 (1955).
[3] H. WOLTER: Ann. Physik (6) **7**, 46, 48 (1950).

Das strenge Dunkelfeldverfahren scheint dem strengen Phasenkontrastverfahren darin überlegen, daß es auch Amplitudenobjekte gut wiedergibt; aber es ist ihm dadurch unterlegen, daß es für die objekttreue Wiedergabe hinreichend weite Isolierung der Objekte voneinander verlangt. Wo Objekte und Zwischenräume nahezu gleiche Breite haben — in Fig. 32b rechts — gibt das Dunkelfeldverfahren kein so eindeutiges Bild wie das Phasenkontrastverfahren (Fig. 32c). Das wurde bei den Gittern in Ziff. 26 näher betrachtet; es war einer der Gründe für die Einführung des Phasenkontrastverfahrens.

36. Reales symmetrisches Dunkelfeldverfahren. Praktisch muß man bei dem symmetrischen Dunkelfeldverfahren ein endlich breites Stück aus der Beugungs-

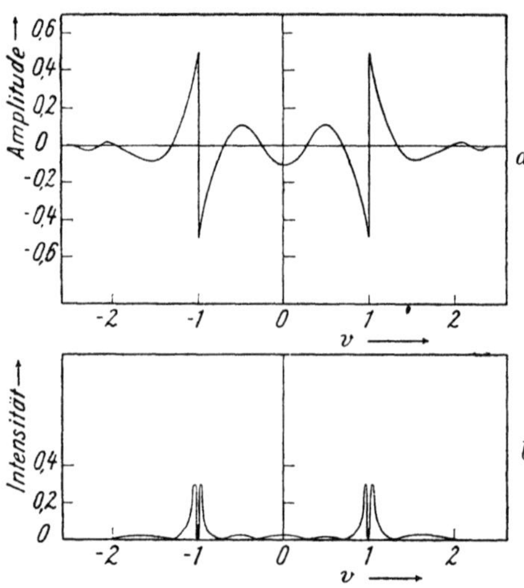

Fig. 37 a u. b. Amplituden (a) und Intensitäten (b) im nicht strengen, aber symmetrischen Dunkelfeldbilde eines Zaponlackstreifens (theoretisch). Vgl. Fig. 32e.

ebene, z.B. für $-\gamma_i < \gamma < \gamma_i$, entfernen. Die Amplitude für diesen Fall erhält man nach Gl. (31.9) durch Differenzbildung zweier Π-Kurven (Fig. 33a). Mit großem Auflösungsvermögen ($u_0 \to \infty$) gibt das für ein mäßig breites Objekt die Werte der Fig. 37. Den zugehörigen experimentellen Befund zeigen Fig. 31e und 32e. Genau an den Objektkanten liegen Intensitätsminima, von hellen Maxima flankiert. Schmalere Objekte werden mehr wie bei dem strengen Dunkelfeldverfahren wiedergegeben. Die Grenzen für die Objekttreue liegen wie bei dem Phasenkontrastverfahren bei Breiten

$$b < 0,1 \cdot \frac{2\lambda f}{Z}. \qquad (36.1)$$

Z bedeudet jetzt die „Drahtdicke" in der Richtungsauswahlebene.

37. Einseitiges Dunkelfeldverfahren. Fast alle als Dunkelfeldverfahren, insbesondere in der Mikroskopie, bezeichneten Schlierenverfahren sind Töpler-Verfahren mit einer Schneide, die eine Seite der Spektren und mindestens auch noch das Spektrum nullter Ordnung (das direkte Licht) abdeckt. Das beschreibt die Eingriffsfunktion nach Gl. (29.6), und das zugehörige Bild ist durch Gl. (31.10) gegeben. Wegen der Unsymmetrie der Abdeckung spielt außer der symmetrischen Funktion $\Pi(u; v)$ auch die unsymmetrische Funktion $\Gamma(u; v)$ mit eine Rolle. Zu dem in Fig. 37b gezeigten Anteil der Intensität addiert sich daher noch ein Intensitätsanteil, der als Norm der Differenz zweier Kurven vom Γ-Typ an jeder Kante des Objektstreifens ein hohes Maximum hat. Dieses füllt das Kantenminimum der Fig. 37b so völlig aus, daß im Bild des einseitigen Dunkelfeldverfahrens die jedem Mikroskopiker geläufigen hellen Kanten der Objekte entstehen. Die Aufnahmen ähneln denen der Fig. 31d und 32d, haben jedoch weniger scharfe Kantenmaxima.

38. Minimumschlierenverfahren. Die Amplitude im Minimumschlierenbild eines Objektstreifens nach Gl. (31.11) wird im wesentlichen durch eine Γ-Funktion (Fig. 33b) dargestellt; die Intensität (Fig. 38a) für hohes Auflösungsvermögen ($\gamma_0 \to \infty$) ist

$$|B_M(x)|^2 = |\mathfrak{a} - 1|^2 \Gamma^2(\infty; v) = |\mathfrak{a} - 1|^2 \cdot \frac{1}{\pi^2} \left(\log \left| \frac{v+1}{v-1} \right| \right)^2 \qquad (38.1)$$

und zeigt an den Objektkanten je ein logarithmisches (extrem schmales) Maximum. Experimentell wird das durch die Fig. 31d und 32d belegt. Das Minimumschlierenverfahren arbeitet auch bei diesen Objekten noch relativ objekttreu insofern, als es Objektorte von gleichem Dickengradienten (Null) ziemlich gleichartig (relativ dunkel) wiedergibt.

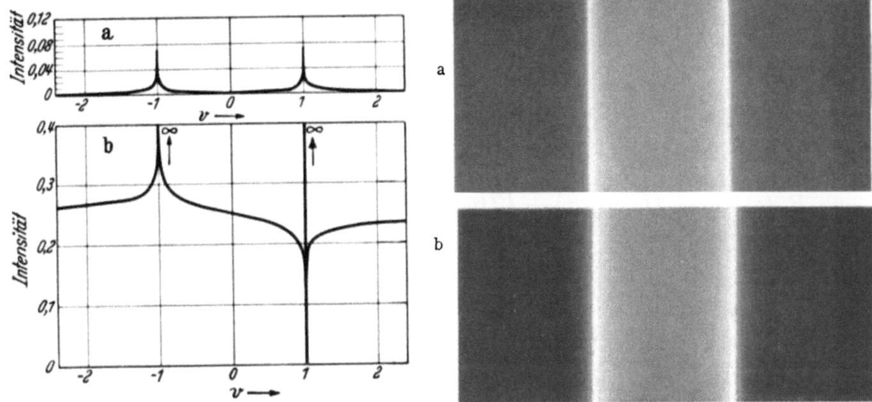

Fig. 38a u. b. Intensitätsverlauf im Bilde eines Zaponlackstreifens, a mit dem Minimumschlierenverfahren, b mit TÖPLERs Schlierenverfahren. Alle Spektren mit $\gamma > 0$ sind abgedeckt. Das Bild mit komplementärer Schneide erhält man, wenn man die Fig. 38b um die Gerade $v = 0$ umklappt.

Fig. 39a u. b. TÖPLERsches Schlierenbild eines Amplitudenobjekts ($\alpha = 1$ in dunkler Umgebung) bei einer durch die optische Achse gehenden Schneidenkante. Die Bildintensität folgt der Formel (39.6). a für extrem hohe Objektivapertur; b für geringere Objektivapertur.

39. TÖPLERsches Schlierenverfahren mit Schneidenkante bei $\alpha = 0$.

Nach Gl. (31.12) ist die Amplitude im TÖPLERschen Schlierenbild eines Streifens mit $\mathfrak{a} = a \cdot e^{-i\varphi}$

$$B_T(x) = \tfrac{1}{2} + \tfrac{1}{2}(a\cos\varphi - 1 - ia\sin\varphi)\,(\Pi(u_0; v) + i\,\Gamma(u_0; v)), \qquad (39.1)$$

und die Intensität ist

$$|B_T(x)|^2 = \tfrac{1}{4}\{1 + (a\cos\varphi - 1)\,\Pi(u_0; v) + a\sin\varphi\,\Gamma(u_0; v)\}^2 + \\ + \tfrac{1}{4}\{-a\sin\varphi\,\Pi(u_0; v) + (a\cos\varphi - 1)\,\Gamma(u_0; v)\}^2. \qquad (39.2)$$

Für ein reines Phasenobjekt ($a = 1$) mit kleiner Phasenverschiebung $|\varphi| \ll 1$ ist in erster Näherung

$$|B_T(x)|^2 \approx \tfrac{1}{4}(1 + \varphi\,\Gamma(u_0; v))^2, \qquad (39.3)$$

und für hohes Auflösungsvermögen ($\gamma_0 \to \infty$) ist das

$$|B_T(x)|^2 \approx \frac{1}{4}\left(1 + \frac{\varphi}{\pi}\log\left|\frac{v+1}{v-1}\right|\right)^2. \qquad (39.4)$$

Fig. 38b zeigt diesen Intensitätsverlauf für $\varphi = -0{,}1$; den experimentellen Befund gibt die Fig. 32f für drei Phasenstreifen. Freilich war hier $\varphi = +0{,}1$, dafür aber auch die TÖPLER-Schneide komplementär zu der in unserer Rechnung vorausgesetzten. Das Bild entspricht dann wieder völlig der Gl. (39.4) und der Fig. 38b. An der linken Kante der Objektstreifen liegt ein „logarithmisches" Maximum (das sich bei endlicher Apertur γ_0 natürlich in ein endliches umwandelt). Genau in der Objektmitte herrscht die Fernabintensität. An der rechten Kante liegt ein Minimum[1].

---

[1] Die Struktur dieses Minimums wurde in Ann. Physik (6) **7**, 51 (1950) untersucht. In seiner Mitte liegt ein gegen ∞ ziehendes steiles Maximum, das jedoch bei endlichem Auflösungsvermögen praktisch stets verschwindet.

Das Töplersche Schlierenbild gibt für Phasen- und Amplitudenobjekte völlig verschiedene Bilder; z.B. hat ein Amplitudenstreifen mit $a = 0$ die Bildamplitude $B_T(x)$ nach Gl. (39.1) und die Intensität

$$|B_T(x)|^2 = \frac{1}{4}\left|1 - \Pi(u_0; v) - i\,\Gamma(u_0; v)\right|^2$$

$$\rightarrow \begin{cases} \dfrac{1}{4\pi}\log^2\left|\dfrac{v+1}{v-1}\right| & \text{für} \quad |x| < \dfrac{b}{2} \\[2ex] \dfrac{1}{4} + \dfrac{1}{4\pi}\log^2\left|\dfrac{v+1}{v-1}\right| & \text{sonst.} \end{cases} \quad \Bigg\} \; \text{für } u_0 \rightarrow \infty. \quad (39.5)$$

Bei einem hellen Objekt $a = 1$ in dunkler Umgebung findet man entsprechend

$$|B_T(x)|^2 = \frac{1}{4}\left(\Pi^2(u_0; v) + \Gamma^2(u_0; v)\right)$$

$$\rightarrow \begin{cases} \dfrac{1}{4} + \dfrac{1}{4\pi}\log^2\left|\dfrac{v+1}{v-1}\right| & \text{für} \quad |x| < \dfrac{b}{2} \\[2ex] \dfrac{1}{4\pi}\log^2\left|\dfrac{v+1}{v-1}\right| & \text{sonst.} \end{cases} \quad \Bigg\} \; \text{für } u_0 \rightarrow \infty. \quad (39.6)$$

Die Bildintensität für reine Amplitudenobjekte ist völlig symmetrisch. Fig. 39 zeigt eine solche Aufnahme für sehr hohes Auflösungsvermögen (a) und darunter (b) für geringere Gesamtapertur des Aufnahmeobjektivs. Entsprechend wird im Töplerschen Schlierenbild auch das leere Objektfeld dargestellt.

Ändert man die Breite des Objektstreifens, so bleibt die relative Intensitätsverteilung in dem Töplerschen Schlierenbilde davon unberührt, sofern nur das Auflösungsvermögen hinreichend groß ist; das zeigt die Gl. (39.4) für Phasenobjekte ebenso wie die Gl. (39.5) mit $u_0 \rightarrow \infty$ für Amplitudenobjekte. Die in Fig. 39 gezeigten „Randerscheinungen" des gesamten ausgeleuchteten Objektfeldes beeinträchtigen daher das Schlierenbild der Objekte in der Mitte des Objektfeldes nicht, da die Objekte als klein gegenüber dem gesamten Objektfeld vorausgesetzt waren.

f) Schlieren- und Phasenkontrastverfahren bei zylindrischen Objekten kreisförmigen oder elliptischen Querschnitts.

40. Objektfunktion, Beugungsfunktion und Bildfunktionen zylindrischer Objekte. Die größte Bedeutung hat das Phasenkontrastverfahren in der Mikroskopie gewonnen. Dort haben sehr viele Objekte, z.B. Bakterien und Chromosomen, annähernd die Gestalt feiner zylindrischer Stäbchen. Für die richtige Deutung beobachteter Phasenkontrastbilder ist es daher zweckmäßig, die Bilder genau bekannter zylindrischer Testobjekte zu beobachten und zu berechnen. Einige Aufnahmen zeigen die Fig. 40 und 46.

Für ein „feines" zylindrisches Objekt vom Radius R und dem Brechungsindex n mit einer Achse in y-Richtung ist die Objektfunktion:

$$O(x) = \begin{cases} e^{-i\beta\sqrt{R^2 - x^2}} \approx 1 - i\beta\sqrt{R^2 - x^2} & \text{für} \quad |x| < R \\ 1 & \text{sonst} \end{cases} \quad \Bigg\} \quad (40.1)$$

mit

$$\beta \approx \frac{4\pi}{\lambda}(n - n_0); \quad |\beta R| \ll 1. \quad (40.2)$$

n_0 ist der Brechungsindex des umgebenden Mediums. Die Objektfunktion (40.1) erfaßt auch den allgemeineren Fall elliptischen Querschnitts bei beliebiger Lage der Ellipsenachsen; Gl. (40.2) wäre dann unwesentlich zu modifizieren. Ist $G \gg R$ die Breite des ausgeleuchteten Objektfeldes, in dessen Mitte das Objekt liegt, so ist nach Gl. (24.4) mit $K = 1$ die Beugungsfunktion

$$E(\gamma) = \int\limits_{-(G/2)}^{G/2} O(x)\, e^{-2\pi i \gamma x}\, dx; \qquad \gamma = \frac{\sin \alpha}{\lambda}, \tag{40.3}$$

$$E(\gamma) = \int\limits_{-(G/2)}^{G/2} e^{-2\pi i \gamma x}\, dx + \int\limits_{-R}^{R} \left(e^{-i\beta \sqrt{R^2 - x^2}} - 1 \right) e^{-2\pi i \gamma x}\, dx, \tag{40.4}$$

$$\left.\begin{aligned} E(\gamma) &= G\,\frac{\sin \pi G \gamma}{\pi G \gamma} - i\beta R^2 \times \\ &\quad \times \int\limits_{-1}^{1} \sqrt{1 - v^2}\, e^{-2\pi i \gamma R v}\, dv; \quad v = \frac{x}{R}, \end{aligned}\right\} \tag{40.5}$$

$$E(\gamma) = G\,\frac{\sin \pi G \gamma}{\pi G \gamma} - i\beta R \cdot \frac{J_1(2\pi \gamma R)}{2\gamma}. \tag{40.6}$$

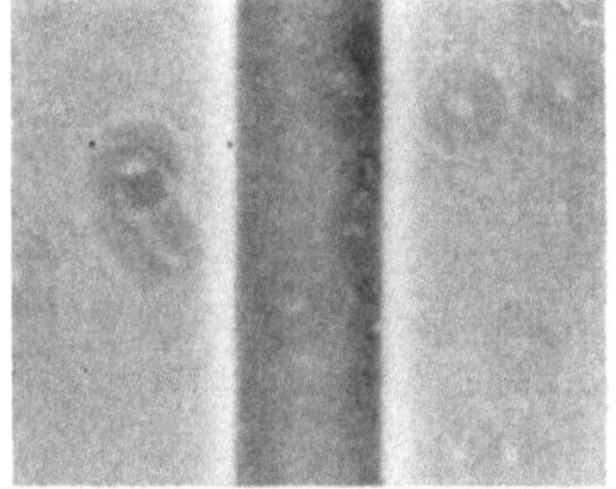

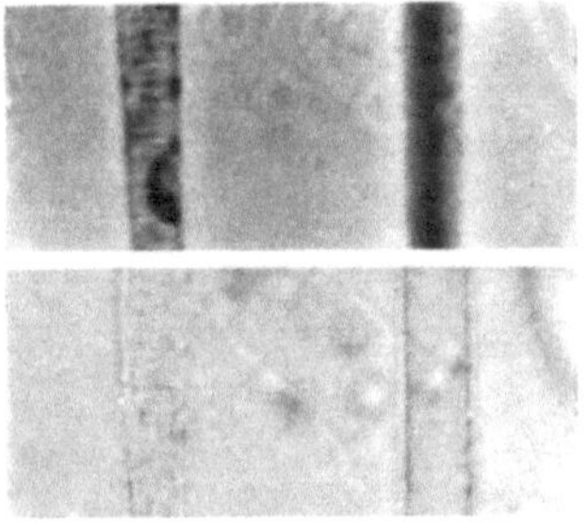

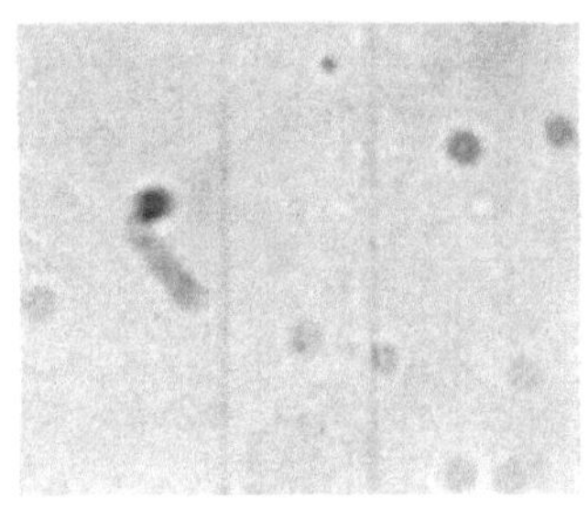

a

b

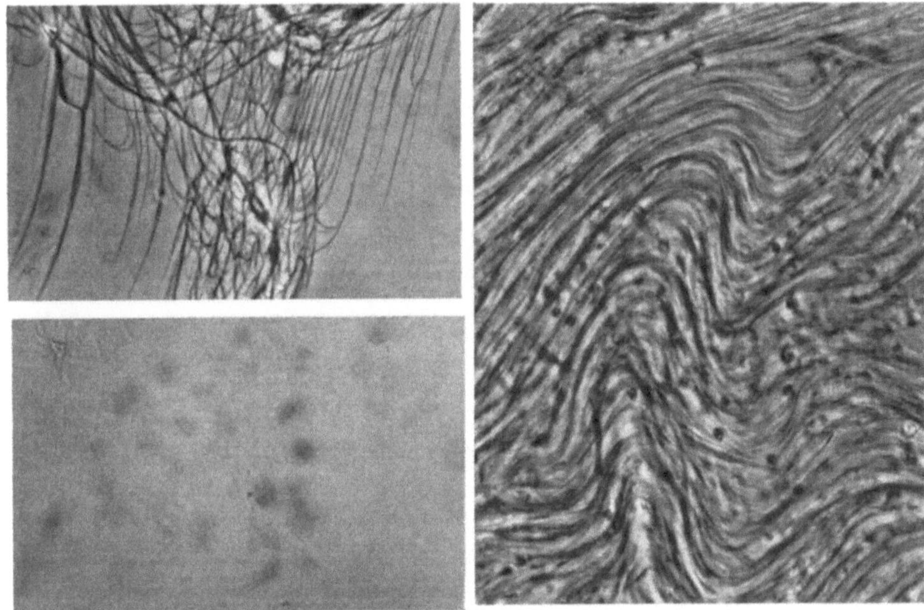

c   d

Fig. 40a—d. Mikroaufnahmen zylindrischer Objekte, oben mit Phasenkontrast, unten im Hellfeld. a Zwei Spinnenfäden (links 2,8 μ Durchmesser) in Xylol; b ein Glasfaden (6,7 μ Durchmesser) in Benzol; beides 1800fach. c Gelfasern einer Natronseife in Wasser, 540mal. d Inneres aus dem Gel einer Natronseife in Wasser, 540mal, Phasenkontrast.

J_1 bezeichnet die erste BESSEL-Funktion. Die Bildamplitude nach Gl. (24.7)

$$B(x) = \int_{-\gamma_0}^{\gamma_0} Q(\gamma)\, E(\gamma)\, e^{2\pi i \gamma x}\, dx \qquad (40.7)$$

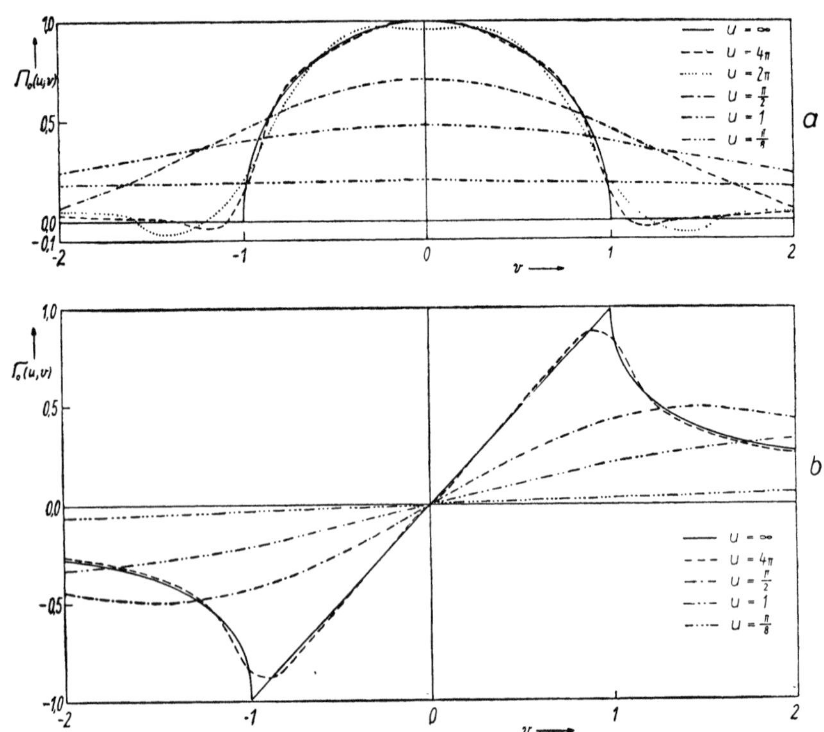

Fig. 41 a u. b. Hilfsfunktionen $\Pi_0(u; v)$ und $\Gamma_0(u; v)$.

mit einer die Objektivapertur kennzeichnenden Größe γ_0 ist dann für die Eingriffsfunktionen $Q(x)$ der verschiedenen Schlierenverfahren, wie sie in den Gl. (29.1) bis (29.8) genannt waren,

$$B_H(x) = 1 - i\beta R \Pi_0(u_0; v), \qquad (40.8)$$

$$B_{sZ}(x) = iA - i\beta R \Pi_0(u_0; v) = B_H(x) - 1 + iA, \qquad (40.9)$$

$$B_Z(x) = iA - i\beta R \Pi_0(u_0; v) - (iA - 1) i\beta R \Pi_0(u_i; v), \qquad (40.10)$$

$$B_{sD}(x) = - i\beta R \Pi_0(u_0; v) = B_H(x) - 1, \qquad (40.11)$$

$$B_D(x) = - i\beta R \{\Pi_0(u_0; v) - \Pi_0(u_i; v)\}, \qquad (40.12)$$

$$B_{eD}(x) = \tfrac{1}{2} B_D(x) + \beta R \{\Gamma_0(u_0; v) - \Gamma_0(u_i; v)\}, \qquad (40.13)$$

$$B_M(x) = + \beta R \Gamma_0(u_0; v), \qquad (40.14)$$

$$B_T(x) = \frac{1}{2} - \frac{i\beta R}{2} \{\Pi_0(u_0; v) + i\Gamma_0(u_0; v)\} = \frac{1}{2}\{B_H(x) + B_M(x)\}. \qquad (40.15)$$

Diese Gleichungen sind ganz analog den Gln. (31.5) bis (31.12) aufgebaut. Es bedeutet jetzt

$$v = \frac{x}{R}; \qquad u_0 = 2\pi R \gamma_0; \qquad u_i = 2\pi R \gamma_i. \qquad (40.16)$$

Der Durchmesser $2R$ tritt also an die Stelle der in Ziff. 31 entsprechend vorkommenden Streifenbreite b. Die reellen Hilfsfunktionen $\Pi_0(u; v)$ und $\Gamma_0(u; v)$ sind jetzt durch

$$\int_0^u \frac{J_1(t)}{t}\, e^{ivt}\, dt = \Pi_0(u; v) + i\, \Gamma_0(u; v) \tag{40.17}$$

definiert und in Fig. 41 für einige Werte von u als Funktionen von v dargestellt; sie sind nur für $u \to \infty$ elementare Funktionen

$$\Pi_0(\infty; v) = \left\{ \begin{array}{cc} \sqrt{1 - v^2} & \text{für} \quad |v| < 1 \\ 0 & \text{sonst}, \end{array} \right\} \tag{40.18}$$

$$\Gamma_0(\infty; v) = \left\{ \begin{array}{cc} v & \text{für} \quad |v| \leqq 1 \\ \dfrac{1}{v + \sqrt{v^2 + 1}} & \text{für} \quad |v| > 1. \end{array} \right\} \tag{40.19}$$

Aus ihnen können die Bildfunktionen bei dem Phasenkontrast- und dem Schlierenverfahren in den nächsten Ziffern abgelesen werden.

41. Hellfeldbild bei Aperturbegrenzung. Im Rahmen unserer Näherung [Vernachlässigung von $(\beta R)^2$ neben 1] hat die Bildintensität $B_H(x)^2$ nach Gl. (40.8) keine Abweichung der Intensität im Bilde des Objekts von der Intensität 1 der Umgebung. Ein Vergleich der Π_0-Kurven (Fig. 41 a) mit den Π-Kurven (Fig. 33 a) zeigt, daß der relativ noch starke Randkontrast des Hellfeldbildes bei Streifenobjekten nur dann auch bei zylindrischen Objekten erreicht werden kann, wenn das Auflösungsvermögen schon recht klein ist. Dem entspricht es, daß zylindrische Objekte — und fast alle mikroskopisch interessanten Objekte haben Ränder ähnlich den Zylinderrändern — bei Hellfeldbeobachtung nur durch eine sehr geringe Randverdunkelung bemerkbar sind, wenn man nicht absichtlich unscharf einstellt (vgl. Fig. 41 oben) oder die Aperturiris stark zuzieht. Beides bedeutet Verlust an Abbildungsschärfe.

Bei absorbierendem Objekt, d. h. z. B. β rein imaginär $\beta = -i\beta'$, ist nach Gl. (40.8)

$$B_H(x) = 1 - \beta'\, R\, \Pi_0(u_0; v). \tag{41.1}$$

Man erhält also die im Hellfeldbild bei Aperturbegrenzung resultierende Bildamplitude im wesentlichen dadurch, daß man die Fig. 41 a auf den Kopf stellt. Wäre die Apertur unbegrenzt $(\gamma_0 \to \infty)$, so besäße die Amplitude am Bildort des Objekts eine „halbkreisförmige" Verdunkelung; wir wollen sie als völlig „objekttreu" bezeichnen, da die Amplitude dann eine lineare Funktion der Objektdicke über das ganze Bild ist. Je stärker die Apertur begrenzt ist, desto unschärfer wird das Objekt durch die Kurven $\Pi_0(u_0; v)$ wiedergegeben. Da $u_0 = 2\pi R\gamma_0$ ist, tritt dasselbe auch bei gleichbleibenden γ_0 ein, wenn R kleiner wird.

42. Strenges Phasenkontrastverfahren nach ZERNIKE. Bei Übergang von $B_H(x)$ zu $B_{sZ}(x)$, also von Gl. (40.8) zu (40.9), wird die 1 im Ausdruck für die Hellfeldamplitude ersetzt durch iA. Da ein rein imaginärer, konstanter Faktor an der Gesamtamplitude belanglos ist, vertauschen also Phasenobjekte (β reell) und Amplitudenobjekte (β imaginär) im wesentlichen ihre Rolle bei den Übergang vom Hellfeld zum Phasenkontrast. Die Amplitude im Bild eines Phasenobjekts ist daher jetzt durch

$$\left| \frac{B_{sZ}(x)}{A} \right| = \left| 1 - \frac{\beta R}{A}\, \Pi_0(u_0; v) \right| \tag{42.1}$$

dargestellt, also durch die auf den Kopf gestellte Fig. 40a. Das strenge Phasenkontrastverfahren stellt also das Objekt bei großem Auflösungsvermögen ($\gamma_0 \to \infty$) völlig „objekttreu", d.h. als „halbkreisartige" Verdunkelung dar; ein geringeres Auflösungsvermögen hat nur die gleichen Beugungsunschärfen zur Folge, die auch im Hellfeldbild an Amplitudenobjekten bekannt sind.

Für optimales $A = \beta R$ nimmt die Amplitude in der Objektmitte auf Null ab.

In Fig. 46e und f können die Spirochäten als objekttreu wiedergegeben angesehen werden, ebenso die Gelfäden in Fig. 40c; aber schon das dicke Bakterium in Fig. 46d zeigt den hellen Hof des nicht strengen Phasenkontrasts, ebenso der Glasfaden in Fig. 40b und die Zellen in Fig. 46e. Die Bildverfälschungen — bedingt durch die zwangsläufig endliche Breite des Zernike-Phasenstreifens — sind Gegenstand der nächsten Ziffer.

43. Reales Phasenkontrastverfahren mit endlich breitem Zernike-Phasenstreifen bei zylindrischen Objekten. Hat der Zernike-Phasenstreifen in der Richtungsauswahlebene eine solche Breite, daß alle „Spektren" [Teile des Beugungsbildes $(E(\gamma))$] mit $|\gamma| < \gamma_i$ um 90° phasengedreht werden und ihre Amplituden auf den Bruchteil A geschwächt werden, so interferiert das resultierende Beugungsbild zu der Bildamplitude nach Gl. (40.10). Wählt man zur Erreichung guten Kontrasts $A = \beta R$, so ist in erster Näherung für $|\beta R| \ll 1$

$$\left| \frac{B_Z(x)}{A} \right| = \left| 1 - \Pi_0\left(2\pi R \gamma_0; \frac{x}{R}\right) + \Pi_0\left(2\pi R \gamma_i; \frac{x}{R}\right) \right| . \tag{43.1}$$

Die Bildfunktion ist in Fig. 42 für fünf verschieden dicke zylindrische Stäbe gleichen Materials (z.B. Glas) in gleichem Medium (z.B. Benzol) dargestellt. Die Amplitude des objektfernen Bildteiles ist als 1 bezeichnet. Das Auflösungsvermögen wird als groß ($\gamma_0 \to \infty$) angesehen. Die „halbkreisartige" Vertiefung im Bild der dünnsten Objekte (unten in Fig. 42) bedeutet objekttreue Wiedergabe. Für

$$R > 0{,}1 \frac{1}{\gamma_i} \tag{43.2}$$

dagegen verfälschen helle „Höfe" und für

$$R > 0{,}3 \frac{1}{\gamma_i} \tag{43.3}$$

auch Aufhellungen im Objektinnern das Bild. Fig. 40 zeigt entsprechende Aufnahmen.

Am auffälligsten und am wichtigsten ist die Tatsache, daß der wahre Rand des Objekts durch das Amplitudenmaximum im Phasenkontrastbild — jedenfalls bei hohem Auflösungsvermögen — angezeigt wird. Spricht man — wie üblich und durch das Aussehen der Phasenkontrastbilder selbst nahegelegt — als Rand den Ort an, an dem zwischen hellem Hof und dunklerem Innenteil die „Fernabintensität" vorliegt, so schätzt man die Objekte in Phasenkontrastbildern zu klein.

Hierin liegt der wahre Grund für die Diskrepanz zwischen Größenmessungen am Phasenkontrastbild und am Hellfeldbild. Als die Phasenkontrastbilder nach der von Zernike[1] veröffentlichten Theorie des Phasenkontrastverfahrens noch als völlig objekttreu galten, schlossen Köhler und Loos[2] aus Phasenkontrastbildern, daß die Durchmesser roter Blutkörperchen um 10 bis 15% kleiner seien,

---

[1] F. Zernike: Physica, Haag **1**, 689 (1934).
[2] A. Köhler u. W. Loos: Naturwiss. **29**, 49 (1941).

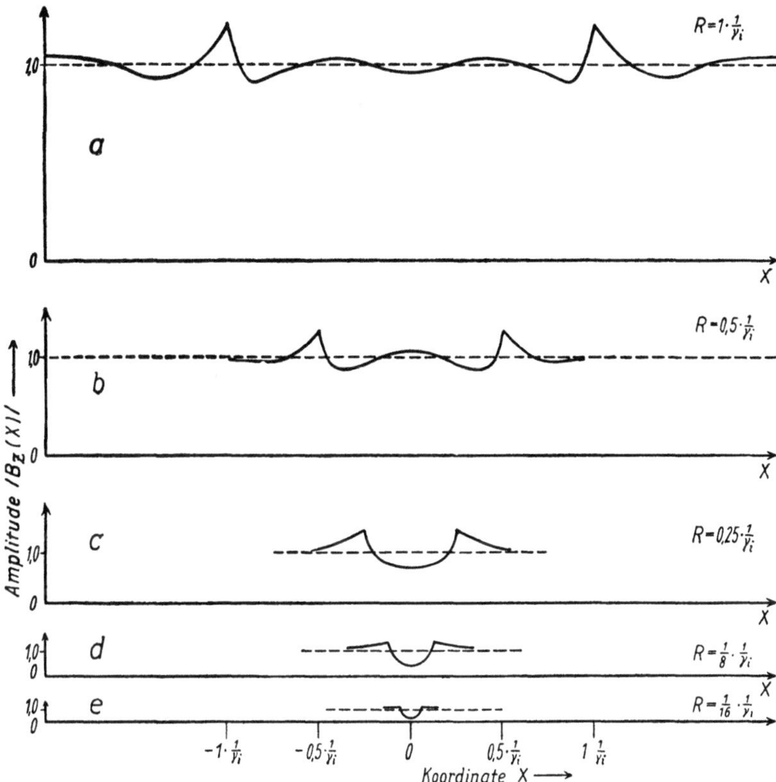

Fig. 42 a—e. Amplituden in Phasenkontrastbildern zylindrischer Objekte gleichen Materials aber verschiedener Radien. Hohes Auflösungsvermögen ($\gamma_0 \to \infty$).

als bis dahin an Hellfeldbildern gemessen. Die Hellfeldmessungen erwiesen sich jedoch als richtig[1]. Auch die Annahme von KÖHLER und LOOS, daß die zentrale Aufhellung im Phasenkontrastbild der roten Blutkörperchen (s. Fig. 46e!) der natürlichen Eindellung entsprechen dürfte, trifft nicht zu. Auch streng zylindrische Glasfäden zeigen bei passender Dicke Aufhellungen, denen nichts Wirkliches entspricht. Die Kurve b in Fig. 42 gibt die theoretische Erklärung hierfür auf Grund der Gl. (43.1)[1].

Leider gibt auch das Intensitätsmaximum in der Praxis nicht den wahren Objektrand; denn das endliche Auflösungsvermögen bewirkt nicht nur eine Abrundung des Maximums sondern auch eine Verlagerung nach außen; das zeigen die theoretischen Kurven der Fig. 43, die abgesehen vom Auflösungsvermögen ganz den Fig. 42b bis e entsprechen. Eine Regel zur Auffindung des wahren Objektrandes am Phasenkontrastbild wäre zwangsläufig kompliziert wegen der vielen Variationsmöglichkeiten.

Fig. 43. Amplituden in Phasenkontrastbildern zylindrischer Objekte gleichen Materials aber verschiedener Radien. Die Kurven b bis d entsprechen den gleich benannten der Fig. 42, gelten jedoch für endliches Auflösungsvermögen $\gamma_0 = 4\gamma_i$. Gesamte Objektivapertur gleich vierfacher Apertur des ZERNIKE-Streifens.

---

[1] H. WOLTER: Ann. Physik (6) **7**, 153 (1950).

44. Dunkelfeldverfahren und TÖPLERS Schlierenverfahren bei zylindrischen Objekten. α) *Das strenge Dunkelfeldverfahren* läßt sich als strenges Phasenkonstrastverfahren mit $A = 0$ auffassen. Sein Bild wird nach Gl. (40.11) durch die Kurven 40a unmittelbar wiedergegeben und ist offenbar so ideal objekttreu wie das Hellfeldbild eines absorbierenden Objekts derselben Gestalt. Erforderlich ist jedoch, daß alle Objekte vereinzelt liegen (vgl. Ziff. 26). Die praktische Durchführung des strengen Dunkelfeldverfahrens setzt außerdem die Verwendung einer sehr schmalen Dunkelfeldblende und eines entsprechend engen Lichtquellenspaltes voraus, hat also dieselben Lichtschwierigkeiten wie das strenge Phasenkontrastverfahren.

β) *Bei dem symmetrischen Dunkelfeldverfahren* mit Abdeckung aller Spektren, deren $|\gamma| < \gamma_i$ ist, ist nach Gl. (40.12) die Bildfunktion einfach gleich der Differenz zweier Π_0-Funktionen (Fig. 40a). Ein Vergleich mit Gl. (42.1) zeigt, daß man die Dunkelfeldbilder aus den Phasenkontrastbildern Fig. 42 und 43 einfach erhalten kann, indem man die mit 1 am linken Rande bezeichnete Horizontale zur Abszissenachse macht und die Kurve jeweils an ihr spiegelt. Im Bilde erhält man offenbar zwei scharfe Linien der Intensität Null. Während diese aber bei Streifenobjekten die Objektränder gaben, würde man hier — wollte man sie als Bilder der Zylinderränder ansprechen — ebenso zu kleine Durchmesser erhalten wie bei der entsprechenden Deutung der Phasenkontrastbeobachtungen.

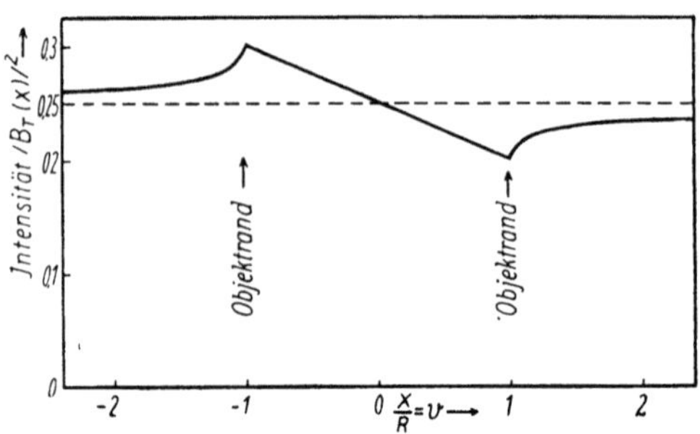

Fig. 44. Intensität in TÖPLERS Schlierenbild zylindrischer Objekte bei großem Auflösungsvermögen ($\gamma_0 \to \infty$) für $\beta R = -0{,}1$.

γ) *Das einseitige Dunkelfeldverfahren,* die in der üblichen Mikroskopie fast nur benutzte Form des Dunkelfeldverfahrens, gibt nach Gl. (40.13) Bilder, die aus den Kurven der Fig. 41 leicht abzuschätzen sind. Die Bilder geben unmittelbar wenig Aufschluß über die Objekte; das Verfahren ist das schlechteste unter den hier diskutierten.

δ) *Das Minimumschlierenverfahren* gibt eine Bildamplitude, die nach Gl. (40.14) besonders einfach als Γ-Funktion nach Fig. 41b aufgebaut ist. Der lineare Verlauf der Amplitude läßt, sobald er auftritt, darauf schließen, daß ein Objekt elliptischen Querschnitts vorliegt.

ε) *Das TÖPLERsche Schlierenbild* des Objekts ist nach Gl. (40.15) im allgemeinen viel komplizierter als das Minimumschlierenbild und erlaubt eine ähnlich einfache Deutung nur bei so dünnen Objekten, daß man statt (40.15) einfacher schreiben kann

$$|B_T(x)| \approx \tfrac{1}{2}\{1 + \beta R\,\Gamma_0(u_0; v)\}. \tag{44.1}$$

Für ein Objekt mit $\beta R = -0{,}1$ gibt Fig. 44 diese Bildamplitude. Bei $\beta R = +0{,}1$ erhielte man das Minimum links und das Maximum rechts. Die vorübergehende Anbringung der TÖPLER-Schneide zusätzlich zur 180°-Phasenplatte ist nützlich, um am qualitativen TÖPLER-Bilde das Vorzeichen von β — also die Entscheidung darüber, ob $n > n_0$ oder $n < n_0$ ist — zu entnehmen. Aus der Intensität des Minimumschlierenbildes allein geht das nicht hervor.

g) Modifikation der Erscheinungen durch endliche Breite und Kreisform des Lichtquellenspaltes.

45. Weil sie die charakteristischen Erscheinungen in Phasenkontrast- und Schlierenbildern am durchsichtigsten zeigen, wurden hier die Rechnungen wiedergegeben, die einen linearen, extrem schmalen Lichtquellenspalt voraussetzen. Eine endliche Breite des Lichtquellenspaltes führt zu einer unwesentlichen Ausbügelung der Wellenlinien in den beiden untersten Kurven der Fig. 36. In den Kurven der Fig. 38 und 44 werden die Maxima und Minima abgerundet. Da man bei den am meisten gebrauchten Beleuchtungsvorrichtungen alle Punkte des Lichtquellenspaltes primär als unkohärent zueinander strahlend ansehen kann, deckt man z. B. bei dem Minimumschlierenverfahren zweckmäßig genau das Bild des Lichtquellenspaltes ab, ohne an der 180°-Phasenverschiebung der oben beschriebenen Phasenplatte etwas zu ändern.

Der lineare Lichtquellenspalt und der lineare ZERNIKE-Phasenstreifen lassen nur Objektstrukturen in einer Koordinatenrichtung senkrecht zum Spalt erkennen. Ein Gitter mit senkrecht zum Spalt liegenden Strichen hätte Spektren, die vom Spektrum nullter Ordnung nicht getrennt wären. Eine sehr kleine Kreisblende als Lichtquellenblende und ein ZERNIKE-Phasenplättchen, das kongruent zum Bild der Kreisblende ist, machen die Strukturen jeder Richtung sichtbar, geben jedoch wegen der endlichen Flächenhelligkeit aller Lichtquellen zu dunkle Bilder z. B. bei der mikroskopischen Anwendung des Phasenkontrastverfahrens.

Nach ZERNIKES Vorschlag benutzt man daher in der Mikroskopie einen ringförmigen[1]

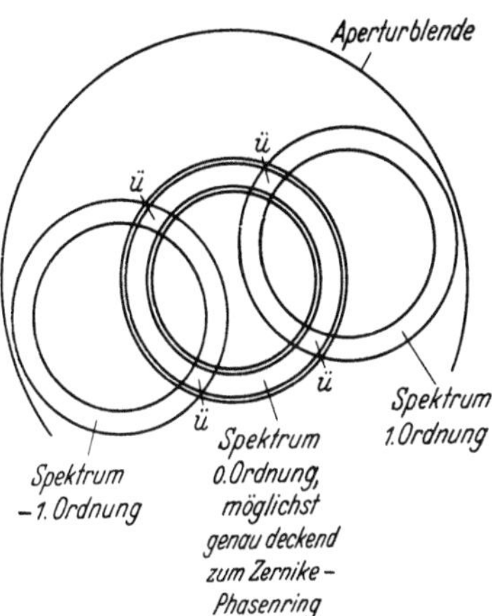

Fig. 45. Skizze der Erscheinung in der oberen Brennebene eines Mikroskopobjektivs, wenn auf dem Objekttisch ein Gitter liegt.

Spalt und einen zu seinem Bilde kongruenten ZERNIKE-Phasenstreifen; er behandelt alle Richtungen gleichmäßig, führt bei gleicher Breite zu viel helleren Bildern und nutzt die Objektivapertur wesentlich besser aus als die Punktblende. Freilich überdecken sich die Spektren gegenseitig auch bei dem Ringspalt ein wenig (in Fig. 45 bei *ü*); doch ist der Flächenanteil der Überdeckung für qualitative Untersuchungen belanglos.

U. SCHMIDT[2] hat kürzlich die mit einem Ringspalt endlicher Breite zu erhaltenden Phasenkontrastbilder für Streifenobjekte berechnet und mit photometrierten Phasenkontrastaufnahmen verglichen, die er mit einem handelsüblichen Phasenkontrastmikroskop von genau bekannten Streifenobjekten gemacht hat. Er konnte die volle Übereinstimmung zeigen und nachweisen, daß gegenüber den hier in den Ziff. 34ff. diskutierten Erscheinungen keine wesentlichen Änderungen auftraten.

h) Zusammenfassung zum Teil C I.

46. α) Unter den zur deutlichen Abbildung nichtabsorbierender Objekte geeigneten Verfahren (Dunkelfeld-, Schlieren- und Phasenkontrastverfahren)

---

[1] F. ZERNIKE: Z. techn. Phys. **16**, 454 (1935). — Phys. Z. **36**, 848 (1935). — A. KÖHLER u. W. LOOS: Naturwiss. **29**, 49 (1941).

[2] U. SCHMIDT: Ann. Physik (6) **16**, 68 (1955). — Eine verwandte theoretische Rechnung, wenn auch ohne Vergleich mit experimentell gewonnenen Photometerkurven, gab bereits BEIER: Jenaer Jb. **1953**, 187, 218.

gibt ein streng durchgeführtes Phasenkontrastverfahren bei hinreichend dünnen und schmalen Objekten das objekttreueste Bild. Das Vorzeichen der Brechungsindexdifferenz gegenüber der Umgebung ist für jedes dünne Phasenobjekt erkennbar. Amplitudenobjekte werden undeutlich.

β) Das strenge Dunkelfeldverfahren ist dem strengen Phasenkontrastverfahren nur dann fast gleichwertig, wenn nur vereinzelte Objekte in großen Abständen voneinander in einem großen Objektfelde liegen. Aber auch dann ist über das Vorzeichen der Brechungsindexdifferenz nichts zu entnehmen. Die Dickenmessung an feinsten Objekten kann durch Dunkelfeldphotometrie geschehen. Phasen- und Amplitudenobjekte werden gleich gut dargestellt.

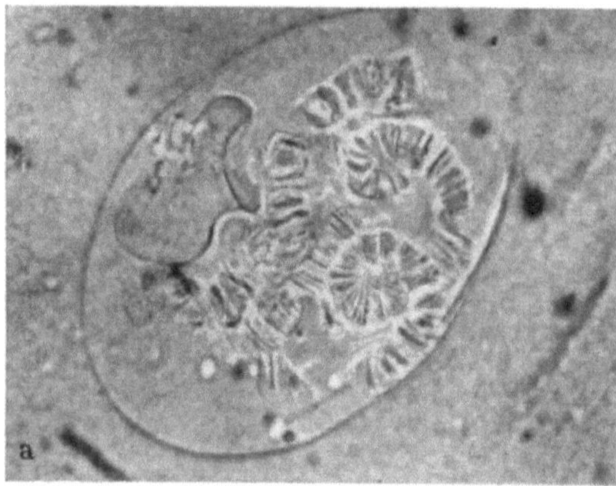
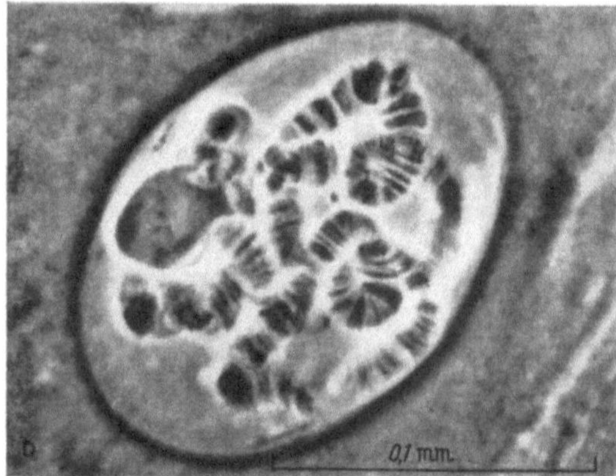

Fig. 46a u. b. Chromosomen einer Speicheldrüsenzelle der Larve einer Chironomide; a Hellfeld, b positiver Phasenkontrast (nach K. MICHEL).

γ) Der Wunsch nach erhöhter Lichtstärke im Bilde zwingt zu breiterem Lichtquellendiaphragma und Abgeben von den strengen Forderungen des Phasenkontrast- und Dunkelfeldverfahrens. Das Dunkelfeldbild verliert dann stark an Objekttreue, das Phasenkontrastbild in geringerem Maße; doch werden bei dem Phasenkontrastverfahren helle Höfe um die Objekte, Innenstrukturen und oft zu kleine Abmessungen vorgetäuscht, sobald die Breite des Objekts den Wert $0,2 \cdot \dfrac{\lambda}{\sin \alpha_i}$ überschreitet (α_i ist die Apertur des ZERNIKE-Phasenstreifens).

δ) TÖPLERs Schlierenverfahren gibt oft anschauliche „Reliefbilder".

ε) Das Minimumschlierenverfahren gibt meist wenig anschauliche Bilder, erlaubt aber die bequemste quantitative Auswertung, wenn eine Vorzeichenentscheidung über die Brechungsindexdifferenz zusätzlich mit der TÖPLER-Schneide gefällt wird.

ζ) Die unsymmetrischen Dunkelfeldverfahren geben schlecht deutbare Bilder.

η) Die große Leistungsfähigkeit des Phasenkontrastverfahrens vor allem ist durch die vielen Phasenkontrastuntersuchungen in der Medizin, der Biologie, Chemie, Technik und der Physik erwiesen[1]. Fig. 46 zeigt einige Beispiele zusätzlich zu den in Fig. 40 mitgeteilten.

---

[1] Siehe die Literaturhinweise in A. H. BENNETT, H. OSTERBERG, H. JUPNIK u. O. W RICHARDS: Phase microscopy, principles and applications. New York u. London 1951. — E. MENZEL: Z. angew. Phys. 3, 308 (1951). — E. INGELSTAM: Progress of research at the optics laboratory, Report Nr. 25 E, IV A 24, 185, 1953. — H. WOLTER: Das Phasenkontrastverfahren und seine Anwendbarkeit auf chemische Untersuchungen. Fortschr. chem. Forsch. 3, 1 (1954).

Hingewiesen sei auch auf die große Leistung des Phasenkontrastverfahrens bei der Prüfung astronomischer Optiken. Zu diesem Zweck hatte ZERNIKE seine erste Phasenkontrastvorrichtung erdacht[1]. LYOT hat an seinem Coronographen, für den besonders streulichtfreie Optik erforderlich ist, das Phasenkontrast-

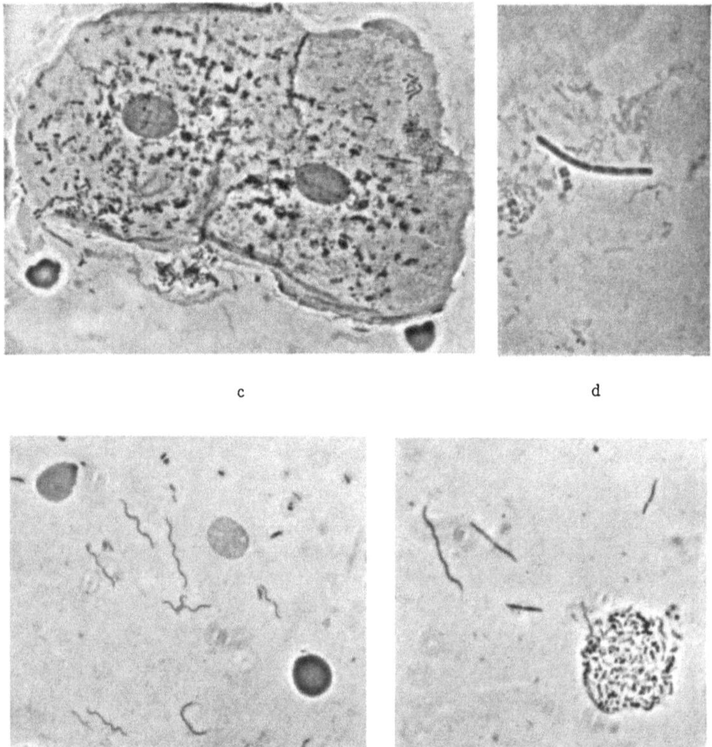

c   d

Fig. 46 c—f. Zellen und Bakterien im Sputum bei positiver Phasenkontrastbeobachtung. c 690fach; d 750fach; e 790fach; f 1000fach vergrößert. Alle Aufnahmen mit Zeiss-Phasenkontrastmikroskop aufgenommen.

verfahren zu Prüfzwecken sehr erfolgreich eingesetzt. Mit stark absorbierendem $\lambda/4$-Plättchen konnte er Abweichungen von der optischen Weglänge in der Größenordnung von 1 Å nachweisen[2].

Nach CAMUS, FRANÇON, INGELSTAM und MARÉCHAL[3] können Beugungsgitter mit Hilfe des Phasenkontrastverfahrens auf die Herkunft der Geister untersucht werden. Das Gitter wird nur im Licht einer Ordnung und zwar in einer Spektrallinie und ihrer Geister auf den Film abgebildet. Die Gitterstriche sind dann an sich nach ABBE unerkennbar, da ja nur *eine* Ordnung zum Bilde beiträgt. Aber die gröbere fehlerhafte (periodische oder unperiodische) Struktur des Gitters wird sichtbar, wenn die Spektrallinie selbst mit einer stark absorbierenden $\lambda/4$-Schicht bedeckt wird, während die Geister ungehindert zum Bilde beitragen. Die Geister spielen die Rolle der Beugungsspektren, die durch Gitterfehler verursacht sind.

---

[1] F. ZERNIKE: Physica, Haag **1**, 689 (1934).
[2] B. LYOT u. M. FRANÇON: Rev. Opt. **29**, 499 (1950).
[3] A. CAMUS, M. FRANÇON, E. INGELSTAM u. A. MARÉCHAL: Rev. Opt. **30**, 121 (1951).

II. Ausführungsformen der Phasenkontrastgeräte; variabler und farbiger Phasenkontrast.

a) Zernikes Grundform und durch geometrisch-optische Maßnahmen variierte Ausführungsformen.

47. Zernikes Grundform der Phasenkontrastgeräte. Zernike hat die 90°-Phasenplatte so in die Richtungsauswahlebene gesetzt, wie das in Teil I beschrieben wurde. Da aber bei starken Mikroskopobjektiven diese Ebene, die beobachterseitige Brennebene, oft zwischen die Linsen selbst fällt, wurde um 1935 während der Entwicklung des Phasenkontrastmikroskops zeitweise auch die Abbildung dieser Ebene in den Okularteil und die Anbringung der Phasenplatte in diesem von Zernike und Köhler in den Kreis der Betrachtungen gezogen. Doch besaßen die ersten in der Firma Carl Zeiss von W. Loos entwickelten Phasenkontrastmikroskope in der Objektivbrennebene einen ringförmigen Phasenstreifen nach Zernikes Vorschlag. Da diese Methode sich sehr bewährte, wurde sie bis heute bei serienmäßig hergestellten Phasenkontrastmikroskopen fast durchweg beibehalten.

Der ringförmige Phasenstreifen besteht z. B. aus einem Lackring, der zwischen zwei Linsen oder Platten mit Kanadabalsam eingekittet ist und wegen seines kleinen Brechungsindexunterschiedes gegenüber dem Kanadabalsam eine beträchtliche Dicke und damit eine gute Reproduzierbarkeit haben kann. Der Ring wird neuerdings auch aus Aufdampfschichten, nichtmetallischen und metallischen, hergestellt, die außer der Phasenverschiebung, die meist ungefähr 90° beträgt, auch die Absorption des direkten Lichtes auf den Bruchteil A bewirken. Der positive Phasenkontrast wird wegen der Ähnlichkeit der Bilder mit denen chemisch gefärbter Präparate bevorzugt.

Die Ringblende in der unteren Kondensorbrennebene wird so einjustiert, daß ihr Bild den Zernike-Ring deckt. Kontrolliert wird das mit einem Hilfsmikroskop, das vorübergehend statt des Okulars in den Mikroskoptubus gesetzt und auf die Objektivbrennebene eingestellt wird.

48. Phasenkontrastplatte im Okularteil. Françon und Nomarski[1] haben die Zwischenabbildung der Richtungsauswahlebene in den Okularteil durchgeführt und damit erstens gewöhnliche Objektive auch für Phasenkontrast verwendbar gemacht und zweitens Platz zu beiden Seiten der Phasenplatte für Variationsmaßnahmen gewonnen (Fig. 47a). Als Phasenplatte dient z. B. ein Glasspiegel, der am Orte des abgebeugten Lichtes metallisiert, im Bereich des direkten Lichtes aber unbelegt ist und vom Licht unter dem Brewsterschen Winkel getroffen wird. Ein Polarisator und ein Analysator erlauben es, das Amplitudenverhältnis zwischen abgebeugtem und direktem Licht willkürlich und die Phasendifferenz in gewissen Grenzen zu variieren.

Die Verlegung der Phasenplatte durch Zwischenabbildung der Richtungsauswahlebene hat den für genaue Messungen wesentlichen Vorteil, daß man die Phasenplatte größer ausführen kann und daß sie dann zugleich nur für kleinere Aperturen gleichartige Phaseneigenschaften zu haben braucht.

49. Phasenkontrast vor dem Objektiv. Durch Zwischenabbildung des Objekts vor dem Objektiv schufen Françon und Nomarski[1] einen Zusatz, der jedes Mikroskop in ein Phasenkontrastmikroskop umwandelt, wenn er zwischen Kondensor und Objektiv angebracht wird. Fig. 47b zeigt einen Schnitt des Geräts. Der Kugelspiegel CQD bildet das Objekt M, das nahe seinem Mittelpunkt liegt, im Maßstab 1:1 bei M' ab; dabei geht das Licht über den halbdurch-

---

[1] M. Françon u. G. Nomarski: Rev. Opt. 29, 619 (1950).

lässig verspiegelten Planspiegel *CRD*. Da der Kugelspiegel bei *M'* nicht metallisiert ist, kann mit dem dort mittels Immersion in Kontakt mit dem massiven Glaskörper des Kugelspiegels gebrachten gewöhnlichen Objektiv das Bild *M'* beobachtet werden. *EF* ist ein Phasenring, der Phasenkontrast verursacht.

Polarisiertes Licht ermöglicht freie Wahl des Amplitudenverhältnisses für direktes und abgebeugtes Licht.

50. Phasenkontrast für Auflicht. Ein Auflichtmikroskop mit Phasenkontrastbeobachtung hat für die Untersuchung von Metalloberflächen großes Interesse. Es gibt daher zur Zeit schon mindestens fünf verschiedene Konstruktionen.

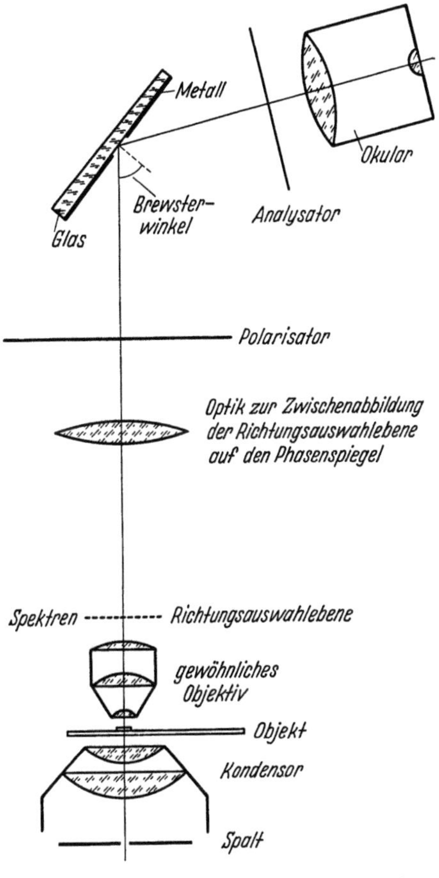

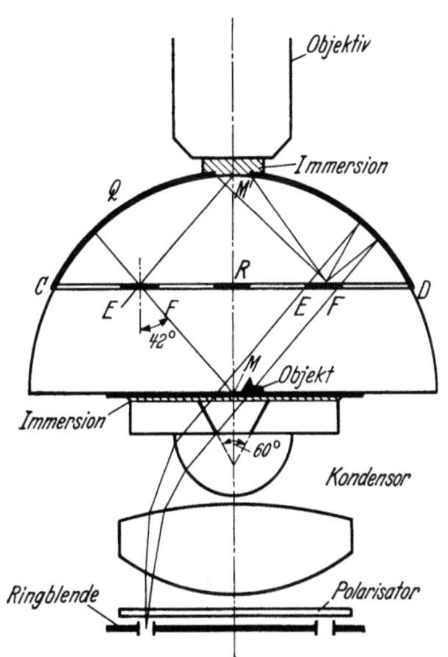

Fig. 47a. Teilweise metallisierter Spiegel als Phasenspiege im Okularteil eines Mikroskops. (Nach FRANÇON und NOMARSKI.)

Fig. 47b. FRANÇONsche Halbkugel als Zusatz zur Umwandlung eines gewöhnlichen Mikroskops in ein Phasenkontrastmikroskop.

Elementar ist die Einfügung der ZERNIKE-Phasenlamelle in ein gewöhnliches Metallmikroskop, in das man das Beleuchtungslicht mit halbdurchlässigem Spiegel oder schräger dünner Platte einschleust. Die Phasenplatte liegt dann in dem bereits wieder einfach gewordenen Teil des Strahlengangs, wenn der einschleusende Spiegel zwischen dem Objektiv und der Objektivbrennebene Platz hat[1].

Bei starken Objektiven ist das nicht der Fall, und dann hilft man sich mit einer Zwischenabbildung der Objektivbrennebene in den Okularteil[2,3] wie in Fig. 47a. FRANÇON und NOMARSKI benutzen auch ihren Phasenspiegel für Auflicht (s. Ziff. 48). Mit Zwischenabbildung läßt sich auch ein Universal-Phasenkontrastmikroskop für Auflicht und Durchlicht schaffen[3].

---

[1] H. JUPNIK, H. OSTERBERG u. G. E. PRIDE: J. Opt. Soc. Amer. **36**, 710 (1946).

[2] J. R. BENDFORD u. R. L. SEIDENBERG: J. Opt. Soc. Amer. **40**, 314 (1950).

[3] COOKE, TROUGHTON u. SIMMS: Siehe M. FRANÇON, Le microscope à contraste de phase et le microscope interférentiel, S. 79. Paris 1954.

b) Variabler Phasenkontrast für optimale Beobachtung und Messungen am Objekt.

51. Verfahren zur Variation des Amplitudenverhältnisses von direktem und gebeugtem Licht. Die „Empfindlichkeit" der Phasenkontrasteinrichtung hängt entscheidend von der Größe A ab (s. Ziff. 33), d.h. dem Amplitudenfaktor, den man außer der Phasenverschiebung dem direkten Licht relativ zum gebeugten aufzwingt. Die Bedeutung dieser Größe für den optimalen Kontrast bei vorgegebenen Objekten, zugleich aber auch für den Eindeutigkeitsbereich, wurde an Hand der Fig. 34 diskutiert. Eine feste Absorption im Zernike-Phasenstreifen, wie sie bei handelsüblichen Objektiven heute noch vorliegt, kann nur einen Kompromiß darstellen[1].

Phasenkontrastvorrichtungen zur Variation der Größe A wurden in den Jahren 1947/48 fast gleichzeitig von Hartley[2], Taylor[3], Osterberg[4], Kaster und Montarnal[5] und Locquin[6] angegeben. Gemeinsam liegt ihnen der Gedanke zugrunde, daß man zwei senkrecht zueinander linear polarisierten Strahlen (Fig. 48) jedes beliebige positive und „negative" Amplitudenverhältnis A geben kann, wenn man mit einem drehbaren Analysator nur jeweils eine Komponente wirken läßt. Sind mit unten diskutierten Mitteln direktes und gebeugtes Licht senkrecht zueinander polarisiert und bildet (nach Fig. 48) die Analysatordurchlaßrichtung den Winkel ϑ_a mit der Polarisationsrichtung für das gebeugte Licht, so ist das Komponentenverhältnis

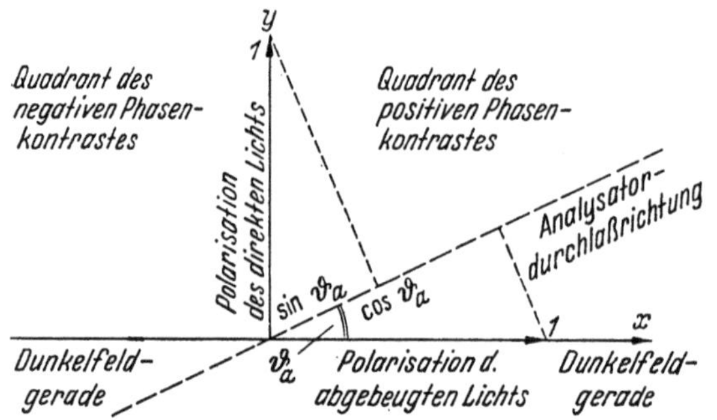

Fig. 48. A-Regelung durch Analysatordrehung, wenn direktes und abgebeugtes Licht senkrecht zueinander polarisiert sind.

$$A = \tan \vartheta_a. \tag{51.1}$$

Es ist

$A = 0$	für $\vartheta_a = 0$;	Dunkelfeld.

$$
\left.\begin{array}{ll}
0 < A \leqq 1 & \text{für} \quad 0 < \vartheta_a \leqq 45° \\
A > 1 & \text{für} \quad 45° < \vartheta_a < 90°
\end{array}\right\}; \quad \text{positiver Phasenkontrast.}
$$

$$
A = \infty \quad \text{für} \quad \vartheta_a = 90°; \quad \text{Bild verschwimmt, da abgebeugtes Licht nicht mehr teilnimmt.}
$$

$$
\left.\begin{array}{ll}
A < -1 & \text{für} \quad 90° < \vartheta_a < 135° \\
-1 \leqq A < 0 & \text{für} \quad 135° \leqq \vartheta_a < 180°
\end{array}\right\}; \quad \text{negativer Phasenkontrast.}
$$

$$
A = 0 \quad \text{für} \quad \vartheta_a = 180°; \quad \text{Dunkelfeld.}
$$

Nach 180° wiederholen sich die Werte von A periodisch.

[1] P. H. Keck u. A. T. Brice: Optik **5**, 31 (1949).
[2] W. G. Hartley: Nature, Lond. **159**, 881 (1947).
[3] E. W. Taylor: Proc. Roy. Soc. Amer. **140**, 422 (1947).
[4] H. Osterberg: J. Opt. Soc. Amer. **36**, 710 (1946); **37**, 726 (1947).
[5] A. Kastler u. R. Montarnal: Nature, Lond. **161**, 357 (1948).
[6] M. Locquin: Mircoscopie (Paris) **1**, M 47 (1948).

Die Verfahren unterscheiden sich in der Realisierung der 90°-Drehung der Polarisationsrichtung des direkten Lichtes zum gebeugten. In allen Fällen freilich wird linear polarisiertes Licht zur Beleuchtung benutzt; der Polarisator liegt unter dem Mikroskopkondensor.

Taylor verwendet zwei Quarzkristallplatten, senkrecht zur Achse geschnitten. Beide Platten sind gleich dick, die Polarisationsrichtung um 45° drehend. Die eine, aus Linksquarz hergestellt, wird so zurechtgeschliffen, daß sie genau am Ort des direkten Lichtes liegt, die andere Platte, aus Rechtsquarz hergestellt, erhält die komplementäre Form und nimmt den Platz des gebeugten Lichtes ein. Zusätzlich wird eine gewöhnliche Zernike-90°-Phasenplatte aufgelegt, um die 90°-Phasenverschiebung zu erreichen.

Kastler und Montarnal schufen eine schon wesentlich leichter realisierbare Vorrichtung gleicher Wirkung. Aus einer quadratischen einachsig doppelbrechenden $\lambda/2$-Platte (Fig. 49a) schnitten sie einen diagonalen Streifen (2) heraus

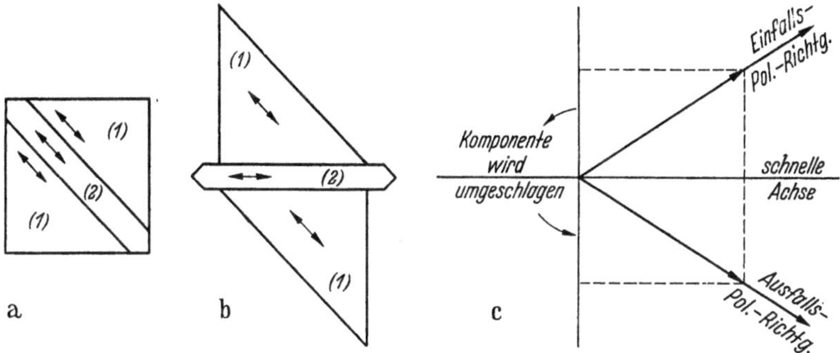

Fig. 49a—c. *a* Doppelbrechende $\lambda/2$-Platte, zerschnitten und b neu zusammengesetzt nach Kastler und Montarnal. c Spiegelung der Polarisationsrichtung an einer Hauptschwingungsrichtung einer doppelbrechenden $\lambda/2$-Platte.

und kitteten ihn so mit den beiden Reststücken (1) des Quadrats zusammen, wie Fig. 49b zeigt. Die Doppelpfeile bezeichnen z. B. die „schnellere" der beiden Hauptschwingungsrichtungen der doppelbrechenden Platte. Läßt man linear polarisiertes Licht einfallen mit einer Schwingungsrichtung parallel zur schnelleren Achse des Streifens (2), so behält das direkte Licht diese Polarisationsrichtung bei. Die des gebeugten Lichtes dagegen wird um 90° gedreht, da eine $\lambda/2$-Platte allgemein linear polarisiertes Licht in linear polarisiertes Licht mit einer neuen Polarisationsrichtung verwandelt, die um 2Θ gedreht ist, wenn Θ der Winkel zwischen einer Kristall-Hauptschwingungsrichtung und der Polarisationsrichtung des einfallenden Lichtes ist (Fig. 49c). Die 90°-Phasenverschiebung zwischen direktem und gebeugtem Licht kann auch hier mit einer zusätzlichen Zernike-Phasenplatte, die der in Fig. 49b skizzierten Platte direkt aufgelegt wird, bewirkt werden. Kastler und Montarnal erreichen das einfacher, indem sie der Platte nach Fig. 49b eine doppelbrechende $\lambda/4$-Platte folgen lassen, deren Hauptrichtungen in die Polarisationsrichtungen des die Platte 49b verlassenden Lichtes gelegt ist.

Nach Locquin bringt man nur an den Ort des direkten Lichtes eine doppelbrechende $\lambda/2$-Schicht (z. B. aus gestrecktem Zellophan). Unter 45° zur Hauptschwingungsrichtung der $\lambda/2$-Schicht ist das einfallende Licht polarisiert; nach Fig. 49c dreht die $\lambda/2$-Schicht die Polarisationsrichtung um 90°, während die Polarisation des gebeugten Lichtes ungeändert bleibt. Die 90°-Phasenverschiebung zwischen direktem und gebeugtem Licht entsteht durch passende Einbettung der $\lambda/2$-Platte wie bei Zernikes Phasenplatte; das Einbettungsmedium ist dafür weniger leicht zu finden, da über die Dicke der $\lambda/2$-Schicht bereits durch die Forderung nach der $\lambda/2$-Doppelbrechung verfügt ist.

Der Phasenspiegel von Françon und Nomarski kann ähnlich zur Regelung der Größe A verwendet werden (s. Ziff. 48).

Hartley sowie Kastler und Montarnal geben noch eine Vorstufe der beschriebenen Verfahren an, bei der die gesamte Phasenplatte aus einer doppelbrechenden $\lambda/4$-Platte geschnitten ist. Im Bereich des direkten Lichtes ist sie um $90°$ gedreht gegenüber dem Rest. Ist das einfallende Licht in der einen Hauptschwingungsrichtung polarisiert, so hat man positiven Phasenkontrast, nach $90°$-Drehung des Polarisators negativen Phasenkontrast.

Die Verfahren nach Taylor, Kastler-Montarnal und Locquin haben besondere Bedeutung, da man aus dem A, das dunkelstes Objektbild gibt, auf die Größe der Objektphasenverschiebung schließen kann. Dunkelstes Objektbild für ein reines Phasenobjekt liegt nach Gl. (33.3) vor, wenn

$$\frac{\partial}{\partial A}\left\{(1 - \cos\varphi)^2 + (A - \sin\varphi)^2\right\} = 0,$$

d.h. wenn

$$A = \sin\varphi \tag{51.2}$$

ist.

Oft reagiert das Auge aber nicht auf absolut dunkelstes sondern auf ein relativ zur Umgebung dunkelstes Objektbild; dann ist

$$\frac{\partial}{\partial A}\ \frac{(1 - \cos\varphi)^2 + (A - \sin\varphi)^2}{A^2} = 0,$$

und hieraus folgt

$$A = 2\ \frac{1 - \cos\varphi}{\sin\varphi}, \tag{51.3}$$

wie Gl. (33.4) schon zeigte. Welcher Fall praktisch vorliegt, hängt von der Bildhelligkeit, vom Beobachter und der Ausgeruhtheit seines Auges ab. Daher ist dieses Verfahren genau nur für kleine φ, bei denen (51.1) und (51.2) übereinstimmend zu

$$A \approx \varphi \approx \vartheta_a$$

führen. Allgemein sind die in den beiden nächsten Ziffern genannten Verfahren für die quantitative Auswertung geeigneter, da bei ihnen die Intensität am Bildort des Objekts auf Null geregelt wird und daher die soeben genannten Fehler fortfallen.

52. Das Polanret-Verfahren. H. Osterberg[1] hat in seinem als „Polanret" bezeichneten Phasenkontrastverfahren außer der Größe A auch die dem direkten Licht gegenüber dem gebeugten aufgezwungene Phasenverschiebung Φ meßbar variabel gestaltet. Man kann dann aus den Werten A und Φ, mit denen die Intensität am Bildort des Objekts zu Null wird, die für das Objekt charakteristischen Größen φ (Phasenverzögerung) und a (Amplitude) durch Auflösen der Gl. (33.7)

$$A \cdot e^{i\Phi} = 1 - a \cdot e^{-i\varphi} \tag{52.1}$$

nach φ und a entnehmen. Die Auflösung gibt

$$a = \sqrt{1 + A^2 - 2A\cos\Phi}; \quad \sin\varphi = \frac{A}{a}\sin\Phi. \tag{52.2}$$

Die A-Regelung bewirkt Osterberg in der gleichen Weise wie Taylor, Kastler und Montarnal durch Analysatordrehung. Direktes und gebeugtes

---

[1] H. Osterberg: J. Opt. Soc. Amer. **37**, 726 (1947).

Licht müssen dafür senkrecht zueinander polarisiert sein; dies erreicht OSTER-BERG ganz unmittelbar dadurch, daß er Polarisationsfolien in die Richtungsaus-wahlebene legt. Die Polarisationsfolie für das direkte Licht habe ihre Durch-laßrichtung in y-Richtung, die Folie am Orte des gebeugten Lichtes in x-Richtung; die Folienkombination heiße „Zonenpolarisator" (s. Fig. 50a).

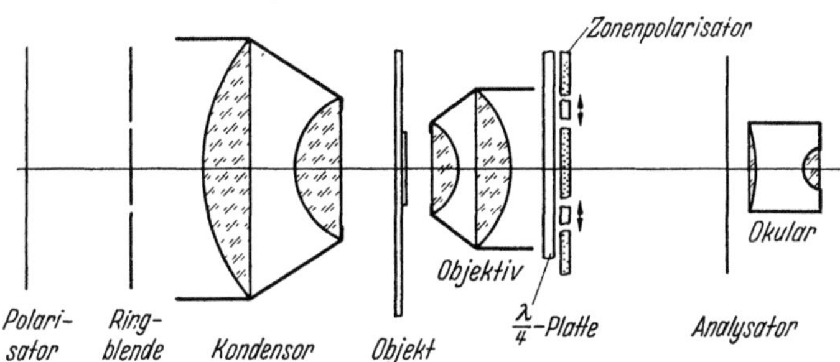

Fig. 50a. „Polanret"-Mikroskop nach OSTERBERG.

Die Phasenverschiebung Φ zwischen direktem und abgebeugtem Licht wird erreicht durch eine vor den Zonenpolarisator gesetzte doppelbrechende $\lambda/4$-Platte, mit ihren Hauptschwingungsachsen (x'- bzw. y'-Achse) unter 45° zu den Polari-sationsrichtungen des Zonenpolarisators orientiert (Fig. 50b).

Das einfallende Licht wird durch einen Polarisator linear polarisiert. Liegt die Polarisatordurchlaßrichtung unter dem Winkel ϑ'_p zur x'-Achse, so macht zunächst die $\lambda/4$-Platte aus der ihr angebotenen Schwingung

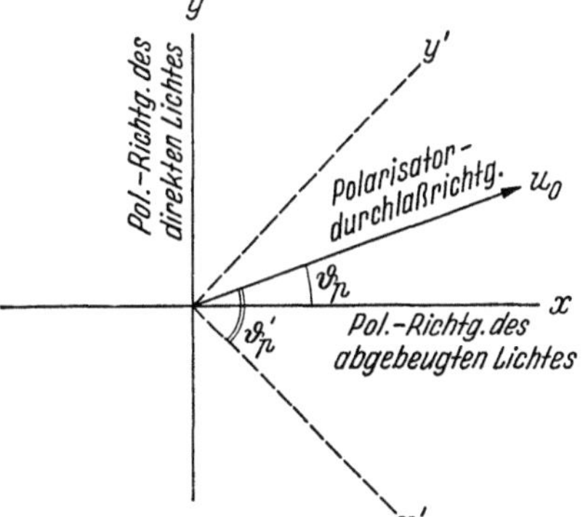

$$u'_x = u_0 \cdot \cos \vartheta'_p; \quad \left. \vphantom{\begin{matrix}a\\b\end{matrix}} \right\} \qquad (52.3)$$
$$u'_y = u_0 \cdot \sin \vartheta'_p$$

die neue Schwingung

$$u'_x = u_0 \cdot \cos \vartheta'_p; \quad \left. \vphantom{\begin{matrix}a\\b\end{matrix}} \right\} \qquad (52.4)$$
$$u'_y = i\, u_0 \cdot \sin \vartheta'_p.$$

Fig. 50b. Zur Lage der Achsen beim Polanret-Verfahren.

Der Faktor i beschreibt die 90°-Phasenverschiebung durch die $\lambda/4$-Platte. Der Zonenpolarisator macht daraus am Orte des direkten Lichts die y-Komponente

$$u_y = \frac{1}{\sqrt{2}} (u'_x - u'_y) = \frac{u_0}{\sqrt{2}} (\cos \vartheta'_p - i \sin \vartheta'_p) = \frac{u_0}{\sqrt{2}} e^{-i \vartheta'_p} \qquad (52.5)$$

und am Orte des gebeugten Lichtes die x-Komponente

$$u_x = \frac{1}{\sqrt{2}} (u'_x + u'_y) = \frac{u_0}{\sqrt{2}} (\cos \vartheta'_p + i \sin \vartheta'_p) = \frac{u_0}{\sqrt{2}} e^{+i \vartheta'_p}. \qquad (52.6)$$

Die Vorrichtung vom Polarisator bis zum Zonenpolarisator einschließlich bewirkt also eine Phasenverschiebung

$$\Phi = -2 \cdot \vartheta'_p \qquad (52.7)$$

und keine Änderung des Amplitudenverhältnisses. Diese tritt allein am Analysator ein und ist nach Gl. (51.1)

$$A = \tan \vartheta_a. \tag{52.8}$$

Darin ist ϑ_a der Winkel zwischen Analysatordurchlaßrichtung und x-Achse. Messen wir auch die Polarisatorstellung statt von der x'-Achse aus von der x-Achse her durch den Winkel

$$\vartheta_p = \vartheta_p' - 45°, \tag{52.9}$$

so ist

$$\Phi = -90° - 2\vartheta_p \tag{52.10}$$

und die Gln. (52.2) gehen über in

$$a = \sqrt{1 + \tan^2\vartheta_a + 2\tan\vartheta_a \sin 2\vartheta_p}; \quad \sin\varphi = -\frac{\tan\vartheta_a \cos 2\vartheta_p}{a}. \tag{52.11}$$

Der Wert der unabhängigen Messung von a und φ liegt auch darin, daß die sonst übliche Annahme $a = 1$ sogar bei nichtabsorbierenden Objekten wegen der Reflexion unzulässig ist und zu etwas verfälschten φ-Werten führt.

Osterberg und Pride (s. Bibliographie) konnten auch sehr kleine und nicht mehr auflösbare Objekte mit dem Polanret-Verfahren vermessen.

Auch abgesehen von den Messungen am Objekt hat das Polanret-Verfahren große Bedeutung, da man für jedes Objekt die beste Beobachtungsmöglichkeit aussuchen kann vom Dunkelfeld über positiven und negativen Phasenkontrast beliebiger Phasenverschiebung bis zum Hellfeld.

53. Einfach realisierbare Phasenkontrastkombinationen für optimale Beobachtung und Messung durch Elliptizitätsanalyse. Das Polanret-Verfahren und die in Ziff. 51 beschriebenen Verfahren erfordern statt des Zernike-Phasenplättchens kompliziert aus Polarisatoren und doppelbrechenden Platten aufgebaute Phasenplatten.

Das Gebiet der Seitenspektren soll von einem Polarisator oder einer doppelbrechenden Platte wohlbestimmter Eigenschaften bedeckt sein, und in das kleine Gebiet des Spektrums nullter Ordnung soll eine andere polarisierende oder doppelbrechende Platte mit ebenfalls sehr genau festgelegten Eigenschaften, insbesondere einer bestimmten Phasenverschiebung, fugenlos eingelassen sein. Dieser Bedingung kann man in der Praxis nur schwer oder ungenau genügen, und daraus resultieren die kostspielige Fertigung bzw. eine erhebliche Ungenauigkeit der Meßwerte.

Insbesondere lassen sich für den Zonenpolarisator des Polanret-Verfahrens keine phasenreinen Polarisationsfolien finden, die hohen Anforderungen genügen könnten.

Deshalb wurde ein Verfahren entwickelt, das die wertvollen Eigenschaften des Polanret-Verfahrens besitzt, aber Polarisatoren nur in phasenunempfindlichen Teilen des Strahlengangs einsetzt, weder fugenloses Passen noch Spezialeinbettung noch enge Toleranzen für doppelbrechende Platten fordert.

Fig. 50c zeigt eine Ausführungsform des Geräts. An die Stelle der Zernike-Phasenplatte tritt z.B. eine auf Glas gekittete doppelbrechende Platte vom Gangunterschied τ ($40° < \tau < 320°$), z.B. eine Gips- oder Glimmerplatte, die am Orte des direkten Lichtes durchbohrt oder bei Ringapertur ringförmig ausgeschliffen ist. Die Platte wird mit irgendeinem schlierenfrei erstarrenden Kitt eingedeckt oder bei gut monochromatischem Licht auch uneingedeckt benutzt. Der Polarisator stehe etwa unter 45° zu den Hauptschwingungsrichtungen der doppelbrechenden Platte.

Dann entsteht auf der Bildebene c die Überlagerung zweier senkrecht zueinander linear polarisierter Phasenkontrastbilder mit einer um den Winkel τ unterschiedlichen Phasenverschiebung der Seitenspektren relativ zum Spektrum nullter Ordnung.

Die Struktur des Objekts nach Amplitude und Phase äußert sich dann in dem resultierenden Bilde durch die von Ort zu Ort, je nach den Objekteigenschaften variierende Elliptizität des Lichtes. Form und Lage der Ellipse am Bildort des Objekts werden mit Viertelwellenlängenplatte h und Analysator i oder anderen hierzu geeigneten Mitteln vermessen und lassen eine genaue Berechnung von Phase und Amplitude des Objekts zu[1].

Stets wird auf Intensität Null am Bildort des Objekts eingestellt, da dann die in Ziff. 51 genannten systematischen Fehler fortfallen.

Gegenüber dem Polanret-Verfahren ist es auch vorteilhaft, daß alle in das Resultat eingehenden Messungen am Analysatorteilkreis geschehen können. Nur er braucht präzise ausgeführt zu sein. Die drehbare Viertelwellenlängenplatte kann mit einem groben Teilkreis oder auch ohne Teilkreis benutzt werden. Die Nullpunkte beider Teilkreise brauchen keinenfalls aufeinander bezogen zu sein. Bedienungsgriffe sind nur am Okularteil nötig; deshalb entfällt hier die bei makroskopischen Polanret-Verfahren erforderliche Fernbedienung des Polarisators.

Die Mannigfaltigkeit der einstellbaren Bilder gleicht der des Polanret-Verfahrens, also der des allgemeinsten Phasenkontrastverfahrens mit beliebig wählbarer Phasenverschiebung und beliebiger Amplitudeneinstellung A des Spektrums nullter Ordnung. Sie umfaßt daher auch das Hellfeldbild und das Dunkelfeldbild.

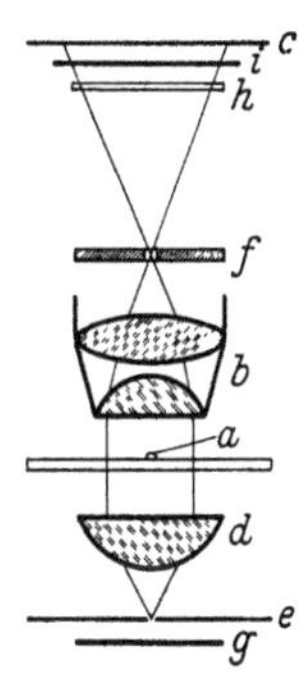

Fig. 50c. Gerät zur Phasen- und Amplitudenmessung durch Elliptizitätsanalyse an Phasenkontrastbildern. a Objekt, b Objektiv, c Bildebene, d Kondensor, e Spalt, Loch- oder Ringblende, f Glimmerplatte (oder dergleichen) mit Ausschnitt im Bereich des direkten Lichtes, g Polarisator, h Kristallplatte, z.B. $\lambda/4$-Platte, i Analysator.

Den Forderungen des strengen Phasenkontrastverfahrens kann man technisch genügen, da man den Bereich des Spektrums nullter Ordnung in der Phasenplatte sehr schmal machen kann. Dazu wird z.B. eine Glimmerfolie oder eine durch Strecken doppelbrechend gewordene Zellophanfolie in zwei Teile geschnitten, die bei der Einbettung auf $^1/_{10}$ mm einander wieder genähert werden. Falls die doppelbrechende Platte absorbiert (gegebenenfalls auch dichroitisch), so kann das bei der Berechnung der Resultate leicht berücksichtigt werden. Die Ausführung der doppelbrechenden Phasenplatte kann in mannigfacher Weise variiert werden. Sogar der Spiegel von FRANÇON und NOMARSKI (s. Fig. 47a) wäre als Ersatz für die doppelbrechende Phasenplatte geeignet. Doch kann man den Einfallswinkel anders als FRANÇON und NOMARSKI wählen. Zweckmäßig ist auch eine schräg vom Licht durchsetzte Glasplatte, die im Bereich der Seitenspektren anders als im Bereich des Spektrums nullter Ordnung dünn metallisiert ist. Bei diesen Lösungen spart man unter Umständen eine schlierenfreie Einbettung, hat aber eine noch etwas kompliziertere Auswertung.

Die Vorteile des Verfahrens werden überhaupt durch eine kompliziertere Rechnung oder Konstruktion in der GAUSSschen Zahlenebene erkauft[1]; sie kann jedoch dem Benutzer durch Tabellen oder Nomogramme abgenommen werden.

Diese variablen Phasenkontrastverfahren erlauben eine Messung der Phase φ auf Bruchteile eines Grad und der Amplitude auf Bruchteile eines Promille.

---

[1] H. WOLTER: Z. Physik **140**, 57 (1955).

Damit sind sie die optimalen Verfahren zur Bestimmung der komplexen Durch-
lässigkeits-Reflexionsfaktoren dünner Schichten, aus denen man die Dicke und
die optischen Konstanten dünner absorbierender oder nichtabsorbierender
Schichten berechnen kann. Die mathematischen Methoden hierzu wurden im
vorstehenden Artikel ,,Optik dünner Schichten" behandelt[1]. Für die Messung
selbst wird aus der zu messenden Schicht ein schmaler Streifen fortgenommen,
der als Objekt im Phasenkontrastgerät in Durchsicht oder Aufsicht betrachtet
wird. Um die Wirkung des strengen Phasenkontrasts zu erhalten, gibt man dem
Streifen keine zu große Breite.

Auch feste Phasenkontrastgeräte sind zu diesem Zweck benutzt worden[2].
Dann ist jedoch eine Photometrie im Bilde erforderlich.

c) Farbige Phasenkontrastverfahren.

54. Die Aufgabe der farbigen Phasenkontrastverfahren. Das Phasenkontrastverfahren nach Zernike mit einer Phasenverschiebung von 90° für das direkte
Licht gegenüber dem abgebeugten macht Phasenobjekte deutlich, läßt zugleich
aber Amplitudenobjekte undeutlich werden. Fig. 31c I zeigt das für ein Amplitudengitter, und das Zeigerdiagramm Fig. 27e in der Gaußschen Zahlenebene macht
das Versagen verständlich; denn Phasen- und Amplitudenobjekte vertauschen
ihre Rollen, wenn man vom Hellfeldbild zum Phasenkontrastbild übergeht.

Um beide Objektarten deutlich sichtbar zu machen, wechselte man zunächst
zwischen Hellfeld- und Phasenkontrastbild. Köhler und Loos[3] gaben schon 1941
ein Verfahren ohne Objektivaustausch hierzu an.

Am einfachsten wechselt man vom Phasenkontrastbild zu einem angenäherten
Hellfeldbild, wenn man die Kondensorblende dejustiert.

Besser reproduzierbare Bilder erhält man, wenn man dicht unter dem Objektträger,
z. B. auf dem Mikroskopkondensor ein winziges Deckglasstückchen schräg mit Caedax auf-
kittet. Der Teil des Gesichtsfeldes über diesem kleinen Prisma erscheint im Hellfeld, der
Rest im Phasenkontrast bei richtiger Justierung[4]. Die Kreuztischverschiebung des Objektträgers läßt den Wechsel für jedes Objektteilchen zu. Befinden sich mehrere Exemplare
z. B. eines Bakteriums im Gesichtsfeld, so erscheinen einige im Hellfeld, andere im Phasenkontrast.

Läßt man eine Glasplatte mit einigen prismatisch aufgeklebten Deckglasstückchen zwischen Kondensor und Lichtquellenblende mit passender Geschwindigkeit rotieren, so wechseln
Phasenkontrastbild und Hellfeldbild 3 bis 8mal pro Sekunde. Beide Objektarten werden
dann durch ein Flimmern erkennbar.

Das Polanret-Verfahren und das Verfahren nach Ziff. 53 geben vollkommenere
Lösungen der Aufgabe. Bei gewissen Phasenverschiebungen Φ (z. B. 45°) werden
Amplituden- und Phasenobjekte zugleich sichtbar. Stets aber gibt es gemischte
Objekte, die Phase und Amplitude ändern und bei einem eingestellten Φ unsicht-
bar bleiben.

Die durch Phase und Amplitude gebildete zweidimensionale Mannigfaltigkeit
kann mit der eindimensionalen Grauskala grundsätzlich nicht eineindeutig in
einem einzigen Bilde zugleich gekennzeichnet werden. Dazu ist es hier wie bei
dem zweidimensionalen Farbschlierenverfahren nötig, die Farben mit heranzuziehen.

Der Farbton, die Farbsättigung und die Helligkeit bilden sogar eine drei-
dimensionale Mannigfaltigkeit, und man kann daher auch noch Polarisationsobjekte, z. B. optisch aktive oder doppelbrechende Stoffe wie Kristalle oder

---

[1] Siehe S. 528ff. dieses Bandes.
[2] M. Françon: Le Contraste de Phase en Optique et en Microscopie. Paris 1950.
[3] A. Köhler u. W. Loos: Naturwiss. **29**, 51 (1941).
[4] H. Wolter: Ann. Physik (6) **9**, 64 (1951).

gewisse Fibrillen in histologischen Präparaten mit Hilfe der Farben zusätzlich nach der Polarisationsrichtung, die sie aus einer ihnen angebotenen machen, kennzeichnen.

Darüber hinausgehende Wünsche aber — etwa nach eineindeutig gekennzeichneter Abbildung auch vielfarbiger Objekte zugleich mit Phasen- und Polarisationsobjekten in einem Bilde — müssen grundsätzlich unerfüllbar bleiben; denn die Dimensionszahl 3 ist die unüberschreitbare Grenze für eine mit unserem Auge in einem Bilde erfaßbare Mannigfaltigkeit.

55. Verfahren zur Darstellung von Phasen- und Amplitudenobjekten in einem einzigen farbigen Bilde. SAYLOR, BRICE und ZERNIKE[1] verstärkten die bei jedem Phasenstreifen bestehende Phasendispersion durch passende Wahl der Dicke. Man kann so erreichen, daß für einen Teil des Spektrums Phasenkontrast, für einen anderen Teil Hellfeld herrschen. Das resultierende Bild ist z. B. eine Kombination eines roten Hellfeldbildes mit einem blauen Phasenkontrastbild. Ein Teil des Spektrums gibt Phasenkontrastbilder anderer Phasenverschiebung. SAYLOR, BRICE und ZERNIKE realisierten für eine Wellenlänge positiven, für eine andere negativen Phasenkontrast und erreichten dadurch stärkere Farbsättigung als bei reiner Phasenkontrast-Hellfeldkombination. Noch größere Möglichkeiten bietet der Gedanke von LOCQUIN[2] und D. KOSSEL[3], den Phasenstreifen als Interferenzfilter auszuführen. (Danach lassen sich auch achromatische Phasenplatten herstellen.)

Ähnliche Wirkung hat ein von GRIGG[4] zuerst veröffentlichtes Verfahren, das in einem Phasenkontrastmikroskop mit handelsüblichem ZERNIKE-Phasenring eine Farbenplatte als Lichtquellenblende benutzt. Die Kondensorringblende der ZERNIKE-ZEISSschen Grundanordnung wird bei einer etwas anderen Ausführung des Verfahrens durch eine Glasplatte ersetzt, die an der Stelle des sonst durchsichtigen Ringes rote Eosingelatine und an den sonst undurchsichtigen Stellen grüne Naphtholgrüngelatine trägt. Licht, das von dem roten Ringgebiet ausgeht, liefert wie im gewöhnlichen Phasenkontrastmikroskop ein Phasenkontrastbild in roter Farbe. Das von dem grün gefärbten Teil ausgehende Licht liefert im wesentlichen ein grünes Hellfeldbild. Um in dem Überlagerungsbild nicht allzusehr das Hellfeldbild überwiegen zu lassen, schwächte GRIGG es durch Polarisatoren, ohne diese selbst aber zur Farbgebung auszunutzen, wie es bei den in den nächsten Ziffern beschriebenen Verfahren geschieht. WOLTER[4] färbte statt dessen nur die unmittelbare Umgebung des roten Ringes grün und beließ weiter entfernte Gebiete undurchlässig. Dadurch geben zugleich die grünen Lichtanteile nicht reines Hellfeld, sondern teilweise negativen Phasenkontrast; das führt zu stärkerer Farbsättigung. Ein Gerät, das nach einem sehr ähnlichen Verfahren arbeitet, wird von der Firma Ernst Leitz gefertigt.

56. Verfahren zur Beobachtung von Amplituden-, Phasen- und Polarisationsobjekten in einem Bilde mit Normalfarben. Als ein „Normalfarbensystem" wird eine echte Teilmenge der Menge aller Farben bezeichnet, wenn jede additive Mischung von Farben der Teilmenge stets wieder zu einer Farbe führt, die einer Farbe der Teilmenge „unbedingt" gleich ist. „Farbe" heißt hier der Quotient

---

[1] C. P. SAYLOR, A. T. BRICE u. F. ZERNIKE: J. Opt. Soc. Amer. **40**, 329 (1950). Dasselbe Verfahren wurde unabhängig im Dezember 1948 (unveröffentlicht) auch in den Optischen Werken Ernst Leitz, Wetzlar entdeckt.

[2] LOCQUIN (1951), zit. bei M. FRANÇON, Le microscope à contraste de phase et le microscope interférential, S. 72. Paris 1954.

[3] D. KOSSEL: Zusatzpatentanmeldung vom 22. 4. 54 zum D.P. 550527.

[4] F. G. GRIGG: Nature, Lond. **165**, 368 (1950). — H. WOLTER: Naturwiss. **37**, 491 (1950). — Ann. Physik (6) **9**, 62 (1951).

des ausfallenden Lichtes zum einfallenden als Funktion z.B. der Frequenz oder Wellenlänge[1]. Die „unbedingte Gleichheit" macht die Normalfarben für quantitative Untersuchungen besonders geeignet, da weder die Beleuchtung noch die Eigenschaften des Beobachterauges oder des für Aufnahmen benutzten Farbfilms die Messung beeinflussen.

Ein Beispiel für Normalfarben sind die Rotationsdispersionsfarben, die entstehen, wenn Licht z.B. durch eine senkrecht zur optischen Achse geschnittene Quarzplatte geschickt wird, die zwischen zwei Polarisatoren liegt. Bequem unterscheidbar werden diese Normalfarben, wenn die Quarzplatte etwa 4 bis 6 mm dick ist. Denn die Größe der Rotationsdispersion des Quarzes bedingt, daß dann ein ganzer Farbenkreis hoher Sättigung durchlaufen wird, während einer der Polarisatoren seinen Winkelbereich 180° durchläuft. Trifft auf eine solche Quarzplatte mit nachfolgendem Analysator eine linear polarisierte Welle, so ist deren Polarisationsrichtung unmittelbar an der Farbe des heraustretenden Lichtes erkennbar. Ist das einfallende Licht elliptisch polarisiert, so wird das an einer Abnahme der Farbsättigung erkennbar.

Die beschriebene Quarzplatte mit nachfolgendem Analysator ist daher ein „Detektor" für Polarisationszustände; er wandelt sie in erkennbare Farben um. Das in Ziff. 53 beschriebene Verfahren, bei dem die Eigenschaften eines Objekts ihren Niederschlag am Bildort des Objekts in der Polarisationsellipse fanden, wird daher zu einem Farbphasenkontrastverfahren, wenn die Platte h des Geräts (s. Fig. 50c) durch die 4 mm dicke Quarzplatte ersetzt wird. Lage und Form der Ellipse — d.h. Amplitude und Phase des Objekts — finden ihren eineindeutigen Ausdruck in der Farbe am Bildort des Objekts, können also aus der Farbe durch Vergleich mit einem „Testobjekt" entnommen werden. Als Testobjekt dient z.B. eine Glashalbkugel, deren verschiedene Oberflächenelemente das Licht einer Mattscheibe in verschiedenen und elementar berechenbaren Polarisationszuständen reflektiert. Wert hat das Farbverfahren vor allem bei veränderlichen Objekten, die für die Polarisationsmessungen nicht hinreichend lange zur Verfügung stehen. Ihr Farbbild kann mit einer Farbfilmaufnahme festgehalten und später durch Vergleich mit dem auf demselben Film aufgenommenen Testobjekt vermessen werden. Als Test genügen drei feste Normalfarben[2].

Weitere verwandte Verfahren wurden 1950/51 entwickelt[3], unter ihnen auch solche, die neben Phasen- und Amplitudenobjekten zugleich Polarisationsobjekte deutlich zeigen. Die Fig. c der Fig. 54 ist nach einem leicht realisierbaren Verfahren aufgenommen, bei dem als Phasenplatte eine handelsübliche Zernike-Platte benutzt wird. Von einem gewöhnlichen Phasenkontrastgerät unterscheidet sich die Apparatur nur dadurch, daß erstens an die Stelle des gewöhnlichen ein Kondensorringspalt tritt, der in aufgeklebte Polarisationsfolie eingedreht ist, und daß zweitens in den Okularteil der Polarisationsdetektor (4 mm-Quarzplatte und Analysator) gesetzt wird. Zwischen Lichtquelle und Kondensorringspalt setzt man zweckmäßig einen weiteren drehbaren Polarisator. Mit ihm kann man unter anderem das Verhältnis ändern, in dem ein Hellfeldbild einer Farbe und ein Phasenkontrastbild anderer Farbe sich zum resultierenden Bilde additiv mischen. Die resultierenden Farben gehören wegen der Eigenschaft der Normalfarben dem Normalfarbenbereich selbst an und sind jeweils mit einer Farbe der Testplatte „unbedingt" gleich.

---

[1] Der Begriff „Farbe" mußte eng an die spektrale Verteilungsfunktion geknüpft werden. Die Sätze über Normalfarben lassen sich mit dem biophysikalischen Farbbegriff nicht formulieren [Ann. Physik (6) **8**, 11 (1950)].

[2] H. Wolter: Ann. Physik **1956** (im Erscheinen).

[3] H. Wolter: Ann. Physik (6) **9**, 57 (1951).

d) Farbige Phasenkontrastverfahren, die zugleich Farbschlierenverfahren sind.

56a. Da das Phasenkontrastverfahren großräumige Objekte nicht wiedergibt (s. Ziff. 56) eignet es sich nicht immer ohne weiteres zur Prüfung von Optiken wie Hochspiegeln und Linsen oder Planparallelplatten. Doch kann man die große Empfindlichkeit des Phasenkontrastverfahrens für feinste Schleifspuren oder andere kleinräumige Fehler vereinen mit der deutlichen Wiedergabe großräumiger „Schlieren", wenn man das Phasenkontrastverfahren mit einem Farbschlierenverfahren kombiniert.

Die meisten Kombinationen dieser Art geben freilich Bilder, die schwer zu analysieren sind. Übersichtliche und leicht deutbare Bilder (z.B. das Bild 54a oben links auf S. 639) erhält man mit einem Phasenkontrastverfahren, das nach Fig. 25 auf S. 590 mit linearem Spalt und einem zu dessen Bild kongruenten linearen ZERNIKE-Phasenstreifen arbeitet und zusätzlich z.B. zwischen Kondensor und Objekt ein ganz schwaches Geradsichtprisma (oder auch eine einfache Keilplatte kleinen Keilwinkels) enthält. Das Spaltbild wird dann zu einem Spektrum auseinandergezogen, von dem nur ein Teil auf den ZERNIKE-Streifen fällt und Phasenkontrast liefert. Der Rest gibt teils negativen Phasenkontrast, teils Hellfeld. Das so entstehende farbige Phasenkontrastbild gibt Amplituden- und Phasenobjekte deutlich wieder, wie die Abbildung 54a (S. 639) zeigt. Darin ist oben links ein Teil eines als Amplitudenobjekt dienenden Tintenpunktes zu sehen, der zur groben Markierung der fehlerhaften Objektstelle angebracht worden war. Außerdem zeigt das Bild im Phasenkontrast Schleifspuren über das ganze Objekt verteilt.

In der Mitte des Bildes ist eine Körnerschliere erkennbar. Sie ist ein großräumiges Objekt, das im reinen Phasenkontrastbild nicht deutlich wurde, hier aber wie in einem Prismenfarbschlierenbild farbig schattiert erscheint. Die Teile dieser Körnerschliere, die das Licht seitlich ablenken, so daß das zum Spektrum auseinandergezogene Spaltbild relativ zum ZERNIKE-Phasenstreifen verschoben wird, geben zwar wie die nicht ablenkenden Objektteile eine Hellfeld-Phasenkontrastkombination; doch ist die spektrale Aufteilung in Hellfeld- und Phasenkontrastanteil hier eine andere, und das führt zu der farbigen Kennzeichnung der Körnerschliere. Die Phasenkontrastwirkung bleibt im Bereich der Schlieren trotzdem erhalten, wenn die Breite des Spektrums nicht zu gering bemessen wird[1].

D. Das Lichtschnittverfahren.

a) Lichtschnittgeräte.

57. Grundsätzliches über Lichtschnittgeräte. Für die Untersuchung von Oberflächen oder dünnen Schichten ist das Lichtschnittverfahren nach G. SCHMALTZ[2] eine wertvolle Ergänzung des Phasenkontrastverfahrens und der Interferenzverfahren. Methodisch ist es den Schlierenverfahren nahe verwandt. Seinen Namen hat es von der elementarsten Anwendung auf die Oberflächenuntersuchung undurchsichtiger Stoffe.

Läßt man von einem schmalen ebenen Lichtband, hergestellt z.B. durch Abbildung eines Spaltes, eine Körperoberfläche „schneiden" (Fig. 51), so kann man die auf dem Körper entstehende helle „Schnittkurve" seitlich beobachten und an ihrer Gestalt unmittelbar die Oberflächenstruktur der betrachteten Körperstelle erkennen. Statt mit dem unbewaffneten Auge beobachtete bereits SCHMALTZ mit einem Mikroskop.

---

[1] H. WOLTER: Ann. Physik (6) **9**, 57 (1951).
[2] G. SCHMALTZ: Technische Oberflächenkunde, S. 73. Berlin 1936.

Fällt das von der Oberfläche reflektierte Licht in das Mikroskop, so spricht man von Hellfeldbeobachtung, sonst von Dunkelfeldbeobachtung bei Licht-schnitt. (Zur Auswertung bei Hellfeld s. Ziff. 61.)

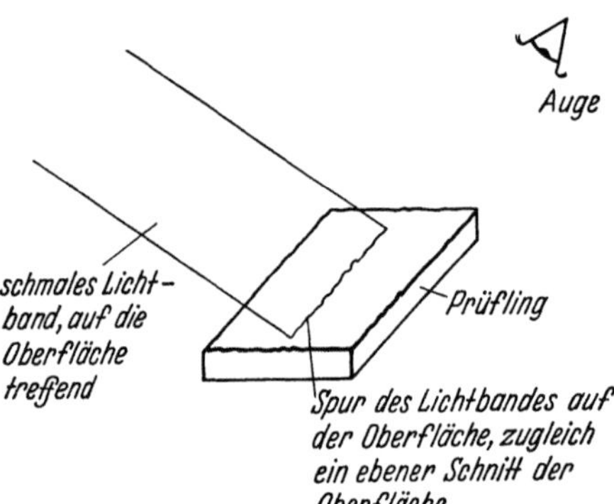

Fig. 51a. Prinzip des Lichtschnittverfahrens nach SCHMALTZ.

58. Die Minimumstrahlkennzeichnung beim Lichtschnitt. Die Anforderungen an die „Schärfe" des Licht-bandes, das als „Schnittebene" dient, ist bei mikroskopischer Beobachtung sehr hoch, und man kann sich fragen, ob nicht eine ebene Schattengrenze dann geeigneter als ein abgebildeter Spalt wirkt. Die Frage nach der optimalen Kennzeichnung einer Ebene oder Kurve im Raum mit Hilfe des Lichtes ist hier dieselbe wie bei den Schlieren-verfahren, und auch hier führt die *Minimumstrahlkennzeichnung* zu der höchsten erreichbaren Genauigkeit[1,2]. Sie wurde von E. MENZEL zuerst auf das mikroskopische Lichtschnittverfahren an-gewandt[3] und erscheint gerade für dieses besonders geeignet. Denn sie ist nicht auf eine Abbildung einer Kante auf das Objekt angewiesen, da das Minimum auf einem großen Teile des Lichtweges sehr scharf bleibt.

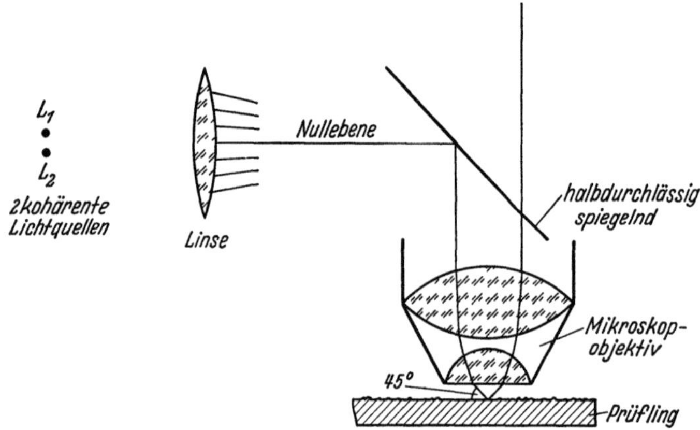

Fig. 51b. Teil einer Apparatur zum mehrfachen mikroskopischen Lichtschnitt. [Siehe auch E. MENZEL: Naturwiss. **38**, 332 (1951).]

Bei hochauflösenden Mikroskopobjektiven[4] wird zweckmäßig ein Strahlen-gang nach Fig. 51b benutzt, bei dem z.B. wie bei gewöhnlichen Auflichtmikro-skopen die eine Hälfte der Objektivapertur für das Einschleusen der durch die Nullebene gekennzeichneten Beleuchtung dient (links) und die andere Hälfte (rechts) für die Beobachtung zur Verfügung steht. Wenn wie in Fig. 51b ein halb-durchlässiger Spiegel (und kein Halbkreisspiegel) zum Einschleusen des Lichtes

---

[1] H. WOLTER: Verfahren zur genauen Kennzeichnung einer Fläche oder Kurve im Raum mittels interferierender Lichtwellen. D.P. 819728 vom 12. 4. 1949.

[2] H. WOLTER: Z. Naturforsch. 5a, 143 (1950). — Phys. Bl. **5**, 234 (1949). Das hier entwickelte Verfahren ist ein Lichtschnittverfahren mit Minimumstrahlkennzeichnung.

[3] E. MENZEL: Mehrfacher mikroskopischer Lichtschnitt. Naturwiss. **38**, 332 (1951).

[4] Außer der MENZELschen Arbeit siehe auch S. TOLANSKY: Nature, Lond. **169**, 445 (1952).

benutzt wird, so steht die volle Objektivapertur dem am Objekt abgebeugten Licht offen, und man erleidet keine Einbuße an Auflösungsvermögen. Zugleich ist die volle Apertur der Beleuchtung zugänglich.

Die Oberfläche wird auf die Netzhaut des Beobachters abgebildet und zeigt (Fig. 51 c) eine dunkle Linie, deren Abweichungen von der Geraden die Abweichungen der zu prüfenden Oberfläche von der Ebene bedeuten. MENZEL benutzte eine Phasenplatte mit mehreren sich je um 180° in Phase unterscheidenden Stufen und erhielt so mehrere „Schnittkurven" mit der zu prüfenden Oberfläche (Fig. 51 c). Statt dessen kann man zur Herstellung mehrerer Nullebenen die Interferenz zweier kohärenter Lichtquellen verwenden, die aus einer spaltförmigen Lichtquelle mit Hilfe des FRESNELschen Biprismas gewonnen werden (Fig. 51 b). Die Nullebenen sind dann exakt äquidistant und haben den kleinsten mit der Apparaturgeometrie überhaupt realisierbaren Abstand voneinander, geben also die feinste Unterteilung und durch den Streifenabstand einen genau berechenbaren Maßstab für die Streifenversetzung[1]. Diese Methode erlaubt eine strenge Auswertung, die wegen der mathematischen Schwierigkeiten bei der Strahlversetzung in anderen Lichtschnittverfahren bis heute nicht allgemein gelungen ist.

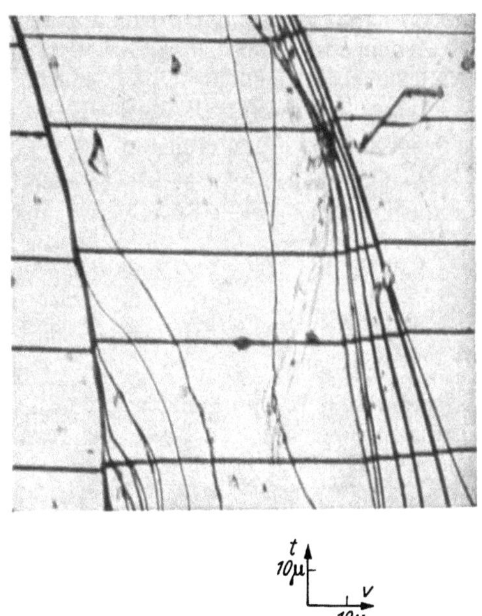

Fig. 51 c. Die horizontalen dunklen Linien sind Minimum-Lichtschnittlinien auf einer Glimmeroberfläche (nach E. MENZEL).

b) Zur Auswertung der Lichtschnittkurven.

59. Strahlversetzung bei dem Lichtschnitt. Bei höchsten Genauigkeitsforderungen muß die Auswertung der Lichtschnittkurven auch die „Strahlversetzung" berücksichtigen. Ein Strahl wird nicht genau von der Oberfläche eines Körpers reflektiert; er wird vielmehr so seitlich versetzt, als ob er von einer Ebene hinter der Oberfläche oder auch in anderen Fällen vor ihr reflektiert würde. Freilich kommt das praktisch nur zur Auswirkung, wenn verschiedene Stoffe die zu prüfende Oberfläche bilden, also z.B. bei der Dickenmessung an Metallschichten auf Glas.

Die Strahlversetzung wurde zuerst bei der Totalreflexion experimentell von GOOS und HÄNCHEN[2] gefunden und von K. ARTMANN[3] berechnet. H. WOLTER[1] entwickelte die Minimumstrahlkennzeichnung an diesem Problem und C. SCHAEFER und C. VON FRAGSTEIN[4,5] haben durch theoretische und experimentelle Arbeiten an der Klärung des ganzen Fragenkreises wesentlichen Anteil. C. SCHAEFER, J. PICHT und F. NOETHER betrachteten bereits vor der Entdeckung von

---

[1] Siehe Fußnoten 1 und 2, S. 634.

[2] F. GOOS u. H. HÄNCHEN: Ann. Physik (6) **1**, 333 (1947); **5**, 251 (1949).

[3] K. ARTMANN: Ann. Physik (6) **2**, 87 (1948).

[4] C. v. FRAGSTEIN: Ann. Physik (6) **4**, 271 (1949) und dort zitierte Arbeiten.

[5] Ausführliche Literatur siehe bei C. v. FRAGSTEIN u. CL. SCHAEFER: Ann. Physik (6) **12**, 84 (1953).

Goos und Hänchen ähnliche Fragen bei der Totalreflexion theoretisch[1] (ähnlich auch König, Voigt, Schuster, Drude und Raman[2]).

Allgemein ist die seitliche (senkrecht zur Nullebene gemessene) Strahlversetzung

$$\mathfrak{D} = \frac{\lambda}{i\,\pi\,n_e}\,\frac{\mathfrak{r}_2 - \mathfrak{r}_1}{\mathfrak{r}_2 + \mathfrak{r}_1}\cdot\frac{1}{\psi_{e2} - \psi_{e1}}\,. \tag{59.1}$$

Darin ist n_e der Brechungsindex des Mediums, aus dem das Licht einfällt, $\psi_{e2} - \psi_{e1}$ der Winkel zwischen den Fortpflanzungsrichtungen der beiden bei der Minimumstrahlkennzeichnung sich kreuzenden Wellen und $\mathfrak{r}_1$ bzw. $\mathfrak{r}_2$ der Fresnelsche Reflexionskoeffizient der Oberfläche für die eine bzw. die zweite Welle.

Der Realteil von $\mathfrak{D}$ bedeutet eine echte Strahlversetzung, der Imaginärteil eine Minimumverbreiterung.

An Oberflächen von Nichtleitern sind $\mathfrak{r}_1$ und $\mathfrak{r}_2$ reell (außer bei Totalreflexion), und es tritt keine reelle Strahlversetzung ein. Bei Metallen kommt z.B. für senkrechte Polarisation (d.h. senkrecht zur Einfallsebene stehende elektrische Feldstärke) die Reflexion aus einer scheinbaren Tiefe unter der Oberfläche

$$\text{Reflexionstiefe} = \left|\frac{\lambda}{2\pi}\,\mathfrak{Re}\,\frac{1}{\sqrt{k^2 - n^2 + n_e^2\sin^2\psi_e + i\,2nk}}\right|\,. \tag{59.2}$$

n und k sind Brechungsindex und Absorptionskoeffizient des Metalls, ψ_e ist der Einfallswinkel der Nullebene. Bei paralleler Polarisation ist die Reflexionstiefe oft negativ, die Reflexion scheint also vor der Oberfläche zu geschehen[2].

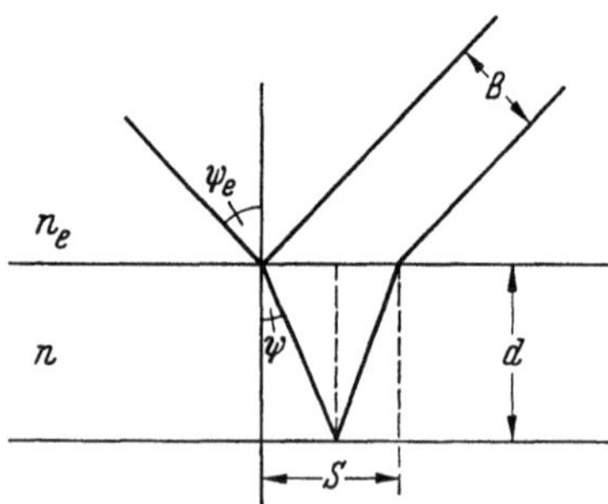

Fig. 52. Reflexion an Vorder- und Rückseite bei dem Lichtschnitt, an einer durchsichtigen Folie.

60. Doppelte Lichtschnittkurven bei durchsichtigen Folien und ihre Ausnutzung zu Dicken- und Brechungsindexmessungen. Eine durchsichtige Folie (Fig. 52) gibt eine Reflexion an der Vorder- und eine an der Rückseite. Beide sind nicht „strahlversetzt" im Sinne der Ziff. 59. Ihre gegenseitige Versetzung B oder die Seitenverschiebung S ist daher nach Fig. 52 strahlenoptisch richtig zu berechnen. Man erhält

$$\frac{S}{2} = \frac{B}{2\cdot\cos\psi_e} = d\tan\psi;\quad \frac{\sin\psi_e}{\sin\psi} = \frac{n}{n_e} \tag{60.1}$$

und daraus die wahre Schichtdicke d aus der Versetzung B bzw. Seitenversetzung S:

$$d = \frac{B}{2\cos\psi_e}\sqrt{\frac{n^2}{n_e^2\sin^2\psi_e} - 1} = \frac{S}{2}\sqrt{\frac{n^2}{n_e^2\sin^2\psi_e} - 1}\,. \tag{60.2}$$

Für einen Einfallswinkel $\psi_e = 45°$ ist speziell

$$d = \frac{B}{\sqrt{2}}\sqrt{\frac{2n^2}{n_e^2} - 1} = \frac{S}{2}\sqrt{\frac{2n^2}{n_e^2} - 1}\,. \tag{60.3}$$

Bei kleinem Einfallswinkel ist

$$d \approx \frac{nB}{2n_e\psi_e} \approx \frac{n\cdot S}{2n_e\psi_e}\,. \tag{60.4}$$

[1] Siehe Fußnote 5, S. 635.
[2] Einen Bericht dazu siehe in H. Wolter: Z. Naturforsch. 5a, 143 (1950).

Während bei dem Phasenkontrastverfahren und Interferenzverfahren effektiv $d(n-n_e)$ gemessen wird, führt die Lichtschnittmessung auf d/n oder allgemeiner nach (60.2) jedenfalls auf eine andere Kombination der Unbekannten d und n. Aus einer Lichtschnittmessung und Phasenkontrastmessung am gleichen Objekt können daher Schichtdicke und Brechungsindex unabhängig voneinander berechnet werden[1].

61. Auswerteformel für einfache Oberflächenstufen an reflektierenden Objekten. Aus (60.2) folgt mit $n = n_e$ zugleich die Formel

$$d = \frac{B}{2 \cdot \sin \psi_e} = \frac{S}{2 \cdot \tan \psi_e} \tag{60.5}$$

für die Berechnung der Höhe d einer Stufe aus der Seitenverschiebung S oder der Versetzung B. Ist $\psi_e = 45°$, so wird $2d = S$; das Oberflächenprofil ist mit Maßstabsverdopplung in die Objektebene optisch umgeklappt. Bei „Dunkelfeldbeobachtungen", bei denen Einfalls- und effektiver Ausfallswinkel verschieden sind, ist die Auswertung ebenfalls sehr einfach (s. G. SCHMALTZ, Fußnote 2 von S. 633). Oft wird aus einer zur Oberfläche senkrechten Richtung beobachtet. Aus dem scheinbaren Abstand $a = S/2$ der Schnittlinien folgt dann wegen (60.5) die Stufenhöhe $d = \dfrac{a}{\tan \psi_e}$. Ist $\psi_e = 45°$, so erscheint nun das Oberflächenprofil maßgerecht in die Objektebene geklappt. Sind verschiedene und darunter auch absorbierende oder totalreflektierende Stoffe beteiligt, so ist die „Strahlversetzung" (Ziff. 59) zu berücksichtigen.

E. Vergleich der Schlieren- und Phasenkontrastverfahren mit den abbildenden Interferenzverfahren.

a) Interferenzverfahren erster, zweiter und dritter Art.

62. Prinzip der abbildenden Interferenzgeräte erster und zweiter Art. Der bei dem Phasenkontrastverfahren und den Schlierenverfahren oben festgestellte Mangel an Objekttreue hat in neuerer Zeit das Interesse an den abbildenden Interferenzverfahren, besonders an der Interferenzmikroskopie, wieder stark belebt. Interferenzmikroskopische Bilder sind frei von inneren Aufhellungen, Höfen um die Objekte und Größenverfälschungen, wie sie an Hand der Fig. 31 und 32 und in den Ziff. 34 und 43 diskutiert wurden. Abbildende Interferenzverfahren können auch beliebig große Objekte völlig objekttreu wiedergeben, und zwar in einer für die quantitative Auswertung besonders geeigneten Form. Sie verwandeln wie das Phasenkontrastverfahren Phasendifferenzen in Intensitätsdifferenzen, um sie dem Auge wahrnehmbar zu machen.

Das Wesen der abbildenden Interferenzverfahren erster Art besteht darin, daß auf das Hellfeldbild gleichmäßiges kohärentes Zusatzlicht gelenkt wird. Dieses interferiert mit dem Licht des Hellfeldbildes zu ganz verschiedenen Intensitäten je nach der jeweils am Bildort eines Objekts vorhandenen Phase des Hellfeldlichtes. Das kohärente Zusatzlicht wirkt also als eine Art Detektor für Phasendifferenzen.

Das kohärente Zusatzlicht wird vor dem Objekt aus der Beleuchtung abgezweigt und hat das Objekt nicht passiert. Im Gegensatz zu den Schlieren- und dem Phasenkontrastverfahren wird in den abbildenden Strahlengang nicht mit beugenden Kanten eingegriffen; damit entfällt hier die Ursache für die aus dem Eingriff resultierenden Bildverfälschungen.

---

[1] H. RAHBEK u. M. OMAR: Nature, Lond. **169**, 1008 (1952).

Neben den älteren Interferenzmikroskopen nach Linnik, der das Michelson-sche Interferometerprinzip auf das Mikroskop übertrug, gibt es neuerdings einige Formen, die einfacher herstellbar und bequemer bedienbar sind. Besonders einfach, aber nur für dünnste Objekte anwendbar ist das Verfahren von Tolansky und Merton[1], die das Objekt zwischen halbdurchlässig verspiegelte Deckgläser und Objektträger legen und unter einem einfachen Hellfeldmikroskop betrachten. Dabei werden die bekannten Farben dünner Blättchen ausgenutzt, die infolge der Reflexerhöhung auch in Durchsicht sehr gesättigt sein können. Lau und Krug[2] bauten ein kleines Mach-Zehndersches Interferenzgerät zwischen Objektiv und Kondensor eines Mikroskops.

Dyson[3] übertrug das Jaminsche Interferometer auf das Mikroskop, und Wolter wandelte dieses so ab, daß statt zweier Planparallelplatten nur eine benötigt wird; ihr Spiegelbild dient als Ersatz der zweiten Platte[4]. Da eine

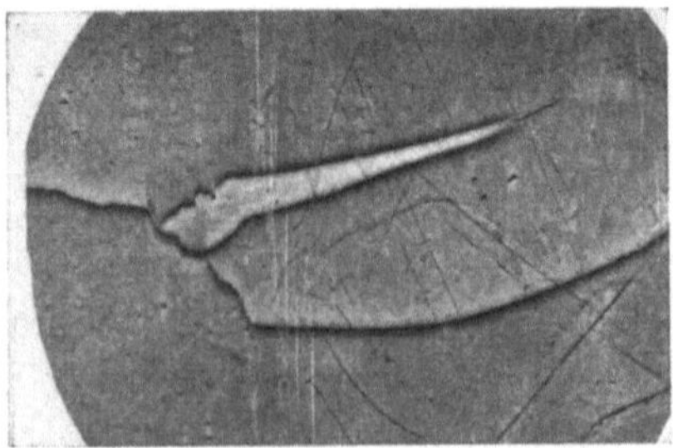

Fig. 53a. Phasenkontrastaufnahme desselben Objekts, das in Bild d der Fig. 54 mit einem Interferenzverfahren erster Art aufgenommen wurde.

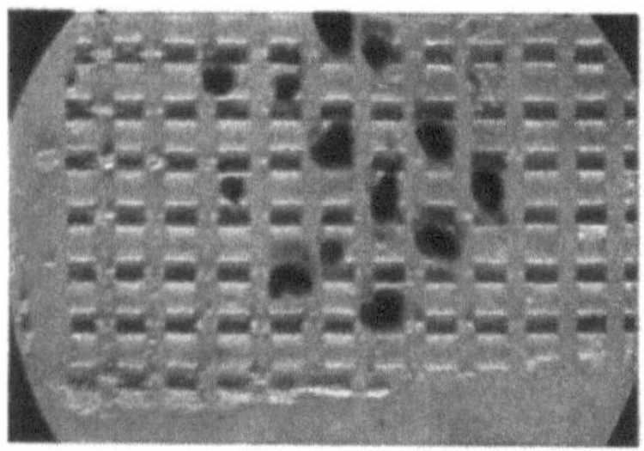

Fig. 53b. Phasenkontrastaufnahme desselben Objekts, das in Bild e der Fig. 54 mit einem Interferenzverfahren erster Art aufgenommen ist.

Planparallelplatte und ihr Spiegelbild exakt gleich dick sind, erhält man automatisch niedrige Ordnungen, die das Objekt in gesättigten Interferenzfarben zeigen (Fig. 54d und e). Sie sind völlig objekttreu im Gegensatz zu den Phasenkontrastbildern derselben Objekte (Fig. 53). Als charakteristisch für alle Interferenzverfahren erster Art wurden sie hier aufgenommen.

„Interferenzverfahren" *erster Art* werden solche abbildenden Interferenzverfahren genannt, die das objektfreie Gesichtsfeld in einheitlicher Färbung zeigen und auch durch die Färbung der Objekte eine bestimmte Phase selbst kennzeichnen wie in Fig. 54d.

Bei älteren Typen von Interferenzgeräten ist eine solche Justierung schwer erreichbar oder kaum festzuhalten. Zwischen Hellfeldbild und dem kohärenten Zusatzlicht besteht vielmehr eine „keilförmig" mit dem Bildort linear sich ändernde Phasendifferenz. Dann ist schon das objektfreie Gesichtsfeld von Interferenzstreifen durchzogen, und die Objekte deformieren und verschieben dieses Streifensystem. Oft ist dieses „Verfahren zweiter Art" trotz der nicht unmittelbar erfüllten Forderung nach Objekttreue für Messungen sehr geeignet. Fast alle bekannten Interferenzbilder sind Bilder zweiter Art.

Bei dem Verfahren erster Art ist neuerdings auch die in Ziff. 53 erwähnte Elliptizitätsanalyse benutzt worden; zu dem Zweck wurden die beiden zur Interferenz kommenden Wellenbündel zuvor senkrecht zueinander polarisiert[4].

[1] S. Tolansky: Multiple-beam interferometry of surfaces and films. Oxford 1948. — T. Merton: Proc. Roy. Soc. Lond., Ser. A **189**, 309 (1947).
[2] W. Krug u. E. Lau: Ann. Physik (6) **8**, 329 (1951).
[3] J. Dyson: Nature, Lond. **164**, 229 (1949).
[4] H. Wolter: Ann. Physik (6) **9**, 65 (1951). — Z. Physik **140**, 565 (1955).

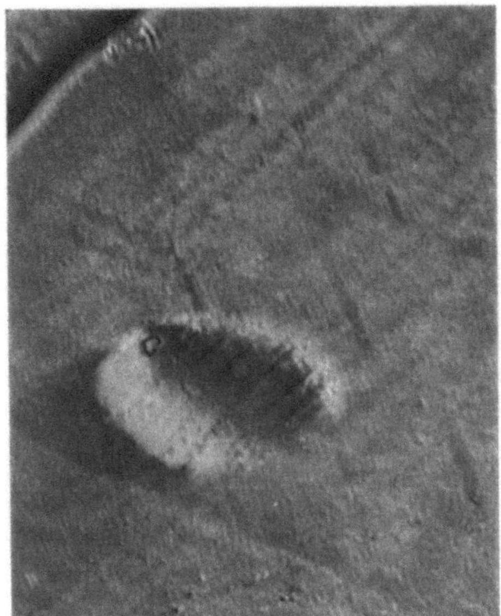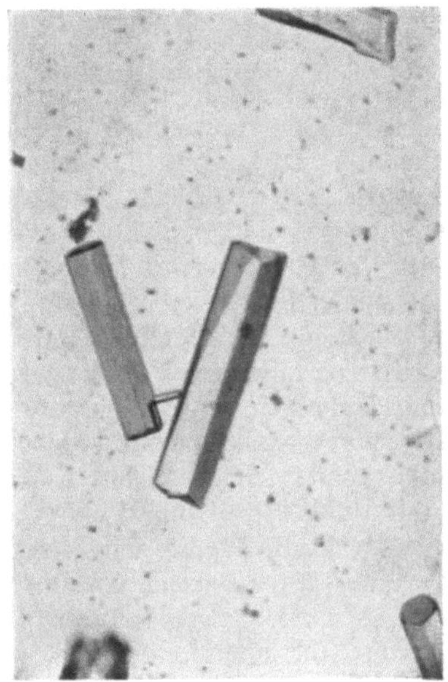

Fig. 54 a—e. a Glasplatte mit kleinräumigen Phasenobjekten (Schleifspuren) und einem großräumigen Phasenobjekt (Körnerschliere). Links oben ein Teil eines Tintenpünktchens als Amplitudenobjekt. Mikroaufnahme (120mal) nach dem in Ziff. 56a beschriebenen Verfahren zur gleichzeitigen deutlichen Abbildung von Phasenobjekten, großräumigen Schlieren und Amplitudenobjekten. b Kaliumsulfatkristalle, in ihrer Mutterlauge wachsend; aufgenommen mit dem zweidimensionalen mikroskopischen (140mal) Farbschlierenverfahren nach Ziff. 9. c Farbige Phasenkontrastaufnahme eines Objekts, das rechts ein Amplitudengitter, links ein Phasengitter und oben zusätzlich Polarisationsobjekte (Glimmerplatten) enthält; aufgenommen mit dem Verfahren nach Ziff. 56. d Nach einem Interferenzverfahren erster Art aufgenommenes interferenzmikroskopisches Bild einer Glimmerplatte mit drei Dickenstufen. Jede Interferenzfarbe zeigt objekttreu die Eigenschaften der betrachteten Objektstelle. (Vgl. die Phasenkontrastaufnahme 53a desselben Objekts). e Nach einem Interferenzverfahren erster Art aufgenommenes interferenzmikroskopisches Bild eines Testobjekts, das aus Phasen- und Amplitudenobjekten auf einer Glasplatte mit schwachem Keilwinkel besteht. (Vgl. die Phasenkontrastaufnahme 53 b desselben Objekts.)

63. Abbildende Interferenzverfahren dritter Art. Statt einer linear veränder-lichen Phasenverschiebung des kohärenten Zusatzlichtes benutzt Françon[1] kohärentes Zusatzlicht, das die gleiche Phasenstruktur wie das Bild selbst hat. Es ist ein um eine kleine Strecke Δx seitlich — wir sagen „in x-Richtung" — verschobenes Bild. Realisiert wird das durch eine Kalkspatplatte im Okular. Damit beide — zunächst automatisch senkrecht zueinander polarisierte — Bilder miteinander interferieren können, wird ein Analysator unter 45° nachgeschaltet. Das Bild zeigt in dieser Form Interferenzerscheinungen höchstens an den Objekt-rändern, da in Gebieten langsamer Phasenänderung das zu dem festen kleinen Δx gehörige $\Delta\varphi$ nicht sehr groß ist. Wird dem Analysator aber noch eine $\lambda/2$-Glimmer-platte vorgeschaltet — mit parallelen Hauptschwingungsrichtungen zu den Polari-sationsrichtungen — so hat das objektfreie Gesichtsfeld ein festes $\Delta\varphi_0 = 180°$ und ist dunkel; die Objekte mit $180° + \Delta\varphi$ erscheinen darin ähnlich wie im Dunkel-feld. Da die Bedingung $\Delta\varphi_0 = 180°$ nicht für alle Lichtwellenlängen streng erfüllt ist, resultiert bei Beleuchtung mit Glühlicht ein farbenprächtiges Interferenzbild, aus dessen z.B. purpurfarbigem Grundton sich alle Stellen farbig abheben, die ein von Null differierendes $\Delta\varphi$ haben.

Objekttreu ist das Bild nicht, aber es erlaubt einen empfindlichen Nachweis von Differenzen $\Delta\varphi$; das bedeutet bei Objekten mit einem gegenüber Δx langen Phasenanstieg die Messung des Gradienten

$$\frac{\Delta\varphi}{\Delta x} \approx \frac{\partial\varphi}{\partial x} \sim \mathfrak{e} - \mathfrak{e}_0. \tag{63.1}$$

Da die Phase eine lineare Funktion der optischen Dicke ist, mißt das Verfahren nahezu die Änderung der Lichtfortpflanzungsrichtung $\mathfrak{e} - \mathfrak{e}_0$ und steht in seiner Wirkung ebenso wie in der Einfachheit seiner Ausführung den Schlierenver-fahren näher als den Interferenzverfahren erster Art, gibt jedoch im Gegensatz zu ihnen unerwünschte Doppelbilder von kleinen Objekten[2].

b) Vorteile des Phasenkontrastverfahrens gegenüber den Interferenzverfahren.

64. Außer der größeren Kostspieligkeit der Interferenzgeräte muß auch ein wesentlicher Unterschied zwischen Interferenz- und Phasenkontrastgeräten betont werden, der sogar die Mängel des Phasenkontrastverfahrens als Vorteile erscheinen läßt. Wie die Fig. 32 und 36 zeigten, kontrastiert das Phasenkontrastverfahren breite Objekte nur an ihrem Rande; das macht zwar den Wunsch nach einein-deutiger Zuordnung zwischen Objekteigenschaften und Bildintensitäten unerfüll-bar; aber es macht auch das Innere des breiten Objekts gleichsam wieder frei für die praktisch unbeeinflußte Kontrastierung kleiner Objekte oder Einzelstrukturen. Insbesondere sind Dickenschwankungen des Objektträgers und des Deckglases solche großräumigen Objekte. Sie stören bei dem Phasenkontrastverfahren nicht, während die Interferenzmikroskopie erster und zweiter Art jede Dickenänderung des Deckglases und des Objektträgers mit voller Empfindlichkeit anzeigt. Gegen-über dem in dieser Hinsicht dem Phasenkontrastverfahren ähnlichen Interferenz-verfahren dritter Art besitzt das Phasenkontrastverfahren bei kleinsten Objekten die größere Objekttreue, da das Interferenzverfahren dritter Art von diesen un-erwünschte Doppelbilder zeigt.

---

[1] M. Françon: Le microscope à contraste de phase et le microscope interférential, S. 113ff. Paris 1954.

[2] Ein ähnlich wirkendes, aber komplizierteres Verfahren, das auf Jamins Kalkspatinter-ferometer [C. R. Acad. Sci. Paris **67**, 814 (1868)] beruht, wurde von F. H. Smith [British Patent Specification 639,014; class 97 (i), J (8b:9:15); Application 1947] zum Patent an-gemeldet. Neugebauer zeigte auf der Physikertagung in Wiesbaden im Herbst 1955 Mikro-aufnahmen, die mit dem Kalkspatinterferometer hergestellt wurden. Siehe auch S. 453 bis 455 dieses Bandes.

Das Phasenkontrastverfahren hat vor allem in der Mikroskopie seine große Bedeutung gerade deshalb erlangt, weil es einen glücklichen Kompromiß darstellt und höchste Empfindlichkeit für die hauptsächlich interessierenden kleinen Objekte mit Unempfindlichkeit gegen die nicht interessierenden großräumigen Phasenänderungen vereint.

65. Übersicht über die optischen Größen, die jeweils ein Verfahren zur Kennzeichnung einer Objektstelle verwertet.

	Verfahren	Objektkennzeichnung durch	Zeichenerklärung in Ziff.
I	Schlierenverfahren	Änderung der Lichtfortpflanzungsrichtung $$\mathfrak{e} - \mathfrak{e}_0 \sim \operatorname{grad} D$$ bzw. $$\mathfrak{e} - \mathfrak{e}_0 \sim \int \frac{1}{n} \operatorname{grad} n\, dz$$	2 und 14 ff.
II	Einfache Phasenkontrastverfahren	Phasenabweichung $\varphi - \varphi$ (einer Umgebung) $$\approx \frac{2\pi d}{\lambda} (n_0 - n)$$	25 ff.
III	Variable und farbige Phasenkontrastverfahren	Phasenabweichung $\varphi - \varphi$ (einer Umgebung) und Amplitudenverhältnis $a : a$ (einer Umgebung)	51 ff.
IV	Lichtschnittverfahren	Stufenhöhe d bzw. d/n bei Folien	59 ff.
V	Interferenzverfahren erster Art	Phase $\varphi -$ Konst.	25 und 62
VI	Interferenzverfahren erster Art mit Elliptizitätsanalyse o. dgl.	Phase $\varphi -$ Konst. Amplitude $a :$ Konst.	62
VII	Interferenzverfahren zweiter Art	$\varphi -$ Konst. $\times x$	62
VIII	Interferenzverfahren dritter Art	Für hinreichend langsamen φ-Anstieg: $$\frac{\partial \varphi}{\partial x} \sim \mathfrak{e}_x - \mathfrak{e}_{0\,x} \quad \text{(siehe I)}$$ sonst $\varphi(x_1) - \varphi(x_2)$ für festes $x_1 - x_2 = \Delta x$	63

Literatur.

Die aufgeführten Originalarbeiten sind ohne Titel zitiert; ihr Inhalt ist jeweils durch einige Stichworte umrissen. Bei Büchern ist der volle Titel angegeben.

ALBER, K. H., and J. T. BRYANT: Ind. Engng. Chem., Analyt. Edit. **12**, 305 (1940). Refractive index of liquids.

ANTWEILER, H. J.: Mikrochemie **36**, 36 (1950). Beugung im PHILPOT-SVENSSON-Verfahren.

—, u. H. KAYER: Angew. Chem. **62**, 412 (1950). Interferenzerscheinungen bei der optischen Aufzeichnung von Konzentrationsgefällen.

ARMBRUSTER, O., W. KOSSEL u. K. STROHMAIER: Z. Naturforsch. **6a**, 510 (1951). Registrierung von Konzentrationsgradienten nach dem Minimumstrahlverfahren.

BARER, R.: Quart. J. Microsc. Sci. **88**, 491—500 (1947). Phase-contrast microscopy.

— Nature, Lond. **164**, 1087 (1949). Variable colour-amplitude phase-contrast microscopy.

BENDFORD, J. R., and R. L. SEIDENBERG: J. Opt. Soc. Amer. **40**, 314 (1950). Phase contrast microscopy for opaque specimens.
BENNETT, A. H., H. JUPNIK, H. OSTERBERG and O. W. RICHARDS: Trans. Amer. Microsc. Soc. **65**, 99 (1946). Ref. Optik **2**, 448 (1947). Phase-microscopy.
— H. OSTERBERG, H. JUPNIK and O. W. RICHARDS: Phase microscopy principles and applications. New York u. London 1951.
DYSON, J.: Nature, Lond. **164**, 229 (1949). A transmission-type interferometer microscope.
EULER, J., u. W. HÜPPNER: Optik **6**, 332 (1950). Quantitatives Schlierenverfahren ohne Photometrie.
FRANÇON, M.: C. R. Acad. Sci. Paris **223**, 504 (1946). Object rectangulaire contraste de phase.
— C. R. Acad. Sci. Paris **228**, 1413 (1949). Limite de séparation du microscope à contraste de phase.
—, and G. NOMARSKI: C. R. Acad. Sci. Paris **230**, 1392 (1950). Lame de phase à contraste variable par réflexion.
— Le contraste de phase en optique et en microscopie. Paris 1950.
— C. R. Acad. Sci. Paris **230**, 1050 (1950). Lame de phase à absorption variable.
— Rev. Opt. **29**, 619 (1950). Phasenkontrast nach dem Okular bzw. vor dem Objektiv.
— Seite 271 im Sammelwerk „La Théorie des Images Optiques". Referat Optik **8**, 191 (1951). Contraste de phase.
— Rev. Opt. **31**, 2 (1952). Interférences par double réfraction en lumière blanche.
— Contraste de phase et contraste par interférences. Paris 1952. Referat Optik **10**, 606 (1953).
— Le microscope à contraste de phase et le microscope interférentiel. Paris 1954.
FREY-WYSSLING, A.: Naturwiss. **39**, 145 (1952). Kontrasteffekt dicker Objekte im Phasenmikroskop.
GOOS, F., u. H. HÄNCHEN: Ann. Phys. (5) **43**, 383 (1943). Über das Eindringen des totalreflektierten Lichtes in das dünnere Medium.
— — Ann Phys. (6) **1**, 333 (1947). Strahlversetzung bei Totalreflexion.
GRIGG, F. C.: Nature, Lond. **165**, 368 (1950). Colour-contrast phase microscopy.
HANSEN, H. G., u. A. ROMINGER: Das Phasenkontrastverfahren in der Hämatologie. Materia Medica Nordmark. Wissenschaftliches Beiblatt Nr. 5, 1953.
HARTLEY, W. G.: Nature, Lond. **159**, 880 (1947). A variable phase-contrast system for microscopy.
HEINEMANN, E.: Optik **9**, 379 (1952). Phasenkontrastverfahren bei Lichtbeugungserscheinungen an Ultraschallwellen.
HOPKINS, H. H.: Proc. Roy. Soc. Lond., Ser. A **217**, 408 (1953). Diffraction theory of optical images.
HORST, U.: Bergbau u. Energiewirtsch. **3**, Sonderbeilage (1950). Phasenkontrasteinrichtung für mikroskopische Kohlenuntersuchungen.
INGELSTAM, E.: Nature, Lond. **168**, 960 (1951). A highly sensitive phase-contrast refractometer for liquids and gases.
— Ark. Fysik **3**, 387 (1951). Phase contrast studies of extremely small refractive differences in liquids.
— Transaction of instruments and measurements conference Stockholm 1952. Phase contrast refractometer.
— Ark. Fysik **6**, Paper 26, 287 (1953). Phase contrast with high accuracy.
—, and E. DJURLE: J. Opt. Soc. Amer. **43**, 572 (1953). Diffraction grating characteristics by phase contrast.
—, and L. JOHANSSON: Appl. Sci. Res. **4**, 100 (1953). Visual receiver for the phase contrast refractometer.
—, E. DJURLE and L. JOHANSSON: J. Opt. Soc. Amer. **44**, 472 (1954). Precision concentration analysis of D_2O/H_2O by means of phase contrast refractometry.
JUPNIK, H., H. OSTERBERG and G. E. PRIDE: J. Opt. Soc. Amer. **36**, 710 (1946). Phase microscopy with vertical illumination.
KÄSEMANN, E.: Optik **3**, 521 (1948). Der Dichroismus des Zellulosefarbstoffkomplexes und seine technische Anwendung als Polarisationsfilter.
— DRP 702026, 749938, 733438, 746663. Verfahren zur Herstellung von Polarisationsfiltern.
KASTLER, A., and R. MONTARNAL: Nature, Lond. **161**, 357 (1948). Phase-contrast in polarized light.
KECK, P. H., u. A. T. BRICE: Optik **5**, 31 (1949). Die günstigste Dimensionierung der Phasenplatte beim Phasenkontrastverfahren.
— J. Opt. Soc. Amer. **39**, 507 (1949). Image contrast in phase-contrast microscopy.
KÖHLER, A., u. W. LOOS: Naturwiss. **29**, 49 (1941). Das Phasenkontrastverfahren und seine Anwendung in der Mikroskopie.

KOSSEL, D.: Zusatzpatentanmeldung vom 22. 4. 54 zum D.P. 550527. Interferenzfilter als Phasenplatte für achromatischen oder farbigen Phasenkontrast.

KOSSEL, W., u. K. STROHMAIER: Z. Naturforsch. 6a, 504 (1951). Zum Elementarvorgang der Lichtbeugung. (Unter anderem Experimente zur Minimumstrahlkennzeichnung.)

KRUG, W., u. E. LAU: Ann. Phys. (6) 8, 329 (1951). Ein Interferenzmikroskop für Durch- und Auflichtbeobachtungen.

LAMM, O.: Nature, Lond. 132, 820 (1933). A new method for determination of the concentration gradient in the ultracentrifuge.

LIPPERT, W.: Optik 8, 543 (1951). Erläuterung des Phasenkontrastmikroskops.

LOOS, W., W. KLEMM u. A. SMEKAL: Naturwiss. 29, 769 (1941). Phasenkontrastmikroskopie zum Poliervorgang an Gläsern.

LOCQUIN, M.: Microscopie (Paris) 1, M 47 (1948). Plaque de phase à absorption réglable en lumière polarisée.

— J. Opt. Soc. Amer. 41, 871 (1951). Phase contrast by the use of an interferential phase plate.

LOHMANN, A.: Optik 11, 478 (1954). Dualitätsprinzip in der Optik.

LONGSWORTH, L. G.: Ann. N.Y. Acad. Sci. 39, 187 (1939) and Chem. Reviews 30, 323 (1942). Elektrophoreseapparatur mit verschiebbarer Schlierenschneide.

LYOT, B.: Optik 8, 238 (1951). Procédés permettant d'étudier les irregularités d'une surface optique bien polie. (La Théorie des Images Optiques.) (Referat.)

MACH, E.: Pogg. Ann. 159, 330 (1876). Momentanbeleuchtung (Funken) bei Schlierenbeobachtungen.

MARX, T., u. F. DIEHL: Naturwiss. 35, 91 (1948). Phasenkontrastverfahren in der Oberflächenmikroskopie.

MENZEL, C., u. E.: Optik 3, 247 (1948). Die Beugungserscheinung optischer Gitter nach Intensität und Phase.

MENZEL, E., u. C.: Optik 4, 22 (1948). Zur Abbildung optischer Gitter.

— Optik 5, 385 (1949). Erhöhter Bildkontrast bei ausgedehnten Objekten.

— Z. angew. Phys. 3, 308 (1951). Phasenkontrastverfahren.

— Naturwiss. 38, 332 (1951). Mehrfacher mikroskopischer Lichtschnitt.

MERTON, T.: Proc. Roy. Soc. Lond., Ser. A 191, 1 (1947). On interference microscopy.

— Proc. Roy. Soc. Lond., Ser. A 189, 309 (1947). On a method of increasing contrast in microscopy.

MICHEL, K.: Naturwiss. 29, 61 (1941). Darstellung von Chromosomen mittels des Phasenkontrastverfahrens.

— Naturwiss. 37, 52 (1950). Das Phasenkontrastverfahren und seine Eignung für zytologische Untersuchungen.

— Die Grundlagen der Theorie des Mikroskops. Physik und Technik, Bd. 1. Stuttgart 1950.

— Jb. Optik u. Feinmechanik 1955. Phasenkontrast.

MOSER, H., u. J. WITTMANN: Z. Physik 131, 48 (1951). Ein experimenteller Beitrag zur Minimumstrahlenkennzeichnung.

—, u. W. SCHMIDT: Z. Physik 134, 546 (1953). Die Anwendung der Minimumstrahlkennzeichnung auf spektroskopische Untersuchungen.

NARATH, A.: Optik 4, 9 (1948). Zur Erklärung des Phasenkontrastverfahrens.

NAUMANN, H.: Bl. Unters.- u. Forsch.-Instr. 16, 25 (1942). Lichtschnittverfahren mit einem Objektiv (s. MENZEL).

OSTERBERG, H.: J. Opt. Soc. Amer. 34, 773 (1944). Phase difference microscopy as a problem in diffraction.

— J. Opt. Soc. Amer. 36, 710 (1946). Polanret microscopy.

— J. Opt. Soc. Amer. 37, 523 (1947) und Optik 2, 326 (1947). The theory of measuring unresolvable particles with the phase microscope.

— J. Opt. Soc. Amer. 37, 726 (1947). The polanret microscope.

— J. Opt. Soc. Amer. 38, 668, 685 (1948). The multipupil in phase microscopy.

— J. Opt. Soc. Amer. 38, 1099 (1948). Phase microscopy with critical illumination.

—, and G. E. PRIDE: J. Opt. Soc. Amer. 40, 64 (1950). The measurement of unresolved, single particles of uniform thickness by means of variable phase microscopy.

— J. Opt. Soc. Amer. 41, 285 (1951). A general theory of fringes formed by multiple reflections in a wedge.

PHILPOT, G.: Nature, Lond. 141, 283 (1938). Direct photography of ultracentrifuge sedimentation curves.

PICHT, J.: Z. Instrumentenkde. 58, 1 (1938). Zum Phasenkontrastverfahren von ZERNIKE.

RAHBECK, H., u. M. OMAR: Nature, Lond. 169, 1008 (1952). Measurement of the thickness of transparent films with the light profile microscope.

RÄNTSCH, K.: Opt. Acta 1, 141 (1954) und Optik 12, 58 (1955). Experimente zur Phasenkontrastmikroskopie.

Reumuth, H.: Kolloid. Z. **115**, 44 (1949). Phasenkontrastmikroskopie, Ergebnisse auf kolloid-chemischen, textiltechnischen und biologischen Anwendungsgebieten.

Rienitz, J.: Optik **9**, 206 (1952). Das Lochkammerinterferometer.

— Optik **9**, 436 (1952). Durchlicht-Interferenzmikroskopie im weißen Licht.

Richter, R.: Optik **2**, 342 (1947). Eine einfache Erklärung des Phasenkontrastmikroskops.

Rösch, S.: Optik **8**, 61 (1951). Drehpolarisationsfarben des Quarzes.

Saylor, C. P., A. T. Brice and F. Zernike: J. Opt. Soc. Amer. **39**, 1053 (1949). Color phase contrast, requirements and applications.

— J. Opt. Soc. Amer. **40**, 329 (1950). Color phasecontrast microscopy, requirements and applications.

Schardin, H.: Ergebn. exakt. Naturw. **20**, 303 (1942). Die Schlierenverfahren und ihre Anwendungen.

—, u. G. Stamm: Glastechn. Ber. **20**, 249 (1942). Prüfung von Flachglas mit Hilfe eines farbigen Schlierenverfahrens.

Schmaltz, G.: Technische Oberflächenkunde. Berlin 1936.

Schuster, K.: Jenaer Jahrbuch 1951. Zur Theorie der Abbildung eines Einzelstreifens nach dem Phasenkontrastverfahren.

Smithson, F.: Nature, Lond. **158**, 621 (1946). Phase-contrast microscopy for mineralogy.

Sperling, J.: Z. Physik **128**, 269 (1950). Das Temperaturfeld im freien Kohlebogen.

Svensson, H.: Kolloid-Z. **87**, 181 (1939). Direkte photographische Aufnahme von Elektrophorese-Diagrammen.

Taylor, E. W.: Proc. Roy. Soc. Lond., Ser. A **190**, 422 (1947). The control of amplitude in phase-contrast microscopy.

—, and B. O. Payne: Nature, Lond. **160**, 338 (1947). A variable phasecontrast system for microscopy.

Tiselius, A.: Naturwiss. **37**, 25 (1950). Elektrophorese und Adsorptionsanalyse als Hilfsmittel zur Untersuchung hochmolekularer Stoffe und ihrer Zerfallsprodukte. (Nobelvortrag.)

Töpler, A.: Beobachtungen nach einer neuen optischen Methode. Bonn 1864.

— Pogg. Ann. **127**, 556 (1866). Über die Methode der Schlierenbeobachtung als mikroskopisches Hilfsmittel, nebst Bemerkungen zur Theorie der schiefen Beleuchtung.

— Pogg. Ann. **128**, 126 (1866). Beobachtungen singender Flammen mit dem Schlierenapparat.

— Pogg. Ann. **131**, 33, 180 (1867). Optische Studien nach der Methode der Schlierenbeobachtung. (Grundlegende Arbeit.)

— Pogg. Ann. **134**, 194 (1868). Schlierenbeobachtung (Funke als Objekt).

Tolansky, S.: Multiple-beam interferometry of surfaces and films. Oxford 1948.

Weber, A. P.: Optik **4**, 213 (1948). Phasenkontrastverfahren.

Weber, P. E.: Optik **7**, 27 (1950). Ein reziproker Beugungseffekt.

Wiener, O.: Wied. Ann. **49**, 105 (1893). Darstellung gekrümmter Lichtstrahlen und Verwertung derselben zur Untersuchung von Diffusion und Wärmeleitung.

Wolter, H.: D.P. Nr. 819728, Kl. 42h, Gr. 34, 11 vom 12. 4. 1949. (Anmelder E. Leitz, Wetzlar.) Genaue Kennzeichnung einer Fläche oder Kurve mittels interferierender Lichtwellen.

— D.P. Nr. 819925, Kl. 42h, Gr. 34, 11 vom 12. 4. 1949. (Anmelder E. Leitz, Wetzlar.) Minimumschlierenverfahren.

— Phys. Bl. **5**, 234 (1949). Minimumstrahlkennzeichnung und Totalreflexion.

— Z. Naturforsch. **5a**, 144 (1950). Untersuchungen zur Strahlenversetzung bei Totalreflexion des Lichtes mit der Methode der Minimumstrahlkennzeichnung.

— Z. Naturforsch. **5a**, 276 (1950). Zur Frage des Lichtweges bei Totalreflexion.

— Ann. Phys. (6) **7**, 182 (1950). Verbesserung der abbildenden Schlierenverfahren durch Minimumstrahlkennzeichnung.

— Ann. Phys. (6) **8**, 1 (1950). Zweidimensionale Farbschlierenverfahren.

— Ann. Phys. (6) **8**, 11 (1950). Physikalische Begründung eines Farbenkreises und Ansätze zu einer physikalischen Farbenlehre.

— Ann. Phys. (6) **7**, 341 (1950). Die Minimumstrahlkennzeichnung.

— Ann. Phys. (6) **7**, 147 (1950). Abbildung zylindrischer Phasenobjekte elliptischen Querschnitts.

— Ann. Phys. (6) **7**, 33 (1950). Objekttreue bei Schlieren-, Dunkelfeld und Phasenkontrastbildern.

— Naturwiss. **37**, 272 (1950). Deutung von Phasenkontrastbildern.

— Naturwiss. **37**, 491 (1950). Farbige Phasenkontrastverfahren.

— Ann. Phys. (6) **9**, 57 (1951). Amplituden-, Phasen- und Polarisationsobjekte in einem einzigen farbigen Bild.

— Ann. Phys. (6) **9**, 65 (1951). Interferenzmikroskop.

— Fortschr. chem. Forsch. **3**, 1 (1954). Das Phasenkontrastverfahren und seine Anwendbarkeit auf chemische Untersuchungen.

Wolter, H.: Z. Physik **140**, 57 (1955). Elliptizitätsanalyse an Phasenkontrastbildern.

— Z. Physik **140**, 565 (1955). Polarisation und Elliptizitätsanalyse im Interferenzmikroskop.

Woodward, L. A., and J. E. Rhodes jr.: Phys. Rev. (2) **87**, 178 (1952). A test and demonstration object for phase microscopy.

Zeiss, Firma Carl: DRP. Nr. 636168/42h/6, 10 vom 26. 11. 1932. Einrichtung zur Verdeutlichung optischer Abbildungen.

Zernike, F.: Mitt. Naturkundig. Labor. Univ. Groningen, Physica, Haag **1**, 43 (1934). Beugungstheorie des Schneidenverfahrens und seiner verbesserten Form, der Phasenkontrastmethode.

— Roy. Astron. Soc. M. N. **94**, 377 (1934). Diffraction theory of the knife edge test and its improved form, the phase contrast method.

— Z. techn. Phys. **16**, 454 (1935). Das Phasenkontrastverfahren bei der mikroskopischen Beobachtung.

— Phys. Z. **36**, 848 (1935). Das Phasenkontrastverfahren bei der mikroskopischen Beobachtung.

— Phase contrast, a new method for the microscopic observation of transparent objects. § 2 in Kap. IV (Physical Optics) des Buches von A. Bouwers „Achievements in Optics". New York u. Amsterdam 1946.

— J. Opt. Soc. Amer. **40**, 329 (1950). Color phase microscopy.

— Optik **8**, 190 (1951) Referat. Observation des amplitudes et des phases au moyen d'un fond cohérent. (Seite 255 im Kapitel CI Contraste des phases des Sammelwerkes „La théorie des images optiques".)

Sachverzeichnis.

(Deutsch-Englisch.)

Bei gleicher Schreibweise in beiden Sprachen sind die Stichwörter nur einmal aufgeführt.

Subject Index.

(English-German.)

Where English and German spelling of a word is identical the German version is omitted.

Table des matières

pour les contributions écrites en français:

A. Maréchal: Optique géométrique générale,
M. Françon: Interférences, diffraction et polarisation.

If you have any concerns about our products,
you can contact us on
ProductSafety@springernature.com

In case Publisher is established outside the EU,
the EU authorized representative is:
Springer Nature Customer Service Center GmbH
Europaplatz 3, 69115 Heidelberg, Germany

Printed by Libri Plureos GmbH
in Hamburg, Germany